High-Frequency Integrated Circuits

A transistor-level, design-intensive overview of high-speed and high-frequency monolithic integrated circuits for wireless and broadband systems from 2GHz to 200GHz, this comprehensive text covers high-speed, RF, mm-wave, and optical fiber circuits using nanoscale CMOS, SiGe BiCMOS, and III-V technologies. Step-by-step design methodologies, end-of-chapter problems, and practical simulation and design projects are provided, making this an ideal resource for senior undergraduate and graduate courses in circuit design. With an emphasis on device–circuit topology interaction and optimization, it gives circuit designers and students alike an in-depth understanding of device structures and process limitations affecting circuit performance, and is accompanied online by supporting lecture slides, student lab assignments and project descriptions, and solutions for instructors.

- Describes step-by-step methodologies to design high-speed circuits as well as layout techniques to maximize both device and circuit performance.
- Contains over 100 end-of-chapter problems and numerous solved examples to aid and test understanding.
- Includes device and circuit simulation and design labs based on real-life scenarios in advanced SiGe BiCMOS and nanoscale CMOS technologies.

Sorin Voinigescu is a Professor at the University of Toronto, where his research and teaching interests focus on nanoscale semiconductor devices and their application in integrated circuits at frequencies beyond 300GHz. The co-founder of Quake Technologies, Inc., he was a recipient of the Best Paper Award at the 2001 IEEE Custom Integrated Circuits Conference, 2005 IEEE Compound Semiconductor IC Symposium (CSICS), and of the Beatrice Winner Award at the 2008 IEEE International Solid State Circuits Conference (ISSCC). His students have won Student Paper Awards at the 2004 VLSI Circuits Symposium, the 2006 RFIC Symposium, and at the 2008 and 2012 International Microwave Symposia.

THE CAMBRIDGE RF AND MICROWAVE ENGINEERING SERIES

Series Editor: **Steve C. Cripps** Distinguished Research Professor, Cardiff University

Peter Aaen, Jaime Plá, and John Wood *Modeling and Characterization of RF and Microwave Power FETs*

Dominique Schreurs, Máirtín O'Droma, Anthony A. Goacher, and Michael Gadringer *RF Amplifier Behavioral Modeling*

Fan Yang and Yahya Rahmat-Samii *Electromagnetic Band Gap Structures in Antenna Engineering*

Enrico Rubiola *Phase Noise and Frequency Stability in Oscillators*

Earl McCune *Practical Digital Wireless Signals*

Stepan Lucyszyn *Advanced RF MEMS*

Patrick Roblin *Non-linear FR Circuits and the Large Signal Network Analyzer*

Matthias Rudolph, Christian Fager, and David E. Root *Nonlinear Transistor Model Parameter Extraction Techniques*

John L. B. Walker *Handbook of RF and Microwave Solid-State Power Amplifiers*

Anh-Vu H. Pham, Morgan J. Chen, and Kunia Aihara *LCP for Microwave Packages and Modules*

Sorin Voinigescu *High-Frequency Integrated Circuits*

FORTHCOMING

David E. Root, Jason Horn, Jan Verspecht, and Mihai Marcu *X-Parameters*

Richard Carter *Theory and Design of Microwave Tubes*

Nuno Borges Carvalho and Dominique Schreurs *Microwave and Wireless Measurement Techniques*

Richard Collier *Transmission Lines*

High-Frequency Integrated Circuits

SORIN VOINIGESCU

University of Toronto

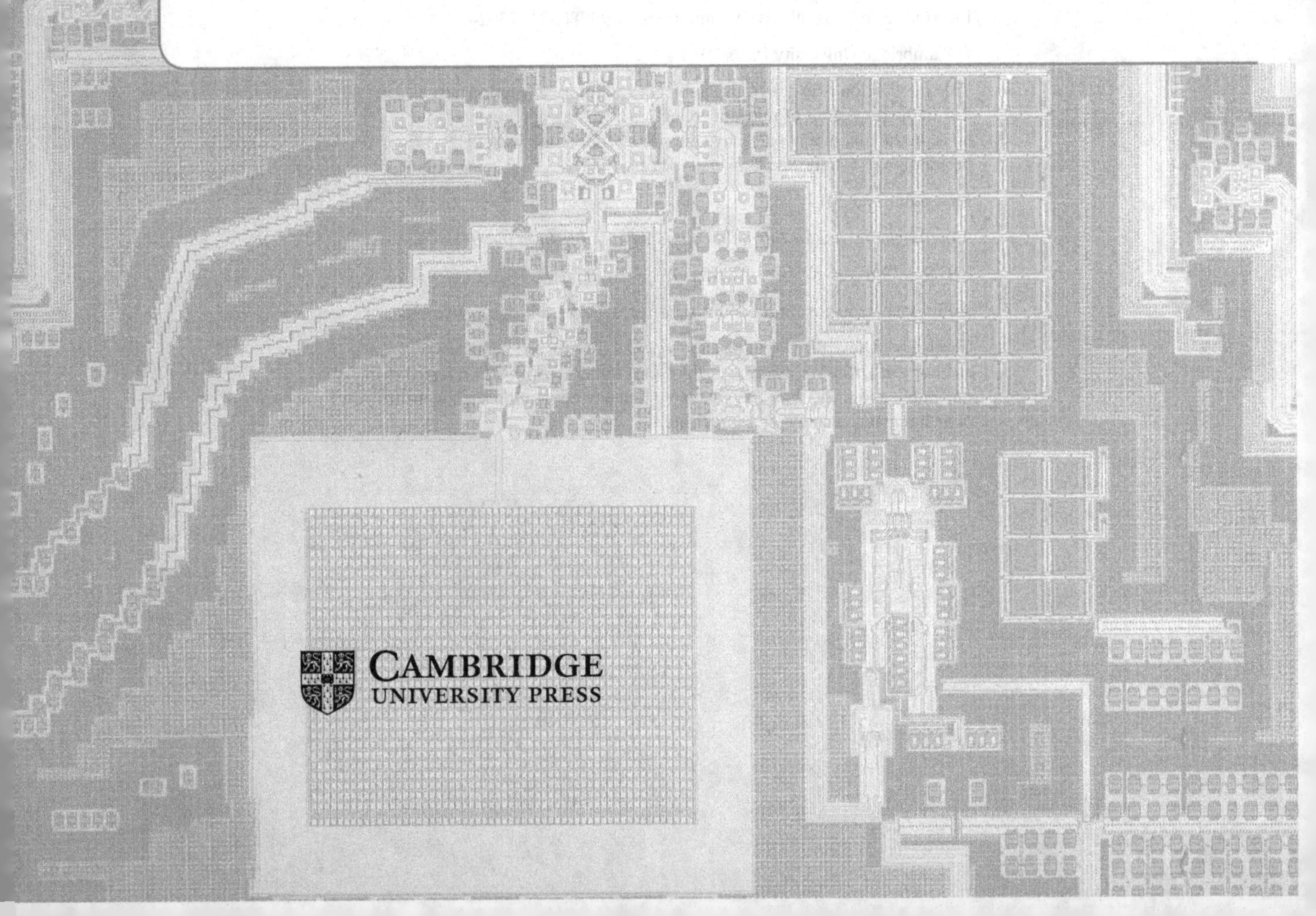

CAMBRIDGE
UNIVERSITY PRESS

CAMBRIDGE
UNIVERSITY PRESS

University Printing House, Cambridge CB2 8BS, United Kingdom

One Liberty Plaza, 20th Floor, New York, NY 10006, USA

477 Williamstown Road, Port Melbourne, VIC 3207, Australia

4843/24, 2nd Floor, Ansari Road, Daryaganj, Delhi - 110002, India

79 Anson Road, #06-04/06, Singapore 079906

Cambridge University Press is part of the University of Cambridge.

It furthers the University's mission by disseminating knowledge in the pursuit of education, learning and research at the highest international levels of excellence.

www.cambridge.org
Information on this title: www.cambridge.org/9780521873024

First published 2013

A catalogue record for this publication is available from the British Library

ISBN 978-0-521-87302-4 Hardback

Additional resources for this publication at www.cambridge.org/voinigescu

"Destined to become a classic reference in high frequency RFICs, Professor Voinigescu's book will certainly become the reference book in the field of high frequency RFIC design. It contains a comprehensive coverage of a vast array of integrated circuits and systems, including fundamental circuit design techniques, system analysis, packaging techniques, and measurements. It has an exceptional tutorial value, and presents the state of the art in microwave and millimeter-wave systems-on-chip."

Gabriel M. Rebeiz, University of California, San Diego

"Sorin Voinigescu has nicely exploited his wide experience in integrated circuits and devices to generate this book. It covers very broadly the field of high-frequency integrated circuits designed in advanced silicon and III–V technologies. Both experienced designers and newcomers in the field will appreciate this book. I am very interested by the many detailed design recipes and tricks – often with a link to the underlying IC technology – that are seldom found in related textbooks."

Piet Wambacq, University of Brussels and IMEC

"With its rational and omni-comprehensive approach encompassing devices, circuits, systems and applications, Sorin Voinigescu's book is a unique encyclopedic 'dictionary' for an in-depth understanding of high-speed and high-frequency microelectronic design.

Original, dense of details, clear and focused on the modern design challenges, it renders at the root several design dogmas that have conditioned research and development of high-frequency integrated circuits, resulting in the first book of a new class with a profound look at the road ahead."

Domenico Zito, University College Cork

"*High-Frequency Integrated Circuits* is the ideal companion for circuit designers wishing to grasp the challenges of circuit design above RF. Professor Voinigescu takes the reader from system specification down to the transistor, and presents the circuit analysis that underlies every RF circuit designer's intuition."

James Buckwalter, University of California – San Diego

ENDORSEMENTS

"Destined to become a classic reference in high frequency RFICs, Professor Voinigescu's book will certainly become the reference book in the field of high-frequency RFIC design. It contains a comprehensive coverage of a vast array of integrated circuits and systems, including fundamental circuit design techniques, system analysis, packaging techniques, and measurements. It has an exceptional tutorial value, and presents the state of the art in microwave and millimeter-wave systems-on-chip."

Gabriel M. Rebeiz, University of California, San Diego

"Sorin Voinigescu has nicely exploited his wide experience in integrated circuits and devices to generate this book. It covers very broadly the field of high-frequency integrated circuits, designed in advanced silicon and III–V technologies. Both experienced designers and newcomers in the field will appreciate this book. I am very interested by the many detailed design recipes and tricks – often with a link to the underlying IC technologies – that are seldom found in related textbooks."

Piet Wambacq, University of Brussels and IMEC

"With its rational and omni-comprehensive approach encompassing devices, circuits, systems, and applications, Sorin Voinigescu's book is a unique encyclopedic 'dictionary' for an in-depth understanding of high-speed and high-frequency microelectronic design.

Original, dense of details, clear and focused on the modern design challenges, it removes at the root several design dogmas that have conditioned research and development of high-frequency integrated circuits, resulting in the first book of a new class with a profound look at the road ahead."

Domenico Zito, University College Cork

"*High-Frequency Integrated Circuits* is the ideal companion for circuit designers wishing to grasp the challenges of circuit design above RF. Professor Voinigescu takes the reader from system specification down to the transistor, and presents the circuit analysis that underlies every RF circuit designer's intuition."

James Buckwalter, University of California – San Diego

CONTENTS

PREFACE

The field of monolithic microwave integrated circuits (MMICs) emerged in the late 1970s and early 1980s with the development of the first GaAs MESFET and pseudomorphic high electron mobility transistor (p-HEMT) IC technologies and started to thrive in the mid-to-late 1990s once the performance of silicon transistors became adequate for radio frequency (RF) applications above 1GHz. Since then, CMOS, SiGe BiCMOS, and III-V HEMT, and heterojunction bipolar transistor (HBT) technology scaling to nanometer dimensions and THz cutoff (f_T) and oscillation frequencies (f_{MAX}) has continued unabated. Despite the increasing dominance of CMOS, each of these technologies has carved its own niche in the high-speed, RF, microwave, and mm-wave IC universe. Today's nanoscale 3-D tri-gate MOSFET is a marvel of atomic-layer and mechanical strain engineering with more "exotic" materials, heterojunctions, and compounds than any SiGe HBT or III-V device. Indeed, InGaAs and Ge are expected to displace silicon channels in "standard" digital CMOS technology sometime in the next five to ten years. More so than in the past, it is very important that high-frequency circuit designers be familiar with all these high-frequency device technologies.

Apart from the general acceptance of CMOS as a credible RF technology, during the last decade "digital-RF" has radically changed the manner in which high-frequency (HF) circuit design is conducted. Traditional RF building blocks such as low-noise amplifiers (LNAs), voltage-controlled oscillators (VCOs), power amplifiers (PAs), phase shifters, and modulators have greatly benefited from this marriage of digital and microwave techniques and continue to play a significant role well into the upper mm-wave frequencies. New, digital-rich, radio transceiver architectures have emerged based on direct RF modulators and IQ power DACs, fully digital PLLs, and digitally calibrated phased arrays.

Increased device speed makes silicon an attractive technology for emerging mm-wave applications, such as 4Gb/s wireless HDTV links at 60GHz, automotive radar at 77GHz, and active and passive imaging at 94GHz. III-V based and, soon, silicon submillimeter wave sensors with integrated antennas are becoming feasible, and undoubtedly more applications will materialize. A similar scenario is unfolding in wireline and fiber-optic links, where higher-frequency applications such as 110Gb/s Ethernet have recently been standardized by the IEEE.

At the same time, the task of designing high-frequency building blocks becomes increasingly difficult in the nanoscale era. While power dissipation, high-frequency noise figure, and phase noise performance improve with scaling, other critical aspects such as maximum output swing, linearity, and device leakage are all degraded. Lower supply voltages constrain the number of devices which can be vertically stacked. From an economics perspective, the high mask costs of these technologies make first-pass design success a must.

Nanoscale MOSFETs do not exhibit the classical behavior described in undergraduate textbooks. Foundries frequently supply high-frequency data such as f_T or minimum noise figure (NF_{MIN}), but how can the design engineer incorporate these data into a design? Is "back-of-the-envelope" design still possible in 32nm CMOS at 77GHz? How about 180GHz? Are there optimal solutions for high-frequency IC building blocks? Can they be seamlessly scaled between technology nodes? Such design methodologies are not covered in any textbook, but are fairly well known – in some circles – to produce robust RF, mm-wave, and wireline IC designs that are generally transferable between technology nodes and even between III-V and silicon technologies. In the nanoscale era, the link between circuit and device performance grows in importance, and successful designers must be well versed in both areas.

This book grew out of a set of notes, assignments, and projects developed and taught during the last nine years as an advanced graduate course at the University of Toronto. Although starting from the basic concepts, the material delves deep into advanced IC design methodologies and describes practical design techniques that are not immediately apparent and which are rooted in over 20 years of microwave and mm-wave silicon and III-V IC design experience. It provides a design intensive overview of high-speed and high-frequency monolithic integrated circuits for wireless and broadband systems with an emphasis on device–circuit topology interaction and optimization. The central design philosophy is that "the circuit is the transistor," and that maximizing transistor and circuit performance go hand-in-hand. The textbook features an extensive treatment of high-frequency semiconductor devices and IC process engineering fundamentals. Properties of CMOS FETs and SiGe HBTs are examined in the context of maximizing transistor performance for high-speed, low-noise, large power, and/or highly linear circuits. Alternative device structures, such as SOI and multi-gate 3-D channels, are presented to keep the reader up-to-date with the latest technology trends. Additionally, compound semiconductor technologies (InP, GaAs, GaAsSb, and GaN), are covered. This topic, which is largely ignored in previous textbooks, is beneficial to a number of practicing engineers as these technologies continue to be incorporated in a number of wireless and wireline production parts today. Furthermore, channel strain engineering, SiGe source–drain heterojunctions, and stacked gate dielectrics speak of the many similarities between state-of-the-art CMOS and III-V heterojunction FETs.

Circuit layout often can make the difference between a successful or unsuccessful high-speed design. For the first time in a textbook in this area, layout techniques that maximize both device and circuit performance are included.

Based on the underlying device fundamentals, step-by-step design methodologies for wireless and wireline building blocks are presented. Circuit design is taught from a current-density centric biasing approach, which relies on biasing transistors at or near their peak f_T, peak f_{MAX}, or optimal NF_{MIN}. Despite the complexity of modern transistors, it is shown through examples that simple design equations and hand analysis are critical and often sufficient for successful designs even at mm-wave frequencies. Numerous practical design examples, validated by fabrication and measurements, are included for bipolar and FET circuits implemented in bulk and SOI CMOS, SiGe BiCMOS, InP, GaAs, and GaN technologies.

Differential, single-ended, and half-circuit noise and impedance matching, stability issues and common misconceptions about differential signaling, noise in feedback circuits, and velocity saturation in nanoscale CMOS are addressed for the first time in this textbook.

RF CMOS designers often complain about model inaccuracies. This book for the first time shows that, through knowledge of CMOS technology scaling rules, engineers can design CMOS high-speed circuits that are robust to bias current and threshold voltage variations, even in the absence of transistor models. Moreover, this understanding of technology scaling allows for designs to be ported between technology nodes, with little-to-no redesign required.

Gate finger segmentation of MOSFETs and Gilbert cells is described as a general technique of providing multi-bit digital control and calibration of high-frequency attenuators, switches, amplifiers, and phase shifters along with on-chip at speed self-test methodologies.

High-frequency IC design is as much an art as it is a science. The art comes from knowing when, what, and how to approximate and simplify, what aspect of a transistor model, simulation or measurement can be trusted and when. It is best acquired by working with experienced designers, learning from their mistakes, and by spending many hours conducting hands-on experiments in the lab. To support the latter aspect, several practical assignments and projects on RF, mm-wave, and optical fiber circuits using nanoscale RF CMOS, SiGe BiCMOS, and III-V technologies are provided as supplementary material on the website.

The book has 13 chapters which are organized in two large sections: (i) foundations of HF IC design, consisting of Chapters 2 through 5, and (ii) HF building block design methodologies, covered in Chapters 6 through 12. Additionally, Chapter 13 surveys representative examples of recent mm-wave silicon systems on chip, SoCs.

Chapter 1 gives a brief history of the field of high-frequency integrated circuits and can be assigned for home reading.

Chapter 2 provides an overview of wireless, fiber-optic, and high-data-rate wireline systems, transceiver architectures, and modulation techniques and explains how they impact the specification of high-frequency ICs.

Chapter 3 briefly reviews multi-port network parameters, *S*-parameters, and the Smith Chart, and introduces the key concepts of noise temperature, noise figure, correlated noise sources, noise parameters, and optimal noise impedance. Analysis techniques for high-frequency linear noisy networks, noise matching bandwidth, optimal noise impedance matching in multi-ports, negative feedback, and differential circuits are treated in depth.

Chapter 4, the largest, consists of five sections, and requires two–three lectures, each two hours long, for complete coverage. It focuses on the small signal, noise, and large signal characteristics, modeling, optimal biasing, sizing and layout of high-frequency field-effect, and heterojunction bipolar transistors fabricated in silicon and III-V technologies. The first section discusses the common high-frequency and noise characteristics of FETs and HBTs, as needed for HF circuit design, and explains the impact of noise correlation and the physical reasons why the noise and input impedances of transistors are different. The last section of Chapter 4 addresses the design and modeling of silicon integrated inductors, transformers, varactor diodes, capacitors, resistors, as well as interconnect modeling. If students have a solid background in advanced semiconductor devices, only Sections 1 and 5 need to be covered. I usually teach Sections 4.2, 4.3, and 4.4 as part of a separate undergraduate/graduate course on advanced electronic devices. These sections discuss the physics, high-frequency, and noise equivalent circuit and layout of nanoscale MOSFETs, HBTs, and HEMTs, respectively, at a fairly detailed and advanced level.

Chapter 5 completes the first half of the textbook with a survey of high-frequency tuned and broadband circuit topologies, impedance matching, bandwidth extension, and circuit analysis techniques. It includes an extensive treatment of stability, common-mode rejection, and single-ended to differential-mode conversion in high-frequency differential circuits. The chapter ends with a description of general topologies and hand analysis methodologies for non-linear circuits, with differential doublers and triplers as main examples.

Chapter 6 introduces another fundamental HF IC concept foreign to both analog and digital circuit designers: that of large signal impedance, which forms the basis of optimal output power matching of transistors in power amplifiers. This chapter deals with the basic principle of operation, classes, topologies, analysis, and step-by-step design methodologies for tuned power amplifiers. Solved PA design problems are provided in CMOS, GaN, and SiGe HBT technologies along with efficiency enhancement and power-combining techniques.

Building up on the optimal noise impedance concept of Chapter 3 and the optimal transistor biasing and sizing techniques developed in Chapter 4, Chapter 7 discusses the specification, design philosophy, topologies, and algorithmic design methodologies for tuned low-noise amplifiers, along with the theory and examples of LNA frequency scaling and design porting from one CMOS technology node to another.

Chapter 8 follows up on the low-noise theme with the design theory and analysis of broadband low-noise transimpedance and transimpedance-limiting amplifiers with digital gain control for fiber-optic and wireline applications.

Chapter 9 examines a variety of non-linear control circuits, largely based on switches, hybrid couplers, and phase shifters, and ranging from mixers to tuned variable gain amplifiers, direct modulators, and tuned high-frequency digital-to-analog converters. The basic concepts of frequency conversion, image frequency, image rejection, analog and digital phase shifting are introduced along with non-linear signal and noise analysis methodologies and simulation techniques. Step-by-step design methodologies for upconvert and downconvert mixers, digital attenuators, and tuned RF DACs and modulators are described along with circuit examples at frequencies as high as 165GHz.

Chapter 10 rounds up the review of non-linear and low-noise circuits with an in-depth treatment of voltage-controlled oscillators. Specification, topologies, analysis, and simulation techniques, and step-by-step design methodologies focusing on minimizing phase noise or power consumption are discussed in detail. For those interested in a deeper understanding of the physical origins of phase noise, the companion Appendix 10 presents an analytical harmonic series formalism to explain how partially correlated noise sidebands, caused by noise sources internal to the oscillator, arise around the carrier in the frequency spectrum of an oscillator.

Chapters 11 and 12 focus on the design of high-speed logic gates and large swing, broadband output drivers, respectively. Algorithmic design methodologies are developed for current-mode logic (CML) FET, HBT, and FET-HBT families with design examples in silicon and III-V technologies. Static, dynamic, and injection-locked divider stages based on CML gates are covered in Chapter 11. The design of laser drivers, optical modulator drivers, and of large swing, broadband power DACs for 40 and 110Gb/s fiber-optic networks with QAM and OFDM modulation formats are discussed in Chapter 12.

Chapter 13 ties it all together with a comprehensive description of a HF design flow, system integration, isolation, simulation, and verification strategies for HF silicon SoCs. Examples of commercial single-chip 60GHz and 77GHz phased array transceivers for short-range, gigabit data-rate wireless communication, automotive radar, and active imaging are discussed in detail, along with a 150GHz SiGe BiCMOS single-chip transceiver with PLL, two receive channels, and integrated transmit and receive antennas.

Although the book is self-contained and each topic is introduced from first principles, it is expected that the reader has an undergraduate level background in semiconductor devices, analog circuits, and microwave circuits.

Chapters 2 through 12 can be taught in twelve two to three hour long lectures. Chapters 6 through 12 can be covered in any order as long as Chapters 8 and 9 are taught after Chapter 7 and Chapter 12 follows Chapter 11. Alternatively, to expose the students to circuit design earlier in the course, the second half of Chapter 3 (noise in circuits with negative feedback) and Section 5 of Chapter 4 can be taught after Chapters 5 and 6. In an RF-only course, Chapters 8, 12 (and even 11) can be left out.

This book was written over the course of six years during which some sections have been updated to keep up with the frantic pace of change in this still fast-evolving field. Despite many efforts to correct mistakes and avoid repetition, errors likely still remain. I shall be thankful to fix them as soon as I receive feedback from readers.

Many people have contributed to this book directly and many more indirectly or unbeknownst to me. I apologize to the latter. Chapter 2 benefited from many discussions on 60GHz wireless system specifications with Nir Sasson and Dr. Magnus Wiklund. The groundwork for the theory of noise in circuits with negative feedback in Chapter 3 were laid out in the late 1980s while collaborating with Dan Neculoiu on developing a low-noise amplifier design project course at the Electronics Department of the Politehnica University in Bucharest. The section on FinFETs and high-frequency parasitics of MOSFETs in Chapter 4 benefited from the discussions with Dr. Ian Young of Intel and Dr. Jack Pekarik of IBM. I am also particularly grateful to Dr. Timothy (Tod) Dickson of IBM for kick-starting Chapters 7, 8, and 11 and for many figures from his Ph.D. thesis. Professor Emeritus Miles Copeland of Carleton University wrote Appendix 10 and painstakingly reviewed Chapter 10, providing many valuable suggestions over lengthy and frequent phone calls. I would like to thank Dr. Pascal Chevalier of STMicroelectronics, Crolles, France, for numerous discussions on SiGe HBTs, SiGe BiCMOS, and nanoscale CMOS physics and process flows. Chapter 13 was enhanced through the gracious help of Dr. Herbert Knapp and Dr. Marc Tiebout of Infineon, and of Dr. Juergen Hasch of Robert Bosch GmbH. who have provided advanced copies of the graphical material used for the description of the 77GHz automotive radar and imaging chip-sets, and who have carefully reviewed the associated text in Chapter 13. Dr. Juergen Hasch also contributed, along with former graduate students Katya Laskin, Ioannis Sarkas, Alexander Tomkins, and Lee Tarnow, to the material describing the 150GHz transceiver in Chapter 13.

Former graduate students Terry Yao, Katya Laskin, Theo Chalvatzis, and Ioannis Sarkas spent long hours carefully reading many of the textbook chapters and provided critical feedback and corrections. I am deeply indebted to them for their dedication and effort.

Most importantly, I would like to thank my present and former graduate and undergraduate students (more or less in chronological order) Tod Dickson, Paul Westergaard, Altan Hazneci, Chihoo Lee, Terry Yao, Michael Gordon, Adesh Garg, Alain Mangan, Katya Laskin, Ken Yau, Keith Tang, Theo Chalvatzis, Sean Nicolson, Mehdi Khanpour, Shahriar Shahramian, Ricardo Aroca, Adam Hart, Alex Tomkins, Ioannis Sarkas, Andreea Balteanu, Eric Dacquay, Lee Tarnow, Valerio Adinolfi, Lamia Tchoketch-Kebir, Olga Yuryevich, George Ng, Pearl Liu, Benjamin Lai, Brian Cousins, David Alldred, Nima Seyedfathi, Nelson Tieu, Stephen Leung, Jonathan Wolfman, Michael Selvanayagam, Danny Li, Kelvin Yu, Ivan Chan, Christopher Yung, Xue Yu, and Rophina Lee for designing, fabricating and/or testing many of the circuits and devices used as examples in this book and for allowing the publication of some of their thesis work.

Most of the experimental data presented in the textbook were made possible by donations, and free access to advanced nanoscale CMOS, SiGe BiCMOS, InP HBT, GaAs p-HEMT, and InP HEMT technology provided over the years by STMicrolectronics, Fujitsu, TSMC, Jazz Semiconductor, Nortel, Ciena, Quake Technologies, the Canadian Microelectronics Corporation, and DARPA. I would also like to acknowledge the Canadian Microelectronics Corporation for providing some of the simulation tools.

I am greatly indebted to my colleagues and collaborators in the industry for their support, discussions, and collaborations over many years. In particular, I'd like to thank Bernard Sautreuil, Rudy Beerkens, Pascal Chevalier, Alain Chantre, Patrice Garcia, Gregory Avenier, Nicolas Derrier, Didier Celi, Andreia Cathelin, and Didier Belot at STMicroelectronics, Paul Kempf (now at RIM), and Marco Racanelli at Jazz Semiconductor, M.T. Yang at TSMC, William Walker and Takuji Yamamoto at Fujitsu, Peter Schvan at Nortel and now Ciena, Petre Popescu, Florin Pera, Douglas McPherson, Hai Tran, Stefan Szilagyi, and Mihai Tazlauanu formerly of Quake Technologies, Gabriel Rebeiz and Peter Asbeck at UC San Diego, Sanjay Raman at DARPA, David Lynch, Ken Martin and Hossein Shakiba at Gennum.

Sorin Voinigescu
Toronto

1 Introduction

1.1 HIGH-FREQUENCY CIRCUITS IN WIRELESS, FIBER-OPTIC, AND IMAGING SYSTEMS

The term **radio frequency integrated circuits** or, in short, **RFICs**, describes circuits operating in the 300MHz to 3GHz range. In the 1990s, RFICs became closely associated with the cellular phone industry. In contrast, **microwave (3–30GHz) and mm-wave (30–300GHz) monolithic integrated circuits**, or **MMICs**, first introduced in the 1970s, have largely been associated with GaAs and III-V technologies and a broader spectrum of commercial and military applications. Today, Si, SiGe, and III-V integrated circuits coexist in commercial products throughout these frequency ranges, each with its own niche market.

In this book, the general term high-frequency integrated circuits, **HF ICs**, is employed to include RFICs, MMICs, as well as high-speed digital and 300+ GHz electronic monolithic integrated circuits. These high-frequency circuits and systems cover the frequency range from 1GHz to 1THz and find their application in:

- wireless,
- backplane,
- optical fiber, and
- other wired communications,

which have become mainstream in the last 20 years. Other, more recent, applications and systems such as:

- road safety and automotive radar,
- security,
- industrial sensors,
- remote sensing and radiometers, and
- radioastronomy

also benefit from the IC concepts and design methodologies developed here. These circuits can be loosely classified in:

- tuned narrowband,
- wideband, and
- broadband.

The first group are encountered in traditional terrestrial (cellular, WiFi, wireless LAN) and satellite wireless communication systems, in automotive radar and radiometers, and operate over bandwidths that do not exceed 20% of the center frequency, f_0.

The second category refers to circuits for applications where the bandwidth is larger than 20% and can sometime extend over one or more octaves. Typical examples are next-generation cellular, 2–10GHz ultra-wideband radios, as well as many military radar systems.

The last category describes circuits which operate from DC (or a few tens of kHz) to a few GHz or tens of GHz, as typically encountered in backplane I/Os, fiber-optic communication systems or in the baseband section of gigabit data-rate 60GHz wireless personal area networks (WPAN) and 80GHz last mile point-to-point radio links.

What all these circuit categories share, and what distinguishes them from traditional analog and digital circuits, is one or several of the following features:

- the presence of low-loss RF passive components such as varactors, MiM capacitors, inductors, transformers, and/or transmission lines;
- the extreme dynamic range requirements which place a strong emphasis on low-noise figure, high linearity, low-phase noise, high-output power, and high-power-added efficiency design;
- operation at frequencies beyond one fifth of the transistor cutoff frequency, f_T, and/or maximum oscillation frequency, f_{MAX}, where the transistor power, voltage, and current gain are 10dB or lower, and where some form of interstage matching may be required.

While the circuit topologies and the system architectures encountered in each category are generally different, they are largely independent of the center frequency. In other words, the circuit topologies and system architectures in each category are widely applicable at center frequencies ranging from a few gigahertz to hundreds of gigahertz. As the application frequency increases, the integration levels also increase since a greater portion of the system, sometimes even the antennas, lend themselves for single-chip integration.

Some circuit topologies scale better with increasing frequency than others. For a given circuit topology, the changes required in the circuit design as the frequency scales affect the technology generation, type, size, and bias current of transistors, size of capacitors, inductors, transformers, transmission lines, and varactor diodes employed. Nevertheless, it should be noted that, with every new generation of transistor technology, the lower boundary, where high-frequency techniques and high-frequency circuit topologies are absolutely necessary, is pushed to higher and higher frequencies. For example, with the 400GHz bipolar and CMOS transistors available in 2011, it is possible to design and manufacture analog-style and conventional CMOS digital circuits operating up to at least 20GHz and 20Gb/s, respectively, without the use of inductors, transformers, or transmission lines.

1.2 A BRIEF HISTORY OF HIGH-FREQUENCY INTEGRATED CIRCUITS

The origins of most of the wireless circuit concepts, topologies, and transceiver architectures encountered in this book can be traced as far back as the second, third, and fourth decades of the twentieth century. Inductive and transformer-coupled tuned amplifiers, inductive degeneration and inductive peaking, distributed amplifiers, out-phasing and Doherty power amplifiers, all oscillator topologies (Colpitts, Hartley, Armstrong, and cross-coupled), static and dynamic frequency dividers, single-balanced and double-balanced mixers, detectors, multipliers, switches and phase shifters, direct detection and heterodyne receiver architectures, direct modulation and upconversion transmitter architectures, radio, radar, and television were all invented before the Second World War and are still being used today. Wireless or "RF" circuits were first demonstrated using vacuum tube diodes, triodes, tetrodes, and pentodes as active elements, and discrete resistors, capacitors, and 3-D coils as inductors and transformers vertically mounted on early versions of the printed circuit board, PCB. In contrast, the microwave circuits of the period were implemented with three-dimensional waveguides as matching elements.

What has kept the high-frequency circuit research booming and rejuvenating, and high-frequency circuit designers in great demand during the last 50 years has been the introduction of new microwave transistor technologies almost every decade, starting with GaAs MESFETs (Metal-Semiconductor Field-effect Transistors) in the 1970s, III-V HEMTs (High Electron Mobility Transistors) and HBTs (Hetrojunction Bipolar Transistors) in the 1980s, Si bipolar and SiGe BiCMOS in the 1990s, planar CMOS and GaN HEMTs in the 2000s, and non-planar MOSFETs in 2011.

While most, if not all, vacuum tube circuit topologies are directly transferable and have, indeed, been transferred to GaAs MESFET, InP HEMT, and silicon MOSFET technologies, and while most high-frequency circuit topologies can be implemented with both bipolar and field-effect transistors, every new transistor technology has enriched the field of high-frequency circuits with new features (e.g. digital control, digital calibration, self-test), new topologies (e.g. current sources, current mirrors, differential pair, Gilbert cell, CMOS inverter), and has facilitated increasing levels of system integration and new system architectures.

Although the first bipolar transistor and the first monolithic integrated circuits were developed in 1947 and 1959, respectively, they were not fast enough to be suitable for high-frequency applications above 1GHz. Throughout the 1960s, the microwave design community had to content itself with thin-film planar microwave integrated circuits (MICs) manufactured on thin (<0.625μm) low-loss insulating substrates, such as alumina and quartz. These circuits relied on surface-mounted semiconductors:

- IMPATT,
- GUNN (also known as transferred-electron devices: TED),
- and tunnel

diodes for signal generation and amplification, on semiconductor:

- Schottky,
- PIN,
- step-recovery, and
- varactor

diodes for detection, modulation, mixing, switching, multiplication, and other control functions, and on:

- microstrip, stripline, and coplanar transmission lines,
- planar spiral inductors,
- transformers, and
- interdigitated capacitors

for matching networks. The lack of a 3-terminal device with cutoff and maximum oscillation frequencies beyond 10GHz was the stumbling block in the development of the first monolithic microwave integrated circuits. However, this did not prevent Gordon Moore from predicting in the last paragraph of his famous 1965 paper: "Even in the microwave area, structures included in the definition of integrated electronics will become increasingly important. The ability to make and assemble components small compared with the wavelengths involved will allow the use of lumped parameter design, at least at the lower frequencies. It is difficult to predict at the present time just how extensive the invasion of the microwave area by integrated electronics will be. The successful realization of such items as phased array antennas, for example, using a multiplicity of integrated microwave power sources, could completely revolutionize radar that in the future, entire radar phased arrays will be integrated on a single-chip" [1].

1.2.1 The early days of GaAs MMICs

With the invention of the GaAs MESFET in 1965 by Carver Mead [2], the development of semi-insulating GaAs substrates and of the first GaAs MESFETs with 1μm gate length, 10GHz cutoff frequency and 50GHz maximum oscillation frequency in 1971 [3], the field was now ready for the first microwave monolithic integrated circuits, MMICs, fabricated on GaAs rather than on silicon wafers. Figure 1.1 shows the genealogical tree of the MMIC. The latter has greatly benefited from earlier progress in both planar thin-film circuits and low-frequency silicon bipolar ICs.

The schematics and die photo of the first high-speed GaAs MESFET logic gate, developed at Hewlett-Packard in Silicon Valley (sic!), California, in 1973 [4], are reproduced in Figure 1.2. It was quickly followed by the first GaAs MMIC low-noise amplifier, LNA, shown in Figure 1.3 and designed and manufactured by Plessey in the UK in 1974 [3]. This single-stage circuit demonstrated 5dB gain in the 8–12GHz band. Shortly thereafter, the first broadband monolithic amplifiers with active transimpedance feedback were introduced by Hewlett-Packard [1], paving the way for the first fiber-optic ICs. Their schematics are depicted in Figure 1.4 and are obviously derived from the NAND/NOR logic gate in Figure 1.2.

What is truly remarkable about these circuits is that, although they operated above 1GHz, they did not feature any of the distributed circuit elements typically encountered in thin-film microwave integrated circuits. Notice the monolithic lumped inductors in the LNA chip realized in the form of a horseshoe. One of the key reasons for the superior performance of GaAs MMICs was the ability to integrate lumped inductors with high-quality factors, exceeding 60 at 10GHz. This was in part due to the low-loss semi-insulating GaAs substrate but also because of the several micron thick, plated-gold metallization. Such high-quality value inductors have yet to be matched even in today's RF silicon processes.

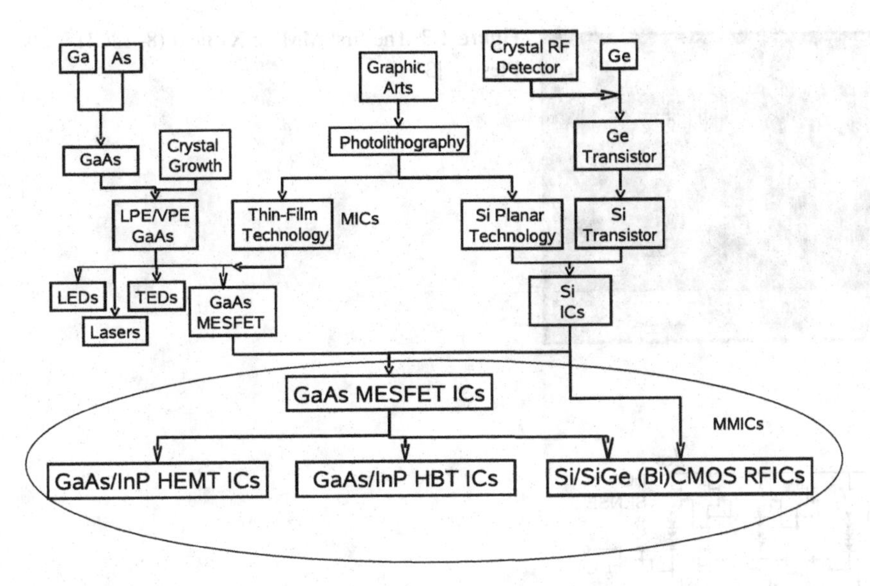

Figure 1.1 The origins of the MMIC. Adapted from [4]

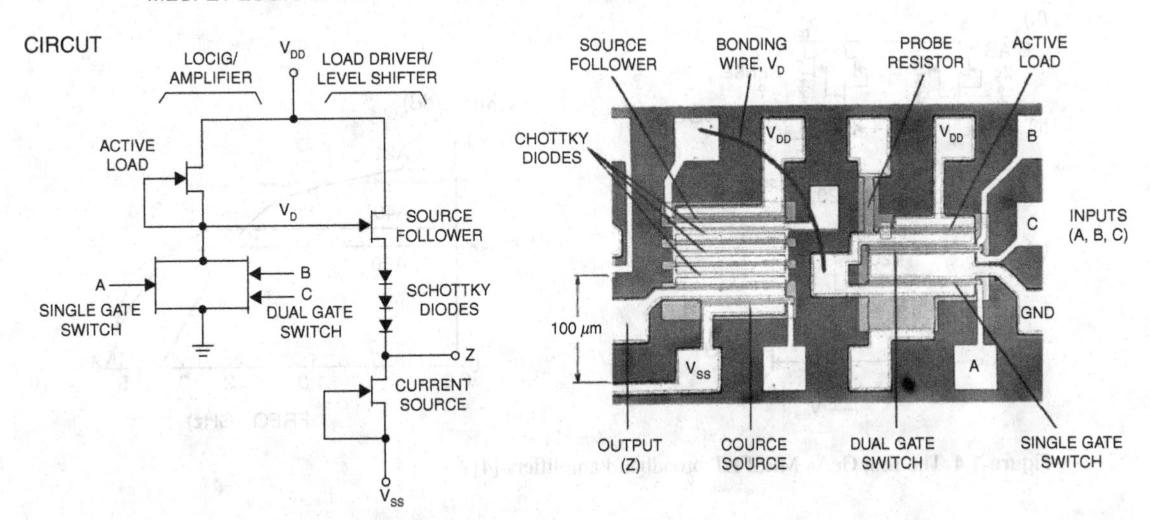

Figure 1.2 The first NAND/NOR GaAs MESFET Logic Gate showing 60ps delay [4]

By the early 1980s, full-blown commercial GaAs MESFET MMIC processes, Figure 1.5, featuring a broad range of active and passive devices, as illustrated in Figure 1.6, were developed by several companies along with multi-functional, small-scale integration RFICs [4], multi-watt power amplifiers, and even transmit/receive modules with digital phase shifters for radar phased arrays [3]. The success of these computer-designed MMICs was facilitated by the development of lumped models for inductors, of a large signal compact model for the GaAs MESFET, and by the introduction of critical microwave test equipment, such as the vector network analyzer, capable of measuring two-port *S*-parameters.

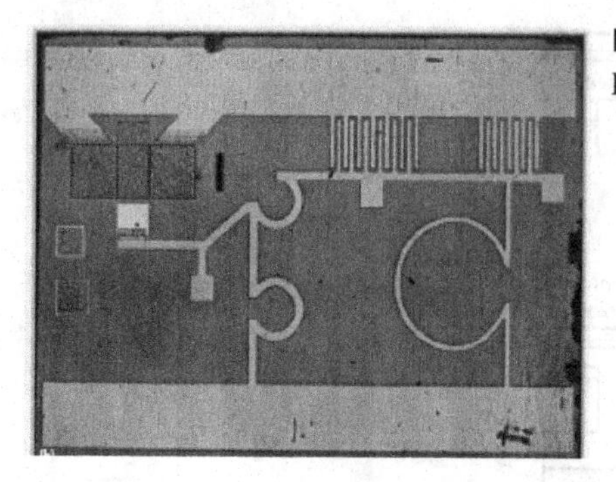

Figure 1.3 The first MMIC: X-Band (8–12GHz) LNA from Plessey [3]

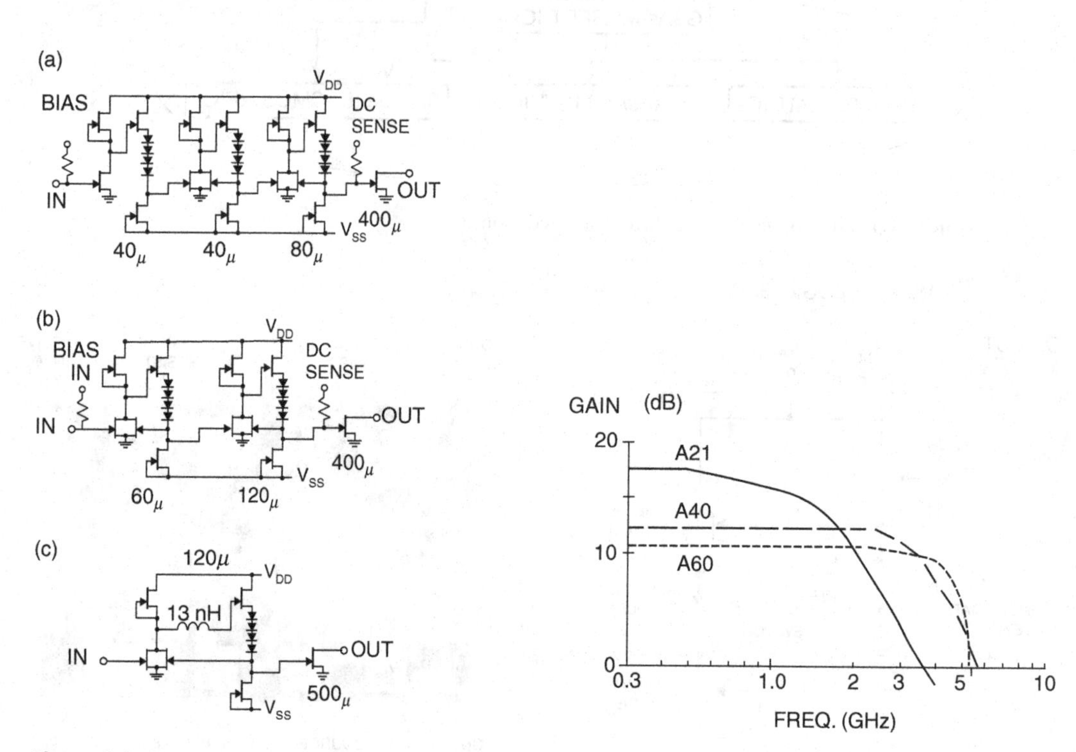

Figure 1.4 The first GaAs MESFET broadband amplifiers [4]

1.2.2 Silicon bipolar, SiGe BiCMOS and CMOS RF, and fiber-optic ICs

At the time of the first MMIC, the fastest silicon bipolar transistors had cutoff frequencies of 2GHz, the physical gate length of silicon MOSFETs was larger than 5μm and complementary CMOS logic was just being introduced. More than 15 years would pass before the first silicon bipolar and BiCMOS processes with transistor f_T and f_{MAX} larger than 10GHz became commercially available. However, the main roadblock in realizing the first silicon RFICs or MMICs was not so much the transistor speed, but rather achieving acceptable quality factors for inductors integrated on silicon. Although for several decades, even today, a large effort has been dedicated to adopting high-resistivity silicon substrates in

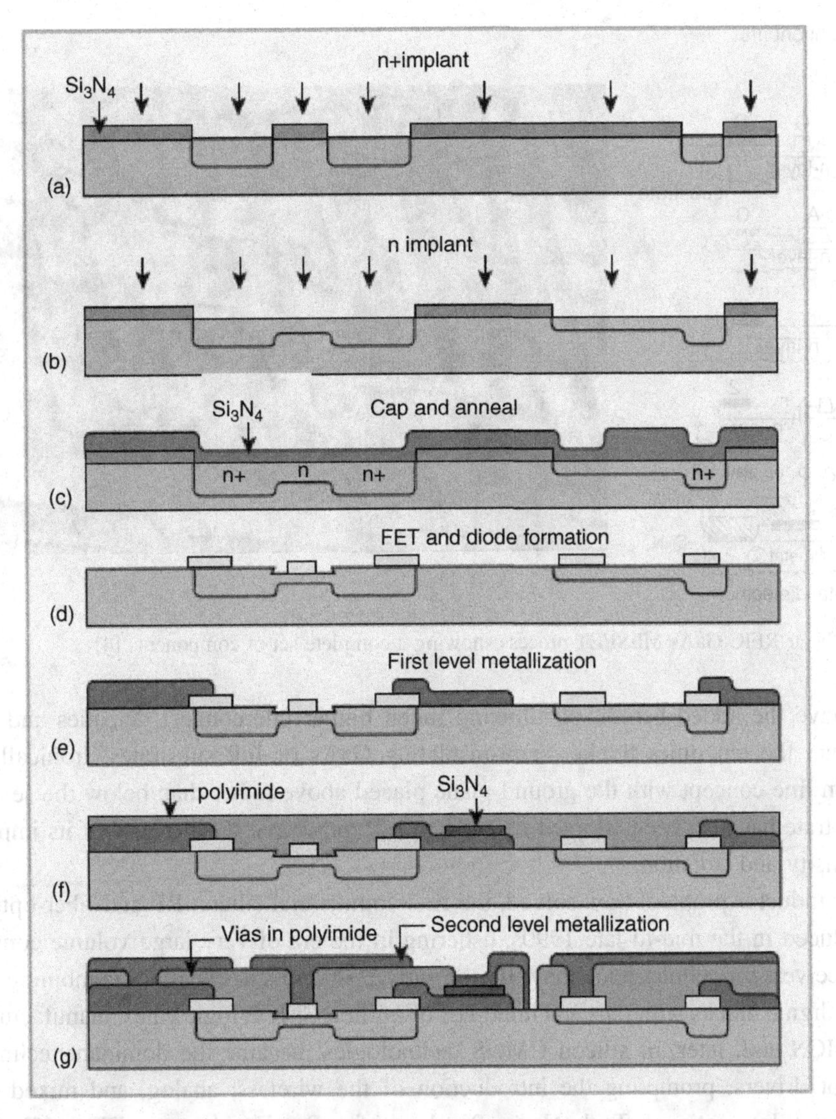

Figure 1.5 Early GaAs MESFET MMIC process flow [3]

high-frequency ICs to reduce substrate losses in transmission lines and inductors at RF and microwave frequencies, the solution was radically different from those in III-V MMICs and occurred without much notice. As the back-end of silicon ICs naturally evolved to thicker dielectric layers and increasing number of metal layers, it became possible to place inductors in the upper metal layers, insulated from the lossy silicon substrate by several microns of low-loss silicon dioxide. Similarly, acceptable microstrip lines were realized using the top metal as the signal line and the lower metal layer, above the silicon substrate, as the ground plane. Thus, the problem of the lossy silicon substrate was avoided for microstrip transmission lines.

Although the metal loss has continued to be higher in silicon interconnect than that achieved in gold-plated III-V MMICs, the above-silicon transmission lines with 5–10μm thick SiO_2

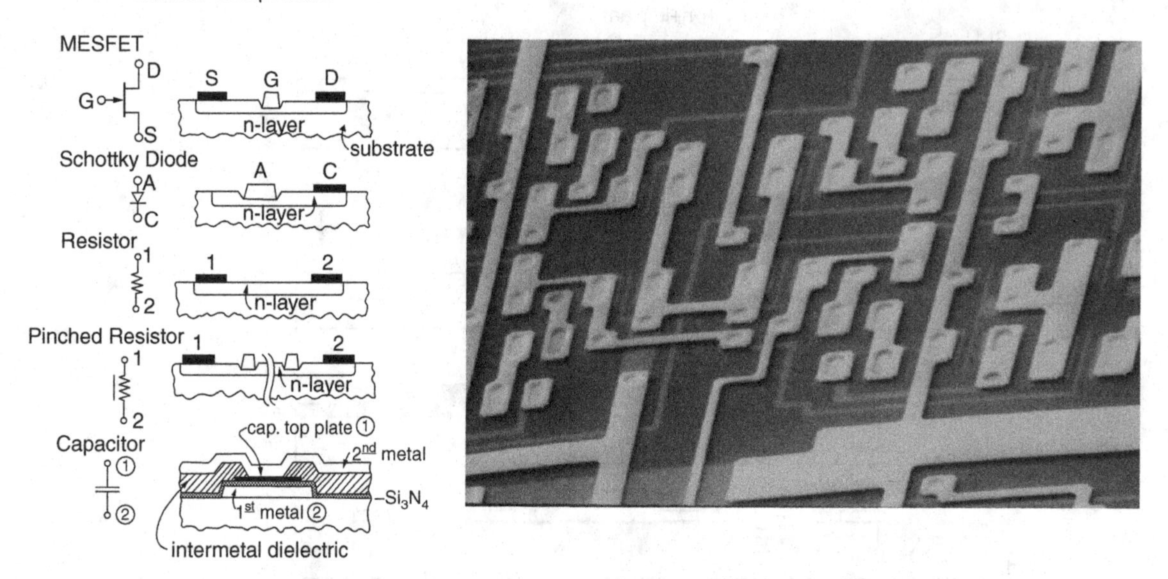

Figure 1.6 Planar RFIC GaAs MESFET process showing a complete set of components [4]

dielectric have the added benefit of allowing much higher interconnect densities and better isolation than the ten times thicker semi-insulating GaAs or InP substrates. Ironically, the transmission line concept with the ground plane placed above rather than below the semiconductor substrate has also been adopted recently in InP mm-wave ICs because of its improved packing density and isolation.

With the inductor problem now solved, the first commercial silicon RF and fiber-optic ICs were introduced in the mid-to-late 1990s, ushering in the era of very large volume consumer RFIC transceivers for cellular telephony. By the early 2000, mixed-signal ICs combining RF or high-speed digital blocks with tens and hundreds of million CMOS logic gates, manufactured in SiGe BiCMOS and, later, in silicon CMOS technologies, became the dominant technology development drivers, prompting the introduction of the wireless, analog, and mixed-signal chapter of the International Technology Roadmap for Semiconductors, ITRS [5]. High-frequency ICs had joined the mainstream!

Some of the more interesting developments in high-frequency ICs that have occurred during the last decade, most of them prompted by silicon becoming the dominant high-frequency technology, include:

- high-resolution, digitally controlled oscillators based on segmented-gate AMOS varactors;
- switching RF power amplifiers and transmitters;
- the emergence of the CMOS inverter and of the cascoded CMOS inverter as popular high-performance power amplifier and low-noise amplifier topologies in nanoscale CMOS;
- RF and mm-wave digital-to-analog converters;
- digital signal-processing (DSP)-based fiber-optic systems with higher-order QAM and OFDM modulation schemes and coherent detection at 40+ Gb/s serial data rates;

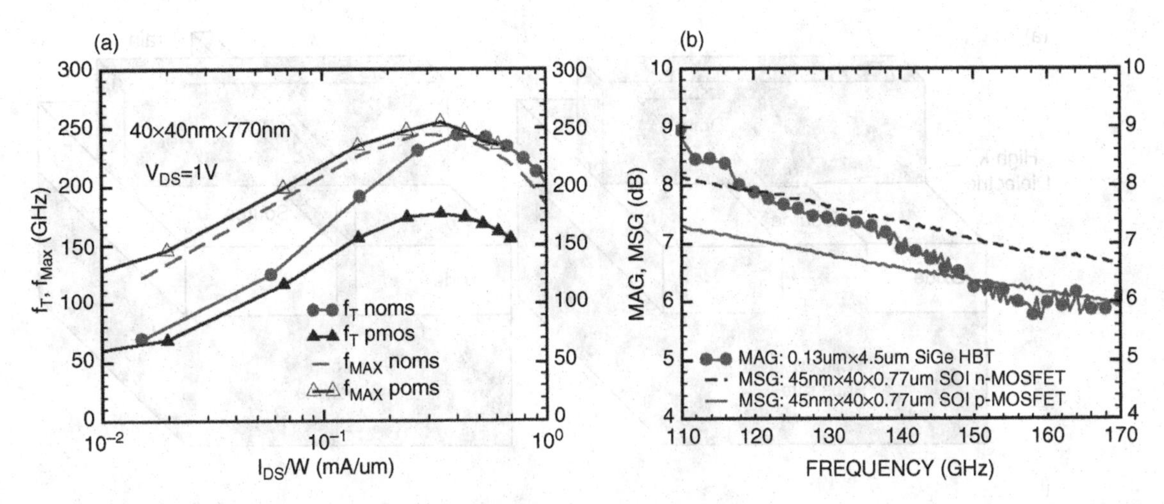

Figure 1.7 (a) Measured f_T and f_{MAX} versus drain current density characteristics of 45nm n-MOS and p-MOS transistors with identical layout geometry (b) Measured maximum available or maximum stable gain in the D-band of state-of-the art 45nm SOI MOSFETs and SiGe HBTs

- the emergence of new, potentially large volume, commercial wireless applications at mm-wave frequencies, such as: automotive radar, passive imaging, and active imaging;
- mm-wave integrated phased array and MIMO (multi-input multi-output) transceivers;
- single-chip mm-wave transceivers with efficient on-die antennas;
- the emergence of free-space power combining as a viable technique.

The last three developments go a long way to fulfilling Gordon Moore's prediction of 45 years earlier.

1.3 WHAT DOES THE FUTURE HOLD?

This is a question that all students and engineers entering, or already active in, this field should ask themselves. Will high-frequency circuits continue to enjoy longevity and the vitality demonstrated over the past four decades?

1.3.1 New high-frequency circuit topologies

Figure 1.7 reproduces the measured f_T and f_{MAX} of fully wired 45nm SOI n-channel and p-channel MOSFETs and compares their maximum stable gains with the maximum available gain of state-of-the-art SiGe HBTs. For all transistors, the power gain is larger than 6dB up to 170GHz. Unlike in other semiconductor technologies, these complementary MOSFETs exhibit almost identical f_{MAX} values of 240GHz and the same peak f_{MAX} current density of 0.3mA/μm, opening the door for a host of new high-frequency circuit topologies which can exploit the symmetrical characteristics of the CMOS inverter to implement existing circuit functions with lower noise, higher bandwidth, higher output power, higher efficiency, and lower power consumption, or to introduce new functions.

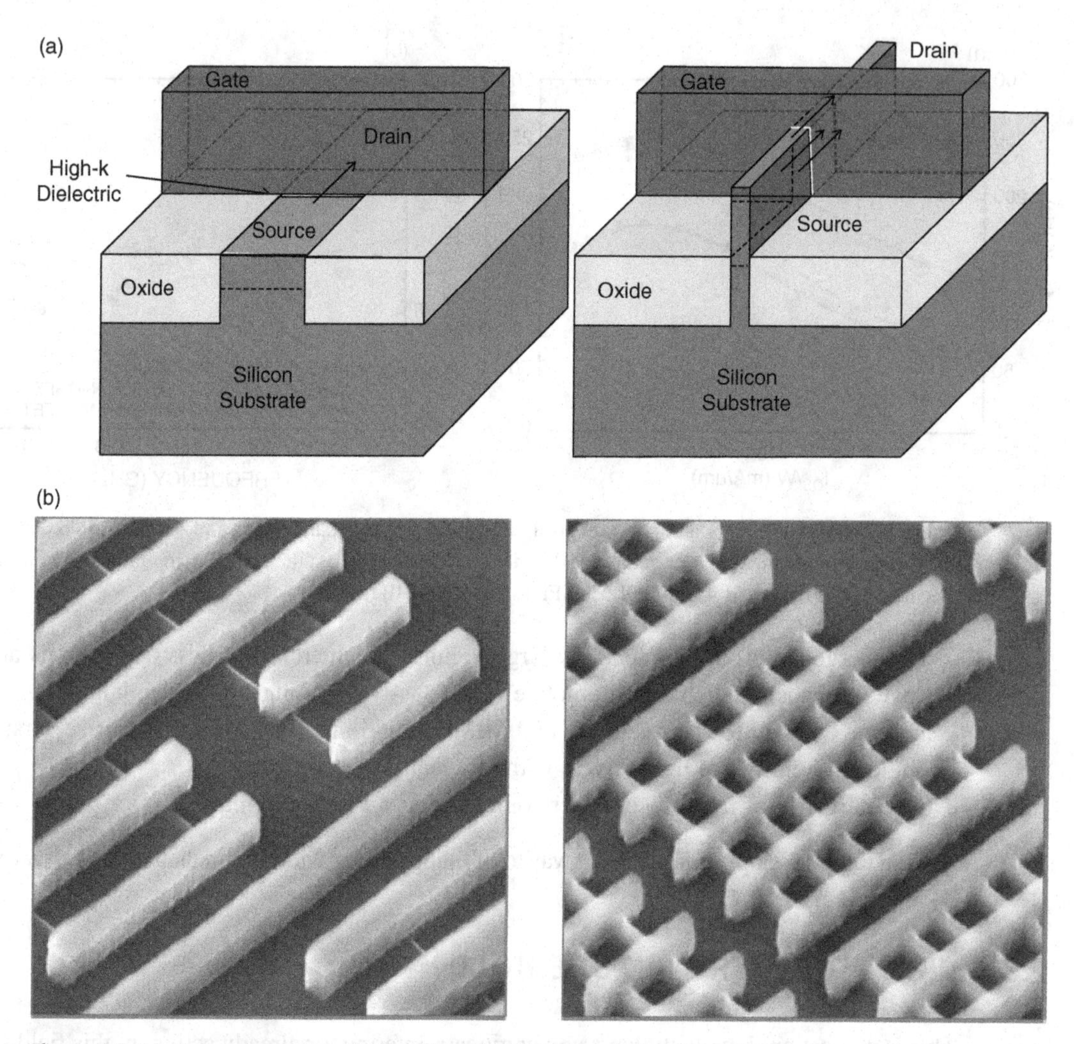

Figure 1.8 (a) Comparison of traditional planar MOSFETs and tri-gate MOSFET structures and (b) scanning electron microscope image of planar 32nm MOSFET digital circuits and of 22nm tri-gate MOSFET circuits from Intel [5]

1.3.2 New transistor architectures

The first 22nm integrated circuit technology based on three-dimensional tri-gate MOSFETs (also known as FinFETs or multi-gate FETs) was announced by Intel in 2011 [5]. A comparison with traditional planar MOSFETs and digital ICs is illustrated in Figure 1.8. Although governed by the classical field-effect principle, the layout design rules of these 3-D transistors are significantly different from those of present-day HF IC technologies. As shown in Figure 1.9, while coexisting in the next few years with planar bulk and silicon-on-insulator (SOI) MOSFETs, the FinFET is expected to become the sole survivor of the MOSFET family by 2020, and to merge with III-V semiconductor FETs in the same time frame. Propelled by progress in atomic layer deposition techniques, higher electron mobility III-V materials and higher hole mobility germanium are predicted to replace silicon in n-channel and p-channel devices, respectively, as early as 2018.

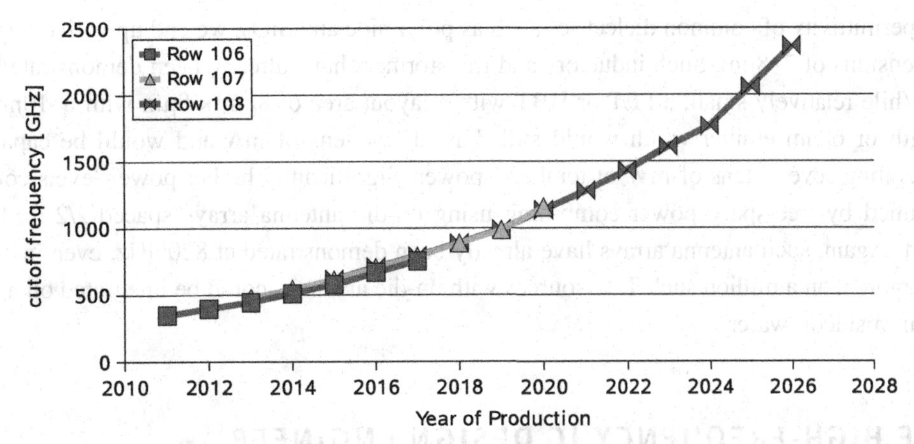

Figure 1.9 The predicted evolution of the cutoff frequency of planar (squares), SOI (up-pointing triangles), and multigate (double triangles) MOSFETs [6]

As with previous generations of transistor technologies, the introduction of multi-gate (MG) FETs will undoubtedly trigger new compact model refinements, new circuit design methodologies, new high-frequency circuit topologies, and, at the very least, a reassessment of existing ones. And then, there is graphene …

1.3.3 What would a 1THz electronic IC look like?

Over the years, circuit designers have established an empirical rule that transistor f_{MAX} values four to five times higher than the frequency of operation of the circuit are needed to develop robust, high-yielding commercial ICs. For example, today's first commercial 77GHz automotive radar and 60GHz radio transceivers are fabricated in technologies with transistor f_{MAX} values in the 250–300GHz range.

It usually takes 10–15 years from the first publication of a new device concept before research results find their way into products. It is reassuring to know that 30nm InP FETs and advanced InP HBTs have already reached cutoff frequencies of 700GHz and maximum oscillation frequencies of 1.2THz, and that MOSFETs with 5nm gate length have been reported as early as 2004. Moreover, three-stage cascode transistor amplifiers operating at 700GHz have been demonstrated using essentially the same design methodology and topologies as those employed in 2GHz cellular phone ICs.

Could lumped transistors, inductors and transformers still be used at 1THz? What would the antenna size be? Could we still employ today's circuit topologies? The answer seems to be a cautious: Yes!

At 1THz, the wavelength, λ, in free space is

$$\lambda = \frac{c}{f} = \frac{3.10^8\text{m/s}}{10^{12}\text{Hz}} = 300\mu\text{m} \tag{1.1}$$

where c is the speed of light and f is the frequency. Applying the empirical rule that lumped components have to be smaller than $\lambda/20$, and allowing for a factor of 2 reduction in size due to the permittivity of common dielectrics such as polyimide and SiO_2, we end up with component dimensions of 7–8μm. Such inductors and transformers have already been demonstrated.

While relatively small, a FET or HBT with a layout area of 8μm × 8μm with 6–10nm gate length or 60nm emitter width would still draw a few tens of mA and would be capable of generating several tens of mW of terahertz power. Significantly higher power levels could be obtained by free-space power combining using on-die antenna arrays spaced $\lambda/2 = 150\mu m$ apart. Again, such antenna arrays have already been demonstrated at 820GHz, even in silicon, and more than a million such THz sources with on-die antennas could be integrated on a single 300mm silicon wafer.

1.4 THE HIGH-FREQUENCY IC DESIGN ENGINEER

The design of high-frequency integrated circuits evolves around four fundamental goals:

- maximizing dynamic range, also characterized as signal-to-noise ratio, SNR;
- maximizing bandwidth, BW;
- minimizing power consumption;
- minimizing component size and cost while increasing functionality.

The first two, together, define Shannon's channel capacity law

$$C = BW \cdot \log_2(1 + SNR) \tag{1.2}$$

The third has always been a priority for circuit designers, while the last is none other than Moore's law. It is only through good, solid, and creative circuit design that these goals can be reached simultaneously. These themes will be pursued throughout the book.

More so than in any other area of electronics, HF circuit designers must learn how to design circuits without wasting transistor speed, transistor power, and by minimizing transistor noise. Like other IC designers, they also have to deal with transistor model inaccuracies and process variation. The most important components that are available to the HF IC designer are the inductors, transformers, and transmission lines which are used to design matching networks, resonators, filters, etc. Fortunately, in modern IC processes these passive elements are photolithographically defined, and, once properly sized, their characteristics vary little, if at all, with process and supply voltage. It is critical for the circuit designer to have full control of the modeling and sizing of these elements.

HF IC design is an art as well as a science. While everything can be accurately simulated in principle given the appropriate amount of resources and time, some circuit and transistor behavior is not well modeled and can only be obtained from measurements. It is often necessary to make simplifying approximations in order to reduce the circuit complexity, and to permit circuit simulations to be carried out in a reasonable amount of time. What and when some high-frequency parameter, component, and/or behavior can be ignored or approximated is a skill that is gained and honed over many years of circuit design and measurement experience.

The twenty-first century HF IC designer must be well versed in:

- microwave circuit and linear noisy network analysis, design, and measurement techniques;
- active and passive semiconductor devices and IC technology;
- inductor, transformer, and transmission line modeling and design;
- electromagnetic field simulation tools;
- analog circuit techniques and topologies for bias and control functions;
- digital circuit techniques for digital control and calibration of HF blocks;

and should have a good understanding of:

- antenna design and of
- wireless/high-speed systems.

With the explosion of wireless data and multi-media devices, HF ICs have already had a profound impact on our everyday life and culture. This is expected to continue for at least several decades. However, unlike in the past, where standalone RF and mm-wave ICs were common-place, in the future, most HF circuits will be developed as intellectual property (IP) blocks for large mixed-signal or digital systems-on-chip (SoCs). This will place even more emphasis on isolation techniques and on digital calibration and self-test.

REFERENCES

[1] Gordon, E. Moore "Cramming more components onto integrated circuits," *Electronics*, **38**(8): 114–117, April 19, 1965.

[2] C. A. Mead, "Schottky barrier field-effect transistor," *Proc. IEEE*, **54**: 307–308, February 1966.

[3] R. Pengelly, "Recalling early GaAs MMIC developments," *Microwaves & RF*, March 2009.

[4] R.L. Van Tuyl, "The early days of GaAs ICs," *IEEE CSICS Digest*, pp. 3–6, October 2010.

[5] M. Bohr, "Evolution of scaling from the homogeneous era to the heterogeneous era," *IEEE IEDM Digest*, pp. 1.1.4, December 2011.

[6] International Technology Roadmap for Semiconductors [ITRS], *Wireless AMS Tables*, 2011, www.itrs.net.

2 High-frequency and high-data-rate communication systems

2.1 WIRELESS AND FIBER-OPTIC COMMUNICATION SYSTEMS

Communication systems transfer information between two points (point-to-point) or from one point to multiple points (point-to-multi-point) located at a distance from each other. The distance may be anywhere from a few centimeters in personal area networks (PAN), to a few thousand kilometers in long-haul optical fiber communication systems. The information can be conveyed using carrier frequencies and energies occupying the audio, microwave, mm-wave, optical, and infrared portions of the electromagnetic spectrum. In this book, we refer to the range spanning GHz to hundreds of GHz as high-frequency. Although optical frequencies do not fall into this category, the baseband information content of most current fiber-optic systems covers the frequency spectrum from DC to tens of GHz. This makes the circuit topologies and design methodologies discussed in this book applicable to the electronic portion of fiber-optic systems.

2.1.1 Wireless versus fiber systems

Figure 2.1 illustrates the block diagrams of typical wireless and fiber-optic communication systems. They both consist of a transmitter and a receiver, a synchronization block, and a transmission medium. The information signal modulates a high-frequency (GHz to hundreds of GHz) or optical (hundreds of THz) carrier which is transmitted through the air, or through an optical fiber, to the receiver. The receiver amplifies the modulated carrier and extracts (demodulates) the information from the carrier. In both cases, an increasing portion of the system is occupied by analog-to-digital converters (ADC), digital to analog converters (DAC), and digital signal processors (DSP), operating with clock frequencies extending well into the GHz domain.

In this chapter, we will review the main system architectures and their specifications, as well as the link between the system-level specifications and the performance targets of the circuits that make up those systems. The circuit topologies and their design will be addressed in detail in the remainder of the book.

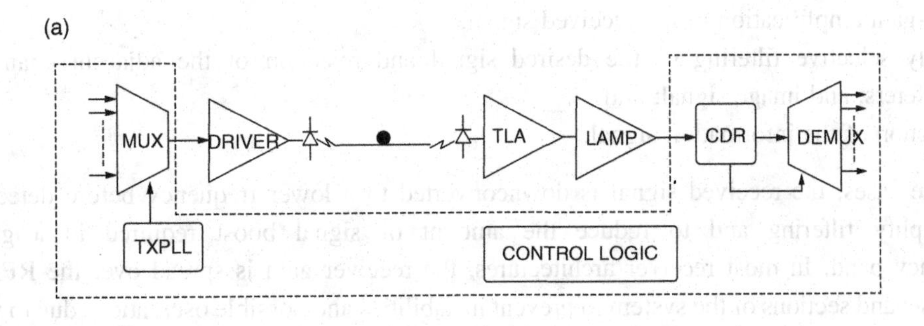

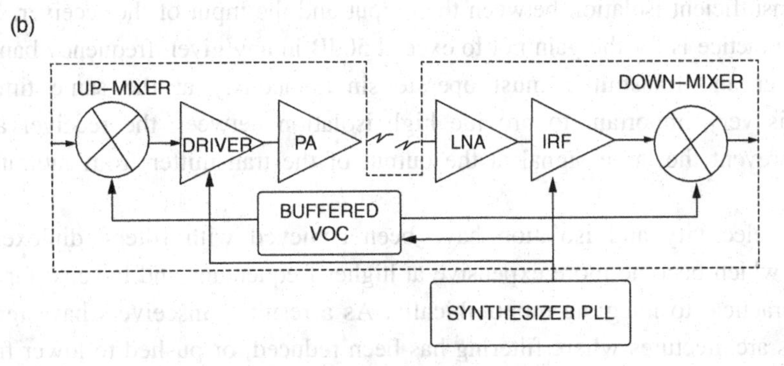

Figure 2.1 (a) Fiber-optic and (b) wireless transceivers

2.2 RADIO TRANSCEIVERS

What is a radio transceiver? A radio transceiver consists of a **trans**mitter and a re**ceiver**, hence its name: **transceiver.**

The role of the transmitter is to modulate a high-frequency carrier with the baseband information signal and to feed it to an antenna which sends it through the air (ether) to its intended destination. In its simplest form, the transmitter consists of a carrier signal generator and a modulator. Very often, a high-frequency power amplifier is also inserted in front of the antenna, after the modulator, to boost the power of the modulated signal before it is transmitted.

The receiver performs the inverse function of the transmitter. Its role is to select the desired modulated carrier signal from a cacophony of transmitting sources, interference signals, and noise, and to recover the information content after amplifying and demodulating it from the carrier. This task is usually more challenging than that of the transmitter because the received signal is very weak, typically −80dBm to −120dBm, and often drowned by much larger interferers. Almost without exception, irrespective of the type of radio application, the receiver must feature *high dynamic range, low noise, high selectivity* (to suppress undesired signals), and very *high gain.* A well-designed receiver must therefore perform several tasks:

- high-gain amplification of the received signal;
- highly selective filtering of the desired signal and rejection of the adjacent channels, interferers, and image signal; and
- detection of the information signal.

In some cases, the received signal is downconverted to a lower frequency before detection to simplify filtering and to reduce the amount of signal boost required in a given frequency band. In most receiver architectures, the receiver gain is spread over the RF, IF, and baseband sections of the system to prevent instabilities and possible oscillations due to very high gain and insufficient isolation between the output and the input of the receiver. A good receiver design practice is for the gain not to exceed 50dB in any given frequency band.

If the receiver and transmitter must operate simultaneously at the same time in a transceiver, it is very important to provide high isolation between the receiver and the transmitter to prevent the large signal at the output of the transmitter from saturating the receiver.

Historically, selectivity and isolation have been achieved with filters, diplexers, and isolators, all of which become more expensive at higher frequencies and have so far proven difficult or impractical to integrate monolithically. As a result, transceivers have tended to migrate towards architectures where filtering has been reduced, or pushed to lower frequencies, and where it can be more readily implemented monolithically, most often in the digital domain.

2.3 MODULATION TECHNIQUES

In communication systems, the sinusoidal carrier is modulated by the data signal.[1] This is accomplished by varying at least one of the 3 degrees of freedom of the carrier: amplitude, frequency, or phase. With the exception of some automotive cruise control radar systems, which employ analog modulation, most modern wireless, fiber-optic, and wireline communication systems rely on digital modulation schemes. Digital modulation is favored because of its improved performance in the presence of noise and fading and ease of implementing error correction and data encryption schemes. The type of modulation directly impacts the SNR, bandwidth, and sensitivity of the communication system. Modulation techniques form the subject of entire books; here we review only the main types of digital modulation schemes and the most important metrics required in specifying system performance.

2.3.1 Types of digital modulation

Digital modulation refers to modulation schemes where the sinusoidal carrier is switched between two states, according to the binary data symbols "1" or "0." If we consider the general expression of the sinusoidal carrier

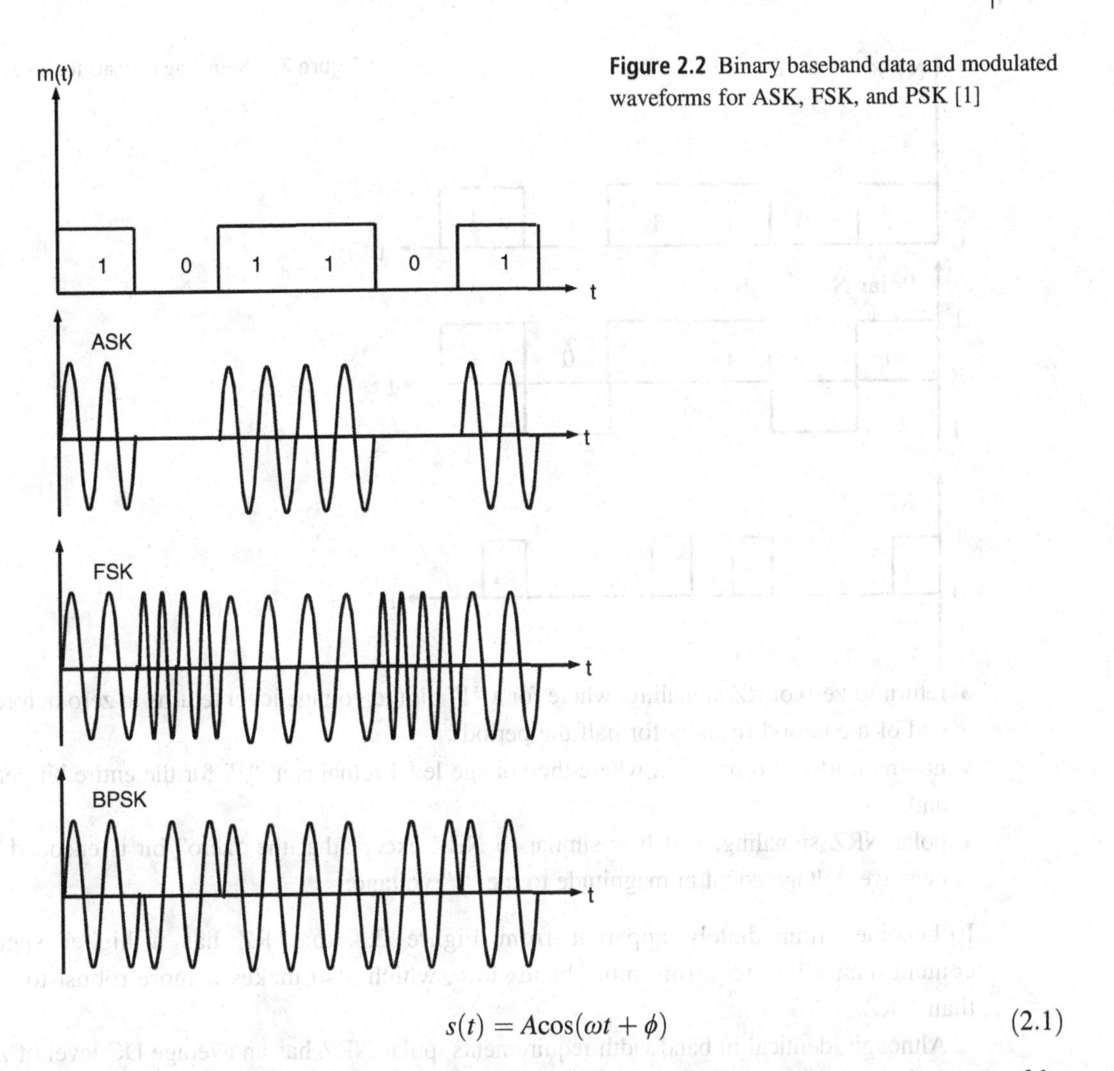

Figure 2.2 Binary baseband data and modulated waveforms for ASK, FSK, and PSK [1]

$$s(t) = A\cos(\omega t + \phi) \tag{2.1}$$

each one of the three variables: the amplitude A, the frequency ω, and the phase ϕ can take on any of the two binary states as a function of time, leading to the three fundamental binary modulation techniques:

- amplitude shift keying (ASK), also known as OOK (on-off keying),
- frequency shift keying (FSK), and
- phase shift keying (PSK).

These fundamental binary modulations schemes are illustrated graphically in Figure 2.2 along with the binary modulating signal *m(t)* which can only take the values "1" or "0." Note that BPSK stands for binary PSK, the most basic digital phase modulation method.

2.3.2 Binary signals

In all three types of modulation, as shown in Figure 2.3, the baseband data signal is a serial bit sequence, which may represent digitized voice, digitized music, digitized video, computer data, or a combination of such binary signals. The binary data are typically encoded in three ways:

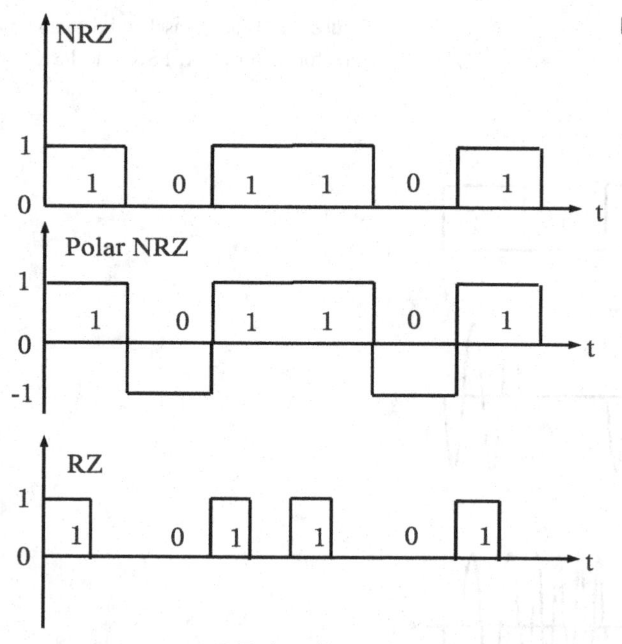

Figure 2.3 Signaling formats for binary data

- return-to-zero or RZ signaling, where for a "1" bit the voltage level returns to zero before the end of the period (usually for half the period);
- non-return-to-zero or NRZ, where the voltage level remains at "1" for the entire bit period; and
- polar NRZ signaling, which is similar to NRZ except that the "zero" bit is encoded as a negative voltage equal in magnitude to the "1" voltage.

It becomes immediately apparent from Figure 2.4 that RZ has a higher spectral content than NRZ, requiring more bandwidth, which also makes it more robust to noise than NRZ.

Although identical in bandwidth requirements, polar NRZ has an average DC level of zero, which offers practical advantages over NRZ in threshold detection.

2.3.3 Amplitude shift keying

In ASK modulated carriers, the carrier amplitude is turned on and off by the binary data. In a radio transmitter, this can be accomplished in two ways:

- by mixing (upconverting) the binary data signal, $m(t)$, with the carrier signal, $A\cos(\omega_0 t)$, or
- by placing an on-off switch after an oscillator that generates the $A\cos(\omega_0 t)$ carrier.

In both cases, the ASK modulated carrier signal is described by

$$s(t) = m(t)\cos(\omega_0 t) \text{ where } m(t) = 0, 1 \tag{2.2}$$

and has a double-sideband spectrum. Note that, throughout this section, we will assume that the carrier has a normalized amplitude $A = 1$.

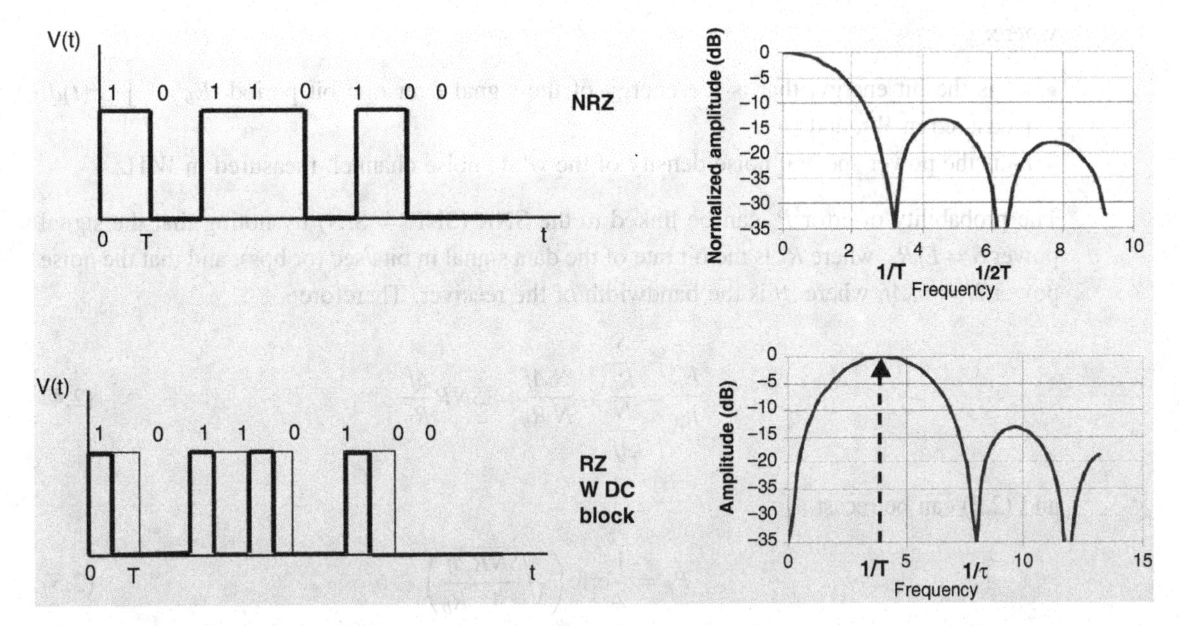

Figure 2.4 Spectral content of RZ and NRZ encoded binary data [2]

In the receiver, an ASK demodulator recovers the original data signal $m(t)$. Demodulation of ASK signals can be performed:

- synchronously (or coherently), by mixing (downconverting) the received signal with a locally generated replica of the carrier $\cos(\omega_0 t)$ and low-pass filtering

$$v(t) = s(t) \times \cos(\omega_0 t) = m(t)\cos^2(\omega_0 t) = \frac{1}{2}m(t)[1 + \cos(2\omega_0 t)] \rightarrow \frac{m(t)}{2}$$

or

- asynchronously, with an envelope or square-law detector

$$v(t) = s^2(t) = m^2(t)\cos^2(\omega_0 t) = \frac{m^2(t)}{2}[1 + \cos(2\omega_0 t)] \rightarrow \frac{m^2(t)}{2} = \frac{m(t)}{2}$$

where we have taken advantage of the fact that, for a binary signal (0,1), $m^2(t) = m(t)$.

Synchronous detection is more expensive to implement because a carrier, synchronized with the one in the transmitter, must be generated in the receiver. However, it leads to higher performance receivers because it requires a lower (by about 1dB) SNR than envelope detection for the same error rate.

It has been shown [1] that the probability of error P_e (also referred to as the bit error rate, BER) in a synchronous ASK detector is described by

$$P_e = \frac{1}{2}\text{erfc}\left(\sqrt{\frac{E_b}{4n_0}}\right) \tag{2.3}$$

where:

- E_b is the bit energy, that is the energy of the signal over one bit period, $E_b = \int_{t=0}^{T} s^2(t)dt$ measured in Ws, and
- n_0 is the power spectral noise density of the white noise channel, measured in W/Hz.

The probability of error P_e can be linked to the SNR (SNR = S/N) by noting that the signal power $S = E_b R_b$, where R_b is the bit rate of the data signal in bits/sec (or bps), and that the noise power $N = n_0 \Delta f$, where Δf is the bandwidth of the receiver. Therefore

$$\frac{E_b}{n_0} = \frac{\frac{S}{R_b}}{\frac{N}{\Delta f}} = \frac{S}{N}\frac{\Delta f}{R_b} = SNR\frac{\Delta f}{R_b} \tag{2.4}$$

and (2.3) can be recast as

$$P_e = \frac{1}{2}\text{erfc}\left(\sqrt{\frac{SNR}{4}\frac{\Delta f}{R_b}}\right). \tag{2.5}$$

Since P_e decreases as the argument of the complementary error function erfc increases, for a given receiver bandwidth and error rate, increasing the bit rate requires an increase in SNR.

2.3.4 Frequency shift keying

In the case of FSK, the carrier frequency switches between two values, ω_1 and ω_2. Such a signal can be generated using a voltage-controlled oscillator whose frequency is changed between ω_1 and ω_2, by applying the binary data signal $m(t)$ to the frequency control pin of the voltage-controlled oscillator (VCO)

$$s(t) = \cos[(\omega_2 + m(t)\Delta\omega)t]. \tag{2.6}$$

where $\Delta\omega = 2\pi\,\Delta f$ and $\Delta f = f_1 - f_2$.

The spectrum of an FSK modulated signal can be shown to have an effective bandwidth

$$B = 2\left(\Delta f + \frac{2}{T}\right) \tag{2.7}$$

where T is the period of the binary data.

As in the case of ASK, demodulation of FSK signals can be performed:

- synchronously, by mixing (downconverting) the received signal with two locally generated carriers $\cos(\omega_1 t)$ and $\cos(\omega_2 t)$, followed by low-pass filtering, as illustrated in Figure 2.5; or
- asynchronously, with two bandpass filters centered on ω_1 and ω_2, respectively, which decompose the FSK signal in two ASK modulated ones, and two envelope detectors, Figure 2.6.

Here, too, coherent detection requires lower transmitter power for the same bit error rate but is more costly to implement than envelope detection.

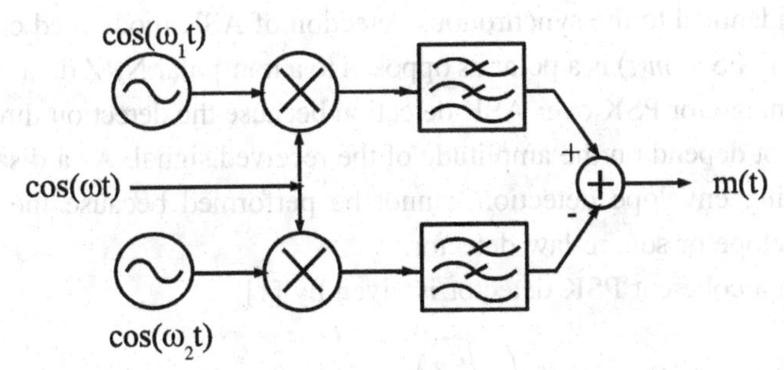

Figure 2.5 Coherent FSK detector [1]

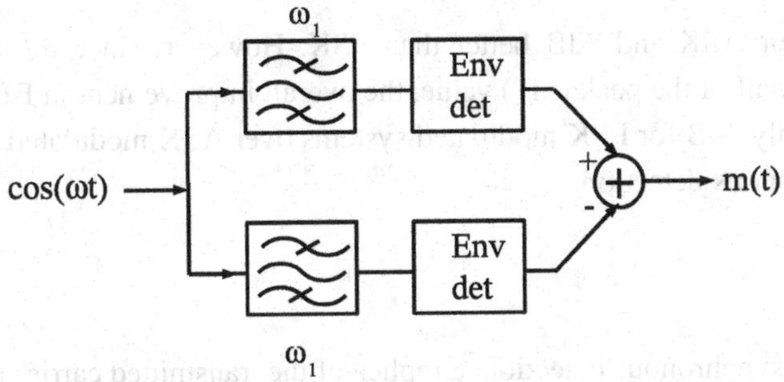

Figure 2.6 Direct FSK detection [1]

The probability of error in a synchronous (coherent) FSK detector is described by

$$P_e = \frac{1}{2}\text{erfc}\left(\sqrt{\frac{E_b}{2n_0}}\right). \tag{2.8}$$

2.3.5 Phase shift keying

In phase shift keying, the phase of the carrier is switched between 0 and 180 degrees by the binary data stream $m(t)$. This can be captured by

$$s(t) = m(t)\cos(\omega_0 t) \tag{2.9}$$

where $m(t)$ is in polar NRZ format, that is $m(t) = 1$ or -1. The generation of a PSK modulated carrier can be performed:

- synchronously by mixing the polar baseband data stream $m(t)$ with the carrier, cos $(\omega_0 t)$, or
- by direct phase (sign) modulation of a differential VCO signal.

Like FSK, PSK modulated signals have constant amplitude, making them more suitable for amplification by non-linear power amplifiers, resulting in systems with better efficiency.

Demodulation of a PSK modulated signal is performed synchronously, by mixing the received signal with a locally generated replica of the carrier, cos $(\omega_0 t)$, and low-pass filtering to obtain $m(t)$

$$v(t) = s(t) \times \cos(\omega_0 t) = m(t)\cos^2(\omega_0 t) = \frac{1}{2}m(t)[1 + \cos(2\omega_0 t)] \rightarrow \frac{m(t)}{2}. \tag{2.10}$$

We note that this process is identical to the synchronous detection of ASK modulated carriers, with the only difference that, in PSK, $m(t)$ is a polar as opposed to a non-polar NRZ data stream. The latter represents an advantage for PSK over ASK detection because the detection threshold can be set to zero and does not depend on the amplitude of the received signal. As a disadvantage, PSK demodulation using envelope detection cannot be performed because the phase information is lost in an envelope or square-law detector.

The probability of error in a coherent PSK detector is given by [1]

$$P_e = \frac{1}{2}\text{erfc}\left(\sqrt{\frac{E_b}{n_0}}\right) \tag{2.11}$$

making it 6dB better than for ASK and 3dB better than FSK. However, since the average transmitted power in ASK is half of the peak ("1") value, the overall improvement in BER for a given transmitted power is only 3dB for PSK modulated systems over ASK modulated or FSK modulated ones with synchronous detection.

2.3.6 Carrier synchronization

We note that in systems with synchronous detection, a replica of the transmitted carrier must be generated locally in the receiver, or recovered from the transmitted signals. This is usually accomplished using a phase lock loop (PLL) and a synthesizer, which are otherwise not needed in receivers with envelope detection.

2.3.7 M-ary digital modulation schemes

Each of the binary modulations schemes discussed above transmit one bit of information during each period and are, therefore, said to have a *bandwidth efficiency* of 1bps/Hz. However, as the communication traffic has continued to increase, driven by the need to transmit data and video signals in addition to voice, the available frequency spectrum has become overcrowded and higher-order modulation methods, with higher bandwidth efficiency, have been devised. Such M-ary modulation methods allow higher data rates to be transmitted over the same bandwidth by packing more than one bit per signal interval. If we transmit $M = 2^n$ bits per signaling interval, theoretically a bandwidth efficiency of n bps/Hz can be achieved. As with binary modulation schemes, all three degrees of freedom of the carrier can be modulated by the M-ary data stream, resulting in M-ary ASK (e.g. 4-PAM), M-ary FSK (e.g. 4-level FSK), M-ary PSK (e.g. QPSK, 8-PSK), and mixed amplitude-phase M-ary QAM (e.g. 16QAM, 64QAM, 256QAM, etc.) modulation schemes.

Due to the lower probability of error, M-ary PSK and QAM modulation methods are the most common modulation techniques encountered in modern digital radio and cable TV communication systems.

A digitally modulated M-PSK signal has M phase states and can be defined as [1]

$$s_i(t) = A\cos(\omega_0 t + \phi_i) \tag{2.12}$$

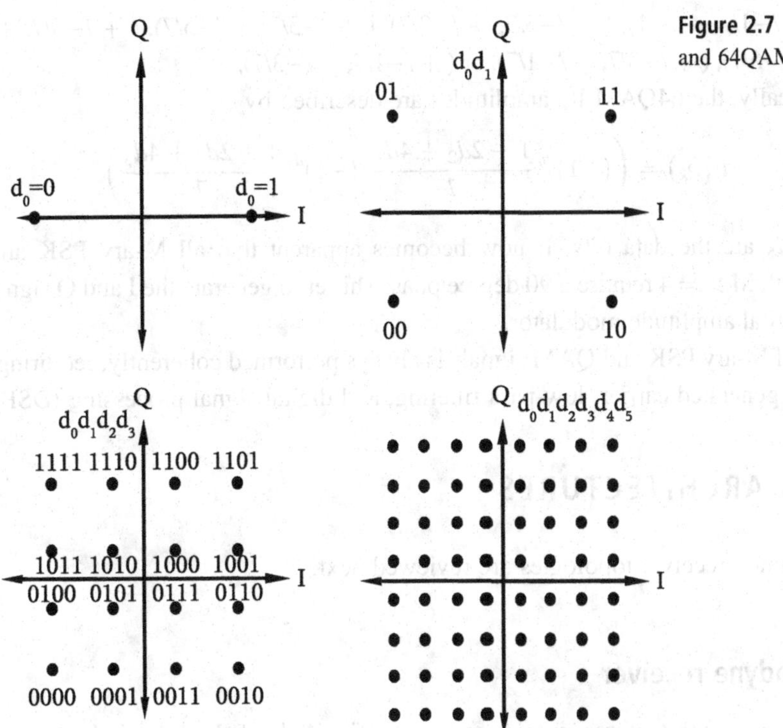

Figure 2.7 BPSK, QPSK, 16QAM, and 64QAM constellations

where
$\phi_i = 2\pi \frac{i}{M}$ for $i = 0, 1, 2 \ldots M-1$, $M = 2^n$ and n is the number of bits per symbol.

BPSK (binary-shift-keying) modulation corresponds to $M = 2$ and $n = 1$, while QPSK modulation is described by $M = 4$ and $n = 2$.

We note that M-ary phase modulation has constant amplitude and does not require very linear power amplifiers to boost the modulated carrier. The M phase states can be represented with phasors whose x-axis projection is the in-phase (I) component and the y-axis projection represents the quadrature component (Q). A QPSK modulated signal can therefore also be described as

$$s(t) = a_I\cos(\omega_0 t) + b_Q\sin(\omega_0 t) \tag{2.13}$$

where a_I and b_Q are polar NRZ binary data bits equal to either 1 or -1.

The M-ary PSK modulation can be generalized to M-ary QAM modulation by allowing the amplitudes of the I and Q components to vary

$$s_k(t) = a_k\cos(\omega_0 t) + b_k\sin(\omega_0 t). \tag{2.14}$$

For a 16QAM signal, $(a_k, b_k) = (+/-1, +/-1), (+/-1/3, +/-1/3), (+/-1, +/-1/3)$, or $(+/-1/3, +/-1)$ and can be mathematically described as a function of the data bits d_0, d_1, d_2, d_3

$$(a_k, b_k) = \left((-1)^{d_0}\frac{1+2d_2}{3}, (-1)^{d_1}\frac{1+2d_3}{3}\right). \tag{2.15}$$

For a 64QAM signal, $(a_k, b_k) = (+/-1, +/-1), (+/-5/7, +/-5/7), (+/-3/7, +/-3/7), (+/-1/7, +/-1/7), (+/-1, +/-5/7)\ (+/-5/7, +/-1), (+/-1, +/-3/7)\ (+/-3/7, +/-1), (+/-1,$

+ /−1/7) (+ /−1/7, + /−1), (+ /−5/7, + /−3/7), (+ /−3/7, + /−5/7), (+ /−5/7, + /−1/7), (+ /−1/7, + /−5/7), (+ /−3/7, + /−1/7), or (+ /−1/7, + /−3/7).

Mathematically, the 64QAM IQ amplitudes are described by

$$(a_k, b_k) = \left((-1)^{d_0} \frac{1 + 2d_2 + 4d_4}{7}, (-1)^{d_1} \frac{1 + 2d_3 + 4d_5}{7} \right) \tag{2.16}$$

where $d_0 - d_5$ are the data bits. It now becomes apparent that all M-ary PSK and QAM modulators with M > = 4 require a 90 degree phase shifter to generate the I and Q signals and a multi-level digital amplitude modulator.

Detection of M-ary PSK and QAM signals is always performed coherently, requiring mixing with a locally generated carrier, low-pass filtering, and digital signal processing (DSP).

2.4 RECEIVER ARCHITECTURES

The most common receiver topologies are reviewed next.

2.4.1 Tuned homodyne receiver

One of the earliest architectures, introduced in the first half of the twentieth century, is the *tuned radio frequency* receiver, also known as the *direct detection* receiver. As illustrated in Figure 2.8, it consists of a series of tunable bandpass RF amplifier stages followed by a square-law detector, all operating at the RF frequency, f_{RF}. At least the first gain stage must be a low-noise amplifier, LNA, whose role is to amplify the weak signal received from the antenna without degrading its signal-to-noise ratio. The bandpass frequency response can be either built in the amplifier stages or realized by placing tunable, narrowband, band-select filters (BSF) before and between the gain stages of the amplifier. Since the detector has relatively poor sensitivity, the overall voltage *gain*, A_V, of the amplifier and of the filter stages preceding it must be large enough to overcome the noise of the detector. It typically exceeds 50–60dB. We note that, to prevent signal distortion before detection, the entire receive chain, up to the decision circuit (indicated with the step function in Figure 2.8), must be linear.

The RF signal at the input of the receiver can be described as

$$s(t) = A(t)\cos(\omega_{RF} t) \tag{2.17}$$

where $A(t)$ is the relatively slow-changing envelope signal which contains the information to be recovered by the receiver, and ω_{RF} is the carrier frequency. We can assume that the spectral content of $A(t)$ extends from DC to ω_B, with $\omega_B << \omega_{RF}$. We typically call this frequency range *baseband*. The detector is realized as an envelope detector or as a square-law device, that is its DC output current or voltage is proportional to the square of the input AC signal, $s(t)$.

After filtering, amplification, and detection, the signal at the output of the detector becomes

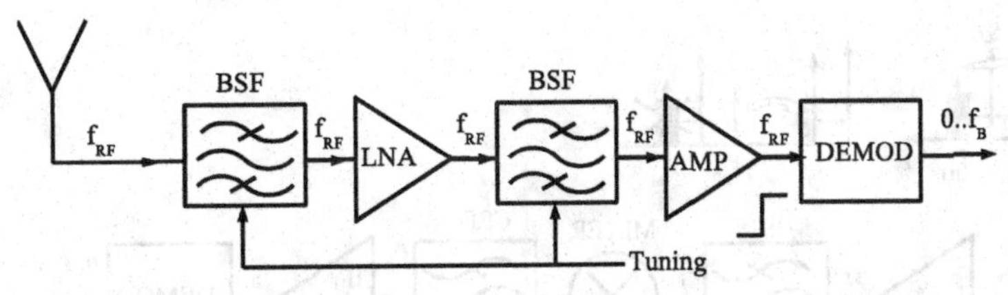

Figure 2.8 Tuned radio frequency receiver architecture

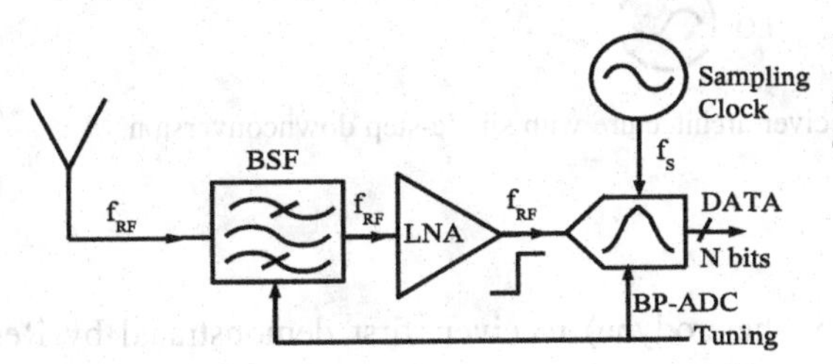

Figure 2.9 Digitally tuned radio frequency receiver architecture and RF bandpass analog-to-digital converter

$$[A_V s(t)]^2 = A_V^2 A(t)^2 \cos^2(\omega_{RF} t) = \frac{A_V^2 A(t)^2}{2}[1 + \cos(2\omega_{RF} t)]. \tag{2.18}$$

By low-pass filtering and taking the square root, we obtain the amplified replica of the information signal

$$\frac{A_V A(t)}{\sqrt{2}}.$$

As discussed earlier, for ASK modulated signals the square root function is not necessary.

The most common example of such a receiver architecture is that of an early twentieth century AM radio, where tuning was mechanical, implemented with variable capacitors or inductors. In a modern AM radio, mechanical tuning has been replaced by electronically tuned filters and, in some cases where either power dissipation or performance are not critical, by a bandpass analog-to-digital converter, BP-ADC, which also performs the detection function, as illustrated in Figure 2.9. The output of the ADC is an n-bit wide digital stream $d_n(t)$ that describes the information signal in the digital domain. Although the complexity and number of active components increases, the digital receiver can be monolithically integrated into a single chip, significantly reducing cost and form factor, and improving reliability.

Another modern example of a tuned radio frequency receiver can be found in upper mm-wave band passive imaging receivers [3–5], which will be discussed in some detail later in this chapter. Nevertheless, the main drawback of this architecture remains that of having to provide a significant amount of gain at a single frequency with extremely high selectivity, a formidable challenge at microwave and mm-wave frequencies. The high gain at RF typically leads to stability problems.

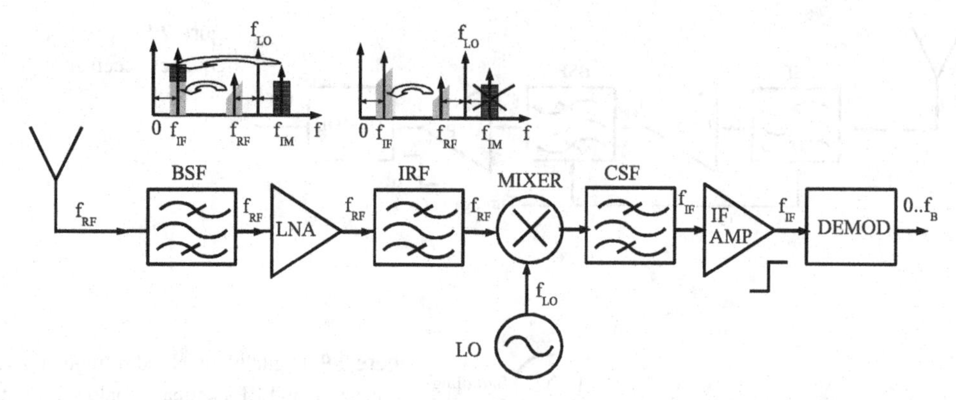

Figure 2.10 (Super) Heterodyne radio receiver architecture with single-step downconversion

2.4.2 Heterodyne receiver

The heterodyne (also known as super-heterodyne) receiver, first demonstrated by Reginald Aubrey Fessenden on Christmas Eve 1906 in Brant Rock, Massachusetts, has been the workhorse of the wireless industry for over a century. Unlike in the tuned radio frequency receiver, the RF frequency is translated to an intermediate frequency, f_{IF}, using either a single or a two-step downconversion process. As a result, the gain requirements imposed on the RF amplifier are relaxed and its stability is improved by distributing the gain across amplifier stages operating in different frequency bands, at RF and IF. As shown in Figure 2.10, following the antenna, the single-step downconversion receiver chain consists of a band-select filter (BSF), which selects the desired frequency band and rejects out-of-band interferers, a low-noise amplifier, an image-reject filter (IRF), a downconvert mixer, a channel select filter (CSF), and a demodulator. The frequency translation from RF to IF is performed by the mixer, a non-linear device, under the control of a local oscillator (LO), which operates at a frequency f_{LO}. The three frequencies encountered along the receive path f_{RF}, f_{LO}, and f_{IF} must satisfy the relationship

$$f_{RF} = f_{LO} - f_{IF} \quad \text{or} \quad f_{RF} = f_{LO} + f_{IF} \tag{2.19}$$

which indicates that signals from two bands, $f_{LO} + f_{IF}$ and $f_{LO} - f_{IF}$, will simultaneously be downconverted to IF.

In most practical systems, only one of these frequency bands contains useful information. We choose to call this band the RF frequency, to distinguish it from the other band, named the image frequency, IM, and whose content must be filtered out. The latter explains the presence of the image-reject filter in the block diagram of Figure 2.10. The IRF allows signals in the RF band to pass unobstructed while strongly attenuating signals at the image frequency f_{IM}. We note that a high f_{IF} relaxes the requirements for the IRF but must be traded off against pushing the CSF and IF amplifier to higher frequencies. If the IF frequency is high enough, then it is possible to integrate the IRF, along with the rest of the transceiver, in silicon [6].

Mathematically, the downconversion process, after the IRF can be described as the product between the RF signal $s(t) = A(t)\cos[\omega_{RF}t + \alpha(t)]$, which can be amplitude and/or phase

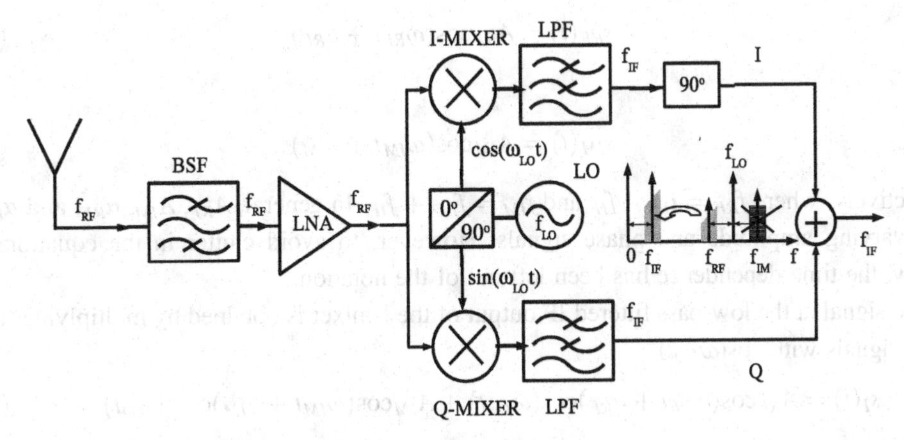

Figure 2.11 Hartley image reject radio receiver architecture

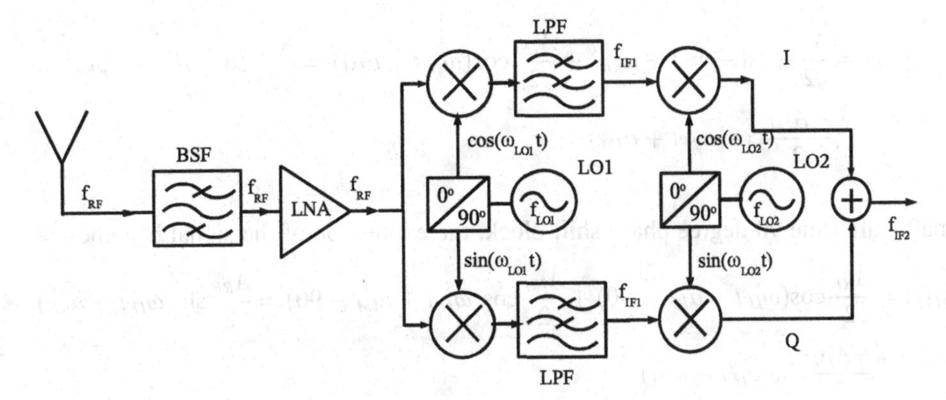

Figure 2.12 Weaver image reject radio receiver architecture

modulated, and the local oscillator signal $\cos(\omega_{LO}t)$. The signal at the input of the demodulator is obtained after filtering

$$s_{IF}(t) = A(t)\cos[\omega_{RF}t + \alpha(t)]\cos(\omega_{LO}t) = \frac{A(t)}{2}\cos[\omega_{IF}t - \alpha(t)] \tag{2.20}$$

where both the amplitude $A(t)$ and the phase $\alpha(t)$ modulating signals are translated to the IF band.

Alternatively, image-reject receiver topologies of the Hartley [7], Figure 2.11, or Weaver [8], Figure 2.12, type can be employed to eliminate the IRF. In both cases, a local oscillator signal is required that generates *in-phase* (I) and *quadrature* (Q) signals, that is signals that are 90 degrees out of phase with respect to each other. The tradeoff is higher complexity in the active circuitry, a relatively small price to pay in an IC where transistors are virtually free.

The image frequency topic and the downconversion equations will be revisited in Chapter 9. Here, we will briefly analyze the image rejection process mathematically for the topologies in Figures 2.11 and 2.12.

For the Hartley architecture, we can start by assuming that the IF and RF signals are sinusoidal and described by

$$s_{RF}(t) = A_{RF}\cos(\omega_{RF}t + \alpha_{RF}) \tag{2.21}$$

and

$$s_{IM}(t) = A_{IM}\cos(\omega_{IM}t + \alpha_{IM}) \tag{2.22}$$

respectively, where $f_{RF} = f_{LO} - f_{IF}$ and $f_{IM} = f_{LO} + f_{IF}$. In general, A_{RF}, A_{IM}, α_{RF}, and α_{IM} are time-varying amplitude and phase signals. However, to avoid clutter in the equations that follow, the time-dependence has been left out of the notation.

The signal at the low-pass filtered IF output of the I-mixer is obtained by multiplying the RF input signals with $\cos(\omega_{LO}t)$

$$s_I(t) = A_{RF}\cos(\omega_{RF}t + \alpha_{RF})\cos(\omega_{LO}t) + A_{IM}\cos(\omega_{IM}t + \alpha_{IM})\cos(\omega_{LO}t). \tag{2.23}$$

Equation (2.23) can be rearranged to separate the low-frequency from the high-frequency terms. The latter are filtered out by the LPF to obtain

$$s_I(t) = \frac{A_{RF}}{2}\cos(-\omega_{IF}t + \alpha_{RF}) + \frac{A_{IM}}{2}\cos(\omega_{IF}t + \alpha_{IM}) = \frac{A_{RF}}{2}\cos(\omega_{IF}t - \alpha_{RF}) + \frac{A_{IM}}{2}\cos(\omega_{IF}t + \alpha_{IM}) \tag{2.24}$$

Finally, after the 90 degree phase shift block, the expression of the signal becomes

$$s_I(t) = \frac{A_{RF}}{2}\cos(\omega_{IF}t - \alpha_{RF} - 90) + \frac{A_{IM}}{2}\cos(\omega_{IF}t + \alpha_{IM} - 90) = \frac{A_{RF}}{2}\sin(\omega_{IF}t - \alpha_{RF}) + \frac{A_{IM}}{2}\sin(\omega_{IF}t + \alpha_{IM}). \tag{2.25}$$

Similarly, for the Q path, we multiply the input signals by $\sin(\omega_{LO}t)$

$$s_Q(t) = A_{RF}\cos(\omega_{RF}t + \alpha_{RF})\sin(\omega_{LO}t) + A_{IM}\cos(\omega_{IM}t + \alpha_{IM})\sin(\omega_{LO}t) \tag{2.26}$$

and after rearranging and low-pass filtering, we get

$$s_Q(t) = \frac{A_{RF}}{2}\sin(\omega_{IF}t - \alpha_{RF}) - \frac{A_{IM}}{2}\sin(\omega_{IF}t + \alpha_{IM}). \tag{2.27}$$

If we add $s_I(t)$ and $s_Q(t)$, the IF components due to the image signal cancel each other while those due to the RF signal add up to obtain

$$s_{IF}(t) = A_{RF}\sin(\omega_{IF}t - \alpha_{RF}) \tag{2.28}$$

which contains both the amplitude and the phase information of the original RF signal, down-converted to the intermediate frequency.

A similar analysis can be conducted for the Weaver architecture in Figure 2.12. We can take advantage of the previous results by noting that we have already derived the expressions of the signals at f_{IF1}. Using (2.24) and (2.27), the signals before the second downconversion on the I and Q paths become

$$s_{I1}(t) = \frac{A_{RF}}{2}\cos(\omega_{IF1}t - \alpha_{RF}) + \frac{A_{IM}}{2}\cos(\omega_{IF1}t + \alpha_{IM}) \quad (2.29)$$

$$s_{Q1}(t) = \frac{A_{RF}}{2}\sin(\omega_{IF1}t - \alpha_{RF}) - \frac{A_{IM}}{2}\sin(\omega_{IF1}t + \alpha_{IM}). \quad (2.30)$$

In the second downconversion, we multiply $s_{I1}(t)$ by $\cos(\omega_{LO2}t)$ and $s_{Q1}(t)$ by $\sin(\omega_{LO2}t)$ and, after summing the results, we obtain

$$s_{IF2}(t) = \frac{A_{RF}}{2}\cos(\omega_{IF2}t - \alpha_{RF}) + \frac{A_{IM}}{2}\cos[(\omega_{IF1} + \omega_{LO2})t - \alpha_{IM}] \quad (2.31)$$

where the second term, representing the image response, is at a much higher frequency than the first term and can be removed with a low-pass filter. We are left with the first term in (2.31), which describes the downconverted RF signal.

The degree of image rejection achieved is sensitive to the amplitude mismatch and phase error between the I and the Q paths. This explains why image rejection topologies have become popular only with the advent of monolithic integration, which facilitates better component matching due to their close proximity on the die.

We wrap up the discussion of the heterodyne architecture by noting that, apart from the relaxed gain and stability, the heterodyne architecture also reduces the number of tuned RF filters by moving the channel filter to IF, where it can be realized at reduced cost and possibly integrated monolithically in the receiver. Furthermore, unlike the tuned radio frequency receiver, which is used with AM modulation, the heterodyne receiver can be deployed in systems with amplitude, frequency, phase, and QAM modulated carriers.

2.4.3 Direct-conversion receiver

The direct-conversion receiver, Figure 2.13, also known as zero-IF or homodyne, was invented by F. M. Colebrooke in 1924 [9]. As the name indicates, the modulated RF carrier is directly downconverted to baseband through a single mixing process. Therefore, it can be regarded as a special case of heterodyne receiver in which $f_{RF} = f_{LO}$ and $f_{IF} = 0$. The image problem disappears since the information signal acts as its own image. Consequently, the IRF is no longer needed while the CSF is replaced by a low-pass filter (LPF), which can be easily integrated in silicon. Moreover, for AM modulated signals, no further detection is required, as can be observed from (2.20) by making $f_{IF} = 0$ and $\alpha(t) = ct$. In practice however, the RF and LO frequencies are not perfectly equal and may drift over time, requiring a tracking loop.

In the case of frequency and phase modulated signals, a quadrature (or IQ) downconversion mixer is needed to recover the phase information, as illustrated in Figure 2.14. This architecture is similar to that of the Hartley receiver with the difference that the I and Q paths are processed independently as the real and imaginary parts of a complex modulation signal $y(t) = a_I + jb_Q$. The real part, a_I, is obtained from the real RF input signal $s(t) = a_I\cos(w_{RF}t) + b_Q\sin(\omega_{RF}t)$ after downconversion and low-pass filtering on the I path, while the imaginary part, b_Q, is recovered in a similar manner on the Q path.

The main advantage of the direct-conversion receiver over a heterodyne architecture is its simplicity, potentially leading to low-cost and low-power consumption. It is not surprising that

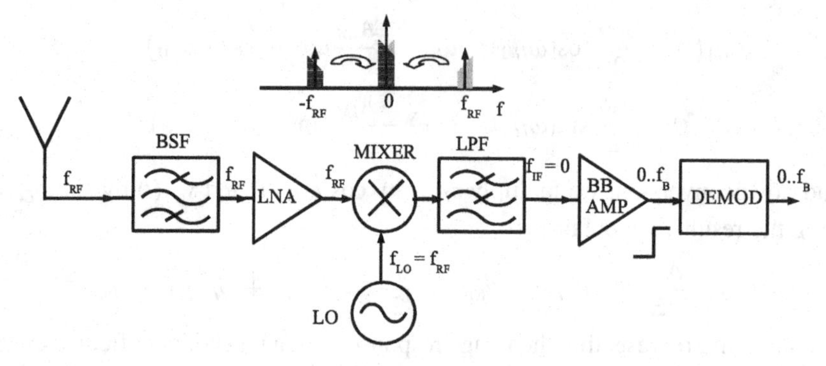

Figure 2.13 Direct conversion radio receiver architecture

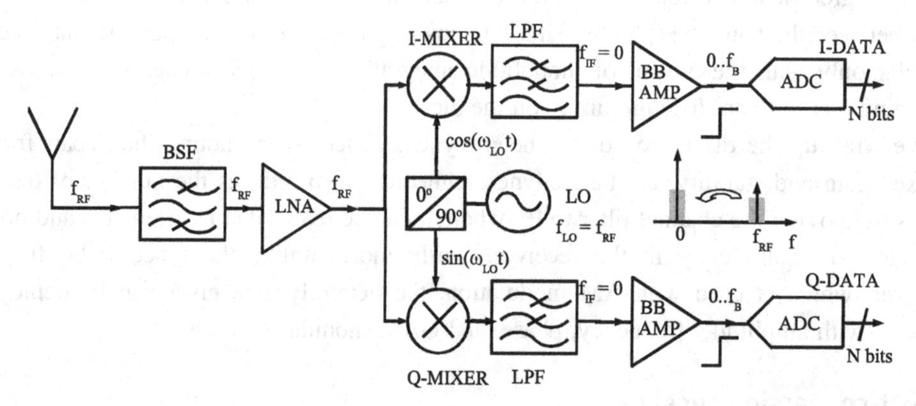

Figure 2.14 Direct IQ conversion radio receiver architecture with baseband digital signal processing

recent wireless and automotive radar standards have been conceived with a direct-conversion architecture in mind, that can be easily integrated in silicon.

Despite its advantages, there are a number of problems that plague the direct-conversion architecture. These include: (i) DC offset, (ii) LO leakage and self-mixing, (iii) LO stability and phase noise, (iv) LO pulling, (v) sensitivity to $1/f$ noise and even-order non-linearity, and (vi) degradation of receiver noise figure and sensitivity in monostatic radar receivers due to transmitter leakage [10]. While some of these issues are present in other architectures, they are exacerbated in the direct-conversion receiver and must be addressed at the circuit level.

2.5 TRANSMITTER ARCHITECTURES

The transmitter largely performs the reverse functions of the receiver. The key components of the transmitter are the modulator and the power amplifier, PA. The modulation of the carrier by the information signal can be performed:

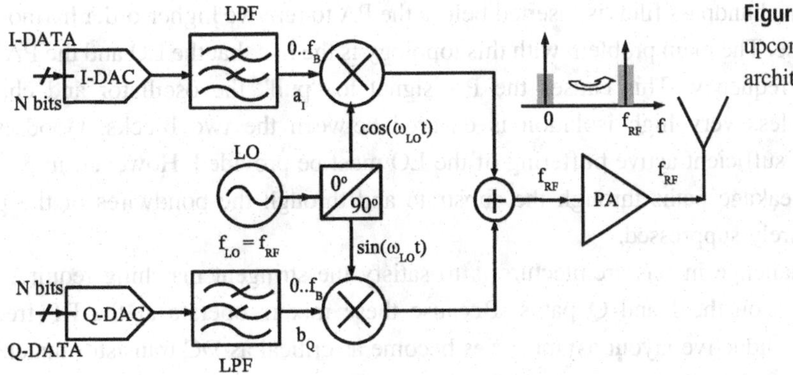

Figure 2.15 Direct IQ upconversion radio transmitter architecture

- linearly by frequency translation using a mixer, also known as an upconverter, or
- by directly modulating the carrier amplitude, frequency, phase, or pulse width.

In turn, direct modulation can be either linear or implemented in a non-linear fashion, most often and most recently by employing some form of high-frequency digital-to-analog converter (DAC).

Depending on the type of modulation, the PA operates in a linear or in saturated regime. The regime in which the PA operates has important ramifications for the efficiency and power consumption of the entire transmitter, as well as for the feasibility of integrating the PA monolithically along with the other blocks of the transmitter. In addition to providing the necessary signal level to the antenna, the role of the PA is also to ensure impedance matching to the antenna.

During the last decade, the transmitter architecture has undergone the most significant transformation from analog to digital content, with some modern transmitters being realized as RF digital-to-analog converters, (RF-DAC) [11].

2.5.1 Direct upconversion transmitter

This architecture, shown in Figure 2.15 in its quadrature version, is the transmit equivalent of the direct-conversion receiver. Although it features several analog blocks, each having to satisfy stringent linearity requirements, its main advantage lies in the elimination of the IF and RF filters, making it ideal for integration in silicon technologies. The baseband section consists of I and Q DACs, which synthesize the real and imaginary parts of the complex information signal $y(t) = a_I + jb_Q$. The analog signals thus obtained are pulse-shaped, to reduce their bandwidth to the minimum necessary, and low-pass filtered to remove sampling products at harmonics of the DAC sampling clock. Pulse-shaping and low-pass filtering are often performed in the analog domain, but, more recently, also in the digital domain [11]. In both situations, a_I and b_Q are analog signals. Next, they are both upconverted to the RF band as a real signal

$$s(t) = (a_I + jb_Q)e^{-j\omega_{LO}t} = a_I\cos(\omega_{LOF}t) + b_Q\sin(\omega_{LO}t). \tag{2.32}$$

In most cases, a bandpass filter is inserted before the PA to remove higher-order harmonics and wideband noise. The main problem with this topology is the fact that the LO and the PA operate at the same frequency. This causes the PA signal to "pull" the oscillator and change its frequency, unless very high isolation is ensured between the two blocks. Good isolation mandates that sufficient active buffering of the LO must be provided. However, in monolithic transmitters, leakage paths through the substrate and through the bondwires of the package cannot be entirely suppressed.

Another challenge in this architecture is to satisfy the stringent matching requirements of the two mixers on the I and Q paths. Because these mixers operate at the RF frequency, capacitive and inductive layout asymmetries become as critical as DC transistor matching and DC offsets.

2.5.2 Single-sideband, two-step upconversion transmitter

The single-sideband, two-step upconversion architecture, shown in Figure 2.16, is perhaps the most common transmitter topology and represents the equivalent of the (super)heterodyne receiver. Although it features additional IF and RF filters when compared to the direct upconversion transmitter, the LO pulling problem is alleviated because the PA and the LO operate at different frequencies. As in the previous transmitter architecture, the complex baseband data signal is directly upconverted. However, the upconversion is to the IF band. A real signal is obtained which is then bandpass filtered to remove higher-order harmonics

$$s_{IF}(t) = (a_I + jb_Q)e^{-j\omega_{IF}t} = a_I\cos(\omega_{IF}t) + b_Q\sin(\omega_{IF}t). \quad (2.33)$$

A second upconversion is performed next to translate $s_{IF}(t)$ to the desired RF band. In this example, $f_{RF} = f_{LO2} - f_{IF}$

$$s(t) = [a_I\cos(\omega_{IF}t) + b_Q\sin(\omega_{IF}t)]\cos(\omega_{LO2}t) = \frac{a_I}{2}\cos(\omega_{RF}t) + \frac{b_Q}{2}\sin(\omega_{RF}t)$$
$$+\frac{a_I}{2}\cos[(\omega_{LO2} + \omega_{IF})t] + \frac{b_Q}{2}\sin[(\omega_{LO2} + \omega_{IF})t]. \quad (2.34)$$

Since a conventional mixer topology, without image rejection, is employed in the second upconversion, an image-reject filter, similar to the one in the receiver, is placed between the mixer and the power amplifier to suppress the terms at $f_{LO2} + f_{IF}$ in (2.34). Apart from the large number of bulky and difficult-to-integrate IF and RF filters, this transmitter topology also demands very high linearity from all the active components in the chain. An advantage of this architecture stems from the fact that the IQ mixers operate at a relatively low frequency, where their characteristics can be more easily matched. Finally, it should be noted that, with a judicious choice of IF frequency, the characteristics of the image-reject filter can be relaxed and the entire transmitter can be integrated monolithically [6],[12].

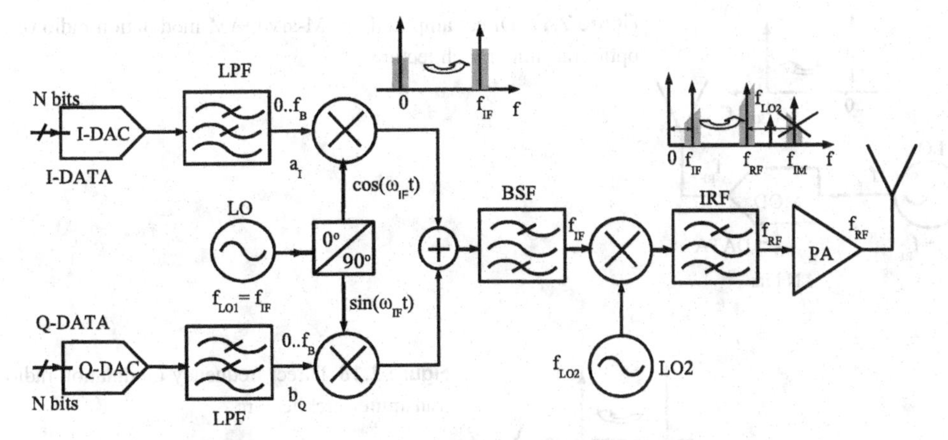

Figure 2.16 Two-step IQ upconversion radio transmitter architecture with baseband digital signal processing

2.5.3 Direct modulation transmitter

The first and arguably also the most modern type of architecture encountered in wireless and fiber-optic transmitters employs direct modulation. Originally, the oscillator signal was modulated in amplitude, frequency, or phase using analog techniques. More recently, the modulation and the modulator itself have been realized with digital techniques. Several direct modulation architectures have been demonstrated where the LO is the only analog signal. However, digitally controlled oscillators are now very common, opening the path for a fully digital transmitter.

Modern digital transmitters with direct modulation employ (i) ASK, BPSK, QPSK [13], or M-ary QAM [14] modulators placed between the oscillator and the PA, as illustrated in Figure 2.17, or (ii) apply a digital word to the control voltage of a VCO (which thus becomes a digitally controlled oscillator or DCO) to modulate its frequency, as in GSM or Bluetooth systems, illustrated in Figure 2.18.

In the case of constant envelope signals, such as those with frequency or phase (BPSK or QPSK) modulation, the PA can operate in non-linear mode with very high efficiency. In transmitters with amplitude, M-ary QAM or OFDM[2] modulation, the PA must be very linear, which inevitably reduces its efficiency because it has to be operated far below its saturated output power bias.

The most recent and most versatile of the transmitters with direct digital modulation, which, theoretically, can satisfy multiple standards, including OFDM, are those based on the IQ RF-DAC architecture shown in Figure 2.19 [11]. The RF-DAC itself, to be discussed in Chapter 9, is composed of binary-weighted direct BPSK modulators. In many ways, the architecture in Figure 2.19 is the fully digital equivalent of the IQ direct upconversion transmitter architecture.

The real RF signal at the output of the transmitter can be described as a function of the complex digital stream $y[k] = a_k + jb_k$

$$s(t) = a_k\cos(\omega_{LO}t) + b_k\sin(\omega_{LO}t) \tag{2.35}$$

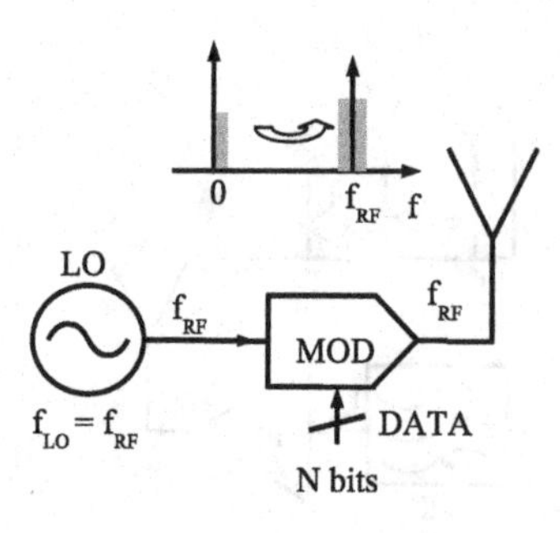

Figure 2.17 Direct amplitude or M-ary QAM modulation radio or fiber-optic transmitter architecture

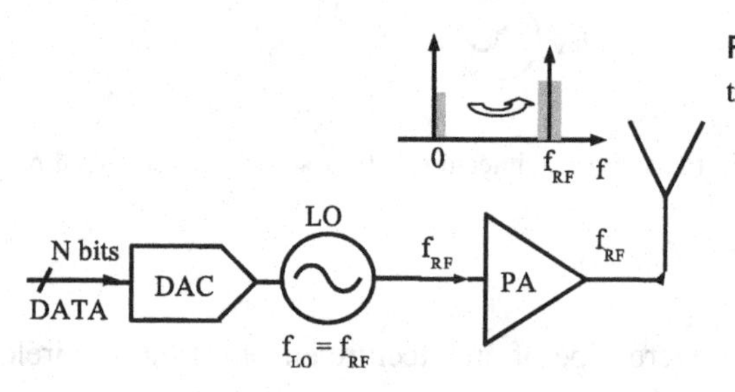

Figure 2.18 Direct frequency modulation radio transmitter architecture

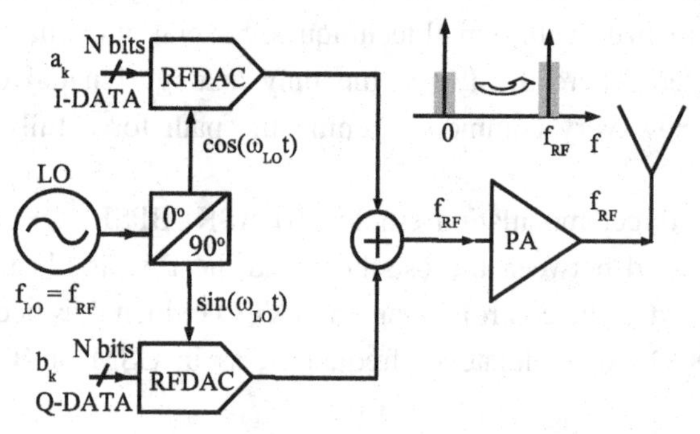

Figure 2.19 IQ RF DAC radio transmitter architecture

where a_k and b_k are the kth samples of the I and Q data streams, respectively.

It is worthwhile to examine the benefits and disadvantages of using a direct digital modulation architecture:

- A direct digital modulator with sufficient output power allows the system to operate in saturated mode, with maximum efficiency, and with the output signal swing constrained only by the reliability limit of the transistors [15].
- It simplifies the baseband circuitry, which can be implemented entirely digitally. In some cases, this can lead to a reduction in power dissipation but also places demands that are more stringent on the transistor switching speed.
- Operation over multiple standards with different modulation schemes becomes possible.
- The pulse-shaping and low-pass filtering can be performed entirely in the digital domain [11],[16].

The main disadvantage for all types of direct modulators is the stability of the local oscillator, a similar problem to that of the direct upconversion transmitter. Pulling of the VCO frequency by the modulator and/or PA is exacerbated. A second drawback is related to the speed of the digital technology and the layout matching and parasitics issues related to distributing many data and, especially, LO signals to the individual cells that make up the modulator. The latter will be addressed in some detail in Chapter 9.

2.6 RECEIVER SPECIFICATION

The main design specifications for a receiver refer to the *frequency of operation*, *dynamic range*, *gain*, *power consumption*, and, depending on the architecture, *image rejection*. The dynamic range is specified in terms of *sensitivity* and *linearity*.

2.6.1 Fundamental limitations of dynamic range

The *dynamic range* of tuned (narrow band) and broadband systems is limited by the *noise floor* and by the *breakdown voltage* of the semiconductor devices employed to realize them.

The former defines the lower end of the dynamic range as the minimum signal level that can be distinguished from background noise and which can still be processed by the electronic system.

The latter defines the maximum signal amplitude that can be processed linearly, without adding distortion. Both degrade with increasing frequency of operation and data rate, as illustrated in Figure 2.20.

Unfortunately, a direct tradeoff exists between transistor speed and transistor breakdown. In any given semiconductor technology, faster transistors exhibit lower breakdown voltage. The breakdown voltage, V_{BR}, is related to the cutoff frequency, f_T, of the transistor through the "Johnson limit"[3]

$$f_T \times V_{BR} = ct. \tag{2.36}$$

The constant in (2.36) is a fundamental limit of the semiconductor material and the type of transistor (FET or HBT). It can be improved by employing transistors fabricated in semiconductor materials with larger bandgap such as GaAs, SiC, GaN, or, ultimately, diamond. This explains why GaN, and to a lesser degree SiC and diamond, are currently being pursued for power amplifiers.

2.6.2 Noise, noise figure, and noise temperature

Noise is critical to the operation and performance of most RF and analog integrated circuits encountered in communication systems and radar sensors because it ultimately determines the threshold for the minimum signal that can be reliably detected by the receiver. Noise power is introduced in the receiver in two ways:

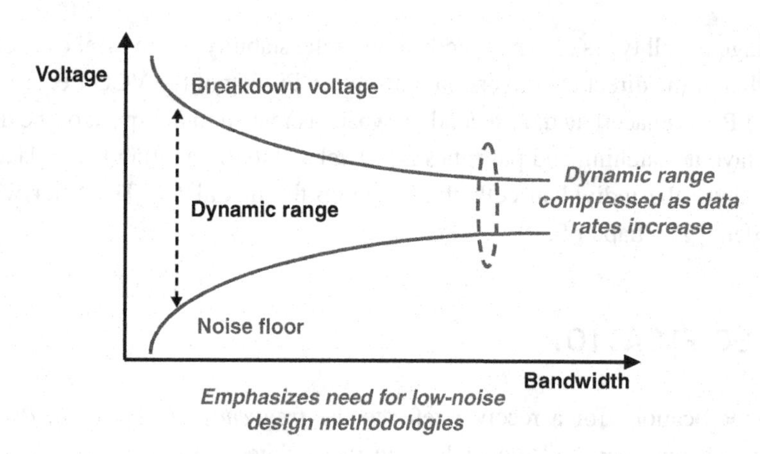

Figure 2.20 Fundamental limitations of dynamic range

- from the *external environment* through the receiving antenna or signal source impedance, and
- by *internal generation* in the receiver's own circuitry.

Noise in electronic circuits is caused by random processes such as the flow of electronic charges through potential barriers or by thermal vibrations in active and passive electronic components at ambient temperatures above absolute zero.

The most important types of noise sources encountered in semiconductor devices and integrated circuits are classified as *thermal*, *shot*, and *flicker* noise.

Thermal noise, also known as Johnson or Nyquist noise, is present in resistive (lossy) components (e.g. resistors, base and emitter resistance of BJTs/HBTs, gate and source resistance of MOSFETs). It is caused by the random vibration and motion of carriers due to their finite temperature, T. Its power, $kT\Delta f$, increases with T and bandwidth. The spectral density, kT, can be considered constant up to at least a few hundred GHz.

Shot noise is caused by the random fluctuations of charge carriers passing through potential barriers (e.g. in diodes, BJTs, and HBTs). It occurs in active devices. Its spectral density, $2qI$, is constant up to extremely high frequencies beyond 300GHz.

Flicker noise, or $1/f$ noise, has a $1/f$ power spectral density and its physical origins are not fully understood. It occurs in active devices and, sometimes, in resistors. It is only relevant at low frequencies, that is $<$ 1MHz. However, in non-linear circuits such as mixers and oscillators, flicker noise is upconverted to very high frequencies and seriously affects the performance of wireless systems.

Thermal noise was first predicted by Albert Einstein in 1906 and was experimentally observed in resistors by Johnson in 1928 and quantified at about the same time by Nyquist [17].

The *available noise power* is defined as the power that can be transferred from a noise source to a *conjugately matched load*, whose physical temperature is 0K, and thus is *unable to reflect* back noise power

$$P_{available} = \frac{\overline{v_n^2}}{4R} = \frac{4kTR\Delta f}{4R} = kT\Delta f \tag{2.37}$$

Although at first glance this may appear counterintuitive, the available noise power from a device/circuit/body/antenna does not depend on its size. In fact, all objects in the universe, whose temperature is larger than 0K, emit broadband thermal radiation whose power depends solely on their temperature and the bandwidth of the observation. This radiation can be detected as thermal noise using broadband low-noise receivers, known as radiometers. For this reason, matched loads, cooled or heated, are often used as noise sources. It should be noted that (2.37) is only an approximation [17]. This topic will be discussed in more detail in Chapter 3.

Noise factor and noise figure

The noise factor F of a two-port (e.g. amplifier, receiver, transistor, etc.) is defined as the signal-to-noise ratio at its input, SNR_i divided by that at its output SNR_o

$$F = \frac{SNR_i}{SNR_o} = \frac{SNR_i}{\dfrac{GP_i}{N_a + GN_i}} = \frac{\dfrac{P_i}{N_i}}{\dfrac{GP_i}{N_a + GN_i}} = 1 + \frac{N_a}{GN_i} \tag{2.38}$$

where:

- G is the power gain of the two-port ($G = A_V A_I$),
- N_i is the input noise power available from the antenna,
- P_i is the input signal power,
- N_a is the noise power added by the two-port.

Noise figure, NF, is the term used to describe the value of the noise factor in dB: $NF = 10\log_{10}(F)$.

Noise temperature

Because of its fundamental nature, thermal noise is used to characterize all other types of noise sources. We can define an equivalent noise temperature, T_e, for a semiconductor device, circuit, or for an entire receiver, as if all the noise generated from that circuit is due to a thermal noise source

$$T_e = \frac{N_a}{kG\Delta f} \tag{2.39}$$

This, in turn allows us to find the relationship between the noise figure and the equivalent noise temperature of the two-port

$$F = 1 + \frac{T_e}{T} \tag{2.40}$$

$$(F - 1) \times T = T_e \tag{2.41}$$

where T is the ambient temperature of the signal source. In satellite systems, T can be much lower than the ambient temperature of the receiver. For example, T may be:

- 290K for a terrestrial antenna,
- 30 … 50K for an antenna pointed at a satellite, and
- > 290K for a noise diode or PIN diode (neither of which are thermal noise sources).

Note that the signal and noise impedance of a signal source are, in general, not equal, precisely because the noise in the signal source is not necessarily of thermal origins.

Since the noise figure of a device or circuit is a function of the ambient temperature, noise figure measurements must specify the ambient temperature at which the measurement occurred.

2.6.3 Noise factor and noise temperature of a chain of two-ports

In the receiver, we are interested in calculating the noise factor of the entire receiver chain as a function of the noise factor and gain of the individual blocks that form the receiver. Such a formula was developed by Friis, at Bell Labs in 1942, and is now known as Friis' cascaded noise figure formula

$$F = F_1 + \frac{F_2 - 1}{Ga_1} + \ldots + \frac{F_n - 1}{Ga_1 \times Ga_2 \times \ldots Ga_{n-1}} \tag{2.42}$$

$$T_e = T_{e1} + \frac{T_{e2}}{Ga_1} + \ldots + \frac{T_{en}}{Ga_1 \times Ga_2 \times \ldots Ga_{n-1}}. \tag{2.43}$$

These formulas can be derived simply by calculating the noise power at the output of the chain due solely to the noise added by the two-ports in the chain

$$N_{out} = (\ldots((kT_{e1}\Delta f Ga_1 + kT_{e2}\Delta f)Ga_2 + kT_{e3}\Delta f)Ga_3 + \ldots + T_{en}\Delta f)Ga_n \tag{2.44}$$

and then dividing it by $k\Delta f\,G$ where G is the power gain of the chain $G = Ga_1 \times Ga_2 \times \ldots\, Ga_n$

$$T_e = \frac{N_{out}}{k\Delta f Ga_1 \times Ga_2 \times \ldots Ga_n} = T_{e1} + \frac{T_{e2}}{Ga_1} + \frac{T_{e3}}{Ga_1 \times Ga_2} + \ldots + \frac{T_{en}}{Ga_1 \times Ga_2 \times \ldots Ga_{n-1}} \tag{2.45}$$

For an infinite cascade of identical two-ports, $F - 1$ converges to the two-port *noise measure*

$$M = \frac{F_i - 1}{1 - \frac{1}{Ga_i}}. \tag{2.46}$$

The latter is an important figure of merit for determining the optimal sequence in which two-ports should be cascaded in a chain, such that the noise factor of the entire chain is minimized. It can be demonstrated that, in this case, the two-ports must be cascaded in increasing order of their noise measure. The one with the lowest noise measure should be placed at the input of the receiver, while the one with the highest noise measure should be placed last.

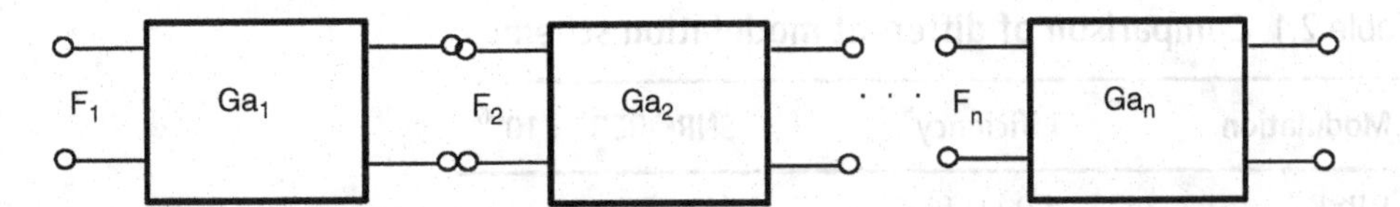

Figure 2.21 Cascade of conjugately matched two-ports in a receiver used to derive Friis' fromula

2.6.4 Receiver noise floor and sensitivity

The noise floor is defined as the noise power measured at the output of the receiver, before the decision circuit or the demodulator, and is expressed as

$$Noise\ Floor = kT\Delta fGF \tag{2.47}$$

where G is the overall power gain of the receive chain and F is its noise factor.

The sensitivity S_i is measured using a bit error rate tester, BERT, and is defined as a function of the receiver noise factor and of the SNR required at the input of the detector to achieve the desired bit error rate (probability of error)

$$S_i = F \times SNR_{RX} \times k \times T \times \Delta f \tag{2.48}$$

or as a function of noise temperature

$$S_i = \left(1 + \frac{T_e}{T}\right) SNR_{RX} \times k \times T \times \Delta f. \tag{2.49}$$

More often, the expression in dBm (decibels relative to 1mW of input power) is preferred

$$S_i(\text{dBm})\,@\,290\text{K} = -174\text{dBm} + NF(\text{dB}) + 10\log\Delta f(\text{Hz}) + SNR(\text{dB}). \tag{2.50}$$

The relationship between the bit error rate and SNR in digitally modulated carriers is described by the error function which, for large values of x, simplifies to

$$\text{erfc}(x) \approx \frac{\exp(-x^2)}{\sqrt{\pi}x} \tag{2.51}$$

For example, for ASK modulation

$$BER = \frac{1}{\sqrt{2\pi}} \frac{\exp\left[-\left(\frac{E_b}{N_o}\right)^2/2\right]}{\frac{E_b}{N_o}} \tag{2.52}$$

Table 2.1 summarizes the bandwidth efficiency and SNR required for proper detection of signals with different types of digital modulation at a bit error rate of 10^{-6}. Typical BER versus carrier-to-noise C/N (i.e. SNR) curves for BPSK, QPSK, and 16QAM are shown in Figure 2.23.

Table 2.1 **Comparison of different modulation schemes.**

Modulation	Efficiency*	SNR@BER = 10^{-6}
BPSK	1.0 (1) bits	12.5dB
4L FSK	1.5 (2) bits	17dB
QPSK	1.6 (2) bits	14dB
8PSK	2.5 (3) bits	19dB
16QAM	3.2 (4) bits	21dB
64QAM	5.0 (6) bits	27dB

Note: *Bandwidth efficiency = data rate/bandwidth = $\frac{R_b}{\Delta f}$ (ideal).

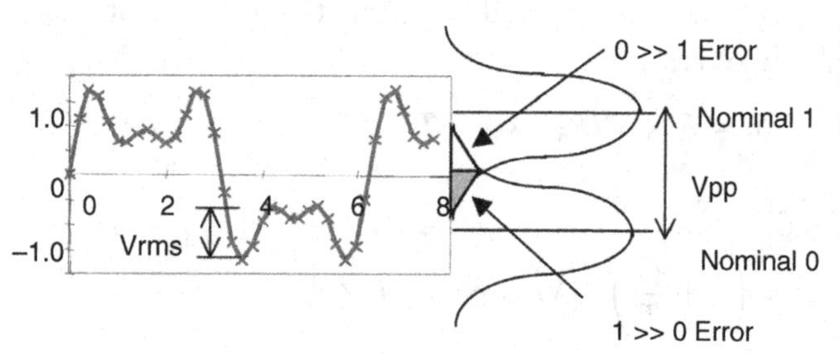

Figure 2.22 Illustration of noisy binary signals and the definition of the rms noise voltage on the "0" and "1" levels

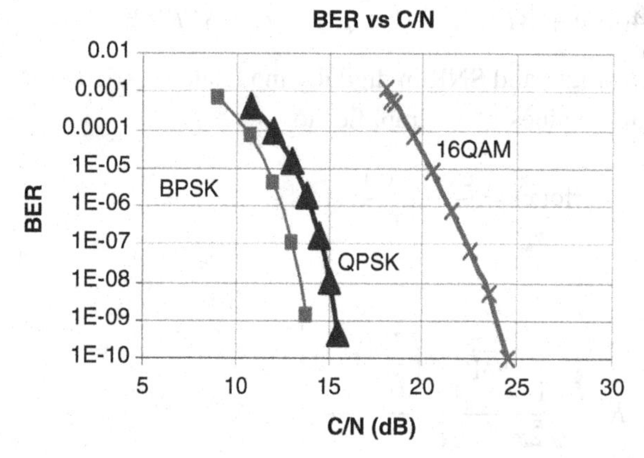

Figure 2.23 Bit error rate vs. SNR for BPSK, QPSK, and 16QAM signals [3]

EXAMPLE 2.1 5GHz wireless LAN system

Let us suppose that we have a receiver with $NF = 6\text{dB}$, $\Delta f = 20\text{MHz}$. We wish to calculate the receiver sensitivity in the case where QPSK modulation with an $SNR_{RX} = 14\text{dB}$ is employed. The latter corresponds to a bit error rate (BER) of 10^{-6}. By applying (2.50), we obtain

$$S_i = -174\text{dBm} + 6 + 10\log\left(\frac{20\text{MHz}}{1\text{Hz}}\right) + 14 = -174 + 6 + 73 + 14 = -81\text{dBm}.$$

EXAMPLE 2.2 60GHz , 1.5Gb/s wireless LAN system

Again, let us assume that $NF = 9\text{dB}$ and $\Delta f = 1\text{GHz}$ and QPSK modulated data with a required SNR of 14dB are received. The receiver sensitivity is calculated as

$$S_i = -174\text{dBm} + 9 + 10\log\left(\frac{1\text{GHz}}{1\text{Hz}}\right) + 14 = -174 + 9 + 90 + 14 = -61\text{dBm}.$$

EXAMPLE 2.3 12GHz satellite receiver

In this case, the antenna temperature T is 30K, the receiver noise figure measured at room temperature is 1dB, $\Delta f = 6\text{MHz}$ and 64QAM modulation with a receiver SNR of 27dB is employed. Since the antenna temperature is different from the temperature at which the receiver noise figure was measured, to determine the receiver sensitivity, we must first calculate the receiver noise temperature T_e and its noise figure at 30K

$$T_e = 300 \times \left(10^{\frac{0.5}{10}} - 1\right) = 300 \times (1.122 - 1) = 36.6\text{K}$$

and

$$NF_{30\text{K}} = 10 \times \log\left(1 + \frac{36.6}{30}\right) = 3.46\text{dB}$$

We can now calculate the receiver sensitivity using (2.49) with $T = 30\text{K}$ and $NF = 3.5\text{dB}$

$$S_i = -184\text{dBm} + 3.5 + 10\log\left(\frac{6\text{MHz}}{1\text{Hz}}\right) + 27 = -184 + 3.5 + 68 + 27 = -85.5\text{dBm}.$$

It is interesting to note that in optical fiber systems, which employ OOK (ASK), a different jargon is used to link the bit error rate to the sensitivity. The equivalent noise current i_n^{rms} at the input of the fiber-optic receiver replaces the noise figure, and Q, the eye quality factor, replaces E_b/N_o. The optical receiver sensitivity, S_i is expressed as

$$S_i = \frac{Qi_n^{rms}}{R} \tag{2.53}$$

where R is the photodiode responsivity (A/W). The relationship between Q and the bit error rate is plotted in Figure 2.24 and given by

$$BER \approx \frac{1}{\sqrt{2\pi}} \frac{\exp\left[-Q^2/2\right]}{Q}. \tag{2.54}$$

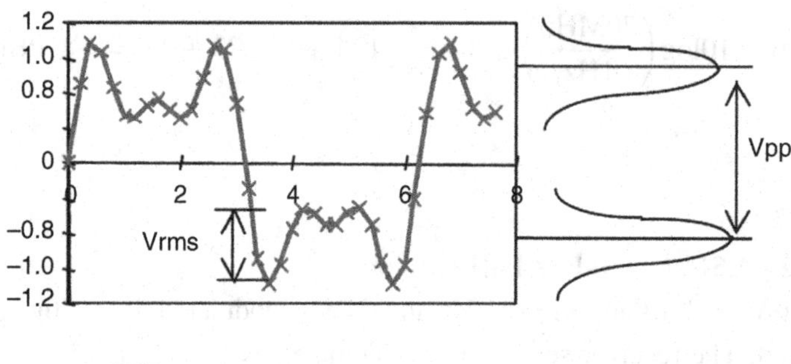

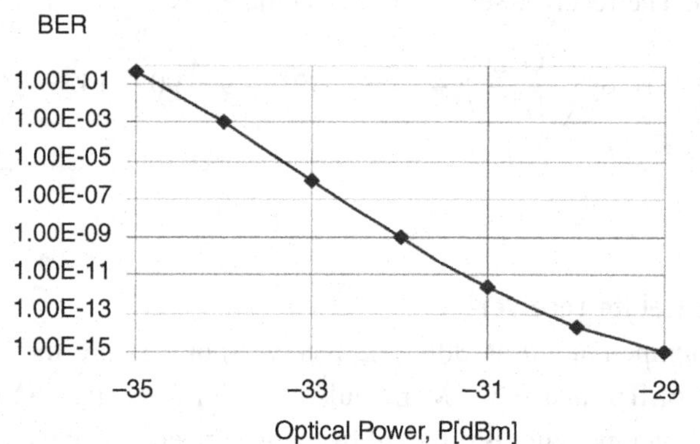

Figure 2.24 Typical variation of the bit error rate as a function of the received optical power in a fiber-optic receiver

2.6.5 Linearity figures of merit

The ideal linear transistor or two-port does not exist in the sense that the signal at its output is always exactly proportional to the signal at its input. In reality, even "linear" electronic devices and two-ports exhibit non-linear output versus input transfer characteristics at very low input power levels when the output signal remains below the noise level (*noise floor*). In addition, at very large input signal levels, all practical devices become non-linear and the output signal level will start to saturate. The latter effect is known as *gain compression*. The noise floor and the onset of gain compression set a minimum and maximum realistic power range, or *dynamic range*, over which a linear component or two-port will operate as desired (i.e. in linear mode).

A graphical method often used to characterize the linearity of a device, or of a circuit which operates over a narrow frequency band, is illustrated in Figure 2.25, also known as the IIP3 plot, where:

- IM3 is the power of the third-order intermodulation products at the output;
- the input 1dB compression point, P_{1dB}, is defined as the input power at which the power gain, G, decreases by 1dB from its small signal value;

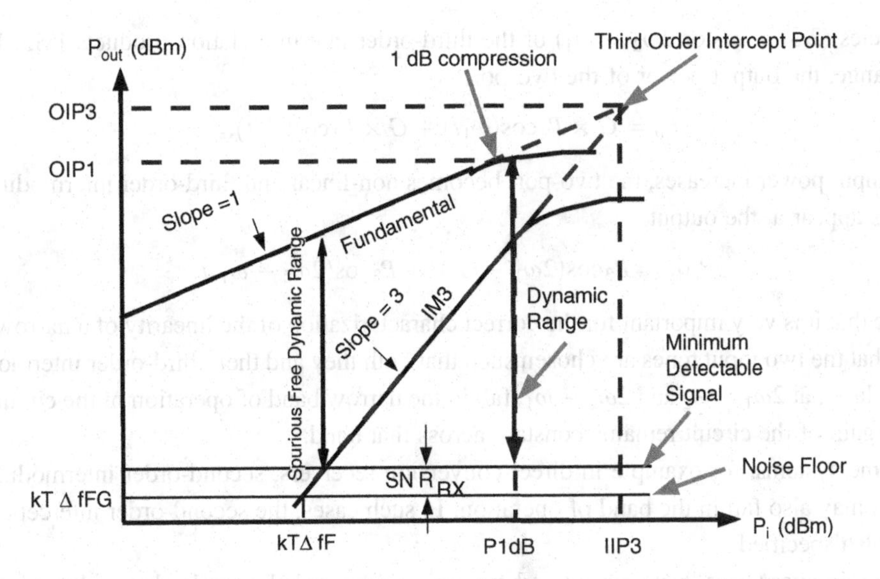

Figure 2.25 Graphical representation of compression points and dynamic range definitions

- the output 1dB compression point, OIP1, is defined as the output power corresponding to the input 1dB compression point;
- the input third-order intercept point, IIP3, is defined as the input power level at which the power of the fundamental signal at the output and the power of the third-order intermodulation components (IM3) at the output become equal;
- the output third-order intercept, OIP3, represents the corresponding output power level for IIP3;
- the minimum detectable signal level is equal to the output noise floor + SNR_{RX};
- the spurious free dynamic range SFDR is defined as the difference in dB between the output power of the fundamental signal and that of the third-order intermodulation product when the third-order product at the output crosses the minimum detectable signal level;
- the dynamic range DR is the difference between the output 1dB compression point OIP1 and the minimum detectable signal level. Often, in the case of analog-to-digital converters, the DR is defined with respect to the input signal rather than the output signal.

How is the IIP3 plot generated?

This is obtained in either simulations or measurements by applying a two-tone input signal

$$P_i = P_1\cos(\omega_1 t) + P_1\cos(\omega_2 t) \tag{2.55}$$

with $\omega_2 - \omega_1 << \omega_1$ and ω_2 such that the power gain of the circuit is practically constant at the two frequencies, and measuring the output power at the fundamental (ω_1 or ω_2) and at the frequencies ($2\omega_1 - \omega_2$ or $2\omega_2 - \omega_1$) of the third-order intermodulation products IM3. In the linear range, the output power of the two-port is

$$P_o = G \times P_1 \cos(\omega_1 t) + G \times P_1 \cos(\omega_2 t). \tag{2.56}$$

As the input power increases, the two-port becomes non-linear and third-order intermodulation products appear at the output

$$P_{IM3} = P_3 \cos(2\omega_1 - \omega_2)t + P_3 \cos(2\omega_2 - \omega_1)t. \tag{2.57}$$

We note that it is very important for the correct characterization of the linearity of a narrowband circuit that the two input tones are chosen such that both they and their third-order intermodulation products, at $2\omega_1 - \omega_2$ and $2\omega_2 - \omega_1$, fall in the narrow band of operation of the circuit and that the gain of the circuit remains constant across that band.

In some systems, for example in direct-conversion receivers, second-order intermodulation products may also fall in the band of operation. In such cases, the second-order intercept point IIP2 is also specified.

Finally, in every broadband circuit and system, non-linear behavior leads to a large number of harmonics being present at the output, even if only a single tone is applied at the input. In this situation, the preferred figure of merit for linearity is the *total harmonic distortion* (THD), which is defined as the sum of the powers of all the harmonics present at the output of the circuit, excluding the fundamental, divided by the power of the fundamental tone at the output.

Equations for calculating the *m*th-order intercept points and dynamic range

We can employ the log-log plot in Figure 2.25 to derive useful system design equations that link the intercept points and the dynamic range of a receiver or radio building block.

The *n*th-order input intermodulation point IIP_n satisfies the following relationship between the power of the fundamental ($P_o = 10\log_{10}G + P_i$) and that of the *n*th-order intermodulation product, IM_n, measured at the output of the receiver/block

$$(n-1)(IIP_n - P_i) = P_i - IM_n \tag{2.58}$$

from which we obtain

$$IIP_n = \frac{nP_i - IM_n}{n-1} \tag{2.59}$$

where $n = 2, 3, 4 \ldots$ and IIP_n, P_i, and IM_n are expressed in dBm. IM_n is the output power of the *n*-intermodulation product when the input tone power is P_i.

Similarly, the output dynamic range, as defined in dB, in Figure 2.25, can be expressed with respect to the *n*th-order spurs as

$$DR_n = \left(1 - \frac{1}{n}\right)(IIP_n - IM_n). \tag{2.60}$$

In a radio receiver, we normally want all spurious signals to be below the receiver sensitivity level, S_i, by a certain margin, C, typically equal to or higher than the receiver SNR

$$IM_n = S_i - C. \tag{2.61}$$

By substituting (2.61) into (2.59), we obtain the required IIP_n for the desired sensitivity

$$IIP_n = \frac{nI - (S_i - C)}{n - 1} \tag{2.62}$$

where I is the power of the interferer in dBm, at the input.

EXAMPLE 2.4

Let us consider a scenario in a 60GHz radio where two interferers $I_1 = I_2 = -38$dBm are present at 64GHz and 62GHz, respectively, along with a weak but desired 60GHz signal at the input of the receiver. Third-order intermodulation products will arise at 60GHz and 66GHz, with the former falling in the same channel as the desired weak signal. Such a situation may occur if the transmitters that produce the interferers are located at a distance of 10cm from the receiver and each has an output power of + 10dBm. Let us assume that the desired sensitivity S_i is -60dBm and that we need a margin C of 14dB for proper reception. The required third-order intercept point of the receiver must satisfy the condition

$$IIP_3 \geq \frac{3 \times (-38) - (-60 - 14)}{2} = \frac{-114 + 74}{2} = -20\text{dBm}.$$

The most relaxed IIP3 requirement is obtained in the absence of interference. A typical scenario would be an OFDM signal with many closely spaced subcarriers, each at the sensitivity level S_i. Assuming $C = -40$dB (SNR requirements are usually tougher for OFDM modulated signals), and $S_i = -60$dBm, we obtain

$$IIP_3 \geq \frac{3 \times (-60) - (-60 - 40)}{2} = -60 + \frac{40}{2} = -40\text{dBm}.$$

2.6.6 Linearity of a chain of two-ports

As in the noise figure case, an important receiver design equation relates the IIP3 of the entire receive chain to those of the individual blocks in the receiver. Consider the chain of two-ports illustrated in Figure 2.26 where Ga_i is the available power gain of stage i (i.e. power gain when its input and output are conjugately matched to the impedance of the preceding and of the following stages). It can be demonstrated that, if conjugate matching exists between each stage, IIP3 and OIP3 satisfy equations (2.63) and (2.64), respectively

$$\frac{1}{IIP3} = \frac{1}{IIP3_1} + \frac{Ga_1}{IIP3_2} + \frac{Ga_1 \times Ga_2}{IIP3_3} + \ldots + \frac{Ga_1 \times Ga_2 \times \ldots Ga_{n-1}}{IIP3_n} \tag{2.63}$$

$$\frac{1}{OIP3} = \frac{1}{OIP3_n} + \frac{1}{Ga_n \times OIP3_{n-1}} + \frac{1}{Ga_n \times Ga_{n-1} \times OIP3_{n-2}} + \ldots + \frac{1}{Ga_2 \times \ldots Ga_n \times OIP3_1}. \tag{2.64}$$

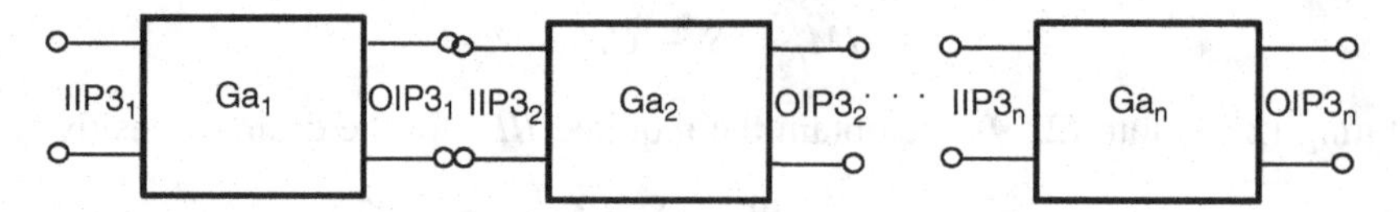

Figure 2.26 Chain of cascaded two-ports employed in the derivation of the overall linearity of the receiver

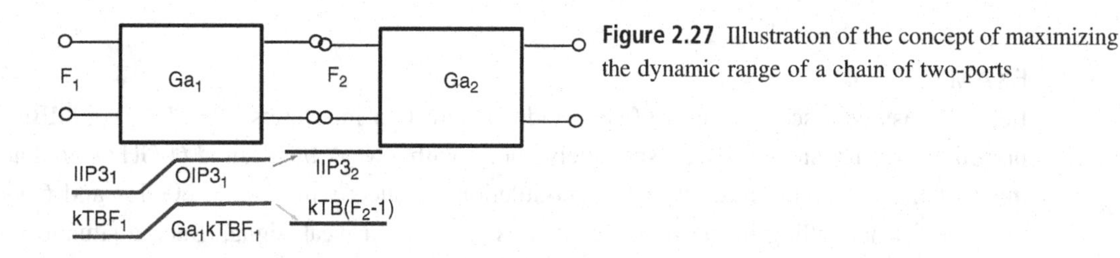

Figure 2.27 Illustration of the concept of maximizing the dynamic range of a chain of two-ports

2.6.7 Optimizing the dynamic range of a chain of two-ports

A likely receiver design scenario pursues the maximization of the dynamic range. Intuitively, the dynamic range of a chain of conjugately matched blocks is maximized when each stage contributes noise and distortion equally. This is accomplished by ensuring that the noise level at the output of the first stage is equal to or higher than the equivalent input noise level of the second stage. The noise level at the output of the second stage equals or is higher than the input equivalent noise of the third stage, and so on. Similarly, the IIP3 of the second stage must be equal or higher than the OIP3 of the first stage, the IIP3 of the third stage must be equal or higher than the OIP3 of the second stage, and so on, as illustrated in Figure 2.27.

In the case of equal noise and distortion contributions from each stage, it can be demonstrated that the gains of each stage must satisfy the condition

$$Ga_i = \sqrt{\left(\frac{F_{i+1} - 1}{F_i}\right)\frac{IIP_{i+1}}{IIP_i}}. \tag{2.65}$$

2.6.8 PLL phase noise

Oscillator and PLL phase noise will be addressed in detail in Chapter 10. Here we limit the discussion to the definition of oscillator phase noise and to its impact on receiver performance. Phase noise manifests itself in a broadening of the spectrum of a local oscillator signal, which, ideally, should have the shape of a Dirac function at the oscillation frequency. In reality, the spectral content of an oscillator exhibits noise "skirts" (see Figure 2.28) caused by the random phase variation of the oscillator signal. The power of the noise decreases as we move away from the oscillation frequency. As a result, phase noise is specified in dBc/Hz (decibels relative to the carrier per hertz) as the ratio of the noise power measured in a 1Hz band at a frequency offset f_m from the oscillation frequency f_{OSC}, and the power of the oscillator signal (or carrier).

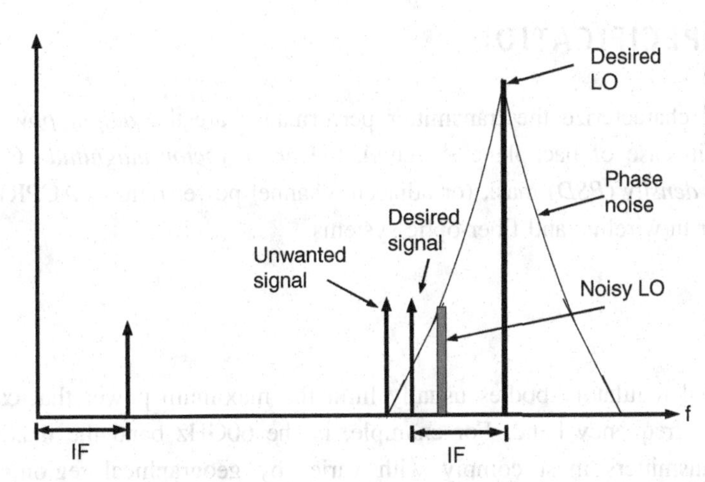

Figure 2.28 Illustration of the impact of LO phase noise on the downconversion of undesired signals in the adjacent channels

In wireless systems, the in-band phase noise of the PLL and the VCO phase noise outside the PLL band affect the sensitivity of the receiver by raising the noise floor.

First, the sensitivity of direct-conversion receivers, such as those used in 77GHz frequency modulated continuous wave (FMCW) automatic cruise control (ACC) radars or in zero-IF radios, is sensitive to the phase noise within the PLL bandwidth.

An example of a situation where phase noise can limit the sensitivity of the receiver is in FMCW radars when detecting the speed of a slow-moving vehicle located at a large distance (> 100m), which produces a very small Doppler shift of 10kHz–100kHz.

Second, in a system with digital phase modulation, the PLL phase noise affects the probability of error in the detection process. A rough estimate of the rms phase error in rads can be obtained from the simplified version of equation (10.22) [1]

$$\theta_{rms} = \sqrt{2L\Delta f} \tag{2.66}$$

where L is the phase noise (as a ratio of powers), and Δf is the radio channel bandwidth in Hz. **For example**, in a radio receiver with OFDM modulation, an average phase noise of −100dBc/Hz over a bandwidth of 5MHz, results in a rms phase error of 1.81 degrees.

Third, the phase noise of a noisy oscillator mixes with undesired nearby signals (interferers) and is downconverted into the IF band, raising the noise floor in super-heterodyne receivers and dictating how closely adjacent channels may be spaced.

The formula which gives the maximum phase noise required to achieve an adjacent channel (at frequency offset f_m from the carrier) rejection of C dB can be derived with the help of Figure 2.28

$$L(f_m) = P(\text{dBm}) - C(\text{dB}) - I(\text{dBm}) - 10\log(\Delta f), (\text{dBc/Hz}) \tag{2.67}$$

where P is the desired signal power (in dBm) typically set equal to the sensitivity of the receiver, I is the undesired (interference) signal level (in dBm), and Δf is the channel bandwidth (in Hz). **For example**, in a 60GHz radio with $P = -60$dBm, $C = 14$dB, $I = -38$dBm, $\Delta f = 1.8$GHz, $f_m = 1$GHz

$$L(1\text{GHz}) = -60\text{dBm} - 14\text{dB} + 38\text{dBm} - 92.55 = -128.55\text{dBc/Hz}.$$

2.7 TRANSMITTER SPECIFICATION

Typical parameters that characterize the transmitter performance are the *output power* (or voltage swing in 50Ω in case of backplane systems), the *error vector magnitude* (*EVM*), transmit *power spectral density* (*PSD*) mask (or adjacent channel power ratio = ACPR), and noise or transmitter jitter in wireline and fiber-optic systems.

2.7.1 Output power

International and regional regulatory bodies usually limit the maximum power that can be transmitted in a specific frequency band. For example, in the 60GHz band the maximum allowed power that transmitters must comply with varies by geographical region, from + 10dBm in Japan and Australia, to + 27dBm in the US. The output compression point $OP_{1\mathrm{dB}}$ is sometimes specified in systems that employ M-ary QAM and OFDM modulation. For systems that employ constant amplitude modulation formats, the saturated output power P_{SAT} is preferred.

2.7.2 EVM

For systems involving M-ary PSK, QAM, and OFDM modulation formats, the distortion in the transmitted signal constellation due to noise and non-linearities in the transmitter is captured by the error vector magnitude (EVM). EVM represents the average rms error over all possible states in the transmitted signal constellation and is typically measured on baseband I and Q data streams after recovery with an ideal receiver. An ideal receiver is a receiver that is capable of converting the transmitted signal into a stream of complex samples at a sufficient data rate, with sufficient accuracy in terms of I/Q amplitude and phase balance, DC offsets, and phase noise. It can perform carrier lock, symbol timing recovery, and amplitude adjustment while making the measurements. Figure 2.29 illustrates how the error vector magnitude is calculated as the Eucledian distance between the coordinates of the received symbol (open circles) and those of the ideal constellation (filled circles).

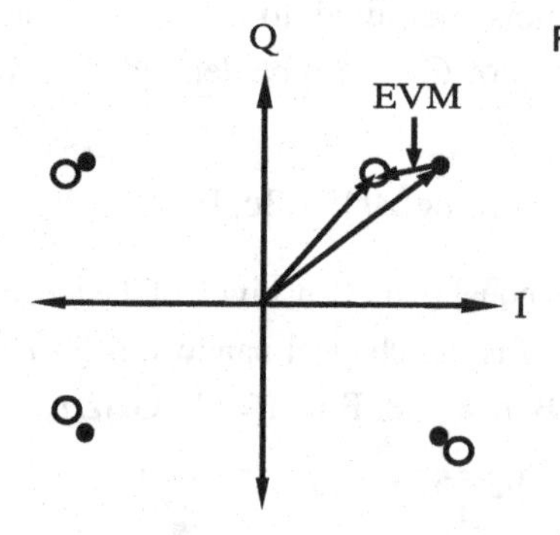

Figure 2.29 Illustration of EVM calculation

EXAMPLE 2.5 EVM specification for the 60GHz WPAN IEEE 802.15–3c standard

For a single-carrier (SC) physical layer (PHY) system, the EVM is calculated by measuring the error of 1000 received symbols at the symbol rate

$$EVM = \sqrt{\frac{1}{1000 \times P_{avg}} \sum_{i=1}^{1000} \left[(I_i - I^*_i)^2 + (Q_i - Q^*_i)^2 \right]} \quad (2.68)$$

where P_{avg} is the average power of the constellation, (I^*_i, Q^*_i) are the complex coordinates of the ith measured symbol, and (I_i, Q_i) are the complex coordinates of the nearest constellation point for the ith measured symbol. The measuring device (receiver) should have an accuracy of at least 20dB better than the EVM value to be measured.

2.7.3 Transmit PSD mask

The role of the PSD mask is primarily to prevent unwanted emissions in adjacent channels. As illustrated in Figure 2.30, it is specified in the form of a plot that describes the maximum signal power allowed at a certain offset frequency from the center of the channel, relative to the signal power in the center of the channel. A related parameter is ACPR which specifies the ratio of the signal power in the adjacent channel to that of the main channel and is measured in dB: 10log10.

2.7.4 Noise

Noise does not play as significant a role in transmitters as it does in receivers. However, transmitter PLL phase noise remains a concern. In fiber-optic and wireline communication systems that employ OOK modulation, the PLL phase noise impacts the jitter of the output eye

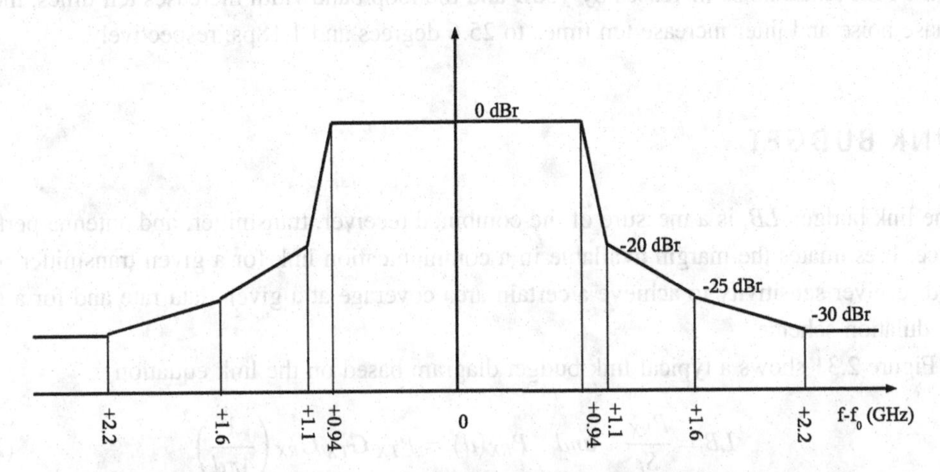

Figure 2.30 PSD mask employed in the 60GHz IEEE 802.15–3c standard

diagram and ultimately the bit error rate of the transmit–receive link. The jitter can be calculated by integrating the phase noise of the carrier from $-\infty$ to $+\infty$ (2.66)

$$t_j^{rms} = \frac{1}{f_o}\sqrt{\int_{-\infty}^{+\infty} S_{\phi n}(f)df}. \quad (2.69)$$

In radio transmitters that operate with digital phase modulation, the PLL phase noise will introduce errors in the transmitted constellation (hence impacting the EVM) in much the same way as in the receiver, affecting the bit error rate and ultimately the link budget. These errors can be calculated using (2.65).

EXAMPLE 2.6

In a 60GHz radio with direct QPSK modulation, the transmitter PLL has a bandwidth of 100KHz and the in-band phase noise is constant at −80dBc/Hz. Beyond 100kHz, the phase noise decreases by 20dB/decade (making the contribution of phase noise above 100kHz negligible). Calculate the rms phase error in degrees and the corresponding jitter in the transmitted constellation. What happens if the in-band PLL phase noise increases to −70dBc/Hz and the PLL loop bandwidth is 1MHz?

Solution

$$\theta_{rms}(\text{degrees}) = \frac{180}{\pi}\sqrt{2L\Delta f} = 57.32\sqrt{2\times 10^{-8}\times 10^{5}} = 2.56^{\circ}$$

$$t_j^{rms} = \frac{1}{f_o}\sqrt{\int_{-\infty}^{+\infty} S_{\phi n}(f)df} = \frac{1}{fo}\sqrt{2L\Delta f} = \frac{1}{60\times 10^{10}}\sqrt{2\times 10^{-8}\times 10^{5}} = 118.5fs.$$

If the PLL phase noise increases by 10dB and the loop bandwidth increases ten times, the rms phase noise and jitter increase ten times to 25.6 degrees and 1.18ps, respectively.

2.8 LINK BUDGET

The link budget, *LB*, is a measure of the combined receiver, transmitter, and antenna performance. It estimates the margin available in a communication link for a given transmitter power and receiver sensitivity to achieve a certain area coverage at a given data rate and for a given modulation scheme.

Figure 2.31 shows a typical link budget diagram based on the link equation

$$LB = \frac{P_{RX}}{S_i} \quad and \quad P_{RX}(d) = P_{TX}\,G_{TX}\,G_{RX}\left(\frac{\lambda}{4\pi d}\right)^2 \quad (2.70)$$

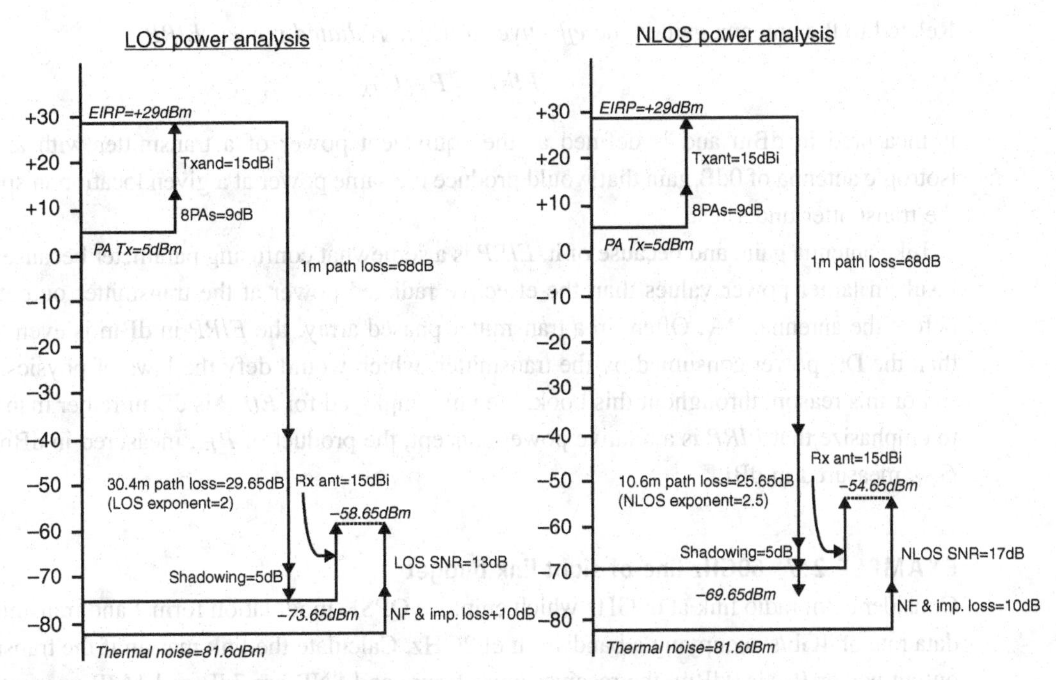

Figure 2.31 Illustration of the link budget in 60GHz LOS and NLOS scenarios [Nir Sason private communication]

where:

- λ is the wavelength,
- d is the distance that must be covered between the transmitter and the receiver (assumed $>> 10\lambda$),
- P_{RX} is the power at the input receiver,
- P_{TX} is the power at the output of the transmitter (or radiated power),
- G_{TX} is the gain of the transmit antenna, and
- G_{RX} is the gain of the receive antenna.

Equation (2.69) is most often used in its dB form

$$LB(\text{dB}) = P_{TX}(\text{dBm}) + G_{TX}(\text{dBi}) + G_{RX}(\text{dBi}) + 20 \cdot \log_{10}\left(\frac{\lambda}{4\pi d}\right)$$

Antenna gain is measured in dBi. The notion of antenna gain can be confusing for circuit designers. Since antennas are passive elements, they cannot amplify the signal from the transmitter, yet the gain of most antennas is higher than 0dB. This can be understood by the fact that the gain of an antenna is normalized to that of an ideal isotropic antenna, which uniformly radiates in all directions and has 100% efficiency, that is no loss.

The gain of the antenna is the product of the antenna efficiency (<1) and of the antenna directivity, D

$$Gain = directivity \times efficiency.$$

Related to the antenna gain is the *effective isotropic radiated power, EIRP*.

$$EIRP = P_{TX} G_{TX}$$

is measured in dBm and is defined as the equivalent power of a transmitter with an ideal isotropic antenna of 0dB gain that would produce the same power at a given location in space as the transmitter under test.

Like antenna gain, and because of it, *EIRP* is a somewhat confusing parameter because it can result in larger power values than the effective radiated power at the transmitter output port, before the antenna, P_{TX}. Often, in a transmitter phased array, the *EIRP* in dBm is even higher than the DC power consumed by the transmitter, which would defy the laws of physics.

For this reason, throughout this book, the unit employed for *EIRP* is dBmi rather than dBm, to emphasize that *EIRP* is a relative power concept, the product of P_{TX}, measured in dBm, and G_{TX}, measured in dBi.

EXAMPLE 2.7 60GHz line-of-sight link budget

Consider a 2m radio link at 60GHz which employs QPSK modulation format and transmits at a data rate of 4Gb/s occupying a bandwidth of 2GHz. Calculate the link margin if the transmitter output power P_{TX} is 0dBm, the receiver noise figure and SNR are 7dB and 14dB, respectively, and the transmit and receive horn antennas each have a gain of 25dB.

Solution

The free-space loss over 2m is

$$LFS = 20 \times \log_{10}\left(\frac{\lambda}{4\pi d}\right) = 74\text{dB where } \lambda = 5\text{mm and } d = 2000\text{mm}.$$

$P_{RX} = P_{TX} + G_{TX} - L_{FS} + G_{RX} = 0 + 25 - 74 + 25 = -24\text{dBm}.$

The receiver sensitivity is:

$$S_i = -174\text{dBm} + 10\log_{10}(2 \times 10^9) + 7\text{dB} + 14\text{dB} = -174 + 93 + 7 + 14 = -60\text{dBm}.$$

That leaves a shadowing and cable + probe loss margin of 36dB. Alternatively, if PCB antennas with 8dBi gain are employed, the loss margin is reduced to only 2dB.

2.9 PHASED ARRAYS

For a fixed signal power, there is a critical value of the data rate, R_b, for which the bit error rate can be made as small as possible. This data rate is known as the channel capacity. It is measured in b/s, and is described by Claude Shannon's formula

$$C = \Delta f \times \log_2(1 + SNR) \tag{2.71}$$

Even when the most bandwidth-efficient modulation formats are employed, present wireless systems operate at only a fraction of the channel capacity. The channel capacity is significantly affected by multi-path fading and interference, both of which degrade the SNR. The latter can be improved by spatial diversity, a technique that relies on processing the signals from multiple transmitters and receivers. Spatial diversity, in the form of MIMO (multiple input, multiple output) or *phased array* systems, allows for several communication channels to be formed, which will, hopefully, be affected in an uncorrelated fashion by noise and interference. By collecting and processing the information transmitted through multiple channels, the SNR can be improved in comparison to that of a single communication channel.

In a MIMO system, the radio transceivers and communication channels are statistically independent, requiring complex, non-linear data processing to improve the SNR.

In phased arrays, where at least a section of the transmitter and of the receiver are shared amongst all transceivers, the data collected from the established communication channels need only be processed in a linear fashion, through delay-and-sum operations. This makes phased arrays less costly than MIMO systems.

The first antenna arrays can be traced as far back as Marconi's transatlantic radio transmission experiment in 1901 between Poldhu in Cornwall, England and St Johns in Newfoundland. The first electronically steered phased arrays were introduced during the second World War and have since been widely used primarily in military applications. More recently, since about 2005, facilitated by the proliferation of silicon-based microwave and mm-wave circuits, phased arrays are finding new applications in low-cost wireless personal networks (WPAN) at 60GHz and in automotive radar at 24GHz and 77–79GHz. The most important properties and specifications for phased arrays are briefly described next. The reader is referred to [18] for a recent and more detailed review of the topic.

2.9.1 Timed versus phased arrays

A **timed or phased array** is a particular set of multiple antenna transmitters, receivers, or transceivers whose signals are independently and electronically delayed, amplified, and summed together such that the transmitted or received electromagnetic wave is steered in a specific direction. A desired pattern is formed by controlling the delay and amplitude of the signal along each path (lane) in the array.

As illustrated in Figure 2.32, a typical phased array consists of N antenna elements placed at a distance d from each other for a total array size $L = (N - 1) \times d$. In most phased arrays, $d = \lambda/2$. Variable delay cells are required in each array element in order to compensate for the different free-space propagation delays of the signal arriving at, or leaving, the array antennas. From Figure 2.32, we can calculate the path-length difference between two adjacent antennas to be equal to $d \times \sin(\theta_{in})$. The latter corresponds to a time delay of $d \times \sin(\theta_{in})/c$, where θ_{in} is the angle of incidence, and $c = 3 \times 10^8$m/s is the speed of light.

The delay elements can operate either in the time (Figure 2.32) or in the phase domain (Figure 2.33). The difference in the behavior of variable phase delay and variable time-delay cells is conceptually illustrated in Figure 2.34. Ideally, what we need in a phased array are true time-delay cells which allow operation over a broad frequency band. However, since true time delay is more difficult to implement, in narrow band applications it is often replaced by phase delay, which, over a narrow band, offers a simpler alternative.

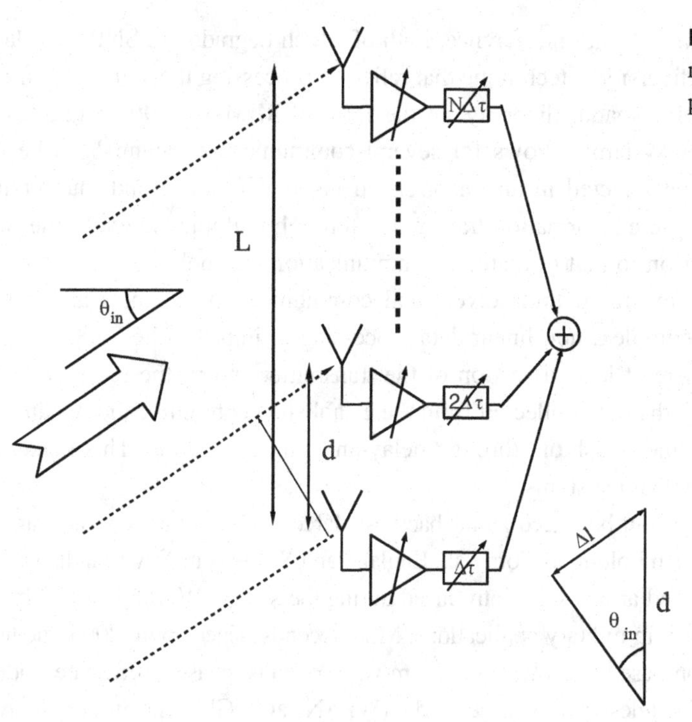

Figure 2.32 Conceptual representation of a true time-delay phased array

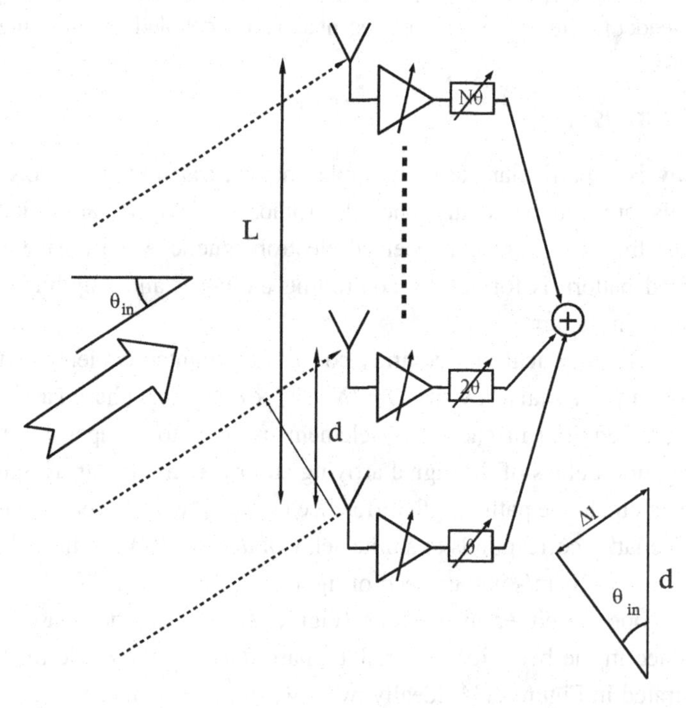

Figure 2.33 Conceptual representation of a phased array with phase shifter elements

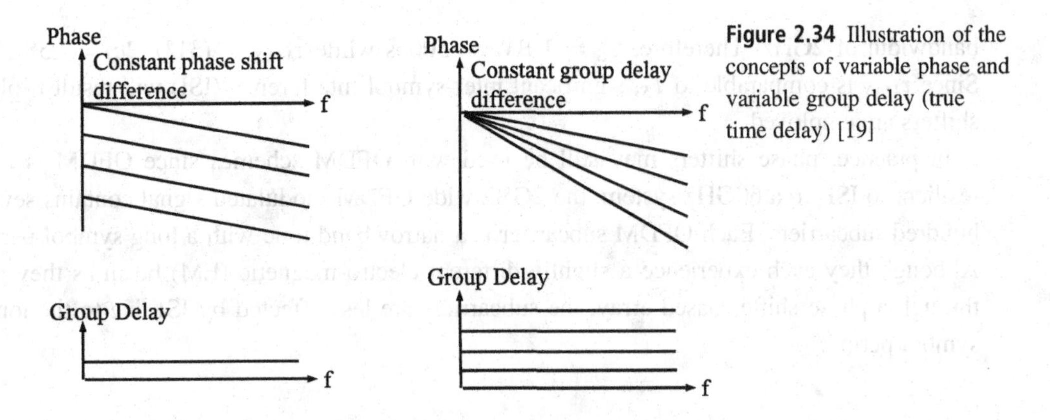

Figure 2.34 Illustration of the concepts of variable phase and variable group delay (true time delay) [19]

Before reaching the delay cell, the signal arriving at antenna i is amplified and can be described by

$$s_i(t) = G_{vi}A\cos\left(\omega\left[t - (i-1)\frac{d}{c}\sin(\theta_{in})\right]\right) \tag{2.72}$$

where G_{vi} is the voltage gain on each path of the array.

To compensate for the propagation delays, the time delays in each array element must be equal to $i(d/c)\sin(\theta_{in}) = i\Delta\tau$ where

$$\Delta\tau = \frac{d}{c}\sin(\theta_{in}) \tag{2.73}$$

and must be arranged in reverse order compared to the linear delay progression of the incident wave. The gain of the amplifier in each path determines the positions of the nulls in the radiation pattern of the phased array.

If the voltage gains on each path are equal to G_v and the delays are arranged as discussed above, the voltage after the summing node becomes

$$s_{out}(t) = \sum_{i=1}^{N} G_v A\cos[\omega[t - (i-1)\frac{d}{c}\sin(\theta_{in}) + i\Delta\tau]\ ] = NG_v A\cos[\omega(t+\Delta\tau)] \tag{2.74}$$

and scales linearly with N.

When can we employ a phase delay cell (phase shifter) instead of a true time-delay cell? A good metric is the ratio of the maximum propagation delay, $\tau_{MAX} = L/c$, between the first and last elements of the array, and the symbol period, T_S, of the data signal transmitted and/or received by the phased array. If $\tau_{MAX} << T_S$, a phase shifter can be safely used.

EXAMPLE 2.8

An eight-element phased array for 77–79GHz collision avoidance radar has a symbol period $T_S = 1/\text{BW} = 500\text{ps}$ while $\tau_{MAX} = (7\lambda)/(2c) = 44.87\text{ps}$. In this case, a phase shifter can be safely employed.

Let us now consider the situation of a 32-element array destined for a 60GHz wireless HDMI video area network (WHD-VAN). It transmits at 4Gb/s and occupies a channel bandwidth of 2GHz. Therefore, $T_S = 1/\text{BW} = 500\text{ps}$ while $\tau_{MAX} = (31\lambda)/(2c) = 258.33\text{ps}$. Since τ_{MAX} is comparable to T_S, significant inter-symbol interference (ISI) will result if phase shifters are employed.

In practice, phase shifters may still be used with OFDM schemes since OFDM is more resilient to ISI. In a 60GHz system, the 2GHz wide OFDM modulated signal contains several hundred subcarriers. Each OFDM subcarrier is a narrowband tone with a long symbol period. Although they each experience a slightly different electro-magnetic (EM) beam as they pass through a phase shifter based array, the subcarriers are less affected by ISI due to the longer symbol period.

2.9.2 Properties of linear phased arrays

The performance of a phased array is described by a number of parameters such as the **array gain**, the **beam-width**, the **beam-pointing angle**, the **nulls**, and the **sidelobes** of the radiation pattern.

The ***array factor*** (or ***gain***) is defined as the ratio of the power gain achieved by the phased array and the gain of a single element of the array. It depends on the number of elements in the phased array and on the unit delay $\Delta\tau$ of the individual array elements [18].

In deriving the maximum gain (when the signals add in phase), we must consider that, in the array case, we have N antennas receiving the same power density per area as a single antenna. Therefore, the total input power is $P_{in} = (N/2)A^2$ whereas the output power is $P_{out} = (N^2 G^2{}_V A^2)/2$. The power gain in the array case becomes $G = P_{out}/P_{in} = NG^2{}_V$ and is N times larger than that of the single element.

Equation (2.73) indicates that, in a true time-delay array, the unit time delay required to maximize the array gain for a given angle of incidence is independent of frequency, explaining why such arrays can be operated over a very broad frequency band.

In practical applications, we are interested in deriving the reverse relationship that gives the angle of incidence for which the array gain is maximized as a function of the unit delay in each element of the phased array. This is also known as the ***beam-pointing angle*** θ_m [18]

$$\theta_m = \arcsin\left(\frac{c}{d}\Delta\tau\right). \tag{2.75}$$

Equation (2.75) describes mathematically the fundamental electronic beam-steering property of phased arrays through $\Delta\tau$. The latter can be controlled either in a continuous fashion (analog beam-steering) or digitally, in discrete steps (digital beam-steering).

Since the array gain decreases for any other incidence angle, (2.75) captures another fundamental property of phased arrays, that of *spatial selectivity*.

We note that, if the distance between the antennas in the array is smaller than $\lambda/2$, the spatial selectivity is reduced, while larger spacing results in multiple main lobes in the radiation pattern.

The array **beam width** (in radians) is approximately equal to λ/L, implying that larger arrays result in a narrower, more focused beam.

EXAMPLE 2.9

Calculate the approximate beam width of an eight-element 60GHz phased array which employs $\lambda/2$ spacing. Calculate the beam width of a similar array operating at 140GHz?

Solution

The beam width is approximately $2/(N - 1) = 2/7 = 0.2857$rad or 16.4 degrees and is independent of frequency.

As can be seen from equation (2.71), for identical path gains G_i, and for a given $\Delta\tau$, the received signal completely vanishes after the summing block for certain incidence angles. These angles are called *nulls*. The nulls can be set arbitrarily at the desired incidence angle by independently controlling the gain in each array element. Furthermore, the radiation patterns exhibit local (or secondary) maxima, known as *sidelobes*, in addition to the main lobe. The number of sidelobes increases with the number of antenna elements.

2.9.3 Benefits of phased arrays

The main advantages of phased array are:

- increased *signal level* in the receiver at the output of the summer for a given power density per area at the receive antennas (the aggregate antenna gain increases N times compared to the single element case) (Figure 2.35);
- increased overall output power, which scales with N (Figure 2.36);
- increased *EIRP*, which scales with N^2;
- immunity to interferers due to beam-forming, but only within the bandwidth of each receiver and only after the signal-summation block;
- immunity to multi-path fading through antenna diversity in both receivers and transmitters;
- improved link reliability, especially in dynamic environments.

EXAMPLE 2.10 60GHz OFDM NLOS + phased arrays

Consider a 10m NLOS link at 60GHz which employs OFDM and 16QAM modulation format and transmits at a data rate of 4Gb/s, occupying a bandwidth of 2GHz.

(a) Calculate the link margin if the transmitter consists of a phased array with 32 elements, each element transmitting with an output power P_{TXi} of 0dBm (10dB back-off from the saturated power, P_{SAT}), the receiver noise figure and detector SNR are 7dB and 21dB, respectively, and the transmit and receive antennas each have a gain of 6dBi.

(b) Recalculate the link margin when the output power of each transmit element increases from 0dBm to + 5dBm.

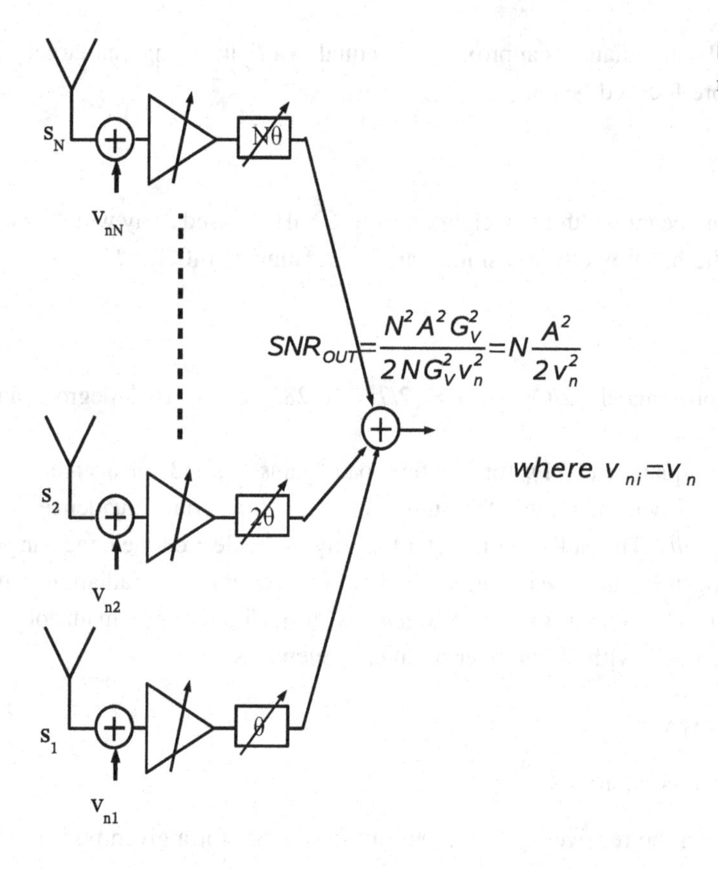

Figure 2.35 Illustration of SNR improvement using a phased array

Solution

The free-space loss over 10m is

$$LFS = 20 \times \log_{10}\left(\frac{\lambda}{4\pi d}\right) = 88\text{dB where } \lambda = 5\text{mm and d} = 10\text{m}.$$

For arrays, $P_{TX} = P_{TXi} + 10\log_{10}(N)$, $G_{TX} = G_{Ai} + 10\log_{10}(N)$,

Hence, $P_{TX} = 0\text{dBm} + 10\log_{10}(32) = 15\text{dBm}$, $G_{TX} = G_{RX} = 6 + 15 = 21\text{dB}$, and $EIRP$ (dBmi) $= P_{TX}$ (dBm) $+ G_{TX}$ (dBi) $= 15\text{dBm} + 21\text{dBi} = 36\text{dBmi}$

$P_{RX} = EIRP - L_{FS} + G_{RX} = 36 - 88 + 6 = -46\text{dBm}$.

The receiver sensitivity is

$$S_i = -174\text{dBm} + 10\log_{10}(2 \times 10^9) + 7\text{dB} + 21\text{dB} - 10\log_{10}(32) = -68\text{dBm}.$$

That leaves a loss margin of 22dB.

(b) If the power transmitted by each element increases from 0dBm to 5dBm, the link margin improves by 5dB to 27dB.

The main advantage of an $N \times$ transmit phased array over a large ($N \times$) transmitter with large ($N \times$) antenna is the fact that the signal from the phased array can be electronically steered to the desired direction, whereas the large transmitter radiates in a fixed direction.

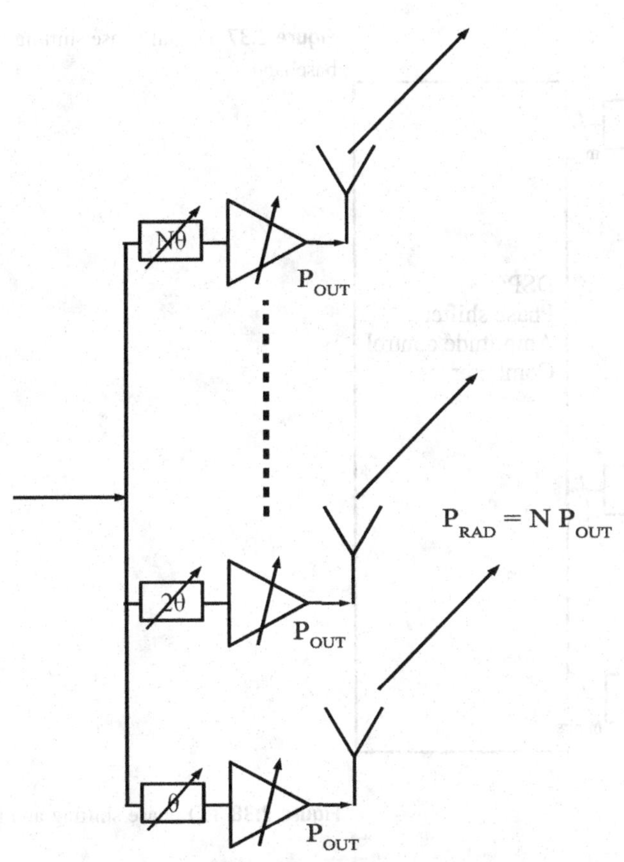

Figure 2.36 Illustration of output power combing using a phased array

2.9.4 Beam-forming transceiver architectures

Depending on where the phase or time-delay cells are inserted in the transceiver array architecture, phased arrays can be classified in arrays with (i) digital, (ii) LO, and (iii) RF phase shifting.

It is important to note that true time-delay cells can only be incorporated in the RF path of a transceiver architecture whereas phase delay cells can be inserted on the RF, LO, and baseband paths of a transceiver.

Digital phase shifting arrays, shown in Figure 2.37, are the most versatile and operate over large bandwidths and at high data rates. However, they occupy the largest area and consume the largest power because, essentially, the entire transceiver, including the ADC and DAC, must be repeated in each array element. Furthermore, they face the most severe linearity demands since the signals are only combined at baseband, after the ADCs. As a result, all components preceding the baseband must be very linear and must survive strong interference signals.

LO phase shifting arrays (Figure 2.38) require a moderate (silicon) area and have moderate power consumption. The linearity requirements on the phase shifters in the LO path are relaxed because the VCO signal is narrow band and is immune to group delay and gain variation. However, since the power combining occurs after the downconvert mixers, the entire RF path, including the downconvert mixers, must be very linear.

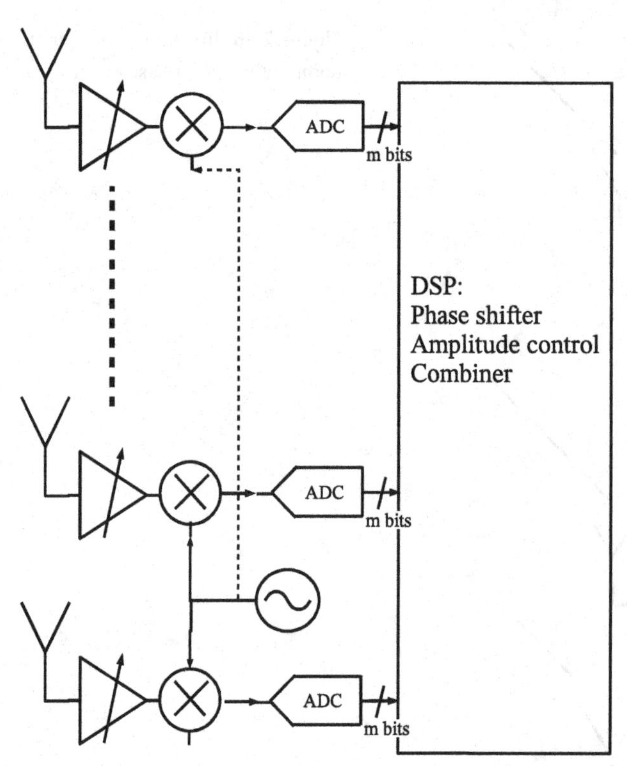

Figure 2.37 Digital phase shifting in baseband

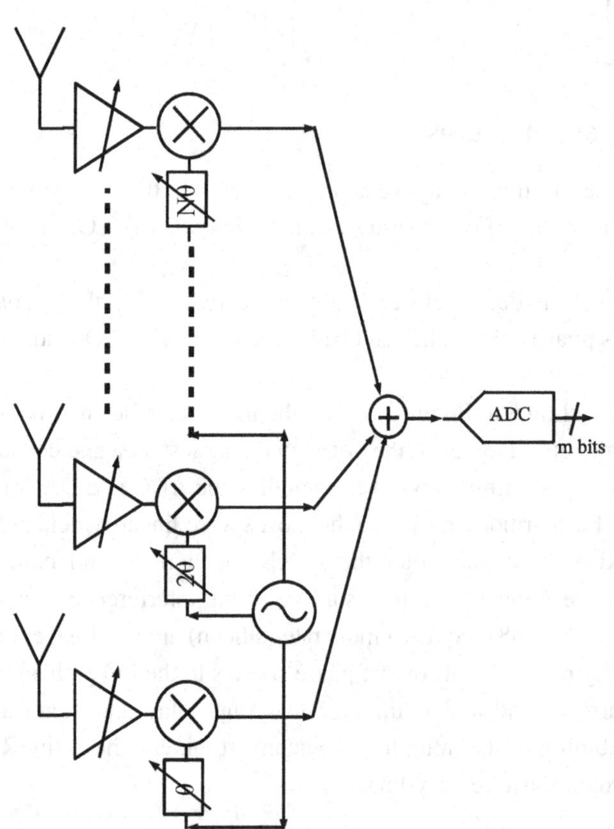

Figure 2.38 LO phase shifting architecture

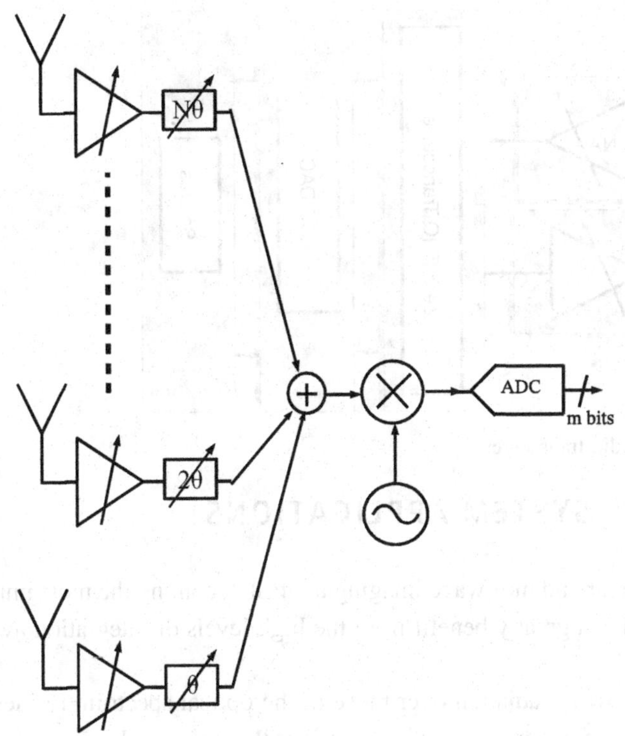

Figure 2.39 RF phase shifting architecture

In terms of power consumption, the **RF phase shifting arrays**, shown in Figure 2.39, are the most efficient because they share the largest number of blocks, including the downconvert and upconvert mixers. However, the linearity, noise figure, and bandwidth demands imposed on the phase or time-delay cells, and on the downconvert/upconvert mixer are the most stringent. Area-wise, they tend to occupy the least area, especially when phase interpolators are employed as phase shifters. They also provide the best architectural choice when interference cancelation is a priority because signal combining occurs before the downconvert mixer. If true time-delay cells are used, the data rate and bandwidth can be as high as in the digital arrays.

2.9.5 Switched-beam systems

In some low-cost, moderate performance applications that do, however, need to consume low power, a phased array can be replaced by a switched antenna system. As illustrated in Figure 2.40, the signal from a single transceiver is switched between antennas with different beam orientations. In this manner, different beams can be scanned sequentially while consuming the power of a single transceiver. Unlike in phased arrays, the SNR and output power remain identical to those of a single antenna element. The only benefit is the wide area coverage and the possibility to avoid obstacles and interferes, in much the same way as a single antenna system with mechanical steering of the antenna. In real implementations, the loss of the antenna switch further degrades the SNR and the output power.

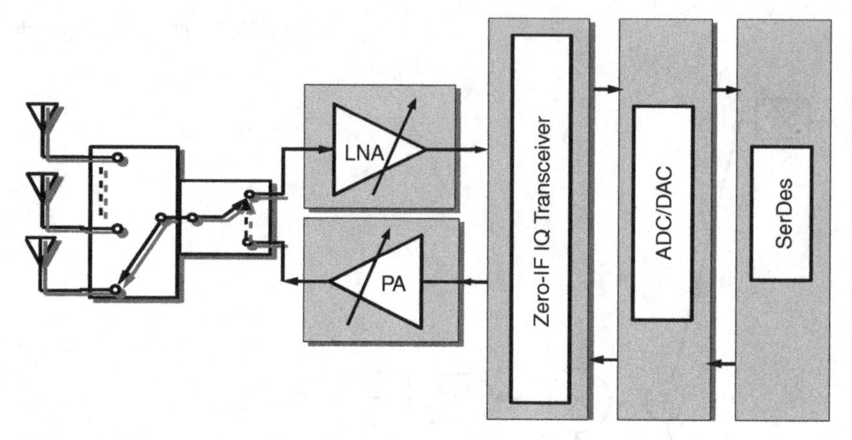

Figure 2.40 A switched antenna radio transceiver

2.10 EXAMPLES OF OTHER SYSTEM APPLICATIONS

Millimeter wave automotive radar and mm-wave imaging are fast becoming the most important markets for mm-wave ICs and can greatly benefit from the high levels of integration available in silicon.

The main advantage of mm-wave radiation over those in the optical spectrum resides in its ability to penetrate through fog, snow, dust, and rain, as well as through a wide range of opaque materials with relatively low water content. The smaller wavelengths also provide higher resolution in imagers than at microwave frequencies.

For many years now, automatic cruise control radars, based on III-V diodes and MMICs have been installed in some luxury brands of cars. The first commercial 77GHz ACC radars employing SiGe BiCMOS transceivers were introduced in some brands of cars in 2008. Millimeter-wave imaging systems based on III-V semiconductors are currently deployed in airports, at security check points, to detect weapons concealed under clothing, as well as in all-weather vision systems.

2.10.1 Doppler radar

Radar (RAdio Detection And Ranging) is one of the earliest applications of microwaves, going back to World War II. The basic principle is based on a transmitter that sends a signal towards a target which partially reflects it back towards a sensitive receiver co-located with the transmitter. The distance to the target is determined from the time required by the signal to travel to the target and back

$$r = \frac{1}{2} c\tau \tag{2.76}$$

where c is the speed of light, and τ is the time of flight delay of the signal to the object and back.

If the target is moving, the radial velocity with respect to the signal direction can be determined from the Doppler frequency shift, Δf_d, between the transmitted and the received reflected signal

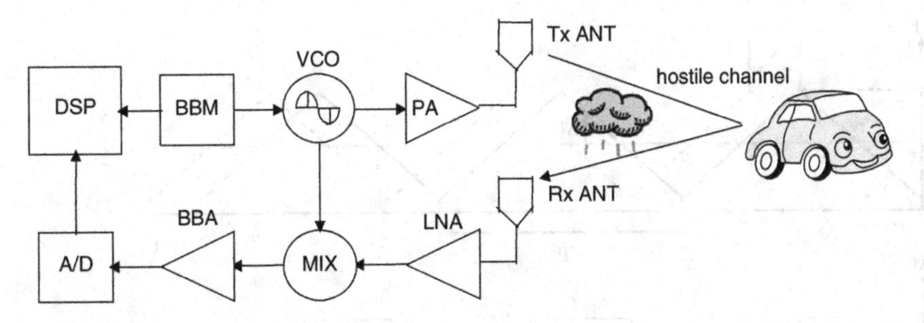

Figure 2.41 Principle of operation and block diagram of a frequency modulated continuous wave (FMCW) automatic cruise control (ACC) radar

$$v = \frac{c\Delta f_d}{2f_{OSC}} \tag{2.77}$$

where f_{OSC} is the transceiver frequency.

Frequency modulated continuous wave (FMCW) radars are likely the most common for automotive applications. They are based on a direct-conversion radio transceiver architecture, Figure 2.41, in which the transmitted signal frequency is linearly modulated by a periodic saw-tooth signal, with period *T*. Both single antenna (monostatic) and dual antenna (bistatic) systems are employed.

As illustrated in Figure 2.42, the range (or position) *r* of the moving object can be extracted solely from frequency domain (rather than time domain) information

$$r = \frac{1}{2}c\tau = \frac{c\Delta fT}{2\Delta f_{OSC}} \tag{2.78}$$

where Δf is the instantaneous difference measured between the transmitted and received FMCW signal frequency, f_{OSC} is the instantaneous VCO frequency and Δf_{OSC} is the maximum tuning (frequency modulation) range of the oscillator.

The target velocity can be determined from the Doppler frequency shift averaged over the modulation period *T*. The average Doppler shift frequency is zero for a stationary target and, depending on the sign convention, positive for targets approaching the transceiver, and negative for vehicles speeding away from the transceiver.

The performance of a radar is described by the radar equation [1]

$$P_{RX} = \frac{\lambda^2 \sigma P_{TX} G_{TX} G_{RX}}{(4\pi)^3 r^4} \tag{2.79}$$

where:

- P_{RX} is the power at the input of the receiver LNA,
- P_{TX} is the power at the output of the transmitters,
- G_{TX} is the gain of the transmit antenna,
- G_{RX} is the gain of the receive antenna (may be the TX antenna),
- r is the distance from the transceiver to the target,

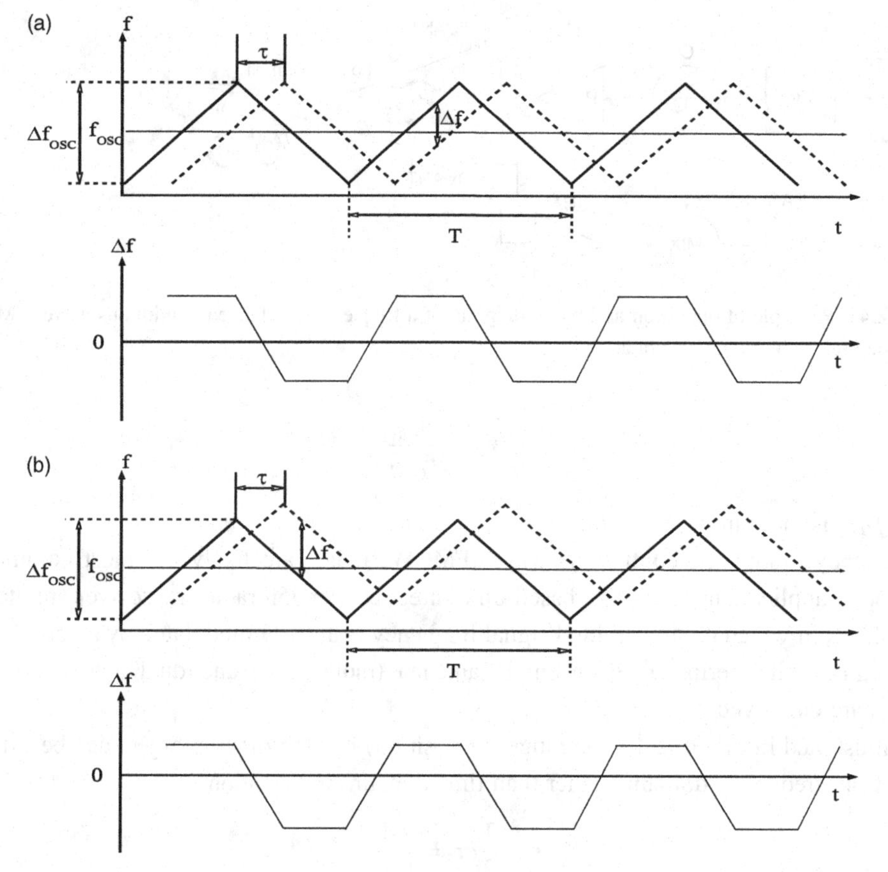

Figure 2.42 Transmitted (solid) and received (dashed) signal frequency and difference frequency as a function of time for (a) stationary target and (b) for a moving (approaching) target illuminated by an FMCW radar

- λ is the wavelength of the radar signal, and
- σ is the radar cross-section of the target, expressed in m^2.

The radar cross-section is target specific and also depends on the angle of incidence when the signal reaches the target.

The maximum detection range r_{MAX} is limited by the received signal power becoming weaker than the sensitivity level S_i of the radar receiver.

Equation (2.79) indicates that the radar performance is primarily dictated by the:

- transmit output power, which directly impacts the range,
- antenna gain,
- receiver noise figure (if the Doppler shift is large enough),
- phase noise and stability of the VCO, which limits the smallest velocity that can be detected,
- the linearity of the VCO tuning characteristics,
- the linearity of the receiver, and
- isolation between receiver and transmitter.

In monostatic systems, where the antenna is shared between the transmitter and the receiver through a coupler or isolator, a relatively large part of the transmitted signal is reflected back into the receiver, potentially driving it in a non-linear regime and significantly degrading its sensitivity.

2.10.2 Remote sensing (passive imaging)

Radar systems collect information about an object by sending a signal (illuminating the target) and measuring the amplitude and phase of the echo received from it. They are therefore considered active remote sensors. Passive imaging sensors or radiometers are high-sensitivity, broadband receivers used to measure the thermal ("black-body") radiation (noise) emitted or reflected by a target. A body in thermal equilibrium at a temperature T emits energy according to Planck's radiation law

$$P = k\Delta f T \tag{2.80}$$

where Δf is the receiver bandwidth over which the black-body radiation is integrated and k is Boltzman's constant.

This equation is strictly valid only for a perfect "black-body" which absorbs all incident energy and reflects none. A non-ideal object partially reflects incident energy and radiates only a fraction of the energy predicted by (2.80). It can therefore be described by a "brightness" (noise) temperature T_B, which is always smaller than its physical temperature and much smaller than the receiver noise temperature T_R.

This radiation is very weak at mm-wave frequencies and, therefore, sensitivity and noise figure are the most important design parameters for a radiometer.

Total power radiometer

Most modern W-Band [3],[5] and D-Band "cameras" [4] employ a direct detection (or tuned homodyne) receiver, also known as a *total power radiometer*, consisting of an LNA, a square-law detector, a post-detector low-bandwidth "video" amplifier, and an integrator, as in Figure 2.43. A 2-D image is formed by mechanically steering the antenna in the X and Y directions and recording the image at each position. This is a slow process. Ideally, if low-cost receivers could be developed, mechanical steering would be replaced by a 2-D receiver array, Figure 2.44, to achieve electronic scanning and image collection at a much faster rate.

The system performance of a total power radiometer is determined by:

- the gain, G, bandwidth, Δf_{RF}, and noise figure (temperature, T_{LNA}) of the LNA,
- responsivity, R, and noise equivalent power, NEP, of the detector, and
- integration time, τ (or bandwidth $\Delta f_{LF} = 1/(2\tau)$) of the integrator.

The system bandwidth Δf is given by the entire receive chain preceding the detector.

The first radiometer systems were developed for radio astronomy [20] and employed a heterodyne receiver architecture to minimize noise figure and improve sensitivity. Since high gain, very low-noise, solid-state amplifiers were not available at mm-wave frequencies, masers and cooled parametric amplifiers were used instead. In a heterodyne radiometer, the system

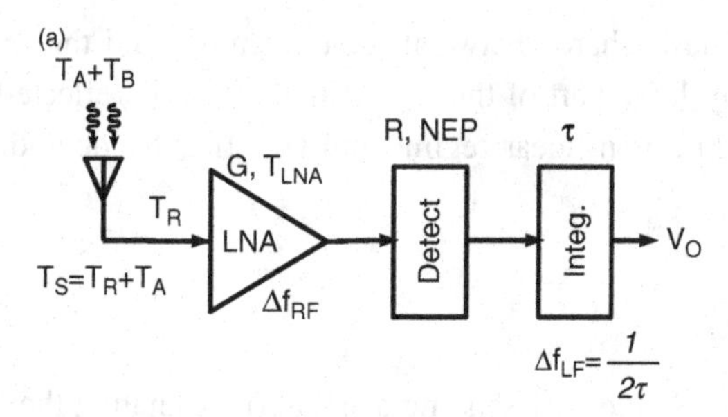

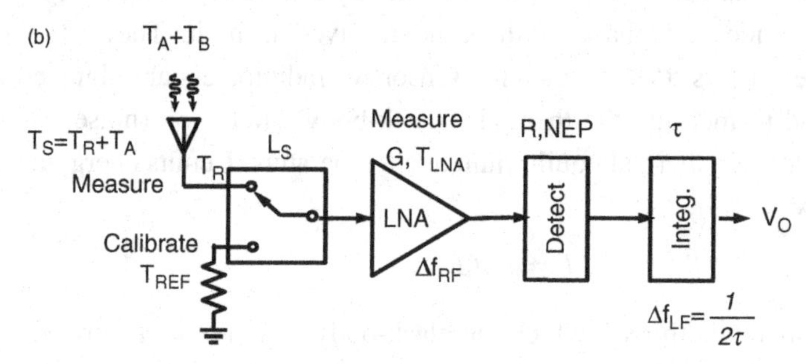

Figure 2.43 (a) Passive imager based on the total power radiometer architecture (b) Radiometer with calibration switch

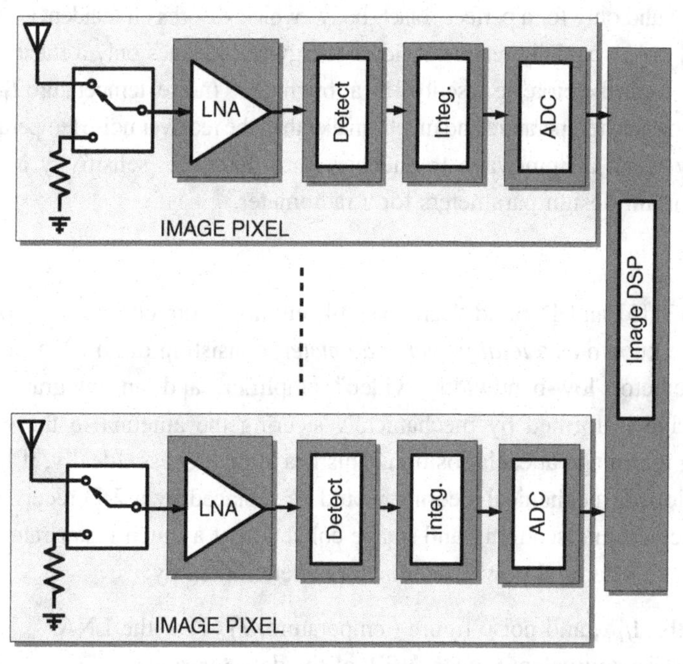

Figure 2.44 Total power radiometer array

bandwidth is typically that of the IF amplifier, while in a direct detection radiometer, the system bandwidth is set by the LNA.

The voltage at the output of the detector in Figure 2.43 is proportional to the input noise power and can be described by [20]

$$V_o = (kT_B \Delta f_{RF} + kT_S \Delta f_{RF})GR \tag{2.81}$$

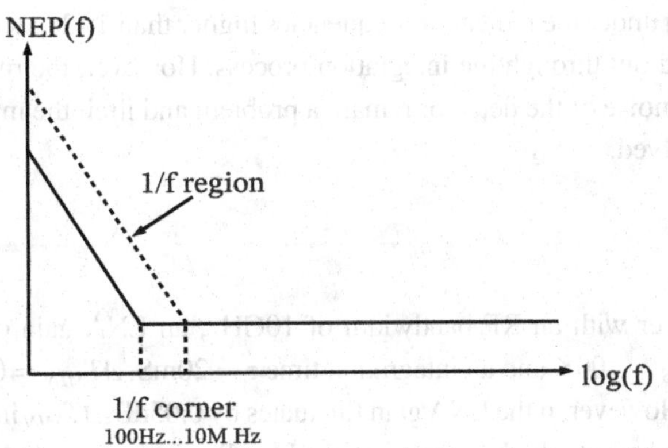

Figure 2.45 Typical NEP vs. frequency curves for semiconductor (III–V, SiGe, and CMOS) detectors

where:

- G is the gain of the LNA,
- $T_S = T_A + T_R$ is the system noise temperature, consisting of
- the receiver noise temperature T_R and
- of the antenna noise temperature T_A.

The latter describes the antenna loss. The receiver noise temperature includes the impact of the noise of the LNA, downconverter, and IF amplifier (if present), and that of the detector diode.

In principle, the system is calibrated by determining the system constants $GRk\Delta f_{FR}$ and $GRk\Delta f_{RF}T_S$ from two measurements in which the antenna is (i) replaced by two known (calibrated) noise sources, or (ii) mechanically steered towards two targets with known temperatures.

The first case is illustrated in Figure 2.43(b) and involves the introduction of an antenna switch. The switch must exhibit low loss (usually below 1dB) and good isolation (> 20dB) to avoid further degrading the overall sensitivity of the radiometer. Two major sources of error occur in a radiometer:

(i) fluctuations in receiver noise (temperature) ΔT_R from one integration interval to another, and
(ii) fluctuations in the overall radiometer gain ΔG.

The imager resolution in degrees Kelvin is given by [20]

$$\Delta T_{MIN} = (2)(T_A + T_R)\sqrt{\frac{1}{\Delta f_{RF}\tau} + \left(\frac{\Delta G}{G}\right)^2} \tag{2.82}$$

where the factor 2 appears in Dicke radiometers only, to be discussed next.

Equation (2.82) indicates that, in order to improve resolution, the system noise temperature must be reduced, and the integration time and bandwidth must be increased. It can be corrected to include $1/f$ and white noise from the detector diode [21]. Indeed, a critical aspect of radiometer design is the choice of a detector with very low $1/f$ noise, Figure 2.45. ΔT_{MIN} is also known as the noise equivalent temperature difference, *NETD* [21].

Fluctuations in noise (first term under the radical) at frequencies higher than $1/(2\tau)$, where τ is the integration time, are smoothed out through the integration process. However, the rms LNA gain fluctuations, ΔG, and the $1/f$ noise of the detector remain a problem and limit the minimum temperature step that can be resolved.

EXAMPLE 2.11

In a 94GHz total power radiometer with an RF bandwidth of 10GHz, an LNA gain of 30dB (1000), a system temperature of $T_S = 400$K and an integration time $\tau = 20$mS, $\Delta T_{MIN} = 0.028$K. This is an outstanding resolution. However, if the LNA gain fluctuates by 0.05dB, ΔT_{MIN} increases dramatically to 4.62K, which is unacceptable. For most practical applications, a resolution of at least 0.5K is considered necessary.

This example shows that overcoming gain variations is more important than averaging out the system noise temperature fluctuations.

Dicke radiometer

Since gain variations have a relatively long time constant (> 1s), it is possible to eliminate their impact by repeatedly calibrating the radiometer at a very fast rate. This is accomplished in the Dicke radiometer, whose conceptual block diagram is illustrated in Figure 2.45. The input of the receiver is switched between the antenna and the reference load with a repetition frequency f_m [20], typically much larger than the $1/f$ corner frequency of the radiometer. If the noise power from the reference load $kT_{REF}\Delta f_{RF}$ is adjusted by the feedback loop such that it becomes equal to the antenna noise power $kT_A\Delta f_{RF}$, then the signal power $k\Delta T\Delta f_{RF}$ is amplitude modulated at frequency f_m and enters the radiometer only during half of the period of the switching frequency. This explains the factor of 2 in (2.82) in the case of the Dicke radiometer. This apparent degradation in sensitivity is more than compensated by the elimination of the gain variation, since f_m is chosen to be much higher (1kHz or higher) than the gain instability frequencies. In Figure 2.46, the multiplier (Mult) switches the detector output in synchronism with the antenna switch in opposite phase to the integrator, such that the voltage corresponding to the reference temperature is subtracted. The DC output voltage from

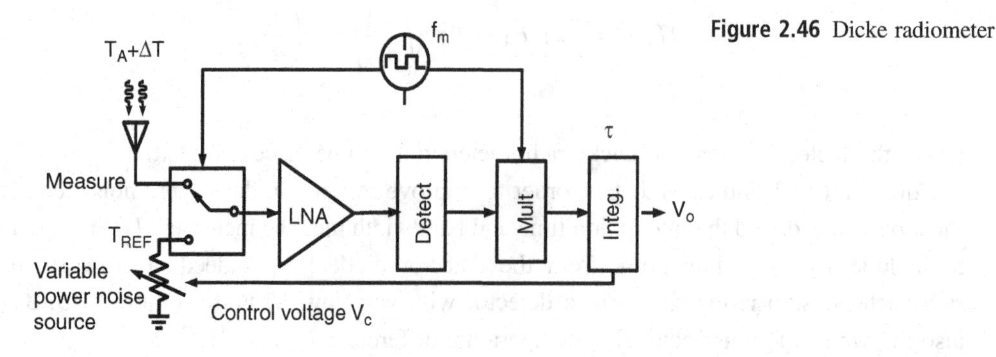

Figure 2.46 Dicke radiometer

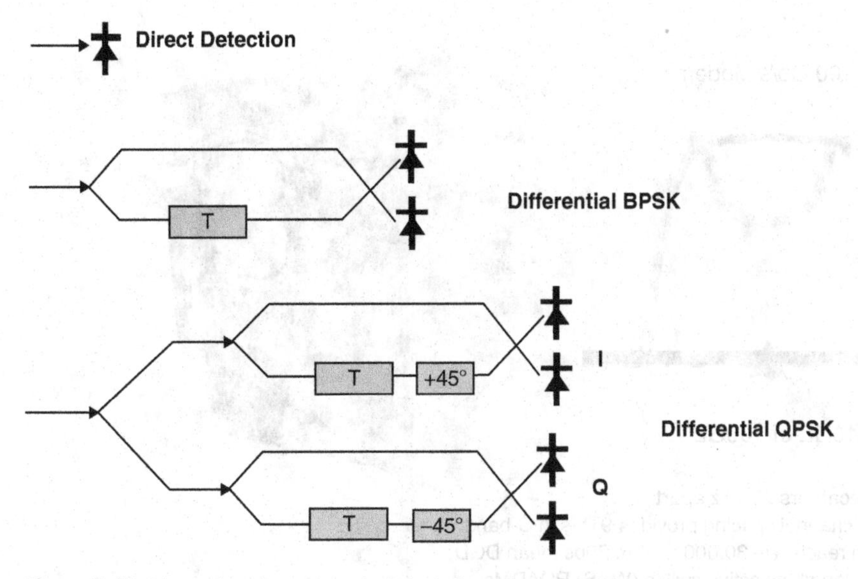

Figure 2.47 Direct (incoherent) optical detection schemes [22]

the integrator thus becomes proportional to the signal temperature ΔT and is zero when no signal is present.

2.10.3 Fiber-optic transceivers

Traditional 2.5Gb/s, 10Gb/s, and 40Gb/s transceivers currently deployed in long-haul fiber-optic communication systems can be analyzed as radio transceivers with direct ASK (OOK) modulation where the carrier is in the optical domain. The modulation is imparted directly through the bias current of the optical oscillator (i.e. laser) or with an amplitude modulator following the laser, as in Figure 2.1, where the modulator is single-bit. The receiver features direct detection schemes, which can be employed for ASK, differential BPSK, and differential QPSK modulation formats, as illustrated in Figure 2.47.

More recently, higher-order modulations schemes such as QPSK and 16QAM variants, and even simple OFDM schemes, have been employed in 40Gb/s and 110Gb/s applications, Figure 2.48. In this case, too, the transmitter and receiver feature direct post-modulation (Figures 2.49–2.50) of a dual-polarized, Figure 2.51) optical carrier by four separate electronic data streams (two for each polarization), and direct downconversion (coherent detection), respectively, in the optical domain, Figure 2.52 [22].

2.10.4 Backplane transceivers

This is a rather unique example where no carrier modulation is employed. Instead, the baseband data are transmitted directly across the backplane using NRZ, or 4-level amplitude modulation (4-PAM).

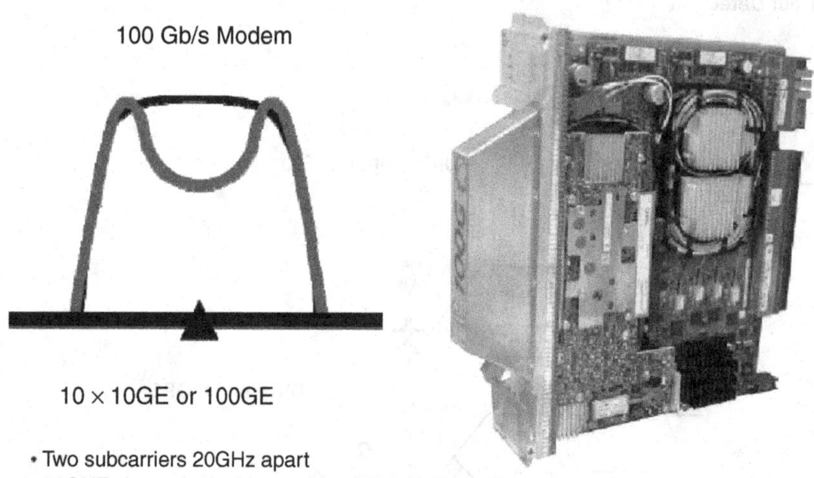

Figure 2.48 100Gb/s modem employing 16QAM and OFDM optical modulation schemes [22]

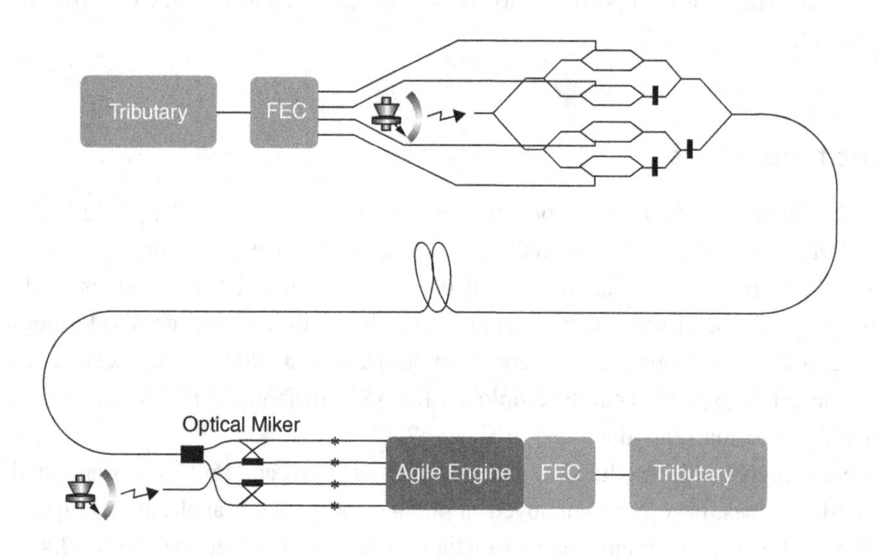

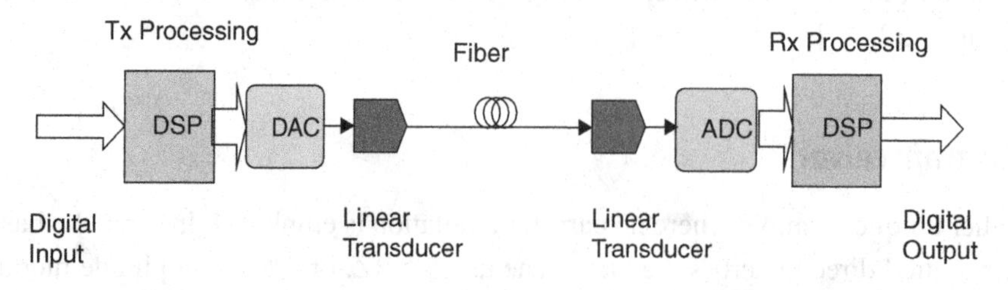

Figure 2.49 100Gb/s direct modulation, direct downconversion DSP-based system architecture [22]

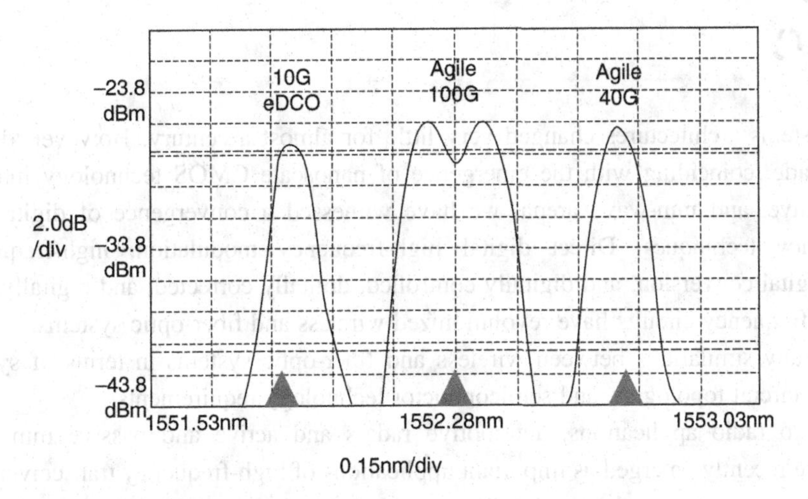

Figure 2.50 100Gb/s OFDM spectrum [22]

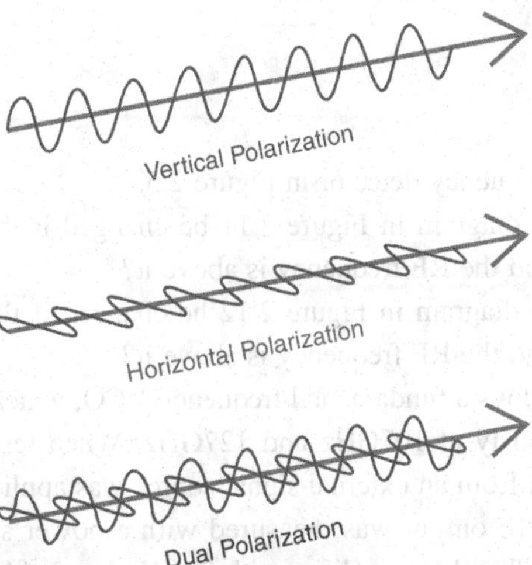

Figure 2.51 Dual optical polarization scheme [22]

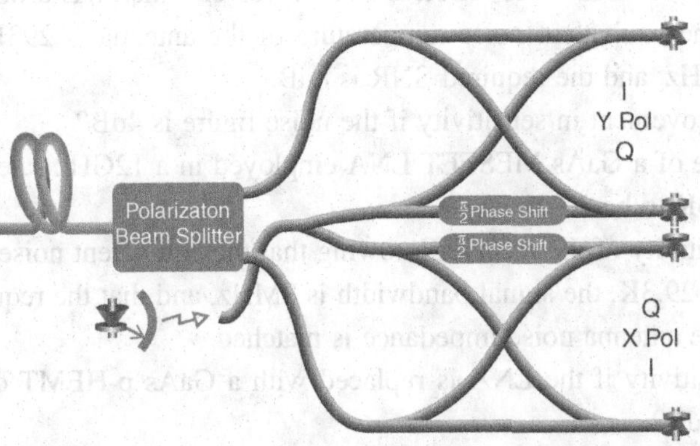

Figure 2.52 Coherent detection scheme [22]

Summary

- Wireless systems architectures changed very little for almost a century. However, during the last decade, coinciding with the emergence of nanoscale CMOS technology into the RF, microwave, and mm-wave arena, we have witnessed a convergence of digital and high-frequency techniques. Direct digital high-frequency modulation, high-frequency analog-to-digital conversion, and digitally controlled, digitally corrected, and digitally calibrated high-frequency circuits have revolutionized wireless and fiber-optic systems.
- There are many similarities between wireless and fiber-optic systems in terms of system architecture, circuit topologies, and semiconductor technology requirements.
- In addition to radio applications, automotive radars and active and passive mm-wave imaging have recently emerged as important applications of high-frequency transceiver ICs.
- In radio and radar systems, phased arrays are employed to improve the SNR, as a form of spatial diversity.

Problems

(1) Derive the equations for the coherent frequency detector in Figure 2.5.

(2) How should the Hartley receiver block diagram in Figure 2.11 be changed if the image frequency is below the LO frequency and the RF frequency is above it?

(3) How should the Weaver receiver block diagram in Figure 2.12 be changed if the image frequency is below the LO frequency and the RF frequency is above it?

(4) A 120GHz distance sensor receiver employs a fundamental frequency VCO, which, due to a design problem, oscillates simultaneously at 107GHz and 127GHz. When testing this receiver in the lab, a 115GHz test signal from an external signal source was applied at the receiver input and the spectrum at the IF output was measured with a power spectrum analyzer. The receiver has a very broadband IF amplifier with 30GHz bandwidth. What frequency components will be seen at the IF output?

(5) (a) Find the sensitivity of a 2.4GHz wireless LAN receiver which has a noise figure of 8dB at 293K. The equivalent noise temperature of the antenna is 293K, the signal bandwidth is 2MHz, and the required SNR is 8dB.

(b) What is the improvement in sensitivity if the noise figure is 4dB?

(6) The 50Ω noise figure of a GaAs MESFET LNA employed in a 12GHz satellite receiver was measured at 293K to be 2.3dB.

(a) What is the sensitivity of the receiver knowing that the equivalent noise temperature of the antenna is 29.3K, the signal bandwidth is 8MHz, and that the required SNR is 6dB? Assume the antenna noise impedance is matched.

(b) What is the sensitivity if the LNA is replaced with a GaAs p-HEMT one having a noise figure of 1dB?

What is the sensitivity if the LNA is replaced with a GaAs p-HEMT one having a noise figure of 0.6dB?

(7) The receiver of a 45GHz radio transceiver with 64QAM modulation at 2GS/s has a receiver noise figure of 8dB. Calculate the required receiver IIP3 if the circuit is to operate correctly with two interferers of −20dBm each in adjacent channels.

(8) A radio receiver has a constant in-band PLL phase noise of −90dBc/Hz and a PLL bandwidth of 1MHz. Calculate the phase error introduced in the received constellation if the PLL phase noise decreases by 20dB per decade outside the PLL bandwidth.

(9) A 60Ω radio link employs two identical single-chip transceivers, one as transmitter, and the other as receiver. The transmit and receive PLL have identical characteristics. The in-band phase noise of the PLL is constant at −86dBc/Hz and the bandwidth of the PLL is 200KHz. Calculate the phase error introduced by the TX-RX link due to the phase noise of the transmit and receive PLLs, assuming that the two PLLs are statistically independent.

(10) Compare the output power and receiver sensitivity of a phased array consisting of 16 transceiver elements and antennas, each with 6dBi antenna gain, 10dB noise figure, and 10dBm output power with those of a single transceiver with 160mW output power, 10dB noise figure, and 18dBi antenna gain. Which one has better SNR and Pout?

(11) Compare power consumption in various 8 × phased array architectures knowing the power consumption is: LNA: 20mW, mixer:10mW, VCO: 30mW, PLL: 60mW, phase shifter: 10mW, VCO buffer: 10mW, summer 20mW (different mixer size for different architectures), ADC = 50mW.

(12) Prove that LO phase shifting by Δt produces the same phase shift at IF as RF phase shifting.

(13) A 140GHz phased array transceiver is fabricated in a SiGe BiCMOS process and consists of 64 transceiver lanes with RF phase shifting, each with its own on-die antenna. The antennas are arranged in an 8 × 8 matrix and are spaced $\lambda/2$ apart. The total transmit power at the antenna port is 500mW.

(a) What is the minimum die area to implement such a system on chip?

(b) What is the transmitted power per element?

(c) If the antenna gain is 5dBi, what is the *EIRP* of the array?

(d) Assuming that each receive lane has a noise figure of 7dB, and a bandwidth of 10GHz, calculate the maximum link range if the phased array is employed for 20Gb/s wireless data transmission with QPSK modulation and an SNR of 14dB.

(14) Prove (2.63) assuming $G = 0$dB.

(15) Calculate the minimum Doppler frequency shift for a velocity and displacement bi-static (separate antennas are employed by the transmitter and by the receiver) sensor which operates at 94GHz over a distance of 20m and can detect velocities as low as 10mm/s for targets of 0.1m^2. The carrier frequency is swept from 93 to 95GHz. If the transmitter output power and antenna gain are + 10dBm and 25dBi, respectively, determine the power of the received signal at the LNA input.

What is the minimum isolation required between the transmitter and the receiver if the receiver IIP3 is −10dBm? Derive the phase noise specification. Is it achievable without range

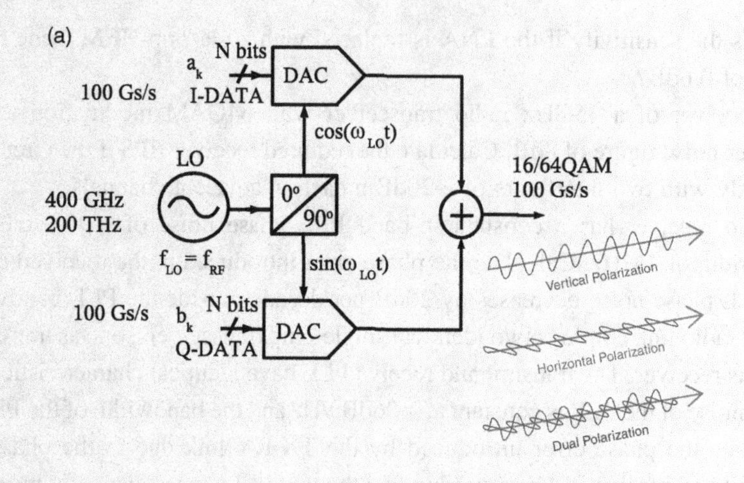

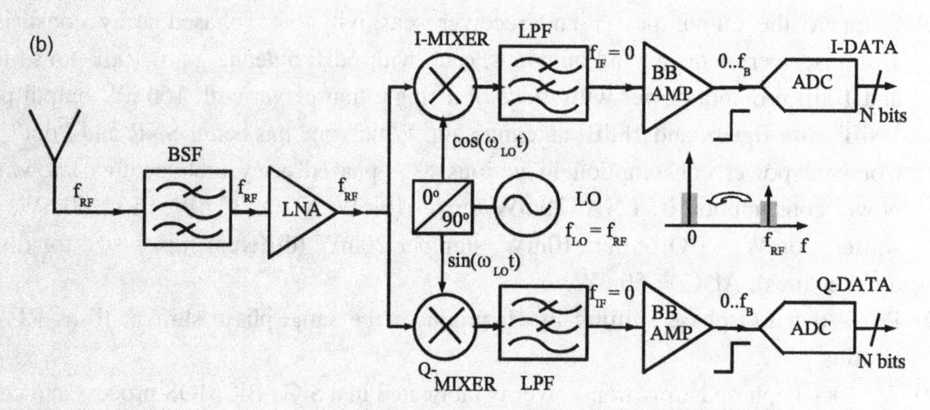

Figure 2.53 (a) Direct 16/64QAM THz or optical transmitter (b) Receiver block diagram

correlation? Assume that the receiver noise figure is 7dB and that the signal bandwidth is 2GHz.

(16) Calculate the temperature resolution if the radiometer in Example 2.11 is realized with a Dicke topology where the antenna switch has a loss of 3dB.

(17) A 112Gb/s Ethernet transmitter with QPSK modulation employs four NRZ 28Gb/s baseband data and modulates an optical carrier with dual polarization. Assuming that the baseband data are not pre-filtered, what is the bandwidth occupied by the main lobe of the modulated optical carrier? Assuming that the bandwidth of each of the four received baseband data paths is 28GHz, calculate the receiver sensitivity if the required SNR is 14dB, the input equivalent noise current of the optical receiver is 15pA/√Hz and the responsivity of the photodiode is $R = 0.7$.

(18) Consider a 16QAM 400GHz wireless link and a 16QAM polarization-multiplexed (vertical and horizontal) 200THz fiber-optic link, each operating at 100Gbaud/sec, as illustrated in Figure 2.53.

(a) What is the aggregate bit rate of each link?

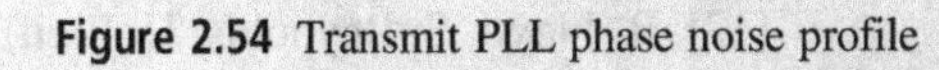

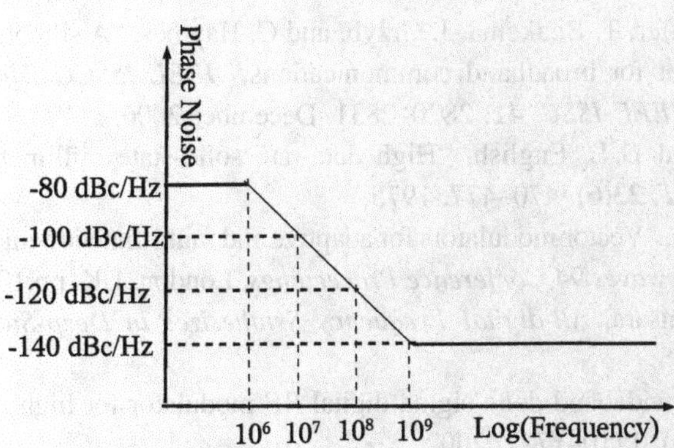

Figure 2.54 Transmit PLL phase noise profile

(b) What is the receiver sensitivity and what is the range of the wireless link if the receiver noise figure is 15dB and the required SNR for a BER of 10^{-6} is 21dB. The transmitter output power is $P_{TX} = 10\text{dBm}$, the transmitter and receiver antenna gains are $G_{TX} = G_{RX} = 18\text{dBi}$, and the receiver bandwidth is 100GHz.

(c) The TX PLL phase noise profile is shown in Figure 2.54. What is the transmitter phase error integrated over the channel bandwidth of 100GHz? What is the best case EVM?

REFERENCES

[1] D.M. Pozar, *Microwave and RF Design of Wireless Systems*, John Wiley & Sons, 2001, Chapter 9.

[2] E. Camargo, "Broadband and mm wave receiver design," *IEEE GaAs IC Symposium Short Course*, Monterey, California, October 2002.

[3] J.J. Lynch, H.P. Moyer, J.H. Schaffner, Y. Royter, M. Sokolich, B. Hughes, Y.J. Yoon, and J.N. Schulman, "Passive millimeter-wave imaging module with preamplified zero-bias detection," *IEEE MTT*, **56**(7): 1592–1600, July 2008.

[4] A. Tessmann, A. Leuther, H. Massler, M. Kuri, M. Riessle, M. Zink, R. Sommer, A. Wahlen, and H. Essen, "Metamorphic HEMT amplifier circuits for use in a high resolution 210GHz radar," *IEEE CSICS Digest*, pp. 248–251, Portland, Oregon, October 2007.

[5] J.W. May and G.M. Rebeiz, "High-performance W-band SiGe RFICs for passive millimeter-wave imaging," *IEEE RFIC Symposium Digest*, Boston, MA, pp. 437–440, June 2009.

[6] M.A. Copeland, S.P. Voinigescu, D. Marchesan, P. Popescu, and M.C. Maliepaard, "5GHz SiGe HBT monolithic radio transceiver with tunable filtering," *IEEE MTT*, **48**(2): 170–181, February 2000.

[7] R. Hartley, "Single-sideband modulator," US Patent no. 1666206, April 1928.

[8] D.K. Weaver, "A third method of generation and detection of single-sideband signals," *Proc. of the IRE*, **44**(12): 1703–1705, December 1956.

[9] F.M. Colebrooke, "Homodyne," *Wireless World and Radio Review*, **13**: 645–648, February 1924.

[10] P.D.L. Beasley, A.G. Stove, B.J. Reits, and B. As, "Solving the problems of a single antenna frequency modulated CW radar," *Proc. IEEE Radar Conference*, pp. 91–395, 1990.

[11] P. Eloranta, P. Seppinen, S. Kallioinen, T. Saarela, and A. Parssinen, "A multimode transmitter in 0.13 μm CMOS using direct-digital RF modulator," *IEEE JSSC*, **42**(12): 2774–2784, December 2007.

[12] B. Floyd, S. Reynolds, U. Pfeiffer, T. Beukema, J. Grzyb, and C. Haymes, "A silicon 60GHz receiver and transmitter chipset for broadband communications," *IEEE ISSCC Digest*, pp. 649–658, February 2006, and *IEEE JSSC*, **41**: 2820–2831, December 2006.

[13] Y.-W. Chang, H.J. Kuno, and D.L. English, "High data-rate solid-state millimeter-wave transmitter module," *IEEE MTT*, **23**(6): 470–477, 1975.

[14] S. Lucyszyn and I.D. Robertson, "Vector modulators for adaptive and multi-function microwave communication systems," *Microwaves 94 Conference Proceedings*, London, UK, pp.103–106.

[15] R.B. Staszewski and P.T. Balsara, *All-digital Frequency Synthesizer in Deep-Submicron CMOS*, John Wiley, 2006.

[16] A. Jerng and C.G. Sodini, "A wideband delta-sigma digital-RF modulator for high data rate transmitters," *IEEE JSSC*, **42**(8): 1710–1722, 2007.

[17] S.A. Mass, *Noise in Linear and Nonlinear Circuits*, Artech House, 2005.

[18] A.M. Niknejad and H. Hashemi, *mm-Wave Silicon Technology 60GHz and Beyond*, Springer, 2008, Chapter 7.

[19] H. Hashemi, "Introduction to antenna arrays," *IEEE VLSI Symposium Workshop*, Kyoto, Japan, June 2009.

[20] M.E. Tiuri, "Radio astronomy receivers," *IEEE AP*, **12**(7): 930–938, December 1964.

[21] A. Tomkins, P. Garcia, and S.P. Voinigescu, "A passive W-band imager in 65nm bulk CMOS," *IEEE CSICS Digest*, Greensboro, NC, October 2009.

[22] K. Roberts, D. Beckett, D. Boertjes, J. Berthold, and C. Laperle, Ciena Corporation, "100G and beyond with digital coherent signal," *IEEE Commun. Mag.*, August 2010.

ENDNOTES

1 The exception is in backplane communications where the baseband data are transmitted over PCB transmission lines, connectors, and cables without modulating a carrier.

2 OFDM stands for orthogonal frequency division multiplexing and was introduced in WLAN systems due to its robustness to multi-path fading. It consists of many frequency-spaced sub-carriers, each modulated by a subset of the data. Most often the modulation method employed is QPSK or M-ary QAM.

3 This is defined in Chapter 4.

3 High-frequency linear noisy network analysis

This chapter reviews the theoretical foundations of small signal and linear noisy network analysis. It starts with the definitions of two-port and multi-port network parameters and matrices, and continues with the introduction of the Smith Chart, *S* – and *ABCD*-parameters, differential *S*-parameters, and a discussion of two-port stability conditions. The coverage of small signal analysis topics ends with the definitions of two-port power gains, which are important in high-frequency amplifier design.

The second part of the chapter provides an introduction to noise power, noise temperature, single-port and *n*-port noise matrix representations, two-port noise figure, and noise parameters. Linear noise analysis and measurement techniques are reviewed. The chapter ends with an extensive treatment of the impact of negative feedback on the optimal noise impedance, minimum noise figure, noise resistance, and noise matching bandwidth of two-ports.

3.1 TWO-PORT AND MULTI-PORT NETWORK PARAMETERS

Just as at lower frequencies, linear network analysis can be conducted at high-frequency using *Y*- or *Z*-parameters which fully describe the behavior of an *n*-port. Additionally, in two-port noise analysis, *H*- and *G*-parameters can also be extended to high frequencies. However, at microwave and mm-wave frequencies, for reasons that will be described shortly, the scattering, or *S*-parameter, and the transmission, or *ABCD*, matrices play central roles.

In the general case of an *n*-port, we can define incident and reflected voltages, V_n^+ and V_n^-, incident and reflected currents, I_n^+ and I_n^-, and a reference terminal plane t_n at port *n* that allows us to synchronize the signals at all ports, as illustrated in Figure 3.1 [1]. The total voltages and currents at port *n* are described as

$$V_n = V_n^+ + V_n^- \tag{3.1}$$

and

$$I_n = I_n^+ + I_n^- \tag{3.2}$$

where

$$I_n = I_n^+ + I_n^- = \frac{V_n^+}{Z_0} - \frac{V_n^-}{Z_0} \tag{3.3}$$

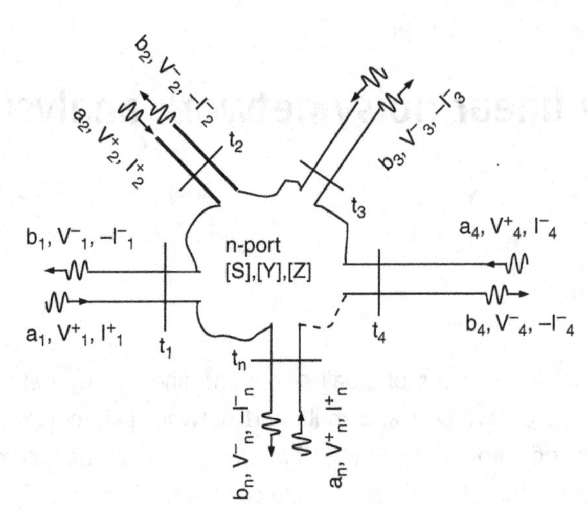

Figure 3.1 Generalized representation of an n-port [1]. Currents which exit a port have a "–" sign

and Z_0 is the reference impedance, for simplicity assumed identical for all ports.

Additionally, we can define normalized incident and reflected power waves $a_n = \frac{V_n^+}{\sqrt{Z_0}}$, $b_n = \frac{V_n^-}{\sqrt{Z_0}}$ at each port, such that the following relations hold at any position x relative to the reference plane t_n at port n

$$a_i(x) = \frac{V_n(x) + Z_0 I_n(x)}{2\sqrt{Z_0}} \tag{3.4}$$

and

$$b_i(x) = \frac{V_n(x) - Z_0 I_n(x)}{2\sqrt{Z_0}}. \tag{3.5}$$

The average power associated with the incident wave at port i becomes

$$P_{in,\,avg,\,i} = \frac{V_n^+ \times I_n^{+*}}{2} = \frac{|a_i|^2}{2} \tag{3.6}$$

while the average reflected (transmitted) power by port i is given by

$$P_{out,\,avg,\,i} = \frac{V_n^- \times I_n^{-*}}{2} = \frac{|b_i|^2}{2}. \tag{3.7}$$

3.1.1 *Y*-parameters

The $n \times n$ *Y*-parameter matrix (or *Y*-matrix) of an n-port describes the relationship between the total currents and the total voltages at each port

$$\begin{bmatrix} I_1 \\ I_2 \\ \cdot \\ \cdot \\ \cdot \\ I_n \end{bmatrix} = \begin{bmatrix} Y_{11} & Y_{12} & \cdot & \cdot & \cdot & Y_{1n} \\ Y_{21} & Y_{22} & \cdot & \cdot & \cdot & Y_{2n} \\ \cdot & \cdot & & & & \cdot \\ \cdot & \cdot & & & & \cdot \\ \cdot & \cdot & & & & \cdot \\ Y_{n1} & Y_{n2} & \cdot & \cdot & \cdot & Y_{nn} \end{bmatrix} \times \begin{bmatrix} V_1 \\ V_2 \\ \cdot \\ \cdot \\ \cdot \\ V_n \end{bmatrix} \tag{3.8}$$

where

$$Y_{ij} = \frac{I_i}{V_j} [V_k = 0 \, for \quad k \neq j]. \tag{3.9}$$

[Y] is also referred to as the *short-circuit admittance matrix* of the *n*-port since its parameters are determined/measured under short-circuit conditions at each port.

3.1.2 *Z*-parameters

Similarly, the dual relationship giving the total voltages at each port as a function of the total currents entering each port is captured by the Z-parameter matrix (in short Z-matrix) of the multi-port

$$\begin{bmatrix} V_1 \\ V_2 \\ \cdot \\ \cdot \\ \cdot \\ V_n \end{bmatrix} = \begin{bmatrix} Z_{11} & Z_{12} & \cdot & \cdot & \cdot & Z_{1n} \\ Z_{21} & Z_{22} & \cdot & \cdot & \cdot & Z_{2n} \\ \cdot & \cdot & & & & \cdot \\ \cdot & \cdot & & & & \cdot \\ \cdot & \cdot & & & & \cdot \\ Z_{n1} & Z_{n2} & \cdot & \cdot & \cdot & Z_{nn} \end{bmatrix} \times \begin{bmatrix} I_1 \\ I_2 \\ \cdot \\ \cdot \\ \cdot \\ I_n \end{bmatrix} \tag{3.10}$$

where

$$Z_{ij} = \frac{V_i}{I_j} [I_k = 0 \, for \quad k \neq j] \tag{3.11}$$

and

$$[Z] = [Y]^{-1}. \tag{3.12}$$

Equation (3.12) shows that, if one of the *Z* or *Y* matrices of a multi-port is known, the other can be immediately obtained from its inverse.

3.1.3 *H*-parameters

In a variety of applications involving two-ports, it is sometimes useful to describe the two-port behavior using the *H*-matrix

$$\begin{bmatrix} V_1 \\ I_2 \end{bmatrix} = \begin{bmatrix} h_{11} & h_{12} \\ h_{21} & h_{22} \end{bmatrix} \times \begin{bmatrix} I_1 \\ V_2 \end{bmatrix} \tag{3.13}$$

where

$$h_{11} = \frac{V_1}{I_1}[\text{when } V_2 = 0]; h_{12} = \frac{V_1}{V_2}[\text{when } I_1 = 0]; h_{21} = \frac{I_2}{I_1}[\text{when } V_2 = 0]; h_{22} = \frac{I_2}{V_2}[\text{when } I_1 = 0]. \tag{3.14}$$

The use of H-parameters is particularly useful in the analysis and design of circuits with series–shunt feedback and in the extraction of the base resistance of bipolar transistors, or of the gate resistance of FETs, from high-frequency measurements.

3.1.4 *G*-parameters

The G-matrix, in many ways the dual of the H-matrix, is a convenient representation in the analysis of circuits with shunt–series feedback and is defined as

$$\begin{bmatrix} I_1 \\ V_2 \end{bmatrix} = \begin{bmatrix} g_{11} & g_{12} \\ g_{21} & g_{22} \end{bmatrix} \times \begin{bmatrix} V_1 \\ I_2 \end{bmatrix} \tag{3.15}$$

with

$$g_{11} = \frac{I_1}{V_1}[\text{when } I_2 = 0]; g_{12} = \frac{I_1}{I_2}[\text{when } V_1 = 0]; g_{21} = \frac{V_2}{V_1}[\text{when } I_2 = 0]; g_{22} = \frac{V_2}{I_2}[\text{when } V_1 = 0]. \tag{3.16}$$

Because of the many possible combinations of port currents and voltages in a multi-port, H- and G-matrices are only defined for two-ports where only two such combinations exist: one for the H-matrix and one for G-matrix.

3.1.5 *ABCD*-parameters

Although Y-, Z-, H-, and G-network parameters can be employed to characterize any two-port, most high-frequency circuits consist of a series of cascaded two-ports. In this case, circuit analysis is considerably simplified if the two-ports are described by their $ABCD$, or transmission, or chain, matrix

$$\begin{bmatrix} V_1 \\ I_1 \end{bmatrix} = \begin{bmatrix} A & B \\ C & D \end{bmatrix} \times \begin{bmatrix} V_2 \\ -I_2 \end{bmatrix} \tag{3.17}$$

where

$$A = \frac{V_1}{V_2}[\text{when } I_2 = 0]; B = \frac{-V_1}{I_2}[\text{when } V_2 = 0]; C = \frac{I_1}{V_2}[\text{when } I_2 = 0]; D = \frac{-I_1}{I_2}[\text{when } V_2 = 0] \tag{3.18}$$

and which relates the input port voltage and current to the output port voltage and current.

Unlike in other network matrices, the $ABCD$-matrix definition assumes that the current at port two flows out of the port, Figure 3.2. This arrangement allows for the overall $ABCD$-matrix of a cascade of two-ports to be calculated as the product of the individual matrices of each two-port

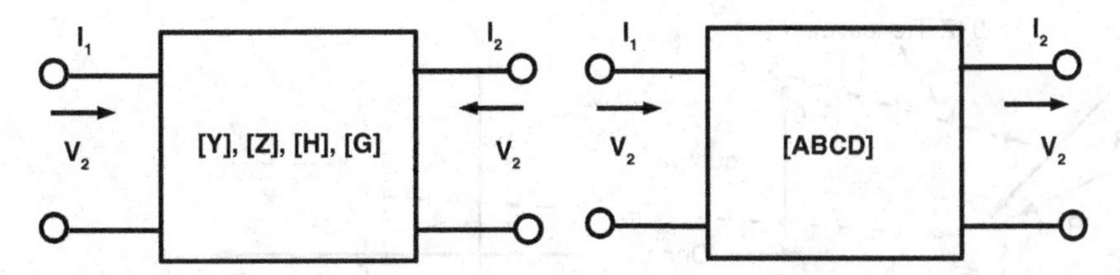

Figure 3.2 Two-port *Y, Z, G, H* and *ABCD*-matrix representations

$$\begin{bmatrix} A & B \\ C & D \end{bmatrix} = \begin{bmatrix} A_1 & B_1 \\ C_1 & D_1 \end{bmatrix} \times \begin{bmatrix} A_2 & B_2 \\ C_2 & D_2 \end{bmatrix} \times \ldots \times \begin{bmatrix} A_n & B_n \\ C_1 & D_n \end{bmatrix} = \prod_{i=1}^{n} \begin{bmatrix} A_i & B_i \\ C_i & D_i \end{bmatrix}. \quad (3.19)$$

Among others, the *ABCD*-matrix is particularly useful in characterizing transmission lines and in de-embedding pad and interconnect parasitics from the test structure measurement data.

3.1.6 The reflection coefficient and the Smith Chart

One of the key differences between standard circuit analysis and high-frequency circuit analysis is electrical size. In low-frequency analysis, the electrical size of the circuit is assumed to be much smaller than the electrical wavelength at the highest frequency of interest. Although it is advisable to minimize the electrical size in high-frequency circuits, inevitably situations arise when the electrical length of the circuit is fractions of, or even several, wavelengths in size. This is especially common in the upper mm-wave frequency range. In such cases, the circuit and the interconnect between circuit blocks must be treated using transmission line theory to account for the possible impedance mismatches between signal source, load, and interconnect impedances, which will result in multiple signal reflections even when the interconnect transmission line is terminated on a matched impedance at one end.

The parameter that describes the impedance mismatch with respect to a reference impedance, Z_0, is the reflection coefficient, Γ, which for an arbitrary impedance, Z, or admittance, Y, is defined as

$$\Gamma = \frac{Z - Z_0}{Z + Z_0} = \frac{Y_0 - Y}{Y + Y_0} \quad (3.20)$$

or, in its normalized form

$$\Gamma = \frac{z - 1}{z + 1} \quad (3.21)$$

where $z = Z/Z_0$. For example, at high frequencies, it is common practice to describe the signal source and load impedances by their respective reflection coefficients Γ_S and Γ_L, both defined with respect to the reference impedance Z_0, typically equal to 50Ω.

A scalar parameter related to the reflection coefficient is the voltage standing wave ratio, *VSWR*, or in shorter form, *SWR*, which is sometimes used to characterize the port impedances when information about the phase is not of concern

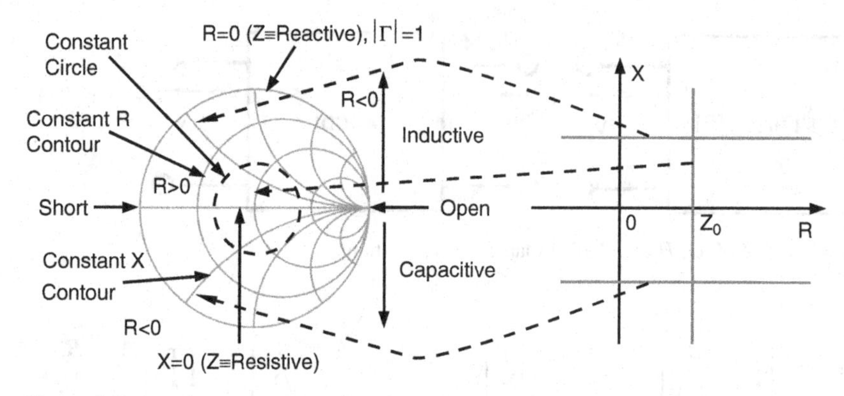

Figure 3.3 Impedance transformation from the complex plane to the Z Smith Chart

$$VSWR = \frac{1 + |\Gamma|}{1 - |\Gamma|}. \tag{3.22}$$

The Smith Chart was developed in 1939 to represent impedances by their reflection coefficient and to solve transmission line problems. It provides a convenient vehicle to graphically display a large range of impedance values in a compact and intuitive manner. It has proven a very useful and intuitive tool in the analytical design of impedance matching networks and has been integrated into the most sophisticated computer-aided design (CAD) software packages, rather than being replaced by them. The Smith Chart is obtained by an impedance transformation from the conventional complex plane to the Γ-plane. As illustrated in Figure 3.3, lines in the complex plane become circles in the Γ-plane. In Smith Chart jargon, we use terms such as *constant resistance*, *constant reactance*, and *constant Γ circles*. Impedances with a positive real part are located inside the unit circle of $|\Gamma| = 1$, while those with a negative real part are outside this circle. The region outside the unit circle is avoided in amplifier design but is useful in the design of negative resistance oscillators. Both impedance, or Z, and admittance, or Y, versions of the Smith Chart are employed in high-frequency circuit design, as shown in Figure 3.4, where a $05 + j0.5$ normalized impedance is represented on both charts.

Techniques employing the Smith Chart in the design of matching networks will be discussed in Chapter 5.

3.1.7 Why *S*-parameters?

The traditional linear network parameters used at low frequencies pose several challenges in high-frequency circuit analysis and measurements. For example, *Y*-, *Z*-, *H*-, and *G*-parameters require open-circuit or short-circuit measurement conditions which are difficult to realize with precision at high frequencies. In contrast, *S*-parameters are measured under controlled, finite impedance conditions. Transmission lines with 50Ω characteristic impedance and broadband 50Ω terminations can be fabricated with excellent performance on alumina or semiconductor substrates, including silicon, at frequencies beyond 300GHz. Other, higher value, precisely controlled broadband impedances can be manufactured as waveguide components.

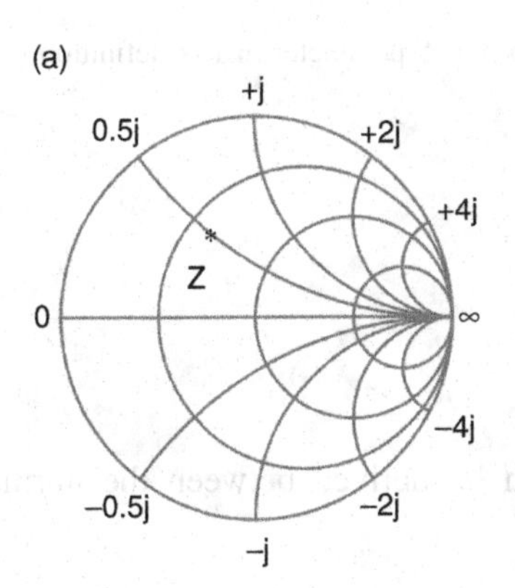

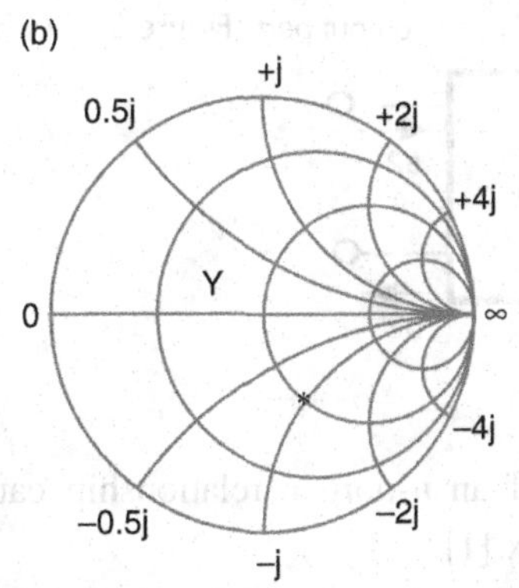

Figure 3.4 (a) Impedance and (b) admittance Smith Chart

While *Y*- and *Z-parameters* are based on voltages and currents, *S*-parameters are defined as ratios of incident and reflected traveling power waves which can be measured accurately using directional couplers and power detectors, typically combined in a *Network Analyzer* [1].

Similar to *Y*- and *Z*-parameters, the scattering matrix can be used to fully characterize the behavior of a linear *n*-port. It describes the relationship between the incident, a_i, and reflected, b_i, voltage waveforms

$$\begin{bmatrix} b_1 \\ b_2 \\ \cdot \\ \cdot \\ \cdot \\ b_n \end{bmatrix} = \begin{bmatrix} S_{11} & S_{12} & \cdot & \cdot & \cdot & S_{1n} \\ S_{21} & S_{22} & \cdot & \cdot & \cdot & S_{2n} \\ \cdot & \cdot & & & & \cdot \\ \cdot & \cdot & & & & \cdot \\ \cdot & \cdot & & & & \cdot \\ S_{n1} & S_{n2} & \cdot & \cdot & \cdot & S_{nn} \end{bmatrix} \times \begin{bmatrix} a_1 \\ a_2 \\ \cdot \\ \cdot \\ \cdot \\ a_n \end{bmatrix} \tag{3.23}$$

where

$$S_{ij} = \frac{b_i}{a_j}[a_k = 0 \text{ for } k \neq j]. \tag{3.24}$$

Equation (3.23) implies that S_{ij}, which represents the transmission coefficient from port *j* to port *i*, is found by driving port *j* with an incident wave a_j and measuring the reflected wave b_i coming out of port *i*, while the incident waves at all other ports than *j* are set equal to zero, and those ports are terminated with matched loads to avoid reflections. S_{ii} represents the reflection coefficient at port *i* when all other ports are terminated on matched loads. The *S*-parameters are normalized with respect to the reference impedance Z_0. In particular, for two-ports (Figure 3.5):

- S_{11} is known as the input reflection coefficient or input return loss,
- S_{12} is known as the reverse gain or the isolation of the two-port,
- S_{21} is known as the transducer power gain of the two-port, and
- S_{22} is known as the output reflection coefficient or output return loss.

S-parameters are most often specified in dB as $20\log_{10}(S_{ij})$.

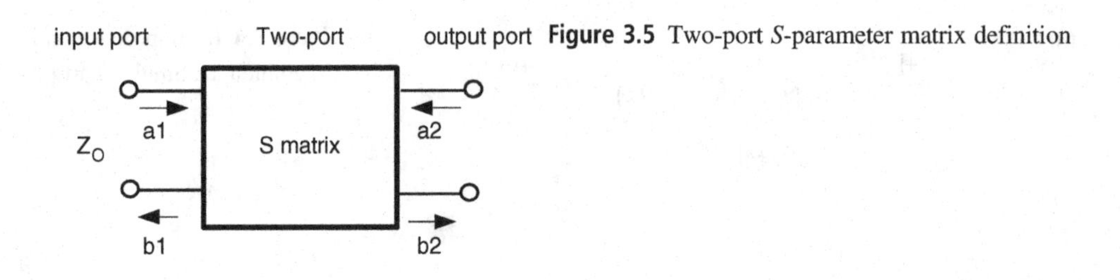

Figure 3.5 Two-port *S*-parameter matrix definition

In the general case of an *n*-port, a relationship can be derived between the normalized *Z*-matrix and the *S*-matrix [1]

$$[S] = ([Z] - [U])([Z] + [U])^{-1} \tag{3.25}$$

$$[Z] = ([U] + [S])([U] - [S])^{-1} \tag{3.26}$$

where *U* is the identity matrix. In the case of a single-port, (3.25) translates to the familiar definition of the reflection coefficient.

Similar to the *ABCD*-matrix, for a two-port we can define a chain scattering transfer matrix, *T*, simply referred to as the transfer matrix, which describes the relationship between the input incident and reflected waves and the incident and reflected waves at the output of the two-port

$$\begin{bmatrix} a_1 \\ b_1 \end{bmatrix} = \begin{bmatrix} T_{11} & T_{12} \\ T_{21} & T_{22} \end{bmatrix} \times \begin{bmatrix} a_2 \\ b_2 \end{bmatrix}. \tag{3.27}$$

Appendix A3 provides matrix transformations between the various representations of a two-port.

3.1.8 Differential *S*-parameters

Differential, common-mode, and single-ended port impedances

Let us consider the differential single-port in Figure 3.6. This circuit can also be regarded as a conventional two-port with input and output voltages and currents v_1, i_1, v_2, and i_2, described by the *Z*-matrix equation

$$\begin{bmatrix} v_1 \\ v_2 \end{bmatrix} = \begin{bmatrix} Z_{11} & Z_{12} \\ Z_{21} & Z_{22} \end{bmatrix} \times \begin{bmatrix} i_1 \\ i_2 \end{bmatrix}. \tag{3.28}$$

For example, two, infinitely long, coupled transmission lines terminated at the far on a matched differential load can be treated as a two-port or as a differential single-port where Z_{ij} are determined by the mutual inductances and capacitances (L_{12}, L_{21}, C_{12}, and C_{21}) and self-inductances and capacitances L_{11}, L_{22}, C_{11}, and C_{22}) of the coupled lines.

We can define the differential- and common-mode voltages and currents of the differential single-port as

$$v_{dm} \stackrel{\text{def}}{=} v_1 - v_2; \; i_{dm} \stackrel{\text{def}}{=} \frac{i_1 - i_2}{2} \tag{3.29}$$

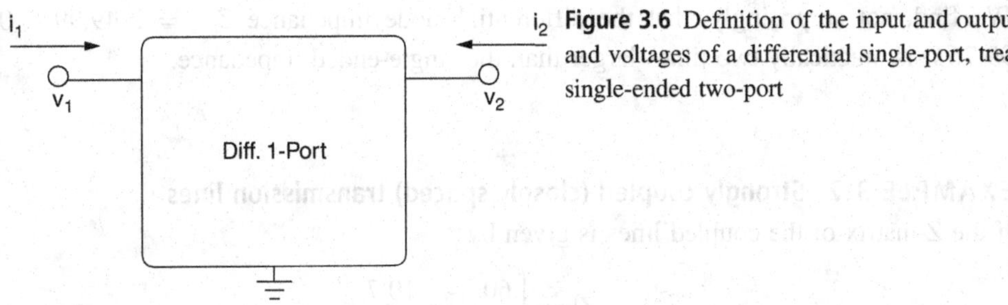

Figure 3.6 Definition of the input and output currents and voltages of a differential single-port, treated as a single-ended two-port

$$v_{cm} \stackrel{\text{def}}{=} \frac{v_1 + v_2}{2}; \ i_{cm} \stackrel{\text{def}}{=} i_1 + i_2. \tag{3.30}$$

Since $i_1 = -i_2$ and $v_1 = -v_2$, from (3.29) and (3.30), we obtain the expression of the differential-mode impedance as

$$z_{dm} \stackrel{\text{def}}{=} \frac{v_{dm}}{i_{dm}} = 2(Z_{11} - Z_{12}). \tag{3.31}$$

Similarly, since $i_1 = i_2$ and $v_1 = v_2$, from (3.30) and (3.28), the common-mode impedance becomes

$$z_{cm} \stackrel{\text{def}}{=} \frac{v_{cm}}{i_{cm}} = \frac{(Z_{11} + Z_{12})}{2}. \tag{3.32}$$

Finally, we can also define a single-ended impedance with one of the differential terminals grounded

$$z_{se} \stackrel{\text{def}}{=} \frac{v_1}{i_1} = Z_{11} + Z_{12}. \tag{3.33}$$

EXAMPLE 3.1 Weakly coupled (widely spaced) transmission lines

Assume that the Z-matrix of the coupled lines is given by

$$[Z] = \begin{bmatrix} 49.99 & 0.11 \\ 0.11 & 49.99 \end{bmatrix}.$$

Calculate the single-ended, common-mode, and differential-mode impedances.

Solution

From the definition (3.32), we can calculate the common-mode impedance as

$$z_{cm} = \frac{(49.99 + 0.11)}{2} = 25.05\text{Ohm}$$

and, from (3.33), the single-ended impedance becomes

$$z_{se} = 49.99 + 0.11 = 50.1\text{Ohm}.$$

We find, not surprisingly, that the differential-mode impedance $Z_{dm} = 2(49.99 - 0.11) = 99.76\,\Omega$ is practically two times larger than the single-ended impedance.

EXAMPLE 3.2 Strongly coupled (closely spaced) transmission lines
If the Z-matrix of the coupled lines is given by

$$[Z] = \begin{bmatrix} 60.24 & 10.7 \\ 10.7 & 60.24 \end{bmatrix}$$

calculate the common-mode, single-ended, and differential-mode impedances of the coupled lines.

Solution

Using the same equations as in the previous example, we can calculate the common-mode, differential-mode, and single-ended impedances as

$$z_{cm} = \frac{(60.24 + 10.7)}{2} = 35.47\text{Ohm}$$

$$z_{dm} = 2\,(60.24 - 10.7) = 99.1\text{Ohm}$$

and

$$z_{se} = 60.24 + 10.7 = 70.94\text{Ohm}.$$

In this case, because of the strong coupling, the differential-mode impedance is no longer two times larger than the single-ended one. This has important implications for circuit design and layout if the coupled lines must be simultaneously matched in differential mode and single-ended mode.

Differential two-port

Just as in low-frequency analog integrated circuits, where fully differential amplifiers are preferred because of their improved immunity to interference and supply noise, fully differential low-noise and power amplifiers, mixers, phase shifters, and modulators have become quite common in high-frequency integrated circuits. A fully differential LNA or PA has four ports (two at the input and two at the output) and is therefore accurately described as a four-port by its associated 4×4 S-, Z-, or Y-parameter matrices. By rearranging the rows in the S-parameter matrix as illustrated in Figure 3.7, we can re-group the four ports into a differential-mode two-port with incident waves a_{dm1} and a_{dm2} and reflected waves b_{dm1} and b_{dm2}, and a common-mode two-port with incident waves a_{cm1} and a_{cm2} and reflected waves b_{dm1} and b_{dm2}. The corresponding S-matrix equation becomes

$$\begin{bmatrix} b_{dm1} \\ b_{dm2} \\ b_{cm1} \\ b_{cm2} \end{bmatrix} = \begin{bmatrix} S_{dd} & S_{dc} \\ S_{cd} & S_{cc} \end{bmatrix} \times \begin{bmatrix} a_{dm1} \\ a_{dm2} \\ a_{cm1} \\ a_{cm2} \end{bmatrix} \tag{3.34}$$

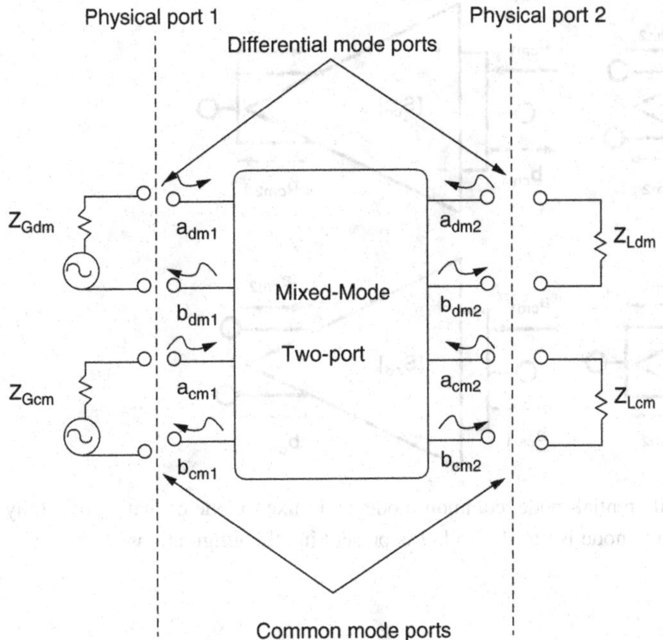

Figure 3.7 Differential *S*-parameter definitions

where S_{dd} and S_{cc} are 2×2 submatrices that describe the differential-mode and common-mode responses, respectively.

Additionally, the off-diagonal submatrices S_{dc} and S_{cm} characterize the (usually) undesired "mixed-mode" responses of the four-port, from the differential-mode input to the common-mode output, and vice-versa. Ideally, $[S_{dc}]$ and $[S_{cd}]$ should be [0].

Fully differential op-amps, low-noise amplifiers and power amplifiers, and coupled transmission lines are the most obvious examples of a differential two-port. Figure 3.8 illustrates how the different mixed-mode responses of a fully differential amplifier are excited and defined.

3.1.9 Two-port stability

The stability analysis of a two-port is central to all amplifier and oscillator design problems. In general, it is possible for oscillation to occur if either the input-port or output-port impedance of the two-port has a negative real part. At the same time, the port impedances are functions of the load and signal source impedances terminating and driving the two-port. We can define the requirements for the unconditional stability of a two-port by imposing that the magnitude of its input and output reflection coefficients be smaller than 1 for any passive load and source impedance. The word "passive" implies that the real part of the load and source impedance is positive, that is $|\Gamma_S| < 1, |\Gamma_L| < 1$.

Mathematically, the unconditional stability condition can be cast as

$$|\Gamma_{in}| = |S'_{11}| = \left| S_{11} + \frac{S_{12}S_{21}\Gamma_L}{1 - S_{22}\Gamma_L} \right| < 1 \tag{3.35}$$

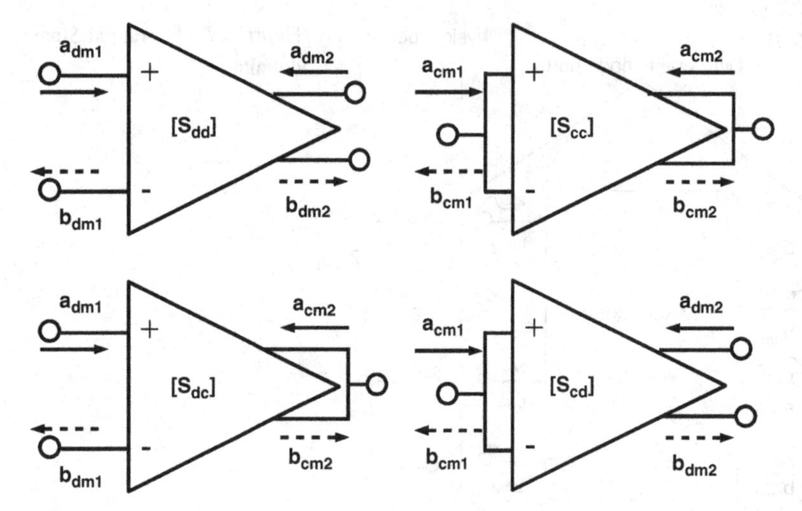

Figure 3.8 Examples of the differential-mode, common-mode, and mixed-mode excitation of a fully differential amplifier. The ground node is not shown but is present in all configurations

and

$$|\Gamma_{out}| = |S'_{22}| = \left|S_{22} + \frac{S_{12}S_{21}\Gamma_G}{1 - S_{22}\Gamma_G}\right| < 1. \tag{3.36}$$

If either (3.35) or (3.36) is violated, the two-port is only conditionally stable. The latter implies that the two-port will be stable for a limited range of passive load and source impedances and it may oscillate for other passive terminations.

Since (3.35) and (3.36) are functions of frequency, the stability of a two-port depends on frequency. For example, transistors are only conditionally stable at low frequencies and, in general, become unconditionally stable at very high frequencies, close to f_T and f_{MAX}.

In practice, the necessary and sufficient test for unconditional stability is defined as

$$k = \frac{1 - |S_{11}|^2 - |S_{22}|^2 + |D|^2}{2|S_{12}||S_{21}|} > 1 \tag{3.37}$$

and

$$|D| = |S_{22}S_{11} - S_{21}S_{12}| < 1 \tag{3.38}$$

which must be satisfied simultaneously. k is known as Rollet's stability factor. Since normally $|D| < 1$, we can show that $k > 1$ is sufficient for unconditional stability.

Although (3.37) and (3.38) are necessary and sufficient conditions to guarantee the stability of a two-port, they cannot be extended to a chain of several cascaded devices/two-ports because they involve constraints on two separate parameters [1]. A new stability criterion, based on a single parameter, μ, was proposed more recently

$$\mu = \frac{1 - |S_{11}|^2}{|S_{22} - DS^*_{11}| + |S_{12}S_{21}|} > 1. \tag{3.39}$$

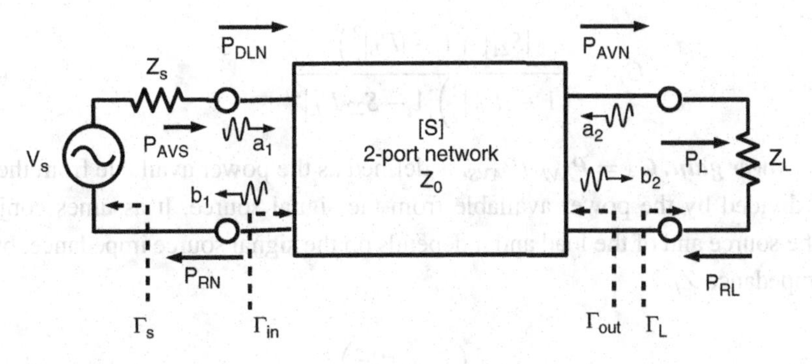

Figure 3.9 Input and output power definitions for a two-port network

Since larger values of μ imply greater stability, it can be a very useful tool in the design of high-frequency amplifiers.

In the analysis of the stability of a differential two-port, we can define stability conditions (k, etc.) for each of the four modes of operation using the corresponding submatrices. However, there are no generalized stability criteria for n-ports where n is larger than 2.

3.1.10 Two-port power gain definitions

While voltages and currents are extremely difficult to measure at high frequencies, it is much easier to measure signal power. At the same time, the interstage impedance levels achievable at high frequencies are neither very high nor very small, making power gain, rather than voltage gain or current gain, critically important in high-frequency circuit design.

Consider the two-port network shown in Figure 3.9, driven by a signal source with impedance Z_S and reflection coefficient Γ_S, and terminated on the load impedance Z_L with a corresponding reflection coefficient Γ_L. The power delivered to the network from the signal source, and to the load from the network, is given by

$$P_{DLN} = P_{in} = P_{AVS} - P_{RN} = \frac{|a_1|^2}{2} - \frac{|b_1|^2}{2} \tag{3.40}$$

and

$$P_L = P_{AVN} - P_{RL} = \frac{|b_2|^2}{2} - \frac{|a_2|^2}{2} \tag{3.41}$$

respectively, where $P_{AVS} = P_{DLN}$ when $\Gamma_{in} = \Gamma^*_S$ and $P_{AVN} = P_L$ when $\Gamma_L = \Gamma^*_{out}$.

Because a variety of load and signal source conditions exist, several power gains can be defined as a function of the two-port S-parameters, Γ_L, Γ_S, and of the input and output reflection coefficients of the two-port: Γ_{in} and Γ_{out}.

- ***The power gain***, $G_P = P_L / P_{DLN}$, is defined as the power dissipated in the load divided by the power delivered to the input of the two-port. It is independent of the signal source impedance Z_S.

$$G_P = \frac{|S_{21}|^2\left(1 - |\Gamma_L|^2\right)}{\left(1 - |\Gamma_{in}|^2\right)|1 - S_{22}\Gamma_L|^2}. \tag{3.42}$$

- ***The available power gain,*** $G_A = P_{AVN}/P_{AVS}$, is defined as the power available from the two-port network divided by the power available from the signal source. It assumes conjugate matching of the source and of the load and it depends on the signal source impedance, but not on the load impedance Z_L

$$G_A = \frac{|S_{21}|^2\left(1 - |\Gamma_S|^2\right)}{|1 - S_{11}\Gamma_S|^2\left(1 - |\Gamma_{out}|^2\right)}. \tag{3.43}$$

- ***The transducer power gain,*** $G_T = P_L/P_{AVS}$, is the ratio of the power delivered to the load to the power available from the source. It depends on both Z_L and Z_S

$$G_T = \frac{|S_{21}|^2\left(1 - |\Gamma_S|^2\right)\left(1 - |\Gamma_L|^2\right)}{|1 - \Gamma_S\Gamma_{in}|^2|1 - S_{22}\Gamma_L|^2} = \frac{|S_{21}|^2\left(1 - |\Gamma_S|^2\right)\left(1 - |\Gamma_L|^2\right)}{|1 - S_{11}\Gamma_S|^2|1 - \Gamma_{out}\Gamma_L|^2}. \tag{3.44}$$

- ***The maximum available power gain*** (MAG – or maximum transducer power gain, $G_{T,MAX}$) represents G_A, G_T when the input and output of the two-port are conjugately matched to the source impedance and to the load impedance, respectively. MAG sets an upper limit on the gain of a single-stage amplifier without risking oscillation

$$MAG = \left|\frac{S_{21}}{S_{12}}\right|\left(k - \sqrt{k^2 - 1}\right) \tag{3.45}$$

where k is Rollet's stability factor.

- ***The unilateral power gain,*** U, is the maximum available power gain, when $S_{12} = 0$. S_{12} can be canceled out externally at each frequency by an appropriate reactive feedback network. Obviously, this is an idealized power gain that is even more difficult to achieve than MAG. In practice, because of its almost ideal 20dB per decade decrease with frequency, U is employed to extract the maximum frequency of oscillation of a transistor from measured S-parameter data.
- ***The associated power gain,*** G_{AS}, is the available power gain when the input of the two-port is driven by a signal source whose impedance is equal to the optimum noise impedance of the two-port (transistor), described by Γ_{SOPT}

$$G_{AS} = \frac{|S_{21}|^2\left(1 - |\Gamma_{OPT}|^2\right)}{|1 - S_{11}\Gamma_{OPT}|^2\left(1 - |\Gamma_{out}|^2\right)}. \tag{3.46}$$

The associated power gain is important in low-noise amplifier design.

3.2 NOISE

Noise sets the limit to the smallest signal that can be processed by a receiver. In electronic circuits, noise manifests itself as a *randomly varying* voltage or current and is treated as a *stationary random process*. This implies that, although noise currents and voltages vary in time randomly, their statistical characteristics do not change with time [2]. Deterministic signals such as interference and signal distortion are mistakenly referred to as noise. However, they are not noise and, therefore, will not be discussed here.

In electronic systems, noise can be generated internally by active and passive devices or it can be received along with the signal. From a circuit designer's perspective, it is critical to be able to predict the noise contribution of the various electronic components that make up the circuit and to be able to minimize the degradation of the signal-to-noise ratio as signals are processed by the electronic circuit. The noise contribution of electronic components is captured by their noise parameters, which in turn, depend on the internal noise sources.

The most important form of noise encountered in electronics is *thermal noise*, also known as Nyquist, or Johnson, noise. Thermal noise is generated by the random motion of charges in lossy components such as resistors, lossy inductors, lossy transmission lines, lossy capacitors, etc., but also by active devices such as diodes and transistors. Thermal noise can also be generated by sources external to the circuit such as the Earth, atmospheric attenuation, the Sun, interstellar background radiation, all of which involve the random motion of thermally excited charges and form the basis of the science of radioastronomy. In this case, thermal noise propagates as a *noise power wave*.

Other important types of noise generated in electronic circuits include:

- *shot noise*, caused to the random motion of charged carriers traversing potential barriers in *pn* junctions and bipolar transistors, and whose power spectral density is $2qI$, where q is the electron charge and I is the DC current passing through the *pn* junction;
- diffusion noise, present in HEMTs;
- *flicker*, or 1/f, noise present in most active and passive electronic components and whose physical origins are still not well understood; and
- generation-recombination, or G-R, noise, encountered in early diodes and transistors, and which is frequency dependent and only important at frequencies below 1GHz.

Although these other sources of noise are different in their origins from thermal noise, they are similar enough that they can be treated in the general framework of thermal noise, which is considered a standard. This situation is prompted by the fact that thermal noise can be precisely predicted from fundamental physics and that it is directly associated with the concept of *noise temperature*, the cornerstone of all noise measurements in electronics.

3.2.1 Thermal noise, noise power, and noise temperature

Two different definitions of noise temperature, T_n, are encountered in the literature [3]. The first, known as the *physical temperature*, states that the noise temperature is "equal to the physical temperature of a resistor whose thermal noise would result in the given spectral density of (available) noise power p_n" [3]. In the second definition, known as the *power definition*, the noise temperature of a resistor is defined as the ratio between the given available noise power of the resistor and Boltzmann's constant (1.38065×10^{-23} J/K)

$$T_n \equiv \frac{p_n}{k} \tag{3.47}$$

The Planck equation for the available noise power density from a resistor

$$p_n^{Planck} = kT \left[\frac{\frac{hf}{kT}}{e^{\frac{hf}{kT}} - 1} \right] \tag{3.48}$$

which is widely considered to be accurate, reduces to the well-known Rayleigh–Jeans form

$$p_n^{R-J} = kT \tag{3.49}$$

when $hf/kT << 1$. The latter is a valid approximation at room temperature from DC up to a few hundred GHz. Here, T is the absolute physical temperature of the resistor in degrees Kelvin, f is the frequency, and h is Planck's constant (6.626068×10^{-34} m^2kg/s).

However, in very low-noise systems or at very high frequencies above 1THz, the zero-point vacuum fluctuations can contribute significantly to the overall noise of an amplifier or mixer [3]

$$p_n^{vac} = \frac{hf}{2}. \tag{3.50}$$

While there is general agreement in the scientific community that this vacuum fluctuation term corresponding to the lowest energy state of the quantum harmonic oscillator must be accounted for, there is disagreement as to how this should be done. One option is to add it to the available noise power spectral density from a resistor, leading to the Callen–Welton form

$$p_n^{C-W} = p_n^{Planck} + p_n^{vac} = kT \left[\frac{\frac{hf}{kT}}{e^{\frac{hf}{kT}} - 1} \right] + \frac{hf}{2}. \tag{3.51}$$

By applying the noise temperature definition (3.46) to the three definitions of the available noise power spectral density from a resistor, we end up with three different expressions for the noise temperature

$$T_n^{Planck} = T \left[\frac{\frac{hf}{kT}}{e^{\frac{hf}{kT}} - 1} \right] \tag{3.52}$$

$$T_n^{R-J} = T \tag{3.53}$$

and

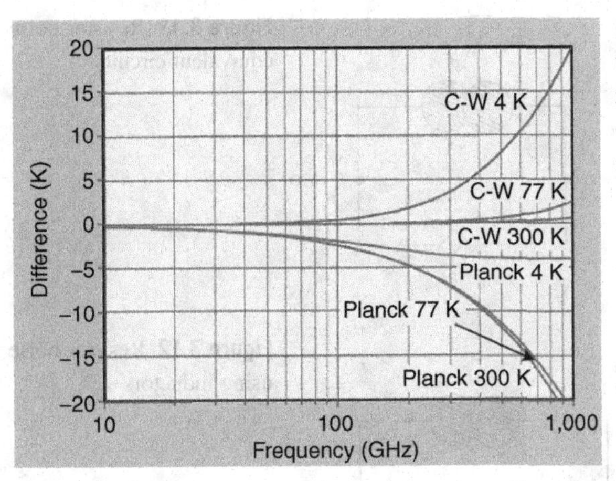

Figure 3.10 Difference between the noise temperature and physical temperature of a resistor for physical temperatures of 4K, 77K, and 300K using the Callen–Welton and Planck forms [3]

$$T_n^{C-W} = T\left[\frac{\frac{hf}{kT}}{e^{\frac{hf}{kT}} - 1}\right] + \frac{hf}{2k}. \tag{3.54}$$

As illustrated in Figure 3.10, for most practical cases below 100GHz, (3.54) and (3.53) agree within 1–2K and T_n can be safely assumed to be equal to the physical temperature, T, of the resistor. In the same frequency range, the spectral power density of thermal noise remains constant, that is *white*. However, in terahertz and optical-amplifier applications, (3.54) must be employed and the thermal noise spectral power density is no longer white. A rule of thumb for estimating when the vacuum fluctuations become significant can be obtained by calculating $hf_{MAX}/k \approx (0.048\text{K/GHz}) \times f_{MAX}$, where f_{MAX} is the maximum frequency of interest for a particular application. If this value is negligible compared to the noise temperature, vacuum fluctuations can be safely ignored.

3.2.2 Noise in RLC single-ports

In view of the previous discussion, for the rest of the book we will assume that the noise temperature of a resistor is approximately equal to its physical temperature, T. As illustrated in Figure 3.11, for the purpose of circuit noise analysis, in conformity with the available noise power definition, the resistor noise can be modeled either as a noise voltage source in series with the noise-free resistor

$$\overline{v_n^2} = \langle v_n, v_n^* \rangle = 4kT_n\Delta fR \approx 4kT\Delta fR \tag{3.55}$$

or as a noise current source in parallel with the noise-free resistor

$$\overline{i_n^2} = \langle i_n, i_n^* \rangle = 4kT_n\Delta fG \approx 4kT\Delta fG \tag{3.56}$$

where $R = 1/G$ and Δf is the bandwidth over which the noise power is integrated.

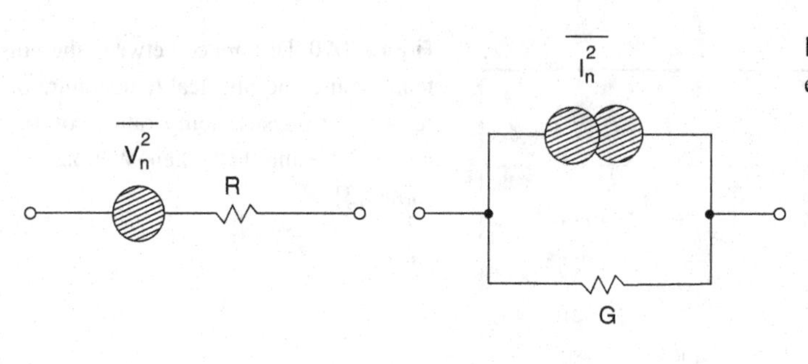

Figure 3.11 Resistor noise equivalent circuits

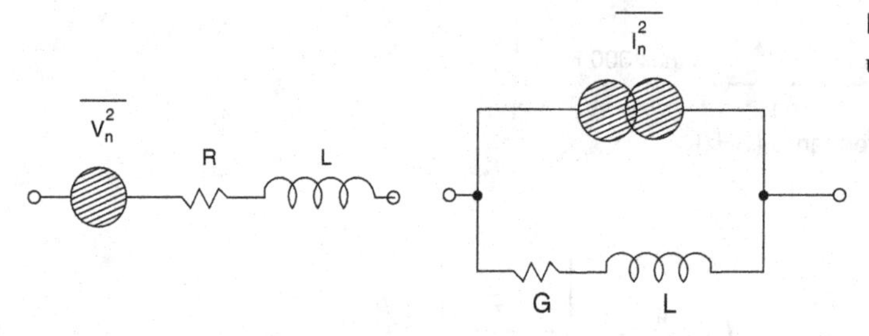

Figure 3.12 Resistor noise filtering using inductors

From (3.55) and (3.56), we can calculate the available noise power, $P_n = p_n \Delta f$, from the resistor as

$$P_n = \frac{\overline{v_n^2}}{4R} = \frac{\overline{i_n^2} R}{4} = kT_n \Delta f \approx kT\Delta f. \tag{3.57}$$

Equation (3.57) reveals, perhaps surprisingly, that **the available noise power from a resistor does not depend on the resistance value**, only on its physical (in the context of the earlier discussion) temperature and bandwidth. However, the equivalent noise voltage of the resistor increases with the square root of the resistance value. A simple rule-of-thumb gives the noise voltage of a resistor at room temperature ($T = 290\text{K}$) as

$$v_n(f) = \sqrt{\frac{R}{1k\Omega}} \times 4.02 \frac{nV}{\sqrt{Hz}}.$$

We might be tempted to conclude from (3.57) that the available noise power from a resistor increases to infinity when the bandwidth increases to infinity and that the available noise power goes to 0 when the physical temperature of the resistor is zero. Neither inferences are correct because, in either case, only the Callen–Welton definition of noise temperature is correct under those extreme conditions.

If an ideal, lossless inductor is added in series with the resistor, as in Figure 3.12, the corresponding series noise voltage source remains the same as in (3.56) while the equivalent noise current connected in parallel with the R-L circuit becomes frequency dependent

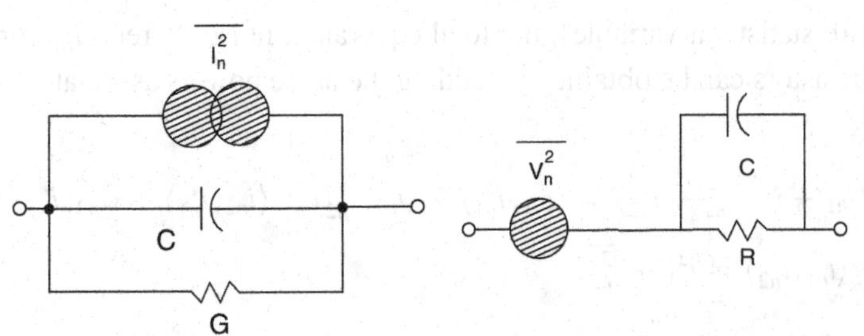

Figure 3.13 Noise filtering at high frequencies using capacitors

$$\overline{i_n^2} = \frac{\overline{v_n^2}}{R^2 + \omega^2 L^2} = \frac{4kT\Delta f G}{1 + \omega^2 L^2 G^2}, \tag{3.58}$$

indicating that the inductor filters out the noise current at high frequencies.

Similarly, for a circuit consisting of an ideal, lossless capacitor in parallel with a resistor, the expression of the parallel noise current source is identical to that given by (3.57), while the noise voltage in series with the parallel RC circuit becomes frequency dependent

$$\overline{v_n^2} = \frac{\overline{i_n^2}}{G^2 + \omega^2 C^2} = \frac{4kT\Delta f R}{1 + \omega^2 C^2 R^2}. \tag{3.59}$$

Both the series and the shunt noise filtering techniques using inductors and capacitors, which, if lossless, do not generate noise, are frequently employed in low-noise amplifiers and oscillators to filter out the noise generated by resistors and by the bias network at high frequency.

EXAMPLE 3.3

Consider a circuit consisting of two resistors, R_1 and R_2, connected in parallel. Knowing that the thermal noise generated in one resistor is statistically independent from the noise generated by the second resistor, find the equivalent noise voltage in series with the parallel combination of R_1 and R_2 and the available noise power of the circuit. How does the noise voltage compare with that generated by a single resistor whose value is $R = \frac{R_1 R_2}{R_1 + R_2}$?

Solution

Because of the parallel combination, it is preferable to represent the noise from each resistor as a noise current source in parallel with the resistor

$$\overline{i_{n1}^2} = \frac{4kT\Delta f}{R_1} \quad \text{and} \quad \overline{i_{n2}^2} = \frac{4kT\Delta f}{R_2}.$$

Since we are dealing with statistical variables, the total equivalent noise current, i_n, connected in parallel with the two resistors can be obtained by adding the noise powers associated with the two noise currents

$$\overline{i_n^2} = \langle i_n, i_n^* \rangle = \langle i_{n1} + i_{n2}, i_{n1}^* + i_{n2}^* \rangle = \langle i_{n1}, i_{n1}^* \rangle + \langle i_{n1}, i_{n2}^* \rangle + \langle i_{n2}, i_{n1}^* \rangle + \langle i_{n2}, i_{n2}^* \rangle$$
$$= \langle i_{n1}, i_{n1}^* \rangle + \langle i_{n2}, i_{n2}^* \rangle = \overline{i_{n1}^2} + \overline{i_{n2}^2}$$

where we have accounted for the fact that the cross-correlation products $\langle i_{n1}, i_{n2}^* \rangle$ and $\langle i_{n2}, i_{n1}^* \rangle$ are zero because the two noise signals are statistically independent. This implies that

$$\overline{i_n^2} = \frac{4kT\Delta f}{R_1} + \frac{4kT\Delta f}{R_2} = 4kT\Delta f\left(\frac{1}{R_1} + \frac{1}{R_2}\right) = 4kT\Delta f\left(\frac{R_1 + R_2}{R_1 R_2}\right)$$

which is identical to the noise power of a resistor whose value is equal to that of the parallel combination of R_1 and R_2.

We can now convert from a Norton to a Thevenin representation of the noise source to obtain

$$\overline{v_n^2} = \langle v_n, v_n^* \rangle = \langle i_n R, i_n^* R \rangle = R^2 \overline{i_n^2} = R^2 \frac{4kT\Delta f}{R} = 4kT\Delta f R$$

where

$$R = \frac{R_1 R_2}{R_1 + R_2}.$$

This demonstrates that the equivalent noise voltage of the parallel combination of two resistors is identical to that of a resistor whose value is equal to that of the parallel combination of the two resistors. The available noise power will remain unchanged: $P = kT\Delta f$.

3.2.3 Noise in diodes

The noise equivalent circuit of a *pn* diode, shown in Figure 3.14, consists of a noise current source, i_{nj}, in parallel with the small signal capacitance and junction resistance of the diode. The noise current has both shot noise and 1/*f* noise terms

$$\overline{i_{nj}^2} = 2q\Delta f I_D + K_1 \Delta f \frac{I_d^a}{f} \tag{3.60}$$

where I_D is the DC current through the diode and K_1 and a are empirical terms which describe the 1/*f* noise and which are obtained from measurements.

Additionally, the parasitic resistance, R_S, associated with the contact and access regions of the diode contributes a thermal noise voltage source in series with the intrinsic diode

$$\overline{v_{ns}^2} = 4kT\Delta f R_s. \tag{3.61}$$

The two noise sources internal to the diode, v_{ns} and i_{nj}, have different physical origins and, therefore, are statistically independent.

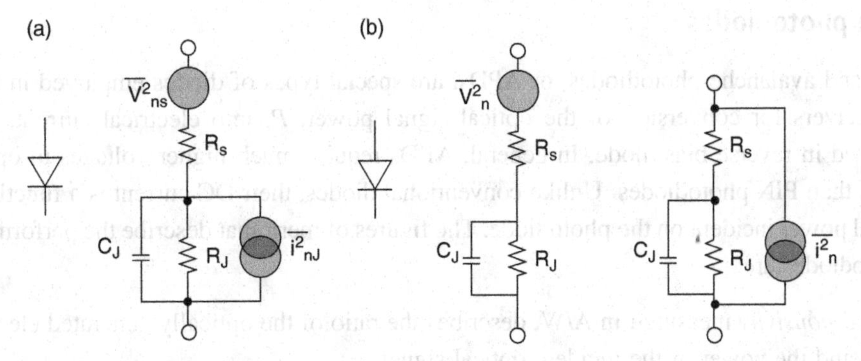

Figure 3.14 Noise equivalent circuit of a diode (a) Internal noise sources (b) Thevenin and Norton representations

Since the diode is a one-port circuit, a single noise source, either a noise voltage source, v_n, in series with the port, or a noise current source, i_n, in parallel with the port, is sufficient to describe its noise behavior. Expressions for v_n or i_n can be easily derived as a function of the "internal" noise sources of the diode, v_{ns} and i_{nj}, using the Thevenin–Norton and Norton–Thevenin transformations. For example, i_{nj} can be converted to a noise voltage source v_{nj} in series with the junction admittance

$$Y_j = j\omega C_J + \frac{1}{R_J} = \frac{1 + j\omega R_J C_J}{R_J}. \tag{3.62}$$

We can now derive the expression of the equivalent noise voltage of the diode as a series connection of two statistically independent noise voltages v_{ns} and v_{nj}

$$\overline{v_n^2} = \overline{v_{ns}^2} + \frac{\overline{i_{nj}^2}}{|Y_J|^2}. \tag{3.63}$$

Alternatively, we can represent the diode noise as a noise current source in parallel with the diode

$$\overline{i_n^2} = \frac{\overline{v_n^2}}{\left|R_s + \frac{1}{Y_j}\right|^2} = \overline{v_{ns}^2}\frac{|Y_J|^2}{|R_s Y_j + 1|^2} + \frac{\overline{i_{nj}^2}}{|R_s Y_j + 1|^2}. \tag{3.64}$$

By ignoring the contributions of the parasitic resistance (i.e. $R_S = 0$) and the 1/*f* noise, we can compare the available noise power from the diode due to its shot noise with the available thermal noise power of an equivalent conductance/resistance. When the diode is forward biased, the equivalent noise conductance of the diode becomes

$$G_e \stackrel{def}{=} \frac{\overline{i_n^2}}{4kT\Delta f} = \frac{qI_d}{2kT} = \frac{1}{2R_j} \tag{3.65}$$

since $R_J = kT/qI_D$. G_e can be interpreted as a conductance which would produce the same amount of thermal noise as the shot noise of the diode.

3.2.4 Noise in photodiodes

The PIN and avalanche photodiodes, or APDs, are special types of diodes employed in fiber-optics receivers for conversion of the optical signal power, P, into electrical current. They are operated in reverse bias mode. In general, APDs require much higher voltages to operate (10–20V) than PIN photodiodes. Unlike conventional diodes, their DC current is a function of the optical power incident on the photodiode. The figures of merit that describe the performance of a photodiodes are:

- R, the *responsivity* measured in A/W, describes the ratio of the optically generated electrical current and the power of the incident optical signal,
- I_{DARK}, the photodiode current in the absence of the optical signal,
- M, the current multiplication factor,
- $F(M)$, the noise factor of the photodiode, and
- P_{MIN}, the optical sensitivity of the photodiode.

The expression of the optically generated electrical current flowing through the photodiode due to an incident optical signal of power, P, is given by

$$I_{ph} = RMP \tag{3.66}$$

from which the shot noise current of the photodiode can be derived as

$$\overline{i_{n,ph}^2} = 2qI\Delta f = 2q\Delta f(RMP + MI_{dark})MF(M) = 2q\Delta f(RP + I_{dark})M^2F(M). \tag{3.67}$$

As in the case of a conventional *pn* junction, we can define an equivalent noise conductance. The noise conductance of a photodiode has the unique feature that it increases with the incident optical signal power

$$G_e = \frac{(RP + I_{dark})M^2F(M)}{\frac{2kT}{q}}. \tag{3.68}$$

From (3.67), we can define the *optical sensitivity* of the photodiode as the minimum optical power that produces an electrical current equal to the dark current of the photodiode

$$P_{min} = \frac{1}{R} \times \sqrt{2q\Delta f I_{dark} F(M)}. \tag{3.69}$$

We note that for PIN photodiodes $M = 1$, $F(M) = 1$, while for APDs $M = 10 \ldots 50$ and $F(M) = 5 \ldots 10$. This leads to the conclusion that, although avalanche photodiodes amplify the optically generated electrical current, they also contribute more noise than a PIN photodiode. In general, optoelectronic receivers with avalanche photodiodes will exhibit higher overall optical sensitivity than those with PIN diodes because the noise contribution of the electronic circuits following the APD is significantly reduced by the gain, M, of the photodiode.

We end the discussion on the noise of passive single-ports with the observation that one noise source, either voltage or current, as in Figure 3.14(b), is sufficient to fully describe the noise

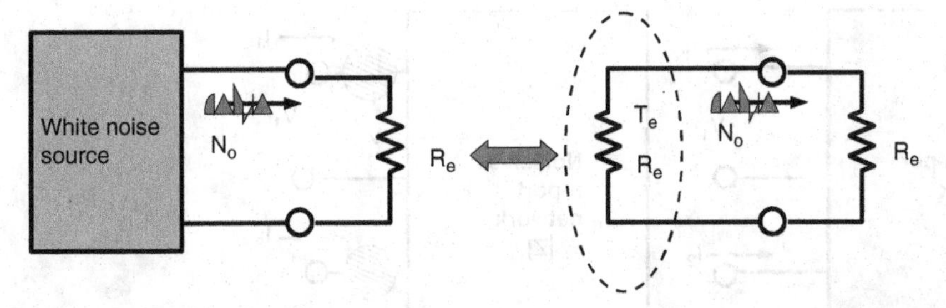

Figure 3.15 Model of an arbitrary white noise source by its equivalent noise temperature T_e

of a single-port. If the noise source is *white* (that means that its power spectral density is constant with frequency), it can be modeled as a thermal noise source and characterized by an equivalent noise resistance, R_e, (or conductance G_e) and noise temperature, T_e. If the noise source is terminated on a matched load of impedance R_e, then the entire available noise power from the noise source, $N_o = kT_e\Delta f$ is transferred to the matched load, as illustrated in Figure 3.15.

Excess noise ratio

In measurement applications, it is important to drive the circuit whose noise performance needs to be characterized with a calibrated noise source. Such noise sources can be realized with resistors placed in liquid nitrogen (77K) or liquid helium (4K). However, for the accurate measurement of most active circuits, much larger noise powers are often necessary. In such situations, active noise generators are employed based on solid-state diodes (e.g. IMPATT diodes) or vacuum tubes which are characterized by elevated noise temperatures of 10,000 to 40,000K. A more common measure of the noise power of these noise generators is the *excess noise ratio*, or ENR, defined as

$$ENR(dB) = 10\log_{10}\left(\frac{T_g - T_0}{T_0}\right) \tag{3.70}$$

where T_g is the noise equivalent temperature of the noise generator and T_0 is a the reference temperature (290K) associated with a passive source (a matched load) at room temperature. Typical ENR values for commercial noise sources are in the 10–20dB range. From (3.70), it is immediately apparent that it is important to indicate the reference temperature T_0 at which the ENR is specified.

3.3 TWO-PORT AND MULTI-PORT NOISE

Large circuits inevitably will contain a large number of noise sources generated by the active and passive devices in the circuit. For circuit noise analysis purposes, these sources can be treated as white noise, at least over some narrow range of frequencies, Δf. In general, they will be at least partially correlated with each other.

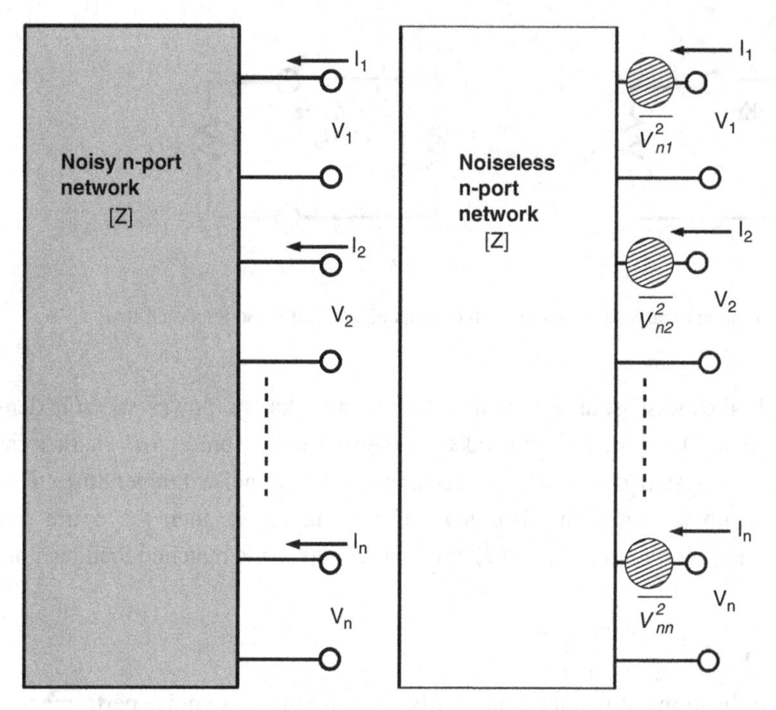

Figure 3.16 *Z*-matrix representation of a noisy multi-port

3.3.1 Two-port and multi-port noise representations

The small signal matrix representation of multi-ports can be extended to include the noise sources of the multi-port. One noise source associated with each port is sufficient to fully capture the behavior of all the internal noise sources in a noisy multi-port. The noise source can be:

- a voltage source,
- a current source exiting the port, or
- a noise wave emanating from the port.

As a result, the noise behavior of an n-port is modeled using:

- an n-dimensional noise-source column vector representing the internal noise sources of the network as seen at its terminals,
- a square $n \times n$ correlation matrix of the above noise sources, and
- the $n \times n$ small signal matrix of the noise-free multi-port.

Just as in the case of linear signal analysis, in the linear analysis of noisy networks, the n-port can be described using impedance [4], admittance or noise wave correlation matrices [5].

In the **impedance formalism**, Figure 3.16, the voltage at each port is described as a function of the signal currents entering each port and of the noise voltage sources connected in series with each port

$$\begin{bmatrix} V_1 \\ V_2 \\ \cdot \\ \cdot \\ \cdot \\ V_n \end{bmatrix} = \begin{bmatrix} Z_{11} & Z_{12} & \cdot & \cdot & \cdot & Z_{1n} \\ Z_{21} & Z_{22} & \cdot & \cdot & \cdot & Z_{2n} \\ \cdot & \cdot & \cdot & & & \cdot \\ \cdot & \cdot & & \cdot & & \cdot \\ \cdot & \cdot & & & \cdot & \cdot \\ Z_{n1} & Z_{n2} & \cdot & \cdot & \cdot & Z_{nn} \end{bmatrix} \begin{bmatrix} I_1 \\ I_2 \\ \cdot \\ \cdot \\ \cdot \\ I_n \end{bmatrix} + \begin{bmatrix} v_{n1} \\ v_{n2} \\ \cdot \\ \cdot \\ \cdot \\ v_{nn} \end{bmatrix} \tag{3.71}$$

where [Z] is the impedance matrix of the noise-free n-port. The noise voltage sources are characterized by the *noise impedance correlation matrix* defined as

$$[C_z] = \begin{bmatrix} \langle v_{n1}, & v_{n1}^* \rangle & \langle v_{n1}, & v_{n2}^* \rangle & \cdot & \cdot & \cdot & \langle v_{n1}, & v_{nn}^* \rangle \\ \langle v_{n2}, & v_{n1}^* \rangle & \langle v_{n2}, & v_{n2}^* \rangle & \cdot & \cdot & \cdot & \langle v_{n2}, & v_{nn}^* \rangle \\ & \cdot & & \cdot & \cdot & & & & \cdot \\ & \cdot & & \cdot & & \cdot & & & \cdot \\ & \cdot & & \cdot & & & \cdot & & \cdot \\ \langle v_{nn}, & v_{n1}^* \rangle & \langle v_{nn}, & v_{n2}^* \rangle & \cdot & \cdot & \cdot & \langle v_{nn}, & v_{nn}^* \rangle \end{bmatrix}. \tag{3.72}$$

For a passive multi-port network at physical temperature T, the impedance correlation matrix is derived from the real part of the Z-matrix, implying that no noise measurements are required

$$[C_z] = 4kT\Delta f\, Re\{[Z]\}. \tag{3.73}$$

Similarly, in the **noise admittance formalism**, Figure 3.17, the current entering each port is described as a function of the voltages at each port and of the noise current sources connected in parallel with each port

$$\begin{bmatrix} I_1 \\ I_2 \\ \cdot \\ \cdot \\ \cdot \\ I_n \end{bmatrix} = \begin{bmatrix} Y_{11} & Y_{12} & \cdot & \cdot & \cdot & Y_{1n} \\ Y_{21} & Y_{22} & \cdot & \cdot & \cdot & Y_{2n} \\ \cdot & \cdot & \cdot & & & \cdot \\ \cdot & \cdot & & \cdot & & \cdot \\ \cdot & \cdot & & & \cdot & \cdot \\ Y_{n1} & Y_{n2} & \cdot & \cdot & \cdot & Y_{nn} \end{bmatrix} \begin{bmatrix} V_1 \\ V_2 \\ \cdot \\ \cdot \\ \cdot \\ V_n \end{bmatrix} + \begin{bmatrix} i_{n1} \\ i_{n2} \\ \cdot \\ \cdot \\ \cdot \\ i_{nn} \end{bmatrix} \tag{3.74}$$

while the noise current sources are characterized by the *noise admittance correlation matrix*

$$[C_y] = \begin{bmatrix} \langle i_{n1}, & i_{n1}^* \rangle & \langle i_{n1}, & i_{n2}^* \rangle & \cdot & \cdot & \langle i_{n1}, & i_{nn}^* \rangle \\ \langle i_{n2}, & i_{n1}^* \rangle & \langle i_{n2}, & i_{n2}^* \rangle & \cdot & \cdot & \langle i_{n2}, & i_{nn}^* \rangle \\ & \cdot & & \cdot & \cdot & & & \cdot \\ & \cdot & & \cdot & & \cdot & & \cdot \\ \langle i_{nn}, & i_{n1}^* \rangle & \langle i_{nn}, & i_{n2}^* \rangle & \cdot & \cdot & \langle i_{nn}, & i_{nn}^* \rangle \end{bmatrix}. \tag{3.75}$$

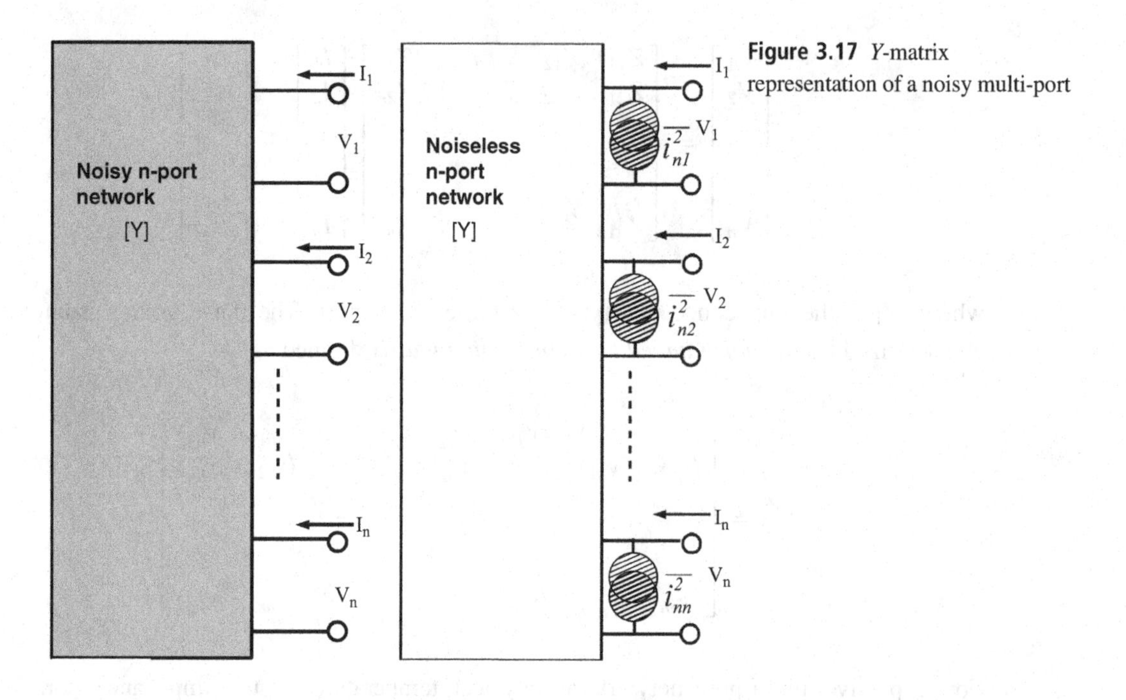

Figure 3.17 *Y*-matrix representation of a noisy multi-port

In (3.74), *[Y]* represents the admittance matrix of the noise-free *n*-port. Again, for a passive multi-port at physical temperature, *T*, the noise admittance correlation matrix is obtained from the real part of [*Y*]

$$[C_y] = 4kT\Delta f\, Re\{[Y]\}. \tag{3.76}$$

Finally, in the **noise wave formalism**, Figure 3.18, the power wave b_i emanating from port i is described as a function of the incident power waves at each port and of the noise waves b_{ni}, emanating from each port

$$\begin{bmatrix} b_1 \\ b_2 \\ . \\ . \\ b_n \end{bmatrix} = \begin{bmatrix} S_{11} & S_{12} & . & . & . & S_{1n} \\ S_{21} & S_{22} & . & . & . & S_{2n} \\ . & . & . & . & . & . \\ . & . & . & . & . & . \\ S_{n1} & S_{n2} & . & . & . & S_{nn} \end{bmatrix} \begin{bmatrix} a_1 \\ a_2 \\ . \\ . \\ a_n \end{bmatrix} + \begin{bmatrix} b_{n1} \\ b_{n2} \\ . \\ . \\ .b_{nn} \end{bmatrix} \tag{3.77}$$

where *[S]* is the *S*-parameter matrix of the noise-free *n*-port. The noise waves are characterized by the *noise wave correlation matrix* and are defined with respect to the same normalization impedance Z_0 as the scattering parameters

$$[C_S] = \begin{bmatrix} \langle b_{n1}, & b_{n1}^*\rangle & \langle b_{n1}, & b_{n2}^*\rangle & . & . & . & \langle b_{n1}, & b_{nn}^*\rangle \\ \langle b_{n2}, & b_{n1}^*\rangle & \langle b_{n2}, & b_{n2}^*\rangle & . & . & . & \langle b_{n2}, & b_{nn}^*\rangle \\ & . & & . & . & . & . & & . \\ & . & & . & . & . & . & & . \\ \langle b_{nn}, & b_{n1}^*\rangle & \langle b_{nn}, & b_{n2}^*\rangle & . & . & . & \langle b_{nn}, & b_{nn}^*\rangle \end{bmatrix}. \tag{3.78}$$

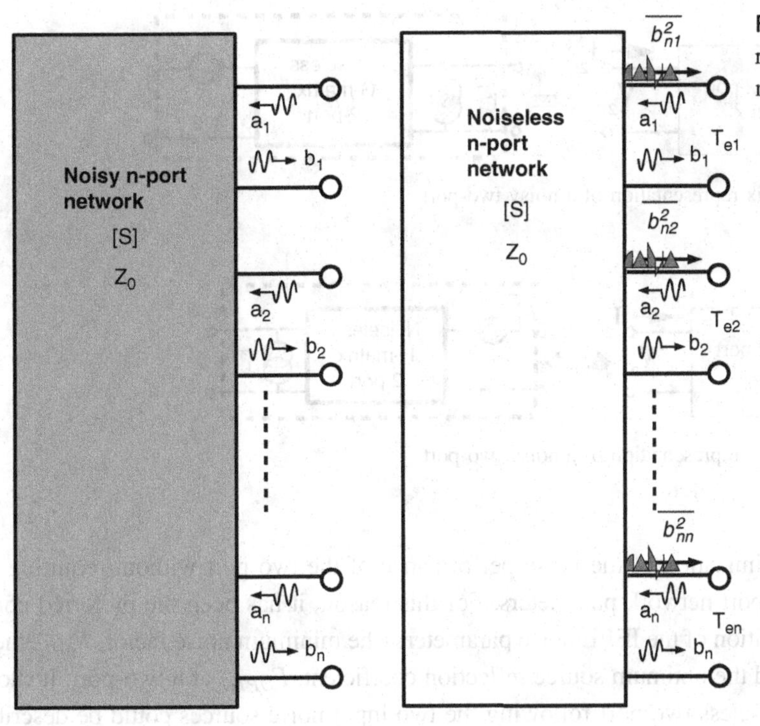

Figure 3.18 *S*-matrix representation of a noisy multi-port

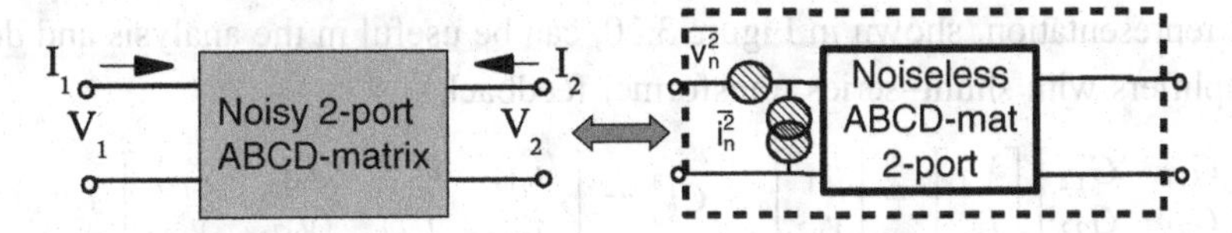

Figure 3.19 *ABCD*-matrix representation of a noisy two-port

The noise wave representation naturally lends itself to the signal flow graph analysis of noisy linear networks. We can now also associate an equivalent noise temperature, T_{ei}, with the noise power emanating from port *i*. For a passive multi-port at physical temperature *T*, the noise wave correlation matrix can be obtained directly from its scattering matrix using Bosma's theorem [7]

$$[C_S] = kT\Delta f\left([U] - [S][S]^{+}\right). \tag{3.79}$$

In the analysis of the noise of circuits consisting of two-port networks, it is possible to extend the number of noise matrix representations to include the *H*-, *G*-, and *ABCD*-matrices, just as in the case of linear networks. The *ABCD*-matrix representation, illustrated in Figure 3.19, gives

$$\begin{bmatrix} V_1 \\ I_1 \end{bmatrix} = \begin{bmatrix} A & B \\ C & D \end{bmatrix}\begin{bmatrix} V_2 \\ -I_2 \end{bmatrix} + \begin{bmatrix} v_n \\ i_n \end{bmatrix} \quad [C_A] = \begin{bmatrix} \langle v_n \;\; v_n^* \rangle & \langle v_n \;\; i_n^* \rangle \\ \langle i_n \;\; v_n^* \rangle & \langle i_n \;\; i_n^* \rangle \end{bmatrix}. \tag{3.80}$$

Apart from the ease of performing matrix calculations in cascaded amplifier stages, this representation is unique in having both noise sources placed at the input of the two-port.

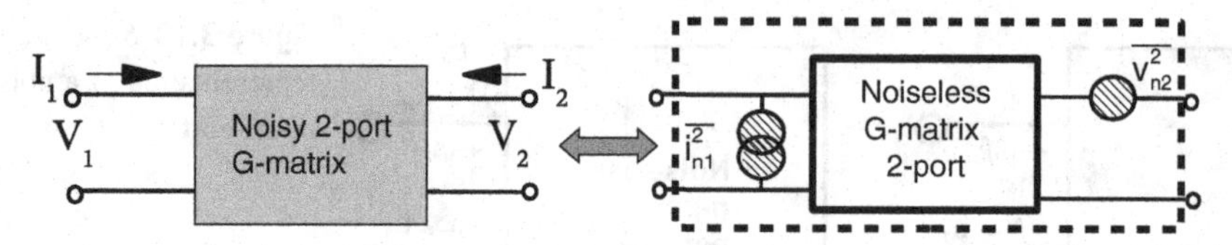

Figure 3.20 *G*-matrix representation of a noisy two-port

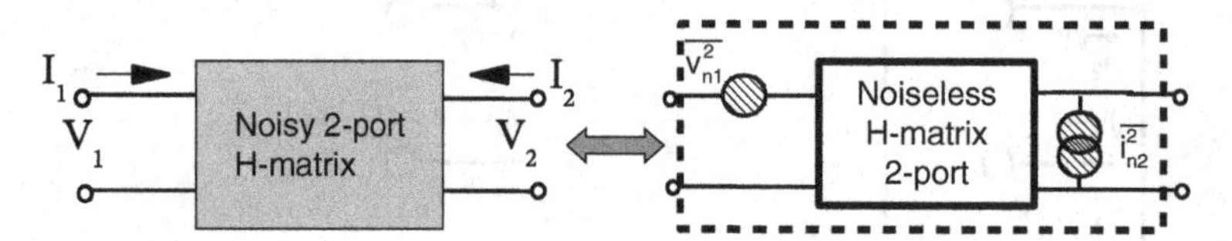

Figure 3.21 *H*-matrix representation of a noisy two-port

It permits the optimization of the noise performance of the two-port without requiring knowledge of the two-port network parameters. For this reason, it has been the preferred configuration for the definition of the IEEE noise parameters (the minimum noise factor, F_{MIN}, the noise resistance, R_n, and the optimum source reflection coefficient, Γ_{OPT},) of a two-port. It should be noted that the noiseless two-port following the two input noise sources could be described by any of the other matrices: *S, Y, Z, G, or H*, without affecting the derivation of the two-port noise parameters.

The *G*-matrix representation, shown in Figure 3.20, can be useful in the analysis and design of low-noise amplifiers with shunt–series transformer feedback

$$\begin{bmatrix} I_1 \\ V_2 \end{bmatrix} = \begin{bmatrix} G_{11} & G_{12} \\ G_{21} & G_{22} \end{bmatrix} \begin{bmatrix} V_1 \\ I_2 \end{bmatrix} + \begin{bmatrix} i_{n1} \\ v_{n2} \end{bmatrix}; \quad [C_g] = \begin{bmatrix} \langle i_{n1}, & i_{n1}^* \rangle & \langle i_{n1}, & v_{n2}^* \rangle \\ \langle v_{n2}, & i_{n1}^* \rangle & \langle v_{n2}, & v_{n2}^* \rangle \end{bmatrix} \tag{3.81}$$

while the *H*-matrix representation of Figure 3.21 is preferred for series–shunt feedback and to describe the noise of a FET in the Pospieszalski noise model, to be discussed in Chapter 4

$$\begin{bmatrix} V_1 \\ I_2 \end{bmatrix} = \begin{bmatrix} H_{11} & H_{12} \\ H_{21} & H_{22} \end{bmatrix} \begin{bmatrix} I_1 \\ V_2 \end{bmatrix} + \begin{bmatrix} v_{n1} \\ i_{n2} \end{bmatrix}; \quad [C_h] = \begin{bmatrix} \langle v_{n1}, & v_{n1}^* \rangle & \langle v_{n1}, & i_{n2}^* \rangle \\ \langle i_{n2}, & v_{n1}^* \rangle & \langle i_{n2}, & i_{n2}^* \rangle \end{bmatrix}. \tag{3.82}$$

The noise correlation matrices have the following properties, which are typical of hermitian matrices

$$Im(c_{nn}) = Im\left(\overline{v_{nn}^2}\right) = Im\left(\overline{i_{nn}^2}\right) = Im\left(\overline{b_{nn}^2}\right) = 0 \tag{3.83}$$

and

$$c_{ki} = c_{ik}^*. \tag{3.84}$$

Prior to performing circuit analysis, the network and noise correlation matrices of all the components that make up the network must be formed and calculated. This can be accomplished for any of the three general *n*-port representations. For example, a set of simple rules can be developed for the noise admittance correlation matrix.

Forming the noise admittance correlation matrix[1]

(i) The matrix element $C_Y(i,i)$ is equal to the sum of all the noise current power spectral densities of each element connected to node i. Thus, the first diagonal element is the sum of noise currents connected to node 1, the second diagonal element is the sum of noise currents connected to node 2, and so on.

(ii) The matrix element $C_Y(i,j)$ (where $j \neq i$) is equal to the negative of all the noise current power densities of the elements connected between nodes i and j. Therefore, a noise current source between nodes 1 and 2 will be included in both entries (1,2) and (2,1).

(iii) If one terminal of a noise current source is grounded, it will only contribute to one entry in the noise correlation matrix, at the appropriate location on the diagonal. Otherwise, if a noise source is connected to nodes i and j, none of which is grounded, it will contribute to four entries in the matrix: two diagonal entries at $C_Y(i,i)$ and $C_Y(j,j)$, and two off-diagonal entries at $C_Y(i,j)$ and $C_Y(j,i)$.

EXAMPLE 3.4

Write the expression of the noise correlation matrix for the circuit in Figure 3.22(a).

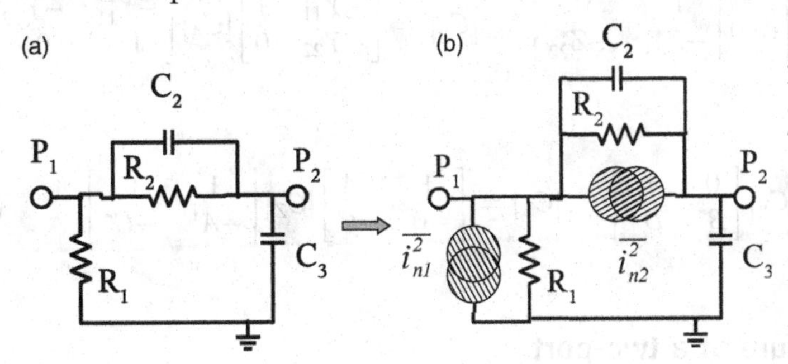

Figure 3.22 Example of noisy two-port used to form the noise admittance correlation matrix (a) Two-port schematic (b) Two-port schematic with thermal noise current sources added for each resistor

Solution

Remembering that we must attach a noise current source in parallel with every resistance/conductance, and that reactive elements such as capacitors and inductors do not contribute noise, we obtain the equivalent circuit in Figure 3.22(b) and

$$[C_Y] = \begin{bmatrix} \overline{i_{n1}^2} + \overline{i_{n2}^2} & -\overline{i_{n2}^2} \\ -\overline{i_{n2}^2} & \overline{i_{n2}^2} \end{bmatrix} = 4kT\Delta f \begin{bmatrix} G_1 + G_2 & -G_2 \\ -G_2 & G_2 \end{bmatrix}$$

which validates Equation (3.76) since the matrix on the right represents the Y-matrix of the passive two-port.

[1] From Lucs.sorceforge.net/tech/nodell.html.

Noise correlation matrix transformations

To simplify circuit noise analysis, to avoid a non-existent representation, or to avoid matrix singularities, it is often necessary to transform the noise representations from one form to another [6]. For n-ports, the relationship between the impedance and admittance noise correlation matrices are given by

$$[C_z] = [Z][C_y][Z]^+; \quad [C_z] = ([U] + [Z])[C_S]([U] + [Z])^+ \tag{3.85}$$

$$[C_y] = [Y][C_z][Y]^+; \quad [C_y] = ([U] + [Y])[C_S]([U] + [Y])^+ \tag{3.86}$$

$$[C_S] = \frac{1}{4}([U] + [S])[C_y]([U] + [S])^+; \quad [C_S] = \frac{1}{4}([U] - [S])[C_Z]([U] - [S])^+ \tag{3.87}$$

where superscript " + " indicates the transpose conjugate of a matrix.

For two-ports, the following relations hold between the noise impedance, noise admittance, and noise $ABCD$-correlation matrices [2]

$$[C_z] = \begin{bmatrix} 1 & -Z_{11} \\ 0 & -Z_{22} \end{bmatrix}[C_A]\begin{bmatrix} 1 & 0 \\ -Z_{11}^* & -Z_{22}^* \end{bmatrix}; \quad [C_y] = \begin{bmatrix} -Y_{11} & 1 \\ -Y_{22} & 0 \end{bmatrix}[C_A]\begin{bmatrix} -Y_{11}^* & -Y_{22}^* \\ 1 & 0 \end{bmatrix} \tag{3.88}$$

$$[C_A] = \begin{bmatrix} 0 & B \\ 1 & D \end{bmatrix}[C_y]\begin{bmatrix} 0 & 1 \\ B^* & D* \end{bmatrix}; \quad [C_A] = \begin{bmatrix} 1 & -A \\ -0 & -C \end{bmatrix}[C_Z]\begin{bmatrix} 1 & 0 \\ -A^* & -C^* \end{bmatrix}. \tag{3.89}$$

3.3.2 The noise temperature of a two-port

As introduced in Chapter 2, a figure of merit for the noise performance of a two-port is the equivalent input noise temperature, T_e, defined in (2.39). For example, let us consider the noisy amplifier in Figure 3.23, characterized by the available power gain, G, and bandwidth, Δf, and matched to the noiseless source and load resistors, R. We can assume that the source resistance is at a hypothetical temperature of 0K and generates no noise. Presumably, N_i, the input noise power, will also be 0. Under these conditions, if we measured the noise power, N_o, at the output of the amplifier, it would be solely due to the added noise by the amplifier itself at its input, N_a.

However, in reality, at 0K, the noise power of the zero-point vacuum fluctuations must be accounted for, such that $N_i = hf\Delta f/2 \neq 0$. At the same time, the lowest possible equivalent input noise power of an amplifier is $N_a = (1\text{–}1/G)hf\Delta f/2$ [3]. For a very high gain amplifier, definition (2.39) results in a minimum possible equivalent input noise temperature

$$T_e = \frac{hf}{2k} \tag{3.90}$$

a minimum output noise power $N_o = Ghf\Delta f$, and a minimum added input noise power of $N_{amin} = hf\Delta f/2$ [3].

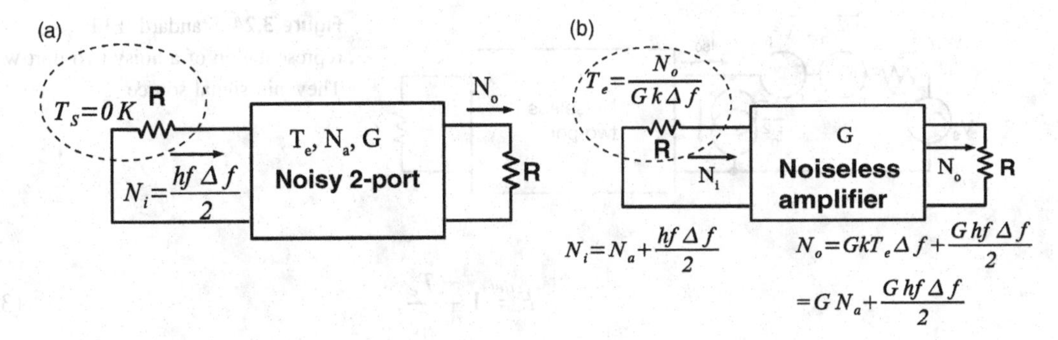

Figure 3.23 Definition of the equivalent input noise temperature, T_e, and the added noise power, N_a, of a noisy amplifier (a) Noisy amplifier (b) Noiseless amplifier

The expression of the equivalent input noise temperature of a noisy two-port with a zero-source reflection coefficient can also be derived from the noise wave formulation and scattering matrix [7]

$$T_e = \frac{\overline{b_{n2}^2}}{k|S_{21}|^2} \tag{3.91}$$

which is identical to (2.39) if we recognize from (3.43) that, when $\Gamma_S = \Gamma_{out} = 0$, $G_A = |S_{21}|^2$. Equation (3.91) can be further extended to the case when $\Gamma_S \neq 0$ [7]

$$T_e = \frac{[\alpha][C_S][\alpha]^+}{k\left(1 - |\Gamma_S|^2\right)}. \tag{3.92}$$

where $[C_S]$ is given by (3.78) and

$$[\alpha] = \left[\Gamma_S, \quad \frac{1 - \Gamma_S S_{11}}{S_{21}}\right]. \tag{3.93}$$

Equation (3.92) indicates that the equivalent noise temperature of a two-port is a function of the source reflection coefficient.

One of the benefits of describing the noise of a matched two-port by its equivalent input temperature is that it leads to simple calculations. Like noise power spectral densities, the noise temperatures of statistically independent noise sources add linearly.

3.3.3 Two-port noise figure

The *noise figure* of an amplifier (two-port) was introduced in Chapter 2, equation (2.38); as the ratio of the *available signal-to-noise ratio* at the signal generator terminals to the available signal-to-noise ratio at the amplifier output. Equation (2.40) underlines the duality between the noise factor and the input equivalent noise temperature, T_e, of a two-port. Given T_e, F is readily calculated.

When he introduced the concept of noise figure (factor), Friis also suggested that the noise figure (factor) be defined with respect to a reference temperature of 290K

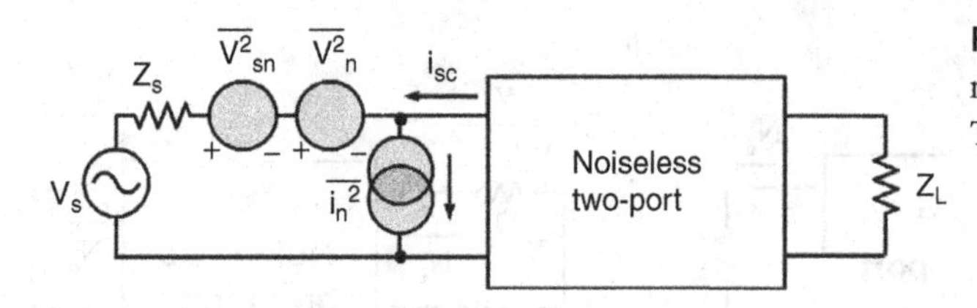

Figure 3.24 Standard IEEE representation of a noisy two-port with Thevenin signal source

$$F \stackrel{def}{=} 1 + \frac{T_e}{T_0}. \tag{3.94}$$

There is still debate in the scientific community as to whether the reference temperature should be the physical temperature $T_0 = 290\text{K}$, as in (3.94), or the noise temperature T_n (290K) of the input termination at a physical temperature of 290K [3]. Within the bounds of the restrictions discussed in Section 3.2.1, the two are practically equal for most of the range of frequencies of interest in electronics. Irrespective of the choice of reference temperature, it is important to recognize that the noise figure of a two-port, like the ENR of a noise source, is not an absolute measure of the two-port noise performance, but rather that it depends on the particular value chosen for the reference temperature. Therefore, in the noise figure definition, N_i must be specified according to a fixed standard. The standard chosen is

$$N_i = kT_0\Delta f. \tag{3.95}$$

The latter also allows us to provide a formula for the noise measure of a two-port as a function of its scattering and noise wave correlation matrices [7]

$$M = \frac{-[\alpha][C_S][\alpha]^+}{kT_0[\alpha]\left([U] - [S][S^+]\right)[\alpha]^+} \tag{3.96}$$

where $[\alpha]$ is given by (3.93).

3.3.4 Deriving the noise figure of a two-port

While it is possible to derive the expression of the IEEE noise parameters of a two-port using any of the two-port noise representations discussed earlier, in measurements and in low-noise amplifier design it is often convenient to represent a noisy two-port with two equivalent noise sources at its input (Figure 3.24): a noise voltage source, v_n, and a noise current source, i_n. The two noise sources, which depend on the internal noise sources present in the two-port, are usually *statistically correlated.* Their correlation coefficient, c, is a complex number defined as

$$c = \frac{\langle i_n, v_n^* \rangle}{\sqrt{\overline{i_n^2}\,\overline{v_n^2}}}. \tag{3.97}$$

Therefore, for a given frequency, there are four real numbers that completely describe the noise of the linear two-port: v_n, i_n, and $c = Re\{c\} + j\,Im\{c\}$.

Let us consider the noisy two-port in Figure 3.24 driven by a signal source with source resistance

$$Z_S = R_S + j\ X_S. \tag{3.98}$$

The thermal noise generated by the signal source resistance R_S can be expressed as a noise voltage source in series with the signal source

$$\overline{v_{sn}^2} = 4kT\Delta f R_S. \tag{3.99}$$

First, for simplicity, we will assume that all noise sources are statistically independent (i.e. $c = 0$). We will remove this restriction later. By calculating the noise power in the load, Z_L, due to v_{sn} alone, and the noise power in the load due to both external (v_{sn}) and internal (v_n, i_n) noise sources, and dividing the latter by the former, we obtain a formula for the signal-to-noise ratio degradation, hence for the noise factor. However, the derivation can be further simplified by noticing that we only need to calculate the short-circuit noise currents at the input of the noiseless two-port since the later will not modify the signal-to noise ratio.

The total input short-circuit noise current is obtained by superposition as

$$i_{sc,\,tot} = \frac{-v_{sn} - v_n}{Z_S} + i_n \text{ and } i_{sc,\,s} = \frac{-v_{sn}}{Z_S} \tag{3.100}$$

which results in the following noise powers at the input of the noiseless two-port

$$\overline{i_{sc,\,tot}^2} = \left\langle \frac{v_{sn} + v_n}{Z_S} + i_n, \frac{v_{sn}^* + v_n^*}{Z_S^*} + i_n^* \right\rangle = \frac{\overline{v_{sn}^2} + \overline{v_n^2}}{|Z_S|^2} + \overline{i_n^2} \text{ and } \overline{i_{sc,\,s}^2} = \frac{\overline{v_{ns}^2}}{|Z_S|^2}. \tag{3.101}$$

The expression of the noise factor becomes

$$F = \frac{\overline{i_{sc,\,tot}^2}}{\overline{i_{sc,\,s}^2}} = 1 + \frac{\overline{v_n^2}}{\overline{v_{sn}^2}} + |Z_S|^2 \frac{\overline{i_n^2}}{\overline{v_{sn}^2}} = 1 + \frac{\overline{v_n^2}}{4kT\Delta f R_S} + \frac{(R_S^2 + X_S^2)\overline{i_n^2}}{4kT\Delta f R_S}. \tag{3.102}$$

The formula above indicates that the noise factor is a function of the two-port noise sources and of the signal source impedance. However, for circuit design purposes, it is desirable to define a set of two-port noise parameters that do not depend on external factors.

3.3.5 The IEEE noise parameters of a two-port

We can derive the expression of the minimum noise factor of the two-port, F_{MIN}, by differentiating (3.102) with respect to R_S, and X_S, respectively, and equating the results to zero

$$\frac{\partial F(R_S, X_S)}{\partial R_S} = 0, \frac{\partial F(R_S, X_S)}{\partial X_S} = 0. \tag{3.103}$$

F_{MIN} is obtained as

$$F_{MIN} = F(R_S = R_{SOPT}, X_S = X_{SOPT}) = 1 + \frac{\sqrt{\overline{v_n^2}\,\overline{i_n^2}}}{2kT\Delta f}. \tag{3.104}$$

The signal source resistance for which the minimum noise factor is reached is called the *optimum noise resistance, R_{SOPT}*

$$R_{SOPT} = \sqrt{\frac{\langle v_n^2 \rangle}{\langle i_n^2 \rangle}} \text{ and } X_{SOPT} = 0 \tag{3.105}$$

This is an intuitive result, similar to the definition of the input resistance of a two-port. We can imagine the two-port as a noisy signal generator. If the input termination of the two-port is equal to its optimum noise impedance, then the maximum transfer of noise from the two-port to the input termination (outside world) occurs without reflections. This, in turn, implies that the least amount of the noise generated inside the two-port is transferred to the load resulting in the least degradation of the signal-to-noise ratio.

It is important to note that F_{MIN} and $Z_{SOPT} = R_{SOPT} + jX_{SOPT}$, depend solely on the internal noise sources of the two-port and not on external factors, such as the signal source impedance. In general, the optimum noise impedance is a complex number and is different from the input impedance of the two-port. As a result, matching for minimum noise is, in general, different from matching for input impedance and for maximum signal power transfer.

Let us now remove the restriction imposed earlier on the two-port noise sources and consider that v_n and i_n are partially correlated.

One option is to describe the correlation with reference to the noise voltage

$$v_n = v_n; i_n = i_u + i_c = i_u + Y_{cor} v_n \tag{3.106}$$

where i_c and i_u represent the correlated and uncorrelated parts, respectively, of the input noise current with respect to the input noise voltage. Y_{cor}, the correlation admittance is defined as

$$Y_{cor} \stackrel{def}{=} \frac{\overline{i_n v_n^x}}{\overline{v_n^2}} = G_{cor} + jB_{cor}. \tag{3.107}$$

Alternatively, the correlation can be described by using the noise current as a reference

$$i_n = i_n; v_n = v_u + v_c = v_u + Z'_{cor} i_n \tag{3.108}$$

where v_c and v_u are the correlated and uncorrelated parts, respectively, of the input noise voltage with respect to the input noise current. Z'_{cor} represents the correlation impedance, defined as

$$Z'_{cor} \stackrel{def}{=} \frac{\overline{v_n i_n^x}}{\overline{i_n^2}} = R'_{cor} + jX'_{cor}. \tag{3.109}$$

Note that $Z'_{cor} \neq 1/Y_{cor}$.

In the **noise admittance formalism**, when the chosen reference source is the input noise voltage, four IEEE noise parameters F_{MIN}, R_n, and $Y_{SOPT} = G_{SOPT} + jB_{SOPT}$ are employed to completely describe the noise of the two-port. Their expressions can be derived with the help of Figure 3.25, where a Norton representation of the signal source has been preferred, as a function of Y_{cor}, the noise resistance R_n, defined as

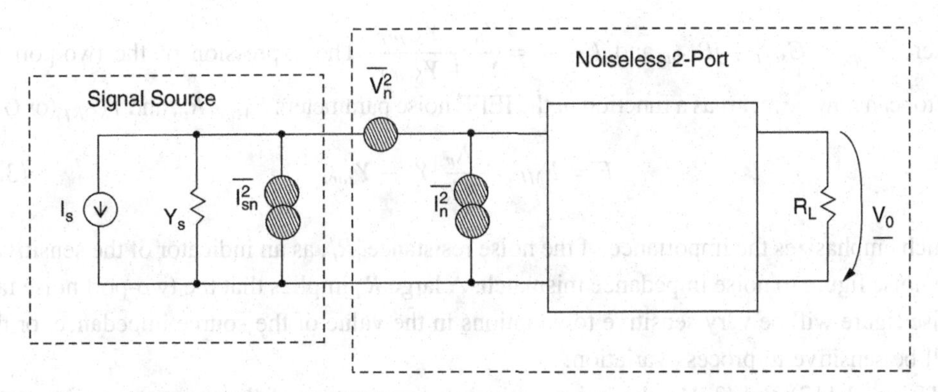

Figure 3.25 Schematic used in the derivation of the noise factor and IEEE noise parameters for a noisy two-port driven by a Norton signal source with finite admittance $Y_s = G_S + jB_S$

$$R_n \overset{def}{=} \frac{\overline{v_n^2}}{4kT\Delta f} \tag{3.110}$$

and of the noise conductance G_u

$$G_u \overset{def}{=} \frac{\overline{i_u^2}}{4kT\Delta f} \tag{3.111}$$

of the two-port.

Using (3.106), the short-circuit currents at the input of the noiseless two-port in Figure 3.25, become

$$i_{sc,\,tot} = i_{sn} + i_n + Y_S v_n = i_{sn} + i_u + (Y_{cor} + Y_S)v_n \text{ and } i_{sc,\,s} = i_{sn} \tag{3.112}$$

where $\overline{i_{sn}^2} = 4kT\Delta f G_S$. By using (3.112) and the definition of the noise factor, we obtain

$$F = \frac{\overline{i_{sc,\,tot}^2}}{\overline{i_{sc,\,s}^2}} = 1 + \frac{\overline{i_u^2}}{\overline{i_{sn}^2}} + \frac{\overline{v_n^2}|Y_{cor} + Y_s|^2}{\overline{i_{sn}^2}}. \tag{3.113}$$

Finally, by inserting (3.110) and (3.111) in (3.113), the noise factor is recast as a function of the noise resistance, noise conductance, and noise correlation admittance of the two-port

$$F = 1 + \frac{R_n}{G_s}|Y_{cor} + Y_s|^2 + \frac{G_u}{G_s}. \tag{3.114}$$

Differentiating equation (3.114) with respect to G_S and B_S yields the minimum noise factor

$$F_{min} = 1 + 2R_n\left(G_{cor} + G_{sopt}\right) \tag{3.115}$$

and the optimum noise admittance

$$G_{sopt} = \sqrt{G_{cor}^2 + \frac{G_u}{R_n}};\, B_{sopt} = -B_{cor} \tag{3.116}$$

where $Y_{sopt} = G_{sopt} + jB_{sopt}$ and $\Gamma_{SOPT} = \frac{Y_0 - Y_{SOPT}}{Y_0 + Y_{SOPT}}$. The expression of the two-port noise factor can now be recast as a function of the IEEE noise parameters F_{MIN}, R_n, and Y_{SOPT} (or G_{SOPT})

$$F = F_{MIN} + \frac{R_n}{G_s}\left|Y_s - Y_{sopt}\right|^2 \tag{3.117}$$

which emphasizes the importance of the noise resistance, R_n, as an indicator of the sensitivity of the noise figure to noise impedance mismatch. A large R_n implies that the two-port noise factor/ noise figure will be very sensitive to variations in the value of the source impedance, or that it will be sensitive to process variation.

Using (3.117) and (3.94), the noise equivalent temperature of the two-port can be cast as a function of R_n, Y_{SOPT}, and the minimum noise temperature of the two-port, T_{MIN}

$$T_e = T_{MIN} + T_0\frac{R_n}{G_s}\left|Y_s - Y_{sopt}\right|^2 \tag{3.118}$$

where

$$T_{MIN} = (F_{MIN} - 1)T_0. \tag{3.119}$$

Similarly, **in the noise impedance formalism,** when the input noise current is considered to be the reference noise source, we can derive a dual set of IEEE noise parameters F_{MIN}, G'_n, and $Z_{SOPT} = 1/Y_{SOPT}$, based on the definition of the noise conductance

$$G'_n \stackrel{def}{=} \frac{\overline{i_n^2}}{4kT\Delta f} \tag{3.120}$$

noise resistance, R'_u, corresponding to the uncorrelated part of the input noise voltage

$$R'_u \stackrel{def}{=} \frac{\overline{v_u^2}}{4kT\Delta f} \tag{3.121}$$

the noise correlation impedance (3.109), and a Thevenin representation of the signal source $Z_S = R_S + jX_S$. The noise factor is first described as a function of the two input equivalent noise sources and of the signal source noise voltage, $\overline{v_{sn}^2} = 4kT\Delta f R_S$,

$$F = 1 + \frac{\overline{i_n^2}\left|Z'_{cor} + Z_s\right|^2}{\overline{v_{sn}^2}} + \frac{\overline{v_u^2}}{\overline{v_{sn}^2}}. \tag{3.122}$$

By plugging the definitions (3.120) and (3.121) into (3.122), we obtain

$$F = 1 + \frac{G'_n}{R_s}\left|Z'_{cor} + Z_s\right|^2 + \frac{R'_u}{R_s} \tag{3.123}$$

which can be differentiated with respect to R_S and X_S to find its minimum F_{MIN}

$$F_{MIN} = 1 + 2G'_n\left(R'_{cor} + R_{sopt}\right) \tag{3.124}$$

corresponding to $Z_{sopt} = R_{sopt} + jX_{sopt}$ where

$$R_{sopt} = \sqrt{R'^2_{cor} + \frac{R_u}{G'_n}}; X_{sopt} = -X'_{cor}. \tag{3.125}$$

Again, we can recast the two-port noise factor as a function of the IEEE noise parameters in the noise impedance formalism

$$F = F_{MIN} + \frac{G'_n}{R_s}\left|Z_s - Z_{sopt}\right|^2 \tag{3.126}$$

noting that, although $R'_u \neq 1/G_u$ and $G'_n \neq 1/R_n$, G'_n plays the same role as R_n as a bell-weather of the sensitivity of the two-port noise factor to noise impedance mismatch.

Equation (3.126) has a dual in terms of the equivalent noise temperature of the two-port

$$T_e = T_{MIN} + T_0\frac{G'_n}{R_s}\left|Z_s - Z_{sopt}\right|^2. \tag{3.127}$$

Expressions that convert the IEEE noise admittance parameters into the IEEE noise impedance parameters, and vice-versa, are derived based on the requirement that, for any noisy two-port, the two formalisms must result in the same minimum noise factor and the same optimum noise impedance/admittance

$$Y_{sopt} = \frac{1}{Z_{sopt}}. \tag{3.128}$$

From these conditions, we obtain

$$Y^x_{cor} = \frac{Z'_{cor}}{\left|Z_{sopt}\right|^2} \tag{3.129}$$

$$R_n = G'_n\left|Z_{sopt}\right|^2 \tag{3.130}$$

and

$$G_u = \frac{R'_u}{\left|Z_{sopt}\right|^2}. \tag{3.131}$$

3.3.6 Noise correlation matrices and the noise parameters of a two-port

In low-noise circuit analysis, it is important to derive relations between the noise correlation matrices and the two-port noise parameters. Given the IEEE noise parameters in the admittance formalism, we can calculate the *ABCD* noise correlation matrix [2]

$$[C_A] = \begin{bmatrix} \langle v_{n1} \; v^*_{n1}\rangle & \langle v_{n1} \; i^*_{n1}\rangle \\ \langle i_{n1} \; v^*_{n1}\rangle & \langle i_{n1} \; i^*_{n1}\rangle \end{bmatrix} = \begin{bmatrix} C_{uu*} & C_{ui*} \\ C_{iu*} & C_{ii*} \end{bmatrix} = \begin{bmatrix} R_n & \frac{F_{MIN}-1}{2} - R_nY^*_{opt} \\ \frac{F_{MIN}-1}{2} - R_nY_{opt} & R_n\left|Y_{opt}\right|^2 \end{bmatrix}. \tag{3.132}$$

Conversely, the IEEE noise parameters can be be obtained from the entries of the *ABCD* noise correlation matrix

$$Y_{opt} = \sqrt{\frac{C_{ii^*}}{C_{uu^*}} - \left[\Im\left(\frac{C_{ui^*}}{C_{uu^*}}\right)\right]^2} + j\Im\left(\frac{C_{ui^*}}{C_{uu^*}}\right) \quad (3.133)$$

$$F_{MIN} = 1 + \frac{C_{ui^*} + C_{uu^*}Y^*_{opt}}{kT} \quad (3.134)$$

and

$$R_n = \frac{C_{uu^*}}{kT} \quad (3.135)$$

Similar relations exist between the noise admittance correlation matrix and the two-port noise parameters [2]

$$[C_y] = \begin{bmatrix} \langle i_{n1} \; i^*_{n1} \rangle & \langle i_{n1} \; i^*_{n2} \rangle \\ \langle i_{n2} \; i^*_{n1} \rangle & \langle i_{n2} \; i^*_{n2} \rangle \end{bmatrix} = \begin{bmatrix} G_u + R_n|Y_{11} - Y_{cor}|^2 & R_n Y^*_{21}(Y_{11} - Y_{cor}) \\ R_n Y_{21}(Y^*_{11} - Y^*_{cor}) & R_n|Y_{21}|^2 \end{bmatrix} \quad (3.136)$$

$$R_n = \frac{C_{y22}}{|Y_{21}|^2} \quad (3.137)$$

$$Y_{cor} = Y_{11} - \frac{C_{y12}}{C_{y22}} Y_{21} \quad (3.138)$$

and

$$G_u = C_{y11} - R_n|Y_{11} - Y_{cor}|^2. \quad (3.139)$$

3.3.7 Constant noise circles

Equations (3.117) and (3.126) are often recast as functions of the optimum noise reflection coefficient and source reflection coefficient, resulting in a set of constant noise factor (noise figure) or constant noise temperature circles

$$F(\Gamma_s) = F_{min} + \frac{4R_n}{Z_0} \frac{|\Gamma_s - \Gamma_{sopt}|^2}{\left(1 - |\Gamma_S|^2\right)|1 + \Gamma_{sopt}|^2} \; ; \; T_e(\Gamma_s) = T_{min} + \frac{4T_0 R_n}{Z_0} \frac{|\Gamma_s - \Gamma_{sopt}|^2}{\left(1 - |\Gamma_S|^2\right)|1 + \Gamma_{sopt}|^2}. \quad (3.140)$$

These can be visually represented on the Smith Chart for easy interpretation during low-noise amplifier design. The radii and centers of these circles are given by

$$r_i = \frac{\sqrt{N_i^2 + N_i\left(1 - |\Gamma_{sopt}|^2\right)}}{1 + N_i} \quad (3.141)$$

and

$$C_i = \frac{\Gamma_{sopt}}{1+N_i} \tag{3.142}$$

respectively, where

$$N_i = \frac{F_i - F_{MIN}}{4r_n}\left|1+\Gamma_{sopt}\right|^2 \quad \text{and} \quad r_n = \frac{R_n}{Z_0}. \tag{3.143}$$

F_i represents the noise factor corresponding to source impedances located on the circle of radius r_i and centered on C_i. For any signal source impedance located inside this circle, the noise factor will be smaller than F_i.

3.3.8 The noise figure of a passive two-port

The equivalent input noise temperature of a passive two-port at physical temperature, T, is obtained by inserting the expression of the noise wave correlation matrix of a passive two-port (3.79) into the equivalent input noise temperature Equation (3.92)

$$T_e = \frac{T[\alpha]\left([U]-[S][S]^+\right)[\alpha]^+}{\left(1-|\Gamma_S|^2\right)}. \tag{3.144}$$

For example, the S-matrix of a lossy transmission line of characteristic impedance Z_0, physical length l and insertion loss L, at physical temperature T, is expressed as $S = \begin{bmatrix} 0 & 1 \\ 1 & 0 \end{bmatrix} \frac{e^{-j\beta l}}{\sqrt{L}}$, while its output reflection coefficient is given by $\Gamma_{out} = \frac{\Gamma_S}{L} e^{-2j\beta l}$ [1]. Then

$$T_e = \frac{T[\alpha]\left([U]-[S][S]^+\right)[\alpha]^+}{\left(1-|\Gamma_S|^2\right)} = \frac{T\left(1-\frac{1}{L}\right)[\alpha][U][\alpha]^+}{\left(1-|\Gamma_S|^2\right)} = \frac{T\left(1-\frac{1}{L}\right)\left(|\Gamma_S|^2+\frac{|1-\Gamma_S S_{11}|^2}{|S_{21}|^2}\right)}{\left(1-|\Gamma_S|^2\right)}$$

which, since $S_{11}=0$ and $|S_{21}|^2 = 1/L$, leads to $T_e = \dfrac{T(L-1)\left(|\Gamma_S|^2+L\right)}{L\left(1-|\Gamma_S|^2\right)}$.

If the t-line is lossless, $L=1$ and $T_e=0$ and $F=1$. In other words, the noise figure of a lossy transmission line is 0dB.

If the line is matched, $\Gamma_S = 0$, then

$$T_e = T(L-1) = \left(\frac{1}{G_A}-1\right)T \text{ and } F = 1+\left(\frac{1}{G_A}-1\right)\frac{T}{T_0} = 1+(L-1)\frac{T}{T_0}. \tag{3.145}$$

where G_A is the available power gain from port 1 to port 2, defined in (3.43), and $1/G_A = L$.

It can be demonstrated that NF_{MIN}, T_{MIN}, $R_n{\times}G_{SOPT}$, $R_n{\times}G_{cor}$, $R_n{\times}G_u$, $G'_n{\times}R_{SOPT}$, $G'_n{\times}R'_{cor}$, and $G'_n{\times}R'_u$ are invariant when connecting lossless reciprocal two-ports (such as an ideal transmission line) at the input and at the output of a noisy two-port [8]. This explains the popularity of reactive matching networks in tuned LNA design.

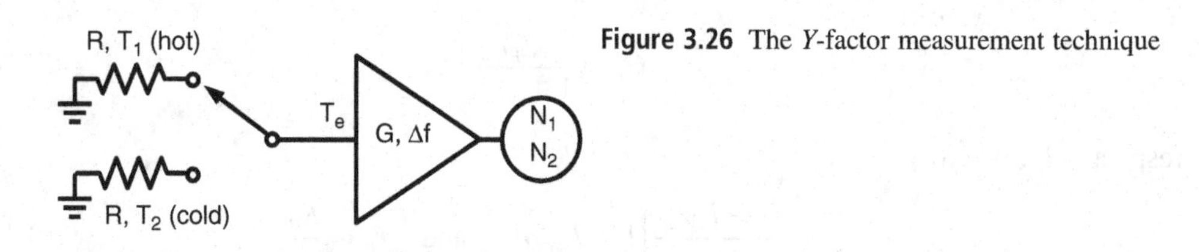

Figure 3.26 The *Y*-factor measurement technique

3.3.9 Two-port noise temperature measurements

The simplest method to measure the equivalent input noise temperature of a two-port is to drive the input of the two-port with a noise source and to measure the noise power at the two-port output. Since, as illustrated in Figure 3.23, the output noise power also includes the contribution of the noise source itself, a ratio technique, known as the *Y*-method, was developed. The technique is summarized in Figure 3.26. The device under test (DUT) of unknown power gain, *G*, and unknown noise temperature, T_e, is first connected to a matched load at temperature T_1, and then to a matched load at temperature T_2. The output noise power is measured in both cases

$$N_1 = GkT_1\Delta f + GkT_e\Delta f \tag{3.146}$$

$$N_2 = GkT_2\Delta f + GkT_e\Delta f. \tag{3.147}$$

The noise temperature (and hence noise factor), and the power gain of the DUT are obtained as

$$Y = \frac{N_1}{N_2} = \frac{T_1 + T_e}{T_2 + T_e}; T_e = \frac{T_1 - YT_2}{Y - 1} \tag{3.148}$$

and

$$G = \frac{N_1 - N_2}{k\Delta f(T_1 - T_2)}. \tag{3.149}$$

3.3.10 Two-port noise parameter measurements

The noise temperature measured under source-matched conditions, like the noise figure, provides only a partial picture of the noise performance of a two-port. In order to fully capture the noise behavior, all four noise parameters of the two-port must be measured. This is accomplished by using a *Noise Figure Meter* to measure the noise figure of the DUT for different signal source impedances. The main idea is to measure the noise figure corresponding to four or more different source impedances and to solve the resulting system of equations with four unknowns: the noise parameters F_{MIN}, R_n, G_{SOPT}, and B_{SOPT}

$$F_k = F_{MIN} + \frac{R_n}{G_{sk}}\left|Y_{sk} - Y_{sopt}\right|^2 \tag{3.150}$$

where $k = 1 \ldots 4$

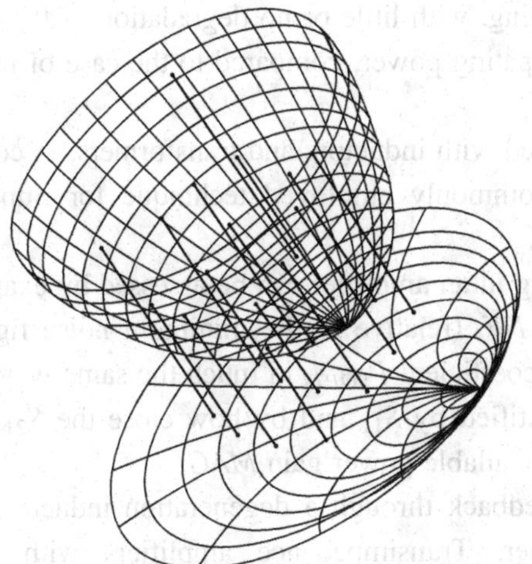

Figure 3.27 3-D representation of the measured noise figure using (3.151) and its projection on the impedance Smith Chart indicating the corresponding source impedance values. Ref. ATN Microwave NP5 System Manual

In theory, four different source impedances and four equations are sufficient to obtain the four noise parameters. In practice, due to measurement error and the difficulty of a priori choosing an appropriate set of source impedances that leads to a unique set of noise parameters, 16 to 32 source impedance states are provided for measurement, as illustrated in Figure 3.27, and a system of 16–32 equations is formed. The choice of Y_{sk} is critical to the measurement accuracy. The challenge of accurately measuring the noise parameters of a two-port, especially of transistors, stems from the fact that noise parameters are not measured directly, but rather that they are extrapolated from the 16 to 32 noise figure measurements. It is only by accident that one of the chosen source impedance values would correspond to the Y_{SOPT} of the DUT

Because of these practical difficulties, methods have been developed to verify the accuracy of the noise parameter measurements. The most common one imposes that the condition (3.151) be satisfied [9]

$$2 \geq \frac{4NT_0}{T_{MIN}} \geq 1 \tag{3.151}$$

where

$$N = R_{SOPT}G'_n = R_n G_{SOPT}. \tag{3.152}$$

3.4 NOISE IN CIRCUITS WITH NEGATIVE FEEDBACK

3.4.1 Feedback networks as noise matching tools

As in low-frequency analog ICs, negative feedback can become a useful vehicle for optimizing the noise impedance of high-frequency amplifiers and the bandwidth over which it can be matched to a desired value. If implemented with purely reactive components, negative feedback can lead to noise and signal impedance matching, with little or no degradation of the overall noise figure of the amplifier and without dissipating power, compared to the case of resistive feedback.

For these reasons, negative feedback, realized with inductors and transformers, in conjunction with transistor sizing, has become a commonly employed technique for impedance matching in high-frequency integrated circuits.

The quality of the noise impedance matching in an amplifier can be assessed by examining both the value of the amplifier noise figure, *NF*, (relative to the minimum noise figure of the transistor NF_{MIN}) and the noise reflection coefficient Γ_{SOPT}, in much the same way as the quality of signal impedance matching is quantified by S_{11} and by how close the S_{21} of the amplifier approaches the transistor maximum available power gain *MAG*.

Most tuned LNAs employ series–series feedback through a degeneration inductor L_S, or series–shunt feedback through a transformer. Transimpedance amplifiers with shunt–shunt feedback are frequently used as broadband low-noise input amplifiers in fiber-optic receivers.

In order to understand how to apply feedback theory to the systematic analysis and design of low-noise amplifiers, the noise parameters of shunt–shunt, series–series, series–shunt, and shunt–series connected two-ports will be derived. Together with the traditional unilateral amplifier approximation, these noise parameters provide the foundations of an intuitive analytical method for the design of low-noise amplifiers. For comparison, the exact computational approach relying on the appropriate two-port *Y*-, *Z*-, *H*-, or *G*-matrix representation and noise correlation matrices will also be provided.

3.4.2 Method for deriving the equivalent noise sources at the input of a two-port

Since the noise parameters of a two-port are defined based on the equivalent noise sources at the input of the two-port, it is important to develop a method for transferring the noise sources (currents or voltages) from their physical location inside the two-port at its input. This technique is generally applicable and can be summarized in two steps:

Step 1: Short circuit the inputs and outputs of the original circuit and of the final circuit, and calculate the short-circuit currents at the output in both cases and force them to be equal. From this equation, you will obtain the expression for v_n.

Step 2: Open circuit the inputs and outputs of the original circuit and of the final circuit and calculate the open-circuit voltage at the output in both cases and force them to be equal. The expression for i_n is obtained from this equation.

EXAMPLE 3.5

Derive the equivalent noise admittance parameters for *T*, *Π*, *I*, and *Γ* resistive networks.

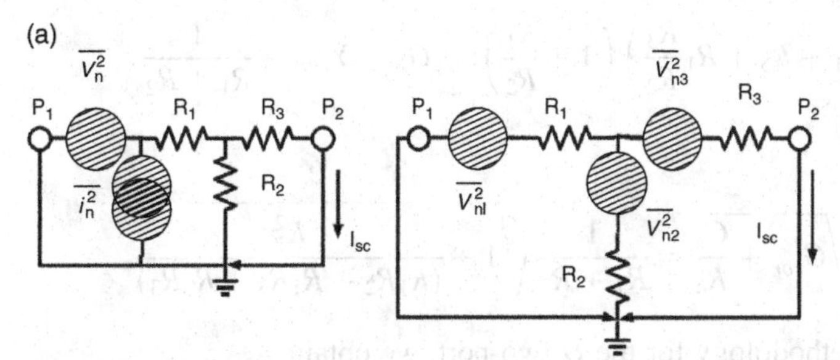

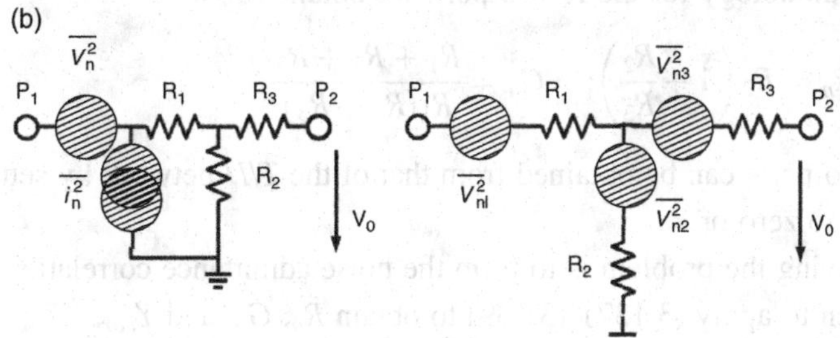

Figure 3.28 Calculation of the noise parameters of a resistive T network (a) Short-circuit output current (b) Open-circuit output voltage

Solution

Let us consider the ***T*** network and label the resistors in order from left to right: R_1, R_2, R_3. The two-step approach for calculating the noise parameters of any two-port is illustrated in Figure 3.28.

(i) Short-circuit the inputs and outputs of the original noisy two-port and of the final noiseless two-port (with equivalent input noise sources) and calculate the output current in both cases. By equating the two output currents, the expression for the equivalent input noise voltage is determined.

The equivalent input noise voltage becomes

$$v_n = v_{n1} + \frac{R_1}{R_2} v_{n2} + \left(1 + \frac{R_1}{R_2}\right) v_{n3}.$$

(ii) Similarly, leaving the input and output of the two-port open, determine the open-circuit output noise voltage in the original noisy two-port and in the final equivalent two-port. By equating the two, the equivalent input noise current source is derived as a function of the internal noise sources

$$i_n = \frac{v_{n2}}{R_2} + \frac{v_{n3}}{R_2}.$$

Applying the definitions for R_n, G_u, and Y_{cor} and considering that

$$\overline{i_n \times v_n^*} = \frac{\frac{R_1}{R_2}\overline{v_{n2}^2} + \left(1 + \frac{R_1}{R_2}\right)\overline{v_{n3}^2}}{R_2}$$

we get

$$R_n = \left(R_1 + R_3 + R_1\frac{R_3}{R_2}\right)\left(1 + \frac{R_1}{R_2}\right); \quad G_u = Y_{cor} = \frac{1}{R_1 + R_2}$$

and

$$Y_{SOPT} = \sqrt{G_{cor}^2 + \frac{G_u}{R_n}} = \frac{1}{R_1 + R_2}\sqrt{1 + \frac{R_2^2}{(R_1R_2 + R_1R_3 + R_2R_3)}}.$$

By following the same methodology for the Π two-port, we obtain

$$R_n = R_2\left(1 + \frac{R_2}{R_3}\right); \quad G_u = \frac{R_1 + R_2 + R_3}{R_1(R_2 + R_3)}.$$

The case of the Γ and I two-ports can be obtained from that of the T/Π network by setting the appropriate resistor values to zero or ∞.

Another approach to solving the problem is to form the noise admittance correlation matrix as in Example 3.4, and then to apply (3.137)–(3.139) to obtain R_n, G_u, and Y_{cor}.

3.4.3 Shunt–shunt (parallel feedback)

Let us consider the case of a network consisting of an amplifier two-port and of a feedback two-port connected in parallel, as in Figure 3.29. The parameters of the main amplifier and of the feedback two-port networks are described by subscripts a and f, respectively. It is convenient to employ Y-parameters and the noise admittance formalism to derive the expressions of the noise parameters R_n, Y_{cor}, and G_u of the overall two-port

$$\begin{bmatrix} Y_{11} & Y_{12} \\ Y_{21} & Y_{22} \end{bmatrix} = \begin{bmatrix} Y_{11f} & Y_{12f} \\ Y_{21f} & Y_{22f} \end{bmatrix} + \begin{bmatrix} Y_{11a} & Y_{12a} \\ Y_{21a} & Y_{22a} \end{bmatrix} \text{ and } [C_y] = [C_{ya}] + [C_{yf}] \tag{3.153}$$

$$v_n = f(v_{nf}, i_{nf}, v_{na}, i_{na}) \tag{3.154}$$

$$i_n = g(v_{nf}, i_{nf}, v_{na}, i_{na}). \tag{3.155}$$

Using the two-step method introduced earlier, we get:

Step 1: Input noise voltage

$$v_n = \frac{Y_{21f}\, v_{nf} + Y_{21a}\, v_{na}}{Y_{21}} = D_f\, v_{nf} + D_a\, v_{na}. \tag{3.156}$$

Step 2: Input noise current

$$i_n = i_{nf} + i_{na} + \frac{Y_{11a}Y_{21f} - Y_{21a}Y_{11f}}{Y_{21}} v_{nf} + \frac{Y_{11f}Y_{21a} - Y_{21f}Y_{11a}}{Y_{21}} v_{na} = i_{nf} + i_{na} + E_f v_{nf} + E_a v_{na} \tag{3.157}$$

And by expanding the noise currents into correlated and uncorrelated term

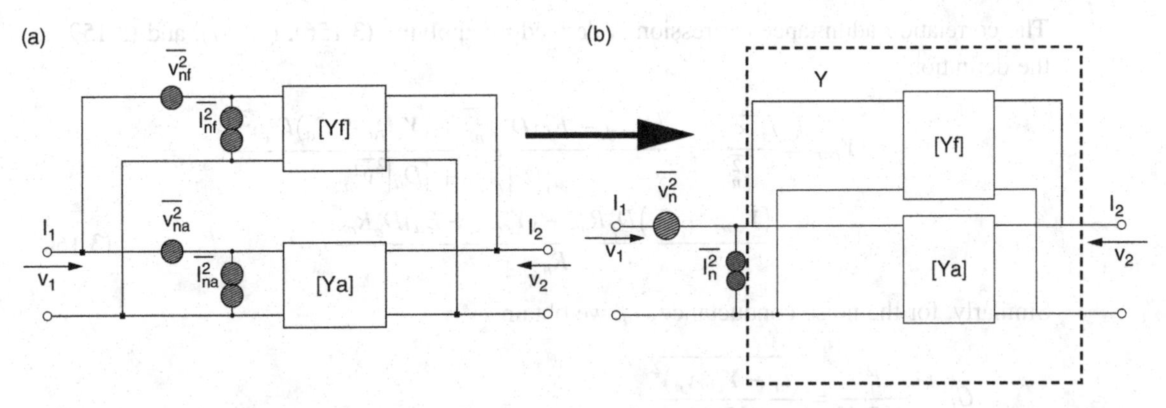

Figure 3.29 (a) Two noisy two-ports connected in parallel (b) Noise equivalent circuit representation of the two two-ports connected in parallel

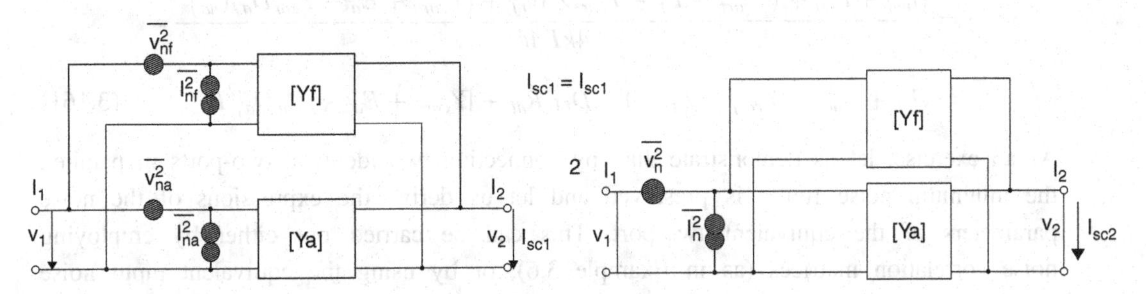

Figure 3.30 Short circuit inputs and outputs

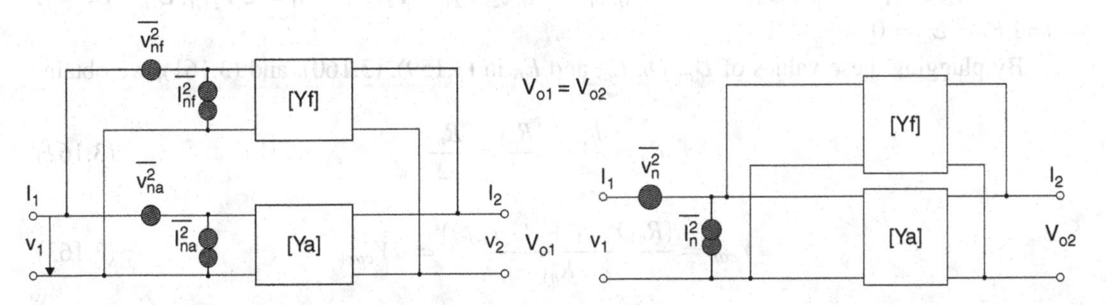

Figure 3.31 Open circuit inputs and outputs

$$i_n = i_{nuf} + i_{nua} + \left(Y_{corf} + E_f\right)v_{nf} + (Y_{cora} + E_a)v_{na}. \tag{3.158}$$

Equations (3.156) and (3.158) indicate that, even if v_{na}, i_{na}, v_{nf}, and i_{nf} are not correlated, i_n and v_n are at least partially correlated simply because their expressions share common terms. These equations can be employed to derive the rather lengthy expressions of the noise parameters of two two-ports connected in parallel as a function of the noise parameters of the individual two-ports [10]. First, we obtain the noise resistance from (3.155)

$$R_n \stackrel{def}{=} \frac{\overline{v_n^2}}{4kT\Delta f} = \frac{\left|Y_{21f}\right|^2 R_{nf} + \left|Y_{21a}\right|^2 R_{na}}{\left|Y_{21}\right|^2} = \left|D_f\right|^2 R_{nf} + \left|D_a\right|^2 R_{na}. \tag{3.159}$$

The correlation admittance expression is derived by applying (3.156), (3.157), and (3.159) to the definition

$$Y_{cor} \stackrel{def}{=} \frac{\overline{i_n \times v_n^*}}{\overline{v_n^2}} = \frac{(Y_{corf} + E_f)D_f^* \overline{v_{nf}^2} + (Y_{cora} + E_a)D_a^* \overline{v_{na}^2}}{|D_f|^2 \overline{v_{nf}^2} + |D_a|^2 \overline{v_{na}^2}}$$

$$= \frac{(Y_{corf} + E_f)D_f^* R_{nf} + (Y_{cora} + E_a)D_a^* R_{na}}{R_n}. \tag{3.160}$$

Similarly, for the noise conductance G_u, we obtain

$$G_u \stackrel{def}{=} \frac{\overline{i_u^2}}{4kT\Delta f} = \frac{\overline{(i_n - Y_{cor}v_n)^2}}{4kT\Delta f}$$

$$= \frac{\overline{(i_{nuf} + i_{nua} + (Y_{corf} + E_f - Y_{cor}D_f)v_{nf} + (Y_{cora} + E_a - Y_{cor}D_a)v_{na})^2}}{4kT\Delta f} =$$

$$= G_{uf} + G_{ua} + |Y_{corf} + E_f - Y_{cor}D_f|^2 R_{nf} + |Y_{cora} + E_a - Y_{cor}D_a|^2 R_{na}. \tag{3.161}$$

As an exercise, let us demonstrate that, by connecting two identical two-ports in parallel, the minimum noise figure is preserved and let us derive the expressions of the noise parameters of the equivalent two-port. This can be carried out either by employing noise correlation matrices (as in Example 3.6), or by using the equivalent input noise sources method. We will use the latter. Because the two two-ports are identical: $[Y]_1 = [Y]_2$, $R_{n1} = R_{n2}$; $Y_{cor1} = Y_{cor2}$; $G_{u1} = -G_{u2}$; $F_{MIN1} = F_{MIN2}$, $Y_{21} = Y_{21a} + Y_{21f} = 2\, Y_{21,1}$, $D_a = D_f = ½$ and $E_f = E_a = 0$.

By plugging these values of D_a, D_f, E_a, and E_f, in (3.159), (3.160), and (3.161), we obtain

$$R_n = \frac{R_{n1}}{4} + \frac{R_{n2}}{4} = \frac{R_{n1}}{2} \tag{3.162}$$

$$Y_{cor} = \frac{(R_{n1}Y_{cor1} + R_{n2}Y_{cor2})}{R_{n1}} = 2Y_{cor1} \tag{3.163}$$

and

$$G_u = G_{u1} + G_{u2} + \left|Y_{cor1} - \frac{Y_{cor}}{2}\right|^2 R_{nf} + \left|Y_{cor2} - \frac{Y_{cor}}{2}\right|^2 R_{na} = G_{u1} + G_{u2} = 2G_{u1}. \tag{3.164}$$

Finally, by inserting (3.162)–(3.163) in the definitions of G_{SOPT} and F_{MIN}

$$G_{sopt} = \sqrt{G_{cor}^2 + \frac{G_u}{R_n}}; \quad F_{MIN} = 1 + 2R_n(G_{cor} + G_{sopt})$$

we find that

$$G_{sopt} = 2G_{sopt1}; \quad F_{MIN} = F_{MIN1}. \tag{3.165}$$

EXAMPLE 3.6

An important application of the shunt–shunt feedback formulas is in the noise analysis of a network consisting of N identical two-ports connected in parallel. This situation occurs often in RF circuits when connecting N identical MOSFET gate fingers in parallel to form a large MOSFET.

Solution

The most elegant solution to derive the noise parameters of such a network is to employ the noise admittance correlation matrix expressions. The noise admittance correlation matrix and the Y-matrix of the resulting two-port becomes

$$[C_y] = \sum_{i=1}^{i=N} [C_y]_i = N \times [C_y]_i \text{ and } Y = \sum_{i=1}^{i=N} [Y]_i = N \times [Y]_i$$

where $[C_y]_i$ and $[Y]_i$ are the noise admittance and Y-parameter matrices, respectively, of two-port i.

From (3.137)–(3.139), we obtain, successively

$$R_n = \frac{C_{y22}}{|Y_{21}|^2} = \frac{N \times C_{iy22}}{N^2 \times Y_{i21}} = \frac{R_{ni}}{N};$$

$$Y_{cor} = Y_{11} - \frac{C_{y12}}{C_{y22}} \times Y_{21} = N \times Y_{i11} - \frac{N \times Ci_{y12}}{N \times Ci_{y22}} \times N \times Yi_{21} = N \times Yi_{cor}$$

$$G_u = C_{y11} - R_n|Y_{11} - Y_{cor}|^2 = N \times Ci_{y11} - \frac{R_{ni}}{N}|N \times Yi_{11} - N \times Yi_{cor}|^2 = N \times G_{ui}$$

$$G_{sopt} = \sqrt{G_{cor}^2 + \frac{G_u}{R_n}} = \sqrt{N^2 G_{cori}^2 + \frac{NG_{ui}}{\frac{R_{ni}}{N}}} = NG_{sopti}$$

$$F_{MIN} = 1 + 2R_n(G_{cor} + G_{sopt}) = 1 + 2\frac{R_{ni}}{N}(NG_{cori} + NG_{sopti}) = 1 + 2R_{¢}(G_{cori} + G_{sopti}) = F_{MINi}.$$

By connecting transistors or transistor gate fingers in parallel, the minimum noise factor is preserved while the optimum source impedance and sensitivity to source mismatch are reduced. This provides the IC designer with a powerful tool for realizing noise matching simply by controlling the size of transistors and without (to first order) compromising the noise factor.

The example above illustrates a practical case in which we can derive the exact noise parameters of a combination of shunt-connected two-ports, without resorting to approximations. However, for most practical applications, as in amplifiers with negative feedback, we need to apply reasonable approximations in order to arrive at simplified noise parameter equations that can be used in the hand design of low-noise amplifiers.

In general, if the unilateral amplifier approximation holds

$$Y_{21} \approx Y_{21a}; \quad Y_{12} \approx Y_{12f}; \quad D_f \approx 0; \quad D_a \approx 1; \quad E_a = -E_f \approx Y_{11f} \tag{3.166}$$

Equations (3.156) and (3.158) simplify to

$$v_n \approx v_{na}; \; i_n = i_{nuf} + i_{nua} + \left(Y_{corf} - Y_{11f}\right)v_{nf} + \left(Y_{cora} + Y_{11f}\right)v_{na} \tag{3.167}$$

Next, by using (3.159)–(3.161) and algebraic manipulations of statistical signals, the noise parameters of the amplifier with feedback can be cast as [11]

$$R_n \approx R_{na}; \; G_u = G_{uf} + G_{ua} + |Y_{corf} - Y_{11f}|^2 R_{nf} \geq G_{ua} \tag{3.168}$$

$$Y_{cor} = Y_{cora} + Y_{11f}; \; G_{cor} = G_{cora} + \Re\left(Y_{11f}\right) \geq G_{cora} \tag{3.169}$$

$$Y_{sopt} = \sqrt{G^2_{sopta} + \frac{G_{uf}}{R_{na}} + 2G_{cora}\Re\left(Y_{11f}\right) + \Re^2\left(Y_{11f}\right) + \frac{|Y_{corf} - Y_{11f}|^2 R_{nf}}{R_{na}}} + j\left[B_{sopta} - \Im\left(Y_{11f}\right)\right] \tag{3.170}$$

$$F_{MIN} = 1 + 2R_{na}\left[G_{cora} + G_{sopt} + \Re\left(Y_{11f}\right)\right] \geq F_{MINa}. \tag{3.171}$$

We note that the noise voltage of the amplifier with parallel feedback is practically equal to that of the main amplifier. The noise currents of the amplifier and feedback networks add, while Y_{SOPT} increases. F_{MIN} increases if the feedback network has resistive elements. If it is purely reactive ($R_{nf} = 0$, $G_{uf} = 0$, $g_{11f} = 0$), it will not degrade the noise figure and G_{SOPT} will remain largely unaffected by feedback. For lossy feedback, we can conclude that:

- Shunt lossy feedback can be used for noise matching in situations where the noise impedance of the original two-port is higher than that of the source impedance. Lossy shunt feedback at the input of an amplifier decreases both the input impedance and the optimal noise impedance.
- Since R_n remains approximately equal to that of the main amplifier, the sensitivity to noise impedance mismatch remains practically the same.
- More importantly, since, according to (3.170), G_{SOPT} increases and B_{SOPT} decreases or remains the same, the quality factor of the optimum noise admittance, $Q = B_{SOPT}/G_{SOPT}$, decreases, implying that the bandwidth over which the noise impedance can be matched increases when negative feedback is applied. This behavior is similar to the signal bandwidth extension property of negative feedback.
- Just as the gain of an amplifier with negative feedback decreases, the minimum noise factor of an amplifier with lossy negative feedback increases compared to that of the main amplifier.

EXAMPLE 3.7 Transimpedance feedback amplifier

Let us consider the shunt–shunt feedback amplifier shown in Figure 3.32. The feedback network consists of resistor R_F while the amplifier network consists of an ideal (but "noisy") inverting voltage amplifier with voltage gain $-A$. The amplifier and feedback networks are described by the Y-parameter matrix Y_a and Y_f, respectively. Find the expressions of the noise parameters of the amplifier with feedback.

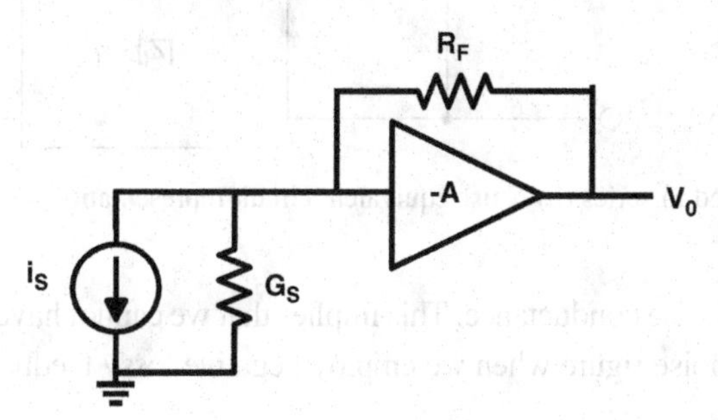

Figure 3.32 Transimpedance feedback amplifier

Solution

By analyzing the feedback network consisting of R_F, we can easily derive its noise and Y-parameters as $R_{nf} = R_F$, $G_{uf} = 0$, $Y_{corf} = 0$; $Y_{11f} = 1/R_F$. The noise admittance correlation matrix of the feedback network is

$$4kT\Delta f \begin{bmatrix} \frac{1}{R_F} & \frac{-1}{R_F} \\ \frac{-1}{R_F} & \frac{1}{R_F} \end{bmatrix}$$

We next plug these expressions into (3.168)–(3.171) to obtain

$$R_n \approx R_{na}\,;\; G_u = G_{ua} + \left| Y_{corf} - \frac{1}{R_F} \right|^2 R_F\,;\; Y_{cor} = Y_{cora} + \frac{1}{R_F}$$

$$Y_{sopt} = \sqrt{G_{sopta}^2 + 2\frac{G_{cora}}{R_F} + \frac{1}{R_F^2} + \frac{1}{R_{na}R_F}} + jB_{sopta}$$

$$F_{MIN} = 1 + 2R_{na}\left(G_{cora} + G_{sopt} + \frac{1}{R_F} \right) > F_{MINa}$$

which demonstrate that G_{SOPT} has increased compared to that of the amplifier. Along with the R_F term in F_{MIN}, this leads to an increase of the minimum noise factor of the amplifier with feedback when compared to that of the amplifier without feedback.

Because G_{SOPT} has increased and B_{SOPT} is unchanged, R_{SOPT} has decreased, confirming that shunt feedback at the input of an amplifier reduces the optimum noise impedance. Note that, if $R_F >> 1/(G_{SOPTa} + G_{cora})$ the feedback resistor, R_F, contributes insignificant noise. However, in that

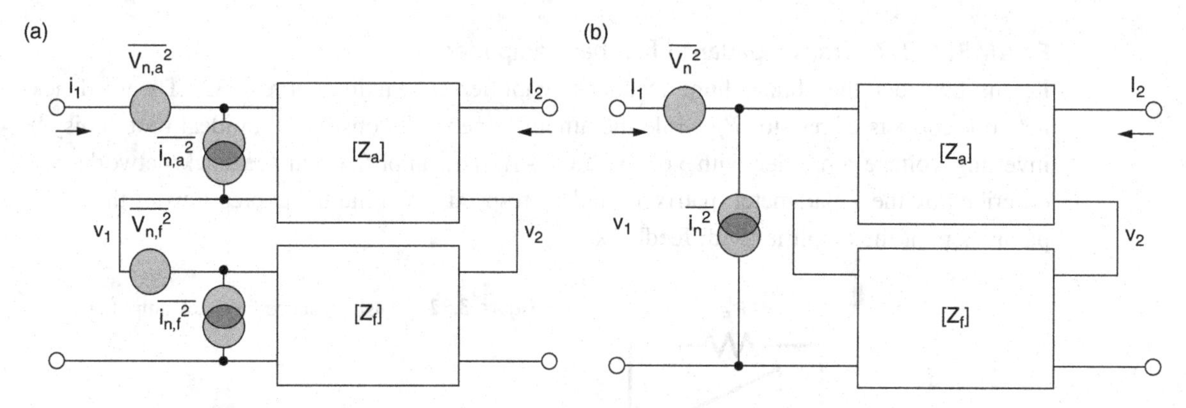

Figure 3.33 (a) Two noisy two-ports connected in series (b) Noise equivalent circuit representation of the two two-ports connected in series

case it also does not change the optimum noise conductance. This implies that we cannot have noise impedance matching without degrading noise figure when we employ negative lossy feedback.

3.4.4 Series–series feedback

Similarly, to analyze circuits consisting of two-ports connected in series, as in Figure 3.33, we can use Z-parameters and the noise impedance formalism

$$\begin{bmatrix} Z_{11} & Z_{12} \\ Z_{21} & Z_{22} \end{bmatrix} = \begin{bmatrix} Z_{11f} & Z_{12f} \\ Z_{21f} & Z_{22f} \end{bmatrix} + \begin{bmatrix} Z_{11a} & Z_{12a} \\ Z_{21a} & Z_{22a} \end{bmatrix} \textit{ and } [C_z] = [C_{za}] + [C_{zf}] \tag{3.172}$$

to find the expressions of the input equivalent noise voltage

$$v_n = v_{nf} + v_{na} + \frac{Z_{11a}Z_{21f} - Z_{21a}Z_{11f}}{Z_{21}} i_{nf} + \frac{Z_{11f}Z_{21a} - Z_{21f}Z_{11a}}{Z_{21}} i_{na} = v_{nf} + v_{na} + B_f i_{nf} + B_a i_{na} \tag{3.173}$$

and noise current

$$i_n = \frac{Z_{21f} i_{nf} + Z_{21a} i_{na}}{Z_{21}} = A_f i_{nf} + A_a i_{na}. \tag{3.174}$$

Equations (3.172)–(3.174) can then be employed to derive the expressions of the noise parameters of two two-ports connected in series as a function of the noise parameters of the amplifier and of the feedback networks. Following an identical approach as for (3.159)–(3.161), we obtain

$$G'_n \stackrel{def}{=} \frac{\overline{i_n^2}}{4kT\Delta f} = \frac{|Z_{21f}|^2 G'_{nf} + |Z_{21a}|^2 G'_{na}}{|Z_{21}|^2} = |A_f|^2 G'_{nf} + |A_a|^2 G'_{na} \tag{3.175}$$

$$Z'_{cor} \stackrel{def}{=} \frac{\overline{v_n \times i_n^*}}{\overline{i_n^2}} = \frac{\left(Z'_{corf} + B_f\right) A_f^* \overline{i_{nf}^2} + (Z'_{cora} + B_a) A_a^* \overline{i_{na}^2}}{|A_f|^2 \overline{i_{nf}^2} + |A_a|^2 \overline{i_{na}^2}}$$
$$= \frac{\left(Z'_{corf} + B_f\right) A_f^* G'_{nf} + (Z'_{cora} + B_a) A_a^* G'_{na}}{G'_n} \tag{3.176}$$

and

$$R'_u \stackrel{def}{=} \frac{\overline{v_u^2}}{4kT\Delta f} = \frac{\overline{(v_n - Z'_{cor} i_n)^2}}{4kT\Delta f}$$

$$= \frac{\overline{\left(v_{nuf} + v_{nua} + \left(Z'_{corf} + B_f - Z'_{cor} A_f\right) i_{nf} + (Z'_{cora} + B_a - Z'_{cor} A_a) i_{na}\right)^2}}{4kT\Delta f} =$$

$$= R'_{uf} + R'_{ua} + \left|Z'_{corf} + B_f - Z'_{cor} A_f\right|^2 G'_{nf} + \left|Z'_{cora} + B_a - Z'_{cor} A_a\right|^2 G'_{na}. \qquad (3.177)$$

If the unilateral amplifier approximation is invoked in (3.172)–(3.177), then

$$Z_{21} \approx Z_{21a}; \quad Z_{12} \approx Z_{12f}; \quad A_f \approx 0; \quad A_a \approx 1; \quad B_a = -B_f \approx Z_{11f} \qquad (3.178)$$

$$i_n \approx \frac{Z_{21a}}{Z_{21}} i_{na} \approx i_{na} \quad v_n = v_{uf} + v_{ua} + (Z'_{cora} + Z_{11f}) i_{na} + \left(Z'_{corf} - Z_{11f}\right) i_{nf} \qquad (3.179)$$

and (3.175)–(3.177) reduce to

$$G'_n \approx G'_{na} \quad Z'_{cor} = Z'_{cora} + Z_{11f} \quad R'_u = R'_{uf} + R'_{ua} + \left|Z'_{corf} - Z_{11f}\right|^2 G'_{nf} \geq R'_{ua} \qquad (3.180)$$

$$Z_{sopt} = \sqrt{R^2_{sopta} + \frac{R'_{uf}}{G'_{na}} + 2R'_{cora}\Re(Z_{11f}) + \Re^2(Z_{11f}) + \frac{\left|Z'_{corf} - Z_{11f}\right|^2 G'_{nf}}{G'_{na}}} + j\left[X_{sopta} - \Im(Z_{11f})\right] \qquad (3.181)$$

$$F_{MIN} = 1 + 2G_{na}\left[R_{cora} + R_{sopt} + \Re(Z_{11f})\right] \geq F_{MINa}. \qquad (3.182)$$

We note that the noise current of the amplifier with series feedback is practically equal to that of the main amplifier. The noise voltages add and Z_{SOPT} increases. F_{MIN} increases if the feedback has resistive elements. If the feedback network is purely reactive, it will not degrade the amplifier noise figure. We can conclude that lossy series feedback should be employed for noise matching in situations where the noise impedance of the original two-port is lower than that of the source impedance.

EXAMPLE 3.8 Inductive degeneration

Let us assume a cascode amplifier with series–series feedback formed by L_G and L_S and with a resonant parallel tank load formed by the output capacitance of the transistor, C_D and L_D, as shown in Figure 3.34. Find the expressions of the noise parameters of the amplifier with feedback.

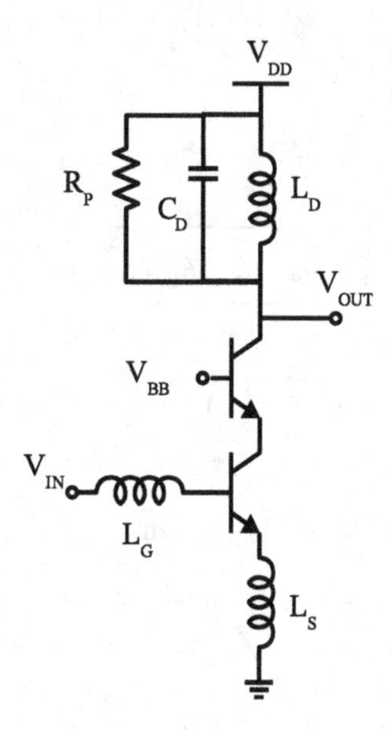

Figure 3.34 Schematics of a cascode LNA stage with inductive degeneration

Solution

The equivalent noise sources at the input of the feedback network are expressed as a function of the loss resistance of each inductor

$$v^2{}_{nf} = 4kTT\Delta f(R_{LG} + R_{LS}) \quad \text{and} \quad i_{nf} = 0.$$

Note that the loss resistances of L_S and L_G, R_{LS} and R_{LG}, can be accounted for by absorbing them in the emitter and the base resistance, respectively, of the common-emitter transistor of the cascode stage. Thus, we can continue the rest of the derivation, assuming ideal inductors. The noise parameters and Z-parameters of the feedback network in the noise impedance formalism become

$$G'_{nf} = 0; R'_{uf} = 0; Z'_{corf} = 0; Z_{11f} = j\omega(L_S + L_G); Z_{12a} = 0, Z_{12f} = Z_{21f} = \ j\omega L_S.$$

From (3.175)–(3.177), we can now derive the expressions of the noise parameters of the entire amplifier with feedback, without invoking the unilateral amplifier approximation

$$G'_n = |A_a|^2 G'_{na} = \frac{|Z_{21a}|^2}{|Z_{21a} + j\omega L_S|^2} G'_{na}$$

$$Z'_{cor} = \frac{Z'_{cora} + B_a}{A_a} \approx Z'_{cora} + j\omega(L_G + L_S)$$

$$R'_u = R'_{ua}$$

$$Z_{sopt} = \sqrt{\Re^2\left(\frac{Z'_{cora} + B_a}{A_a}\right) + |A_a|^2 \frac{G'_{na}}{R'_{ua}}} + j\left[\Im\left(\frac{Z'_{cora} + B_a}{A_a}\right) - \omega(L_G + L_S)\right]$$

$$\approx R_{sopta} + j[X_{sopta} - \omega(L_G + L_S)]$$

$$F_{MIN} = 1 + \frac{2|Z_{21a}|^2}{|Z_{21a} + j\omega L_S|^2} G'_{na}[R'_{cora} + R_{sopta}] \leq F_{MINa}.$$

This somewhat unexpected result shows that, if a purely inductive feedback network is employed, it is possible for the minimum noise figure of the amplifier with feedback to become smaller than that of the main amplifier as L_S (the amount of feedback) increases. However, as demonstrated theoretically by Haus and Adler [4], when L_S increases, the gain with feedback decreases and the noise measure, $M = F_{MIN}/(1-1/G)$, is preserved. In reality, though, the real part of the optimum noise impedance and the minimum noise figure increase slightly compared to those of the main amplifier, due to the losses in the feedback network described by R_{LG} and R_{LS}. Nevertheless, these results explain why the amplifier topology with inductive degeneration shown in Figure 3.34 is the most popular for HF tuned low-noise amplifiers.

For a connection of N identical two-ports in series, we obtain $G'_n = G'_{ni}/N$; $Z_{SOPT} = NZ_{SOPTi}$; $F_{MIN} = F_{MINi}$. By connecting N identical two-ports in series, the minimum noise figure is preserved while the optimum source impedance increases and sensitivity to source mismatch is reduced. An example of such a circuit is the differential pair, shown in Figure 3.35, which can be represented as a series connection of two identical half circuits, indexed with 1 and 2.

EXAMPLE 3.9 Differential stage

Prove directly using the noise representation in Figure 3.35, and without resorting to (3.175)–(3.177), that the minimum noise factor of the differential amplifier is equal to that of the differential-mode half circuit and that the optimum noise impedance is twice as large.

Solution

Assuming that the noise sources in the two half circuits are not statistically correlated, we can derive the expressions for the noise parameters of the differential pair in Figure 3.35 [12] from the definition of the noise factor in the noise impedance formalism (3.120)

$$F = 1 + \frac{\overline{v_{u1}^2} + \overline{v_{u2}^2}}{\overline{v_{ns1}^2} + \overline{v_{ns2}^2}} + \frac{|Z'_{cor1} + Z_{s1}|^2 \overline{i_{n1}^2} + |Z'_{cor2} + Z_{s2}|^2 \overline{i_{n2}^2}}{\overline{v_{ns1}^2} + \overline{v_{ns2}^2}}$$

$$F = 1 + \frac{R'_{u1} + R'_{u2}}{R_{s1} + R_{s2}} + \frac{|Z'_{cor1} + Z_{s1}|^2 G'_{n1} + |Z'_{cor2} + Z_{s2}|^2 G'_{n2}}{R_{s1} + R_{s2}}.$$

Since the two differential half circuits are identical, the following relations hold

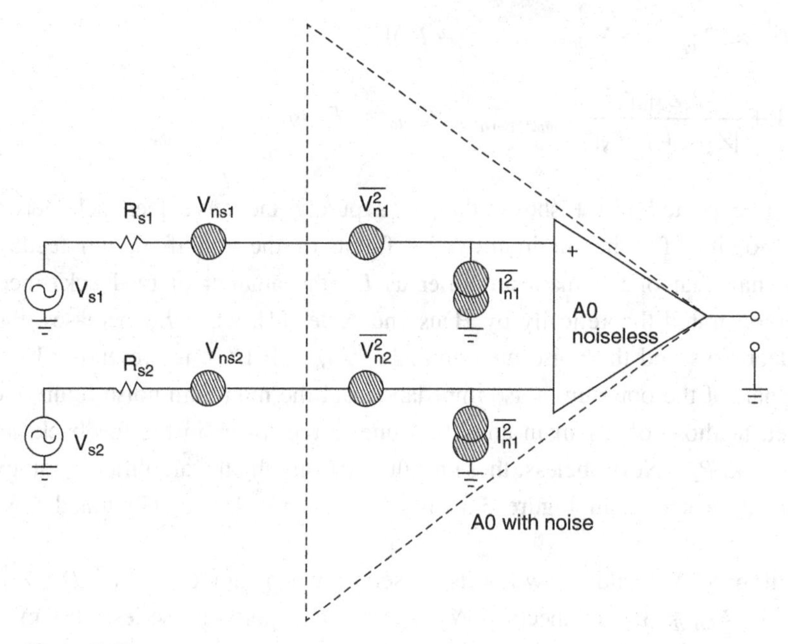

Figure 3.35 Noise equivalent circuit of a noisy differential amplifier showing the statistically independent noise sources associated with each half circuit

$$Z_{s1} = Z_{s2} = \frac{Z_s}{2}; R'_{u1} = R'_{u2}; G'_{n1} = G'_{n2}; Z'_{cor1} = Z'_{cor2}.$$

As a consequence, the expressions of the noise parameters become

$$F = 1 + \frac{2R'_{u1}}{R_s} + \frac{|2Z'_{cor1} + Z_s|^2 \frac{G'_{n1}}{2}}{R_s}; \; R'_u = 2R'_{u1}; G'_n = \frac{G'_{n1}}{2}; Z'_{cor} = 2Z'_{cor};$$

$$X_{soptdif} = 2X_{sopt}; R_{soptdif} = \sqrt{4R'^2_{cor1} + 4\frac{R'_{u1}}{G'_{n1}}} = 2R_{sopt}\,; \; F_{MINdif} = 1 + |2R_{sopt} + 2R'_{cor1}|\frac{G'_{n1}}{2} = F_{MIN}.$$

3.4.5 Series–shunt feedback

Similarly, to analyze circuits consisting of two-ports connected in series at the input and in parallel at the output, as illustrated in Figure 3.36, we can use H-parameters and the noise impedance formalism

$$\begin{bmatrix} h_{11} & h_{12} \\ h_{21} & h_{22} \end{bmatrix} = \begin{bmatrix} h_{11f} & h_{12f} \\ h_{21f} & h_{22f} \end{bmatrix} + \begin{bmatrix} h_{11a} & h_{12a} \\ h_{21a} & h_{22a} \end{bmatrix} \text{ and } [C_h] = [C_{ha}] + [C_{hf}] \tag{3.183}$$

to obtain the input equivalent noise voltage and noise current of the combined two-port

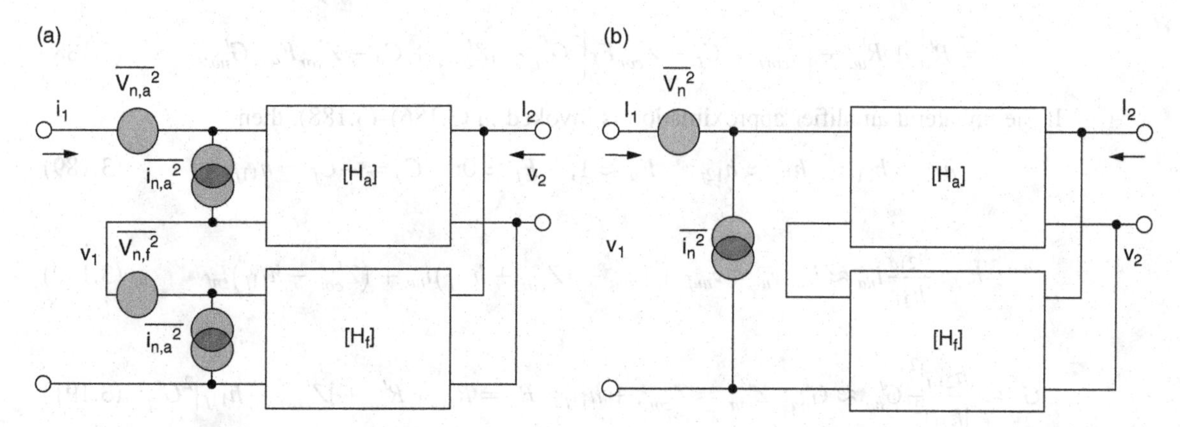

Figure 3.36 (a) Two, series–shunt connected noisy two-ports (b) Noise equivalent circuit representation of the series–shunt two-ports

$$v_n = v_{nf} + v_{na} + \frac{h_{11a}h_{21f} - h_{21a}h_{11f}}{h_{21}} i_{nf} + \frac{h_{11f}h_{21a} - h_{21f}h_{11a}}{h_{21}} i_{na} = v_{nf} + v_{na} + C_f i_{nf} + C_a i_{na} \tag{3.184}$$

$$i_n = \frac{h_{21f} i_{nf} + h_{21a} i_{na}}{h_{21}} = F_f i_{nf} + F_a i_{na} \tag{3.185}$$

and the expressions of the noise parameters in the noise impedance formalism

$$G'_n \stackrel{\text{def}}{=} \frac{\overline{i_n^2}}{4kT\Delta f} = \frac{|h_{21f}|^2 G'_{nf} + |h_{21a}|^2 G'_{na}}{|h_{21}|^2} = |F_f|^2 G'_{nf} + |F_a|^2 G'_{na} \tag{3.186}$$

$$Z'_{cor} \stackrel{\text{def}}{=} \frac{\overline{v_n \times i_n^*}}{\overline{i_n^2}} = \frac{\left(Z'_{corf} + C_f\right) F_f^* \overline{i_{nf}^2} + (Z'_{cora} + C_a) F_a^* \overline{i_{na}^2}}{|F_f|^2 \overline{i_{nf}^2} + |F_a|^2 \overline{i_{na}^2}}$$

$$= \frac{\left(Z'_{corf} + C_f\right) F_f^* G'_{nf} + (Z'_{cora} + C_a) F_a^* G'_{na}}{G'_n} \tag{3.187}$$

and

$$R'_u \stackrel{\text{def}}{=} \frac{\overline{v_u^2}}{4kT\Delta f} = \frac{\overline{\left(v_n - Z'_{cor} i_n\right)^2}}{4kT\Delta f}$$

$$= \frac{\overline{\left(v_{nuf} + v_{nua} + \left(Z'_{corf} + C_f - Z'_{cor} F_f\right) i_{nf} + \left(Z'_{cora} + C_a - Z'_{cor} F_a\right) i_{na}\right)^2}}{4kT\Delta f} =$$

$$= R'_{uf} + R'_{ua} + \left|Z'_{corf} + C_f - Z'_{cor}F_f\right|^2 G'_{nf} + \left|Z'_{cora} + C_a - Z'_{cor}F_a\right|^2 G'_{na}. \quad (3.188)$$

If the unilateral amplifier approximation is invoked in (3.186)–(3.188), then

$$h_{21} \approx h_{21a}; \quad h_{12} \approx h_{12f}; \quad F_a \approx 1; \quad F_f \approx 0; \quad C_a = -C_f = h_{11f} \quad (3.189)$$

$$i_n \approx \frac{h_{21a}}{h_{21}} i_{na} \approx i_{na}; v_n = v_{nuf} + v_{nua} + \left(Z'_{cor} + h_{11f}\right) i_{na} + \left(Z'_{cor} - h_{11f}\right) i_{nf} \quad (3.190)$$

$$G_n \approx \frac{|h_{21a}|^2}{|h_{21}|^2} G'_{na} \approx G'_{na}; \; Z'_{cor} = Z'_{cora} + h_{11f}; \; R'_u = R'_{uf} + R'_{ua} + |Z'_{corf} - h_{11f}|^2 G'_{nf} \quad (3.191)$$

$$Z_{sopt} = \sqrt{R^2_{sopta} + \frac{R'_{uf}}{G'_{na}} + 2R'_{cora}\Re\left(h_{11f}\right) + \Re^2\left(h_{11f}\right) + \frac{\left|Z'_{corf} - h_{11f}\right|^2 G'_{nf}}{G'_{na}}} + j\left[X_{sopta} - \Im\left(h_{11f}\right)\right] \quad (3.192)$$

$$F_{MIN} = 1 + 2G'_{na}\left[R'_{cora} + R_{sopt} + \Re\left(h_{11f}\right)\right]. \quad (3.193)$$

EXAMPLE 3.10 Common-emitter amplifier with transformer feedback

A transformer-feedback LNA is shown in Figure 3.37 [13]. The transformer turns ratio is $n = n_s/n_p$, and the H-parameters of the ideal transformer are given by

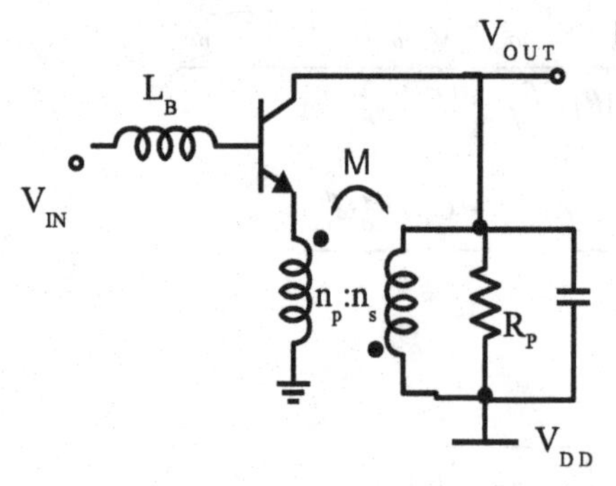

Figure 3.37 Series–shunt transformer feedback amplifier

$$h_{11} = j\omega\left(1 - k^2\right)L_{11}; h_{12} = \frac{-k}{n}; h_{21} = \frac{k}{n}; h_{22} = \frac{1}{j\omega L_{22}}$$

where

$$n = \sqrt{\frac{L_{22}}{L_{11}}} \text{ and } k = \frac{M}{\sqrt{L_{11}L_{22}}} = 0.8.$$

Find the expression of the noise parameters of the amplifier in the noise impedance formalism.

Solution

In this circuit with series–shunt feedback (voltage feedback), the transformer forms the feedback network and the feedback factor $\beta = h_{12f} = -k/n$. The H-matrix of the amplifier with feedback becomes

$$h_{11} = h_{11a} + h_{11f} = h_{11a} + j\omega\left[L_B + (1-k^2)L_{11}\right]$$

$$h_{21} = h_{21a} + h_{21f} = h_{21a} + \frac{k}{n};\ h_{12} = h_{12a} + h_{12f} = h_{12a} - \frac{k}{n} = \frac{1}{1+\frac{C_\pi}{C_\mu}} - \frac{k}{n}$$

$$h_{22} = h_{22a} + h_{22f} = h_{22a} + \frac{1}{j\omega L_{22}}.$$

We notice that the amplifier can be neutralized (such that $h_{12} = 0$) if $n/k - 1 = C_p/C_m \cong 10$. In general, this condition, which makes the amplifier with feedback unilateral, is easier to accomplish in MOSFETs [14] than in HBTs because $C_p/C_\mu >> C_{gs}/C_{gd} = 2$.

Since the transformer is ideal with infinite Q, $v_{nf} = i_{nf} = 0$; $R'_{uf} = 0$; $G'_{nf} = 0$; $Z'_{corf} = 0$ and Real $(h_{11f}) = 0$. From (3.192) and (3.193), we obtain

$$G_n = \frac{|h_{21a}|^2}{\left|h_{21a} + \frac{k}{n}\right|^2} G'_{na};\ Z'_{cor} = Z'_{cora} + j\omega\left[L_B + (1-k^2)L_{11}\right];\ R'_u = R'_{ua}$$

$$Z_{sopt} = R_{sopta} + jX_{sopta} - j\omega\left[L_B + (1-k^2)L_{11}\right]$$

$$F_{MIN} = 1 + \frac{2|h_{21a}|^2}{\left|h_{21a} + \frac{k}{n}\right|^2} G'_{na}\left[R_{cora} + R_{sopt}\right] \leq F_{MINa}.$$

Again, we find that ideal reactive feedback can result in slightly smaller minimum noise factor than that of the main amplifier. Since the minimum noise measure must be preserved, the power gain also decreases. In practical implementations, the transformer loss will actually lead to a degradation of the minimum noise figure compared to that of the main amplifier.

Another important observation is that, to first-order, ideal reactive feedback, either series–shunt, shunt–shunt, or series–series, does not affect the real part of the optimum noise impedance or admittance, only their imaginary part. This aspect has important ramifications in developing strategies for low-noise amplifier topology selection and design.

3.4.6 Shunt–series feedback

Finally, to analyze circuits consisting of two-ports connected in shunt at the input and in series at the output, we can use G-parameters and the noise admittance formalism to find the expressions of the input equivalent noise voltage and noise current sources

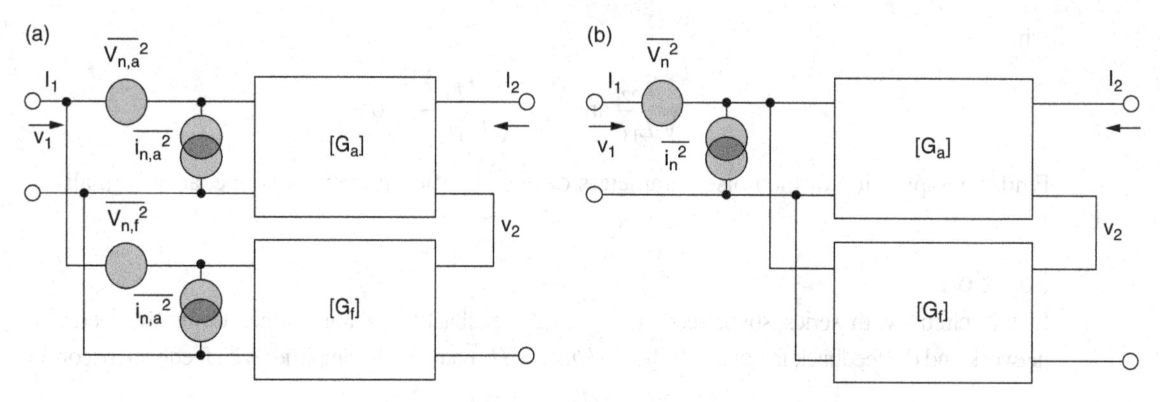

Figure 3.38 (a) Two noisy two-ports connected in parallel at the input and series at the output (b) Noise equivalent circuit representation of the two shunt–series connected two-ports

$$\begin{bmatrix} g_{11} & g_{12} \\ g_{21} & g_{22} \end{bmatrix} = \begin{bmatrix} g_{11f} & g_{12f} \\ g_{21f} & g_{22f} \end{bmatrix} + \begin{bmatrix} g_{11a} & g_{12a} \\ g_{21a} & g_{22a} \end{bmatrix} \textit{ and } [C_g] = [C_{ga}] + [C_{gf}]. \tag{3.194}$$

Following a similar approach as in the shunt–shunt feedback topology, and replacing *Y*- with *G*-parameters, we find

$$v_n = \frac{g_{21f}v_{nf} + g_{21a}v_{na}}{g_{21}} = P_f v_{nf} + P_a v_{na} \tag{3.195}$$

$$i_n = i_{nf} + i_{na} + \frac{g_{11a}g_{21f} - g_{21a}g_{11f}}{g_{21}} v_{nf} + \frac{g_{11f}g_{21a} - g_{21f}g_{11a}}{g_{21}} v_{na} = i_{nf} + i_{na} + Q_f v_{nf} + Q_a v_{na}. \tag{3.196}$$

From (3.195) and (3.196) and their definition, the expressions of the noise parameters become

$$R_n \stackrel{\text{def}}{=} \frac{\overline{v_n^2}}{4kT\Delta f} = \frac{|g_{21\text{f}}|^2 R_{nf} + |g_{21\text{a}}|^2 R_{na}}{|g_{21}|^2} = |P_f|^2 R_{nf} + |P_a|^2 R_{na}. \tag{3.197}$$

$$\begin{aligned} Y_{cor} &\stackrel{\text{def}}{=} \frac{\overline{i_n \times v_n^*}}{\overline{v_n^2}} = \frac{(Y_{corf} + Q_f)P_f^* \overline{v_{nf}^2} + (Y_{cora} + Q_a)P_a^* \overline{v_{na}^2}}{|P_f|^2 \overline{v_{nf}^2} + |P_a|^2 \overline{v_{na}^2}} \\ &= \frac{(Y_{corf} + Q_f)P_f^* R_{nf} + (Y_{cora} + Q_a)P_a^* R_{na}}{R_n} \end{aligned} \tag{3.198}$$

and

$$\begin{aligned} G_u &\stackrel{\text{def}}{=} \frac{\overline{i_u^2}}{4kT\Delta f} = \frac{\overline{(i_n - Y_{cor}v_n)^2}}{4kT\Delta f} \\ &= \frac{\overline{\left(i_{nuf} + i_{nua} + (Y_{corf} + Q_f - Y_{cor}P_f)v_{nf} + (Y_{cora} + Q_a - Y_{cor}P_a)v_{na}\right)^2}}{4kT\Delta f} = \\ &= G_{uf} + G_{ua} + \left|Y_{corf} + Q_f - Y_{cor}P_f\right|^2 R_{nf} + |Y_{cora} + Q_a - Y_{cor}P_a|^2 R_{na}. \end{aligned} \tag{3.199}$$

If the unilateral amplifier approximation holds

$$g_{21} \approx g_{21a}; \quad g_{12} \approx g_{12f}; \quad P_a \approx 1; \quad P_f \approx 0; \quad Q_a = -Q_f = g_{11f} \tag{3.200}$$

and (3.197)–(3.199) simplify to

$$v_n \approx v_{na}; \; i_n = i_{nf} + i_{na} + (Y_{cora} + g_{11f})v_{na} + (Y_{corf} - g_{11f})v_{nf} \tag{3.201}$$

$$R_n \approx R_{na}; \; G_u = G_{uf} + G_{ua} + |Y_{corf} - g_{11f}|^2 R_{nf}; \; Y_{cor} = Y_{cora} + g_{11f} \tag{3.202}$$

$$Y_{sopt} = \sqrt{G_{sopta}^2 + \frac{G_{uf}}{R_{na}} + 2G_{cora}\Re(g_{11f}) + \Re^2(g_{11f}) + \frac{|Y_{corf} - g_{11f}|^2 R_{nf}}{R_{na}}} + j[B_{sopta} - \Im(g_{11f})] \tag{3.203}$$

$$F_{MIN} = 1 + 2R_{na}[G_{cora} + G_{sopt} + \Re(g_{11f})]. \tag{3.204}$$

We note that the noise voltage of the amplifier with shunt–series feedback is equal to that of main amplifier. The noise currents of the amplifier and feedback networks add, while Y_{SOPT} increases. We can conclude that shunt–series feedback can be used for broadband noise matching in situations where the noise impedance of the original two-port is higher than that of the source impedance.

EXAMPLE 3.11 Common-source amplifier with shunt–series transformer feedback

Consider the wideband LNA proposed in [15]. Its first stage features lossless shunt–series feedback using a transformer as shown in Figure 3.39. Knowing that the *G*-parameters of the transformer can be expressed as

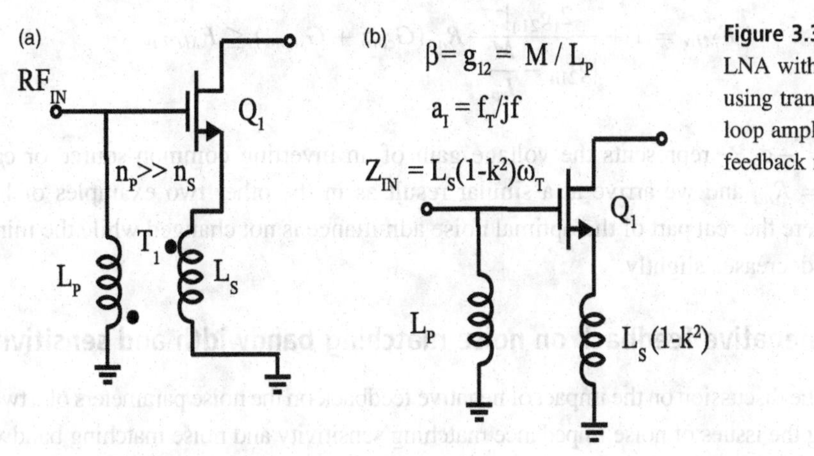

Figure 3.39 (a) CS MOSFET LNA with shunt–series feedback using transformer T_1 (b) Open loop amplifier with loading from feedback network

$$g_{11f} = \frac{-j}{\omega L_P} + G_P; \quad g_{12f} = \frac{M}{L_P}; \quad g_{21f} = \frac{-M}{L_P}; \quad g_{22f} = j\omega L_S(1-k^2) + R_{SEC}$$

derive the expressions of the noise parameters of the amplifier with feedback.

Solution

The noise sources at the input of the transformer-feedback network can be obtained using the two-step open-circuit voltage and short-circuit current method

$$v_{nf} = \frac{g_{12f}}{g_{11f}\,g_{22f} - g_{12f}\,g_{21f}} v_{n2f} \approx \frac{M}{L_S} v_{n2f} \text{ and } i_{nf} = i_{n1f}$$

where v_{n2f} is the noise voltage source associated with the series loss resistance of the transformer secondary

$$\bar{v}^2_{n1f} = 4kT\Delta f \frac{M^2}{L_S^2} R_{SEC}$$

and i_{n1f} is the noise current source associated with the shunt loss conductance of the transformer primary

$$\bar{i}^2_{nf} = 4kT\Delta f G_P.$$

Since the two noise sources are not statistically correlated, we obtain

$$R_{nf} = \frac{M^2}{L_S^2} R_{SEC}; \; G_{uf} = G_P; \; Y_{corf} = 0 \text{ and Real}\left(g_{11f}\right) = G_P.$$

If the transformer is lossless, $G_P = 0$ and $R_{SEC} = 0$, leading to

$$R_n = \frac{|g_{21a}|^2}{\left|g_{21a} - \frac{M}{L_p}\right|^2} R_{na}, \; G_u = G_{ua} \text{ and } Y_{cor} = Y_{cora} + g_{11f} = Y_{cora} + \frac{1}{j\omega L_P}$$

$$Y_{sopt} = G_{sopta} + j\left(B_{sopta} + \frac{1}{\omega L_P}\right)$$

$$F_{MIN} = 1 + \frac{2|g_{21a}|^2}{\left|g_{21a} - \frac{M}{L_p}\right|^2} R_{na}(G_{cora} + G_{sopta}) \leq F_{MINa}.$$

Since $g_{21a} = -g_m R_L$ represents the voltage gain of an inverting common-source or cascode stage, $R_n <= R_{na}$ and we arrive at a similar result as in the other two examples of lossless feedback where the real part of the optimal noise admittance is not changed while the minimum noise figure decreases slightly.

3.4.7 Impact of negative feedback on noise matching bandwidth and sensitivity

We wrap up the discussion on the impact of negative feedback on the noise parameters of a two-port, by addressing the issues of noise impedance matching sensitivity and noise matching bandwidth.

When using negative feedback for noise impedance matching, one is typically interested in controlling the variation of the noise matching condition with respect to frequency and with respect to the signal source impedance. The former is described by the *noise matching bandwidth*, which plays a similar role as the 3dB bandwidth of a two-port, while the latter is known as the *noise mismatch sensitivity* and is fully described by R_n in the noise impedance formalism, or by G'_n in the noise admittance formalism.

As with the bandwidth over which signal impedance matching is achieved, the noise matching bandwidth of a two-port is maximized by reducing the quality factor of the optimum noise impedance/admittance B_{SOPT}/G_{SOPT} or X_{SOPT}/R_{SOPT}. In addition, to reduce the sensitivity of the noise figure of the two-port to noise impedance mismatch and to frequency, we need to minimize R_n (G'_n).

3.4.8 Impact of external network transformations on the noise measure of noisy linear networks

We have studied the impact of all negative feedback configurations and of cascaded lossless two-ports on the IEEE noise parameters of an active (transistor or amplifier) two-port. We learn (Problem 7) that, when cascading the input and output of an amplifier two-port with lossless passive two-ports, the minimum noise figure, maximum available gain, as well as several products of noise parameters (R_nG_{cor}, R_nG_{SOPT}, R_nG_u) remain invariant. We have also demonstrated that it is possible to slightly improve the minimum noise figure of an amplifier two-port by applying purely reactive negative feedback.

All these examples are particular cases of an exhaustive theoretical investigation conducted by Haus and Adler [4] on the impact of an arbitrary passive interconnection of amplifiers on the amplifier noise measure. For an amplifier with arbitrarily large gain, the minimum noise measure directly relates to the minimum noise figure of the amplifier. The most important conclusions of this study are summarized below:

1. The lower bound, $M_{MIN} \stackrel{\text{def}}{=} \frac{F_{MIN}-1}{1-\frac{1}{G}}$ of the noise measure of an amplifier can be achieved by appropriate lossless embedding and this is accomplished in such a way that subsequent cascading of identical amplifier units realizes M_{MIN} at arbitrarily high gain.
2. An arbitrary passive interconnection (encompasses all types of feedback and cascaded two-ports) of independently noisy amplifiers with different M_{MIN} cannot yield a new two-port amplifier with lower M_{MIN} than that of the best amplifier.
3. The use of passive dissipative embedding networks for a given two-port amplifier driven with a positive source impedance cannot achieve a noise measure less than its M_{MIN}.

From 1 above, it is obvious that the slight improvement in the minimum noise figure of an amplifier with lossless feedback comes with the penalty of a decrease in the available power gain since the minimum noise measure remains invariant to a lossless network transformation.

These fundamental properties of noisy linear networks will help us understand the noise performance limitations of each circuit topology and to judge how close to optimal a low-noise amplifier design is.

Summary

- Linear network analysis at high frequencies is usually conducted using *S*-parameters because they can be measured with great precision.
- Matrix transformations can be employed to convert *S*-parameters to *Z*-, *Y*-, *G*-, *H*-, or *ABCD*-parameters, and vice-versa.
- The noise behavior of linear noisy *n*-ports is fully described by *n* noise sources, and their $n \times n$ correlation matrix, in addition to the signal matrix. The noise sources can be voltages, currents, or noise waves.
- The noise figure is introduced as a figure of merit for two-ports.
- *NF* is a function of signal source impedance and of the four noise parameters of the two-port.
- For a two-port, there is an optimum signal source impedance which minimizes *NF*.
- To achieve NF_{MIN} the two-port must be "noise-matched."
- Noise matching can be achieved using negative feedback or (lossless) impedance transformation techniques.
- The noise figure of a passive two-port is equal to its insertion loss.
- Feedback can be used for noise impedance matching in much the same way as it is used for input/output impedance matching.
- Lossy feedback increases the overall noise figure of an amplifier.
- Purely reactive negative feedback, typically implemented with inductors or transformers, does not degrade the amplifier noise figure. In some cases, it may even improve it. However, lossless feedback has little impact on the real part of the optimum noise impedance.
- To reduce the noise impedance matching sensitivity at a given frequency, R_n or G'_n must be minimized.
- To realize noise impedance matching over a wide frequency bandwidth we must minimize B_{SOPT}/G_{SOPT} or X_{SOPT}/R_{SOPT}.
- As with its impact on gain, input/output impedance and bandwidth, negative feedback helps noise impedance matching robustness and noise matching bandwidth but at the expense of increased noise figure.

Problems

(1) In the Raleigh–Jeans approximation, find the expression of the equivalent noise voltage source and of the available spectral noise power from a series combination of two resistors R_1 and R_2. Assume that both resistors are at the physical temperature *T*.

(2) For a bandwidth of 1MHz and room temperature (293K), calculate the rms noise voltage v_n for the following resistors: (a) $R = 50\Omega$ (b) $R = 10\text{k}\Omega$.

(3) Find the expression of the equivalent noise conductance of a PIN photodiode which, in addition to the shot noise current source, has a series parasitic resistance R_s.

(4) A noise source has an ENR of 13dB and an ideal reflection coefficient equal to zero.

(a) What is the equivalent noise temperature of the noise source if the reference noise temperature is 290K?

(b) Calculate the equivalent noise temperature seen by the LNA if the noise source is connected to the LNA through a coaxial cable whose loss is 1dB?

(5) Find the noise parameters of a two-port consisting of a series R_f, L_f network connected between port 1 and port 2.

Solution

$R_n = R_f$; $G_u = 0$; $Y_{cor} = 0$

$$Y_{11} = \frac{1}{R_f + sL_f}.$$

(6) (a) Find the equivalent noise parameters of a parallel connected $R_f - C_f$ network connected in series with the signal path. (b) If this network is used as a shunt–shunt feedback, how does this case compare with the situation described of a series R-L circuit connected in shunt–shunt feedback?

Solution

(a)

$$R_n = \frac{R_f}{1 + (sC_fR_f)^2}; \; G_u = 0; \; Y_{cor} = 0; Y_{11} = \frac{1 + sC_fR_f}{R_f}.$$

(b) By using the shunt–shunt feedback theory and the approximations presented

$$R_n \approx R_{na}; G_u = G_{ua} + \frac{1}{R_f}; Y_{cor} = Y_{cora} + \frac{1}{R_f} + sC_f.$$

The minimum noise figure degrades with increasing frequency (the terms R_nG_{cor} and G_n/G_{cor} increase by adding the feedback).

(7) (a) Find the expression of the noise parameters at the input of two cascaded two-ports in which the first two-port is passive while the second one is active (i.e. a transistor) as a function of the noise parameters of each two-port (*Hint:* you may want to use *ABCD*-matrix parameters).

(b) Find the necessary condition for the minimum noise factor of the overall network to be the same as that of the second (active) two-port. (*Solution:* the terms R_nG_{cor} and G_n/G_{cor} must be preserved).

(c) Think of a practical case when (b) may be used.

(8) (a) Derive the expressions of the $[C_z]$ matrix entries of a resistive T network formed by resistors R_1, R_2, and R_3.

(b) What happens if the arm resistors R_1 and R_2 are replaced by lossless inductors L_1 and L_2?

(9) Derive the noise parameters of the amplifier with shunt–shunt feedback in the case where R_F is replaced by an ideal, lossless inductor L_F. How does the impact of lossless feedback on G_{SOPT} and F_{MIN} compare to that of Examples 3.8, 3.10, and 3.11?

(10) What is the minimum system noise figure for a cascade of transistors with 1dB noise figure and 8dB gain? What is the minimum noise temperature?

(11) A one-stage low-noise amplifier has NF_{MIN}= 1.4dB and $G = 7$dB. By adding source feedback, the noise figure is improved until $NF = 1.2$dB and $G = 5$dB. Show that the noise measure has not changed.

(12) (a) Find the equivalent noise conductance of a PIN photodiode with a responsivity of 0.8 A/W operating at 10Gb/s when the incident optical power is 10 μW average. Assume the noise bandwidth to be 10GHz and $I_{DARK} = 10$ nA.

(b) What is the value of the equivalent noise conductance if the optical power is 1 μW?

(c) What is the optical sensitivity of a PIN-preamplifier receiver if the equivalent input noise current of the preamplifier is 1.2 μA rms?

(13) Find the best order in which to cascade the following two-ports to obtain the highest overall gain with the minimum overall noise figure. Each two-port can only be used once. Does gain depend on order?

$F1 = 2$; $G1 = 20$dB

$F2 = 1.5$; $G2 = 15$dB

$F3 = 1$dB; $G3 = 10$dB.

(14) For a passive element, show that the noise figure is given by $F = 1/G_a$ where G_a is the available power gain. For an ideal generator (matched to the two-port), show that

$$F = \frac{1 - |S_{22}|^2}{|S_{21}|^2}.$$

What is the noise temperature and noise figure of an ideal (perfectly input/output matched) 3dB attenuator?

(15) Consider the T resistive network with R_1 on the left and right, R_2 in the center. Derive the expression for the noise factor (you can use problem (8)) and find the values of R_1 and R_2 if (a) $NF = 6$dB; (b) $NF = 0$dB. Assume in both cases that the reference source and load impedance is 50Ω.

Answer (a) $R_1 = Z_o/3$ and $R_2 = 4Z_o/3$; (b) $R_1 = 0$ and $R_2 = \infty$.

(17) The following noise parameters were measured for a GaAs MESFET in a 50Ω system at 12GHz: NF_{MIN}= 2.2dB, $R_n = 10\Omega$; $\Gamma_{SOPT} = 0.65/120°$.

(a) Determine F_{MIN}, G_n, and Y_{cor}.

(b) What is the noise figure for a 50Ω source impedance?

(18) (a) Given the following measured noise parameters for the *Si* bipolar transistor 2N6617 at 2 GHz: NF_{MIN}= 1.72dB, R_n = 15.73Ω; Γ_{SOPT} = 0.41/88° Determine F_{MIN}, G_n, and Y_{cor}.

(b) What is the noise figure for a 50Ω source impedance?

REFERENCES

[1] David M. Pozar, *Microwave and RF Design of Wireless Systems*, John Wiley & Sons, 2001, Chapter 2.

[2] S. A. Maas, *Noise*, Boston: Artech House, 2005, Chapter 1.

[3] A. R. Kerr and J. Randa, "Thermal noise and noise measurements – a 2010 update," *IEEE Micro. Mag.*, October 2010, pp. 40–52.

[4] H. A. Haus and R. Adler, *Circuit Theory of Linear Noisy Networks*, New York: Wiley, 1959, Chapter 5.4.

[5] P. Penfield, "Wave representation of amplifier noise," *IRE Trans. Circuit Theory*, vol. CT-9, pp. 581–590, October 1976.

[6] H. Hillbrand and P. H. Russer, "An efficient method for computer-aided noise analysis of linear amplifier networks," *IEEE Trans. Circuits Syst.*, Vol. CAS-23, pp.235–238, April 1976.

[7] S. W. Wedge and D. B. Rutledge, "Wave techniques for noise modeling and measurement," *IEEE MTT*, **40**: 2004–2012, November 1992.

[8] J. Lange, "Noise characterization of linear two-ports in terms of invariant parameters," *IEEE JSSC*, SC-**2**: 37–40, June 1967.

[9] M. W. Pospieszalski, "Interpreting transistor noise," *IEEE Micro. Mag.*, pp. 61–69, October 2010.

[10] S. Iversen, "The effect of feedback on noise figure," *Proc. IEEE*, **63**: 540–542, March 1975.

[11] T. O. Dickson, K. H. K. Yau, T. Chalvatzis, A. Mangan, R. Beerkens, P. Westergaard, M. Tazlauanu, M. T. Yang, and S. P. Voinigescu, "The invariance of the characteristic current densities in nanoscale MOSFETs and its impact on algorithmic design methodologies and design porting of Si(Ge) (Bi)CMOS high-speed building blocks," *IEEE JSSC*, **41**(8): 1830–1845, August 2006.

[12] S. P. Voinigescu and M. C. Maliepaard, "5.8GHz and 12.6GHz Si Bipolar MMICs," *IEEE ISSCC Digest*, pp. 372–373, 1997.

[13] J. R. Long, M. Copeland, S. Kovacic, D. Malhi, and D. Harame, "RF analog and digital circuits in SiGe technology," *IEEE ISSCC Digest*, pp. 82–83, February 1996.

[14] D. J. Cassan and J. R. Long, "A 1-V transformer-feedback low-noise amplifier for 5GHz wireless LAN in 0.18-μm CMOS," *IEEE JSSC*, **38**: 427–435, March 2003.

[15] M. T. Reiha, J. R. Long, and J. J. Pekarik, "A 1.2V reactive-feedback 3.1–10.6 GHz ultrawide-band low-noise amplifier in 0.13μm CMOS," *Proc. IEEE RFIC Symposium Digest*, pp. 55–58, 2006.

4 High-frequency devices

This chapter reviews the DC, high-frequency, and noise characteristics of field-effect and heterojunction bipolar transistors and discusses the figures of merit and design methodology of high-frequency passive devices such as inductors, transformers, transmission lines, and fixed and variable capacitors (varactors).

4.1 HIGH-FREQUENCY ACTIVE DEVICES

4.1.1 Definition of an active device

We define as ***active*** an electronic device whose power gain is larger than 1 or 0dB. The power gain is made possible by the conversion of DC power into time-varying power [1].

As illustrated in Figure 4.1, high-frequency active devices can be divided into two large families: **field-effect** and **bipolar** devices. In the field-effect category, we include the MOSFET and its derivatives (LDMOS, SOI, FinFET, nanowire FET) and the high electron mobility transistor (HEMT) with its pseudomorphic (p-HEMT) and metamorphic (m-HEMT) derivatives. MOSFETs are currently fabricated in silicon with some silicon-germanium present in the source and drain regions of p-MOSFETs in advanced technology nodes. Commercial HEMTs are realized in several III-V material systems, the most popular being GaAs/InGaAs, InP/InGaAs, and AlGaN/GaN.

The most widely deployed high-frequency bipolar devices are the SiGe HBT, the GaAs HBT, and the InP HBT, where HBT stands for heterojunction bipolar transistor.

Figure 4.1 also includes legacy high-frequency devices such as (i) the junction(*j*-FET) and (ii) metal-semiconductor (MESFET) field-effect transistors, and (iii) the silicon bipolar transistor, all of which have already been or are about to be phased out from high-frequency ICs.

It is interesting to note that all state-of-the-art high-frequency transistors, even the MOSFET, include some heterojunctions and compound materials.

4.1.2 FETs versus HBTs

Before analyzing each device type in detail, it is instructive to compare their fundamental features, pictorially illustrated in Figure 4.2. In both cases, the principle of operation is that of controlled charge transfer between two terminals: source and drain in the case of FETs, and

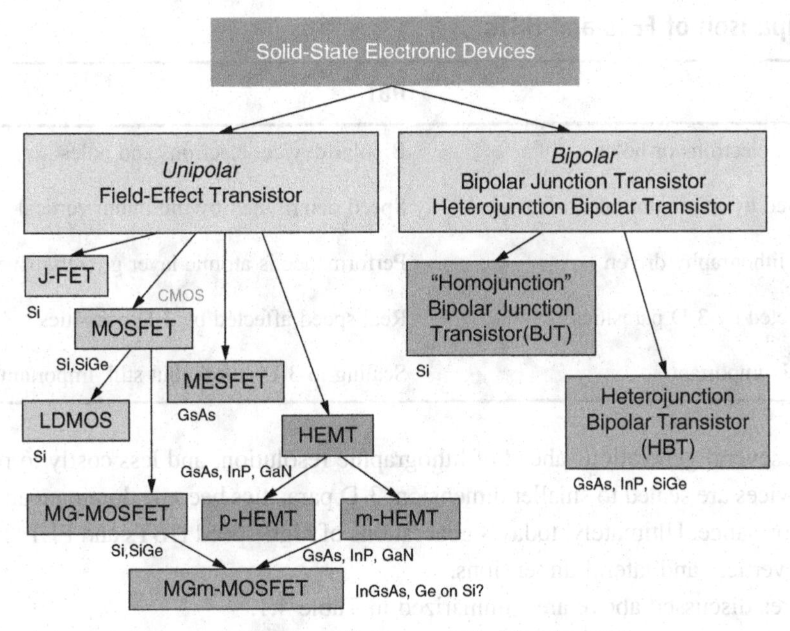

Figure 4.1 Main high-frequency device types. Updated from [1]

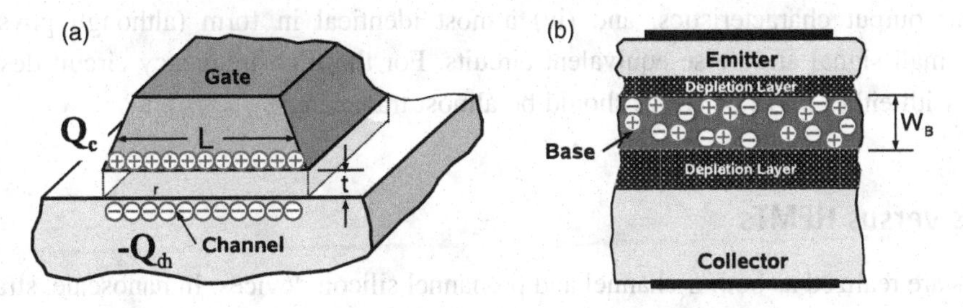

Figure 4.2 Charge control principle in (a) FETs and (b) HBTs [1]

emitter and collector in the case of HBTs. In field-effect transistors, the controlling charge resides on the gate, is of opposite sign, and is physically separated from the controlled charge which travels through the channel between the source and drain. Since only one type of carrier (electrons or holes) contributes to current flow in the active mode of operation typically employed in HF circuits, FETs are also described as *unipolar* devices.

On the contrary, in HBTs, the controlling charge (holes in npns, electrons in pnps) is co-located in the base with the controlled charge (electrons in npns, and holes in pnps). This explains the *bipolar* nature of these devices.

Other differences between the two structures include the direction of current flow – lateral in FETs, vertical in HBTs – and the technological control of the minimum feature size: gate length, L, in FETs and vertical distance between emitter and collector in HBTs. As a result, the speed of the intrinsic FET is driven by progress in lithography. In contrast, HBT speed is determined by the precision with which thin semiconductor layers can be grown vertically, for example by atomic layer deposition techniques. Historically, vertical control of semiconductor

Table 4.1 **Comparison of FETs and HBTs.**

FET	HBT
Unipolar device: electrons or holes	Bipolar device: electrons and holes
Speed determined by minimum lateral feature: L	Speed determined by minimum vertical feature
Performance is lithography driven	Performance is atomic layer growth driven
Real speed affected by 3-D parasitics	Real speed affected by 3-D parasitics
Scaling in 3-D is important	Scaling in 3-D is less, but still, important

layers has been several generations ahead of lithographic resolution, and less costly to realize. However, as devices are scaled to smaller dimension, 3-D parasitics become dominant and limit real device performance. Ultimately, today's generations of high-speed HBTs and FETs require scaling in both vertical and lateral dimensions.

All the features discussed above are summarized in Table 4.1.

Perhaps ironically, nanoscale MOSFETs show many bipolar-like features such as: (i) gate leakage current not unlike the base current of the HBTs, (ii) exponential subthreshold behavior, (iii) similar output characteristics, and (iv) almost identical in form (although physically different) small signal and noise equivalent circuits. For the high-frequency circuit designer, designing with either FETs or HBTs should be almost transparent.

4.1.3 MOSFETs versus HEMTs

MOSFETs are realized as both n-channel and p-channel silicon devices. In nanoscale, strained-channel technologies, p-MOSFETs and n-MOSFETs exhibit almost symmetrical DC and high-frequency characteristics. Si MOSFETs can be realized on bulk silicon and on insulating SiO_2 (SOI-MOSFETs) or sapphire (SOS-MOSFETs) substrates. The latter two are ideal for operation at high frequencies due to the presence of the insulating, low-loss substrate, which reduces junction capacitances and facilitates the realization of low-loss transmission lines. Although over the past four decades Ge, GaAs, InP, and GaN MOSFETs have been repeatedly attempted, a high-quality native oxide/semiconductor interface has so far proven an insurmountable hurdle to overcome. In contrast and as an alternative, HEMTs are fabricated in III-V material systems, on a semi-insulating substrate, and have been restricted primarily to n-channel devices. While p-channel HEMTs have been manufactured successfully, the poor mobility of holes in III-V semiconductors has so far prevented complementary HEMTs from being used in high-frequency circuits.

As illustrated in Figure 4.3, in both MOSFETs and HEMTs, a quantum-well (QW) channel arises at the insulator-channel interface where the mobile carriers form a two-dimensional gas. By choosing a material for the channel with high intrinsic electron mobility, such as InGaAs, and by leaving the channel undoped, very large mobilities and short transit times are achieved

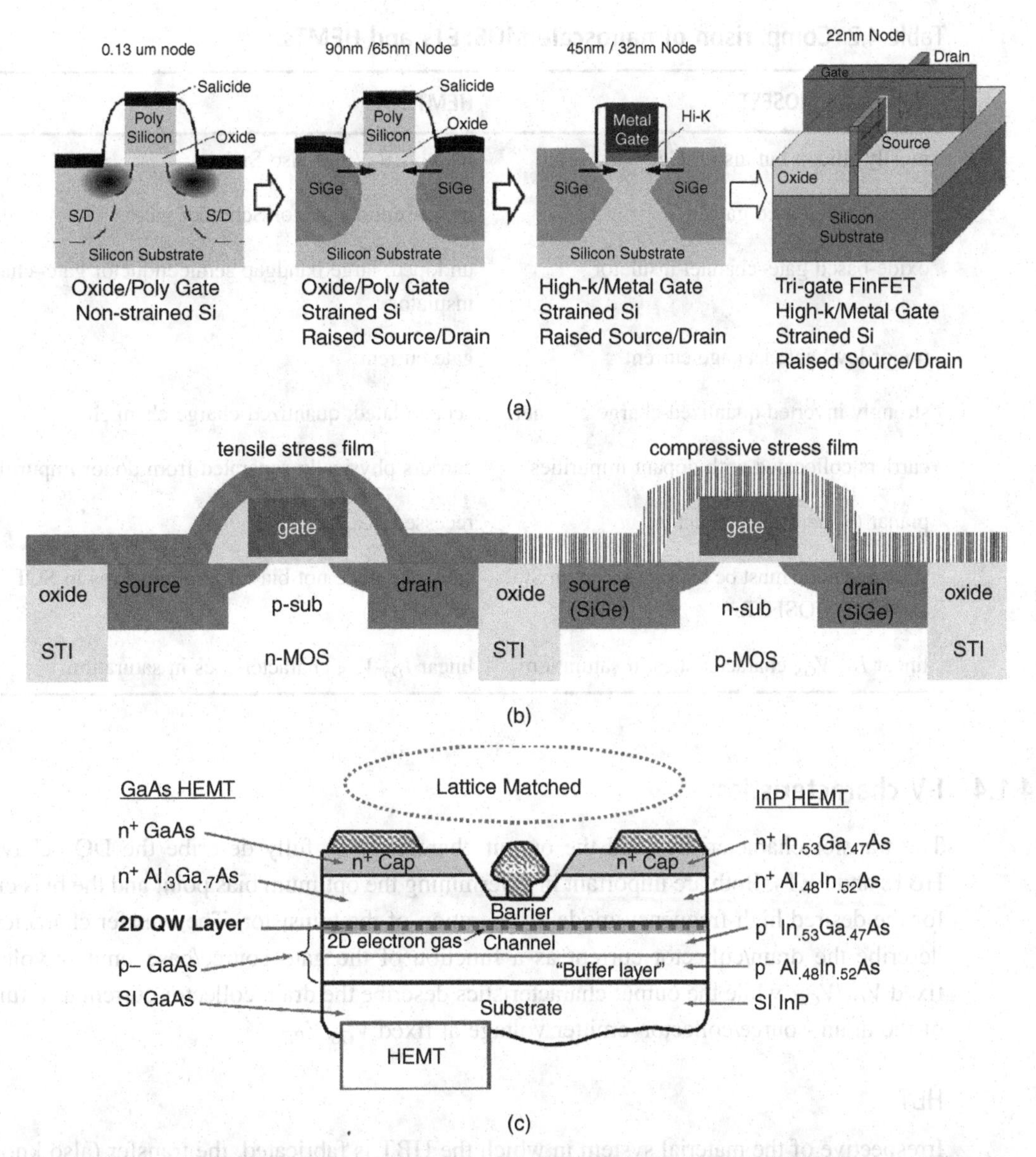

Figure 4.3 (a) MOSFFET evolution [2] (b) strained n/p-MOSFETs and (c) III–V HEMT structure [1]

in a HEMT. To compensate for the poor mobility of electrons and holes in silicon, mechanical strain is applied through special capping films and selectively grown SiGe source and drain regions in silicon MOSFETs to enhance the carrier mobility in the 2-D electron or hole gas that forms the channel. As in HEMTs, metal gate electrodes have become the norm in 45nm CMOS and beyond. The similarities between HEMTs and MOSFETs go beyond the use of metal gates and strained heterojunctions, to the linear I-V characteristics in the saturation region, to the small signal equivalent circuit, and to the internal noise sources.

Table 4.2 summarizes the main similarities and differences between MOSFETs and HEMTs.

Table 4.2 **Comparison of nanoscale MOSFETs and HEMTs.**

Nanoscale MOSFET	HEMT
mostly silicon but also SiGe	mostly III-V but also SiGe
metal-oxide-based gate	metal-semiconductor Schottky gate
oxide-based gate–channel insulator	undoped, large bandgap semiconductor gate–channel insulator
no, or low, gate leakage current	gate current
strongly inverted quantized-charge channel	accumulated, quantized-charge channel
carriers collocated with dopant impurities	carriers physically separated from donor impurities
planar (to date) gate structure	recessed gate structure
substrate node must be biased (except in SOI and SOS MOSFETs)	substrate node not biased, grounded (as in SOI MOSFETs)
linear I_{DS}-V_{GS} characteristics in saturation	linear I_{DS}-V_{GS} characteristics in saturation

4.1.4 I-V characteristics

The transfer characteristics and the output charcateristics fully describe the DC behavior of HBTs and FETs. Both are important in determining the optimum bias point and the bias circuits for the desired high-frequency mode of operation of the transistor. The transfer characteristics describe the drain/collector current as a function of the gate–source/base–emitter voltage at fixed V_{DS}/V_{CE}, while the output characteristics describe the drain/collector current as a function of the drain–source/collector–emitter voltage at fixed V_{GS}/V_{BE}.

HBTs

Irrespective of the material system in which the HBT is fabricated, the transfer (also known as **Gummel**) and output characteristics of the transistor in the active region are captured by the classical silicon bipolar transistor equations

$$I_C = J_S A_E \exp\left(\frac{V_{BE}}{VT}\right)\left(1+\frac{V_{CE}}{V_A}\right), I_B = \frac{I_C}{\beta} \tag{4.1}$$

where:

- J_S is the saturation current density, a technology constant subject to process variation,
- A_E is the emitter area,
- $V_T = kT/q$ is the thermal voltage, and
- V_A is the Early voltage, also a technology constant subject to process variation.

The circuit designer has access to only two levers through which he or she can control the collector current: V_{BE} and A_E. However, for bias stability reasons, V_{BE} is never used as a circuit design variable. Instead, the collector current density, $J_C = I_C/A_E$, is set first, typically with a current mirror, and next the emitter area is sized to obtain the desired collector current, input or noise impedance, or output power.

FETs

The traditional undergraduate textbook defines three regions of operation for FET: triode, active (or saturation), and subthreshold. However, in nanoscale MOSFETs and HEMTs, the saturation region can be divided into two distinct parts. Thus, three distinct regions can be identified in the transfer characteristics of an n-channel MOSFET:

(i) the subthreshold region (for $V_{GS} < V_T$ where V_T is the threshold voltage), characterized by an exponential variation of the drain current with respect to V_{GS}, similar to that of an HBT,
(ii) the well-known square-law region (for $V_{GS} >= V_T$ and $I_{DS}/W <= 0.15\text{mA}/\mu\text{m}$), where the drain current is described by a second-order polynomial in V_{GS}, and
(iii) the linear-law region (for $V_{GS}>V_T$ and $I_{DS}/W > 0.15\text{mA}/\mu\text{m}$), where the drain current is a linear function of V_{GS}.

Remarkably, as detailed in Sections 4.2 and 4.4, the same regions and shape of the transfer characteristics apply to III-V HEMTs and MOSFETs. In the linear portion of the saturation region, where FETs are biased in most high-frequency and high-speed circuits, the drain–source current can be approximated by a simple equation

$$I_{DS} = g'_{mp} N_f W_f (V_{GS} - V_T)(1 + \lambda V_{DS}) \tag{4.2}$$

where:

g'_{mp} is the peak transconductance per unit gate width, a technology constant that also depends on gate length and which is best obtained from measurements in the $I_{DS}/W =$ 0.2–0.4mA/μm range,

N_f is the number of gate fingers,
W_f is the gate finger width,
$W = N_f W_f$ is the total gate width, and
λ is the channel length modulation parameter.

V_{GS}, V_{DS}, and N_f are the only MOSFET or FET variables over which the HF circuit designer has control. As in HBT circuits, the drain current density, rather than V_{GS}, is fixed by the circuit designer to establish the desired operation regime, with N_f and W_f sized to achieve the desired drain current, input or noise impedance, output power, etc. Although HEMTs exhibit similar transfer characteristics to nanoscale MOSFETs, most HEMTs have negative threshold voltage, with V_{GS} being negative for most of the useful bias range.

The output characteristics of nanoscale MOSFETs and HEMTs are similar to those of the HBT but the current densities are typically one order of magnitude smaller.

Table 4.3 **Comaprison of the DC characteristics of FETs and HBTs.**

Characteristics	FET	HBT
Transfer	Exponential, square law, linear	Exponential
	Low turn-on voltage: $V_T = 0.3$–0.4 V	High turn-on voltage: 0.7–0.9 V
Output	$V_{DS,SAT} = 0.1$–0.5 V	$V_{CESAT} = 0.2$–0.7 V
	Low output resistance	High output resistance

4.1.5 What dictates high-frequency device performance?

Although technology, minimum feature size, and proper design play a critical role, ultimately the high-speed and high-frequency performance is determined by the physical properties of the semiconductor materials used to fabricate the transistor. These can be grouped into:

- transport properties:
  electron mobility, μ_n (InSb, GaSb, InGaAs >InP > SiGe > Si),
  hole mobility, μ_p (Ge >SiGe > Si > InP),
  saturation velocity, v_{sat} (GaN > InP > Si/SiGe),
- intrinsic breakdown field, E_B (GaN > SiC> GaAs>InP > Si>Ge), and
- thermal conductivity, K (C >SiC > Si > SiGe > InP > GaAs).

Several material figures of merit (FoMs) have been defined [3] which capture the suitability of a semiconductor material for high-frequency transistors. These include the Johnson limit, which rewards high-frequency operation through v_{sat} and large voltage swing through the E_B term

$$\text{JFoM} = \left(\frac{E_B v_{sat}}{2\pi}\right)^2 \tag{4.3}$$

and the combined figure of merit

$$\text{CFoM} = K\varepsilon\mu v_{sat} E_B^2 \tag{4.4}$$

which, in addition, includes the switching loss through ε and μ, and the thermal conductivity as a measure of the material's capacity to dissipate heat. High thermal conductivity is desirable in large power transistors because self-heating degrades high-frequency power gain, switching speed and voltage swing.

Table 4.4 summarizes the material properties for several semiconductor materials used in high-frequency transistors. A quick inspection of these data reveals that Si and Ge have the lowest JFoM and CFoM, with GaN exhibiting the highest CFoM. The latter explains the strong interest in GaN as a material suitable for high-frequency power amplifiers.

Historically, III-V FETs were the first HF devices to be manufactured using electron beam lithography and 0.1μm gate length. However, today it is silicon-based devices that tend to employ the most advanced lithography and smallest feature size. The leading-edge lithography

Table 4.4 **Properties of the main semiconductor materials used in high-frequency transistors [4].**

Material	Bandgap (eV)	Mobility (cm^2/Vs)	ε_r	E_B (V/cm)	v_{sat} (cm/s)	K (W/cm-K)
n-SiC (4H)	3.26	300	9.66	2.2×10^6	2×10^7	3.0–3.8
n-GaN	3.39	1500	9	3×10^6	2.5×10^7	2.2
n-Si	1.12	1300	11.7	2.5×10^5	0.8×10^7	1.5
n-GaAs	1.4	5000	13.1	3×10^5	0.8×10^7	0.5
n-InP	1.35	4500	12.4	5×10^5	1×10^7	0.68
n-Ge	0.66	3900	15.8	2×10^5	0.6×10^7	0.58
$In_{0.53}Ga_{0.47}As$	0.78	11000	13.9	2×10^5	0.8×10^7	0.05

and complex processing partially make up for the poorer transport properties of silicon compared to InP or InGaAs.

4.1.6 High-frequency equivalent circuit

The high-frequency small signal equivalent circuits of the intrinsic HBT in the common-emitter configuration and of the intrinsic HEMT/MOSFET in the common-source configuration are shown in Figure 4.4. The corresponding *Y*-parameter matrices are given by (4.5) and (4.6), respectively

$$[y] = \begin{bmatrix} g_\pi + j\omega(C_{be} + C_{bc}) & -j\omega C_{bc} \\ g_m e^{-j\omega\tau} - j\omega C_{bc} & g_o + j\omega(C_{cs} + C_{bc}) \end{bmatrix} \tag{4.5}$$

$$\left[\frac{y}{W}\right] = \begin{bmatrix} \dfrac{R'_i\omega^2 C'^2_{gs} + j\omega C'_{gs}}{1 + \omega^2 R'^2_i C'^2_{gs}} + j\omega C'_{gd} & -j\omega C'_{gd} \\ \dfrac{g'_m e^{-j\omega\tau}}{1 + j\omega R'_i C'_{gs}} - j\omega C'_{gd} & g'_o + j\omega\left(C'_{db} + C'_{gd}\right) \end{bmatrix} \tag{4.6}$$

where τ represents the transconductance delay, g'_m, g'_o, C'_{gs}, C'_{gd}, and C'_{db} represent technology parameters per unit gate width, and R'_i, the channel access resistance, is specified in Ohm times unit gate width to emphasize that all intrinsic y -parameters of a FET scale linearly with W.

A simplified high-frequency equivalent circuit valid for both types of devices in the common-emitter/source configuration is illustrated in Figure 4.5. This approximate circuit, which includes the "Miller" capacitance C_{gd}/C_{bc}, but ignores the base–emitter resistance, negligible at HF, is useful in first-order circuit analysis and design. The channel access resistance of the FET, R_i, is also ignored but will be considered in the more accurate equivalent circuits in the MOSFET and HEMT chapter sections.

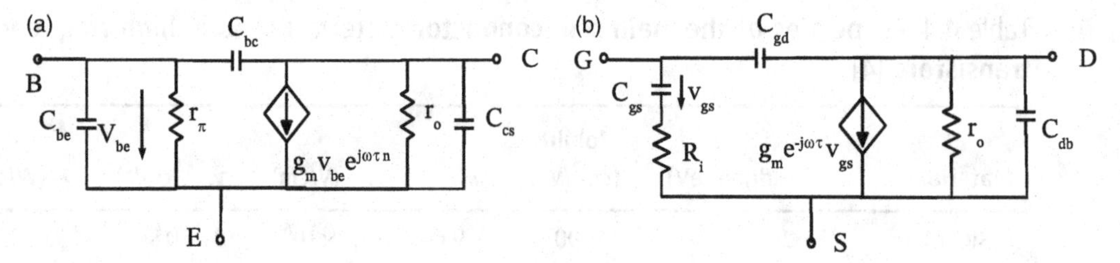

Figure 4.4 High-frequency equivalent circuits for the intrinsic (a) HBT and (b) FET

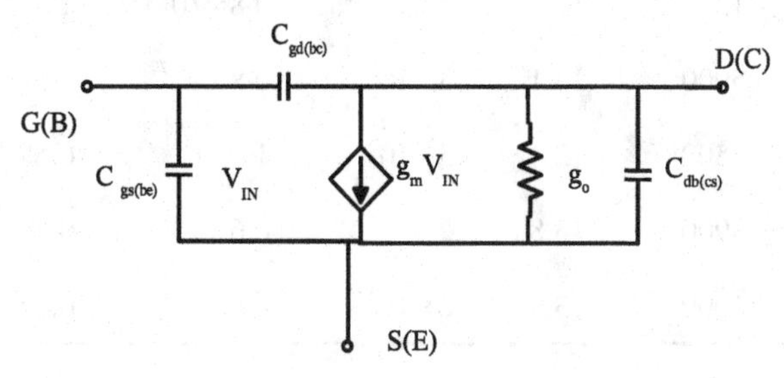

Figure 4.5 Simplified HF circuit describing the intrinsic transistor

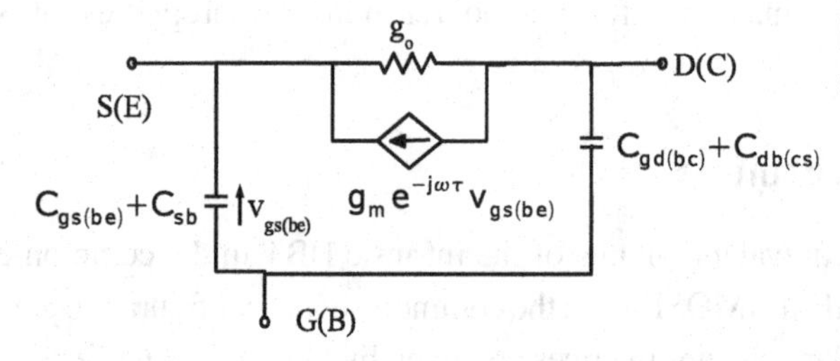

Figure 4.6 Common-gate(base) small signal equivalent circuit

The input admittance and voltage gain of a CE/CS stage loaded by the admittance Y_L can be directly obtained from the Y-parameter matrix using

$$Y_{IN} = Y_{11} - \frac{Y_{12}Y_{21}}{Y_L + Y_{22}} \text{ and } A_v = \frac{-Y_{21}}{Y_L + Y_{22}}. \tag{4.7}$$

Similarly, with the help of the small signal equivalent circuit in Figure 4.6, the Y-parameter matrix of the intrinsic transistor in the common-base/gate configuration (ignoring R_i) is expressed as

$$[y] = \begin{bmatrix} g_m e^{-j\omega\tau} + g_o + g_\pi + j\omega C_{be} & -g_o \\ -g_m e^{-j\omega\tau} - g_o & g_o + j\omega(C_{cs} + C_{bc}) \end{bmatrix} \tag{4.8}$$

for HBTs, and

$$\left[\frac{y}{W}\right] = \begin{bmatrix} g'_m e^{-j\omega\tau} + g'_o + j\omega\left(C'_{gs} + C'_{sb}\right) & -g'_o \\ -g'_m e^{-j\omega\tau} - g'_o & g'_o + j\omega\left(C'_{db} + C'_{gd}\right) \end{bmatrix} \tag{4.9}$$

for FETs.

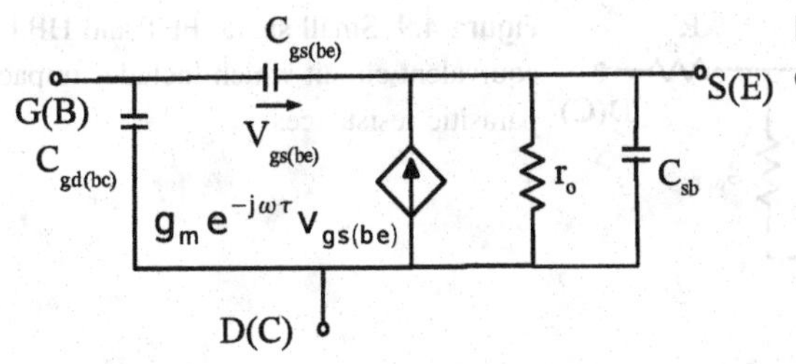

Figure 4.7 Common-collector(drain) small signal equivalent circuit

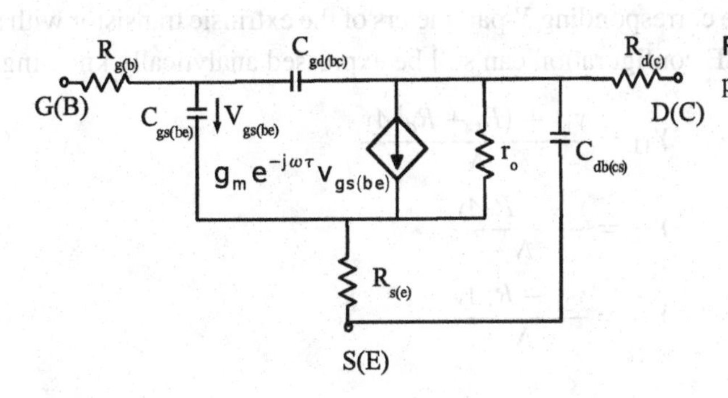

Figure 4.8 HF quivalent circuit with series parasitic resistances

In the common-collector or common-drain configuration, the intrinsic *Y*-parameter matrix is obtained from the equivalent circuit in Figure 4.7

$$[y] = \begin{bmatrix} g_\pi + j\omega(C_{be} + C_{bc}) & -g_\pi - j\omega C_{be} \\ -g_m e^{-j\omega\tau} - g_\pi - j\omega C_{be} & g_m e^{-j\omega\tau} + g_o + j\omega C_{bc} \end{bmatrix} \tag{4.10}$$

$$\left[\frac{\mathrm{y}}{\mathrm{W}}\right] = \begin{bmatrix} \mathrm{j}\omega\left(\mathrm{C}'_{gs} + \mathrm{C}'_{gd}\right) & -\mathrm{j}\omega\mathrm{C}'_{gs} \\ -\mathrm{g}'_{\mathrm{m}}\mathrm{e}^{-\mathrm{j}\omega\tau} - \mathrm{j}\omega\mathrm{C}'_{gs} & \mathrm{g}'_{\mathrm{m}}\mathrm{e}^{-\mathrm{j}\omega\tau} + \mathrm{g}'_{\mathrm{o}} + \mathrm{j}\omega\left(\mathrm{C}'_{\mathrm{sb}} + \mathrm{C}'_{gd}\right) \end{bmatrix}. \tag{4.11}$$

Finally, the input admittance and voltage gain of a cascode stage, another often-encountered gain stage in high-frequency circuits, are derived from the *Y*-parameters of the common-source, CS, and common-gate, CG, stages as

$$Y_{IN} = Y_{11,\mathrm{CS}} - \frac{Y_{12,\mathrm{CS}} Y_{21,\mathrm{CS}}}{Y_{22,\mathrm{CS}} + Y_{11,\mathrm{CG}} - \dfrac{Y_{12,\mathrm{CG}} Y_{21,\mathrm{CG}}}{Y_{\mathrm{L}} + Y_{22,\mathrm{CG}}}} \tag{4.12}$$

$$A_v(\mathrm{casc}) = \frac{-Y_{21,\mathrm{CS}}}{Y_{12,\mathrm{CG}} - \left[\dfrac{Y_{11,\mathrm{CG}} + Y_{22,\mathrm{CS}}}{Y_{21,\mathrm{CG}}}\right]\left[Y_L + Y_{22,\mathrm{CG}}\right]}. \tag{4.13}$$

To accurately capture the power gain of the transistor, especially at mm-wave frequencies, we must consider the series resistive parasitics, as illustrated in Figure 4.8. In particular, the source resistance, R_s, or emitter resistance, r_E, and the gate/base resistance, $R_{g(b)}$, play an increasingly significant role in determining the input impedance and power gain of the transistor as the frequency increases.

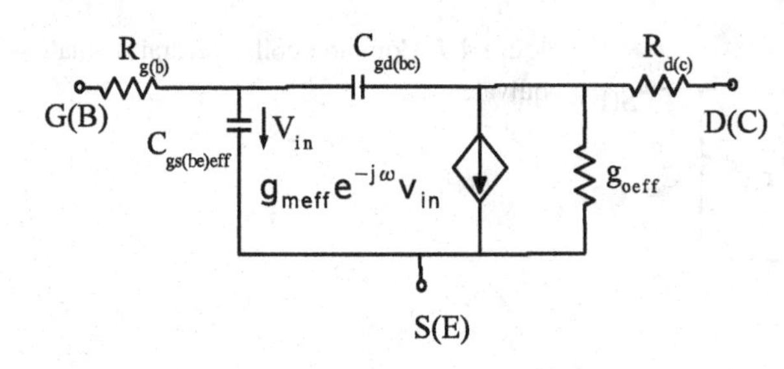

Figure 4.9 Small signal FET and HBT equivalent circuit which includes impact of parasitic resistances

Although more complex, the corresponding Y-parameters of the extrinsic transistor with series resistive parasitics in the CS/CE configuration can still be expressed analytically knowing that

$$Y_{11} = \frac{y_{11} + (R_s + R_d)\Delta y}{N} \tag{4.14}$$

$$Y_{12} = \frac{y_{12} - R_s\Delta y}{N} \tag{4.15}$$

$$Y_{21} = \frac{y_{21} - R_s\Delta y}{N} \tag{4.16}$$

and

$$Y_{22} = \frac{y_{22} + (R_s + R_d)\Delta y}{N} \tag{4.17}$$

where

$$N \approx 1 + R_s(y_{11} + y_{12} + y_{21} + y_{22}) + R_d y_{22} + R_g y_{11} + \Delta y R_s^2 \tag{4.18}$$

and

$$\Delta y = y_{11}y_{22} - y_{21}y_{12}. \tag{4.19}$$

In the equations above, the Y-parameters of the equivalent circuit with series parasitics are represented with capitals.

Simplified CS/CE HF circuit with series parasitics useful for hand analysis and design

By applying series–series feedback theory, the equivalent circuit in Figure 4.8 can be further simplified to that in Figure 4.9. The impact of $R_s(r_E)$ is included in $R_{G(B)}$, g_{meff}, g_{oeff}, $C_{gs(be)eff}$ as follows

$$g_{meff} = \frac{g_m}{1 + g_m R_s} \text{ or } g_{meff} = \frac{g_m}{1 + g_m r_E} \tag{4.20}$$

$$g_{oeff} = \frac{g_o}{1 + g_m R_s} \text{ or } g_{oeff} = \frac{g_o}{1 + g_m r_E} \tag{4.21}$$

$$C_{gseff} = \frac{C_{gs}}{1 + g_m R_s} \text{ or } C_{beeff} = \frac{C_{be}}{1 + g_m r_E} \tag{4.22}$$

where

$$R_G = R_g + R_s \text{ and } R_B = R_b + r_E. \tag{4.23}$$

4.1.7 High-frequency figures of merit

The most relevant high-frequency device figures of merit (FoM) for wireless applications are the cutoff frequency, f_T, the maximum oscillation frequency, f_{MAX}, and the minimum noise factor, F_{MIN}. The analog figure of merit $g_m/I_{C(DS)}$ is also relevant at high frequency because, through g_m, it reflects how much current is consumed to achieve a certain optimum noise impedance, R_{SOPT}, or a certain noise resistance value, R_n. In HBTs, g_m/I_C reaches a theoretical maximum of 38 V^{-1} at room temperature, while for n-MOSFETs and HEMTs it lies between 1 and 2 V^{-1}.

In addition, for high-speed digital applications, the intrinsic slew rate is important in judging the switching speed of large swing CML or distributed output drivers, as well as the maximum operation frequency of static dividers.

All of these FoMs will be discussed in more detail next.

4.1.8 Unity current gain (cutoff) frequency

The unity current gain frequency, f_T, as the name indicates, is the frequency at which the current gain becomes equal to 1. Mathematically, this is expressed as

$$|H_{21}(f = f_T)| = 1 \text{ or } 20 \times \log_{10}|H_{21}(f = f_T)| = 0 \tag{4.24}$$

where $|H_{21}(f)|$ represents the magnitude of the current gain as a function of frequency. It can be obtained from the measured or the simulated Y-parameters of the transistor

$$H_{21}(f) = \frac{Y_{21}(f)}{Y_{11}(f)}. \tag{4.25}$$

It is immediately apparent from (4.25) that H_{21} and f_T do not depend on the output impedance of the transistor. Measured $|H_{21}(f)|$ characteristics are shown in Figure 4.10 for a SiGe HBT at several V_{BE} bias points and $V_{CE} = 1.2$V. Similar characteristics can be generated by simulation. We note that for most of the frequency range where $20\log_{10}(|H_{21}(f)|)$ is larger than 0dB, the characteristics have an ideal 20dB/decade slope which allows the f_T to be obtained by linear extrapolation as the x-axis intercept of the $20\log_{10}|H_{21}(f)|$ curves. As in the case of some of the curves in Figure 4.10, the f_T of advanced HBTs or FETs is often larger than the maximum frequency range of available measurement equipment. Measured Y- or S-parameters are not always available up to f_T. In these situations, extrapolation is the only method available to extract f_T.

For device and circuit design purposes, it is useful to derive the expression of f_T as a function of the small signal equivalent circuit parameters in Figure 4.9 which includes the impact of the series parasitic resistances. First the expression of the output short-circuit curent, i_{sc}, is obtained in several steps as a function of the input current, i_{in}, by ignoring g_{oeff} and C_{db} which are shunted by R_d

$$i_{sc} = -g_{meff}v_{in} + (v_{in} - i_{sc}R_d)sC_{gd} \tag{4.26}$$

$$i_{in} = sC_{gs}v_{in} + (v_{in} - i_{sc}R_d)sC_{gd} \tag{4.27}$$

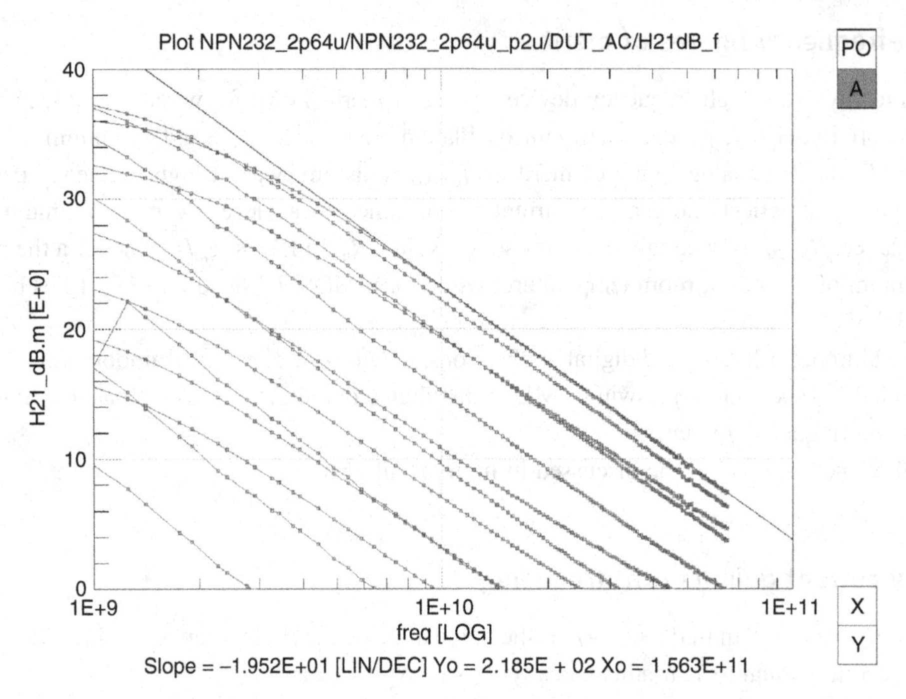

Figure 4.10 Measured current gain vs. frequency characteristics of a SiGe HBT for different base emitter voltages and V_{CE} = 1.2 V

$$v_{in} = \frac{1 + sC_{gd}R_d}{-g_{meff} + sC_{gd}} i_{sc} \tag{4.28}$$

$$H_{21}(s) = \frac{i_{sc}(s)}{i_{in}} = \frac{g_{meff} - sC_{gd}}{s(C_{gseff} + C_{gd} + g_{meff}R_dC_{gd}) - s^2R_dC_{gd}C_{gseff}}. \tag{4.29}$$

From the definition of f_T

$$|H_{21}(f_T)| = |\frac{Y_{21}(f_T)}{Y_{11}(f_T)}| = \frac{|g_{\text{meff}} - j\omega_T C_{gd}|}{|j\omega_T(C_{gseff} + C_{gd} + g_{meff}R_dC_{gd}) + \omega_T^2 R_dC_{gd}C_{gseff}|} = 1 \tag{4.30}$$

assuming that $g_{meff} > 3\omega_T C_{\text{gd}}$ and igoring the $\omega_T{}^2$ term in (4.30)

$$\frac{1}{\omega_T} = \frac{1}{2\pi f_T} = \frac{C_{gseff} + C_{gd}}{g_{meff}} + C_{gd}R_d. \tag{4.31}$$

By expanding g_{meff} and C_{gseff} using (4.20) and (4.22), (4.31) can also be rearranged as

$$\frac{1}{2\pi f_T} = \frac{\dfrac{C_{gs}}{1 + g_mR_s} + C_{gd}}{\dfrac{g_m}{1 + g_mR_s} + C_{gd}R_d} = \frac{C_{gs} + C_{gd}}{g_m} + C_{gd}(R_s + R_d). \tag{4.32}$$

The corresponding expression for HBTs is

$$\frac{1}{2\pi f_T} = \frac{(C_{be} + C_{bc})}{g_m} + (r_E + r_C)C_{bc}. \tag{4.33}$$

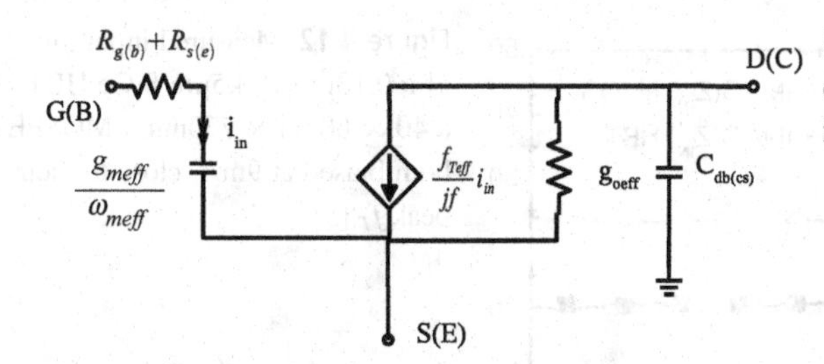

Figure 4.11 Simplified, unilateral, CE/CS and cascode stage equivalent circuit based on frequency-dependent current source

Although still missing higher-order terms due to g_o and C_{db}, (4.32) and (4.33) account for the f_T degradation caused by the parasitic source/emitter and drain/collector resistances. In advanced FETs and HBTs, the contribution of the second term in (4.32) and (4.33) can be as high as 15–20% of the intrinsic f_T, while that of the ignored higher-order terms remains below 5%. Unlike R_s and R_d, the gate/base resistance has no impact on f_T.

Note that f_T and g_{meff} represent the actual simulated or measured values. The intrinsic g_m and f_T of a transistor can only be obtained indirectly from the measured or simulated transconductance and cutoff frequency values after correcting for the measured parasitic resistances.

It is possible and useful in circuit design to define the cutoff frequency of a cascode stage or of a CMOS inverter stage. For the cascode stage, the effective f_T can be approximated by

$$\frac{1}{2\pi f_{Teff}(\text{casc})} \approx \frac{(C_{gs}+2C_{gd})}{g_m} + (R_s + R_d)C_{gd} \tag{4.34}$$

and

$$\frac{1}{2\pi f_{Teff}(\text{casc})} = \frac{(C_{be}+2C_{bc})}{g_m} + (r_E + r_C)C_{bc}. \tag{4.35}$$

Because of its reduced Miller effect and improved isolation, the unilateral high-frequency equivalent circuit with a frequency-dependent current source shown in Figure 4.11 can be employed to derive the expression of the input impedance of a cascode stage as a function of the effective transconductance and cutoff frequency of the cascode

$$Z_{in} = R_G - j\frac{f_{Teff}(\text{casc})}{fg_{meff}} = R_g + R_s - j\frac{f_{Teff}(\text{casc})}{fg_{meff}} \tag{4.36}$$

$$Z_{in} = R_B - j\frac{f_{Teff}(\text{casc})}{fg_{meff}} = R_b + r_E - j\frac{f_{Teff}(\text{casc})}{fg_{meff}}. \tag{4.37}$$

Figure 4.12 shows the measured input impedance of a 0.13μm × 4.5μm SiGe HBT and of a 40 × 60nm × 770nm n-MOSFET, both biased at 9mA, close to their peak f_T. The real part of the input impedance remains practically constant from 110 to 170GHz while the imaginary part is capacitive and decreases with increasing frequency. These measurements confirm that the simplified equivalent circuit, with a resistance in series with a capacitor, is adequate up to at least 170GHz, even for the CE/CS stage.

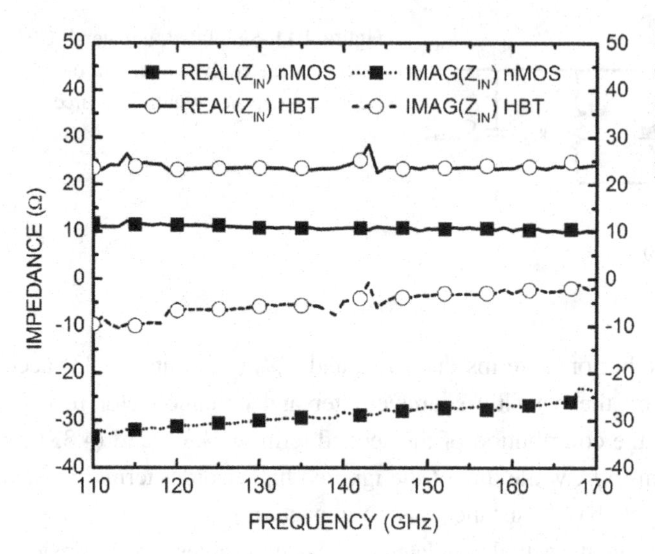

Figure 4.12 Measured input mpedance of a 0.13μm × 4.5μm SiGe HBT and of a 40 × 60nm × 770nm n-MOSFET, both biased at 9mA, close to their peak f_T [5]

4.1.9 Unity power gain (maximum oscillation) frequency f_{MAX}

In high-frequency circuits, power gain, rather than voltage or current gain, plays the most significant role. We have already seen that power gain defines whether a device is active or not.

The maximum frequency of oscillation of a transistor is defined as the frequency at which the maximum available power gain is equal to 1 (or 0dB)

$$MAG(f = f_{MAX}) = 1 \text{ or}$$
$$MAG_{dB}(f = f_{MAX}) = 10\log_{10} MAG(f = f_{MAX}) = 0\text{dB}. \tag{4.38}$$

Therefore, f_{MAX} can be obtained as the x-axis intercept of the maximum available power gain versus frequency characteristics. However, as for f_T, it is often the case in modern transistors that measured $MAG(f)$ characteristics are not available up to the frequencies where $MAG_{dB}(f)$ intercepts the frequency axis. As shown in Figure 4.13, unlike $20 \times \log_{10}|H_{21}(f)|$, the $MAG_{dB}(f)$ characteristics do not exhibit a region of constant slope and cannot be extrapolated. Fortunately, as illustrated in Figure 4.13, for all types of transistors, $MAG(f)$ and the unilateral power gain $U(f)$ intercept the x-axis at the same point. The $10 \times \log_{10}U(f)$ characteristics do exhibit a region of constant slope, approximately 20dB/decade which can be employed to extract f_{MAX}.

U can be expressed as a function of the Y-parameters

$$U = \frac{|y_{21} - y_{12}|^2}{4[\Re(y_{11})\Re(y_{22}) - \Re(y_{12})\Re(y_{21})]} \tag{4.39}$$

or Z-parameters of the transistor

$$U = \frac{|z_{21} - z_{12}|^2}{4[\Re(z_{11})\Re(z_{22}) - \Re(z_{12})\Re(z_{21})]}. \tag{4.40}$$

From (4.39) and (4.40), it can be seen that f_{MAX} (unlike f_T) contains information about the real part of the input and output impedance of the transistor, making it a more meaningful figure of

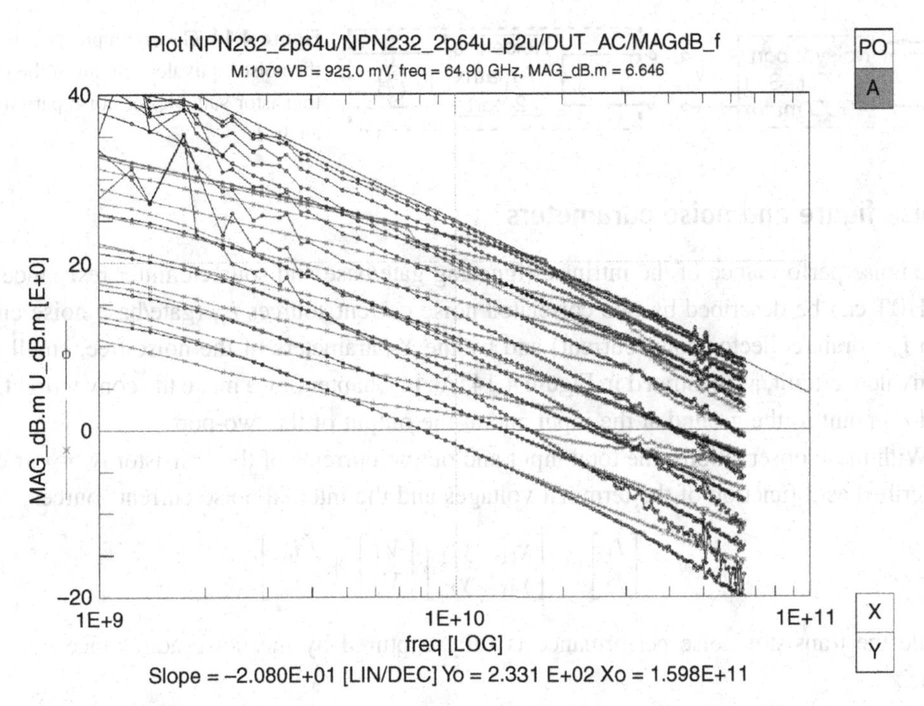

Figure 4.13 Measured unilateral (*U*) and maximum available gain (*MAG*) as a function of frequency for a SiGe HBT at different base–emitter voltages

merit for the high-frequency performance of a transistor. It is possible, though painful, to derive an exact analytical expression of f_{MAX} as a function of all the small signal parameters of the transistor with series parasitics. Simplified expressions for f_{MAX} are provided in the MOSFET, HBT, and HEMT subchapters. Here we will just derive an approximate expression of f_{MAX} based on the unilateral equivalent circuit in Figure 4.11, and the observation that the definition of *MAG* implies matched input and output impedances. Therefore, the power gain can be calculated simply as the product between current gain and voltage gain, which is equal to the square of the current gain multiplied by the ratio of the output and input resistances and divided by 4 to account for the current flowing through the matched signal source and load impedances

$$MAG(f = f_{MAX}) \approx \frac{1}{4}|H_{21}(f = f_{MAX})|^2 \frac{r_{oeff}}{R_g + R_s} = \frac{f_T^2}{4f_{MAX}^2(R_g + R_s)g_o} = 1. \tag{4.41}$$

From (4.41), the approximate expression for f_{MAX} becomes

$$f_{MAX} = \frac{f_T}{2\sqrt{(R_s + R_g)g_{oeff}}} = \frac{f_T}{2\sqrt{(R_b + r_E)g_{oeff}}} \tag{4.42}$$

whose primary merit is that it illustrates the dependence of f_{MAX} on the gate, source, and output resistances of the transistor. However, because of the unilateral nature of the circuit used to derive it, (4.42) does not capture the impact of C_{gd} and C_{bc}.

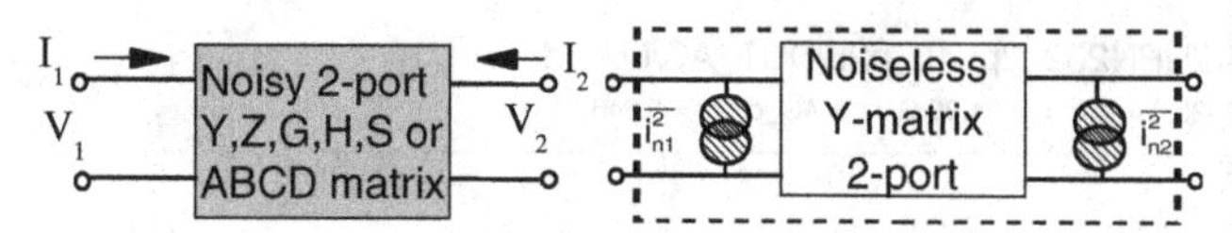

Figure 4.14 General representation of the noise equivalent circuit of the intrinsic transistor suitable for noise parameter analysis

4.1.10 Noise figure and noise parameters

The noise performance of the intrinsic (ignoring gate/base and source/emitter resistance) FET or HBT can be described by two correlated noise current sources i_{n1} (gate/base noise current) and i_{n2} (drain/collector noise current) and by the *Y*-parameters of the noise-free, small signal equivalent circuit, as illustrated in Figure 4.14. As in Chapter 3, we make the convention that i_{n1} and i_{n2} point to the ground at the input and at the output of the two-port.

With these observations, the total input and output currents of the transistor two-port can be described as a function of the terminal voltages and the internal noise current sources

$$\begin{bmatrix} I_1 \\ I_2 \end{bmatrix} = \begin{bmatrix} y_{11} & y_{12} \\ y_{21} & y_{22} \end{bmatrix} \begin{bmatrix} V_1 \\ V_2 \end{bmatrix} + \begin{bmatrix} i_{n1} \\ i_{n2} \end{bmatrix} \tag{4.43}$$

while the transistor noise performance is fully captured by the noise admittance correlation matrix

$$\left[C_y\right] = \begin{bmatrix} \langle i_{n1}\, i^*_{n1}\rangle & \langle i_{n1}\, i^*_{n2}\rangle \\ \langle i_{n2}\, i^*_{n1}\rangle & \langle i_{n2}\, i^*_{n2}\rangle \end{bmatrix}. \tag{4.44}$$

In the CS/CE configuration, the equivalent input noise voltage, V_n, noise current, I_n

$$V_n = \frac{-i_{n2}}{y_{21}}, I_n = \frac{-y_{11}}{y_{21}} i_{n2} + i_{n1} \text{ and } I_u = \frac{Y_{cor} - y_{11}}{y_{21}} i_{n2} + i_{n1} \tag{4.45}$$

and the two-port noise parameters in the noise admittance formalism R_n, Y_{cor}, and G_u can be derived from their definitions in Chapter 3 as

$$R_n = \frac{\langle V^2_n\rangle}{4kT\Delta f} = \frac{\langle i^2_{n2}\rangle}{4kT\Delta f|(y_{21})|^2} \tag{4.46}$$

$$Y_{cor} = \frac{\langle I_n V^*_n\rangle}{\langle V^2_n\rangle} = \frac{y_{11}\langle i^2_{n2}\rangle - y_{21}\langle i_{n1} i^*_{n2}\rangle}{4kT\Delta f R_n|(y_{21})|^2} = y_{11} - \frac{\langle i_{n1} i^*_{n2}\rangle}{4kT\Delta f R_n y^*_{21}} \tag{4.47}$$

$$G_u = \frac{\langle i^2_{n1}\rangle}{4kT\Delta f} + \frac{\langle i^2_{n2}\rangle}{4kT\Delta f}\left|\frac{y_{11} - Y_{cor}}{y_{21}}\right|^2 - 2\Re\left[\frac{y^*_{11} - Y^*_{cor}}{y^*_{21}} \frac{\langle i_{n1} I^*_{n2}\rangle}{4kT\Delta f}\right] = \frac{\langle i^2_{n1}\rangle}{4kT\Delta f} - \left|\frac{\langle i_{n1} i^*_{n2}\rangle}{4kT\Delta f R_n |y_{21}|^2}\right|^2 \tag{4.48}$$

where the *y*-matrix entries are those of the intrinsic transistor, given by (4.5) or (4.6).

From R_n, G_u, and Y_{cor}, we can obtain the IEEE two-port noise parameters (R_n, Y_{SOPT}, F_{MIN}) of the CS/CE FET/HBT in the noise admittance formalism

$$Y_{sopt} = \sqrt{G^2_{cor} + \frac{G_u}{R_n}} - jB_{cor} \tag{4.49}$$

$$F_{MIN} = 1 + 2R_n(G_{cor} + G_{sopt}) \text{ and } F = F_{MIN} + \frac{R_n}{G_s}|Y_s - Y_{sopt}|^2. \tag{4.50}$$

Note from (4.47) that, for both FETs and HBTs, B_{cor} is approximately equal to the imaginary part of y_{11}, a property that is essential to achieving simultaneous noise and input impedance matching in CS/CE and cascode LNA stages, as will be discussed in detail in Chapter 7.

If the noise contributions from the gate(base) and source(emitter) resistances of the transistor are considered, the noise parameters in the CS(CE) configuration become [6]

$$R_n = R_{s(E)} + R_{g(b)} + \frac{\langle i_{n2}^2 \rangle}{4kT\Delta f|(y_{21})|^2} \tag{4.51}$$

$$Y_{cor} = y_{11} - \frac{\langle i_{n1} i_{n2}^* \rangle}{4kT\Delta f R_n y_{21}^*}. \tag{4.52}$$

We underscore that, to facilitate comparison between HBTs and FETs, expressions for i_{n1} and i_{n2} were not substituted in equations (4.46)–(4.48) and (4.51)–(4.52). This also allows us to account for correlation between the base and collector noise current sources in an HBT, and between the gate and drain noise current sources in a FET. The particular expressions of the noise currents in FETs and HBTs are provided in Chapters 4.2 and 4.3. For circuit design purposes, only the simplified expressions of R_{SOPT} and F_{MIN} are provided here for CS/CE stages

$$Z_{SOPT}(\text{FET}) \approx \frac{1}{\omega(C_{gs} + C_{gd})}\left[\sqrt{\frac{g_m(R_s + R_g)}{k_1}} + j\right] = \frac{f_{Teff}}{f g_{meff}}\left[\sqrt{\frac{g'_m\left(R'_s + W_f R'_g(W_f)\right)}{k_1}} + j\right] \tag{4.53}$$

$$Z_{SOPT}(\text{HBT}) \approx \frac{1}{\omega(C_{be} + C_{bc})}\left[\sqrt{\frac{g_m}{2}}(r_E + R_b) + j\right] = \frac{f_{Teff}}{f g_{meff}}\left[\sqrt{\frac{g_m}{2}}(r_E + R_b) + j\right] \tag{4.54}$$

$$F_{MIN}(\text{FET}) \approx 1 + \frac{2f}{f_{Teff}}\sqrt{k_1}\sqrt{g_m(R_s + R_g) + 1} \tag{4.55}$$

$$F_{MIN}(\text{HBT}) \approx 1 + \frac{1}{\beta} + \frac{f}{f_{Teff}}\sqrt{2g_m(R_b + r_E)} \tag{4.56}$$

where k_1 is a technology constant that depends on the degree of correlation between the input and output noise current sources and is obtained by fitting the measured noise parameters. Equations (4.53)–(4.56) can also provide a reasonable approximation for a cascode stage if f_{Teff} represents the cutoff frequency of the cascode stage.

It is instructive to derive the expressions of the transistor noise parameters in the common-gate(base) configuration. Following the same approach, ignoring the series parasitics, and using the equivalent circuit in Figure 4.6 and the common-base/gate Y-parameter matrix (4.8), (4.9), we obtain

$$R_n = \frac{\langle i_{n2}^2 \rangle}{4kT\Delta f|(y_{21})|^2} \tag{4.57}$$

$$G_u = \frac{\langle i_{n1}^2 \rangle}{4kT\Delta f} - \left| \frac{\langle i_{n1} i_{n2}^* \rangle}{4kT\Delta f R_n |y_{21}|^2} \right|^2 \tag{4.58}$$

and

$$Y_{cor} = y_{21} + y_{11} + \frac{\langle i_{n1} i_{n2}^* \rangle}{4kT\Delta f R_n y_{21}^*}. \tag{4.59}$$

Noting that $y_{21} + y_{11}$ in CG/CB configuration is equal to y_{11} in the CS/CE configuration, we can conclude that, for the intrinsic transistor

$$R_n(CG) = R_n(CS), G_u(CG) = G_u(CS), G_{cor}(CG) = G_{cor}(CS), \text{ and } F_{MIN}(CG) \cong F_{MIN}(CS). \tag{4.60}$$

Finally, we note that G_u, G_{SOPT} decrease with increasing correlation between i_{n1} and i_{n2}, while R_{SOPT} increases and F_{MIN} decreases with increasing correlation. This implies that transistors with strongly correlated input and output noise sources will tend to exhibit lower minimum noise figure.

4.1.11 f_T, f_{MAX}, and NF_{MIN} versus I_C /I_D characteristics and characteristic current densities

Historically, unlike in analog design, high-frequency circuit designers have relied on the measured or simulated g_m versus I_{DS}, f_T versus I_C, and f_T versus I_{DS} characteristics to choose the best operating point at which to bias a FET or an HBT for a certain HF circuit function. For example, in power amplifier design, the f_{MAX} versus I_C or f_{MAX} versus I_{DS} curves are employed, while for low-noise amplifiers, the NF_{MIN} versus I_C or NF_{MIN} versus I_{DS} characterstics, at the frequency of interest, provide the most useful design information. Examples of such characteristics, measured for a SiGe HBT, are illustrated in Figure 4.15 while Figure 4.16 shows the measured f_T versus J_{DS} characteristics for several recent generations of strained channel n-MOSFETs biased at V_{DS} = 1.1V [2]. J_{pfT} and J_{pfMAX} represent the current densities (in mA/μm of gate width in FETs, and mA/μm^2 of emitter area for HBTs) at which f_T and f_{MAX}, respectively, reach their maximum values. As shown in Figure 4.15, J_{OPT} represents the current density at which the minimum noise figure reaches its optimum value as a function of bias current density. In general, J_{OPT} is a function of frequency, typically increasing with frequency. However, in MOSFETs, in a given technology, J_{OPT} appears to remain largely invariant with frequency and with the circuit topology [7]. Finally, it should be noted that, due to the channel length modulation effect in FETs and due to the Kirk effect in HBTs, J_{OPT}, J_{pfT}, and J_{pfMAX} all increase with V_{DS}/V_{CE}.

The characteristic current densities are obtained either from measurements or from simulations of f_T, f_{MAX}, and NF_{MIN}, respectively, at different bias current densities. Even in the absence of transistor models, the measured $f_T - J_C/J_{DS}$, $f_{MAX} - J_C/J_{DS}$, and $NF_{MIN} - J_C/J_{DS}$ characteristics

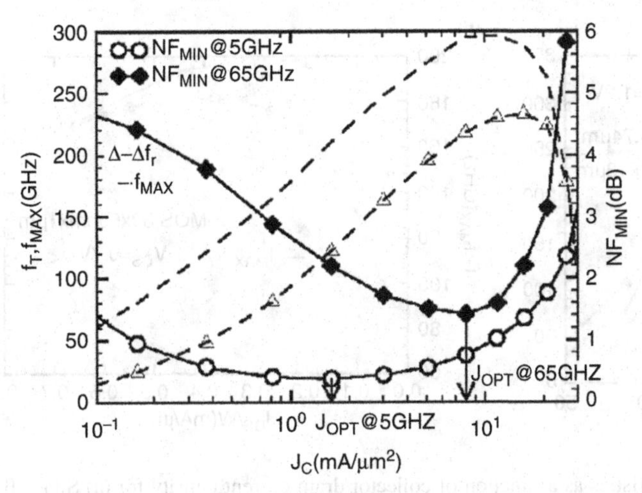

Figure 4.15 Measured f_T, f_{MAX}, and NF_{MIN} characteristics of a SiGe HBT

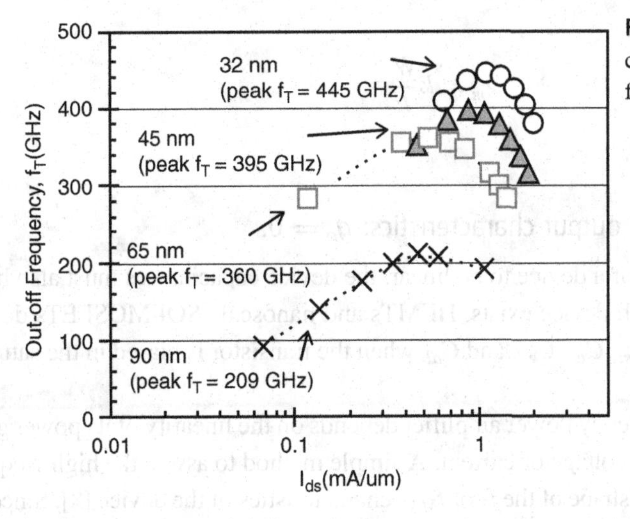

Figure 4.16 Measured f_T vs. J_{DS} characteristics of n-channel MOSFETs in four technology nodes [2]

are sufficient to allow for circuit design of the basic circuit blocks such as LNAs, PAs, and VCOs with first-time success.

4.1.12 The link between linearity and the f_T and f_{MAX} characteristics

The ideal linear device should have constant small signal equivalent circuit parameters that do not change with V_{DS}/V_{CE} and V_{GS}/V_{BE}. At low frequencies this implies

Linear transfer charactersistics

$$I_D = k\frac{W}{L}(V_{GS} - V_T) \text{ for } V_{GS} > V_T \text{ and } I_D = 0 \text{ for } V_{GS} \le V_T \tag{4.61}$$

which results in constant g_m as a function of V_{GS} and bias current

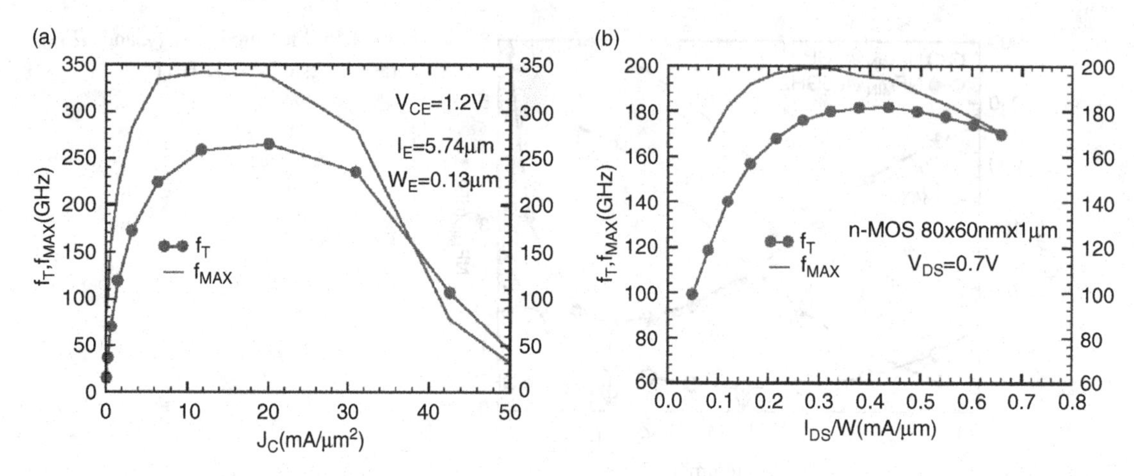

Figure 4.17 f_T and f_{MAX} characteristics as a function of collector/drain current density for (a) SiGe HBTs and (b) 65nm MOSFETs

$$g_m = k\frac{W}{L} \tag{4.62}$$

and

Linear (preferably constant) output characteristics: $g_o = 0$.

However, at high frequency, for a device to be linear, the device capacitances must also be bias independent. Although no such device exists, HEMTs and nanoscale SOI MOSFETs do come close to the ideal of constant g_m, C_{gs}, C_{gd}, and C_{db} when the transistor is biased in the saturation region at fixed V_{DS}.

The linearity of a high-frequency power amplifier depends on the linearity of its power gain as a function of the applied input voltage or current. A simple method to assess the high-frequency non-linearity is to examine the shape of the f_T or f_{MAX} characteristics of the device [8]. Since f_{MAX} is obtained directly from the maximum available power gain of the transistor, it is most suitable for estimating PA linearity. Transistor linearity depends on the flatness of the $f_{MAX}(V_{DS}/V_{CE})$, $f_{MAX}(I_{DS}/I_C)$ characteristics in the region around the peak. The transistor remains linear over the range of bias currents and drain–source or collector–emitter voltages for which f_{MAX} changes by less than 10%, which largely corresponds to the 1dB compression point. Typical f_T and f_{MAX} characteristics versus collector–drain current density are shown in Figure 4.17 for a SiGe HBT and for a 65nm n-MOSFET, respectively. In both cases, there is at least a 4:1 range of collector–drain currents over which f_{MAX} and f_T remain within 10% of their peak values.

A generic linearity figure of merit (to which IIP3 is proportional) can be employed for LNAs using the second-order derivative of g_m, f_T, or f_{MAX} with respect to V_{GS} or I_{DS} [8]

$$\mathrm{IIP3} \propto 20\log_{10}\left[\frac{g_m}{\frac{\partial^2 g_m}{\partial V_{GS}^2}}\right], \text{ or } \mathrm{IIP3} \propto 20\log_{10}\left[\frac{f_{MAX}}{\frac{\partial^2 f_{MAX}}{\partial V_{GS}^2}}\right] \text{ or } \mathrm{IIP3} \propto 20\log_{10}\left[\frac{f_T}{\frac{\partial^2 f_T}{\partial V_{GS}^2}}\right] \tag{4.63}$$

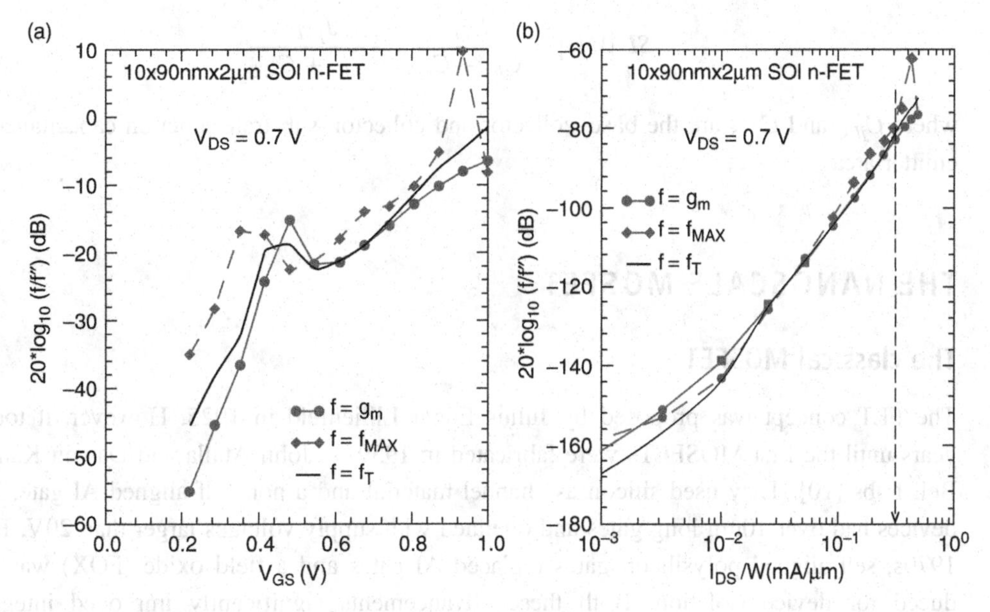

Figure 4.18 Linearity as a function of (a) V_{GS} and of (b) I_{DS}

$$\text{IIP3} \propto 20\log_{10}\left[\frac{f_{MAX}}{\frac{\partial^2 f_{MAX}}{\partial I_{DS}^2}}\right],\ \text{IIP3} \propto 20\log_{10}\left[\frac{f_T}{\frac{\partial^2 f_T}{\partial I_{DS}^2}}\right],\ \text{or IIP3} \propto 20\log_{10}\left[\frac{g_m}{\frac{\partial^2 g_m}{\partial I_{DS}^2}}\right]. \tag{4.64}$$

Figure 4.18 compares various linearity figures of merit as functions of V_{GS} and I_{DS}. It is important to note that, since the derivatives are taken with respect to different variables, the *Y*-axes in Figure 4.18(a) and 4.18(b) have different values. A linearity "sweet spot" can be identified in Figure 4.18(a) for a narrow range of gate voltages, just above the transistor threshold voltage. However, it occurs at different V_{GS} values depending on whether it is derived from f_{MAX}, f_T, or g_m. More interestingly, if we use the derivative with respect to bias current, the "sweet spot" vanishes altogether, indicating that it is most likely a mathematical artifact, and thus of little practical value. If one is to ensure that the transistor is biased for the best possible linearity, V_{GS} should be large, corresponding to a current density of 0.3mA/μm or higher.

4.1.13 Intrinsic slew rate

The ratio of the collector–drain current and the output capacitance of the transistor, known as the intrinsic slew rate, SL_i, is measured in *V/s* and is an indicator of the switching speed of the transistor in a current-mode-logic (CML) gate, in a 50Ω output driver, or in laser- or optical-modulator drivers. When considering that both the current and the output capacitance scale with the gate width or emitter area of the transistor, the slew-rate expressions become

$$SL_i[V/s] = \frac{I_{pfT}}{C_{gd}+C_{db}} = \frac{J_{pfT}}{C_{gd}+C_{db}} \tag{4.65}$$

and

$$SL_i[V/s] = \frac{I_{pfT}}{C_{bc} + C_{cs}} = \frac{J_{pfT}}{C_{jBC} + C_{jCS}} \quad (4.66)$$

where C_{jBC} and C_{JCS} are the base–collector and collector–substrate junction capacitances per emitter area.

4.2 THE NANOSCALE MOSFET

4.2.1 The classical MOSFET

The FET concept was proposed by Julius Edgar Lillienfeld in 1925. However, it took 34 years until the first MOSFETs were fabricated in 1959 by John Atalla and Dawon Kahng of Bell Labs [10]. They used silicon as channel material and a non-self-aligned Al gate. These devices had over 10μm long gates and operated with supply voltages larger than 20V. By the 1970s, self-aligned polysilicon gates replaced Al gates and a field oxide (FOX) was introduced for device isolation. Both these advancements significantly improved integration density and, along with the complementary n-MOS/p-MOS inverter, paved the way for CMOS to become the dominant logic family and to break into some low-end analog applications. At this time, the long-channel, square-law I-V model became the standard tool in the analysis and design of CMOS analog circuits. In the 1990s, with gate lengths shrunk well beyond the 1μm mark, further structural refinements included shallow-trench isolation (STI) which replaced FOX to improve surface planarity, self-aligned silicides to reduce source and drain contact resistance, and dual p-type and n-type polysilicon gates to allow for surface channel p- and n-MOSFET devices with symmetrical characteristics. Simultaneously, CMOS transistors were being considered for RF applications. The most recent steps in the evolution of the MOSFET, coincidental with the scaling of the gate length into the nanoscale (sub-100nm) region, are (i) the introduction of strain engineering techniques at the 90nm node to enhance carrier mobility in the channel and (ii) the (perhaps ironically) replacement of silicon dioxide as the gate dielectric, and of the polysilicon gate with a high-k dielectric stack and metal gates, respectively, at the 45nm node. By 2011, 99% of the semiconductor ICs, including most analog and some high-frequency circuits, were fabricated in CMOS technology.

As shown in Figure 4.19, the MOSFET is a four-terminal structure consisting of the intrinsic device, source–bulk and drain–bulk diodes, and parasitic resistances associated with each of the four terminals:

- R_g (the resistance of the polysilicon gate),
- R_s (the resistance of the source contact region),
- R_d (the resistance of the drain contact region), and
- the substrate resistance network consisting of three resistors placed between the drain and bulk nodes, R_{db}, the source and bulk nodes, R_{sb}, and the resistance of the substrate region below the channel, R_{dsb}.

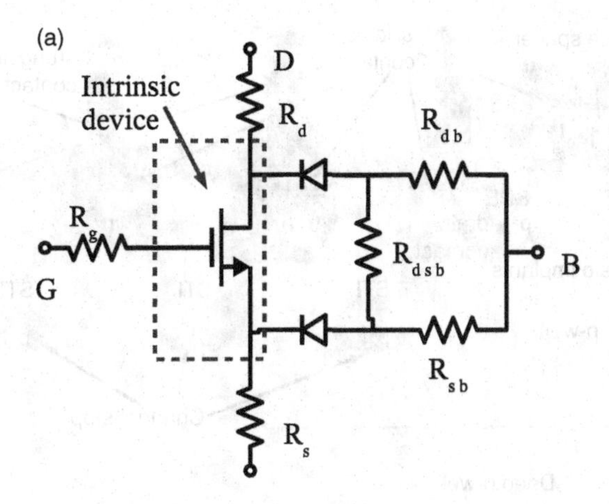

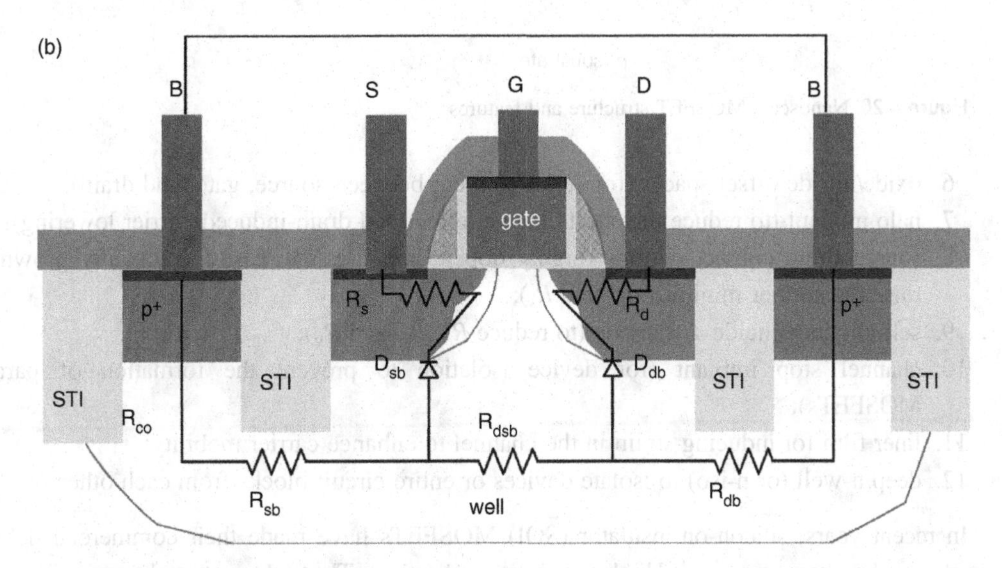

Figure 4.19 (a) Large signal equivalent circuit of a MOSFET including device parasitics and (b) device cross-section indicating the location of the parasitic diodes and resistors

4.2.2 The nanoscale MOSFET structure

Figure 4.20 sketches a vertical cross-section through a bulk planar MOSFET. This structure is representative of both p-channel and n-channel devices fabricated in the 90nm, 65nm, 45nm, and 32nm technology nodes. Its main features are indicated in Figure 4.20 and are listed below:

1. shallow trench isolation STI (to isolate devices and reduce $C_{db/sb}$);
2. retrograde well with V_T-adjust doping profile (V_T, punch-through and latch-up control, reduce $C_{db/sb}$);
3. SiON or SiON-HfO gate dielectric stack (gate leakage is a concern);
4. gate material (polysilicon or metal stack beyond the 45nm node);
5. source–drain extensions (SDE) (shallow and highly doped to reduce $R_{d,s}$);

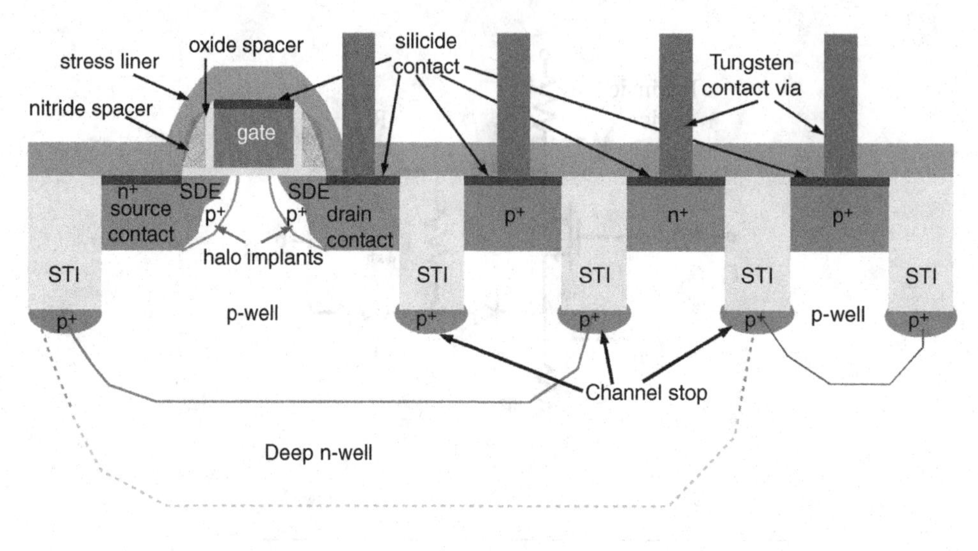

Figure 4.20 Nanoscale MOSFET structure and features

6. oxide/nitride offset spacer (for self-alignment between source, gate, and drain);
7. halo implant (to reduce the short-channel-effect and drain-induced barrier lowering);
8. source–drain contact regions (highly doped and, in some cases, epitaxially grown and raised to further minimize R_d and R_s);
9. self-aligned-silicide = salicide (to reduce R_d, R_s, and R_g);
10. channel stop implant (for device isolation to prevent the formation of parasitic MOSFETs);
11. liner film for inducing strain in the channel to enhance carrier mobility;
12. deep n-well (or n-iso) to isolate devices or entire circuit blocks from each other.

In recent years, silicon-on-insulator (SOI) MOSFETs have made their commercial debut in advanced microprocessors and high-frequency applications. The SOI structure shown in Figure 4.21 is similar to the bulk device, except that a buried oxide (BOX) layer is employed between the silicon substrate (handle wafer) and the thin silicon film in which the MOSFETs are formed. Although the cost of SOI wafers is higher than that of conventional silicon wafers, SOI allows for a simpler CMOS fabrication process. The buried oxide provides a number of benefits over a bulk process:

- excellent isolation between devices at low and moderate frequencies, without the need for deep n-wells and channel stop implants, reducing die area;
- no latch up;
- reduced source–bulk and drain–bulk junction and periphery capacitances when a floating body (no body contact) MOSFET is employed;
- no bulk diodes and no need for substrate contacts;
- compressive strain which helps improve hole mobility.

Before discussing the behavior of the MOSFET in detail, we must first clarify the various definitions of gate length, as captured in Figure 4.22:

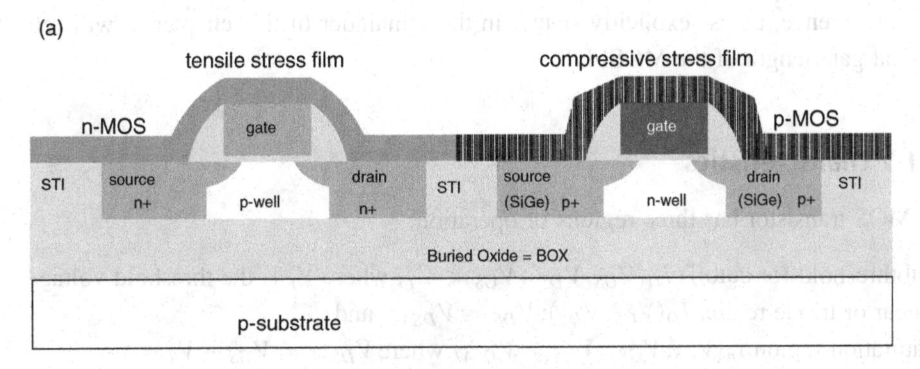

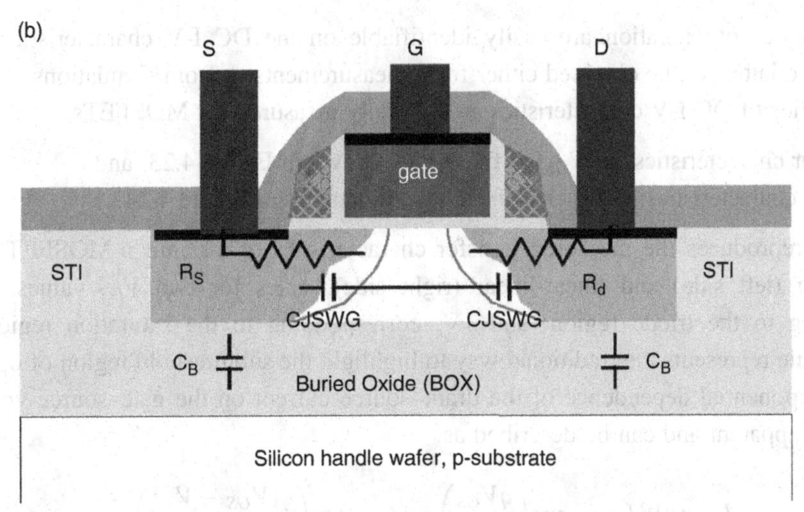

Figure 4.21 (a) Cross-section through a 65/45nm SOI CMOS wafer with n-MOS and p-MOS devices separated by STI regions and by the buried oxide (b) Cross-section through the n-MOSFET indicating the location of the parasitic source and drain resistances and of the source–bulk and drain–bulk parasitic capacitances

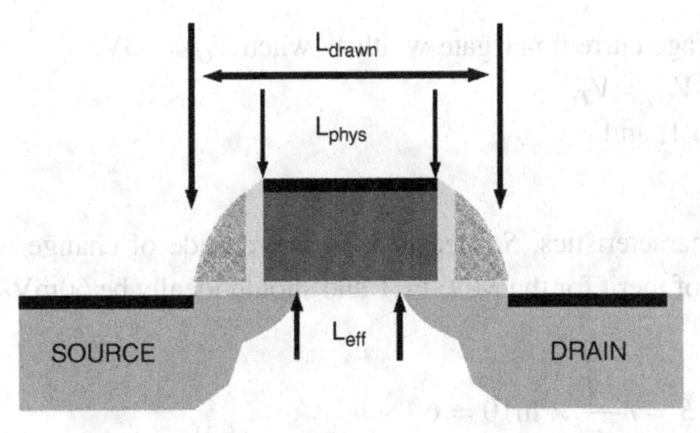

Figure 4.22 Definitions of the MOSFET gate length

- the *printed* (*drawn*) *gate length* is the photo-lithographically defined length of the (polysilicon) gate electrode;
- the *physical gate length* is the final, as-etched, length of the bottom of the gate electrode;
- the *effective* (*metallurgical*) *gate length* is the electrical gate length after the lateral out-diffusion of the source and drain contact regions.

For convenience, unless explicitly stated, in the remainder of this chapter, L will refer to the physical gate length of the MOSFET.

4.2.3 DC I-V characteristics

The MOS transistor has three regions of operation:

- subthreshold (or cutoff) $I_D(V_{GS}, V_{DS}); V_{GS} < V_T$, where V_T is the threshold voltage,
- linear or triode region $I_D(V_{GS}, V_{DS}); V_{DS} < V_{DSAT}$, and
- saturation region $I_n(V_{GS}, V_{GS}); V_{DS} \geq V_{DSAT}$ where $V_{DSAT} = V_{GS} - V_T$.

The three modes of operation are easily identifiable on the DC I-V characteristics of the transistor. The latter can be obtained either from measurements or from simulations.

Two families of DC I-V characteristics are typically measured for MOSFETs:

- the transfer characteristics: $I_D(V_{GS})$ at fixed V_{DS}, shown in Figure 4.23, and
- the output characteristics: $I_D(V_{DS})$ at fixed V_{GS}, illustrated in Figure 4.24.

Figure 4.23 reproduces the measured transfer characteristics of a 65nm n-MOSFET plotted on linear-log (left side) and linear-linear (right side) scales for two V_{DS} values: 40mV, corresponding to the triode region, and 1V, corresponding to the saturation region. The linear-log scale represents the traditional way to highlight the subthreshold region of operation where the exponential dependence of the drain–source current on the gate–source voltage is immediately apparent and can be described as

$$I_{DS} = WJ_{SDleak}exp\left(\frac{qV_{GS}}{nkT}\right) = I_{DSO}exp\left(\frac{q(V_{GS} - V_T)}{nkT}\right) \tag{4.67}$$

where:

- W is the total gate width,
- J_{SDleak} is the subthreshold leakage current per gate width W when $V_{GS} = 0$V,
- I_{DSO} is the drain current when $V_{GS} = V_T$,
- n is the ideality factor, close to 1, and
- k is Boltzmann's constant.

The slope of the subthreshold characteristics, S, measured in mV/decade of change in drain current, is an often-quoted figure of merit for the MOSFET and should ideally be 60mV/decade at room temperature

$$S = n\frac{kT}{q} \times \ln 10 = 60 \times n. \tag{4.68}$$

Values as large as 80mV/decade are tolerated in nanoscale technologies where short channel effects degrade the steepness of the subthreshold characteristics.

For most part of the transfer characteristics, I_{DS} is a linear function of V_{GS}, a significant departure from the long-channel MOSFET, where the dependence is governed by the square law. In its simplest form, the drain current expression in the saturation region can be cast as

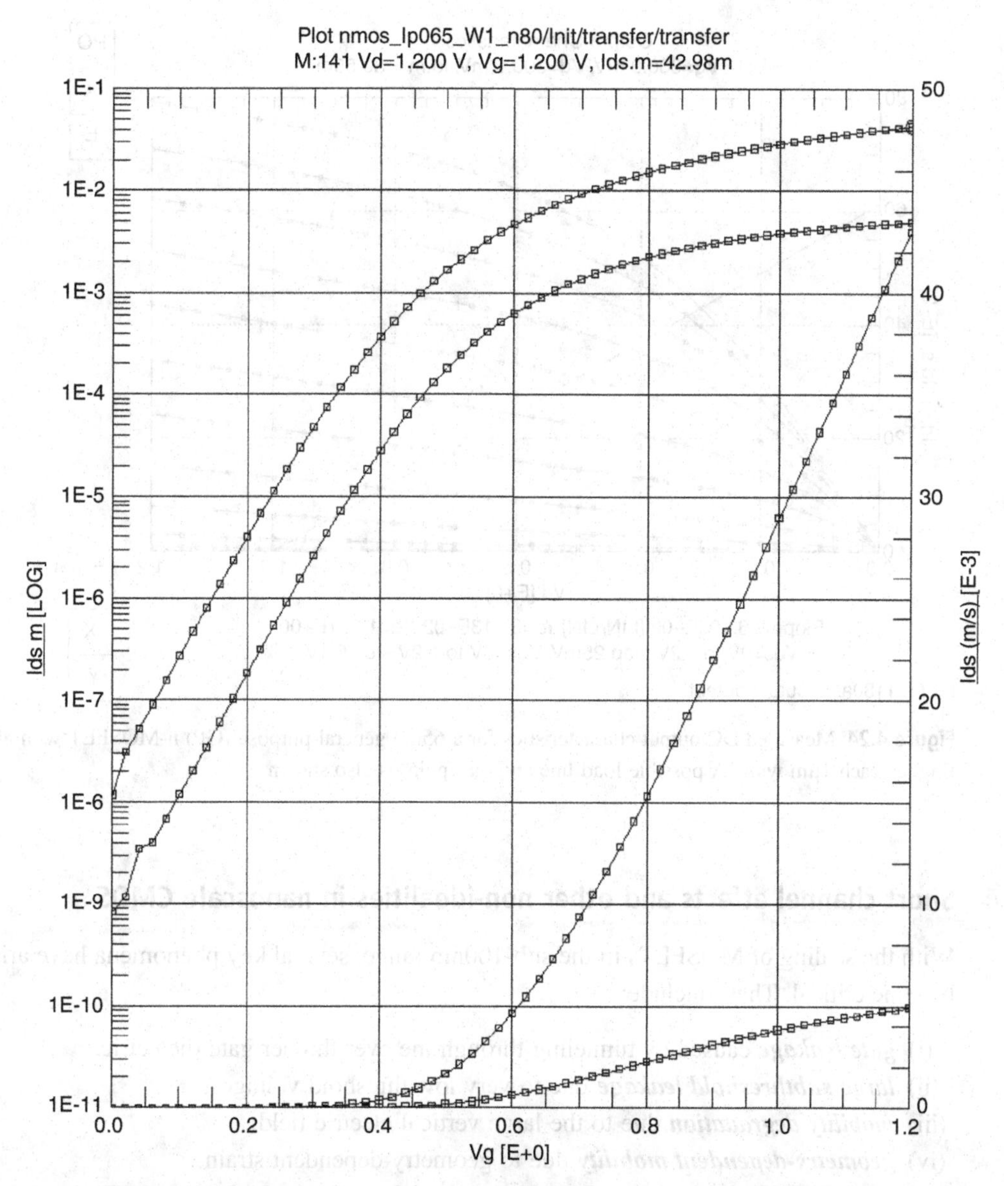

Figure 4.23 Measured transfer characteristics of a 65nm low-power (LP) MOSFET at $V_{DS} = 1.2$V and $V_{DS} = 40$mV illustrating the subthreshold region on the logarithmic-scale *Y*-axis, at the left, and on a linear-scale *Y*-axis at the right

$$I_{DS} = W v_{xo} C_{OX}^{inv}(V_{GS} - V_T)(1 + \lambda V_{DS}) \tag{4.69}$$

where

- v_{x0} is the effective carrier velocity at the intrinsic source of the device (SDE-channel junction below gate) and depends on the carrier mobility in the inversion channel at that location and on the gate length,
- C_{OX}^{inv} is the effective capacitance of the channel at inversion, and
- λ is the channel length modulation parameter.

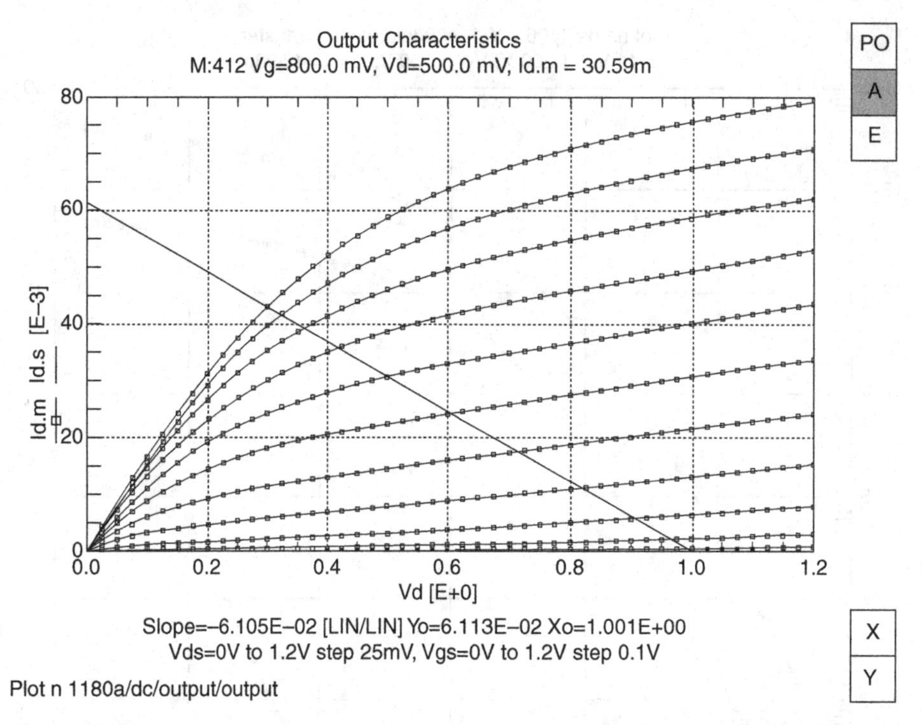

Figure 4.24 Measured DC output characteristics for a 65nm general-purpose (GP) n-MOSFET with 80 gate fingers, each 1μm wide. A possible load line and bias point is also shown

4.2.4 Short channel effects and other non-idealities in nanoscale CMOS

With the scaling of MOSFETs in the sub-100nm range, several key phenomena have arisen or become critical. These include:

(i) ***gate leakage*** caused by tunneling through the ever thinner gate dielectric,
(ii) ***large subthreshold leakage*** due to very low threshold voltages,
(iii) ***mobility degradation*** due to the large vertical electric field,
(iv) ***geometry-dependent mobility*** due to geometry-dependent strain,
(v) greatly diminished scaling of the capacitance effective thickness (*CET*) due to charge quantization in the channel at the silicon-gate oxide interface,
(vi) increasing contribution of the source and drain ***series resistances*** to the overall channel resistance and to the degradation of the extrinsic g_m, f_T, f_{MAX}, and *NF*, and
(vii) increased $\boldsymbol{V_T}$ ***variation*** due to larger fluctuations in the number of dopant atoms in very short channels.

However, before discussing the nanoscale-related phenomena, we will first review the short channel effect, the reverse short channel effect, and drain induced barrier lowering (DIBL), all of which were already present in older deep-submicron CMOS technologies.

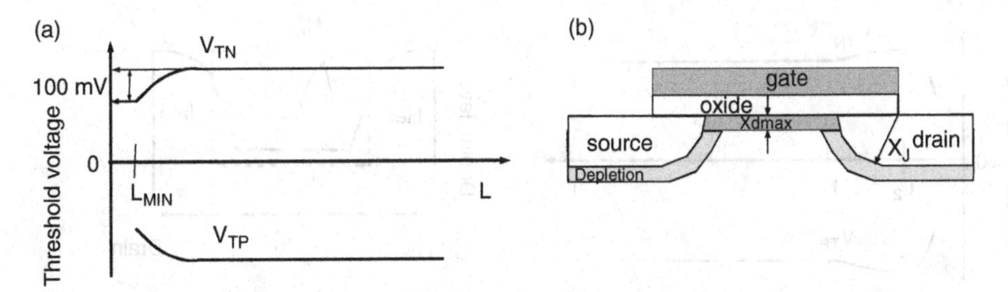

Figure 4.25 Illustration of (a) the V_T roll off and (b) the charge-control in the channel and in the source/drain regions

Short channel effect (SCE)

As illustrated in Figure 4.25, the short-channel effect refers to the decrease of V_T in a given technology as L is reduced. This phenomenon is also known as *threshold roll-off.* To avoid significant subthreshold leakage in a deeply scaled CMOS process, it is important to limit the threshold voltage reduction in devices with minimum channel length. Typically, a variation by no more than 100mV is tolerated compared to the V_T of a long channel device.

The V_T roll-off can be understood by invoking the charge sharing model [11] to explain how the gate loses control of the channel charge as the gate length is reduced. This model considers the ionized acceptors or donors in the channel depletion region which determine the bulk charge, Q_B, that is the first square-root term in the threshold voltage expression (4.70). As L decreases, more of these charges become associated with the source and drain depletion regions and, therefore, as illustrated in Figure 4.25, the gate sees a diminished Q_B. This implies that, when compared to long-channel devices, in short-channel n-MOSFETs the threshold voltage is reduced. The effect is exacerbated as the gate length decreases. It can be described analytically by

$$V_T = V_{FB} + \frac{\sqrt{4\varepsilon_s q N_B \phi_F}}{C_{OX}}\left[1 - \frac{x_J}{L}\sqrt{1 + \frac{2x_{dmax}}{x_J}}\right] - 2\phi_F. \tag{4.70}$$

where V_{FB} is the flat-band voltage, C_{OX} is the oxide capacitance, and $\phi_F = (kT/q)\ln(N_B/n_i)$ is the quasi-Fermi level in the well where the transistor is formed.

Typically, to alleviate the impact of the SCE, both the depletion region width below the channel, x_{dmax}, and the junction depth of the source and drain contact regions, x_J, must be smaller than one third of the physical gate length. The former is accomplished with a retrograde doping profile in the channel (i.e. the doping level is lower at the surface and increases with depth), while the latter is implemented by employing shallow, yet highly doped, source–drain extension regions. We note that, if $L >> x_J$, (4.70) reduces to the expression of the threshold voltage of a long-channel MOSFET.

The reverse short channel effect (RSCE)

As the gate length decreases, the source–bulk and drain–bulk depletion regions become physically closer to each other and may even merge, causing "punch-through." To prevent punch-through (an extreme manifestation of the SCE), the doping in the channel must be

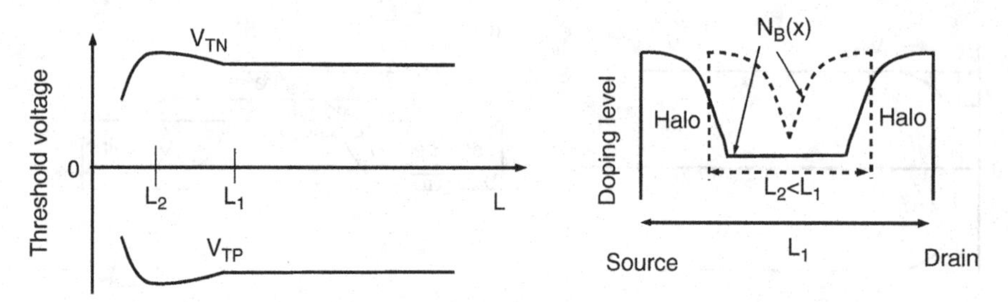

Figure 4.26 Ilustration of the reverse short channel effect

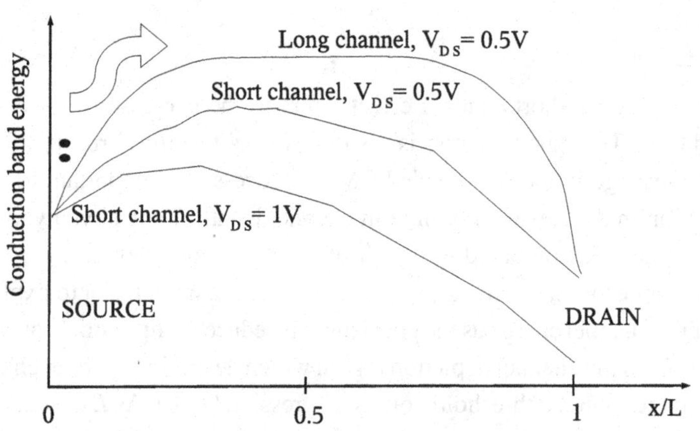

Figure 4.27 The lowering of the conduction band barrier between source and drain

increased. Unfortunately, the higher doping degrades carrier mobility. A good compromise is to use halo (or pocket) implants to raise the doping level in the channel only in the vicinity of the source and of the drain diffusion regions. The halo implants are of the same polarity as the well doping and increase the average effective doping, N_B, in the channel. As shown in Figure 4.26, N_B becomes a function of channel length and raises to higher values in short gate length devices than in long-channel devices. Since in a n-MOSFET V_T increases with N_B, the n-channel threshold voltage first increases as the gate length shrinks, before finally succumbing to short-channel effects. As its name indicates, the RSCE delays the onset of the SCE. In p-MOSFETs, the $V_T(L)$ dependence is a mirror image of that of n-MOSFETs with respect to the X-axis.

Important note: One of the ramifications of the RSCE is that, in a given process, the mobility, output impedance, and source–bulk and drain–bulk capacitances become functions of gate length, impacting the manner in which CMOS analog and RF circuit design is conducted. Rather than employing the W/L ratio as the main device sizing parameter for circuit design, the gate length must be fixed, and only the gate width, or better still, the number of gate fingers, N_f, should be varied.

Drain-induced barrier lowering (DIBL)

As the name suggests, in an n-channel MOSFET, the drain-induced-barrier-lowering effect is caused by the reduction of the potential barrier seen by the electrons (or holes in a p-MOSFET) in the source towards the drain as the drain–source voltage increases. The

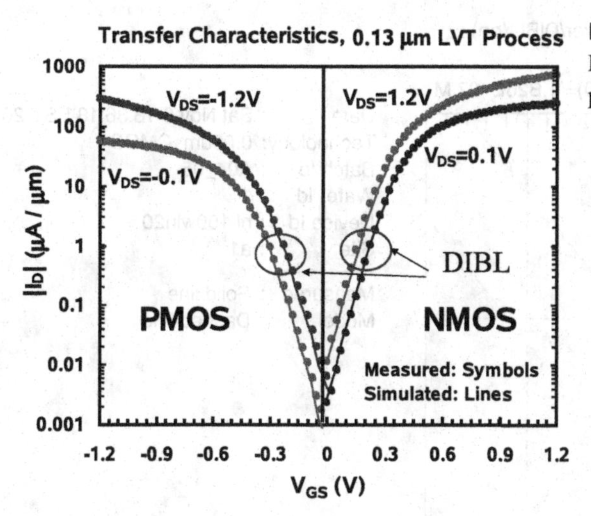

Figure 4.28 Illustration of the impact of the DIBL effect on the I-V characteristics of 130nm p- and n-channel MOSFETs

phenomenom, only present in short channels devices, is sketched in Figure 4.27 and manifests itself in a reduction of the threshold voltage of an n-channel MOSFET as the drain–source voltage increases

$$V_T = V_{T0} - \eta V_{DS}. \tag{4.71}$$

Equation (4.2.5) also applies to a p-MOSFET.

The DIBL effect impacts both the subthreshold and the saturation regions (Figure 4.28) and, for a constant V_{GS}, causes the drain current in the subthreshold region to increase in comparison to that of a device where V_T does not depend on V_{DS}

$$I_{DS_{subth}} = I_{DS0} \exp\left(\frac{V_{GS} - V_{T0} + \eta V_{DS}}{S} \ln 10\right). \tag{4.72}$$

The DIBL effect parameter, η, can be easily extracted from the transfer characteristics in the subthreshold region, as illustrated in Figure 4.29.

EXAMPLE 4.1

Consider the cross-section in Figure 4.20 through a 65nm n-MOSFET in a CMOS process with retrograde p- and n-well, and deep n-well. Assume that the substrate is doped 10^{14}cm^{-3}:

(a) If the physical gate length of the general purpose (GP) device is 45nm, what should be the depth of the junction in the extension region?

$Xj = 15\text{nm}$.

(b) What should be the average effective doping below the gate such that $x_d <= L/3 = 15\text{nm}$?

(c) What is the minimum average doping in the halo implants needed to avoid punch-through between the source and the drain at the maximum allowed voltage of 1.5V? Assume a doping level of 10^{21} cm^{-3} in the S/D contact region.

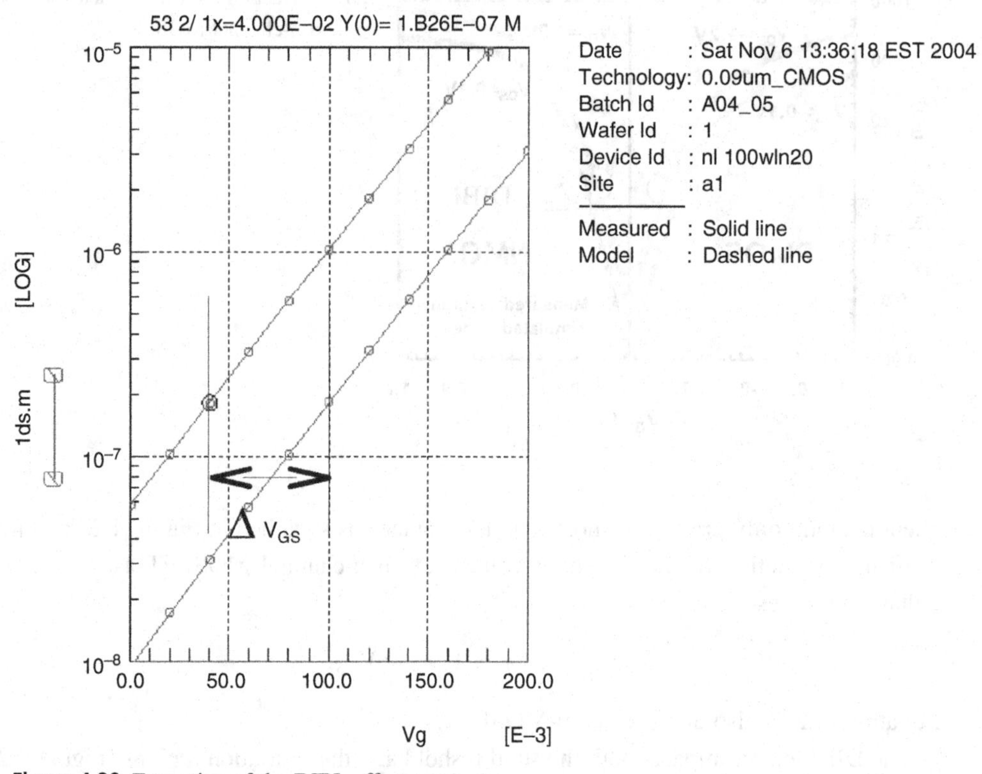

Figure 4.29 Extraction of the DIBL effect parameter η

Solution

Using the equations that give the depletion region at inversion and the surface potential [11], we obtain

$$x_d = \sqrt{\frac{2\varepsilon_s\varepsilon_0|(\phi_S)|}{qN_A}} < L/3 = 15\text{nm where } \phi_S = 2\frac{kT}{q}\ln\left(\frac{N_A}{n_i}\right) \approx 1V$$

$$N_A = \frac{2\varepsilon_S\varepsilon_0\Phi_S}{qx_d^2} = 4.1 \times 10^{18}\text{cm}^{-3}.$$

(c) From the equation of the depletion region width at a reverse-biased hyperabrubt pn junction [11], and from the built in voltage equation [11], we obtain

$$x_d = \sqrt{\frac{2\varepsilon_s\varepsilon_0|(\phi_{bi}+V)|}{qN_A}} < L/2 = 22.5 \text{ nm where } \phi_{bi} = \frac{kT}{q}\ln\left(\frac{N_A N_D}{n_i^2}\right) => N_A > 6.4 \times 10^{18}\text{cm}^{-3}.$$

Gate leakage

The gate leakage current has become a problem starting with the high-performance (HP) 90nm node due to electron *tunneling* through the thin-gate oxide. It is a function of the *physical thickness* of the gate dielectric and of the potential barrier between the channel and the gate, but does not depend on the dielectric constant of the gate oxide stack.

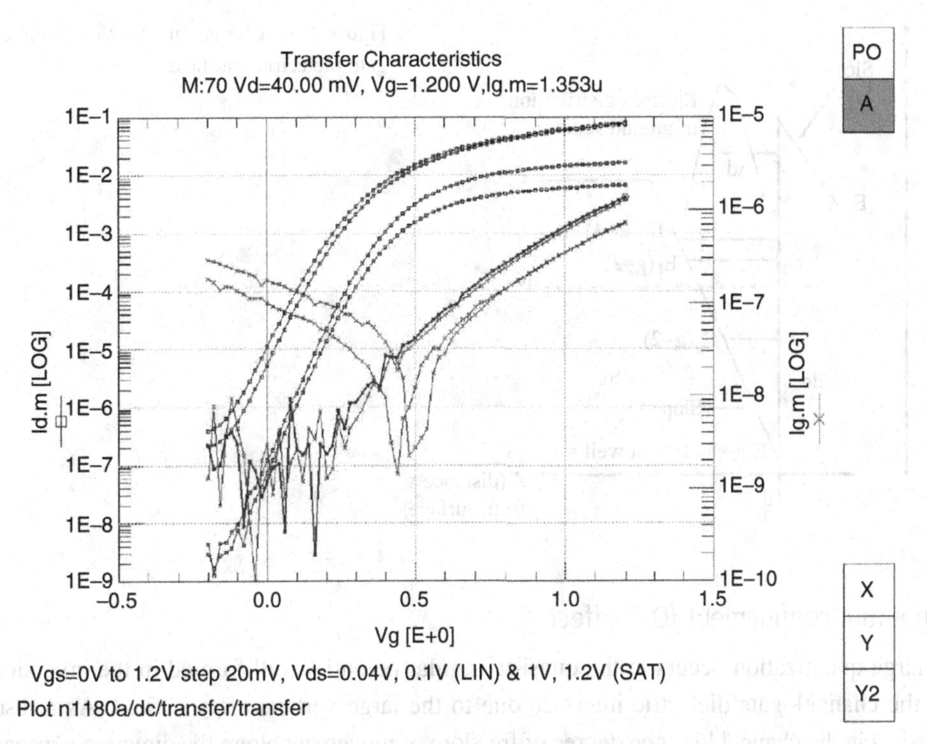

Figure 4.30 Measured drain current (black squares) and gate leakage current (blue crosses) vs. gate voltage in a 65nm 80×1mm GP n-MOSFET at V_{DS} = 40mV, 100mV, 1 V and 1.2 V

An empirical gate leakage model was developed by C. Osborne of NCSU to describe the gate current density [12]

$$J_G\left[\frac{A}{\text{cm}^2}\right] = 1.44e5 \times \left(\exp\left(-4.02 \times V_{GS}^2[V] + 13.05 \times V_{GS}[V]\right)\right) \times \exp(-1.17 \times T_d[A]) \quad (4.73)$$

where T_d is the thickness of the gate dielectric in Angstrom.

Measurements of the gate leakage current $I_G = J_G \times W \times L$ of a 65nm n-MOSFET with 1.25nm thick oxide are reproduced in Figure 4.30 as a function of V_{GS} for several V_{DS} values. In this device, note that the gate current is only 4 orders of magnitude smaller than the drain current. By deploying a thicker, high-k dielectric gate stack, it is possible to reduce J_G without simultaneously reducing gate capacitance and transconductance.

Important note: Gate and subthreshold leakage does not pose insurmountable problems for high-speed and high-frequency circuit design. GaAs p-HEMTs, InP HEMTs, InP HBTs, and SiGe HBTs exhibit much larger gate or base currents than 65nm MOSFETs, yet they have been used successfuly for over two decades to build microwave and mm-wave products. In fact, in many CMOS current-mode-logic CML and RF circuits, low-V_T devices (with increased subthreshold leakage) are preferred to achieve high-speed operation with a low-voltage power supply and thus minimize power consumption.

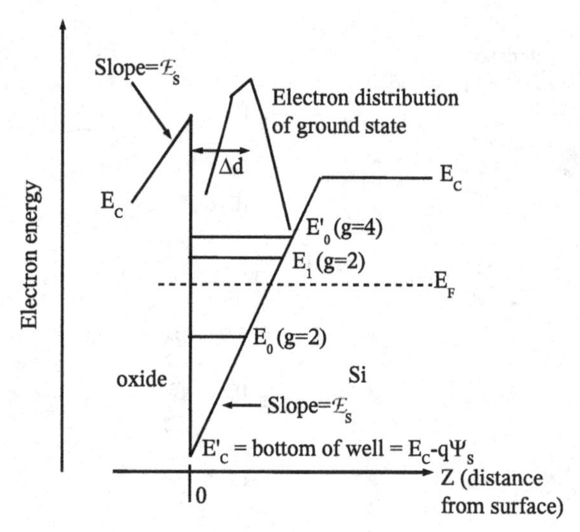

Figure 4.31 Charge quantization at the channel–gate dielectric interface

Quantum confinement (QC) effect

Charge quantization occurs in the quasi-triangular potential well formed in the inversion layer at the channel–gate dielectric interface due to the large vertical electric field. As a result, the carriers in the channel lose one degree of freedom of movement along the direction perpendicular to the gate and form a two-dimensional electron or hole gas. This effect impacts the MOSFET behavior in several ways. First, because the 2-D density of states of the free carriers in the channel is smaller than the 3-D one, the inversion charge in the channel and the drain current are smaller than 3-D Fermi–Dirac statistics predict [11]

$$Q_i^{QM} = -qn = \frac{-4\pi qkT}{h^2}\left[2m_t\sum_j \ln\left(1+e^{\frac{E_f-E_C-E_j+q\Psi_S}{kT}}\right) + 4\sqrt{m_t m_l}\sum_j \ln\left(1+e^{\frac{E_f-E_C-E'_j+q\Psi_S}{kT}}\right)\right] \tag{4.74}$$

where, as illustrated in Figure 4.31:

- E_C is the bottom of the conduction band;
- E_F is the Fermi level;
- Ψ_S is the surface potential;
- E_j and E'_j are the allowed energy states in the potential well, referenced to the bottom of the conduction band energy at the silicon/gate oxide interface, corresponding to the Δ_2 and Δ_4 conduction valleys, respectively [11];
- m_t and m_l are the effective transversal and longitudinal electron masses, respectively, in a <100> silicon wafer with MOSFET channels along the <110> direction; and
- h is Planck's constant.

Second, the quantum confinement of the free carrier population in the inversion channel pushes the channel charge away from the channel–gate dielectric interface, reducing the effective gate capacitance and increasing the threshold voltage compared to the situation when QC is ignored. The spatial offset, Δd, shown in Figure 4.31, is a function of the gate voltage (smaller for larger V_{GS}) and of the gate dielectric thickness, and can be approximated by the following empirical equation [11]

$$\Delta d = 6.2 \times 10^{-5} \left(\frac{V_{GS} + V_T}{t_{ox}} \right)^{-0.4} \tag{4.75}$$

where t_{ox} and Δd are in cm.

Although the quantum-confinement effect in MOSFETs was investigated for the first time in the late 1960s [13], it is only in recent years, when t_{ox} has became comparable to Δd, that it has started to play a significant role.

The polysilicon gate depletion effect

In very thin oxide MOSFETs with polysilicon gates, a depletion region of depth X_{gd} forms in the polysilicon gate because of the strong vertical electric field and finite doping concentration of the polysilicon gate. It can be approximated by [11]

$$X_{gd} \approx \frac{C_{OX}(V_{GS} - V_T + \gamma\sqrt{2\phi_F})}{qN_{gate}} \tag{4.76}$$

where N_{gate} is the doping in the polysilicon gate and γ is the body effect parameter. For MOSFETs with metal gates, $X_{gd} = 0$.

The combined impact of the gate poly depletion and of the quantum confinement effect is to increase the effective thickness of the gate dielectric and thus to reduce the gate capacitance

$$\frac{1}{C_{gate}} = \frac{1}{C_{gd}} + \frac{1}{C_{ox}} + \frac{1}{C_{ac}} = \frac{X_{gd}}{\varepsilon_0 \varepsilon_{Si}} + \frac{t_{ox}}{\varepsilon_0 \varepsilon_{ox}} + \frac{\Delta d}{\varepsilon_0 \varepsilon_{Si}}. \tag{4.77}$$

A capacitive equivalent thickness, CET, of the gate dielectric is defined as

$$CET = t_{inv} \approx EOT + \frac{\varepsilon_{Si}}{\varepsilon_{ox}}(X_{gd} + \Delta d) \tag{4.78}$$

where EOT is the effective oxide thickness in case high-k dielectrics are employed for the gate oxide

$$EOT = t_{ox} \frac{\varepsilon_{SiO_2}}{\varepsilon_{ox}} \tag{4.79}$$

and

$$C_{gate} = \frac{\varepsilon_{SiO_2}}{CET}. \tag{4.80}$$

Note that in some publications t_{inv} is used instead of CET.

EXAMPLE 4.2

If EOT = 2nm, N_{gate} = 2 $\times 10^{20}$ cm^{-3}, V_T = 0.3V, V_{GS} = 0.5V, $\phi_s = 2\phi_F$ = 0.96V, γ = 0.42 V^{-1}, C_{OX} = 17.26mF/m^2, calculate CET.

Solution

$$\Delta d = 6.2 \times 10^{-5}\left(\frac{0.8}{2 \times 10^{-7}}\right)^{-0.4} = 1.41\text{ nm}$$

$$X_{gd} \approx \frac{0.01726(0.2 + 0.42\sqrt{0.96})}{1.6 \times 10^{-19} \times 2 \times 10^{26}} = 0.33\text{nm}$$

$$CET \approx 2\text{nm} + \frac{1.74\text{nm}}{3} = 2.58\text{nm}.$$

This value is 30% larger than the effective oxide thickness of 2nm and will result in smaller g_m and I_{DS} than expected from the effective oxide thickness.

Now, let us assume that we scale the oxide thickness to 1.2nm. $N_{gate} = 2{\times}10^{20}\ \text{cm}^{-3}$, $V_T = 0.3\text{V}$, $V_{GS} = 0.5\text{V}$, $\phi_s = 2\phi_F = 1\text{V}$, $\gamma = 0.3\ \text{V}^{-1}$, $C_{OX} = 28.76\text{mF/m}^2$

$$\Delta d = 6.2 \times 10^{-5}\left(\frac{0.8}{1.2 \times 10^{-7}}\right)^{-0.4} = 1.15\text{nm}$$

$$X_{gd} \approx \frac{0.02876(0.2 + 0.3\sqrt{1})}{1.6 \times 10^{-19} \times 2 \times 10^{26}} = 0.44\text{nm}$$

$$CET \approx 1.2\text{nm} + \frac{1.59\text{nm}}{3} = 1.73\text{nm}.$$

The gate capacitance has increased by a factor of 1.5 when C_{OX} was scaled by 1.67.

We note that if metal gates were to be used, $Xgd = 0$ and $CET = 1.583$nm.

What happens when EOT is reduced below 1.25nm for the next technology node?

Mobility degradation due to the vertical electric field

The main physical phenomenon responsible for the linear I-V characteristics of the MOSFET at moderate and high V_{GS} is the carrier mobility degradation due to the vertical electric field. The latter can be as high as 4–5MV/cm [14]. The effect can be understood by examining the experimental universal surface mobility curves for electrons and holes, which are independent of the doping in the channel. The term "universal" is used because these curves are independent of substrate doping and apply to silicon MOSFETs fabricated in any technology node.

The universal surface mobility curve for electrons in silicon can be expressed as [11]

$$\mu_n = \frac{\mu_{ac}\mu_{sr}}{\mu_{ac} + \mu_{sr}} \tag{4.81}$$

where the acoustic-phonon-limited mobility μ_{ac} and surface scattering-limited mobility μ_{sr} terms

$$\mu_{ac}\left[\frac{cm^2}{Vs}\right] \approx 330E_{eff}^{-1/3}\left[\frac{MV}{cm}\right],\ \mu_{sr}\left[\frac{cm^2}{Vs}\right] \approx 1450E_{eff}^{-2.9}\left[\frac{MV}{cm}\right] \tag{4.82}$$

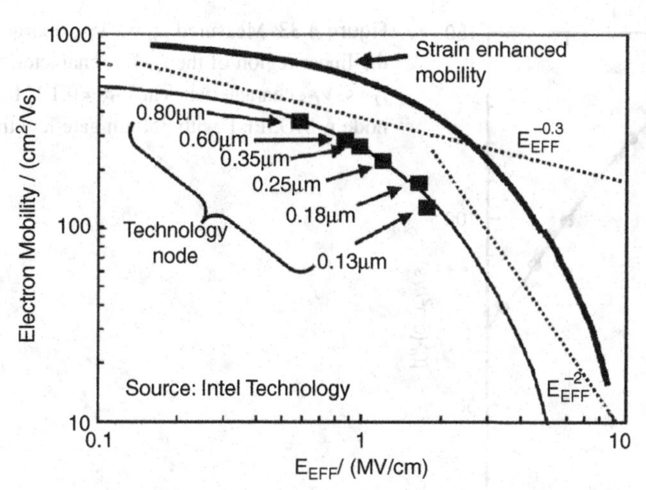

Figure 4.32 Mobility versus technology scaling for Intel process technologies[14] (a) Electron mobility and (b) Hole mobility

are functions of the effective vertical field, E_{eff}. For devices fabricated in the 180nm, 130nm, and 90nm nodes, E_{eff} falls in the 1 to 2MV/cm range where μ_{sr} dominates over μ_{ac} and leads to a rapid reduction in electron mobility as the gate voltage increases [14]. Recent data [14], shown in Figure 4.32, indicate that, notwithstanding larger overall values, the electron mobility in strained silicon channels has the same dependence on the vertical electric field, with the onset of surface scattering occurring in at slightly higher vertical fields. A similar behavior is observed for strain enhanced hole mobilities when the SiGe concentration in the source and drain regions of p-MOSFETs increase from one technology node to another.

The mobility degradation due to the vertical (normal) electric field is a stronger effect than velocity saturation due to the lateral drain–source field [7]. The latter reaches a value of at most 0.2MV/cm in a 50nm n-MOS channel biased at $V_{DS} = 1$V. Figure 4.33 illustrates that, for gate–source voltages larger than the threshold voltage, the dependence of mobility on V_{GS} can be empirically described as

$$\mu_{eff}(V_{GS}) = \frac{\mu_0}{1 + \theta(V_{GS} - V_T)} \tag{4.83}$$

where θ is a fitting parameter that captures mobility degradation due to the vertical electric field [7], and μ_0 is the low field mobility when $V_{GS} = V_T$.

Velocity saturation due to the lateral field

At low lateral electric fields, E_{lat}, between the source and drain, the carrier velocity varies linearly with the electric field

$$v = \mu E_{lat} = \mu V_{ds}/L. \tag{4.84}$$

At high electric fields, this ceases to be true. Carriers scatter off atoms more frequently and velocity saturates at approximately $6 - 10 \times 10^6$ cm/s and, $4 - 8 \times 10^6$ cm/s, for electroncs and holes, respectively.

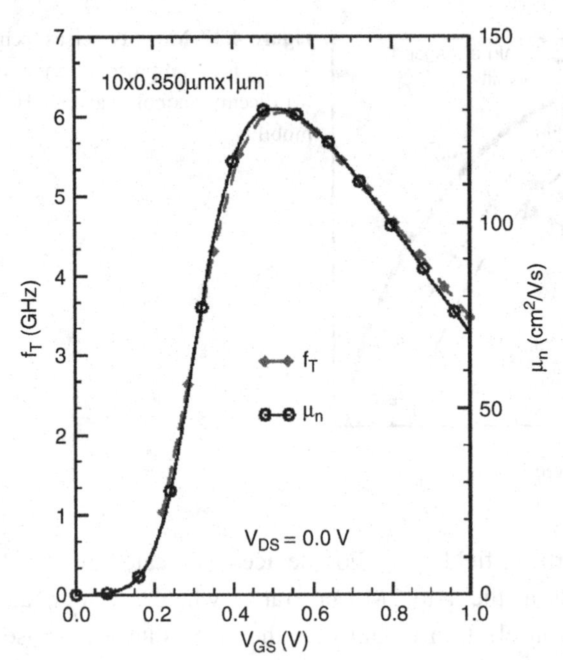

Figure 4.33 Measured μ_n vs. V_{GS} (extracted from the linear region of the I_D-V_{GS} characteristics) and f_T vs. V_{GS} characteristics at $V_{DS} = 0.1$ V for a 90nm node n-MOSFET with 350nm gate length [7]

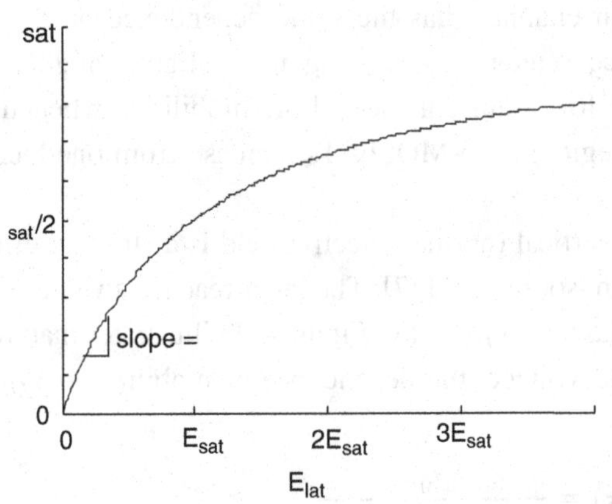

Figure 4.34 Illustration of the velocity saturation effect [11]

A general, empirical velocity-field dependence model that captures this effect is given by

$$v(E_{LAT}) = \frac{\mu_{eff} E_{lat}}{\left[1 + \left(\frac{E_{lat}}{E_{sat}}\right)^n\right]^{1/n}} \tag{4.85}$$

where $n = 1$ for holes, $n = 2$ for electrons

$$E_{sat} = \frac{v_{sat}}{\mu_{eff}} \tag{4.86}$$

and μ_{eff} is a function of the vertical electrical field through V_{GS}, described by (4.83), as discussed earlier.

What is the impact of velocity saturation on the I-V characteristics?

In the absense of velocity saturation and vertical field mobility degradation, the MOSFET drain current in the saturation region increases with $(V_{GS} - V_T)^2$. If the channel were fully velocity-saturated, then the drain current would simply be expressed as

$$I_{DS} = WC_{OX}^{inv}(V_{GS} - V_T)v_{sat}. \quad (4.87)$$

However, real transistors are only partially velocity-saturated and their I-V characteristics feature a combination of quadratic and linear behavior.

A closed form expression for drain current that lumps together the vertical and lateral field dependence of mobility is present in all advanced MOSFET models [15]. However, this has made it rather difficult in the past to actually pinpoint which effect is dominant. There is increasing evidence in recent literature that, in deep-submicron and nanoscale MOSFETs, mobility degradation due to the applied vertical field prevails [7]. Typically, the vertical field is one order of magnitude larger than the lateral electric field. In addition, the critical lateral field $E_C = v_{sat}/\mu_n$ for velocity saturation is not constant. It increases as the carrier mobility μ_n, μ_p degrades [11]. This phenomenon helps to delay the onset of velocity saturation in deep-submicron and nanoscale MOSFETs to ever larger lateral electric fields and to larger V_{DS} values as V_{GS} increases.

EXAMPLE 4.3

Let $L = 120\text{nm}$, $t_{ox} = 2\text{nm}$, $V_T = 0.4\text{V}$, $E_\perp(V_T) = (0.4 + 0.2)/6\text{nm} = 1\text{MV/cm}$, $V_{DS} = 1.2\text{V}$, $E_{LAT} = 1.2\text{V}/120\text{nm} = 0.1\text{MV/cm}$. Find the low lateral-field electron–hole mobilities after accounting for the large vertical field and calculate the values of the saturation field.

Solution

From (4.71) and (4.72), we obtain

$$\mu_{neff} = 250\,\text{cm}^2/\text{Vs},\ \mu_{peff} = 60\,\text{cm}^2/\text{Vs},\ E_{satn} = v_{nsat}/\mu_{neff} =$$

$8\text{E}6/250 = 3.2\times10^4\text{V/cm}$, and

$$E_{satp} = v_{psat}/\mu_{peff} = 6\text{E}6/60 = 10^5\text{V/cm}.$$

EXAMPLE 4.4

Repeat the calculations in Example 4.3 for a technology with $L = 45\text{nm}$, $t_{ox} = 1\text{nm}$, $V_T = 0.4\text{V}$, $E_\perp(V_T) = (0.4 + 0.2)/3\text{nm} = 2\text{MV/cm}$, $V_{DS} = 0.9\text{V}$, $E_{LAT} = 0.9\text{V}/45\text{nm} = 0.2\text{MV/cm}$.

Solution

$\mu_{neff} = 100\ \text{cm}^2/\text{Vs}$, $\mu_{peff} = 30\ \text{cm}^2/\text{Vs}$, $E_{satn} = v_{nsat}/\mu_{neff} = 8\text{E}6/100 = 8\times10^4\text{V/cm}$, and $E_{satp} = v_{psat}/\mu_{peff} = 6\text{E}6/30 = 2\times10^5\text{V/cm}$.

From these two examples, we notice that the low-field mobility of electrons and holes has degraded significantly from the 130nm node to the 45nm node. At the same time the electric field at which velocity saturation occurs has doubled. This explains why even 45nm devices are not operating under full saturation regime and why an increasing amount of channel strain has to be introduced in new technology nodes to compensate for mobility degradation due to larger doping levels in the channel and larger vertical electric fields.

I-V model for the saturation region that accounts for both the vertical and lateral fields

Using a piecewise linear $\mu(E_{lat})$ model where v reaches v_{sat} at $E_{lat} = 2v_{sat}/\mu_{eff}$ [11]

$$v(E_{lat}) = \frac{\mu_{eff}E_{lat}}{1 + \frac{\mu_{eff}|E_{lat}|}{2v_{sat}}} \text{ where } E_{lat} = \frac{-dV}{dx} \tag{4.88}$$

and (4.83) to describe the μ_{eff} dependence on the vertical (transversal) field $E_{\perp}$, the drain current, which is constant at any position x in the channel between the source and drain, can be expressed as

$$I_{DS}(x) = WQ_i(V)v(E) = -WQ_i(V)\frac{\mu_{eff}(V)\frac{dV}{dx}}{1 + \frac{\mu_{eff}(V)}{2v_{sat}}\left|\frac{-dV}{dx}\right|} = ct. \tag{4.89}$$

where Q_i is the inversion charge in the channel and varies with x through $V(x)$.

Equation (4.89) can be recast as

$$I_{DS}dx = -\mu_{eff}(V)\left[WQ_i(V) + \frac{I_{DS}}{2v_{sat}}\right]dV \tag{4.90}$$

and integrated from 0 to L on the left-hand side with respect to dx and from 0 to V_{DS} on the right-hand side with respect to the channel voltage V at location x

$$\int_0^L I_{DS}dx = -\mu_{eff}\left[W\int_0^{V_{DS}} Q_i(V)dV + \frac{I_{DS}}{2v_{sat}}V_{DS}\right]. \tag{4.91}$$

The inversion charge can be expressed as a function of the local voltage in the channel

$$Q_I(V) = \frac{\varepsilon_{OX}}{CET}(V_{GS} - V_T - mV) \tag{4.92}$$

where m is a correction factor, larger than 1, which accounts for the capacitance of the depletion region in the channel

$$m = 1 + \frac{\sqrt{\frac{\varepsilon_{Si}qN_B}{4\Phi_F}}}{C_{OX}} = 1 + \frac{C_{dm}}{C_{OX}} = 1 + \frac{\varepsilon_{Si}}{\varepsilon_{OX}}\frac{t_{ox}}{X_{dm}}. \tag{4.93}$$

By inserting (4.92) in (4.91), we obtain sequentially

$$I_{DS} = -\mu_{eff}\frac{W}{L}\frac{\int_0^{V_{DS}} Q_i(V)dV}{1+\frac{\mu_{eff}V_{DS}}{2v_{sat}L}} = \mu_{eff}\frac{W}{L}\frac{\varepsilon_{OX}}{CET}\frac{\left(V_{GS}-V_T-\frac{mV_{DS}}{2}\right)V_{DS}}{1+\frac{\mu_{eff}V_{DS}}{2v_{sat}L}} \tag{4.94}$$

and

$$I_{DS} = \mu_0\frac{W}{L}\frac{\varepsilon_{OX}}{CET}\frac{\left(V_{GS}-V_T-\frac{mV_{DS}}{2}\right)V_{DS}}{1+\theta(V_{GS}-V_T)+\frac{\mu_0 V_{DS}}{2v_{sat}L}}. \tag{4.95}$$

V_{DSAT}, the drain–source saturation voltage, is obtained from the condition that the carrier velocity saturates at the source end of the channel, implying that I_{DSAT} must satisfy the equation

$$I_{DSAT} = -WQ_i v_{sat} = W\frac{\varepsilon_{OX}}{CET}v_{sat}(V_{GS}-V_T-mV_{DSAT}) \tag{4.96}$$

as well as equation (4.95) when $V_{DS} = V_{DSAT}$. Thus, the expression for V_{DSAT} becomes

$$V_{DSAT} = \frac{(V_{GS}-V_T)}{m}\frac{[1+\theta(V_{GS}-V_T)]}{1+\left[\theta+\frac{\mu_0}{2mv_{SAT}L}\right](V_{GS}-V_T)} \tag{4.97}$$

and that of I_{DSAT} follows by setting $V_{DS} = V_{DSAT}$ in (4.95)

$$I_{DSAT} = \mu_0\frac{W}{L}\frac{\varepsilon_{OX}}{CET}\frac{\left(V_{GS}-V_T-\frac{mV_{DSAT}}{2}\right)V_{DSAT}}{1+\theta(V_{GS}-V_T)+\frac{\mu_0 V_{DSAT}}{2v_{sat}L}}. \tag{4.98}$$

In the derivation above, it is assumed that the velocity of the carriers at the drain end reaches v_{sat}. By inserting (4.97) in (4.98), we obtain

$$I_{DSAT} = v_{sat}W\frac{\varepsilon_{OX}}{CET}\left(V_{GS}-V_T-\frac{(V_{GS}-V_T)[1+\theta(V_{GS}-V_T)]}{1+\theta(V_{GS}-V_T)+\frac{\mu_0(V_{GS}-V_T)}{2mv_{sat}L}}\right) \tag{4.99}$$

which reduces to

$$I_{DSAT} = \frac{\mu_0}{2m}\frac{W}{L}\frac{\varepsilon_{OX}}{CET}\frac{(V_{GS}-V_T)^2}{1+\left[\theta+\frac{\mu_0}{2mv_{SAT}L}\right](V_{GS}-V_T)}. \tag{4.100}$$

Finally, by adding the channel length modulation term to (4.100), the expression of the drain current in the saturation region is cast as

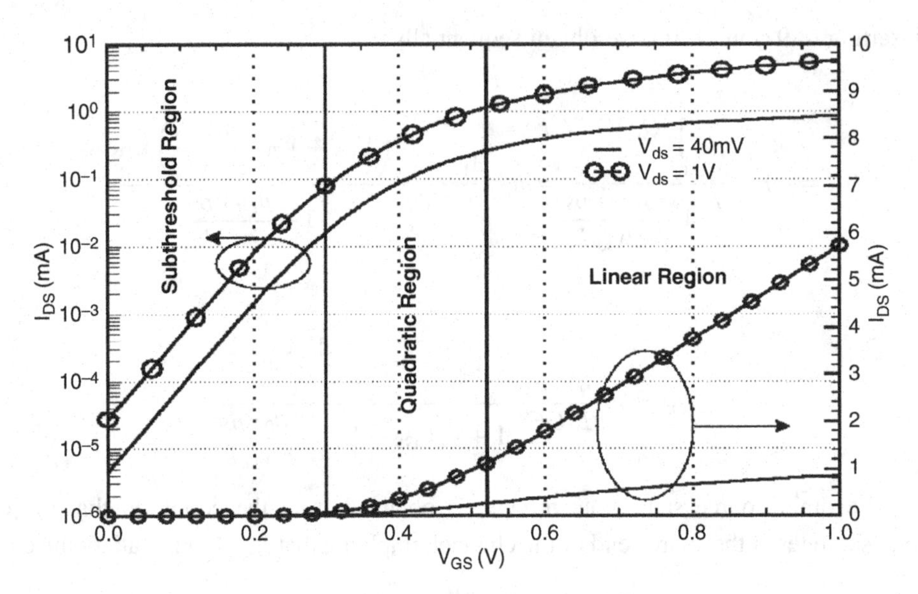

Figure 4.35 Measured transfer and subthreshold characteristics of a 10 × 1μm × 90nm n-MOSFET indicating the narrow voltage range over which the square law is still valid

$$I_{DS} = \frac{\mu_0}{2m}\frac{W}{L}\frac{\varepsilon_{OX}}{CET}\frac{(V_{GS}-V_T)^2}{1+\left[\theta+\frac{\mu_0}{2mv_{SAT}L}\right](V_{GS}-V_T)}(1+\lambda V_{DS}). \tag{4.101}$$

We note that the mobility degradation terms due to vertical and lateral electrical fields are grouped together in the denominator as a single multiplication constant for V_{GS}–V_T. Equation (4.101) reduces to the traditional square law if $m = 1$, $\theta = 0$ and $v_{sat} \rightarrow \infty$.

The validity of the square law for the I-V characteristics of nanoscale MOSFETs

The bias voltage range over which the quadratic law correctly describes the DC characteristics of the transistor shrinks with every new technology generation. For example, as illustrated in Figure 4.35, in 90nm n-channel MOSFETs, the effective gate voltage range is smaller than 250 mV. At the same time, due to the DIBL effect, V_T can vary by at least 50–100 mV when the drain bias voltage is changed between 0 and 1V. A similar variation in threshold voltage occurs as a result of the reverse short-channel effect when the gate length is changed from the minimum feature size to two to three times the minimum feature size, a common practice among analog designers. Obviously, designing circuits with a certain effective gate overdrive voltage, V_{eff}, and V_T in mind is no longer feasible. Adding to the predicament is the impact of ever-increasing process variation on V_T at nanoscale gate lengths and, with the advent of strain engineering [14], the gate length and gate finger-width dependence of carrier mobility. It has become apparent that CMOS RF design techniques must be overhauled to keep pace with nanoscale CMOS phenomena. Fortunately, the change is not that difficult to make. Constant-current-density design techniques, which have long been in practice in the industry for high-speed and RF circuits using bipolar transistors, can also be reliably applied to nano-CMOS

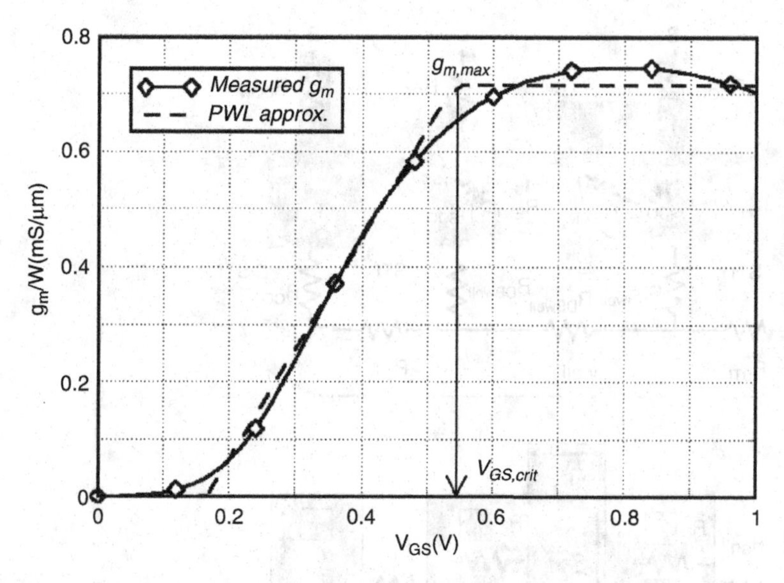

Figure 4.36 Linear piecewise approximation of the transconductance in a 130nm MOSFET biased in saturation

circuits. In this proposed design methodology, once the gate length has been set (typically to the minimum size allowed) the transistors are:

(i) first biased at the desired constant current density per unit gate width I_{DS}/W and
(ii) then the desired current or transconductance is achieved by fixing the gate finger width, W_f, and varying the number of gate fingers, N_f, connected in parallel.

The actual DC V_{GS} value is of secondary importance, but, should one need to calculate it, an estimate can be obtained in the saturation region using a piecewise approximation of (4.101)

$$I_{Dsat} = \frac{\mu_0}{2}\frac{\varepsilon_{ox}}{CET}\frac{W_f}{L}N_f(V_{GS} - V_T)^2(1 + \lambda V_{DS}) \tag{4.102}$$

$$g_m = \mu_0\frac{\varepsilon_{ox}}{CET}\frac{W_f}{L}N_f(V_{GS} - V_T)(1 + \lambda V_{DS}) \text{ if } \frac{I_{Dsat}}{W} < 0.15\frac{\text{mA}}{\mu\text{m}} \tag{4.103}$$

and

$$I_{Dsat} = g_{mp}(L)W_fN_f(V_{GS} - V_T)(1 + \lambda V_{DS}) \tag{4.104}$$

$$g_m = g_{mp}(L)W_fN_f(1 + \lambda V_{DS}) \text{ if } \frac{I_{Dsat}}{W} \geq 0.3\frac{\text{mA}}{\mu\text{m}} \tag{4.105}$$

$$V_{eff}@\frac{0.3\text{mA}}{\mu\text{m}} \approx \frac{0.3\text{mA}}{\mu\text{m}}\frac{1}{g'_{mp}} \tag{4.106}$$

where g'_{mp} is the measured peak transconductance per unit gate width in saturation, as illustrated in Figure 4.36. It can be easily obtained from DC I-V measurements or from device simulation using a compact model.

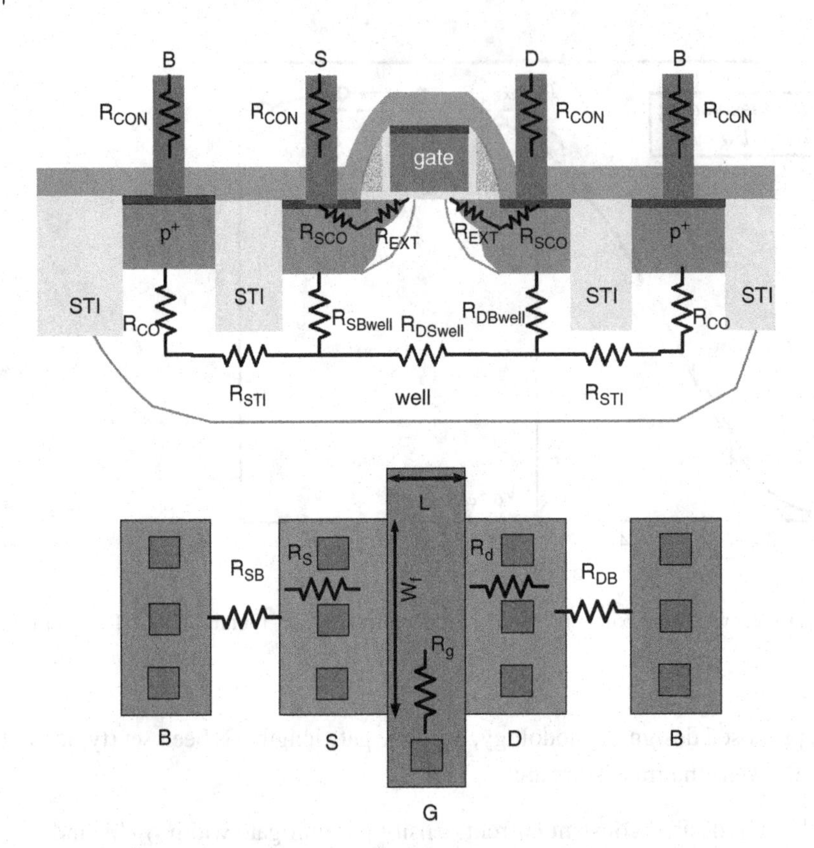

Figure 4.37 MOSFET cross-section and top view of layout showing the various components of the gate, substrate and source/drain resistances

For example, in a 130nm n-MOSFET, g'_{mp} = 0.72mS/μm and the corresponding V_{eff} = 0.375V. In a 90nm n-MOSFET, g'_{mp} = 1mS/μm and the corresponding V_{eff} = 0.25V, while in a 65nm GP n-MOSFET, g'_{mp} = 1.2mS/μm and the corresponding V_{eff} = 0.20V.

4.2.5 Parasitic resistances

Having reviewed the main phenomena responsible for the DC I-V characteristics of the intrinsic nanoscale MOSFET, we can now address the extrinsic parasitics and the small signal behavior. The most important parasitic resistances are illustrated in Figure 4.37.

Source and drain resistance

The source, R_s, and drain, R_d, resistances degrade the I-V characteristics of the transistor. For example, when the transistor is biased fully on, at $V_{GS} = V_{DD}$, typically 10% to 30% of the channel resistance, $R_{sat} = V_{DD} / I_{Dsat}$, is due to R_s and R_d. This contribution increases with every new technology generation. R_s, in particular, is very important because it also degrades the g_m, f_T, f_{MAX}, and F_{MIN}, of the transistor.

Both the source and drain resistance are inversely proportional to the total gate width W

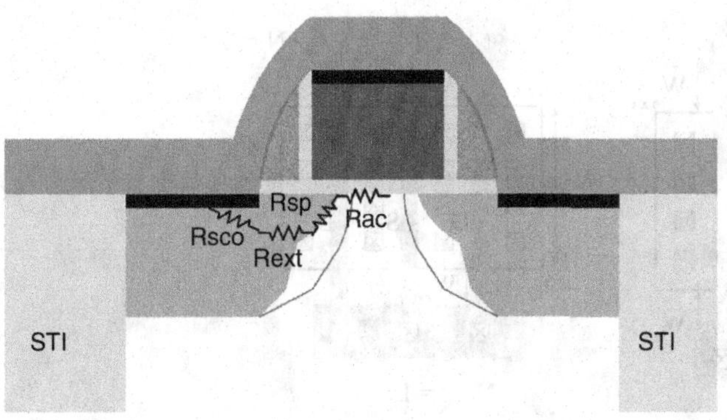

Figure 4.38 Components of the source resistance

$$R_s = R'_s/W \tag{4.107}$$

$$R_d = R'_d/W \tag{4.108}$$

R'_s and R'_d are extracted from DC and/or high-frequency measurements and, in nanoscale technologies, are in the 150–200Ω × μm range.

As shown in Figure 4.38, there are several internal transistor components that contribute to R_s and R_d:

- R_{SCO}, the resistance of the source–drain contact diffusion region,
- R_{ext}, the resistance of the source–drain extension region (also known as tip resistance),
- R_{sp}, the current spreading resistance, and
- R_{ac}, the tip resistance of the carrier accumulation layer.

To reduce the tip and spreading resistance, together forming the "link resistance," the doping profile at the channel end of the SDE region must be very abrupt.

The number of contact vias on the source–drain stripes of the MOSFET is also going to affect the overall source and drain resistances by an additional component, R_{CON}/N_{CON}, where R_{CON} is the resistance of one via contact and N_{CON} is the number of contacts per source–drain diffusion stripe. In a well-designed MOSFET layout, this component of the source–drain resistance should not exceed 10–15% of the total source–drain resistance. This condition becomes increasingly difficult to satisfy in deeply scaled nano-CMOS technologies because R_{CON} typically increases with the square of the technology scaling factor, doubling from one technology node to another.

Gate resistance

Figure 4.39 illustrates several possible gate geometry layouts of single- and multi-gate finger MOSFETs contacted on one side or on both sides of the gate. Since the sheet resistance of the polysilicon gate is non-zero, its contribution to the AC and transient behavior of the transistor must be accounted for. Ignoring the DC gate leakage current, only AC current, coupled from the MOSFET channel through the gate oxide capacitance, flows through this resistance, as shown in Figure 4.39. Because the gate is contacted outside the active region of the transistor, the AC current flows through the gate poly in a direction perpendicular to the source–drain current flow.

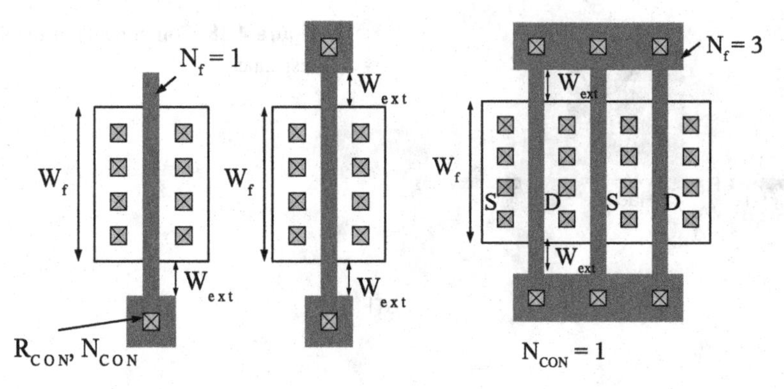

Figure 4.39 Single and multiple-gate-finger layouts

If the gate is contacted on one side only, the distributed nature of the current flow from the gate contact into the channel through the gate oxide leads to an effective gate resistance which is approximately one third of the DC resistance of the gate stripe: $R_{SHG} \times W_f/(3L)$.

If the gate is contacted on both sides, the AC gate current is divided into two equal parts, each flowing into half of the channel through polysilicon gate stripes $W_f/2$ long and L wide. Since the two sides of the gate finger are externally shunted together through metal interconnect (see Figure 4.40(b)), the effective AC resistance of the intrinsic gate finger contacted on both sides becomes $R_{SHG} \times W_f/(12 \times L)$. For a complete treatment of the gate resistance, we must also include the contributions of the gate contact vias and of the gate extension region between the gate contact and the edge of the active region of the transistor. When the gate fingers are contacted on one side, the expression of the gate resistance becomes

$$R_g = \frac{\left[\frac{R_{CON}}{N_{CON}} + \frac{R_{SHG}}{L}\left(W_{ext} + \frac{W_f}{3}\right)\right]}{N_f}. \tag{4.109}$$

while the gate resistance for transistors where the gate fingers are contacted on both sides is described by

$$R_g = \frac{\left[\frac{R_{CON}}{N_{CON}} + \frac{R_{SHG}}{L}\left(W_{ext} + \frac{W_f}{6}\right)\right]}{2N_f}. \tag{4.110}$$

In (4.109) and (4.110):

- $W = N_f \times W_f$ is the total gate width,
- N_f is the number of gate fingers,
- W_f is the gate finger width,
- L is the physical gate length,
- W_{ext} is the distance between the gate contact and the active region,
- R_{SHG} is the sheet resistance per square of the polysilicon or metal gate, typically 10Ω/↔ in 90nm CMOS,
- N_{CON} is the number of metal1-to-poly vias per gate contact, and
- R_{CON} is the metal1-to-poly via resistance, typically 15–20Ω in 90nm CMOS.

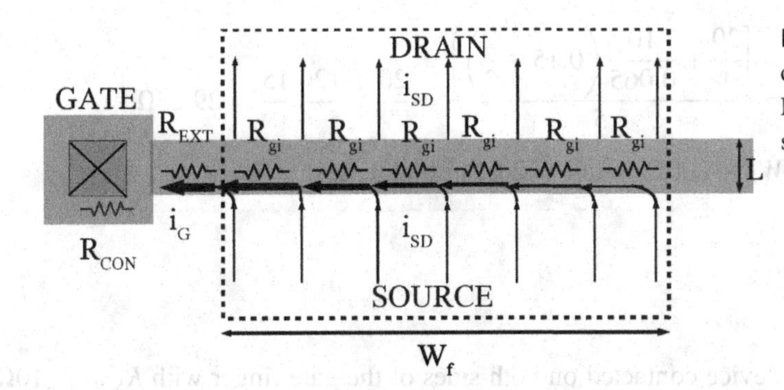

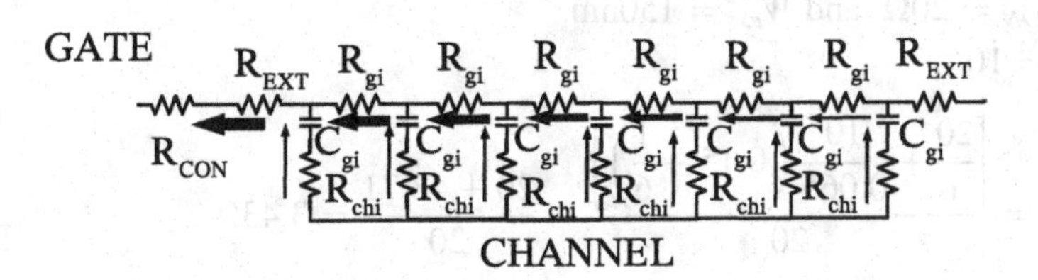

a)

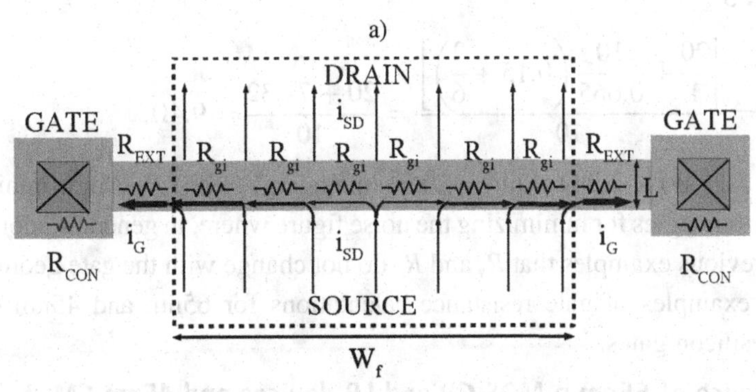

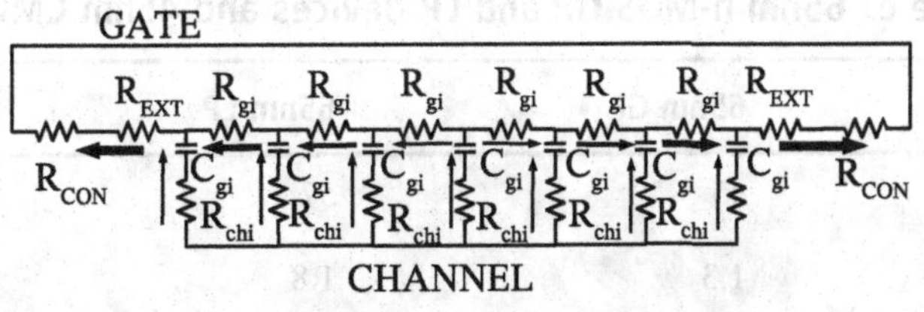

b)

Figure 4.40 Distributed AC current flow in the gate of MOSFETs contacted on (a) one side and (b) both sides of the gate

EXAMPLE 4.5

For a 10μm × 90nm device contacted on one side (physical gate length is 65nm) with $R_{SHG} = 10\Omega$, $L = 65\text{nm}$, $N_{CON} = 1$, $R_{CON} = 20\Omega$, and $W_{ext} = 150\text{nm}$.

(a) If $W_f = 1\mu\text{m}$, $N_f = 10$

$$R_g = \frac{\left[\frac{20}{1} + \frac{10}{0.065}\left(0.15 + \frac{1}{3}\right)\right]}{10} = \frac{20 + 74.3}{10} = 9.4\Omega.$$

(b) If $W_f = 2\mu\text{m}$, $N_f = 5$

$$R_g = \frac{\left[\frac{20}{1} + \frac{10}{0.065}\left(0.15 + \frac{2}{3}\right)\right]}{5} = \frac{20 + 126.15}{5} = 29.23\Omega$$

while $R_s = R_d = (1/W) \times 200\Omega \times \mu m = 20\Omega$ in both cases.

EXAMPLE 4.6

For a 10μm × 90nm device contacted on both sides of the gate finger with $R_{SHG} = 10\Omega$, $L =$ 65nm, $N_{CON} = 1$, $R_{CON} = 20\Omega$, and $W_{ext} = 150$nm.

(a) If $W_f = 1\mu m$, $N_f = 10$

$$R_g = \frac{\left[\frac{20}{1} + \frac{10}{0.065}\left(0.15 + \frac{1}{6}\right)\right]}{20} = \frac{20 + 48.71}{20} = 3.43\Omega.$$

(b) If $W_f = 2\mu m$; $N_f = 5$

$$R_g = \frac{\left[\frac{20}{1} + \frac{10}{0.065}\left(0.15 + \frac{2}{6}\right)\right]}{10} = \frac{20 + 74.32}{10} = 9.4\Omega.$$

In both cases, $R_S = R_d = (1/W) \times 200\Omega \times \mu m = 20\Omega$ remain large, while R_g can be minimized. This has implications in strategies for minimizing the noise figure where, in general R_s dominates R_g. We see from the previous examples that R_s and R_d do not change with the gate geometry.

Table 4.5 provides examples of gate resistance calculations for 65nm and 45nm CMOS technologies with polysilicon gates.

Table 4.5 **Gate resistance of 65nm n-MOS GP and LP devices and 45nm CMOS.**

Parameter	65nm GP	65nm LP	45nm
Physical L (nm)	45	57	35
EOT (nm)	1.3	1.8	1.3
W_f (μm)	1	1	0.7
N_{CON}	1	1	1
Contact on both sides	No	No	No
R_{CON} (Ω)	40	40	60
R_{SHG} (Ω/□)	15	15	20
W_{ext} (nm)	120	120	100
N_f	1	1	1
R_G (Ω)	191	159	250.5

In technologies where metal gates are used, the gate sheet resistance and the contact via resistance should be significantly lower and result in different optimal gate geometries than for transistors with polysilicon gates.

Substrate resistance

The substrate resistance is important in high-frequency and high-speed applications because it degrades f_{MAX} and F_{MIN}. It can be calculated based on the geometry of the transistor layout and on the doping and sheet resistance of the well in which the transistor is formed. The various components of the susbtrate resistance for a bulk MOSFET layout are illustrated in Figure 4.41.

4.2.6 Gate and junction parasitic capacitance

Figure 4.42 compiles all the capacitive components associated with the MOSFET gate and with the source–bulk and drain–bulk junctions.

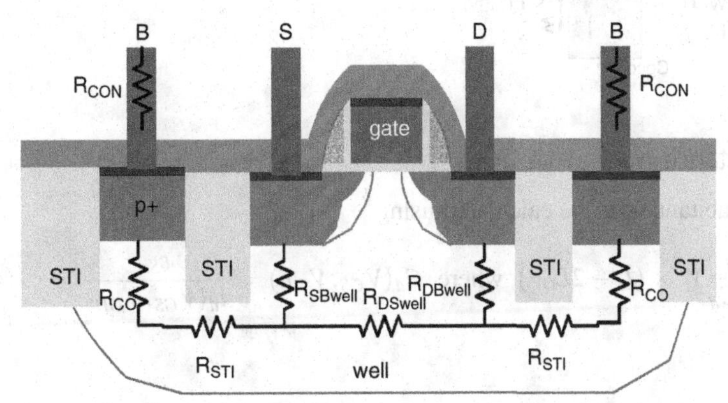

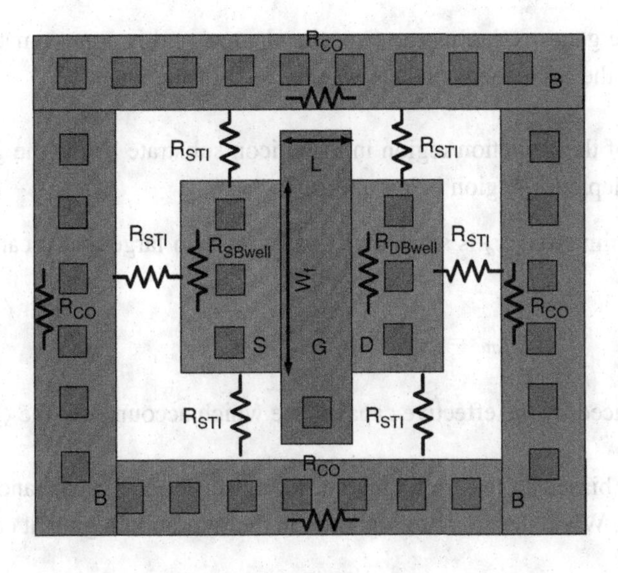

Figure 4.41 Illustration of the various components that make up the source–bulk and drain–bulk

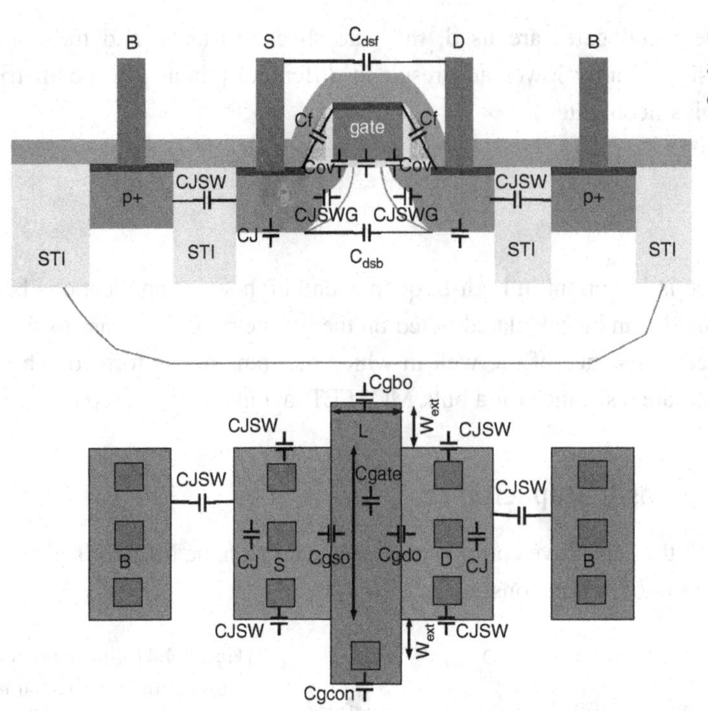

Figure 4.42 Illustration of the gate and junction capacitance components

Intrinsic gate capacitance

The intrinsic gate capacitance can be calculated using

$$C_{gate} = \frac{C_{ox}C_d}{C_{ox}+C_d} W \times (L - 2L_{OV}) \text{ where } C_d(V_{GS}, V_{GD}) = \frac{\varepsilon_{Si}}{x_d(V_{GS}, V_{GD})} \tag{4.111}$$

where:

- L is physical gate length,
- L_{OV} is the length of the gate overlap region over the channel and is approximately equal to the lateral diffusion of the S/D extensions below each side of the gate,
- W is the gate width,
- C_d is the capacitance of the depletion region in the silicon substrate below the gate, and
- x_d is the width of the depletion region below the gate.

If the channel is strongly inverted, x_d is small and C_d becomes so large that it can be ignored such that

$$C_{gate} = \frac{\varepsilon_{OX}}{CET} W(L - 2L_{OV}) \tag{4.112}$$

where C_{OX} has been replaced by the effective capacitance which accounts for the QC, and gate poly depletion effects.

When the transistor is biased in the triode region, the intrinsic gate capacitance is equally split between C_{gs} and C_{gd}. When the transistor operates in the saturation region, it can be shown that the intrinsic gate–source capacitance is approximately two thirds of the gate capacitance, while the intrinsic gate–drain capacitance is approximately 0.

Gate–drain overlap, gate–source overlap, and gate–bulk overlap capacitances

The gate–source overlap capacitance, C_{GSO}, is located on the source side of the gate along the gate finger width. Similarly, the gate–drain overlap capacitance, C_{GDO}, extends on the drain side of the gate, along the gate finger width. C_{GSO} and C_{GDO} are only weakly dependent on the gate bias voltage

$$C_{GSO} = C_{GDO} = C_{ox} \times L_{OV} + C_f \text{ per unit gate width (F/m)} \quad (4.113)$$

where

$$C_f = \frac{2\varepsilon_{SION}}{\pi}\left[\ln\left(1 + \frac{t_{poly}}{t_{ox}}\right) + \ln\frac{\pi}{2} + 0.308\right] \text{in F/m} \quad (4.114)$$

is the fringing capacitance and t_{poly} is the thickness of the polysilicon (or metal) gate.

The gate–bulk overlap capacitance is described by

$$C_{GB0} = 2\frac{\varepsilon_{ox}}{t_{STI}}\Delta W + C_{fGB0} \text{ per unit gate length (F/m)} \quad (4.115)$$

where $\Delta W = W_{EXT}$ is the length of the polysilicon gate overlap of the STI region, C_{fGBO} is the fringing component

$$C_{fGBO} = \frac{2\varepsilon_{ox}}{\pi}\left[\ln\left(1 + \frac{t_{poly}}{t_{STI}}\right) + \ln\frac{\pi}{2} + 0.308\right] \text{in F/m} \quad (4.116)$$

and t_{STI} is the thickness of the STI region, typically 200–400 nm.

Taking the overlap and fringing capacitances into consideration, the total gate–source and gate–drain capacitances become:

(a) in the triode region

$$C_{GS} = \frac{C_{OX}}{2}(L - 2L_{OV})W + C_{GSO}W + N_f C_{GBO}L \text{ and } C_{gd} = \frac{C_{OX}}{2} W \times (L - 2L_{OV}) + C_{GDO}W \quad (4.117)$$

(b) in the saturation region

$$C_{GS} = \frac{2}{3}C_{OX}(L - 2L_{OV})W + C_{GSO}W + N_f C_{GBO}L \text{ and } C_{gd} = C_{GDO}W. \quad (4.118)$$

EXAMPLE 4.7

If $L = 65\text{nm}$, $L_{OV} = 10\text{nm}$, $EOT = 16\text{nm}$, $t_{poly} = 130\text{nm}$, $t_{STI} = 300\text{nm}$, $\Delta W = 120\text{nm}$, and $N_f = 1$, then

$C_f = 0.223\ \text{fF/µm}$, $C_{OV} = \varepsilon_{OX} L_{OV}/EOT = 0.022\ \text{fF/µm}$, $C_{fGBO} = 0.025\ \text{fF/µm}$, $C_{GBO} = 0.06\ \text{fF/µm}$.

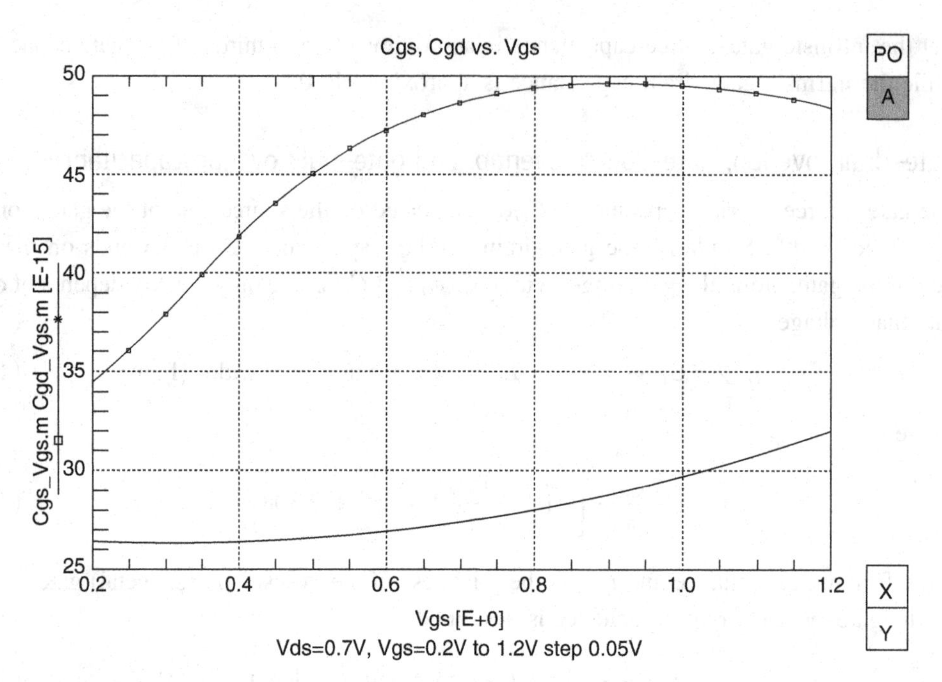

Figure 4.43 Measured $C_{gs}(V_{GS})$ and $C_{gd}(V_{GS})$ characteristics for a 80×1µm, 65nm GP n-MOSFET at $V_{DS} = 0.7$ V

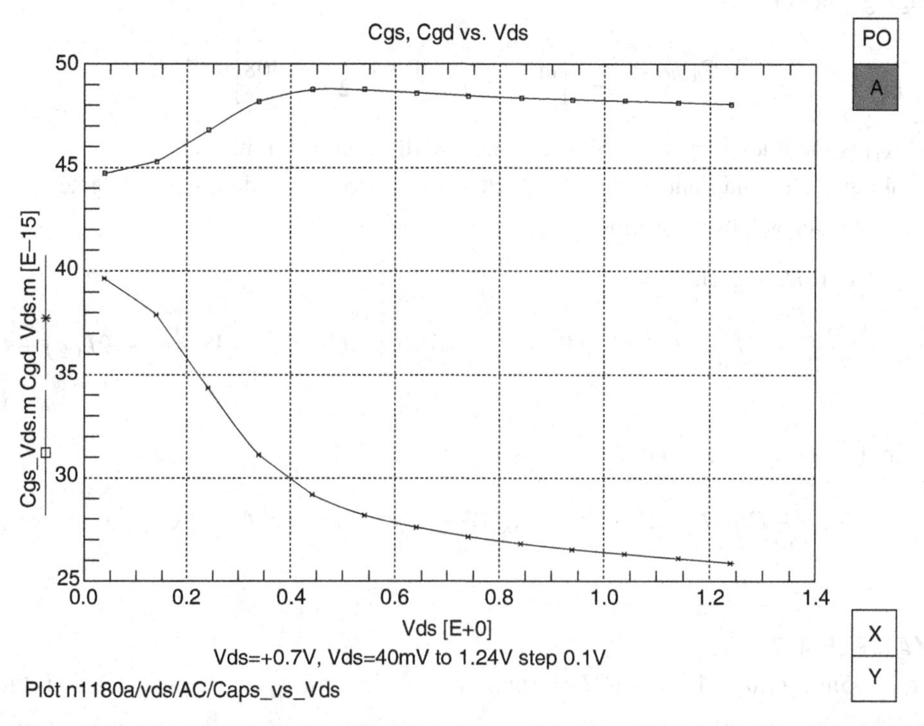

Figure 4.44 Measured $C_{gs}(V_{DS})$ and $C_{gd}(V_{DS})$ characteristics for a 80 × 1µm, 65nm GP n-MOSFET at $V_{GS} = 0.7$ V

Note that the fringing capacitance is typically larger than the value calculated based on the permittivity of SiO_2 because the spacer is made of a combination of oxide and nitride which has a higher effective dielectric constant than pure silicon dioxide.

Figures 4.43 and 4.44 show the measured V_{GS} and V_{DS} dependence of C_{gs} and C_{gd} for a 65nm GP n-MOSFET with 80 gate fingers, each 1µm wide, illustrating that, for most of the saturation region, the two capacitances remain largely constant.

Drain–bulk and source–bulk parasitic capacitances

Like the gate and substrate resistance, the drain–bulk and source–bulk capacitances are layout-dependent. Different layouts are needed in different transistor configurations (CS, CD) in order to minimize circuit capacitance. There are three components that make up the source–bulk and drain–bulk capacitances and they can be calculated separately:

- an area component, C_j, when the junction is zero-biased and which depends on the doping in the retrograde well (or buried oxide thickness in SOI) and which scales with the gate finger width and the contact width, l_S,
- two sidewall components,
- C_{JSWG} on the channel side of the source–drain which depends on the halo and channel well doping and scales with the gate finger width and the number of gate fingers, and
- C_{JSW} on the STI side of the source–drain (on two or three sides of the source–drain stripe and depends on the layout),

where:

- l_S is the width of the source contact region,
- l_D is the width of the drain contact region,
- N_S is the number of source contact stripes, and
- N_D is the number of drain contact stripes.

As shown in Figure 4.45, if the layout has an odd number of gate fingers, then

$$C_{S(D)B} = \frac{C_j l_{S(D)} W_f (N_f + 1)}{2\left(1 - \frac{V_{BS(D)}}{\phi_{bi}}\right)^{mA}} + \frac{C_{JSWG} W_f N_f}{\left(1 - \frac{V_{BS(D)}}{\phi_{bi}}\right)^{mP}} + C_{JSW}\left[W_f + l_{S(D)}(N_f + 1)\right] \quad (4.119)$$

If the layout has an even number of gate fingers, then, depending on which of N_S or N_D is larger

$$C_{S(D)B} = \frac{C_j l_{S(D)} W_f N_{S(D)}}{\left(1 - \frac{V_{BS(D)}}{\phi_{bi}}\right)^{mA}} + \frac{C_{JSWG} W_f N_f}{\left(1 - \frac{V_{BS(D)}}{\phi_{bi}}\right)^{mP}} + C_{JSW}\left(2W_f + 2l_{S(D)} N_{S(D)}\right) \quad (4.120)$$

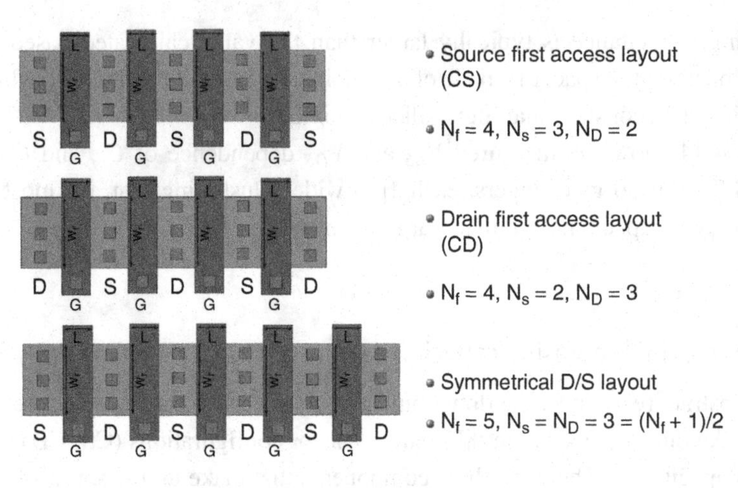

Figure 4.45 Layouts with different source–bulk and drain–bulk capacitances

or

$$C_{S(D)B} = \frac{C_j l_{S(D)} W_f N_{S(D)}}{\left(1 - \frac{V_{BS(D)}}{\phi_{bi}}\right)^{mA}} + \frac{C_{JSWG} W_f N_f}{\left(1 - \frac{V_{BS(D)}}{\phi_{bi}}\right)^{mP}} + 2C_{JSW} l_{S(D)} N_{S(D)}. \tag{4.121}$$

4.2.7 The small signal equivalent circuit

The complete small signal circuit, including substrate parasitics, series gate, drain, and source resistances, and non-quasi-static (NQS) effects represented by the channel resistance, R_i, and by the transconductance delay, τ, is shown in Figure 4.46. The most important of the small signal equivalent circuit parameters are g_m, g_o, C_{gs}, C_{gd}, and C_{db} while R_i and τ become relevant at frequencies above $f_T/2$.

The negative feedback through the source series resistance can be exploited to simplify the equivalent circuit to that illustrated in Figure 4.46(b) where C_{gseff}, g_{meff}, and g_{oeff} now include the impact of R_s. The expression of the f_T of the MOSFET in the common-source configuration can be derived by calculating the short-circuit current at the drain terminal, as shown in Figure 4.47. The substrate network can be ignored in this calculation because it is short circuited by R_s and R_d, respectively, both much smaller than the substrate resistance. Since the impedance of C_{ds} is also much larger than R_d, even at frequencies close to f_T, it, too, can be neglected. Using the Y' matrix of the intrinsic MOSFET (which does not include the parasitic source, gate and drain resistances), we obtain [11]

$$H_{21} = \frac{I_O}{I_{in}} = \frac{Y'_{21}}{Y'_{11}}\left[1 - \frac{\Delta Y'}{Y'_{11}}(R_s + R_d) - \frac{\Delta Y'}{Y'_{21}} R_s\right] \tag{4.122}$$

where

$$\Delta Y' = Y'_{11}Y'_{22} - Y'_{12}Y'_{21}. \tag{4.123}$$

Finally, from (4.122), by setting $H_{21} = 1$ at $f = f_T$, and after several algebraic manipulations, the f_T formula, including the impact of the resistive parasitics, can be cast as [16]

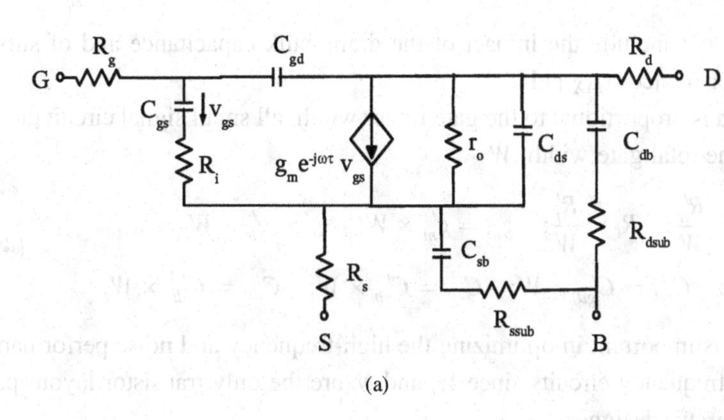

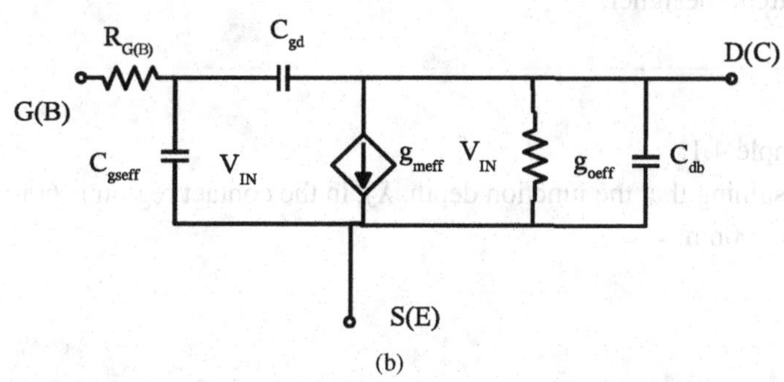

Figure 4.46 (a) Small signal circuit model for Si MOSFETs (b) Simplified equivalent circuit in the common-source configuration which includes the impact of R_s

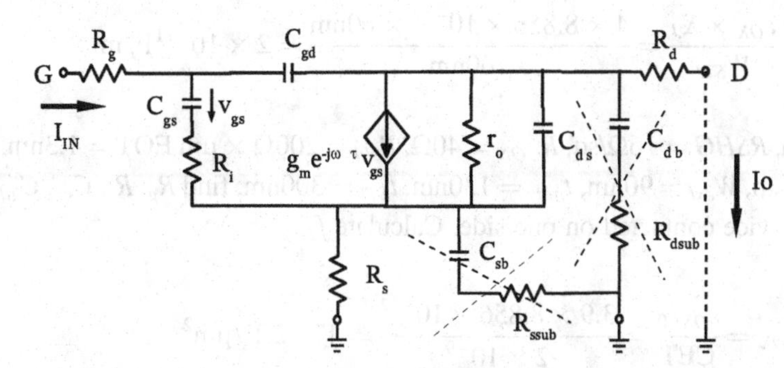

Figure 4.47 Small signal equivalent circuit of a common-source MOSFET employed to derive the f_T expression

$$\frac{1}{2\pi f_T} = \frac{(C_{gs} + C_{gd})}{g_m} + (R_s + R_d)C_{gd} + (C_{gs} + C_{gd})(R_s + R_d)\frac{g_o}{g_m}. \tag{4.124}$$

In the 90nm node, the first term in the right-hand side of (4.124) accounts for approximately 80% of the cutoff frequency of an n-channel MOSFET while the second term contributes about 15% [7]. However, the contributions of the second and third terms increase in more advanced technology nodes.

Similarly, but a much lengthier derivation, it can be shown that the f_{MAX} of a MOSFET can be approximated by [17]

$$f_{MAX} \approx \frac{f_T}{2\sqrt{(R_i + R_s + R_g)g_o + 2\pi f_T R_g C_{gd}}} \tag{4.125}$$

Equation (4.125) does not include the impact of the drain–bulk capacitance and of substrate resistance which further reduce f_{MAX} [11].

Except for R_g, which is proportional to the gate finger width, all small signal circuit elements scale with respect to the total gate width, W

$$R_s = \frac{R_s'}{W};\quad R_d = \frac{R_d'}{W};\quad R_i = \frac{R_i'}{W};\quad g_m = g_m' \times W;\quad g_o = g_o' \times W$$
$$C_{gs} = C_{gs}' \times W;\quad C_{gd} = C_{gd}' \times W;\quad C_{sb} = C_{sb}' \times W;\quad C_{db} = C_{db}' \times W. \tag{4.126}$$

The dependence on W is important in optimizing the high-frequency and noise performance of Si MOSFETs in high-frequency circuits since W_f and N_f are the only transistor layout parameters available to the circuit designer.

EXAMPLE 4.8

For the device in Example 4.1:

(a) Calculate $CJSW$ assuming that the junction depth, X_J, in the contact region is 60nm and the STI width, W_{STI}, is 100nm.

Solution

$$CJSW = \frac{\varepsilon_{OX} \times X_J}{W_{STI}} = \frac{4 \times 8.856 \times 10^{-12} \times 60\text{nm}}{100\text{nm}} = 2 \times 10^{-11}\text{F/m}.$$

(b) If $g_m = 1.5\text{mS}/\mu\text{m}$, $RSHG = 16\Omega/\text{sq}$, $R_{CON} = 40\Omega$, $R_{s/d} = 200\Omega \times \mu\text{m}$ EOT = 1.3nm, CET = 2nm, $L_{OV} = 10\text{nm}$, $W_{ext} = 90\text{nm}$, $t_{poly} = 130\text{nm}$, $t_{STI} = 300\text{nm}$, find R_g, R_s, C_{gs}, C_{gd} for a 1µm gate finger device contacted on one side. Calculate f_T.

Solution

$$C_{OX} = \frac{\varepsilon_{OX}\varepsilon_0}{\text{CET}} = \frac{3.9 \times 8.856 \times 10^{-12}}{2 \times 10^{-9}} = 17.22\text{fF}/\mu\text{m}^2.$$

$C_{gs} = 0.66 \times C_{ox} \times (L - 2 \times L_{OV}) + C_{ov} + C_{GBO} \times L = 0.66 \times 17.22 \times 25 \times E - 5\ \text{fF}/\mu\text{m} + 0.172\ \text{fF}/\mu\text{m} + 0.121\ \text{fF}/\mu\text{m} + 0.001\ \text{fF}/\mu\text{m} = (0.28 + 0.293 + 0.01)\ \text{fF}/\mu\text{m} = 0.574\text{fF}\ \mu\text{m}$.
$g_{meff} = g_m/(1 + 0.3) = 1.15\text{mS}/\mu\text{m}$

$$\frac{1}{2\pi f_T} = \frac{C_{gs} + C_{gd}}{g_m} + (R_s + R_d)C_{gd} = \frac{0.867\text{fF}}{1.5\text{mS}} + 400 \times 0.293\text{fF} = 738fs;\ f_T = 215.8\text{GHz}$$

4.2.8 The bias dependence of the f_T and f_{MAX} characteristic current densities

The measured f_T of a 90nm n-MOSFET is plotted versus drain current density in Figure 4.48 at different V_{DS} values. Except at very low drain–source voltages when the device operates close to, or in the triode region, the peak f_T current density, J_{pfT}, is only

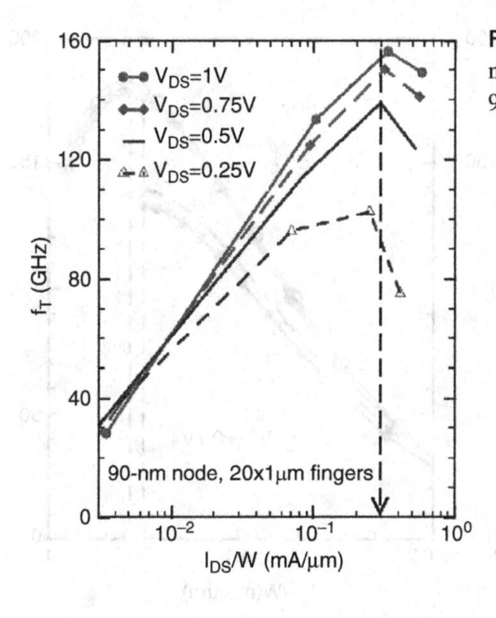

Figure 4.48 Measured f_T as a function of drain current per micron of gate width for different drain-source voltages in a 90nm n-MOSFET [7]

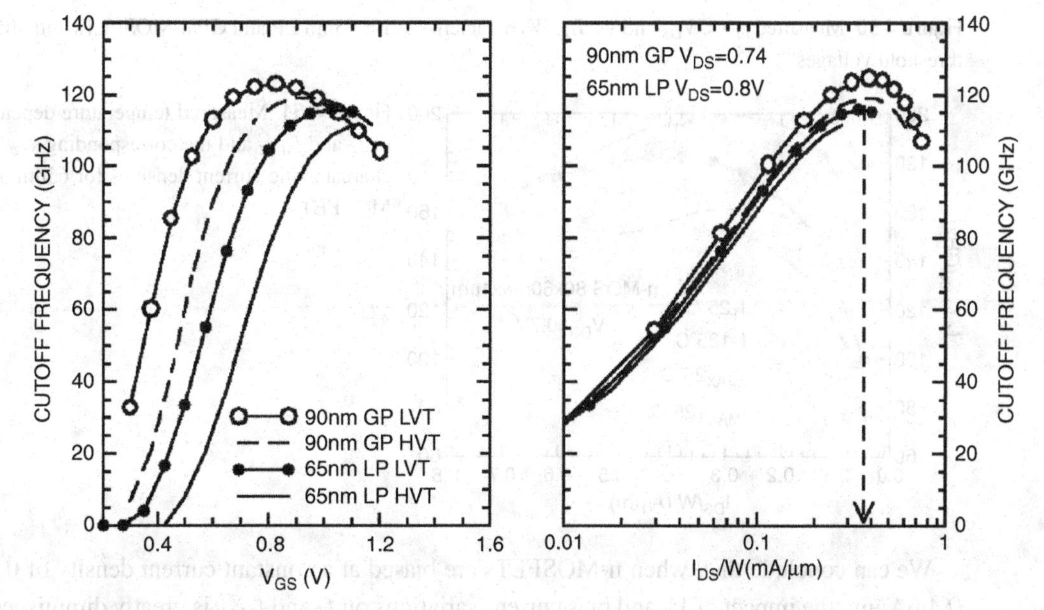

Figure 4.49 Measured f_T vs. V_{GS} and vs. I_{DS}/W characteristics for 90nm GP and 65nm LP n-MOSFETs with different threshold voltages [18]

a weak function of V_{DS} and, in the saturation region, it remains practically constant at 0.3–0.4 mA/μm [7].

Figures 4.49 and 4.50 indicate that, even in devices with different threshold voltages (by more than 0.35V), J_{pfT} remains practically constant while the V_{GS} bias corresponding to the peak f_T value varies widely. The same is true of J_{pfMAX}, except that, as shown in Figure 4.51, its value is always smaller than J_{pfT} at approximately 0.25–0.3 mA/μm in an n-MOSFET.

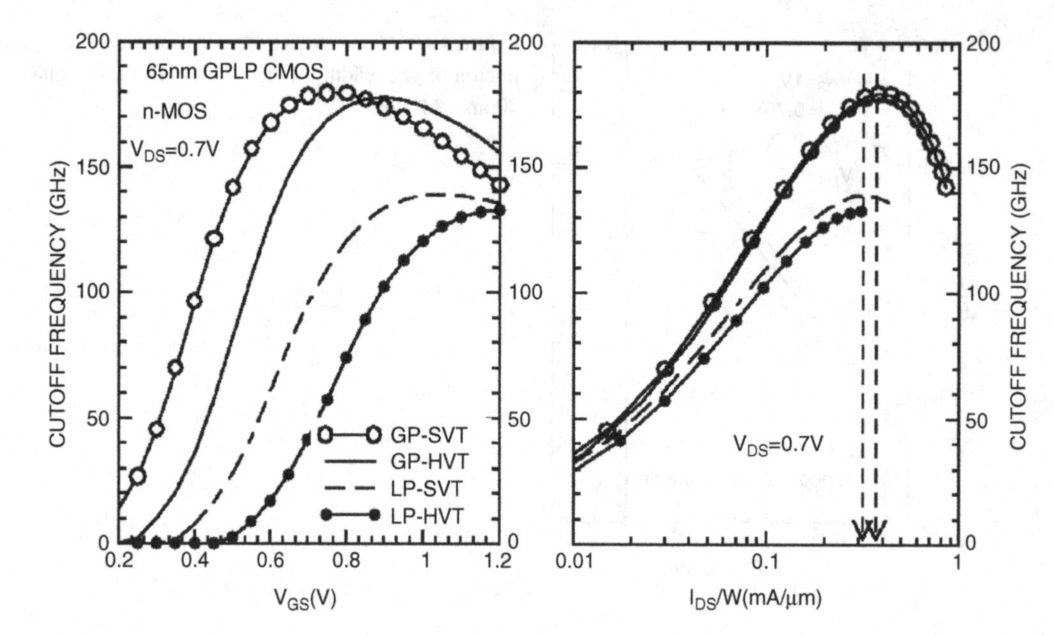

Figure 4.50 Measured f_T vs. V_{GS} and vs. I_{DS}/W characteristics for 65nm LP and GP n-MOSFETs with different threshold voltages

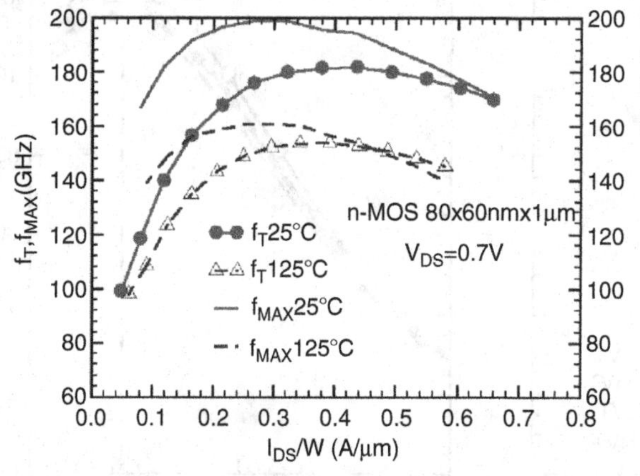

Figure 4.51 Measured temperature dependence of f_T and f_{MAX} and the corresponding charcateristic current densities for 65nm n-MOSFETs

We can conclude that, when n-MOSFETs are biased at a constant current density of 0.15 to 0.4mA/μm, the impact of V_T and bias current variations on f_T and f_{MAX} is greatly diminished. By comparison, biasing at constant V_{GS}, at low effective gate voltages, or, even worse, in the subthreshold region, leads to significant variability in f_T and f_{MAX}. Furthermore, in a given technology node, the characteristic current densities remain invariant with gate length [7] and, as shown in Figure 4.51, do not change with temperature.

It becomes apparent from Figure 4.51 that, at any temperature and drain–source voltage in the saturation region, f_T and f_{MAX} remain within 10% of their peak values for drain current densities varying between 0.2mA/μm and 0.5mA/μm.

Finally, Figures 4.52 and 4.53 reproduce the measured dependence of f_T, f_{MAX}, and g_m on the drain–source voltage for 65nm n-MOSFETs from two different foundries and for general

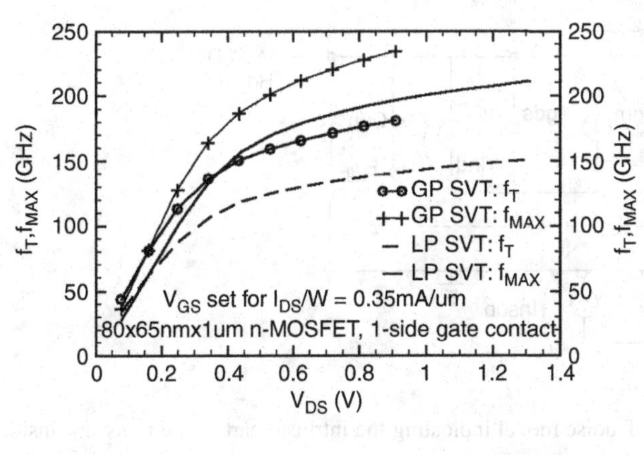

Figure 4.52 Measured V_{DS} dependence of f_T and f_{MAX} for 65nm LP and GP n-MOSFETs

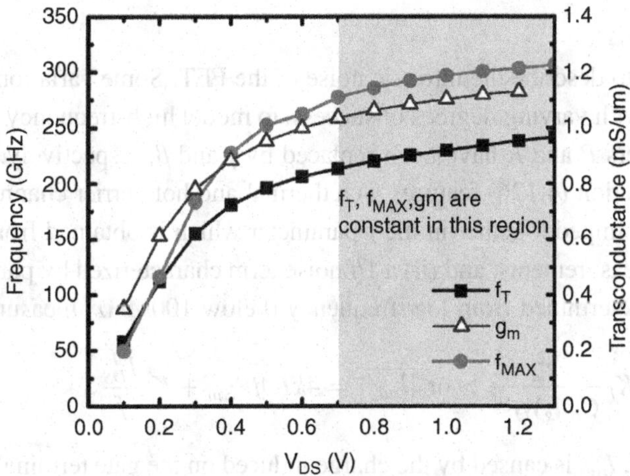

Figure 4.53 Measured V_{DS} dependence of g_m, f_T, and f_{MAX} for 65nm n-MOSFETs, second foundry [19]

purpose and low-power flavors of a 65nm CMOS process. In both cases, the V_{DS} dependence in saturation follows the channel length modulation law $(1 + \lambda V_{DS})$. The saturation region, where f_T, f_{MAX}, and g_m reach their peak values and change by less than 20%, is highlighted in Figure 4.53. In HF circuits, MOSFETs should be biased in this region because their perfomance becomes immune to I_{DS}, V_{DS}, and V_T variations. We note that, because of their smaller physical gate length, GP MOSFETs exhibit higher f_T and f_{MAX} than their LP counterparts.

4.2.9 Noise parameters

The first high-frequency noise model for field-effect transistors was developed by Van der Ziel [20]. This model was later adapted and enhanced by Pucel, Haus, and Statz to make it suitable for MESFETs and HEMTs operating at microwave frequencies [21]. The model is considered quasi-physical because it is derived based on the physical analysis of the transistor but its noise parameters are determined empirically. Because the noise sources are explicit functions of the bias currents, it is also suitable for non-linear noise analysis. As illustrated in Figure 4.54, it employs two noise current sources, one for the drain, I_{nd}, and one for the gate, I_{ng}, and three

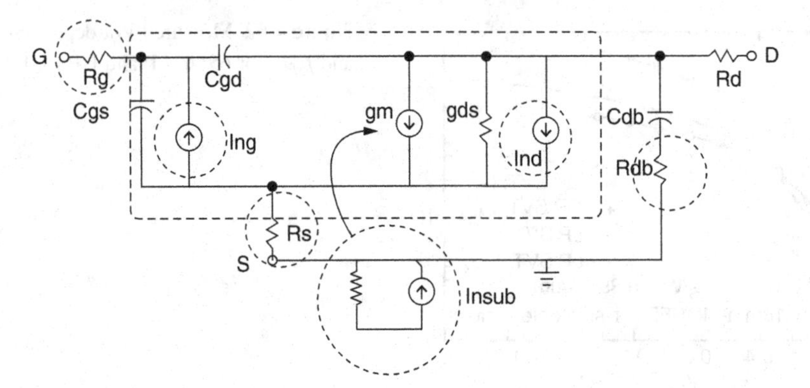

Figure 4.54 Pucel and Enz MOSFET noise model indicating the intrinsic part of the transistor inside the dashed box

noise parameters P, R, and C, to describe the intrinsic noise of the FET. Some variations of the Pucel model have been used, with varying degrees of success, to model high-frequency noise in Si MOSFETs, where parameters P and R have been replaced by γ and β, respectively.

The drain noise current equation (4.126) features (i) a thermal and hot-carrier channel noise term whose contributions are lumped together in the γ parameter which is obtained from direct high-frequency noise figure measurements, and (ii) a $1/f$ noise term characterized by parameters K_f and af, both of which are determined from low-frequency (below 100MHz) measurements

$$\overline{|I_{nd}{}^2|} = 4kT\Delta f\gamma g_m + K_f \frac{g_m^2}{C_{ox} l_g W f^{af}} > \text{or } \overline{|I_{nd}{}^2|} = 4kT\Delta f P g_m + K_f \frac{I_{DS}^{af}}{f}. \tag{4.127}$$

The induced gate noise current, I_{ng}, is caused by the charge induced on the gate terminal at high frequency by the randomly fluctuating carriers in the channel. Since its origins are in the channel noise current, it is correlated to the drain noise current souce. The coupling of the channel noise to the gate occurs through the gate–channel capacitance. As a result, the induced gate noise current power spectral density increases quadratically with frequency. Its magnitude is matched to measurements using the fitting parameter R (or β)

$$\overline{|I_{ng}{}^2|} = 4kT\Delta f\beta \frac{(\omega C_{gs})^2}{g_m} = 4kT\Delta f\beta g_m \frac{f^2}{f_T^2} \text{ or}$$
$$\overline{|I_{ng}{}^2|} = 4kT\Delta f R g_m \frac{f^2}{f_T^2}. \tag{4.128}$$

The correlation between the gate and drain noise current sources is assumed to be purely imaginary, described by the third noise parameter, C

$$jC = \frac{\overline{I_{ng} I_{nd}^x}}{\sqrt{\overline{|I_{nd}{}^2|}\,\overline{|I_{ng}{}^2|}}}. \tag{4.129}$$

Note that, in Figure 4.54, the orientation of I_{ng} is reversed from the standard definition adopted in this book such that the sign of C, defined in (4.129), remains positive.

The Pucel model is augmented with thermal noise sources associated with the parasitic resistances of the transistor: R_g, R_s, R_d, R_{ssub}, and R_{dsub} of which only the first two are actually important. In (4.127)–(4.129), T represents the physical temperature of the transistor. The bias and layout geometry dependence of the noise current sources are captured primarily through g_m, C_{gs}, and f_T. The three noise parameters P, R, and C (or γ, β, and C) have been shown to exhibit some bias and temperature dependence. Although estimates of their values can be derived analytically from transistor physics, they are most often obtained by fitting experimental noise data for specific bias conditions and temperatures. In fact, the lack of validity over a wide range of temperatures is the main limitation of this model. Nevertheless, it has been successfully used over the last four decades to model and design low-noise amplifiers in III-V MESFET, HEMT as well as Si CMOS technologies.

The Pospieszalski noise model

The Pospieszalski model [22] has become popular because of its simplicity and its scalability with temperature and frequency. In this model, originally developed for HEMTs, the noise of the intrinsic MOSFET is described by two parameters: T_g the physical temperature of the intrinsic channel resistance R_i, and T_d the elevated temperature of the output conductance g_o. Except in the triode region, T_d is always higher than the physical device temperature because of the high lateral electric field at the drain end of the FET channel. In contrast, T_g is almost always equal to the physical temperature of the MOSFET because it is associated with the source-end of the channel, where the lateral electric field is low.

The corresponding noise equivalent circuit is shown in Figure 4.55 and consists of two intrinsic noise sources: a voltage noise source

$$\overline{V_{ngs}^2} = 4kT_g \Delta f R_i \tag{4.130}$$

in series with R_i, and a noise current source in parallel with r_o

$$\overline{I_{nd}^2} = 4kT_d \Delta f g_o. \tag{4.131}$$

A shot noise current source

$$\overline{I_{gs}^2} = 2q\Delta f I_g, \tag{4.132}$$

where I_g is the DC gate current, is added to the gate to capture the impact of the gate leakage current in nanoscale MOSFETs. Unlike in the Pucel model, the intrinsic noise sources I_{gs}, I_{nd}, and V_{ngs} are not correlated to each other because they have different physical origins. Finally, thermal noise voltage sources are added in series with each parasitic resistance.

Starting from the equivalent circuit in Figure 4.55 and ignoring the noise of the parasitic resistances and the gate leakage, we can derive the expressions of the four IEEE noise parameters T_{MIN} (F_{MIN}), g_n, R_{SOPT}, and X_{SOPT} in the noise impedance formalism as a function of T_g, T_d, the reference temperature T_o, and the small signal parmeters g_m, g_o, C_{gs}, R_i, and f_T

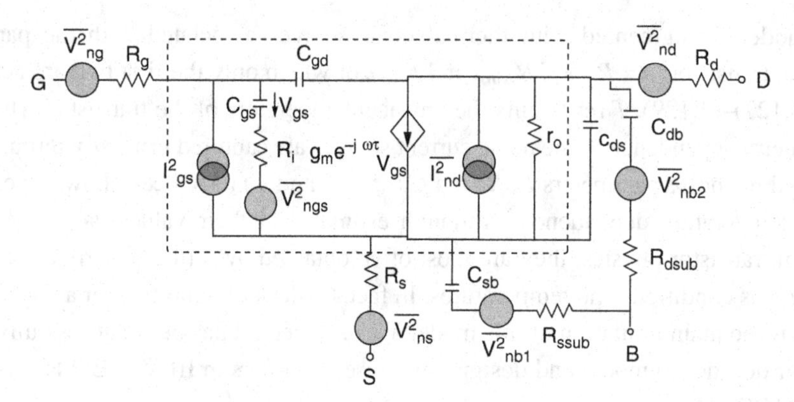

Figure 4.55 The Pospieszalski MOSFET noise model showing the intrinsic MOSFET in the dashed box

$$T_{MIN} = 2\frac{f}{f_T}\sqrt{g_o R_i T_g T_d + \left(\frac{f}{f_T}\right)^2 R_i^2 g_o^2 T_d^2 + 2\left(\frac{f}{f_T}\right)^2 R_i g_o T_d} \tag{4.133}$$

$$R_{SOPT} = \sqrt{\left(\frac{f_T}{f}\right)^2 \frac{R_i}{g_o}\frac{T_g}{T_d} + R_i^2} \tag{4.134}$$

$$X_{SOPT} \approx \frac{1}{\omega C_{gs}} \tag{4.135}$$

$$g_n = \left(\frac{f}{f_T}\right)^2 g_o \frac{T_d}{T_0} \tag{4.136}$$

where $T_0 = 290$ K, $F_{MIN} = 1 + T_{MIN}/T_0$, and $f_T = g_m/(2\pi\, C_{gs})$.

Equations (4.133)–(4.136) can be modified to account for the gate leakage curent [22]

$$T_{MIN}^L = T_{MIN} + 2g_n R_{SOPT} T_0 \left[\sqrt{1 + A\left(\frac{f_T}{f}\left|\frac{Z_{SOPT}}{R_{SOPT}}\right|\right)^2} - 1\right] \tag{4.137}$$

$$R_{SOPT}^L = \frac{R_{SOPT}}{1 + A\left(\frac{f_T}{f}\right)^2}\sqrt{1 + \left(\frac{f_T}{f}\left|\frac{Z_{SOPT}}{R_{SOPT}}\right|\right)^2 A} \tag{4.138}$$

$$X_{SOPT}^L = \frac{X_{SOPT}}{1 + A\left(\frac{f_T}{f}\right)^2} \tag{4.139}$$

$$g_n^L = g_n\left[1 + A\left(\frac{f_T}{f}\right)^2\right] \tag{4.140}$$

where $A = \frac{\overline{i_{ngs}^2}}{\overline{i_{nd}^2}}$, $Z_{SOPT} = R_{SOPT} + jX_{SOPT}$, and superscript L indicates the noise parameters with gate leakage. If $A = 0$, (4.137)–(4.140) reduce to (4.133)–(4.136).

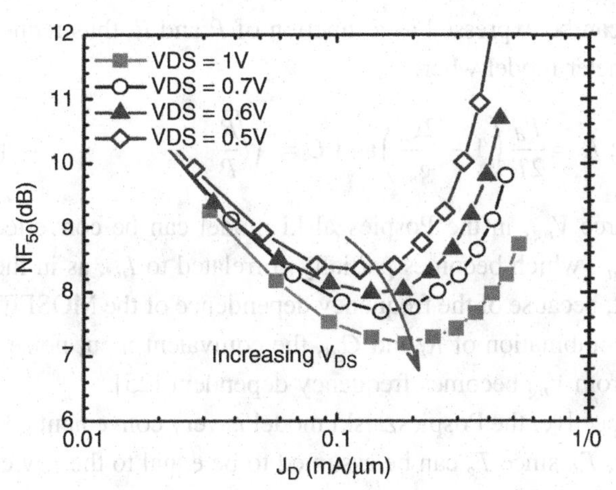

Figure 4.56 Measured V_{DS} dependence of NF_{50} and J_{OPT} at 84GHz for a 65nm GP n-MOSFET [23]

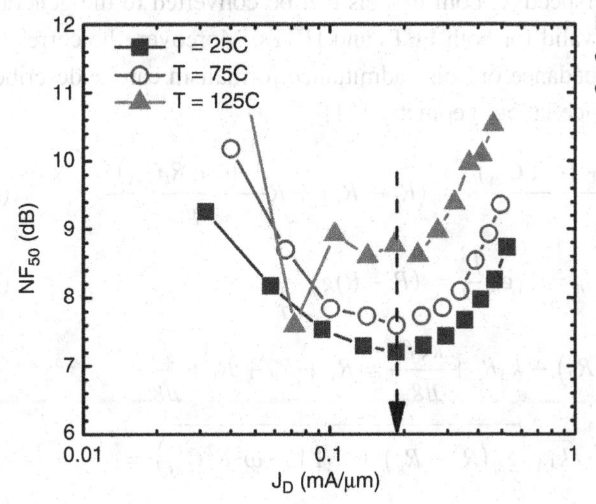

Figure 4.57 Measured temperature dependence of NF_{50} at 84GHz as a function of drain current density for a 65nm GP n-MOSFET [23]

One of the shortfalls of the Pospieszalski model is that it is strictly a linear model. To apply it to non-linear problems such as VCO phase noise and mixer noise calculations, T_d will have to be linked to the drain current. Another problem is related to the fact that R_i, which features prominently in the expressions of T_{MIN} and R_{OPT}, is particularly difficult to extract from small signal measurements, except at frequencies very close to the f_T of the transistor. Most foundries do not yet measure the S-parameters of Si MOSFETs at frequencies above 110GHz and most MOSFET compact models do not include R_i and τ in their small signal and noise equivalent circuits.

Figures 4.56 and 4.57 illustrate the measured 50Ω noise figure, NF_{50}, of 65nm n-MOSFETs at 84GHz as a function of the drain current density, V_{DS} and temperature [23]. As can be observed, J_{OPT} increases with V_{DS} from 0.15 to 0.2mA/μm but remains invariant with temperature. At the same time, NF_{50} is improved as V_{DS} increases and the temperature decreases.

It is important to note that, in terms of describing the linear noise behavior of MOSFETs, as long as their parameters are accurately extracted, the Pucel and Pospieszalski models are equivalent. It has been shown theoretically and experimentally [24] that the correlation coefficient C in the Pucel model can be expressed as a function of P and R, thus reducing the Pucel model to a two-noise parameter model where

$$P = \frac{T_d}{2T};\ R = \frac{T_d}{2T}\left(1 - \frac{2g_o}{g_m}\right) \text{and } C = \sqrt{\frac{R}{P}}. \tag{4.141}$$

The equivalent noise voltage source V_{ngs} in the Pospieszalski model can be converted to an equivalent noise current source, I_{ng}, which becomes partially correlated to I_{nd}, as in the Pucel model. It can be demonstrated that, because of the frequency dependence of the MOSFET input impedance formed by the series combination of R_i and C_{gs}, the equivalent input noise current power spectral density obtained from V_{ng} becomes frequency dependent [25].

From the device modeling perspective, the Pospieszalski model is very convenient because it requires only one noise parameter, T_d, since T_g can be assumed to be equal to the device noise temperature.

From the circuit designers' perspective, both models can be converted to the general noise model described in Chapter 4.1, valid for both FETs and HBTs. Moreover, the corresponding noise parameters in the noise impedance or noise admittance formalism can be described as a function of bias currents and device layout geometry [21]

$$R_u = (R_s + R_g) + k_2 \frac{1 + (\omega R_i C_{gs})^2}{g_m} = (R_s + R_g) + R \frac{1 + (\omega R_i C_{gs})^2}{g_m} \tag{4.142}$$

$$G_n = k_1 g_m \frac{f^2}{f_T^2} = (P - R) g_m \frac{f^2}{f_T^2} \tag{4.143}$$

$$Z_{cor} = (R_s + R_g) + k_3 R_i + \frac{k_3 f_T}{jf g_m} = R_s + R_g + R_i + \frac{f_T}{jf g_m} \tag{4.144}$$

$$F_{min} \approx 1 + \frac{2f}{f_T}\sqrt{k_1}\sqrt{g_m (R_s + R_g) + k_2\left(1 + \omega^2 R_i^2 C_{gs}^2\right)} = 1 + \frac{2f}{f_T}\sqrt{P - R}\sqrt{g_m (R_s + R_g) + R\left(1 + \omega^2 R_i^2 C_{gs}^2\right)} \tag{4.145}$$

$$k_1 = P + R - 2C\sqrt{PR} = P - R;\ k_2 = \frac{PR(1 - C^2)}{k_1} = R;\ k_3 = \frac{P - C\sqrt{PR}}{k_1} = 1 \tag{4.146}$$

$$Z_{SOPT} = \frac{f_T}{f g_m}\sqrt{\frac{g_m (R_s + R_g) + R\left(1 + \omega^2 R_i^2 C_{gs}^2\right)}{P - R}} + j\frac{f_T}{f g_m}. \tag{4.147}$$

4.2.10 Layout optimization

The performance of a MOSFET of given gate length, finger width and total gate width can vary significantly depending on the geometry of its layout. Both resistive and capacitive layout parasitics degrade the high-frequency figures of merit of the transistor, typically by more than

15–20%. This is one of the main reasons why schematic-level simulations without extracted layout parasitics do not provide a useful estimate of high-frequency circuit performance. In high-speed and high-frequency applications, a multi-gate finger layout geometry is employed in order to reduce the gate resistance and to accommodate the relatively large transistor current. The latter typically exceeds 1mA and, therefore, requires a total MOSFET gate width larger than 3μm. In some situations, unless multiple stacked metals, shunted together, are employed to extract the current from the channel, the minimum gate finger width is dictated by the electromigration rules for the metal used in the source and drain contact stripes.

For example, in a 65nm CMOS process, if a minimum-width metal-1 contact stripe is employed on the source and/or on the drain, no more than 0.15–0.2 mA could be extracted per 1 μm wide gate finger. In a multi-finger transistor, where a source or drain contact stripe collects current from two gate fingers, either shunted metal-1 and metal-2 (or even higher-level metals) stripes must be used, or the contact stripe must be widened. In the first case, C_{gd} and C_{ds} increase while in the second case C_{db} increases. In either scenario, f_{MAX} and, often, f_T and F_{MIN} are degraded. If the finger width is made too narrow, the electromigration problems are avoided but the gate–bulk fringing capacitance becomes a significant fraction of C_{gs} and C_{gd}, and f_T is severely reduced.

Layout schemes, with the gate contacted on one side or on both sides, and techniques to minimize parasitic capacitances associated with the multi-layer interconnect on top of the source and drain are illustrated in Figure 4.58. A corresponding 3-D view of the transistor layout is shown in Figure 4.59.

In addition to minimizing capacitive parasitics, the series resistance associated with the source and drain contact resistance, and with the via stacks on top of the gate, source and drain must be painstakingly reduced. For example, in the top part of Figure 4.58, five contact vias are employed to reduce the contact resistance per source–drain stripe to $40/5 = 8\Omega$. Similarly, the M1–M6 via stack contacting the gate in the bottom part of the figure has an equivalent resistance of 0.4Ω. At first glance this does not appear significant, but if we consider that the transistor has 56 gate fingers, each with one gate contact, and and that all 56 gate fingers are connected together through the via stack, the 0.4Ω vias stack resistance contributes $0.4 \times 56 = 22.4\Omega$ to the overal resistance of each gate finger. This figure is almost 60% of the nominal gate poly contact resistance of 40Ω.

The transistor layouts and the associated parasitics must be tailored to suit the configuration in which the transistor will be used in a circuit. For example, a different choice of interconnect between gate and drain is preferred for common-source than for common-gate devices. In the former case, gate–drain capacitance is critical but is less important in a common-gate stage.

As illustrated in Figure 4.60, the optimal MOSFET layout which maximizes the *MAG* can change with the operating frequency while Figure 4.61 shows that the best layout for minimizing the noise figure is different from the one best suited for maximum power gain. These measurements and simulations after layout extraction indicated that different transistor layouts must be employed in power amplifiers than in LNAs or oscillators.

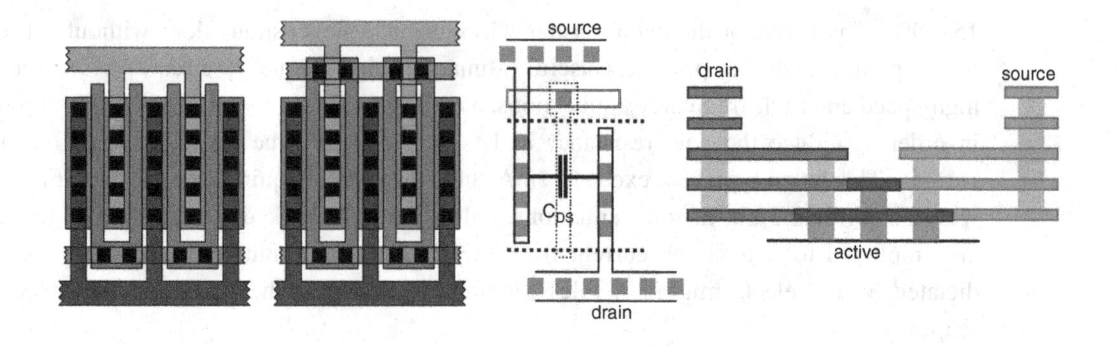

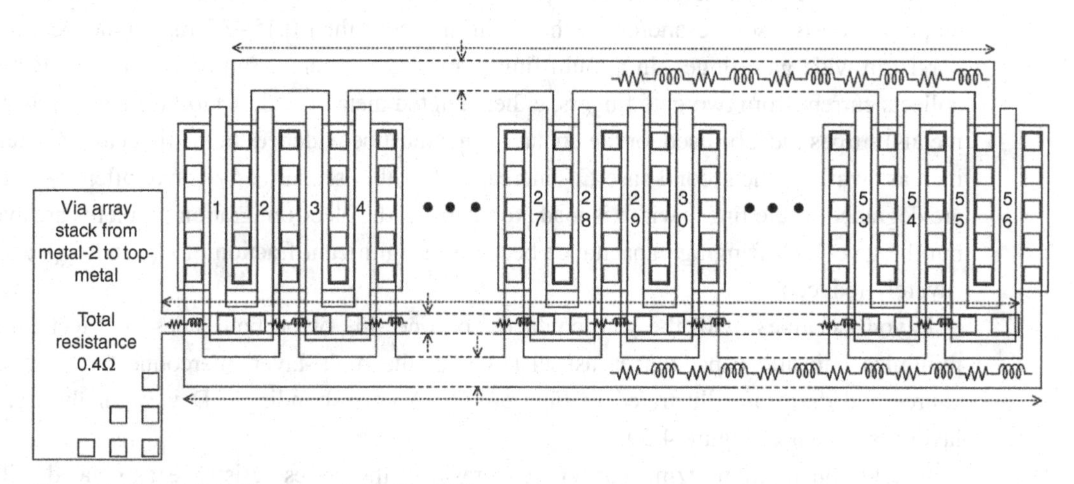

Figure 4.58 MOSFET layout optimization [26]

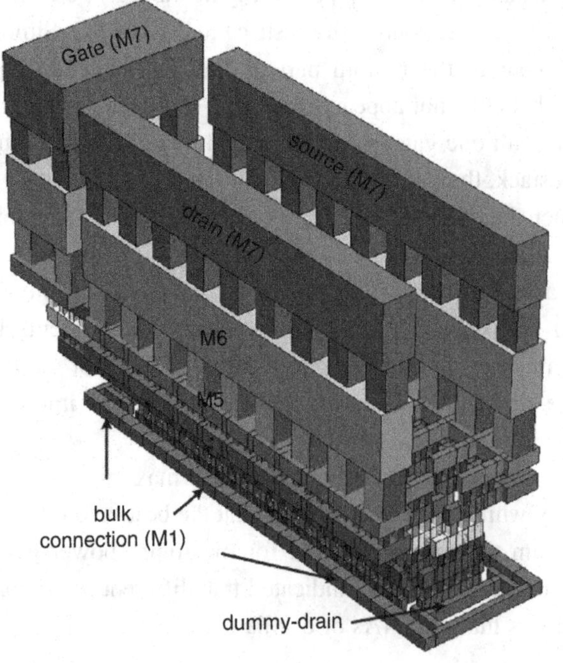

Figure 4.59 3-D view of a MOSFET layout showing 7 metal layers and intermetal vias [27]

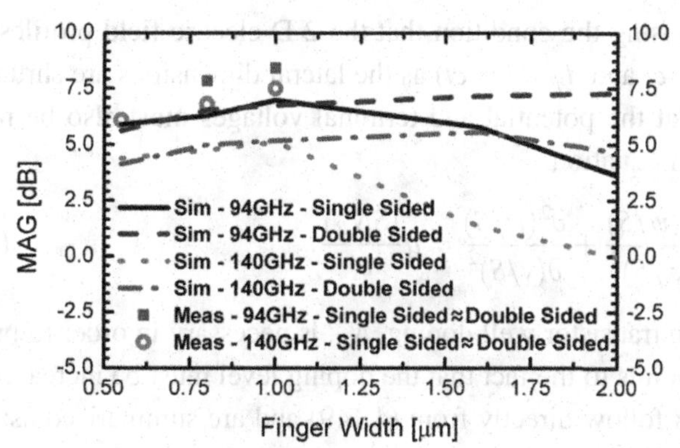

Figure 4.60 MOSFET layout optimization for MAG at different frequencies [28]

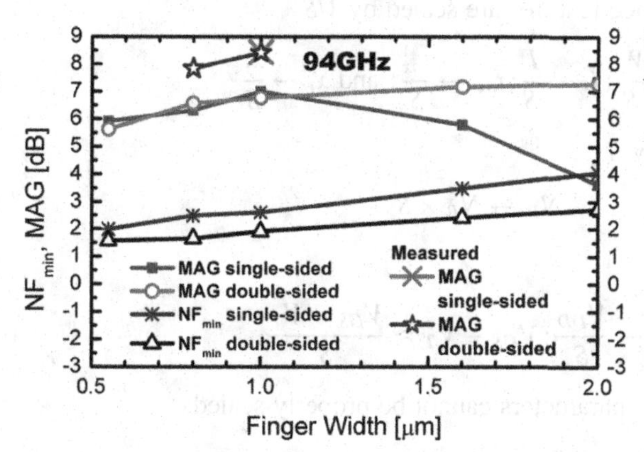

Figure 4.61 MOSFET layout optimization for *MAG* vs. NF_{MIN} at 94GHz [28]

4.2.11 MOSFET scaling

Ever since the first commercial LSI CMOS circuits appeared in the early 1970s, the minimum feature size of CMOS technology, usually represented by the physical gate length of the MOSFET, has shrunk by a scaling factor $S = \sqrt{2}$ every 2–3 years. As a result, transistors have become faster and the cost per electronic function has decreased at every new technology node.

What changes between technology nodes depends on the scaling rules. If *constant field scaling* is employed, all dimensions ($x, y, z => W, L, t_{ox}$), the supply voltage, V_{DD}, and the doping levels in the channel change by a factor S.

When *constant voltage scaling* (also known as lateral scaling) rules are applied, only the gate length changes. Constant voltage scaling is often done as a quick gate shrink ($S = 1.05$) between technology nodes.

Modeling MOSFET scaling

Theoretical equations for MOSFET scaling were derived by Dennard in 1974 [29] starting from the 2-D Poisson equation

$$\frac{\partial^2 \psi}{\partial x^2} + \frac{\partial^2 \psi}{\partial y^2} = \frac{qN_B}{\varepsilon_s} \quad (4.148)$$

where ψ is the potential, and imposing the condition that the 2-D electric field profiles in the device are preserved (i.e. $d\psi/dx = ct$ and $d\psi/dy = ct$) as the lateral dimensions are shrunk by a factor S. This, in turn, implies that the potential and terminal voltages must also be reduced by S, resulting in a scaled Poisson equation

$$\frac{\partial^2(\psi/S)}{\partial(x/S)^2} + \frac{\partial^2(\psi/S)}{\partial(y/S)^2} = q\frac{(SN_B)}{\varepsilon_s}. \tag{4.149}$$

The scaling factor S in front of the transistor well doping, N_B, is necessary in order to preserve the integrity of the equation and points to the fact that the doping level must be increased by S.

The constant field scaling rules follow directly from (4.149) and are summarized as:

(i) both lateral and vertical device features are scaled by $1/S$

$$W \rightarrow \frac{W}{S}; L \rightarrow \frac{L}{S}; t_{ox} \rightarrow \frac{t_{ox}}{S}; \text{ and } x_j \rightarrow \frac{x_j}{S} \tag{4.150}$$

(ii) substrate doping is scaled by S

$$N_B \rightarrow N_B \times S \tag{4.151}$$

(iii) voltages are scaled by $1/S$

$$V_{DD} \rightarrow \frac{V_{DD}}{S}; V_{GS} - V_T \approx \frac{V_{GS} - V_T}{S}. \tag{4.152}$$

At the same time, some transistor parameters cannot be properly scaled:

(i) because of their non-linear dependence on doping, V_T and Φ_F do not scale linearly;
(ii) the source, drain, gate doping concentrations do not scale because they are already at the solid solubility limit of the impurity species in silicon;
(iii) the effective gate oxide thickness stopped scaling beyond 1.2nm due to the quantum confinement effect.

Table 4.6 summarizes the constant field ($\alpha = 1$, $x = 1.4$) and generalized scaling ($\alpha > 1$, $k = 1.4$) rules for the most important DC, digital, analog and HF parameters of the MOSFET. Several observations can be derived:

- gate capacitance per micron of gate width is nearly independent of process:
  - ON resistance × micron improves with process scaling,
    - logic gates get faster with scaling (good),
  - dynamic power goes down with scaling (good),
- current density goes up with scaling (bad).

The impact of constant field scaling on analog and HF circuits can be distilled as:

- The V_{eff} corresponding to the square-law regime becomes smaller with every new node.
- The current density at which the square law breaks down remains constant over nodes at about 0.15mA/μm in n-MOSFETs and at 0.08mA/μm in p-MOSFETs.
- The peak g'_m, value improves with scaling.

Table 4.6 **MOSFET generalized scaling rules: electric field increases by *a* and all physical dimensions reduce by a factor *k*, where *a*,*k* > = 1.**

Depletion width X_d	$1/k$	
Gate capacitance	$1/k$	
Inversion-layer charge Q_i	a	
Carrier velocity	$1/k$	
Current, I_{ON}	a/k	
Circuit delay time	$1/k$	
Power-dissipation per circuit	a^2/k^2	Bad
Power-delay product per circuit	a^2/k^3	
Circuit density	k^2	
Power per unit area (density)	a^2	Very bad
Transconductance	1	
Cutoff frequency	k	
Source drain resistance	k	
Gate resistance	1	
f_{MAX}	$k^{-1/2}$	
$F_{MIN}-1$	$k^{-1/2}$	

- The peak f_T, *peak* f_{MAX}, and F_{MIN} values improve with scaling, as illustrated in Figures 4.62–4.66.
- J_{pfT}, J_{pfMAX}, and J_{OPT} remain constant if V_{DS} is set to approximately $V_{DD}/2$ in each node.
- J_{pfT}(NMOS) = 0.3–0.4mA/μm (Figure 4.62), I_{DSfT}(PMOS) = 0.15–0.3mA/μm (Figure 4.63):
  - J_{pfMAX}(NMOS) = 0.2–0.3 mA/μm (Figure 4.64), $I_{DSfMAX,NF}$(PMOS) = 0.125–0.2mA/μm (Figure 4.65)
  - J_{OPT} (NMOS) = 0.15 = 0.2mA/μm (Figure 4.66).
- Reactances per gate width remain largely constant and so does the optimum noise impedance if the gate finger width is scaled.

It should be noted though, that because of the significant amount of channel strain introduced in sub-65 nm SOI p-MOSFETs, their peak f_T and peak f_{MAX} current densities have increased dramatically, approaching those of the n-type MOSFETs. At the same time, the improvememnt with scaling in the f_{MAX} of devices contacted on one side of the gate is not as dramatic as in

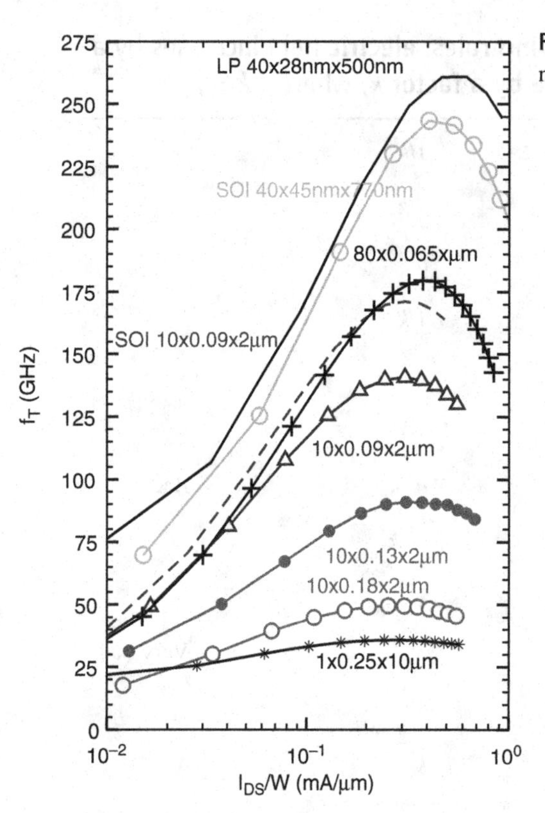

Figure 4.62 Scaling of the f_T and J_{pfT} of n-MOSFETs across technology nodes

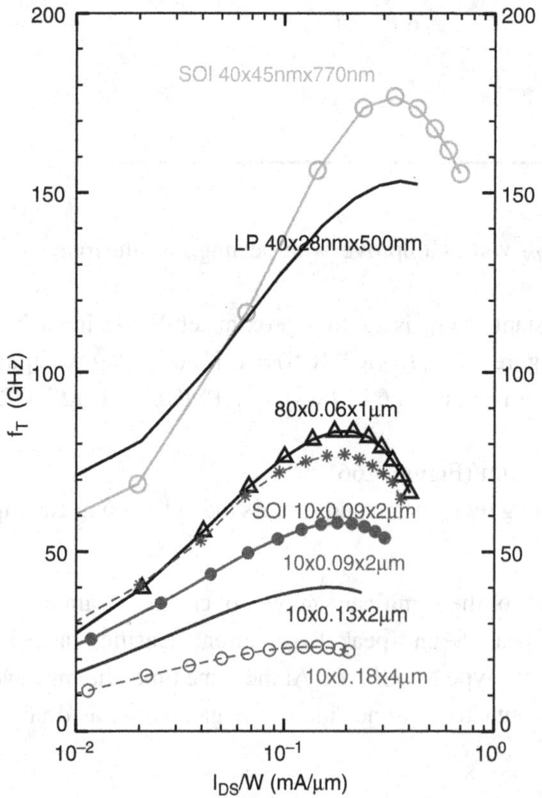

Figure 4.63 Scaling of the f_T and J_{pfT} of p-MOSFETs across technology nodes

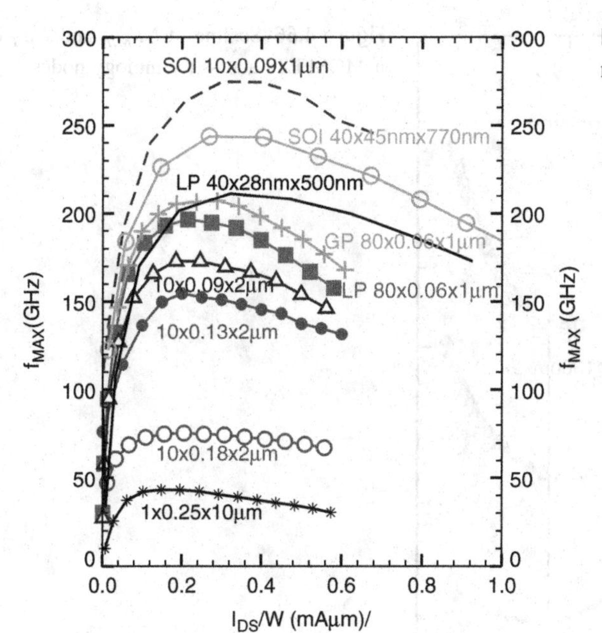

Figure 4.64 Scaling of the f_{MAX} and J_{pfMAX} of n-MOSFETs across technology nodes

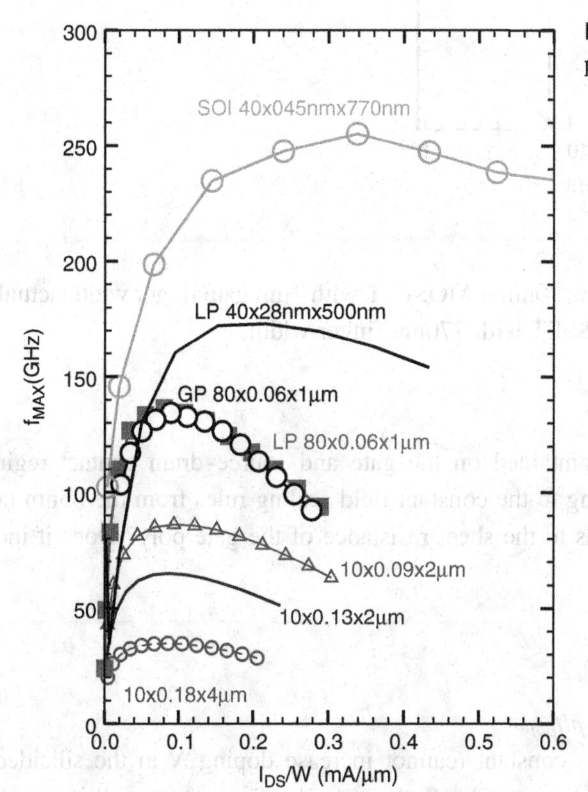

Figure 4.65 Scaling of the f_{MAX} and J_{pfMAX} of p-MOSFETs across technology nodes

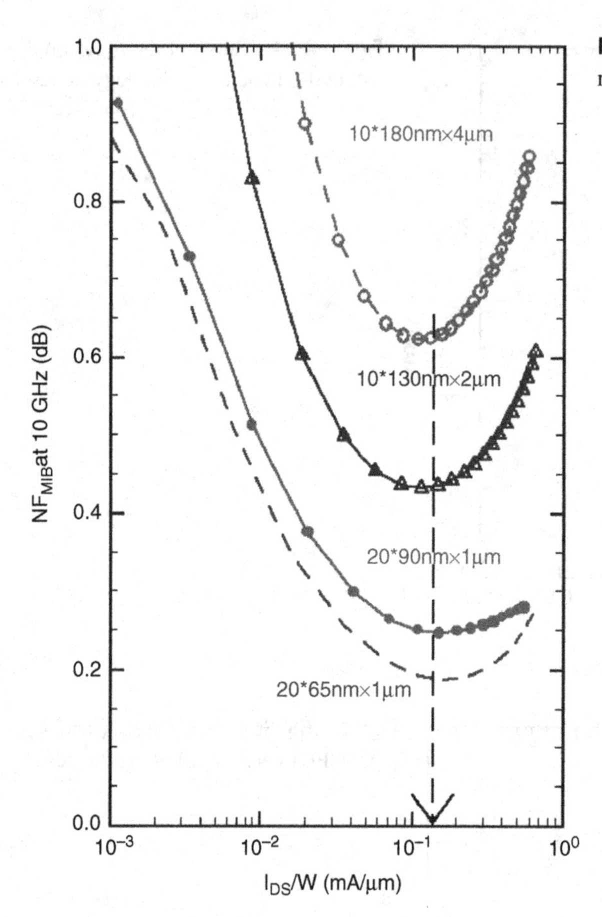

Figure 4.66 Scaling of NF_{MIN} and J_{OPT} of n-MOSFETs across technology nodes

earlier nodes. In the case of SOI, the 90nm n-MOSFET with 1µm gate finger width actually has higher f_{MAX} than the 45nm n-MOSFET with 770nm finger width.

EXAMPLE 4.9

(a) The same NiSi silicide is maintained on the gate and source–drain contact regions as MOSFETs are scaled according to the constant field scaling rules from the 65nm node to the 45nm node. What happens to the sheet resistance of the gate poly? Does it increase, decrease, or remain constant?

Solution

The gate sheet resistance $R_{SHG} = \rho/t_{poly}$

Since $\rho = 1/(q\ \mu_n\ N)$ remains constant (cannot increase doping N in the silicided poly anymore) and the poly thickness decreases by S, the gate sheet resistance will increase by S. That is one of the reasons why metal gates have been introduced in high-performance (HP) 45nm and 32nm CMOS processes.

(b) Derive the constant field scaling rules for the peak f_T value and for the optimum noise resistance R_{SOPT} of a MOSFET, knowing that the g_m/W at peak f_T increases with the scaling factor S. Ignore the contribution of the source and drain resistances to f_T and R_{SOPT}.

Solution

$$f_T = \frac{g_m}{2\pi(C_{gs}+C_{gd})} = \frac{g'_m W}{2\pi\left(C'_{gs}+C'_{gd}\right)W} \rightarrow \frac{g'_m (W/S)}{2\pi\left(C'_{gs}+C'_{gd}\right)(W/S)} = \frac{g'_m}{2\pi\left(C'_{gs}+C'_{gd}\right)}.$$

Since C'_{gs} and C'_{gd} do not change with scaling and g'_m ($= g_m/W$) scales by S, f_T also scales by S

$$R_{SOPT} = \frac{0.5}{\omega(C_{gs}+C_{gd})} = \frac{0.5}{\omega\left(C'_{gs}+C'_{gd}\right)W} \rightarrow \frac{0.5}{\omega\left(C'_{gs}+C'_{gd}\right)(W/S)} = \frac{0.5S}{\omega\left(C'_{gs}+C'_{gd}\right)W} \rightarrow SR_{Sopt}.$$

Both f_T and R_{SOPT} scale by S. However, R_{SOPT}/W remains constant. Since I_{DS}/W at peak f_T, peak f_{MAX} or minimum noise figure bias also remains constant with scaling, LNAs can be scaled from one technology node to another without changing the size and bias current of the transistors in the LNA, while the LNA noise figure and gain (proportional to f_T) improve.

Problem: Derive the expression for gate resistance scaling knowing that the sheet resistance of the gate increases by S and that the silicides and contact via materials remain the same while their dimensions (height and width) scale according to constant field scaling rules.

4.2.12 Multi-gate MOSFETs

The ultimate scaling of planar FETs to shorter and shorter gate lengths is limited by the electric field lines from the source and drain encroaching into the channel and competing with the gate field lines to control the channel charge. This leads to strong short channel effects which degrade the subtheshold slope and exacerbate the DIBL effect. Furthermore, in a planar MOSFET, the main conduction channel is formed in a thin layer at the silicon–gate oxide interface, with most of the silicon body acting as a source of leakage current when the transistor is off. As illustrated in Figure 4.67, the leakage current can be reduced by employing an ultra-thin silicon body while short channel effects can be delayed by surrounding the channel with the gate, thus improving the electrostatic control of the FET channel by the gate.

The first commercial three-dimensional channel device, a "tri-gate" MOSFET was introduced in 2011 at the 22nm CMOS node. Its main features are compared in Figure 4.68 with those of a 32nm planar MOSFET for a device with two gate fingers and three fins [31]. They include:

- fully depleted, undoped fins to reduce the leakage current, short channel effects, and increased channel mobility and substhreshold slope;
- high-k metal gates on three sides of the fin, reducing gate–bulk fringing capacitance;
- strained silicon channel to increase hole and electron mobility;
- raised, epitaxial source and drain regions, contacting each fin on three sides for reduced source and drain resistances;

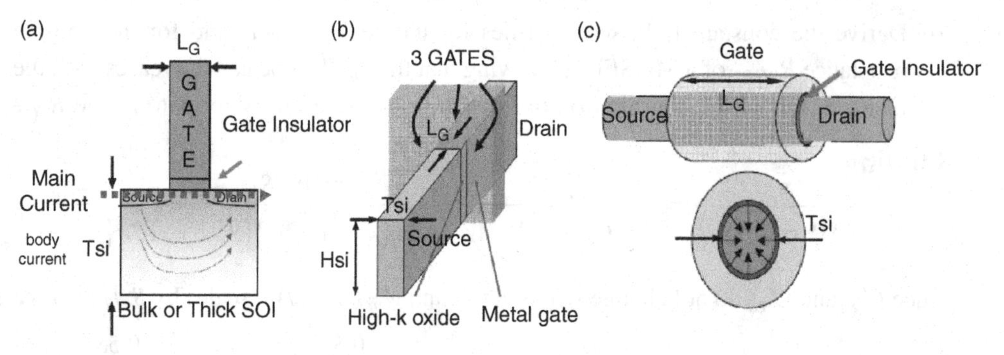

Figure 4.67 (a) Planar MOSFET, (b) tri-gate, and (c) nanowire FinFETs [30]

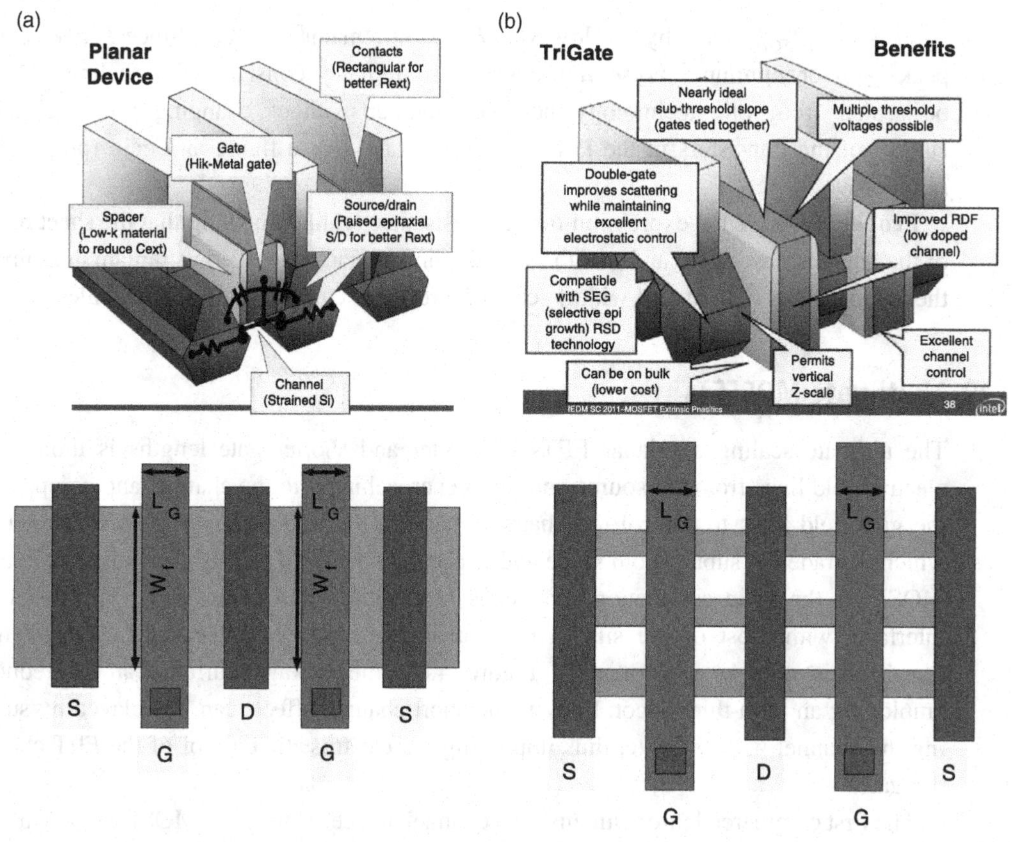

Figure 4.68 Comparison of planar MOSFETs and tri-gate FinFETs (a) main structural features [31] and (b) top level layout views

- rectangular metal contacts to minimize contact resistance;
- fin base is directly connected to the silicon substrate, to avoid using SOI wafers for reduced cost.

An important observation is that tri-gate and nanowire FinFETs are governed by similar DC characteristics, small signal and noise equivalent circuits as 32nm planar MOSFETs. The transistor layout seen by the IC designer is only slightly changed. In FinFETs, the active area of the transistor is replaced by one or several fins connected together by the source and drain contacts.

As illustrated in Figure 4.67, the fin is described by its height, H_{si}, width, T_{si}, and by the gate length, L_G. The width of the fin is determined by the minimum feature size, and must be sized such that the depletion regions on each side of the fin merge under the control of the gate. Since the fin is undoped and fully depleted, its width determines the threshold voltage. The fin height is typically chosen to be comparable to the fin width, as a good compromise between reducing the source and drain contact resistance, which demands a tall fin, and minimizing gate to source–drain contact side capacitance, asking for a short fin. Tight control of the fin width and fin height is critical to achieve acceptable yields in production.

The unit finger width is determined by the height of the fin

$$W_f = 2H_{si} + T_{si} \tag{4.153}$$

and is not available as a design variable to the circuit designer. In circuit design, the total gate width can be increased by connecting fins in parallel, and/or by increasing the number of gate fingers, N_f, per fin. By solving Poisson's equation for planar and multi-gate devices with up to N gates ($N = 1 \ldots 4$), it is possible to calculate a generalized minimum gate length, $L_{gN} \approx 6 \cdot \sqrt{\frac{\varepsilon_{Si}}{N\varepsilon_{ox}} t_{ox} T_{si}}$ beyond which short-channel effects can be contained [11]. For the same effective gate oxide thickness, t_{ox}, and silicon body or fin thickness, T_{si}, a tri-gate FET ($N = 3$) can have 3 times smaller gate length than a planar ($N = 1$) MOSFET. The improvement from a tri-gate to a full surround gate nanowire FET is a mere 2/3. For continued MOSFET switching and high-frequency performance scaling beyond the 14nm node, the silicon channel is expected to be replaced by InGaAs in n-MOSFETs (Figure 4.69) and Ge in p-MOSFETs.

Irrespective of the channel material, the parasitic capacitances and resistances of FinFETs, shown in Figure 4.70, are going to dominate their high-frequency and noise performance. In general, the resistances improve because of the increased number of contacted surfaces. The gate–substrate fringing capacitance and source–drain–bulk capacitances, too, improve. However, the parasitic gate–drain–source capacitances increase with additional components that do not exist in planar FETs [31]. Some of these capacitances are illustrated in Figure 4.70 and include:

- C_{OV}, the overlap capacitance, between the gate and the source and drain extension regions under the gate, similar to its planar FET equivalent;
- C_1, between the top of the gate and the top of the fin contact, similar to the top fringing gate capaciatnce in a planar FET
- C_2, between the gate sidewall and the top surfaces of the fins, which depends in the fin height similar to the gate side fringing capacitance in planar FETs;
- C_3, unique to the FinFET, between the gate in the inter-fin (dead-space) region and the source–drain contact bars.

In general, as in planar nanoscale FETs, the various components of the parasitic side capacitance between the gate and the source–drain contact bars can be reduced by increasing the contact pitch. Fortunately, the analog performance, determined by r_o and $g_m' r_o$, improves because of better electrostatics [31].

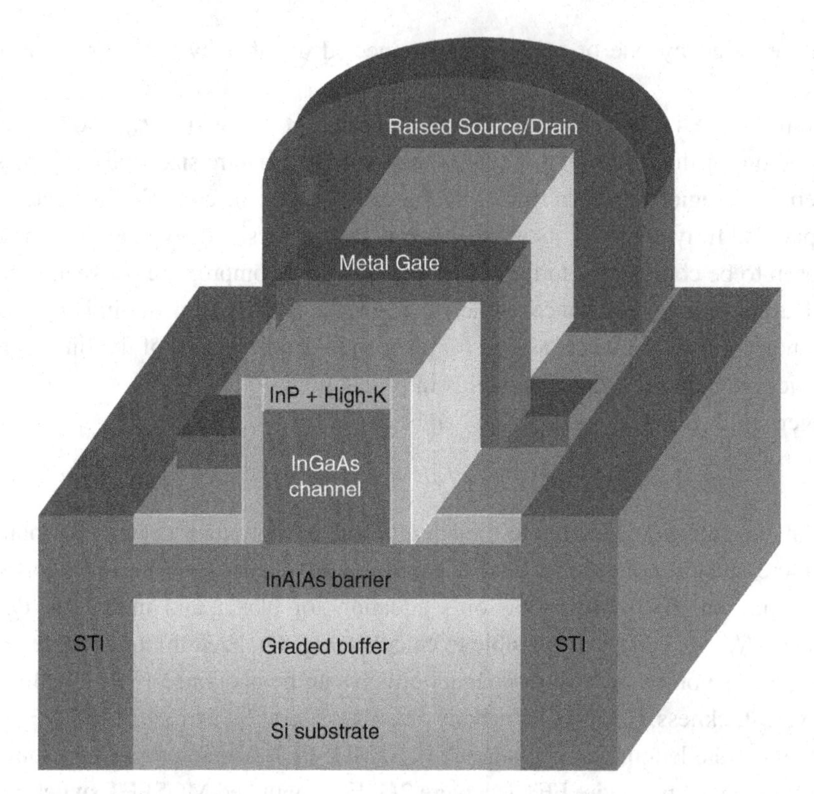

Figure 4.69 Multi-gate InGaAs quantum-well channel MOSFETs [32]

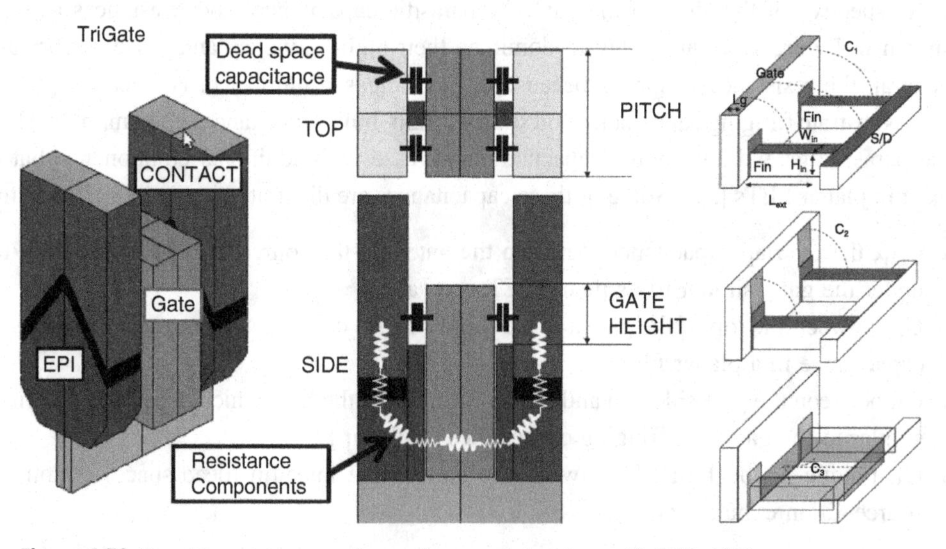

Figure 4.70 Parasitic resistances and capacitances of the tri-gate FinFETs [31]

4.3 THE HETEROJUNCTION BIPOLAR TRANSISTOR

4.3.1 The heterojunction bipolar transistor concept

The idea was first explained by Herbert Kroemer in 1957 and won him the Nobel Prize in physics in 2000. The main goal was to improve the high-frequency performance of a non bipolar transistor by:

- suppressing hole injection from the base into the emitter to increase the DC current gain, β, and
- speeding up transport across the base by reducing the base transit time, τ_B.

Kroemer had silicon and germanium in mind as the materials that would form the emitter/base heterojunction. However, its first practical realization had to wait until the late 1980s.

4.3.2 InP or GaAs HBT structure

The first experimental heterojunction bipolar transistor was fabricated in the AlGaAs/GaAs system by Peter Asbeck of Rockwell in 1980, with an InP-based HBT being announced by Bell Labs a few years later. From the outset, these III-V HBTs were developed for microwave applications by employing materials with much larger electron mobilities than either silicon or germanium. The most common III-V material combinations, summarized in Table 4.7, are AlGaAs/GaAs, InGaP/GaAs, InP/InGaAs, InAlAs/InGaAs/InP, InP/InGaAs/InP, and InP/InGaAsSb/InP. More so than in Si/SiGe HBTs, the primary focus on III-V HBT design engineers was on:

- increasing β by selecting a large bandgap difference between the emitter and the base,
- reducing the base resistance by doping the base at much higher levels than the emitter,
- choosing a material with high breakdown field and high saturation velocity in the collector, and
- selecting low bandgap materials with high electron mobility for the base.

The basic structure is illustrated in Figure 4.71 and consists of a **semi-insulating InP substrate,** typically 50μm to 500μm thick, on which a highly doped n + InP **subcollector** layer is grown where the collector contact is placed. The role of this layer is to provide a low-resistance access to the intrinsic **n-doped collector** formed either in InGaAs, InAsP, or InAlAs. The narrow bandgap, **p+ base** is typically realized in InGaAs or GaAsSb, and is etched away outside the emitter and base contact regions to expose the collector layer and thus to minimize the base–collector junction area and base–collector capacitance. Finally, an InP or InAlAs n-type emitter layer is grown on top of the base along with an n+ InGaAs layer which forms a low-resistance contact with the emitter metal alloy, typically Au-Ge. The emitter–base area, too, is minimized by mesa etching to reduce base–emitter capacitance. Since the mobility of holes in III-V materials is significantly lower than that of electrons, the base must be highly doped ($> 5 \times 10^{19}\text{cm}^{-3}$) in order to ensure that the base resistance remains low. In compensation, the emitter

Table 4.7 **Material systems used for HBT fabrication.**

Substrate	Emitter	Base	Collector
Si	Si/poly silicon	graded SiGe	Si
GaAs	AlGaAs	GaAs or graded AlGaAs	GaAs or AlGaAs (DHBT)
GaAs	InGaP	GaAs or graded AlGaAs	GaAs or InGaP (DHBT)
InP	InP or InAlAs	InGaAs or graded InAlAs	InGaAs or InP (DHBT)
InP	InP	GaAsSb	InP (DHBT)
SiC or sapphire	AlGaN	GaN	GaN

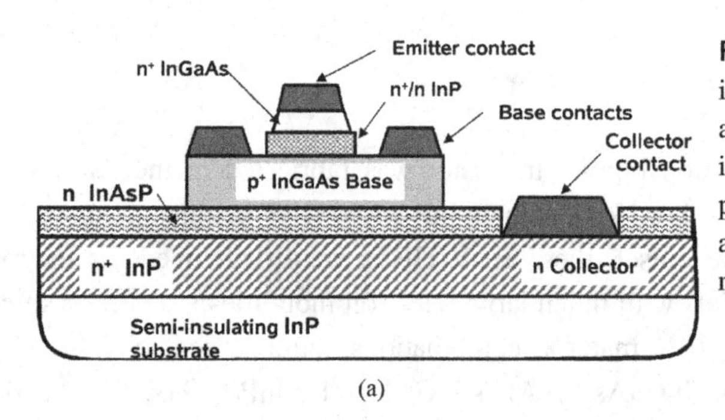

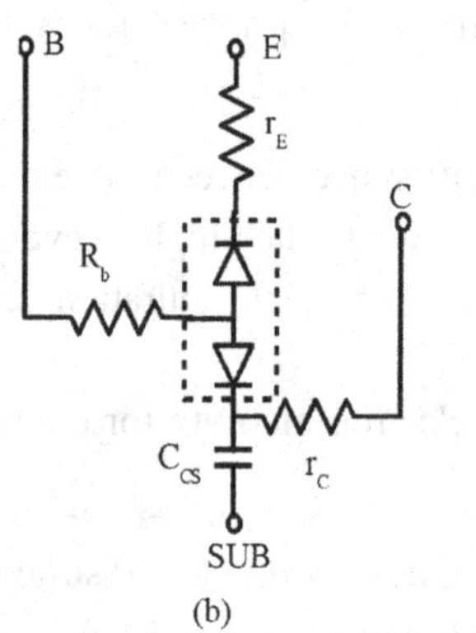

Figure 4.71 (a) InP HBT cross-section indicating the various semiconductor materials and regions (b) Large signal equivalent circuit illustrating the BE and BC diodes and the parasitic base, emitter and collector resistances associated with the access regions between the metal contacts and the intrinsic device

doping is set 10 to 100 times lower, to maintain a low base–emitter junction capacitance. This doping arrangement is different than in SiGe HBTs where the emitter doping is higher than the base doping.

The equivalent circuit in Figure 4.71(b) includes the resistances r_E and r_C associated with the highly doped emitter and collector access regions to the intrinsic transistor. The intrinsic transistor is formed strictly below the emitter contact and consists of the BE and BC junctions and of the intrinsic part of the base resistance, r_{bi}. The parasitic collector–substrate capacitance, C_{CS}, is usually negligible unless the semi-insulating substrate is thinned down to less than 50μm.

The energy band diagrams, reproduced in Figure 4.72(a) for single-heterojunction AlGaAs/GaAs or InP/InGaAs HBTs [33], illustrate that the injection of electrons from the emitter into the base is favored over that of holes from the base into the emitter because of the significantly lower potential barrier ($\Delta\phi_n < \Delta\phi_p$) they have to overcome. In addition, in the case of the InP/GaAsSb/InP HBT with staggered line-up double heterojunctions [34] illustrated in Figure 4.72(b), the electrons experience a significant boost in energy, caused by the conduction band offset ΔE_C of approximately 0.15 eV, as they are injected from the base into the collector. ΔE_C describes the difference between the conduction band energy in the base and in the collector.

4.3.3 The SiGe HBT structure

Figure 4.73 sketches a vertical cross-section through a SiGe HBT. The main features are indicated in the figure and are listed below:

- $n+$ polysilicon emitter,
- self-aligned external $p+$ polysislicon base,
- SiGe base with graded Ge concentration to create a quasi-electric field aiding electron transport,
- selectively implanted collector (SIC) to reduce base–collector capacitance,
- shallow trench isolation, STI, to isolate the base from the collector and to reduce C_{bc},
- collector sinker contact region to reduce the extrinsic collector resistance while contacting the collector at the silicon surface,
- highly doped buried layer (also known as subcollector) to isolate the transistor from the substrate and to reduce the extrinsic collector resistance,
- deep trench isolation, DTI, to provide electrical isolation from other devices while reducing the device area and the collector–substrate capacitance, and
- p-well with $p+$ substrate contact.

The main goal in the design of the SiGe HBT is to reduce the base transit time. Because the electron mobility is not as high as in III-V semiconductors, for the same base transit time, the base region must be made thinner in SiGe HBTs. Additionally, as shown in Figure 4.74, the Ge concentration in the base is graded, increasing from the base–emitter towards the base–collector junction. This creates a quasi-electric field which speeds up the movement of electrons across the base. The Ge profile in the base is quantified by two parameters:

- the bandgap difference at the base–emitter junction, $\Delta E_{gGe}(0)$, which reflects the bandgap reduction in the base with respect to the bandgap of *Si* used in the emitter, and
- $\Delta E_{gGe}(\text{grade})$ which describes the amount of change in bandgap from the edge of the base–emitter junction to the edge of the base–collector junction, due to *Ge* concentration grading

(a)

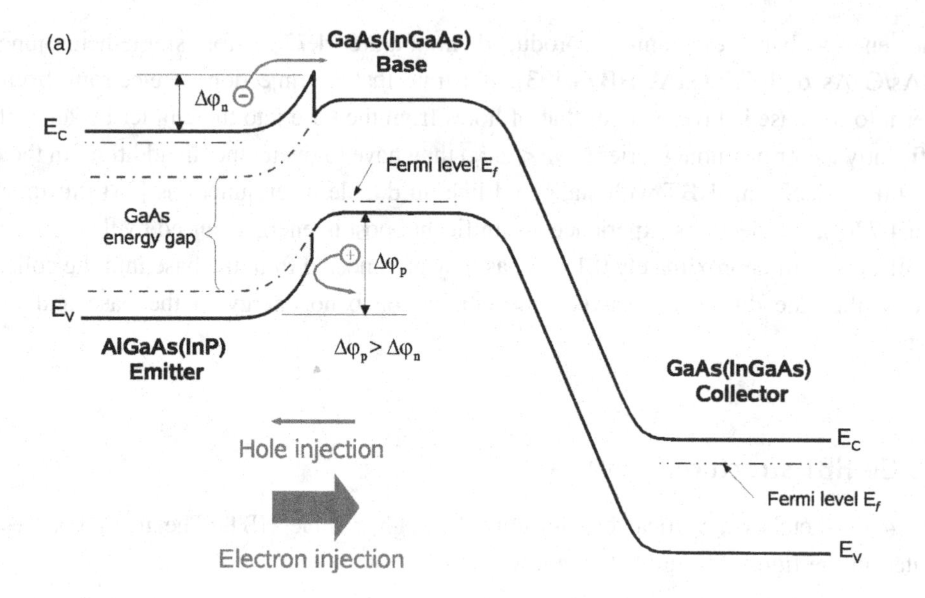

(b)

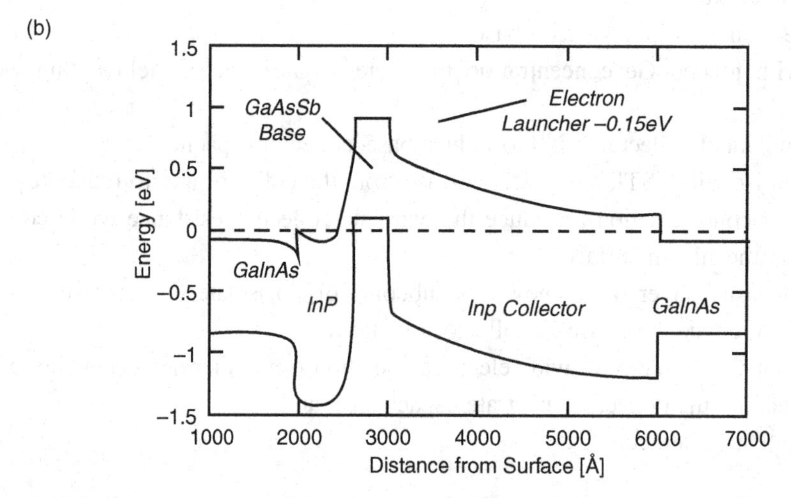

Figure 4.72 Band diagrams of (a) single heterojunction AlGaAs/GaAs or InP/InGaAs [33], and (b) of double-heterojunction, type II InP/GaAsSb/InP HBTs [34]

$$\Delta E_{gGe}(\text{grade}) = \Delta E_{gGe}(x = W_B) - \Delta E_{gGe}(x = 0) \tag{4.154}$$

where W_B is the width of the neutral base region, as defined in Figure 4.75.

Why was germanium introduced in the base of a silicon BJT? The main reasons are summarized below:

- *Ge* in base increases β;
- allows to significantly increase the base doping, N_B, to reduce r_B without significantly degrading β,

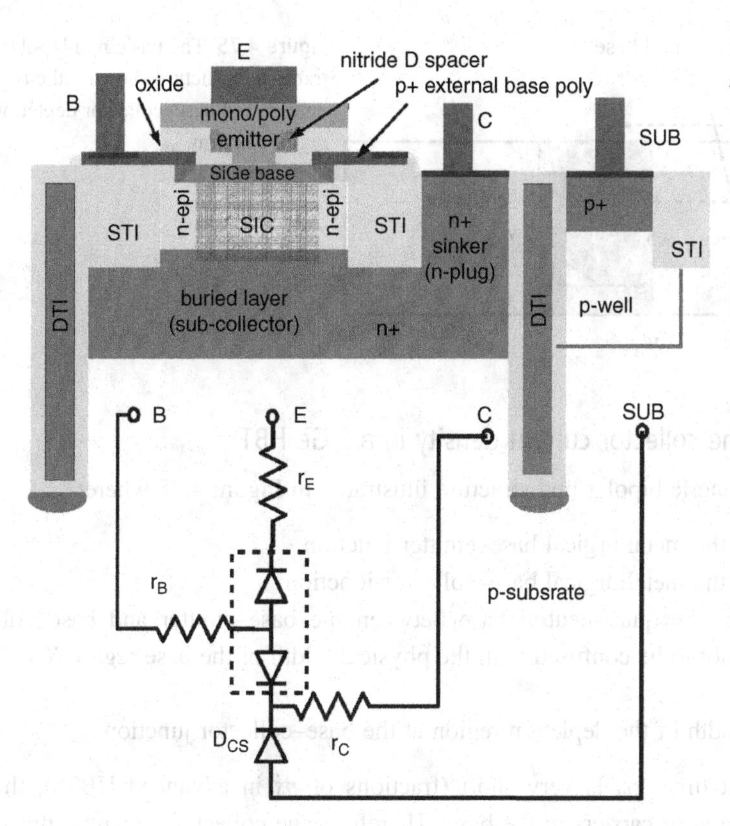

Figure 4.73 SiGe HBT structure and parasitics

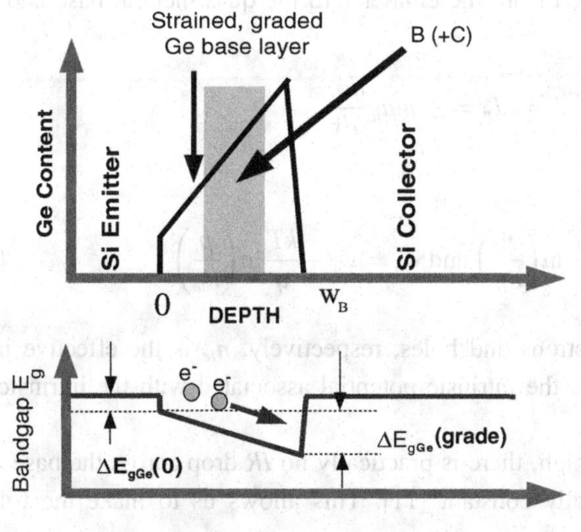

Figure 4.74 Energy band diagrams of a graded base SiGe HBT

- lower base resistance increases f_{MAX} and decreases NF_{MIN};
- *Ge* grading across the base reduces τ_b and thus increases f_T;
- *Ge* grading across the base increases the Early voltage, V_A;
- increases βV_A which is desirable in analog applications where a large intrinsic voltage gain is required.

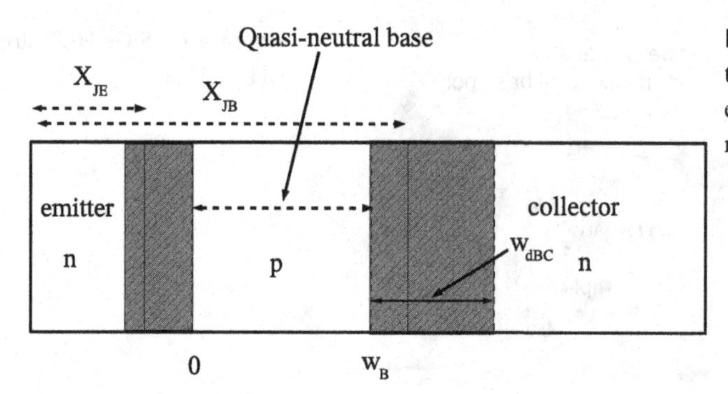

Figure 4.75 The basic npn bipolar transistor structure showing the base–emitter and base–collector depletion regions in gray

The expression of the collector current density in a SiGe HBT

Let us consider the generic bipolar npn structure illustrated in Figure 4.75 where:

- X_{JE} is the depth of the metallurgical base–emitter junction,
- X_{JB} is the depth of the metallurgical base–collector junction,
- W_B is the width of the quasi-neutral base between the base–emitter and base–collector depletion regions (not to be confused with the physical width of the base region $X_{JB} - X_{JE}$), and
- W_{dBC} is the total width of the depletion region at the base–collector junction.

Since the base transit time, τ_b, is very short (fractions of *ps* in advanced HBTs), there is negligible recombination of carriers in the base. Therefore, the collector current is practically equal to the electron current injected from the emitter into the quasi-neutral base and can be expressed as [11]

$$J_c = J_n = -qn\mu_n \frac{d\phi_n}{dx} \tag{4.155}$$

where

$$\phi_n = \psi_i - \frac{kT}{q}\ln\left(\frac{n}{n_{ie}}\right) \text{ and } \phi_p = \psi_i + \frac{kT}{q}\ln\left(\frac{p}{n_{ie}}\right) \tag{4.156}$$

are the quasi-Fermi levels for electrons and holes, respectively, n_{ie} is the effective intrinsic concentration in the base, and ψ_i is the intrinsic potential associated with the intrinsic Fermi level.

Since the doping in the base is high, there is practically no *IR* drop across the base and the quasi-Fermi level remains practically constant [11]. This allows us to make the following approximation

$$\frac{d\phi_p}{dx} \approx 0 \tag{4.157}$$

which, combined with

$$\phi_p - \phi_n = \frac{kT}{q}\ln\left(\frac{p_p n_p}{n_{ie}^2}\right) \tag{4.158}$$

allows for the electron current to be expressed as

$$J_n = q n_p \mu_n \frac{d\left(\phi_p - \phi_n\right)}{dx} = q D_n \frac{n_{ie}^2}{p_p}\frac{d}{dx}\left(\frac{n_p p_p}{n_{ie}^2}\right) \tag{4.159}$$

where $D_n = kT\mu_n/q$ is the diffusion constant of the minority electrons in the base region.

By rearranging (4.159) and integrating across the base from 0 to W_B, we get

$$J_n \int_0^{W_B} \frac{p_p}{q D_n n_{ie}^2} dx = \frac{n_p(w_B) p_p(w_B)}{n_{ie}^2(w_B)} - \frac{n_p(0) p_p(0)}{n_{ie}^2(0)} \approx \frac{-n_p(0) p_p(0)}{n_{ie}^2(0)}. \tag{4.160}$$

The last approximation in (4.160) is possible because $n_p(W_B) << n_p(0)$, where

$$n_p(0) = n_{p0}(0) e^{\frac{qV_{BE}}{kT}}. \tag{4.161}$$

As a result, the expression of the collector current density becomes

$$J_c = J_n = \frac{q e^{qV_{BE}/kT}}{\int_0^{W_B} \frac{P_P(x) dx}{D_{nB}(x) n_{ieB}^2(x)}} \tag{4.162}$$

where $p_p \approx N_B$, n_{ieB} is the position-dependent intrinsic concentration in the SiGe base, n_i is the reference intrinsic concentration in silicon, and the integral in the denominator represents the scaled base Gummel number $G_B = \int_0^{W_B} \frac{P_p(x) n_i^2 dx}{D_{nB}(x) n_{ieB}^2(x)}$.

We note from (4.162) that the collector current is dependent solely on the properties of the base region.

For a trapezoidal *Ge* profile in the base, which is the most common, the intrinsic concentration becomes position dependent

$$n_{iB}^2(x) = \gamma n_{io}^2 e^{\Delta E_{gB}^{app}/kT} e^{\Delta E_{gGe}(\text{grade}) \frac{x}{W_B kT}} e^{\Delta E_{gGe}^{(0)}/kT} \tag{4.163}$$

where n_{io} is the intrinsic concentration at the emitter side of the base and

$$\gamma = \left[(N_C N_V)_{SiGe}/(N_C N_V)_{Si}\right] \approx 1 \tag{4.164}$$

$$\eta = (D_{nB})_{SiGe}/(D_{nB})_{Si} = 1 \ldots 2 \tag{4.165}$$

are material parameters that depend on the Ge profile in the base.

By inserting (4.163) into (4.162), we obtain the expression of the collector current as a function of the parameters that describe the trapezoidal germanium profile in the base

$$J_{CSiGe} = \frac{qD_{nB}n_{io}^2}{W_B N_B}\left(e^{\Delta E_{gB}^{app}/kT}e^{qV_{BE}/kT} - 1\right)\left[\gamma\eta\frac{\Delta E_{gGe}(\text{grade})}{kT}\frac{e^{\Delta E_{gGe}(0)/kT}}{1 - e^{-\Delta E_{gGe}(\text{grade})/kT}}\right]. \quad (4.166)$$

The base current density expression

The base current, J_B, is due to the holes injected from the base into the emitter and depends only on parameters that describe the emitter region. In SiGe HBTs, the latter remains the same as in a Si BJT. Therefore, the base current expression is identical to that of a silicon BJT [11]

$$J_B = \frac{qe^{qV_{BE}/kT}}{\int_0^{X_E}\frac{N_E(x)dx}{D_{pE}(x)n_{iE}^2(x)}} \quad (4.167)$$

where

$$G_E = \int_0^{X_E}\frac{N_E(x)n_i^2 dx}{D_{pE}(x)n_{iE}^2(x)} \quad (4.168)$$

is known as the emitter Gummel number, D_{pE} is the diffusion constant of holes in the emitter, n_{iE} is the intrinsic concentration in the emitter, and N_E represents the doping in the emitter region.

Next, we will derive expressions for the relative improvement in β, V_A, and τ_b, compared to those of a Si BJT, when a trapezoidal Ge profile is used in the base region of a SiGe HBT.

4.3.4 Current gain enhancement

From the definition of the current gain

$$\beta = \frac{J_C}{J_B} = \frac{G_E}{G_B} = \frac{\int_0^{X_E}\frac{N_E(x)dx}{D_{pE}(x)n_{iE}^2(x)}}{\int_0^{W_B}\frac{N_B(x)dx}{D_{nB}(x)n_{ieB}^2(x)}} \quad (4.169)$$

and (4.166), we can derive an expression that quantifies the improvement in current gain relative to a silicon BJT as

$$\frac{\beta_{SiGe}}{\beta_{Si}} \approx \frac{J_{CSiGe}}{J_{CSi}} = \gamma\eta\frac{\Delta E_{gGe}(\text{grade})}{kT}\frac{e^{\Delta E_{gGe}(0)/kT}}{1 - e^{-\Delta E_{gGe}(\text{grade})/kT}}. \quad (4.170)$$

We note that, for HBTs with uniform composition and constant doping in the emitter and in the base (as in most III-V HBTs) $\beta \approx \frac{N_E D_{nB} X_E n_{ieB}^2}{N_B D_{pE} W_B n_{iE}^2}$. The only difference from the expression of the current gain of a BJT is due to the presence of the ratio of the intrinsic concentrations in the base and in the emitter. This provides the additional degree of freedom to optimize the transistor performance. The intrinsic concentration in a specific region of the transistor depends exponentially on the bandgap of the semiconductor material employed in that region. This observation was first made by H. Kroemer in 1954 and is critical in allowing the base doping to become much larger than the emitter doping, without degrading β.

4.3.5 Early voltage improvement

Similarly, starting from the definition of the Early voltage [11]

$$V_A \approx \frac{J_C}{\frac{dJ_C}{dV_{CB}}} \tag{4.171}$$

the improvement due to the graded Ge profile in the base is described by [11]

$$(V_A)_{SiGe}/(V_A)_{Si} \approx e^{\Delta E_{gGe}(\text{grade})/kT}\left[\frac{1-e^{-\Delta E_{gGe}(\text{grade})/kT}}{\Delta E_{gGe}(\text{grade})/kT}\right]. \tag{4.172}$$

We note that the Early voltage increases exponentially with the amount of Ge grading across the base region.

4.3.6 Base transit time

The high-frequency operation of the intrinsic HBT is dominated by the total delay encountered by the electrons (in an npn) as they travel from the emitter to the collector. The total delay is the sum of the transit times through the emitter, τ_E, neutral base, τ_b, and the collector, τ_C, as well as parasitic RC delays. Along with the collector transit time, identical to that of a BJT, which scales with the width of the base–collector space charge region, W_{dBC}, and is inversely proportional to the saturation velocity of carriers, v_{sat} [11]

$$\tau_C = \frac{W_{dBC}}{2v_{sat}} \tag{4.173}$$

the base transit time, τ_b, is the most important parameter that characterizes the intrinsic high-frequency performance of a bipolar transistor. By definition [11]

$$\tau_b = \frac{|Q_B|}{|J_C|}, Q_B = -q\int_0^{W_B}[n_p(z) - n_{p0}(z)]dz \tag{4.174}$$

where Q_B represents the excess minority charge in the base.

It can be demonstrated [11] that

$$\tau_b = \int_0^{W_B} \frac{n_{iB}^2(z)}{N_B(z)} \left[\int_z^{W_B} \frac{N_B(x)dx}{D_{nB}(x)n_{iB}^2(x)} \right] dz = \frac{W_B{}^2}{\zeta D_B} \tag{4.175}$$

where $\zeta = 2$ for a uniformly doped Si BJT and $\zeta = 2 \ldots 20$, depending on the germanium profile in the base, for SiGe HBTs.

Following lengthier mathematical manipulations, the impact of a trapezoidal Ge profile in the base in reducing the base transit time, compared to a uniform silicon base, is given by [11]

$$\frac{\tau_{bSiGe}}{\tau_{bSi}} = \frac{2}{\zeta} = \frac{2}{\eta} \frac{kT}{\Delta E_{gGe}(grade)} \left[1 - \frac{kT}{\Delta E_{gGe}(grade)} \left(1 - e^{-\Delta E_{gGe}(grade)/kT} \right) \right]. \tag{4.176}$$

Again, as in the case of the Early voltage, the improvement in τ_b is solely due to the grading of the Ge concentration across the base. In very thin base transistors, another term, W_B/v_{exit}, is added to the base transit time to capture the finite velocity of carriers as they exit the neutral base at the base–collector junction, v_{exit}.

EXAMPLE 4.10

Consider an early generation of SiGe HBTs where the Germanium composition varies from 5% at the base–emitter junction to 10% at the base–collector junction. Therefore, $\Delta E_{gGe}(0) =$ 37.5meV, $\Delta E_{gGe}(\text{grade}) = 75\text{meV}$, $kT = 25\text{meV}$, $\eta = 1.2$, and $\gamma = 1$. By inserting these values in (4.170), (4.172), and (4.176), we obtain

$$\frac{\tau_{bSiGe}}{\tau_{bSi}} = \frac{2}{\zeta} = \frac{2}{1.2} \frac{1}{3} \left[1 - \frac{1}{3} \left(1 - e^{-3} \right) \right] = 0.379$$

$$\frac{\beta_{SiGe}}{\beta_{Si}} \approx 1.2 \times 3 \frac{e^{1.5}}{1 - e^{-3}} = 17$$

$$\frac{(V_A)_{SiGe}}{(V_A)_{Si}} \approx e^3 \left[\frac{1 - e^{-3}}{3} \right] = 6.36.$$

These values emphatically illustrate the benefit of grading the Ge composition in the base to reduce transit time to less than 40% of that in a silicon BJT of identical base width and to improve β by more than an order of magnitude.

EXAMPLE 4.11

Let us now assume a more advanced SiGe HBT where the Germanium composition varies from 10% at the base–emitter junction to 20% at the base–collector junction. As a result, $\Delta E_{gGe}(0) =$ 75meV, $\Delta E_{gGe}(\text{grade}) = 150\text{meV}$, $kT = 25\text{meV}$, $\eta = 1.2$, and $\gamma = 1$. This leads to a sharper reduction in base transit time

$$\frac{\tau_{bSiGe}}{\tau_{bSi}} = \frac{2}{\zeta} = \frac{2}{1.26}\frac{1}{6}\left[1 - \frac{1}{6}\left(1 - e^{-6}\right)\right] = 0.23.$$

However, we note that doubling the Ge concentration and doubling the grading slope results in less than 2x reduction in base transit time. This implies that there are diminishing returns to further Ge grading in terms of base transit time reduction. Reducing the base thickness is also needed to continue the scaling of SiGe HBTs to higher speeds.

4.3.7 DC characteristics

In high-frequency applications, bipolar transistors are biased in the active region. As for any BJT, the DC performance of SiGe and InP HBTs biased in the active region is described by:

(a) the *Gummel or transfer characteristics* ilustrated in Figure 4.76, where the base/collector currents are plotted versus V_{BE}, and
(b) by the *output characteristics*, shown in Figure 4.77.

Both are captured by equation (4.177), where A_E represents the emitter area, J_S is the collector saturation current and is inversely proportional to the base Gummmel number, n_F (normally $= 1$) is the ideality factor of the collector current, while J_{BO} describes the base saturation current which is inversely proportional to the emitter Gummel number, G_E

$$I_C = A_E J_S\left(e^{\frac{qV_{BE}}{kTn_F}} - 1\right)\left(1 + \frac{V_{CE}}{V_A}\right) I_B \approx \frac{I_C}{\beta} = A_E J_{B0}\left(e^{\frac{qV_{BE}}{kT}} - 1\right). \quad (4.177)$$

We note that two distinct regions can be identified in the Gummel characteristics of Figure 4.76. At low and moderate V_{BE}, the collector and base currents show the expected exponential dependence on the base–emitter voltage. However, at V_{BE} values exceeding 0.85V, high current effects become important, the collector and base currents start to saturate, leading to β degradation. It is at the boundary between these two regions that HBTs are biased for high-frequency operation.

The output characteristics in Figure 4.77 show that the edge of the saturation region, corresponding to low V_{CE} values where the collector current increases linearly with V_{CE}, extends to V_{CE} values larger than 0.5V when the transistor is biased at large currents, as needed in high-frequency applications.

In the active region, there is clear evidence of self-heating, caused by the large power density dissipated in a small emitter and collector area. This manifests itself as an apparent negative output resistance. This effect becomes more pronounced at large currents and large V_{CE}. The junction temperature in modern HBTs can rise by as much as 50–80°C with respect to the ambient when the transistor is biased for maximum speed [35].

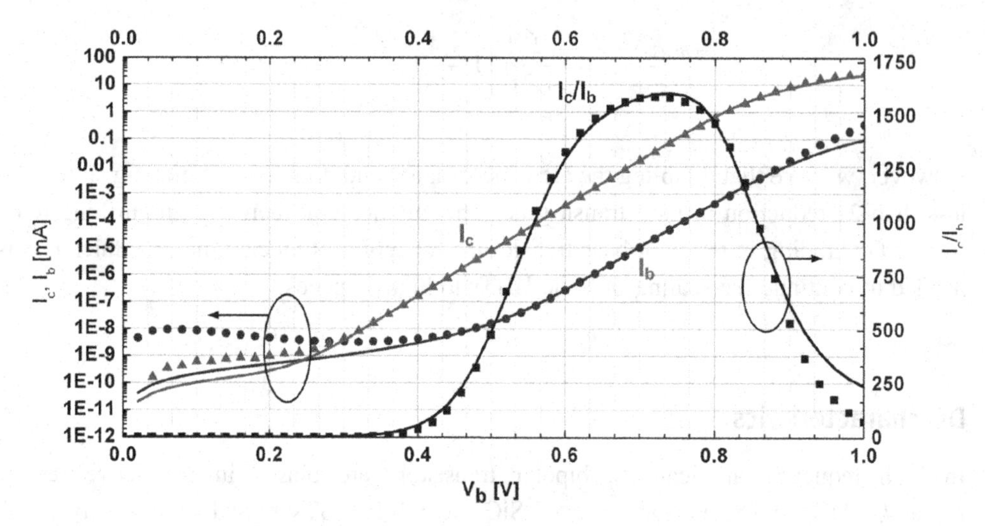

Figure 4.76 Gummel (transfer characteristics) of an HBT

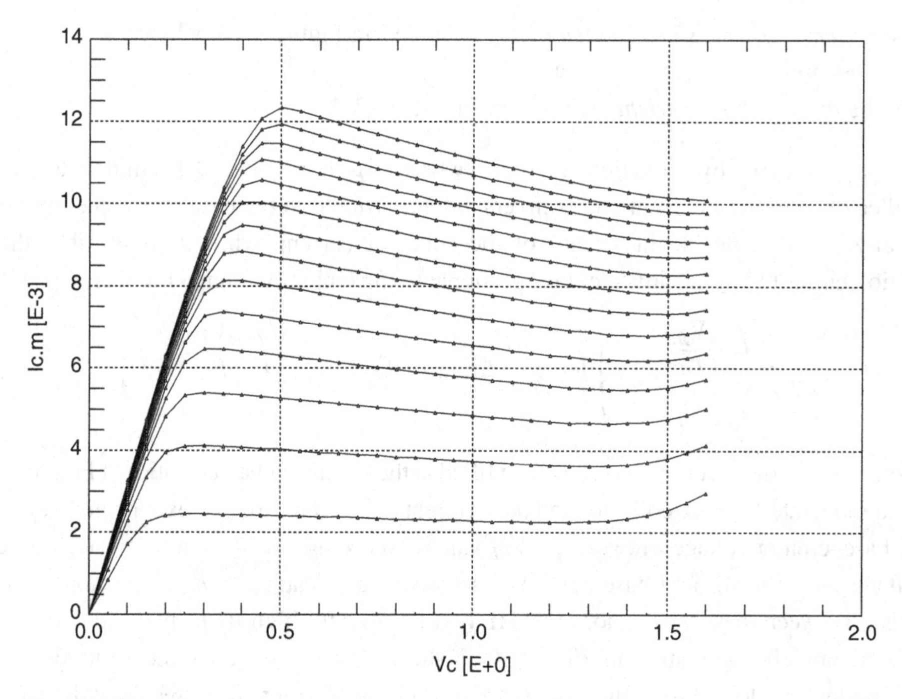

Figure 4.77 DC output characteristics when I_B varies from 0 to 28 μA in 2 μA steps

4.3.8 Avalanche breakdown

At large BC voltages, the electric field in the BC junction exceeds the breakdown field, E_B, of the semiconductor material used in the collector, causing a sudden increase in I_C. This regime must be avoided. BV_{CEO}, which represents the collector–emitter breakdown voltage measured when the base is left open ($I_B = 0$), as in Figure 4.78(a), and BV_{CBO}, describing the collector–base breakdown voltage when the emitter is left open ($I_E = 0$), are the Figures of Merit, FoMs,

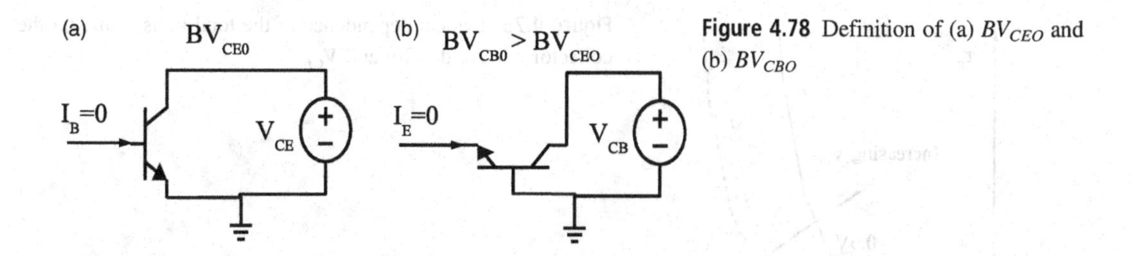

Figure 4.78 Definition of (a) BV_{CEO} and (b) BV_{CBO}

that characterize the transistor performance at large voltage. In addition to setting the safety bounds for the DC voltages applied to the HBT, they also limit the maximum voltage swing that can be safely supported by an HBT at high frequencies.

The transistor should not be operated with V_{CB} larger than BV_{CBO} because it will be destroyed. It is, however, possible to safely operate SiGe HBTs at collector–emitter voltages larger than BV_{CEO} but only under special conditions when the resistance seen by the transistor between base and emitter is small and the $I_C \times V_{CE}$ product does not exceed the thermal limits (a few tens of mW/μm^2 of emitter area) for safe operation.

4.3.9 The Kirk effect and other high current effects

One of the main advantages of HBTs over MOSFETs and HEMTs is that they operate at much higher current densities. In general, at low and moderate currents, the high-frequency performance of the HBT improves as the collector current density increases, primarily because the transconductance, g_m, increases almost linearily with the collector current. However, at high collector current densities, intrinsic "high-current effects" and the extrinsic parasitic resistances become important, limiting g_m and severely degrading β, f_T, f_{MAX}, and F_{MIN}.

The most important intrinsic effect is the base-pushout effect, also known as the Kirk effect. As the collector current increases, the concentration of minority carriers leaving the base and entering the collector increases significantly, reaching and exceeding the background doping concentration in the collector, N_C. When that happens, the depletion charge on the collector side of the base–collector space charge region is neutralized by the large number of free carriers entering the collector, effectively eliminating the depletion region and extending the base region into the collector, hence the name "base pushout." The outcome is a sudden increase of the effective base width, W_B, and, therefore, of τ_b, triggering a drop in β and f_T. The critical collector current density, J_{CMAX}, at which this phenomenon occurs is a function of the collector doping N_C

$$J_{CMAX} = qv_{sat}N_C \tag{4.178}$$

and constitutes one of the main design specifications for the intrinsic collector region of an HBT.

The typical dependence of the total intrinsic device transit time, ($\tau_F = \tau_b + \tau_c$), through the base and through the collector, on the collector current density and collector–emitter voltage, V_{CE}, is sketched in Figure 4.79. We note that τ_F improves (decreases) as V_{CE} increases. J_{CMAX}, too, improves (increases) with V_{CE}. The latter behavior can be explained by the fact that the

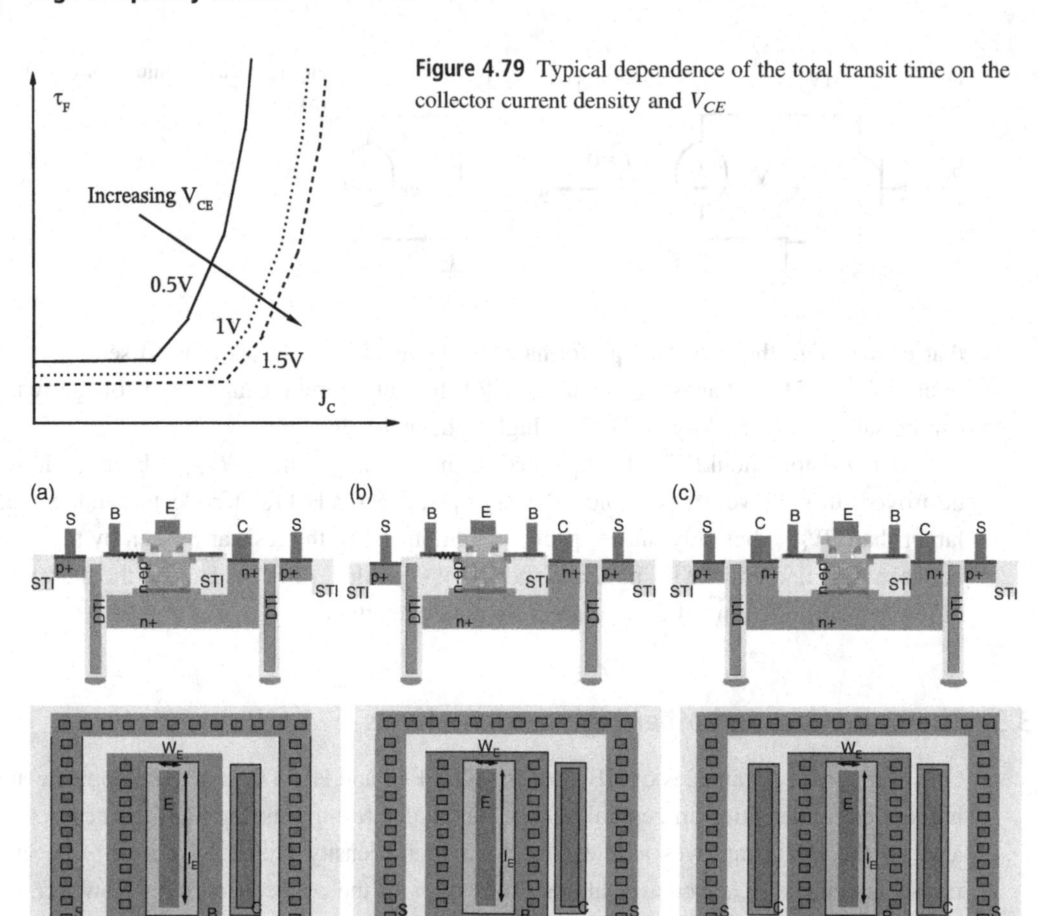

Figure 4.79 Typical dependence of the total transit time on the collector current density and V_{CE}

Figure 4.80 SiGe HBT layouts (a) CBE, (b) CBEB, (c) CBEBC

doping concentration increases from the base–collector junction towards the substrate. Therefore, as V_{CE} becomes larger, the space charge region extends deeper into the collector where the effective doping is higher.

4.3.10 HBT layout

The high-frequency circuit designer is often faced with having to choose between several HBT layouts available in the design kit. Figure 4.80 illustrates cross-sections and top level views of SiGe HBT layouts with single base, single emitter and single collector contact stripes (BEC), double base, single emitter and single collector contact stripes (BEBC), and double base, single emitter and double collector contact stripes (CBEBC). In addition to the number of emitter, base, and collector contact stripes, which impact the high-frequency and noise performance of the transistor, the circuit designer is usually allowed to vary the emitter length, l_E, and, sometimes (but rarely), the emitter width, w_E. At the time of writing (spring 2011), the most

(a)

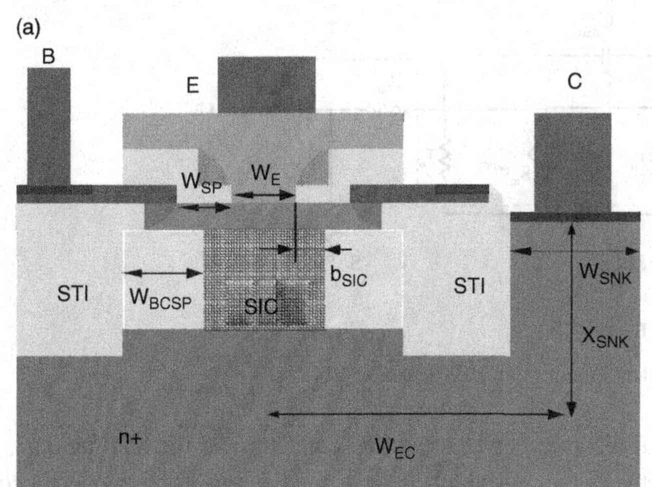

(b)

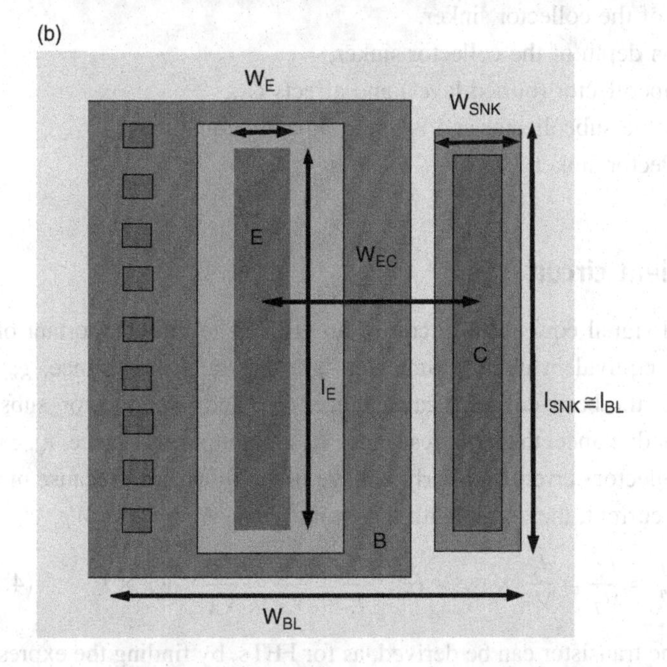

Figure 4.81 Definition of the vertical and lateral dimensions of a BEC SiGe HBT layout using (a) the vertical cross-section and (b) the top view of the layout, respectively

advanced SiGe BiCMOS processes employ 130nm emitter width and lithography, with 55nm lithography being considered for the next generation BiCMOS technology. The length of the emitter stripe typically varies between 0.5μm and 15μm, with 1–5 μm being the most common range. If larger emitter lengths and currents are desired, 2 – or, at most, 3 – emitter stripe layouts can be employed. Beyond that, transistors must be connected in parallel to avoid performance degradation due to current-crowding, electromigration, and/or self-heating.

The 3-D geometrical dimensions of the SiGe HBT are defined in Figure 4.81 as follows:

- W_{SP} represents the distance between the external base poly and the emitter contact and affects the external base resistance,
- b_{SIC} is the lateral spread of the SIC implant in the collector,

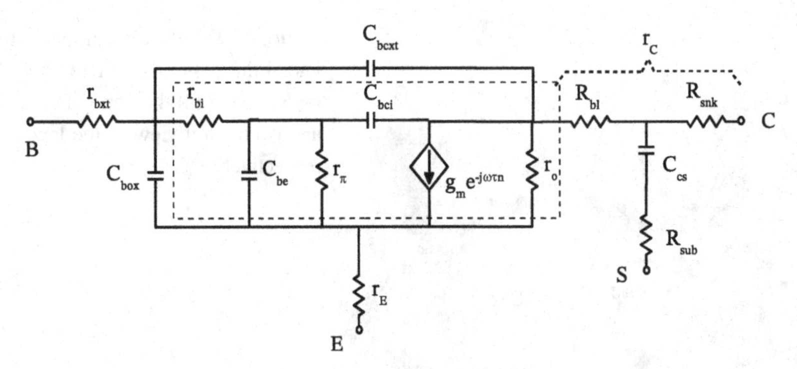

Figure 4.82 Small signal equivalent circuit of an HBT (SiGe or III–V)

- w_{BCsp} represents the lateral distance between the SIC collector and the STI region,
- W_{EC} defines the spacing between the center of the emitter and the center of the collector contacts,
- W_{SNK} describes the width of the collector sinker,
- X_{SNK} describes the junction depth of the collector sinker,
- W_{BL} is the width of the subcollector (buried layer) and affects C_{cs},
- l_{BL} describes the length of the subcollector and is approximately equal to,
- l_{SNK}, the length of the collector sinker.

4.3.11 The small signal equivalent circuit

Figure 4.82 shows the small signal equivalent circuit of an HBT. The most important of the high-frequency small signal equivalent circuit parameters are the transconductance, g_m, the base–emitter capacitance C_{be}, the base–collector capacitance, C_{bc}, and the collector–substrate capacitance, C_{cs}. The transconductance, the input resistance r_π, and output resistance, r_o, can be obtained directly from the collector current and Early voltage of the transistor. Because of their dependence on the collector current, they scale with the emitter area, $A_E = l_E \times W_E$

$$I_C = J_C \times w_E \times l_E; \quad g_m = \frac{I_C}{V_T} = \frac{J_C}{V_T} \times w_E \times l_E; \quad \frac{1}{r_o} = \frac{I_C}{V_A} = \frac{J_C}{V_A} \times w_E \times l_E. \tag{4.179}$$

The expression of the f_T of the transistor can be derived, as for FETs, by finding the expression of the short-circuit current at the collector terminal, in the common-emitter configuration when the emitter and substrate terminals are shorted together in Figure 4.82.

4.3.12 Parasitic resistances

Figure 4.83 summarizes the parasitic resistances of SiGe and InP HBTs with single emitter, two base and two collector contact stripes. Because of the large emitter and collector currents, the metal contacts to the emitter and to the collector are formed as continuous metal bars. Their contribution to the total emitter and collector resistance can be ignored. However, in SiGe HBTs, the contacts to the substrate and to the external base poly are formed as minimum size

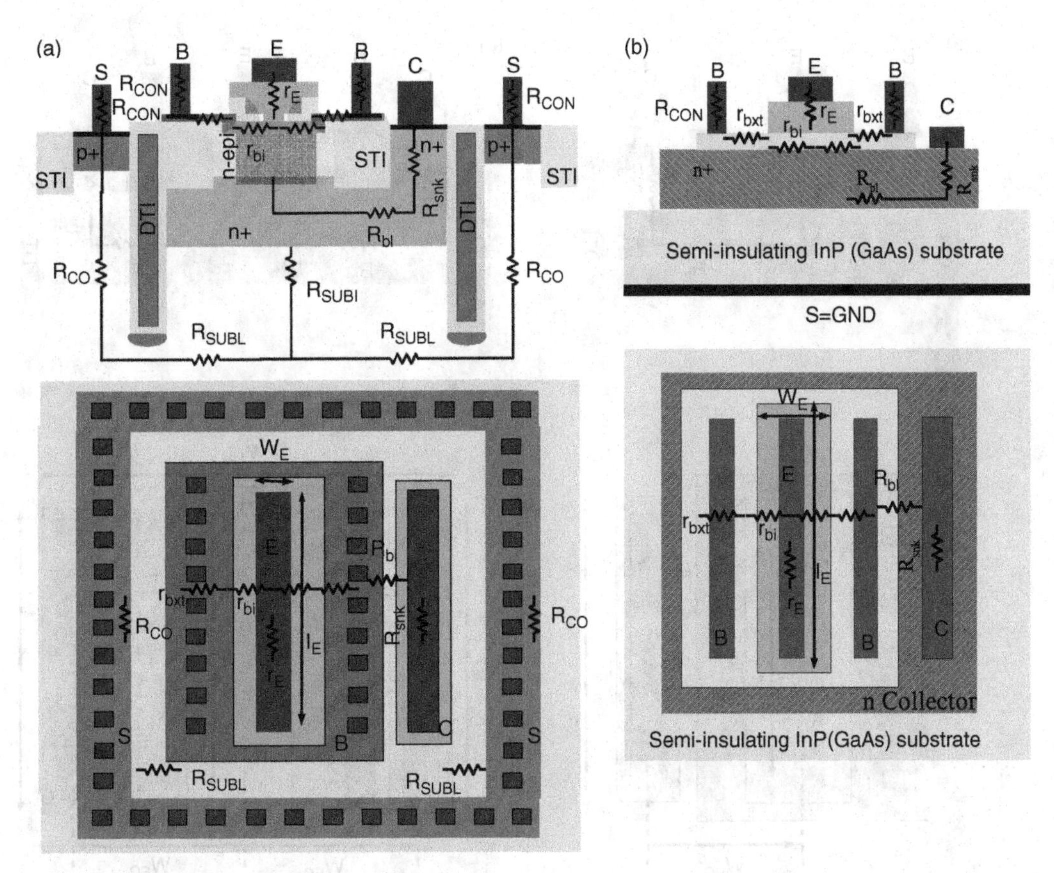

Figure 4.83 (a) SiGe and (b) III–V HBT cross-section and top layout view indicating the regions associated with the various components of the emitter, base, collector, and substrate resistances

tungsten squares, identical to those employed for contacting the MOSFET source, gate, and drain regions. The resistance, R_{CON}, of such a minimum size tungsten contact can be as high as 10–15Ω in a 130nm SiGe BiCMOS process. The components of the emitter, collector, base, and substrate resistances are discussed in detail next for several HBT layouts.

Emitter resistance

The emitter resistance affects the DC I-V characteristics by increasing the base–emitter voltage at which the peak device performance is achieved. The associated DC voltage drop on the parasitic emitter resistance can be as high as 20–50 mV. Additionally, and more importantly, the g_m, f_T, f_{MAX} and the minimum noise factor, F_{MIN}, of the transistor are all degraded by r_E. Its contribution, similar to that of the source resistance in MOSFETs, becomes increasingly important in more advanced lithographic nodes. Its expression as a function of process parameters and emitter geometry is

$$r_E = \frac{R_{Se}}{l_E w_E} \tag{4.180}$$

where the emitter sheet resistance, R_{Se}, is a process parameter specified in Ohm × μm^2.

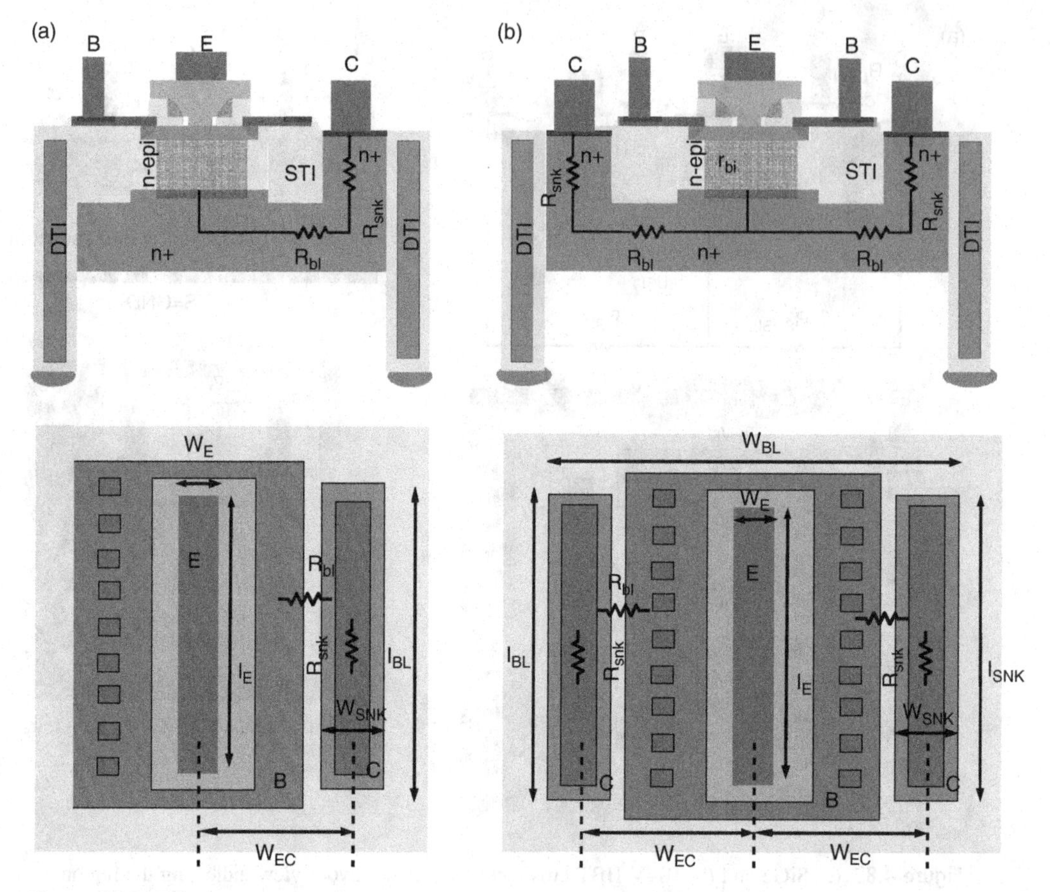

Figure 4.84 Components of the collector resistance for (a) single- and (b) double-collector contact layouts

Collector resistance

The collector resistance depends primarily on the doping levels and geometry of the sinker and buried layer regions. It affects the DC output charactersictics by increasing V_{CEsat} and reduces the f_T and f_{MAX} of the transistor. For single collector stripe (BEC or BEBC) structures (Figure 4.84 left side)

$$r_C = R_{BL} + R_{Sinker} = \frac{R_{Sbl} \times w_{EC}}{l_{BL}} + \frac{R_{Ssnk} \times x_{SNK}^2}{l_{SNK} w_{SNK}} \tag{4.181}$$

For the double collector stripe (CBEBC) structure (Figure 4.84 right side)

$$r_C = \frac{R_{BL}}{2} + \frac{R_{Sinker}}{2} = \frac{R_{Sbl} \times w_{EC}}{2 l_{BL}} + \frac{R_{Ssnk} \times x_{SNK}^2}{2 l_{SNK} w_{SNK}} \tag{4.182}$$

Here, R_{Sbl} represents the sheet resistance of the buried layer (10–20Ω/sq) and R_{Ssnk} is the sheet resistance in the sinker region.

As a general guideline, for an HBT layout with a single emitter stripe shorter than 2μm, the benefit of using two collector stripes to reduce the collector resistance and improve f_T and f_{MAX}

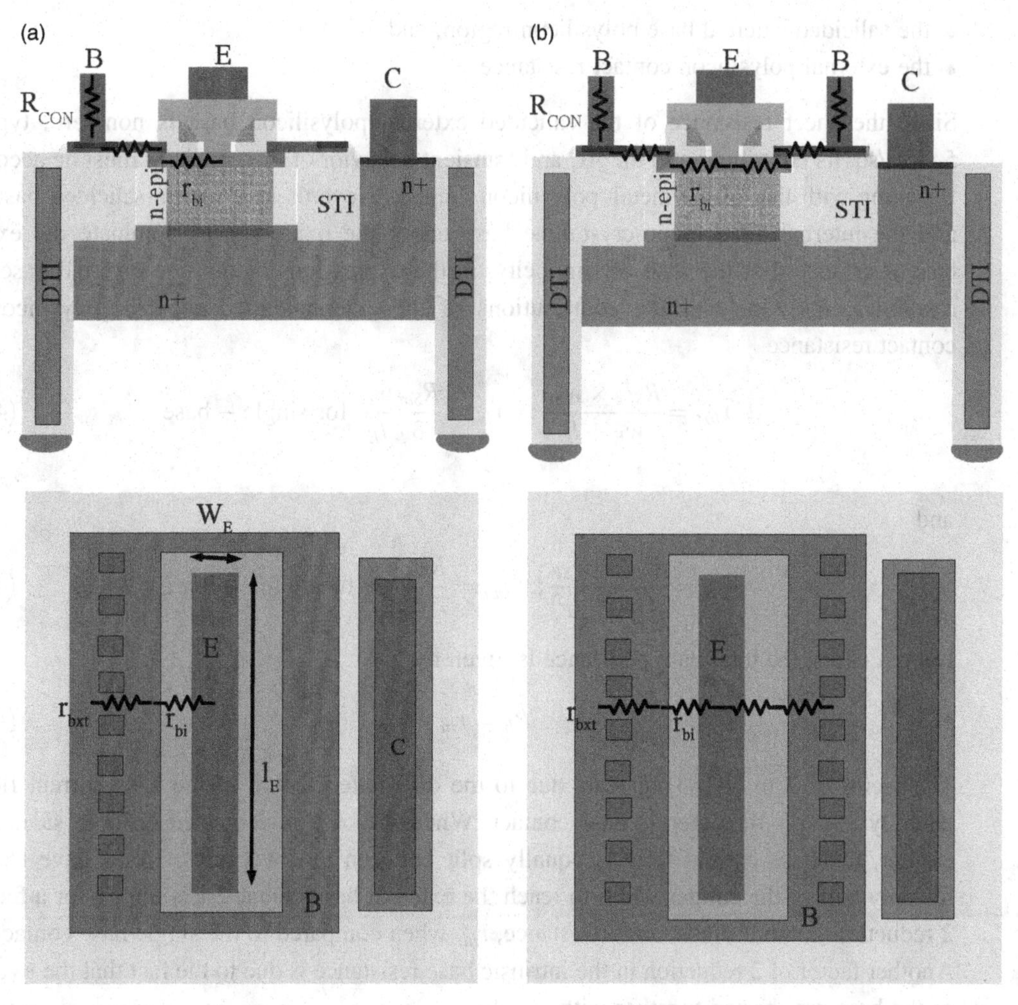

Figure 4.85 Components of the base resistance for (a) single- and (b) double-base contact layouts

is lost by the increase in the collector substrate capacitance. For longer emitters, layouts with two or more collector stripes should always be used.

Base resistance

Figure 4.85 illustrates the contributors to the internal and external base resistances in single emitter HBT layouts with single and two base contact stripes. The intrinsic base resistance is determined by the base doping, N_B, the thickness, W_B, of the intrinsic SiGe base region and by the emitter width, W_E. Most often, the base width and base doping are lumped together into the pinched-base sheet resistance, R_{sbi}, an important process parameter. The latter is obtained directly from transistor test structure measurements and is specified as a sheet resistance in Ω/sq.

There are four contributors to the extrinsic base resistance:

- the mono-crystalline SiGe base region below the base–emitter spacer, W_{SP},
- the un-salicided external base polysilicon region below the emitter oxide,

- the salicided external base polysilicon region, and
- the external polysilicon contact resistance.

Since the sheet resistance of the salicided external polysilicon base is non-zero, typically 5–15Ω/sq, its contribution to the *AC* and transient behavior of the transistor must be accounted for along with that of the metal–polysilicon contact. Nevertheless, the un-salicided base poly and the internal SiGe mono-crystalline base under the oxide spacer dominate the external base resistance. For the sake of simplicity, in the expressions below, the external base sheet resistance, R_{sbx}, includes the contributions of the external base salicided polysilicon and contact resistance

$$r_{bxt} = \frac{R_{Sbx} \times w_{SP}}{w_E + l_E}; \quad r_{bi} = \frac{R_{Sbi}}{3}\frac{w_E}{l_E} \text{ for single} - \text{base} \tag{4.183}$$

and

$$r_{bxt} = \frac{R_{Sbx} \times w_{SP}}{2(w_E + l_E)}; \quad r_{bi} = \frac{R_{Sbi}}{12}\frac{w_E}{l_E} \text{ for double} - \text{base.} \tag{4.184}$$

In both cases, the total base resistance is given by

$$R_b = r_{bi} + r_{bxt}. \tag{4.185}$$

The factor of 3 in (4.183) appears due to the distributed nature of the base current flowing laterally towards the external base contact. When the base is contacted on both sides of the emitter, the base current flow is equally split between the two sides. Holes have to travel for only half of the emitter width to reach the external base contact, accounting for a factor of 2 reduction in the intrinsic base resistance, r_{bi}, when compared to the single base contact case. Another factor of 2 reduction in the intrinsic base resistance is due to the fact that the two sides of the base are shorted together with metal.

Substrate resistance

The substrate resistance is important in high-frequency and high-speed applications because it degrades f_{MAX} and, to a smaller extent, F_{MIN}. It can be calculated based on layout geometry, the location and shape of the substrate contact, and on the sheet resistance in the substrate. In III-V HBTs, the substrate resistivity is so high (10^7 to $10^8 \Omega \times$cm) that the semi-insulating substrate can be approximated by a pure capacitance. In a SiGe BiCMOS process, the substrate resistivity is typically 10–20Ω × cm. In Figure 4.86, R_{CO} represents the component of the substrate resistance corresponding to the region of the substrate located immediately below the substrate contact, while R_{SUBL} and R_{SUBI} describe the contributions of the substrate region below the transistor. To ensure that their values are well-modeled and remain independent of layout style, most SiGe HBT foundries strictly enforce a continuous ring of substrate contacts surrounding the transistor.

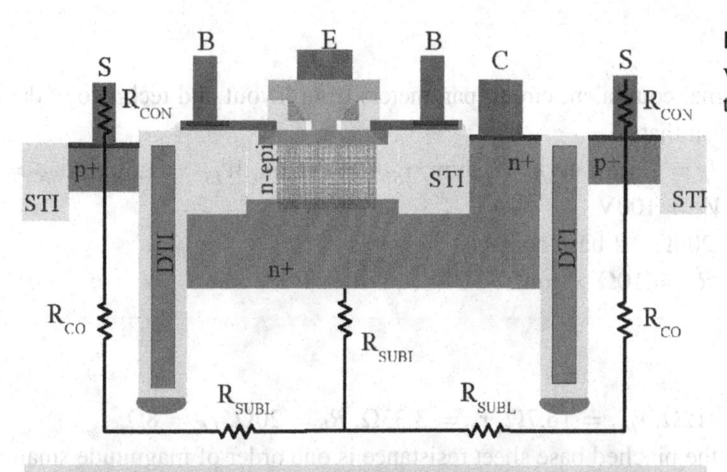

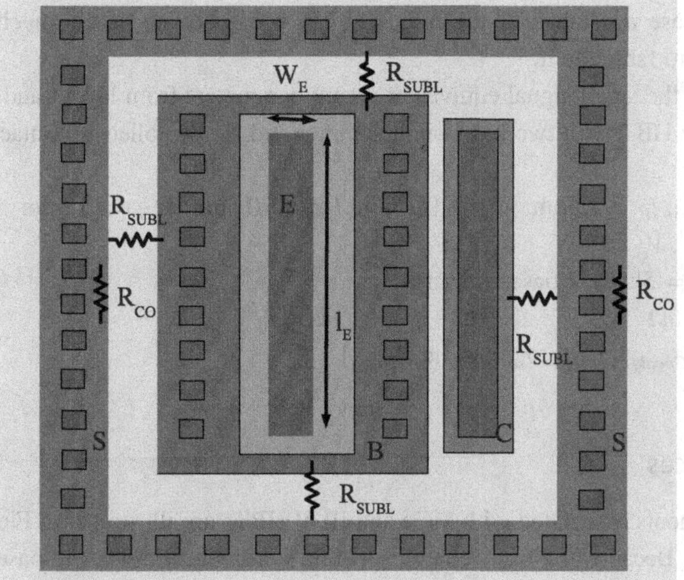

Figure 4.86 Illustration of the various components that make up the substrate resistance

EXAMPLE 4.12

Calculate the small signal equivalent circuit parameters from layout and technology data for a 300GHz SiGe HBT with two base, single emitter and single collector contact stripes, knowing that $w_E = w_{SNK} = 0.1\mu m$, $l_E = 2.5\mu m$, $w_{SP} = 0.1\mu m$, $x_{SNK} = 0.25\mu m$, $W_{EC} = 1.5\mu m$, $b_{SIC} = 10nm$

$J_{CpeakfT} = 20mA/\mu m^2$, $V_A = 100V$

$R_{sbx} = 200\Omega$ $R_{Sbi} = 2k\Omega$ 2 base contacts

$R_{sbl'} = R_{sink'} = 20\Omega$ $R_{Se} = 2\Omega \times \mu m^2$.

Solution

$I_{CpeakfT} = 5mA$, $r_C = 17\Omega$, $r_{bxt} = 3.8\Omega$, $r_{bi} = 6.7\Omega$, $R_b = 10.5\Omega$, $r_E = 8\Omega$.

We note that r_C is the largest of the parasitic resistances.

EXAMPLE 4.13

Calculate the small signal equivalent circuit parameters from layout and technology data for a 300GHz InP HBT, given that

$w_E = w_{SNK} = 0.5\mu m$, $l_E = 2.5\mu m$, $w_{SP} = 0.5\mu m$, $x_{SNK} = 0.1\mu m$, $W_{EC} = 1.5\mu m$, $b_{SIC} = 0$

$J_{CpeakfT} = 8mA/\mu m^2$, $V_A = 100V$

$R_{sbx} = 200\Omega$ $R_{Sbi} = 200\Omega$ 2 base contacts

$R_{sbl'} = R_{sink'} = 20\Omega$ $R_{se} = 10\Omega \times \mu m^2$.

Solution

$I_{cpeakfT} = 10mA$, $r_C = 12\Omega$, $r_{bxt} = 16.7\Omega$, $r_{bi} = 3.33\Omega$, $R_b = 20\Omega$, $r_E = 8\Omega$.

Note that, although the pinched base sheet resistance is one order of magnitude smaller than in a SiGe HBT, the base resistance of the InP HBT is larger because of the much coarser lithography employed to fabricate it.

Problem: Calculate the small signal equivalent circuit parameters from layout and technology data for a 150GHz HBT with two base, single emitter and single collector contact stripes, knowing that

$w_E = w_{SNK} = 0.2\mu m$, $l_E = 2.5\mu m$, $w_{SP} = 0.15\mu m$, $l_{SNK} = 0.5\mu m$, $W_{EC} = 1.5\mu m$

$J_{CpeakfT} = 6mA/\mu m^2$, $V_A = 80V$

$R_{sbx} = 500\Omega$ $R_{Sbi} = 5k\Omega$ 2 base contacts

$R_{sbl} = 20\Omega$ $R_{se} = 5\Omega$

Find the values of $I_{CpeakfT}$, g_{mpfT}, r_o, r_C, and R_b.

4.3.13 Junction capacitances

The capacitive components associated with SiGe and III-V HBTs are illustrated in Figure 4.87 and 4.88, respectively. Because the base–emitter junction is forward biased, the base–emitter capacitance consists of the diffusion capacitance, $C_{be,dff}$, and of the junction depletion capacitance C_{JE}. The latter comprises a periphery, C_{JEP}, and an area component, C_{JEA}. The diffusion capacitance is equal to the product between the total transit time of the HBT, τ_F, and the transconductance. Since the transconductance scales with the emitter–base junction area, so should the base–emitter capacitance, if it were not for the periphery component of the depletion capacitance. Additionally, since the transconductance is proportional to the collector current, the total base–emitter capacitance increases with increasing bias current and is typically dominated by the diffusion component

$$C_{be} = C_{be,diff} + C_{JE} = \tau_F g_m + C_{JEA} w_E l_E + 2C_{JEP}(l_E + W_E) = C_{beA} w_E l_E + 2C_{JEP}(l_E + W_E). \quad (4.186)$$

In the case of SiGe HBTs, because the emitter polysilicon and metal layers overhang the oxide above the base, an oxide capacitance, C_{EOX}, which is proportional the emitter perimeter, is also included in the high-frequency equivalent circuit between the external base and the emitter.

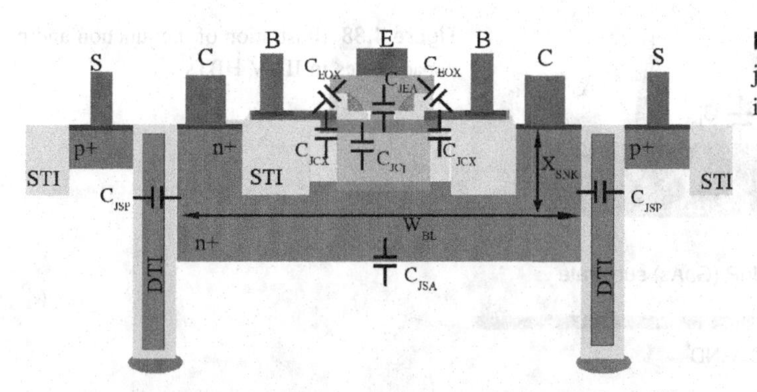

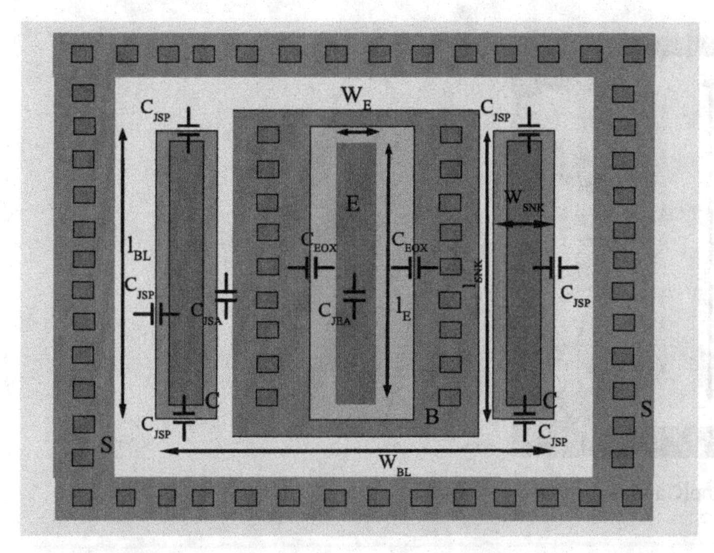

Figure 4.87 Illustration of the junction and parasitic capacitances in SiGe HBTs

The base–collector capacitance, similar to the base resistance, has a distributed nature and is typically approximated by two components, an intrinsic one, C_{bci}, corresponding to the area below the emitter

$$C_{bci} = C_{jCi} \times A_{BCi} = C_{jCi} \times l_E \times (w_E + 2b_{sic}) \tag{4.187}$$

and an extrinsic one, C_{bcxt}, as illustrated in Figure 4.87

$$C_{bcxt} = C_{jCX} \times A_{BCext} \approx C_{JCX} \times 2 \times w_{BCsp} \times \left(w_E + l_E + 2w_{BCsp}\right). \tag{4.188}$$

In advanced SiGe HBTs, the intrinsic collector doping is higher than the doping in the extrinsic region of the collector. However, in III-V HBTs, the two regions usually have identical doping concentrations (Figure 4.88).

The collector–substrate capacitance, C_{cs}, is largest for single-emitter HBTs with two collector and two base contact stripes and smallest for layouts with one base and one collector contact stripe. It, too, can be broken into area, C_{jSA}, and periphery, C_{jSP}, components

$$C_{CS} = C_{jSA} \times l_{BL} \times w_{BL} + C_{jSP} \times 2 \times (l_{BL} + w_{BL}). \tag{4.189}$$

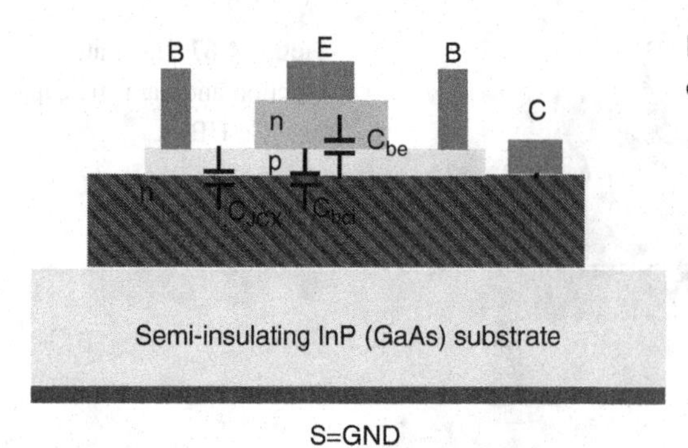

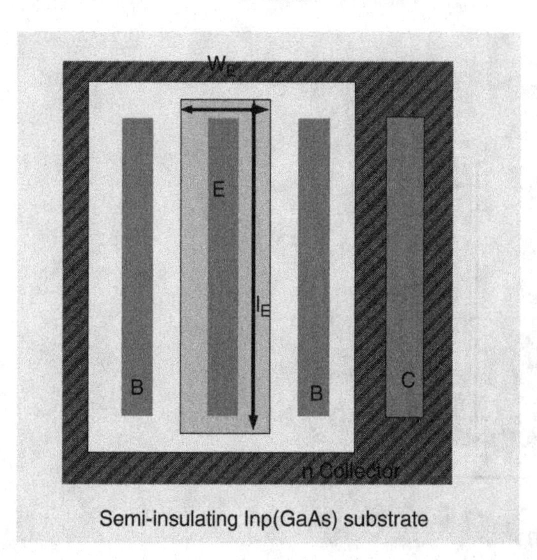

Figure 4.88 Illustration of the junction and parasitic capacitances in III–V HBTs

4.3.14 f_T and f_{MAX} characteristic current densities

From its inception in the late 1980s, the design of HBT-based high-frequency integrated circuits has relied on the (experimental) f_T and f_{MAX} versus J_C characteristics to select the proper bias current density of the transistor that maximizes the performance for a particular circuit function. SiGe BiCMOS and III-V HBT foundries typically provide, at the very least, experimental f_T and f_{MAX} versus J_C characteristics, as illustrated in Figure 4.89 for an experimental 130nm SiGe HBT process.

The shape of the characteristics in Figure 4.89 can be understood by examining the expression of the overall emitter–collector transit time of the HBT, τ_{EC}, which includes components due to the base-collector transit time, charging time of the base–emitter and base–collector capacitances, as well as the external RC parasitics

$$\frac{1}{2\pi f_T} = \tau_{EC}(I_C) = \tau_F(J_C A_E) + \frac{kT}{qJ_C A_E}(C_{be} + C_{bc}) + (r_E + r_C)C_{bc}. \tag{4.190}$$

The last term in (4.190) is a weak function of the collector current density and, in a first-order approximation, can be considered constant. As illustrated in Figure 4.89, at low collector current densities of 0 to 5mA/μm^2, τ_F is constant and the middle term in (4.190) dominates, leading to a fast reduction in τ_{EC} and to a sharp increase in f_T as J_C increases. At moderate current densities, in the 5–20 mA/μm^2 range for this SiGe HBT, the middle term becomes smaller than the sum of the other two terms, corresponding to the relatively flat region of the f_T - J_C characteristics in Figure 4.89, with J_{pfT} = 15–20 mA/μm^2. Finally, at larger current densities, beyond 20mA/μm^2, J_{CMAX} is reached, resulting in a sharp rise of τ_F (Figure 4.79), and a corresponding sharp fall in f_T.

The measured data in Figure 4.89 show that f_{MAX} closely follows the behavior of f_T. This can be explained by the equation that links f_{MAX} to f_T, R_b and C_{bc}

$$f_{MAX} \approx \sqrt{\frac{f_T}{8\pi R_b C_{bc}}} \tag{4.191}$$

where only f_T is a strong function of the collector current density.

In general, for most III-V and SiGe HBTs, the peak f_T and peak f_{MAX} current densities coincide. Furthermore, J_{pfT} and J_{pfMAX} remain practically constant as the temperature varies from 25°C to 125°C.

Most interesting is the behavior of the effective transconductance, $g_{me} = g_m/(1 + r_E\, g_m)$, which, in the range of current densities of interest for high-frequency circuits, 5–25 mA/μm^2, departs sharply from the ideal $g_m = (qJ_C A_E)/kT$ law, reaching a peak value at 25mA/μm^2 and then decreasing as J_C increases. The lower effective transconductance can be explained by the negative feedback through the parasitic emitter resistance, by the increase in the ideality factor of the collector current, n_F, in equation (4.177) from 1 to 2 at high injection levels, and by self-heating.

The dependence of f_T, f_{MAX}, and of the characteristic current densities on the HBT layout geometry

To understand the dependence of the f_T and f_{MAX} characteristics of the HBT on the width and length of the emitter, the two layout parameters available to the circuit designer, we must start from the expression that links f_T to the structural parameters of the HBT

$$\frac{1}{2\pi f_T} = \frac{1}{\zeta}\left[\frac{W_B^2}{D_B} + \frac{W_B}{v_{exit}}\right] + \frac{W_{dBC}}{2v_{SAT}} + \frac{kT}{qI_C}(C_{be} + C_{bc}) + (r_C + r_E)C_{bc} \tag{4.192}$$

where a second term has been added to the base transit time in the square bracket to capture the exit velocity in base regions with graded material composition [37].

The transit time depends primarily on the vertical structure of the transistor and is only weakly dependent on the layout geometry. However, the last two terms in (4.192) are expected to vary with the emitter length and width. This becomes apparent when the formulae of the transistor capacitances and parasitic resistances are inserted in (4.192) to obtain

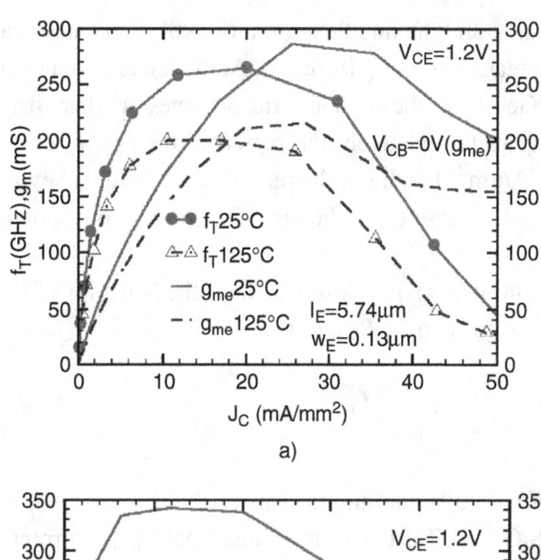

Figure 4.89 Measured (a) g_m, f_T, and (b) f_{MAX} vs. J_C characteristics for 130nm SiGe HBTs at 25°C and 125°C [36]

$$\frac{1}{2\pi f_T} = \tau_F + \frac{KT}{qJ_C}\left[C_{JEA} + C_{JCi}\left(1 + \frac{2b_{SIC}}{w_E}\right) + (C_{JEP} + W_{BCsp}C_{JCX})\left(\frac{2}{w_E} + \frac{2}{l_E}\right) + \frac{4C_{JCX}W_{BCsp}^2}{w_E l_E}\right] + (r_C l_E w_E + R_{sE})C_{JCi}\left(1 + \frac{2b_{SIC}}{w_E}\right) \tag{4.193}$$

To first order, f_T is independent of emitter length if $l_E >> w_E$. For example, in a 130nm process, this corresponds to $l_E > 1.3\mu m$. As a second-order effect, f_T varies slightly with l_E due to peripheral and oxide components of the BE capacitance and due to r_C, which does not scale linearly with the inverse of the emitter area. This behavior is captured in the measured and simulated $f_T - I_C$ characteristics of 250GHz SiGe HBTs with variable emitter lengths shown in Figure 4.90 [35]. We note that devices with very short emitters have 10–15% lower cutoff frequencies. The current density at which the peak f_T occurs tends to increase for shorter emitter lengths, resulting in imperfect scaling with emitter length.

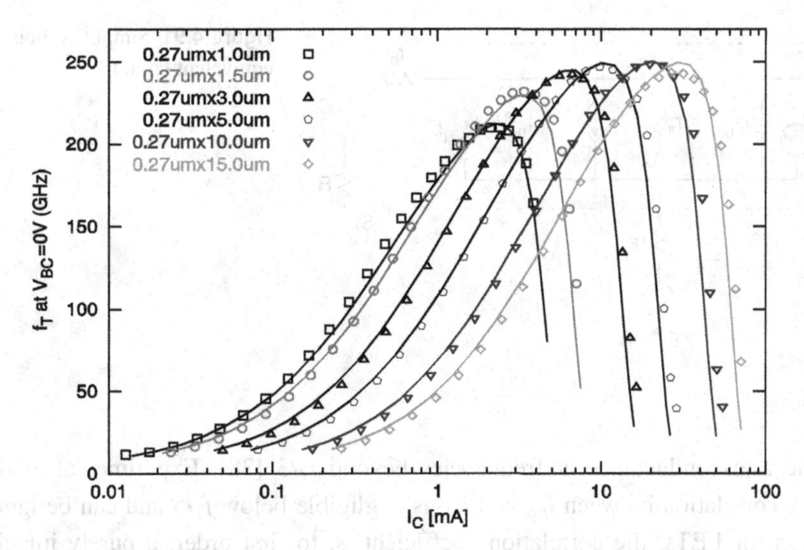

Figure 4.90 Cutoff frequency as a function of collector current density in SiGe HBTs with different emitter lengths [35]

Because of the sensitivity of R_b and C_{bc} to the transistor layout geometry, the maximum frequency of oscillation is somewhat more sensitive than f_T to the transistor layout. However, this sensitivity is mild compared to that encountered in Si MOSFETs.

Problem: Which of the HBT layouts CEB, CBEB, or CBEBC has the highest cutoff frequency and why?

4.3.15 Noise parameters

The simplified noise equivalent circuit of the HBT is illustrated in Figure 4.91 where the distributed R-C network describing the base–collector region has been simplified to a single resistor, R_b, and a single capacitor, C_{bc}. The input noise current source is dominated by the base current shot noise, with contributions from the collector current only at frequencies approaching the f_T of the transistor

$$\langle i_{n1} i_{n1}^* \rangle = \langle i_{nB} i_{nB}^* \rangle = 2q\Delta f \left(I_B + |1 - e^{-j\omega\tau_n}|^2 I_C \right). \tag{4.194}$$

The output noise current is due to the collector shot noise current

$$\langle i_{n2} i_{n2}^* \rangle = \langle i_{nC} i_{nC}^* \rangle = 2q\Delta f I_C \tag{4.195}$$

where I_B and I_C are the DC base and collector currents of the HBT and q is the electron charge.

The statistical corellation between the input and output noise currents is described by

$$\langle i_{n1} i_{n2}^* \rangle = \langle i_{nB} i_{nC}^* \rangle = 2qI_C[exp(j\omega\tau_n) - 1] \approx j2qI_C\omega\,\tau_n \tag{4.196}$$

where τ_n, always smaller than the device transit time, τ_F, is believed to be related to the transit time through the base–collector space charge region. Experimentally, τ_n can be extracted from

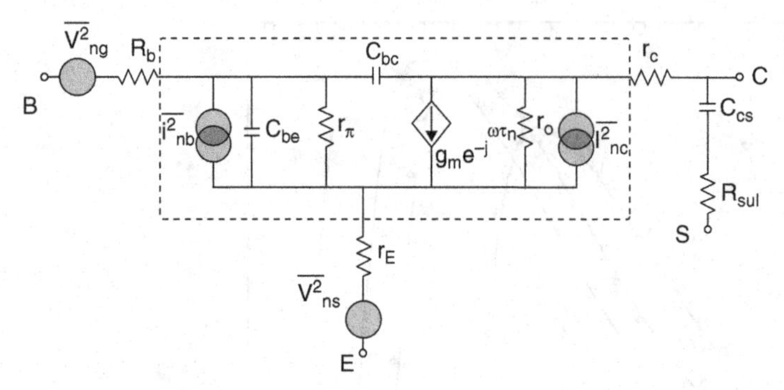

Figure 4.91 Simplified noise equivalent circuit

the phase of the transconductance at frequencies beyond $f_T/3$ [38]. Experimental evidence indicates that the correlation between i_{nB} and i_{nC} is negligible below $f_T/5$ and can be ignored. Note also that, as for FETs, the correlation coefficient is, to first order, a purely imaginary number.

In addition to the two intrinsic shot noise currents, thermal noise voltage sources associated with the parasitic emitter and base resistances are also significant contributors to the noise behavior of the HBT

$$\langle v_{nB}v_{nB}^{*}\rangle = 4kT\Delta f R_b \tag{4.197}$$

$$\langle v_{nE}v_{nE}^{*}\rangle = 4kT\Delta f r_E. \tag{4.198}$$

By inserting (4.194)–(4.195) into (4.38)–(4.40), and after some lengthy algebraic manipulations [39], we obtain

$$R_n = R_b + r_E + \frac{\langle I_{n2}^2\rangle}{4kT\Delta f|(y_{21})|^2} \approx R_b + r_E + \frac{n_F^2 V_T}{2I_C} \tag{4.199}$$

$$Y_{sopt} \approx \frac{f}{f_T R_n}\left[\sqrt{\frac{I_C}{2V_T}(r_E + R_b)\left(1 + \frac{f_T^2}{\beta f^2}\right) + \frac{n_F^2 f_T^2}{4\beta f^2}} - j\frac{n_F}{2}\right] \tag{4.200}$$

$$F_{MIN} \approx 1 + \frac{n_F}{\beta} + \frac{f}{f_T}\sqrt{\frac{2I_C}{V_T}(r_E + R_b)\left(1 + \frac{f_T^2}{\beta f^2}\right) + \frac{n_F^2 f_T^2}{\beta f^2}} \tag{4.201}$$

where n_F is the collector current ideality factor, close to 1.

By examing (4.199), we notice that the noise resistance, R_n, is independent of frequency, as confirmed by the experimental data in Figure 4.92. From (4.200) and Figure 4.93, the optimum noise resistance, R_{SOPT}, is seen to be inversely propotional to frequency.

Finally, (4.201) shows that the minimum noise figure increases linearly with frequency with a slope that is inversely proportional to f_T. As a result, transistors with higher f_T exhibit a slower degradation of F_{MIN} as the frequency increases. This can be observed in Figure 4.94, where the measured noise figure and associated power gain, G_A, at the optimal noise figure match are plotted versus frequency up to 170GHz for two generations of SiGe HBTs.

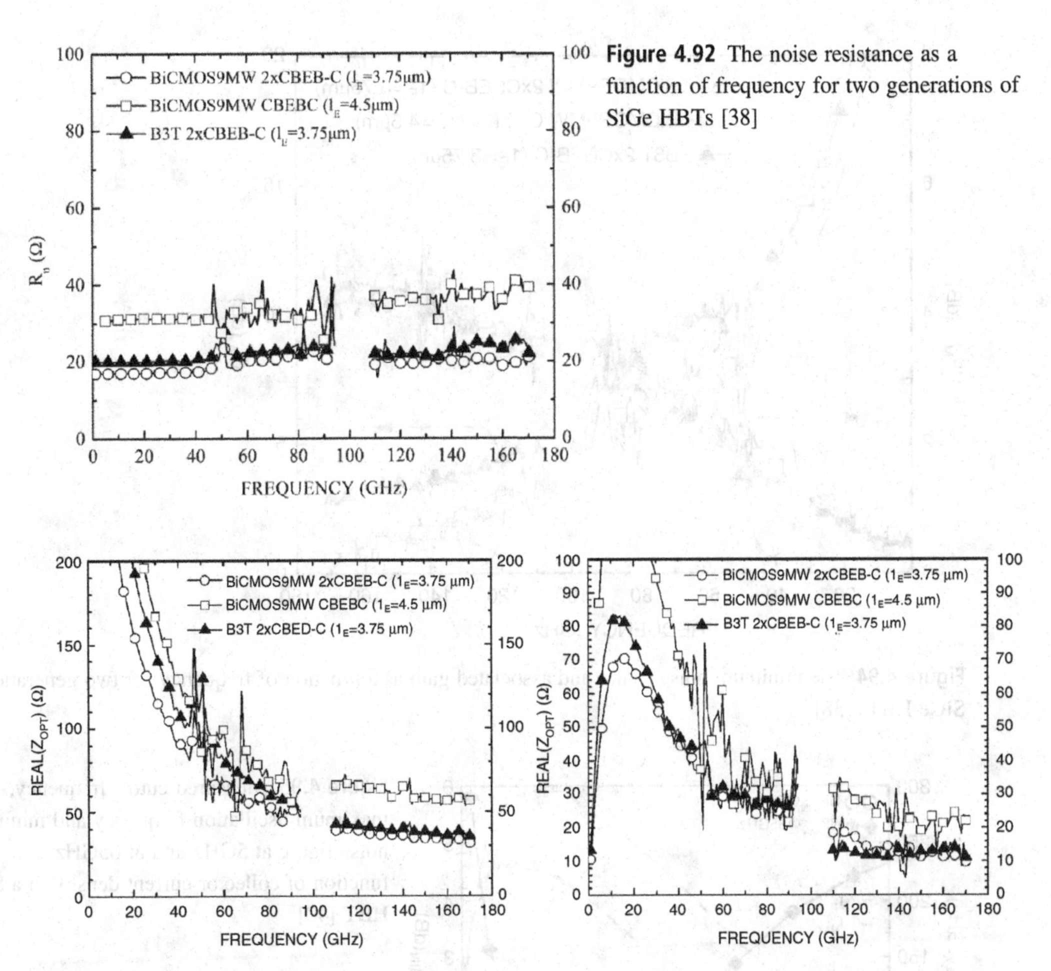

Figure 4.92 The noise resistance as a function of frequency for two generations of SiGe HBTs [38]

Figure 4.93 The optimum noise resistance, R_{SOPT}, and reactance, X_{SOPT}, as a function of frequency for two generations of SiGe HBTs [38]

Bias dependence of the HBT noise parameters

(a) NF_{MIN} The typical dependence of the minimum noise figure of an HBT on collector current is illustrated in Figure 4.95 at 5GHz and at 65GHz. At low-current densities, NF_{MIN} first decreases as the collector current density increases. This is caused by the decrease in the base resistance with increasing collector current. In this region of operation, the thermal noise of the base and emitter resistance is dominant. As the collector current density continues to increase, the collector shot noise current contribution becomes comparable to that of the thermal noise, and the minimum noise figure reaches its optimum value. The corresponding optimum noise bias current density, J_{OPT}, occurs below the peak f_T/f_{MAX} current densities. At even higher collector current densities, the collector shot noise current dominates and causes a significant degradation in the minimum noise figure. This region should be avoided in low-noise circuits. We note that J_{OPT} first increases with frequency but remains practically constant at frequencies beyond $f_T/5$.

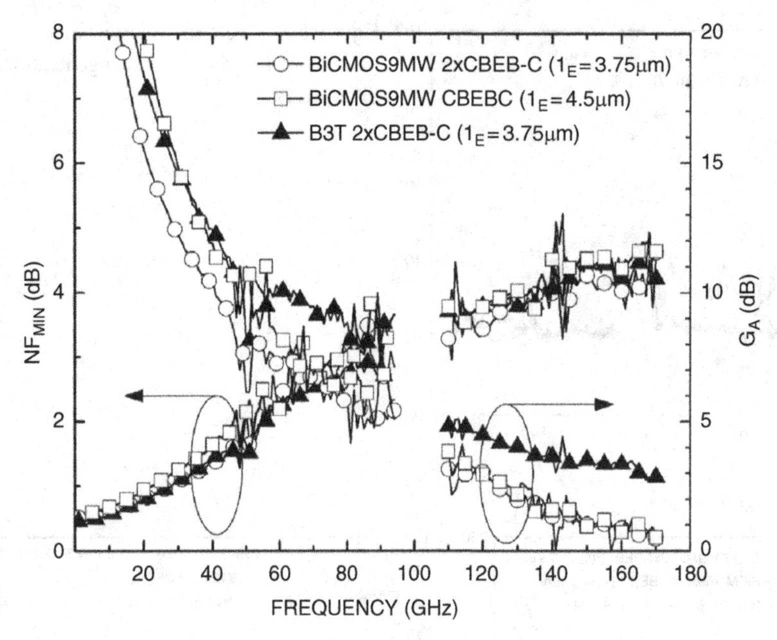

Figure 4.94 The minimum noise figure and associated gain as a function of frequency for two generations of SiGe HBTs [38]

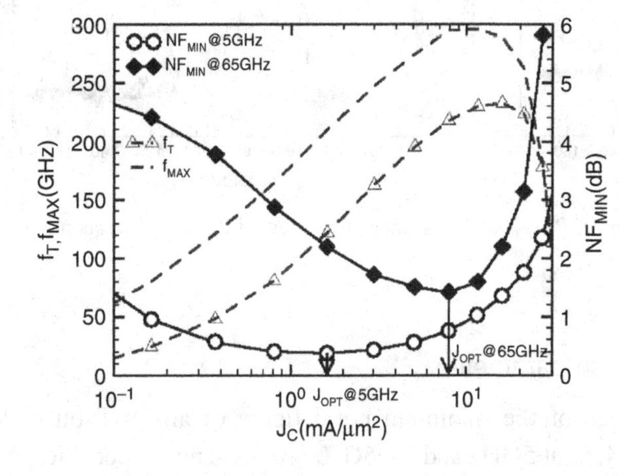

Figure 4.95 Measured cutoff frequency, maximum oscillation frequency and minimum noise figure at 5GHz and at 65GHz as a function of collector current density in a SiGe HBT [40]

It should be noted that J_{OPT} increases if a series resistance is added to the emitter or the base terminal. This is a common occurrence in circuits where lossy inductors, transformers, or transmission lines are employed for impedance matching. In such circuits, J_{OPT} should be determined for the specific LNA stage topology and not just for the transistor employed in the LNA stage.

(b) $\boldsymbol{R_n}$ From (4.199), it is easy to conclude that the last term is inversely proportional to current. However, r_E is constant and R_b decreases weakly with increasing current. This implies that R_n decreases as the collector current increases, making the device noise figure less sensitive to noise impedance mismatch at higher currents.

(c) $\boldsymbol{Y_{SOPT}}$, $\boldsymbol{Z_{SOPT}}$ Similarly, from (4.200) it is apparent that Y_{SOPT} scales with I_C while R_{SOPT} is inversely proportional to I_C.

Dependence on emitter geometry w_E and l_E

As can be observed from (4.202), in devices with two base contact stripes for each emitter stripe, R_n is inversely proportional to the emitter length and has a more complicated dependence on the emitter width. However, for low noise and maximum speed, the latter should always be fixed at the minimum value allowed by the technology

$$R_n \approx R_b + r_E + \frac{V_T}{2I_C} \approx \frac{R_{Se} + \dfrac{V_T}{2J_C}}{l_E w_E} + \frac{R_{sbi}}{12}\frac{w_E}{l_E} + \frac{R_{sbx} W_{SP}}{2l_E\left(1 + \dfrac{w_E}{l_E}\right)}. \tag{4.202}$$

Although the corresponding formula is more complex, it can be shown that R_{SOPT}, too, scales with $1/l_E$ [39].

F_{MIN} is a weak function of l_E but a (strong) function of w_E. This dependence is due primarily to the emitter geometry dependence of the base resistance. HBT layouts that minimize R_b also minimize F_{MIN}. The minimum emitter width should always be used to produce the minimum possible noise figure.

Finally, we note that the statistical correlation between the base and the collector shot noise currents is often not captured by currently available commercial HBT models. This can lead to an underestimation of the transistor bias current and of the transistor size required for optimum noise matching, especially at mm-waves.

4.3.16 Self-heating

Due to the very high current and high-power densities at which they operate, HBTs suffer from significant self-heating. Self-heating describes the rise in the local junction temperature of the device with respect to the ambient temperature. For example, a 130nm emitter SiGe HBT operating at the peak f_T current density of 15mA/μm^2 and biased at a V_{CE} of 1.4V ($V_{CB} = 0.5$V) dissipates 21mW/μm^2. As illustrated in Figure 4.96 [35], this can lead to a local temperature rise of up to 50°C. The increased junction temperature lowers the f_T and f_{MAX} and increases the minimum noise figure. Even more critical, if the ambient temperature reaches 75°C, which can often happen in packaged ICs, the HBT junction temperature rises above 125°C. The latter is the accepted limit for reliable operation of the transistor.

Self-heating is captured in modern bipolar transistor compact models such as HICUM, MEXTRAM, and VBIC with the help of the thermal subcircuit shown in Figure 4.97. This consists of a voltage-controlled current source, which represents the DC power, P_D (in *W*), consumed in the device, in parallel with a thermal resistance, R_{TH} (in degrees Celsius per *W*), and a thermal capacitance, C_{TH} (in Joules per degree Celsius). The voltage developed across the thermal resistance describes the local temperature rise, from the ambient temperature, T_A, to the junction temperature, T_J, while the thermal time constant $C_{TH} \times R_{TH}$, on the order 0.1–10μs, describes the delay associated with the self-heating process. Note that in the thermal subcircuit, currents correspond to watts of thermal power while net voltages represent degrees celsius.

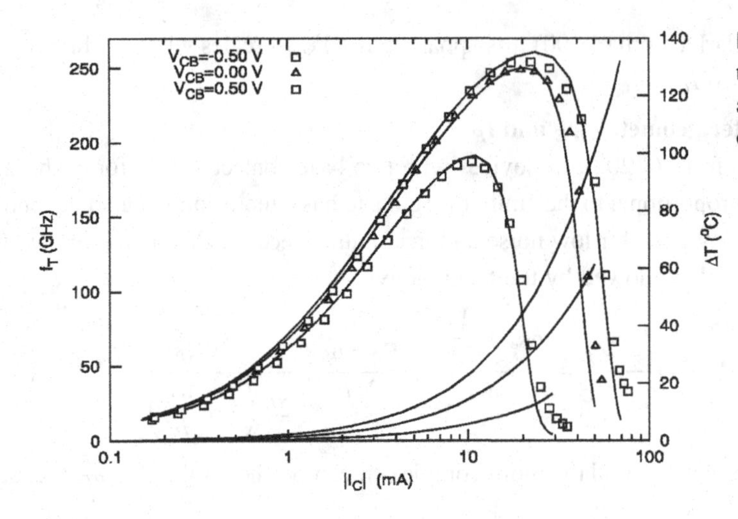

Figure 4.96 f_T and junction temeperature rise relative to the ambient as a function of collector current in SiGe HBTs [35]

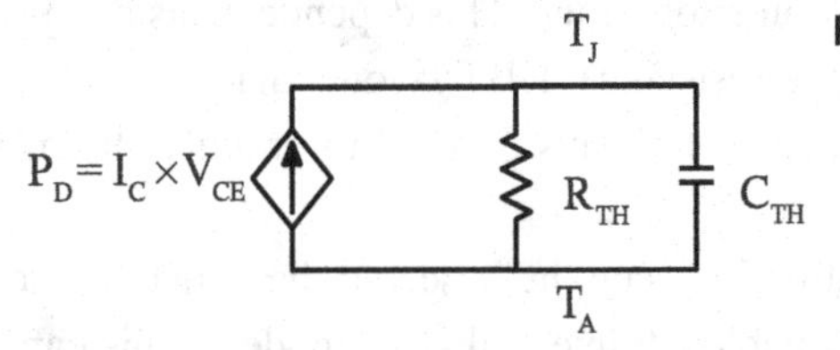

Figure 4.97 Self-heating subcircuit

4.3.17 HBT scaling

As in FETs, the HBT physical dimensions and doping levels can be scaled to increase their f_T and f_{MAX}, and to decrease their minimum noise figure from one technology generation to another. Although initially only the vertical structure of the HBT and the doping levels were scaled, nowadays, both lateral and the vertical scaling rules are applied in order to maximize device performance.

Why vertical scaling? Vertical scaling is employed to:

- reduce transit time and thus increase f_T,
- push the Kirk-effect to higher current density, allowing for higher f_T, and to
- achieve the optimal tradeoff between f_T and breakdown voltage.

How is vertical scaling realized in practice? This is accomplished by:

- decreasing the base and the collector thickness by a factor S,
- increasing the base and the collector doping by approximately S^2,
- increasing the Ge concentration and the Ge grading slope by S.

The process of vertical scaling is graphically summarized in Figure 4.98. We note that, although desired, as in MOSFETs, the emitter, subcollector, and sinker doping levels do not scale because they are already at the maximum values realizable in silicon or InGaAs, or in whatever material is used for those low sheet resistance regions.

Why lateral scaling? The goal is to reduce the base resistance and the base–collector capacitance which are degraded in the vertical scaling steps. Both R_b and C_{bc} affect (i) the

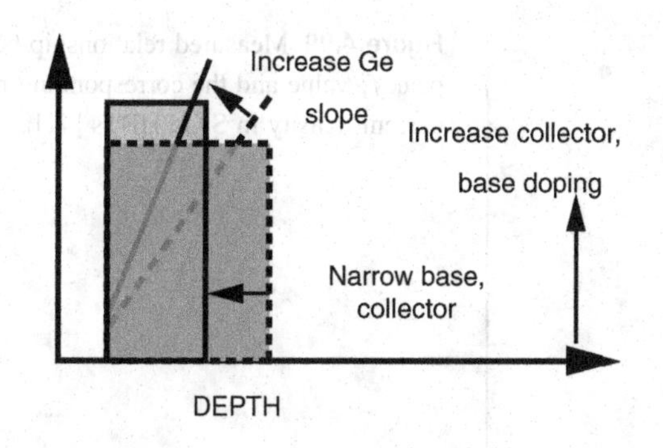

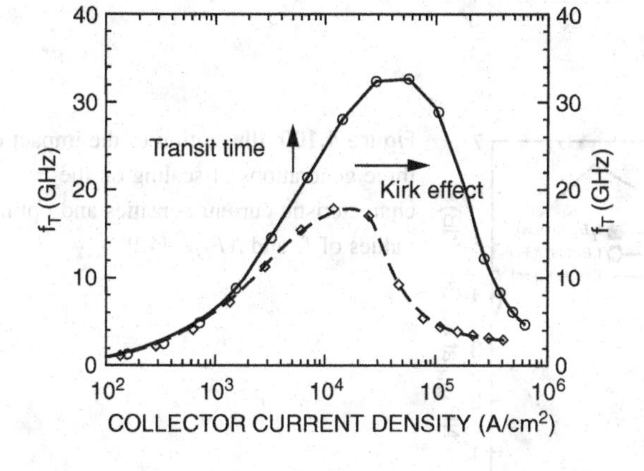

Figure 4.98 Illustration of HBT scaling methodology and its impact on the peak f_T value and on the peak f_T current density

maximum oscillation frequency, (ii) the minimum noise figure, and (iii) the delay in ECL and CML digital circuits.

How is lateral scaling carried out in practice? This is implemented:

- by shrinking the emitter width by a factor *S*, thus reducing the internal r_{bi} and C_{bci},
- by shrinking the emitter–base spacer width, W_{SP}, to reduce the external base resistance, and
- by shrinking the emitter–lateral spacing to reduce r_C and C_{cs}.

From a high-frequency perspective, the beneficial outcome of scaling is that f_T, f_{MAX}, NF_{MIN} improve at each new node. A somewhat undesired but tolerable result is that J_{pfT} and J_{OPT} increase by S to S^2 (Figure 4.99) [41], implying that the collector current of the minimum size HBT at peak f_T remains practically the same from one technology node to another.

Scaling of SiGe HBTs to 500GHz and of InP HBTs to 1THz is possible and already underway in several laboratories around the world. An example of typical f_T, f_{MAX}, and NF_{MIN} scaling is illustrated in Figure 4.100 with measured data for two existing SiGe HBT technology nodes and with simulation results for a would-be future node [40].

Table 4.8 summarizes the structural and high-frequency performance parameters of state-of-the-art SiGe BiCMOS and InP HBTs.

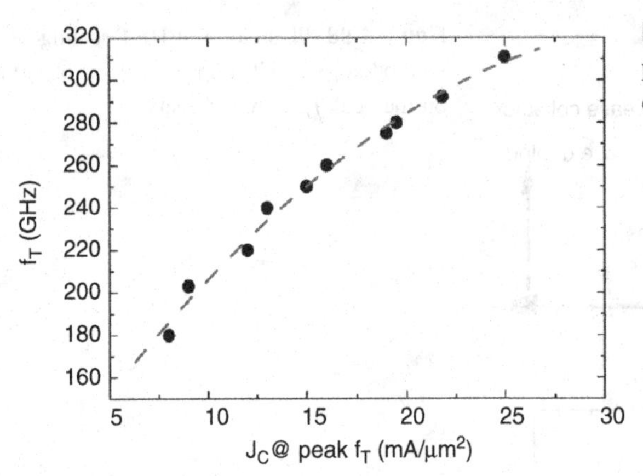

Figure 4.99 Measured relationship between the peak f_T value and the corresponding peak f_T current density in SiGe HBTs [41]

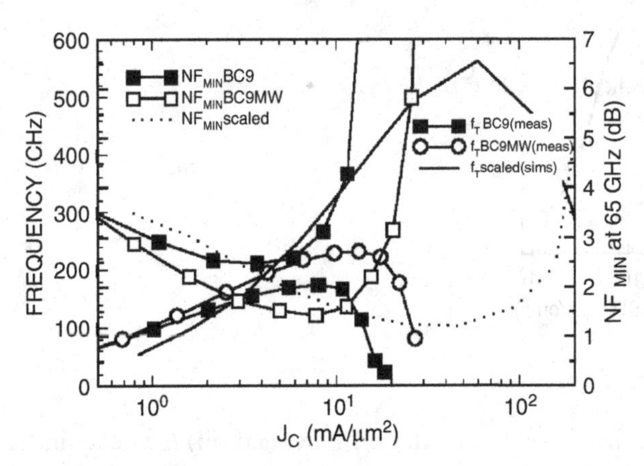

Figure 4.100 Illustration of the impact of three generations of scaling on the characteristic current densities and optimal values of f_T and NF_{MIN} [40]

Implications of HBT scaling rules for circuit design. Since the characteristic current densities increase from one technology node to another, circuits will have to be redesigned if their performance is to be maximized in new technology nodes. Both the transistor sizes and the bias currents must be re-sized.

4.3.18 Biasing HBTs

The bias network must ensure the stability of the DC operating point of the transistor over temperature, supply voltage, and process variation. In SiGe BiCMOS technologies, the temperature and supply voltage stability is secured by employing a bandgap circuit from which stable reference currents are distributed using p-MOSFETs or pnps to local npn-current mirrors in the high-frequency blocks in an IC. P-type devices are usually not available in III-V technologies. Instead, in III-V high-frequency circuits, the stable reference current is often provided by a separate, silicon chip.

The current mirrors realized with III-V and SiGe HBTs are similar to those implemented with Si BJTs in low-frequency analog circuits. However, several measures must be taken to ensure:

- that the current mirror does not short circuit the HF input of the HF transistor,
- that it does not contribute prohibitive capacitive parasitics to the HF path, and
- that it does not contribute noise.

Table 4.8 **Example of state-of-the art SiGe HBT and InP HBTs.**

Parameter	SiGe HBT [35]	InP/InGaAs DHBT [42]
w_E (nm)	120	250
N_E (cm^{-3})	5×10^{20}	$5–7\times10^{17}$
E_{GE} (eV)	1.12	1.35
N_B (cm^{-3})	2×10^{19}	$(5 \ldots 15)\times10^{19}$
D_n (cm^2/s)	3 … 5	10 … 15
E_{GB} (eV)	0.97	0.78 (or 0.72eV)
W_B (nm)	20	30
N_C (cm^{-3})	$10^{18}–5\times10^{18}$	$5\times10^{16} + 3\times10^{18}$
W_{dBC} (nm)	70	120–150
v_{sat} (cm/s)	0.8	>1.0
f_T/f_{MAX} (GHz)	230/290	390/800
J_{pfT} (A/μm^2)	12	13
BV_{CEO} (V)	1.6	2

The latter is particularly important in low-noise amplifiers and oscillators. Typical low-voltage bias networks for high-frequency single-ended and (often) differential HBT circuits are illustrated in Figure 4.101. In all schematics, the role of capacitor C_1 is to prevent noise from the bias network to reach the HF input node. In Figure 4.101(a), L_B plays a similar role as C_1 in filtering out noise from the bias network. It can also be employed as matching element to tune out the input capacitance of Q_2. The advantage of using a bias filter consisting only of capacitors and inductors is that they do not contribute noise. However, L_B consumes a large area, and like L_E (needed to perform the desired HF function), due to its non-zero parasitic resistance, affects the accuracy of the current mirroring ratio. To solve this problem, a resistor R_B is used instead of L_B in Figures 4.101(b) and (c) The value of R_B must be at least 20 times larger than the input impedance of the circuit such that it does not load the input and it does not contribute noise. To ensure perfect mirroring, a n times larger base resistance must also be added to the base of Q_1.

The current mirror can be simple, as in Figures 4.101(a) and (b), or with resistive emitter degeneration, as in Figure 4.101(c). More so than in MOSFET or HEMT circuits, because of the exponential dependence of the collector current on the base–emitter voltage, the bipolar current mirror must employ emitter resistive degeneration to ensure that the mirrored current is insensitive to ground resistance or to the parasitic resistance of L_E. Both would cause a V_{BE}

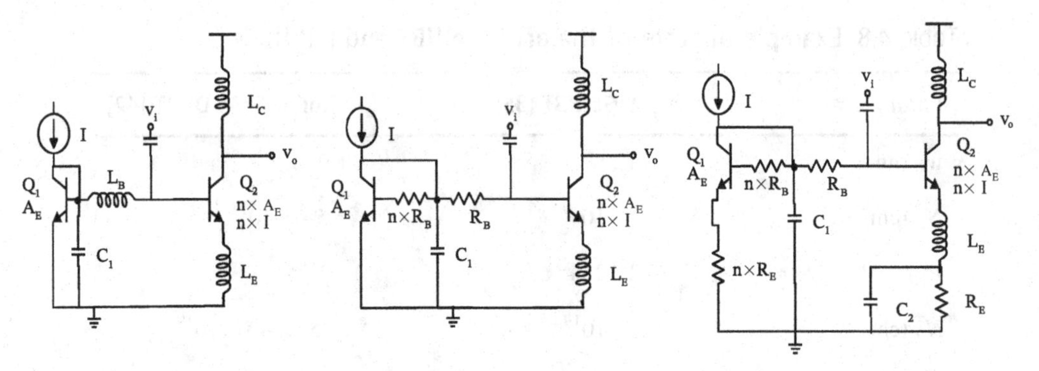

Figure 4.101 Bias circuits for tuned HBT amplifier stages

mismatch of a few mV between the transistors in the current mirror. A DC drop of 150–200 mV on the emitter resistor R_E is sufficient to suppress this negative effect without taking up prohibitive voltage headroom from the HF transistor Q_2. Capacitor C_2 is added across R_E to suppress its noise. The capacitance value, the self-resonant frequency, and the layout of C_2 are critical in ensuring that C_2 behaves like an ideal short circuit at the frequency of interest, and that it does not cause Q_2 to oscillate at any frequency.

Finally, it should be noted that imperfect mirroring ocurrs over process and temperature variations when using devices with unequal emitter length and width for Q_1 and Q_2, rather than realizing Q_2 as a parallel connection of n Q_1 devices. However, the latter scheme increases interconnect parasitics and degrades circuit performance at mm-wave frequencies. A compromise between bias current stability and high-frequency performance must be reached. Usually, because HF performance is not very sensitive to bias current variation when the device is biased at J_{OPT} or J_{pfT}, imperfect mirroring is tolerated in HF circuits.

4.4 THE HIGH ELECTRON MOBILITY TRANSISTOR

4.4.1 HEMT structure and operation principle

The first high electron mobility transistor or HEMT was fabricated by Fujitsu, in Japan, in 1980 [43]. It is a direct descendant of the GaAs Metal-Semiconductor FET, MESFET, Figure 4.102(a), proposed by Carver Mead in 1966 at Caltech [44]. By the 1970s, the MESFET had emerged as the dominant microwave semiconductor device, the workhorse of the new microwave monolithic integrated circuits, known under the acronym MMICs.

As illustrated in Figure 4.102, in all types of FETs, the basic principle of operation is the lateral current flow from the source to the drain due to the mobile channel charge, as a result of the applied drain–source voltage, V_{DS}. The mobile channel charge (electrons, typically) and, therefore, the source–drain current, are electrostatically controlled by the applied gate–source voltage, V_{GS}.

In constrast to MOSFETs, in the absence of a high-quality native oxide for III-V semiconductors, in MESFETs and HEMTs a Schottky (also known as metal-semiconductor) contact is

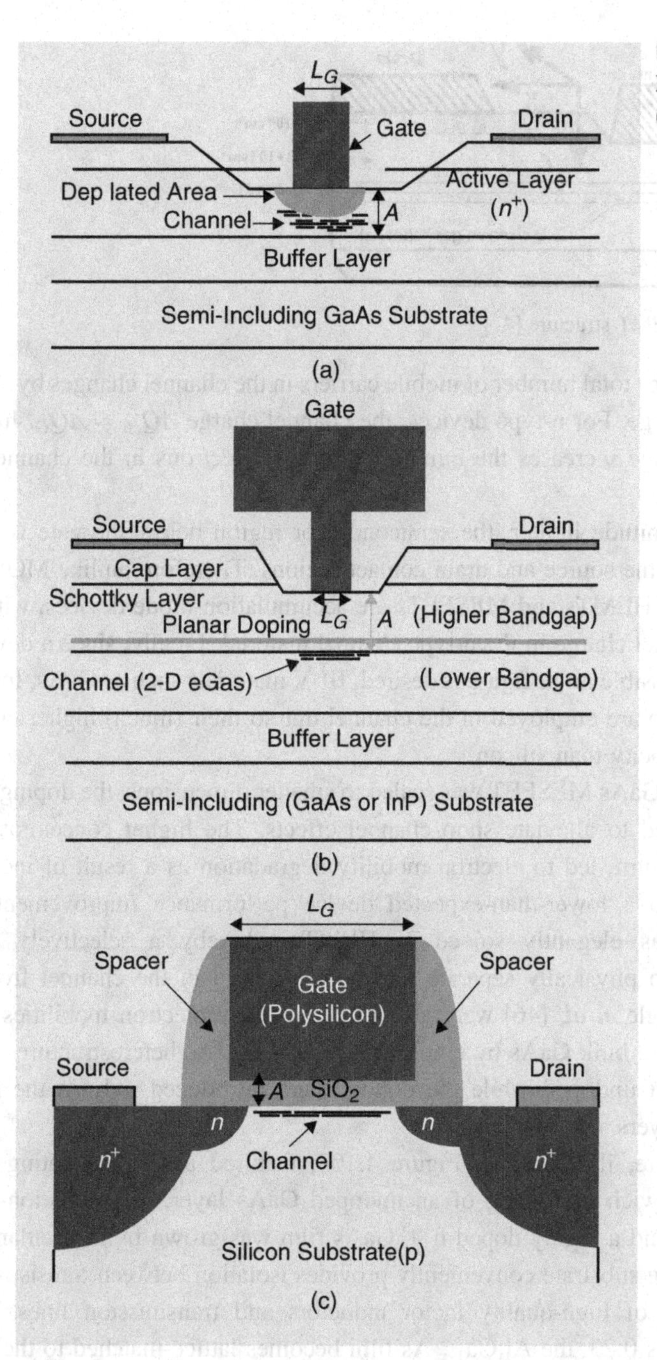

Figure 4.102 Cross-sections through FET devices (a) MESFET, (b) HEMT, (c) MOSFET [45]

placed directly on the semiconductor material above the channel to form the gate electrode. In the normal (active) mode of operation, the gate is reverse-biased, creating a depletion region which extends into the channel, modulating the flow of carriers between the source and drain. A change in gate voltage ΔV_G causes an increase in the charge accumulated on the gate by ΔQ_G. To maintain charge neutrality, the latter, in turn, causes a change $-\Delta Q_G$ in the depletion region

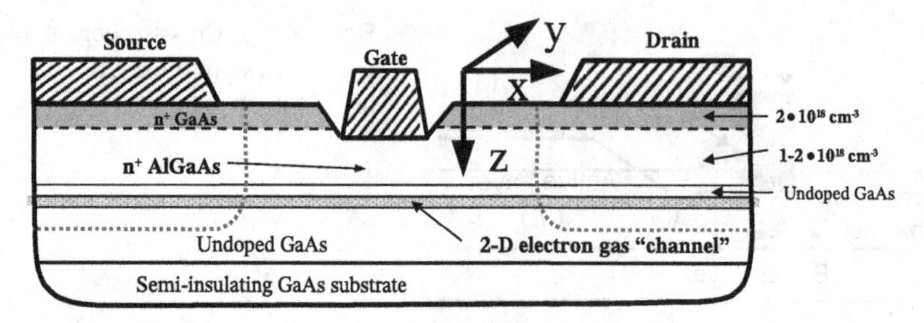

Figure 4.103 AlGaAs/GaAs HEMT structure [47]

charge. As a consequence, the total number of mobile carriers in the channel changes by $\Delta Q_G/q$, where q is the electron charge. For n-type devices, the channel charge $\Delta Q_{ch} = \Delta Q_G$. In other words, when the gate voltage increases the number of mobile electrons in the channel also increases.

Although orders of magnitude lighter, the semiconductor region below the gate is of the same type as the doping in the source and drain contact regions. Therefore, unlike MOSFETs which operate in inversion, HEMTs and MESFETs are accumulation-mode devices, with free electrons forming the channel charge in the n-type channel material. Finally, since a device of utmost speed and lowest possible noise figure is desired, III-V materials such as GaAs, InGaAs, GaN, InAs, GaSb, and InSb are employed in the channel due to their (much) higher electron mobility and saturation velocity than silicon.

As the gate length of the GaAs MESFET was scaled to smaller dimensions, the doping in the channel had to be increased to alleviate short-channel effects. The higher concentration of dopants in the channel, in turn, led to electron mobility degradation as a result of increased impurity scattering, and to a lower-than-expected device performance improvement with scaling. This problem was elegantly solved in HEMTs whereby a selectively doped heterostructure was used to physically separate the mobile charge in the channel from the donor atoms. In 1978, Dingle *et al.* [46] were able to demonstrate electron mobilities much higher than those in undoped bulk GaAs by employing AlGaAs/GaAs heterostructures where the GaAs regions were left undoped while the donors were introduced only in the higher bandgap AlGaAs barrier layers.

The first HEMT structure, illustrated in Figure 4.103, featured a semi-insulating GaAs substrate on which a sandwich consisting of an undoped GaAs layer, a modulation-doped n-type Al_xGa_{1-x} As layer, and a highly doped n + GaAs film was grown by molecular beam epitaxy. The semi-insulating substrate conveniently provides isolation between transistors and allows for the fabrication of high-quality factor inductors and transmission lines. If the aluminum mole fraction x is 0.23, the Al_xGa_{1-x} As film becomes lattice-matched to the GaAs substrate, forming a high-quality heterojunction, which explains why this material system was preferred. The channel is formed in the undoped GaAs layer, thus maximizing its mobility, at the heterointerface with the AlGaAs layer. The latter acts as an insulator between the gate and the channel, and as a source of electrons for the channel. To minimize the scattering between the free electrons in the channel and the donor atoms in the AlGaAs barrier, a thin (2–5 nm) undoped AlGaAs spacer is placed immediately above the GaAs channel. The role of the top

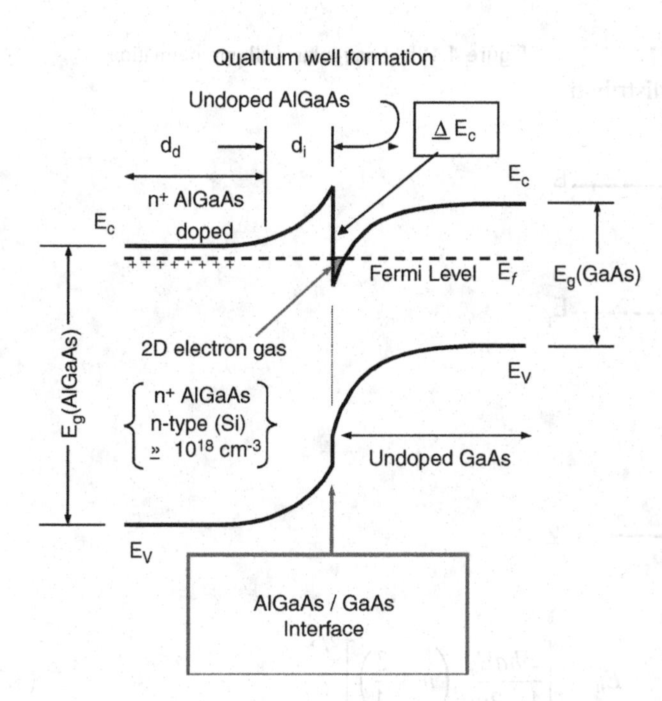

Figure 4.104 Energy bands and material composition in a modulation-doped heterostructure used in AlGaAs/GaAs HEMTs [47]

n + GaAs layer is to provide low-resistance ohmic contacts to the source and drain regions. The n + GaAs film is removed from the gate region, resulting in a "recessed" gate geometry which ensures that the gate forms a Schottky rather than an ohmic contact with the AlGaAs layer, and that short-channel effects are minimized.

Figure 4.104 reproduces the energy bands diagram at the channel–barrier interface in a cross-section perpendicular to the gate and channel. The conduction band forms a narrow potential well with a height equal to the conduction band offset, ΔE_C, between the AlGaAs and GaAs regions. When a sufficiently large voltage is applied on the gate, the bottom of the conduction band dips below the Fermi level, leading to a large number of free electrons accumulating at the interface into a thin sheet. The latter is known as a two-dimensional electron gas, or 2DEG, and is characterized by its 2-D charge density, n_s, measured in cm^{-2}. Since the drain–source current, I_{DS}, and the transconductance, g_m, are proportional to n_s and to the mobility, μ_n, of the channel charge, the main design goal is to maximize the $\mu_n \times n_s$ product, an important figure of merit. A large conduction band offset is needed to increase n_s and a material with large electron mobility must be chosen for the unintentionally doped channel.

Due to the positive gate–channel voltage, a large vertical electric field, ε_S, arises perpendicular to the hetrointerface and produces a sharp bending of the conduction band on the channel side, leading to charge quantization. The channel charge loses a degree of freedom along the z-axis (perpendicular to the gate) with motion allowed only in the x and y directions. As shown in Figure 4.105, only discrete energy subbands E_0, E_1, E_2, etc. (or eigen-energies) can be occupied by the free electrons in the channel. In a triangular quantum well, typical of HEMT accumulation and Si MOSFET inversion channels, an analytical solution exists for the eigen energies [13]

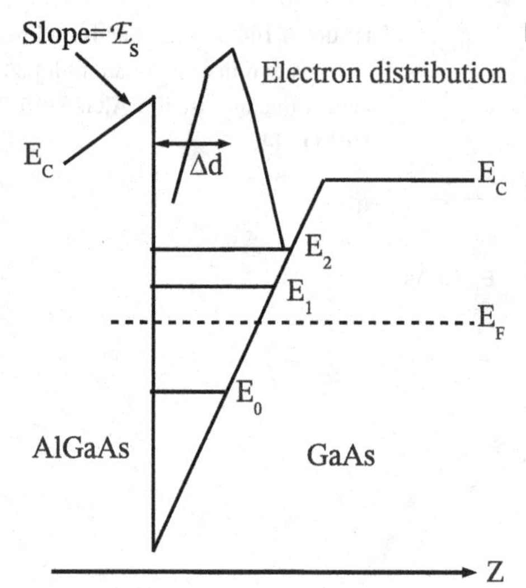

Figure 4.105 Triangular well approximation

$$E_n = \left[\frac{3hqE_S}{4\sqrt{2m_e^*}}\left(n+\frac{3}{4}\right)\right]^{2/3} \tag{4.203}$$

where m_e^* is the effective electron mass along the z-axis. The centroid of the resultant 3-D electron distribution in the channel is located several nanometers away from the interface, farther separated from the donors in the doped AlGaAs barrier layer, as illustrated in Figure 4.105.

4.4.2 Charge control model

In the one-dimensional case where only the influence of the potential from the gate is considered and that of the electric field between source and drain is ignored, the steady-state electron charge distribution in the channel can be obtained by iteratively solving the coupled Schrödinger and Poisson equations along the z-direction, assuming that there is no current flow from the gate into the channel

$$\frac{-h^2}{2\pi^2 m_e{}^*}\frac{d^2\zeta_n}{dz^2} + E_C\zeta_n = E_n\zeta_n \tag{4.204}$$

$$\frac{d}{dz}\left(\varepsilon_S\frac{dE_C}{dz}\right) = q(N_D - n_e - N_A). \tag{4.205}$$

In (4.204) and (4.205), ζ_n represents the nth electron wavefunction associated with the eigen-energy E_n, $E_C(z)$ is the position-dependent conduction band energy, assumed as the main variable (rather than the electron potential) in Poisson's equation, N_D is the ionized dopant concentration in the AlGaAs barrier layer, N_A is the background doping in the undoped p – GaAs buffer that forms the channel, and

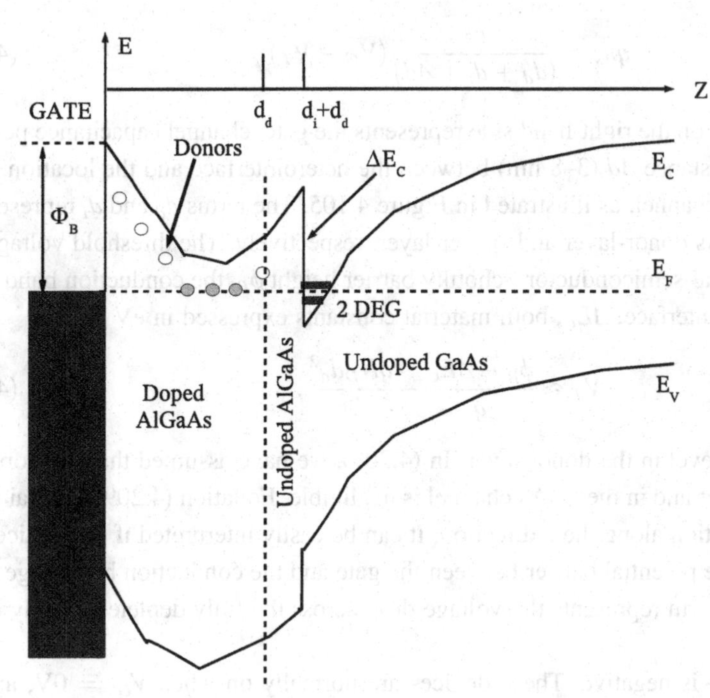

Figure 4.106 Band diagrams used to derive the V_T and n_s expression

$$n_e = \frac{4\pi kTm_e^*}{h^2}\sum_{n=0}\zeta_n^2 ln\left(1 + exp\frac{E_F - E_n}{kT}\right) \quad (4.206)$$

is the 3-D free electron concentration integrated over all energy bands, E_n.

The three equations above are solved iteratively at discrete points in the one-dimensional vertical cross-section through the device, along the *z*-direction. Starting from an assumed initial charge distribution $\rho(z) = qN_D(z) - qn_e(z) - qN_A(z)$, Poisson's equation is solved first to determine $E_C(z)$. The latter is then plugged into Schrodinger's equation to solve for E_n and ζ_n. In the third step, a new free electron profile $n_e(z)$ is calculated from (4.206). This three-step process is iterated until the solution converges and the changes from one iteration to the next become smaller than a predetermined error.

When $E_F > E_n$, the logarithmic term in (4.206) becomes a linear function of $E_F - E_n$. This implies that, at large gate voltages, when the channel is fully turned on and the Fermi level in the channel rises above the conduction band edge and above the first allowed energy level E_0, the electron charge in the channel will increase linearly with the applied gate voltage. In contrast, for small gate voltages, the Fermi level remains below the conduction band edge and the exponential term in the logarithm becomes much smaller than 1, such that the 3-D channel charge becomes proportional to $exp\left(\frac{E_n - E_F}{kT}\right)$. This translates into an exponential dependence of the channel charge on the applied voltage on the gate, characteristic of the subthreshold region of operation for the HEMT.

As in MOSFETs, the gate charge must satisfy the equation

$$Q_G = CV \quad (4.207)$$

where C is the gate–channel capacitance. V can be approximated by the voltage difference between the gate and the channel, $V_{GS} - V_T$. The two-dimensional channel charge per unit gate width is given by

$$qn_s = \frac{\varepsilon_S}{(d_d + d_i + \Delta d)}(V_G - V_T) \qquad (4.208)$$

where the fraction term on the right-hand side represents the gate–channel capacitance per area, and accounts for the distance Δd (3–8 nm) between the heterointerface and the location of the charge centroid in the channel, as illustrated in Figure 4.105. The terms d_d and d_i represent the thickness of the AlGaAs donor-layer and spacer-layer, respectively. The threshold voltage, V_T, is a function of the metal-semiconductor Schottky barrier height ϕ_B, the conduction band offset at the channel–barrier interface, ΔE_C, both material constants expressed in eV

$$V_T \approx \frac{\phi_B - \Delta E_C}{q} - \frac{qN_D d_d^2}{2\varepsilon}. \qquad (4.209)$$

N_D is the 3-D doping level in the donor layer. In (4.209), we have assumed that the doping in the AlGaAs spacer layer and in the GaAs channel is negligible. Equation (4.209) is obtained by solving Poisson's equation along the z-direction. It can be easily interpreted if one notices that $\phi_B - \Delta E_C$ represents the potential barrier between the gate and the conduction band edge in the channel, and the third term represents the voltage drop across the fully depleted, highly doped barrier layer.

In most HEMTs, V_T is negative. These devices are normally on when $V_{GS} = 0$V, and are known as depletion-mode HEMTs, or simply HEMTs. In some applications, where a negative supply is not available, or in logic circuits, an enhancement-mode device, or E-HEMT, with $V_T > 0$, is desirable. As indicated by (4.209), the threshold voltage can be adjusted by changing the thickness and doping level in the barrier layer. In all HEMTs, the voltage excursion on the gate is limited to the range between V_T and ϕ_B/q. For gate voltages larger the ϕ_B/q, the gate–channel junction turns on and the gate current becomes too large for the device to operate properly. Once the material system is chosen, the only knob available to the device designer for adjusting the threshold is the doping, N_D, in the barrier layer. The thickness of the barrier layer below the gate is dictated by the condition that short channel effects be suppressed. The rule of thumb is to set the gate–channel thickness $d_d + d_i = L_G/10$, where L_G is the physical gate length, as defined in Figure 4.102(b). This corresponds to the vertical electric field being apprximately ten times larger than the lateral drain–source electric field.

4.4.3 DC characteristics

From (4.208), the drain–source current at a location x along the channel can be expressed as

$$I_{DS}(x) = qn_s(x)v(\varepsilon)W \qquad (4.210)$$

where W is the gate width, $\varepsilon(x)$ is the lateral (drain–source) electric field, and $v(\varepsilon)$ is the electron velocity in the channel.

Assuming that $V_S = 0$ and $V_{GS} = V_G$, the potential $V(x)$ at a given postion x in the channel must satisfy the condition

$$V_S = 0 < V(x) < V_{DS}. \qquad (4.211)$$

We can now modify (4.208) to describe the local charge at location x along the channel as

$$qn_s(x) = \frac{\varepsilon_S}{(d_d + d_i + \Delta d)}[V_G - V_T - V(x)]. \tag{4.212}$$

Knowing that there is no current flow into the gate or into the substrate, the drain–source current (4.210) must satisfy the condition

$$I_{DS}(x) = -Wqn_s(x)v(\varepsilon(x)) = ct \tag{4.213}$$

at any position in the channel. By inserting (4.212) into (4.213), we obtain

$$I_{DS}(x) = \frac{\varepsilon_S W}{d_d + d_i + \Delta d}[V_G - V_T - V(x)]v(\varepsilon) \tag{4.214}$$

where, by definition, the lateral electric field in the channel at location x is given by

$$\varepsilon(x) = \frac{-dV(x)}{dx}. \tag{4.215}$$

For the expression of the drain–source current in the general case of a realistic velocity-field dependence, the reader is referred to the MOSFET section. Two useful extreme cases, which result in simple analytical solutions, are described below:

(a) Long channel approximation ($V_{DS}/L << \varepsilon_{sat}$) Assuming a linear dependence of the electron velocity on the lateral electric field, $v(\varepsilon) = \mu_{neff}\,\varepsilon$, valid at small V_{DS}, integrating the drain–source current along the channel, and knowing that $V(x = L) = V_{DS}$ and that $V(x = 0) = V_S = 0$V, we obtain

$$\int_0^L I_{DS}(x)dx = W\mu_{neff}\int_0^{V_{DS}} qn_s(V)\frac{dV}{dx}dx = LI_{DS} \tag{4.216}$$

$$I_{DS} = \frac{\mu_{neff}\varepsilon_s}{d_d + d_i + \Delta d}\frac{W}{L}\int_0^{V_{DS}}(V_G - V_T - V)dV = \frac{\mu_{neff}\varepsilon_s}{d_d + d_i + \Delta d}\frac{W}{L}\left[(V_G - V_T)V_{DS} - \frac{V_{DS}^2}{2}\right]. \tag{4.217}$$

As expected, equation (4.217) is identical to the standard long-channel MOSFET equation.

(b) Short channel approximation ($V_{DS}/L >> \varepsilon_{sat}$) Assuming that the lateral electric field is large enough that $v(V_{DS}/L) = v_{sat}$ throughout the channel, for $V_{DS} > V_{DSAT}$, I_{DS} remains constant at

$$I_{DS} = \frac{\varepsilon_s v_{sat} W}{d_d + d_i + \Delta d}(V_{GS} - V_T) \tag{4.218}$$

and varies linearly, rather than quadratically with $V_{GS} - V_T$. In real devices, both these approximations are rather crude, and a model similar to that developed for MOSFETs, accounting for realistic $v(\varepsilon)$ characteristics can be derived.

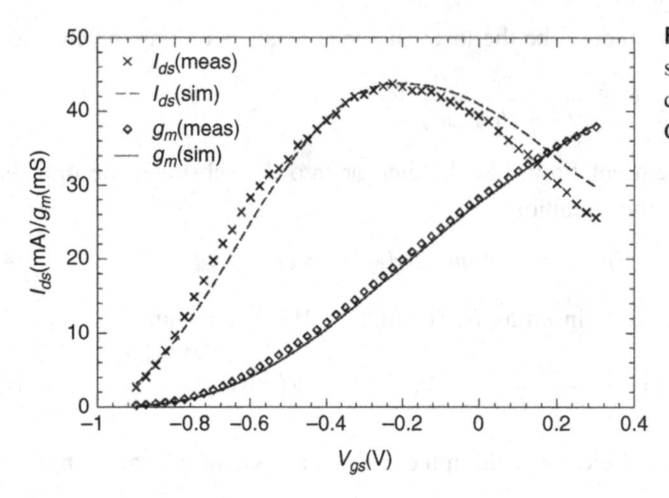

Figure 4.107 Measured (symbols) and simulated (lines) transfer and g_m-V_{GS} characteristics of a 0.15μm 2 × 40μm GaAs p-HEMT [48]

Transfer characteristics: $I_{DS}(V_{GS})$

In the subthreshold region, based on (4.206), an exponential dependence of the drain curent on V_{GS} is assumed, as in MOSFETs: $I_{DS} \sim \exp(V_{GS})$

In the saturation region of the transfer characteristics, at large V_{GS}, the following relationship holds

$$I_{DS} = k_n W(V_{GS} - V_T)(1 + \lambda V_{DS}) \quad (4.219)$$

where k_n [*V/m*] is a technology constant which depends on the effective mobility, physical gate length and barrier layer thickness, while λ captures the channel length modulation effect. As in nanoscale MOSFETs, the threshold voltage is a function of V_{DS}

$$V_T = V_{T0} - \eta V_{DS} \quad (4.220)$$

with η extracted as a fitting parameter from DC measurements.

While (4.219) and (4.220) can be used to get an intuitive understanding of device characteristics, for the purpose of device modeling in circuit design, compact models based on quasi-physical device equations fitted to experimental data are employed. One of the most commonly employed compact models is Agilent's EEHEMT model.

Figure 4.107 illustrates measured and simulated transfer charcateristics for a GaAs p-HEMT with 150nm gate length and two gate fingers, each 40μm wide [48]. We observe that this is a depletion-mode device with a threshold voltage of around −0.8V. The V_{GS} dependence of the drain current starts off as quadratic between −0.8V and −0.6V and becomes linear above −0.6V up to about + 0.1V. Above 0.1V, parasitic conduction starts to occur in the barrier layer, which has lower carrier mobility, leading to transconductance degradation. The transconductance is obtained as $g_m = \frac{\partial I_{DS}}{\partial V_{GS}}$.

Output characteristics: $I_{DS}(V_{DS})$

The measured output characteristics in the saturation region at different V_{GS} values are described by (4.219), as shown in Figure 4.47 for the same device as in Figure 4.107. The measurements are well matched by simulations using the EEHEMT model [48]. The drain

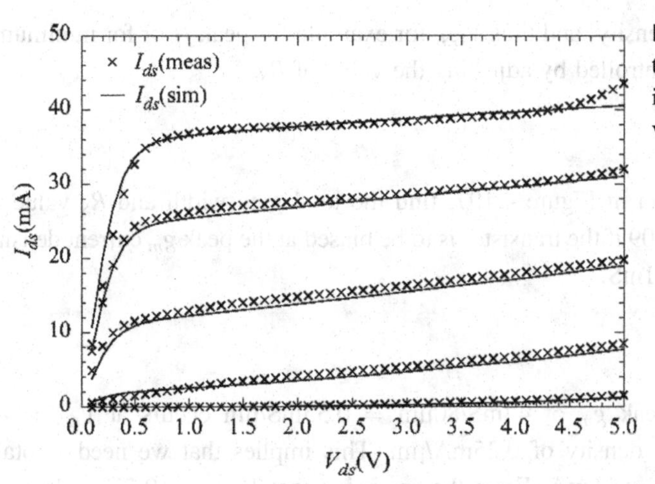

Figure 4.108 Output characteristics of the 0.15μm 2 × 40μm GaAs p-HEMT in Figure 4.4.6 for five different V_{GS} values [48]

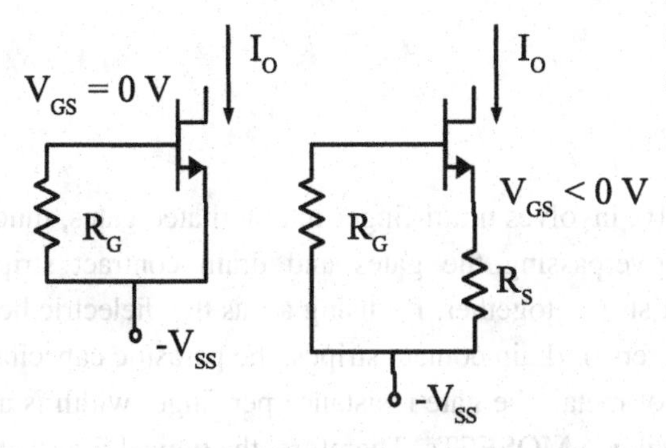

Figure 4.109 Current sources realized with depletion-mode HEMTs

saturation voltage, V_{DSAT}, remains at around 0.4V, irrespective of V_{GS}, indicating that saturation of the drain current occurs due to velocity saturation. The onset of avalanche multiplication and breakdown becomes apparent at large currents for $V_{DS} > 4.5$V.

Figure 4.109 shows two possible depletion-mode HEMT current sources that can be employed as the tail current source of a differential pair. The biasing scheme on the right can also be used to self-bias a HEMT in a tuned high-frequency amplifier, at the desired current density. In that situation, a capacitor can be placed across R_S to remove its impact from the small signal response.

How is the current source designed? Since the gate current is negligible, the DC voltage drop on the resistance connected between the gate and V_{SS}, R_G is approximately 0V. The output current of the current source, I_O, is determined by the total gate width W of the HEMT, and by the value of the resistance R_S, by solving the system of two equations with two unknowns (V_{GS} and R_S)

$$V_{GS} = -I_O R_S \text{ and } I_O = k_n W(V_{GS} - V_T)(1 + \lambda V_{DS})$$

where V_{DS} is set to a desired value in the saturation region and k_n and λ are technology constants. In the circuit with $R_S = 0\Omega$, V_{GS} becomes 0V and the current density of the HEMT is simply determined by its threshold voltage. This arrangement may not be optimal for all applications. The current density, and thus V_{GS}, for example for peak f_T or for minimum noise figure operation, can be controlled by adjusting the value of R_S.

EXAMPLE 4.14

Using the experimental data in Figure 4.107, find the total gate width and R_S value for the current source in Figure 4.109 if the transistor is to be biased at the peak g_m current density for a total transconductance of 11mS.

Solution

From Figure 4.107, the peak g_m of 44mS/80μm = 0.55mS/μm occurs at $V_{GS} = -0.2$V, corresponding to a current density of 0.25mA/μm. This implies that we need a total W of 20μm and a total current I_O of 5mA. From the condition that $V_{GS} = -0.2\text{V} = I_O \times R_S$, we obtain $R_S = 40\Omega$.

4.4.4 HEMT layout

A traditional HEMT layout, Figure 4.110, involves multi-finger interdigitated gates, much like MOSFETs [49]. A metal airbridge, overpassing the gates and drain contract stripes, is employed to connect the source contact stripes together. By using air as the dielectric between the source metal bridge and the gate fingers or drain contact stripes, the parasitic capacitance is minimized. Since the gate is made out of metal, the gate resistance per finger width is at least one order of magnitude smaller than in silicon MOSFETs. Therefore, the typical finger width is about one order of magnitude larger than in MOSFETs of identical gate length: $W_f = 10$–100μm for $L_G = 0.1$μm to 0.25μm. Smaller gate finger widths tend to be dominated by the gate fringing capacitance to the substrate, resulting in f_T degradation. Today, the most aggressively scaled HEMTs have 30nm gate lengths and finger widths as small as 1.2μm [50].

The circuit designer has the freedom to adjust the finger width, W_f, and the number of fingers, N_f, to control the total gate width $W = N_f \times W_f$, drain current, and the transconductance of the device.

4.4.5 HF equivalent circuit

Figure 4.111 shows a cross-section through one of the fingers of a HEMT indicating the location the various parameters that make up the small signal high-frequency equivalent circuit of the device. The corresponding schematic is reproduced in Figure 4.112 which also includes the noise sources. In Figure 4.111, R_{CON} represents the source and drain contact resistance. In Figure 4.112, R_{CON} is included in r_s and r_d, respectively.

The small signal equivalent circuit parameters are similar to those of a MOSFET. The expressions for f_T and f_{MAX} are also identical if R_i is replaced by R_{gs}. The back-side of the

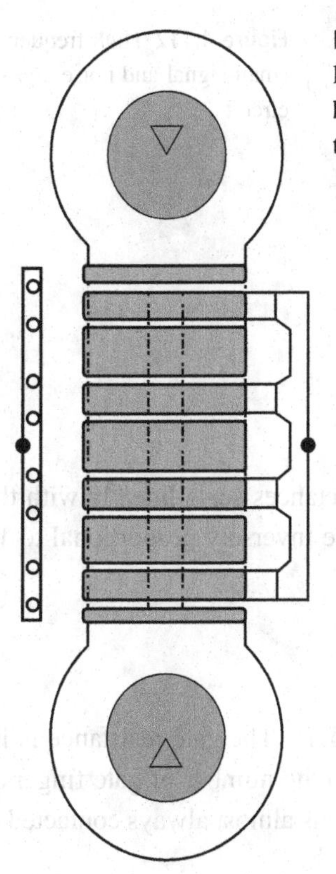

Figure 4.110 Interdigitated 8-gate-finger HEMT layout with the gate on the left, the drain on the right and the source grounded through two circular via holes, at the top and at the bottom, to the backside. The wider contact stripes to the active region of the device represent the local source contacts [49]

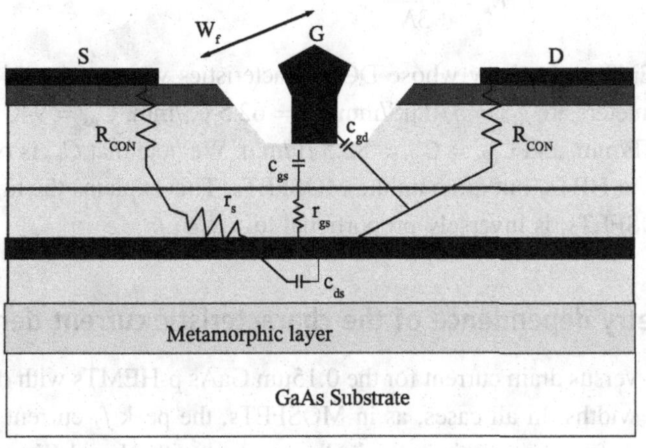

Figure 4.111 Location of the resistive and capacitive parasitics in a HEMT structure

semi-insulating wafer is metallized and grounded. In most practical cases, even if the source is not connected to the back-side ground, the parasitic capacitance from any terminal to ground is negligible because the wafer is a few hundred micrometers thick. However, in some power amplifier or terahertz technologies, the GaAs or InP substrate is thinned to 50μm or less and the parasitic drain and source capacitances to the backside ground must be accounted for.

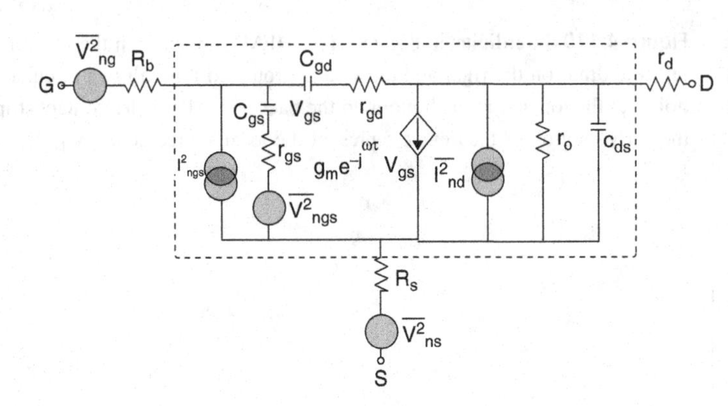

Figure 4.112 High frequency small-signal and noise equivalent circuit

Again, as in MOSFETs, all capacitances and transconductances scale linearly with the total gate width of the HEMT while all resistances, except r_g, are inversely proportional to W

$$C_{gs}, C_{gd}, C_{ds}, g_m \propto W \tag{4.221}$$

$$r_{gs}, r_{gd}, R_s, r_d, r_o \propto \frac{1}{W}. \tag{4.222}$$

Another, more common, notation for r_{gs} is R_i, as in MOSFETs. The gate resistance is linearly dependent on the finger width and inversely proportional to the number of gate fingers. Since the metal gate sheet resistance R_{sq} is less than $1\Omega/\square$, the gate is almost always contacted on one side only

$$R_g = \frac{W_f R_{sq}}{3N_f}. \tag{4.223}$$

For the 0.15µm GaAs p-HEMT technology whose DC characteristics were presented earlier, the typical small signal parameters are $g_m = 530$ µS/mm, $g_o = 62.5$ µS/mm, $C_{gs} = 940$ fF/mm, $C_{gd} = 90$ fF/mm, $C_{ds} = 73$ fF/mm, and $C_{db} = C_{sb} = 82.5$ fF/mm. We note that C_{gs} is typically ten times larger than C_{gd}, as in HBTs, but quite unlike MOSFETs. This explains the high f_{MAX} of HEMTs which, as in MOSFETs, is inversely proportional to $\sqrt{R_g C_{gd}}$.

4.4.6 f_T, f_{MAX}, and the geometry dependence of the characteristic current densities

Figure 4.113 compiles the f_T versus drain current for the 0.15µm GaAs p-HEMTs with different finger widths and total gate widths. In all cases, as in MOSFETs, the peak f_T current density $I_{DSpfT} = 0.3$mA/µm. The peak f_T value remains practically constant with N_f and W_f, except at small W_f, when the gate fringing capacitance degrades f_T.

Remarkably, experimental evidence collected from various publications indicates that scaling across technology nodes and different material systems does not change the peak f_T (approximately 0.3mA/µm), peak f_{MAX} (approximately 0.2mA/µm) and minimum noise figure current densities (0.15–0.2 mA/µm). The measured data in Figure 4.114 include 0.18µm n-MOSFETs, 0.15µm GaAs p-HEMTs, and 0.1µm InP HEMTs.

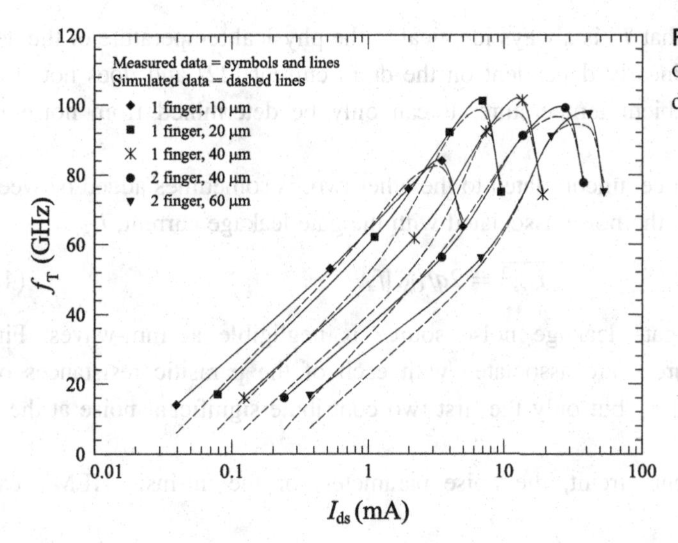

Figure 4.113 Measured f_T as a function of drain current for GaAs p-HEMTs with different gate width [48]

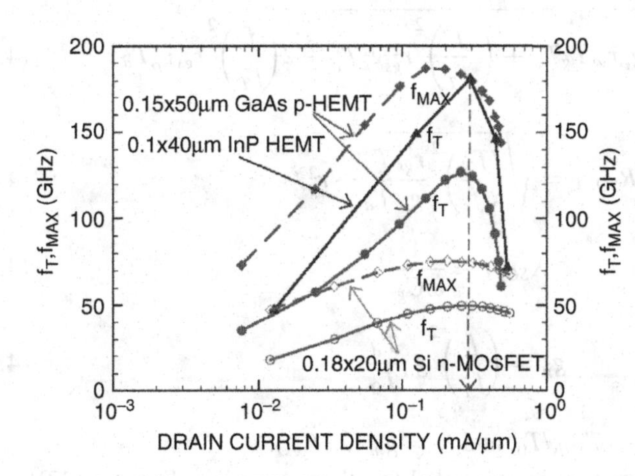

Figure 4.114 Measured f_T-I_{DS}/W and f_{MAX}-I_{DS}/W characteristics for Si MOSFETs, GaAs p-HEMTS and InP HEMTs with different gate lengths [51]

4.4.7 Noise parameters

The noise behavior of the intrinsic HEMT, shown in the dashed box in Figure 4.112, is described by two uncorrelated noise sources [52]. One source is associated with the thermal noise of the channel resistance, r_{gs}, and is described by the noise temperature T_g

$$\overline{v_{ngs}{}^2} = 4kT_g r_{gs} \Delta f, \tag{4.224}$$

while the second describes the drain current noise

$$\overline{i_{nds}{}^2} = 4kT_d g_o \Delta f \tag{4.225}$$

where k is Boltzmann's constant and T_d is the noise equivalent temperature of the drain current.

It is generally accepted that T_g is always identical to the physical temperature of the device. In contrast, T_d is almost linearly dependent on the drain current [22] and does not strongly depend on the device ambient temperature. It can only be determined from noise figure measurements.

A shot-noise current source, uncorrelated to the other two, is sometimes added between the gate and source to describe the noise associated with the gate leakage current, I_{GS}

$$\overline{i_{ngs}{}^2} = 2qI_{GS}\Delta f. \tag{4.226}$$

The contribution of the gate leakage noise source is negligible at mm-waves. Finally, thermal noise voltage sources are associated with each of the parasitic resistances of the extrinsic circuit, r_g, r_s, and r_d, but only the first two contribute significant noise at the input of the transistor.

Based on this equivalent circuit, the noise parameters of the intrinsic HEMT can be expressed as [22]

$$T_{MIN} = 2\frac{f}{f_T}\sqrt{g_o r_{gs} T_g T_d + \left(\frac{f}{f_T}\right)^2 r_{gs}^2 g_o^2 T_d^2 + 2\left(\frac{f}{f_T}\right)^2 r_{gs} g_o T_d} \tag{4.227}$$

$$R_{SOPT} = \sqrt{\left(\frac{f_T}{f}\right)^2 \frac{r_{gs}}{g_o}\frac{T_g}{T_d} + r_{gs}^2} \tag{4.228}$$

$$X_{SOPT} \approx \frac{1}{\omega C_{gs}} \tag{4.229}$$

$$g_n = \left(\frac{f}{f_T}\right)^2 g_o \frac{T_d}{T_0} \tag{4.230}$$

where $T_0 = 290$ K, $NF_{MIN} = 1 + T_{MIN}/T_0$, and $f_T = g_m/(2\pi C_{gs})$.

When the gate leakage noise current is accounted for, the expressions become [22]

$$T_{MIN}^L = T_{MIN} + 2g_n R_{SOPT} T_0 \left[\sqrt{1 + A\left(\frac{f_T}{f}\left|\frac{Z_{SOPT}}{R_{SOPT}}\right|\right)^2} - 1\right] \tag{4.231}$$

$$R_{SOPT}^L = \frac{R_{SOPT}}{1 + A\left(\frac{f_T}{f}\right)^2}\sqrt{1 + \left(\frac{f_T}{f}\left|\frac{Z_{SOPT}}{R_{SOPT}}\right|\right)^2 A} \tag{4.232}$$

$$X_{SOPT}^L = \frac{X_{SOPT}}{1 + A\left(\frac{f_T}{f}\right)^2} \tag{4.233}$$

$$g_n^L = g_n\left[1 + A\left(\frac{f_T}{f}\right)^2\right] \tag{4.234}$$

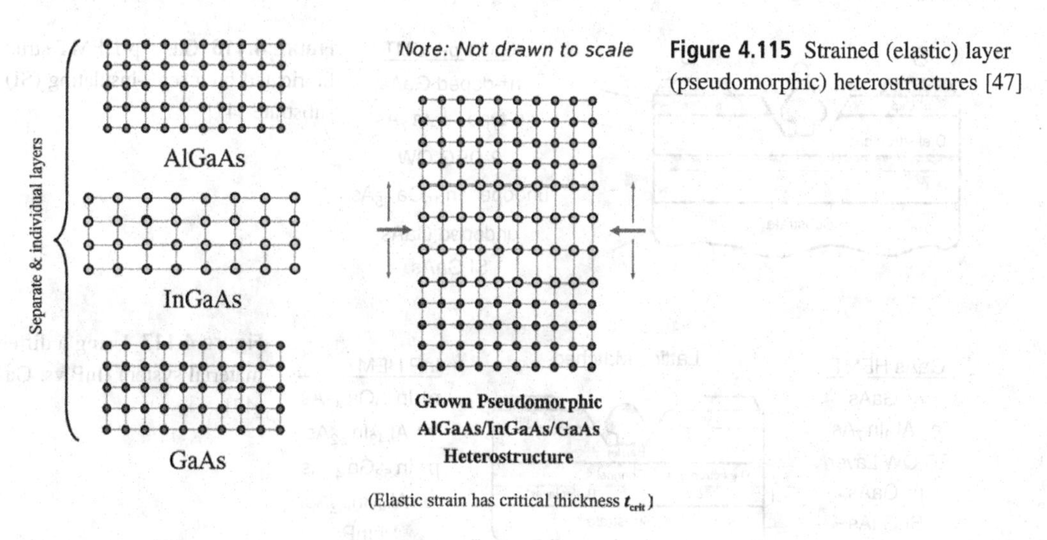

Figure 4.115 Strained (elastic) layer (pseudomorphic) heterostructures [47]

where $A = \frac{\overline{i_{ngs}^2}}{\overline{i_{nd}^2}}$ and superscript L indicates the noise parameters with gate leakage. If $A = 0$, (4.231)–(4.234) reduce to (4.227)–(4.230).

It should be noted that the same equivalence between the Pucel and the Pospieszalski models exists as for MOSFETs and the same noise parameter equations apply.

4.4.8 Other HEMT structures and material systems

The original lattice-matched AlGaAs/GaAs HEMT suffered from several problems:

- I-V or drain collapse at lower temperatures,
- sharply peaked transconductance, and
- persistent photoconductivity,

all of which were due to the presence of donor-complex (DX) centers in $Al_xGa_{1-x}As$ layers with $x > 0.23$.

A search for new material systems was launched early on to eliminate these problems and to further improve device performance. Apart from the goal of eliminating the DX centers, the research aimed:

(a) for higher mobility-channel charge product:
  - deeper quantum well by materials choice,
  - double quantum well structures,
  - higher mobility material for channel region;

(b) for higher cutoff frequency:
  - use of "strained" layers with higher *In* mole fraction (e.g. the pseudomorphic HEMT in Figure 4.115, 4.116 with $In_x\ Ga_{1-x}$ As channel where $x > 0.5$),
  - materials with higher saturated velocity (e.g., InP, Figure 4.117);

(c) for higher power: AlGaN/GaN HEMTs.

The use of compositionally graded layers (Figures 4.118 and 4.119) allows for high-mobility indium-based compound semiconductors and for these layers to be grown on GaAs

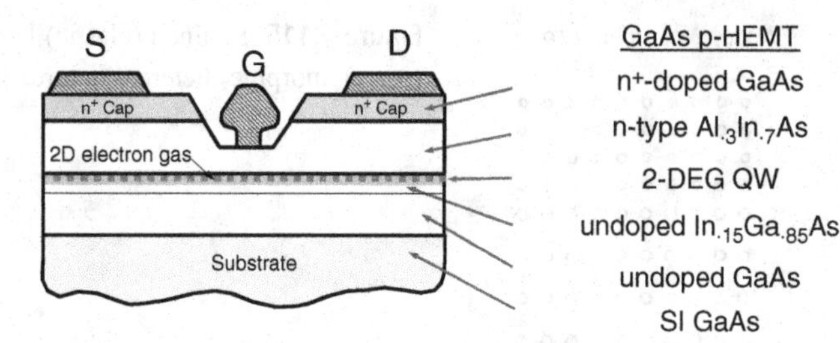

Figure 4.116 GaAs p-HEMT structure fabricated on a semi-insulating (SI) GaAs substrate [47]

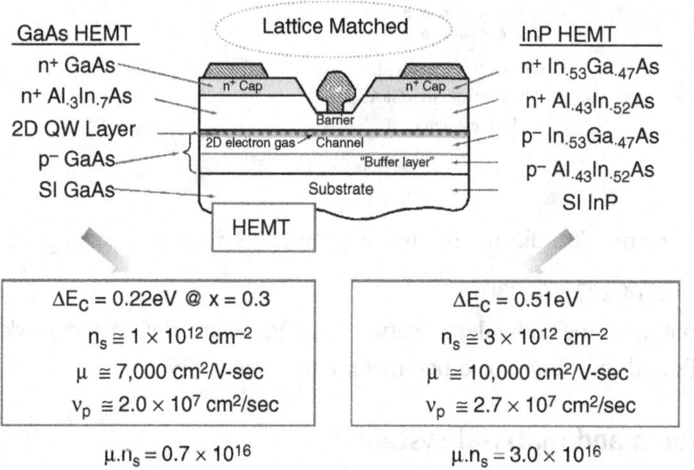

Figure 4.117 Using a different material system (InP vs. GaAs) [47]

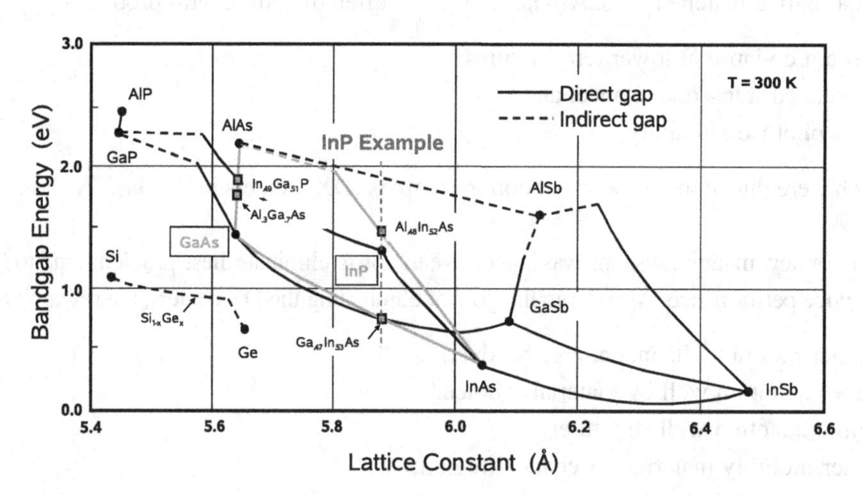

Figure 4.118 Bandgap Engineering: Bandgap Energy vs. Lattice Constant showing the AlInAs/InGaAs on InP example [47]

substrates for lower cost. Compared to InP substrates, GaAs ones are considerably less expensive, less brittle, and available in larger sizes (up to 6″) [47]. The resulting transistor is called a metamorphic HEMT (Figure 4.120) and is almost as fast and as low noise as an InP HEMT.

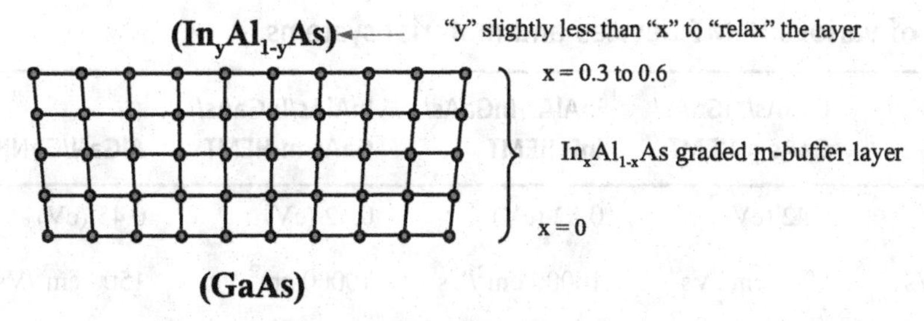

Figure 4.119 Metamorphic growth [47]

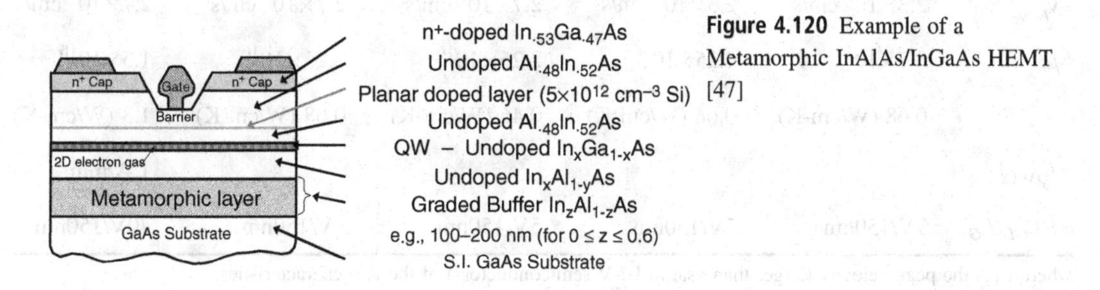

Figure 4.120 Example of a Metamorphic InAlAs/InGaAs HEMT [47]

The InP-based structures with InGaAs/InAlAs channels and barrier layers, and the metamorphic HEMTs give the highest frequency and lowest noise devices. The current record for the fastest transistor in any semiconductor technology is held by a 30nm InP HEMT which has an f_{MAX} of 1.2THz [50].

EXAMPLE 4.15

Design the doping level in the barrier layer of a double-heterostructure (DH) InAlAs/InGaAs/InP p-HEMT with a gate length of 150nm, and $V_T = -0.8$V knowing that $d_i = 5$nm, $\Delta d =$ 5 nm, $\Phi_B = 0.7$eV, $\varepsilon_S = 12 \times 8.84 \times 10^{-14}$ F/m, $\Delta E_C = 0.52$ eV. Draw the energy band diagrams at $V_{GS} = 0$V and $V_{GS} = V_T$ and calculate the sheet charge n_s in the channel at $V_{GS} - V_T = 0.8$V and $V_{DS} = 0$V.

Solution

The total thickness of the barrier layer must be approximately $L_G/10$ to prevent short channel effects $=> d_d + d_i <= 15$nm. Assuming that $\Delta d = 5$nm, and allowing for a thin undoped barrier layer $d_i = 5$nm, this leaves $d_d = 10$nm. We can now apply the threshold voltage equation to determine N_D, the doping in the barrier layer

$$\frac{qN_D d_d^2}{2\varepsilon_s} = \frac{\phi_B}{q} - \frac{\Delta E_C}{q} - V_T;\ N_D = \frac{2\varepsilon_s\left(\frac{\phi_B}{q} - \frac{\Delta E_C}{q} - V_T\right)}{q d_d^2}$$

$$N_D = \frac{2 \times 12 \times 8.856 \times 10^{-12}(0.7 - 0.52 + 0.8)}{1.6 \times 10^{-19}\left(1.5 \times 10^{-8}\right)^2} = 5.8 \times 10^{24}\text{m}^{-3} = 5.8 \times 10^{18}\text{cm}^{-3}$$

Table 4.9 **Comparison of various HEMT devices and material systems.**

Param.	AlGaAs/ GaAsHEMT	InAlAs/InGaAs/ GaAs p-HEMT	InAlAs/InGaAs/ InP HEMT	InAlAs/InGaAs/ GaAs m-HEMT	AlGaN/GaNHEMT
ΔE_C	0.22 (eV)	0.42 (eV)	0.52 (eV)	0.52 (eV)	0.45 (eV)
μ_n	7000 cm^2/Vs	9000 cm^2/Vs	10000 cm^2/Vs	10000 cm^2/Vs	1500 cm^2/Vs
n_s	1×10^{12} cm^{-2}	2.5×10^{12} cm^{-2}	3×10^{12} cm^{-2}	3×10^{12} cm^{-2}	1×10^{13} cm^{-2}
v_p	2.3×10^7 cm/s	2.3×10^7 cm/s	2.7×10^7 cm/s	2.7×10^7 cm/s	2.7×10^7 cm/s
$\mu_n \times n_s$	0.7×10^{16}	2.25×10^{16}	2.25×10^{16}	2.25×10^{16}	1.5×10^{16}
K	0.68 (W/cm-K)	0.68 (W/cm-K)	0.46 (W/cm-K)	0.68 (W/cm-K)	1.3 (W/cm-K)
I_{DMAX}					1 A/mm
BV_{GD}/L_G	7V/150nm	7V/150nm	5V/150nm	7V/150nm	40V/150nm

where v_p is the peak velocity (larger than vsat in III-V semiconductors) of the $v(\varepsilon)$ characteristics.

$$n_s = \frac{\varepsilon_S}{q(d_d + d_i + \Delta d)}[V_G - V_T] = \frac{12 \times 8.856 \times 10^{-12} \times 0.8}{1.6 \times 10^{-19} \times 2 \times 10^{-8}} = 2.65 \times 10^{12}\text{cm}^{-2}.$$

The corresponding energy-band diagrams are shown in Figure 4.121.

RF and microwave power devices require large voltage and current swings and a substrate with good thermal conductivity in order to minimize self-heating. For such devices, in addition to the $\mu_n \times n_s$ product, v_{sat}, the gate–drain breakdown voltage BV_{GD} and the maximum DC current, I_{MAX}, are important parameters to evaluate the merits of a given material system and transistor performance. Figure 4.122 reproduces a typical AlGaN/GaN HEMT structure, the most recent device in the HEMT family. GaN is used as the channel material while $Al_{0.25}Ga_{0.75}N$ is employed for the barrier layer. Historically, because until very recently GaN substrates were not available, GaN HEMTs have been grown on sapphire, SiC, or silicon substrates. The sapphire substrates provide excellent isolation but have poor thermal conductivity, while silicon substrates, the most cost-effective, suffer from inadequate thermal conductivity and a large lattice mismatch with GaN. SiC substrates are semi-insulating and have the highest thermal conductivity but are only available in small sizes.

The advantages of GaN over GaAs for microwave and mm-wave applications can be summarized as:

- larger bandgap (3.45 eV versus 1.43 eV) leading to higher breakdown voltage BV_{GD} (40–100V) and higher temperature operation;
- higher thermal conductivity: 1.3 W/cm-K (4.9 for SiC substrates) versus 0.46 W/cm-K;
- higher breakdown field (4×10^6V/cm versus 5.5×10^5V/cm) allowing for larger voltage swing;

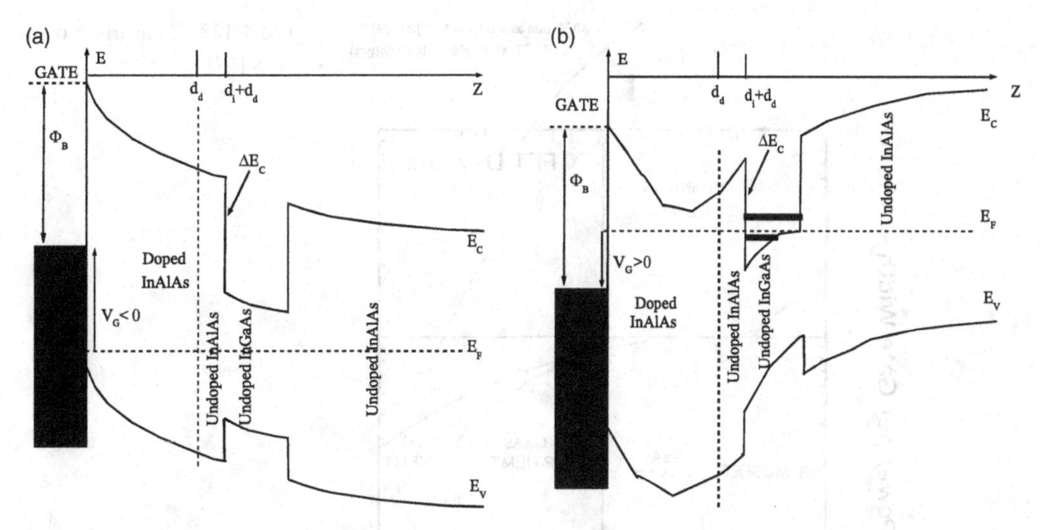

Figure 4.121 Pictorial band diagrams (a) at threshold and (b) for $V_{GS} - V_T > 0$ V for an InP HEMT

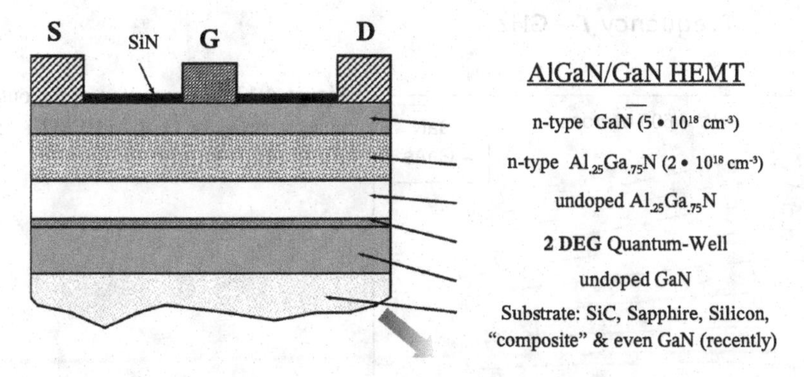

Figure 4.122 Example of AlGaN/GaN HEMT structure [43]

- higher v_{sat} (2.3×10^7 cm/s versus 1×10^7 cm/s) leading to faster operation: high f_T and high I_{MAX},
- higher electron charge density, n_s ($>10^{13}$ cm^{-2} versus 2×10^{12} cm^{-2}) resulting in larger I_{MAX}.

All of the above, combined, result in much larger output power for GaN HEMTs than for GaAs and InP-based devices, as illustrated in Figure 4.123. The Johnson FoM for GaN HEMTs is approximately 5THz × V.

The fastest devices reported at the time of writing feature 40nm gate length with f_T/f_{MAX} = 220/260GHz [53].

4.4.9 Applications

Today HEMTs are used in niche applications where very low noise, very high-power and/or very high-frequency gain are critical requirements and cannot be matched by other semiconductor devices. For example, the first stage of all satellite TV receivers, the antenna switches in cellphones and most microwave and mm-wave power amplifiers for space and military applications use HEMTs.

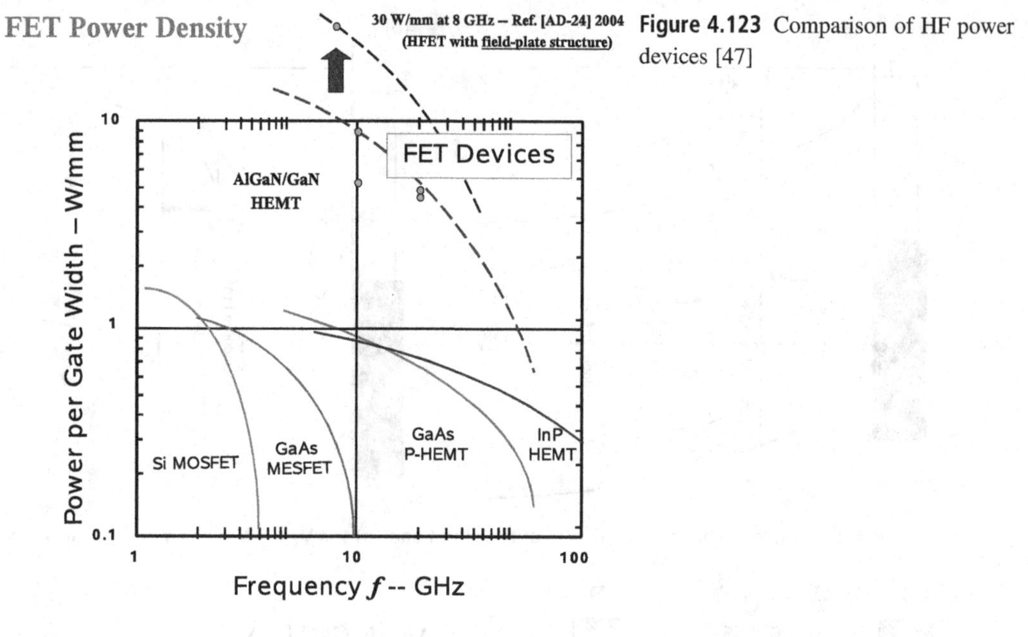

Figure 4.123 Comparison of HF power devices [47]

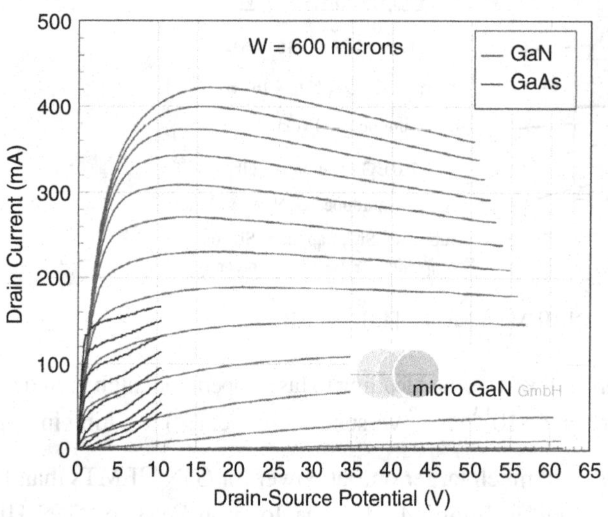

Figure 4.124 Comparison of the output characteristics of GaAs p-HEMT and AlGaN/GaN HEMT devices [47]

4.5 HIGH-FREQUENCY PASSIVE COMPONENTS

4.5.1 Inductors

Inductors are the most important passive components used in high-frequency and high-speed integrated circuits. Their realizability usually defines whether a semiconductor process can be classified as an RF process or not. Historically, inductors and transformers have been the last devices to be integrated in silicon, in the early 1990s. They require a BEOL (back-end-of-line) with a relatively thick dielectric and metal layers with excellent conductivity, positioned at least 4–5 μm above the lossy silicon substrate. Over the years, an increasingly larger number of metal

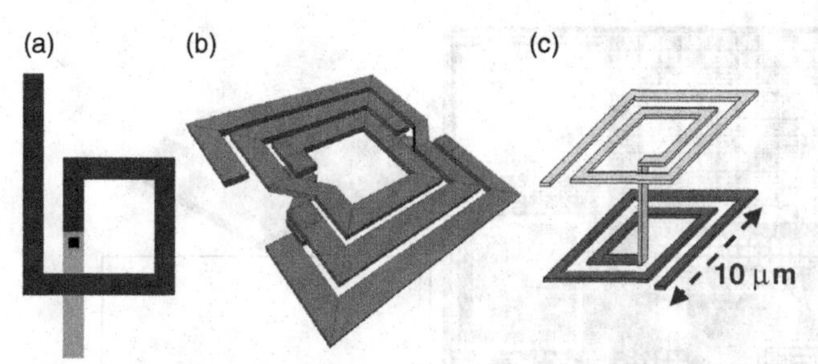

Figure 4.125 Inductors encountered in hybrid and monolithic ICs: (a) rectangular spiral, (b) symmetrical rectangular spiral, (c) 3-D vertically stacked rectangular spiral

layers and thicker overall dielectric stack have become the norm in advanced nanoscale CMOS and SiGe BiCMOS technologies, even in processes not specifically developed for high-frequency applications.

Inductors are as indispensable to HF IC designers as resistors are to analog designers. Although most RF design kits provide scalable models for some limited inductor geometries, a high-frequency circuit designer will inevitably end up using custom inductor and interconnect structures which he or she will have to design and model.

Figure 4.125 illustrates different types of inductors commonly encountered in monolithic integrated circuits. These include:

- spiral (the most common in ICs),
- symmetrical spiral,
- multi-layer shunted (not shown in Figure 4.125),
- 3-D vertically stacked spiral.

The last two inductor types require a BEOL with at least three metal layers, whereas the simpler planar spirals can be realized with just two metal layers, or with a single metal layer and an air-bridge, as often encountered in the early III-V MESFET and HEMT MMICs.

Integrated inductors can also be classified based on the number of terminals in:

- 2-terminal and
- 3-terminal or T-coil inductors.

The most important figures of merit that describe the inductor performance are:

- the inductance, L,
- the quality factor, Q,
- the self-resonance frequency, *SRF*, and
- the peak Q frequency, *PQF*.

In addition to the inductance, physical inductors exhibit parasitic resistive and capacitive elements which limit their performance at very high frequencies and impact the effective inductance and quality factor at any given frequency.

Figure 4.126 shows a top layout view of a planar, rectangular spiral inductor test structure fabricated in the top two metal layers of an RF CMOS process. Due to constraints imposed by manufacturing yield, the inductor is surrounded by three concentric square rings of floating metal pieces. To facilitate S-parameter measurements using special on-wafer probes, the

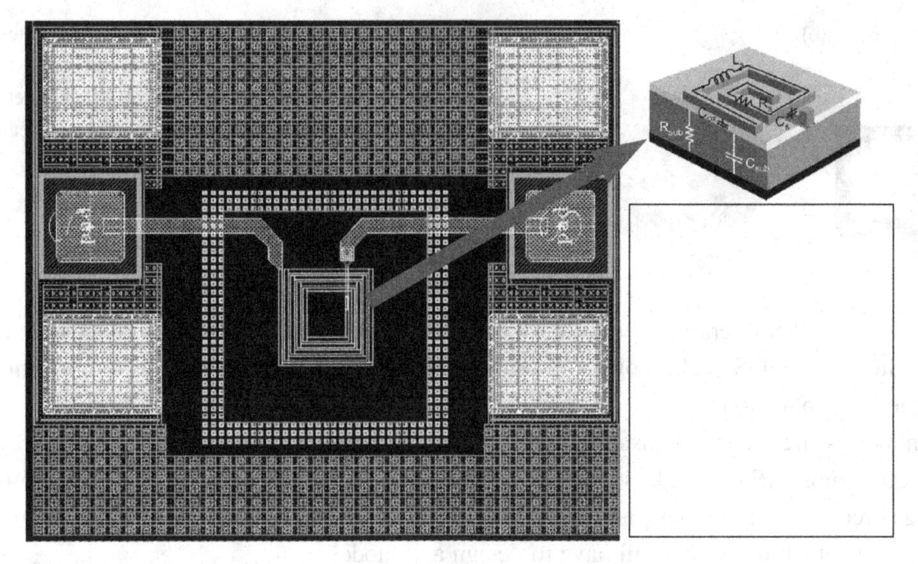

Figure 4.126 Inductor test structure layout and cross-section

inductor is placed as a two-port device between two sets of ground–signal–ground test pads whose contributions are later removed from the measured data by special de-embedding techniques. The inset of Figure 4.126 reproduces a 3-D view of that inductor indicating the most important parasitic components: R_s, C_{OX}, C_{sub}, R_{sub}, and C_p.

Inductance

An inductor induces and stores magnetic field. Its inductance value, L, is defined as the ratio between the total magnetic flux and the electrical current flowing through the inductor. For a planar spiral inductor, the inductance L can be calculated using Greenhouse's equation

$$L = \left(\sum_{j=1}^{n} L_j\right) + \left(\sum_{i=1}^{n}\sum_{j=2(j\neq i)}^{n} M_{ij}\right) \tag{4.235}$$

where L_j and M_{ij} represent the self-inductance of line segment j, and the mutual inductance between segments i and j, respectively. As illustrated in Figure 4.127, if two segments are perpendicular to each other, their mutual inductance is zero. If the current flows in the same direction in two parallel, adjacent line segments, their mutual inductance is positive, while if the current flows in opposite directions, the mutual inductance is negative.

It becomes immediately apparent from (4.235) and Figure 4.127 that, in order to maximize the inductance per area and per volume (an important goal in inductor design which ensures the lowest area and cost), the layout of the inductor must be designed in such a way that currents always flow in the same direction in neighboring parallel segments of the inductor. This rule is important in both side-coupled and vertically coupled adjacent segments in planar and vertically stacked inductors.

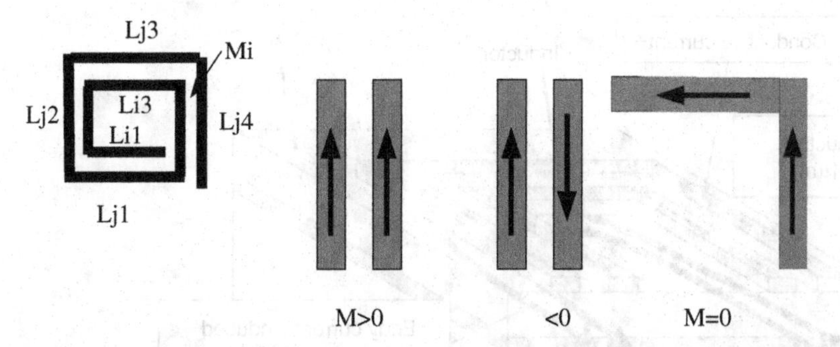

Figure 4.127 Illustration of the mutual inductance between segments of a spiral inductor

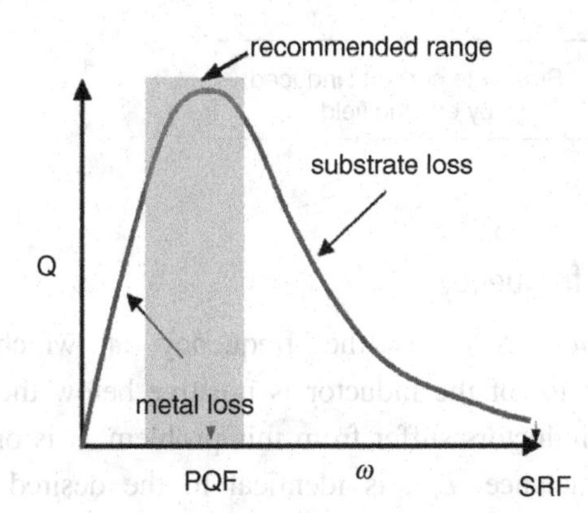

Figure 4.128 Typical inductor quality factor as a function of frequency

Inductor *Q*

Figure 4.128 sketches the typical dependence of the quality factor of an inductor, integrated on a silicon substrate, as a function of frequency, and graphically defines *SRF* and *PQF*. Ideally, *Q*, *SRF*, and *PQF* should be infinite. The inductor quality factor, *Q*, describes the efficiency with which the inductor stores energy. *Q* is defined as the ratio between the energy stored in the inductor and the energy dissipated within the inductor. An ideal inductor has no losses and, therefore, its *Q* is infinite. In a real inductor, several loss mechanisms associated with the metal segments and with the semiconductor substrate lead to the $Q(f)$ characteristics shown in Figure 4.128, where the optimal frequency range of operation for the inductor is highlighted.

Equations (4.236) and (4.237) provide a convenient way to estimate the effective quality factor and the effective inductance of a monolithic inductor from its measured or simulated *Y-parameters*

$$Q_{eff} = \frac{\Im(-Y_{11})}{\Re(Y_{11})} \tag{4.236}$$

$$L_{eff} = \frac{\Im\left[(Y_{11})^{-1}\right]}{\omega}. \tag{4.237}$$

It is important to note that, due to the parasitic capacitances of the monolithic inductor, the effective inductance becomes a function of frequency.

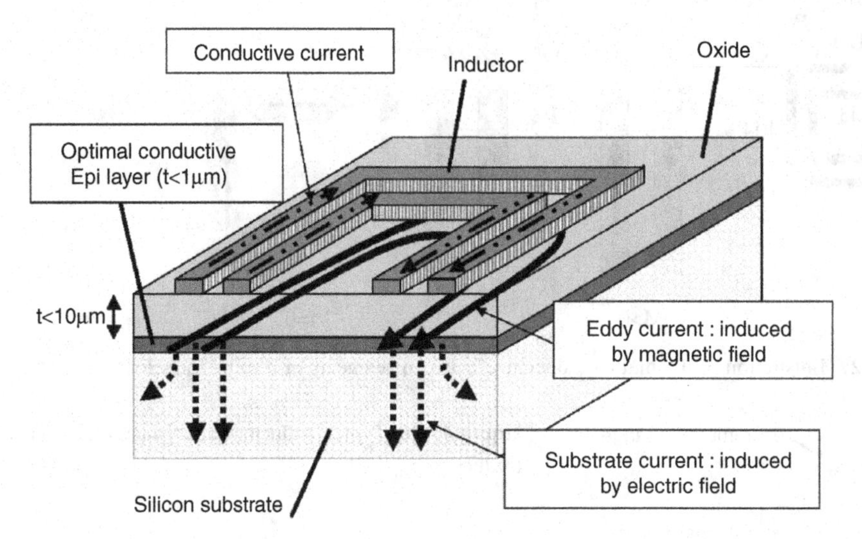

Figure 4.129 Loss mechanisms in inductors [54]

The self-resonance frequency and peak *Q* frequency

By definition, the *self-resonance frequency*, *SRF*, is the frequency at which the inductor becomes a capacitor. The quality factor of the inductor is positive below the *SRF* and negative above the *SRF*. All monolithic inductors suffer from this problem. It is only at very low frequencies that the effective inductance, L_{eff}, is identical to the desired ideal inductance value.

Similarly, *the peak Q frequency*, *PQF*, represents the frequency at which the inductor *Q* reaches its maximum value. It marks the frequency at which the monolithic inductor most closely resembles an ideal inductor, and where it should be used.

Loss mechanisms in inductors

The loss mechanisms that degrade the performance of a monolithic inductor are illustrated graphically in Figure 4.129. Inductor losses occur either:

- in the metal or
- in the substrate.

At frequencies below the *PQF*, the losses in the metal interconnect dominate while above the *PQF*, the substrate losses are responsible for the decrease in the quality factor with increasing frequency.

Metal losses are due to the conductive current flowing through the metal strips of the inductor. As illustrated in Figure 4.130, they can be classified in:

- DC losses due to the finite resistance of the thin metal strips that form the inductor, and
- frequency-dependent losses as a result of the reduction in the conductor cross-section of the metal strips.

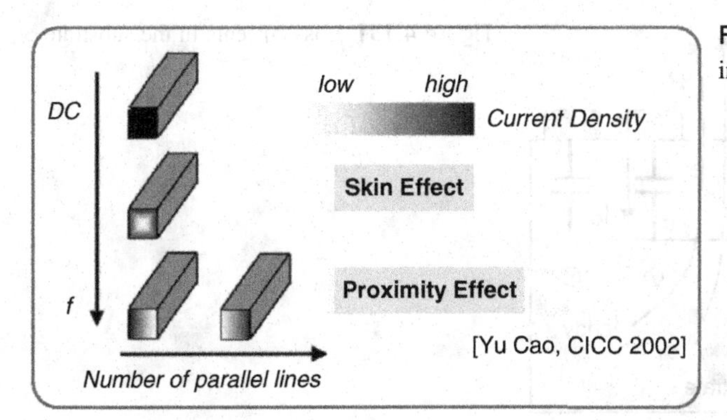

Figure 4.130 Metal loss in monolithic inductors [55]

The latter are caused by:

- the self-induced magnetic field B, known as the skin effect, and
- by the magnetic field induced by adjacent segments, the proximity effect.

Conductive losses can be modeled as a frequency-dependent resistance, R_s, placed in series with the inductance in the high-frequency equivalent circuit of the inductor.

Several techniques are regularly employed to reduce metal loss:

- shunting multiple metal layers together to reduce the DC resistance (which, unfortunately also reduces the *PQF* and *SRF*);
- using narrow inner coils and wider outer coils to minimize the impact of the proximity effect while reducing DC loss and inductor area for a given inductance value;
- employing thicker metal layers (which requires a process change and is not normally an option available to circuit designers); and
- using metals with higher conductivity, for example Cu or Au instead of AlCu, also requiring process change.

However, it should be noted that Cu interconnect is now used in all advanced silicon CMOS and SiGe BiCMOS processes while Au is employed in all III-V processes. Both Cu and Au have excellent conductivity and further improvement is difficult to envision at the moment. Finally, in deeply scaled nanoscale CMOS technologies, the resistance of the inter-metal vias can be a significant contributor to the DC resistance of multi-layer inductors.

Losses in the semiconductor substrate are due to undesirable currents flowing through the semiconductor substrate at high frequencies. The total substrate current, J

$$J = J_p + J_e \tag{4.238}$$

consists of the potential current, J_p, induced by the electric field, ε, and of the Eddy current, J_e, induced by the magnetic field, B. The two are perpendicular to each other and must satisfy equation (4.239)

$$\nabla \times J_p = 0; \quad \nabla J_e = 0 \tag{4.239}$$

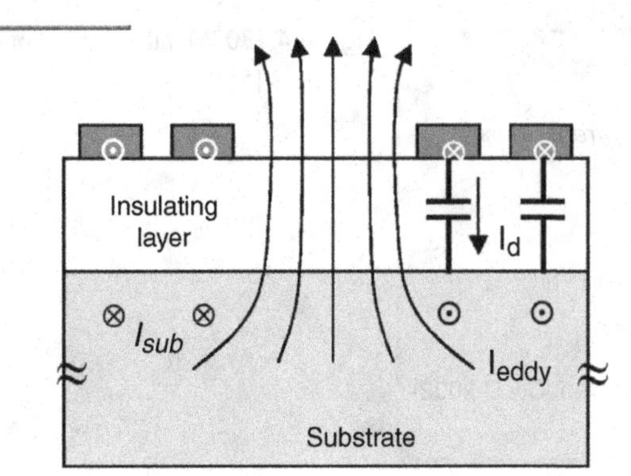

Figure 4.131 Loss currents in the substrate

where

$$J_p = \sigma\varepsilon = -\sigma\nabla\phi \tag{4.240}$$

$$\nabla^2\phi = 0 \tag{4.241}$$

$$\nabla \times J_e = j\omega\sigma B. \tag{4.242}$$

σ is the substrate conductivity, and ϕ is the electric potential.

The Eddy current flows parallel but in opposite direction to the current flow in the metal windings of the inductor, causing inductance and resistance loss.

As shown in Figure 4.131, in silicon ICs, the potential current triggers two loss mechanisms:

- dielectric losses due to the displacement current, J_d, induced by the vertical electric field through the oxide;
- substrate losses due to the substrate current, J_p, also induced by the vertical electric field.

At all frequencies of interest, the silicon substrate losses dominate the dielectric losses, which can be neglected. The frequency dependence of the substrate and Eddy current losses in silicon are illustrated in Figure 4.132.

Equations (4.241) and (4.242) clearly indicate that all substrate losses can be suppressed if the substrate is an insulator, that is $\sigma = 0$. This explains why semi-insulating GaAs and InP substrates have been the first to be used in MMICs, starting in the 1970s.

Techniques to reduce substrate loss

These can be classified in techniques which:

- minimize electric coupling to the substrate (also increase the *SRF* and *PQF*):
  - choose small inductor area,
  - increase dielectric thickness (can imply process change), and
  - select low-k dielectric (process change);
- reduce Eddy currents in substrate:
  - an appropriately patterned ground shield is placed below the inductor (reduces *PQF*, *SRF*);
- reduce both Eddy and potential currents in the substrate:

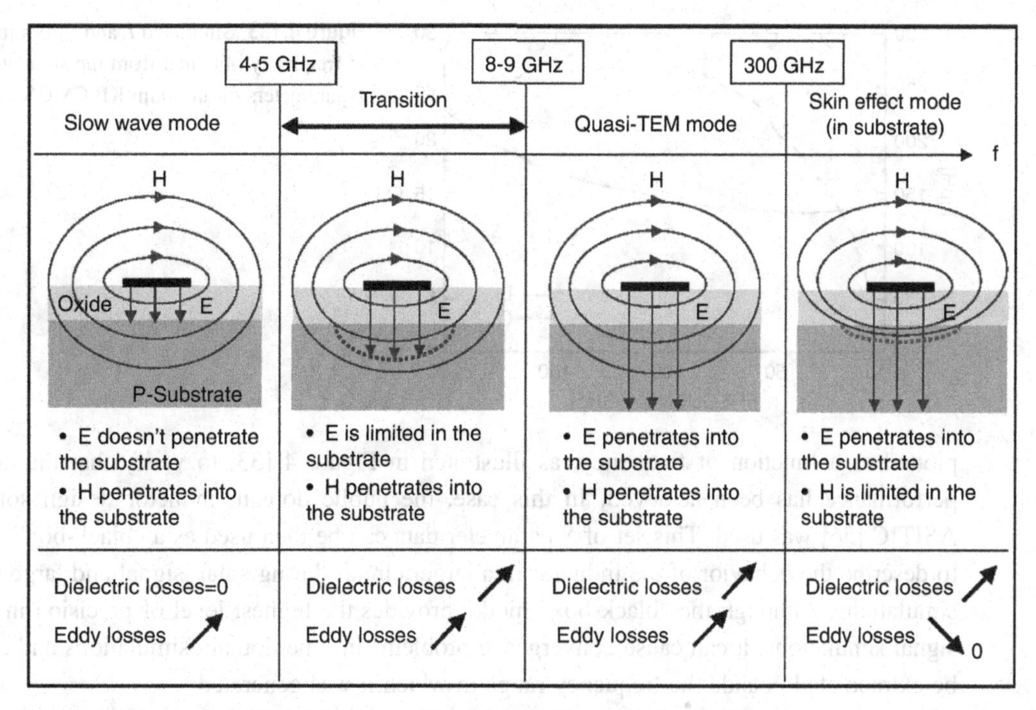

Figure 4.132 Modes of EM field propagation into the silicon substrate [54]

- employ narrower inner inductor turns to reduce Eddy current loss and wider outer turns to reduce DC resistance (increases PQF and SRF),
- increase substrate resistivity (process change), and
- place substrate contacts at an adequately short distance from inductor, typically one quarter of the spiral's external diameter.

Despite the successful integration of inductors in silicon RF processes in recent years, several issues remain. These include:

- low-quality factor which leads to inadequate performance in low-phase noise and in high-power efficiency circuits;
- large chip area (high cost);
- limited application frequency range:
  - lower bound limited by size and Q,
  - upper bound (SRF) limited by dielectric thickness and permittivity.

In general, for a given inductance, the goal is to aim for high *SRF* and high Q, although, in some high-speed digital and broadband applications, a moderate Q is acceptable.

High-frequency equivalent circuit and compact modeling of inductors

Inductors have a truly distributed nature and their performance must be simulated using a 2.5-D or 3-D electromagnetic field simulator which generates the two-port or multi-port S-parameters of the inductor as a function of frequency. The effective inductance and quality factor can be

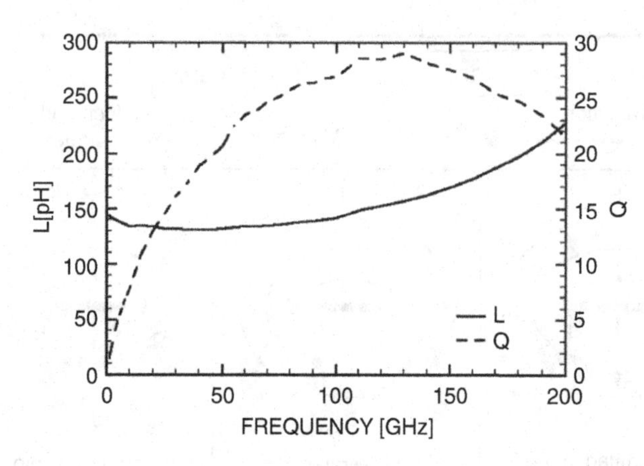

Figure 4.133 Simulated L and Q as a function of frequency obtained from the simulated Y-parameters of an 90nm RF CMOS inductor

plotted as a function of frequency, as illustrated in Figure 4.133, to verify that the desired performance has been achieved. In this case, the public domain inductor design software ASITIC [56] was used. This set of S-parameter data can be then used as a "black-box" model to describe the behavior of the inductor in a larger circuit during small signal and large signal simulations. Although the "black-box" model provides the highest level of precision in small signal simulations, it can cause convergence problems in time domain simulations and cannot be extrapolated outside the frequency range in which it was generated.

In order to simulate and design complex high-frequency circuits that use tens or hundreds of inductors, it is convenient to develop a compact model which relies on a lumped-element, small signal equivalent circuit. The compact model must satisfy a minimum of requirements:

- must be applicable from DC (including for calculating the DC operating point) to frequencies beyond the *SRF* of the inductor;
- must contain a sufficient number of lumped-element sections to accurately capture the distributed nature of the inductor at frequencies significantly larger than the *SRF*;
- must not provide any DC path to the substrate/ground, not even through 1MΩ resistors, because this incorrectly alters the DC operating point of bipolar transistors;
- must be physical in the sense that its elements should be easily derivable from the inductor geometry and BEOL technology data;
- all circuit elements should be constant over the entire range of frequencies of interest; and,
- ideally, it should be scalable.

In practice, single-π or double-π high-frequency equivalent circuits have been successfully employed.

Single-π equivalent circuit

The single-π equivalent circuit has the smallest number of circuit elements. For a planar spiral inductor, formed in the top metal layer and using the second topmost metal layer for the underpass, all circuit elements can be calculated from the physical dimensions of the inductor layout and from the BEOL data typically provided by the foundry. The inductor cross-section and the relationship between the single-π equivalent circuit elements and specific regions of the inductor metalization, dielectric, and silicon substrate are illustrated in Figure 4.134, where T

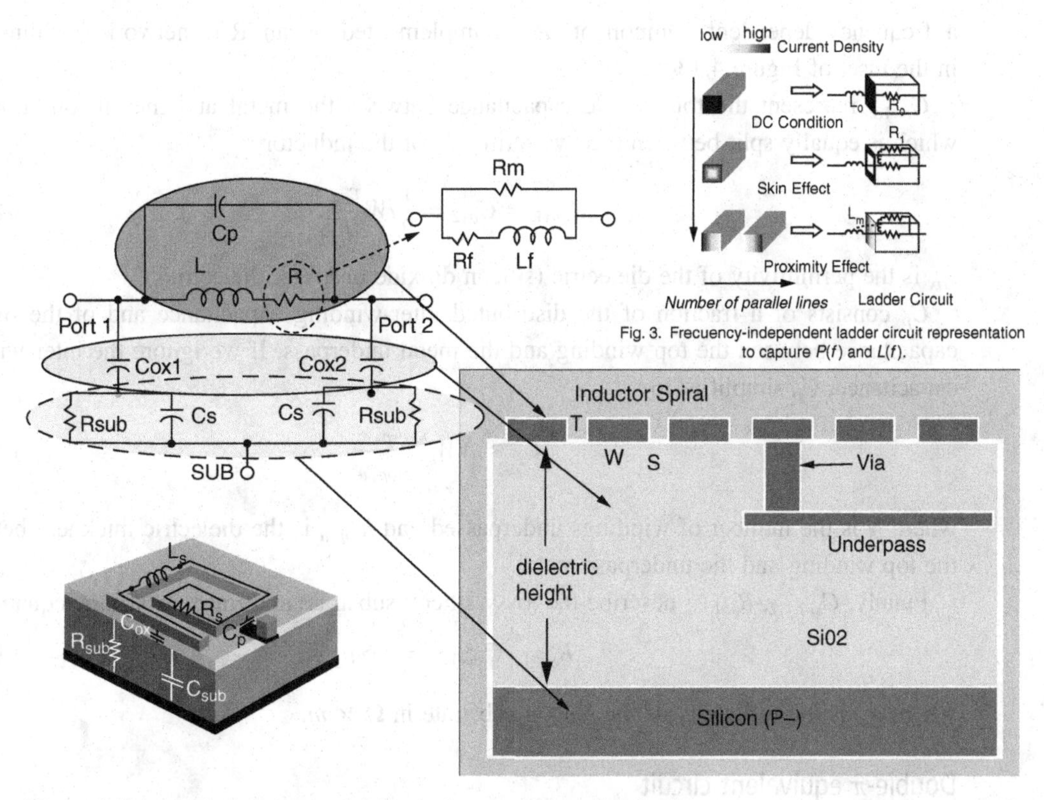

Figure 4.134 Physical derivation of the single-π equivalent circuit [55]

represents the thickness of the top metal layer, W is the width of the metal winding, S is the spacing between adjacent windings, and h is the thickness of the dielectric layer between the top metal and the silicon substrate.

The expression for L is approximated by [55]

$$L \approx \frac{6\mu_0 n^2 d_{avg}^2}{11d - 7d_{avg}} \tag{4.243}$$

where n is the number of windings of the inductor, d is the outer diameter of the inductor, d_{avg} is the arithmetic mean of the inner and outer diameters, and $\mu_0 = 4\pi 10^{-13}$ H/μm is the permeability of vacuum.

The series resistance, R_s, represents the frequency-dependent metal loss and can be expressed as

$$R_s = \frac{\rho l}{W\delta\left[1 - exp\left(\frac{-t}{\delta}\right)\right]} \tag{4.244}$$

where

$$\delta = \sqrt{\frac{1}{\pi f \mu \sigma}} \tag{4.245}$$

l is the total length of the inductor windings, ρ is the resistivity of the metal, f is the frequency, μ is the permeability, and σ is the conductivity of the metal strip. To avoid using a frequency-dependent component, R_s is implemented as an R-L network, as illustrated in the inset of Figure 4.134.

$C_{ox1,2}$ represent the total oxide capacitance between the metal and the silicon substrate, which is equally split between the two terminals of the inductor

$$C_{ox1} = C_{ox2} = \frac{1}{2} lW \frac{\varepsilon_{ox}}{h}. \tag{4.246}$$

ε_{ox} is the permittivity of the dielectric (silicon dioxide or low-k dielectrics).

C_p consists of a fraction of the distributed inter-winding capacitance and of the overlap capacitance between the top winding and the metal underpass. If we ignore the inter-winding capacitance, C_p simplifies to

$$C_p \approx NW^2 \frac{\varepsilon_{ox}}{h_{top,u}} \tag{4.247}$$

where N is the number of windings underpassed and $h_{top,u}$ is the dielectric thickness between the top winding and the underpass metal.

Finally, $C_{sub1,2}$, $R_{sub1,2}$ describe the lossy silicon substrate and must satisfy the equation

$$R_{sub1,2} C_{sub1,2} = \varepsilon_r \varepsilon_0 \rho_{su} \tag{4.248}$$

where ρ_{su} is the resistivity of the silicon substrate in $\Omega \times m$.

Double-π equivalent circuit

The simplicity of the single-π circuit is also its weakness. Since the inductor is a distributed element, in most situations a more elaborate equivalent circuit is needed. For example, if one terminal of the inductor is connected to ground or to V_{DD}, half of the capacitance is artificially removed from the equivalent circuit. Also, the single-π circuit is unable to predict the change of sign in the real part of the inductor Y_{12} at high frequency. For these reasons, the double-π circuit should be preferred. The parameters of the double-π circuit can be obtained directly from those of the single-π circuit, as illustrated in Figure 4.135. where

$$R_{sub} = \frac{R_{sub1} + R_{sub2}}{2} \tag{4.249}$$

and

$$C_s = \frac{C_{sub1} + C_{sub2}}{2}. \tag{4.250}$$

Parameter extraction

A straightforward methodology can be developed to extract the equivalent circuit parameters of the inductor from the simulated or measured Y-parameters as a function of frequency. To ensure that the extracted parameters are physical, the extraction must be direct, ideally without resorting to computer optimization. A direct extraction methodology has been described in [57] and is summarized below.

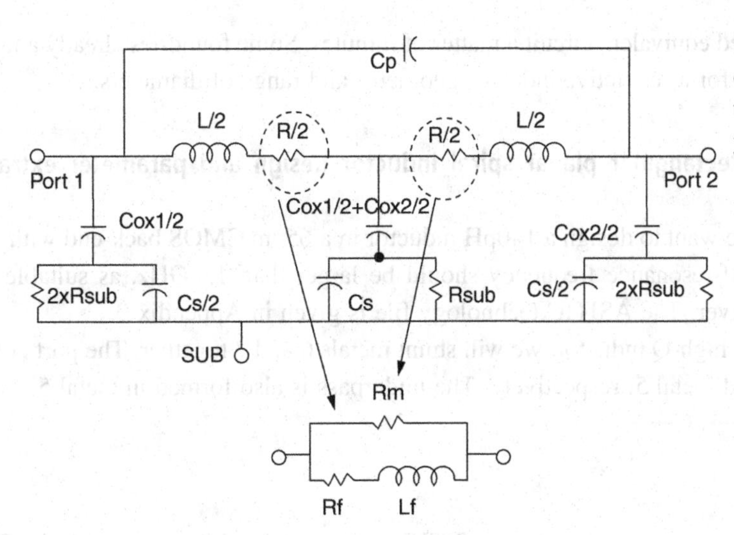

Figure 4.135 Double-π equivalent circuit as function of the single-π equivalent circuit parameters

From the low-frequency range, that is 0.1–0.5 GHz, we can extract directly, without optimization

$$L = \frac{\Im[-Y_{12}{}^{-1}]}{\omega} \tag{4.251}$$

$$R = \Re[-Y_{12}{}^{-1}] \tag{4.252}$$

$$R_{sub1} = \Re\left[(Y_{11} + Y_{12})^{-1}\right] \tag{4.253}$$

$$R_{sub2} = \Re\left[(Y_{22} + Y_{12})^{-1}\right] \tag{4.254}$$

$$C_{OX1} = \frac{\left[\Im\left[(Y_{11} + Y_{12})^{-1}\right]\right]^{-1}}{\omega} \tag{4.255}$$

$$C_{OX2} = \frac{\left[\Im\left[(Y_{22} + Y_{21})^{-1}\right]\right]^{-1}}{\omega} \tag{4.256}$$

and

$$C_s = \frac{\varepsilon_r \varepsilon_0 \times \rho_{su}}{R_{sub}}. \tag{4.257}$$

Next, from the high-frequency range, if possible above the *SRF*, we can obtain the overlap capacitance from the frequency at which imag(Y_{12}), changes sign, becoming capacitive.

Finally, the skin effect parameters R_f, L_f are obtained by fitting real(Y_{12}) as a function of frequency. As a rule of thumb, an R-L pair, as shown in the inset of Figure 4.134, is needed for every decade of frequencies over which the skin effect loss must be modeled.

It should be noted that, when designing inductors using the public domain ASITIC software, $C_{ox1,2}$, R, L, $R_{sub1,2}$, and $C_{sub1,2}$ can be obtained with a simple command at a single frequency, or at two frequencies if the skin effect is to be captured, making the inductor design and the generation of a lumped equivalent circuit a matter of minutes. Some foundries already automate this process but only for a restrictive inductor geometry and range of diameters.

EXAMPLE 4.16 Rectangular planar spiral inductor design and parameter extraction using ASITIC

Let us assume that we want to design a 140pH inductor in a 65nm CMOS back-end with seven metal layers. The self-resonance frequency should be larger than 130GHz, as suitable for a 60GHz radio transceiver. The ASITIC technology file is given in Appendix 9.

In order to create a high-Q inductor, we will shunt metals 6 and 7 together. The port contacts will be in metal 7 and metal 5, respectively. The underpass is also formed in metal 5.

Solution

Inductor definition commands (Note that ASITIC commands and responses are in BOLD letters.)

ASITIC> sq (square spiral)
Name? a
Note: **To specify inner hole dimensions, enter −1.**
Outer dimension(s) of spiral (edge to edge)? 26
Metal width? 2.6 (*W*)
Spacing (metal edge to metal edge)? 1 (*S*)
Turns (fractions okay)? 2.75 (*n*)
Metal layer? 6
Add exit segment (y/n)?n (no underpass added)
Origin of spiral center (x y)? 20 20
ASITIC> cp a b 7 (copy inductor a from metal 6 to b in metal 7)
ASITIC> shunt a b (shunt metal 6 and metal 7 together to reduce series resistance)
ASITIC> wire (create the underpass connection to the inductor in M5)
Name? b
Length? 15
Width? 2.6
Metal layer? 5
Spiral Electrical phase (+ 1 or −1)? 1
Origin of spiral center (x y)? 21 19
Spiral orientation in degrees? 90
ASITIC> cp b c 7 (create connection for the first port in M7)
ASITIC> mv c −7.2 33 (and move it to the appropriate location)
ASITIC> join a b (join the M5 underpass with one port of the inductor)
ASITIC> join a c (join the M7 exit segment with the other port of the inductor)
ASITIC> mv a 0 −5 (move the inductor in the center of the simulation area for better accuracy)

We have now created the inductor whose layout is shown in Figure 4.136.

Figure 4.136 Layout of 140pH inductor designed in Example 4.6.1

The following commands are employed for the extraction of the single-π model parameters.

ASITIC > pix a 0.1 (This gives directly R, L, $C_{ox1} = Cs_1$, $C_{ox2} = Cs_2$, $R_{sub1} = R_{s1}$ and $R_{sub2} = R_{s2}$. This command also provides an estimate of SRF = 160GHz. Note that the calculated SRF is optimistic because ASITIC does not capture inter-winding side capacitance. Leave yourself at least 20% margin). If L and SRF are not acceptable, the design should be modified before proceeding any further. (Note that you can also get an estimate for Cs from (4.257) but you can best extract Cs from the pix command at very high frequency.)

Pi Model at f = 0.10GHz: Q = 0.067,0.067,0.067

L = 0.144 nH R = 1.34

Cs1 = 6.89 fF Rs1 = 1.53e + 04

Cs2 = 3.54 fF Rs2 = 1.26e + 04 f_res = 159.99GHz

ASITIC> pix a 62.5 (This command gives R at 62.5GHz and Q at 62.5GHz. These are not model parameters but rather performance Figures of Merit at the frequency where the inductor will be used. They help us to determine whether the design is acceptable or not and provide a data point for extracting the skin effect network.)

Pi Model at f = 62.50GHz: Q = 12.903,13.064,13.297

L = 0.142 nH R = 4.1

Cs1 = 0.681 fF Rs1 = 757

Cs2 = 0.481 fF Rs2 = 1.21e + 03 f_res = 511.60GHz

ASITIC> pix a 160 (This command gives $C_{sub1} = C_{s1}$ and $C_{sub2} = C_{s2}$)

Pi Model at f = 160.00GHz: Q = 14.337,15.292,16.585

L = 0.158 nH R = 8.58

Cs1 = 0.662 fF Rs1 = 120

Cs2 = 0.458 fF Rs2 = 199 f_res = 507.25GHz

ASITIC> pix a 200 (This command confirms the C_{sub1} and C_{sub2} values obtained at 160 GHz)

Pi Model at f = 200.00GHz: Q = 12.063,13.373,15.160
L = 0.177 nH R = 12.4
Cs1 = 0.662 fF Rs1 = 77.5
Cs2 = 0.46 fF Rs2 = 128 f_res = 469.86GHz

From the commands above, all the single-π equivalent circuit parameters are determined, except C_p and the skin effect R-L network. C_p can be estimated simply by assuming that, along with L, it determines the *SRF*. This turns out to be a rather pessimistic value but makes up for the lack of inter-winding side capacitance in ASITIC.

The skin-effect network parameters can be obtained from the three values of R_s provided by ASITIC at 0.1GHz, 62.5GHz, and 160GHz. In the early phase of the design of a circuit, it is often sufficient simply to use a constant value for R_s since the inductor values and designs may still change. In that case, it is best to pick the value of R_S at the frequency of interest. In Example 4.16, this would be 62.5GHz.

3-Terminal inductors

Three-terminal inductors or t-coils are typically employed in differential circuits, where the common terminal is connected to a bias supply, and as impedance matching elements. The layout and the 2-π equivalent circuit of a 3-terminal inductor are illustrated in Figures 4.137 and 4.138, respectively. The equivalent circuit is similar to that of a 2-terminal inductor but with an additional inductor and resistor associated with the third terminal, and a coupling coefficient to capture the mutual inductance of the two halves of the symmetrical coil.

The equivalent circuit parameters can be calculated in much the same way as those of the 2-terminal inductor. In fact, if the third terminal is left floating, the parasitic capacitances $C_{ox1,2}$, $C_{sub1,2}$ and the substrate resistance values can be estimated or extracted with the same equations as those used for the equivalent 2-terminal inductor. To derive the inductances of each port and the mutual inductance and coupling, we can use the Z-parameters of the 3-terminal inductor obtained from measurements or simulations when the center tap (third terminal) is grounded. Again, from the low-frequency range (0.1–0.5 GHz), we obtain directly, without fitting or optimization

$$L_{11} = \frac{\Im[Z_{11}]}{\omega} \tag{4.258}$$

$$L_{22} = \frac{\Im[Z_{22}]}{\omega} \tag{4.259}$$

$$M = \frac{-\Im[Z_{12}]}{\omega} = \frac{-\Im[Z_{21}]}{\omega} \tag{4.260}$$

$$k = \frac{M}{\sqrt{L_{11} \times L_{22}}} \tag{4.261}$$

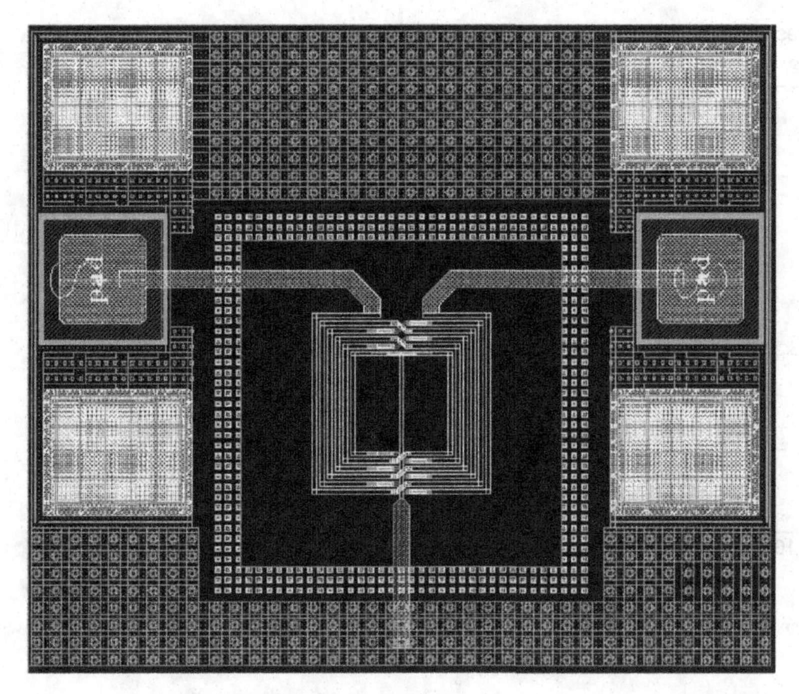

Figure 4.137 Layout of 3-terminal inductor test structure

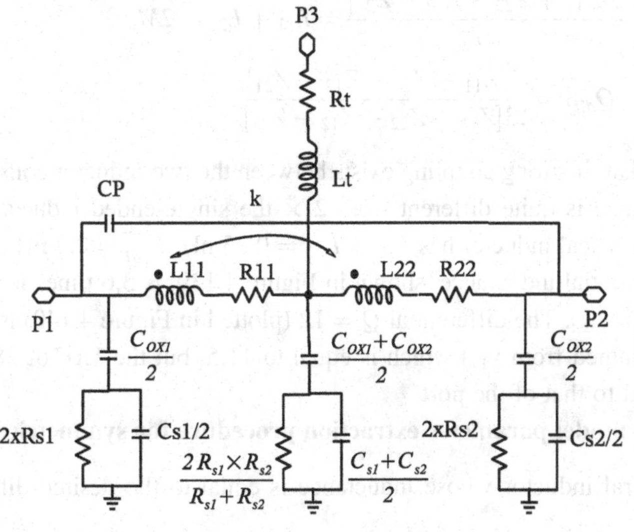

Figure 4.138 Equivalent circuit of a 3-terminal inductor

and

$$R_{11} = \Re[Z_{11}] \tag{4.262}$$

$$R_{22} = \Re[Z_{22}] \tag{4.263}$$

$$R_t = \Re[Z_{12}] = \Re[Z_{21}]. \tag{4.264}$$

The differential inductance and the differential Q (with the center terminal floating) are given by

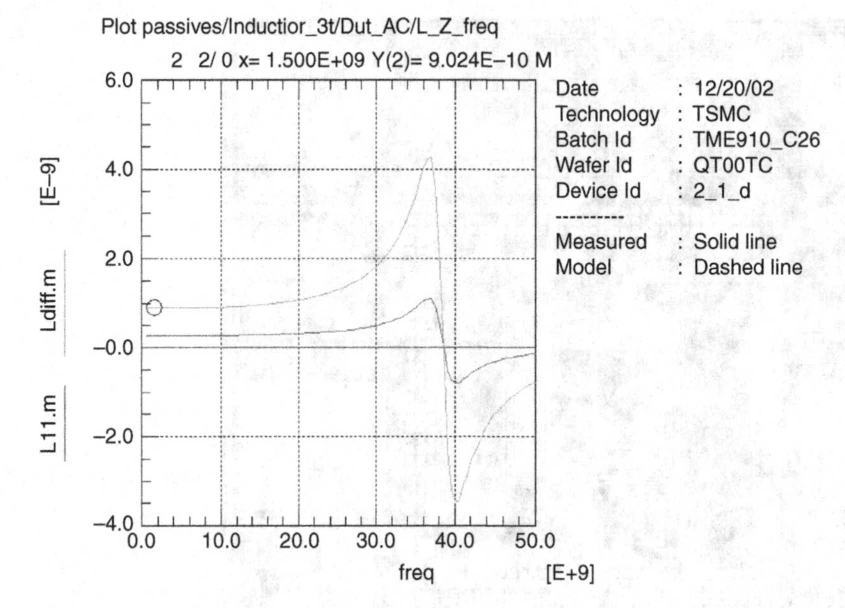

Figure 4.139 Measured differential and port inductance of the 3-terminal inductor test structure shown in Figure 4.137

$$L_{diff} = \frac{\Im[Z_{11} + Z_{22} - Z_{12} - Z_{21}]}{\omega} = L_{11} + L_{22} + 2M \tag{4.265}$$

$$Q_{diff} = \frac{\Im[Z_{11} + Z_{22} - Z_{12} - Z_{21}]}{\Re[Z_{11} + Z_{22} - Z_{12} - Z_{21}]}. \tag{4.266}$$

It is important to note here that, if strong coupling exists between the two inductor coils L_{11} and L_{22}, the differential inductance is quite different from 2 × the single-ended inductance. For example, a 3-terminal symmetrical inductor has $L_{11} = L_{22} = 0.25$ nH, $L_{diff} = 0.9$ nH and $M = 0.2$ nH. The measured differential inductance, shown in Figure 4.139, is 3.6 times larger than the inductance per side $L_{11} = L_{22}$. The differential $Q = 12$ (plotted in Figure 4.140) is slightly better than the port1 Q (obtained from y_{11}) which is equal to 11.5, but the *SRF* of 38GHz in differential mode is identical to that of the port.

ASITIC-based compact model parameter extraction procedure for symmetrical t-coils

- Create a symmetrical spiral inductor whose inductance is equal to the desired differential inductance:
  - Use the "pix" command at 0.1GHz to find C_{OX}, R_{SUB}, R_{diff}, L_{diff}, and *SRF*;
  - Use the "pix" command at *SRF* or higher frequency to find C_{SUB}:
    - Split the inductor in its two symmetrical halves.
- Use the "pix" command on the each half coil to get R_{ii}, L_{ii}.
- Use the "C" command with one of the two halves grounded to get the capacitance between the two coils C_p:
  - Use the "k" command at 0.1GHz to get the coupling, k, between the two coils.

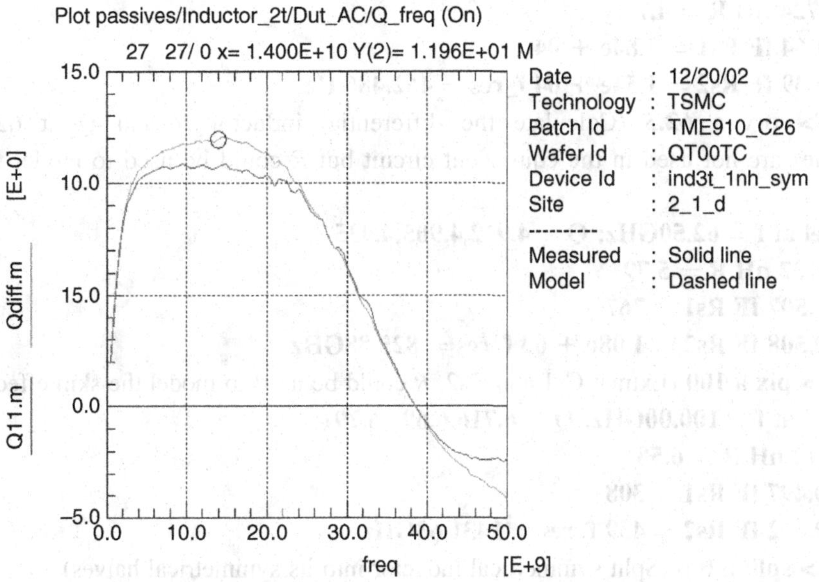

Plot passives/Inductor_2t/Dut_AC/Q_freq (On)

27 27/ 0 x= 1.400E+10 Y(2)= 1.196E+01 M

Date : 12/20/02
Technology : TSMC
Batch Id : TME910_C26
Wafer Id : QT00TC
Device Id : ind3t_1nh_sym
Site : 2_1_d

Measured : Solid line
Model : Dashed line

Q11.m Qdiff.m [E+0]

freq [E+9]

Figure 4.140 Measured differential and port Q of the 3-terminal inductor test structure shown in Figure 4.137

EXAMPLE 4.17 Three-terminal inductor design

In the same 65nm BEOL as in Example 4.16, design a symmetrical spiral inductor with center tap, such that it can be used as a 3-terminal inductor, and extract its equivalent circuit. The differential inductance should be 75pH and its self-resonance frequency should be suitable for operation in a 60GHz radio transceiver.

Solution

ASITIC> symsq (a symmetrical square spiral inductor, with two turns, formed in metal 7, is employed)

Name? a
Outer Length Edge to Edge: 18
Metal width? 1.5
Metal layer? 7
Transition metal layer? 6
Spacing (metal edge to metal edge)? 1
Space for Transitions: 2.5
Turns? 2
Origin of spiral center (x y)? 20 20

ASITIC> pix a 0.1 (Calculate the differential inductance, the SRF, R, extract C_{ox1}, C_{ox2}, R_{sub1}, and R_{sub2})

Pi Model at f = 0.10GHz: Q = 0.010,0.010,0.010

L = 0.0724 nH R = 4.7

Cs1 = 1.64 fF Rs1 = 1.84e + 04

Cs2 = 2.39 fF Rs2 = 1.54e + 04 f_res = 462.48GHz

ASITIC> pix a 62.5 (Calculate the differential inductor *R* and *Q* at 62.5GHz. These values are not used in the equivalent circuit but *R* could be used to model the skin effect)

Pi Model at f = 62.50GHz: Q = 4.912,4.905,4.935

L = 0.0732 nH R = 5.79

Cs1 = 0.507 fF Rs1 = 767

Cs2 = 0.508 fF Rs2 = 1.08e + 03 f_res = 825.88GHz

ASITIC> pix a 100 (Extract Cs1 and Cs2, *R* could be used to model the skin effect)

Pi Model at f = 100.00GHz: Q = 6.716,6.694,6.791

L = 0.072 nH R = 6.58

Cs1 = 0.497 fF Rs1 = 308

Cs2 = 0.492 fF Rs2 = 439 f_res = 1431.63GHz

ASITIC> split a 0 b (Split symmetrical inductor into its symmetrical halves)

ASITIC> pix a 0.1 (Extract the single-π model parameters: R_{11}, L_{11} of inductor half *a*)

Pi Model at f = 0.10GHz: Q = 0.004,0.004,0.004

L = 0.027 nH R = 3.83

Cs1 = 1.72 fF Rs1 = 1.55e + 04

Cs2 = 1.28 fF Rs2 = 1.91e + 04 f_res = 736.90GHz

ASITIC> pix b 0.1 (Extract the single-π model parameters: R_{22}, L_{22} of inductor half *b*)

Pi Model at f = 0.10GHz: Q = 0.019,0.019,0.019

L = 0.0269 nH R = 0.875

Cs1 = 1.5 fF Rs1 = 1.62e + 04

Cs2 = 1.52 fF Rs2 = 1.8e + 04 f_res = 791.58GHz

ASITIC> pix a 62.5 (Find the *Q* and R_{11} at 62.5GHz for half *a*)

Pi Model at f = 62.50GHz: Q = 2.463,2.464,2.467

L = 0.0258 nH R = 4.1

Cs1 = 0.477 fF Rs1 = 1.19e + 03

Cs2 = 0.459 fF Rs2 = 640 f_res = 1435.48GHz

ASITIC> pix b 62.5 (Find the *Q* and R_{22} at 62.5GHz for half *b*)

Pi Model at f = 62.50GHz: Q = 8.881,8.882,8.902

L = 0.0264 nH R = 1.16

Cs1 = 0.474 fF Rs1 = 894

Cs2 = 0.46 fF Rs2 = 957 f_res = 1422.35GHz

ASITIC> k a b (Calculate the coupling coefficient between the two halves of the 3-terminal inductor from *a* to *b*)

Coupling coefficient of A and B: k = 0.34348 and M = 0.00923 (nH).

ASITIC> k b a (…and from *b* to *a*)

Coupling coefficient of B and A: k = 0.34348 and M = 0.00923 (nH).

ASITIC> c a 0.1 b (Calculate the capacitance C_p between the two halves for the 3-terminal inductor)

At 0.100GHz:
Total Capacitance = 4.316 (fF)
Total Resistance = 4567.870.

When should one use t-coils and differential-mode inductors? In differential mode, t-coils have the same inductance, *Q*, *SRF*, and occupy the same area as the 2-terminal inductor obtained by leaving the center tap of the t-coil floating. However, in very high-speed MOS CML or BiCMOS CML differential circuits, where the self-resonance frequency of the peaking inductor is critical, one should use two 2-terminal inductors rather than a t-coil. Although this choice will result in slightly larger area for the CML gate, the *SRF* in both single-ended and differential mode will be higher than in the case where a single t-coil is used differentially. Furthermore, in circuits that employ 3-terminal inductors as differential loads, the single-ended and differential impedances cannot be simultaneously impedance-matched due to the large mutual inductance. As a result, it is not recommended to place t-coils in differential output drivers. However, they can be used inside the chip, as long as they do not drive long differential transmission lines.

The following are useful design tips for high-frequency inductors:

- when a high *SRF* is desired: choose narrow metal windings (small *W*), wide spacing, minimum diameter;
- for high *Q*, low *SRF* inductors: choose wide, thick metal, shunted metal layers, patterned ground shield;
- for inductors with large *L*: choose large diameter, narrow spacing, and vertically stacked, helical geometry with series-connected metal layers;
- for large *L*, high *SRF*, moderate-to-low *Q* inductors used in high-speed current-mode-logic (CML) cells: choose small diameter, narrow metal, vertically stacked helical geometry with series-connected metal layers;
- place grounded substrate contact taps at a distance of approximately *d*/4 from the edge of the inductor to improve the inductor *Q* and to reduce coupling to adjacent inductors in a large circuit with many inductors;
- to minimize the layout area of differential circuits: use one center-tapped differential inductor rather than two inductors.

Below are design tips useful for the design of high-*SRF*, high-*Q* inductors:

- small diameter (*d*),
- narrow (narrower in inner turns) (*W*),
- thick (*T*),
- widely spaced (*S*),
- top metal only,
- windings on,
- thick dielectric (*h*),
- with low permittivity (ε_{OX}).

Inductor scaling to higher frequencies

It is important to note that, like MOSFETs, spiral inductors also follow scaling laws. Therefore, inductor scaling can be accomplished by shrinking all lateral and vertical dimensions of the inductor and of the BEOL by a factor S [58]. The outcome of scaling can best be demonstrated using the simple equations presented earlier for a rectangular planar spiral inductor.

From (4.243), it is observed that, if the diameter is scaled by a factor S, the inductance also reduces by the same factor

$$L \approx \frac{6\mu_0 n^2 d_{avg}^2}{11d - 7d_{avg}} => \frac{L}{S} \approx \frac{6\mu_0 n^2 \left[\frac{d_{avg}}{S}\right]^2}{11\frac{d}{S} - 7\frac{d_{avg}}{S}}. \tag{4.267}$$

Moreover, the oxide capacitance between the bottom metal layer of the inductor and the semiconductor substrate is a function of the total metal line length in one layer (l), the trace width (W), and the distance from the bottom metal layer to the substrate (h). If the length and width are scaled by the same factor S, the inductor area is reduced by a factor of S^2, while the total parasitic capacitance to ground

$$C_{ox} = \frac{1}{2} lW \frac{\varepsilon_{ox}}{h} => \frac{C_{ox}}{S} = \frac{1}{2}\frac{1}{S}\frac{W}{S}\frac{\varepsilon_{ox}}{\frac{h}{S}} \tag{4.268}$$

and the cross-capacitance between the two inductor terminals

$$C_p = nW^2 \frac{\varepsilon_{ox}}{h_{M9-M8}} => \frac{C_p}{S} = n\left(\frac{W}{S}\right)^2 \frac{\varepsilon_{ox}}{\frac{h_{M9-M8}}{S}} \tag{4.269}$$

decrease S times.

As a result, despite the fact that the thickness of the dielectric between the top metal and the silicon substrate has been reduced S times (typical of the trend in nanoscale CMOS technologies), the self-resonance frequency increases S times

$$SRF \approx \frac{1}{2\pi\sqrt{L(C_{OX} + C_p)}} => S \times SRF = \frac{1}{2\pi\sqrt{\frac{L}{S}\left(\frac{C_{OX}}{S} + \frac{C_p}{S}\right)}}. \tag{4.270}$$

For completeness, it is instructive to see how Q scales with the inductor size. The DC resistance in a single metal layer can be determined from the metal thickness, t, and resistivity, ρ.

$$R_{DC} = \frac{\rho l}{Wt} => R_{DC} = \frac{\rho \frac{l}{S}}{\frac{W}{S} t}. \tag{4.271}$$

While l and W are both reduced by S [1], the top metal thickness and sheet resistance usually do not change from one technology node to another. Therefore, the total series resistance of the inductor remains constant. The same is true at high frequencies when the skin depth, δ, of the conductor is considered

$$R_{AC} = \frac{\rho \frac{l}{S}}{\frac{W}{S}\delta\left[1 - \exp\left(\frac{-t}{\delta}\right)\right]}. \tag{4.272}$$

Consequently, the peak value of the quality factor of the scaled inductor remains unchanged, while the frequency at which the Q peaks increases roughly by a factor of S. It follows naturally that spiral structures are as suitable for use at mm-wave frequencies as they were for RF applications at 2–5 GHz.

4.5.2 Transformers

Transformers can be analyzed, modeled, and designed in much the same way as 3-terminal inductors. Two non-inverting transformer structures, one relying on vertical coupling between the primary and secondary, while the other using lateral coupling between the primary and secondary, are illustrated in Figure 4.141.

Figure 4.142 shows the schematic of an inverting transformer and one of the possible physical implementations using vertically coupled windings. Finally, two symmetrical transformer structures with laterally and vertically coupled primary and secondary coils are illustrated in Figure 4.143 for a conventional "digital" BEOL and for a mm-wave BEOL.

Transformer modeling

The operation of an ideal transformer as in Figure 4.142(a) is described in the time domain by the following equations

$$v_1 = L_1 \frac{di_1(t)}{dt} - M\frac{di_2(t)}{dt} \tag{4.273}$$

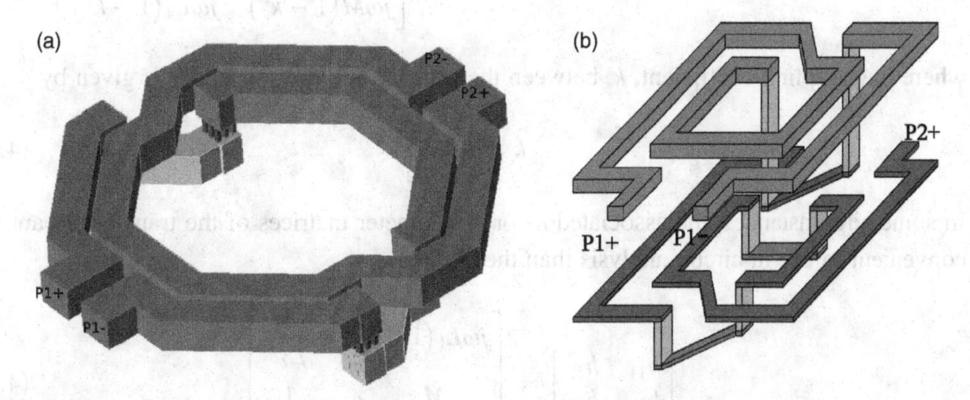

Figure 4.141 (a) Side-coupled symmetrical transformer or balun and (b) vertically stacked transformer/balun with symmetrical primary and secondary

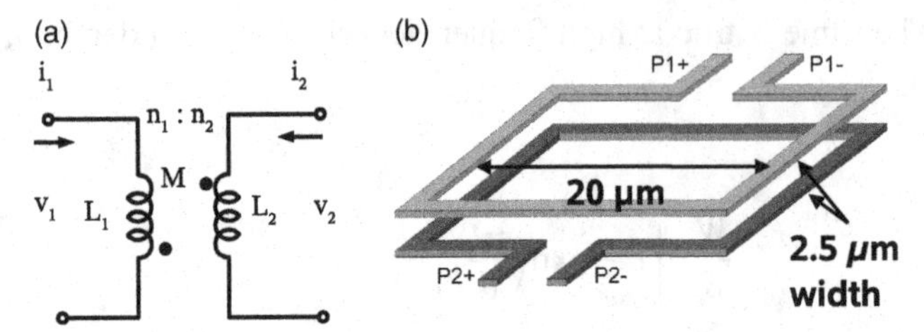

Figure 4.142 (a) Schematic representation of an inverting transformer and (b) possible implementation as a vertically stacked structure with symmetrical primary and secondary. Note that the current is flowing into opposite directions in the primary and in the secondary windings

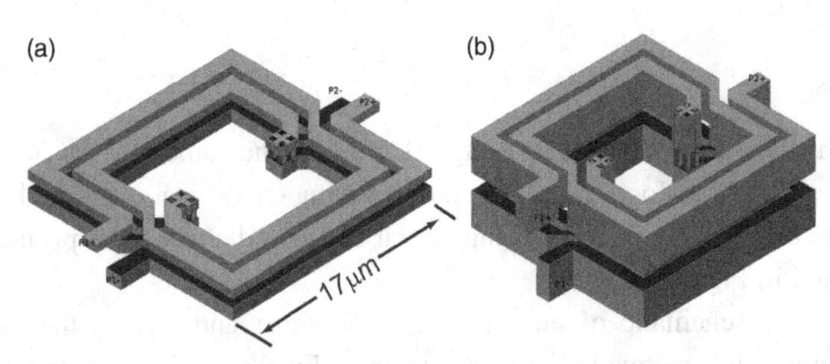

Figure 4.143 Symmetrical transformer (balun) with vertically and laterally coupled two-layer primary and secondary in (a) "digital" BEOL and (b) thick-metal, mm-wave BEOL [59]

$$v_2 = -M\frac{di_1(t)}{dt} + L_2\frac{di_2(t)}{dt}. \tag{4.274}$$

These can be readily represented in the frequency domain by the equivalent Z- or Y-parameter matrices

$$\begin{bmatrix} z_{11} & z_{12} \\ z_{21} & z_{22} \end{bmatrix} = \begin{bmatrix} j\omega L_1 & -j\omega M \\ -j\omega M & j\omega L_2 \end{bmatrix}; \begin{bmatrix} y_{11} & y_{12} \\ y_{21} & y_{22} \end{bmatrix} = \begin{bmatrix} \dfrac{1}{j\omega L_1(1-k^2)} & \dfrac{k^2}{j\omega M(1-k^2)} \\ \dfrac{k^2}{j\omega M(1-k^2)} & \dfrac{1}{j\omega L_2(1-k^2)} \end{bmatrix} \tag{4.275}$$

where the coupling coefficient, k, between the primary and the secondary is given by

$$k = \frac{M}{\sqrt{L_1 L_2}}. \tag{4.276}$$

In some circumstances, the associated h- or g-parameter matrices of the transformer are more convenient to use in circuit analysis than the Y-matrix

$$\begin{bmatrix} h_{11} & h_{12} \\ h_{21} & h_{22} \end{bmatrix} = \begin{bmatrix} j\omega L_1(1-k^2) & \dfrac{-M}{L_2} \\ \dfrac{M}{L_2} & \dfrac{1}{j\omega L_2} \end{bmatrix} \tag{4.277}$$

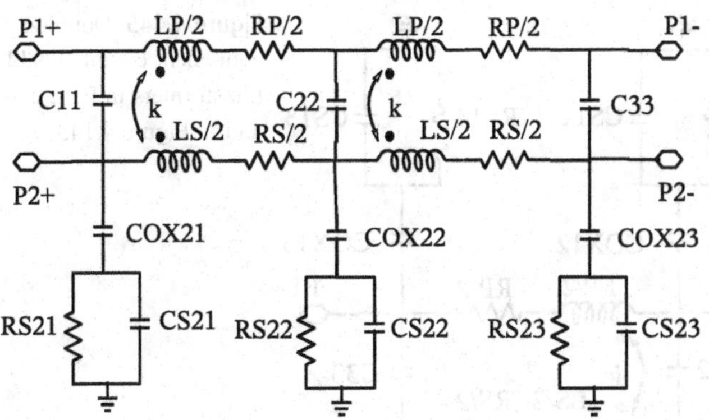

Figure 4.144 Double-π equivalent circuit suitable for the transformers in Figure 4.141(b) and in Figure 4.142

$$\begin{bmatrix} g_{11} & g_{12} \\ g_{21} & g_{22} \end{bmatrix} = \begin{bmatrix} \dfrac{1}{j\omega L_1} & \dfrac{M}{L_1} \\ \dfrac{-M}{L_1} & j\omega L_2\left(1-k^2\right) \end{bmatrix}. \tag{4.278}$$

The latter two matrices have the interesting property that the forward and the reverse transfer functions are equal in magnitude but have opposite signs.

3-D stacked transformer modeling

Just like an inductor, a real, integrated transformer has parasitic elements that need to be captured by a compact model. Depending on the layout of the transformer, on the vertical or lateral coupling, custom 2-π high-frequency equivalent circuits can be developed similarly to those employed for inductors. For example, for vertically coupled transformers where the primary is placed directly above the secondary, only the primary coil is effectively shielded from the substrate by the secondary coil. As a result, the equivalent circuit shown in Figure 4.144 can be used along with the following extraction methodology.

ASITIC-based compact model parameter extraction procedure for vertically stacked transformers

- Create two separate inductors in different metal layers.
- Use the "pix" command on bottom coil alone to find its complete 2-π equivalent circuit C_{OX}, C_{SUB}, R_{SUB}, R_2, L_2.
- Use the "pix" command on the top coil alone to find R_1 and L_1 for the top coil.
- Use the "c" command at 0.1GHz on the top coil with the bottom coil grounded to get the parasitic capacitance, C_p, between the two coils and distribute it among $C_{11} = C_{33} = C_{p/4}$ and $C_{22} = C_{p/2}$.
- Use the "k" command at 0.1GHz to find the coupling coefficient. The inverting or non-inverting nature of the coupling must be reflected in the lumped equivalent circuit.

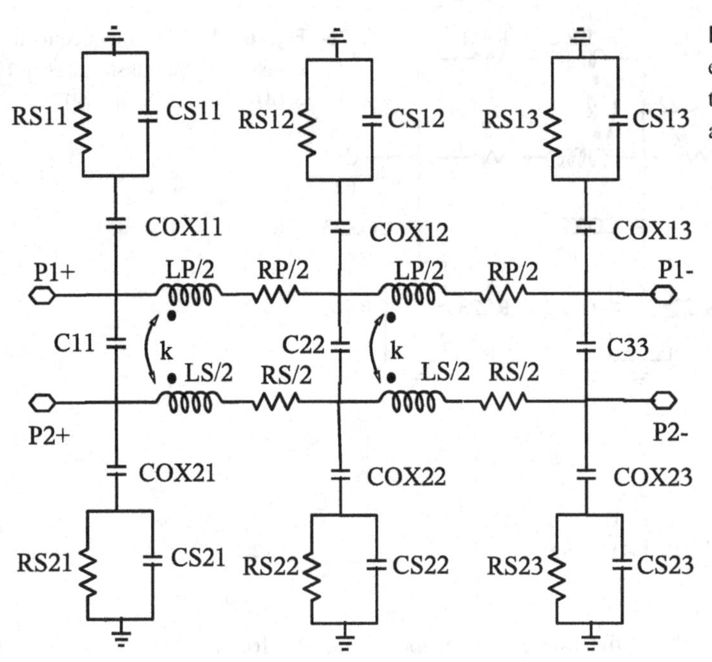

Figure 4.145 Double-π equivalent circuit suitable for the transformers in Figure 4.141(a) and in Figure 4.143

For baluns (symmetrical transformers) as in Figure 4.143 or Figure 4.141(a), where the primary and secondary coils are symmetrical, side-coupled, and equally exposed to the silicon substrate, the more complex equivalent circuit shown in Figure 4.145 must be used along with the following parameter extraction methodology.

Compact model parameter extraction procedure for symmetrical, side-coupled transformers

- Create two separate inductors one for the primary, one for the secondary, or create a transformer and split it into its primary and secondary coils.
- Use the "pix" command on each coil to find its complete 2-π equivalent circuit C_{OX}, C_{SUB}, R_{SUB}, R, L.
- Use the "c" command at 0.1GHz on the primary coil with the secondary coil grounded to get the parasitic capacitance, C_p, between the two coils and distribute it among $C_{11} = C_{33} = C_{p/4}$ and $C_{22} = C_{p/2}$.
- Use the "k" command at 0.1GHz to find the coupling coefficient between the two coils.

Figure 4.146 shows the die photo and measured S_{21} of a vertically coupled transformer fabricated in a 90nm RF-CMOS back-end [9] where the primary and secondary each have two turns. The transmission coefficient remains better than −2dB from 60GHz up to 94GHz.

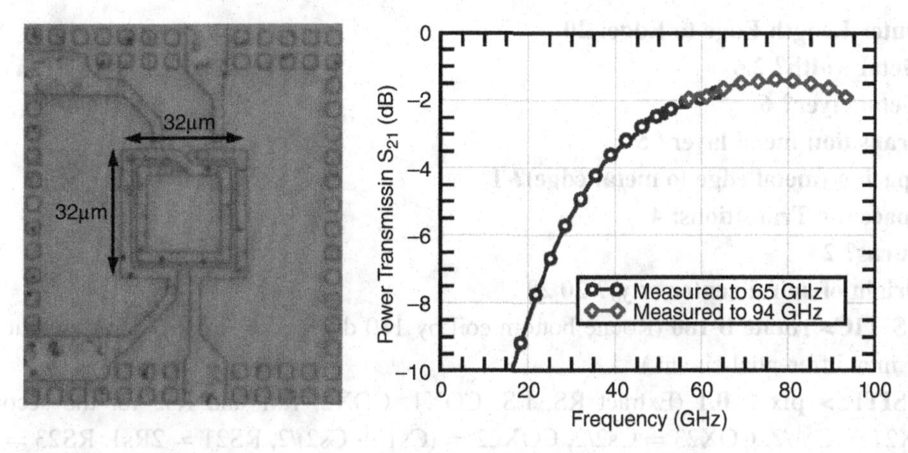

Figure 4.146 Die photo and measured S_{21} of a vertically stacked transformer fabricated in a 90nm, RF-CMOS process with $W = S = 2\mu m$, $t = 3\mu m$ [9]

EXAMPLE 4.18 Transformer design and compact model extraction using ASITIC

Design a vertically coupled, non-inverting, symmetrical transformer (balun) with a 1:1 turn ratio. The primary and secondary each have 2 turns. The self-resonance frequency should be over 150GHz and the outer diameter should be 20μm. Use the same 65nm back-end as in Examples 4.16 and 4.17. Use the layout in Figure 4.141(b) and the equivalent circuit in Figure 4.144. Extract the high-frequency equivalent circuit.

Solution

ASITIC> symsq (Create the top coil with two turns)

Name? a

Outer Length Edge to Edge: 20

Metal width? 2.6

Metal layer? 7

Transition metal layer? 6

Spacing (metal edge to metal edge)? 1

Space for Transitions: 4

Turns? 2

Origin of spiral center (x y)? 20 20

ASITIC> pix a 0.1 (Extract R_P, L_P for the primary)

Pi Model at f = 0.10GHz: Q = 0.021,0.021,0.021

L = 0.0581 nH R = 1.71

Cs1 = 2.15 fF Rs1 = 1.64e + 04

Cs2 = 2.54 fF Rs2 = 1.69e + 04 f_res = 450.54GHz

ASITIC> symsq (Create the bottom coil with two turns)

Name? b

Outer Length Edge to Edge: 20
Metal width? 2.6
Metal layer? 6
Transition metal layer? 5
Spacing (metal edge to metal edge)? 1
Space for Transitions: 4
Turns? 2
Origin of spiral center (x y)? 20 20

ASITIC> rotate b 180 (Rotate bottom coil by 180 degrees to simplify input-output signal flow in a differential circuit.)

ASITIC> pix b 0.1 (Extract RS, LS, COX1, COX2, Rs1 and Rs2 for the secondary. COX21 = Cs1/2, COX23 = Cs2/2, COX22 = (Cs1 + Cs2)/2, RS21 = 2Rs1, RS23 = 2Rs2, RS22 = 2Rs1Rs2/(Rs1 + Rs2))

Pi Model at f = 0.10GHz: Q = 0.044,0.044,0.044
L = 0.064 nH R = 0.924
Cs1 = 2.9 fF Rs1 = 1.62e + 04
Cs2 = 2.96 fF Rs2 = 1.77e + 04 f_res = 369.80GHz

ASITIC> k a b (Extract k)

Coupling coefficient of A and B: k = 0.68421 and M = 0.04177 (nH).

ASITIC> c a 0.1 b (Calculate the total capacitance C between the primary and secondary and distribute it C11 = C33 = C/4, C22 = C/2)

At 0.100GHz:
Total Capacitance = 10.668 (fF)
Total Resistance = 1459.215.

ASITIC> pix a 62.5 (Find Q and R_P of the primary at 62.5GHz)

Pi Model at f = 62.50GHz: Q = 7.002,7.002,7.037
L = 0.0526 nH R = 2.92
Cs1 = 0.538 fF Rs1 = 865
Cs2 = 0.52 fF Rs2 = 972 f_res = 946.06GHz

ASITIC> pix b 62.5 (Find Q and R_S of the secondary at 62.5GHz)

Pi Model at f = 62.50GHz: Q = 14.181,14.187,14.308
L = 0.0557 nH R = 1.52
Cs1 = 0.553 fF Rs1 = 909
Cs2 = 0.537 fF Rs2 = 945 f_res = 906.74GHz

ASITIC> pix a 300 (Find R_P of primary at 300GHz for skin effect modeling)

Performing Analysis at 300.00GHz
Pi Model at f = 300.00GHz: Q = 13.313,13.340,14.509
L = 0.0551 nH R = 6.59
Cs1 = 0.521 fF Rs1 = 40.1
Cs2 = 0.5 fF Rs2 = 45.3 f_res = 946.44GHz

ASITIC> pix b 300 (Find R_S of secondary at 300GHz for skin effect modeling, extract Cs1 and Cs2. Set CS21 = Cs1/2 CS23 = Cs2/2, CS22 = (Cs1 + Cs2)/2))

Pi Model at f = 300.00GHz: Q = 29.316,29.476,34.232
L = 0.0633 nH R = 2.98
Cs1 = 0.533 fF Rs1 = 42
Cs2 = 0.52 fF Rs2 = 43.6 f_res = 872.85GHz
--

4.5.3 Transmission lines

Transmission lines have always played a significant role in high-frequency integrated circuits. They show up by design, as circuit matching elements, or inevitably, as interconnect between circuit blocks. The most common types of transmission lines encountered in monolithic integrated circuits are microstrip lines and coplanar waveguides, CPWs, as illustrated in Figure 4.147. Historically, because accurate models have been more readily available for microstrip lines, this has led to their proliferation. If a separate ground plane is easy to implement, as on the back side of semi-insulating GaAs and InP wafers, or above the silicon wafer using the lower metal layers, microstrip has the advantage of occupying less area than CPWs which require relatively large metal ground planes in the same metal layer as the signal lines. However, in situations where a ground plane on top of the semiconductor wafer is not feasible or has high loss, coplanar waveguides can have (significantly) lower loss than microstrip lines realized directly over the silicon substrate. This is especially the case on high-resistivity SOI substrates.

The loss mechanisms of silicon microstrip and coplanar waveguides are similar to those encountered in inductors, except for the proximity effect which is not present in isolated metal lines. In nanoscale CMOS BEOLs, a significant part of the metal loss occurs in the ground plane. This can be alleviated by shunting several lower-level metal layers together to reduce the resistance of the ground plane.

Figure 4.148 reproduces the measured attenuation in dB/mm for two microstrip lines realized in standard CMOS BEOL and in a thick metal, mm-wave BEOL and compares it to the attenuation of a coplanar waveguide fabricated on a low-loss alumina substrate (CS15 CPW) [60].

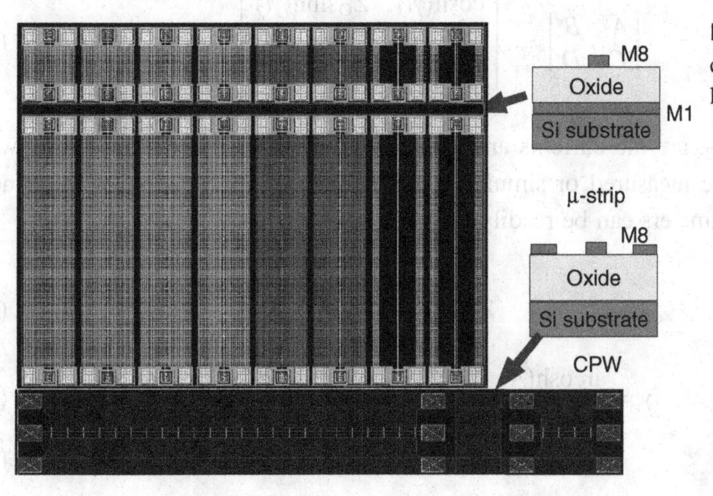

Figure 4.147 Microstrip and coplanar waveguide transmission line test structures

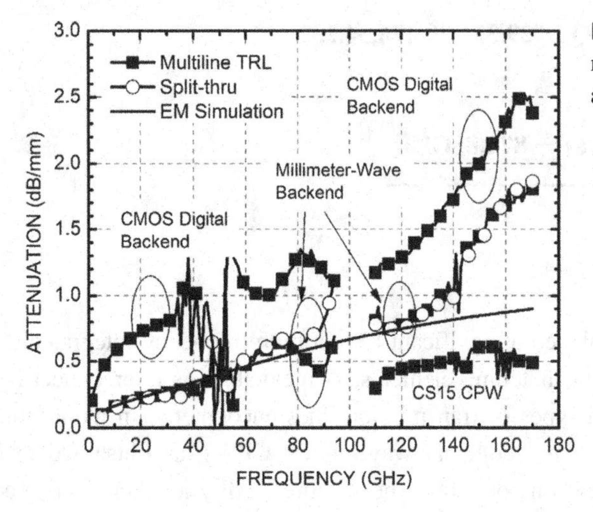

Figure 4.148 Measured attenuation of several microstrip transmission lines on silicon and on alumina substrates [60]

Transmission line model

The most important parameters that describe the high-frequency performance of a transmission line are the characteristic impedance, Z_O, and the propagation constant, $\gamma = \beta + j\alpha$, from which the attenuation constant, α (in dB/mm), and the delay, τ (in ps/mm) are obtained.

The simplest and most common analytical model of the transmission line as a distributed circuit element describes its two-port *ABCD*-matrix as a function of Z_O, γ, and the transmission line length, l

$$\begin{bmatrix} v_1 \\ i_1 \end{bmatrix} = \begin{bmatrix} \cosh(\gamma l) & Z_O \sinh(\gamma l) \\ \frac{\sinh(\gamma l)}{Z_O} & \cosh(\gamma l) \end{bmatrix} \times \begin{bmatrix} v_2 \\ -i_2 \end{bmatrix} \tag{4.279}$$

$$\begin{bmatrix} A & B \\ C & D \end{bmatrix} = \begin{bmatrix} \cosh(\gamma l) & Z_O \sinh(\gamma l) \\ \frac{\sinh(\gamma l)}{Z_O} & \cosh(\gamma l) \end{bmatrix} \tag{4.280}$$

where i_1, v_1 and i_2, v_2, are the currents and voltages at the input and output of the two-port, respectively. From the measured or simulated *ABCD*-matrix as a function of frequency, the transmission line parameters can be readily obtained as

$$Z_O = \sqrt{\frac{B}{C}} \tag{4.281}$$

$$\gamma = \frac{\operatorname{arcosh}(A)}{l} \quad \text{or} \quad \gamma = \frac{\operatorname{arcsinh}\left(\sqrt{BC}\right)}{l} \tag{4.282}$$

$$\alpha(\text{in dB/mm}) = 8.688\Re(\gamma) \tag{4.283}$$

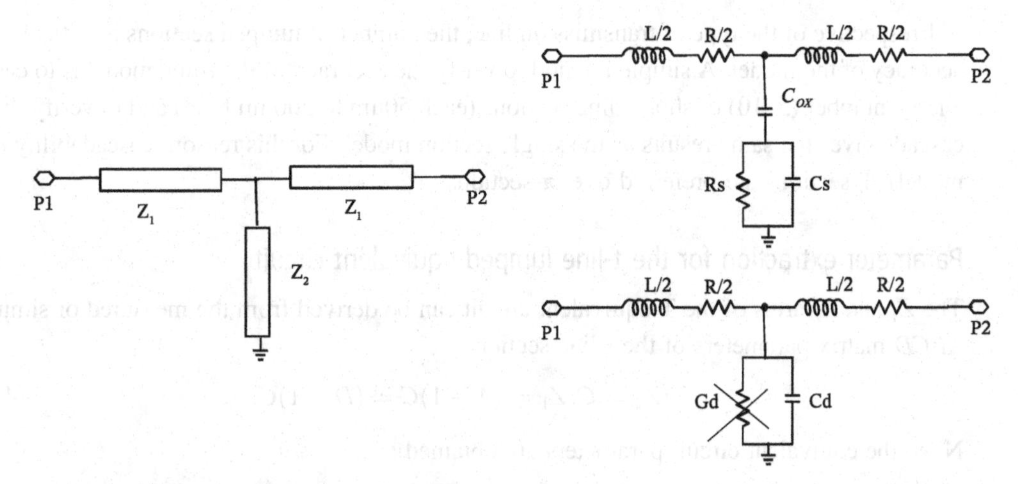

Figure 4.149 Lumped model equivalent circuit of a short section of microstrip transmission line as a T-section. The circuit in the upper right corner is suitable for microstrip lines over the lossy silicon substrate. The circuit in the bottom right is suitable for microstrip lines over semi-insulating. III–V substrates or for metal-oxide-metal microstrip lines on top of a silicon substrate.

$$\tau = \frac{\partial \beta}{\partial \omega} \times l. \tag{4.284}$$

Likewise, the capacitance, C_d, and inductance, L, per unit length of the transmission line can be obtained directly from the *ABCD*-parameters as

$$C_d = \frac{1}{\omega} \Im\left(\frac{\gamma}{Z_0}\right) \tag{4.285}$$

$$L = C_d \frac{B}{C}. \tag{4.286}$$

For circuit design purposes, a lumped model, similar to the ones developed for inductors and transformers, is desirable. Most simulators have built-in frequency-domain models for GaAs, InP, and Si metal-oxide-metal (M-O-M) microstrip lines. However, we should verify that the built-in simulator models are adequate for transient simulations. In some cases, even the propagation delay is not properly accounted for. In any event, lumped scalable RLC models must be developed for metal-over-silicon lines since they are not available in simulators.

The symmetrical T-equivalent circuit, as illustrated in Figure 4.149, is the most convenient topology for representing the behavior of a short transmission line section of length l ($l < \lambda/10$). Specific implementations of this equivalent circuit for microstrip t-lines over the lossy silicon substrate and over a semi-insulating substrate are shown in the right-hand side of Figure 4.149. Note that frequency-dependent losses can be included in the resistance R using the same approach as for inductors.

Important note: To prevent the unintentional alteration of the DC operating point of a bipolar transistor whose base is connected to a transmission line, the t-line equivalent circuit must not allow DC paths between any of its ports and ground. For this reason, the loss conductance of the insulating dielectric should be removed from the bottom right equivalent circuit. This solution only marginally affects the accuracy of the transmission line model.

Irrespective of the type of transmission line, the number of lumped sections is critical for the accuracy of the model. A simple method to verify the accuracy of the t-line model is to cascade a large number (5–10) of short t-line sections (each 50μm to 200μm long) and to verify that the cascade gives the same results as the single section model. For this reason (cascadability of the model), T-sections are preferred over π-sections.

Parameter extraction for the t-line lumped equivalent circuit

The Z_1 and Z_2 arms of the T-equivalent circuit can be derived from the measured or simulated *ABCD*-matrix parameters of the t-line section

$$Z_2 = C; Z_1 = (A - 1)C = (D - 1)C. \tag{4.287}$$

Next, the equivalent circuit parameters are obtained as

$$\frac{L}{2} = \frac{\Im(Z_1)}{2\pi f} \tag{4.288}$$

$$\frac{R(f)}{2} = \Re(Z_1) \tag{4.289}$$

$$C_d = \frac{\Im\left(Z_2^{-1}\right)}{2\pi f} \tag{4.290}$$

and

$$G_d = \Re\left(Z_2^{-1}\right). \tag{4.291}$$

For the special case of a microstrip line placed over the silicon substrate, with the ground plane located on the back of the substrate, we can use the equations and methodology developed in the inductor section, where the inductor is a simple wire.

4.5.4 MIM and MOM capacitors

Metal-insulator-metal (MiM) or Metal-Oxide-Metal) MoM capacitors are preferred at high frequencies over polysilicon capacitors because of their linearity and high-quality factors. While MoM capacitors (Figure 4.150) come for free, MiM capacitors require a special, very thin, insulator layer and additional process steps and fabrication masks compared to a MoM capacitor. On the other hand, they typically exhibit higher Q and area capacitance, better reliability and repeatability.

Irrespective of the type of capacitor, their high-frequency behavior can be modeled using the Γ-equivalent circuit shown in Figure 4.151. In addition to the desired capacitance, C, the latter includes the parasitic loss resistance, R, and inductance, L, associated with the capacitor plates, the parasitic capacitance, C_{OX}, to the silicon substrate, and the R_s,C_s loss of the substrate itself, much like in an inductor.

When isolation and crosstalk are major concerns, a deep n-well may be placed below the MiM or MoM capacitor in order to minimize the injection of noise to and from the substrate,

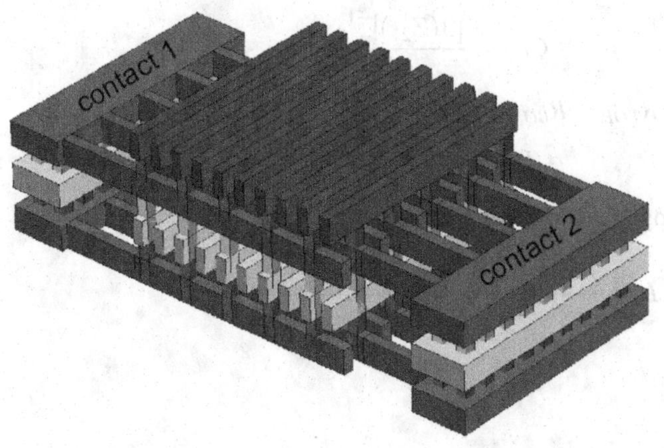

Figure 4.150 MoM capacitor layout and 3-D view [27]

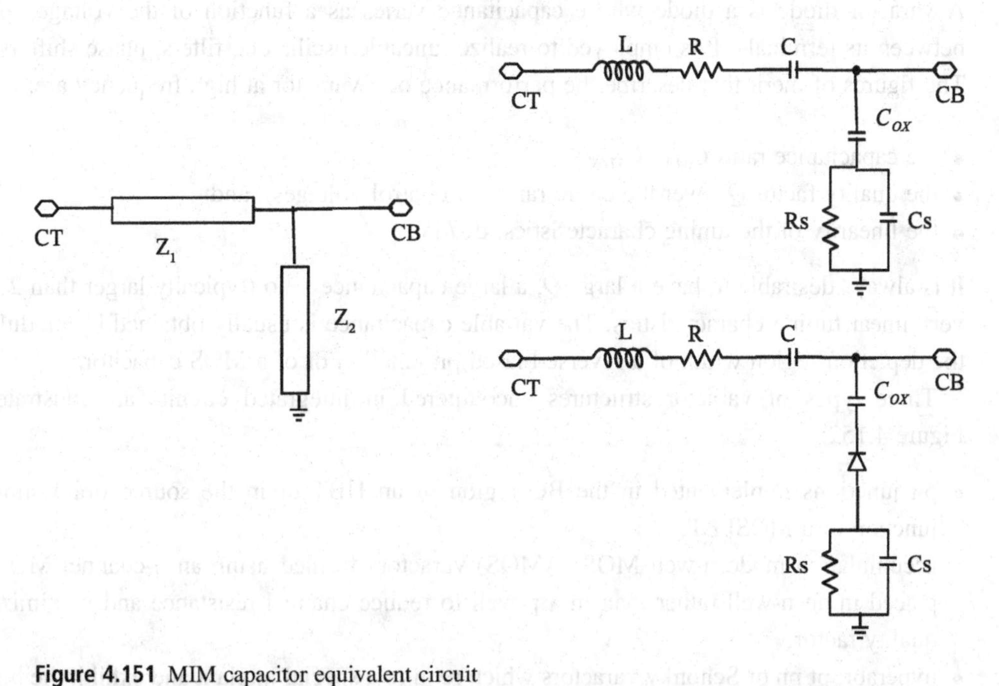

Figure 4.151 MIM capacitor equivalent circuit

which further reduces the capacitance to the substrate. This is captured by the equivalent circuit at the bottom right side of Figure 4.151.

The parameter of the HF equivalent circuit of a MiM or MoM capacitor are extracted directly, without the need for optimization, from the measured or simulated *S*-parameters, using a combination of *Y*- and *Z-parameters*.

From the low-frequency range (0.1GHz to 0.5GHz)

$$C_{mim} = \frac{[\Im[-Y_{12}{}^{-1}]]^{-1}}{\omega} \tag{4.292}$$

$$C_{OX} = \frac{[\Im(Z_{21})]^{-1}}{\omega} \quad (4.293)$$

$$R = R_{TOP} + R_{BO} = \Re(Z_{11} + Z_{22} - Z_{12} - Z_{21}) \quad (4.294)$$

$$R_{sub} = \Re(Z_{21}). \quad (4.295)$$

From the high-frequency range (>100GHz)

$$L = \frac{\Im(Z_{11} + Z_{22} - Z_{12} - Z_{21}) + \frac{1}{\omega C}}{\omega}. \quad (4.296)$$

4.5.5 Varactors

A varactor diode is a diode whose capacitance varies as a function of the voltage applied between its terminals. It is employed to realize tuneable oscillators, filters, phase shifters, etc. The figures of merit that describe the performance of a varactor at high frequency are:

- the capacitance ratio C_{MAX}/C_{MIN},
- the quality factor Q (over the entire range of control voltages), and
- the linearity of the tuning characteristics: dC/dV.

It is always desirable to have a large Q, a large capacitance ratio (typically larger than 2), and very linear tuning characteristics. The variable capacitance is usually obtained by modulating the depletion region width of a reverse-biased pn junction or of a MOS capacitor.

Three types of varactor structures encountered in integrated circuits are illustrated in Figure 4.152:

- pn junctions implemented in the BC region of an HBT or in the source–drain substrate junction of a MOSFET;
- accumulation-mode n-well MOS (AMOS) varactors formed using an n-channel MOSFET placed in an n-well rather than in a p-well to reduce channel resistance and maximize the quality factor;
- hyperabrupt pn or Schottky varactors which require a special implant and exhibit the highest capacitance ratio but are only available in specialized processes.

During the last decade, due to the proliferation of nanoscale CMOS technologies, the AMOS varactor has become the most popular variable capacitance device.

The high-frequency equivalent circuit of an AMOS varactor is illustrated in Figure 4.153 while the cross-section through the device is shown in Figure 4.154. The equivalent circuit is similar to that of the MiM/MoM capacitor with three notable differences. First, the fixed capacitance C is now replaced by the variable capacitor C_{var}. Second, the n-well-to-substrate diode is represented as a variable capacitance to illustrate its variation as a function of the applied voltage between the n-well and the substrate. Finally, since short finger widths are employed, the parasitic inductance can be ignored.

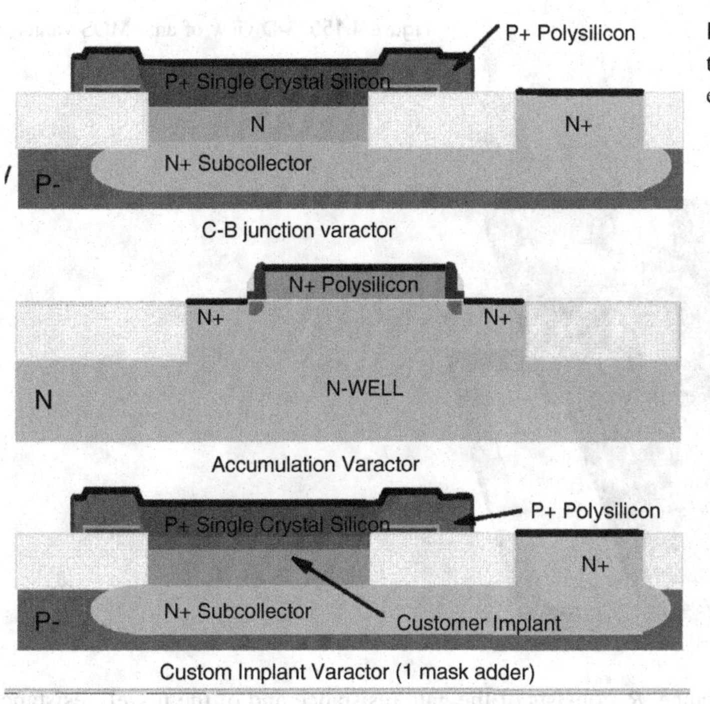

Figure 4.152 Cross-sections through typical varactor diodes encountered in silicon IC [61]

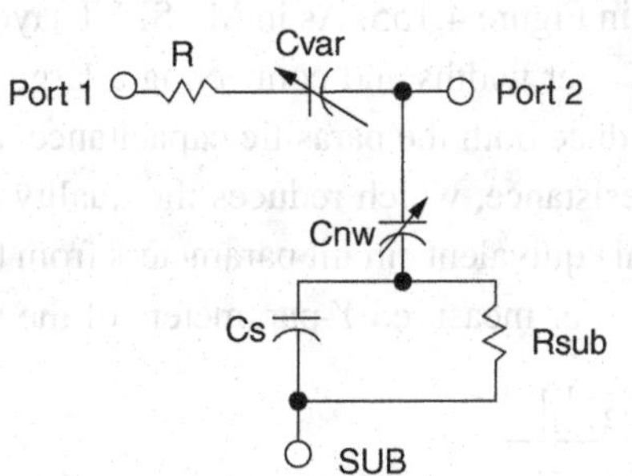

Figure 4.153 Simplified small-signal equivalent circuit adequate for parameter extraction

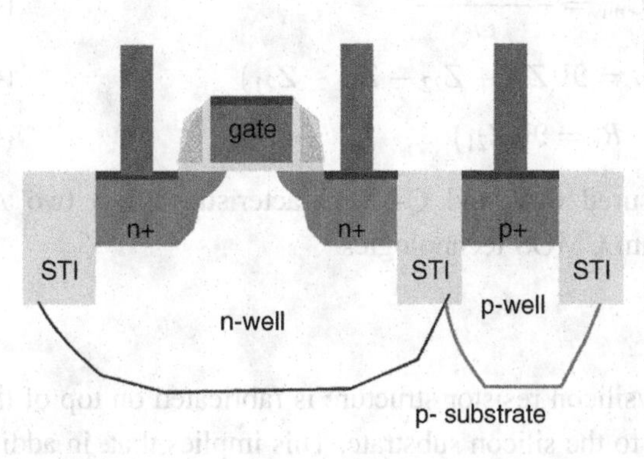

Figure 4.154 AMOS n-type varactor

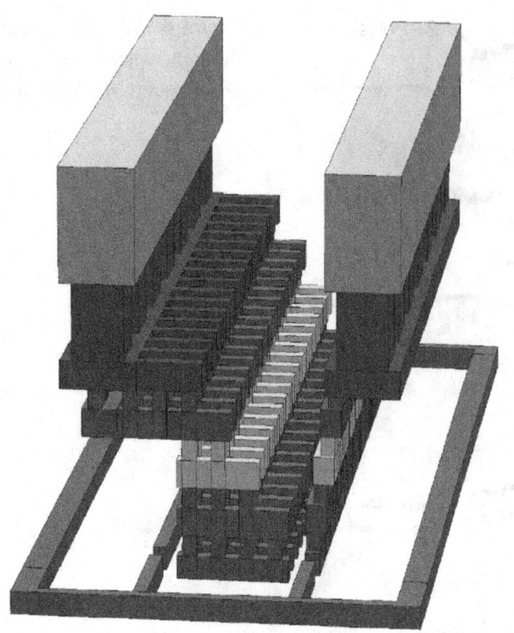

Figure 4.155 3-D view of an AMOS varactor layout [27]

The parasitic resistance, R, consists of the gate resistance and of the n-well resistance below the gate. Both must be minimized by reducing the gate length and the finger width of each varactor finger. A typical varactor layout is shown in Figure 4.155. As in MOSFET layouts, the gate resistance is minimized by employing small finger widths and connecting a large number of gate fingers in parallel. Care must be taken to reduce both the parasitic capacitance, to avoid the degradation of the capacitance ratio, and via resistance, which reduces the quality factor.

As with inductors, we can extract the small signal equivalent circuit parameters from the low-frequency range (0.1GHz to 1GHz) of the simulated or measured Y-parameters of the varactor

$$C_{var} = \frac{[\Im[-Y_{12}{}^{-1}]]^{-1}}{\omega} \tag{4.297}$$

$$C_{nw} = \frac{[\Im(Z_{21})]^{-1}}{\omega} \tag{4.298}$$

$$R = R_G + R_{CH} = \Re(Z_{11} + Z_{22} - Z_{12} - Z_{21}) \tag{4.299}$$

$$R_s = \Re(Z_{21}). \tag{4.300}$$

Figures 4.156 and 4.157 show measured C-V and Q-V characteristics from two AMOS varactors fabricated in 130nm and 45nm CMOS technologies.

4.5.6 Resistors

As illustrated in Figure 4.158, the polysilicon resistor structure is fabricated on top of the STI (shallow-trench-isolation) oxide, close to the silicon substrate. This implies that, in addition to its resistance, given by (4.301)

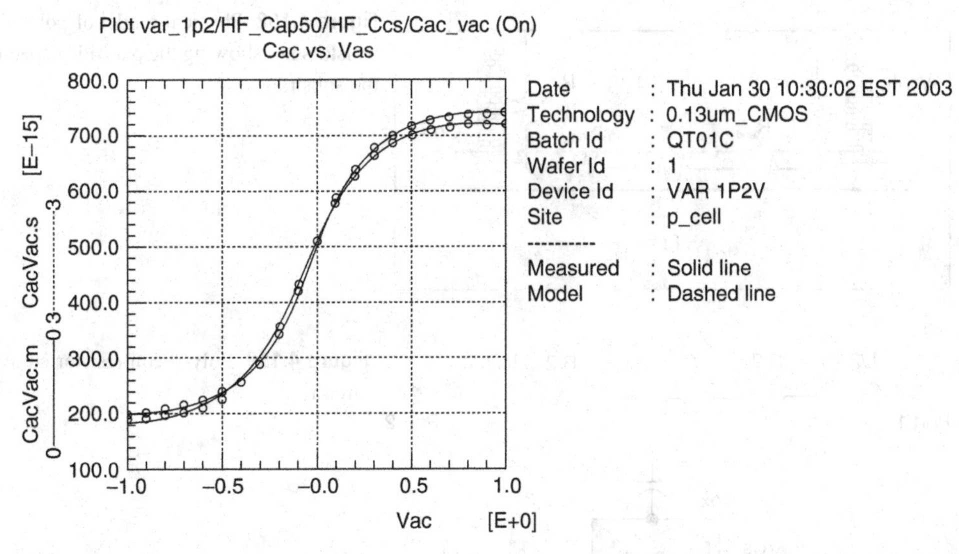

Figure 4.156 Measured and simulated C-V characteristics of an AMOS varactor realized with the thin gate oxide n-MOSFET in a 130nm RF CMOS process. The capacitance ratio exceeds 3.5

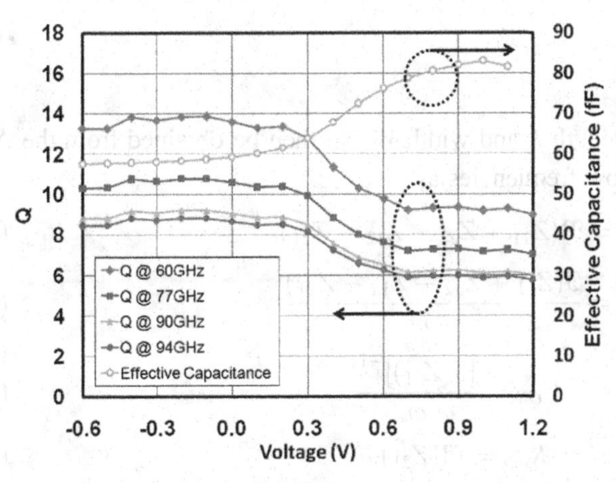

Figure 4.157 Measured C-V and Q-V characteristics of an AMOS varactor realized with the thin gate oxide n-MOSFET in a 45nm CMOS process. *Ldrawn* = 45nm, *Wf* = 0.7μm, Wtotal = 78.4μm, C_{VAR} = 1.07fF/μm, *C* variation: 58fF – 84fF, *Q* = 6–8.5 at 94GHz [62]

$$R = 2\frac{R_{CON}}{N_{CON}} + \frac{2R_{end}}{W} + R_{SH}\frac{L}{W} \tag{4.301}$$

the equivalent high-frequency model of the resistor must include the parasitic oxide capacitance to the silicon substrate as well as the parasitic RC network describing the silicon substrate losses. As in the case of the MiM capacitor, an inductance must be associated with the polysilicon stripe. The resultant symmetrical T-equivalent circuit is shown in Figure 4.159.

In case such a high-frequency resistor model is not available in the design kit, it can be easily obtained based on the BEOL information. For example, the polysilicon layer can be defined as the first metal layer in the ASITIC technology file. The equivalent circuit parameters *R*, *L*, C_{ox},

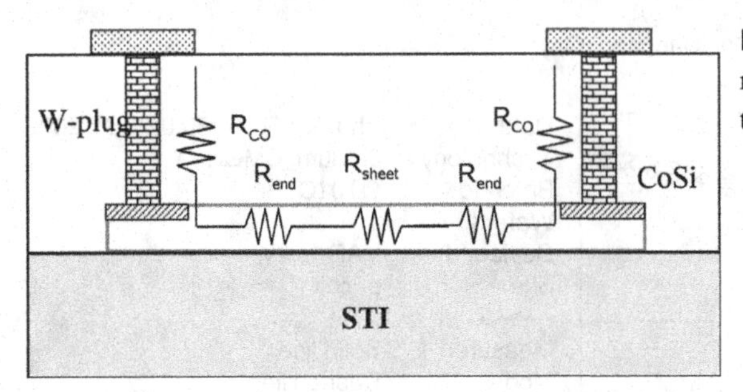

Figure 4.158 Physical model of polysilicon resistors not showing the parasitic capacitance to the substrate

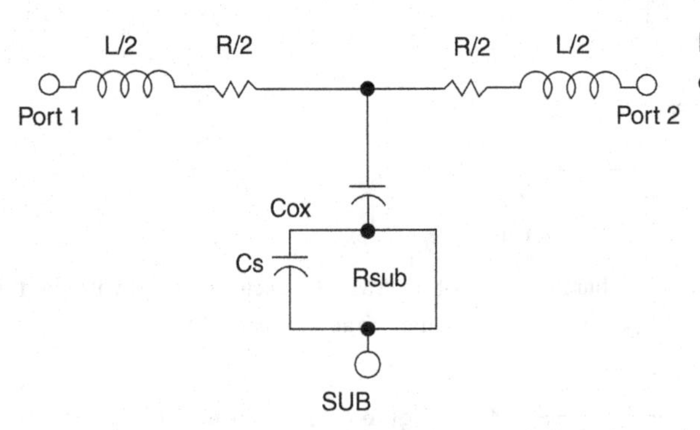

Figure 4.159 Polysilicon resistor equivalent circuit

and R_{sub} of a polysilicon line of length L and width W can then be obtained from the ASITIC simulated Y- or Z-parameters at low frequencies as

$$R = \Re(Z_{11} + Z_{22} - Z_{12} - Z_{21}) \tag{4.302}$$

$$L = \frac{\Im(Z_{11} + Z_{22} - Z_{12} - Z_{21})}{\omega} \tag{4.303}$$

$$C_{OX} = \frac{[\Im(Z_{21})]^{-1}}{\omega} \tag{4.304}$$

$$R_{sub} = \Re(Z_{21}). \tag{4.305}$$

C_{sub} is obtained from the high-frequency range, as in the case of inductors.

4.5.7 Modeling interconnect parasitics

If the design kit does not have RF models, we can use ASITIC to calculate the HF parasitics of all components as follows:

- MIM caps, poly resistors, varactors have identical substrate network (R_{sub}, C_{sub}).
- Use ASITIC to model the substrate network for a metal line of similar width and length.
- MIM caps and poly resistors are realized in the oxide, above the silicon substrate.
- Use ASITIC to model C_{OX}, L, R for a metal line of similar width and length and located at the same distance from the substrate as the MIM cap or polysilicon resistor.

Summary

- Today's silicon MOSFET is significantly different from the classical planar MOSFET. It features strain engineering in the channel introduced at the 90nm node, high-k dielectric gate stack and metal gates introduced at the 45nm node, selectively grown source and drain contact regions for reduced resistance, and 3-D fin channels, surrounded by the gate on three sides, at the 22nm node.
- The real MOSFET has parasitic source, drain, and substrate resistances which are important in high-frequency and high-speed logic circuits.
- The parasitic resistances can be calculated as a function of layout geometry and of the sheet resistance in the source–drain and well regions.
- The parasitic capacitances of the MOSFET limit its switching speed. They, too, can be calculated from the MOSFET layout geometry and doping concentration in the well, source–drain contact regions, and halo implants.
- In a given technology node, nanoscale MOSFETs can be fabricated in many flavors determined either by:
  - the effective gate oxide thickness EOT and physical gate length L: GP, HP, LP;
  - V_T: low-V_T (LVT), standard-V_T (SVT), and high-V_T (HVT), native V_T;
  - oxide thickness, channel doping profile, and gate length for high voltage I/Os: 1.8V, 2.5 V or 3.3V);
  - nanoscale CMOS technologies come in both bulk and SOI flavors. Starting at the 22nm node 3-D multi-gate FinFETs have been introduced, likely to be followed later by nanowire (surround gate) FETs with III-V and Ge channel materials.
- Due to the large vertical electric field between the gate and the channel, the I-V characteristics and high-frequency FoMs of nanoscale MOSFETs are determined by a combination of mobility degradation and, to a smaller extent, by velocity saturation.
- The $I_D - V_{GS}$, $g_m - V_{GS}$, $f_T - V_{GS}$, and $f_{MAX} - V_{GS}$ characteristics in saturation can be described by three regions:
  - subthreshold (exponential dependence on V_{GS}, much as in a bipolar transistor);
  - square (classical) I_D - V_{GS} law, which corresponds to a linear increase of g_m, f_T, and f_{MAX} with V_{GS} beyond the threshold;
  - linear $I_D - V_{GS}$ law which, to first order, corresponds to g_m, f_T, and f_{MAX} being approximately constant with V_{GS} beyond the threshold. This region has been rarely exploited in analog and HF circuits before the 180nm node because of the corresponding large effective gate voltage V_{eff} required.
  - Due to the constant field scaling rules being applied rather rigorously by all foundries between the 0.5μm and 65nm nodes, and which more-or-less preserve the 2-D electric field profile in the device, the transistion between the square and linear I-V regions occurs at approximately 0.15mA/μm in n-MOSFETs, largely independent of technology node and foundry. The corresponding value for p-MOSFETs is about 0.08mA/μm.

- The effective voltage corresponding to the boundary between the square and linear law regions decreases with every new technology generation, making it increasingly challenging to design analog and HF circuits using the square law and the traditional constant V_{eff} biasing rules in nanoscale MOSFETs.
- For current densities beyond 0.2mA/μm, g_m, f_T, and f_{MAX} remain practically constant with V_{GS}.
- The V_{DS} dependence of g_m, f_T, and f_{MAX} follows that of I_{DS} through the $(1 + \lambda V_{DS})$ channel length modulation term.
- The characteristic current densities of MOSFETs remain invariant across technology nodes at least down to the 65nm node. However, the characteristic current densities have started to increase in newer nodes, 45nm and 32nm, because of significant strain engineering, reduced effective oxide thickness scaling, and because the supply voltage has not been scaled.
- Biasing MOSFETs in HF circuits at constant current density, as in bipolar circuits, solves the dwindling V_{eff} problem and improves circuit performance robustness to process, bias current, and temperature variation.
- The "RF" MOSFET is a "digital" MOSFET to which a subcircuit has been added to include gate, source and substrate resistance, and non-quasi-static effects (NQS) effects through R_i, τ, and R_{gd}.

- HBTs can be analyzed using the semiconductor equations with position-dependent n_{ie}. Their I-V charactersitcs are described by the same equations as those employed for Si BJTs.
  - The double-poly, self-aligned SiGe HBT with shallow trench and deep trench isolation and selectively implanted collector is the most common HF silicon bipolar transistor today.
  - The GaAs HBT is the dominant device used in cell-phone power amplifiers today and InP HBTs hold the record for the highest frequency bipolar transistors.
  - Scaling of HBTs is based on shrinking lateral and vertical dimensions, increasing doping, bandgap offsets, and bandgap grading.
- HEMTs were the first commercial heterojunction transistors.
  - Charge is confined to a 2-D electron gas.
  - $n_s \times \mu_n$ product dictates performance and material choice.
  - HEMTs are used in most critical antenna interface circuits: LNAs, PAS, and antenna switches.
  - InP HEMTs and metamorphic GaAs p-HEMTs are the most common devices today for very mm-wave low-noise applications.
  - GaN HEMTs are most actively pursued for microwave and mm-wave power applications.

Problems

(1) Based on the measured data in Figure 4.12 and the size of the transistor and bias current, find out the gate width and bias current of the n-MOSFET whose real part of the input impedance is 50Ω? Knowing that the impedance of the HBT scales with the emitter

length, calculate the emitter length and bias current required to achieve an input resistance equal to 50Ω.

(2) A 32nm CMOS technology has the following transistor data: $L = 22$nm, $J_{OPT} = 0.3$mA/μm; $J_{pfT} = 0.5$mA/μm for $0.3\text{V} < V_{DS} < 0.9\text{V}$. $g'_{mn} = g'_{mp} = 2.6$mS/μm; $g'_{dsn} = g'_{dsp} = 0.4$mS/μm.

For both n-channel and p-channel devices: $C'_{gs} = 0.55$ fF/μm; $C'_{gd} = 0.3$ fF/μm; $C'_{sb} = C'_{db} = 0.55$ fF/μm; R'_s (n-MOS) $= R'_d = 200\Omega \times$ μm; R'_s (p-MOS) $= R'_d = 200\Omega \times$ μm; Total R_g/1μm finger contacted on one side is 96Ω. Noise parameter: $k_1 = 0.5$.

(a) Calculate the f_T of the n-MOSFET and p-MOSFET using the small signal equivalent circuit parameters.

(b) Calculate the corresponding f_T of the n-MOSFET cascode.

(c) Assuming the transistor is biased at J_{OPT}, what is the total gate width and bias current of a n-MOSFET whose R_{SOPT} is 50Ω? What is the corresponding R_n?

(d) Calculate the intrinsic slew-rate of the n-MOS and p-MOS transistor.

(3) Repeat problem 2 for a 65nm CMOS process with $L = 45$nm, J_{OPT} (n-MOSFET) $= 0.2$mA/μm; $g'_{mn} = 2g'_{mp} = 1.3$mS/μm; $g'_{dsn} = 2g'_{dsp} = 0.18$mS/μm; J_{pfT} (n-MOSFET) $= 0.4$mA/μm for $0.3\text{V} < V_{DS} < 1\text{V}$; J_{pfT} (p-MOSFET) $= 0.2$mA/μm for $0.3\text{V} < V_{SD} < 1\text{V}$.

For both n-channel and p-channel devices: $C'_{gs} = 0.7$ fF/μm; $C'_{gd} = 0.4$ fF/μm; $C'_{sb} = C'_{db} = 0.7$ fF/μm;

R'_s (n-MOS) $= R'_d = 160\Omega \times$ μm; R'_s (p-MOS) $= R'_d = 320\Omega \times$ μm; total R_g/0.7μm finger contacted on one side is 158Ω. Noise parameter: $k_1 = 0.5$.

(4) (a) Using the model parameters given for 0.13μm n-MOSFETs and assuming a fringing gate capacitance of 0.3 fF at each side of a gate finger, irrespective of its width, calculate:

the peak f_T, f_{MAX} for a device with 20 fingers each 1μm wide,

the peak f_T, f_{MAX} for a device with five fingers each 4μm wide.

(b) Which of the two devices has the better minimum noise figure?

(c) Between a device with ten fingers, each 2μm wide, and one with four fingers, each 2μm wide, which one has the better f_{MAX}? What about NF_{MIN}?

(d) Calculate the minimum noise figure and optimum noise impedance at 5GHz for a 0.13μm x 2μm MOSFET with 50 gate fingers biased at peak f_T current density and knowing that the gate is contacted on one side only.

(e) Calculate the minimum noise figure if a 1.5 nH inductor, having a series resistance of 2Ω, is added in series with the gate of the MOSFET.

(5) (a) Using the model parameters given for the 230GHz, 130nm SiGe HBT and $\beta = 200$, calculate:

the peak f_T, f_{MAX} for a device with a 5μm long emitter,

the minimum noise figure and optimum noise impedance at 6GHz for four transistors connected in parallel if they are biased at one tenth of the peak f_T current density.

(b) Calculate the minimum noise figure and optimum noise impedance at 6GHz.

(6) A 65nm CMOS process allows you to create an accumulation-mode MOS varactor using the same gate oxide as that of the n-channel MOSFET.

(a) Write an expression for extracting from measured *Y*/*Z* data the variable gate capacitance as well as the parasitic $n+n$ junction capacitance of the MOS accumulation-mode varactor whose equivalent circuit is shown in Figure 4.153.

(b) If the series resistance R is 5Ω and the variable capacitance C_{var} varies between 200 fF and 600 fF, calculate the quality factor Q at 5GHz of the intrinsic varactor (consisting of R and C_{var}) for the maximum and minimum capacitance value. Ignore substrate effects.

REFERENCES

[1] Donald Estreich, Device Chapter, *IEEE CSICS-2009 Primer Course*, October 2009.

[2] C.-H. Jan, M. Agostinelli, H. Deshpande, M. A. El-Tanani, W. Hafez, U. Jalan, L. Janbay, M. Kang, H. Lakdawala, J. Lin, Y.-L. Lu, S. Mudanai, J. Park, A. Rahman, J. Rizk, W.-K. Shin, K. Soumyanath, H. Tashiro, C. Tsai, P. VanDerVoorn, J.-Y. Yeh, and P. Bai, "RF CMOS technology scaling in high-k/metal gate era for RF SoC (system-on-chip) applications," *IEEE IEDM Digest*, pp. 604–607, December 2010, and M. Bohr, "Evolution of scaling from the homogeneous era to the heterogeneous era," *IEEE IEDM Digest*, pp.1.1.4, December 2011.

[3] Stephen Long, *IEEE CSICS-2009 Primer Course*, October 2009, Materials chapter.

[4] Semiconductor materials data base Ioffe Institute, http://www.ioffe.ru/SVA/NSM/

[5] S.P. Voinigescu, E. Laskin, I. Sarkas, K.H.K. Yau, S. Shahramian, A. Hart, A. Tomkins, P. Chevalier, J. Hasch, P. Garcia, A. Chantre, and B. Sautreuil, "Silicon D-band wireless transceivers and applications," *IEEE APMC*, pp. 1857–1864, December 2010.

[6] G. Dambrine, H. Happy, F. Danneville, and A. Cappy, "A new method for on-wafer noise measurement," *IEEE MTT*, **41**(3): 375–381, March 1993.

[7] T.O. Dickson, K.H.K. Yau, T. Chalvatzis, A. Mangan, R. Beerkens, P. Westergaard, M. Tazlauanu, M.-T. Yang, and S.P. Voinigescu, "The invariance of the characteristic current densities in nanoscale MOSFETs and its impact on algorithmic design methodologies and design porting of Si(Ge) (Bi)CMOS high-speed building blocks," *IEEE JSSC*, **41**(8): 1830–1845, August 2006.

[8] S.P. Voinigescu, T.O. Dickson, M. Gordon, C. Lee, T. Yao, A. Mangan, K. Tang, and K. Yau, "RF and millimeter-wave IC design in the nano-(Bi)CMOS Era," in Will Z. Cai (ed.), *Si-based Semiconductor Components for Radio-Frequency Integrated Circuits (RF IC)*, Transworld Research Network, 2006, pp. 33–62.

[9] T. Yao, M.Q. Gordon, K.K.W. Tang, K.H.K. Yau, M.-T. Yang, P. Schvan, and S.P. Voinigescu, "Algorithmic design of CMOS LNAs and PAs for 60GHz Radio," *IEEE JSSC*, **42**(5): 1044–1057, May 2007.

[10] John Atalla and Dawon Kahng, US Patent no. 3102230, filed in 1960, issued in 1963.

[11] Y. Taur and T.H. *Ning in Fundamentals of Modern VLSI Devices*, Cambridge University Press, 2nd Edition, 2009.

[12] International Technology Roadmap for Semiconductors (ITRS) 2003.

[13] F. Stern and W.E. Howard, "Properties of semiconductor surface inversion layers in the electric quantum limit," *Phys. Rev.* **163**: 816–835, 1967.

[14] S.E. Thompson, M. Armstrong, C. Auth, M. Alavi, M. Buehler, R. Chau, S. Cea, T. Ghani, G. Glass, T. Hoffman, C.-H. Jan, C. Kenyon, J. Klaus, K. Kuhn, Z. Ma, B. Mcintyre,

K. Mistry, A. Murthy, B. Obradovic, R. Nagisetti, P. Nguyen, S. Sivakumar, R. Shaheed, L. Shifren, B. Tufts, S. Tyagi, M. Bohr, and Y. El-Mansy, "A 90nm logic technology featuring strained-silicon," *IEEE ED*, **51**: 1790–1797, 2004.

[15] W. Liu, *MOSFET Models for Spice Simulation*, John Wiley, 2001

[16] P. J. Tasker and B. Hughes, "Importance of source and drain resistance to the maximum f_T of millimeter-wave MODFETs," *IEEE EDL*, **10**(7): 291–293, July 1989.

[17] P. H. Ladbrooke, *MMIC Design: GaAs FETs and HEMTs*, Boston: Artech House, 1989, Chapter 6.

[18] S. P. Voinigescu, S. T. Nicolson, M. Khanpour, K. K. W. Tang, K. H. K. Yau, N. Seyedfathi, A. Timonov, A. Nachman, G. Eleftheriades, P. Schvan, and M. T. Yang, "CMOS SOCs at 100GHz: system architectures, device characterization, and IC design examples," *IEEE ISCAS*, pp. 1971–1974, May 2007.

[19] A. Tomkins, R. A. Aroca, T. Yamamoto, S. T. Nicolson, Y. Doi, and S. P. Voinigescu, "A Zero-IF 60GHz 65nm CMOS transceiver with direct BPSK modulation demonstrating up to 6Gb/s data rates over a 2m wireless link," *IEEE JSSC*, **44**(8): 2085–2099, August 2009.

[20] A. Van Der Ziel, *Noise: Sources Characterization, Measurement*, Englewood Cliffs, NJ: Prentice-Hall, 1970.

[21] R. A. Pucel, H. A. Haus, and H. Statz, "Signal and noise properties of gallium arsenide microwave field effect transistors," *Advances in Electronics and Electron Physics (AEEPS)*, 38, New York: Academic Press, 1975.

[22] M. W. Pospieszalski, "Interpreting Transistor Noise," *IEEE Microwave Mag.*, **11**(6): 61–69, 2010.

[23] K. H. K. Yau, M. Khanpour, M.-T. Yang, P. Schvan, and S. P. Voinigescu, "On-die source-pull for the characterization of the W-band noise performance of 65nm general purpose (GP) and low power (LP) n-MOSFETs," *IEEE IMS, Digest*, pp. 773–776, June 2009.

[24] F. Danneville, H. Happy, G. Dambrine, J.-M. Belquin, and A. Cappy, "Microscopic noise modeling and macroscopic noise models: how good a connection?" *IEEE ED*, **41**(5): 779–786, May 1994.

[25] S. A. Maas, *Noise*, Boston: Artech House, Boston, 2005, Chapter 4.

[26] S. Nicolson, Ph.D. Thesis, *ECE Department*, University of Toronto, September 2008.

[27] M. Kraemer, "Design of a low-power 60GHz transceiver front-end and behavioral modeling and implementation of its key building blocks in 65nm CMOS," Ph.D. thesis, LAAS, Toulouse, France, December 2010.

[28] E. Laskin, M. Khanpour, S. T. Nicolson, A. Tomkins, P. Garcia, A. Cathelin, D. Belot, and S. P. Voinigescu, "Nanoscale CMOS transceiver design in the 90–170 GHz Range," *IEEE MTT*, **57**: 3477–3490, December 2009.

[29] R. H. Dennard, F. H. Gaensslen, H. N. Yu, V. L. Rideout, E. Bassous, and A. R. LeBlanc, "Design of ion-implanted MOSFETs with very small physical dimensions," *IEEE JSSC*, **9**: 256–258, 1974.

[30] M. Mayberry, "Intel's tri-gate transistor to enable next era in energy efficient performance," Intel Press Release Presentation Slides, *June 6*, 2006.

[31] I. Young, *IEEE IEDM Short Course*, December 2011.

[32] M. Bohr, "Evolution of scaling from the homogeneous era to the heterogeneous era," *IEEE IEDM Digest*, pp. 1.1.4, December 2011.

[33] P.M. Asbeck *et al.*, "Heterojunction bipolar transistor technology," in C.T. Wang (ed.), *Introduction to Semiconductor Technology: GaAs and Related Compounds*, New York: Wiley-Interscience, 1990, pp. 170–230.

[34] C.R. Bolognesi, N. Matine, M.W. Dvorak, X.G. Xu, and S.P. Watkins, "InP/GaAsSb/InP DHBT's with high f_T and f_{MAX} for wireless communication applications," *IEEE GaAs IC Symposium*, pp. 63–66. October 1999.

[35] G. Avenier, P. Chevalier, S. Pruvost, J. Bouvier, G. Troillard, L. Depoyan, M. Buczko, S. Montusclat, A. Margain, S. T Nicolson, K.H.K. Yau, D. Gloria, M. Diop, N. Loubet, N. Derrier, C. Leyris, S. Boret, D. Dutartre, S.P. Voinigescu, and A. Chantre, "0.13μm SiGe BiCMOS technology fully dedicated to mm-wave applications," *IEEE JSSC*, **44**(8): 2312–2321, September 2009.

[36] S.P. Voinigscu, E. Laskin, S.T. Nicolson, M. Khanpour, K. Tang, K. Yau, A. Tomkins, P. Chevalier, and P. Garcia, "80–160 GHz CMOS and SiGe BiCMOS building block design," IEEE IMS/RFIC 2008 Workshop WSB, "Advances in Circuit Design for Wideband Millimeter Wave Applications," Atlanta, June 15, 2008.

[37] H. Kroemer, "Two integral relations pertaining to the electron transport through a bipolar transistor with non-uniform energy bandgap in the base region," *Solid State Electron*, **28**: 1101–1103, 1985.

[38] K.H. K Yau, P. Chevalier, A. Chantre, and S.P. Voinigescu, "Characterization of the noise parameters of SiGe HBTs in the 70–170 GHz range," *IEEE MTT*, **59**: 1983–2001, August 2011.

[39] S.P. Voinigescu, M.C. Maliepaard, J.L. Showell, G. Babcock, D. Marchesan, M. Schroter, P. Schvan, and D.L. Harame, "A scalable high-frequency noise model for bipolar transistors with application to optimal transistor sizing for low-noise amplifier design," *IEEE JSSC*, **32**(9): 1430–1438, 1997.

[40] S.P. Voinigescu, T. Chalvatzis, K.H.K. Yau, A. Hazneci, A. Garg, S. Shahramian, T. Yao, M. Gordon, T.O. Dickson, E. Laskin, S.T. Nicolson, A.C. Carusone, L. Tchoketch-Kebir, O. Yuryevich, G. Ng, B. Lai, and P. Liu, "SiGe BiCMOS for analog, high-speed digital and millimetre-wave applications beyond 50GHz," *IEEE BCTM Digest*, pp. 223–230, October 2006.

[41] P. Chevalier, B. Barbalat, M. Laurens, B. Vandelle, L. Rubaldo, B. Geynet, S.P. Voinigescu, T.O. Dickson, N. Zerounian, S. Chouteau, D. Dutartre, A. Monroy, B. Sautreuil, F. Aniel, G. Dambrine, and A. Chantre, "High-speed SiGe BiCMOS technologies: 120nm status and end-of-roadmap challenges," *Si Monolithic Integrated Circuits in RF Systems*, pp. 18–23, January 2007.

[42] M. Urteaga, R. Pierson, J. Hacker, M. Seo, Z. Griffith, A. Young, P. Rowell, and M.J.W. Rodwell, "InP HBT integrated circuit technology for terahertz frequencies," *IEEE CSICS Digest*, pp. 1–4, October 2010.

[43] T. Mimura et al., "A new field-effect transistor with selectively doped GaAs/n-$Al_xGa_{1-x}As$ heterojunctions," *Jpn J. Appl. Phys.*, **19**: L225–L227, 1980.

[44] C.A. Mead, "Schottky barrier field-effect transistor," *Proc. IEEE*, **54**: 307–308, February 1966.

[45] F. Danneville, "Microwave noise and FET devices," *IEEE Microwave Mag.*, **11**(6): 53–60, October 2010.

[46] R. Dingle, H.L. Stormer, A.C. Gossard, and W. Wiegmann, "Electron mobilities in modulation doped GaAs-AlGaAs heterojunction superlattices," *Appl. Phys. Lett.*, **33**: 665, 1978.

[47] D. B. Estreich, "Basics of compound semiconductor ICs," *IEEE CSICS, Primer Course*, October 2010.

[48] D. S. McPherson, F. Pera, M. Tazlauanu, and S. P. Voinigescu, "A 3V fully differential distributed limiting driver for 40Gb/s optical transmission systems," *IEEE JSSC*, **38**(9): 1485–1496, 2003.

[49] S. Tabatabaei, *IEEE CSICS Short Course*, October 2009.

[50] W. Deal, K. Leong, X. B. Mei, S. Sarkozy, V. Radisic, J. Lee, P. H. Liu, W. Yoshida, J. Zhou, and M. Lange, "Scaling of InP HEMT cascode integrated circuits to THz frequencies," *IEEE CSICS Digest*, pp. 195–198, October 2010.

[51] S. P. Voinigescu, D. S. McPherson, F. Pera, S. Szilagyi, M. Tazlauanu, and H. Tran, "A comparison of silicon and III-V technology performance and building block implementations for 10 and 40Gb/s optical networking ICs," *IJHSES*, **13**(1), and book chapter in *Compound Semiconductor Integrated Circuits*, pp. 27–58, 2003.

[52] M. W. Pospieszalski, "Modeling of noise parameters of MESFETs and MODFETs and their frequency and temperature dependence," *IEEE MTT*, **37**: 1340–1350, 1989.

[53] J. D. Albrecht and T. Chang, "DARPA's nitride electronic next generation technology program," *IEEE CSICS Digest*, pp. 7–10, October 2010.

[54] D. Dubuc, T. Tournier, I., Telliez, T. Parra, C. Boulanger, and J. Graffeuil, "High quality factor and high self-resonant frequency monolithic inductor for millimeter-wave Si-based IC's," *IEEE MTT-S Digest*, pp.193–196, June 2002.

[55] Y. Cao, R. A. Groves, X. Huang, N. D. Zamdmer, J.-O. Plouchart, R. A. Wachnik, Tsu-Jae King, and C. Hu, "Frequency-independent equivalent-circuit model for on-chip spiral inductors," *IEEE JSSC*, **38**: 419–426, March 2003.

[56] http://rfic.eecs.berkeley.edu/~niknejad/doc-05-28-01/faq.html#expo–t

[57] T. O. Dickson, M.-A. LaCroix, S. Boret, D. Gloria, R. Beerkens, and S. P. Voinigescu, "30–100GHz inductors and transformers for millimeter-wave (Bi)CMOS integrated circuits," *IEEE MTT*, **53**(1): 123–133, 2005.

[58] T. O. Dickson and S. P. Voinigescu, "Low-power circuits for a 10.7-to-86 Gb/s serial transmitter in 130nm SiGe BiCMOS," *IEEE JSSC*, **42**(10): 2077–2085, October 2007.

[59] S. P. Voinigescu, E. Laskin, I. Sarkas, K. H. K. Yau, S. Shahramian, A. Hart, A. Tomkins, P. Chevalier, J. Hasch, P. Garcia, A. Chantre, and B. Sautreuil, "Silicon D-band wireless transceivers and applications," *IEEE, APMC*, pp.1857–1864, December 2010.

[60] Kenneth H. K. Yau, "On the metrology of nanoscale silicon transistors above 100GHz," Ph.D. Thesis, ECE Department, University of Toronto 2011.

[61] D. Harame, *IEEE IEDM Short Course*, December 2002.

[62] R. A. Aroca, A. Tomkins, Y. Doi, T. Yamamoto, and S. P. Voinigescu, "Circuit performance characterization of digital 45nm CMOS technology for applications around 110GHz," *IEEE VLSI Circuits Symposium Digest*, pp.162–163, Honolulu, June 2008.

ENDNOTES

1 A lower limit on the trace width W is usually determined by the current through the inductor and electromigration reliability rules for the technology.

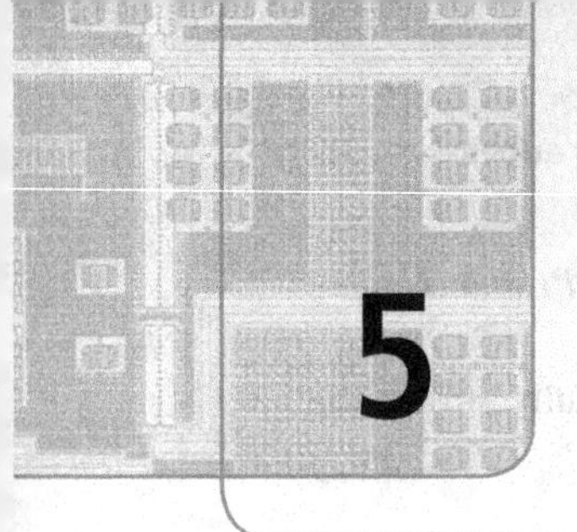

5 Circuit analysis techniques for high-frequency integrated circuits

5.1 ANALOG VERSUS HIGH-FREQUENCY CIRCUIT DESIGN

Analog ICs are characterized by the common use of:

- differential stages with active loads,
- DC-coupled broadband amplifiers (i.e. no DC-blocking capacitors are present between stages),
- strong impedance mismatch between stages,
- input and output impedance matching based on negative feedback or resistors.

At the same time, analog designers work with:

- currents,
- voltages,
- transistors,
- capacitors, and
- resistors,

while employing small signal AC, noise, and transient large signal simulations to analyze the performance of their circuits. In contrast, traditional microwave circuit design deals with:

- single-ended stages with a small number of transistors,
- AC-coupled, tuned narrowband or broadband gain stages,
- reactive components such as inductors, transformers, capacitors, and transmission lines, and
- lossless impedance matching to maximize power gain and minimize noise figure.

Microwave circuit designers frequently rely on:

- *S*-parameters and the Smith Chart,
- noise figure and noise parameters,
- *P*1dB and IIP3,
- harmonic Fourier series analysis,

and conduct:

- small signal *S*-parameter and noise figure simulations,
- periodic steady-state (*PSS*) or harmonic balance (*HB*) large signal simulations

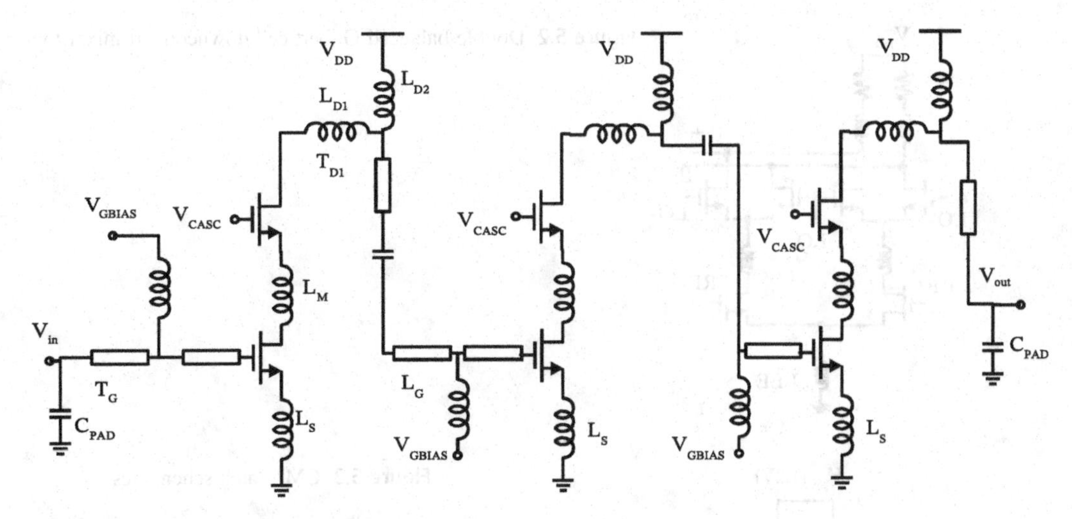

Figure 5.1 Three-stage cascode LNA schematic

to analyze their circuits. Furthermore, especially at mm-wave frequencies, an electromagnetic-field simulator is employed to analyze, model, and design inductors, transformers, and interconnect and parasitic coupling in circuit layout.

High-frequency and high-speed IC design borrows from both analog and microwave circuit topologies and analysis techniques. It deals with:

- differential tuned narrowband and/or broadband circuits;
- resistive, active, and reactive input, output, and interstage matching;
- differential-mode, common-mode, and single-ended analysis for gain, impedance, and noise matching and stability;
- small signal *S*-parameter, noise, transient, *PSS* or *HB* large signal simulation techniques; and
- electromagnetic field simulators for inductor, transformer, and interconnect design and modeling.

Figures 5.1 through 5.4 show typical examples of topologies used in HF and high-speed ICs. They are all characterized by the presence of inductors, transformers, and/or transmission lines.

The first example, shown in Figure 5.1, is a low-noise-amplifier which consists of three single-ended cascode stages, each with inductive series–series feedback (source degeneration inductors L_S), broadbanding inductors, L_M, and tapped inductor loads, L_{D1}/L_{D2}. The metal interconnect between stages is realized and modeled as transmission lines, which in Figure 5.1 are represented as open rectangles. Inductors are also employed to provide the bias voltage to the gate of the common-source transistors, without contributing noise. The DC gain of the LNA is zero since inductors and transmission lines act as short circuits at DC. Furthermore, the entire supply voltage is distributed across the two vertically stacked transistors in the cascode, maximizing their drain–source voltage, and therefore the voltage swing, for a given supply.

The second example is that of a double-balanced Gilbert cell downconversion mixer, shown in Figure 5.2. It operates as a four quadrant multiplier at high frequency, with the

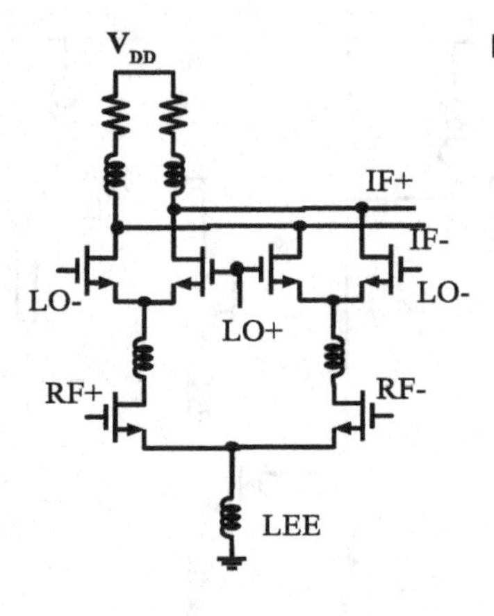

Figure 5.2 Double balanced Gilbert cell downconvert mixer topology

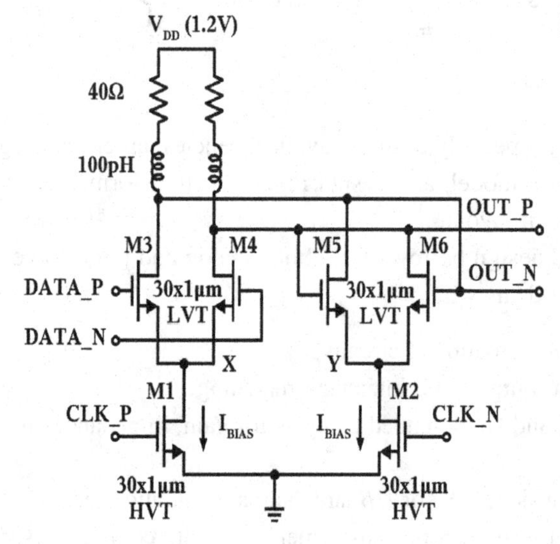

Figure 5.3 CML latch schematics

(low-frequency) differential output voltage being a product of the differential signals applied at the RF and LO ports. It is an essential block in most radio transceivers. Like the LNA, it features inductive broadbanding. The complex load impedance consists of a resistor in series with an inductor, providing gain from DC up to moderately high frequencies. Since, at high frequencies, the output impedance of current sources is capacitive, and since a current source contributes noise, the preferred method to achieve common-mode and even-mode harmonic rejection at high frequency is through the use of a common-mode inductor, L_{EE}. We will cover more of the latter topic towards the end of this chapter.

The third example, a high-speed current-mode-logic (CML) latch with a similar topology as the Gilbert cell downconverter, is described in Figure 5.3. This circuit is meant to operate at low, as well as at high data rates. For this reason, resistive loads in series with inductors are employed to compensate for the transistor capacitance at the output nodes.

The final example of a HF topology is that of a differential, distributed optical-modulator driver, shown in Figure 5.4, where inductors or a relatively high-impedance transmission lines

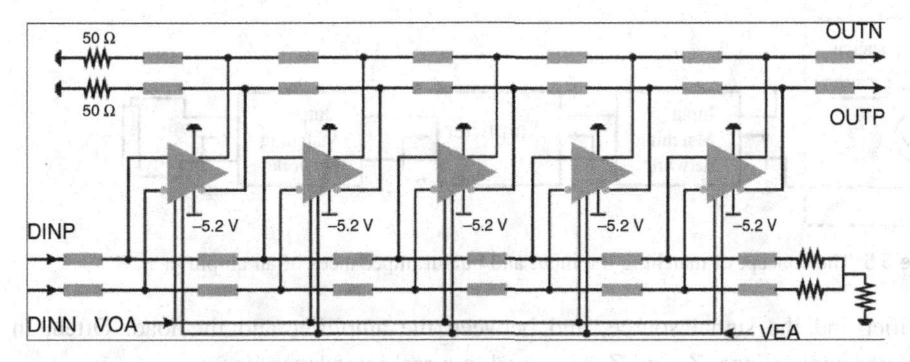

Figure 5.4 Block diagram of a distributed amplifier (DA)

form artificial transmission lines with the input and output capacitances of several, typically 3 to 7, gain cells, thus maximizing the bandwidth.

Techniques to analyze and design these circuits will be presented in the following chapters.

5.2 IMPEDANCE MATCHING

Traditional analog circuit design relies on strongly impedance-mismatched circuit stages to maximize voltage or current gain. As an example, for maximum voltage transfer, the impedance of the load must be significantly larger than that of the driving stage. In contrast, for maximum current transfer, the load impedance must be much smaller than the impedance of the driving stage. The question then arises: **why do we need matched inputs and outputs at high frequency?**

First, unlike at low frequencies where the signal wavelength exceeds one meter, at high frequency the electrical distance between circuit pads and external terminations becomes comparable to the signal wavelength. Impedance mismatches at either the near or the far end cause reflections that can significantly reduce the signal power arriving at the destination circuit.

The typical matching impedance value imposed by standard test equipment and standard bodies is 50Ω, and in some (mostly older) applications, 75Ω. In differential high-frequency circuits, three matching conditions must be met simultaneously:

- 100Ω in differential mode,
- 50Ω in single-ended mode, and
- 25Ω in common-mode.

Secondly, to ensure maximum power transfer between the signal source and an amplifier, the input impedance of the amplifier, Z_{IN}, must be conjugately matched to that of the signal source, Z_S. The same applies for the maximum power transfer from the amplifier to its load. The output impedance of the amplifier, Z_{OUT}, must be conjugately matched to the load impedance, Z_L. Therefore, as illustrated in Figure 5.5, matching networks must be inserted between the

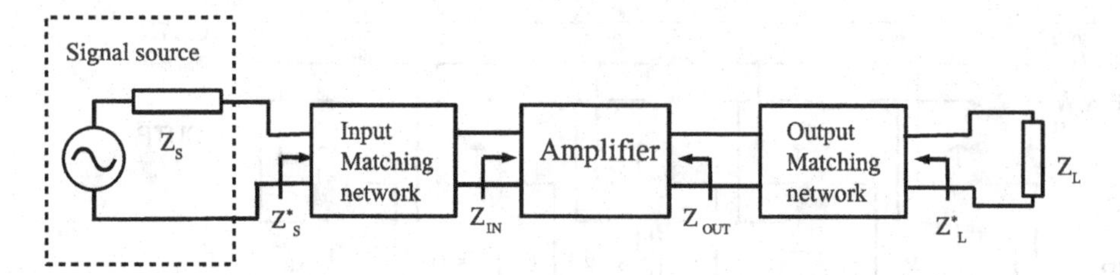

Figure 5.5 The concept of matching the input and output impedances of an amplifier

amplifier and the signal source, and between the amplifier and the load. Often, in high-frequency applications, Z_S and Z_L are equal to a real impedance Z_0.

Thirdly, in high sensitivity receivers, to minimize the noise contributed by the receiver to the degradation of the SNR, the receiver must see an impedance equal to its optimum noise impedance looking towards the antenna.

Finally, in transmit high-power amplifiers, the antenna, which acts as a load to the amplifier, must provide a conjugately matched impedance to the optimal large signal output impedance of the power amplifier. In both of the last two cases, a lossless passive matching network is employed to convert the antenna impedance to the desired value.

Why is impedance matching rarely employed in low-frequency analog design? The simple answer is that in most cases the transistor power gain is so high that we can afford to lose some signal power, especially if by doing so we avoid using inductors, transformers, and transmission lines which occupy significantly larger die area (and therefore are much more expensive) than transistors. However, even in low-frequency circuits, impedance matching using negative feedback is often employed, but rarely between the stages of an amplifier.

5.2.1 Narrow band matching networks

As we have seen in Chapter 4, the input and output nodes of transistors and active circuits typically behave like parallel or series R-C circuits. In high-frequency circuit design, the input and/or output impedance of a transistor often needs to be matched to a real impedance $Z_0 (= 1/Y_O)$. Ideally, the matching network should be lossless to avoid noise figure degradation and unnecessary power loss.

As long as the real part of the load impedance, Z_L, is non-zero, a matching network can always be found to convert Z_L to a real value Z_0. In general, many solutions exist and, usually, the ones leading to the simplest physical implementation, or the widest bandwidth, are chosen.

5.2.2 *L*-section matching

The simplest type of matching network is the *L*-section (also known as *L*-network) which consists of two lumped elements, a reactance jX and a susceptance jB. Two possible topologies exist, as shown in Figure 5.6 [1].

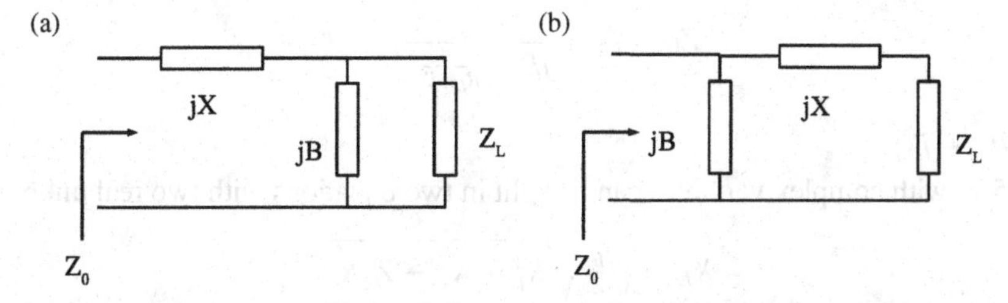

Figure 5.6 *L*-section matching networks: (a) for Z_L located inside the $1+jx$ circle and (b) for Z_L located outside the $1+jx$ circle

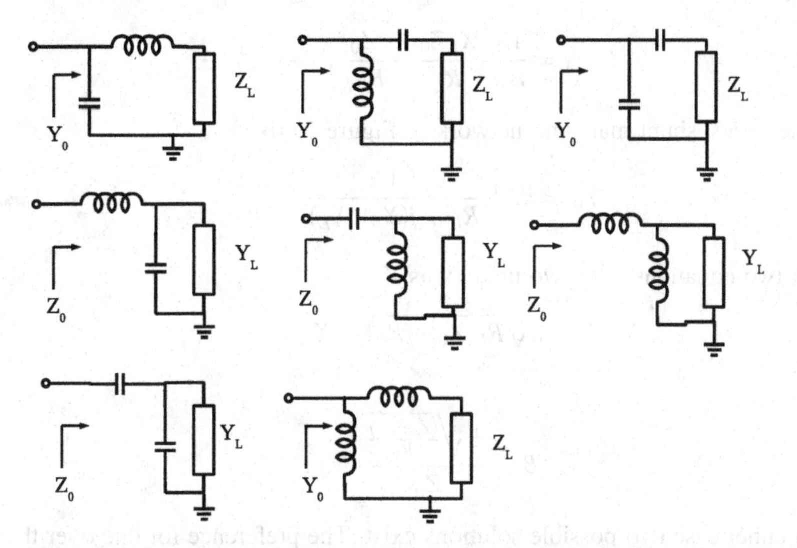

Figure 5.7 *L*-section matching networks based on lumped inductors and capacitors

(a) if the normalized load impedance $z_L = Z_L/Z_0$ is located inside the $1 + jx$ circle of the Smith Chart (i.e. $r_L > 1$), a susceptance, jB, is placed in parallel with Z_L, followed by a series reactance, jX;
(b) if the normalized load impedance $z_L = Z_L/Z_0$ is located outside the $1 + jx$ circle of the Smith Chart (i.e. $r_L < 1$), a reactance, jX, is placed in series with Z_L, followed by a shunt susceptance, jB.

Since, in both type a and type b networks, X and B can be either positive or negative, eight possible 2-element *L*-networks exist that can be realized with lumped inductors and capacitors, as illustrated in Figure 5.7. As long as the physical dimensions of L and C are smaller than $\lambda/10$, *L*-networks can be (and have been) used in ICs at frequencies as high as 170 GHz [2].

We note that tapped inductors or tapped capacitor matching networks represent special cases of the network in Figure 5.6(b) where $X = \omega L_1$, $B = -1/\omega L_2$, and $X = -1/\omega C_1$, $B = \omega C_2$, respectively. Tapped inductors and capacitors are employed when $R_L < Z_0$.

5.2.3 Analytical solutions

We can now proceed to derive exact analytical solutions for the *L*-networks. In the case of the shunt–series topology in Figure 5.6(a) the following equality holds

$$Z_0 = jX + \frac{1}{jB + \frac{1}{R_L + jX_L}} \quad (5.1)$$

where $Z_L = R_L + jX_L$.

Equation (5.1) with complex variables can be split in two equations with two real unknowns B

$$B = \frac{X_L \pm \sqrt{\frac{R_L}{Z_0}}\sqrt{R_L^2 + X_L^2 - Z_0 R_L}}{R_L^2 + X_L^2} \quad (5.2)$$

and X

$$X = \frac{1}{B} + \frac{X_L Z_0}{R_L} - \frac{Z_0}{BR_L}. \quad (5.3)$$

Similarly, for the series–shunt matching network in Figure 5.6(b)

$$Y_0 = jB + \frac{1}{R_L + j(X + X_L)} \quad (5.4)$$

which results in two equations with two unknowns X

$$X = \pm\sqrt{R_L(Z_0 - R_L)} - X_L \quad (5.5)$$

and B

$$B = \frac{\pm\sqrt{\frac{(Z_0 - R_L)}{R_L}}}{Z_0} \quad (5.6)$$

We note that in either case two possible solutions exist. The preference for one over the other depends on:

- ease of implementation: some inductor/capacitor values are easier to fabricate with adequately low parasitics in a given technology;
- need for bias decoupling (i.e. a series capacitance is required between stages and therefore an L-network with a series-connected capacitor is preferred);
- stability considerations: since transistor gain is higher at lower frequencies, a high-pass matching network (series capacitor, shunt inductor) will suppress low-frequency gain;
- lower noise: smaller series inductors and smaller shunt capacitors have lower loss and, therefore, generate less noise;
- harmonic termination/filtering: e.g. low-pass matching networks (series L, shunt C) suppress higher-order harmonics in the output matching network of a power amplifier.

5.2.4 Graphical solutions using the normalized impedance/admittance (ZY) Smith Chart

The L matching network can also be determined graphically, in two steps, using the impedance and the admittance Smith Charts and the impedance-to-admittance conversion. The latter is obtained by superimposing the Z-Smith Chart with an admittance Smith Chart rotated by

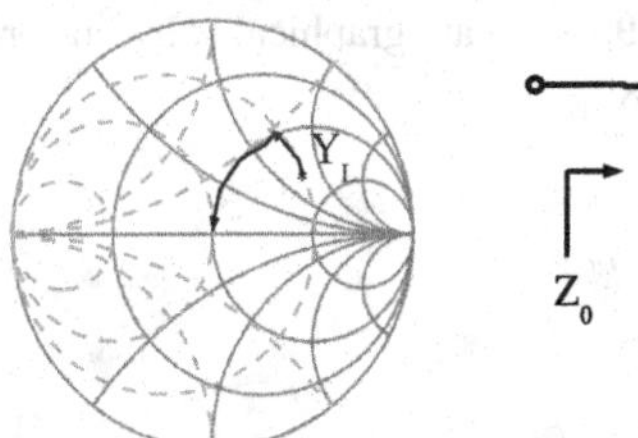

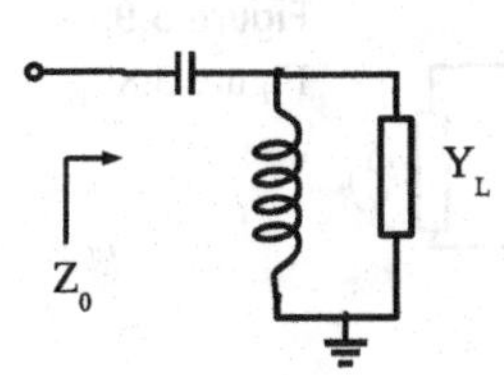

Figure 5.8 Graphical solution for L-network matching using the *ZY* Smith Chart when z_L is inside the $1+jx$ circle. A typical practical example would be matching to the optimal noise impedance of a transistor

180 degrees, as illustrated in Figure 5.8. This combined chart is known as a normalized impedance and admittance Smith Chart, or simply *ZY* Smith Chart [3]. The impedance circles are drawn with solid lines while the rotated admittance circles are shown with dashed lines. In the *ZY* Smith Chart, for a given value of z, the corresponding normalized admittance value y is read directly from the admittance coordinates. For example, in Figure 5.8, $y_L = Y_L Z_0 = 0.3 - j0.3$ and corresponds to $z_L = 1.67 + j1.67$.

In the upper half of the *ZY* Smith Chart, the susceptance is negative while the reactance is positive. The roles are reversed in the lower half, with the reactance being negative and the susceptance positive. A series-connected reactance, jX, produces a movement of the impedance along a constant resistance circle R_L/Z_0. The movement is upwards if the reactance is positive (inductive) or downwards if the reactance is negative (capacitive). A shunt-connected positive (capacitive) susceptance, jB, moves the impedance downwards along a constant conductance circle $G_L Z_0$, while a negative (inductive) susceptance moves the impedance upwards along the same constant conductance circle.

When one designs a matching network using the *ZY* Smith Chart, one moves along constant resistance and/or constant conductance circles from one value of impedance or admittance to another. Each movement along a constant resistance or constant conductance circle gives the value of the reactive element (*L* or *C*) in the *L*-network, which corresponds to that particular movement [3].

EXAMPLE 5.1

Let us consider the situation illustrated in Figure 5.8, where the normalized load impedance is $z_L = 1.67 + j1.67$ and where we want to design a matching network from Z_L to $Z_0 = 50\Omega$. The exact values of B and X are calculated from (5.2) and (5.3)

$$B = \frac{83.5 - \sqrt{1.67}\sqrt{6972.25 + 6972.25 - 4175}}{6972.25 + 6972.25} = -0.00317S \text{ or } b_L = -j0.158$$

$$X = \frac{1}{B} + \frac{X_L Z_0}{R_L} - \frac{Z_0}{BR_L} = -315.46 + 50 + 188.896 = -76.56\Omega \text{ or } X_L = -j1.53$$

At 94GHz, the corresponding L and C values are 534pH and 22.4fF. The inductor value is rather high for the back-end of a typical silicon technology, resulting in a SRF that may not exceed 90GHz.

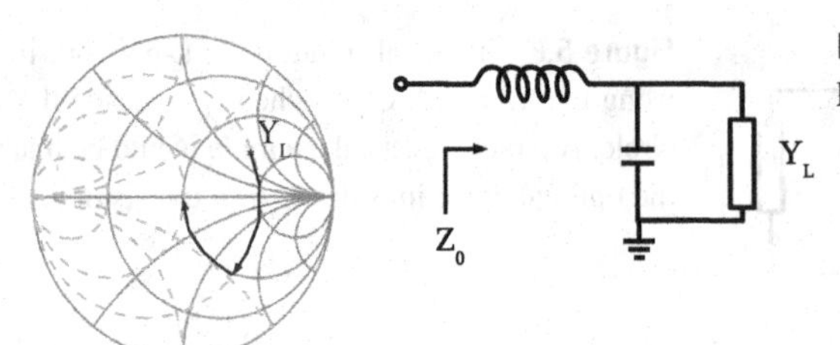

Figure 5.9 Alternate graphical solution for Figure 5.8

The other possible solution is

$$B = \frac{83.5 + \sqrt{1.67}\sqrt{6972.25 + 6972.25 - 4175}}{6972.25 + 6972.25} = 0.015S \text{ or } b_L = j0.757$$

$$X = \frac{1}{B} + \frac{X_L Z_0}{R_L} - \frac{Z_0}{BR_L} = 66.66 + 50 - 39.92 = 76.74\Omega \text{ or } x_L = j1.535$$

which, at 94GHz, results in C and L values of 25.4fF and 130pH, respectively. These values are more suitable for fabrication in silicon technology because they can be designed such that their SRF exceeds 200GHz. The corresponding graphical solution is illustrated in Figure 5.9.

EXAMPLE 5.2

Let us now match the load impedance $z_L = 0.25 + j0.2$, located outside the $1 + jx$ circle, to 50Ω, as illustrated in Figure 5.10.

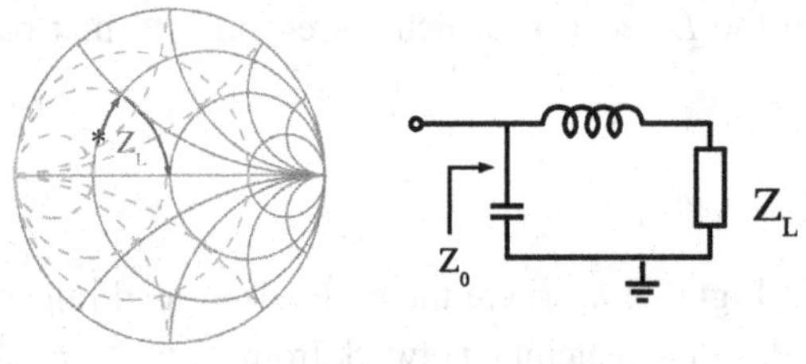

Figure 5.10 Example of graphical solution to L-network matching when Z_L is outside the $1+jx$ circle

The analytical solution results in

$$X = \sqrt{R_L(Z_0 - R_L)} - X_L = \sqrt{12.5(50 - 12.5)} = 21.65\Omega \text{ or } x_L = j0.433.$$

$$B = \frac{\sqrt{\frac{(Z_0 - R_L)}{R_L}}}{Z_0} = \frac{\sqrt{\frac{(50 - 12.5)}{12.5}}}{50} = 0.03464S \text{ or } b_L = \text{j}1.73.$$

At 94GHz, the corresponding L and C values are 36.67pH and 58.68fF, respectively, suitable for implementation in silicon.

5.2.5 The quality factor and bandwidth of matching networks

An L-network can be regarded as a two-pole (second-order) low-pass or high-pass filter. At each node of the L-network there is a series ($R_S + jX_S$) or parallel ($G_P + jB_P$) resonant circuit for which we can define a quality factor Q_n, as

$$Q_n = \frac{|X_S|}{R_S} \text{ or } Q_n = \frac{|B_P|}{G_P}. \tag{5.7}$$

For a narrow frequency band around the resonant frequency f_0 where the L-network achieves matching between the load and the desired source impedance, we can describe the combination of the signal source impedance, L-network, and its load as a second-order bandpass filter. For such a second-order bandpass filter, we can define the loaded quality factor, Q_L, as a function of the 3dB bandwidth BW_{3dB}

$$Q_L = \frac{f_0}{BW_{3dB}}. \tag{5.8}$$

Since at every node in the L-network the impedances to the left and to the right are conjugately matched, and since the equivalent series (parallel) resistance (conductance) at resonance is $2R_S$ ($2G_P$), Q_L is equal to $Q_n/2$. We immediately realize that the bandwidth over which the L-network can achieve matching depends on the center frequency and the loaded quality factor. Also, in order to obtain high Q_L, the circuit node Q must be very high.

EXAMPLE 5.3 Input matching bandwidth of a SiGe HBT at 120GHz

A SiGe HBT biased at the peak f_T current density of 15mA/μm^2 has a *MAG* of 6dB, an input impedance $Z = 31.5–j29.4\Omega$ ($z = 0.6303–j0.588$), and an output impedance $Z_{out} = 113.8–j93.7\Omega$ ($z_{out} = 2.27–j1.875$) at 120GHz. Design the input matching network to 50Ω at 120GHz using a single-stage L-network. Assume that the transistor is already matched at the output and that (ideally) the output matching network does not affect the input impedance of the transistor.

Solution

We note that Z is located outside the $1 + jx$ circle. As illustrated in Figure 5.11, we first employ a series inductor L_1 of 7.5pH (still realizable in silicon) to move the impedance to the $g = 1$ circle in the admittance Smith Chart. Next, we move the impedance to the center of the Smith Chart by adding an inductor L_2 of 86pH in parallel with the input.

To separate the bias circuit from the signal source, two bias decoupling capacitors are added: C_{DC}, in series with the input, and C_{BIAS}, at the other end of the shunt inductor, L_2, through which the base is biased. This arrangement is not optimal because each capacitor contributes parasitics and increases the loss of the matching network. Note that the Q of the input and output impedances of the HBT is lower than 1, making it relatively easy to obtain good matching over a broad bandwidth. In fact, after conducting S-parameter computer simulations, we obtain $S_{11} < -20$dB from 102 to 145GHz, and $S_{11} < -15$dB from 90 to 190GHz.

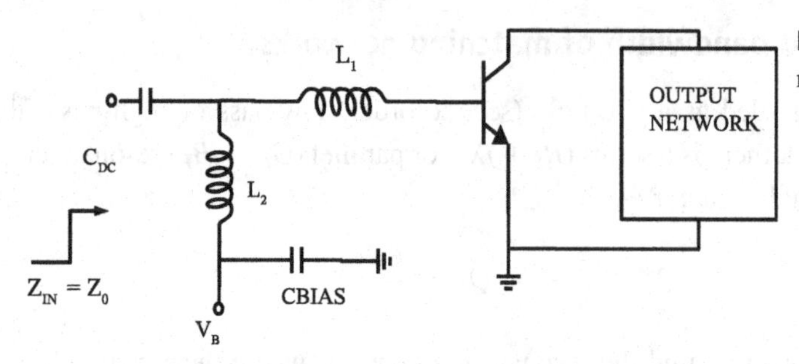

Figure 5.11 Input matching and bias network for Example 5.3

The main disadvantage of the *L*-network is that it only has two design variables. These are employed to set the impedance transformation ratio and the frequency at which the impedance is matched. The Q and bandwidth of the matching network, usually adequate for narrow band application, are set from the other design considerations. For a single-step *L*-match, the Q is reasonably well approximated by the square root of the impedance transformation ratio [4].

To provide an extra degree of freedom to control the bandwidth and Q, more sophisticated matching networks and techniques have been developed. Two frequent design scenarios arise:

(a) How can the bandwidth of the matching network be increased (broadband matching scenario)?
(b) How can we achieve high Q_L (PAs are important applications of this scenario)?

Higher bandwidths or loaded Qs than those obtained with the *L*-networks can be realized using matching circuits with three or more elements. For example, three *L* and/or *C* elements can form lossless π or *T* networks [4]. The reader is referred to [4] for an in-depth treatment of the π or *T* networks.

5.2.6 Transmission line matching networks

The input impedance into a lossless transmission line of physical length, l, characteristic impedance, Z_0, and terminated on a load impedance Z_L is given by

$$Z_{IN}(l) = Z_0 \left[\frac{Z_L + jZ_0 \tan\left(\frac{2\pi f\sqrt{\varepsilon_{eff}}}{c} l\right)}{Z_0 + jZ_L \tan\left(\frac{2\pi f\sqrt{\varepsilon_{eff}}}{c} l\right)} \right] \tag{5.9}$$

where c is the speed of light and f is the frequency. Depending on the value of the load impedance, Z_L, on the physical length and characteristic impedance, purely reactive impedances, either inductive or capacitive can be realized. If, at a given frequency, the electrical line length corresponds to 90 degrees, the line acts as an impedance transformer and the input impedance becomes

$$Z_{IN}\left(l = \frac{\lambda}{4}\right) = \frac{Z_0^2}{Z_L} \tag{5.10}$$

where $\lambda = \frac{c}{f\sqrt{\varepsilon_{eff}}}$ is the wavelength at frequency f along the transmission line whose effective permittivity is ε_{eff}.

The quarter-wave transformer represents a simple circuit element that can be used, in theory, to match any real load impedance to any real source or line impedance over a relatively narrow band simply by choosing the characteristic impedance of the line such that $Z_0 = \sqrt{Z_{IN}Z_L}$. In practice, the range of characteristic line impedance values is limited by the back-end of the semiconductor technology. In the case where good matching is required over wider bandwidth, cascaded (multi-section) quarter-wave transformers or tapered impedance transformers can be used, obviously with a penalty in size and (in the real world) loss.

Because of their ease of manufacturing and modeling, shunt and series transmission line stub matching is the most used technique in microwave and mm-wave integrated circuits fabricated on PCB, ceramic substrates and on semi-insulating GaAs, SiC, or InP substrates. In silicon ICs, because t-lines have higher loss and because the area occupied by such networks is larger than that of the equivalent inductor-and-capacitor-based L-section, it is only employed in mm-wave circuits, at extremely high frequencies [3]. Most recently, 30nm InP HEMT amplifiers with microstrip line matching have been demostrated well into the terahertz region, at frequencies as high as 700GHz [5].

5.2.7 Lossless broadband matching

The matching techniques discussed so far are valid over a narrow bandwidth, typically 10–15%. In some applications, the bandwidth over which a certain load impedance must be matched to a desired real impedance is significantly larger than 10% and can extend over decades of frequency. Recognizing that L-networks are filters, we can employ higher-order filters and the associated synthesis and analysis theory to increase the matching bandwidth. Higher-order matching networks are typically realized as:

- a cascade of L-networks (π-, T-, tapped inductor and tapped capacitor networks),
- transformers and/or t-coils,
- distributed LC matching networks.

Transformer matching

Although more difficult to synthesize, a transformer matching network has several advantages over an L-section. First, it provides a very compact solution to broadband matching, applicable up to at least 170GHz [2]. Second, the primary and secondary coils are isolated at DC, thus simplifying bias decoupling. In a simplistic view, where the coupling coefficient $k = 1$ and the coil parasitic capacitances are ignored, the impedance looking into the primary of an ideal transformer whose secondary is terminated on a load resistance R_L is a factor of L_P/L_S times larger than R_L. Here L_P and L_S are the inductances of the primary and secondary coils, respectively. However, at high frequency, we must account for the non-ideal coupling coefficient, $k < 1$, and for the parasitic capacitance and resistance of the coils of a real transformer, all of which affect the impedance transformation. A common matching technique is to tune out the parasitic inductances of the primary and of the secondary by placing a shunt capacitor at the primary or at the secondary port. As illustrated in Figure 5.12, if we represent

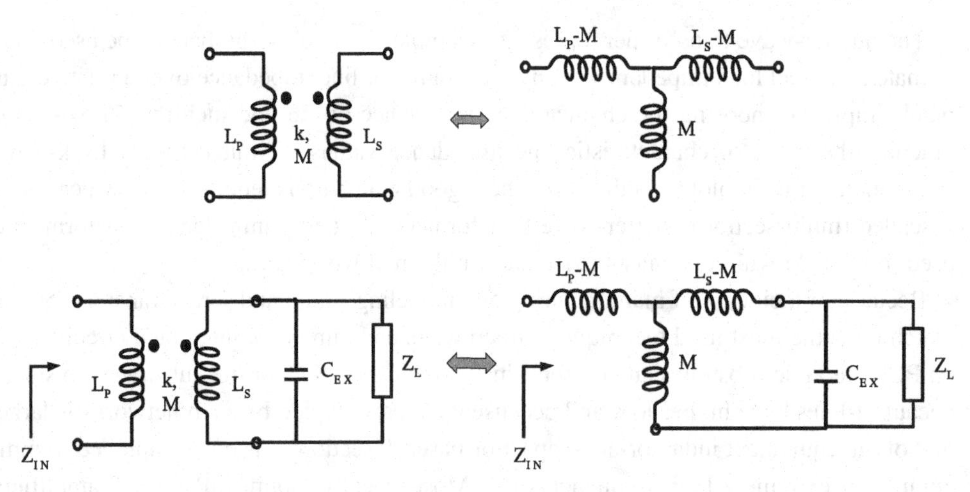

Figure 5.12 Transformer matching networks

the transformer by its T equivalent model with three inductors, the transformer and the shunt capacitor C_{EX} form a two-stage matching network consisting of two cascaded L-sections. Intuitively, this explains why transformers provide broader bandwidth matching than a simple LC network.

EXAMPLE 5.4

For the transistor in Example 5.3, design the output matching network to a load impedance of 50Ω at 120GHz using (a) a single L-section, (b) two cascaded L-sections, and (c) a transformer.

Solution

(a) The output impedance locus is inside the $1 + jx$ circle. Therefore, as illustrated in Figure 5.13, we start with a shunt impedance (inductance L_3 of 90pH) on the admittance Smith Chart, followed by a series – capacitor, C_4, of 17fF (also needed for separating the load from the circuit and supply at DC). Simulating this circuit in the design kit, we obtain a very good S_{22}, but S_{11} has been degraded! Why is that happening? The simple answer is that the transistor does not provide adequate isolation at 120GHz. We need to use a gain stage with higher isolation or else we will be forced to iterate several times between the design of the input and output matching networks. A possible solution is a bipolar cascode stage which normally provides sufficient isolation to allow for the relatively independent design of the input and output matching networks.

Note that we could have also used an L-section consisting of a shunt capacitor followed by a series inductor. Although the latter solution represents a shorter movement on the Smith Chart, it would not have provided DC-blocking.

Finally, simulations indicate that the output matching bandwidth is much narrower than that at the input, with S_{22}<–20dB from 111 to 128GHz.

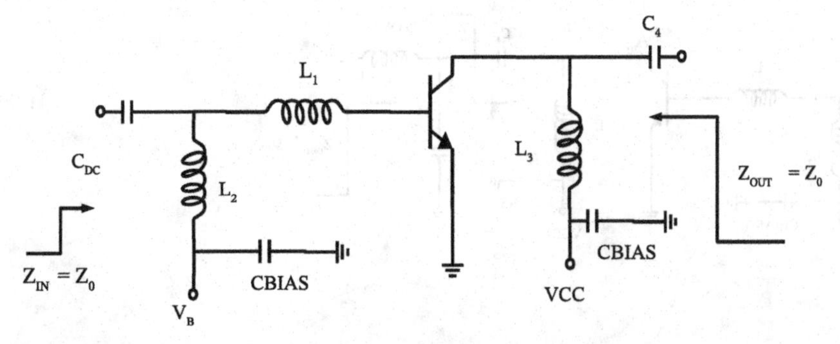

Figure 5.13 Single *L*-section output matching network for Example 5.4

(b) 2-step *L*-section matching: a popular approach is to match first from 113Ω to 75Ω (approximately $\sqrt{113 \times 50}$) and then from 75Ω to 50Ω. This reduces the quality factor of the intermediary matching networks and increases the bandwidth.

The *L*-section that matches the transistor output impedance to 75Ω is synthesized using a shunt $L_3 = 106$pH and a series $C_4 = 16$fF. At this point (*B* in Figure 5.14), the output is perfectly matched to 75Ω at 120GHz. Next, a second *L*-section is added which consists of a shunt $C_5 = 13.4$fF (to point *C* on the Smith Chart in Figure 5.14) and a series $L_6 = 46.44$pH, with an equivalent reactance of 35Ω at 120GHz. The output is perfectly matched to 50Ω at 120GHz with $S_{11} < -20$ dB from 106.8 to 145GHz. The matching bandwidth is significantly larger than that obtained with a single *L*-section.

(c) Transformer matching: let us first assume an ideal 1:1 transformer with $k = 1$, $L_P = L_S$. One minimalist approach is to select L_P such that it resonates with the output capacitance of the transistor at 120GHz. This results in $L_P = L_S = 230$pH. However, if the transformer impedance ratio is equal to 1 (i.e. if $L_P = L_S$), the normalized output impedance becomes 3.43, much higher than desired. Therefore, a transformer impedance ratio of 3.43 is chosen, which matches the output impedance of the transistor to 50Ω. The inductance of the primary remains 230pH and the inductance of the secondary is set to 67pH = 230pH/3.43. Computer simulation shows that the bandwidth over which $S_{22} < -20$dB extends from 100 to 160GHz, which is significantly larger than that obtained with a single *L*-section and even larger than that of the cascaded 2-step *L*-section network. As mentioned earlier, in a real transformer, $k = 0.7$–0.8 and the primary and secondary windings exhibit parasitic capacitances which will affect the optimal L_P and L_S values and further reduce the bandwidth.

Can we do even better in terms of broadband matching than with the previously described networks? The ultimate limitation for broadband matching with "lumped" elements was derived theoretically by Bode and Fano and is known as the *Bode–Fano limit* [6],[7]. For a given impedance described by a parallel R-C network, the following inequality holds

$$\int_0^{\infty} \ln\left(\frac{1}{|S_{11}(\omega)|}\right) d\omega \leq \frac{\pi}{RC} \tag{5.11}$$

which embodies the tradeoff between impedance matching and bandwidth. One can only achieve good matching, characterized by the return loss S_{11}, over a small bandwidth.

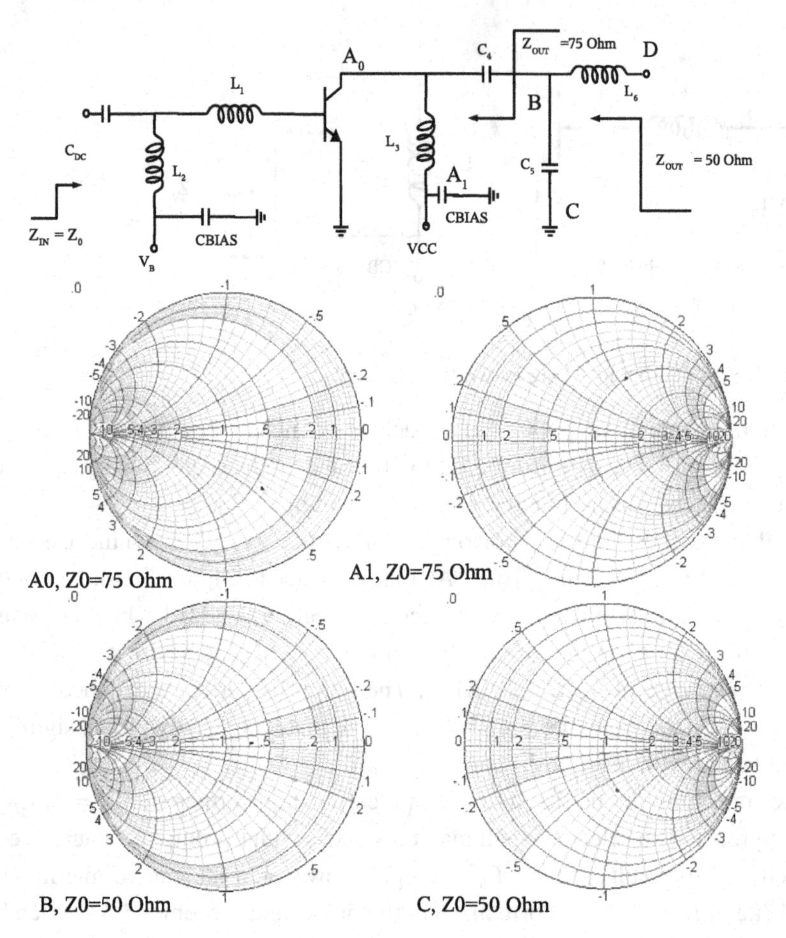

Figure 5.14 Two-step *L*-section output matching network for Example 5.4

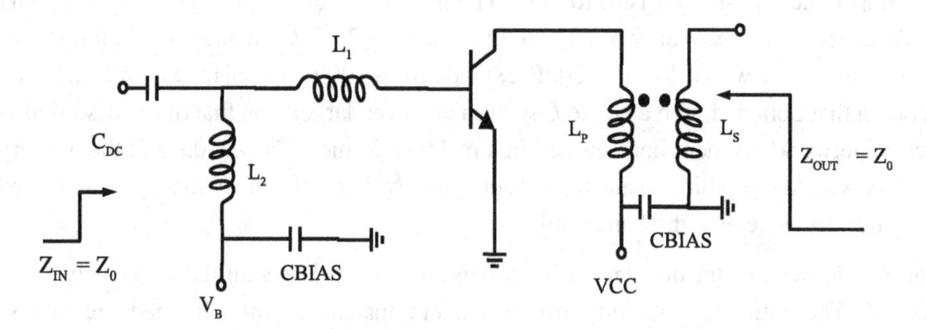

Figure 5.15 Transformer-based output matching network for Example 5.4

Similar ultimate limits exist for series R-C

$$\int_0^{\infty} \frac{1}{\omega^2} \ln\left(\frac{1}{|S_{11}(\omega)|}\right) d\omega \leq \pi RC \tag{5.12}$$

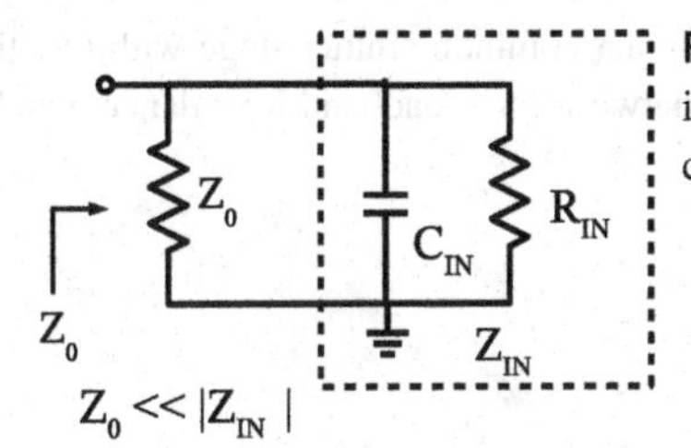

Figure 5.16 "Brute force" impedance matching by connecting a Z_0 impedance in parallel with the input of a very high input impedance circuit

parallel R-L

$$\int_0^\infty \frac{1}{\omega^2} \ln\left(\frac{1}{|S_{11}(\omega)|}\right) d\omega \leq \frac{\pi L}{R} \quad (5.13)$$

and series R-L loads

$$\int_0^\infty \ln\left(\frac{1}{|S_{11}(\omega)|}\right) d\omega \leq \frac{\pi R}{L}. \quad (5.14)$$

Using t-coils and distributed amplifier topologies, we can overcome this limit by reducing L or C, respectively, for a given R.

5.2.8 Lossy matching

As we have seen in the previous section, reactive lossless matching can be accomplished only over a limited bandwidth. It is particularly difficult to realize at low frequencies because of the need for extremely large L and C values, which cannot be integrated. Lossy matching, which implies adding resistors in series with inductors or transmission lines to broaden the bandwidth, can be used in those situations. Indeed, lossy matching is applied to match the input and output of the DC-to-HF amplifiers encountered in fiber-optic and high-data-rate wireline communication systems. The penalty is noise figure degradation and power loss, a compromise that is accepted in some practical cases.

Lossy matching is typically implemented by brute force or with negative feedback. In the brute force scenario, a resistor of value Z_0 is connected in parallel with a high input/output impedance stage (Figure 5.16) or in series with a low input/output impedance stage. The latter scheme is rarely used at high frequency, and only at the output of a circuit.

The most common examples of high input impedance circuits are MOSFET differential stages at low and moderate frequencies and bipolar emitter-follower or bipolar differential pair stages with resistive emitter degeneration. The drawback of such a brute force approach is that half of the power received from the signal source is wasted on the shunt-connected Z_0 resistor and that the noise figure is degraded.

The equivalent circuit in Figure 5.16 applies to all cases discussed above. It also describes the typical output matching network of a broadband output buffer. The expression of the input/output impedance as a function of frequency can be cast as

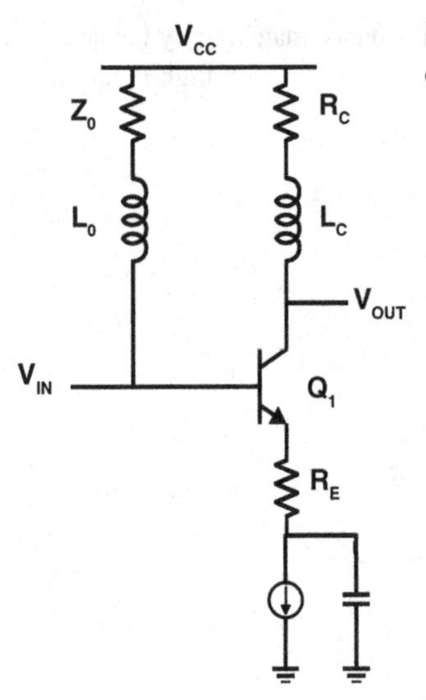

Figure 5.17 Example of a bipolar common emitter stage with resistive degeneration and shunt *R-L* network for broadband impedance matching

$$Z(\omega) = \frac{1}{\frac{1}{Z_0} + \frac{1}{R_{IN}} + j\omega C_{IN}} \approx \frac{Z_0}{1 + j\omega C_{IN} Z_0} = \frac{Z_0(1 - j\omega C_{IN} Z_0)}{1 + \omega^2 C_{IN}^2 Z_0^2} \tag{5.15}$$

and indicates that matching is perfect at low frequencies (i.e. $Z(\omega = 0) = Z_0$) and degrades at high frequencies. In (5.15), R_{IN} represents the input resistance of the transistor at DC and can be ignored at high frequencies if it was properly designed to be large. $Z(\omega)$ is a single-pole function of frequency with the pole at $1/(C_{IN}Z_0)$. To further increase the matching bandwidth, a zero can be introduced in the input impedance expression. This is accomplished by adding an inductor in series with the input or in series with the shunt-connected Z_0, as in Figure 5.17. The corresponding input impedance formula, ignoring R_{IN}, thus becomes

$$Z(\omega) \approx \frac{1}{\frac{1}{Z_0 + j\omega L} + j\omega C_{IN}} = \frac{Z_0\left(1 + j\omega \frac{L}{Z_0}\right)}{1 - \omega^2 C_{IN} L + j\omega Z_0 C_{IN}} \tag{5.16}$$

and the zero is at Z_0/L.

The inductor increases the bandwidth over which matching remains better than −20dB or −15dB. Typically L is chosen such that $\frac{Z_0^2 C_{IN}}{3} < L < Z_0^2 C_{IN}$ with the lower value preferred to avoid ringing.

Negative feedback Transimpedance amplifiers are typically employed as the input low-noise, low-current, large-bandwidth amplifier of high-data-rate communication receivers. The transimpedance feedback resistor (or resistor in series with inductor as in Figure 5.18) provides broadband input matching.

If for the circuits in Figures 5.18(a) and (b) we assume that the input capacitance of the amplifier is C_{IN}, the input impedance can be expressed as

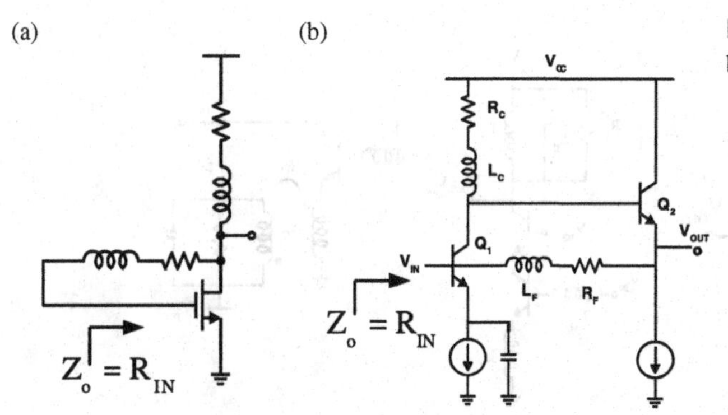

Figure 5.18 Typical (a) FET and (b) bipolar transimpedance amplifier stages

$$Z(\omega) = \frac{Z_F}{A+1} \frac{1}{1 + j\omega \frac{Z_F C_{IN}}{A+1}} \tag{5.17}$$

where A is the open loop voltage gain of the transimpedance amplifier, assumed constant for the sake of simplicity, and $Z_F = R_F + j\omega L_F$.

By setting $Z_0 = R_F/(1 + A)$, good matching can be achieved from DC up to very high frequencies. Again, as in the "brute force" approach, by adding an inductor, L_F, in series with the feedback resistor, R_F, the matching bandwidth can be increased further.

5.3 TUNED CIRCUIT TOPOLOGIES AND ANALYSIS TECHNIQUES

Tuned circuits are an important class of circuits encountered in all wireless transceivers. As we will see shortly, they can be described as second-order bandpass filters whose performance can be analyzed and designed in much the same way as that of low-frequency amplifiers.

5.3.1 Tuned circuit topologies

The most common tuned amplifier topologies are:

- the common-emitter (source) or common-base (gate) stage with parallel resonant tanks at input and output (encountered in PAs and LNAs, Figure 5.19(a);
- the common-emitter (source) stages with series resonance at input and parallel resonance at output (a traditional topology employed in the first stage of LNAs with inductive degeneration, Figure 5.19(b));
- the common-emitter (source) or common-base (gate) stage with selective negative feedback (Figure 5.19(c)).

Because of the relatively modest voltage, current, and power gain per stage at high frequencies, it is often necessary to cascade a number of tuned amplifier stages to achieve the desired overall gain. As illustrated in Figure 5.20, these stages can be either tuned at the same frequency,

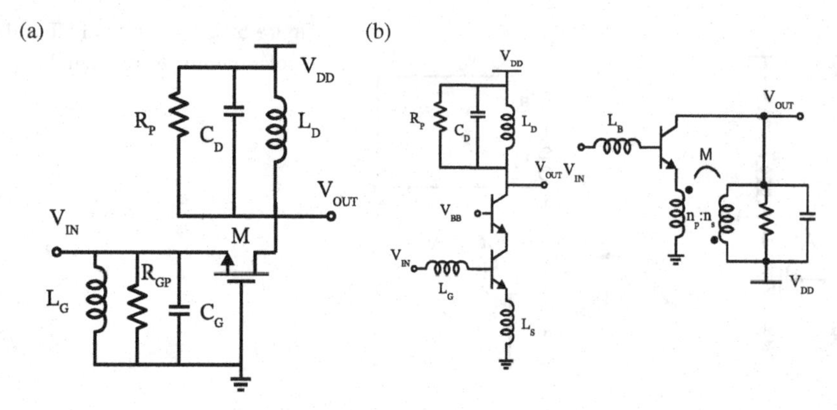

Figure 5.19 Typical tuned amplifier stages: (a) common gate with parallel input and parallel output resonant circuits, (b) casocode stage with inductive degeneration and parallel output tank, and (c) common-emitter stage with transformer feedback

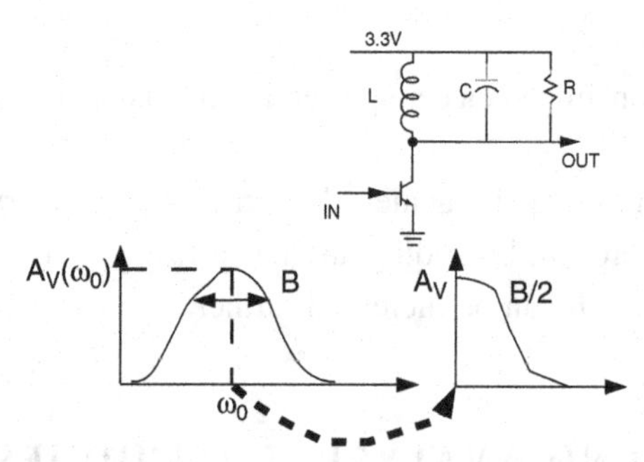

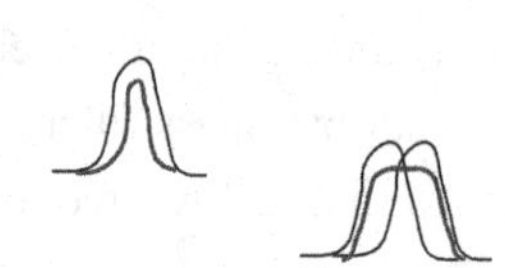

Figure 5.20 Pictorial illustration of the conversion of a tuned amplifier stage to a baseband stage and the concept of stagger-tuning to increase bandwidth

typically resulting in large gain over a narrow band, or can be stagger-tuned to achieve moderate gain over a broader bandwidth.

5.3.2 Tuned circuit analysis techniques

Next we will derive analytical expressions for the voltage gain and power gain of the most common-tuned amplifier stages. These relatively simple expressions provide useful insight into the operation of tuned amplifiers and a fairly accurate starting point for the computer design of tuned amplifiers.

Voltage gain of a tuned single-stage amplifier

Let us consider the most basic topology, shown in Figure 5.20, of a CE or CS stage with a resonant parallel RLC load (single-tuned circuit). To simplify the analysis, the transistor output resistance and capacitance are incorporated in the load impedance, Z_L.

Just like at low frequencies, and ignoring the Miller capacitance, the voltage gain can be expressed as the product of the effective transistor transconductance, g_{meff}, and Z_L, which is now frequency dependent

$$A_v(s) = -g_{meff}Z_L(s) = \frac{-g_{meff}}{C}\frac{s}{s^2 + \left(\frac{1}{CR}\right)s + \frac{1}{LC}} \tag{5.18}$$

$$A_v(s = j\omega_0) = -g_{meff}R \tag{5.19}$$

where the resonance frequency ω_0, the quality factor Q, and bandwidth, B, of the amplifier are given by

$$\omega_0 = \frac{1}{\sqrt{LC}}; Q = \omega_0 CR \, and \, B = \frac{1}{CR}. \tag{5.20}$$

A two-step analysis technique, using the simplified (no Miller capacitance) transconductance-based high-frequency equivalent circuit of the transistor is summarized as follows:

- assume that the circuit has been conjugately matched at its input and at its output at the "resonant" frequency f_o to the signal source and load impedance, respectively;
- analyze the circuit at the resonant frequency f_o by shifting the entire frequency response to DC and applying the same methodology as in DC amplifiers to calculate midband gain, input and output impedance.

For common-emitter (source) or common-base (gate) stages with tuned output, this process is illustrated in Figure 5.21. The input and output impedance at resonance, the short-circuit output current, and the open load voltage gain can then be expressed as in baseband amplifiers:

- $R_{OUT}(f_o) = R_P$ (includes r_o and loss resistance of L_D)
- $R_{IN}(f_o) = R_{GP}$ (includes impact of transistor input resistance and loss resistance of L_G)
- $I_{sc}(f_o) = (+/-)g_{meff}V_{in}$
- $A_v(f_o) = I_{sc} \times R_{out} = (+/-)g_{meff}R_P$

where the " + " sign applies to a common-base (gate) stage and the "−" sign corresponds to a common-emitter (source) stage.

In high-frequency amplifiers, power gain is often more relevant than voltage gain. If the circuit is driven by a matched impedance signal source, $Z_S = R_{IN}$, and terminated on a matched load, $Z_L = R_P$, we can easily derive the expression of the power gain as described next for the most common tuned amplifier stages.

CE/CS or (ac-coupled) cascode stage

First, from the expressions of the short-circuit output current

$$I_{sc} = -g_{meff}V_{in} \tag{5.21}$$

of the output voltage

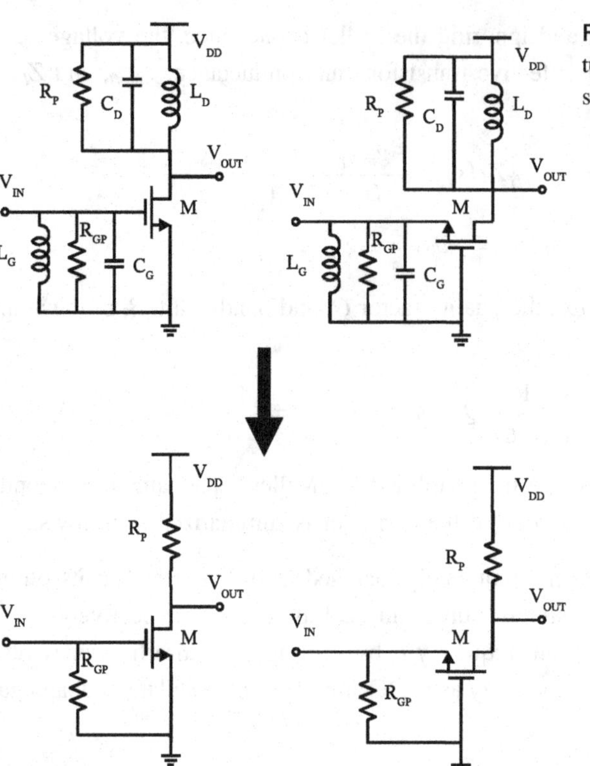

Figure 5.21 Illustration of the conversion of tuned CS and CG stages to baseband amplifier stages at resonance

$$V_{out} = \frac{I_{sc} \times R_P}{2} = \frac{-g_{meff} \times R_P}{2} V_{in} \tag{5.22}$$

and of the current through the load

$$I_{load} = \frac{I_{sc}}{2} = \frac{-g_{meff} \times V_{in}}{2} \tag{5.23}$$

we can derive the formula for the power delivered by the circuit to the load

$$P_{load} = V_{out} \times I_{load} = \frac{g_{meff}^2 \times R_P}{4} V_{in}^2 \tag{5.24}$$

and for the power delivered by the signal source to the circuit

$$P_{in} = V_{in} \times I_{in} = \frac{V_{in} \times V_{in}}{R_{GP}} = \frac{V_{in}^2}{R_{GP}}. \tag{5.25}$$

Hence, the power gain becomes

$$G = \frac{P_{load}}{P_{in}} = \frac{g_{meff}^2 \times R_P \times R_{GP}}{4}. \tag{5.26}$$

It is interesting to relate the power gain expression to the f_{MAX} of the transistor. If ideal (i.e. infinite Q) inductors are assumed for tuning the input and output impedances, we recognize that

$R_P \approx r_{oeff}$ and that the only contribution to R_{GP} is the input resistance of the transistor (i.e. $R_b + r_E$ for HBT, $R_g + R_s$ for MOSFETs). From the HF equivalent circuit of the transistor, we obtain

$$R_{GP} \approx \frac{1}{(R_g + R_s)\omega_0^2 C_{IN}^2} \quad or \quad R_{GP} \approx \frac{1}{(R_b + R_e)\omega_0^2 C_{IN}^2}. \tag{5.27}$$

By inserting (5.27) into (5.26), the expression of the power gain can be recast as

$$G(f) = \frac{g_m^2 r_o}{4\omega^2 C_{IN}^2 (R_g + R_s)} = \frac{f_T^2}{4f^2} \frac{r_o}{R_g + R_s} = \frac{f_T^2}{4f^2} \frac{r_o}{R_b + R_e} = \frac{f_{MAX}^2}{f^2} \tag{5.28}$$

where

$$f_{MAX} = \frac{f_T}{2} \sqrt{\frac{R_{OUT}}{R_{IN}}}. \tag{5.29}$$

Equation (5.28) shows that the power gain of a reactively tuned, input and output-matched, single-stage amplifier decreases with the square of frequency and that it scales with the f_{MAX} (not f_T) of the transistor technology.

CB/CG stage

Similarly to the common-source stage, we can sequentially derive the expressions of the output short-circuit and load currents, of the voltage on the load and of the output and input power delivered to the circuit

$$I_{sc} = g_m V_{in} \tag{5.30}$$

$$V_{out} = \frac{I_{sc} \times R_P}{2} = \frac{g_m \times R_P}{2} V_{in} \tag{5.31}$$

$$I_{load} = \frac{I_{sc}}{2} = \frac{g_m \times V_{in}}{2} \tag{5.32}$$

$$P_{load} = V_{out} \times I_{load} = \frac{g_m^2 \times R_P}{4} V_{in}^2 \tag{5.33}$$

$$P_{in} = V_{in} \times I_{in} = \frac{V_{in} \times V_{in}}{R_{GP}} = \frac{V_{in}^2}{R_{GP}}. \tag{5.34}$$

From these, the power gain of the input- and output-matched tuned common-gate stage is described by

$$G = \frac{P_{load}}{P_{in}} = \frac{g_m^2 \times R_P \times R_{GP}}{4}. \tag{5.35}$$

As a homework problem, try to link (5.35) to the transistor f_T and f_{MAX}. Are there similarities with the common-source stage?

Tuned CE/CG or cascode stage with inductive degeneration

The analysis can be more easily conducted if we employ the current gain-based high-frequency equivalent circuit for the transistor, as in Figure 5.22. The input capacitance of the transistor and the degeneration inductor form a series resonant circuit at the input. The parallel resonance at

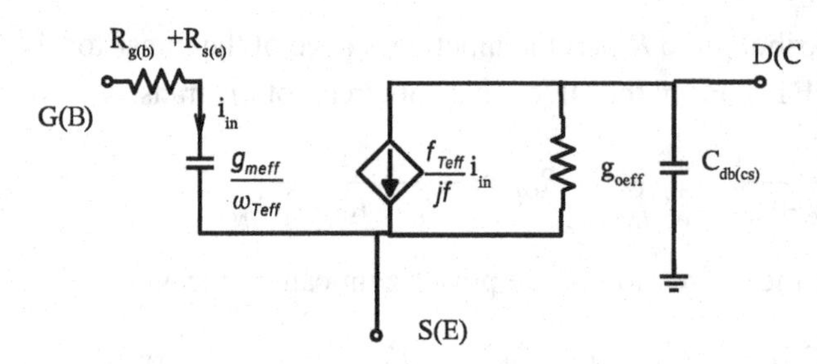

Figure 5.22 Simplified current-gain based high-frequency equivalent circuit of a FET or HBT

the output and the effective transistor transconductance $g_{meff} = g_m/(1 + g_m R_S)$ remain the same as in the previously analyzed CE/CB stages.

How do we deal with the Miller effect?

One option is to express the equivalent circuit parameters of the transistor as a function of the effective cutoff frequency, f_{Teff}, and of the effective transconductance, g_{meff}, of the transistor or of the cascode stage itself (if the stage employs a cascode). Both are readily available from measurements or from simulations and f_{Teff} includes, at least partially, the impact of C_{gd}. Although this equivalent circuit is still unilateral and does not account for the reverse feedback through the gate–drain or base–collector capacitance, it is simple and sufficiently precise, typically with less than +/−15% error, even at the upper millimetre-wave frequencies beyond 100GHz.

Let us consider the tuned CS stage with inductive degeneration shown in Figure 5.23(a). By replacing the FET with the current gain-based high-frequency equivalent circuit in which the impact of g_{oeff} is lumped into the output loss resistance, R_P, we obtain the simplified circuit shown in Figure 5.23(b).

A step-by-step approach is summarized below:

Step 1: Calculate the input impedance at resonance by deriving the expression of the input voltage as a function of the input current

$$v_{in} = i_{in}\left(j\omega L_G + R_g + R_s + \frac{\omega_{Teff}}{j\omega g_{meff}} + j\omega L_S\right) + j\omega L_S \frac{\omega_{Teff}}{j\omega} i_{in} \tag{5.36}$$

from which the formula for the input impedance is obtained by dividing both sides of (5.36) by i_{in}

$$Z_{in} = R_g + R_s + \omega_{Teff} L_S + j\left[\omega(L_G + L_S) - \frac{\omega_{Teff}}{\omega g_{meff}}\right]. \tag{5.37}$$

Step 2: Calculate the short-circuit current i_{sc}

$$i_{sc} = \frac{-f_{Teff}}{jf} i_{in} = j\frac{f_{Teff}}{f} i_{in}. \tag{5.38}$$

Step 3: To match the input to a signal source with real impedance Z_0, the imaginary part of the input impedance, Z_{in}, is canceled by proper choice of L_G, while the real part, R_{IN}, is made equal to the source impedance, Z_0, by selecting L_S such that $R_{IN} = Z_0$.

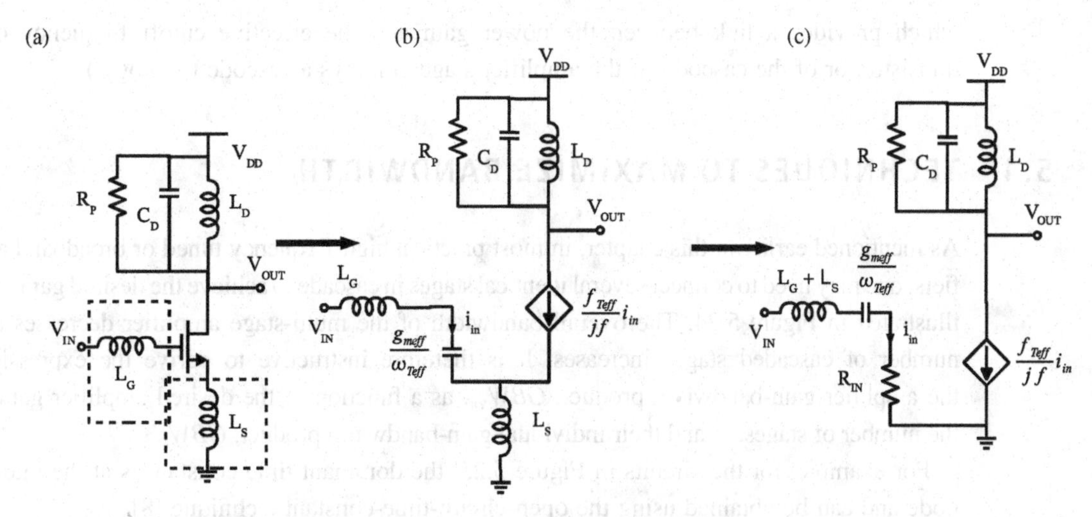

Figure 5.23 (a) Tuned common-source stage with inductive degeneration, (b) simplified equivalent circuit, and (c) equivalent circuit after transferring the impact of the degeneration inductor at the input

In Figure 5.23

$$C_{IN} = \frac{g_{meff}}{\omega_{Teff}} \text{ and } R_{IN} = R_s + R_g + \omega_{Teff} \times L_S. \tag{5.39}$$

Step 4: Derive the expression of the power gain for the circuit in Figure 5.23(a) by noting that the circuit can now be represented as in Figure 5.23(c). When terminated by a matched load, R_P, at resonance, only half of the short-circuit current ends up in the load, with the other half flowing through R_P.

$$V_{out} = \frac{I_{sc} \times R_P}{2} = j\frac{f_{Teff} \times R_P}{jf} i_{in} \tag{5.40}$$

$$I_{load} = \frac{I_{sc}}{2} = j\frac{f_{Teff}}{2f} i_{in} \tag{5.41}$$

$$P_{load} = V_{out} \times I^*_{load} = \frac{f^2_{Teff} \times R_P}{4f^2} |i_{in}|^2 \tag{5.42}$$

$$P_{in} = v_{in} \times i^*_{in} = i_{in} \times R_{IN} \times i^*_{in} = R_{IN}|i_{in}|^2. \tag{5.43}$$

Hence, the power gain in a load matched to R_P can be expressed as

$$G = \frac{P_{load}}{P_{in}} = \frac{f^2_{Teff}}{4f^2} \times \frac{R_P}{R_{IN}}. \tag{5.44}$$

Normally, $R_{IN} = \omega_{Teff} L_S$ is matched to the signal source impedance Z_0.

We note that (5.44) reduces to (5.28) if the degeneration inductance $L_S \rightarrow 0$.

For a single-stage amplifier, matched both at the input and at the output to $Z_0 = R_P = R_{IN}$, the power gain simplifies to

$$G = \frac{f^2_{Teff}}{4f^2} \tag{5.45}$$

which provides a link between the power gain and the effective cutoff frequency of the transistor, or of the cascode (if the amplifier stage employs a cascode topology.)

5.4 TECHNIQUES TO MAXIMIZE BANDWIDTH

As mentioned earlier in this chapter, in most practical high-frequency tuned or broadband amplifiers, one may need to connect several identical stages in cascade to achieve the desired gain. This is illustrated in Figure 5.24. The overall bandwidth of the multi-stage amplifier decreases as the number of cascaded stages increases. It is therefore instructive to derive the expression of the amplifier gain-bandwidth product, GBW_{tot}, as a function of the desired amplifier gain, A_{tot}, the number of stages, n, and their individual gain-bandwidth product, GBW_s.

For example, for the circuits in Figure 5.25, the dominant time constant is at the interstage node and can be obtained using the open-circuit-time-constant technique [8].

$$\tau_p = (R_C||r_o)[C_{cs} + C_{be} + (2 + |A_0|)C_{bc}] + R_b[C_{be} + (1 + |A_0|)C_{bc}] \approx R_C[C_{cs} + C_{be} + (2 + |A_0|)C_{bc}]$$
$$\tau_p = (R_D||r_o)\left[C_{db} + C_{gs} + (2 + |A_0|)C_{gd}\right] + R_g\left[C_{gs} + (1 + |A_0|)C_{gd}\right] \approx R_D\left[C_{db} + C_{gs} + (2 + |A_0|)C_{gd}\right] \quad (5.46)$$

where R_b (R_g) is the base (gate) resistance and

$$A_0 = -g_{meff}(R_C||r_o) \approx -g_{meff}R_C \text{ or } A_0 = -g_{meff}(R_D||r_o) \approx -g_{meff}R_D. \quad (5.47)$$

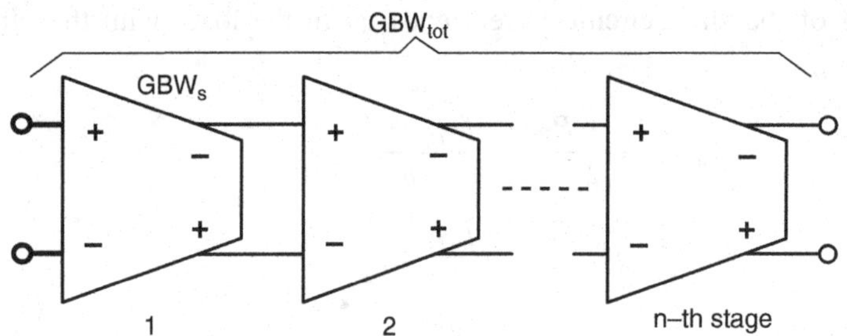

Figure 5.24 A multi-stage amplifier consisting of a chain of identical gain stages described by the individual gain-bandwidth product GBW_s

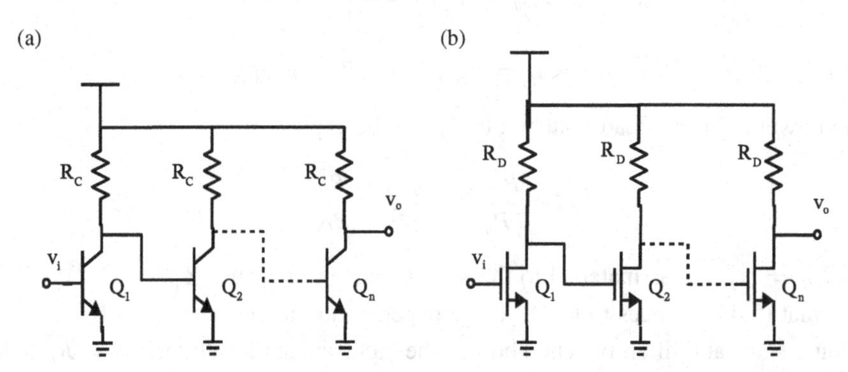

Figure 5.25 Single-ended implementation of a chain of identical broadband: (a) CE stages with resistive loads, R_C, and (b) CS stages with resistive loads, R_D. In both cases, bias circuits and bias decoupling capacitors have been left out

If one assumes that the voltage gain of each stage is described by a single-pole function

$$A(s) = \frac{A_0}{\left(1 + \frac{s}{\omega_p}\right)} \tag{5.48}$$

where A_0 is the voltage gain at low frequency, $\omega_p = 1/\tau_p$ is the pole, and the gain bandwidth product of the stage is given by $GBW_s = A_0\omega_p$, for a cascade of n identical single-pole stages, we obtain

$$A_{tot}(s) = \frac{A_0^n}{\left(1 + \frac{s}{\omega_p}\right)^n}. \tag{5.49}$$

The 3dB bandwidth of the chain, ω_{3dB}, can be calculated from the condition

$$|A_{tot}(s = j\omega_{3dB})| = \frac{|A_0|^n}{\sqrt{2}} \tag{5.50}$$

which is equivalent to

$$\left|1 + \frac{j\omega_{3dB}}{\omega_p}\right| = 2^{\frac{1}{2n}} \; or \; 1 + \frac{\omega_{3dB}^2}{\omega_p^2} = 2^{\frac{1}{n}}. \tag{5.51}$$

Hence, the 3dB bandwidth and gain-bandwidth product of the amplifier chain become

$$\omega_{3dB} = \omega_p\sqrt{2^{\frac{1}{n}} - 1} \tag{5.52}$$

and

$$GBW_{tot} = A_0^n\omega_p\sqrt{2^{\frac{1}{n}} - 1} = A_{tot}\omega_p\sqrt{2^{\frac{1}{n}} - 1}. \tag{5.53}$$

Finally, the ratio of the *GBW* of the entire cascade of stages and that of an individual stage is given by

$$\frac{GBW_{tot}}{GBW_S} = A_{tot}^{1-1/n} \times \sqrt{2^{1/n} - 1}. \tag{5.54}$$

This result provides guidance on how to maximize the GBW of a chain of identical first-order stages for a desired total gain when the unit stage has fixed GBW_s [9]. Such a situation occurs commonly in broadband amplifiers used in fiber-optics and backplane receivers, or in the baseband amplifiers of gigabit data-rate radios. By differentiating (5.54) with respect to n, and equating the result to 0, the optimal number of cascaded stages that leads to the maximum GBW_{tot} is found to be approximately [9]

$$n_{opt} \approx 2 \times lnA_{tot}. \tag{5.55}$$

EXAMPLE 5.5

A broadband amplifier stage with GBW_S = 100GHz is realized in a SiGe BiCMOS process with SiGe HBT f_{MAX} of 300GHz. Find the optimal number of cascaded identical stages that leads to the highest bandwidth amplifier with 40dB gain.

Solution

The gain of 40dB corresponds to A_{tot} = 100. From (5.55), we find the optimal number of stages to be n_{OPT} = 9.2. Indeed, if we use $n = 9$, we obtain

$$\frac{GBW_{tot}}{GBW_S} = 100^{1-1/9} \times \sqrt{2^{1/9} - 1} = 16.95.$$

If $n = 10$, we obtain:

$$\frac{GBW_{tot}}{GBW_S} = 100^{0.9} \times \sqrt{2^{0.1} - 1} = 16.90$$

and, if $n = 5$

$$\frac{GBW_{tot}}{GBW_S} = 100^{0.8} \times \sqrt{2^{0.2} - 1} = 15.34$$

These results indicate that the best overall GBW_{tot} is 1.695 THz (much higher than the f_{MAX} of the transistor in this BiCMOS process), corresponding to an amplifier bandwidth of 17GHz. Each of the 9 stages has a gain A_0 = 1.668 (4.44dB) and a 3dB bandwidth of 60GHz. However, 9 stages will occupy a large die area and the solution may prove uneconomical. Could we design an amplifier with a smaller number of stages and acceptable bandwidth? It turns out that even if we use only 5 stages, GBW_{tot} is 1.534 THz, only 10% smaller than the maximum possible. The amplifier will have a bandwidth of 15.34GHz, and each of the 5 stages would have a gain of 2.51 (or 8dB) and a bandwidth of 39.8GHz.

Equations (5.54) and (5.55) are strictly valid for single-pole gain stages. Similar expressions can be derived for amplifier stages with second- or higher-order frequency response functions. For the frequently encountered case of a chain of second-order Butterworth stages with $Q = \sqrt{2}$ and no zeros, we obtain [9]

$$\frac{GBW_{tot}}{GBW_S} = A_{tot}^{1-1/n}\left(2^{1/n} - 1\right)^{1/4}. \tag{5.56}$$

Finally, it should be noted that the analysis above also applies to the gain-bandwidth product of a chain of cascaded tuned amplifier stages.

How can we improve the gain-bandwidth product of the unit baseband amplifier stage?

The GBW_S tends to be a technology constant and cannot be increased unless we employ a faster transistor technology or higher bandwidth topologies for the individual amplifier stage. Some of the most commonly encountered topologies and techniques to maximize bandwidth include:

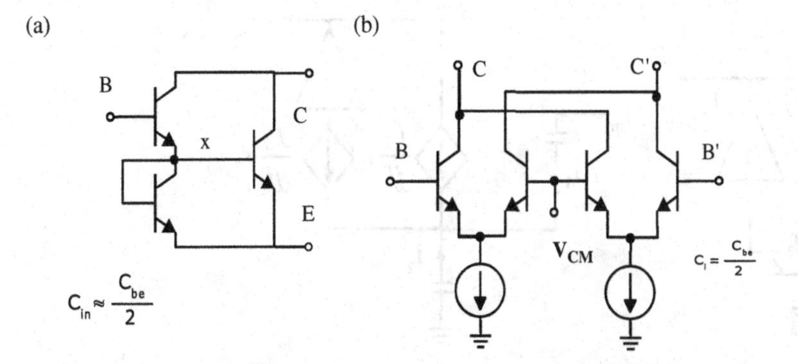

Figure 5.26 Schematics of (a) single-ended and (b) differential bipolar f_T-doubler topologies

- reducing the input capacitance of the amplifier stage,
- the Cherry–Hooper stage,
- buffering and scaling,
- using inductive peaking and distributed amplifier topologies to compensate for node capacitance.

Next we will briefly analyze each one of these concepts.

5.4.1 Reducing the input capacitance

In common-emitter (common-source) amplifier stages with voltage gain A, the frequency response is often dominated by the input capacitance. Ultimately, even for a very low impedance signal source, the input pole is limited by the base + emitter (gate + source) resistance of the input transistor and by the input capacitance, C_{in}. The latter consists of the base–emitter (gate–source) capacitance and the Miller capacitance which is represented by the base–collector or gate–drain capacitance multiplied by the voltage gain of the stage $C_{in} = C_{be} + C_{bc}(1 + |A|)$ *or* $C_{in} = C_{gs} + C_{gd}(1 + |A|)$.

The f_T doubler, shown in Figure 5.26, acts on reducing C_{be} or C_{gs} as important contributors to the input capacitance, by connecting two approximately equal capacitances in series, such that

$$C_{in} \approx \frac{C_{be}}{2} \; or \; C_{in} \approx \frac{C_{gs}}{2}. \tag{5.57}$$

From the outset, we note that this technique does not reduce the contribution of C_{bc} or C_{gd} to the input capacitance and, therefore, the f_T doubler label is somewhat misleading.

The concept can be realized in both single-ended and differential variants with the differential version consuming twice the DC power.

To understand why the topology is called the f_T doubler, it is instructive to derive the expressions of the short-circuit output current and of the equivalent unity current gain frequency for the circuit in Figure 5.26(b). This can be accomplished by resorting to the current gain HF equivalent circuit of the transistor and the half circuits in Figure 5.27, where the "−1" gain blocks represent signal sign inversion.

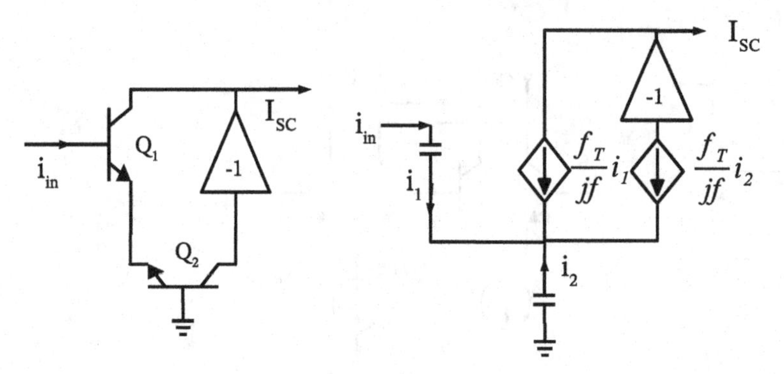

Figure 5.27 Half-circuit and corresponding high-frequency equivalent circuit for the schematic in Figure 5.26(b)

The input current is seen to be equal to the base current of Q_1

$$i_{in} = i_1 \tag{5.58}$$

and the short-circuit output current is obtained as

$$i_{sc}(f) = \frac{f_T}{jf}(i_2 - i_1). \tag{5.59}$$

The condition that the sum of all the currents entering the common-emitter node of Q_1 and Q_2 is equal to zero

$$i_1 + i_2 = \frac{-f_t}{jf}(i_1 + i_2) \tag{5.60}$$

implies that $i_1 = -i_2$ and

$$i_{sc}(f) = \frac{f_T}{jf}(i_2 - i_1) = 2\frac{f_T}{jf}i_{in}. \tag{5.61}$$

The current gain can now be expressed as a function of frequency as

$$h_{21}(f) = \frac{i_{sc}(f)}{i_{in}} = \frac{2f_T}{jf} \tag{5.62}$$

proving that the unity current gain frequency of the circuit is two times that of the transistor itself.

In practice, the speed improvement is not as emphatic as (5.62) predicts because, by employing the simplified HF equivalent circuit of the transistor, we have ignored the resistive parasitics R_b/R_g, r_o, as well as the Miller capacitance.

The f_T doubler technique has proven most effective in applications where current gain rather than power gain is important. It works well if C_{be} (C_{gs}) >> C_{bc}, C_{cs} (C_{gd}, C_{db}). This is generally true of bipolar circuits. MOSFET implementations suffer from the source–bulk capacitance of the additional transistor and because, in nanoscale technologies, C_{gd} is approximately $C_{gs}/2$. Therefore, its impact cannot be ignored.

As illustrated in Figure 5.28, the input capacitance can also be minimized by:

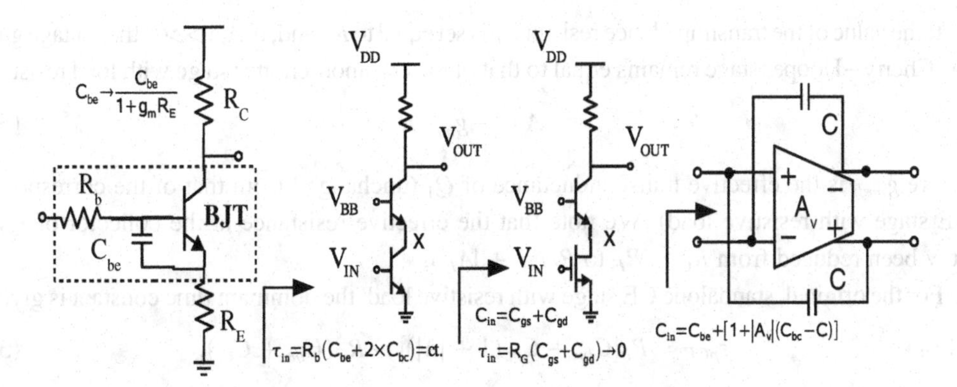

Figure 5.28 Techniques to reduce the input capacitance of an amplifier stage

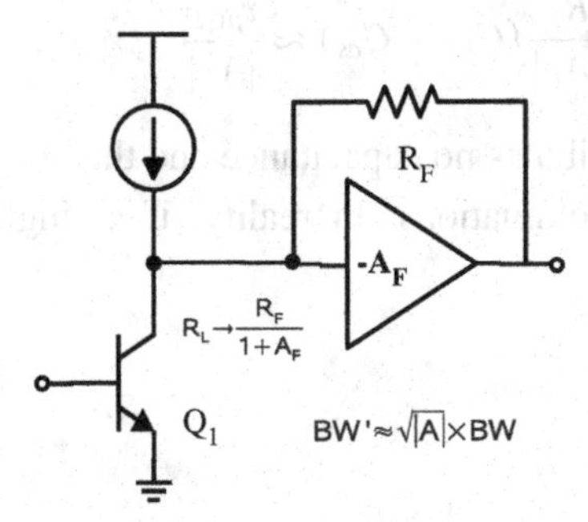

Figure 5.29 Cherry–Hooper stage

(a) using series resistive feedback (emitter degeneration) which reduces C_{be} and C_{gs} by a factor of $(1 + g_m r_E)$, increasing the bandwidth at the expense of the gain (no improvement in *GBW*);
(b) reducing the Miller capacitance with an HBT–HBT or MOS–MOS cascode stage (HBT implementations generally leading to better results than MOSFET ones);
(c) reducing both the Miller capacitance and the $R_g \times C_{in}$ time constant with a MOS-HBT cascode (since R_g, unlike R_b, can be made very small through layout techniques) [10];
(d) canceling out the Miller capacitance with positive feedback (also known as neutralization and which can lead to instability if overdone), and
(e) through combinations of all of the above.

5.4.2 Cherry–Hooper stage

This topology was introduced by Cherry and Hooper in 1963 [11] and, as the cascode stage, reduces the Miller effect on input capacitance. Additionally, it decreases the output resistance while maintaining the overall gain. The conceptual schematic is shown in Figure 5.29. The load resistor of a common-emitter or common-source stage is replaced by an active load, formed by the TIA stage. The latter consists of an inverting amplifier with voltage gain $-A_F$ and infinite input impedance, and a feedback resistor R_F.

If the value of the transimpedance resistor R_F is set equal to R_C and, if $|A_F|>>1$, the voltage gain of the Cherry–Hooper stage remains equal to that of the common-emitter stage with load resistor R_C

$$A = -g_{meff}R_F \tag{5.63}$$

where g_{meff} is the effective transconductance of Q_1 (unchanged from that of the corresponding CE stage with resistive load). We note that the effective resistance at the collector of Q_1 has now been reduced from $R_C = R_F$ to $R_C/(1 + |A_F|)$.

For the original, standalone CE stage with resistive load, the dominant time constant is given by

$$\tau_{pCE} = R_b[C_{be} + C_{bc}(1 + |A|)] + R_C(C_{be} + C_{cs}). \tag{5.64}$$

Similarly, for the Cherry–Hooper stage, we get

$$\tau_{pCH} = R_b\left[C_{be} + \left(1 + \left|\frac{A}{A_F}\right|\right)C_{bc}\right] + \frac{R_F}{|A_F|}(C_{bc} + C_{cs}) \approx \frac{\tau_{pCE}}{|A_F|}. \tag{5.65}$$

In deriving (5.65), we have assumed that the TIA contributes no capacitance and that its gain is constant with frequency. Both are rather crude approximations. In reality, if a single-pole frequency response is assumed for the TIA stage

$$A_F(s) = \frac{A_{F0}}{1+\dfrac{s}{\omega_F}} \tag{5.66}$$

then the expression of the load impedance $Z_L(s)$ at the collector of Q_1 becomes

$$Z_L(s) = \frac{R_F}{1+\dfrac{A_{F0}}{1+\frac{s}{\omega_F}} + sR_FC_{out}} = \frac{R_F\left(1+\dfrac{s}{\omega_F}\right)}{(1+A_{F0})\left[1+\dfrac{s}{1+A_{F0}}\left(\dfrac{1}{\omega_p}+\dfrac{1}{\omega_F}\right)+\dfrac{s^2}{(1+A_{F0})\omega_p\omega_F}\right]} \tag{5.67}$$

where $C_{out} = C_{bc1} + C_{cs1} + C_{inTIA}$ is the capacitance at the collector of Q_1, including the input capacitance of the TIA stage, and $\omega_p = 1/(R_FC_{out})$.

The voltage gain of the Cherry–Hooper stage can now be expressed as a second-order function of the complex frequency s

$$A(s) = -g_{meff}Z_L(s)A_F(s) \approx \frac{-g_{meff}R_F}{\left[1+\dfrac{s}{1+A_{F0}}\left(\dfrac{1}{\omega_p}+\dfrac{1}{\omega_F}\right)+\dfrac{s^2}{(1+A_{F0})\omega_p\omega_F}\right]} = \frac{A}{1+\dfrac{s}{Q\omega_0}+\dfrac{s^2}{\omega_0^2}} \tag{5.68}$$

where $\omega_0 = \sqrt{(1+A_{F0})\omega_p\omega_F}$ and $Q = \dfrac{\omega_0}{\omega_p+\omega_F} = \dfrac{\sqrt{(1+A_{F0})\omega_p\omega_F}}{\omega_p+\omega_F}$.

It can be shown that, if Q in (5.68) is set to 0.707, the bandwidth, ω_0, is improved by a factor of approximately $\sqrt{|A|}$ compared to ω_p [9], and the gain response versus frequency is flat, without exhibiting any peaking.

The main disadvantage of the Cherry–Hooper topology is its increased power consumption since it features a cascade of two stages. Moreover, because at least 3 transistors must be stacked

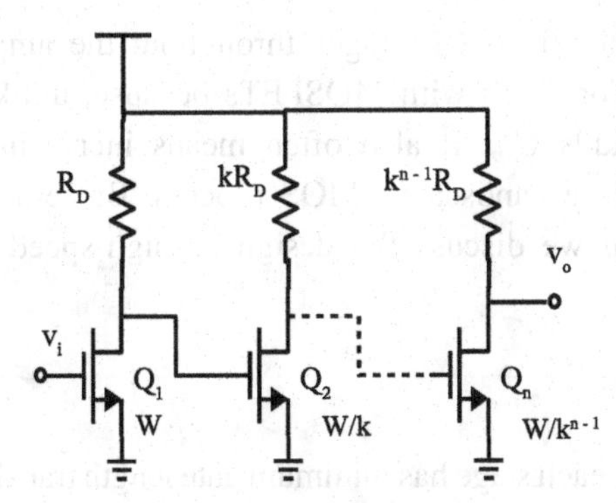

Figure 5.30 A multi-stage amplifier consisting of a chain of gain stages whose transistor size and bias current decrease by a factor k from input to output. The voltage gain of each stage is the same

vertically between the supply and ground in differential versions of the circuit, the supply voltage is also rather high. Bipolar implementations are most common, but they require relatively high (3.3V and up) power supplies. It is possible to reduce the power supply voltage to 1.8V by employing MOS-HBT topologies in a BiCMOS process. In nanoscale CMOS technologies, supplies as low as 1.5V are feasible if low-VT MOSFETs are employed, see Problem 5.7.

5.4.3 Buffering and size scaling

The capacitance at the output node of the gain stage can be reduced by decreasing the size and bias current in each gain stage, starting from the input towards the output. Although the topology of each stage remains the same, the stages are no longer identical in component size and bias current. Such an approach can be employed in the input amplifier of a broadband receiver, where the large transistor and large bias current of the input stage help to reduce the receiver noise figure and make the circuit less sensitive to noise impedance mismatch. To understand why the bandwidth improves, let us rewrite the expression of the dominant time constant (5.46) for a chain, as in Figure 5.30, where all transistor sizes and bias currents are reduced by a factor $k > 1$ from stage to stage, and the load resistance R_D is increased by the same factor k from stage to stage, such that the voltage gain per stage remains unchanged

$$\tau_p = (R_D || r_o)\left[C_{gd} + C_{db} + \frac{C_{gs}}{k} + (1 + |A_0|)\frac{C_{gd}}{k}\right] + R_g\left[C_{gs} + (1 + |A_0|)C_{gd}\right]. \quad (5.69)$$

We note that the term in the first square bracket has been reduced compared to (5.46), leading to increased bandwidth.

Ultimately, the size scaling is limited by the minimum transistor size and by the input capacitance of the first stage which, if too large, along with the 50Ω signal source impedance, will reduce the bandwidth. This technique is particularly popular in CMOS amplifiers. Often, the problem of the large input capacitance is addressed by employing a large-swing, high-bandwidth preamplifier realized in SiGe BiCMOS or III-V technologies to drive the CMOS chip [12].

Alternatively, the topology of the individual stage can be changed from common-emitter (common-source) to emitter-follower + common-emitter. This is known as *buffering and scaling* because the EF acts as a buffer between two CE stages and its size and input capacitance is scaled down compared to the CE stage. By making the EF size and current smaller than that of the CE stage, the capacitance seen at the output node of the CE stage is reduced while still having identical composite EF + CE stages throughout the amplifier chain [9]. This alternate approach does not work well with MOSFETs because, unlike the emitter-follower, the source-follower stage adds C_{sb}. It also often means increasing the supply voltage above that typically allowed in a nanoscale CMOS process. Buffering and scaling will be revisited in Chapter 11 when we discuss the design of high-speed logic circuits.

EXAMPLE 5.6 65nm CMOS chain scaling

Consider a 6-stage CMOS amplifier chain where each stage has minimum gate length transistors, with $W = 20\mu m$ and 20 unit fingers, each 1µm wide and contacted on both sides of the gate. $R_D = 200\Omega$. Each stage is biased at 0.15mA/µm and the corresponding small signal parameters of the MOSFET are $R_g = 10\Omega$, $(200\Omega/N_f)$ $C'_{gs} = 0.7$fF/µm, $C'_{sb} = C'_{db} = 0.7$fF/µm, $C'_{gd} = 0.4$fF/µm, $g'_{meff} = 0.9$mS/µm, $g'_{oeff} = 0.18$mS/µm. (a) Calculate the DC gain, the 3dB bandwidth and the gain-bandwidth product of the individual stage and of the amplifier chain. (b) Re-design the amplifier chain using size and bias scaling by a factor of 2 from the input to output, knowing that the first stage has $R_D = 62.5\Omega$ and $W = 64\mu m$. Assume the same bias current density, and small signal parameters per unit gate width as in the original amplifier.

Solution

(a) The low-frequency voltage gain per stage is

$$A_S = -g_{meff}(r_o||R_D) = -0.9e - 3 \times 20 \times 116 = -2.088$$

$$\tau_p = (R_D||r_o)[C_{db} + C_{gs} + (2 + |A_0|)C_{gd}] + R_g[C_{gs} + (1 + |A_0|)C_{gd}] = 7.55ps$$

$$BW_S = \frac{1}{2\pi\tau_p} = 21.1GHz$$

$GBW_S = 44.05$GHz.

For the chain, $A_{tot} = 2.088^6 = 82.86 = 38.36$dB, $BW_{tot} = \left(\sqrt{2^{1/6} - 1}\right)21.1e9 = 7.38$GHz and $GBW_{tot} = 632.9$GHz.

An exact computer simulation using the design kit models gives $A_S = 6.4$dB (2.089), $BW_s = 18.96$GHz, and $A_{tot} = 38.37$dB, $BW_{tot} = 6.2$GHz. The error between the hand calculations and exact simulation is 10% for the stage and about 16% for the entire chain.

(b) If we use scaling by $k = 2$: $W_1 = 64\mu m$, $R_{D1} = 62.5\Omega$, $W_2 = 32\mu m$, $R_{D2} = 125\Omega, \ldots,$ $W_6 = 2\mu m$, $R_{D6} = 2$ kΩ.

The gain per stage remains unchanged, but the time constant and bandwidth of the stage change to

$$\tau_p = (R_D||r_o)\left[C_{gd} + C_{db} + \frac{C_{gs}}{k} + (1 + |A_0|)\frac{C_{gd}}{k}\right] + R_g\left[C_{gs} + (1 + |A_0|)C_{gd}\right]$$
$$= 36.25[25.6f + 44.8f + 22.4f + 3.1 \times 12.8f] + 3.125[44.8f + 3.1 \times 25.6f]$$
$$= 4.9ps + 0.388ps = 5.288ps$$

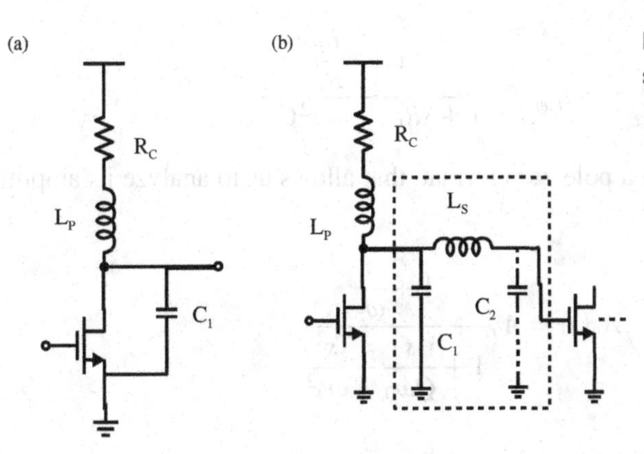

Figure 5.31 (a) Shunt peaking and (b) shunt–series peaking concepts

We obtain $BW'_S = 30.1\text{GHz}$, almost a 50% increase in stage bandwidth compared to the original amplifier, and $BW'_{tot} = \left(\sqrt{2^{1/6}-1}\right) 30.1e9 = 10.53\,\text{GHz}$.

Using computer simulation, we obtain $BW'_S = 28.17\text{GHz}$, $BW'_{tot} = 8.73\text{GHz}$, $GBW'_S = 59.15\text{GHz}$, $GBW'_{tot} = 748.7\text{GHz}$.

Again, the error between hand calculations and computer simulations using design kit models is within 10% for the single stage but increases to 17.1% for the entire chain.

Note that, if we consider interconnect capacitance and the parasitic capacitance of the load resistor, the calculated bandwidth will be smaller because the interconnect capacitance is comparable to that of the smallest size stage.

5.4.4 Inductive peaking and distributed topologies

The idea of using an inductive element to tune out the node capacitance in a circuit to increase its bandwidth dates back to the 1930s. By placing an inductor or short transmission line in series with the load resistance at the offending node, a zero is introduced that partly compensates the original pole. The inductor also introduces a second pole, resulting in a second-order voltage gain response function. The technique can be applied to both the input (as in the impedance matching section) and the output node of the amplifier stage.

In its simplest implementation, when only one inductive element is used, as in Figure 5.31(a), the method is known as *shunt peaking*. To understand how it works, let us first derive the expression of the voltage gain of the circuit in Figure 5.31(a) as a function of frequency, ignoring the Miller effect

$$A_V(s) = -g_{meff} Z_L(s) \tag{5.70}$$

where

$$Z_L(s) = \frac{R_C + sL_P}{1 + sR_C C_1 + s^2 C_1 L_P} \tag{5.71}$$

is the frequency-dependent load impedance at the collector of Q_1. By inserting (5.71) into (5.70), we obtain

$$A_v(s) = -g_{meff}R_C \frac{1 + s\dfrac{L_P}{R_C}}{1 + sR_C C_1 + s^2 C_1 L_P}. \tag{5.72}$$

Equation (5.72) can be recast in a pole-zero format, that allows us to analyze its amplitude and phase response

$$A_v(s) = A_0 \frac{1 + \dfrac{s}{\omega_z}}{1 + \dfrac{s}{Q\omega_0} + \dfrac{s^2}{{\omega_0}^2}} \tag{5.73}$$

where

$$A_0 = -g_{meff}R_C; \omega_0 = \frac{1}{\sqrt{L_P C_1}}; \omega_z = \frac{R_C}{L_P} = \frac{\omega_0}{Q} \text{ and } Q = \frac{1}{R_C}\sqrt{\frac{L_P}{C_1}}. \tag{5.74}$$

Unlike (5.68) which has a Butterworth response, (5.73) also features a zero. The presence of the zero implies that lower pole-Q values are needed for a flat gain response. The following design equation can be used to find the value of the shunt peaking inductor L_P for a given load resistor, node capacitance, and pole Q

$$L_P = Q^2 {R_C}^2 C_1. \tag{5.75}$$

If in (5.73) $Q = 0.63$, a maximally flat gain response is obtained which imposes that

$$L_P = 0.4{R_C}^2 C_1 \tag{5.76}$$

and results in a 70% increase in bandwidth compared to the case where $L = 0$.

However, in high-data-rate fiber-optic receivers and in radio transceivers with phase or QAM modulation, a maximally flat group delay response is desired, and an even smaller pole-Q is needed, with the optimal shunt peaking inductor value being given by [4]

$$L_P = 0.32{R_C}^2 C_1 = \frac{{R_C}^2 C_1}{3.1}. \tag{5.77}$$

In this case, the bandwidth improvement is 60%.

EXAMPLE 5.7 65nm n-MOS common-source stage with shunt peaking

For the circuit in Example 5.6, find the value of the peaking inductor and bandwidth extension if a maximally flat gain response is desired.

Solution

The load resistance R_D is 200Ω and the total capacitance at the output node is

$$C = \left[C_{db} + C_{gs} + (2 + |A_0|)C_{gd}\right] = 14\text{fF} + 14\text{fF} + 32.8\text{fF} = 60.8\text{fF}.$$

The peaking inductance value for flat gain response is obtained from (5.76)

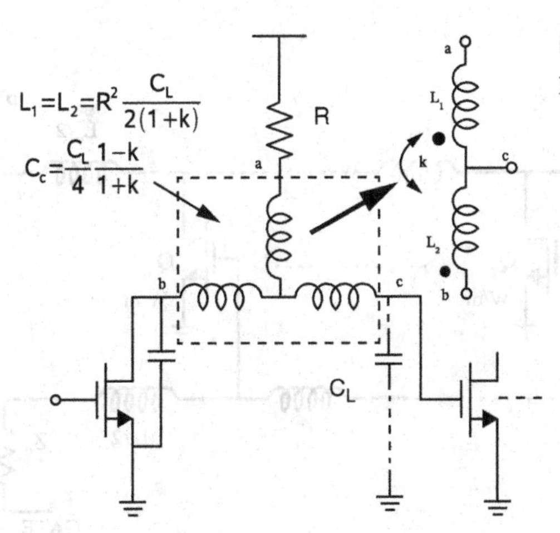

Figure 5.32 Bandwidth extension using t-coils. C_c is placed between nodes a and b [4]

$$L_P = 0.4R_D{}^2C = 0.97nH$$

which results in a bandwidth of $1.7 \times 21.1\text{GHz} = 35.87\text{GHz}$.

Computer simulation using the design kit and BSIM4 MOSFET models shows that $A'_S = 6.4\text{dB}$ (2.089), $BW'_S = 32.96\text{GHz}$, $A'_{tot} = 38.38\text{dB}$ (83), $BW'_{tot} = 15.5\text{GHz}$, $GBW'_S = 68.85\text{GHz}$, $GBW'_{tot} = 1.286\text{THz}$.

We note that, compared to the case without peaking, the simulated bandwidth has improved by a factor of 32.96/18.96 = 1.74, very close to the theoretical value. This is a significantly larger GBW than that obtained in the case where only scaling, without inductive peaking is applied. The two amplifier chains consume approximately the same current 18mA (inductive peaking), and 19mA (scaling), respectively, from a 1.2V supply. A disadvantage of the stage with inductive peaking is that it requires a larger die area because of the area occupied by the peaking inductor. However, in a typical multi-metal CMOS back-end, the size of a 600pH inductor can be made as small as 10μm × 10μm.

More complicated inductive peaking schemes, with increasing number of inductors, can lead to even larger bandwidth extension:

shunt–series peaking, employing two inductors, Figure 5.31b, results in a doubling of the bandwidth compared to that of the original amplifier stage without inductive elements;

shunt and double-series peaking using a t-coil as in Figure 5.32, producing a 2.8x increase in bandwidth;

distributed amplifier (DA) topology, leading to the largest possible bandwidth for a given transistor technology.

The benefits of the distributed amplifier topology, invented by Parcival in 1936 [13], can be explained by first considering the expression of the characteristic impedance and 3dB bandwidth of a transmission line with line impedance L and line capacitance C

$$Z_0 = \sqrt{\frac{L}{C}} \text{ and } BW_{3\text{dB}} = \frac{1}{\pi\sqrt{LC}}. \quad (5.78)$$

If the original lumped amplifier stage with load resistance R_D, input capacitance C_I, and output capacitance C_O, is split into m identical stages, each with an input capacitance C_I/m and an

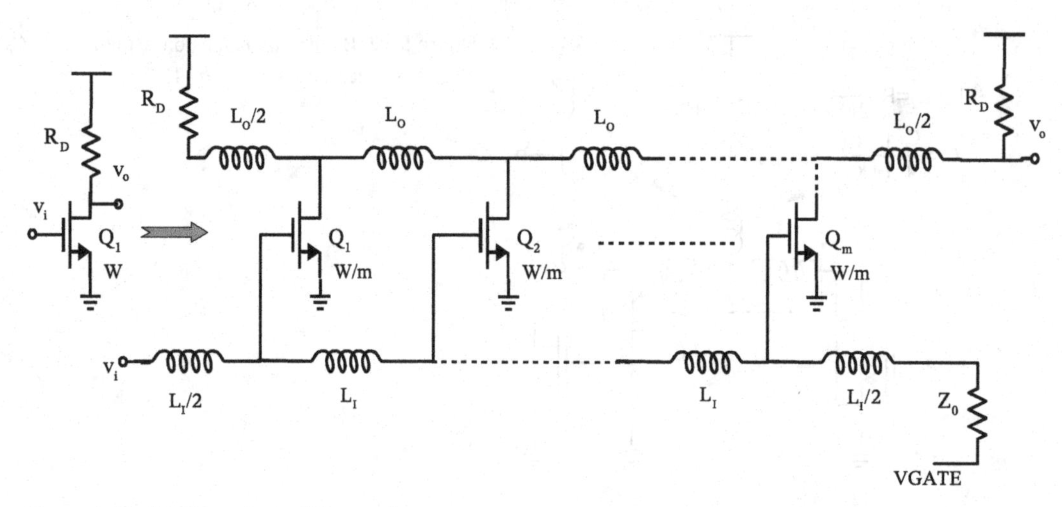

Figure 5.33 Distributed amplifier topology

output capacitance C_O/m, the overall DC gain remains the same. If we now connect the inputs to a distributed input transmission line of inductance L_I, and the outputs to a distributed output transmission line of inductance L_O, such that

$$L_I = Z_0^2 \frac{C_I}{m} \text{ and } L_O = R_D^2 \frac{C_O}{m} \tag{5.79}$$

the input line will have a characteristic impedance Z_0 and must be terminated on Z_0, while the output line will have a characteristic impedance equal to R_D and must be terminated on R_D at both ends. The resulting distributed amplifier schematic is shown in Figure 5.33. We note that, at DC, all inductors behave as short circuits and that the voltage gain is reduced in half compared to that of the lumped stage because of the presence of the second load resistance, R_D, needed to match the output transmission line

$$A = -m \frac{g_{meff}}{m} \frac{R_D}{2} = -g_{meff} \frac{R_D}{2}. \tag{5.80}$$

The bandwidth is given by the smaller of the 3dB bandwidths of the input and output transmission lines

$$BW_{3\text{dB}} = \min\left(\frac{1}{\pi \sqrt{L_I \frac{C_I}{m}}}, \frac{1}{\pi \sqrt{L_O \frac{C_O}{m}}} \right) \tag{5.81}$$

Usually, $C_i = C_O$ (if not, extra capacitance is added in parallel with the smaller of the two) to ensure that the output and input transmission lines have identical delays and cutoff frequencies. If we set $Z_O = R_D$, then

$$BW_{3\text{dB}} = \min\left(\frac{m}{\pi R_D C_I}, \frac{m}{\pi R_D C_O} \right) \tag{5.82}$$

which, in theory, can be made infinite if $m \to \infty$.

Since the GBW of the original lumped stage is

$$GBW_S = \frac{g_{meff}}{2\pi(C_I + C_O)} \tag{5.83}$$

the gain-bandwidth product of the distributed amplifier

$$GBW'_S = \min\left(m\frac{g_{meff}}{2\pi C_I}, m\frac{g_{meff}}{2\pi C_O}\right) \tag{5.84}$$

becomes at least 2 × m larger, because the node capacitance has been split in two (C_I and C_O) and both C_I and C_O have been reduced m-times compared to those of the original lumped stage.

Ultimately, the GBW of the distributed amplifier is limited by the f_{MAX} of the transistor. The parasitic base/gate resistance and output resistance of the transistor, the loss of the transmission lines, all ignored in the previous analysis, contribute increasing loss to the input and output t-lines as the number of sections in the DA increases. Therefore, in practice the number of sections is between 4 and 7.

The gain cell in the distributed amplifier can be a transistor (usually a FET due to its high-Q input impedance), a cascode stage, other more advanced amplifier stages such as EF + cascode, and even a multi-stage lumped amplifier [14]. In each situation, the bandwidth of the distributed amplifier far exceeds that of the unit gain cell. In the case of HBT amplifiers, the gain cell usually employs an emitter-follower at its input in order to increase the input impedance and thus minimize the loss along the input transmission line. In SiGe BiCMOS technologies, the MOS-HBT cascode provides both high-Q input and output impedances, large output voltage swing, and large gain cell bandwidth [15].

It should be noted that, although most of these topologies have "peaking" in their name, the design is conducted in such a manner that peaking is avoided in both the amplitude and in the group delay response of the transfer characteristics. When in doubt, it is always wise to design on the side of caution, to avoid peaking in all process corners, by choosing smaller values for the peaking inductors.

EXAMPLE 5.8 65nm CMOS distributed amplifier

Consider a lumped 1.2V CS stage with resistive load where $W = 180\mu m$ and $R_D = 25\Omega$. Find the bandwidth extension if a distributed topology is used.

Solution

(a) For the lumped stage, biasing each transistor at 0.15mA/μm, $g_{meff} = 162mS$, $r_{oeff} = 30.86\ \Omega$, $A_S = -0.162 \times 13.81 = -2.23$.

We assume that this stage is in a chain of identical stages

$$\tau_P = (R_D||r_o)[C_{db} + C_{gs} + (2 + |A_O|)C_{gd}] + R_g[C_{gs} + (1 + |A_O|)C_{gd}]$$
$$= (25||30.86) \times 556.56f + 1.11 \times 358.6f = 8.073ps$$

$$BW_S = \frac{1}{2\pi\tau_p} = 19.72GHz.$$

(b) In the distributed amplifier case: $W = 36$ um, 5 stages. Again, each transistor is biased at 0.15mA/μm. The total current is 26mA, $W_{total} = 180\mu m$. $V_{DD} = 1.2V$.

The calculated input capacitance per stage is (3.23×0.4f + 0.7f) × 36 = 71.76fF.

To match the input line to $Z_0 = 50\Omega$, $L_{gate} = 179$pH (from $L = CZ_0^2$) and results in a line cutoff frequency of 88.76GHz. To match the output line impedance and delay to that of the input, we add 0.55fF×36 = 19.8fF at the drain of each stage and select the same inductance value $L_D = 179$pH. We obtain in simulation $A_S = 6.9$–7.3dB (ripple), $BW_S = 73$GHz, 3.7 times larger than that of the lumped stage.

5.5 CHALLENGES IN DIFFERENTIAL CIRCUITS AT HIGH FREQUENCY

The importance of using differential topologies in integrated circuits at microwave and mm-wave frequencies cannot be overstated. In addition to the well-known benefit of providing immunity to common-mode noise and crosstalk, differential topologies alleviate the negative impact of the bonding wire inductance, or of the flip-chip bump inductance, on the gain, output power, and stability of amplifiers. High-frequency single-ended circuits are particularly vulnerable to ground impedance. The latter increases with frequency and introduces undesired feedback within and between circuit blocks. On the contrary, as we will see, in differential topologies, if judiciously placed at strategic nodes, ground and supply inductance can improve common-mode rejection at high frequencies. Nevertheless, some of the traditional benefits of differential circuits are diminished at high frequencies. These include:

- little or no common-mode rejection due to the capacitive nature of current sources, especially the "advanced" ones;
- capacitive degeneration of differential pairs in common-mode, potentially leading to negative common-mode resistance and even oscillations;
- increased coupling between the differential inputs due to the capacitive nature of the input impedance at high frequency, causing the differential input impedance to no longer equal 2 times the single-ended impedance, as at low frequencies, and making it difficult to simultaneously match the single-ended, differential-, and common-mode impedances of the differential pair.

5.5.1 Stability of differential circuits at high frequencies

In the absence of a generalized stability theory in circuits with more than two-ports, we can ensure that single-stage differential circuits are stable by:

- verifying that each of the 2 × 2 differential-mode, common-mode, and mixed-mode sub-matrices are stable;
- checking that $|S_{ii}| < 1$ at any port i;
- avoiding peaking larger than 1dB in the common-mode, single-ended, and differential-mode gain versus frequency characteristics;
- checking that the closed-loop phase margin of amplifiers with feedback is larger than 60° in all modes of operation: differential, common-mode, and single-ended.

As in single-ended amplifiers, there is no guarantee that a multi-stage differential amplifier is stable if its 2×2 mixed-mode submatrices are stable. As a precaution, in multi-stage amplifiers one must analyze the stability of each stage separately.

5.5.2 Common reasons for instability in differential circuits at high frequencies

Differential-mode and/or common-mode negative resistance can occur at high frequency at various nodes in a differential amplifier stage. Some of the most common cases are listed below:

- Capacitively loaded emitter/source nodes give rise to negative resistance at the base/gate nodes.
- In amplifier stages with inductive degeneration, the parasitic capacitance of the monolithic inductor causes its impedance to become capacitive beyond the self-resonance-frequency, SRF. As a result, the gain stage becomes capacitively degenerated at frequencies above the SRF and, if the transistor still has gain in that frequency range, the real part of the impedance looking into its base/gate becomes negative.
- Current sources (tails) employed to bias a differential pair become capacitive at high frequency, causing negative resistance to appear at the input of the differential pair in common-mode.
- A bipolar or FET cascode stage can become unstable if a non-zero parasitic layout inductance appears at the base/gate of the common-base/gate transistor.
- Inductive power supply lines, if not properly decoupled to ground locally in the amplifier stage, can cause negative resistance to arise at the common-mode input of a differential pair.

In all the situations above, adding a resistive element in series or in parallel with the offending parasitic inductor or capacitor reduces its Q and the associated negative resistance. The penalty is often a somewhat degraded amplifier gain and noise figure in LNAs, increased DC voltage drop on resistive parasitics and lower power-added efficiency in power amplifiers. However, these are usually less catastrophic than having an oscillating amplifier.

5.5.3 Single-ended, differential-mode, and common-mode input impedances of a differential pair

As at low frequencies, the differential-mode input impedance of the circuit in Figure 5.34(a) is two times that of the differential-mode half circuit, Figure 5.34(b). The latter is simply a common-source stage with inductive load. Using the simplified high-frequency FET equivalent circuit in Figure 5.22, where the Miller capacitance is ignored, we obtain the following expression

$$Z_{indiff} = 2(R_g + R_s) + \frac{2\omega_{Teff}}{j\omega g_{meff}}. \tag{5.85}$$

It is apparent from (5.85) that the real part of the differential-mode impedance is very small, due solely to the parasitic source/emitter and gate/base resistances of the transistor. However, especially for common-source stages with high-Q inductive loads, the Miller capacitance cannot be ignored. As discussed in Chapter 6.6, at high frequencies, the differential input impedance becomes a function of the load inductor causing its real part to become negative, potentially leading to oscillations in differential-mode.

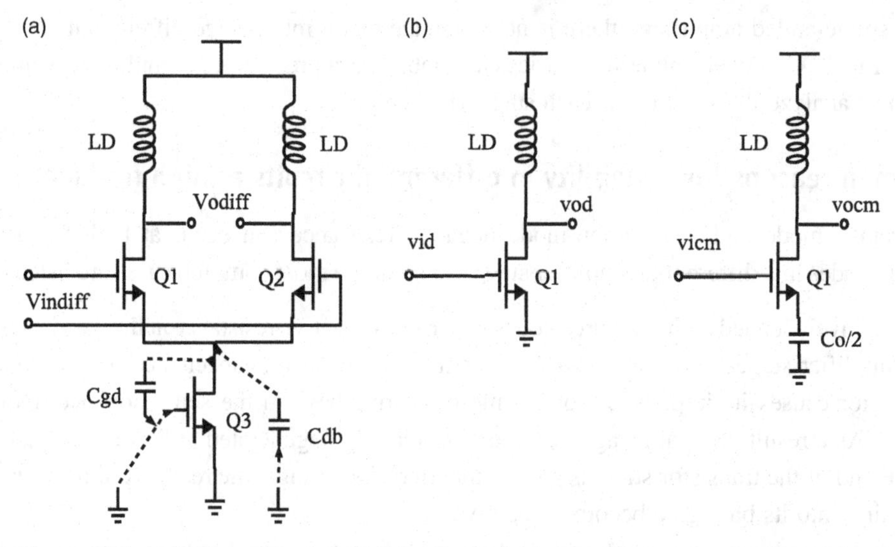

Figure 5.34 (a) Differential pair with inductive loads showing the common mode parasitic capacitance, (b) differential-mode half circuit, and (c) common-mode half circuit

Similarly, the common-mode impedance of the differential amplifier in Figure 5.34(a) is half that of its associated common-mode half circuit, shown in Figure 5.34(c). We can lump together C_{db3}, C_{gd3}, C_{sb1}, and C_{sb2} into a single capacitor, C_o, which represents the total capacitance at the common-source/emitter node of the differential pair. To first order, C_o remains largely the same if a conventional or a cascode current source is employed. The output impedance of the current source can now be expressed as

$$z_{OCS} = \frac{r_o}{1 + j\omega r_o C_o}. \tag{5.86}$$

It becomes capacitive beyond the zero frequency, f_z, of the common-mode gain response. For example, in a 65nm CMOS technology

$$f_z = \frac{1}{2\pi r_o C_o} \approx \frac{1}{2\pi \frac{1.8\text{fF}/\mu\text{m}}{0.2\text{mS}/\mu\text{m}}} = \frac{100\text{GHz}}{1.8 \times 3.14} = 17.7\text{GHz}.$$

For a bipolar differential pair, or for a differential pair with cascode current source, f_z can be five to 50 times lower than that of a general purpose, minimum gate length MOSFET current source. It is also important to note that the frequency of the common-mode zero is invariant to the transistor size (gate width or emitter length) and bias current, as long as the current source operates at the same current density (or $V_{GS} - V_T$).

Two important problems plague the common-mode half circuit due to the capacitance of the current source:

(a) decreasing common-mode rejection as the operation frequency increases beyond f_z and
(b) negative input resistance in common-mode which can lead to instability and oscillation.

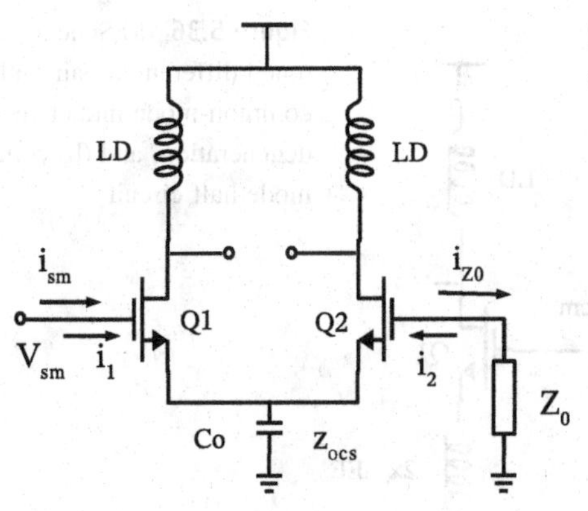

Figure 5.35 Equivalent circuit for calculating the single-ended input impedance of the differential pair

Using the simplified current gain-based HF equivalent circuit of the MOSFET and ignoring the output resistance of the current source, r_o (an acceptable approximation at $f > f_z$) we can derive the expression of the common-mode input impedance from the common-mode half circuit in Figure 5.34.c:

$$Z_{incm} = \frac{v_{icm}}{2i_{icm}} = \frac{R_g + R_s}{2} + \frac{1 + \frac{\omega_{Teff}}{j\omega}}{2j\omega\frac{C_o}{2}} + \frac{\omega_{Teff}}{2j\omega g_{meff}} = \frac{R_g + R_s}{2} - \frac{\omega_{Teff}}{\omega^2 C_o} + \frac{1}{j\omega C_o} + \frac{\omega_{Teff}}{2j\omega g_{meff}} \tag{5.87}$$

which clearly shows the onset of negative resistance if $\frac{\omega_{Teff}}{\omega^2 C_o} > \frac{R_g + R_s}{2}$.

The resistance of the current source, neglected in the derivation above, actually will help cancel a portion of the negative resistance. From this point of view, a bipolar or a cascode current source is not as effective as a simple MOSFET current source because of their larger output resistance.

The single-ended input impedance of the differential stage can be calculated using the equivalent circuit in Figure 5.35, and connecting one of the inputs through a resistor Z_0 to ground (or directly to ground). The current through Z_0 is obtained as

$$i_{zo} = -i_2 \frac{\left(R_s + R_g + Z_0 + \frac{\omega_{Teff}}{j\omega g_{meff}}\right)}{z_{ocs}} = i_{ism}\left(1 + \frac{\omega_{Teff}}{j\omega}\right) + i_2\left(1 + \frac{\omega_{Teff}}{j\omega}\right) \tag{5.88}$$

$$i_2 = -i_{ism} \frac{\left(1 + \frac{\omega_{Teff}}{j\omega}\right)}{1 + \frac{R_s + R_g + Z_0}{z_{ocs}} + \frac{\omega_{Teff}}{j\omega}\left(1 + \frac{1}{g_{meff} z_{ocs}}\right)} \tag{5.89}$$

$$V_{ism} = \left(R_g + R_s + \frac{\omega_{Teff}}{j\omega g_{meff}}\right) i_{ism} + \frac{\left(R_s + R_g + Z_0 + \frac{\omega_{Teff}}{j\omega g_{meff}}\right)\left(1 + \frac{\omega_{Teff}}{j\omega}\right)}{1 + \frac{R_s + R_g + Z_0}{z_{ocs}} + \frac{\omega_{Teff}}{j\omega}\left(1 + \frac{1}{g_{meff} z_{ocs}}\right)} i_{ism} \tag{5.90}$$

From this, we obtain the single-ended mode impedance when the other input is terminated on Z_0 as

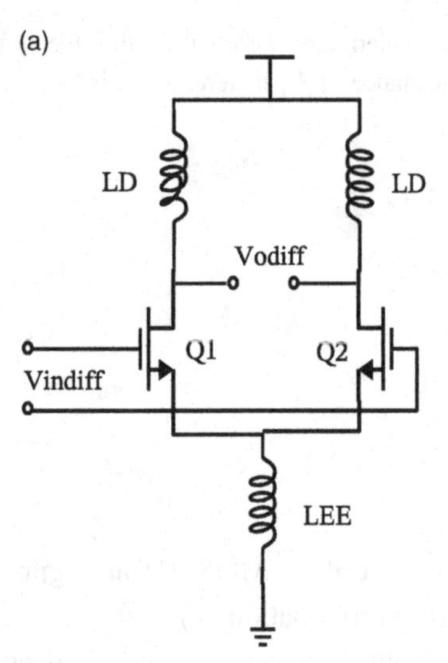

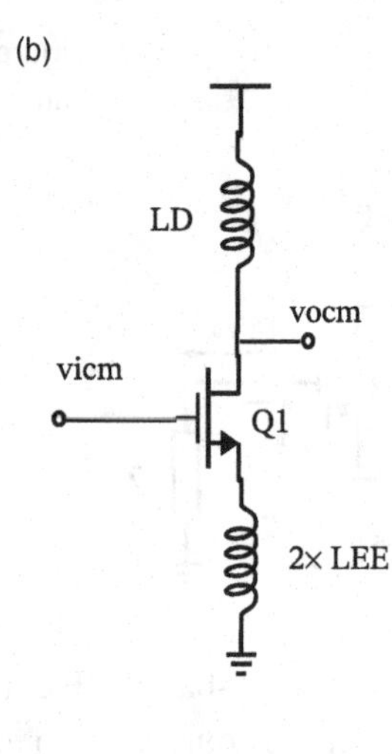

Figure 5.36 (a) Schematics of a tuned differential pair with common-mode inductive degeneration, and (b) common-mode half circuit

$$Z_{ism} = \frac{V_{ism}}{i_{ism}} = \left(R_g + R_s + \frac{\omega_{Teff}}{j\omega g_{meff}}\right) + \frac{\left(R_s + R_g + Z_0 + \frac{\omega_{Teff}}{j\omega g_{meff}}\right)\left(1 + \frac{\omega_{Teff}}{j\omega}\right)}{1 + \frac{R_s + R_g + Z_0}{z_{ocs}} + \frac{\omega_{Teff}}{j\omega}\left(1 + \frac{1}{g_{meff} z_{ocs}}\right)} \tag{5.91}$$

If z_{ocs} is very large and $Z_0 = 0$, then the single-ended mode impedance simplifies to (5.85), the differential-mode input impedance.

If z_{ocs} is vanishingly small (valid at large frequencies), the single-ended mode impedance reduces to

$$Z_{ism} = \left(R_g + R_s + \frac{\omega_{Teff}}{j\omega g_{meff}}\right) \tag{5.92}$$

which is identical to the input impedance of the differential-mode half circuit.

For finite z_{ocs}, the single-ended mode impedance varies between these two extreme cases.

5.5.4 Techniques to improve the common-mode rejection and the single-ended-to-differential conversion at high frequencies

Both of these problems can be alleviated by increasing the common-mode degeneration impedance at high frequencies. The solutions include:

- resistor, instead of current source, placed at the common-emitter/source node,
- resistive added in series with the current source at the common-emitter/source node,
- inductor placed at the common-emitter/source node, as in Figure 5.36,
- a series R-L circuit placed at the common-emitter/source node, and
- a resistor in series with a parallel RLC resonant circuit at the common-emitter/source node.

Each of these solutions has its drawbacks. The inductive degeneration is helpful at high frequencies and the parallel RLC resonant circuit solution is effective in tuned amplifiers over a narrow band. However, neither is useful in a broadband DC-to-HF amplifier where good

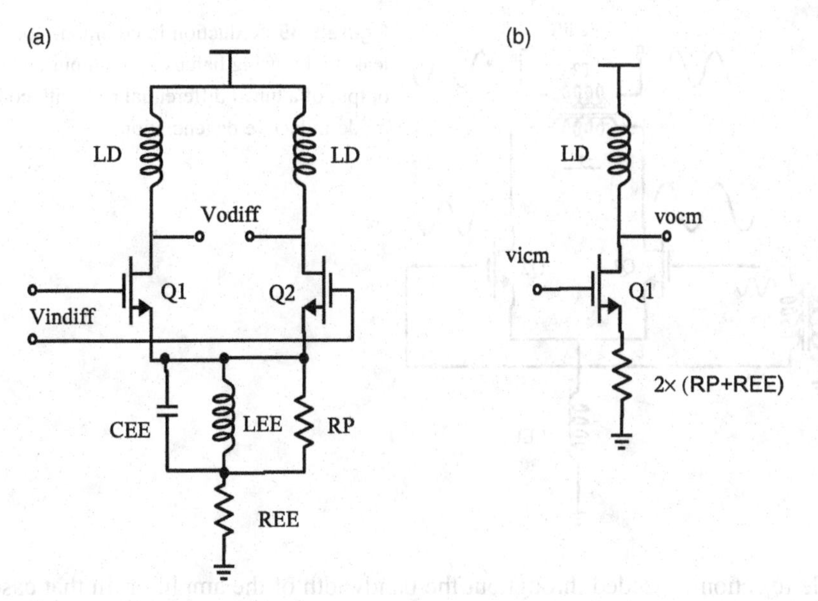

Figure 5.37 (a) Schematics of a tuned differential pair with common-mode resistive and resonant tank degeneration, and (b) common-mode half-circuit at resonance

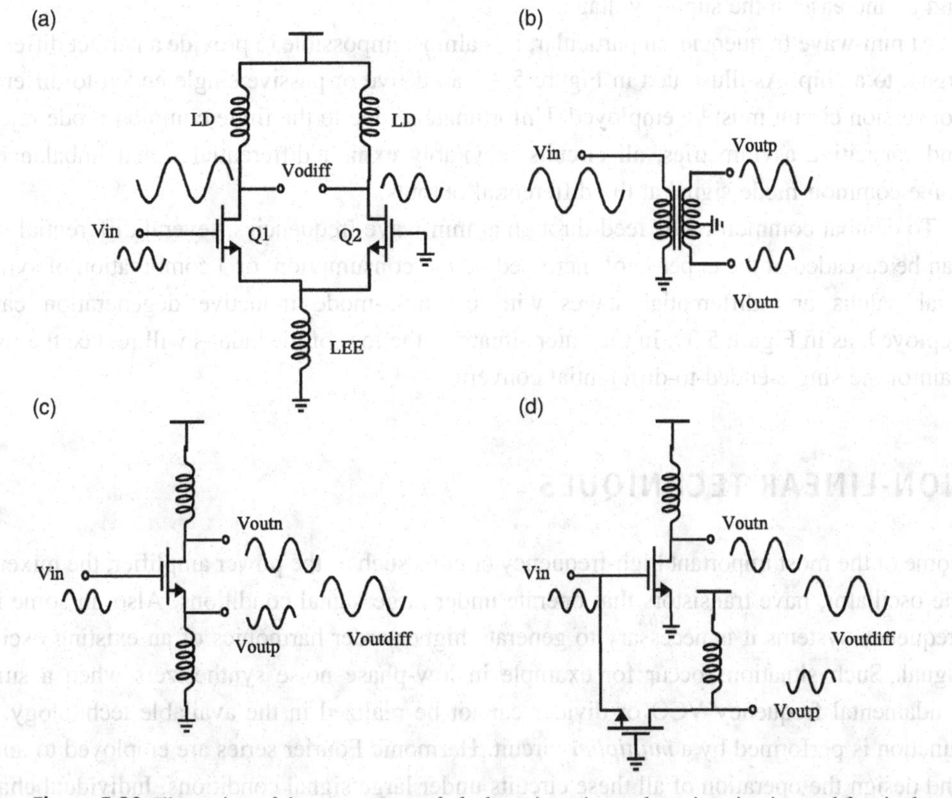

Figure 5.38 Illustration of the common-mode leakage in active and passive circuits used for single-ended to differential-mode conversion at high frequencies: (a) tuned differential pair with common-mode inductive degeneration, (b) balun, (c) common-source with emitter degeneration, and (d) common-source/common-gate stage

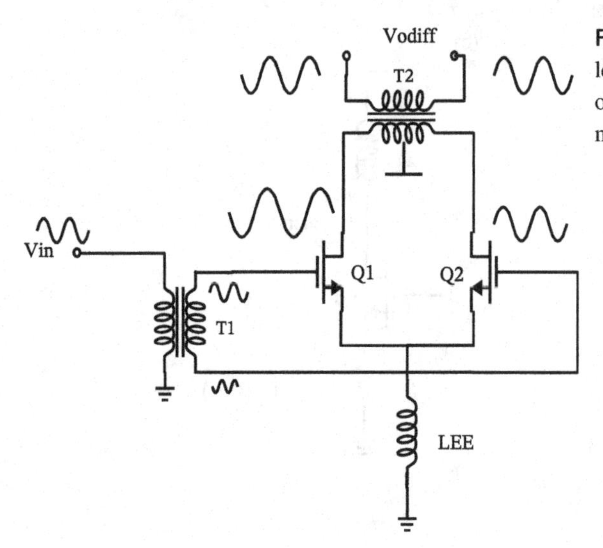

Figure 5.39 Reduction in common-mode leakage by using baluns at the input and at the output of a tuned differential pair with common-mode inductive degeneration

common-mode rejection is needed throughout the bandwidth of the amplifier. In that case, the resistor in series with the current source provides acceptable common-mode rejection without stability problems but it results in significant DC voltage drop across the degeneration resistor and an increase in the supply voltage.

At mm-wave frequencies in particular, it is almost impossible to provide a perfect differential signal to a chip. As illustrated in Figure 5.38, an active or passive single-ended-to-differential conversion circuit must be employed. Unfortunately, due to the finite common-mode rejection and capacitive asymmetries, all circuits inevitably exhibit differential signal imbalance and some common-mode signal at the differential outputs.

To combat common-mode feed-through at mm-wave frequencies, several differential stages can be cascaded at the expense of increased power consumption, or a combination of symmetrical baluns and differential stages with common-mode inductive degeneration can be deployed, as in Figure 5.39. In the latter situation, the loss of the baluns will reduce the overall gain of the single-ended-to-differential converter.

5.6 NON-LINEAR TECHNIQUES

Some of the most important high-frequency circuits such as the power amplifier, the mixer, and the oscillator, have transistors that operate under large signal conditions. Also, in some high-frequency systems it is necessary to generate higher-order harmonics of an existing oscillator signal. Such situations occur for example in low-phase noise synthesizers when a suitable fundamental frequency VCO or divider cannot be realized in the available technology. This function is performed by a *multiplier* circuit. Harmonic Fourier series are employed to analyze and design the operation of all these circuits under large signal conditions. Individual chapters are dedicated to power amplifiers, mixers, and oscillators. Frequency multipliers will be briefly discussed here.

High-frequency multipliers are realized by taking advantage of the non-linear characteristics of diodes or transistors, and typically produce integer harmonics of the input signal. The most common multiplier stages either double (doubler) or triple (tripler) the input frequency and often employ a differential pair driven into non-linear regime by a very large differential input signal. While second harmonic signals are relatively easy to generate using push–push oscillators, without the need for a separate doubler circuit, a tripler is particularly useful because it allows us to generate harmonics much beyond the f_{MAX} of a given transistor technology. By cascading several doubler and/or tripler stages, higher-order (by 4, 6, 8, 9, 18, etc.) multipliers can be realized [16]. The most important figures of merit for multipliers are the conversion gain (or loss), defined as the ratio of the output power at the nth harmonic and the input power at the fundamental, and the DC-to-RF efficiency.

The differential-stage multiplier can be understood by examining the circuits in Figures 5.40 and 5.41. Transistors Q_1 and Q_2 (which can be either HBTs or FETs) are biased at the onset of conduction, in class B [17]. If the input signal power is low, there is practically no current passing through Q_1 and Q_2. However, when a large signal is applied at the input, Q_1 and Q_2 will each alternately conduct current, during the positive and negative halves of the input excitation cycle, respectively. Therefore, the currents through Q_1 and Q_2 will resemble a train of rectified cosine pulses with a duty cycle of at most 50%. In this respect, a multiplier stage operates in much the same way as a class B push-pull power amplifier stage. The latter will be discussed in more detail in Chapter 6.

The even harmonics of the input signal appear at the common nodes of the tuned differential pair while only the odd harmonics show up at the differential outputs. Based on this observation, a doubler can be implemented by collecting the output signal at the common collector–drain node of the differential pair. To maximize the conversion gain of the multiplier, the common-mode output node is loaded with a parallel resonant tank tuned to the second harmonic of the input signal, as illustrated in Figure 5.40. All odd harmonics of the input signal and the fourth- and higher-order even harmonics see a short circuit at the output node and are, therefore, suppressed.

To further improve the conversion gain, second harmonic reflectors are placed at the input to prevent leakage of the output signal. These reflectors are typically realized with open-circuit/l/4 t-line stubs at the second harmonic, or with short circuited/l/4 t-line stubs at the fundamental frequency.

It is also possible to collect the output from the doubler differentially, if a common-mode inductor or transmission line is placed at the common-emitter node where the second harmonic of the input signal also appears. This arrangement allows for the second harmonic output signal to be collected differentially between the common-collector output and the common-emitter output. Ideally the two nodes are 180 degrees out of phase. In practice, because of the capacitive loading at the various internal nodes of the circuit, the phase difference will not be exactly 180°, and some tuning will be necessary.

Similarly, a tripler can be implemented by collecting the third harmonic currents differentially at the output of the differential pair, using a resonant load tuned at the third harmonic, as in Figure 5.41. A series resonant trap for the fundamental frequency is placed at the differential

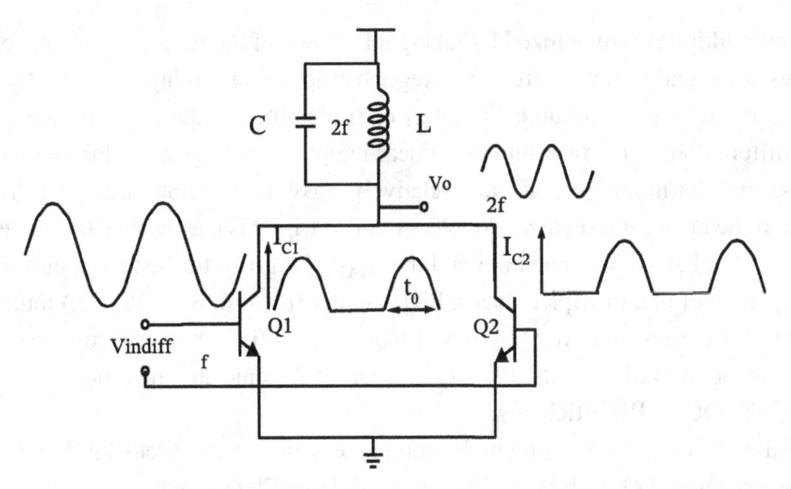

Figure 5.40 Conceptual schematic of a differential pair used as a frequency doubler

output to prevent the input signal from leaking at the output. A parallel resonant circuit centered on the second harmonic of the input signal is inserted at the common-emitter/source node.

Note that the common-emitter node presents an AC-ground for the fundamental and third harmonic signals but not for the second harmonic signal. By placing a high impedance at the second harmonic of the input signal at this node, the second harmonic current flowing through the differential pair is reduced and, therefore, it does not propagate to the output node.

To understand and optimize the design of a multiplier based on a bipolar or FET differential pair, we have to employ the Fourier series expansion of the collector–drain current through Q_1 (or Q_2) [17]

$$I_C(t) = I_0 + I_1\cos(\omega t) + I_2\cos(2\omega t) + \ldots + I_n\cos(n\omega t) \quad (5.93)$$

where the amplitude of the nth harmonic of the transistor output current is given by

$$I_n = I_{MAX}\frac{4t_0}{\pi T}\left|\frac{\cos\left(n\pi\frac{t_0}{T}\right)}{1-\left(2n\frac{t_0}{T}\right)^2}\right|. \quad (5.94)$$

Here, I_{MAX} represents the value at which the collector/drain current peaks, $t_0 < T/2$ describes the duration of each current pulse (or duty cycle), and T is the period of the fundamental signal at the input [17]. For MOSFET multipliers, I_{MAX} cannot exceed I_{ON}. In HBT multipliers, I_{MAX} should be kept smaller than 1.5–2J_{pfT} to avoid significant degradation of conversion gain and efficiency.

When $n = 0$, we obtain the expression of the rectified DC current

$$I_0 = I_{MAX}\frac{2t_0}{\pi T} \quad (5.95)$$

and, if $t_0/T = 1/(2n)$ and $n \neq 0$

$$I_n = I_{MAX}\frac{t_0}{T}. \quad (5.96)$$

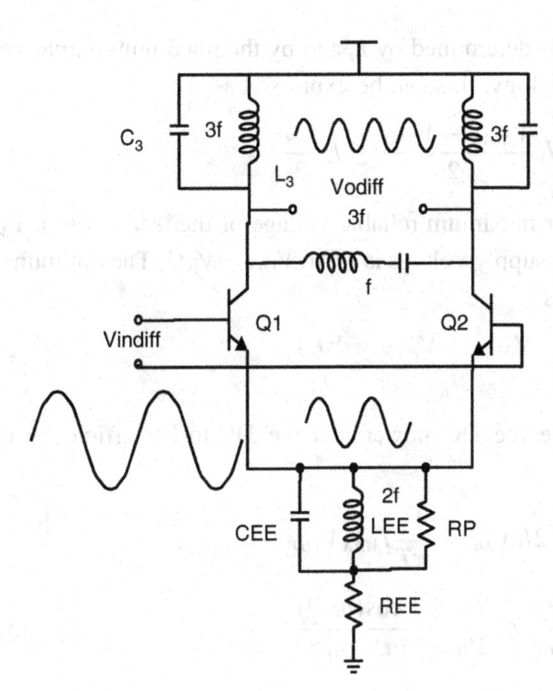

Figure 5.41 Conceptual schematic of a tripler circuit based on a bipolar differential pair with 3rd harmonic output tank and 2nd harmonic common-mode rejection

In the case of the tripler circuit in Figure 5.41, the output current passes through the resonant tank at the output and only the I_3 term is amplified and converted to a differential voltage at the output

$$V_{odiff}(3\omega) = -2R_{P3}I_{MAX}\frac{4t_0}{\pi T}\left|\frac{\cos\left(3\pi\frac{t_0}{T}\right)}{1-\left(6\frac{t_0}{T}\right)^2}\right|. \tag{5.97}$$

R_{P3} represents the output resistance at resonance at the third harmonic. A factor of 2 was added in (5.97) to account for the fact that the currents from both transistors of the differential pair add in the resonant load.

In the case of the doubler circuit in Figure 5.40, the common collector/drain output node provides a short circuit for the fundamental and third harmonic components of the currents through the differential pair, while the second harmonic currents add up and are converted to voltage

$$V_o(2\omega) = -2R_{P2}I_{MAX}\frac{4t_0}{\pi T}\left|\frac{\cos\left(2\pi\frac{t_0}{T}\right)}{1-\left(4\frac{t_0}{T}\right)^2}\right|. \tag{5.98}$$

Equations (5.94), (5.97), and (5.98) indicate that, in order for the multiplier to achieve maximum conversion gain and power efficiency, the conduction angle t_0/T must be separately optimized for each harmonic. In all cases where $n>1$, the optimal conduction angle is smaller than 80° [17].

The output power of the multiplier is determined by I_n and by the maximum output voltage swing. For a differential multiplier topology, this can be expressed as

$$P_{out} = I_n \frac{V_{MAX} - V_{MIN}}{2} \leq I_n \frac{V_B}{2} \tag{5.99}$$

where V_B is the breakdown voltage (or maximum reliable voltage of the technology), V_{MIN} is approximately V_{CEsat} or V_{DSAT}, and the supply voltage is set to $V_{DD} = V_B/2$. The optimum large signal load resistance per side becomes

$$R_{LOPT} = \frac{V_{MAX} - V_{MIN}}{2I_n} \approx \frac{V_{DD}}{I_n}. \tag{5.100}$$

Finally, using (5.95) we can calculate the DC power and the DC-to-RF efficiency of the differential multiplier as [17]

$$P_{DC} = 2I_0 V_{DD} = \frac{4t_0}{\pi T} I_{MAX} V_{DD} \tag{5.101}$$

$$\eta_{DC} = \frac{P_L}{P_{DC}} \leq \frac{I_n}{2I_0} = \left| \frac{\cos\left(n\pi \frac{t_0}{T}\right)}{1 - \left(2n \frac{t_0}{T}\right)^2} \right|. \tag{5.102}$$

Summary

In this chapter, we have reviewed the fundamental circuit concepts, topologies, and analysis techniques encountered in high-frequency and high-speed integrated circuits.

First, we discussed the concept of impedance matching and techniques to accomplish it using L-sections, transmission lines, and lossy networks.

Next, the concept of narrowband tuned amplifiers was introduced along with a simple method to analyze their gain by "downconverting" their "band-pass" response to a low-pass one centered at DC.

Thirdly, we reviewed techniques and circuit topologies commonly employed to maximize the bandwidth of multi-stage amplifiers.

We concluded the chapter with an overview of the challenges associated with the deployment of differential topologies in high-frequency circuits and of modern approaches to exploit the non-linearity of the differential pair to realize doublers and triplers. Harmonic Fourier series analysis was introduced as a powerful technique to analyze and design non-linear circuits and circuits operating under large signal conditions. This will be employed throughout the remainder of the textbook.

Problems

(1) The input impedance of a MOSFET can be approximated by a 15Ω resistance in series with a 30fF capacitance. Design the simplest L matching network that will match the MOSFET to 50Ω at 60GHz.

(2) The optimal noise impedance of a MOSFET at 60GHz is $Z_{SOPT} = 75\Omega + j225\Omega$. Design a transformer matching network to 50Ω.

(3) Match the output impedance of a MOSFET, consisting of a shunt R-C network (200Ω, 15fF), to 100Ω at 90GHz using a two-step *L*-section.

(4) The input impedance of a 0.13μ × 2μm SiGe HBT is 40Ω-*j*10Ω at 150GHz. Design a matching network to 50Ω at 150GHz. What is the bandwidth over which S_{11} is better than −10dB? Hint: the input impedance of the HBT can be described as a series R-C circuit.

(5) What is the bandwidth improvement if a CE stage with gain *A* is replaced by two cascaded identical stages, each with a gain of $\sqrt{|A|}$? Assume a single-pole response for the CE stage.

(6) The circuit in Figure 5.42 is used as a broadband LO signal distribution amplifier in a 10Gbs fiber-optics transceiver.

(a) What type of stages and what broadbanding techniques are employed for the amplifier in Figure 5.42?

(b) Draw the small signal differential half circuit and identify the type of negative feedback employed. Find the expression of the open-loop gain.

(c) Using the simplified HF equivalent circuit of the transistor with current gain shown in Figure 5.43 below, derive the expression of the 3dB bandwidth of the second stage of the amplifier.

Calculate the common-mode rejection ratio.

(7) Compare a chain of Cherry–Hooper stages with a chain of stages with shunt inductive peaking realized in the same technology. They both have a second-order frequency response. What are their advantages and disadvantages in terms of bandwidth, area consumption, power consumption, robustness to process and temperature variation?

(8) For a chain of three Cherry–Hooper stages with pole-*Q* of 0.707 (maximally flat response), find the required stage bandwidth and gain if the amplifier gain is 40dB and its bandwidth is 10GHz. Use the same SiGe BiCMOS technology with $GBWs = 100$GHz.

(9) Design a distributed amplifier based on the 65nm n-MOSFET-SiGe HBT cascode with a gain of 10dB and a maximum linear voltage swing at the output of $3V_{pp}$. The 65nm MOSFETs will be biased at a current density of 0.2mA/μm and have $g'_{meff} = 0.9$mS/μm, $C'_{gs} = C'_{db} = C'_{sb} = 0.7$fF/μm, $C'_{gd} = 0.4$fF/μm. The 0.1μm emitter width SiGe HBTs are biased at the peak f_T current density of 1.5mA/μm of emitter length and have $C'_{bc} = 1.6$fF/μm and $C'_{cs} = 1$fF/μm per emitter length, $R'_E = 20\Omega$/μm, $R'_c = 100\Omega$/μm, and $R'_b = 70\Omega \times$ μm.

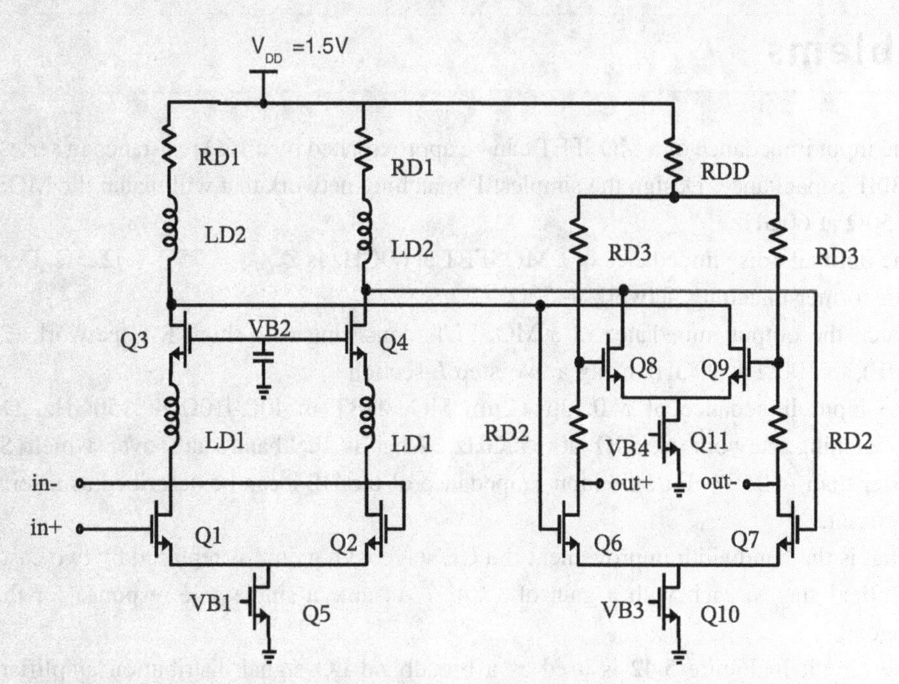

Figure 5.42 CMOS broadband amplifier. All MOSFETs have $L = 130$ nm

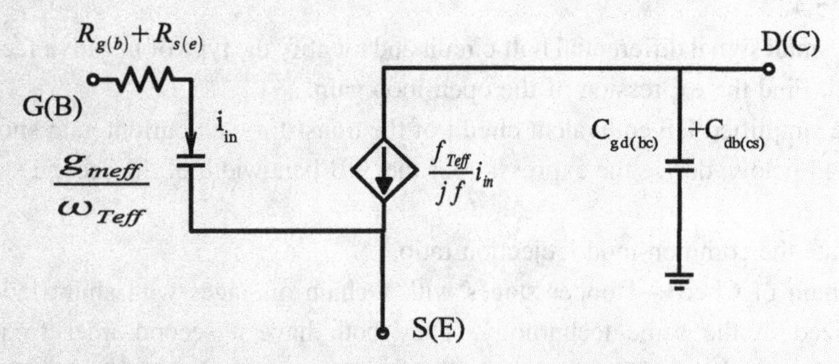

Figure 5.43 MOSFET HF current-gain based equivalent circuit

(10) A 65nm CMOS distributed amplifier consists of seven stages. Each stage is formed using three cascaded, scaled stages, as in Figure 5.30, where $k = 0.5$. The total gate width of the first transistor in the chain is 10μm and its load resistance is 400Ω. The third stage in each DA cell drives the output transmission line directly (i.e. no load resistor R_D is employed in this stage). The transistors are biased as in Example 5.6. What is the total DC current drawn by the circuit? What is the DC gain if the amplifier drives 50Ω loads and its input impedance is 50Ω. What is the value of the unit inductance of the input and output line, respectively? What is the bandwidth of the unit DA cell and of the entire DA? How does the bandwidth of the unit cell and of the DA change if constant group-delay inductive peaking is employed in the first two stages of each DA cell?

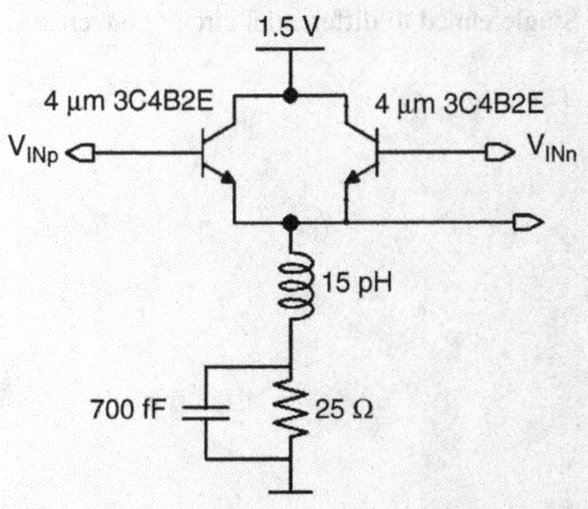

Figure 5.44 SiGe HBT 150-to-300 GHz doubler

(11) For each of the active baluns in Figure 5.38, find the expression of the transfer function from the input to the in-phase and out-of-phase outputs.

(12) Using the SiGe HBT parameters in problem 9, design a differential tripler from 120 to 360GHz tripler to provide 0dBm output power at 360GHz in a 100Ω differential antenna load. Use the topology in Figure 5.41. Find the optimal bias current, transistor size, and the output L-section matching network. Limit the maximum current to 2.25mA/μm of emitter length. Assume that the Q of the load inductors is 10 and that the maximum voltage swing at the transistor output cannot exceed $2V_{pp}$ per transistor. Hint: Plot equation (5.97) to find the optimum conduction angle for the tripler.

(13) A 100–200GHz 45nm differential CMOS doubler operates from a 1.1V supply. The minimum gate length MOSFETs have $V_T = 0.3$V, $I_{ON} = 1$mA/μm, and $W = 30$μm. Calculate the maximum output power and DC-to-RF efficiency assuming that $I_{MAX} = I_{ON}$. What is the optimum load resistance? Assume that the optimal conduction angle is $t_0/T = 0.35$.

(14) The schematic of a 150–300GHz SiGe HBT doubler is shown in Figure 5.44.

(a) Knowing that the optimal conduction angle for maximum efficiency is 118 degrees ($t_o/T = 0.33$), find the maximum theoretical efficiency of the doubler.

(b) Knowing that a 4μm 3C4B2E transistor has 0.12μm emitter width and two 4μm long emitters, and that $J_{MAX} = 24\text{mA}/\mu\text{m}^2$, calculate I_{DC} and I_2 for maximum efficiency.

(c) If $V_2 = 0.3$V, find R_{LOPT} and the output power at 300GHz.

(15) (a) For the circuit in Figure 5.45, derive the expression of the input impedance using the simplified HF equivalent circuits of the MOSFET in Figures 5.46 and 5.47 and assuming $R_s = 0$ and $R_g = 0$.

(b) Derive the expressions of V_{outn}/V_{in} and V_{outp}/V_{in}. Size Q_1 and Q_2 such that the input resistance is 50Ω and the voltage gains between the input and each output are equal in magnitude. Assume that $R_s = R_g = 0$ and that both outputs are impedance matched to the load impedance Z_0.

(c) As illustrated in Figure 5.48, the tri-port circuit in Figure 5.45 can be regarded as a shunt–series connection of two two-ports. Derive the expressions of the IEEE noise parameters of the equivalent two-port. Assume that V_{ng} and I_{nd} are not correlated and that

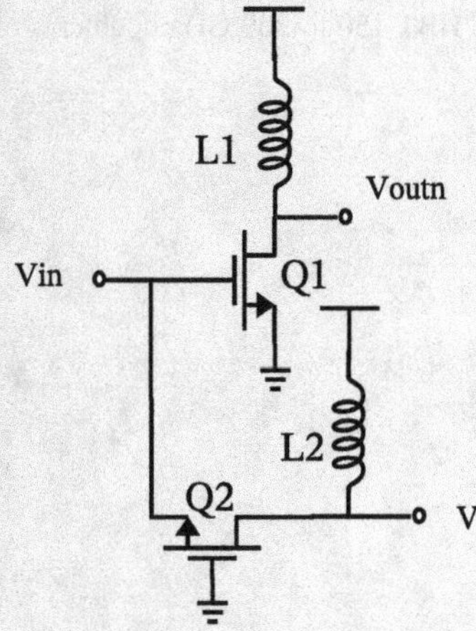

Figure 5.45 Single-ended to differential circuit converter

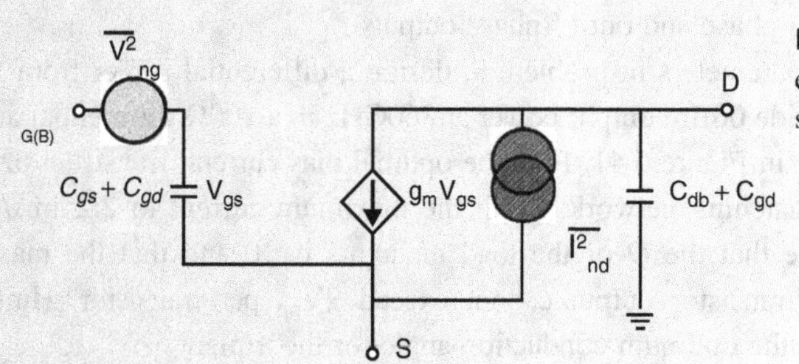

Figure 5.46 Simplified HF and noise equivalent circuit of the common-source stage

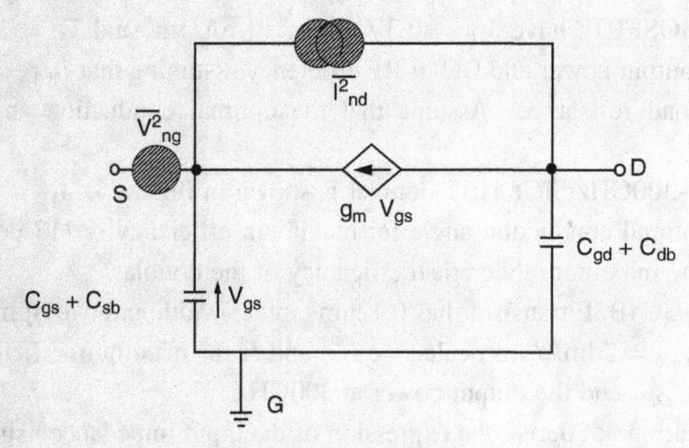

Figure 5.47 Simplified HF and noise equivalent circuit of the common-gate stage

$\overline{v_{ng}^2} = 4kT\Delta f\left(R_g + R_s + R_i\right)$ $\overline{i_{nd}^2} = 4\mathrm{k}\mathrm{T}\Delta f P g_m$ and $\langle i_n, v_n \rangle = 0$. Assume that Q_1 and Q_2 have the same L and W.

(16) The distributed amplifier in Figure 5.49 has seven stages and is fabricated in a 32nm CMOS technology where the CS FETs are replaced by the cascoded CMOS inverters shown in Figure 5.50. Assuming that $R_D = Z_0 = 50\Omega$, find L_I, L_O, and W for each transistor such that the bandwidth of the amplifier is 100GHz. The transistors are biased at the minimum noise figure current density. What is the low-frequency voltage gain?

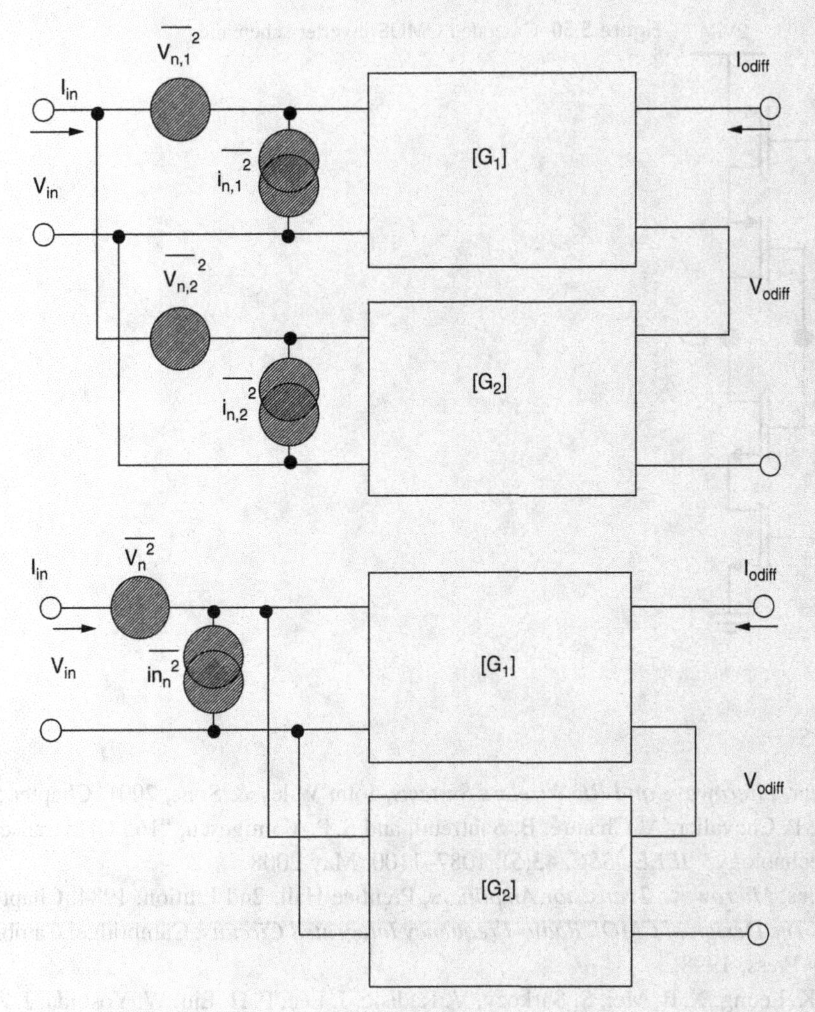

Figure 5.48 Noisy two-port representation of the circuit in Figure 5.45

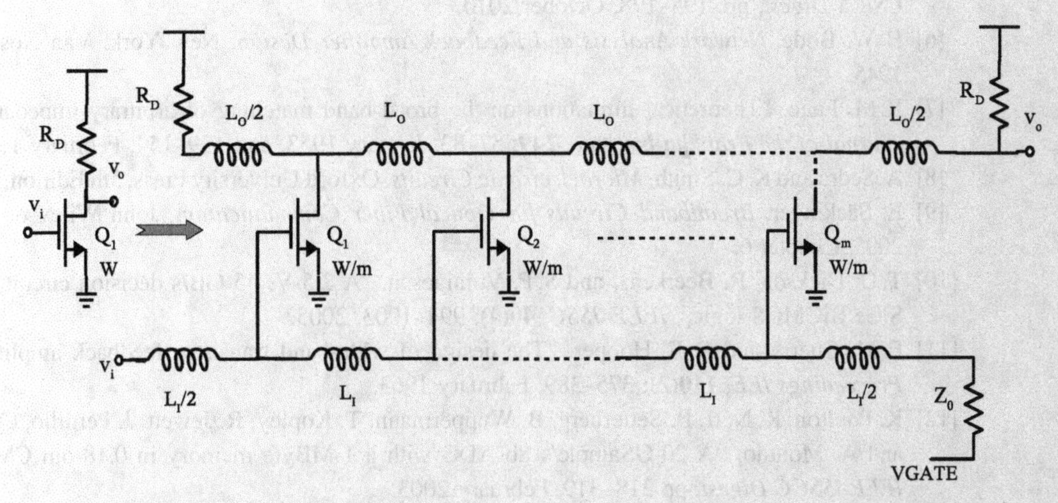

Figure 5.49 Distributed amplifier schematic

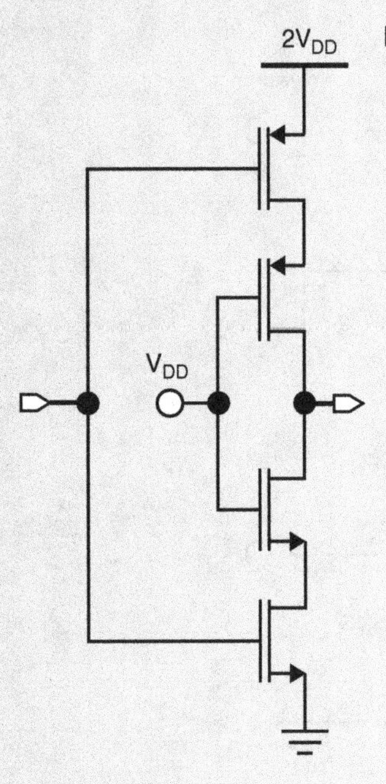

Figure 5.50 Cascoded CMOS inverter schematic

REFERENCES

[1] D. M. Pozar, *Microwave and RF Wireless Systems*, John Wiley & Sons, 2001, Chapter 5.

[2] E. Laskin, P. Chevalier, A. Chantre, B. Sautreuil, and S. P. Voinigescu, "165 GHz transceiver in SiGe technology," *IEEE JSSC*, **43**(5): 1087–1100, May 2008.

[3] G. Gonzales, *Microwave Transistor Amplifiers*, Prentice Hall, 2nd Edition, 1984, Chapter 2.

[4] T. H. Lee, *The Design of CMOS Radio-Frequency Integrated Circuits*, Cambridge: Cambridge University Press, 1998.

[5] W. Deal, K. Leong, X. B. Mei, S. Sarkozy, V. Radisic, J. Lee, P. H. Liu, W. Yoshida, J. Zhou, and M. Lange, "Scaling of InP HEMT cascode integrated circuits to THz frequencies," *IEEE CSICS Digest*, pp. 195–198, October 2010.

[6] H. W. Bode, *Network Analysis and Feedback Amplifier Design*, New York: Van Nostrand, 1945.

[7] R. M. Fano, "Theoretical limitations on the broad-band matching of arbitrary impedances," *Journal of the Franklin Institute*, **249**: 57–83, January 1953, and 139–154, February 1950.

[8] A. Sedra and K. C. Smith, *Microelectronic Circuits*, Oxford University Press, 5th Edition, 2004.

[9] E. Säckinger, *Broadband Circuits for Optical Fiber Communication*, John Wiley & Sons, 2005, Chapter 6.

[10] T. O. Dickson, R. Beerkens, and S. P. Voinigescu, "A 2.5-V, 45 GB/s decision circuit using SiGe BiCMOS logic," *IEEE JSSC*, **40**(4): 994–1003, 2005.

[11] E. M. Cherry and D. E. Hooper, "The design of wide-band transistor feedback amplifiers," *Proceedings IEE*, **110**(2): 375–389, February 1963.

[12] K. Poulton, R. Neff, B. Setterberg, B. Wuppermann, T. Kopley, R. Jewett, J. Pernillo, C. Tan, and A. Montijo, "A 20-GSample/s 8b ADC with a 1-MByte memory in 0.18-um CMOS," *IEEE ISSCC Digest*, pp 318–319, February 2003.

[13] W. S. Percival, "Thermionic valve circuits," British Patent no. 460562, filed 24 July 1936, granted January 1937.

[14] J.-C. Chien and L.-H. Lu, "40 Gb/s high-gain distributed amplifiers with cascaded gain stages in 0.18μm CMOS," *IEEE JSSC*, **42**: 2715–2725, December 2007.

[15] R. A. Aroca and S. P. Voinigescu, "A large swing, 40 GB/s SiGe BiCMOS driver with adjustable pre-emphasis for data transmission over 75Ω coaxial cable," *IEEE JSSC*, **43**(10): 2177–2186, October 2008.

[16] E. Öjefors, B. Heinemann, and U. Pfeiffer, "A 325GHz frequency multiplier chain in a SiGe HBT technology," *IEEE RFIC Symposium Digest*, 91–94, May 2010.

[17] S. A. Maas, *Nonlinear Microwave and RF Circuits*, Norwood, MA: Artech House, 2003, Chapter 10.

6 Tuned power amplifier design

WHAT IS A TUNED POWER AMPLIFIER?

Tuned power amplifiers are key components in the transmission path of wireless communication systems and in automotive radar. They typically deliver the power required for transmitting information to the antenna with high efficiency and, usually, high linearity, over bandwidths of 10% to 20% relative to the center frequency of the amplifier. For battery-operated applications in particular, minimum DC power consumption at a specified output power level is required.

As shown in Figure 6.1, in its idealized representation, a tuned power amplifier consists of a common-emitter or common-source (very large) transistor operating under large signal conditions with large output voltage swing. The transistor drain/collector is biased through a bias T (which presents an infinite inductor to the power supply and an infinite capacitor towards the load) and is loaded with a parallel resonant tank (formed by R_L, C_1, L) at the frequency of interest. For the sake of simplicity, we will assume that the transistor output capacitance, $C_{out} = C_{ds} + C_{gd}$ or $C_{bc} + C_{cs}$, is absorbed in the load capacitance C_1. In a similar manner, R_L includes the loss resistance of inductor L_1 and capacitor C_1. The circuit draws DC power from the supply to amplify the input signal power and deliver it to the load. Ideally all of the DC power and the input signal power should be converted to output signal power. In practice, at least some of the DC power is dissipated as heat.

Although only recently emerging, CMOS PAs have advantages in system-on-chip applications where a relatively small transmit power is needed. However, at the time of the writing of this book chapter (late 2011), the main cellular phone PA market continues to be overwhelmingly dominated by GaAs HBT PAs. In discrete PAs with higher transmit power, e.g. for cellular base station applications, other technologies such as LDMOS (laterally diffused drain MOSFET) or compound semiconductor technologies have prevailed. At mm-wave frequencies, commercial PAs are typically manufactured using GaAs pseudomorphic and metamorphic HEMT, GaN HEMT, and, more recently, SiGe HBT and nanoscale CMOS technologies.

From the outset, the reader is advised that, more so than in most other high-frequency circuit design topics, because of the large signal nature of the PA stages and because of their strong non-linearities, the analytical expressions commonly derived to explain their behavior are fairly approximate. A precise computer simulation analysis and a computer-driven design methodology are compulsory in understanding the operation of these amplifiers. This chapter

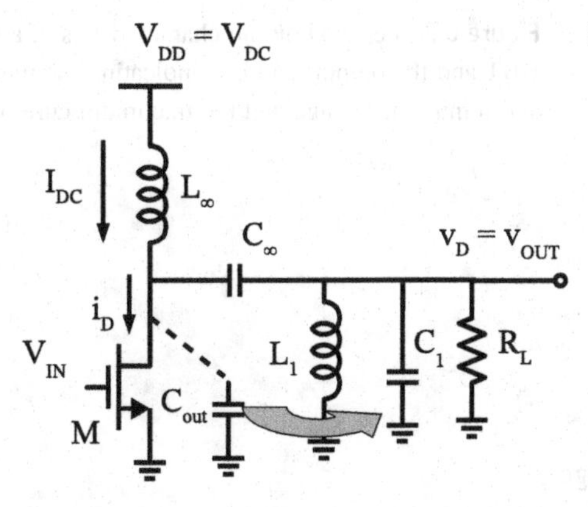

Figure 6.1 Idealized power amplifier stage

introduces the basic high-frequency power amplifier concepts. For an in-depth treatment of power amplifiers and their design methodologies, the reader is referred to [1].

6.1 TUNED PA FUNDAMENTALS

To establish a performance figure of merit, several key parameters must be taken into account. These include output power P_{out}, power gain G, carrier frequency f, linearity (in terms of $OP1$dB), and power-added-efficiency (PAE). Unfortunately, linearity strongly depends on the operating class of the amplifier, making it difficult to compare amplifiers of different classes. Moreover, the most recent developments increasingly focus on highly non-linear, saturated output power switching PA topologies because of their high efficiency. To remain independent of the design approach and the specifications of different applications, the linearity is often not included in the PA figure of merit. To compensate for the 20dB/decade roll-off of the transistor power gain as a function of frequency, a factor of f^2 is included into the figure of merit. This results in

$$FoM_{PA} = P_{OUT} \cdot G \cdot PAE \cdot f^2 \tag{6.1}$$

where:

- $PAE = (P_{OUT} - P_{IN}) / P_{DC}$ is the power-added efficiency of the amplifier,
- P_{OUT} is the output power,
- P_{IN} is the input power, and
- P_{DC} is the power drawn by the PA from the power supply.

Sometimes, the drain (or collector) efficiency, $\eta_{coll,drain} = P_{OUT}/P_{DC}$ is employed, instead of PAE, to describe the efficiency of the amplifier. For large power gains, PAE asymptotically approaches $\eta_{coll,drain}$.

What is important to maximize the output power and the PAE of a PA stage? Understanding the physical limits of the transistor, captured by the output characteristics of Figure 6.2, is critical for optimizing the PA output power and efficiency.

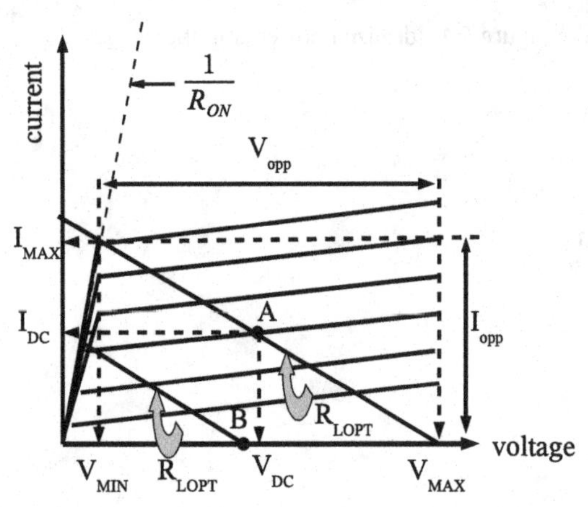

Figure 6.2 Idealized output characteristics of a FET or HBT and the output load line indicating the maximum and minimum voltage and the maximum current

The output voltage swing is limited by V_{MAX}, which represents the maximum allowed drain–source or collector–emitter voltage, and by V_{MIN}. V_{MAX} is normally determined by the transistor breakdown voltage or by some other maximum allowed voltage beyond which the reliability of the transistor is degraded. In FETs, V_{MIN} is determined by the carrier mobility and the saturation voltage, while in HBTs it is approximated by V_{CESAT}. Another transistor figure of merit equivalent to V_{MIN} is R_{ON} ($V_{MIN} \cong I_{MAX} \times R_{ON}$) which is typically employed, instead of V_{MIN}, to describe the losses of the transistor when operated as a switch.

The output current swing of the PA is limited by I_{MAX}. I_{MAX} scales with the total gate width or with the emitter area. In FETs, I_{MAX} depends on the saturation velocity and mobility of the carriers, and is approximately equal to I_{ON} in MOSFETs. In HBTs, I_{MAX} is more difficult to define precisely. It is linked to the f_T degradation due to high injection effects and does not exceed $2J_{pfT}$. From Figure 6.2, it is apparent that

$$P_{OUT} \propto (V_{MAX} - V_{MIN}) \cdot I_{MAX} \tag{6.2}$$

and

$$PAE \propto (1 - V_{MIN}/V_{MAX}). \tag{6.3}$$

These equations naturally lead to the most important PA design criterion. For an ideal PA, given the choice, we must select the transistor technology with the largest V_{MAX} and I_{MAX} and, ideally, $V_{MIN} = 0$ and $R_{ON} = 0$.

A quick glance at Figure 6.1 allows us to calculate the small signal voltage gain at the center frequency f as $A_V = -g_m R_L$, identical to that of a low-frequency amplifier [2]. L_1 and C_1 must satisfy the resonance condition

$$f = \frac{1}{2\pi\sqrt{L_1 C_1}} \tag{6.4}$$

while the quality factor Q and bandwidth Δf of the amplifier are given by (6.5) and (6.6), respectively

$$Q = 2\pi f C_1 R_L \tag{6.5}$$

$$\Delta f = \frac{f}{2Q}. \tag{6.6}$$

If the Q of the output tank is high, the output voltage and current through the load will be sinusoidal. All other harmonics will be attenuated by the bandpass characteristics of the parallel RLC tank. In a rather crude approximation where $V_{MIN} = V_{DSAT} = 0$, the voltage at the output of the transistor will swing between 0 and 2 V_{DD}. This is possible because an inductor, rather than a resistor, is connected between the drain/collector and V_{DD}.

At resonance, the sinusoidal output voltage waveform $V_{DD} \times \cos(\omega t)$ produces a sinusoidal current $(V_{DD}/R_L) \times \cos(\omega t)$ through the load resistor R_L. The power delivered to the load can then be expressed as

$$P_{OUT} = \frac{1}{T}\int_0^T V_{DD}\cos(\omega t)\frac{V_{DD}\cos(\omega t)}{R_L}dt = \frac{V_{DD}^2}{2T.R_L}\int_0^T [1 + \cos(2\omega t)]dt = \frac{V_{DD}^2}{2R_L}. \tag{6.7}$$

If the transistor is biased such that $I_{DC} = V_{DD}/R_L$, then $P_{OUT} = V_{DD} \times I_{DC}/2$ and the drain/collector efficiency, P_{OUT}/P_{DC}, becomes as high as 50%. We note that, because of the choke inductor, the output voltage is allowed to swing to 2 V_{DD} and the maximum efficiency of a class A RF PA is two times larger than that of a low-frequency class A amplifier, which is limited to 25% [2].

6.2 CLASSES OF TUNED PAs AND THE ASSOCIATED VOLTAGE WAVEFORMS

As shown in Figure 6.3 [3], PAs can be classified in: (i) PAs in which the transistor operates as a voltage-controlled current source and (ii) PAs in which the transistor acts as a switch. In both cases, the transistor drain/collector output is terminated on a high-Q resonant tank which acts as a highly selective bandpass filter at the frequency of interest. The PA classes are defined based on the shape of the voltage and current waveforms and on the angle of current conduction, in much the same way as output stages operating at audio frequencies [2].

In the first group, we have:

- class A PAs with 100% current conduction, that is the device is always "on" and the conduction angle is 2π,
- class B PAs with a conduction angle of π, that is the transistor only conducts for half of the period of the input voltage waveform,
- class AB PAs where the angle of conduction can be anywhere between π and 2π, and
- class C PAs with a conduction angle smaller than π.

The switching PA group includes:

- class D,
- class E,
- class F, and
- inverse F (F^{-1})

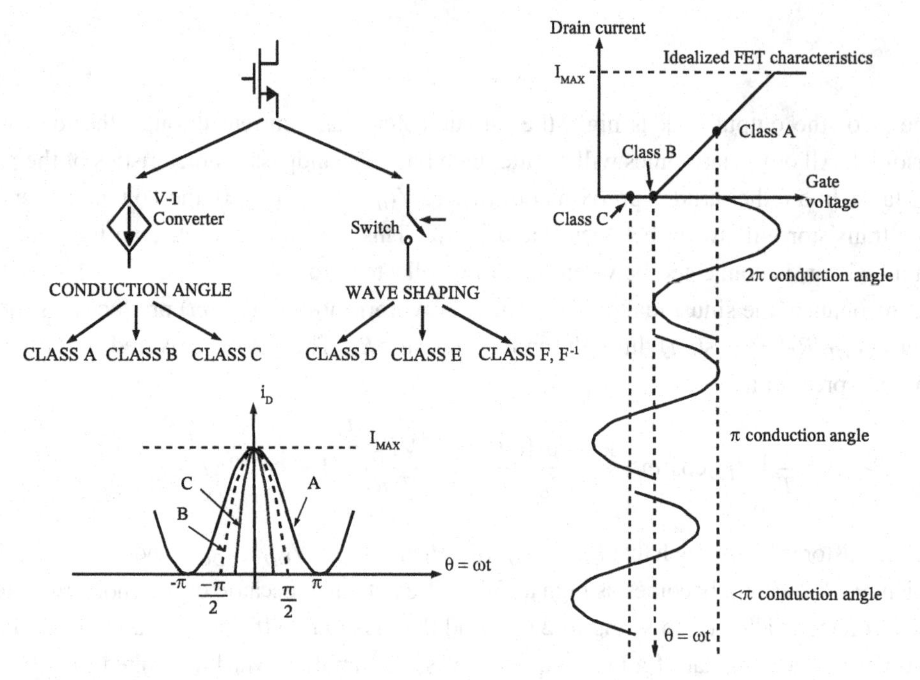

Figure 6.3 The PA family tree and conduction angle definitions for class A, B, and C operation [3]

PAs. In general, the switching PAs achieve the highest efficiency but they also exhibit the most non-linear behavior. The key idea in maximizing the efficiency in all PA classes is to minimize the overlap of the current and voltage waveforms in the transistor.

6.2.1 PA design fundamentals

R_{LOPT}

The PA output stage operates with the largest voltage and current swing. Since the large signal output impedance of the transistor is very different from the small signal value, small signal *S*-parameters obtained from measurements or from simulations are not useful in the design of the output matching network. Instead, as shown in Figure 6.2, the load line impedance

$$R_{LOPT} = \frac{V_{opp}}{I_{opp}} \approx \frac{V_{MAX}}{I_{MAX}} = \frac{V_{MAX}}{2 \cdot I_{DC}} \tag{6.8}$$

is usually chosen as a first guess of the impedance on which to terminate the output stage [1]. V_{opp} and I_{opp} represent the peak-to-peak swing of the output voltage and of the load current, respectively. This optimal load impedance maximizes the output power delivered to the load when the amplifier is operated at the 1dB compression point and beyond. Although it leads to about 1dB lower gain at low input power than the corresponding small signal S_{22}-matched output stage, it was found experimentally to produce 2dB higher output power [1]. Matching the output stage impedance to R_{LOPT} is sometimes called *power matching* to distinguish it from conventional *small signal* impedance matching.

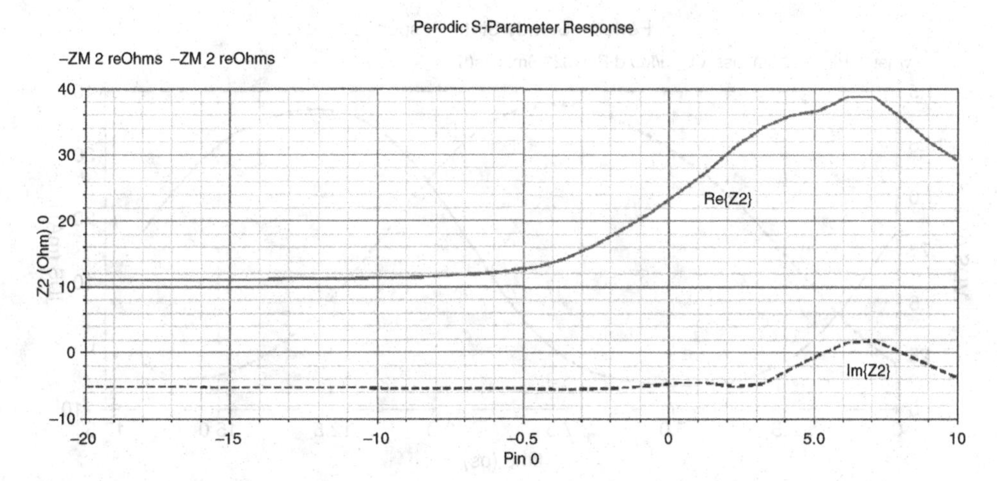

Figure 6.4 Simulated output impedance of a class A PA stage as a function of input power

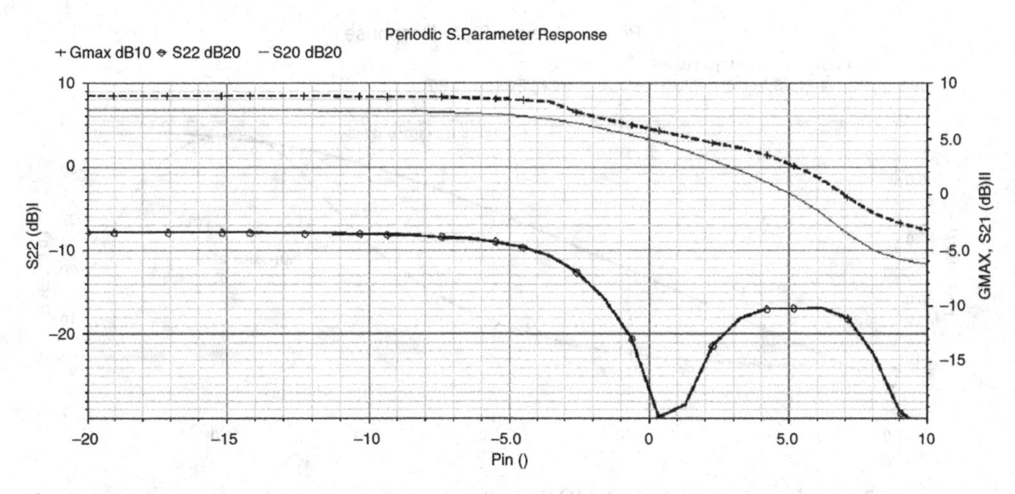

Figure 6.5 Simulated maximum available gain (G_{MAX}), S_{21} and S_{22} of a class A PA as a function of input power

Simulation techniques have been developed to allow for the prediction of the large signal output impedance, return loss, and constant output power contours of transistors and amplifiers as a function of the load impedance. Figure 6.4 reproduces the simulated output impedance of a 65nm CMOS output stage at 60GHz as a function of the power applied at its input. The supply voltage is $V_{DD} = 0.7$V. Large signal periodic-steady-state (pss) S-parameter simulations were employed (Figure 6.5), confirming that the output power at the 1dB compression point is maximized when the output stage is conjugately matched to its LARGE SIGNAL output impedance. In a narrowband PA output stage, the large signal output resistance is well approximated by the ratio of the amplitudes of the fundamental components of output voltage and current. In class A PAs, the latter values are very close to those of the DC operating voltage

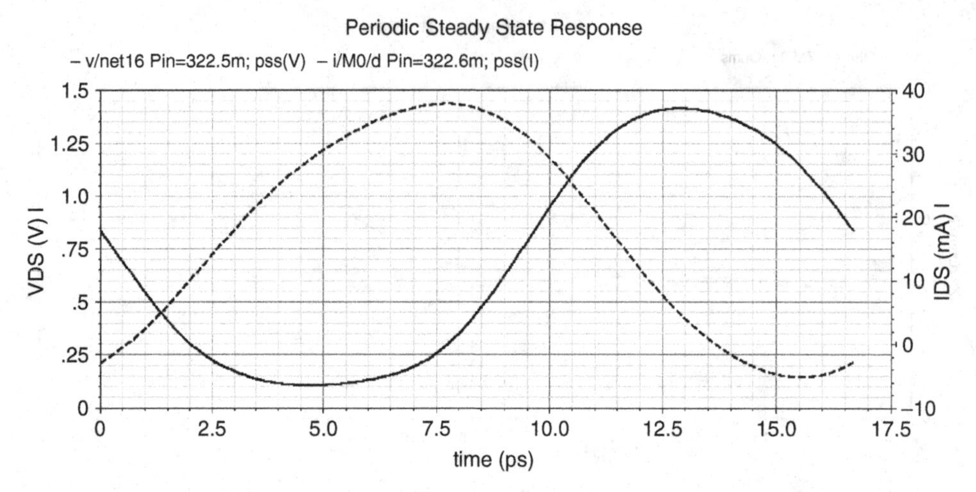

Figure 6.6 Simulated instantaneous drain voltage (solid line) and drain current (dashed line) of the class A 65nm CMOS PA for an input power of 0.323dBm

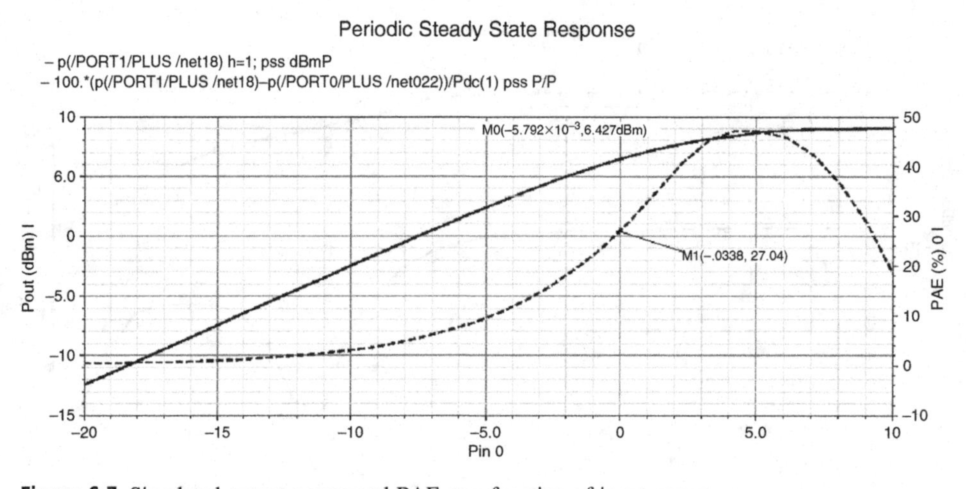

Figure 6.7 Simulated output power and PAE as a function of input power

and current of the transistor, explaining why a good approximation for R_{LOPT} can be obtained from the DC output characteristics of the transistor.

The output matching network must be designed to transform the 50Ω (or, in general, Z_0) system impedance to R_{LOPT} at the transistor output. Since the input power is relatively small, the input matching network is typically designed to achieve conjugate matching between the small signal input impedance of the transistor and the impedance of the previous stage or signal source.

For example, as illustrated in Figures 6.6 and 6.7, if the output voltage swing at the 1dB compression point is 1.25V peak-to-peak and the corresponding current swing in the transistor is 42mA peak-to-peak, the optimum load resistance becomes 1.25V/42mA = 29.7Ω. The corresponding output power is 6.5mW (8.1dBm). The discrepancy between simulation and simple hand analysis results for R_{LOPT} and P_{out} is about 25%, part of it being due to the fact that

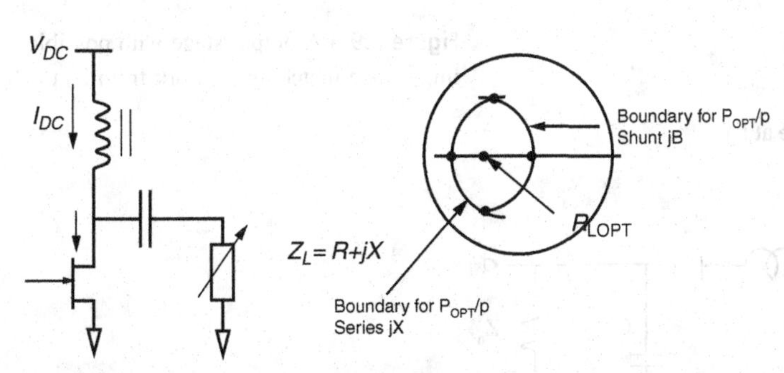

Figure 6.8 Illustration of the load pull technique and of the constant output power contours on the Smith Chart [1]

the voltage and current waveforms are no longer perfect sinusoids at the 1dB compression point. The output matching network must be designed to transform the 50Ω load impedance to 29.7Ω at the transistor output. We note that V_{MAX} is 1.4V, which may already be unsafe for the long-term reliable operation of a 65nm MOSFET.

Load pull contours

If $V_{MIN} = 0$, the maximum output power

$$P_{LOPT} = \frac{V_{opp} \cdot I_{opp}}{8} \approx \frac{V_{MAX} \cdot I_{MAX}}{8} = \frac{V_{DC} \cdot I_{DC}}{2} \tag{6.9}$$

is obtained when $R_L = R_{LOPT}$. However, the exact value of R_{LOPT} can only be obtained by measuring or simulating the output power for different load impedance values and selecting the load impedance corresponding to the highest output power. The measurement or simulation technique by which the load impedance locus for optimal output power is determined is known as load pull and the resulting constant output power contours are plotted on the Smith Chart, as illustrated in Figure 6.8. In addition to the optimum load impedance, the figure includes constant resistance ($R_L = R_{LOPT}/p$) and constant conductance circles ($G_L = 1/(pR_{LOPT})$) along which the power is $10 \times \log(p)$ dB smaller than P_{LOPT} [1]. Since the current in the load resistance remains constant as the series reactance is changed, the power along the constant resistance circle is given by

$$P_{OUT} = \frac{I_{DC}^2 R_L}{2} = \frac{I_{DC}^2 R_{LOPT}}{2p} = \frac{P_{LOPT}}{p}. \tag{6.10}$$

Similarly, the power along the constant conductance circle can be calculated as

$$P_{OUT} = \frac{V_{DC}^2 G_L}{2} = \frac{V_{DC}^2 G_{LOPT}}{2p} = \frac{P_{LOPT}}{p}. \tag{6.11}$$

For load impedances located within the boundaries of these closed contours, the output power will be within $10 \times \log_{10}(p)$dB of the maximum possible. The source-pull contours play a much more significant role in the design of power amplifiers than the constant gain circles used in high-frequency linear amplifier design.

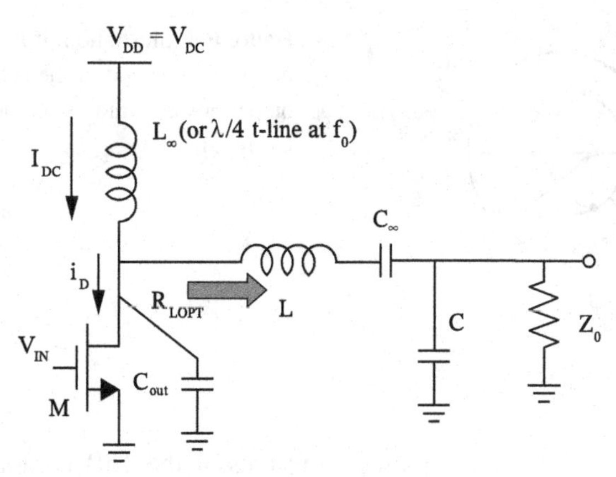

Figure 6.9 PA output stage with possible impedance matching network from Z_0 to R_{LOPT}

Output matching network design

The output matching network performs several functions:

- transforms the load impedance to R_{LOPT} (Z_{LOPT}) for maximum output power,
- provides suitable terminations to all harmonics so that only the power of the fundamental is transferred to the load, and
- maintains the impedance transformation over the bandwidth of interest.

In theory, any of the LC matching networks discussed in Chapter 5 can be employed in the impedance transformation network. In practice, as shown in Figure 6.9, a series L followed by a shunt capacitor across the load impedance is the preferred implementation, with a series bypass capacitor added to separate the AC and DC paths.

In general, the impedance transformation ratio between the 50Ω load impedance and R_{LOPT} leads to a high-Q matching network, which makes it relatively narrow band. To maintain constant power across the band of interest, a low-Q matching network is needed. This can be accomplished with a two-step impedance transformation, as discussed in Chapter 5.

EXAMPLE 6.1

Let us assume that we want to design the output matching network of a 60GHz PA with R_{LOPT} of 2Ω and $Z_0 = 50\Omega$. The corresponding Q of the output impedance transformation network can be calculated as [4]

$$Q = \sqrt{\frac{Z_0}{R_{LOPT}} - 1} = 4.9 \approx 5.$$

The DC-blocking capacitor, C_∞, is typically chosen such that its impedance is 100 times smaller than Z_0 at 60GHz

$$C_\infty = \frac{100}{\omega Z_0} = \frac{100}{6.28 \times 60 \times 10^9 \times 50} = 5.3pF.$$

This value is quite large and, depending on the technology, may result in a bottom plate capacitance as high as 100fF, as well as some parasitic inductance of several picoHenry.

L can be calculated as [4]

$$L = \frac{Q \times R_{LOPT}}{\omega} = \frac{4.9 \times 2}{6.28 \times 60 \times 10^9} = 26\text{pH}.$$

Finally, C is obtained from [4] as

$$C = \frac{L}{R_{LOPT} \times Z_0} = \frac{26p}{100} = 260\text{fF}.$$

C can absorb the bottom plate capacitance of the DC-blocking capacitor while the value of L can be corrected to account for the series parasitic inductance of the DC-blocking capacitor. If we employ a two-step transformation, first from $Z_0 = 50\Omega$ to $R_1 = \sqrt{Z_0 \cdot R_{LOPT}} = 10\Omega$ and then from R_1 to $R_{LOPT} = 2\Omega$, the new matching network will have a lower Q of

$$Q = \sqrt{\frac{50}{10} - 1} + \sqrt{\frac{10}{2} - 1} = 2 + 2 = 4.$$

Input matching and source pull

Once the output stage has been matched to R_{LOPT}, the impedance provided by the signal source to the PA stage can be varied until the best PAE is obtained over the desired bandwidth. The corresponding signal source impedance is Z_{SOPT} and the associated technique is known as *source pull*. The input matching network can then be synthesized to transform the signal source impedance from Z_0 to Z_{SOPT}.

6.2.2 Voltage waveforms

The class A, B, and C PAs have identical drain/collector voltage waveforms in the shape of a full sinusoid. As illustrated in Figure 6.3, the difference between the three classes lies in the angle of conduction and in the shape of the current waveforms.

Class A

In a class A PA, shown in Figure 6.10, the transistor is normally biased at the peak f_T current density. The output matching network is designed such that the transistor is loaded by R_L, at the fundamental frequency f_0 and is terminated on a short circuit at all of its harmonics.

$$Z_L(f_0) = R_L; Z_L(nf_0) = 0 \text{ for } n > 1. \tag{6.12}$$

Conditions (6.12) are ensured if the parallel resonant circuit formed by L_1 and C_1 (Figure 6.1) is designed to resonate at $f_0 = \frac{1}{2\pi\sqrt{L_1 C_1}}$. At $2f_0$ and higher frequencies, C_1 produces a short circuit at the output. In some cases, inductors L_∞ and L_1 can be replaced by a single inductor L_D

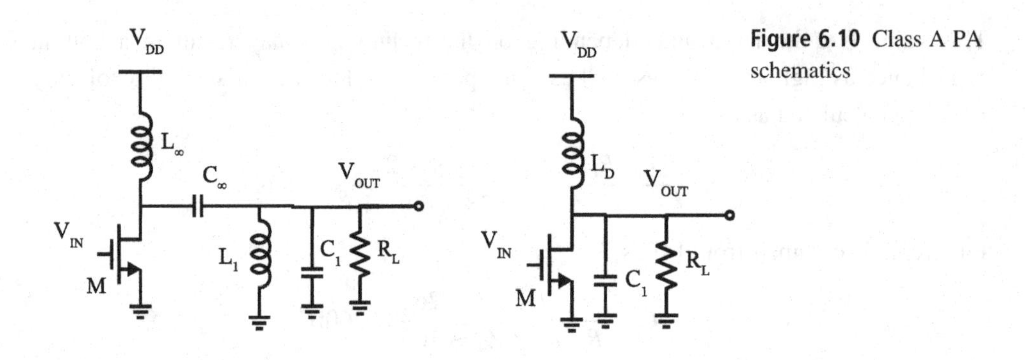

Figure 6.10 Class A PA schematics

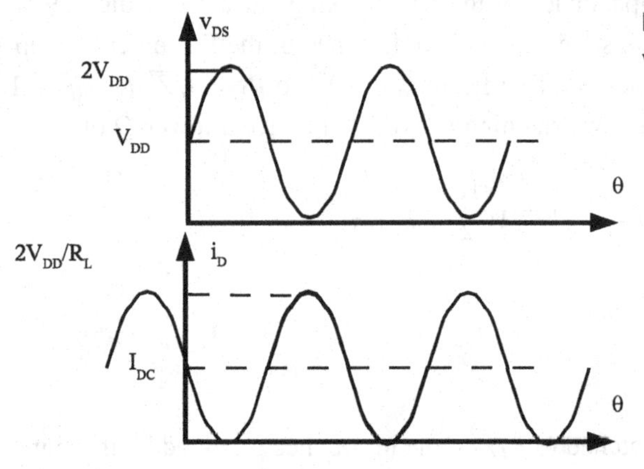

Figure 6.11 Output voltage and current waveforms through the load resistance R_L

connected directly between the transistor drain–collector terminal and V_{DD}, as shown in Figure 6.10. As mentioned, the parasitic output capacitance of the transistor is absorbed in C_1.

The drain efficiency of a class A power amplifier stage depends on the signal amplitude at the fundamental frequency and is given by

$$\eta_{drain} = \frac{P_L}{P_{DC}} = \frac{V_1 \cdot I_1}{2 \cdot I_{DC} \cdot V_{DD}} = \frac{(V_{DD} - V_{MIN})(I_{DC} - I_{MIN})}{2 \cdot V_{DD} \cdot I_{DC}} \tag{6.13}$$

where I_{MIN} is the minimum instantaneous value of the current through the load, $I_1 = I_{DC} - I_{MIN}$ is the amplitude of the fundamental current in the load, $V_1 = V_{DD} - V_{MIN}$ is the amplitude of the fundamental voltage in the load, $V_{DC} = V_{DD}$ and $I_{DC} = V_{DC}/R_L$. As illustrated in Figure 6.11, the maximum values of the output voltage and current waveforms are $V_{MAX} = 2\ V_{DD}$ and $I_{MAX} = 2\ V_{DD}/R_L$, respectively. From (6.13), it follows that the maximum drain efficiency of 50% occurs at the maximum output swing when $V_{MIN} = 0$ and $I_{MIN} = 0$ while the minimum efficiency is 0% when the input signal is 0V.

How big is the impact of V_{MIN} on efficiency? As indicated by (6.3), it depends on the value of V_{MAX}. Let us first consider a 65nm CMOS process with $V_{MAX} = 1.5$V and $V_{MIN} = 0.1$V as in Figure 6.6. This implies that $V_1 = (V_{MAX} - V_{MIN})/2 = 0.7$V, instead of the ideal 0.75V, reducing the drain efficiency from the theoretical maximum of 50% to $0.7 \times 50 / 0.75 = 46.6\%$. In contrast, in a class A GaN PA with $V_{MAX} = 80$V and $V_{MIN} = 0.5$V, the maximum efficiency drops to 49.7%.

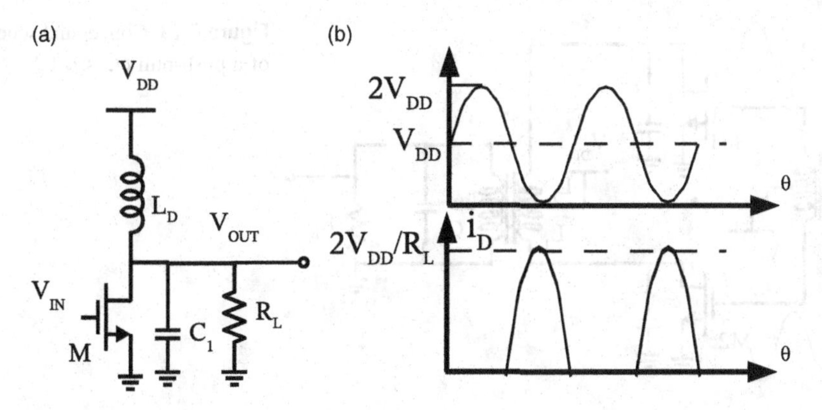

Figure 6.12 (a) Schematics and (b) voltage and current waveforms of a class B PA

Class B

Although the peak drain efficiency of a class A stage is 50%, it decreases sharply at lower output power levels. Unfortunately, modern wireless transmitters employ digital modulation formats with large peak-to-average output power ratios and most of the time the PA operates in backoff from the peak efficiency point. Thus, the average drain efficiency is smaller than 50% because the transistor draws the same DC current, even when the output power is backed off from its maximum value.

How can we improve efficiency? A solution is to **allow the transistor to enter the cutoff** region, for at least part of the signal period. This is the idea behind the class B stage.

The peak current of the class B stage is similar to that of class A, however, as shown in Figure 6.12, the transistor conducts current for only half the period of the fundamental frequency. Biasing the transistor at the edge of cutoff reduces the conduction angle to π and minimizes the transistor current when the output voltage is high. As a consequence, the power dissipation in the transistor is reduced compared to the class A stage. Several important and somewhat unwelcome, yet manageable, developments result from this arrangement:

1. The transconductance and power gain are reduced to half those of the class A stage.
2. The amplifier becomes highly non-linear.
3. Larger input voltage swing is required for the same output power.

Nevertheless, the strong non-linearity helps to increase the efficiency which reaches a theoretical maximum of 78%.

Assuming $V_{MIN} = 0$, for the same fundamental voltage amplitude, $V_1 = V_{DD}$, the maximum allowed voltage, $V_{MAX} = 2\,V_{DD}$, and the maximum current, I_{MAX}, the DC current of the class B stage can be obtained as the average of the current waveform in Figure 6.12(b) over one signal period, knowing that the transistor conducts only for half of the period (from π to 2π)

$$I_{DC} = \frac{-I_{MAX}}{2\pi}\int_{\pi}^{2\pi} \sin\theta d\theta = \frac{I_{MAX}}{\pi} = \frac{2\cdot V_{DD}}{\pi\cdot R_L}. \tag{6.14}$$

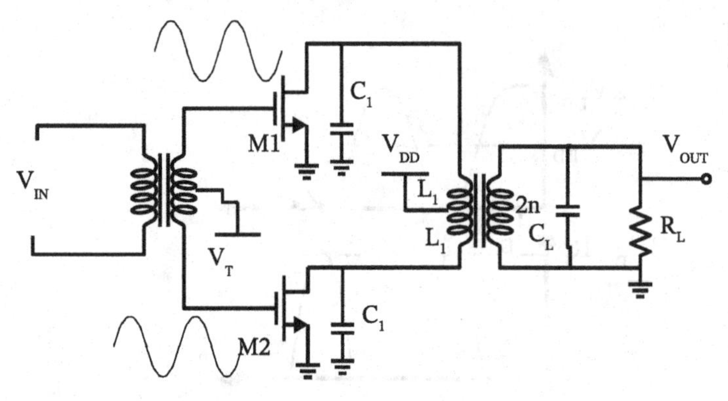

Figure 6.13 Conceptual schematic of a push-pull class B PA

Similarly, the amplitude of the first and second harmonic of the current through the load can be calculated from the Fourier series of the current waveform as $I_1 = I_{MAX}/2$ and $I_2 = 2I_{MAX}/3\pi$. This result indicates that the amplitude of the fundamental current and the amplitude of the fundamental voltage waveforms of the class B stage are the same as those of the class A stage. Therefore, if designed for the same output power, the optimal load resistance R_{LOPT} and the output matching network of the class B stage remain the same as those of the class A stage, where

$$R_{LOPT} = \frac{V_1}{I_1} = \frac{V_{DD}}{\frac{I_{MAX}}{2}} = \frac{2V_{DD}}{I_{MAX}} = \frac{V_{MAX}}{I_{MAX}}. \tag{6.15}$$

However, assuming $V_{MIN} = 0$, the peak drain efficiency increases to

$$\eta_{drain} = \frac{P_L}{P_{DC}} = \frac{V_1 \cdot I_1}{2 \cdot I_{DC} \cdot V_{DD}} = \frac{V_{DD} \cdot \frac{I_{MAX}}{2}}{\frac{2 \cdot V_{DD} \cdot I_{MAX}}{\pi}} = \frac{\pi}{4}. \tag{6.16}$$

Unlike class A, in which the DC current remains constant irrespective of the amplitude of the input signal, in class B and class C PAs the DC current increases with increasing signal power, thus explaining their improved efficiency.

In the case of the class C stage, which has the highest theoretical efficiency among the PAs in the voltage-controlled current source group, the strong non-linearity which accompanies the high efficiency, the low output power, and the power gain are responsible for its rare use.

A significant contributor to the non-linearity of the class B stage is the gate–source or base–emitter capacitance, which varies significantly when the transistor is driven into the cutoff region. This can lead to deviations from the 50% duty cycle of the stage and to degradation of efficiency. As discussed at the end of Chapter 5 in the design of multiplier stages, one technique to reduce this input non-linearity and improve efficiency is to place a short circuit at the second harmonic of the input signal at the gate or base of the transistor.

Similar to audio frequency output stages [2], a push-pull topology consisting of two class B stages/180 degrees out of phase, can be employed to improve the linearity, as indicated in Figure 6.13. The transformer forms a parallel LC resonant tank at the output, cancels the even harmonics, making the output current waveform sinusoidal, Figure 6.14.

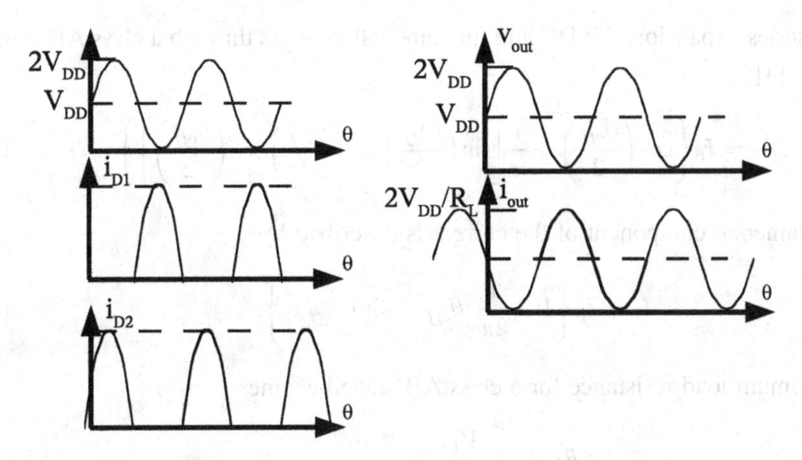

Figure 6.14 Current and voltage waveforms of the class B push-pull PA in Figure 6.13

Class AB

The class AB PA has been the workhorse of the linear PA industry for many decades because it reaches a compromise between the linearity of the class A and the efficiency of the class B power amplifiers. As the name indicates, the transistor is biased between class A and class B operation, with the angle of conduction larger than π. An experimentally determined bias condition that produces a linearity sweet spot in FET-based class AB amplifiers appears to be the current at which the transconductance is about one half of the peak transconductance of the transistor [3].

The drain–collector current waveforms $i_D(t)$ can be described by a train of chopped sinusoids with a conduction angle larger than 180 degrees

$$i_D(t) = I_{DC} + I_p \sin(\omega t) \text{ for } \omega t \le \frac{3\pi}{2} - \frac{\theta_{off}}{2} \tag{6.17}$$

$$i_D(t) = 0 \; for \frac{3\pi}{2} - \frac{\theta_{off}}{2} \le \omega t \le \frac{3\pi}{2} + \frac{\theta_{off}}{2} \tag{6.18}$$

$$i_D(t) = I_{DC} + I_p \sin(\omega t) \text{ for } \omega t \ge \frac{3\pi}{2} + \frac{\theta_{off}}{2} \tag{6.19}$$

where θ_{off} is the angle during which the transistor is off and $I_{MAX} = I_{DC} + I_p$.

The relationship between the current amplitude I_p and the quiescent current I_{DC} can be obtained from the condition

$$I_{DC} + I_p \sin\left(\pi + \frac{\pi - \theta_{off}}{2}\right) = 0 \tag{6.20}$$

which leads to

$$\theta_{off} = 2\cos^{-1}\left(\frac{I_{DC}}{I_p}\right). \tag{6.21}$$

Using Fourier series expansion, the DC and fundamental currents through a class AB stage can be expressed as [4]

$$I_{DC} = I_p\left\{\cos\left(\frac{\theta_{off}}{2}\right) + \frac{1}{\pi}\left[\sin\left(\frac{\theta_{off}}{2}\right) - \left(\frac{\theta_{off}}{2}\right)\cos\left(\frac{\theta_{off}}{2}\right)\right]\right\} \tag{6.22}$$

while the fundamental component of the current is described by

$$I_1 = I_p\left\{1 - \frac{1}{2\pi}\left[\theta_{off} - \sin(\theta_{off})\right]\right\}. \tag{6.23}$$

Finally, the optimum load resistance for a class AB stage becomes

$$R_{LOPT} = \frac{V_{DC} - V_{MIN}}{I_1}. \tag{6.24}$$

Using (6.22) and (6.23), the drain efficiency can be calculated as a function of θ_{off} assuming that all harmonics, except the fundamental, are shorted at the output

$$\eta_{drain} = \frac{P_L}{P_{DC}} = \frac{I_1}{2I_{DC}}\frac{V_{DC} - V_{MIN}}{V_{DD}} = \frac{1}{2}\frac{1 - \frac{1}{2\pi}\left[\theta_{off} - \sin(\theta_{off})\right]}{\cos\left(\frac{\theta_{off}}{2}\right) + \frac{1}{\pi}\left[\sin\left(\frac{\theta_{off}}{2}\right) - \frac{\theta_{off}}{2}\cos\left(\frac{\theta_{off}}{2}\right)\right]}\left(\frac{V_{DD} - V_{MIN}}{V_{DD}}\right). \tag{6.25}$$

This expression can be used to calculate the drain efficiency of all class A, B, AB, and C power amplifiers.

Switching PAs

As we have mentioned, the key idea behind maximizing the efficiency is to avoid the overlap of the current and voltage waveforms in the transistor. In theory, it is possible to achieve 100% efficiency by operating the transistor as a switch. This can be accomplished through waveform engineering, a process by which the voltage and current waveforms at the drain/collector are shaped such that the current is zero when the voltage is high and the voltage is zero when the current is high.

In practice, the desired non-overlapping waveforms for the drain current or voltage can be obtained by properly terminating the higher-order harmonics of the output signal, as in the class F and class F^{-1} stages where either the current or the voltage waveforms are square waves. The class E stage uses a switch without a specific harmonic termination and analog-like waveforms, whereas the class D stage employs a complementary switch with square waveforms. In all cases, the magnitude of the output current depends on how hard the transistor is driven into conduction. Ideally, for fast switching, the transistor should be driven as in a CML gate (see Chapter 11), swinging up to J_{pfT} in FETs and up to 1.5–2 times J_{pfT} in HBTs. These values are typically smaller than I_{MAX}. If the transistor is driven too hard or too weakly, f_T and f_{MAX} degrade, reducing the switching speed and the frequency at which the transistor can be used as a switch. Obviously, for a switching PA implementation to be even

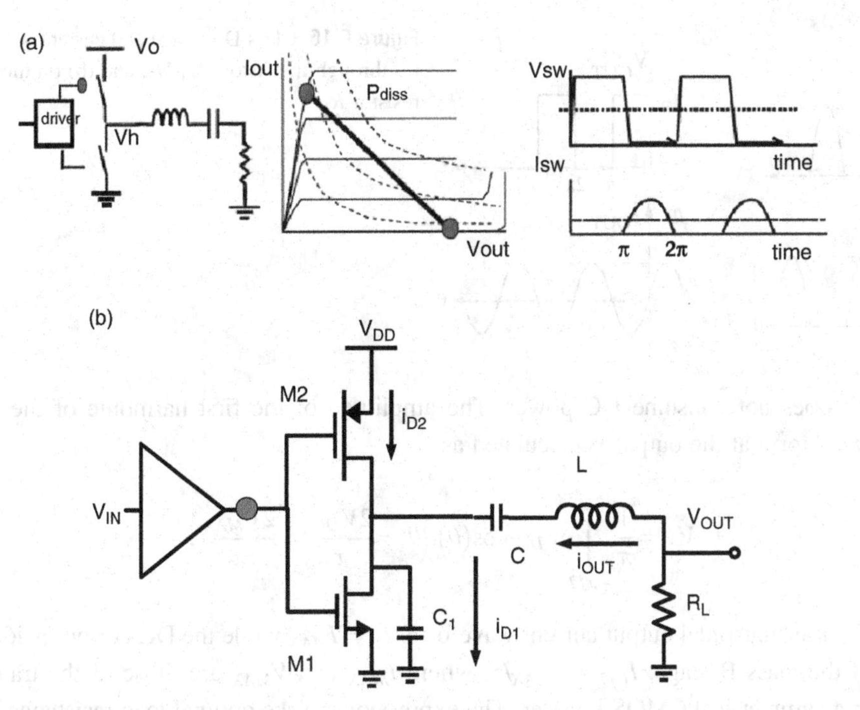

Figure 6.15 Class D PA (a) concept and (b) CMOS inverter schematics [4]

possible, the transistor has to exhibit useful gain at the higher-order harmonics. Therefore, switching PAs will have superior performance to a class AB or class B PA, only at frequencies much smaller than the f_T/f_{MAX} of the transistor.

Class D

Figure 6.15 illustrates the class D PA concept along with a practical implementation with a nanoscale CMOS inverter. Although it looks very similar to the push-pull class B PA in Figure 6.8, the class D switching amplifier differs from it in several important ways:

- the input voltage is a square rather than a sinusoidal waveform,
- the transistors act as switches with a square drain voltage waveform,
- the output load forms a high-Q series RLC filter, forcing the load current to become sinusoidal and, ideally, offering an infinite impedance at all harmonics higher than 1

$$Z_L(f_0) = R_L; Z_L(nf_0) = \infty \text{ for } n > 1. \tag{6.26}$$

The circuit draws current from the supply only when M_2 conducts. Thus, $V_{DD} = 2V_{DC} = V_{MAX}$. Although the output voltage waveform is digital and the current waveforms through each switch can be approximated by a train of half sinusoids, due to the selectivity of the R_L, L, C circuit and the push-pull operation of the complementary switches, the output current waveform through the load resistor is sinusoidal, as shown in Figure 6.16.

If $R_{ON} = 0$ and the number of harmonics is infinite, the theoretical efficiency is as high as 100% because the drain voltage and current waveforms of each transistor do not overlap, and

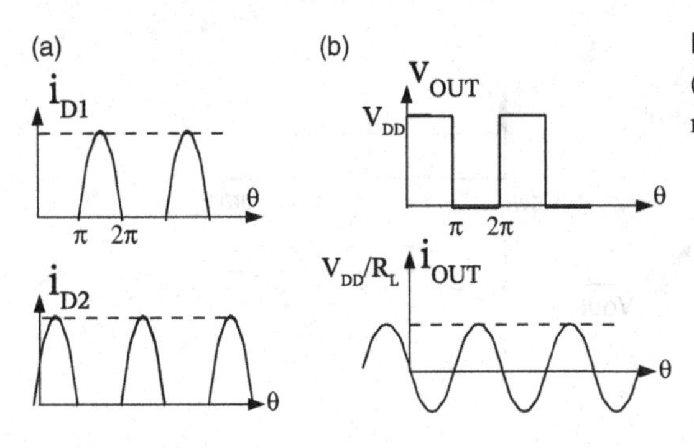

Figure 6.16 Class D voltage and current waveforms (a) through transistors M_1,M_2, and (b) on the load resistor R_L

the circuit does not consume DC power. The amplitude of the first harmonic of the digital voltage waveform at the output is calculated as

$$V_1 = \frac{1}{\pi}\int_{-\pi/2}^{\pi/2} V_{DD}\cdot\cos(\theta)d\theta = \frac{2V_{DD}}{\pi} = \frac{2V_{MAX}}{\pi}. \tag{6.27}$$

Because of the sinusoidal output current waveform, $I_1 = I_{MAX}$ while the DC current is identical to that of the class B stage: $I_{DC} = I_{MAX}/\pi$, where I_{MAX} and V_{MAX} are those of the transistor, assuming a symmetrical CMOS inverter. The expressions of the optimal load resistance and of the maximum efficiency thus become

$$R_{LOPT} = \frac{V_1}{I_1} = \frac{2V_{MAX}}{\pi I_{MAX}} = \frac{2\ V_{DD}}{\pi\ I_{MAX}} \tag{6.28}$$

$$P_L = \frac{V_1 I_1}{2} = \frac{2V_{DD}I_{MAX}}{2\pi} = \frac{V_{DD}I_{MAX}}{\pi}, \eta = \frac{P_L}{P_{DC}} = \frac{V_{DD}\cdot I_{MAX}}{\frac{\pi\cdot V_{DD}\cdot I_{MAX}}{\pi}} = 1. \tag{6.29}$$

However, like in the other PA stages, in a real class D stage, the non-zero R_{ON} (V_{MIN}) leads to power dissipation and degradation of efficiency. The charging and discharging of the output capacitance C_1 have also been ignored. They too, lead to increasing power dissipation with frequency. Additionally, asymmetries between the p-MOS and n-MOS transistors in the complementary switch lead to power dissipation. Nevertheless, since the performance of p-MOSFETs is now almost identical to that of n-MOSFETs, this topology is becoming popular in nanoscale CMOS technologies with good results demonstrated at GHz and even mm-wave frequencies.

Class E

The challenges of realizing the ideal square and half sinewave current-voltage waveforms of the class D stage are somewhat relaxed in the class E PA shown in Figure 6.17 [5]. The output network, consisting of the shunt capacitor C_1 (which includes the output capacitance of the transistor) and the series resonant circuit formed by C_2 and L is designed so that, when the switch closes and the current starts to flow, not only is the drain (switch) voltage $v_D = 0$, but also its derivative $dv_D/dt = 0$. This voltage and current waveform arrangement also makes the

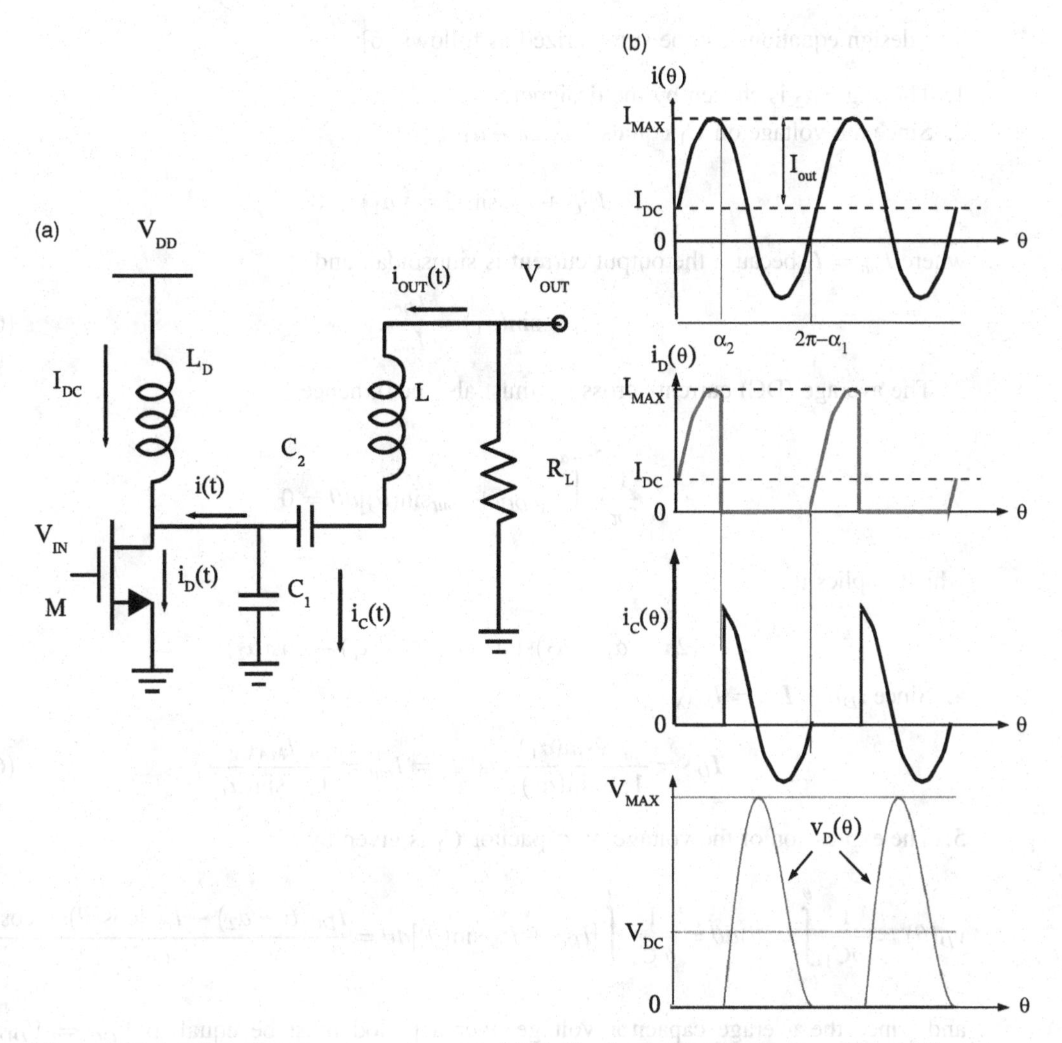

Figure 6.17 (a) Class E output stage schematics (b) Idealized current and voltage waveforms [1]

operation relatively insensitive to the rise time of the input signal [4]. The higher efficiency that ensues is accompanied by a very strong non-linearity. The maximum drain voltage V_{MAX} becomes as high as $3.6 \times V_{DD}$ and the maximum current, I_{MAX}, is $1.7 \times V_{DD}/R_L$ [1].

The current $i(t)$ is equal to the sum of the current through the switch, $i_D(t)$, and the current through the capacitor C_1, $i_C(t)$. The operation of the class E stage can be summarized as follows:

- The switch closes at $\theta = 2\pi - \alpha_1$. At this moment, the switch starts supplying the current i_D but the voltage v_D across the capacitor C_1 and the switch is 0.
- The switch opens at $\theta = \alpha_2$. At this time, capacitor C_1 supplies the current $i_C(t)$. The voltage across the switch and C_1 reaches $V_{MAX} > 2V_{DD}$.

Capacitor C_1 (which includes the transistor output capacitance) is sized such that $v_D(2\pi - \alpha_1) = 0$ and $i_C(2\pi - \alpha_1) = 0$.

The design equations can be summarized as follows [6]:

1. The angle α_2 is chosen by the designer.
2. Since the voltage on C_1 equals 0 at $2\pi - \alpha_1$

$$I_{DC} + I_{out}\sin(2\pi - \alpha_1) = 0$$

where $I_{out} = I_1$ because the output current is sinusoidal, and

$$\sin(\alpha_1) = \frac{I_{DC}}{I_{out}}. \tag{6.30}$$

3. The average (DC) current across C_1 must also be 0, hence

$$\frac{1}{2\pi}\int_{\alpha_2}^{2\pi-\alpha_1}[I_{DC} + I_{out}\sin(\theta)]d\theta = 0$$

which implies that

$$(2\pi - \alpha_1 - \alpha_2)\sin(\alpha_1) = \cos(\alpha_1) - \cos(\alpha_2).$$

4. Since $I_{DC} + I_{out} = I_{MAX}$

$$I_{DC} = \frac{I_{MAX}\sin(\alpha_1)}{1+\sin(\alpha_1)} \text{ and } I_1 = I_{out} = \frac{I_{MAX}}{1+\sin(\alpha_1)}. \tag{6.31}$$

5. The expression of the voltage on capacitor C_1 is given by

$$v_D(\theta) = \frac{1}{\omega C_1}\int_{\alpha_2}^{\theta} i_C(\theta)d\theta = \frac{1}{\omega C_1}\int_{\alpha_2}^{\theta}[I_{DC} + I_{out}\sin(\theta)]d\theta = \frac{I_{DC}(\theta - \alpha_2) - I_{out}[\cos(\theta) - \cos(\alpha_2)]}{\omega C_1}$$

and, since the average capacitor voltage over a period must be equal to $V_{DC} = V_{DD}$, the following condition must be satisfied

$$V_{DD} = V_{DC} = \frac{1}{2\pi}\int_{\alpha_2}^{2\pi-\alpha_1}\frac{I_{DC}(\theta - \alpha_2) - I_{out}\,[\cos(\theta) - \cos(\alpha_2)]}{\omega C_1}d\theta$$

from which, after integration, an equation for sizing C_1 is obtained

$$C_1 = \frac{I_{out}}{4\pi\omega V_{DC}}\left\{(2\pi - \alpha_1 - \alpha_2)[\cos(\alpha_1) + \cos(\alpha_2)] + 2[\sin(\alpha_1) + \sin(\alpha_s)]\right\}$$

$$C_1 = \frac{I_{out}}{4\pi\omega V_{DC}}\frac{[\sin(\alpha_1) + \sin(\alpha_2)]^2}{\sin(\alpha_1)} = \frac{I_{MAX}}{4\pi\omega V_{DC}}\frac{[\sin(\alpha_1) + \sin(\alpha_2)]^2}{\sin(\alpha_1)[1 + \sin(\alpha_1)]} = \frac{I_{DC}}{4\pi\omega V_{DC}}\frac{[\sin(\alpha_1) + \sin(\alpha_2)]^2}{\sin^2(\alpha_1)} \tag{6.32}$$

From this equation, several observations can be made:

- As the frequency increases, C_1 decreases. At some frequency f_m, the device output capacitance needed to achieve the desired output power will become too large to implement a class E stage.
- C_1, and thus f_m, can be increased if the supply voltage $V_{DD} = V_{DC}$ is decreased.
- Though not immediately apparent, these results and design equations are indirectly dependent on transistor f_T and f_{MAX} through V_{MAX}, which is inversely proportional to f_T and f_{MAX}.

Finally, the output power in the load can be calculated by integrating the voltage across the capacitor C_1 and the load current (including a "–" sign to account for the direction of i_{out})

$$P_L = \frac{-1}{2\pi}\int_{a_2}^{2\pi-a_1} v_D(\theta)I_{out}\sin(\theta)d\theta = \frac{V_{DC}I_{MAX}\sin(a_1)}{\sin(a_1)+1} = V_{DC}I_{DC}$$

(i.e. $\eta_d = 100\%$) and the optimum load impedance becomes

$$R_{LOPT} = \frac{2{\cdot}P_L}{I_{out}^2} = \frac{2V_{DC}I_{MAX}\sin(\alpha_1)[\sin(\alpha_1)+1]^2}{[\sin(\alpha_1)+1]I_{MAX}^2} = \frac{2V_{DD}}{I_{MAX}}\sin(\alpha_1)[\sin(\alpha_1)+1] = \frac{2V_{DC}}{I_{DC}}\sin^2(\alpha_1). \tag{6.33}$$

The class E topology has been successfully employed in 2GHz PAs realized in 130nm and 90nm CMOS technologies [7], in GaN HEMT PAs [6] and in mm-wave SiGe HBT PAs [8]. It should be noted that, compared to class AB and B stages with the same V_{MAX}, V_{DD} is reduced. The popularity of the class E stage has increased as MOSFETs have become faster in advanced technology nodes.

Class F

The class F topology, shown in Figure 6.18, has the same square drain voltage waveform as the class D stage, while the drain current waveform can be approximated by a periodic half sinusoid. Unlike class D or E, the load forms a parallel resonant tank, with $R_L = Z_0$. By placing a $\lambda/4$ transmission line at the fundamental frequency f_O between the load and the drain/collector, the transistor is terminated on R_L at f_O, on a short circuit at even harmonics of f_O, and on an open circuit at odd harmonics

$$Z_L(f_0) = R_L; Z_L(2{\cdot}k{\cdot}f_0) = 0; Z_L[(2k+1){\cdot}f_0] = \infty. \tag{6.34}$$

Using the waveforms in Figure 6.18, the equations for the DC and maximum current, DC and maximum voltage, and for R_{LOPT} become

$$I_{DC} = \frac{I_{MAX}}{\pi},\quad I_1 = \frac{I_{MAX}}{2},\quad V_{DC} = V_{DD} = \frac{V_{MAX}}{2},\quad V_1 = \frac{4V_{DD}}{\pi} = \frac{2V_{MAX}}{\pi} \tag{6.35}$$

$$P_L = \frac{V_1I_1}{2} = \frac{V_{MAX}I_{MAX}}{2\pi}, R_{LOPT} = \frac{V_1}{I_1} = \frac{4V_{MAX}}{\pi I_{MAX}} = \frac{8V_{DD}}{\pi^2 I_{DC}}. \tag{6.36}$$

This multiple harmonic output termination network helps to concentrate most of the power in the load at the fundamental frequency and to maximize the PA efficiency. The efficiency increases with the number of harmonic terminations designed to satisfy (6.34). For example, if the first three harmonics are properly terminated, the theoretical drain efficiency is as high as 81%. It increases to 90% if the first five harmonics are appropriately terminated. The maximum drain voltage of 2.55 × V_{DD} is higher than in classes A, B, C, and D, but not as high as in the

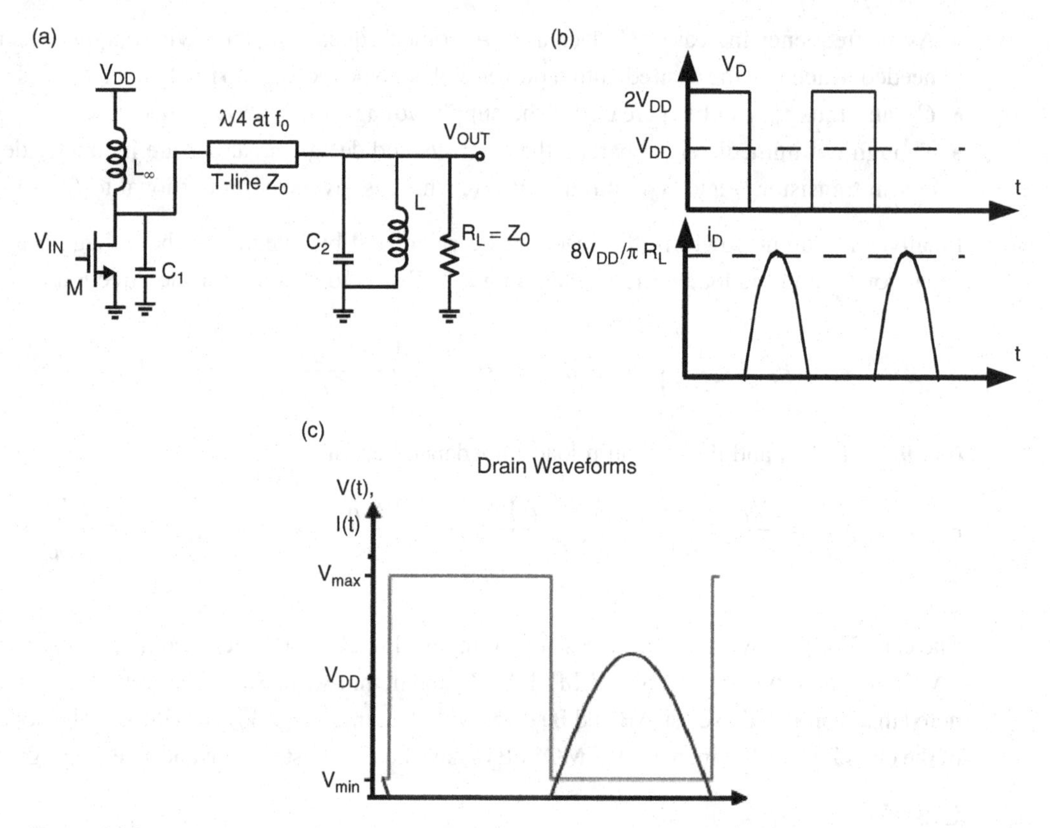

Figure 6.18 (a) Class F output stage schematics (b) Idealized drain voltage and current waveforms (c) Real drain voltage and current waveforms

class E PA. Like the class D amplifier, the class F PA suffers from switching loss. Furthermore, the output capacitance of the transistor ultimately limits the number of odd harmonics than can be terminated on a high impedance.

Inverse class F (F^{-1}) stage

As shown in Figure 6.19, this is the dual of the class F stage with the load being a series resonant tank. The ideal current waveforms are rectangular while the ideal voltage waveforms are a periodic train of half sinusoids. The corresponding optimal harmonic termination, I_{DC}, V_{MAX}, R_{LOPT} are given by

$$Z_L(f_0) = R_L; Z_L(2kf_0) = \infty; Z_L[(2k+1)f_0] = 0 \tag{6.37}$$

$$V_{DC} = \frac{V_{MAX}}{\pi},\ V_1 = \frac{V_{MAX}}{2},\ I_{DC} = \frac{I_{MAX}}{2},\ I_1 = \frac{4I_{DC}}{\pi} = \frac{2I_{MAX}}{\pi} \tag{6.38}$$

$$P_L = \frac{V_1 I_1}{2} = \frac{V_{MAX} I_{MAX}}{2\pi},\ R_{LOPT} = \frac{V_1}{I_1} = \frac{\pi V_{MAX}}{4I_{MAX}} = \frac{\pi^2 V_{DC}}{8I_{DC}}. \tag{6.39}$$

Figure 6.20 summarizes the optimal termination conditions for the highly non-linear switching amplifier stages [10] while Table 6.1 compares the theoretical R_{LOPT}, P_L, and drain efficiencies of the main amplifier classes.

Table 6.1 **Comparison of the theoretical performance of the various PA classes.**

param/class	A	B	D	E (one case)	F	F^{-1}
V_{DD}	$V_{MAX}/2$	$V_{MAX}/2$	V_{MAX}	$V_{MAX}/3.6$	$V_{MAX}/2$	V_{MAX}/π
I_{DC}	$I_{MAX}/2$	I_{MAX}/π	I_{MAX}/π	$I_{MAX}/2.06$	I_{MAX}/π	$I_{MAX}/2$
$P_L/(V_{MAX}I_{MAX})$	0.13	0.13	0.16	0.16	0.16	0.16
R_{LOPT}	V_{MAX}/I_{MAX}	V_{MAX}/I_{MAX}	$2V_{MAX}/\pi I_{MAX}$	$1.075V_{MAX}/I_{MAX}$	$4V_{MAX}/\pi I_{MAX}$	$\pi V_{MAX}/4I_{MAX}$
η_d	50.00%	78.00%	100.00%	100.00%	100.00%	100.00%

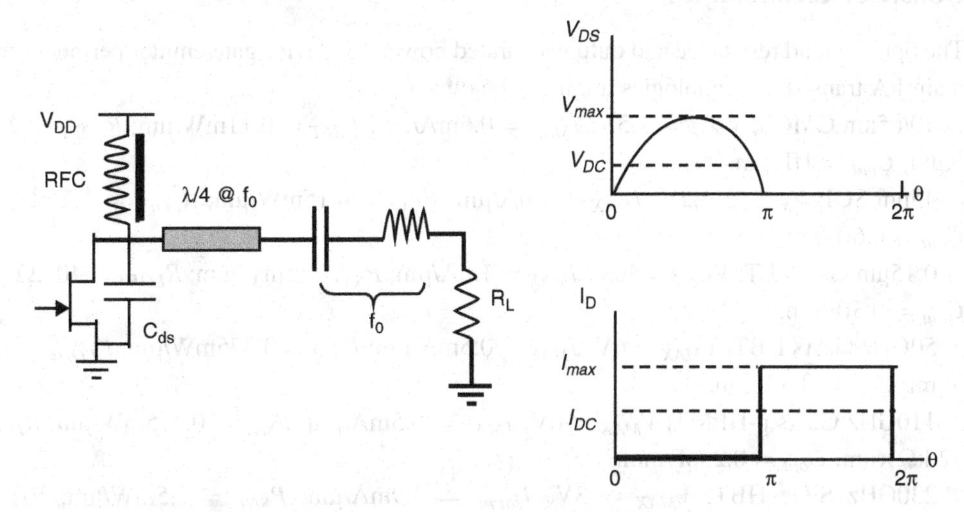

Figure 6.19 Class F^{-1} stage schematics and idealized waveforms [9]

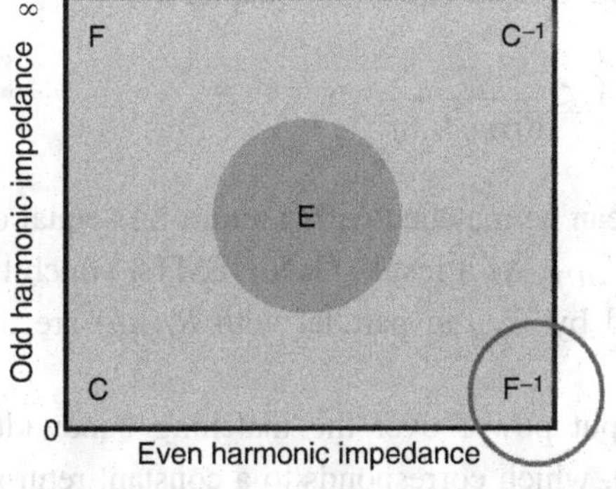

Figure 6.20 Ideal harmonic terminations for class C, E, and F stages [10]

Final notes on switching PAs

- As in the case of the class B stage, because the transistor is operated as a switch, the power gain is at least 6dB lower than that of a class A stage.
- Theoretically, the improvement in output power and efficiency compared to a class B stage is 1dB or less.

- A drain efficiency of 100% is impossible to achieve in reality because of:
  - non-zero V_{MIN} and R_{ON},
  - losses in the output matching network, and
  - transit and RC delays in the transistor itself [9].

Also, a perfect periodic square waveform requires an infinite number of odd harmonics while a perfect periodic half sinewave requires an infinite number of even harmonics. In practice, close to 90% efficiency has been demonstrated with a finite number of harmonics (at least 5) using GaN HEMT technology.

6.2.3 Typical class A PA saturated output power densities and R_{LOPT} in different transistor technologies

The optimal load resistance and output saturated power per device gate/emitter periphery for the main PA transistor technologies are listed below:

90/65nm CMOS: $V_{MAX} = 1.5$V, $J_{MAX} = 0.6$mA/μm; $P_{SAT} = 0.11$mW/μm; $R_{LOPT} = 2.5$ kΩ ×μm, $C_{out} = 1$fF/μm.

45nm SOI: $V_{MAX} = 1.2$V, $J_{MAX} = 1$mA/μm; $P_{SAT} = 0.15$mW/μm; $R_{LOPT} = 1.2$ kΩ ×μm, $C_{out} = 0.6$fF/μm.

0.15μm GaN FET: $V_{MAX} = 40$V, $J_{MAX} = 1$mA/μm; $P_{SAT} = 5$mW/μm; $R_{LOPT} = 40$ kΩ ×μm, $C_{out} = 0.3$fF/μm.

50GHz GaAs HBT: $V_{MAX} = 6$V, $J_{MAX} = 0.5$mA/μm; $P_{SAT} = 0.375$mW/μm; $R_{LOPT} = 12$ kΩ ×μm, $C_{out} = 0.5$fF/μm.

110GHz GaAs p-HEMT: $V_{MAX} = 6$V, $J_{MAX} = 0.5$mA/μm; $P_{SAT} = 0.375$mW/μm; $R_{LOPT} =$ 12 kΩ×μm, $C_{out} = 0.25$pF/mm.

230GHz SiGe HBT: $V_{MAX} = 3$V, $J_{sat,pp} = 3.9$mA/μm; $P_{SAT} = 1.5$mW/μm; $R_{LOPT} =$ 770Ω×μm, $C_{out} = 3.5$fF/μm.

It should be noted that, according to the Bode–Fano equation (Chapter 5)

$$\int_0^{\infty} \ln\left|\frac{1}{\Gamma(\omega)}\right| d\omega \leq \frac{\pi}{R_{LOPT}C_{out}}$$

the bandwidth, BW, over which the PA stage can be matched with a return loss equal or better than Γ is inversely proportional to $R_{LOPT} \times C_{OUT}$. As a result, GaN HEMTs, which have the highest Q optimal output impedance (formed by C_{out} in parallel with R_{LOPT}) are the most difficult to match over a broad bandwidth

For example, for a 1dB reduction in output power over the matching bandwidth (3dB bandwidth is totally inadequate in PA design), which corresponds to a constant return loss of 20 dB (Γ = 0.1), the `1dB matching bandwidths of the above technologies are given by

$$BW_{1\text{dB}} < \frac{4.343}{20 R_{LOPT} C_{out}}.$$

The GaN HEMT has the smallest bandwidth, while the nanoscale CMOS PAs exhibit the largest bandwidth.

6.2.4 Examples of 1-W class A, AB and B PA designs in various technologies

Based on the PA technology parameters in the previous section, it is instructive to compare the calculated values of the device periphery and of optimal load resistance for 1-W output stages in different technologies:

90/65nm n-MOS: $W = 9.09$mm; $R_{LOPT} = 0.27\Omega$ (impossible to match from 50Ω).
45nm SOI n-MOS: $W = 6.7$mm; $R_{LOPT} = 0.18\Omega$ (impossible to match from 50Ω).
150nm GaN FET: $W = 0.2$mm; $R_{LOPT} = 200\Omega$ (easy to match from 50Ω).
50GHz GaAs HBT: $l_E = 2.66$mm; $R_{LOPT} = 4.5\Omega$ (possible to match from 50Ω).
110GHz GaAs p-HEMT: $W = 2.66$mm; $R_{LOPT} = 4.5\Omega$ (possible to match from 50Ω).
230GHz SiGe HBT: $l_E = 0.67$mm; $R_{LOPT} = 1.16\Omega$ (possible but difficult to match from 50Ω).

These power densities, transistor sizes, and optimal load resistance values are independent of the frequency of the PA as long as the transistor has adequate power gain at the frequency of interest and at the corresponding bias conditions. For example, in a class B PA the gain is approximately 6dB lower than in a class A PA, and the class B stage will be the first to run out of gain at high frequencies. Although, in theory, any of these technologies could be used to produce 1W PAs at 1–5GHz, it is obvious that a high voltage device is needed to deliver >1 W of power with realistic on-chip matching networks.

The reduction in breakdown voltage with continued scaling and the saturation of I_{ON} at levels of 1–1.2mA/μm in newer technology nodes dictates the need for larger bias currents to achieve the same output power as in earlier technology nodes. This leads to larger MOSFET sizes and hence smaller R_{LOPT}, which complicates the output matching process. It gives III-V technologies and SiGe HBTs a clear advantage in PA design, since their higher breakdown voltage permits much lower bias currents and hence smaller devices, simplifying the output matching to 50Ω.

In nanoscale CMOS technologies, this limitation can be overcome at GHz frequencies by employing cascode stages which combine thin-oxide, short channel MOSFETs and thick-oxide, higher breakdown MOSFETs, or by series stacking, as will be discussed later.

EXAMPLE 6.2 10-W, 3GHz , GaN HEMT class E, F and inverse F PA designs

Let us consider the following 0.25μm GaN on SiC HEMT process from Triquint with the following transistor parameters: $J_{MAX} = 1$mA/μm, $C'_{out} = 0.3$fF/μm, $R'_{ON} = 2$kΩ ×μm [6]. We will contrast the class E, F, and F^{-1} designs assuming the same $V_{MAX} = 40$V and output power $P_L = 10$W.

Class E

According to the design equations, the class E design will reach the highest V_{MAX} with respect to VDD, when the conduction angle $\alpha_1 + \alpha_2 = 180°$. Although the maximum efficiency does not depend on the choice of α_1 and the conduction angle, according to the technology parameters, a good choice for α_2 is 110°. Thus $\alpha_1 = 70°$.

Assuming $V_{MAX} = 40$V, the design starts by calculating the highest possible $V_{DD} <= V_{MAX}/3.6 = 11.11$V.

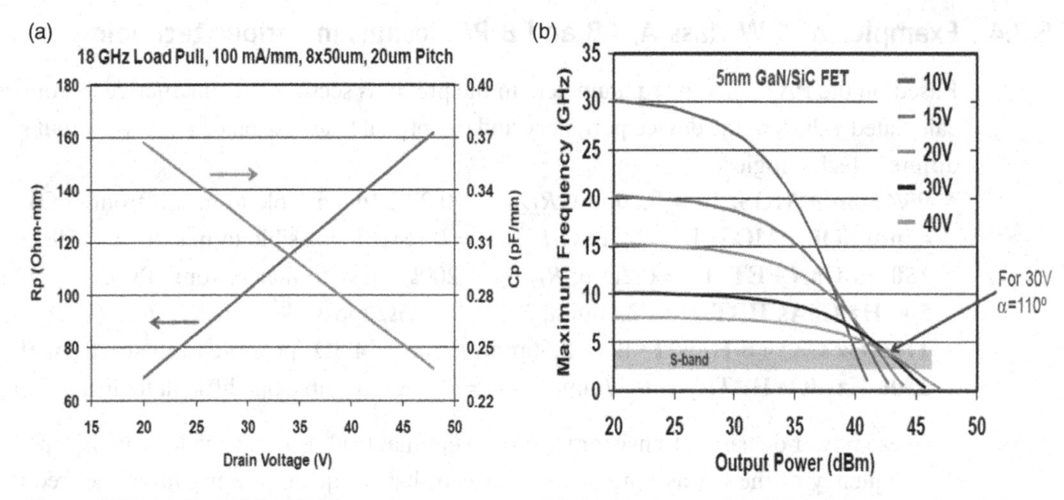

Figure 6.21 (a) GaN/SiC HEMT transistor parameters [6] (b) Maximum class E PA frequency plotted parametrically vs. PA output power for a 5mm gate periphery GaN HEMT with different V_{DD} [6]

We select $V_{DD} = 11$V, which results in a maximum drain voltage $V_{MAX} = 39.6$V.

Assuming that 80% efficiency is achievable, we budget the DC current to account for the 20% loss in power.

$I_{DC} = P_L/(V_{DD} \times 0.8) = 10/(11\times0.8) = 1.136$ A

$$R_{LOPT} = \frac{2V_{DD}}{I_{DC}}\sin^2(\alpha_1) = \frac{22}{1.136}\sin^2(\alpha_1) = 19.36\sin^2(\alpha_1).$$

Based on the experimental data in Figure 6.21(b), we select a_1.

The earlier choice of $\alpha_2 = 110°$ and $\alpha_1 = 70°$ leads to

$$I_{MAX} = I_{DC}\frac{\left(1+\sin(\alpha_1)\right)}{\sin(\alpha_1)} = \frac{1.136\cdot1.93969}{0.93969} = 2.3449A \text{ and } R_{LOPT} = 17.1\Omega.$$

This leads to a total device periphery of $W = I_{MAX}/J_{MAX} = 2.35$mm. In practice, the device size will be somewhat larger because we have not accounted for the impact of R_{ON}.

For a 3GHz design, we obtain $C_1 = 1.74$pF which includes the output capacitance of the transistor $C_{out} = 0.3$(fF/μm) × 2350μm = 705fF.

Indeed, Figure 6.21 reproduces measurements of the output power and power-added efficiency (G >13dB) of such a class E GaN HEMT PA designed for the 3GHz band [6]. The gate periphery is 2.55mm and the peak drain efficiency is approximately 70%, 10% lower than the value assumed in the design.

Class F

We assume the same $V_{MAX} = 40$V and $J_{MAX} = 1$mA/μm.

Hence, we obtain $V_{DD} = V_{MAX}/2 = 20$V.

Assuming 80% efficiency, because we expect to be able to properly terminate only the first three harmonics of the output signal, $I_{DC} = 10\text{W}/(0.8 \times V_{DD}) = 625$mA.

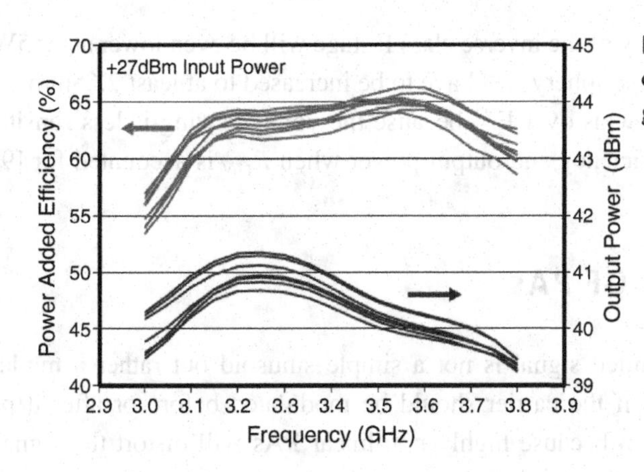

Figure 6.22 Measured output power and PAE of a 3GHz class E GaN HEMT PA. The total gate periphery is 2.55mm [6]

$I_{MAX} = \pi I_{DC} = 1.9625$ A. Hence the total gate periphery becomes $W = I_{MAX}/J_{MAX} =$ 1.9625 mm

$$R_{LOPT} = \frac{4 \cdot V_{MAX}}{\pi \cdot I_{MAX}} = \frac{160}{6.16} = 26\Omega.\ R_{ON} = 1.019\Omega.$$

If we account for R_{ON}, since $V_{MIN} = I_{MAX} \times R_{ON}$ [9]

$$P_L = \left(\frac{4}{\pi}\right)^2 \frac{V_{MAX}^2}{8R_{LOPT}} \left(1 + \frac{8}{\pi}\frac{R_{ON}}{R_{LOPT}}\right)^{-2} = 10.3W \text{ and } \eta = \left(1 + \frac{8}{\pi}\frac{R_{ON}}{R_{LOPT}}\right)^{-1} = 0.909.$$

Note that this will have to be multiplied by 0.81 if we are to account for the fact that only the first three harmonics will be terminated. Clearly, we need to increase the gate periphery by another 20–30% to end up with 10W of output power and a drain efficiency of $0.81 \times 0.909 <= 73\%$. Therefore, the practical transistor gate periphery for 10W will be closer to 2.5mm.

Inverse class F

$V_{MAX} = 40$V and $J_{MAX} = 1$mA/μm remain unchanged.

$V_{DD} = V_{MAX}/\pi = 12.7$V. Assuming 80% efficiency because only the first three harmonics will be matched, $I_{DC} = 10\text{W}/(0.8 \times V_{DD}) = 982$mA.

$I_{MAX} = 2 \times I_{DC} = 1.964$ A. Hence the total gate periphery becomes $W = I_{MAX}/J_{MAX} =$ 1.964mm

$$R_{LOPT} = \frac{\pi \cdot V_{MAX}}{4 \cdot I_{MAX}} = \frac{125.66}{7.856} = 16\Omega \cdot\ R_{ON} = 1.018\Omega.$$

If we account for R_{ON} [9]

$$P_L = \frac{V_{MAX}^2}{8R_{LOPT}} \left(1 + \frac{\pi^2}{4}\frac{R_{ON}}{R_{LOPT}}\right)^{-2} = 9.33W \text{ and } \eta = \left(1 + \frac{\pi^2}{4}\frac{R_{ON}}{R_{LOPT}}\right)^{-1} = 0.864.$$

The impact of R_{ON} reduces the efficiency and the output power significantly. If we were to account for the 81% ideal efficiency because only the first three harmonics are matched, then the overall output power and efficiency of the inverse class F stage will be even lower at 7.55W and 70%, respectively. The transistor periphery will have to be increased to at least 2.65mm.

The conclusion is that class F wins over F^{-1} because the voltage swing is less sensitive to R_{ON}, resulting in higher drain efficiency and output power when R_{ON} is accounted for [9].

6.3 LINEAR MODULATION OF PAs

In most applications, the transmitted signal is not a simple sinusoid but rather a modulated carrier. The question then arises if the carrier should be modulated before or after it passes through the PA. This is a concern because highly non-linear PAs will distort the signal and corrupt the amplitude information that needs to be transmitted.

Modulation of the carrier in the transmitter is typically accomplished in three ways:

- the signal arriving at the PA input is already modulated (typically in frequency, phase, or pulse width in the case of some digital PAs),
- the modulating signal is introduced through the supply node of the PA, or
- the modulating signal is applied at the output of the PA.

The latter technique is also known as direct RF modulation. In the case of radio systems with class A, B, or AB PAs, the signal is already modulated by the data as it arrives at the input of the PA.

In the highly non-linear class C, D, E, or F PAs, the modulating data signal is normally separated from the carrier. The carrier signal is applied at the input of the PA while the modulating data signal is typically introduced at its output through the supply node, as illustrated in Figure 6.23. This approach dates back to the radio transmitters of the 1930s [11] and can be employed to provide amplitude modulation. Another solution that has been applied in pulse modulated radars is to simply place a switch after the PA. The switch turns the transmitted signal on and off, producing amplitude modulation. Direct BPSK, QPSK, or 16/64 QAM modulation can also been introduced in a similar manner, as described in Chapter 9.

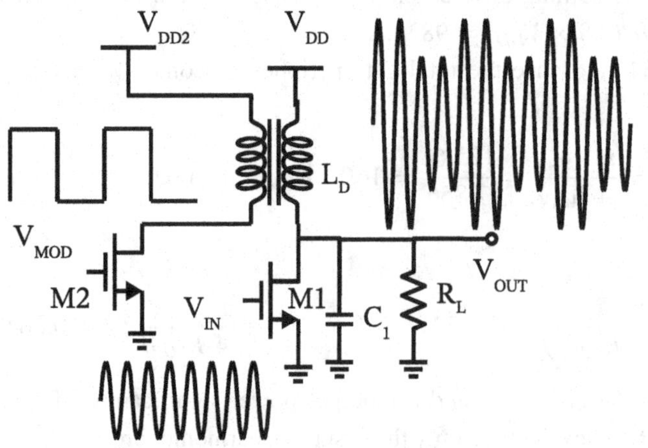

Figure 6.23 Principle of linear modulation of PAs through the supply node

More recently, as discussed in Section 6.6.2, digital signal processing has been employed to encode the amplitude information into the pulse width of a digital signal that is directly applied to the input of the PA [7].

6.4 CLASS A PA DESIGN METHODOLOGY

6.4.1 Design equations

The linear swing of a class A PA depends on the flatness of the f_{MAX} versus J_{DS}/J_C characteristics. The OP_{1dB} is maximized when the transistor (HBT or FET) is biased at the peak f_T current density J_{pfT}. As discussed in Chapter 4, in SiGe HBTs, J_{pFT} increases almost with the square of the technology scaling factor. This implies that, despite the reduction in the breakdown voltage of the transistor at higher f_T, large output power can be obtained in newer technology nodes at increasingly higher frequencies, with roughly the same transistor size as in older nodes, by increasing the collector current density to compensate for the smaller voltage swing.

As illustrated in Figure 6.24, the situation is quite different in MOSFETs, where the linear voltage swing at input/output decreases with each new node while J_{pFT} remains largely constant at 0.3–0.4mA/μm. The lower value corresponds to $V_{DS} <= 0.6$V and the upper bound corresponds to $V_{DS} = 1.2$V. Therefore, in MOSFET PAs the OP_{1dB} current swing $J_{swing,pp}$ remains largely constant across nodes at about 0.4mA$_{pp}$/μm, to avoid transistor cutoff.

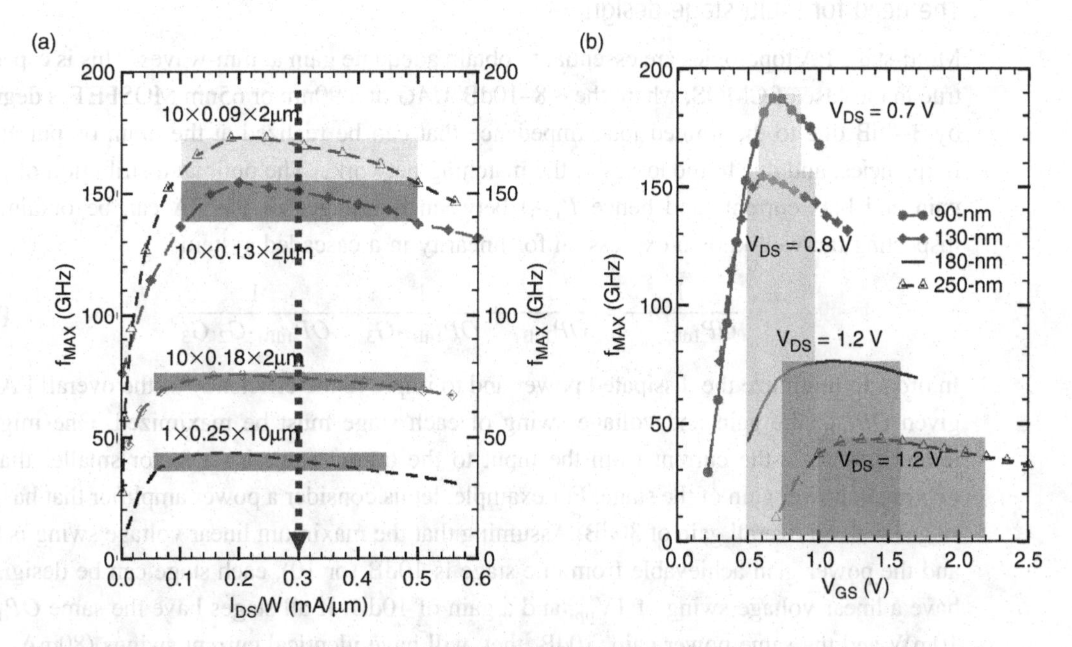

Figure 6.24 Measured f_{MAX} of n-channel MOSFETs from different technology nodes as a function of (a) drain current density per unit gate width and (b) as a function of V_{GS}. The range of voltages and current over which f_{MAX} changes by less than 10% from its peak value are highlighted

With these observations, and the definitions of current and voltage swings illustrated on the output transistor characteristics in Figure 6.24, the output compression point can be directly linked to the technology data and transistor size

$$OP_{1\text{dB}} = W{\cdot}J_{swing,pp}\frac{(V_{DD} - V_{dsat})}{4} = W{\cdot}J_{swing,pp}\frac{(V_{MAX} - V_{dsat})}{8}. \tag{6.40}$$

We note that (6.40) is valid when the output stage is terminated on R_{LOPT}. Similarly, to achieve the saturated output power P_{SAT}, the corresponding current swing $J_{sat,pp}$ is 0.6–0.7mA$_{pp}$/μm for n-FETs and $2 \times J_{pfT}$ for HBTs. The maximum current that can pass through a MOSFET is limited by I_{ON}. The latter is expected to continue to increase in future CMOS nodes. I_{ON} values as high as 1.3mA/μm and 1mA/μm have been reported for production 45nm n-MOSFETs and p-MOSFETs, respectively [12].

As we have seen, for class A, B, and AB PAs, the maximum voltage swing $V_{sat,pp}$ of the fundamental at P_{SAT} is approximately $2 \times V_{DD}$, which is limited by V_{MAX}. Similarly, in all three classes, the maximum peak-to-peak current swing at the fundamental frequency is limited by $J_{sat,pp}$, therefore the saturated output power for these PA classes is limited by

$$P_{sat} = \frac{W{\cdot}J_{sat,pp}{\cdot}V_{DD}}{4} = \frac{W{\cdot}J_{sat,pp}{\cdot}V_{MAX}}{8}. \tag{6.41}$$

6.4.2 mm-Wave class A PA design methodology

The need for multi-stage design

Multi-stage PA topologies are essential to obtain adequate gain at mm-waves. This is especially true in the case of CMOS, where the ~8–10dB *MAG* of a 90nm or 65nm MOSFET is degraded by 3–4dB due to the limited load impedance that can be realized at the drain output at high frequencies, and due to the losses of the matching networks. The optimal distribution of power gain and bias current (and hence $P_{1\text{dB}}$) between the stages of the PA can be obtained by inspecting the well-known expression for linearity in a cascaded system

$$\frac{1}{OP_{1\text{dB}_{cascade}}} = \frac{1}{OP_{1\text{dB}_3}} + \frac{1}{OP_{1\text{dB}_2}{\cdot}G_3} + \frac{1}{OP_{1\text{dB}_1}{\cdot}G_2{\cdot}G_3}. \tag{6.42}$$

In order to minimize the dissipated power and to improve the efficiency of the overall PA for a given $OP_{1\text{dB}}$, the gain and voltage swing of each stage must be maximized. One might be tempted to scale the current from the input to the output stage by a factor smaller than the maximum power gain of the stage. For example, let us consider a power amplifier that has three stages with an overall gain of 30dB. Assuming that the maximum linear voltage swing is 1V$_{pp}$, and the power gain achievable from one stage is 10dB (or 10), each stage can be designed to have a linear voltage swing of 1V$_{pp}$, and a gain of 10dB. If all stages have the same $OP_{1\text{dB}}$ = 10mW and the same power gain, 10dB, they will have identical current swings (80mA$_{pp}$) and the overall output compression point will be 9mW, approximately equal to that of the output stage (10mW).

Alternatively, we could scale the currents by a factor of 10. The OP_{1dB} of each stage will increase by a factor of 10 from stage to stage. To achieve the same overall $OP_{1dB} = 9$mW, the output stage will have a three times larger OP_{1dB} of 27mW, with those of the first two stages being 0.27mW and 2.7mW, respectively. The corresponding peak-to-peak current swings in each stage will be 216mA$_{pp}$, 21.6mA$_{pp}$, and 2.6mA$_{pp}$, respectively. Interestingly, both designs will have the same overall gain (30dB), power dissipation, PAE, and output compression point, but the second design might require an output stage that is too large to allow for its current to be carried by on-chip inductors and to be easily matched to 50Ω.

An algorithmic design methodology can be developed for a linear mm-wave class A CMOS PA based on the load line theory [1] and constant current-density biasing for optimal linearity. This is summarized in the seven steps outlined below [13]:

Step 1 Starting at the output stage, determine the maximum allowed voltage swing for the given technology. From load line theory, the optimal linearity and output power are obtained when the transistor (with inductive load) is biased such that the drain voltage swings equally between $V_{DS,sat}$ and V_{MAX} (dictated by the device breakdown or reliability limit), centered at V_{DD}. Thus, the maximum voltage swing is $V_{swing} = V_{MAX} - V_{DSAT}$, where $V_{MAX} = 2V_{DD} - V_{DSAT}$

Step 2 Set the bias current density to 0.3–0.4mA/μm to maximize linearity.

Step 3 Determine the bias current that meets the P_{1dB} requirements and, from that find the transistor width. An expression for P_{1dB} can be derived from load line theory

$$P_{1dB} = I_{swing} \times \frac{(V_{DD} - V_{DSAT})}{4} \tag{6.43}$$

where $I_{swing} = (0.4\text{mA}_{pp}/\mu\text{m}) \times W$ is the maximum current swing before 1dB compression, instead of $2 \times I_{DC}$.

In most nanoscale CMOS technologies, the value of V_{DSAT} corresponding to the optimal linearity bias point of 0.3mA/μm is about 0.3V. From (6.21) $W = 4\times P_{1dB}/[0.4\text{mA}_{pp}/\mu\text{m} \times (V_{DD} - V_{DSAT})]$ and $I_{DC} = W \times 0.3\text{mA}/\mu\text{m}$.

Step 4 Add a degeneration inductor L_S to improve stability, Figure 6.25. Otherwise, the input resistance can become negative for an inductively loaded CS/CE stage. Iterations may be needed since L_S also changes *MAG*. The degeneration inductor also improves linearity.

Step 5 Add an output matching network (from Z_0 to R_{LOPT}) for the last stage and (if necessary) interstage matching networks for intermediate stages to maximize power transfer.

Step 6 Repeat steps 1–5 for each preceding stage with the V_{swing} determined by V_{input} for the subsequent stage to avoid gain compression.

Step 7 Design the first stage to be input-matched to 50Ω. A cascode topology may be used in the first stage for higher gain.

In multi-stage PAs, the size and bias current of transistors increase towards the output.

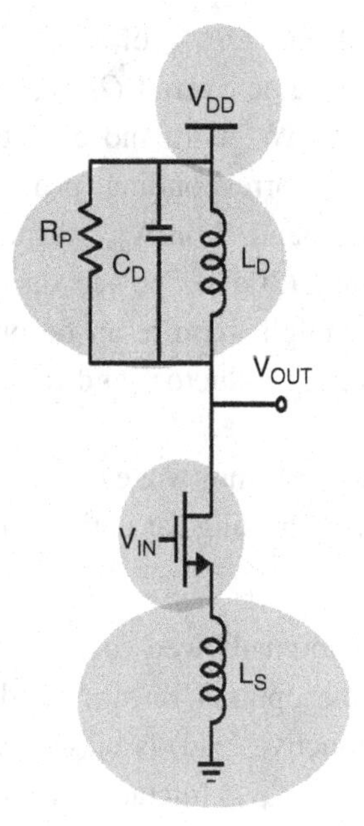

Figure 6.25 Schematic of output stage used in the design methodology

EXAMPLE 6.3 Design a 65nm CMOS output stage with P_{1dB} of 7dBm at 60GHz

Given: $V_{DD} = 0.7V$, $V_{DSAT} = V_{MIN} = 0.2V$ at 0.3mA/μm and the goal of $OP_{1dB} = 7dBm = 5mW$, one determines $V_1 = V_{DD} - V_{DSAT} = 0.5V$ ($1V_{pp}$) and $V_{MAX} = 1.5V$.

From $OP_{1dB} = 5mW = V_1 \times I_{swing}/4 => I_{swing} = 40mA_{pp}$ and $W = I_{swing}/0.4mA_{pp}/\mu m = 100\mu m$.

$R_{LOPT\,(P1dB)} = 2V_1/I_{swing} = 1000/40 = 25\Omega$. The corresponding DC bias current in the output transistor is $W \times 0.3mA/\mu m = 30mA$. $P_{DC} = I_{DC} \times V_{DD} = 30mA \times 0.7V = 21mW$, $\eta = 22.2\%$.

The output matching network is simplified because it only needs to tune out the reactance of the output transistor.

In the design methodology described above, we have ignored the losses in the output matching network due to its finite Q.

EXAMPLE 6.4

Design a SiGe HBT output stage with OP_{1dB} of 10dBm at 60GHz using a supply voltage of 1.5V. The peak f_T current density is 2mA/μm (of emitter length), the linear current swing at peak f_T bias is $3mA_{pp}/\mu m$, and $V_{CESAT} = V_{MIN} = 0.5V$.

Given: $V_{CC} = 1.5V$ and the goal of $OP_{1dB} = 10dBm = 10mW$, one determines $V_1 = V_{CC} - V_{CESAT} = 1V$ ($2V_{pp}$) and $V_{MAX} = 2.5V$.

From $OP_{1dB} = 10mW = V_1 \times I_{swing}/4 => I_{swing} = 40mA_{pp}$ and $l_E = I_{swing}/(3mA_{pp}/\mu m) = 13.3\mu m$.

$R_{LOPT(P1dB)} = 2V_1/I_{swing} = 2000/40 = 50\Omega$. The corresponding DC bias current in the output transistor is $l_E \times 2mA/\mu m = 26.7mA$. $P_{DC} = I_{DC} \times V_{CC} = 26.7mA \times 1.5V = 40\ mW$, $\eta = 25\%$.

6.4.3 Computer-based algorithmic class A PA design methodology to obtain a target OP_{1dB}

1. Start with a periodic S-parameter simulation at the desired frequency of operation. Bias the transistor at the maximum linearity bias current density. Use the transistor size calculated by hand analysis and 50Ω input and output ports. Plot MAG as a function of P_{in} and find the 1dB compression point at the input, IP_{1dB}. Since we need OP_{1dB} output power, the input power at 1dB compression will be at least ($OP_{1dB} - MAG$) dBm. By the same reasoning as above, the best possible OP_{1dB} is $IP_{1dB} + MAG(IP_{1dB})$. Set this OP_{1dB} power level to be at least 1–2dBm larger than the target. Otherwise, we need to increase the transistor size while maintaining the optimal linearity bias current density.
2. Plot the real and imaginary parts of Z_{in} and Z_{out}. Examine their values at the IP_{1dB} input power level and conjugate match the input and the output at this input/output power level.
3. Re-run the periodic S-parameter simulation on the matched amplifier and plot P_{out} versus P_{in}, Gain versus P_{in} and PAE versus P_{in}. A load pull simulation should be conducted to further tweak the output impedance and to obtain the maximum possible output power. If you still cannot achieve the desired output power, repeat steps 1–2 after further increasing the transistor sizes.

6.4.4 Additional PA design considerations

Output power control

The transmitted power level is usually prescribed by the various IEEE standards. Most of these standards mandate at least some capability to dynamically control the output power level. Furthermore, it is desirable that the efficiency of the PA does not degrade as the output power is reduced. Most of the efficiency-enhancement techniques to be discussed in Section 6.7 deal with the important issue of improving efficiency when the output power is backed off from the maximum PAE bias. All of these techniques support power control.

PA linearity, and the associated methods to improve it, remains a very active area of research, prompted by the proliferation of wireless standards that employ some form of amplitude modulation. Two transmitter specification parameters address the linearity performance of the PA:

- the spectral mask and
- the error vector magnitude (EVM) of the output signal.

The spectral mask is concerned with preventing the excessive noise generated by transmitters from leaking into adjacent communication channels and corrupting them.

The EVM specification deals with the signal-to-noise ratio (SNR) budget of a communication link and allocates a maximum SNR degradation to the transmitting PA.

Depending on the type of modulation format mandated by a particular wireless standard, the PA design is limited either by the spectral mask or by the EVM specification. In general, in higher data-rate standards, which require more complex modulations and higher bandwidth efficiency, the EVM specification is more difficult to satisfy.

6.5 NON-IDEALITIES IN PAs

6.5.1 Instability in CS/CE stages due to the drain–collector inductance

The input impedance of a common-emitter or common-source stage with inductive load Z_L can be derived using Miller's theorem to account for both the source/emitter degeneration impedance, Z_E, and for the Miller capacitance C_{gd} or C_{bc} [4]

$$Z_{in} = \frac{V_{in}}{I_{in}} = \frac{\frac{1}{g_m} + \left(1 + j\frac{\omega}{\omega_T}\right) Z_E}{\frac{1}{g_m (Z_f + Z_L)} + \frac{Z_E + Z_L}{Z_f + Z_L} + j\frac{\omega}{\omega_T}\left(1 + \frac{Z_E}{Z_f + Z_L}\right)} \tag{6.44}$$

where
$Z_L = j\omega L_L, Z_f = \frac{1}{j\omega C_{gd(bc)}}, Z_E = j\omega L_E, \omega_T = \frac{g_m}{C_{gs}}$, L_L is the load inductor and L_E is the emitter degeneration inductor. If no inductive degeneration is present ($Z_E = 0$)

$$Z_{in} \approx \frac{1}{g_m}\frac{Z_f}{Z_L} \approx \frac{-1}{\omega^2 g_m C_{gd(bc)} L_L}. \tag{6.45}$$

It should be noted from (6.44) that, in the absence of inductive degeneration, this negative resistance can cause instability and even oscillation of the output stage. By adding L_E, an additional positive real term appears in (6.45) that stabilizes the amplifier. The value of L_E must be chosen as small as possible so as not to degrade the power gain.

6.5.2 Thermal runaway in class A bipolar PAs

In large output stages realized with bipolar transistors, the transistor layout will consist of many emitter stripes connected in parallel. Due to the heat dissipated in the transistor, the stripes in the center of the layout will tend to have higher local temperature than those on the periphery. The higher temperature leads to a local increase in current. The stripes in the center will effectively "steal" the current from the other stripes, resulting in additional heat and further increase in current. This positive feedback phenomenom is one of the major disadvantages of bipolar PAs and can destroy the device. It is known as thermal runaway.

The practical solution to thermal runaway, illustrated in Figure 6.26, is to add "ballast" resistors in each emitter stripe, which provide local negative feedback. Thermal runaway is normally not a problem in FET-based PAs.

6.5.3 Avalanche breakdown

As we have seen, in order to maximize the output power and the power-added efficiency, all PAs operate with large voltage swings, close to the collector–emitter or drain–gate breakdown of the transistor. Avalanche multiplication occurs when electron–hole pairs are generated at high electric field in the reversed-biased base–collector junction of a bipolar transistor. Unless

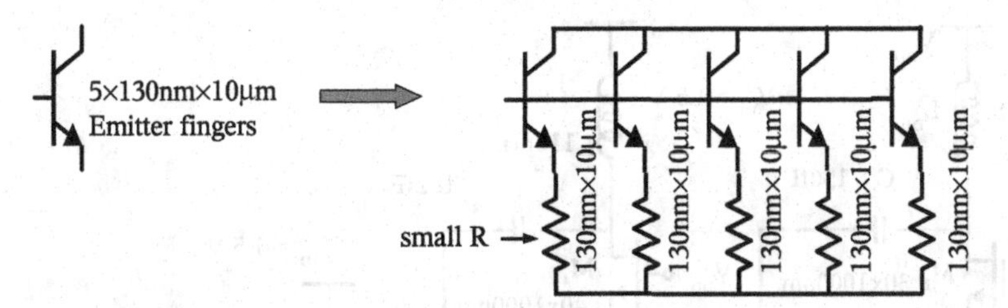

Figure 6.26 Emitter ballasting technique to alleviate thermal runaway in bipolar transistor PAs

a very low impedance is provided at the base to facilitate their removal, these holes will enter the base–emitter junction and cause an increase in the number of electrons injected in the collector. In turn, these electrons will generate more electron–hole pairs in the collector. This positive feedback quickly leads to avalanche breakdown and is accompanied by a rapid rise in device temperature which can permanently damage the transistor.

The avalanche multiplication process can be contained if the base is connected to AC ground through a low impedance. It is quite common to bias SiGe HBT (but not III-V HBT) PA stages safely beyond BV_{CEO} without destroying the transistor, as long as the instantaneous collector–emitter voltage remains below BV_{CBO} (collector–base breakdown voltage when the emitter is left open). In all these situations, the HBT must be biased with a low-impedance voltage source on the base, rather than with a current mirror as in traditional analog design.

In FET PAs, breakdown occurs due to the large electric field between the gate and drain leading to gate oxide breakdown in MOSFETs and to Schottky-diode breakdown in MESFET and HEMT PAs.

6.6 IMPLEMENTATION EXAMPLES OF CMOS AND SiGe HBT MM-WAVE PAs

6.6.1 2-stage – cascode PA with common-source output stage in 90nm CMOS

Based on the technique described above, the 60GHz PA whose schematic is reproduced in Figure 6.27 has been designed and fabricated in a 90nm CMOS technology [14]. All transistors have the minimum gate length of 90nm. The number of gate fingers and the finger width are indicated in the figure for each MOSFET. The amplifier consists of two single-ended cascode stages, and a common-source output stage biased in class A mode. Although the cascode topology has higher gain, larger output impedance, and flat I_{DS}–V_{DS} characteristic, the single transistor CS configuration is advantageous for the moderate power PA implementation because of the lower supply voltage required, leading to higher efficiency and good linearity. The main drawback of the single transistor topology is the reduced reverse isolation, which complicates the input/output matching process.

The PA was designed for nominal 1.2V operation and a saturated power of 7dBm. With $V_{DSAT} = 0.3$V, the voltage amplitude at the output is $V_{DD} - V_{DSAT} = 0.9$V. The required current

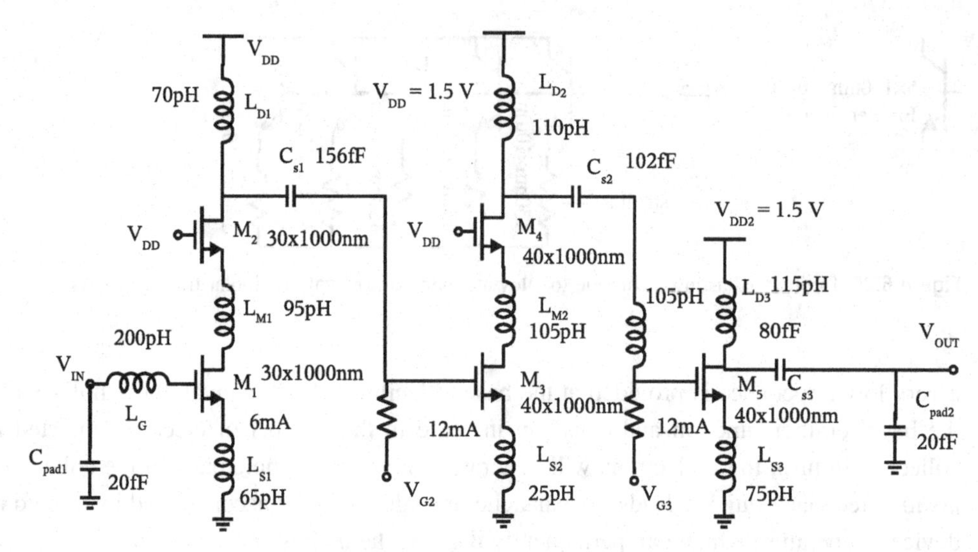

Figure 6.27 Schematic of 60GHz PA fabricated in a 90nm CMOS technology [14]

in the final stage for 7dBm (5mW) output power is then 11mA. This leads to an output transistor gate width of (11mA)/(0.28mA/μm) = 39μm, which was rounded off to 40μm in the design. With an output swing of 2 × 0.9V = 1.8V_{pp} and a transistor *MAG* of 8dB at 60GHz, the gate voltage swing becomes 1.8V_{pp}/2.5 = 0.7V_{pp}, exceeding the P_{1dB} limit of 0.45V_{pp} at the gate of a 90nm MOSFET. Hence, inductive degeneration was employed to prevent the G–S junction from becoming non-linear. The first and second stages were designed in a similar manner given the required output voltage swing and *MAG*. The first stage is biased at the peak f_{MAX} bias of 0.2mA/μm to maximize gain. The second and third stages are biased at the optimal linearity current density of 0.3mA/μm. The transistor widths are 30μm (M1), 40 μm (M3), and 40μm (M5) (Figure 6.27).

The input matching network consists of the inductor L_{S1}, which sets the real part of the input impedance to 50Ω, and L_{G1} which cancels the imaginary part of Z_{IN}, at 60GHz, as discussed in Chapter 5. A 1-stage *L*-network consisting of the load inductor L_{D3} and capacitor C_{C3} is employed at the output. Simplicity in the matching network is critical in minimizing series parasitics, whose effects are more pronounced at 60GHz. Interstage matching is used between the second and third stages to maximize power transfer and consists of the 1-stage *L*-networks (L_D, C_C) at the output of the second stage and the gate (L_G) and source (L_S) inductors of the third stage, where the source inductor also improves linearity. The schematic of the PA is shown in Figure 6.27. Inductors L_{M1} and L_{M2} improve the power gain of the cascode stage, as will be explained in more detail in the following chapter.

Figures 6.28–6.32 summarize the experimental results for this power amplifier in the 40–65GHz range over dies, temperature, and power supply variation. Figure 6.28 reproduces the *S*-parameters indicating a peak gain of 15dB at 55GHz with a 3dB bandwidth extending from 48–61GHz. Figure 6.29 illustrates the degradation of S_{21} by 2.5dB as V_{DD} is reduced from 1.5V to 1.2V.

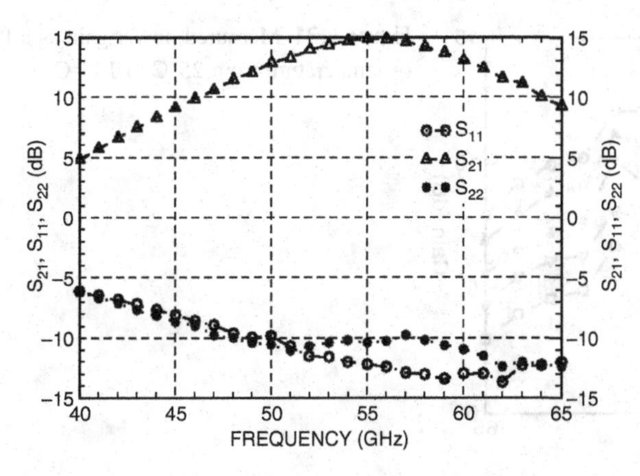

Figure 6.28 Measured S-parameters versus frequency

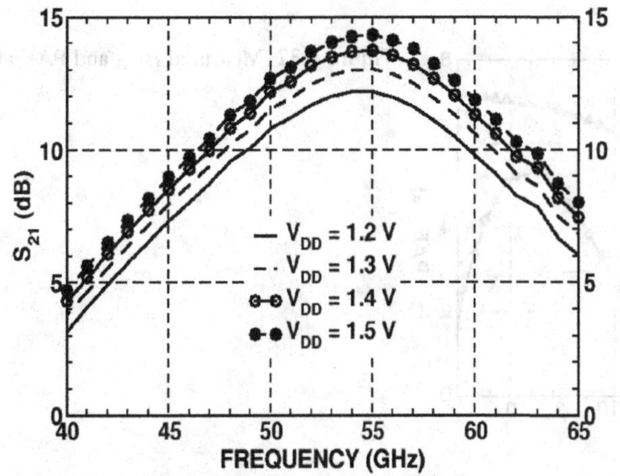

Figure 6.29 Measured S_{21} as a function of V_{DD}

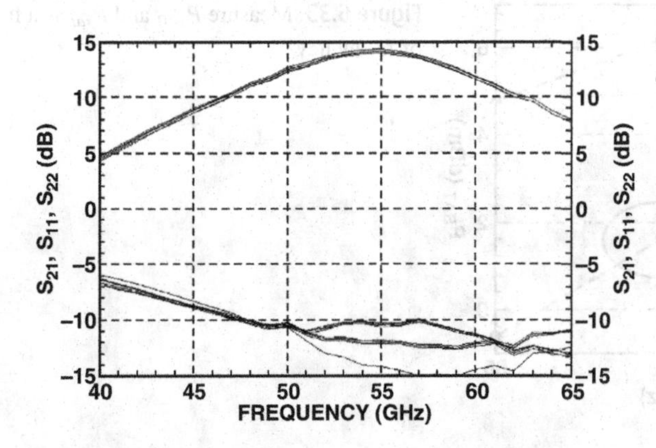

Figure 6.30 Measured *S*-parameters for five dies

Figure 6.30 compiles the measured *S*-parameters across 5 dies, indicating excellent peak gain repeatability and that S_{11} and S_{22} are lower than −10dB from 48GHz beyond 65GHz.

While supply variation and process variation can be easily dealt with through constant current density biasing schemes, the strong degradation of power gain versus temperature

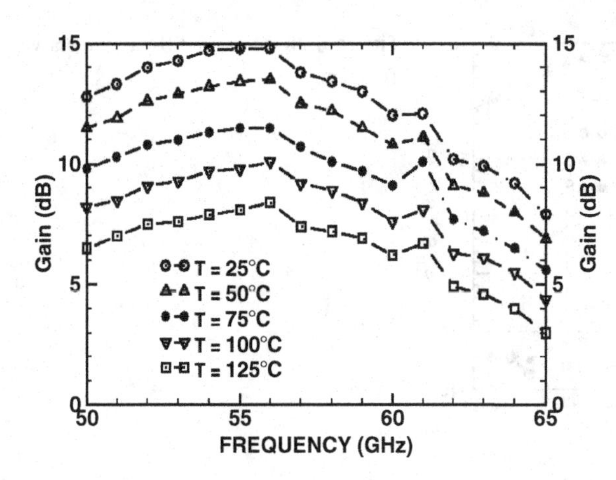

Figure 6.31 Measured power gain as a function of temperature from 25°C to 125°C

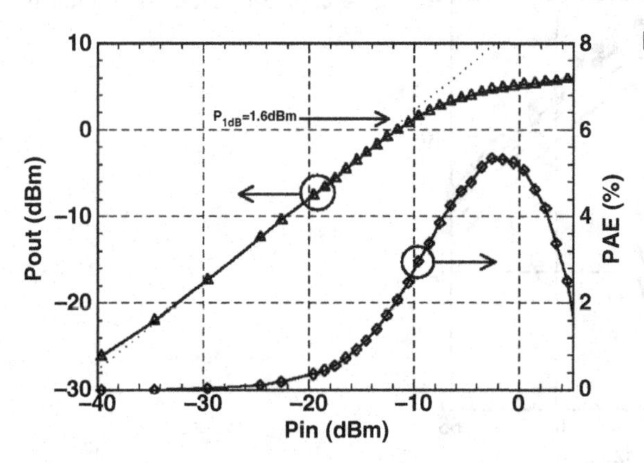

Figure 6.32 Measured $P_{1\mathrm{dB}}$ and PAE at 60GHz

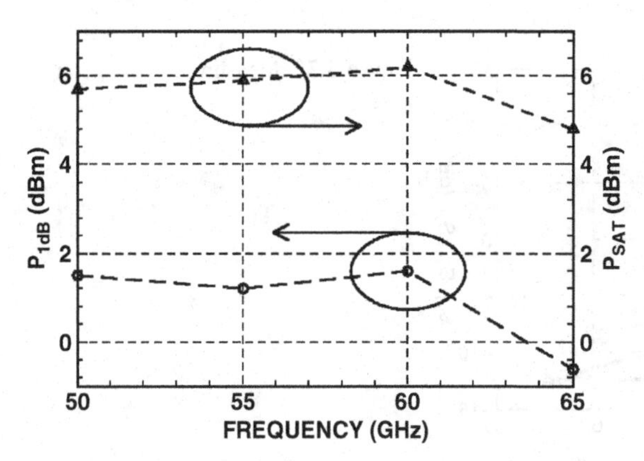

Figure 6.33 Measure $P_{1\mathrm{dB}}$ and P_{sat} as a function of frequency

is illustrated in Figure 6.31 and cannot be compensated without compromising gain at room temperature.

The large signal power measurements versus frequency, temperature and bias current density are compiled in Figures 6.32–6.35. These results indicate that in MOSFET PAs the bias current density for maximum linearity (about 0.3–0.35mA/μm) differs from that for

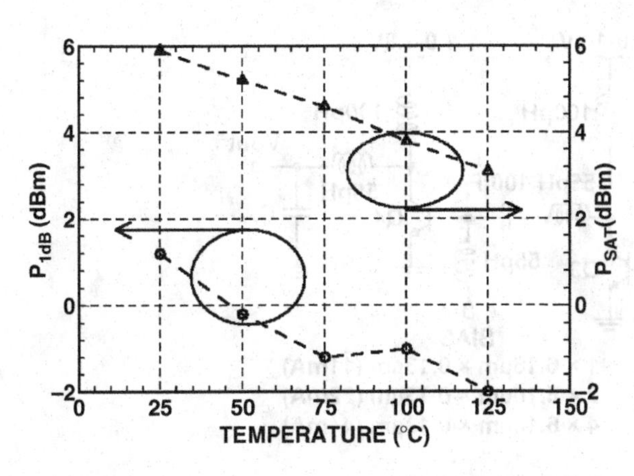

Figure 6.34 Maximum output power and 1dB compression point versus temperature at 55GHz

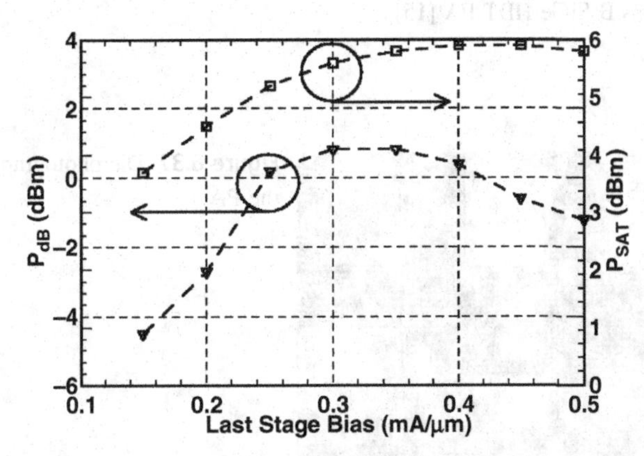

Figure 6.35 Measured saturated output power and 1dB compression point at 60GHz as a function of the bias current density in the output stage

maximum output power (about 0.4–0.45mA/μm). Note that the maximum output power is limited to + 6dBm, lower than the designed value of 7dBm. As mentioned earlier, this discrepancy can be explained by the fact that the losses of the matching networks have not been accounted for in the hand analysis, and by uncertainty in de-embedding the setup and probe losses in large signal power measurements. This uncertainty can be as high as + /−1dB in a heterogeneous interface waveguide-to-coaxial setup, such as the one employed in these 60GHz measurements.

Figure 6.35 shows maximum output power and 1dB compression point versus frequency at room temperature.

6.6.2 Class AB 77GHz PA in SiGe HBT technology [15]

Figures 6.36 and 6.37 illustrate the schematic and die photo, respectively, of a 77GHz SiGe HBT PA. As can be seen in Figure 6.36, the last two stages are common-emitter to minimize power consumption and maximize PAE. The latter was measured to be 15% (Figure 6.38). The HBT sizes are scaled by a factor of 2 from stage to stage. The optimal bias current density for the HBTs in the first stage of the PA is the peak f_T/f_{MAX} current density because the first stage is meant to behave as a linear class A amplifier. The last two stages however operate

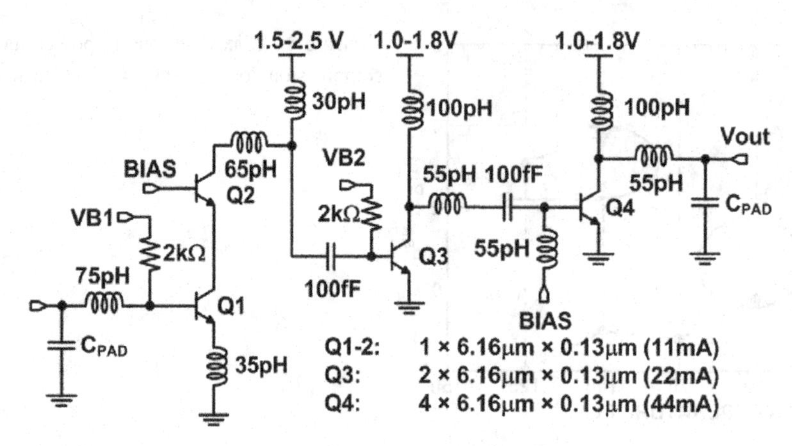

Figure 6.36 Schematics of 77GHz, class B SiGe HBT PA [15]

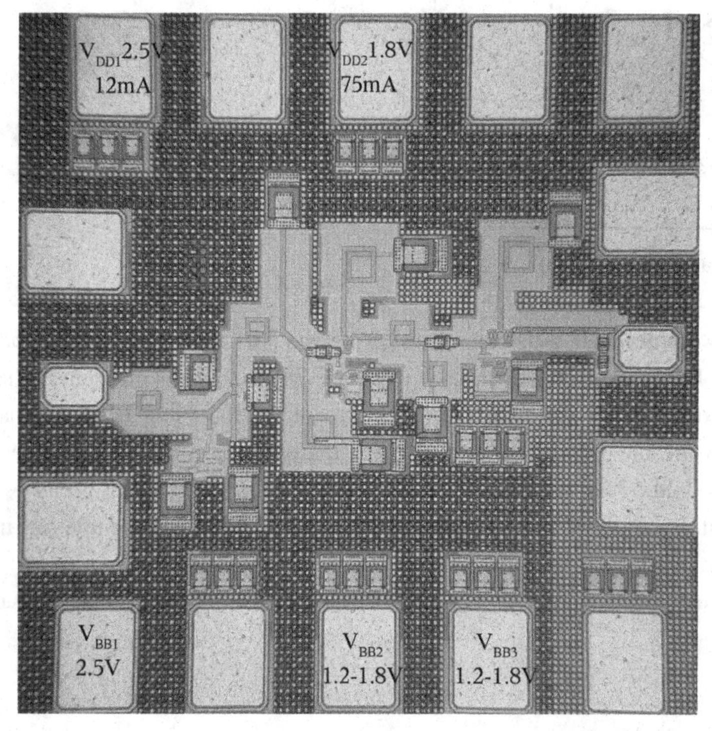

Figure 6.37 Die photograph of the PA

as class AB amplifiers. Using the measurement results illustrated in Figure 6.39, the bias current density of Q_3 and Q_4 is set to 8mA/μm^2, approximately $J_{pfT}/2$, to maximize the performance of the PA. The measurements show that the OP_{1dB}, P_{SAT}, and PAE all peak at the same current density.

The saturated output power can be calculated by ignoring V_{CESAT} and the losses in the output matching network as

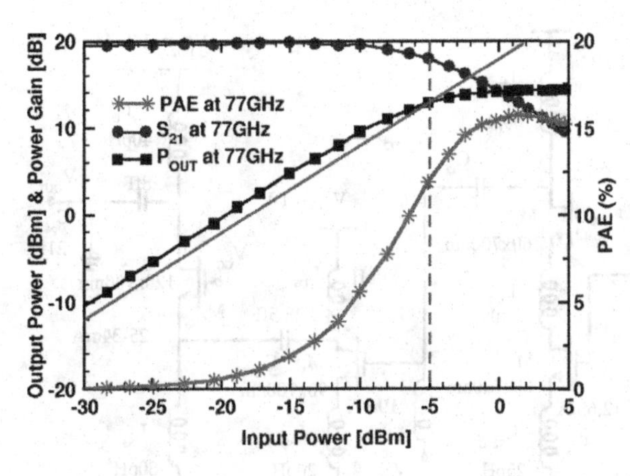

Figure 6.38 Measured PA S_{21}, OP_{1dB}, and P_{SAT} at 77 GHz

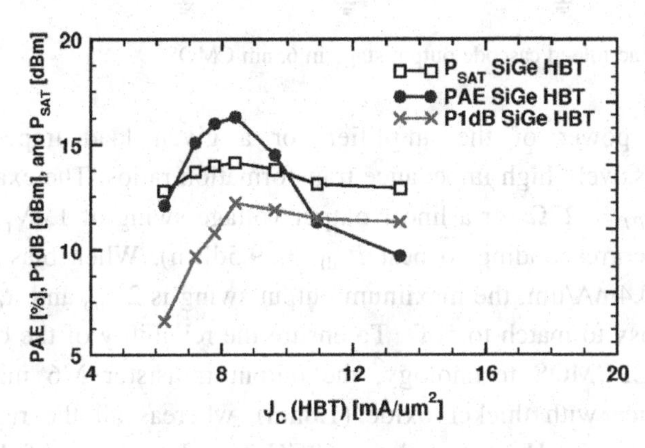

Figure 6.39 Measured P_{SAT}, OP_{1dB}, and PAE versus bias current density in the final stage of the PA

$$P_{sat} \leq \frac{I_{DC}V_{CC1}}{2} = \frac{44\text{mA} \times 1.5\text{V}}{2} = 33\text{mW} = 15.1\text{dBm}.$$

This value is not that different from the measured P_{SAT} of 14.5dBm at 77GHz. The measured output compression point, PAE, and the small signal gain are 12dBm, 15.7% and 19dB, respectively.

6.6.3 Common-source, common-gate mm-wave output stage

The schematics of a PA consisting of a common-source stage followed by a common-gate stage are illustrated in Figure 6.40. All transistors have a minimum gate length of 60nm. The number of gate fingers and the finger width are indicated for each MOSFET. This circuit benefits from the:

- good isolation,
- high gain, and
- the high output impedance

of a cascode stage, while maximizing the V_{DS} drop across each transistor to V_{DD}.

The large V_{DS} maximizes the power gain at mm-wave frequencies while the higher output impedance for a given bias current allows us to critically increase the size, bias

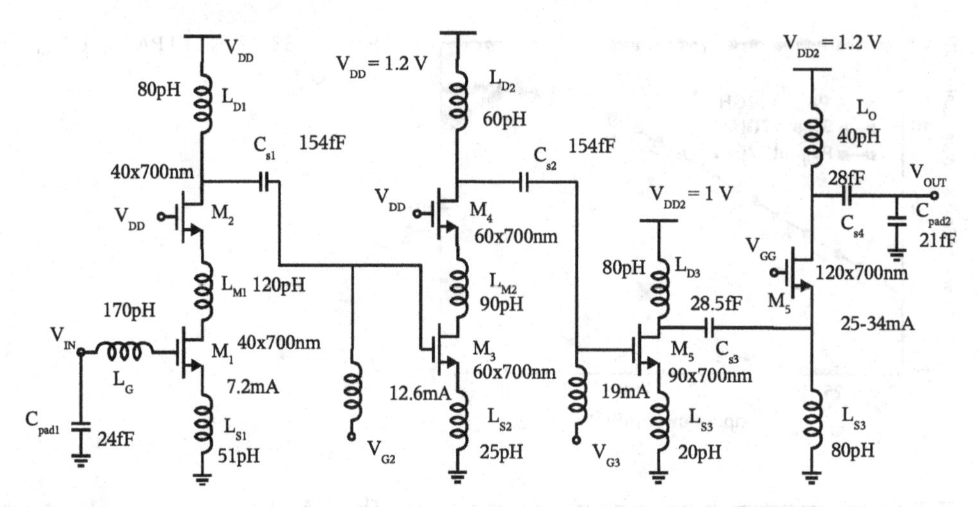

Figure 6.40 94 GHz class-A PA with ac-folded cascode output stage in 65nm CMOS

current, and therefore output power of the amplifier for a given load impedance without having to resort to excessively high impedance transformation ratios. The example below is designed such that $R_{OPT} = 27\Omega$ for a linear output voltage swing of $1.4V_{pp}$ at a current density of 0.3mA/μm (corresponding to best $P_{1dB} = 9.5$dBm). When biased for maximum $P_{SAT} = 12.3$dBm at 0.4mA/μm, the maximum output swing is $2V_{pp}$ and $R_{OPT} = 29.4\Omega$. These R_{OPT}values are easy to match to 50Ω. To ensure the reliability of the circuit when implemented in 65nm GPCMOS technology, the output transistor M6 may be realized with an LVT LP device with thicker oxide (1.8nm), whereas all the rest are LVT GP MOSFETs to maximize gain. However, above 60GHz, the lower gain of the LP device can significantly affect the overall PA gain and output power, despite its larger voltage swing.

6.6.4 Other mm-wave CMOS PAs

70–77GHz PA

This circuit whose schematic is shown in Figure 6.41, was realized in 90nm CMOS and employs four common-source stages with microstrip transmission lines as impedance matching networks [16]. The last stage is two times larger than the third stage by connecting two identical stages, with their matching networks, in parallel. The circuit operates with $V_{DD} = 1$V, has a saturated output power of + 7dBm and a gain of 12dB, or 3dB/stage.

Transformer-coupled 60GHz differential PA

Another 90nm CMOS common-source design, shown in Figure 6.42, features a 3-stage differential topology where transformers are employed for single-ended-to-differential conversion at the input, for interstage impedance matching, and for power combining and differential to single-ended conversion at the output, as illustrated in Figure 6.42 [17]. This 60GHz power

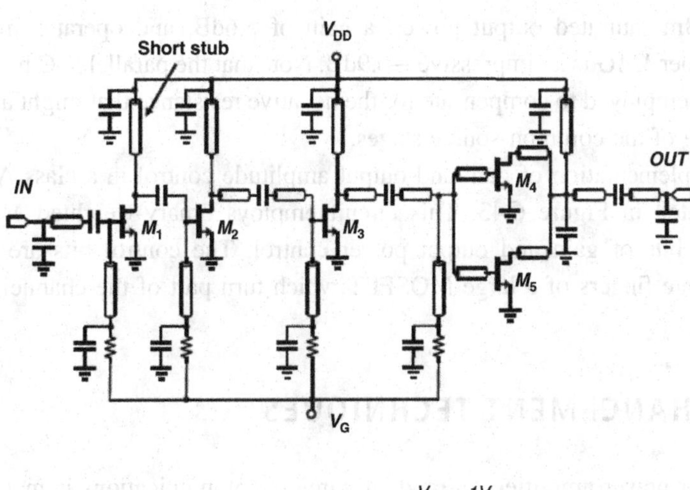

Figure 6.41 Schematics of a 77 GHz CMOS PA in 90nm CMOS [16]

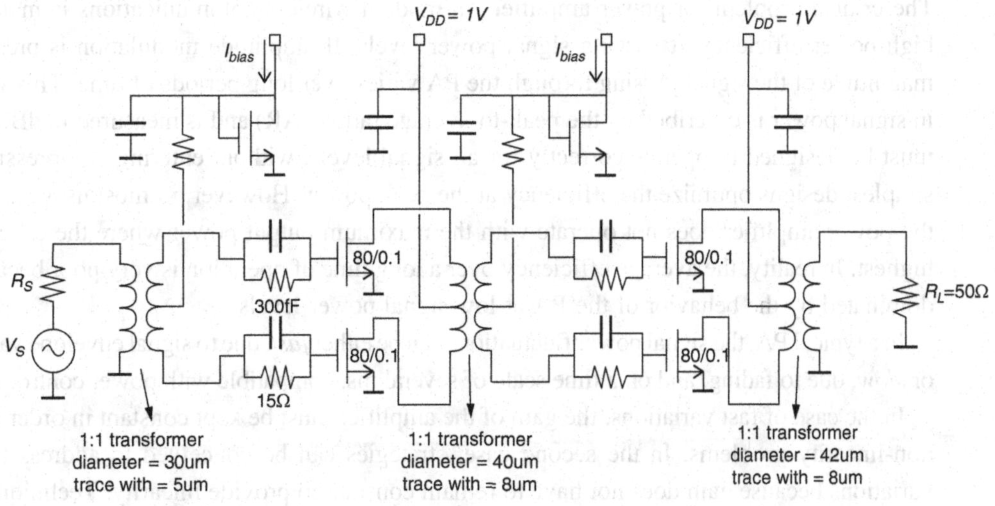

Figure 6.42 Schematics of a 60GHz differential PA fabricated in 90nm CMOS [17]

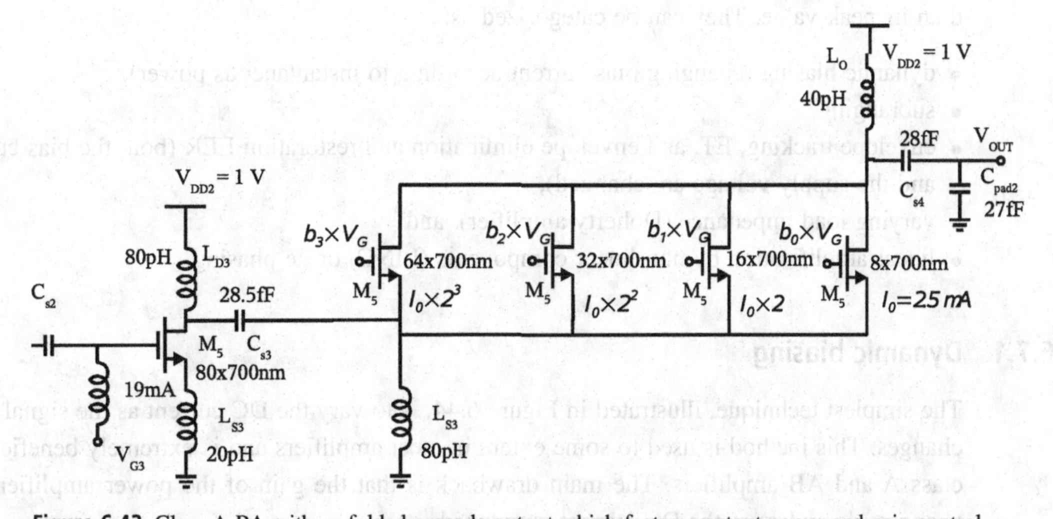

Figure 6.43 Class A PA with ac-folded cascode output which features output power and gain control

amplifier has +12.3dBm saturated output power, a gain of 5.6dB, and operates from 1V. The measured transformer *MAG* is an impressive −0.9dB. Note that the parallel R-C network at the gate of each FET is employed to compensate for the negative resistance that might arise due to the inductive loading of the common-source stages.

Finally, a simple implementation of gain and output amplitude control in a class A or AB output stage is illustrated in Figure 6.43. This circuit employs binary-weighted MOSFET fingers to realize four bits of gain and output power control. The control bits are applied directly to groups of gate fingers of a large MOSFET which turn part of the channel current on and off.

6.7 EFFICIENCY ENHANCEMENT TECHNIQUES

The critical problem for power amplifiers in modern wireless communications is maintaining high-power efficiency at varying signal power levels. If amplitude modulation is present, the magnitude of the signal passing through the PA varies over long periods of time. This variation in signal power is described by the peak-to-average ratio (PAR) and is measured in dB. The PA must be designed to operate correctly for all signal levels, without entering compression. The simplest designs optimize the efficiency at the peak power. However, in most usage scenarios, the power amplifier does not operate with the maximum output power where the efficiency is highest. In reality, the average efficiency over a long time of operation is very poor because it is dominated by the behavior of the PA at low signal power levels.

In a typical PA, the signal power fluctuations can be either *fast*, due to signal envelope variations, or *slow*, due to fading, and on a time scale of several ms, compatible with power control loops.

In the case of fast variations, the gain of the amplifier must be kept constant in order to avoid non-linearity problems. In the second case, strategies can be conceived to address the slow variations because gain does not have to remain constant to provide linearity. Techniques were therefore developed to maintain high efficiency when the output power is significantly lower than its peak value. They can be categorized as:

- dynamic biasing (changing bias current according to instantaneous power),
- subranging,
- envelope tracking, ET, and envelope elimination and restoration EER (both the bias current and the supply voltage are changed),
- varying load impedance (Doherty amplifier), and
- linear amplification by non-linear components (LINC) or de-phasing.

6.7.1 Dynamic biasing

The simplest technique, illustrated in Figure 6.44, is to vary the DC current as the signal level changes. This method is used to some extent in most amplifiers and is extremely beneficial in class A and AB amplifiers. The main drawback is that the gain of the power amplifier also changes (degrades) as the DC current is reduced.

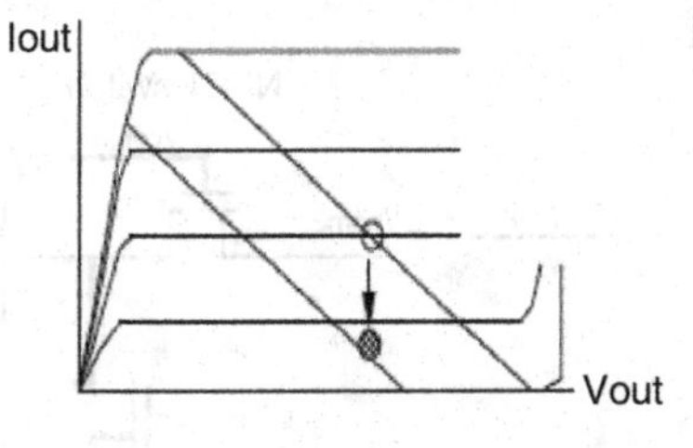

Figure 6.44 Illustration of the dynamic biasing concept [4]

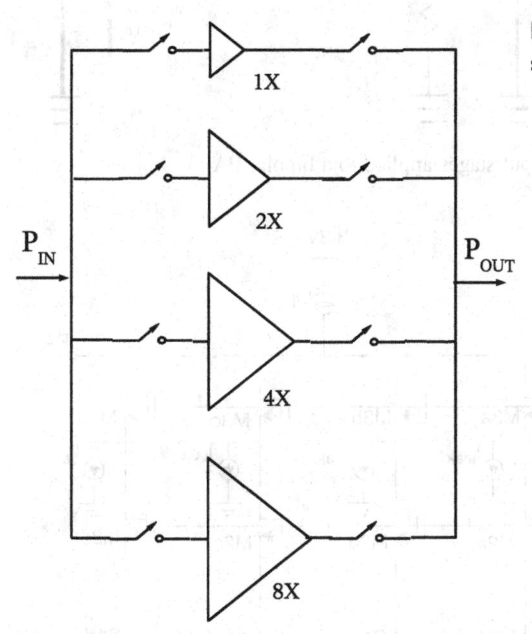

Figure 6.45 Concept of binary-weighted PA output stages connected in parallel [3]

6.7.2 PA subranging

A better, yet still brute-force, solution is to control the output power by connecting several binary-weighted PA sections in parallel and switching them on and off as needed by the transmitter system. The advantage of this approach, shown in Figure 6.45 [3], is that the gain and efficiency of each PA section are maximized, thus increasing the average efficiency and gain of the overall transmitter. However, the challenge comes from the design of the input and output matching networks which must accommodate the change in the input and output impedances of the PA as sections are switched on and off. Ideally, at the very least, the output matching network should also be adaptable to provide different R_{LOPT} terminations for different PA configurations.

Practical implementations of this approach in bipolar and MOSFET PAs are illustrated in Figures 6.46 [4] and 6.47 [17], respectively.

The first example employs MOSFET switches to select a binary-weighted number of emitter fingers, thus controlling the output power. Each emitter finger is biased optimally for gain and drain efficiency.

The second example involves 2.4GHz and 5GHz digitally controlled class A WLAN PAs which feature a 4-bit, digitally programmable output power level with a peak output power of

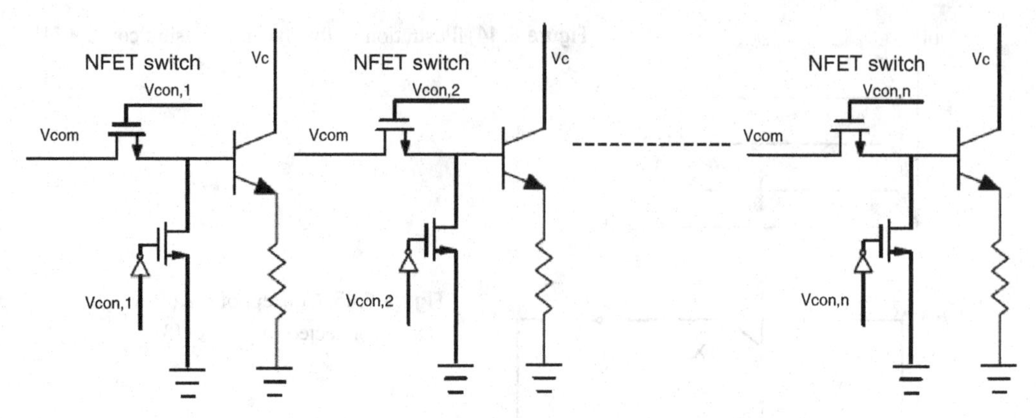

Figure 6.46 Concept of binary weighted output stages applied to a bipolar PA [4]

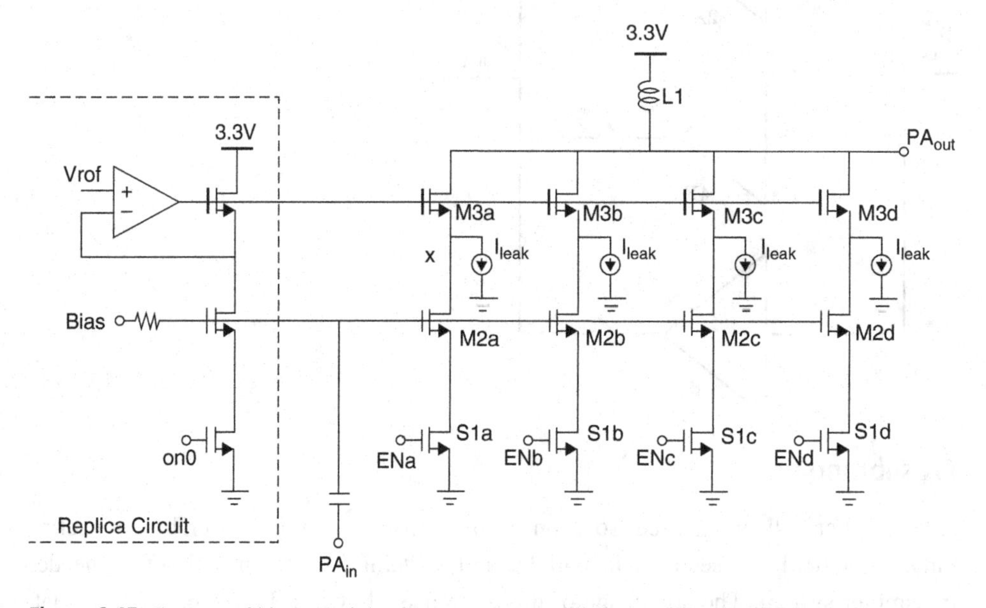

Figure 6.47 Concept of binary weighted output stages applied to a 5GHz CMOS PA [18]

10mW [18]. The design employs common-gate devices with thick-gate oxide for reliability and to maximize the output voltage swing, and thin-gate oxide MOSFETs in common-source configuration for large power gain and efficiency. A replica bias scheme allows for turning off the power to the entire PA. The I_{leak} current sources ensure that the common-gate devices are always on, and exhibit a sufficiently large V_{GS} to avoid overstress of the thin-oxide CS devices when a particular PA section is turned off.

Solved problems

The schematic of a class A 80GHz PA realized in a 130nm SiGe BiCMOS process is shown in Figure 6.45. The HBTs have 130nm wide emitters and are biased for maximum saturated output power at 14mA/μm^2. V_{REF} is 1.5V, $V_{BIAS} = 1.2$V, and $V_{DD} = 3$V. $C_{je} = 17$fF/μm^2;

$C_\mu = 15\text{fF}/\mu\text{m}^2$; $C_{cs} = 1.2\text{fF}/\mu\text{m}$. The numbers next to each HBT in the schematic indicate the emitter length in micrometers. The PA operates in saturated output power mode. MOSFETs M0, M1, M2, and M3 have 150nm gate length and are biased at 0.4mA/μm and sized to provide the appropriate current for the HBTs. The gain of the opamp can be assumed to be very large:

(a) What is the total bias current through L_C when all control bits b_0, b_1, b_2, b_3 are set to "1"?

Solution

$$I_{Ctotal} = J_{pfT}(A_{E0} + A_{E1} + A_{E2} + A_{E3}) = 14\frac{\text{mA}}{\mu\text{m}^2} \times 15 \times 0.13 \times 1\mu\text{m}^2 = 27.3\text{mA}.$$

(b) Find the maximum peak-to-peak saturated output voltage swing assuming that the output voltage swings symmetrically below and aboveV_{DD}.

Solution

Since this is the SATURATED output swing, we ignore $V_{CESAT} = V_{MIN}$ (typically 0.3–0.5V). The voltage at the output swings down to the DC voltage at the emitter of the CB transistors which is set by V_{REF} to 1.5V. Therefore $V_{opp} = 2V_1 = 2\ (V_{DD}\ \text{-}V_{REF}) = 3V_{pp}$.

(c) Find R_{LOPT} when the output voltage swing is maximized and all bits are "1."

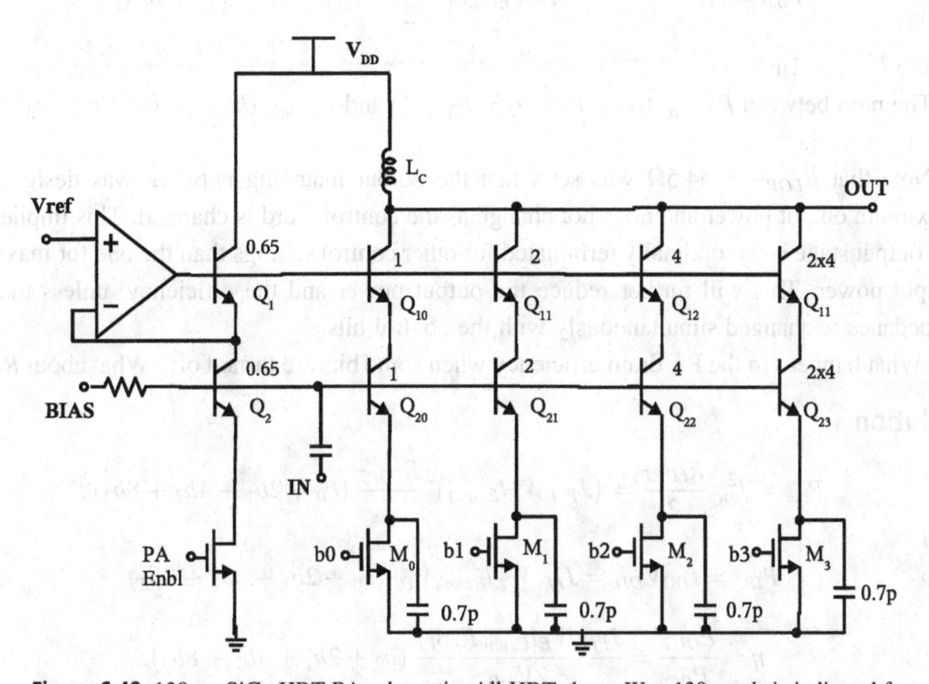

Figure 6.48 130nm SiGe HBT PA schematic. All HBTs have W_E =130nm. l_E is indicated for each HBT

Solution

$$R_{LOPT} = \frac{V_{MAX}}{I_{MAX}} = \frac{V_{opp}}{2I_{DC}} = \frac{3V}{54.6\text{mA}} = 54.5\Omega.$$

(d) Calculate L_C such that the output of the PA resonates at 80GHz.

Solution

$$C_{OUT} = A_{Etotal} \times C'_{bc} + l_{Etotal} \times C'_{cs} = 1.95 \times 15\text{fF} + 15 \times 1.2\text{fF} = 47.25\text{fF}$$

$$L_C = \frac{1}{(2\pi f)^2 C_{out}} = 84\text{pH}.$$

(e) [1p] Assuming that the circuit is terminated on R_{LOPT}, calculate the maximum output power delivered to the load.

Solution

$$P_{out} = \frac{V_{opp} I_{opp}}{8} = I_{DC}^2 \frac{R_{LOPT}}{2} = 20.3\text{mW}.$$

(f) [2p] Write the expression of the output power as a function of the control bits b_0, b_1, b_2, b_3. What is the ratio between the maximum output power and the output power when $b_0 = 1$, $b_1 = b_2 = b_3 = 0$?

Solution

$$P_{out} = I_{DC}^2 \frac{R_{LOPT}}{2} = J_{PfT} W_E l_{Eunit} \frac{R_{LOPT}}{2} (b_0 + 2b_1 + 4b_2 + 8b_3)^2$$

where $l_{Eunit} = 1\mu\text{m}$.

The ratio between P_{outmax} ($b_0 = b_1 = b_2 = b_3 = 1$) and P_{outmin} ($b_0 = 1, b_1 = b_2 = b_3 = 0$) is $(15/1)^2 = 225$.

Note that $R_{LOPT} = 54.5\Omega$ was set when the output matching network was designed for maximum output power and does not change as the control word is changed. This implies that the output stage is not optimally terminated for other control settings than the one for maximum output power. This will further reduce the output power and the efficiency, unless the load impedance is changed simultaneously with the control bits.

(g) What happens to the PA drain efficiency when some bits are turned off? What about R_{LOPT}?

Solution

$$P_{out} = I_{DC}^2 \frac{R_{LOPT}}{2} = (J_{PfT} W_E l_{E,unit})^2 \frac{R_{LOPT}}{2} (b_0 + 2b_1 + 4b_2 + 8b_3)^2$$

and

$$P_{DC} = I_{DC} V_{DD} = J_{PfT} W_E l_{E,unit} V_{DD} (b_0 + 2b_1 + 4b_2 + 8b_3)$$

$$\eta = \frac{P_{OUT}}{P_{DC}} = \frac{J_{PfT} W_E l_{E,unit} R_{LOPT}}{2V_{DD}} (b_0 + 2b_1 + 4b_2 + 8b_3).$$

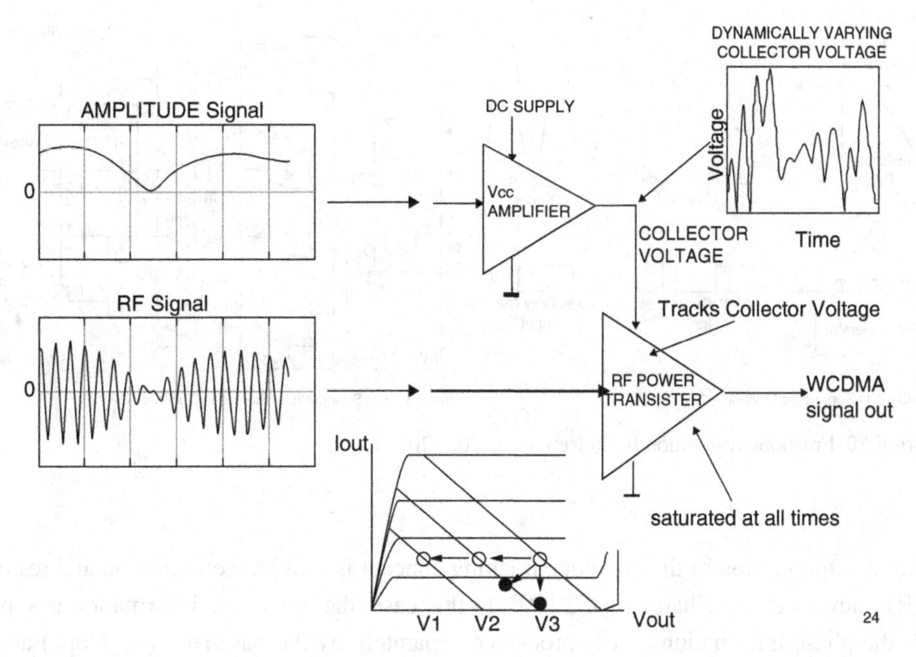

Figure 6.49 The principle of envelope tracking and restoration [4]

Unlike the output power which depends on the square of the digital word, the drain efficiency is linearly dependent on the digital word.

As the output power is reduced, the efficiency DOES NOT remain constant. However, it degrades linearly, rather than exponentially. For the efficiency to remain constant at lower output power, V_{DD} and V_{REF} must also be reduced simultaneously, not just I_{DC}. This approach is pursued in the envelope restoration techniques discussed next. Alternatively, if V_{REF} and V_{DD} remain constant, R_{LOPT} will have to be adjustable to maintain the same efficiency as the output power is reduced. This problem is partially addressed by the Doherty amplifier.

6.7.3 Envelope tracking techniques

The envelope tracking technique maximizes the PA efficiency by keeping the transistor at the edge of the active region. It is based on the observation that the back-off efficiency of a class AB PA can be significantly improved by dynamically reducing the supply voltage when the amplitude of the input signal decreases. This is accomplished, as illustrated in Figure 6.49, where, in addition to the main, class AB PA, a second amplifier is employed to track the envelope of the input signal and to adjust the supply voltage to the main amplifier. Two conditions must be satisfied:

1. The envelope tracking amplifier, though operating at low frequency, must have sufficient bandwidth to track the baseband signal that modulates the carrier.
2. The overall efficiency must be better than that of the class AB PA.

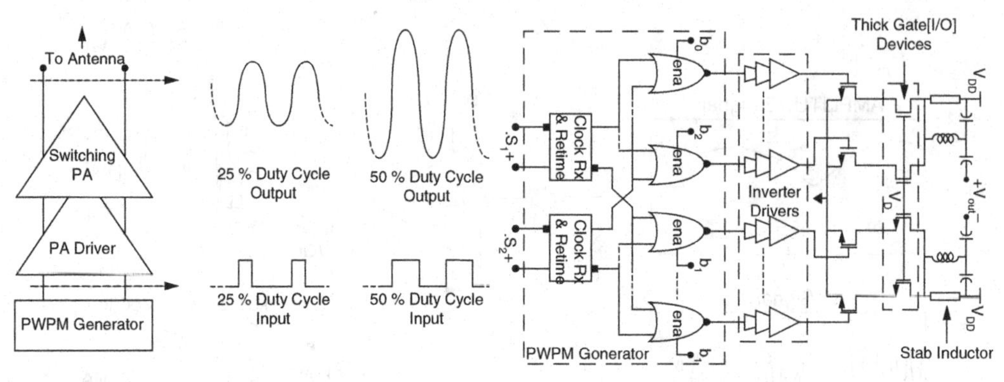

Figure 6.50 Envelope restoration by pulse modulation [20]

An extreme version of the envelope tracking concept is envelope elimination and restoration (EER), introduced by Khan in 1952 [19]. In this case, the amplitude information is separated from the phase information and is processed separately by the baseband (envelope) amplifier before it is re-introduced in the main amplifier through the power supply. The main amplifier now only processes constant amplitude information and can therefore be operated in a highly non-linear regime as a class E switching PA, and made more efficient. The main challenge in this approach is the accurate synchronization between the amplitude and phase information along two different paths.

6.7.4 Pulse-width modulated class-E PA with ER

A somewhat different, very recent 65nm CMOS implementation of the envelope restoration concept is illustrated in Figure 6.50 [7], [20]. Here, the amplitude information is coded in the pulse width and a purely digital signal is applied at the PA gate. Since a high-Q tank is present at the PA output, the output voltage remains sinusoidal and its amplitude scales with the pulse width of the input signal. The gain and output power are digitally controlled by 4 bits. We note that, for a 10:1 pulse-width modulation, the frequency spectrum of the input signal can extend over a ten times larger bandwidth than the fundamental frequency. This no longer poses an implementation challenge at 2–5GHz if nanoscale CMOS technologies with transistor f_T exceeding 150GHz are employed [7].

6.7.5 Doherty amplifier

The Doherty amplifier, whose block diagram is shown in Figure 6.51, can be regarded as a clever variation on the subranging concept, or as an amplifier with variable load impedance. The original version [21] relies on splitting the input signal between two parallel paths. The main path consists of a class AB amplifier and a $\lambda/4$ transmission-line transformer, while the

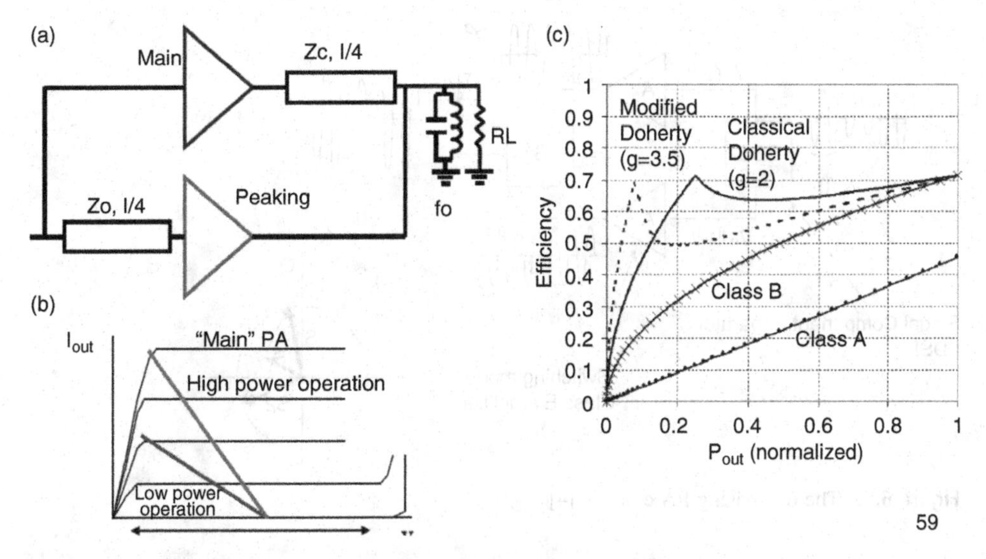

Figure 6.51 The Doherty amplifier (a) concept, (b) load lines in low-power and high-power modes, and (c) efficiency as a function of normalized output power [4]

second path features a $\lambda/4$ transmission-line transformer first, followed by an auxiliary class C peaking amplifier which only becomes active when the main amplifier reaches about one quarter of the maximum output power (P_{OMAX}) of the entire amplifier. The key idea is that the drain currents of the transistors on the two paths are summed on the output inductor. The quarter-wave transformers are typically realized as transmission lines and their role is to convert voltage at one port to current at the second port.

The peaking amplifier acts as an active load for the main path, reducing the load impedance seen by the main path and, therefore, limiting its output voltage swing while allowing its drain current and output power to continue to grow as the input signal power increases beyond $P_{OMAX}/4$. This arrangement ensures that the main transistor remains in the active region and improves linearity.

The main advantage of the Doherty amplifier is its power-added efficiency. At power levels below $P_{OMAX}/4$, only the main amplifier is on. Its drain current is converted linearly to voltage applied to the load, in much the same way as in a conventional PA. The PAE corresponds to that of a class AB stage and can be as high as 70% when the output power reaches $P_{OMAX}/4$ and the output voltage swing at the drain of the main transistor is approximately $2 \times V_{DD}$. The main transistor is sized to reach this voltage swing at about half its maximum current, which corresponds to 6dB back-off from the maximum overall output power. As the peaking transistor turns on, the system is adjusted such that the voltage swing at the output of the main transistor remains constant while the current swing in the load continues to increase. The peaking transistor supplies exactly the necessary amount of power to maintain a linear relationship between the overall input power and output power of the amplifier up to P_{OMAX}, at approximately the maximum power-added efficiency of a class AB stage [3].

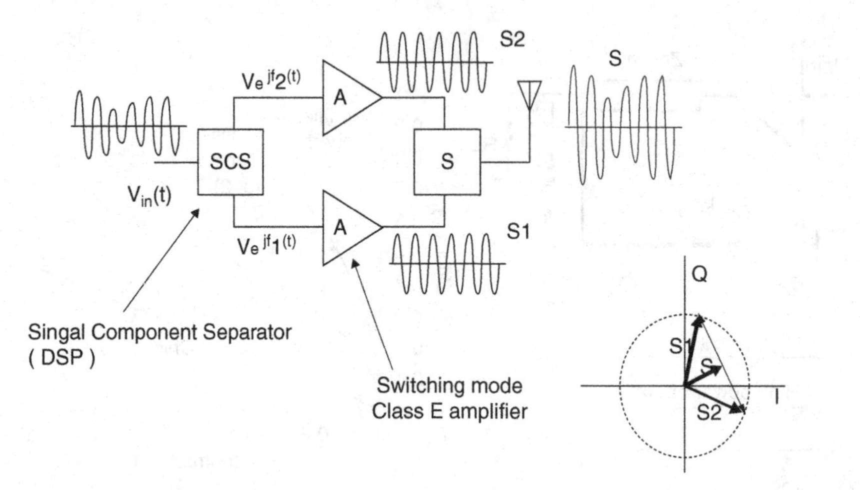

Figure 6.52 The de-phasing PA concept [4]

The Doherty amplifier concept has regained popularity in the last few years and is intensely investigated for CMOS implementations, even at 60GHz [22].

6.7.6 Linear amplification with non-linear components (de-phasing)

This concept, which is even older than the Doherty amplifier, was invented by Chireix in 1935 [23] and is illustrated in Figure 6.52. It is based on pure geometrical considerations which allow for any RF signal (typically represented as a rotating vector with time-varying amplitude and phase) to be decomposed in two new rotating vectors of constant amplitude. All the information of the original RF signal is now contained in two constant envelope signals. The latter two signals can now be amplified with a highly non-linear, high efficiency PA (typically a class E) and linearly combined to recreate an amplified version of the original RF signal. Today, the signal decomposition is realized with digital-signal-processing techniques. The key challenge remains in efficiently combining two constant envelope vectors through simple vector addition if the two vectors are not in phase [24]. For example, in the worst case, the two vectors can be almost 180 degrees out of phase. When they are combined by conventional vector addition, their entire power is dissipated in the process, resulting in 0% efficiency.

6.7.7 Digital predistorsion

With the proliferation of SiGe BiCMOS and, more recently, CMOS technologies as viable options for RF PAs, analog-to-digital converters (ADCs) and digital signal-processing (DSP) have been employed to linearize the output power versus input power transfer function of PAs, especially for wireless standards that use QAM and OFDM modulation

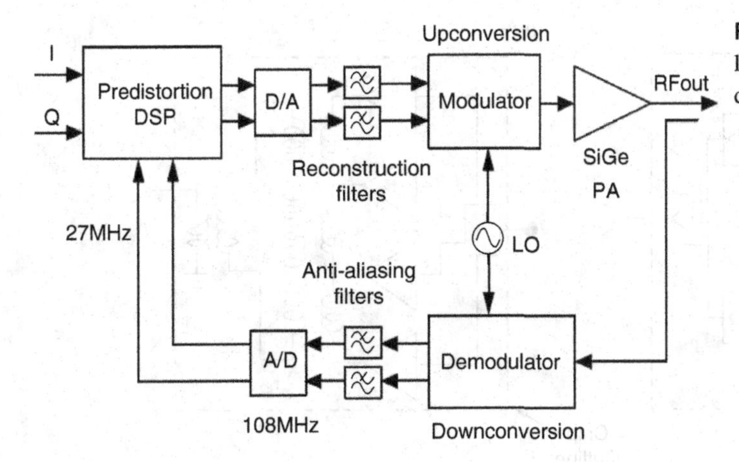

Figure 6.53 Transmitter linearization scheme based on digital pre-distortion [4]

formats. The key idea here is to intentionally pre-distort the signal going into the PA such that its output satisfies the EVM and spectral mask specifications imposed by the standard, even when the PA operates close to P_{SAT}. One such IQ transmitter topology with loop-back from the PA output to the IQ upconverter input is illustrated in Figure 6.53. In this case, a replica of the inverse of the non-linear transfer function of the PA is obtained by sampling the PA output and storing it in memory. This information is then fed to the DSP to pre-distort the low-frequency IQ data signals before being applied to the IQ modulator/ upconverter.

Some of the implementation issues are related to the size of the memory table and the speed with which the data provided to DSP are updated. However, most of these are disappearing as more advanced CMOS nodes, with faster switching transistors are employed.

6.8 POWER COMBINING TECHNIQUES

Traditional power amplifier design, irrespective of class, relies on the common-source or common-emitter configuration with the device periphery (gate width or emitter area) determined by the desired maximum output power. In all PAs, the layout consists of many transistor gate or emitter fingers connected in parallel to a main gate or base bus. For efficient power combining, the signal must arrive in phase to all gate/emitter fingers and without attenuation. Since attenuation increases with gate finger width and emitter finger length, with the overall transistor footprint, and with frequency, this power-combining technique has limitations which are exacerbated at higher frequencies and for technologies with lower breakdown voltages. The limitation is severe in nanoscale CMOS where the safe operating output voltage swing is low and current and device periphery must be increased in compensation, resulting in extremely low port impedances and lossy matching networks. The solution to this problem relies on combining the power output from several PAs into a single output.

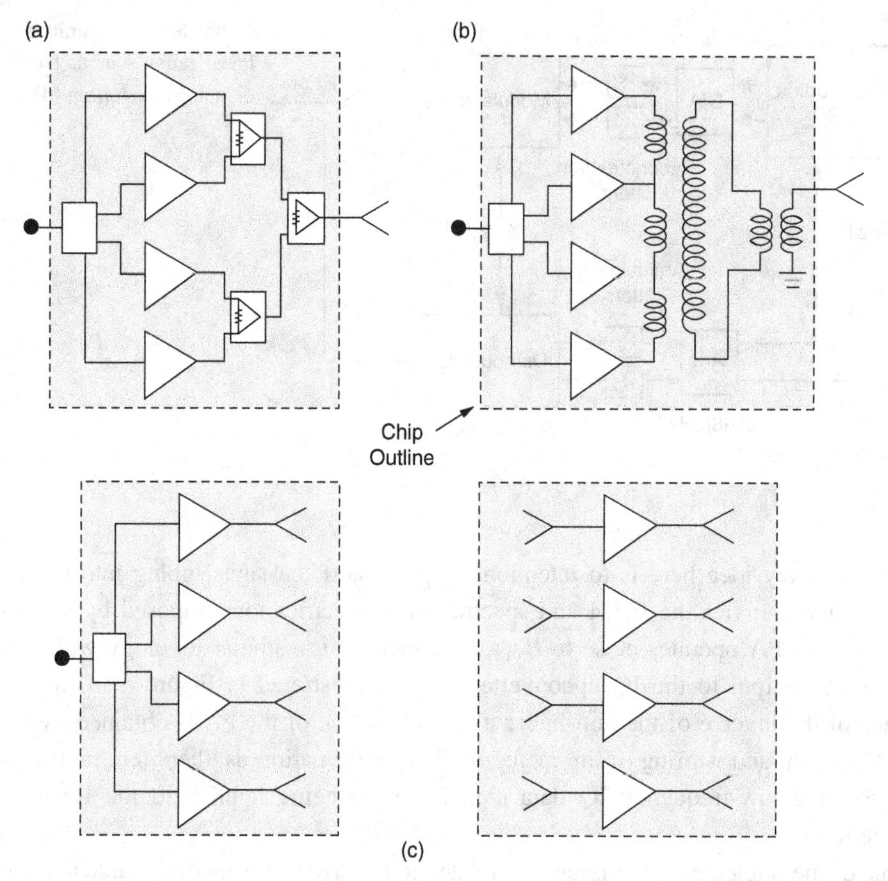

Figure 6.54 Block diagram of (a) Wilkinson, (b) transformer, and (c) quasi-optical power-combining techniques [26]

Power combining techniques, some of which are illustrated in Figure 6.54, can be classified in:

- two-way:
  - push-pull,
  - in-phase power combining using Wilkinson couplers Figure 6.54(a),
  - balanced (two identical amplifiers are combined using input and output 90 degree couplers, Figure 6.55);
- multi-way power combining:
  - series connection of transistor output voltage using transformers [25], Figure 6.54(b), or vertical stacking,
  - half-wavelength paralleling:
    - simple,
    - Bagley polygon,

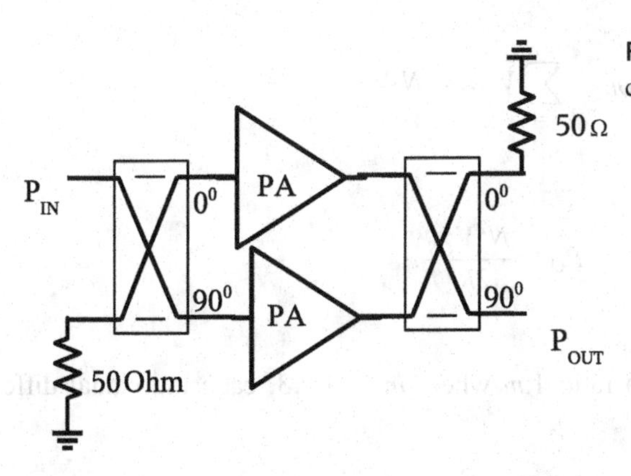

Figure 6.55 A balanced amplifier based on power combining through 90-degrees hybrid couplers

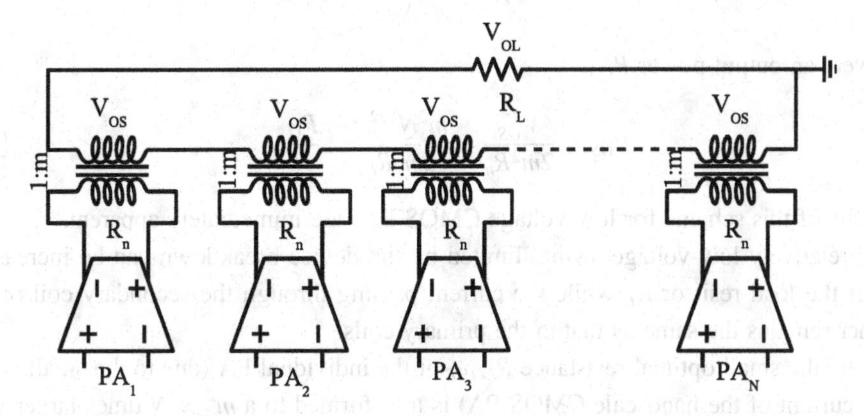

Figure 6.56 Principle of power combining using series-connected transformers

- spatial (through the air, as in phase arrays) $\lambda/2$ power combining also known as quasi-optical power combing, Figure 6.54(c) [26].

In general, with the exception of spatial power combining, all these techniques suffer from the losses associated with the power-combining process itself. Power combining efficiency, and the PAE of the overall PA, degrades as the number of power amplifiers increases. Typical (good) values are in the 70–80% range.

In the case of the balanced amplifier, the isolation provided by the hybrid couplers ensures matching to 50Ω even if the individual amplifiers are not matched to 50Ω. Practical implementations of hybrid couplers will be discussed in more detail in Chapter 9.

Transformer-based power combining has become the most popular technique for the implementation of silicon PAs at RF, microwave, and mm-wave frequencies. The principle is illustrated in Figure 6.56, where N identical differential power amplifiers are coupled through 1:m transformers to the load R_L. The secondary coils of the N transformers are connected in series. As a result, when the entire system is impedance-matched, the voltage developed across the load resistor, V_{oL}, becomes the sum of the individual voltages V_{os} in the secondary

$$V_{OL} = \sum_{i=0}^{i=N} V_{OSi} = N \cdot V_{OS} \tag{6.46}$$

and the output power, P_O, becomes

$$P_O = \frac{N^2 V_{OS}^2}{2R_L}. \tag{6.47}$$

Accounting for the transformer turn ratio 1:m where $m = 1 \ldots 3$, each individual differential amplifier sees a load impedance R_n

$$R_n = \frac{R_L}{m^2 N} \tag{6.48}$$

and delivers an output power P_{OS}

$$P_{OS} = \frac{V_{OS}^2}{2m^2 R_n} = \frac{m^2 N V_{OS}^2}{2m^2 R_L} = \frac{P_O}{N}. \tag{6.49}$$

The benefits of this scheme for low-voltage CMOS PAs are immediately apparent.

First, a relatively low-voltage swing, limited by the device breakdown can be increased N times over the load resistor R_L while the current passing through the secondary coil of each transformer remains the same as that in the primary coils.

Secondly, the small optimal resistance R_{LOPT} of the individual PA (due to the small voltage and large current of the nanoscale CMOS PA) is transformed to a $m^2 \times N$ times larger value, which is otherwise impractical to accomplish with a regular matching network.

For example, a 2GHz , 3-W CMOS PA based on this concept was reported in [25]. It consists of eight differential cascode PA stages coupled through a 3-coil transformer to the 50Ω load impedance, as illustrated in Figure 6.57. The transformer has two concentric primary coils with the secondary sandwiched in-between.

Assuming a transformer power transfer efficiency of 80% and a 1:1 coil ratio between each primary and the secondary, the individual differential stages would have to generate 0.468W into a 12.5Ω differential load. This implies that each stage sees a 3.42V differential voltage amplitude, or $3.42V_{pp}$ per side. This can be reliably sustained if the common-gate devices in the cascode stages are thick-oxide 2.5V MOSFETs.

Series-stacking of transistors An alternate series-stacking scheme which forms a composite transistor [27], also known as a high-voltage/high-power transistor [28] or a super-cascode [29], and which does not suffer from the transformer loss, is illustrated in Figure 6.58. This concept was first applied to a GaAs MESFET PA [27] but works equally well with HEMTs, HBTs, and SOI, [29] or Silicon on Sapphire (SOS) MOSFETs.

The circuit consists of N large transistors connected in series between the supply and ground. By appropriately designing the resistive ladder that biases the gates/bases of the transistors in the super-cascode, and by loading the gates/bases of the transistors with appropriately sized shunt capacitors, C_i, to ground, each transistor operates under the same DC and RF conditions,

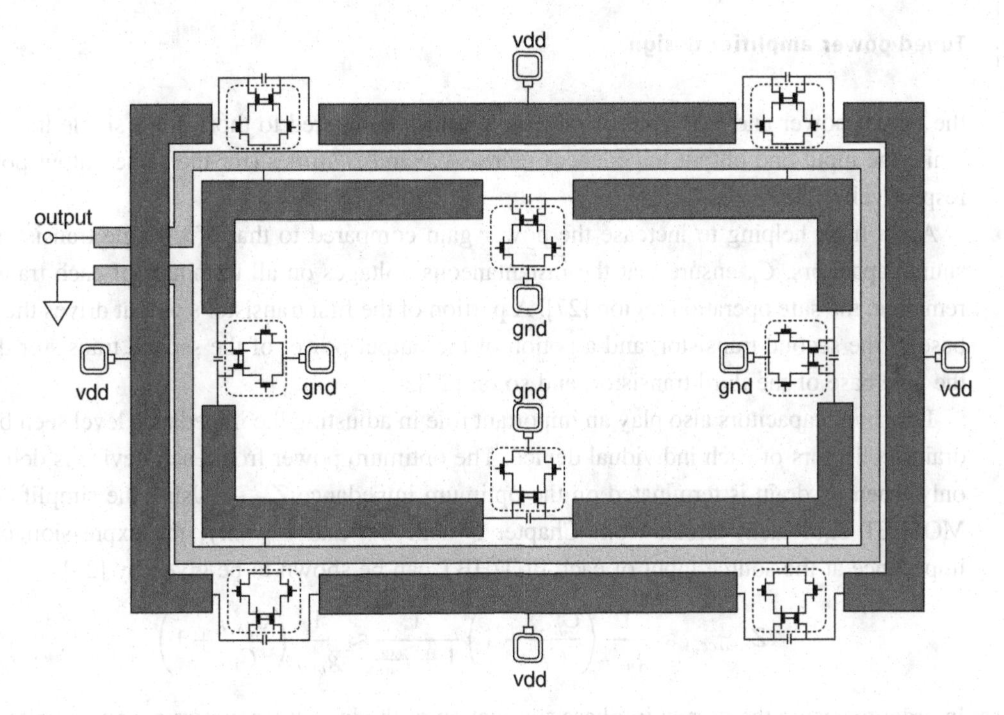

Figure 6.57 Pictorial sketch of a 3W power amplifier in 130nm CMOS realized using transformer-based power combining techniques [25]

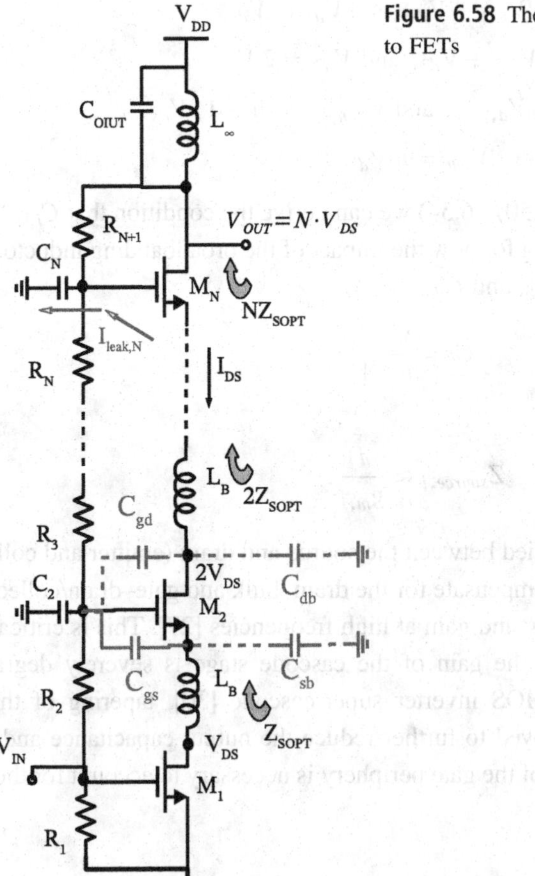

Figure 6.58 The concept of series-stacking (super-cascode) applied to FETs

the output power and bias voltage increase N times compared to those for a single transistor, while the input and output impedances increase N and N^2 times (for the same output power), respectively [27].

Apart from helping to increase the power gain compared to that of a single transistor, the shunt capacitors, C_i, ensure that the instantaneous voltages on all terminals of each transistor remain in the safe operation region [27]. A portion of the first transistor's output drives the gate/base of the second transistor, and a potion of the output power of the second transistor drives the gate/base of the third transistor, and so on [27].

The shunt capacitors also play an important role in adjusting the impedance level seen by the drains/collectors of each individual device. The optimum power from each device is delivered only when its drain is terminated on the optimum impedance Z_{SOPT}. Using the simplified HF MOSFET equivalent circuit from Chapter 5 with C_{gs} and g_m only, the expression of the impedance at the source input of each FET/HBT can be shown to be given by [28]

$$Z_{source,i} \approx \frac{1}{g_{m,i}}\left(\frac{C_{gs,i}}{C_i}+1\right)\frac{1}{1+j\frac{\omega}{\omega_T}} \approx \frac{1}{g_{m,i}}\left(\frac{C_{gs,i}}{C_i}+1\right). \tag{6.50}$$

In order to ensure the correct in-phase summation of the instantaneous drain–source voltages at the output of the PA, the following arrangement must hold [28]

$$V_{g2} = V_{gs2} + V_{d1}, \text{ and } V_{d1} = V_{ds} \tag{6.51}$$

$$V_{g3} = V_{gs3} + V_{d2}, \text{ and } V_{d2} = 2 \cdot V_{ds} \tag{6.52}$$

$$V_{gn} = V_{gsn} + V_{d,n-1}, \text{ and } V_{d,n-1} = (n-1) \cdot V_{ds} \tag{6.53}$$

$$V_{out} = n \cdot V_{ds}. \tag{6.54}$$

For example, if $C_2 >> C_{gs2}$, from (6.50)–(6.54) we can derive the condition that C_i ($i > 2$) must satisfy using Figure 6.58 and ignoring for now the impact of the broadbanding inductors L_B and of the parasitic capacitances C_{gd}, C_{sb}, and C_{db}

$$C_i = \frac{C_{gs,i}}{i-1} \tag{6.55}$$

which leads to

$$Z_{source,i} \approx \frac{i}{g_{m,i}}. \tag{6.56}$$

Series inductive peaking can be applied between the source and drain (emitter and collector) of adjacent transistors in the stack to compensate for the drain–bulk and gate–drain/collector–base capacitance and to improve matching and gain at high frequencies [27]. This is critical at mm-wave frequencies in CMOS where the gain of the cascode stage is severely degraded. As illustrated in Figure 6.59 for a CMOS inverter super-cascode [30], tapering of the device periphery in the stack can be employed to further reduce the output capacitance and increase the output impedance. The tapering of the gate periphery is necessary to account for the leakage

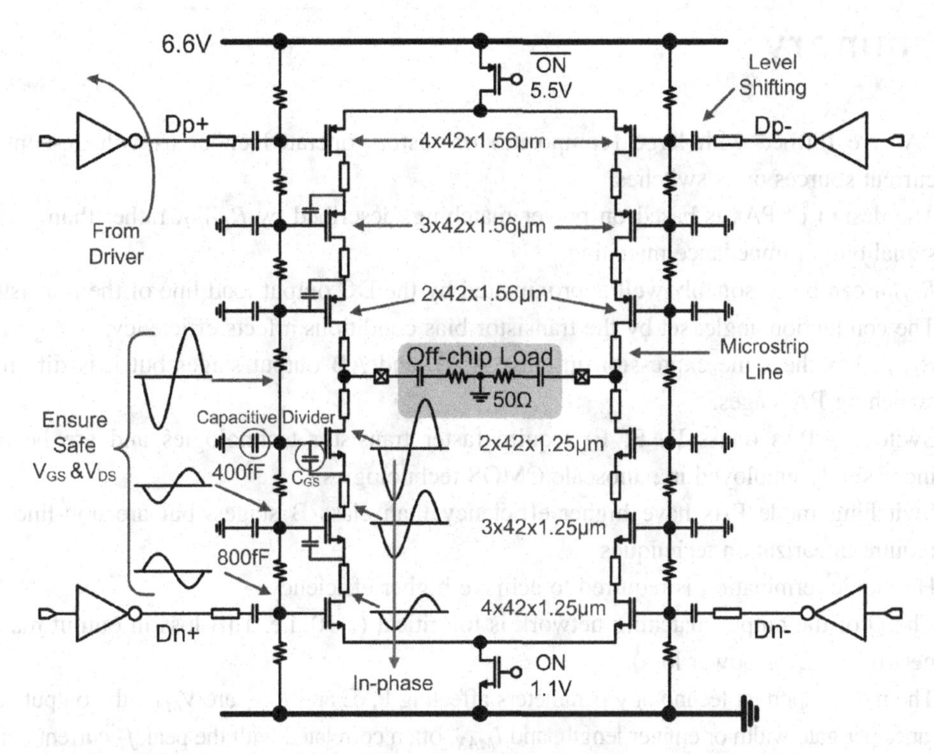

Figure 6.59 Class D SOI CMOS series-stacked PA schematics [30]

current to ground through C_{gsi} and C_i (see Figure 6.58), which reduces the total current flowing through the upper transistors in the stack and the current reaching the load.

Series stacking has several advantages over parallel stacking:

- N times lower current and periphery are needed for the same output power;
- a higher optimum impedance by a N^2 factor is obtained compared to a regular large FET with the same output power;
- simpler, lower-cost, lower-loss matching networks can be implemented at the input and at the output; and
- higher power and higher gain are achieved than in a conventional design using a very large transistor.

However, it should be noted that in bulk CMOS technologies stacking is limited to two to three transistors because of the limited breakdown voltage of the deep-nwell-to-substrate junction. This is not a problem for III-V FETs and HBTs, fabricated on a semi-insulating substrate, or for SOI CMOS where the buried oxide can safely operate with 10V or higher supplies.

Similarly, as well-known from bipolar operational amplifiers, the improvement of the output impedance in a bipolar cascode saturates if more than two transistors are stacked, making the super-cascode HBT concept rather ineffective beyond the regular cascode.

Summary

- PAs are formed with large (composite) transistors operated either as voltage-controlled current sources or as switches.
- The design of PAs is based on power matching, described by R_{LOPT}, rather than on small signal output impedance matching.
- R_{LOPT} can be reasonably well approximated by the DC output load line of the transistor.
- The conduction angle, set by the transistor bias conditions affects efficiency.
- R_{LOPT} has the same expression in class A, B, and AB output stages but it is different in switching PA stages.
- Switching PAs (class D, E, F) require faster transistor technologies and are becoming increasingly employed in nanoscale CMOS technologies.
- Switching mode PAs have higher efficiency than class B stages, but are non-linear and require linearization techniques.
- Harmonic termination is required to achieve higher efficiency.
- The Q of the output matching network is to critical (>10, i.e. 1dB loss in output matching network = 21% power loss).
- The most important technology parameters affecting PAE and P_{SAT} are V_{MAX}, the output capacitance per gate width or emitter length, and I_{MAX}, often correlated with the peak f_T current density.
- The bandwidth over which the PA stage can be matched is inversely proportional $R_{LOPT} \times C_{OUT} \times \omega_0$. As a result, GaN HEMTs, which have the highest Q optimal output impedance, are the most diffciult to match over a broad bandwidth.
- In class A MOSFET PAs, biasing at 0.3mA/μm ensures maximum linearity and efficiency.
- For >1W PAs, special device technologies are typically employed, such as GaAs HBTs, LDMOS, and GaN HEMTs, in conjunction with power-combining techniques.

Problems

(1) Your manager has assigned you the task of evaluating the merits of integrating a 60GHz power amplifier in a 130nm SiGe BiCMOS process or in a 65nm GP CMOS process.

(a) Assume that the last stage of the class A power amplifier is implemented as a differential pair with an inductive load and a common-mode inductor instead of a tail current source. Draw the schematic of such a stage implemented with SiGe HBTs.

(b) Calculate the size and bias current of the SiGe HBTs and of the 65nm MOSFETs in the last power amplifier stage such that the saturated output power in differential-mode is 10dBm. Assume that the maximum voltage allowed on the HBT is $V_{MAX} = 3$V and that the maximum voltage allowed on the 65nm GP MOSFETs is 1.2V.

(c) What is the optimal load resistance that must be presented per side to each transistor in the two implementations?

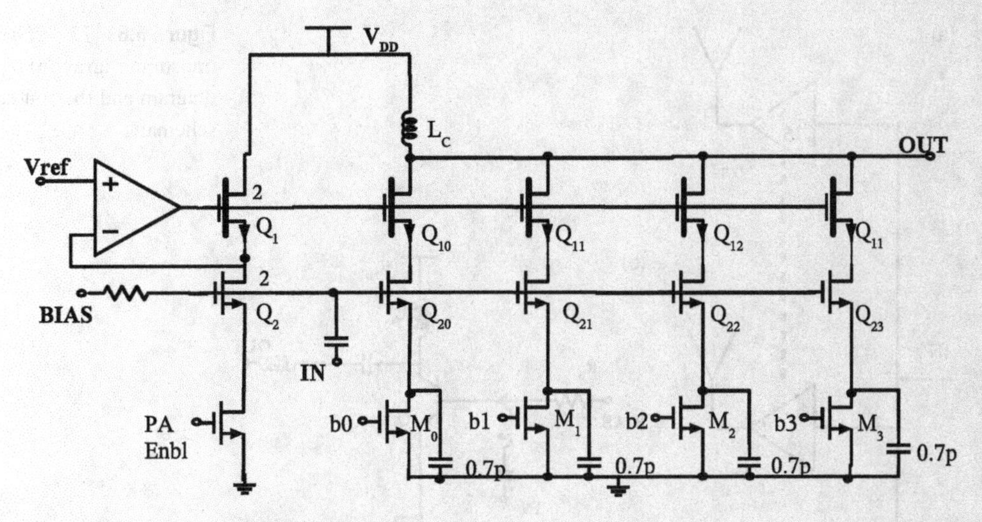

Figure 6.60 45nm CMOS PA schematic

(d) If the common-mode inductor L_{EE} has an inductance of 300pH and a DC resistance of 10Ω, calculate its equivalent impedance at 60GHz and 120GHz. Assume a self-resonance frequency of 240GHz. What is the DC voltage drop on L_{EE} in the SiGe PA and in the CMOS PA?

(e) What is the best possible drain efficiency of the last stage of the power amplifier in the SiGe HBT and in the 65nm CMOS implementations? Assume that the voltage swing in each amplifier is 0V to V_{MAX}. What are the corresponding power supply voltages V_{CC} and V_{DD} of the two amplifiers?

(2) Using the 150nm GaN HEMT technology, design a 2-W class A power amplifier at 10GHz.

(a) Assuming that the maximum transistor power gain follows the f^2_{MAX}/f^2 rule when the device is unconditionally stable and MSG $= g_m/\omega C_{gd}$ otherwise, estimate the maximum gain of a common-source output stage at 10GHz.

(b) Determine the optimum load resistance and the gate width of the GaN HEMT. What is the value of the maximum allowed power supply voltage?

(c) Redesign the 10GHz output stage as a class B stage with the same output power. What is different from the class A case?

(3) A 45nm CMOS implementation of a 5GHz PA is shown in Figure 6.60. The circuit employs 45nm thin-gate oxide CS MOSFETs for high speed and high gain, and thick-oxide 180nm MOSFETs in common-gate for large output voltage swing. All transistors are biased at the saturated power current density of 0.4mA/μm. The thick-oxide MOSFETs have $V_{GS} = 1.2$V, 180nm gate lengths and a V_{MAX} (the maximum allowed instantaneous V_{DS}) of 3V.

(a) Find the size and bias current of each MOSFET such that the saturated output power is 36mW when the PA is terminated on R_{LOPT}. $V_{REF} = 1.5$V, $V_{BIAS} = 1.3$V and $V_{DD} = 3$V.

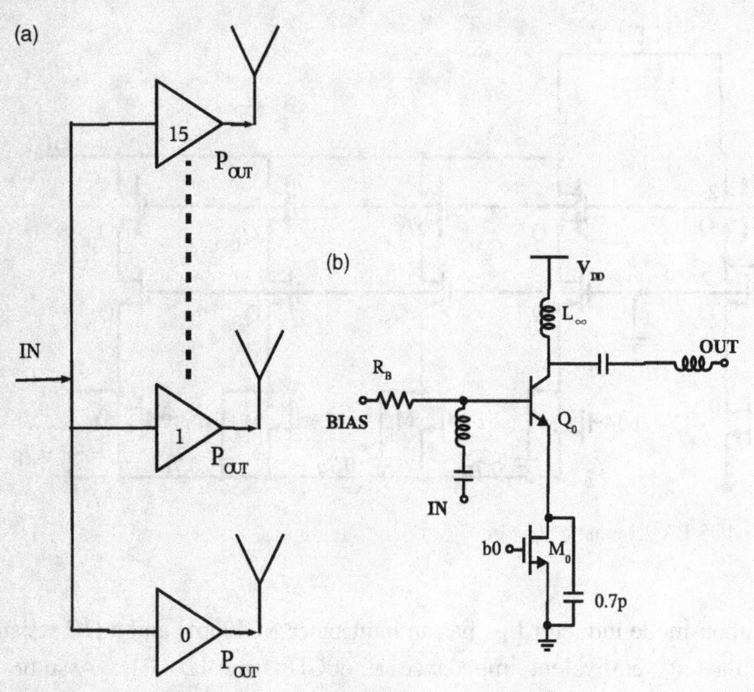

Figure 6.61 77–79GHz radar transmitter array (a) block diagram and (b) unit cell schematic

The gain of the opamp is very large. Assume that the gate widths of the MOSFETs are binary-weighted.

(b) How would you implement digital amplitude modulation in this circuit?

(4) Figure 6.58 shows the block diagram of a 77–79GHz automotive radar transmitter array for collision avoidance (ACC). The radar array consists of 16 identical PAs, each with its own integrated antenna, and each antenna has a gain of 3dB. Each PA is operated in saturated output power mode, with a PAE of 30%. The bias current can be turned fully off or fully on at a data rate of up to 2 Gb/s using the control bit.

(a) Calculate the P_{SAT} of each PA such that the total effective isotropic radiated power (*EIRP*) of the array at saturation is 29dBmi. ($EIRP = G_{TX}P_{TXSAT}$).

(b) Calculate the link budget at 77GHz for a target located 2m away which has a radar cross-section of 0.1m^2. The receiver consists of 16 elements, each with an antenna gain of 3dB and a noise figure of 7dB. The receiver bandwidth is 2GHz and the detector SNR is 10dB. What is the link budget if the TX and RX antennas have gains of 7dB?

(c) The unit cell of the SiGe HBT PA array is implemented as a single-ended CE stage, as shown in Figure 6.61(b). The HBT has $W_E = 0.13\mu m$. $V_{BIAS} = 1V$ and $V_{DD} = 1.5V$. When $b_0 = 1$, the V_{DS} of M_0 is 100 mV. Calculate the emitter length and collector current when the unit cell PA is designed to have an OP_{1dB} of 7dBm. The optimal linearity bias current density of the SiGe HBT is 10mA/µm^2. Assume that the corresponding transistor f_T is 230GHz and that $V_{CESAT} = 0.7V$.

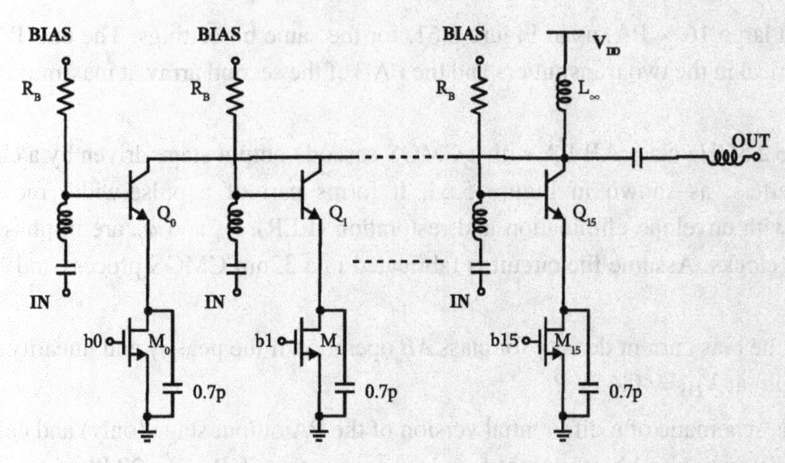

Figure 6.62 Transmitter schematic with 16x PA and on-die (ideal) power combining

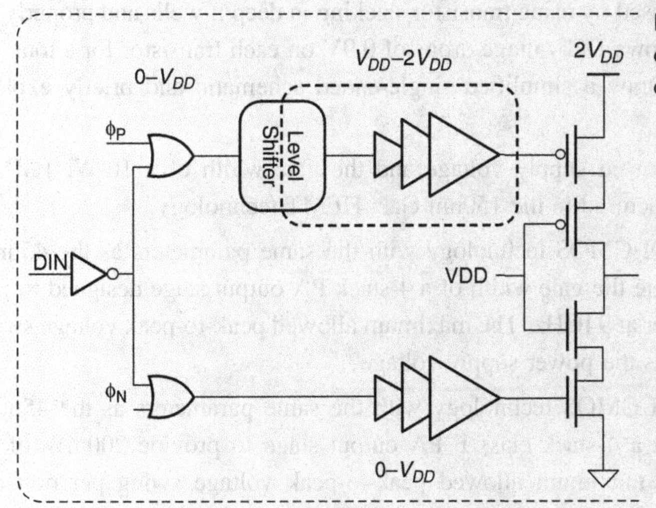

Figure 6.63 32nm CMOS class AB output stage driven by CMOS-inverter chains [31]

(d) The inputs of the 16 PAs are connected together. If each PA has an input impedance of 16Ω, a power gain of 10dB, and an output power of 10dBm, what is the peak-to-peak voltage swing at the common input node? Assume that the input is impedance-matched to the signal source.

(e) How would you implement 4-bit gain control using the 16 PAs knowing that at least one element of the array is always on?

(f) Write the expression of the output power of the array as a function of the 16 control bits $b_0, b_1, b_2, b_3, \ldots b_{15}$ and the output power of the individual PA, P_{OUT}.

(g) What is the total PAE of the array if the PAE of each individual PA is 30% and $b_0, b_1, b_2, b_3, \ldots b_{15}$ are all set to "1"? What about the total PAE when b_0, b_1, b_2, b_3 are set to 1 and all other bits are 0? How does the PAE variation with the control bits compare to that of a large 16 × PA, as in Figure 6.61, for the same bit settings. The unit PA cells are identical in the two transmitters and the PAE of the second array at maximum power is 30%.

(5) Consider the 2.5GHz class AB PA with a CMOS cascode output stage driven by a chain of CMOS inverters, as shown in Figure 5.63. It forms part of a pulse-width modulated transmitter with envelope elimination and restoration (EER). φ_n and φ_p are in-phase non-overlapping clocks. Assume the circuit is fabricated in a 32nm CMOS process and $V_{DD} =$ 0.9V.

(a) What is the bias current density for class *AB* operation if the peak f_T and linearity bias is 0.4mA/μm at $V_{DS} = 0.45$V ?

(b) Draw the schematic of a differential version of the PA output stage (only) and calculate the transistor size and bias current for class *A* operation if $P_{SAT} =$ 20dBm.

(c) Find R_{LOPT} for the differential PA in (b) above.

(d) R_{LOPT} can be increased by using transistor stacking in deep n-wells and properly placed capacitors. This allows DC voltage drops of 0.9V on each transistor for a total supply voltage of 3.6V. Draw a simplified single-ended schematic and briefly explain the concept.

(6) Find the maximum allowed supply voltage and the gate width of a 10-W, 1GHz class E switching PA manufactured in the 150nm GaN HEMT technology.

(7) Assuming a 45 nm SOI CMOS technology with the same parameters as the 45nm bulk CMOS process, calculate the gate width of a 4-stack PA output stage designed to provide 100mW of output power at 94GHz. The maximum allowed peak-to-peak voltage swing per transistor is 1V. What is the power supply voltage?

(8) Assuming a 45nm SOI CMOS technology with the same parameters as the 45nm bulk CMOS process, design a 4-stack class F PA output stage to provide 200mW of output power at 45GHz. The maximum allowed peak-to-peak voltage swing per transistor is 1V. Calculate the transistor size, the power supply voltage, and the output matching network for the first three harmonics.

REFERENCES

[1] S. Cripps, *RF Power Amplifiers for Wireless Communications*, Artech House, 2nd Edition, 2006.

[2] A. Sedra and K. C. Smith, *Microelectronic Circuits*, Oxford University Press, 5th Edition, 2003.

[3] V. Prodanov and M. Banu, "Power amplifier principles and modern design techniques," in *Wireless Technologies, Systems, Circuits Devices*, edited by K. Iniewski, CRC Press, 2008, pp. 349–381.

[4] L. Larson, "Bipolar-based power amplifier design beyond handsets and into MMIC: design, modeling, characterization, packaging, reliability," *IEEE BCTM Short Course*, October 2007.

[5] N. O. Sokal and A. D. Sokal, "Class E, a new class of high-efficiency tuned single-ended power amplifiers," *IEEE JSSC*, **10**: 168–176, June 1975.

[6] C. Campbell, "Advanced PA design in GaN," *IEEE CSICS Short Course*, October 2009.

[7] R. B. Staszewski and P. T. Balsara, *All-digital Frequency Synthesizer in Deep-Submicron CMOS*, John Wiley, 2006.

[8] A. Valdes-Garcia, S. Reynolds, and U. Pfeiffer, "A class-E power amplifier in SiGe," *IEEE SiRF Meeting Digest*, pp.199–202, January 2006.

[9] Stephen Long, "Basics of power amplifier design," *IEEE CSICS Short Course*, October 2009.

[10] F. Raab, "Class C, E, and F power amplifiers based upon a finite number of harmonics," *IEEE MTT*, **49**(8): 1462–1468, August 2001.

[11] T. H. Lee, *The Design of CMOS Radio-Frequency Integrated Circuits*, Cambridge, 2nd Edition, 2004, Chapter 15.

[12] K. Mistry *et al.*, "A 45 nm logic technology with high-k metal gate transistors, strained silicon, 9 Cu interconnect layers, 193 nm dry patterning, and 100% Pb-free packaging," *IEEE IEDM Digest*, pp. 247–250, December 2007.

[13] T. Yao, M. Q. Gordon, K. K. W. Tang, K. H. K. Yau, M.-T. Yang, P. Schvan, and S. P. Voinigescu, "Algorithmic design of CMOS LNAs and PAs for 60GHz radio," *IEEE JSSC*, **42**(5): 1044–1057, May 2007.

[14] M. Khanpour, S. P. Voinigescu, and M. T. Yang, "A high-gain, low-noise, + 6dBm PA in 90 nm CMOS for 60GHz radio," *IEEE CSICS Digest*, pp. 121–124, October 2007.

[15] S. T. Nicolson, K. H. K. Yau, S. Pruvost, V. Danelon, P. Chevalier, P. Garcia, A. Chantre, B. Sautreuil, and S. P. Voinigescu, "A low-voltage SiGe BiCMOS 77GHz automotive radar chipset," *IEEE MTT*, **56**: 1092–1104, May 2008.

[16] T. Suzuki, Y. Kawano, M. Sato, T. Hirose, and K. Joshin, "60 and 77GHz power amplifiers in standard 90 nm CMOS," *IEEE ISSCC Digest*, pp.562–563, February 2009.

[17] D. Chowdhury, P. Reynaert, and A. M. Niknejad, "A 60GHz 1V + 12.3dBm transformer-coupled wideband PA in 90 nm CMOS," *IEEE ISSCC Digest*, pp.560–561, February 2008.

[18] L. Nathawad *et al.*, "A dual-band CMOS MIMO radio SoC for IEEE 802.11n wireless LAN," *IEEE ISSCC Digest*, pp. 358–359, February 2008.

[19] L. Khan, "Single-sided transmission by envelope elimination and restoration," *Proc. IRE*, pp. 803–806, July 1952.

[20] J. Walling, H. Lakdawala, Y. Palaskas, A. Ravi, O. Degani, K. Soumyanath, and D. Allstot, "A 28.6dBm 65 nm Class-E PA with envelope restoration by pulse-width and pulse-position modulation," *IEEE ISSCC Digest*, pp. 566–567, February 2008.

[21] W. H. Doherty, "A new high efficiency power amplifier for modulated waves," *Proc. IRE*, **24**: 1163–1182, September 1936.

[22] B. Wicks, E. Skafidas and R. Evans, "A 60GHz, fully integrated Doherty power amplifier based on a 0.13-μm CMOS process," *IEEE RFIC Symposium Digest*, pp. 69–72, 2008.

[23] H. Chireix, "High power outphasing modulation," *Proc. IRE*, **23**: 1370–1393, November 1935.

[24] S. Moloudi, K. Takinami, M. Yaoussef, M. Mikhemar, and A. Abidi, "An outphasing power amplifier for a software-defined radio transmitter," *IEEE ISSCC Digest*, pp. 568–569, February 2008.

[25] I. Aoki, S. Kee, R. Magoon, R. Aparicio, F. Bohn, J. Zachman, G. Hatcher, D. McClymont, and A. Hajimiri, "A fully integrated quad-band GSM/GPRS CMOS power amplifier," *IEEE ISSCC Digest*, pp. 570–571, February 2008.

[26] Y. A. Atesal, B. Cetinoneri, M. Chang, R. Alhalabi, and G. M. Rebeiz, "Millimeter-wave wafer-scale silicon BiCMOS power amplifiers using free-space power combining," *IEEE MTT*, **59**: 954–965, April 2011.

[27] M. Shifrin, Y. Ayasli, and P. Kazin, "A new power amplifier topology with series biasing and power combining of transistors," *IEEE 1992 MWSYM Digest*, pp. 39–41, 1992.

[28] A. K. Ezzedine and H. C. Huang, "The high voltage/high power FET (HiVP)," *IEEE RFIC Symposium Digest*, pp. 215–218, June 2003.

[29] S. Pornpromlikit, J. Jeong, C. D. Presti, A. Scuderi, and P. M. Asbeck, "A watt-level stacked-FET linear power amplifier in silicon-on-insulator CMOS," *IEEE MTT*, **58**: 57–64, January 2010.

[30] I. Sarkas, A. Balteanu, E. Dacquay, A. Tomkins, and S. P. Voinigescu, "A 45 nm SOI CMOS class D mm-wave PA with >10Vpp differential swing," *IEEE ISSCC Digest*, pp. 24–25, February 2012.

[31] Sang-Min Yoo, J. S. Walling, Eum Chan Woo, and D. J. Allstot, "A switched-capacitor power amplifier for EER/polar transmitters," *IEEE ISSCC Digest*, pp. 428–429, February 2011.

7 Low-noise tuned amplifier design

What is a low-noise amplifier? Digital-processing systems require interfaces to the analog world. The first amplification stage in a receiver is typically called a low-noise amplifier or, in short, LNA. The LNA amplifies the input signal to a level that makes further signal processing insensitive to noise. The key performance issue for an LNA is to deliver the undistorted but amplified signal to the signal-processing unit without adding further noise.

In this chapter, we will examine the design of narrowband, tuned low-noise amplifiers. Tuned amplifiers have a bandpass filter response with positive gain. By "narrowband," we mean that the 3dB bandwidth of the amplifier is less than 20% of its center frequency. Such LNAs are common in radio receiver front-ends where the desired input signal is confined to a specific frequency range. Broadband LNAs will be discussed in Chapter 8.

First, the basic design philosophy and design theory will be introduced. Next, design methodologies for several different LNA topologies will be developed followed by design examples. Finally, the impact of temperature, supply and process variation on LNA performance will be discussed along with bias techniques to combat it.

7.1 LNA SPECIFICATION AND FIGURE OF MERIT

The first amplification stage in a wireless receiver plays a critical role in determining the overall receiver sensitivity. To understand why, let us re-visit the noise performance of N cascaded stages, as discussed in Chapter 2. Such a system is depicted in Figure 7.1, where the gain and noise factor of the Nth stage are described by the available power gain, $G_{a,N}$, and the noise factor, F_N, or noise temperature, T_{eN}, respectively. The blocks in Figure 7.1 could represent the LNA, the mixer, and any variable or fixed gain amplifiers in the IF path of a radio receiver. Alternatively, they could describe individual stages of a multi-stage LNA.

Assuming maximum power transfer (i.e., the output of one stage is impedance matched to the input of the subsequent stage), the overall noise factor and noise temperature of the cascaded system are given by Friis' formulas (2.42) and (2.43), respectively, and reproduced here for convenience [1]

$$F = F_1 + \frac{F_2 - 1}{G_{a,1}} + \ldots + \frac{F_N - 1}{G_{a,1}G_{a,2}\ldots G_{a,N-1}} \text{ and } T_e = T_{e,1} + \frac{T_{e,2}}{G_{a,1}} + \ldots + \frac{T_{e,N}}{G_{a,1}G_{a,2}\ldots G_{a,N-1}}$$

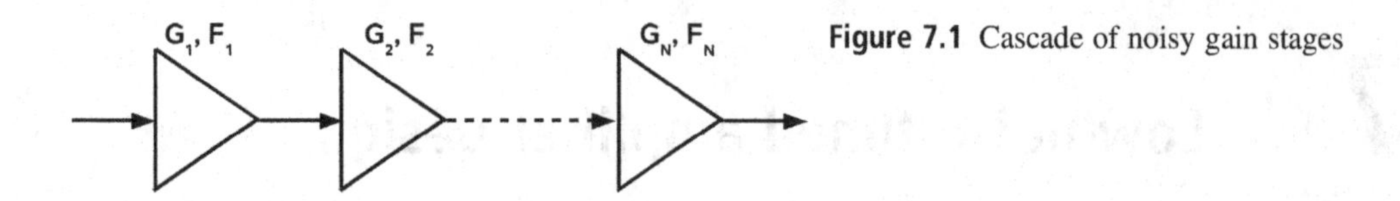

Figure 7.1 Cascade of noisy gain stages

We note that the noise contribution of each stage is reduced by the aggregate gain of all the preceding stages. Thus, the first stage should be designed to have a low-noise factor along with adequate gain to suppress the noise contribution of subsequent stages.

The **power gain** and **the noise** factor are not the only metrics to consider in the design of this first stage. LNA applications (4G wireless, WiFi, GPS, Bluetooth, satellite links, automotive radar, etc.) operate in many frequency bands. The operating frequency, f, and, in wireline and fiber-optic applications, the operating bandwidth of the LNA will impact the maximum achievable performance. Non-linearity must also be considered to meet the specifications of many applications. These parameters must be included in the figure of merit, FoM. On the other hand, different systems may not be directly comparable, and in fact have diverging requirements. For example, very wide bandwidth is needed for high-performance wired applications, but this increases power consumption. Low power consumption is an important design attribute for low-bandwidth wireless applications. For wide bandwidth systems, bandwidth is often more important than linearity to describe the performance of an LNA. In this chapter, we focus on tuned LNAs in the context of *wireless* communication, automotive radar, and imaging systems, all of which tend to be relatively narrowband.

The **dynamic range** should be high such that the receiver can process both large and small input signal levels. In radio systems where out-of-band tones are present, the mixing of these tones with the desired signal often results in intermodulation terms which lie in-band. These scenarios dictate a highly linear LNA, whose linearity is measured in terms of its third-order intercept point, IP_3. This intercept point can be described as an input-referenced quantity, IIP3, or as an output-referenced quantity (OIP3), where the two are related through the power gain, G, of the LNA

$$OIP_3 = G \times IIP_3. \tag{7.1}$$

The metrics described so far – gain, linearity, and noise factor – lead us to a figure of merit for a tuned LNA (FoM_{LNA}). This figure of merit, defined by the ITRS roadmap [2], captures the dynamic range of an amplifier versus its power consumption. This performance measure is independent of circuit topology and of frequency and thus is independent of the specific application

$$FOM_{LNA} = \frac{G \times IIP_3 \times f}{(F-1) \times P} = \frac{T_0}{T_e}\frac{G \times IIP_3 \times f}{P} = \frac{OIP_3 \times f}{(F-1) \times P} = \frac{T_0}{T_e}\frac{OIP_3 \times f}{P}. \tag{7.2}$$

None of the parameters in (7.2) is expressed in dB. The normalized noise temperature, (F-1), is used in the figure of merit, rather than the noise factor. F is a measure of the minimum signal that is correctly amplified by the LNA. However, T_e/T_0 = (F-1) is a better measure of the contribution of the amplifier to the total noise, since it allows the ratio between the noise of the amplifier and the noise already present at its input to be directly evaluated. Moreover, the noise temperature of a transistor improves with technology scaling as $1/S$, rather than logarithmically, and is easier to quantify in small increments.

Important note: In a well-designed LNA stage, OIP_3 depends only on the supply voltage and $F-1$ increases linearly with frequency. Consequently, (7.2) captures the fact that the dynamic range of the LNA, described by $OIP_3/(F\text{-}1)$, is expected to decrease with increasing application frequency, f.

Typical LNA design specifications also include other parameters not present in the FoM_{LNA}. Simple microwave circuit theory tells us that the power gain is maximized if both the input and the output of the LNA are impedance matched. In most cases, the input and output return loss, described by S_{11} and S_{22} respectively, should both be better than -15dB over the desired bandwidth. Other design considerations in both narrowband or broadband LNAs include the reverse isolation and the stability factor. The latter tells us if the amplifier does not turn into an oscillator.

7.2 DESIGN GOALS FOR TUNED LNAs

Maximizing the FoM of the LNA is usually the ultimate goal. This implies simultaneously:

- minimizing the noise factor,
- increasing the gain beyond a minimum required value,
- maximizing linearity,
- minimizing power dissipation,
- reducing the input reflection coefficient below a given value over the specified bandwidth, and
- in standalone LNAs only, reducing the output reflection coefficient below a given value over the specified bandwidth.

When minimizing the noise factor of the LNA, the clearly defined target is the minimum noise figure of the chosen transistor technology. Based on Haus and Adler's theoretical study discussed in Chapter 3, this is theoretically achievable by optimally biasing the input transistor in the LNA and perfectly matching its noise impedance to the signal source impedance.

7.3 LOW-NOISE DESIGN PHILOSOPHY AND THEORY

Traditional low-noise amplifier design at microwave and mm-wave frequencies employed lossless reactive matching networks to transform the signal source impedance to the optimum noise impedance of a discrete transistor. The transistor itself was biased at its minimum noise figure sweet spot. Since, in general, the real part of the optimum noise impedance and the real part of the input impedance of transistors are quite different, this approach compromises the input reflection and the gain of the amplifier in order to minimize its noise factor. The designer would use the measured S-parameters of the transistor at the optimum noise bias and the measured constant noise figure circles to carry out the design. Typical measured S-parameter and noise data would be provided by the transistor manufacturer in the data sheet or would have to be collected by the designer in the lab.

Before the emergence of monolithic microwave integrated circuits in the 1980s, this LNA design technique was "the only game in town" for hybrid circuits using discrete transistors and microstrip matching networks. Life was easy for the designer because the solution was always unique and optimal. Business was booming for the microwave transistor manufacturers and millions of X-band satellite receivers with GaAs p-HEMT LNAs were sold worldwide.

With the advent of monolithic microwave integrated circuits (MMICs), first in GaAs in the 1980s, and later in silicon in the 1990s, two significant changes occurred. First, the geometry and optimal bias current of the transistor were freed up as design variables. Second, the passive matching networks, at least ten times larger in area than the transistors themselves, dictated the size and cost of the LNA while their performance degraded – especially on silicon substrates – due to higher substrate and metal loss.

The need for a radical change in the design methodology for silicon LNAs became apparent. The size and number of passive components had to be significantly reduced and "active matching" using the transistor itself, the lowest cost and highest performance component in an IC, had to somehow be harnessed. Such a design methodology will be described in the rest of this chapter. It is independent of the transistor type – bipolar or FET [3] – and of the semiconductor material – silicon, silicon-germanium, GaAs, InP, GaN, etc. The key elements that provide its foundations were introduced in Chapters 3 and 4 and are summarized below:

- Rather than transforming the signal source impedance to the optimum noise impedance of the transistor, the transistor itself is designed to have an optimum noise impedance equal to the signal source impedance.
- The optimum noise impedance is inversely proportional to frequency, the transistor size (emitter length/gate width), number of transistors, number of transistor fingers, or MOSFET fins connected in parallel.
- The real part of the optimum noise impedance is different from the real part of the input impedance, except by accident and at a single frequency.
- If the bias current density is fixed, the minimum noise factor is invariant with respect to transistor size (emitter length/gate width), number of transistors, number of transistor gate fingers, or fins connected in parallel.
- The imaginary part of the input impedance of a transistor (FET or HBT) is typically within 15% of the imaginary part of its optimum noise impedance. Consequently, they can be tuned out simultaneously (with at most 10% error margin) using series inductors.
- The real part of the optimum noise impedance and the real part of the input impedance of a transistor (FET or HBT) can be independently tuned using transistor sizing and reactive feedback, respectively.
- The power consumption of a minimum noise, noise-matched LNA stage in a given technology is dictated by the bias current required to achieve a certain transistor noise

impedance, and is proportional to a device technology figure of merit Power_Noise_FoM = $J_{OPT}V_{DD}f_T/g'_{meff}$.
- Saving power comes at the expense of noise and linearity.

Based on these properties, the LNA design can be accomplished in two major steps:

(i) active matching (transistor sizing) to make the real part of the optimum noise impedance equal to the source impedance, and
(ii) reactive matching with lossless feedback to match the input impedance and tune out the imaginary part of the optimum noise impedance.

Following this approach, the designed LNA is optimal in terms of noise factor or noise equivalent temperature, and consumes the least die area with the lowest untested die cost.

Part 1: Active device matching

As discussed in Chapter 3, a low-noise amplifier can be represented as a two-port network driven by a signal source with a (complex) admittance $Y_S = G_S + jB_S$, and loaded with an impedance Z_L.

In the noise admittance formalism, the noise factor of this amplifier can be expressed as

$$F = F_{MIN} + \frac{R_n}{G_S}|Y_S - Y_{sopt}|^2 \tag{7.3}$$

where F_{MIN} is the minimum noise factor of the amplifier, ideally identical or close to the minimum noise factor of the transistor itself, and R_n is the noise resistance of the amplifier. The second term in (7.3) becomes zero when the complex signal source impedance Y_S equals the optimum source admittance Y_{SOPT}. For this unique admittance, the noise factor of the two-port equals its minimum noise factor, F_{MIN}. Therefore, two steps must be taken to minimize the noise figure/factor of an LNA:

- First, the minimum noise figure must be made as low as possible to reduce the contribution of the first term in (7.3). The lowest limit is the minimum noise figure of the transistor itself.
- Second, the optimal source admittance(impedance) must be made equal to the signal source admittance(impedance)$Y_S = 1/Z_S$.

In most radio systems, $Z_S = Z_0 = 50\Omega$, so our design goal is to set $Z_{SOPT} = 1/Y_{SOPT} = 50\Omega$. This can be elegantly and most effectively achieved through a matching network, of which the transistor is an integral part, as we shall see.

If we are to minimize the F_{MIN} of the amplifier, it follows that the F_{MIN} of the transistor and of the LNA topology should be minimized first. This is accomplished by properly biasing and sizing the transistor as discussed in Chapter 4. Let us re-visit the transistor high-frequency and noise characteristics introduced Chapter 4. Figure 7.2 plots the NF_{MIN} of a SiGe HBT at 5GHz and 65GHz as a function of the collector current per emitter area, J_C.

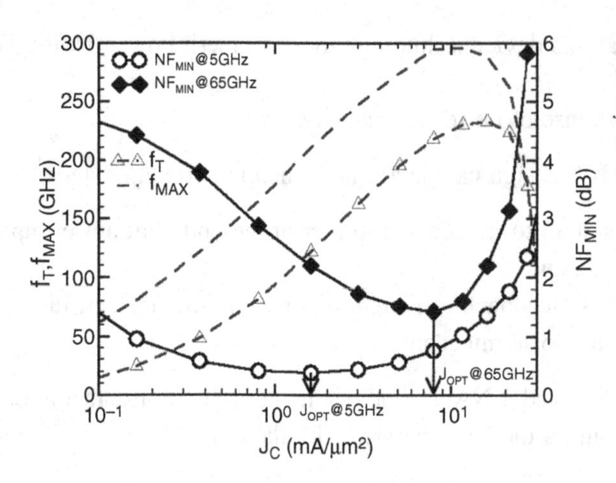

Figure 7.2 Measured SiGe HBT cutoff frequency, f_T, maximum oscillation frequency, f_{MAX}, and minimum noise figure, NF_{MIN}, at 5 and 65GHz, plotted as a function of collector current density [4]

From this plot, we note that NF_{MIN} reaches a minimum at an optimal noise current density, J_{OPT}. ***For low-noise designs, the amplifier stage should be biased at*** J_{OPT}. In HBTs, this current density increases with frequency and is also dependent on the amplifier configuration – J_{OPT} of a single HBT is typically smaller than J_{OPT} of a cascode stage. Therefore, the NF_{MIN} versus J_C curve should be measured or simulated at each frequency of interest. Moreover, at all frequencies, this current density is lower than the peak f_T or peak f_{MAX} current densities. For design purposes, it is also useful to plot the f_T and f_{MAX} dependencies on the collector current density, as shown in Figure 7.2. It is apparent that J_{OPT}increases from 1.5mA/μm^2 at 5GHz to 8mA/μm^2 at 65GHz. The corresponding f_T values are 120GHz and 220GHz, respectively, while f_{MAX} is 200GHz and 290GHz, respectively. This leads to the conclusion that, when biasing an HBT for low-noise operation, a penalty in power gain of at least 1–2 dB is to be expected.

The same low-noise biasing concept can be applied to field-effect transistors, as evidenced in Figure 7.3. As discussed in Chapter 4.2, in MOSFETs, J_{OPT} does not change with frequency. Even more interestingly, J_{OPT} in a silicon n-MOSFET has remained fairly constant at about 0.15mA per micron of gate width even as CMOS technologies have been scaled according to constant field scaling rules from the 0.5μm technology generation to the 65nm node. In more recent nodes, it has increased to 0.2–0.3mA/μm due to strain engineering and increasing vertical electric field between the gate and the channel. No experimental data have been published for FinFETs but similar optimum noise current densities of 0.15–0.2mA/μm have been observed in GaAs, InP, and GaN HEMTs.

Once the optimal bias current density has been selected, the next step is to size the transistor such that the optimal noise impedance of the LNA, including the matching network, becomes equal to the signal source admittance/impedance Y_S/Z_S. At a given bias, the transistor noise parameters – R_n, G_u, G_{cor}, and B_{cor} – all scale with emitter length or gate width. Hence, the HBT noise parameter equations from Chapter 4.3, can be recast as functions of frequency, emitter length, l_E, and number of emitter stripes, N

$$R_n = \frac{R_{HBT}}{Nl_E} \tag{7.4}$$

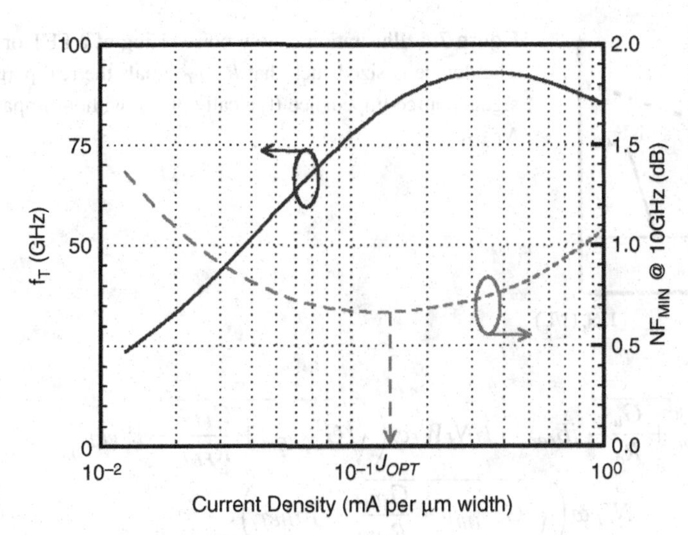

Figure 7.3 130nm silicon n-channel MOSFET cutoff frequency (f_T) and minimum noise figure at 10GHz (NF_{MIN}) as a function of drain current per micron of gate width

$$G_u = G_{HBT}\omega^2 NI_E \tag{7.5}$$

$$G_{cor} = G_{C,HBT}\omega NI_E \tag{7.6}$$

$$B_{cor} = B_{HBT}\omega NI_E \tag{7.7}$$

where R_{HBT} [$\Omega \times \mu$m], $G_{u,HBT}$ [Ω/µm × Hz^2], $G_{C,HBT}$ [Ω/µm × Hz], and B_{HBT} [Ω/µm × Hz] are technology constants. They depend mainly on the collector current density and, when the HBT is biased in the active region, are weak functions of V_{CE}. They can also be obtained from simulations using the compact model provided by the foundry with the design kit.

For FETs, the total gate width, W, is controlled by connecting a number of gate fingers, N_f, in parallel, each with a fixed gate finger width, W_f, and by connecting planar MOSFET active areas (transistors) or FinFET fins in parallel. Under these conditions, the noise parameters of a FET can be expressed in terms of the number of gate fingers and number of active areas or fins, N

$$R_n = \frac{R_{N,FET}}{N_f N} \tag{7.8}$$

$$G_u = G_{N,FET}\omega^2 N_f N \tag{7.9}$$

$$G_{cor} = G_{C,FET}\omega N_f N \tag{7.10}$$

and

$$B_{cor} = B_{FET}\omega N_f N. \tag{7.11}$$

Again, $R_{N,FET}$ [Ω], $G_{N,FET}$ [Ω/Hz^2], $G_{C,FET}$ [Ω/Hz], and B_{FET} [Ω/Hz] represent technology constants which depend on the drain current density per unit gate width, on the gate finger width, and are only weak functions of V_{DS}. They can be obtained from measurements or simulations.

From the two-port noise theory presented in Chapter 3 and (7.4)–(7.11), the optimal source admittance can be described as a function of frequency, gate width, and number of transistors or fins

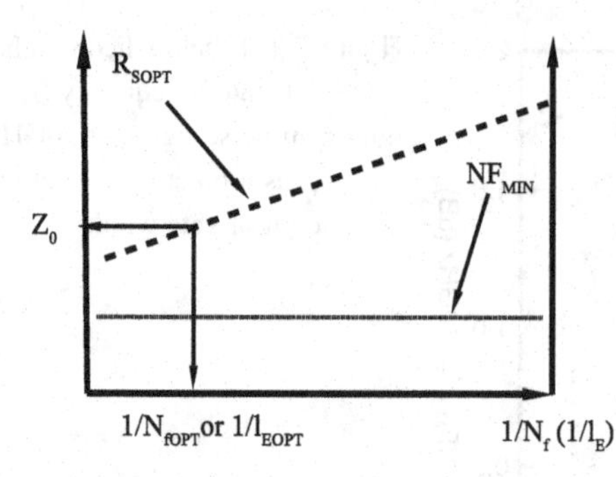

Figure 7.4 Illustration of low-noise sizing of a FET or HBT. The device is sized such that R_{SOPT} equals the real part of the signal source impedance (typically 50Ω), without impacting NF_{MIN}

$$Y_{sopt} = \sqrt{G_{cor}^2 + \frac{G_u}{R_n}} - jB_{cor} = NN_f W_f \omega \left(\sqrt{G_{C,FET}^2 + \frac{G_{FET}}{R_{FET}}} - jB_{FET} \right)$$

$$\text{or } NI_E \omega \left(\sqrt{G_{C,HBT}^2 + \frac{G_{HBT}}{R_{HBT}}} - jB_{HBT} \right). \tag{7.12}$$

Alternatively, from (4.53) and (4.54), the optimum noise impedance is expressed as a function of the small signal parameters and layout geometry of FETs and HBTs. We thus obtain critical design equations that give the transistor size, in terms of number of transistors, fins, gate fingers or emitter length, as a function of the desired noise impedance, Z_0,

$$Z_{sopt}(FET) \approx \frac{f_{Teff}}{N \cdot N_f \cdot W_f \cdot f \cdot g'_{meff}} \left[\sqrt{\frac{g'_m \cdot R'_s + W_f \cdot g'_m \cdot R'_g(W_f)}{k_1}} + j \right] = Z_0 + jX_{sopt} \tag{7.13}$$

$$NN_f = \frac{f_{Teff}}{Z_0 \cdot W_f \cdot f \cdot g'_{meff}} \sqrt{\frac{g'_m \cdot R'_s + W_f \cdot g'_m \cdot R'_g(W_f)}{k_1}} \tag{7.14}$$

where $k_1 = 0.5 \ldots 1$, depending on the choice of gate finger width, W_f, or fin height, and for HBTs

$$Z_{sopt}(HBT) \approx \frac{f_{Teff}}{f \cdot N \cdot l_E \cdot g'_{meff}} \left[\sqrt{\frac{g'_m}{2}(r'_E + R'_b)} + j \right] = Z_0 + jX_{sopt} \tag{7.15}$$

$$Nl_E = \frac{f_{Teff}}{Z_0 \cdot f \cdot g'_{meff}} \sqrt{\frac{g'_m}{2}(r'_E + R'_b)} \tag{7.16}$$

where the small signal equivalent circuit parameters with primes are given per unit gate width or per unit emitter length. It should be noted that (7.15) is not valid at frequencies below $\frac{1}{2\pi r_\pi C_\pi}$ because, in deriving (7.15), we have ignored r_π in the noise equivalent circuit of the HBT.

As illustrated in Figure 7.4, (7.14) and (7.16) permit the optimization of the transistor geometry as part of the noise matching network. This can be accomplished with little or no impact on the minimum noise figure of the transistor, since, to first order, NF_{MIN} and J_{OPT} are largely independent of emitter length, number of gate fingers, transistors, or FinFET fins connected in parallel.

Note that the best possible noise figure that can be attained for the LNA at a given frequency is the minimum noise figure of the transistor in the input stage at that operating frequency. Also, since the minimum noise figure degrades with increasing emitter width in HBTs, and with increasing gate length in MOSFETs, only the minimum emitter width and the minimum gate length devices are employed in low-noise amplifiers.

EXAMPLE 7.1

(a) The measured R_{SOPT} of a 3×130nm×2.5μm HBT (3 emitter stripes, each 130nm wide and 2.5μm long) is 466.6Ω at 6GHz and J_{OPT} = 2mA/μm^2. Find the total emitter length of a SiGe HBT with 130nm emitter width such that its R_{SOPT} = 50Ω at 6GHz. Calculate the bias current knowing that the HBT is biased at the optimum noise figure current density.

(b) Find the total gate width of a 90nm planar n-MOSFET such that its R_{SOPT} = 50Ω at 6GHz knowing that the optimum noise resistance of a 20μm wide device with 2μm fingers is 275Ω at 5GHz when biased at the optimum noise figure current density J_{OPT} = 0.15mA/μm.

Solution

(a) If, for an HBT with a total emitter length of 3 × 2.5μm = 7.5μm R_{SOPT} is 466.6Ω, then we need an emitter length of 466.6/50 × 7.5 = 70μm to satisfy the noise matching condition. This could be realized with N = 28 emitters stripes each 2.5μm long. More realistically, to minimize interconnect resistance and die area, we could have five transistors connected in parallel, each with 2 × 7μm emitters, i.e. l_E = 7μm and N = 10 emitter stripes. The corresponding bias current (at the optimum noise current density) becomes 2 × 0.13 × 70 = 18mA. In an older technology, with lower f_T of 45GHz and 0.5μm wide emitters, the optimum noise current was only about 3mA [3]. This significant increase in device size and current is partly due to the narrower emitter and partly due to the larger f_T of new generation HBTs.

(b) Using a 2μm finger width contacted on one side, and taking advantage of the frequency dependence of the optimum source resistance, we obtain a total gate width W_{OPT} of

$$W_{OPT}(6\text{GHz}) = W(5\text{GHz}) \times \frac{5\text{GHz}}{6\text{GHZ}} \times \frac{R_{SOPT}(20\mu\text{m}@5\text{GHz})}{50\Omega} = (20\mu\text{m}) \times \frac{5}{6} \times \frac{275}{50} = 91.7\mu\text{m}.$$

The total gate width is rounded off to 92μm and corresponds to an optimal noise bias current of 13.8mA.

This completes the "active matching" part of the design. At this stage, the transistor is biased at the minimum noise figure current density, and exhibits the lowest possible minimum noise factor (figure) at that particular frequency. At the same time, the real part of its optimum noise impedance is equal to the real part of the signal source impedance, Z_0.

Part II: Passive component matching

The second part of the design process relies on traditional impedance matching and is specific to the topology that is chosen for the LNA. Adding resistive components will degrade the noise figure and should be avoided. Purely reactive components do not contribute noise, so ideal inductors, transformers, transmission lines, and/or capacitors are preferred as matching elements in an LNA. These components have a finite (and often quite poor) Q and occupy considerably more die area than transistors. Therefore, using as few passive elements as possible is advisable. On the positive side, inductors do not dissipate DC power, do not contribute noise and, because of their smaller size in comparison with distributed structures, have become the predominant passive device in an LNA, even at millimeter wave frequencies.

A minimalist design, would just tune out the imaginary part of the optimum noise impedance to achieve the minimum noise figure of the LNA, as in traditional LNA design with discrete transistors, accepting non-optimal input return loss and non-optimal power gain. However, in most integrated LNAs, negative reactive feedback is applied in order to independently match the real part of the input impedance, without modifying the real part of the optimum noise impedance or degrading the minimum noise figure.

Let us briefly summarize this LNA design philosophy:

- Select the LNA topology with the lowest number of transistors since each device added to the circuit will contribute noise. Avoid active loads and active inductors because they increase noise and degrade the amplifier linearity.
- Bias the transistors at the minimum noise figure current density of the topology to achieve the lowest noise figure. In planar FETs, the optimal finger width should be chosen at this stage. In FinFETs, the designer has no control over the fin height and width.
- Calculate the transistor size (either the emitter length, the number of gate fingers, or the number of transistors or fins connected in parallel) that corresponds to the desired optimum noise impedance, Z_0.
- Add reactive (lossless) feedback to transform the real part of the input impedance to the desired value, Z_0.
- Add reactive matching network to tune out the imaginary parts of the optimum noise impedance and of the input impedance. Passive devices should be used only sparingly. Their quality factor is fairly low, which undermines the overall noise figure, and their area can be significantly larger than that of transistors.

We note that CE/CS, CB/CG, and cascode topologies with all four types of negative reactive feedback, or without feedback, have been employed in LNAs. Perhaps because of their lack of voltage gain and because of higher noise, common-collector, CC, and common-drain, CD, topologies have been avoided in tuned LNAs.

Next, we will analyze and develop design methodologies for the most common tuned LNA topologies.

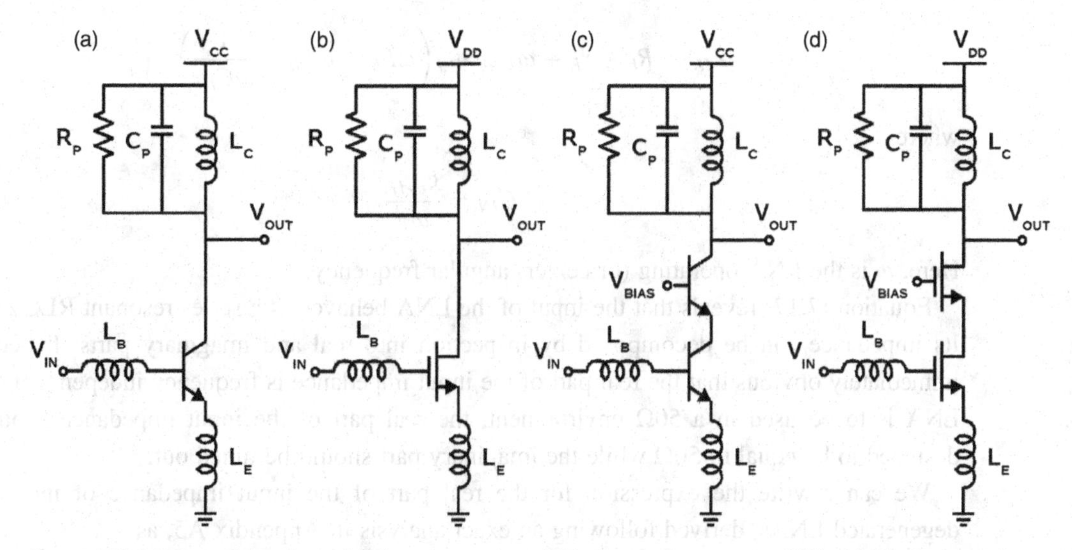

Figure 7.5 LNA configurations: (a) Common-emitter (CE), (b) common-source (CS), (c) bipolar cascode, and (d) MOS cascode

7.4 LNAs WITH INDUCTIVE DEGENERATION

The common-emitter (CE), common-source (CS), and cascode amplifier topologies, shown in Figure 7.5, are the most widely used in LNA designs. The secret behind their popularity is simplicity and that they permit simultaneous noise and input impedance matching, making it possible to achieve noise figure values closely approaching the NF_{MIN} of the transistor itself, at least over a narrow 5–10% bandwidth.

7.4.1 Design equations

The LNA topologies in Figure 7.5 can be represented as a series–series feedback network in which the amplifier network is formed by the CE/CS transistor or by the cascode, and the feedback network is purely reactive, consisting of L_B and L_E. We recall from Chapter 3 that a series–series feedback network is best described in terms of the Z-parameters of the forward amplifier, Z_a, and those of the feedback network, Z_f. We also established that a lossless inductor series–series feedback network does not alter the real part of the optimum noise impedance and that the overall minimum noise figure of the amplifier with feedback remains largely the same or is marginally better than that of the CE/CS or cascode topology.

If it does not change the NF_{MIN} and the R_{SOPT}, what does the feedback network do? As we will see next, the role of the reactive feedback network is to transform the real part of the input impedance to Z_0 and to cancel out most of the imaginary parts of the input and optimum noise impedance.

Input impedance matching

The expression of the input impedance of an amplifier with inductive degeneration (i.e. inductive series–series feedback) was derived in Chapter 5 using a simplified equivalent circuit

$$Z_{IN} \approx R_b + r_E + \omega_T L_E + j\left(\omega L_E + \omega L_B - \frac{1}{\omega C_{IN}}\right) \tag{7.17}$$

where

$$C_{IN} = \frac{g_{meff}}{\omega_T} \tag{7.18}$$

Here, ω is the LNA operating (or center) angular frequency.

Equation (7.17) reveals that the input of the LNA behaves as a series resonant RLC circuit. Its impedance can be decomposed by inspection into real and imaginary parts. It becomes immediately obvious that the real part of the input impedance is frequency independent. If the LNA is to be used in a 50Ω environment, the real part of the input impedance should be designed to be equal to 50Ω while the imaginary part should be tuned out.

We can rewrite the expression for the real part of the input impedance of inductively degenerated LNAs, derived following an exact analysis in Appendix A5, as

$$\Re\{Z_{IN,HBT}\} = R_b + r_E + R_{LB} + R_{LE} + \omega_{T,HBT} L_E \tag{7.19}$$

$$\Re\{Z_{IN,MOS}\} = R_g + r_s + R_{LB} + R_{LE} + \omega_{T,MOS} L_E \tag{7.20}$$

$$\Re\{Z_{IN,casc}\} = R_b + r_E + R_{LB} + R_{LE} + \omega_{T,casc} L_E \tag{7.21}$$

where we have accounted for the series loss resistance of the feedback inductors L_B and L_E.

These equations illustrate the role of the degeneration inductor. By adding L_E, the real part of the input impedance can be increased to achieve a 50Ω impedance match. The amount of degeneration needed depends only on the f_T of the topology and is independent of the LNA operating frequency, transistor size, and bias current. It only depends on the bias current density which sets the value of f_T and, therefore, of L_E

$$L_E = \frac{Z_0 - r_E - R_b - Z_{LE} - R_{LE}}{2\pi f_T}. \tag{7.22}$$

In most practical cases, the emitter/source inductance is small and its associated series resistance can be ignored in (7.22). However, L_B is always large enough that its series resistance cannot be neglected. Since the value of L_B is not known at this stage in the design process, we are forced to ignore its series resistance and make corrections to the L_E value obtained from (7.22) later.

Note that for cascode LNAs, the cutoff frequency of the cascode stage is used in (7.17) and (7.22) rather than that of the transistor. As can be seen in Figure 7.6, the peak f_T current density and the minimum noise figure density of the MOSFET cascode stage are identical to those of the transistor.

Equation (7.22) has implications for scaled CMOS technologies – as we move to smaller feature sizes and faster devices, the required amount of inductive degeneration in an LNA stage decreases.

Once the value of L_E has been determined, we can use the imaginary term in (7.17) to find the value of the input inductor L_B. L_B tunes the imaginary part of the input impedance to zero at the LNA operating frequency. Its value is given by

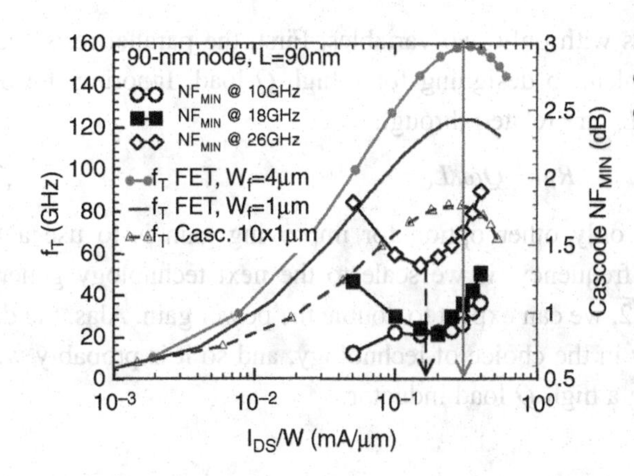

Figure 7.6 Measured f_T and NF_{MIN} vs. current density for 90nm n-MOSFET common-source and cascode stages [5]

$$L_B \approx \frac{\omega_T}{\omega^2 g_{meff}} - L_E. \tag{7.23}$$

It must also cancel out X_{SOPT} in (7.13) and (7.15). As mentioned, this is possible because, for both FETs and HBTs, the imaginary parts of the noise and input impedances are within 20% of each other. For MOSFET common-source and cascode topologies

$$L_S = \frac{Z_0 R_g - R_s}{2\pi f_T} \text{ and } L_G \approx \frac{\omega_T}{\omega^2 g_{meff}} - L_S. \tag{7.24}$$

At this stage in the design process, the input matching network is complete.

Output matching to maximize gain

In an IC, the load of the LNA is usually provided by another gain stage, a phase shifter, or by the mixer. As discussed in Chapter 5, the output of the LNA is capacitive. The simplest (yet narrowband) matching network can be realized with an *L*-section consisting of a shunt inductor, L_C, connected to V_{DD}, followed by a series capacitor toward the load. This arrangement automatically provides a DC block to the next stage. Unlike the series resonance at the input of the LNA, L_C forms a parallel resonant circuit with the output capacitance of the transistor and whatever parasitic capacitance may be presented by the next stage. An added benefit of using an inductive load is that it allows the output voltage to swing above the supply rail, thus improving linearity compared to the case when a resistive load is employed.

At resonance, the input- and output-matched LNA simplifies to a resistively loaded CE/CS or cascode stage whose maximum power gain expression was derived in the tuned circuit analysis from Chapter 5

$$G \leq \frac{1}{4}\left(\frac{f_T}{f_0}\right)^2 \frac{R_P}{Z_0}. \tag{7.25}$$

Equation (7.25) assumes that the amplifier is matched to Z_0 at the input and to R_P at the output. It gives insight into how to maximize the LNA power gain. Larger power gain is achieved for a lower signal source impedance, although Z_0 is usually not a variable over which the design engineer has control. This leaves us with only two variables. First, the parallel resistance R_P must be increased, which is equivalent to designing for a high-Q load. Ignoring the output resistance of the amplifier, R_P and L_C are related through

$$R_P = Q\omega_0 L_C \tag{7.26}$$

Once Q has been maximized, the only other option for improving gain is to use a faster transistor. For the same operating frequency, if we scale to the next technology generation where ω_T improves by a factor of $\sqrt{2}$, we can expect to double the power gain. Alas, the design engineer usually has little to no say in the choice of technology, and so it is probably wise to direct your efforts towards realizing a high-Q load inductor.

Design for linearity

As in the case of power amplifiers discussed in Chapter 6, the linearity of the LNA stage is determined by either the gate–source or base–emitter junction becoming non-linear at large input voltages, or by the maximum output voltage swing that causes the transistor to exit the active region. The latter condition is avoided or delayed by maximizing the V_{DS}/V_{CE}, of the transistor for a given supply voltage. In a cascode stage, the V_{DS}/V_{CE} of the common-gate/common-base transistor should be set larger than that of the common-source/common-emitter transistor since the top transistor always experiences a larger signal swing.

The linearity of the input junction is proportional to the amount of feedback provided by the source/emitter degeneration inductor

$$1 + j\omega L_S g_{meff} = 1 + j\omega L_S N_f W_f g'_{meff}(J_{OPT}) = 1 + j\frac{\omega}{\omega_T} Z_0 N_f W_f g'_{meff}(J_{OPT})$$

for FETs, or

$$1 + j\omega L_E A_E \frac{J_{OPT}}{V_T} = 1 + j\frac{\omega}{\omega_T} Z_0 \frac{J_{OPT} A_E}{V_T}$$

for HBTs. Therefore, to maximize linearity, the LNA transistor(s) must be biased at high current density, typically where the transconductance peaks, and large emitter/source degeneration must be employed. However, the overriding requirement for low-noise operation imposes that the current density is set to J_{OPT}. Similarly, the input impedance matching condition sets the value of the degeneration inductance to Z_0/ω_T.

If the linearity of the optimally noise-matched LNA is limited by the input junction and does not meet specification, the best compromise between noise figure and linearity is to increase the bias current by making transistors larger while still biasing them at J_{OPT}. Although this approach reduces R_{SOPT} below Z_0, R_n (G'_n), too, is reduced such that the noise impedance mismatch does not lead to significant departure from F_{MIN}.

7.4.2 Step-by-step design methodology for CE, CS, and cascode LNAs

The design equations discussed earlier, give rise to a simple, algorithmic design methodology for a CE, CS, or cascode LNA stages that has a unique optimal solution. The design can be conducted by hand, and/or, for higher precision, using small signal computer simulation:

Step 1 Set the V_{CE} or V_{DS} of transistors for maximum linearity, such that clipping at the output is avoided. In the case of a CE/CS stage, $V_{DS}/V_{CE} = V_{DD}$.

Step 2 Determine the J_{OPT} of the amplifier, noting that this current density might be different for an HBT-cascode than for a single transistor. Maintain this current density throughout the rest of the design steps. Mathematically, this is equivalent to solving: $\frac{\partial F_{MIN}(J)}{\partial J} = 0$.

Step 3 Select the best FET finger width and HBT emitter length that leads to the best minimum noise figure without degrading f_{MAX}. In FinFETs, the finger width is already set by the process. Mathematically this is described as: $\frac{\partial F_{MIN}(W_f)}{\partial W_f} = 0$ or $\frac{\partial F_{MIN}(l_E)}{\partial l_E} = 0$.

Step 4 Size the transistor such that $\Re\{Z_{sopt}\} = Z_0$ which is equivalent to $\frac{\partial F_{Z_0}(N_f)}{\partial N_f} = 0$ or $\frac{\partial F_{Z_0}(N)}{\partial N} = 0$. Step 4 is carried out by connecting gate fingers, FinFET fins, or HBTs in parallel. At this point, the bias current and size of all transistors in the LNA stage are known. If the design is carried out by computer, the transistors should be laid out and the layout parasitics should be extracted, before proceeding to the next design step, otherwise, the matching network will have to be re-designed after layout extraction.

Step 5 Add the degeneration inductance L_E, L_S to set the real part of the input impedance equal to Z_0 without affecting Z_{SOPT}: $L_E = \frac{Z_0 - R_b - r_E}{2\pi f_T}$ $L_S = \frac{Z_0 - R_g - R_s}{2\pi f_T}$.

Step 6 Add the base, L_B, or gate inductance, L_G, such that the imaginary parts of Z_{IN} and Z_{SOPT} become equal to zero.

Step 7 Design the output matching network using the techniques introduced in Chapter 5 to maximize the power delivered to the load over the bandwidth of interest.

Step 8 Add bias circuitry without degrading the noise figure.

7.4.3 Frequency scaling of CMOS LNAs

As illustrated in Figure 7.7, the minimum noise figure current density of n-channel MOSFETs does not change with frequency and, in technology nodes prior to the 45nm CMOS generation, with the technology node. Consequently, we can apply the simple design equations developed earlier to scale an existing LNA design in frequency from f_o to $f'_o = \alpha f_0$. The design scaling can be accomplished algorithmically in four steps.

Step 1 Bias transistors for minimum noise at $J_{OPT} = 0.15$mA/µm.

Step 2 Device sizing: W_f remains unchanged but $N'_f = N_f/\alpha$ and $W' = W/\alpha$.

Step 3 Input impedance matching: L_S remains largely unchanged since f_T has not changed but the source and gate resistances, of secondary importance, are now larger because of the smaller overall MOSFET gate width

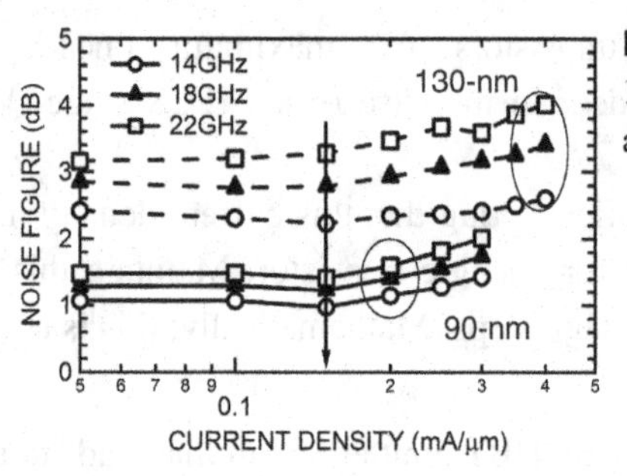

Figure 7.7 The measured minimum noise figure in 90nm and 130nm n-channel MOSFET as a function of drain current density and frequency [6]

$$L_S^{'} = \frac{Z_0 - \alpha(R_g + R_s)}{2\pi f_T} \approx L_S$$

and

$$L_G^{'} = \frac{1}{\alpha^2 \omega^2 \frac{C_{IN}}{\alpha}} - L_S^{'} \approx \frac{L_G}{\alpha}$$

Step 4 Output matching: $L_D^{'} = L_D/\alpha$; $C_D^{'} = C_D/\alpha$.

Note that this method does not apply to HBT LNAs.

EXAMPLE 7.2 6GHz and 12GHz LNA design in 180nm CMOS

You have just been hired as an RF circuit designer by a *fabless* semiconductor company focusing on developing wireless products over a wide range of frequencies. Your boss has assigned you the task of designing a n-MOS cascode LNA (like the one in Figure 7.5(d) with a center frequency of 6GHz in a 180nm RF CMOS process. While this may not be the most advanced technology, it has adequate performance for frequencies up to 20GHz and is very cost effective. At the same time, there is a pressing need to design a 12GHz LNA for a new satellite receiver, which also falls among your design responsibilities. You have the following data for the 180nm RF CMOS process (Table 7.1).

C'_{db}, C'_{sb}, C'_{gs}, C'_{gd}, and g'_{meff} represent the device capacitances and transcoductance per micron of gate width. Note that g'_{meff}, g'_o, C'_{gs}, and C'_{db} already include the impact of R_s. Assume that $k_1 = 0.5$. Furthermore, assume that X_{SOPT} is equal to $1.15\omega_T/\omega g_{meff}$ and that the load inductor has a Q of 10, irrespective of frequency and that the output of the LNA is terminated on a matched load.

Determine the transistor sizes and bias current, the required inductor values, and estimate the gain of the 180nm CMOS cascode LNA.

Solution

Let us start with the 6GHz LNA. From device simulations, we find that, in this 180nm process, the best finger width for minimum noise is $W_f = 2\mu m$. We can calculate $g'_m = g'_{meff}/(1-R'_s \times g'_{meff}) = 0.434 mS/\mu m$. The first step is to size the transistors in the cascode such that $R_{SOPT} = 50\Omega$. At 6GHz, we determine the gate width $W_G = N_f \times W_f$:

Table 7.1 **180nm RF-CMOS process data.**

Parameter	**180nm**
k_n	0.2mA/V2
f_T (of n-MOS cascode stage) @ 0.15mA/μm	35GHz
$C'_{db/sb}$	1.1fF/μm
C'_{gs}	1.0fF/μm
C'_{gd}	0.4fF/μm
NF_{MIN} @ 6GHz, J_{OPT}	0.30dB
NF_{MIN} @ 12GHz, J_{OPT}	0.58dB
g'_{meff} @ 0.15mA/μm	0.4mS/μm
g'_o @ 0.15mA/μm	0.04mS/μm
R'_s	200Ω×μm
W_f	2μm
R'_g (W_f = 2μm)	75Ω
V_T	0.5V
V_{DD}	1.8V

$$N_f = \frac{f_{Teff}}{fZ_0 W_f g'_{meff}} \sqrt{\frac{g'_m R'_s + W_f g'_m R'_g (W_f)}{k_1}}$$

$$= \frac{35}{6 \times 50 \times 2 \times 0.4 \times 10^{-3}} \sqrt{\frac{0.43 \times 0.2 + 2 \times 0.43 \times 10^{-3} \times 75}{0.5}} \approx 71.$$

$W = N_f \times W_f =$ 142μm. If we bias at J_{OPT}, the drain current is now fixed.

$$I_{DS} = J_{OPT} W = 0.15 \frac{\text{mA}}{\mu\text{m}} \times 142\mu\text{m} = 21.3\text{mA and } g_{\text{meff}} = 0.4\text{mS } 142 = 56.8\text{mS}.$$

Fixing the gate width also sets the input capacitance, which can be used to determine the gate inductance needed to cancel out the imaginary part of the input impedance

$$L_G + L_S = \frac{\omega_T}{(2\pi f)^2 W g'_{meff}} = \frac{6.28 \times 35 \times 10^9}{(6.28 \times 6 \times 10^9)^2 \times 142\mu\text{m} \times 0.4 \frac{\text{mS}}{\mu\text{m}}} = 2.72\text{nH}.$$

Additionally, the degeneration inductance is a function of the cascode f_T. The inductance required to achieve a 50Ω input impedance match in the 180nm design is calculated as

$$L_S = \frac{Z_0 - R_g - R_s}{2\pi f_T} = \frac{50 - 75/77 - 100/77}{6.28 \times 35 \times 10^9} = 217\text{pH}.$$

which leaves $L_G = 2.72\text{nH} - 217\text{pH} = 2.5\text{nH}$.

One option for the output network is to select the output inductor, L_D, such that it resonates with the transistor output capacitance at 6GHz. This output capacitance can be calculated as a function of the device size.

$$C_{OUT} = \left(C'_{db} + C'_{gd}\right)W = (1.1 + 0.4)\frac{\text{fF}}{\mu\text{m}} \times 142\mu\text{m} = 213\text{fF}.$$

At 6GHz, the load inductor becomes

$$L_D = \frac{1}{(2\pi f)^2 C_{OUT}} = \frac{1}{(6.28 \times 6 \times 10^9)^2 \times 213\text{fF}} = 3.3\text{nH}.$$

If $Q = 10$, then the equivalent parallel loss resistance of the inductor becomes

$$R_{PL} = 2\pi f_0 Q L_D = 6.28 \times 6 \times 10^9 \times 10 \times 3.3 \times 10^{-9} = 1244\Omega.$$

However, we must also consider the loading due to the output resistance of the cascode stage itself, r_{out}, which is approximated by $g_m/g_o{}^2 = 1760\Omega$ for a 180nm MOSFET cascode with a transconductance of 56.8mS. The output impedance then becomes $R_{PL}//r_{out} = 728\Omega$. The resulting maximum power gain is

$$|G| = \frac{f_T^2}{f_0^2}\frac{R_p}{4Z_0} = \frac{5.833^2 \times 728}{200} = 126 = 21\text{dB}.$$

We can apply the same methodology to scale the design to 12GHz. As the frequency has doubled, the gate width required for noise matching is half that of the 6GHz design, or 71μm. Since the unit finger width is 2μm, we will round the total gate width up to 72μm and 36 fingers. It follows that the current consumption, transconductance, and capacitances will also scale by a factor of 2

$$I_{DS} = J_{OPT}W = 0.15\frac{\text{mA}}{\mu\text{m}} \times 72\mu\text{m} = 10.8\text{mA and } g_{meff} = 0.4\text{mS} \times 72 = 28.8\text{mS}.$$

The total inductance required to tune out the input capacitance is now

$$L_G + L_S = \frac{\omega_T}{(2\pi f)^2 W g'_{meff}} = \frac{6.28 \times 35 \times 10^9}{(6.28 \times 12 \times 10^9)^2 \times 72\mu\text{m} \times 0.4\frac{\text{mS}}{\mu\text{m}}} = 1.46\text{nH}$$

nearly half the value in the 6GHz design. The degeneration inductance depends only on the cascode f_T, and hence should remain unchanged from the previous design (217pH). The input inductance for the 12GHz LNA becomes

$$L_G = 1.46\text{nH} - 217\text{pH} = 1.24\text{nH}.$$

Similarly, the output capacitance reduces in half and is now only 108fF, resulting in a load inductance of

Table 7.2 **6GHz and 12GHz LNA parameters.**

Design parameter	6GHz LNA	12GHz LNA	Comment
Gate width (W)	142μm	72μm	R_{SOPT} scales inversely with frequency
Bias current (I_{DS})	21.3mA	10.8mA	Current density remains unchanged
Max. power gain (G)	21dB	15.8dB	5.2dB reduction due to lower transistor gain
Degeneration (L_S)	217pH	217pH	Depends only on technology
Input inductance (L_G)	2.5nH	1.24nH	Lower C_{IN}, higher ω_0
Load inductance (L_D)	3.3nH	1.63nH	Reduced by half

$$L_D = \frac{1}{(2\pi f)^2 C_{OUT}} = \frac{1}{(6.28 \times 12 \times 10^9)^2 \times 108\text{fF}} = 1.63\text{nH}$$

which, not surprisingly, is half that of the 6GHz design. The equivalent output parallel resistance is unchanged, as the doubling of frequency and halving of load inductance cancel each other

$$R_P = (2\pi f_0)QL_D = (2\pi \times 12\text{GHz}) \times 10 \times 1.63\text{nH} = 1228\Omega.$$

Now, after accounting for the output resistance of the amplifier (3472Ω), the maximum power gain of the 12GHz LNA is calculated as

$$|G| = \frac{f_T^2}{f_0^2}\frac{R_P}{4Z_0} = \frac{2.91^2 \times 907}{200} = 38.4 = 15.8\text{dB}.$$

This value is almost 6dB lower than that of the 6GHz LNA and reflects the decreasing available gain of the transistor at higher frequencies. The key parameters of the 6GHz and 12GHz designs are summarized in Table 7.2.

In the previous design examples, we have ignored the parasitic resistances of L_G and L_S since they are not known before the inductor values are found and the inductor layouts are designed. In practice, especially at frequencies below 10GHz, because of the relatively large inductor values, the series resistances R_{LG} and R_{LS} are comparable or larger than R_g and R_s. This leads to design iterations and larger transistors sizes than originally calculated.

Indeed, 6GHz and 12GHz CMOS cascode LNAs based on these design results were fabricated in a 180nm RF-CMOS process and using the generic schematic from Figure 7.8 [6]. In each case, the LNAs were optimized for best possible noise performance. The optimized transistor widths were 160μm and 80μm for the 6GHz and 12GHz designs, respectively. L_G was 2.32nH for the 6GHz LNA, and 1.1nH for the 12GHz design. In both cases, L_S was 350pH. Excepting L_S, these values are within 10% of the hand analysis, verifying that simple design equations can be applied to CMOS LNA designs which can easily be scaled in frequency.

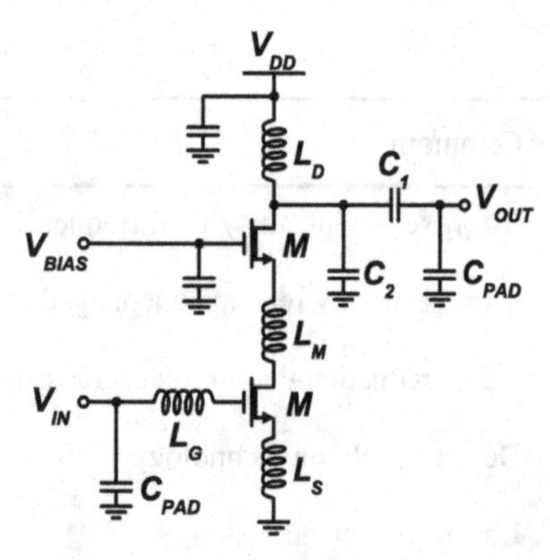

Figure 7.8 Generic scalable LNA schematic

7.4.4 Design porting of CMOS LNAs from one technology node to another

Based on Figure 7.7, we can also attempt to port an existing CMOS LNA, with the schematic of Figure 7.8, from one technology node to another, while keeping the center frequency unchanged.

Step 1 Bias for minimum noise at J_{OPT} which remains unchanged from one technology node to another.

Step 2 Device size remains practically unchanged since Z_{SOPT} is almost invariant between technology nodes. To avoid the dramatic increase in gate resistance and degradation of F_{MIN} in new technology nodes, the finger width must be scaled: $W'_f = W_f/S$ and $N'_f = N_f \times S$ such that $W' = W$. In practice, and especially in more recent CMOS nodes, 45nm and below, the total transistor gate width will increase due to the fact that R_s and R_g do not decrease by a factor S from one technology node to another.

Step 3 Input matching: L_S scales roughly as $1/f_T$

$$L_S^{'} = \frac{\Re(Z_0) - (R_g + R_s)}{2\pi(Sf_T)} = \frac{L_S}{S}.$$

Because the transistor size has not changed, $L_S + L_G$ and $R_g + R_s$ remain the same.

Finally, because scaling improves f_{MAX} and g'_{meff}, the power gain and the noise figure also improve when porting the LNA to a more advanced node.

The following example highlights the impact of CMOS scaling on LNA designs, emphasizing how they can be ported from one technology node to the next.

EXAMPLE 7.3 Porting a 12GHz LNA from 180nm to 90nm CMOS

Your company has decided to switch to a 90nm CMOS technology due to the savings in the die area for the digital baseband signal processor that is incorporated into the 12GHz satellite receiver. You have been asked to port your 12GHz LNA design from the 180nm CMOS process into the 90nm technology. Assume that the optimum noise figure current density in n-MOSFETs is approximately 0.15mA/μm, irrespective of technology generation and foundry. The technology parameters are provided in Table 7.3 below.

Table 7.3 **180nm and 90nm CMOS technology parameters.**

Parameter	180nm	90nm
k_n	0.2mA/V2	0.4mA/V^2
f_T (of n-MOS cascode stage) @ 0.15mA/μm	35GHz	70GHz
$C'_{db/sb}$	1.1fF/μm	0.8fF/μm
C'_{gs}	1fF/μm	1fF/μm
C'_{gd}	0.4fF/μm	0.4fF/μm
NF_{MIN} @ 12GHz	0.58dB	0.42dB
g'_{meff} @ 0.15mA/μm	0.4mS/μm	0.8mS/μm
g'_{oeff} @ 0.15mA/μm	0.04mS/μm	0.08mS/μm
R'_s	200Ω × μm	160Ω × μm
W_f	2μm	1μm
R'_g (W_f = 2μm)/R'_g (W_f = 1μm)	75Ω	37Ω
V_T	0.5V	0.3V
V_{DD}	1.8V	1.2V

Solution

From the given g'_{meff} and R'_s, the calculated $g'_m = 0.95\text{mS}/\mu\text{m}$ and has more than doubled compared to the 180nm node. Note that the source resistance has not scaled fully. Hence, because of the parasitic resistances, R_{SOPT} does not remain exactly the same from one technology node to another, and the gate width W increases slightly. From the condition that the $R_{SOPT} = 50\Omega$, we calculate the number of fingers as

$$N_f = \frac{f_{Teff}}{fZ_0 W_f g'_{meff}}\sqrt{\frac{g'_m R'_s + W_f g'_m R'_g(W_f)}{k_1}} =$$

$$= \frac{70}{12 \times 50 \times 0.8 \times 10^{-3}}\sqrt{\frac{0.95 \times 0.16 + 1 \times 0.95 \times 10^{-3} \times 37}{0.5}} \approx 89.$$

Therefore, $W = N_f \times Wf = 89 \times 1\mu\text{m} = 89\mu\text{m}$ instead of 72μm as in the 180nm node. The drain current becomes

$$I_{DS} = J_{OPT} \times W = 0.15\frac{\text{mA}}{\mu\text{m}} \times 89\mu\text{m} = 13.35\text{mA}.$$

Both the size and current have increased by about 23% from the 180nm design. This change could be made smaller with further reduction of the gate finger width in the 90nm node beyond that imposed by the scaling factor $S = 2$. However, there is not that much that the designer can do about the source resistance.

Since the cutoff frequency of the FET has doubled, we will need a smaller degeneration inductance to make the real part of the input impedance equal to 50Ω

$$L_S = \frac{Z_0 - R_g - R_s}{2\pi f_T} = \frac{50 - 0.415 - 1.79}{2\pi \times 70\text{GHz}} = 109\text{pH}.$$

The gate inductor, L_G, should be slightly smaller than in the 180nm design due to the 20% higher input capacitance

$$L_G + L_S = \frac{\omega_T}{(2\pi f)^2 W g'_{meff}} = \frac{6.28 \times 70 \times 10^9}{(6.28 \times 12 \times 10^9)^2 \times 89\mu\text{m} \times 0.8 \frac{\text{mS}}{\mu\text{m}}} = 1.087\text{nH}$$

and $L_G = 1.087\text{nH} - 0.109\text{nH} = 0.98\text{nH}$.

We notice that the drain/source–bulk capacitance per unit gate width has decreased from the 180nm to the 90nm node. This compensates for the larger gate width. Thus, the output capacitance, C_{OUT}, and L_D remain practically unchanged at 107fF and 1.63nH, respectively

$$L_D = \frac{1}{(2\pi f)^2 C_{OUT}} = \frac{1}{(6.28 \times 12 \times 10^9)^2 \times 107\text{fF}} = 1.63\text{nH}.$$

Assuming the same inductor quality factor, load impedance, and intrinsic voltage gain for the transistor, g_m/g_o, the parallel tank resistance remains the same while the LNA power gain increases by 6dB because of the doubling of f_T

$$|G| = \frac{f_T^2}{f_0^2}\frac{R_P}{4Z_0} = \frac{5.833^2 \times 907}{200} = 154 = 21.9\text{dB}.$$

More importantly, the minimum noise figure of the amplifier improves in the new design because the transistor minimum noise figure improves. The component sizes and performance of the 180nm and 90nm LNA designs are compared in Table 7.4.

These calculations were confirmed in an experimental study conducted on LNAs designed in 130nm and 90nm CMOS processes [6].

The previous example illustrates that by porting an LNA to a more advanced CMOS node, a design engineer should be able to easily improve the power gain and the noise figure of the amplifier with only trivial design modifications. This trend is unclear beyond the 45nm node. Recent 32nm data indicate that the minimum noise figure of MOSFETs is degraded compared to the 45nm node [7].

A final note: In the previous two examples, we have ignored the series resistance of the gate and source inductors which will dominate over the source and gate resistance of the MOSFET at frequencies below 20GHz. Ignoring them will underestimate the total gate width and the needed bias current. We have also assumed that the amplifier stage is unilateral, which implies that the output matching network does not affect the input noise and signal impedance matching. This

Table 7.4 **180nm and 90nm LNA designs.**

Design parameter	180nm	90nm
Gate width (W_G)	72µm	89µm
Bias current (I_{DS})	10.8mA	13.35mA
Max. power gain (G)	15.8dB	21.9dB
Degeneration (L_E)	217pH	109pH
Input inductance (L_B)	1.24nH	0.98nH
Load inductance (L_C)	1.63nH	1.63nH

assumption is acceptable in cascode stages at lower frequencies but becomes problematic at mm-wave frequencies, especially in CMOS, because of the reduced transistor isolation. In all these situations, design iterations will be required. Lastly, the parasitic capacitance of L_B, together with the pad capacitance, will reduce the signal source resistance seen by the input transistor in the LNA. This parasitic effect is particularly important in mm-wave LNAs and will be addressed next.

7.4.5 Accounting for the parasitic capacitance of the pad and of the input inductor

To compensate for the capacitive loading due to the pad and due to the gate/base inductor parasitic capacitance, we can employ the shunt–series transformation of the input pad and signal source impedance, as illustrated in Figure 7.9, and then follow the usual procedure to match the noise and input impedance of the amplifier to the new complex signal source impedance $R_1 + 1/j\omega C_1$ [9]. We note that the source impedance seen by the amplifier is now complex, and its real part is frequency dependent

$$R_1 = \frac{Z_O}{1 + \omega^2 C_{pad}^2 Z_O^2} = \frac{Z_O}{k} \tag{7.27}$$

$$C_1 = \frac{k}{k-1} C_{pad} \tag{7.28}$$

where

$$k = 1 + \omega^2 C_{pad}^2 Z_O^2. \tag{7.29}$$

Equations (7.27)–(7.28) can be extended to the situation where a bondwire is attached to the pad and the impact of its inductance must be accounted for in the design process [9]. This technique was applied in the design the mm-wave SiGe HBT and CMOS LNAs.

Figure 7.10 shows the schematic of a 65GHz LNA with two cascode stages, implemented in a commercial 180nm SiGe BiCMOS process [10] with f_T and f_{MAX} of approximately 160GHz. The component sizes and bias currents are indicated in the figure. In this technology, the capacitance of the 60µm × 60µm pad was measured to be 20fF. The cascode is biased at the

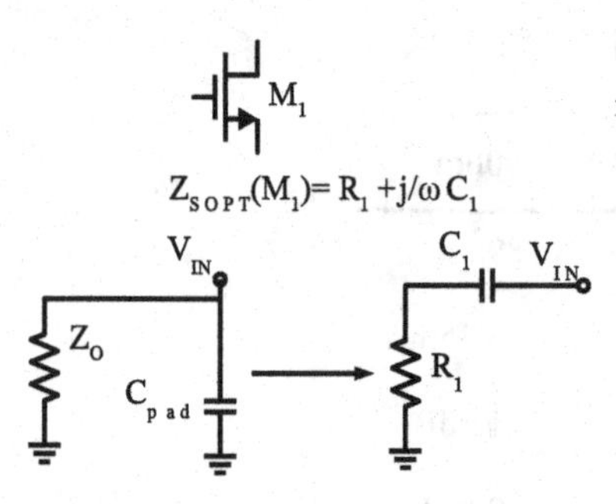

Figure 7.9 Schematics illustration of converting the shunt-pad source impedance circuit to a series one

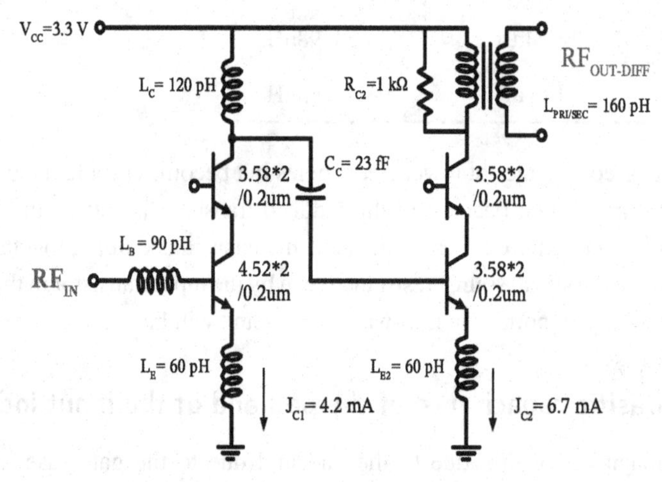

Figure 7.10 65GHz 2-stage SiGe HBT cascode LNA schematic [10]

optimum noise figure current density of 2.4mA/μm^2, which is higher than the J_{OPT} of the transistor itself. The corresponding f_T is 125GHz for the transistor, and about 12% lower, or 110GHz, for the cascode stage $f_T \approx \frac{g_m}{2\pi(C_\pi + C_\mu)}$ and $f_{T,cascode} \approx \frac{g_m}{2\pi(C_\pi + 2C_\mu)}$ with $C_\pi \approx 10C_\mu$.

Given the pad capacitance and the 65GHz center frequency, the signal source impedance correction factor, k, becomes equal to 1.166. As a result, the size of the input transistor, 9μm × 0.2μm, was chosen to match R_{SOPT} to $50/1.166 = 42.85\Omega$. The value of L_E, 60pH, results in the real part of the input impedance being equal to 41.5Ω, very close to 42.85Ω. In fact, measurements show that the real part of the input impedance is 40.35Ω and that the imaginary part of the input impedance is canceled out at 78GHz. The small error of only 3% in the input resistance is remarkable for hand analysis at 65GHz. However, a systematic larger error is observed for the base inductance. A 15–20% larger value for L_B would have shifted the frequency for best S_{11} from 78 to 65GHz.

7.4.6 Design methodology refinements for mm-wave LNAs with inductive degeneration

Unlike HEMTs and HBTs, where $C_{gd}(C_{bc})$ is about one tenth of $C_{gs}(C_{be})$, in MOSFETs C_{gd} is approximately 50% of C_{gs} and the f_T of the cascode stage is at least 33% smaller than that of the transistor. As a consequence, the CMOS cascode LNA stage, in particular, requires bandwidth

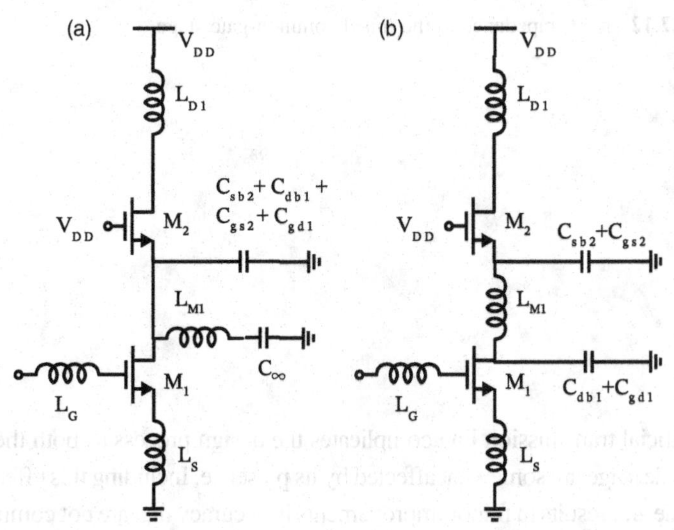

Figure 7.11 (a) Tuning out the middle node capacitance with a shunt inductor L_{M1}. (b) Creating an artificial transmission line in the CMOS cascode by using a broad-banding inductor L_{M1}

extension techniques in order to achieve acceptable gain at mm-wave frequencies. One approach [11], illustrated in Figure 7.11(a), is to place a shunt inductor to the AC ground (no DC path) at the middle node between the CS and CG transistors. However, this resonance is narrow band. Another technique is to form an artificial transmission line by introducing a series inductor between the drain of the CS FET and the source of the CG FET, as depicted in Figure 7.11. This approach has the added benefit of improving the noise figure of the cascode stage at mm-waves.

The characteristic impedance of this artificial transmission line can be expressed as

$$Z_{01} = \sqrt{\frac{L_{M1}}{C_{gs2} + C_{sb2}}} \tag{7.30}$$

while its 3dB bandwidth, when matched at both ends (difficult but possible with FETs) becomes

$$f_{3dB} = \frac{1}{\pi\sqrt{L_{M1}(C_{sb2} + C_{gs2})}}. \tag{7.31}$$

As illustrated in Figure 7.12, the design of the MOS cascode with artificial interstage transmission line proceeds from the output of the cascode. First, the impedance $Z_{M2,in}$ of the common-gate transistor M_2 is calculated based on its load resistance at resonance. Next, the value of L_{M1} is chosen so that the impedance of the interstage transmission line satisfies the condition

$$Z_{01} = Z_{M2,in} = \frac{1}{g_{m2}} + \frac{\omega L_{D1} Q}{2(1 + g_{m2} r_{o2})}. \tag{7.32}$$

The 1/2 factor in the second term of (7.32) originates from the assumption that the cascode stage is loaded with an impedance which is matched to that of the inductor L_{D1} at resonance. In other words, the loaded Q of the inductor L_{D1} under matched conditions is at best 50% of the unloaded Q.

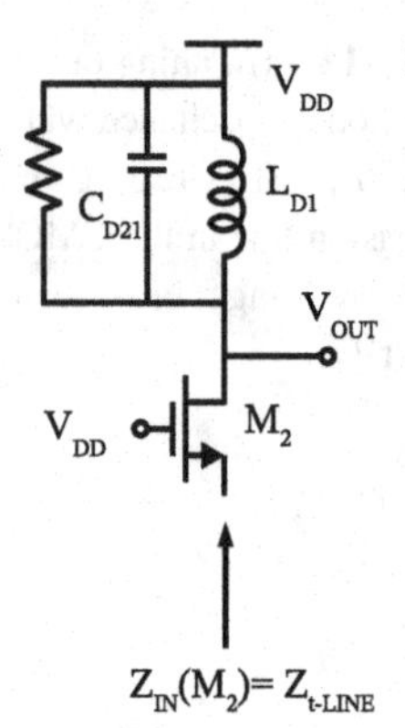

Figure 7.12 Input impedance in the tuned common-gate stage

Note: The interstage artificial transmission line complicates the design process as both the Z_{SOPT} and the f_T of the MOS cascode stage are somewhat affected by its presence. Including this effect in the hand analysis is cumbersome and results in minor improvements in accuracy that are not commensurate with the effort entailed. Suffice it to mention that this effect is readily captured during simulation.

After calculating the effective signal source resistance seen by the amplifier across its bond pad, the design method can be reduced to seven steps:

Step 1 Set the bias to the optimum NF_{MIN} current density (J_{OPT}) to minimize the transistor noise figure.

Step 2 Choose the optimal W_f to minimize NF_{MIN}. For 65/90nm CMOS, W_f is 0.7–1.5µm.

Step 3 Find the best L_M value for the cascode biased at J_{OPT} by plotting the f_T of the cascode versus L_M. Note that L_M scales with W^{-1} (N_f^{-1}). This step is best carried out by simulation rather than hand analysis, as illustrated in Figure 7.13.

Step 4 With all devices biased at J_{OPT}, scale the number of fingers (N_f), number of transistors or fins connected in parallel, and L_M to match the optimal noise impedance, R_{SOPT}, to the source impedance at the frequency of operation.

Step 5 After layout and parasitic extraction, with f_T from Step 4 and according to (7.24), find $L_S = (Z_0 - R_s - R_g)/\omega_T$. Note that ω_T corresponds to the cascode stage with L_M, not just to the transistor, and after extraction of layout parasitics.

Step 6 Add L_G to tune out Im$\{Z_{IN}\}$ and Im$\{Z_{SOPT}\}$.

Step 7 Add output matching network to maximize gain.

Following this design methodology, it is possible to achieve the lowest noise performance in the given technology. In the case of mm-wave SiGe HBT LNAs, adding L_M is rarely necessary.

EXAMPLE 7.4 Single-stage cascode, 60GHz 90nm CMOS LNA

Consider a single stage with $C_{PAD} = 20$fF, $W_{OPT} = 20$µm, $W_f = 1$µm, $N_f = 20$, $g_{meff} = 16$mS, $R_S = 12.5$, $R_g = 100\Omega/20 = 5\Omega$. Determine L_M, L_S, and L_G using Figure 7.13 and $k = 1.142$.

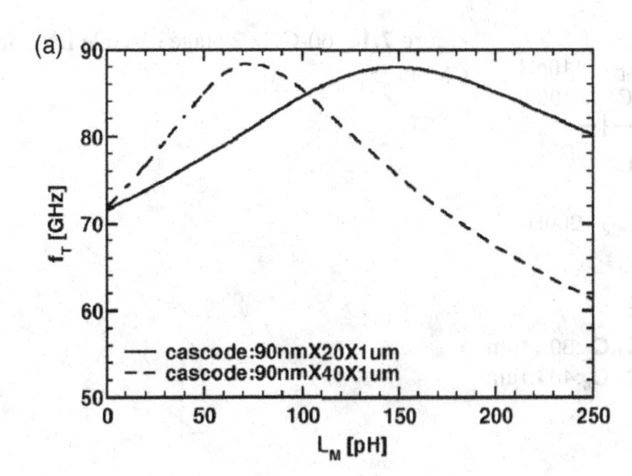

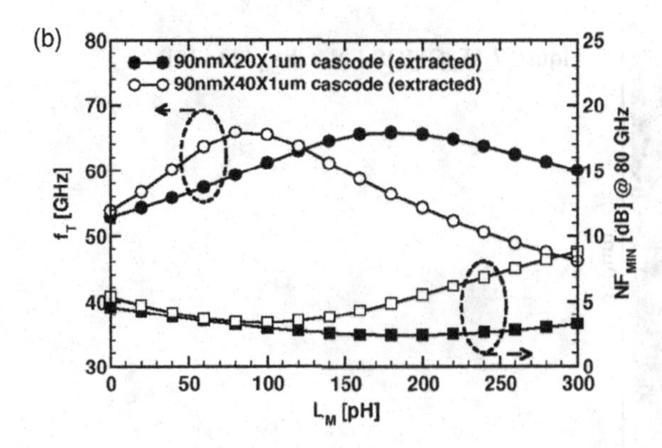

Figure 7.13 Impact of broad-banding inductor L_M on the cascode f_T and NF_{MIN} simulated for two different 90nm n-MOSFET cascode sizes (a) before layout parasitic extraction and (b) after extraction [5]

Solution

From Figure 7.13, the highest $f_{T,casc}$ of 66GHz for a cascode with 20μm MOSFETs is obtained when $L_M = 95$pH. By combining (7.24) and (7.27), we obtain $L_S = \frac{50/k-17.5}{6.28\times 66\text{GHz}} = 63.4$pH.

From the condition that $\frac{1}{j\omega C_1} = j\omega(L_G + L_S) + \frac{\omega_{T,casc}}{j\omega g_{meff}}$, we obtain $L_G = \frac{\omega_{T,casc}}{\omega^2 g_{meff}} - \frac{1}{\omega^2 C_1} - L_S$ where $C_1 = 160.8$fF. $L_G = 182.6\text{pH} - 43.82\text{pH} - 63.4\text{pH} = 75.38$pH.

Note that C_{in} and L_G will be affected by L_M. Therefore, they, too, should be obtained by simulations rather than hand analysis.

Even with the bandwidth and gain extension technique described above, single-stage MOS cascodes fabricated in 90nm and 65nm CMOS technology do not exhibit more than 4–5dB gain above 60GHz if they are designed to be simultaneously noise and input/output impedance matched. For practical power gains of 10dB or higher, at least two stages are needed, as in the 60GHz 90nm CMOS LNA whose schematics and die photo are shown in Figures 7.14 and 7.15, respectively. The circuit operates with a 1.5V power supply and consumes 14mA. All transistors are biased at 0.2mA/μm, a "painless" compromise between linearity and noise. Normally, the current and transistor size in the second stage are set to satisfy the linearity condition (OIP3). The bias current density should be either 0.15mA/μm if noise is the biggest concern, or 0.3mA/μm if linearity is a critical goal [5]. The L-network

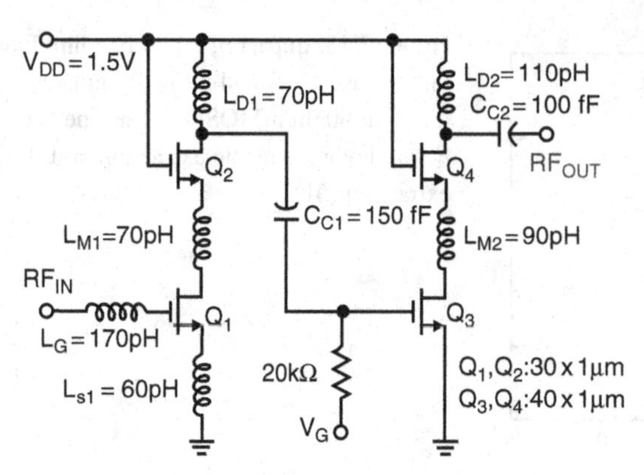

Figure 7.14 60-GHz 2-stage cascode LNA in 90nm CMOS [5]

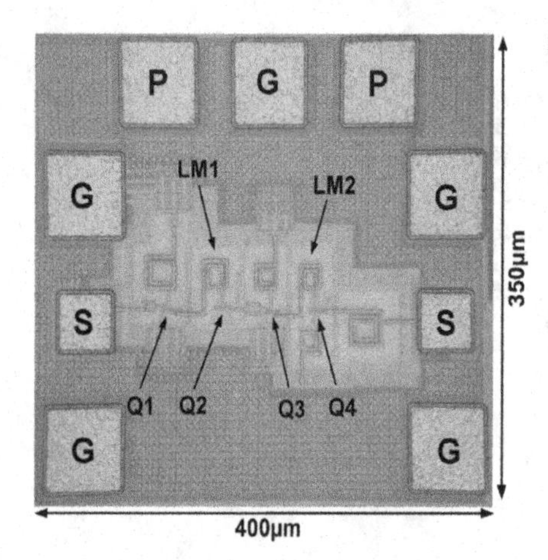

Figure 7.15 CMOS LNA die photo [5]

at the output, formed by L_{D2} and $C_{C2,}$ ensures matching to 50Ω. The noise figure of this LNA was about 4.5dB.

Figure 7.16 reproduces the schematics of an 80GHz cascode LNA implemented in a 230/300GHz f_T/f_{MAX} SiGe HBT technology with 130nm emitter width. It consists of two CE stages followed by a cascode stage [12]. The CE stages are chosen because they have lower noise than the cascode stage. They are biased at about 6mA/μm^2 which, in this technology, corresponds to the HBT J_{OPT} at 80GHz. Because of the standard digital BEOL, the pad capacitance is 30fF, which results in the correction factor k at 80GHz being 1.56. As a result, the circuit was designed to be noise and input impedance-matched to 31.9Ω. Note that, when this circuit was ported in the same process with a mm-wave BEOL which features thick dielectrics and thick metals, the pad capacitance was reduced to 8fF, k decreased to less than 1.1, and the input transistor size was reduced to 4.5mm in order to ensure matching to an optimal noise impedance of 48Ω [13]. Both versions of this LNA, achieved a noise figure of 3.2dB in the 80GHz band, very close to the minimum noise figure of the transistor.

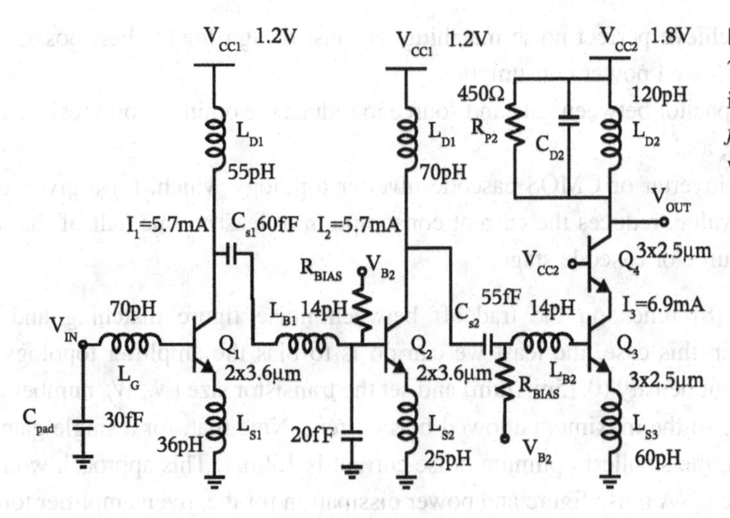

Figure 7.16 Schematics of a 77-GHz cascode LNA implemented in a 230/300GHz f_T/f_{MAX} SiGe HBT technology with 130nm emitter width [12]

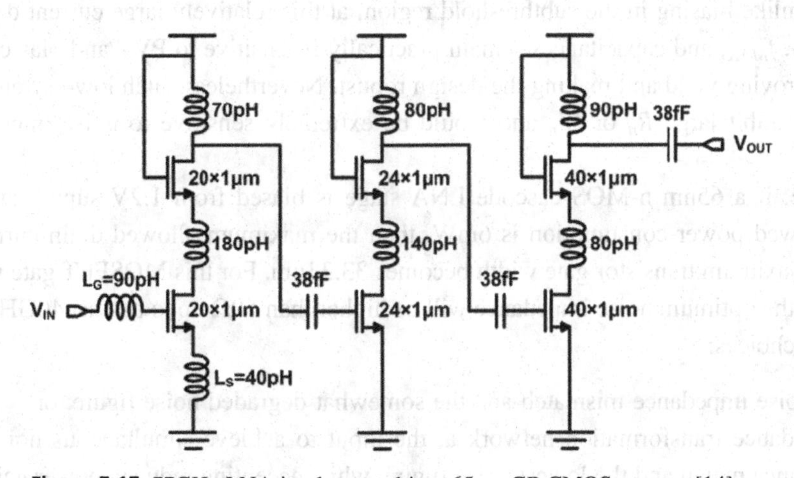

Figure 7.17 77GHz LNA implemented in a 65nm GP CMOS process [14]

Finally, Figure 7.17 reproduces the schematics of a 77GHz, 3-stage cascode LNA with inductive degeneration implemented in a 65nm GP CMOS process. Since the pad capacitance is 34fF, at 80GHz, $k = 1.73$, and the optimum noise impedance should be 28.9Ω.

7.5 POWER-CONSTRAINED CMOS LNA DESIGN

The popularity of the CE, CS, and cascode LNA stages with inductive degeneration lies first in the simplicity with which simultaneous noise and impedance matching can be performed and, second, in the fact that the amplifier noise figure closely approaches the minimum noise figure of the transistor. A potential drawback in CMOS implementations of these topologies at frequencies below 20GHz is that they require large device sizes and, hence, high bias currents. To overcome this problem, three solutions have been proposed:

- ignore trying to achieve perfect noise matching and just design for the best possible noise figure given the allowed power consumption;
- add an external capacitor between gate and source to reduce the optimal noise resistance for a given bias current; and
- employ a CMOS inverter or CMOS cascode inverter topology which, for a given optimal noise impedance value, reduces the current consumption at least to one half of that of a n-MOS common-source or cascode stage.

The first approach [8] relies on the tradeoff between noise figure matching and power consumption. Even in this case, the least we can do is to bias the amplifier topology at the minimum noise current density (0.15mA/μm) and set the transistor size (N_f, W_f, number of fins) to that corresponding to the maximum allowed bias current. Note that, for a single gate finger FET with $W_f = 1\mu m$, the smallest optimum noise current is 150μA. This approach would lead to the lowest possible LNA noise figure and power dissipation for the given amplifier topology. Furthermore, unlike biasing in the subthreshold region, at this relatively large current density, the transistor f_T, f_{MAX}, and capacitances remain practically insensitive to PVT and bias current variations, improving yield and making the design robust. Nevertheless, such low-current FET LNAs would exhibit large R_n or G'_n and would be extremely sensitive to noise impedance mismatch.

For example, if a 65nm n-MOS cascode LNA stage is biased from 1.2V supply and the maximum allowed power consumption is 6mW, then the maximum allowed drain current is 5mA and the maximum transistor gate width becomes 33.33μm. For this MOSFET gate width, the real part of the optimum noise impedance will be higher than 50Ω up to at least 40GHz. We now have two choices:

- accept the noise impedance mismatch and the somewhat degraded noise figure, or
- use an impedance transformation network at the input to achieve simultaneous noise and input impedance match and the lowest noise figure, while accepting reduced power gain due to the higher Z_0 in (7.25).

To understand the idea behind the second approach, let us assume that the added capacitance C_1 in Figure 7.18 is simply part of the gate–source capacitance of the transistor. Since the real part of the optimum noise impedance is inversely proportional to the input capacitance, it follows that we will need a smaller transistor gate width and bias current to achieve a given R_{SOPT}. However, there is a significant price that we must pay. The cutoff frequency of this artificial transistor becomes

$$f'_T \rightarrow \frac{f_T}{1 + \frac{C_1}{C_{gs} + 2C_{gd}}} \tag{7.33}$$

and the required degeneration inductance for input impedance matching also increases

$$L'_S \rightarrow L_S\left(1 + \frac{C_1}{C_{gs} + 2C_{gd}}\right). \tag{7.34}$$

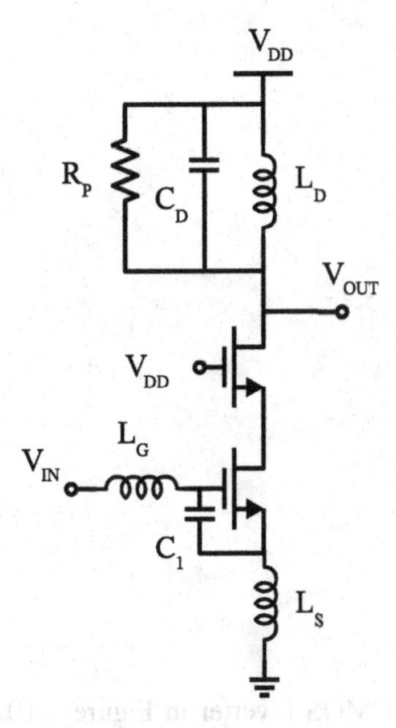

Figure 7.18 Adding a capacitance C_1 across the gate–source junction

Since G and F_{MIN} depended on f_T, both will be degraded. In fact, simulations show that even a noise-mismatched LNA with the same bias current will have higher gain and lower noise than the one with capacitor C_1 strapped across the gate–source junction of the MOSFET. The FoM of the LNA is actually degraded and the noise resistance R_n increases, making it more sensitive to noise impedance mismatch.

It could be argued that in the power consumption budget of a radio transmit–receive link, saving power in the LNA is not as efficient as saving power in the PA for the same overall system dynamic range. Aiming for the minimum noise figure in the LNA should always be the priority, even if more power is expended. Exceptions may be justified for applications where only a receiver is needed (like GPS) and where a tradeoff between the power consumption of the LNA and that of the PA is not possible.

To save current in CMOS LNAs, the best approach is to employ a CMOS inverter topology with higher g_m/I_{DS} ratio than a simple n-MOSFET or n-MOSFET cascode. This topology will be analyzed next.

7.6 LOW-CURRENT CMOS INVERTER LNAs

A CMOS inverter with inductive degeneration can be used to reduce the power dissipation of a noise-impedance matched stage to less than half that of a n-MOSFET only implementation [15],[16]. However, part of the power savings can be lost if the supply voltage must be doubled to compensate for the voltage headroom required by the p-channel MOSFET. Nevertheless, there are many practical systems where the supply voltage is fixed and is sufficiently large to allow for a CMOS inverter to operate with good linearity and low noise at high frequencies.

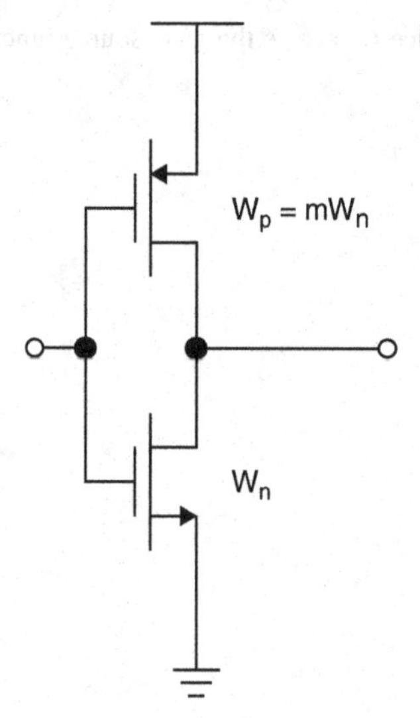

Figure 7.19 CMOS inverter.

From the small signal and noise point of view, the CMOS inverter in Figure 7.10 can be represented as a parallel connection of the small signal equivalent circuits of the n-MOSFET and p-MOSFET devices. Let us consider a minimum gate length CMOS inverter in which the p-MOSFET gate width is sized to be *m* times as large as that of the n-MOSFET. To ensure symmetrical I-V characteristics and transconductance, *m* is set approximately equal to $I_{ON,n}/I_{ON,p}$. While in older CMOS nodes *m* was 2–3, due to the significantly larger hole mobility enhancements in advanced CMOS nodes, *m* is typically 1.2–1 in the 32nm and 22nm nodes.

To derive the small signal and noise parameters of the CMOS inverter, the gate width and small signal parameters of the p-MOSFET can be described as a function of the corresponding values of the n-MOSFET, scaled by *m*

$$W_p = mW_n, g'_{mp} = g'_{mn}/m, R'_{sp} = mR'_{sn}, R'_{dp} = mR'_{dn}, g_{mn} = g_{mp}; I_{dn} = I_{Dp};$$
$$mg_{op} = g_{on}, R_{sp} = R_{sn}, R_{dp} = R_{dn}, C'_{gsp} = C'_{gsn}; C'_{gdp} = C'_{gdn}, C'_{dbp} = C'_{dbn};$$
$$C_{gsp} = mC_{gsn}; C_{gdp} = mC_{gdn}, C_{dbp} = mC_{dbn} \tag{7.35}$$

By appropriately selecting the finger width of the n-MOS and p-MOS device, it is also possible to ensure that $R_{gn} = R_{gp}$. Starting from their definitions, it is now relatively straightforward to arrive at the following expressions for the equivalent f_T, F_{MIN}, and noise parameters of the CMOS inverter relative to those of the n-MOSFET

$$\frac{1}{2\pi f_{T,CMOS}} = \frac{m+1}{2}\frac{C_{gs,nMOS} + C_{gd,nMOS}}{g_{m,nMOS}} + \frac{m+1}{2}C_{gd,nMOS}\left(R_{s,nMOS} + R_{d,nMOS}\right);$$
$$f_{T,CMOS} = \frac{2}{m+1}f_{T,nMOS}. \tag{7.36}$$

$$F_{MIN,CMOS} - 1 = \frac{m+1}{2}\left[F_{MIN,n-MOS} - 1\right] \tag{7.37}$$

$$\frac{g_m}{I_{DS}} = \frac{2g_{m,nMOS}}{I_{DS}}; \frac{R_n}{I_{DS}} = \frac{R_{n,nMOS}}{2I_{DS}}; \frac{R_{sopt}}{I_{DS}} = \frac{R_{sopt,nMOS}}{(m+1)I_{DS}}; X_{sopt} = \frac{X_{sopt,nMOS}}{m+1}; X_{IN} = \frac{X_{IN,nMOS}}{m+1}. \quad (7.38)$$

The first thing to notice is that, if the p-MOS and n-MOS devices are identical, that is $m = 1$, the CMOS inverter f_T, f_{MAX}, and F_{MIN} are identical to those of the MOSFET. In general, compared to an n-MOSFET-only implementation, in a CMOS inverter LNA, the transistor size and the bias current can be made $m + 1$ times smaller because the optimum noise impedance is inversely proportional to the input capacitance of the CMOS inverter, which is $m + 1$ times larger than that of the n-channel MOSFET.

$$R_{sopt} \propto \frac{1}{C_i} = \frac{1}{C_{i,pMOS} + C_{i,nMOS}} \approx \frac{1}{(1+m)C_{i,nMOS}} \quad (7.39)$$

The slight degradation in noise figure due to the p-MOSFET will be offset by the lower loss of the matching inductors which are now $m + 1$ times smaller.

Please see problem 7.6. for a cascode CMOS inverter implementation of this type of LNA with inductive degeneration in a 32nm high-k metal gate CMOS process.

7.7 LOW-VOLTAGE LNA TOPOLOGIES

In advanced CMOS nodes, supply voltages are reduced to 0.9V and below. This is a relatively serious problem in high-frequency and mm-wave circuits because the f_T, f_{MAX}, and F_{MIN} of MOSFETs quickly degrade as the drain–source voltage of the transistor decreases below 0.5V. We can no longer stack MOSFETs without significantly reducing V_{DS} and, hence, f_T and f_{MAX}. Moreover, if the V_T of the common-gate device varies in a telescopic cascode, the V_{DS} of the CS device changes and so do the gain and noise figure of the stage.

A possible solution to this problem is to employ only single transistor CS and/or common-gate topologies and learn to deal with the reduced isolation and the required iterations in the design of a multi-stage amplifier. An example of a single-stack, multi-stage CS-CS-CG-CS 60GHz LNA schematic, implemented in a 65nm CMOS technology is illustrated in Figure 7.20. Figure 7.21 reproduces the measured gain at different supply voltages from 0.8V to 1.2V, demonstrating the relative robustness of this topology to large V_{DD} variations.

Such an approach is manageable in an amplifier but it becomes problematic in control circuits like mixers, modulators, and latches where cascode topologies are needed. A second option is to use ac-coupled cascode stages which reduce the number of transistors in the vertical stack and improve V_{DD} and V_T variation as illustrated in Figure 7.22, but double the bias current. Although the doubling of the bias current per cascode stage is rather wasteful, the robustness to process variation and model uncertainty associated with the ac-coupled cascode topology is often an acceptable price to pay at mm-wave frequencies where the transistor

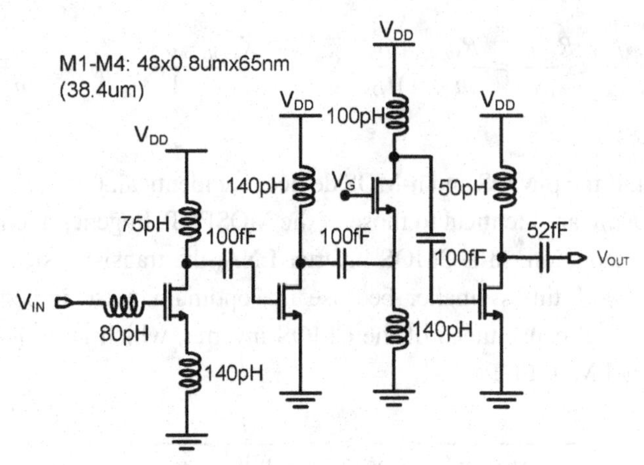

Figure 7.20 Low-voltage CS + AC cascode + CS CMOS LNA schematic [17]

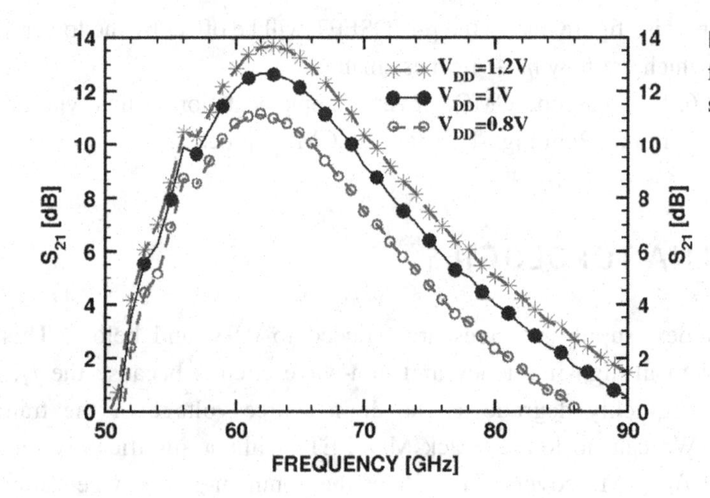

Figure 7.21 Measured gain vs. frequency characteristics at different supply voltages

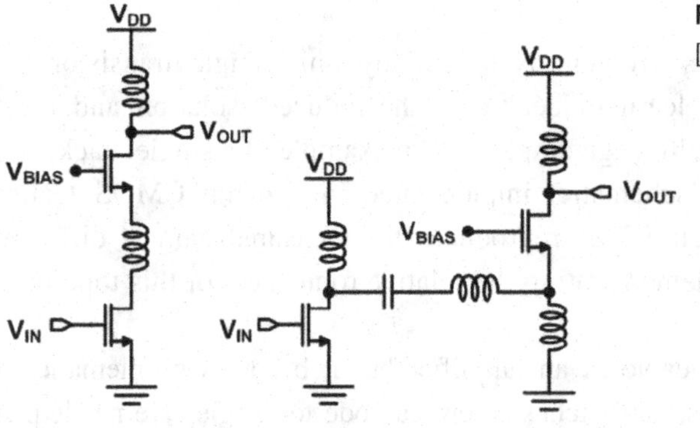

Figure 7.22 AC-coupled cascode topology [18]

models may not be very accurate. The improved isolation also simplifies the design methodology in multi-stage amplifiers [18].

Figure 7.23 depicts the schematics of a three-stage ac-coupled cascode (or six-stage CS-CG-CS-CG-CS-CG) LNA [18] designed for operation at 140GHz in a 65nm CMOS

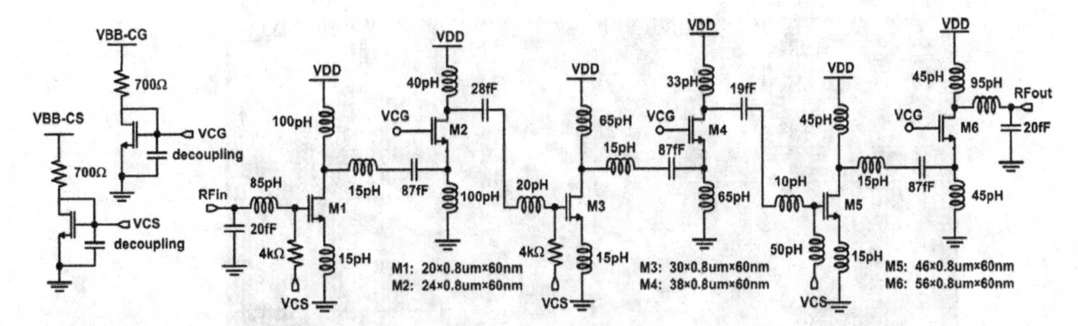

Figure 7.23 140GHz 65nm CMOS LNA schematic [18]

process. The circuit was intended to be portable, without modifications in a future 32nm CMOS technology. Each transistor employs 0.8μm finger width with the gate finger contacted on both sides to reduce gate resistance. The choice of ac-coupled cascode topology was dictated by the fact that no transistor models were available for the 32nm technology, and that the maximum allowed supply voltage was going to be 0.9V or lower. In addition to the higher current consumption, the price that is paid for design ease is an increased number of interstage passive components: three inductors and one capacitor instead of just a broadbanding inductor. The drain inductor of M1 and the source inductor of the M2 are chosen to be equal and, in parallel, to resonate with the total capacitance at the drain of M1 and at the source of M2. The series LC network between M1 and M2 is designed to resonate at 140GHz. Exactly the same approach is taken for the design of the interstage matching networks between M3 and M4 and M5 and M6 with the shunt inductor values being scaled according to the size scaling of the MOSFETs. Only lumped inductors, designed using ASITIC, were employed. The die photograph of the chip fabricated in the standard 65nm CMOS process is reproduced in Figure 7.24 while its measured S_{21} and S_{11} are shown in Figure 7.25 as a function of frequency and supply voltage.

7.8 OTHER LNA TOPOLOGIES

Common-base (CB) [19] and common-gate (CG) [11] topologies have also been employed in LNAs, especially in applications where high-power consumption and wider bandwidths are more important than achieving minimum noise figure.

Transformer feedback in both common-source and common-gate configurations, some of them with gm-boosting, and feed-forward common-source and common-gate topologies connected in parallel (sometimes rather misleadingly called noise-cancelation topologies) have also become popular in recent years. All of them will be discussed next.

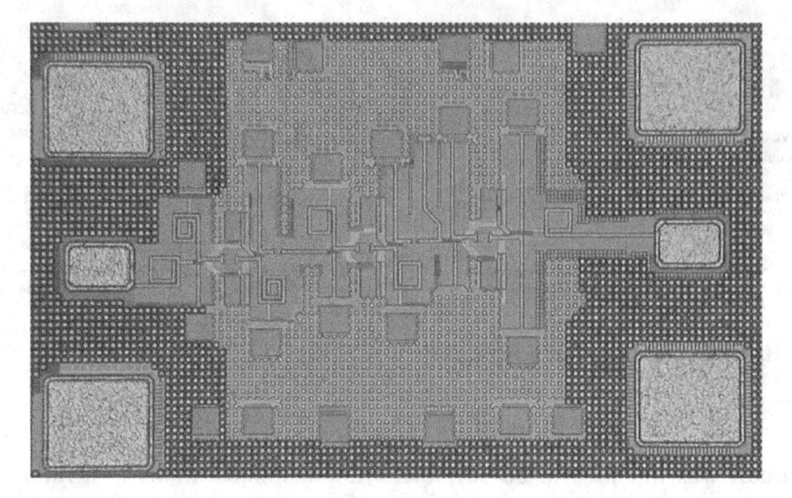

Figure 7.24 Layout of 140GHz 65nm CMOS LNA: 0.3mm × 0.5mm [18]

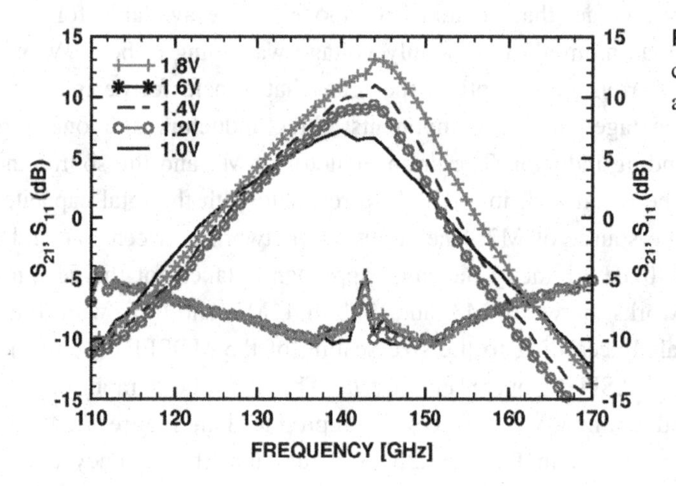

Figure 7.25 Measured S_{21} and S_{11} characteristics as a function of frequency and supply voltage

7.8.1 Common-base and common-gate LNAs

If lower power and bandwidth extension down to DC are desired and low-noise is not the prime goal, a common-base (CB) or common-gate (CG) LNA configuration should be considered. It is stated from the outset that these topologies will yield higher noise and poorer linearity than the CE, CS, or cascode LNAs. However, if these drawbacks are acceptable for the intended application, it is possible to design a CB or CG LNA with small bias currents and hence very low power.

Common-base (CB) and common-gate (CG) LNAs topologies are depicted in Figure 7.26. As discussed in Chapter 5, they can be analyzed as tuned amplifiers with two parallel resonant circuits: one at the input and one at the output. The input resonant circuit consists of L_E and of the total capacitance between the emitter (source) and ground, which includes the pad capacitance. As with the CE, CS, and cascode amplifiers, the output node at the collector or

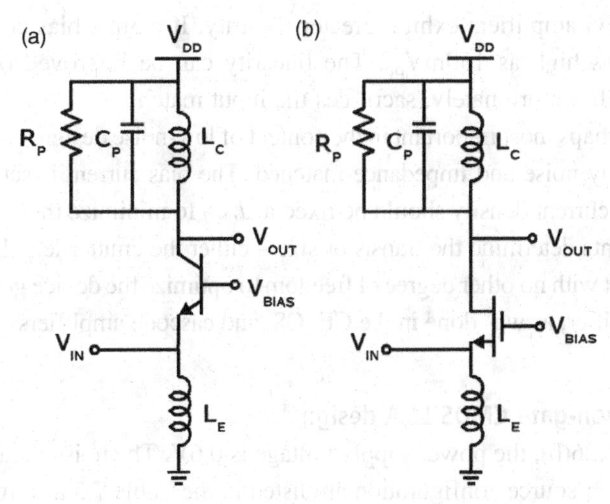

Figure 7.26 (a) Common-base (CB) and (b) common-gate (CG) LNA topologies

drain of the transistor is loaded with a tuned tank which can be represented as a parallel RLC network designed to resonate at ω_0.

It is relatively simple to obtain a good input impedance match with the CB and CG topologies. The real part of the input impedance is largely determined by the DC parameters of the transistor. For the bipolar implementation, this is simply given as

$$\Re\{Z_{IN,CB}\} = \frac{1}{g_{m,HBT}} = \frac{V_T}{I_C}. \tag{7.40}$$

To achieve a 50Ω input impedance match at room temperature (25°C), the bias current is fixed at approximately 0.5mA (without counting for the parasitic emitter resistance). The input capacitance, set by C_{be} and by the parasitic capacitance of the bonding pad, can be canceled by the shunt inductor, L_E,.

Impedance matching in a MOS CG amplifier is somewhat more complicated due to the large output conductance, g_o, which degrades the isolation between input and output. In this case, the real part of the input impedance becomes a function of the bias condition as well as of the load resistance at resonance, R_P

$$\Re\{Z_{IN,CG}\} = \frac{1}{g_{m,MOS}} + \frac{R_P}{1+\frac{g_{m,MOS}}{g_o}} \tag{7.41}$$

Still, in nanoscale CMOS technologies, the bias current needed to achieve a 50Ω impedance is less than 3mA, yielding a fairly low-power LNA.

The maximum voltage gain of the CB/CG stage *is* $2R_P/Z_0$. However, as we saw in Chapter 5, the maximum power gain is limited by $(g^2_{meff}\, Z_0\, R_P)/4$ and by the maximum available gain of the transistor at the LNA operation frequency.

The previously mentioned drawbacks of the CB and CG topologies result from the fixing of the bias current to set the input impedance.

First, the maximum input linear range is set by the 50Ω input impedance and by the bias current. For a HBT CB LNA biased at 0.5mA, only 25mV$_{pp}$ input swing can be tolerated in the linear operating region. MOS CG amplifiers exhibit greater linearity. If a 3mA bias current is used, the input swing can be as high as $150mV_{pp}$. The linearity can be improved only by increasing the bias current which, unfortunately, sacrifices the input match.

The second drawback, and perhaps most important in the context of low-noise design, is that the amplifier cannot be independently noise and impedance matched. The bias current is set by the input matching criterion and the current density should be fixed at J_{OPT} to minimize the amplifier noise figure. These two constraints determine the transistor size – either the emitter length or the gate width – and hence we are left with no other degree of freedom to optimize the device geometry in order to noise-match the amplifier, as was done in the CE, CS, and cascode amplifiers.

EXAMPLE 7.5 60GHz common-gate CMOS LNA design

For the 60GHz LNA in Figure 7.26(b), the power supply voltage is 0.6V. The noise parameters of the n-MOSFET in the common-source configuration are listed in the Table 7.5 as a function of frequency. Assume $g'_{meff} = 1.1\text{mS}/\mu\text{m}$, $g'_{oeff} = 0.11\text{mS}/\mu\text{m}$, $C'_{gs} = 1\text{fF}/\mu\text{m}$.

(a) Determine the size and bias current of the transistor such that the real part of the optimum noise impedance is equal to 50Ω. The minimum noise figure and optimum noise impedance of a common-gate stage can be assumed to be identical to those of the common-source stage.

(b) Calculate L_S, the transistor gate width and the bias current such that input impedance is matched to 50Ω for $R_P = 110\Omega$. What is the power gain if the output resonates at 60GHz and the input is matched to 50Ω? The real part of the input impedance in the common-gate stage can be approximated by $R_{IN} = \frac{1}{g_{meff}} + \frac{R_p}{1+\frac{g_{meff}}{g_{oeff}}}$?

(c) If the measured *MAG* (maximum available power gain) of the transistor is 9dB at 60GHz, how large can the effective R_p at the output node really be? Is a 700Ω value realistic? What would the power gain and R_{IN} be in the latter case?

Table 7.5 Noise parameters of a 90nm 20 × 1μm n-MOSFETs at 0.15mA/μm, $f_T = 120$GHz.

F(GHz)	Rn (Ω)	R_{SOPT} (Ω)	X_{SOPT} (Ω)	X_{IN} (Ω)	F_{MIN}	NF_{MIN} (dB)
5	130.6	275	1237	−979.6	1.05	0.2
30	46.7	79.5	206	−164	1.4	1.46
60	38.26	45	103.5	−83.5	1.75	2.43

Solution

(a) We can use the data in the table and the fact that $R_{SOPT} = R'_{SOPT}/W$.

At 60GHz R_{SOPT} for a 20μm device is 45Ω. Hence a device with $W = 20 \times 45/50 = 18\mu\text{m}$ will have $R_{SOPT} = 50\Omega$.

$I_{DS} = W \times 0.15\text{mA}/\mu\text{m} = 2.7\text{mA}$.

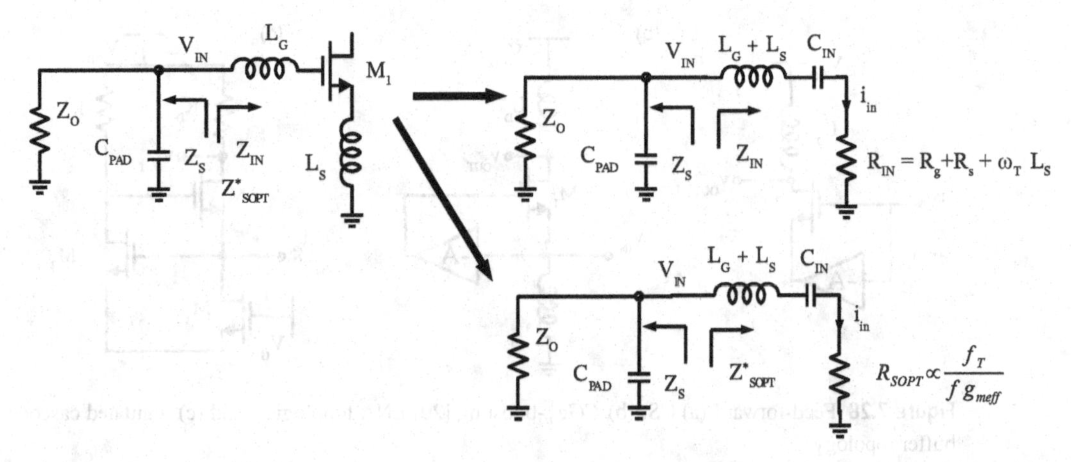

Figure 7.27 The equivalent circuit of the input impedance and noise impedance of a tuned LNA stage with inductive degeneration illustrating the frequency dependence of the real part of the optimum noise impedance R_{SOPT}. This fundamentally prevents broadband noise impedance matching

(b)
$$R_{IN} = \frac{1}{g'_{meff}W} + \frac{R_P}{1+\frac{g_{meff}}{g_{oeff}}} = \frac{1}{g'_{meff}W} + 10 = 50\,\text{Ohm}$$

$W = \{25\text{mS}\}/\{1.1\text{mS}/\mu\text{m},\} = 22.7\mu\text{m}$. We pick $W = 23\mu\text{m}$ and $I_{DS} = 3.45\text{mA}$ and $g_{meff} =$ 25.3mS.

$$C_{gs} = W \times C'_{gs} = 23\text{fF}.\ C_{sb} = W \times C'_{sb} = 25.3\text{fF}.$$

$$L_S = \frac{1}{\omega^2(C_{gs}+C_{sb})} = 150\text{pH}$$

This device size corresponds to a noise matching impedance of 40Ω. A compromise between noise and input impedance matching would be to select $W = 20\mu\text{m}$ and thus end up with a noise impedance of 45Ω and an input impedance of 55Ω.

(c) $G <= 1/4\, g^2_{meff} R_P R_{IN}$ for $R_P = 700\Omega$; $R_{IN} = 40 + 700/11 = 103.63\Omega$. $G = 11.6 = 10.6$dB. This value is larger than the transistor *MAG*. That means that R_P, which includes the output resistance of the MOSFET and that of the drain inductor, cannot be as high as 700Ω.

7.8.2 Gm-boosting and "noise-canceling" LNAs

In the early 2000s, with the emergence of the new ultra-wideband (UWB) radio standard which covered the 2–12 GHz band, RF designers were faced with the formidable challenge of implementing high-power, low-noise amplifiers with good input impedance matching over octave bandwidths. As illustrated in Figure 7.27, the input and noise impedance of a LNA stage with inductive degeneration form a series tuned resonant circuit. The real part of the input impedance can be made constant and equal to Z_0 over a large frequency range. However, because R_{SOPT} is frequency-dependent, simultaneous noise and input impedance matching is not possible except at a single frequency where $R_{SOPT} = Z_0$. Furthermore, at low frequencies (GHz-range), large inductors are required in CMOS LNAs to tune out the imaginary parts of the

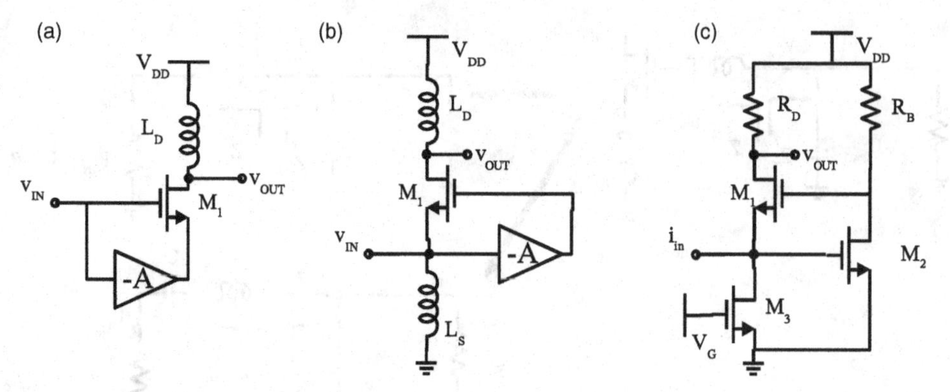

Figure 7.28 Feed-forward (a) CS, (b) CGg_m-boosting [20] LNA topologies, and (c) regulated cascode current buffer topology

input and noise impedances. That, too, can only be accomplished over a relatively narrow band unless complicated and area-intensive matching networks are employed. As we have already mentioned, a third problem of the inductively degenerated LNA is the relatively large current needed for noise impedance matching in low-frequency CMOS LNAs.

One idea that was put forward to overcome these problems is to take advantage of the low-current (and hence low input capacitance) required for input-impedance matching in a common-gate LNA, and to find some other technique to compensate for its poor noise-impedance matching and, therefore, poor noise figure. In principle, this could be achieved if one could simultaneously increase the transconductance and the input capacitance of the transistor in the LNA stage without concomitantly increasing the bias current. The larger transconductance and input capacitance would reduce both the noise resistance, R_n, and the real part of the noise impedance and bring it closer to Z_0, without sacrificing bias current. As a result, for a given signal source of impedance Z_0, the LNA noise figure is expected to improve and to approach the minimum noise figure of the transistor, itself.

Three such circuits are illustrated in Figure 7.28. In each case, a secondary shunt–series feed-forward path, featuring an ideal inverting voltage amplifier, is employed between the gate and source, or between the source and gate, of the main CS or CG stage, respectively. As we know from the theory of noise in feedback networks developed in Chapter 3, using a shunt connection at the input always reduces the noise resistance and the optimum noise impedance. The key (and rather sobering) restriction here is that the ideal inverting voltage amplifier should consume no power and should contribute no noise.

How can the inverting gain, noiseless amplifier be implemented in practice? In theory, it can be realized with an ideal transformer, as explained in the next section. In that case, the voltage gain A cannot be larger than 1. Moreover, in practice, the transformer exhibits some loss and, therefore, the feed-forward path will contribute some noise.

A second option is to implement the inverting amplifier as a common-source stage, like that formed by M_2 and R_B in the regulated cascode current buffer of Figure 7.28(c). Alas, M_2 and R_B contribute noise and require extra bias current. Furthermore, M3, which acts as a current source for M1, also contributes noise.

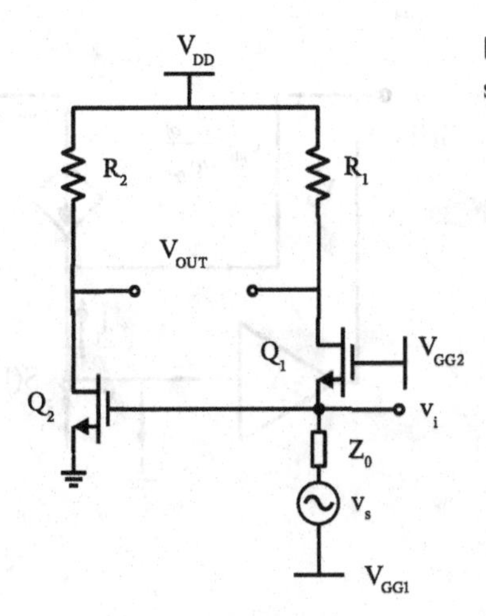

Figure 7.29 Example of a "noise cancellation" topology or shunt-connected impedance matching and voltage sensing topology

A third solution is to connect a CG and a CS stage in parallel at the input and to collect the output differentially, as in the (rather misleadingly called) "noise-canceling" topology [21] reproduced in Figure 7.29. A closer inspection unveils that the circuit in Figure 7.29 is similar but not identical to that in Figure 7.28(c).

Interestingly, by using cross-coupled connections [20] or large capacitors, as in problem 7.7, the circuits in Figures 7.28(a), 7.28(b), and 7.29 evolve into the same differential amplifier, with the inputs simultaneously driving the gate and the source of MOSFETs.

In fact, all the circuits in Figure 7.28 and Figure 29 can be described, and their small signal and noise behavior can be analyzed, using the shunt–series connection of two two-ports. It seems that we have stumbled upon a circuit which combines the g_m-boosting and the noise cancelation concepts into one. However, according to the linear noisy network analysis developed by Adler and Haus and discussed in Chapter 3, the noise measure of these circuits cannot be smaller than the noise measure of the individual transistors. Interestingly, only the circuit in Figure 7.28(b) was recognized as "g_m-boosting" [20]. However, close inspection will convince us that all four topologies in Figures 7.28 and 7.29 exhibit this property. Additionally, their input and noise impedance are reduced by the same scaling factor by which the transconductance is boosted.

Small signal analysis

Let us first try to derive the expressions of the transconductance and of the input impedance (capacitance) using the simplified unilateral equivalent circuit of the MOSFET with C_{in} and g_{meff}, as illustrated in Figure 7.30. The effective transconductance and input impedance of the circuit in Figure 7.28(a) become

$$G_m \stackrel{\text{def}}{=} \frac{-i_{sc}}{v_i} = \frac{g_{meff}v_{gs}}{v_i} = \frac{g_{meff}[v_i - (-Av_i)]}{v_i} = (1+A)g_{meff} \tag{7.42}$$

$$Z_{in} \stackrel{\text{def}}{=} \frac{v_i}{i_i} = \frac{v_i}{j\omega C_{in}v_{gs}} = \frac{v_i}{(1+A)v_i j\omega C_{in}} = \frac{1}{j\omega(1+A)C_{in}} = \frac{\omega_{Teff}}{j\omega(1+A)g_{meff}} = \frac{\omega_{Teff}}{j\omega G_m} \tag{7.43}$$

where we have assumed that the inverting voltage amplifier is ideal, with infinitely large input impedance (no input current flows in it) and infinitely small output impedance. Note that the

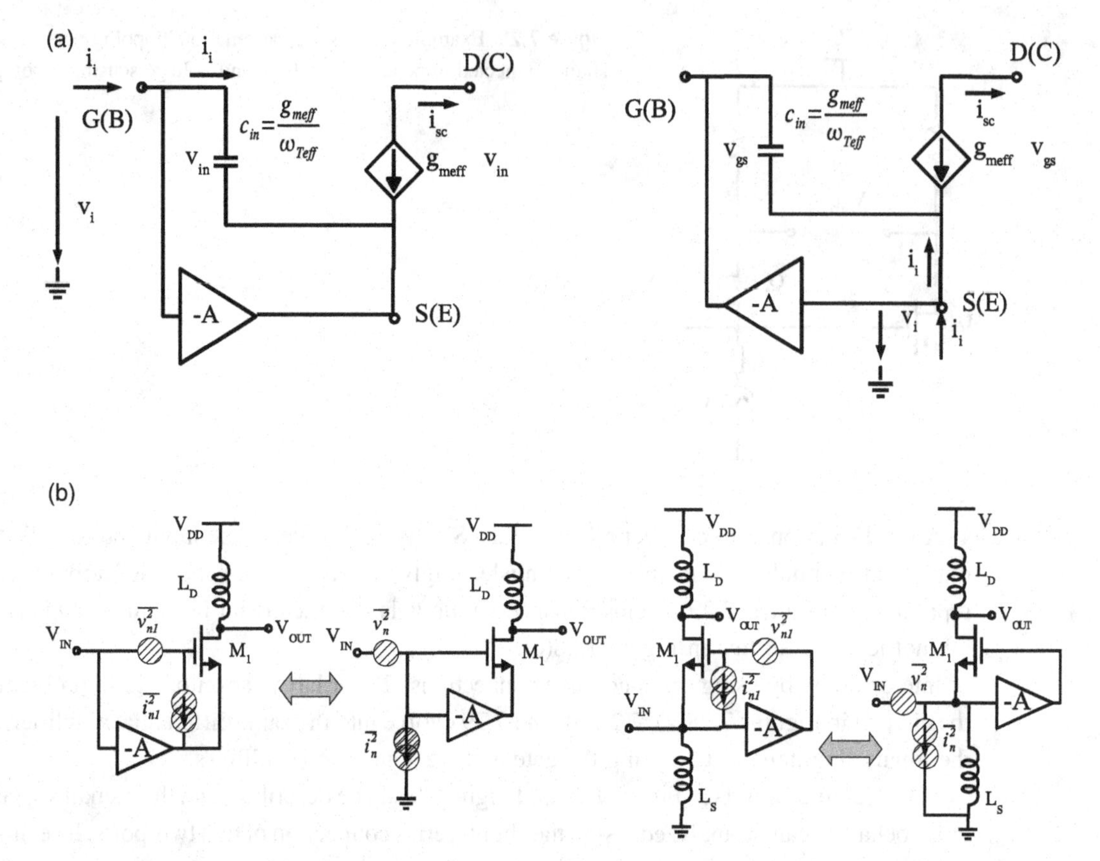

Figure 7.30 (a) Simplified small signal equivalent circuits and (b) noise equivalent circuits for the g_m-boosting topologies in Figures 7.28(a) and 7.28(b)

product between the transconductance and the input impedance (equivalent to current gain) has not changed. We cannot get something for nothing!

Similarly, for the circuit in Figure 7.28(b), we obtain that $G_m = (1 + A)\, g_{meff}$ and $Y_{in} = (1 + A)g_{meff} + j\omega[(1 + A)\, C_{gs} + C_{sb}]$. (Note that here we have accounted for the source–bulk capacitance of the MOSFET.)

The equivalent G_m of the "noise-cancelation" circuit in Figure 7.29 is obtained from the short-circuit current at the differential output. The short-circuit current at the drain of Q_1 is $i_{sc1} = g_{m1}v_i$. Similarly, the short-circuit current at the drain of Q_2 is $i_{sc2} = -g_{m2}v_i$. The differential short-circuit current then becomes $i_{sc} = i_{sc1} - i_{sc2} = (g_{m1} + g_{m2})v_i$ which proves g_m-boosting with $G_m = g_{m1} + g_{m2}$. The equivalent input capacitance has also been "boosted." Ignoring the Miller effect, we obtain $C_{in} = C_{gs1} + C_{gs2} + C_{sb1}$. However, in this case the current consumption also doubles.

Noise analysis for the ideal case when the feed-forward amplifier is noiseless

The *G*-matrices of the feed-forward path, main amplifier path, and of the overall amplifier can be described as a function of the *Y*-parameters of the main path transistor and of the voltage gain of the feed-forward path

$$[G_f] = \begin{bmatrix} 0 & 0 \\ -A & 0 \end{bmatrix}, [G_a] = \begin{bmatrix} y_{11a} - \frac{y_{12a}y_{21a}}{y_{22a}} & \frac{y_{12a}}{y_{22a}} \\ \frac{-y_{21a}}{y_{22a}} & \frac{1}{y_{22a}} \end{bmatrix} \text{ and } [G] = \begin{bmatrix} y_{11a} - \frac{y_{12a}y_{21a}}{y_{22a}} & \frac{y_{12a}}{y_{22a}} \\ \frac{-y_{21a}}{y_{22a}} - A & \frac{1}{y_{22a}} \end{bmatrix} \quad (7.44)$$

where the index f describes the blocks in the feed-forward path while index a designates the main amplifier path.

Assuming that the shunt–series connected feed-forward network is noiseless, i.e. $R_{nf}=0$, $G_{uf}=0$, $Y_{corf}=0$, from (3.197)–(3.199), we obtain that

$$R_n = |P_a|^2 R_{na} = \frac{|g_{21a}|^2}{|g_{21}|^2} R_{na} \quad (7.45)$$

$$Y_{cor} = \frac{Y_{cora} + Q_a}{P_a} = \frac{g_{21}Y_{cora} + Ag_{11a}}{g_{21a}} \quad (7.46)$$

$$G_u = G_{ua}. \quad (7.47)$$

Alternatively, using the simplified noise equivalent circuits in Figure 7.30(b), we can obtain the expressions of the input equivalent noise voltage, v_n, and noise current, i_n, as $v_n = \frac{v_{n1}}{1+A}$ and $i_n = i_{n1}$ for the circuit in Figure 7.28(a), and $v_n = \frac{-v_{n1}}{1+A}$ and $i_n = -i_{n1}$ for the circuit in Figure 7.28(b). The expressions of the noise parameters in the noise admittance formalism become

$$R_n = \frac{\overline{v_n^2}}{4kT\Delta f} = \frac{R_{na}}{(1+A)^2} \quad (7.48)$$

$$Y_{cor} = \frac{\overline{i_n v_n^*}}{\overline{v_n^2}} = \frac{(1+A)^2}{1+A}\frac{\overline{i_{n1}v_{n1}^*}}{\overline{v_{n1}^2}} = (1+A)Y_{cora} \quad (7.49)$$

and $G_u = G_{ua}$

$$G_{sopt} = \sqrt{G_{cor}^2 + \frac{G_u}{R_n}} = \sqrt{(1+A)^2 G_{cora}^2 + (1+A)^2 \frac{G_{ua}}{R_{na}}} = (1+A)G_{sopta} \quad (7.50)$$

$$\begin{aligned} F_{MIN} &= 1 + 2R_n(G_{cor} + G_{sopt}) = 1 + \frac{2R_{na}}{(1+A)^2}(1+A)(G_{cora} + G_{sopta}) \\ &= 1 + \frac{2R_{na}}{1+A}(G_{cora} + G_{sopta}) < F_{MINa}. \end{aligned} \quad (7.51)$$

Equations (7.48)–(7.51) prove that, in the ideal case, the noise resistance and the optimum noise impedance are (drastically) reduced and that, if the feed-forward amplifier is noiseless, the minimum noise figure can be smaller than that of the main amplifier. As a note, the noise figure is still not smaller than that of the feed-forward amplifier, so there is no contradiction with Haus and Adler's linear noisy network theory.

Noise analysis for the real case when the feed-forward amplifier is noisy

Let us try to derive the expression of the noise figure for the differential version of these topologies, shown in Figure 7.31, and demonstrate that it is not smaller than that of the individual transistor. For simplicity, but without reducing generality, let us assume that

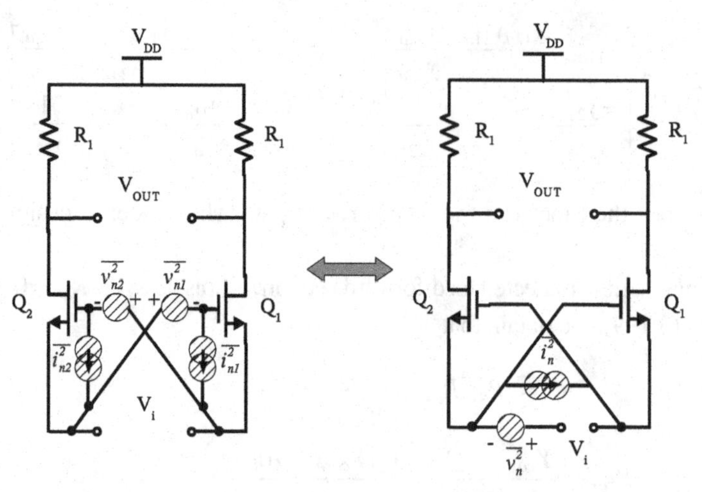

Figure 7.31 Noise equivalent circuit of the differential version of the amplifiers in Figures 7.28 and 7.30

$Q_1 = Q_2$, $R_1 = R_2$ and that the noise sources of Q_1 and Q_2 are statistically independent since they originate in different transistors.

As explained in Chapter 3, from the short-circuit current condition, we obtain the expression of the equivalent differential noise voltage source, v_n, as a function of the noise voltage and noise current sources of the two differential-mode half circuits v_{n1}, v_{n2}, i_{n1}, and i_{n2}

$$v_n = \frac{g_{m1}v_{n1} - g_{m2}v_{n2}}{g_{m1} + g_{m2}} = \frac{v_{n1}}{2} - \frac{v_{n2}}{2}. \tag{7.52}$$

Similarly, from the open-circuit voltage condition, the formula of the equivalent input differential noise current is derived

$$i_n = i_{n1} - i_{n2}. \tag{7.53}$$

The noise parameters in the noise admittance formalism are obtained from (7.52)–(7.53)

$$R_n = \frac{R_{n1}}{4} + \frac{R_{n2}}{4} = \frac{R_{n1}}{2} \tag{7.54}$$

$$Y_{cor} = \frac{\overline{i_n v_n^*}}{\overline{v_n^2}} = \frac{\frac{Y_{cor1}}{2}R_{n1} + \frac{Y_{cor2}}{2}R_{n2}}{R_n} = 2Y_{cor1} \tag{7.55}$$

$$i_u = i_n - Y_{cor}v_n = i_{u1} + Y_{cor1}v_{n1} - i_{u2} - Y_{cor2}v_{n2} - Y_{cor}\frac{v_{n1}}{2} + Y_{cor}\frac{v_{n2}}{2}$$

$$= i_{u1} - i_{u2} + \left(Y_{cor1} - \frac{Y_{cor}}{2}\right)v_{n1} - \left(Y_{cor2} - \frac{Y_{cor}}{2}\right)v_{n2} = i_{u1} - i_{u2} \tag{7.56}$$

$$G_u = G_{u1} + G_{u2} = 2G_{u1}$$

$$G_{sopt} = \sqrt{G_{cor}^2 + \frac{G_u}{R_n}} = \sqrt{4G_{cor1}^2 + 4\frac{G_{u1}}{R_{n1}}} = 2\sqrt{G_{cor1}^2 + \frac{G_{u1}}{R_{n1}}} = 2G_{sopt1} \tag{7.57}$$

$$F_{MIN} = 1 + 2R_n\left(G_{cor} + G_{sopt}\right) = 1 + \frac{2R_{n1}}{2}\left(2G_{cor1} + 2G_{sopt1}\right) = F_{MIN1}. \tag{7.58}$$

Equations (7.57) and (7.58) can also be directly obtained from the general noise parameter expressions of the shunt–series connection of two identical noisy two-ports derived in Chapter 3. The optimum noise admittance of the differential g_m-boosting amplifier is actually two times larger than

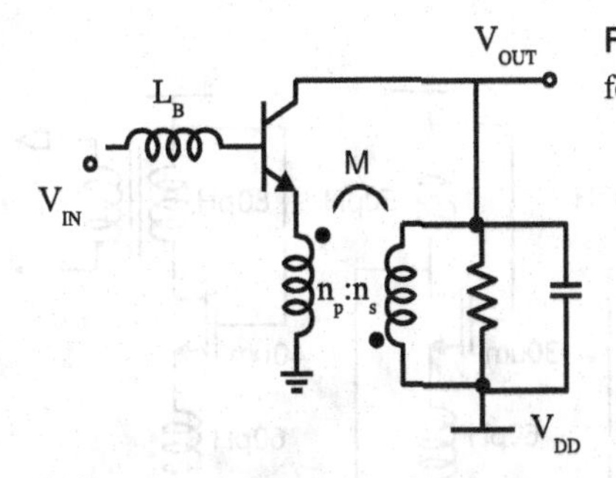

Figure 7.32 Schematics of LNA with shunt-series transformer feedback [22]

that of the transistor. It implies that the optimum noise impedance per side has been reduced four times while the bias current has increased only two times. The minimum noise figure has remained the same as that of the transistor. This brings back memories of the CMOS inverter. Indeed, the noise of Q_1 has not been canceled by Q_2. If the CS transistor were to cancel the noise of the CG transistor, then we could make the same statement about the noise of a multi-finger MOSFET. One gate finger cancels the noise of another gate finger and the overall minimum noise figure remains the same as that of the unit finger. That IS NOT thermal noise cancelation. It is simply noise impedance tailoring to provide noise matching, while preserving the minimum noise figure. It is also R_n reduction to make the LNA less sensitive to noise impedance mismatch over a broad bandwidth.

Although it does not cancel noise, the circuit topology does an excellent job at providing wideband input impedance matching with very low-noise resistance and, therefore, with less degradation of noise figure when compared to the minimum noise figure of the transistor itself. The main advantage of these topologies, like the CMOS inverter, is that the transconductance, input admittance and noise admittance per bias current increase roughly $(1 + A)$ times.

7.8.3 Transformer-feedback LNAs

Instead of inductors, we can employ transformers in the feedback network. As described in Chapter 3, the transformer turns ratio can be used to control the gain of the amplifier with feedback [22], while the inductance of the primary is employed, in much the same fashion as the source/emitter degeneration inductor in the previously discussed LNAs, to match the real part of the input impedance to the desired value. Figure 7.32 illustrates the case where the transformer forms a series–shunt feedback network around either a bipolar [22] or a MOS transistor [23]. Its small signal and noise analysis is described in detail in Appendix A6. Here, we will dwell on the dual shunt–series case [24], adapted in Figure 7.33 for a W-Band LNA designed in a 65nm CMOS technology. This circuit can also be described by the feed-forward topology in Figure 7.28(a) where the role of the inverting voltage amplifier is played by the transformer. Note that a common-gate version of the shunt–series transformer-feedback amplifier is also possible.

As in the case of the inductor feedback LNA, the bias current and size of the transistor in the LNA are chosen to set the real part of the optimum noise impedance to the desired value. Besides controlling the gain, the role of the transformer is to provide broadband input impedance matching and to tune out the imaginary part of the noise impedance. Unlike the inductive degeneration LNA, which had a series and a parallel resonant circuit at its input, the input and

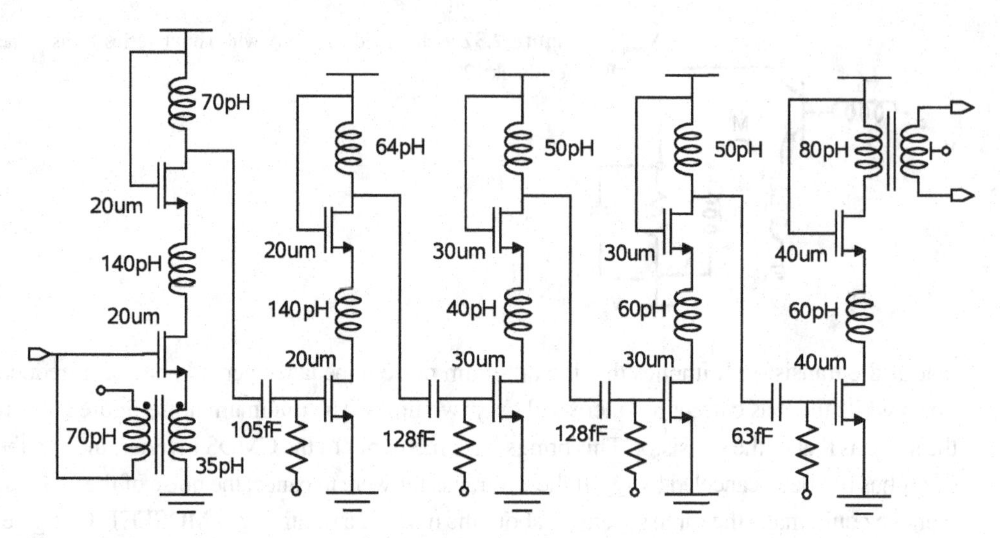

Figure 7.33 W-Band transformer-feedback LNA implemented in a 65nm CMOS process [25]

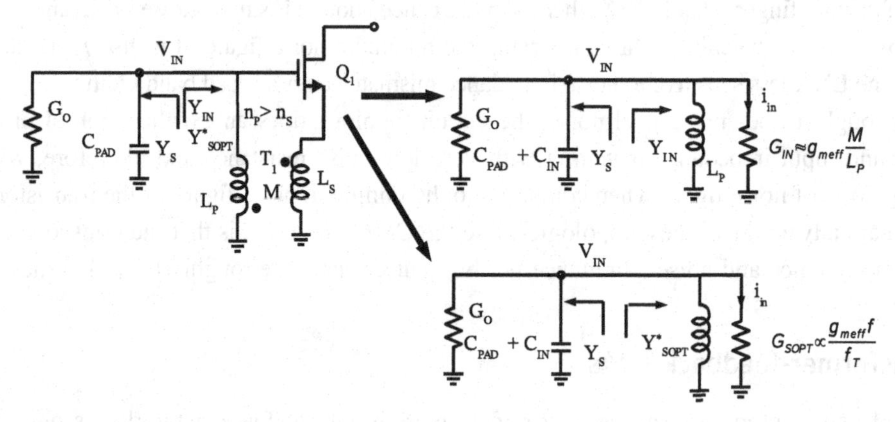

Figure 7.34 The equivalent circuit of the input impedance and noise impedance of a tuned shunt–series transformer-feedback LNA stage

noise impedance equivalent circuit of this topology, illustrated in Figure 7.34, features a single-parallel resonance which also includes the pad capacitance. This explains its much broader input matching bandwidth, apparent from the measurements and simulations shown in Figure 7.35. The noise figure (less than 7dB) and the input matching bandwidth (S_{11} <-10dB for over 40GHz) are significantly better than those of the W-Band LNA with inductive degeneration fabricated in the same technology [14]. Note that, because the pad capacitance is resonated out by the inductance of the transformer primary, the real part of the signal source impedance remains equal to Z_0 and $k = 1$ at the frequency of interest.

7.8.4 Design methodology for shunt–series transformer-feedback LNAs

Using the equations developed in Appendix A7, which account for the loss of the transformer, a step-by-step algorithmic LNA design methodology can be derived for the transformer-feedback LNA [14], similar to the one developed for the series–series inductor feedback one. The

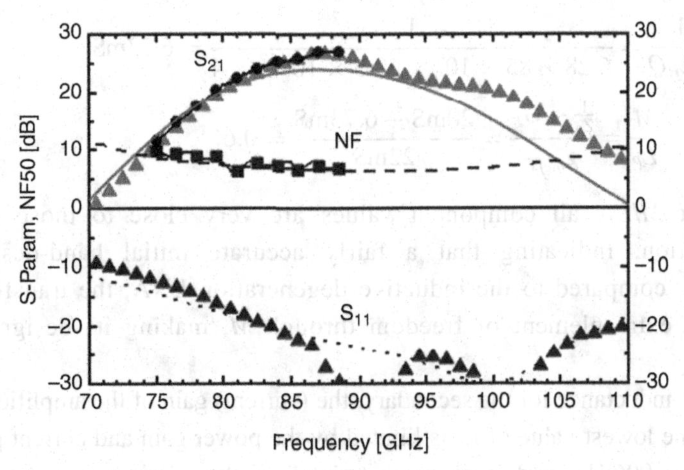

Figure 7.35 Measured vs. simulated S-parameters and noise figure of W-Band CMOS LNA [25]

numerical example below is for the cascode LNA in Figure 7.33, at 85GHz. The 65nm GP CMOS process in which the circuit was fabricated features minimum gate length n-MOSFETs with $g'_m = 1.41\text{mS}/\mu\text{m}$, $R'_S = 160\Omega \times \mu\text{m}$, $g'_{meff} = 1.1\text{mS}/\mu\text{m}$. The gate resistance of a 1μm wide finger contacted on both sides is 68Ω. The f_{Teff} of the cascode stage with inductive broadbanding is 115GHz, $J_{OPT} = 0.25\text{mA}/\mu\text{m}$ and $k_1 = 0.5$.

Step 1 Transistor sizing for $R_{SOPT} = 50\Omega$ and assuming 1μm wide gate fingers contacted on both sides of the gate

$$N_f = \frac{f_{Teff}}{fZ_0 W_f g'_{meff}} \sqrt{\frac{g'_m R'_s + W_f g'_m R'_g (W_f)}{k_1}} =$$

$$= \frac{115}{85 \cdot 50 \cdot 1.1 \cdot 10^{-3}} \sqrt{\frac{1.41 \cdot 0.16 + 1 \cdot 1.41 \cdot 10^{-3} \cdot 68}{0.5}} \approx 20$$

where $R'_S = 160\Omega \times \mu\text{m}$ is the source resistance times unit gate width.

Step 2 Cascode bias current calculation assuming $J_{OPT} = 0.25\text{mA}/\mu\text{m}$

$$I_{DS} = J_{OPT} W_G = 0.25 \frac{\text{mA}}{\mu\text{m}} \cdot 20\mu\text{m} = 5\text{mA}.$$

Step 3 Determine L_P for input susceptance cancelation from the imaginary part of (A7.10)

$$L_P = \frac{1}{\omega^2 \left(C_{PAD} + \frac{g_{meff}}{\omega_{Teff}} \right)} = \frac{1}{(2 \cdot \pi \cdot 85 \cdot 10^9)^2 \cdot \left(20fF + \frac{22mS}{6.28 \cdot 115 \cdot 10^9} \right)} = 69.54\text{pH} \approx 70\text{pH}$$

where a pad capacitance of 20fF was assumed.

Step 4 Find M/L_P from the real part of (A7.10) for input conductance matching, (assuming a Q of 4 for the primary and a pad capacitance of 20fF)

$$G_P = \frac{1}{\omega L_P Q} = \frac{1}{6.28 \times 85 \times 10^9 \times 69.5 \times 10^{-12} \times 4} = 6.73\text{mS}$$

$$\frac{M}{L_P} = \frac{\frac{1}{Z_0} - G_P}{g_{meff}} = \frac{20\text{mS} - 6.73\text{mS}}{22\text{mS}} = 0.6.$$

With the exception of M/L_P, all component values are very close to those finally arrived at by simulation, indicating that a fairly accurate initial hand-design is possible. We note that, compared to the inductive degeneration LNA, the transformer-feedback LNA has an extra element of freedom through M, making its design more complicated.

By choosing a smaller inductance for the secondary, the (current) gain of the amplifier stage is increased. However, the lowest value of L_S is limited by the power gain and current gain of the transistor itself at 80–90GHz, and is also constrained by the inductance value of the primary, the coupling coefficient k, and layout realizability. The power gain and the peak gain of both the inductor-feedback and transformer-feedback stages is set by the Q and inductance, respectively, of the inductor connected between V_{DD} and the drain of M_2.

7.9 DIFFERENTIAL LNA DESIGN METHODOLOGY

It is often common in a single-chip transceiver implementation for the LNA to be realized as a differential topology. As demonstrated in Chapter 3, the differential-mode optimum noise impedance is simply equal to two times that of the differential-mode half circuit [26]. In the case of the differential LNA with inductive degeneration in Figure 7.36, the differential-mode optimum noise impedance and input impedance are give by

$$Z_{soptdiff} = 2Z_{sopt}(half\_ckt) \tag{7.59}$$

$$Z_{INdiff} = 2\omega_T L_{S1} \tag{7.60}$$

where L_{S1} is the degeneration inductance per side and ω_T is the cutoff frequency of the differential-mode half circuit.

7.10 PROCESS VARIATION IN TUNED LNAs

One of the most under-appreciated benefits of transformer or inductor-feedback LNAs is their robustness to process variation. This is not merely the result of the most important property of negative feedback networks: that of desensitizing the gain of the amplifier to transistor parameter variability. Primarily this is due to the fact that the main parameters of the feedback network itself (i.e. those of the transformer and inductor: winding inductance and coupling coefficient) are solely a function of lithography and, therefore, they remain largely process and temperature independent. The only important passive device parameter that is process – and

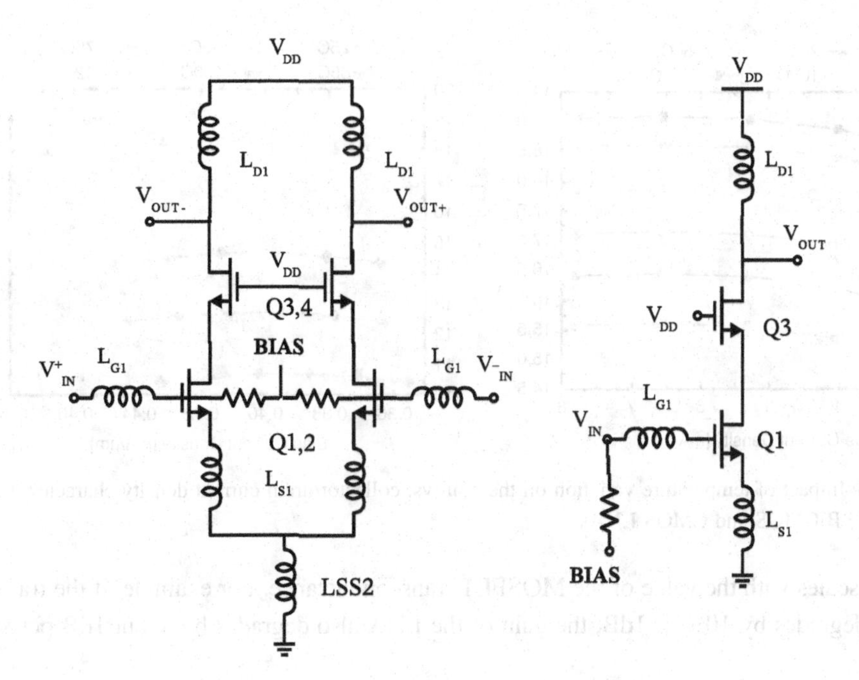

Figure 7.36 Differential LNA and differential-mode half circuit

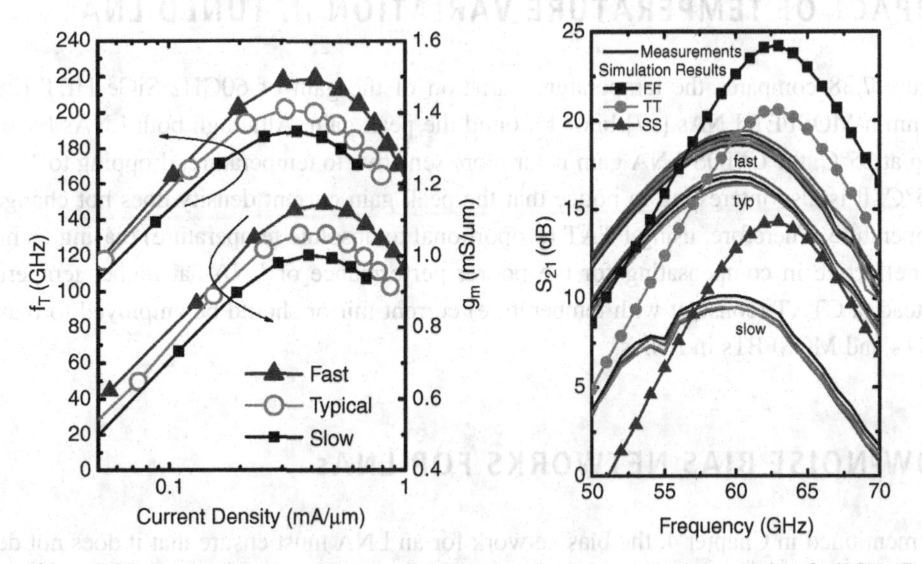

Figure 7.37 Impact of transistor process variation on the gain vs. frequency characteristics of a 60GHz CMOS LNA [27]

temperature sensitive is the quality factor. It affects the gain of the LNA and, to far less extent, the bandwidth and port impedances of the LNA.

Figure 7.37 illustrates the measured g_m and f_T of 65nm n-MOSFETs and the gain versus frequency characteristics of an entire LNA across transistor process corners [27]. These measured data indicate that the transistor parameter variation affects the peak gain (and minimum noise figure) values, but not the 3dB bandwidth or the frequency at which the gain peaks. The LNA

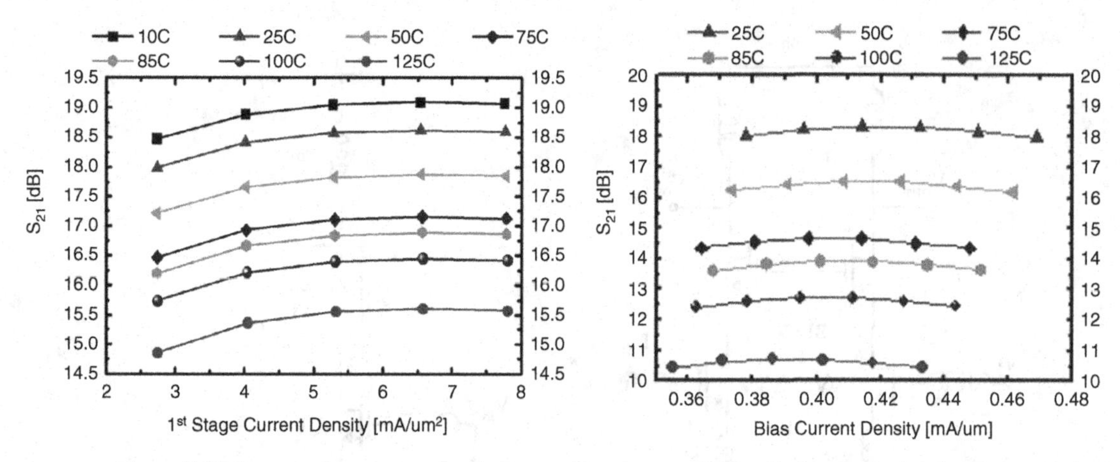

Figure 7.38 Impact of temperature variation on the gain vs. collector/drain current density characteristics of 60GHz SiGe BiCMOS and CMOS LNAs

peak gain scales with the value of the MOSFET transconductance. For example, if the transconductance degrades by 10% or 1dB, the gain of the LNA also degrades by about 1dB per stage.

7.11 IMPACT OF TEMPERATURE VARIATION IN TUNED LNAs

Figure 7.38 compares the temperature variation of the gain of 60GHz SiGe HBT [28] and 65 nm n-MOSFET LNAs [27] biased around the peak gain. Although both LNAs have 18dB gain at 25°C, the CMOS LNA gain is far more sensitive to temperature, dropping to 10.5dB at 125°C. It is also interesting to notice that the peak gain current density does not change with temperature. Therefore, using PTAT (proportional to absolute temperature) biasing techniques is ineffective in compensating for the poorer performance of LNAs at higher temperatures. Instead, a CTAT (constant with temperature) current mirror should be employed to bias SiGe HBTs and MOSFETs in LNAs.

7.12 LOW-NOISE BIAS NETWORKS FOR LNAs

As mentioned in Chapter 4, the bias network for an LNA must ensure that it does not degrade the HF and noise performance of the LNA and it does not cause stability problems (oscillation) either in or outside the bandwidth of interest. The bias circuit of the 2-stage cascode SiGe HBT LNA discussed earlier is shown in Figure 7.39 in gray. It consists of current mirrors $Q5$ and $Q6$, bias resistors and de-coupling capacitors.

Carefully sized decoupling capacitors, are placed in proximity of the base of each common-base transistor and at the output of each current mirror. These capacitors should provide very low capacitive impedance in the bandwidth of interest and their self-resonance frequency should be higher than the highest frequency of operation. Any inductive behavior of the de-coupling capacitors

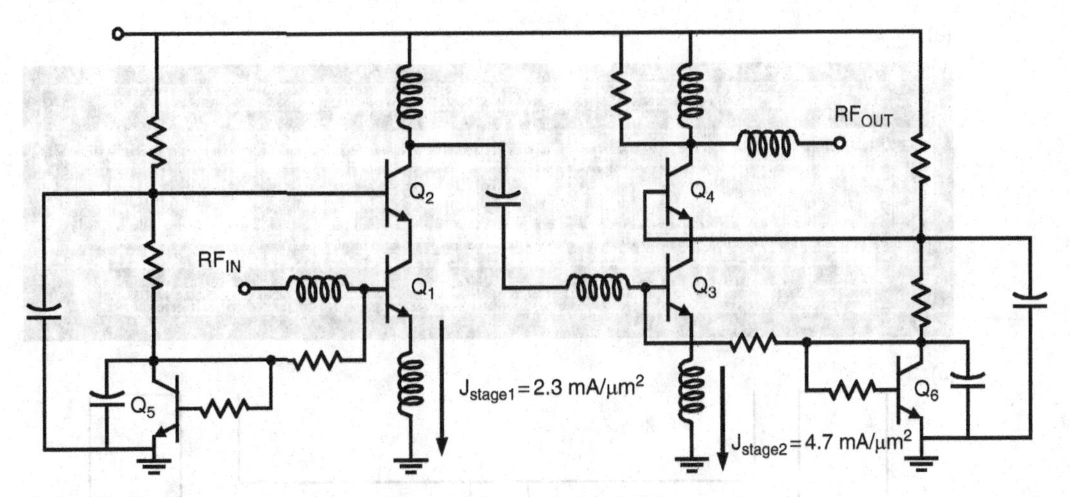

Figure 7.39 Bias circuit for 60GHz 2-stage HBT-cascode LNA [29]

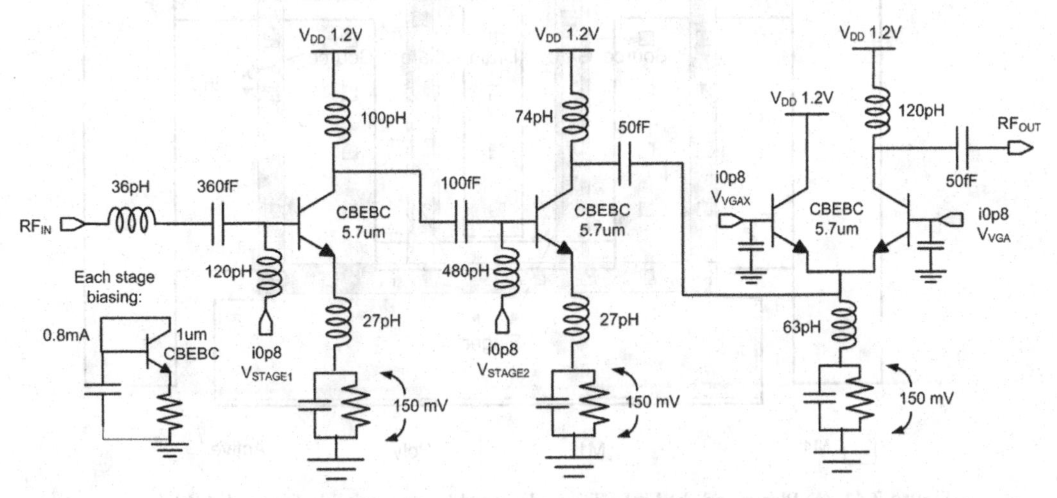

Figure 7.40 Example of HBT LNA biasing scheme using resistive degeneration to stabilize the bias point over temperature and process variation and capacitive decoupling to eliminate bias resistor noise [27]

connected to the base of a HBT can cause the circuit to oscillate. At the same time, the impedance of the capacitor should not be larger than 5Ω to avoid gain and noise figure degradation.

The reference current to the current mirror(s) usually comes from a bandgap circuit. The resistor connected between the current mirror and the base of the LNA HBT is chosen to be at least 0.5–1KΩ, and should have low parasitic capacitance. The series bias resistor can sometimes be replaced by an inductor. The latter then becomes part of the LNA matching network.

In the LNA layout, the current mirror HBT (Q_5, Q_6) must be placed very close (less than 50μm) to the LNA HBT (Q_1, Q_3) with a wide, solid metal ground connection between them. For an even more robust biasing scheme, emitter degeneration resistors can be employed, as illustrated in Figure 7.40. The DC voltage drop on the emitter degeneration resistor should be set to at least 150mV and an appropriately sized capacitor must be placed in parallel with the resistor to filter out its noise.

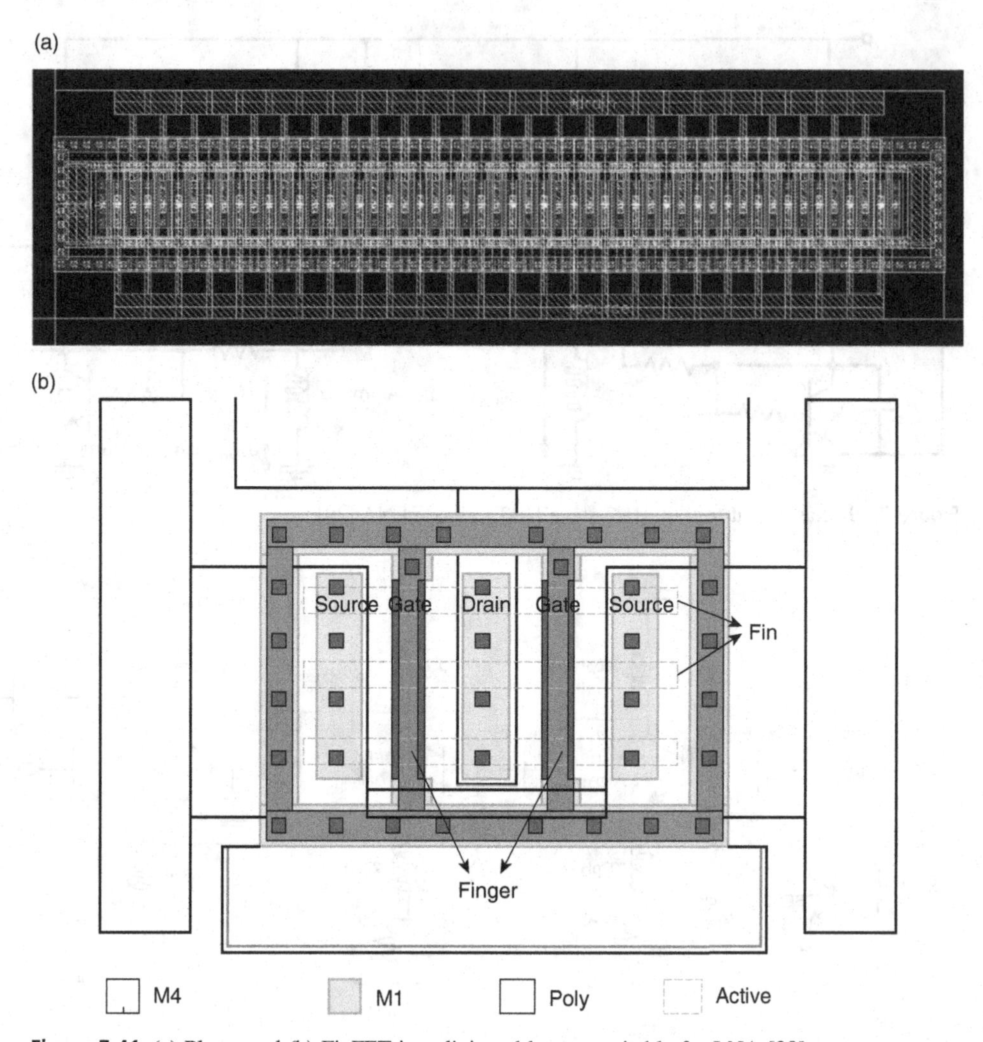

Figure 7.41 (a) Planar and (b) FinFET interdigitated layouts suitable for LNA [30]

The same bias de-coupling schemes are employed in MOSFET LNAs as in HBT LNAs. However, a matching gate resistance is not needed in the MOSFET current mirror and source degeneration is not that critical. In the case of III-V HEMT LNAs, since the threshold voltage is negative, only the bias circuits discussed in Chapter 4.4 can be employed.

7.13 MOSFET LAYOUT IN LNAs

The gate resistance of the MOSFET is critical in achieving low-noise figure in a CMOS LNA. Layouts with interdigitated gate fingers, contacted on both sides of the gate, should be used as illustrated in Figure 7.41 for a planar FET and for a FinFET. The planar MOSFET layout features 50 gate fingers, in this case 0.8μm wide, contacted on both sides of the gate and two dummy gate fingers, one on each side. The FinFET layout has two gate fingers and three fins for a total of 6 MOSFET unit cells connected in parallel.

Summary

- The LNA figure of merit aims to minimize noise figure and power dissipation, and to maximize gain and linearity.
- An optimal solution and an algorithmic design methodology exist for the design of noise – and impedance matched tuned low-noise amplifiers. This topology and design method are scalable to at least 200GHz.
- In the case of CMOS LNAs, J_{OPT} and Z_{SOPT} are invariant between technology nodes and the design equations are scalable across technology nodes and across frequencies.
- Single-stack transistor topologies are needed in sub 65nm CMOS nodes.
- Feed-forward (or g_m-boosting) topologies have been employed to realize broader bandwidth noise matching while also achieving relatively broadband input impedance matching.
- The bipolar cascode with inductive degeneration continues to be the LNA topology of choice because it combines the excellent isolation of the CB stage with the high-power gain, good linearity, and ease of simultaneous noise and input impedance matching, typical of the CE/CS stage. Furthermore, the high g_m/I_C ratio and small noise resistance make this stage, unlike a MOSFET one, insensitive to impedance mismatch and model inaccuracy.

Problems

(1) A folded cascode stage is typically employed in nanoscale CMOS technologies to improve the output resistance, voltage gain and the frequency response of opamps while operating from 1.2V supply. Similarly, an ac-folded cascode can be employed at 60GHz in 65nm GP CMOS LNAs and power amplifiers allowing them to be biased from a supply voltage $V_{DD} = 1$V, as shown in Figure 7.42. Inductors L_1 and L_2 are chosen to resonate at 60GHz with the total capacitance at the drain of Q_1 and gate of Q_2. In order to maximize the f_{Teff} of the ac-folded cascode, transistor Q_2 has two times the number of gate fingers of Q_1. When biased at the minimum noise bias, The transistor characteristics are $g'_{meff} = 0.95$mS/µm, $g'_{oeff} = 0.115$mS/µm, $C'_{gs} = 0.7$fF/µm, $C'_{gd} = 0.35$fF/µm and $C'_{db} = C'_{sb} = 0.6$fF/µm. The gate resistance of a 1µm wide gate finger is 200Ω. Its source resistance is also 200Ω. All inductors have a quality factor Q of 15 at 60GHz.

(a) Calculate the size and bias currents of Q_1 and Q_2 when biased at the optimal noise figure current density knowing that the effective cutoff frequency of the ac-folded cascode is 80% that of a transistor, and that the LNA is optimally noise-matched to a 50Ω signal source impedance. $k_1 = 0.5$.

(b) Determine L_1 and L_2 such that they resonate at 60GHz with the total capacitance at the drain of Q_1 and at the gate of Q_2.

(c) Calculate L_S such that the real part of the input impedance of the LNA is 50Ω.

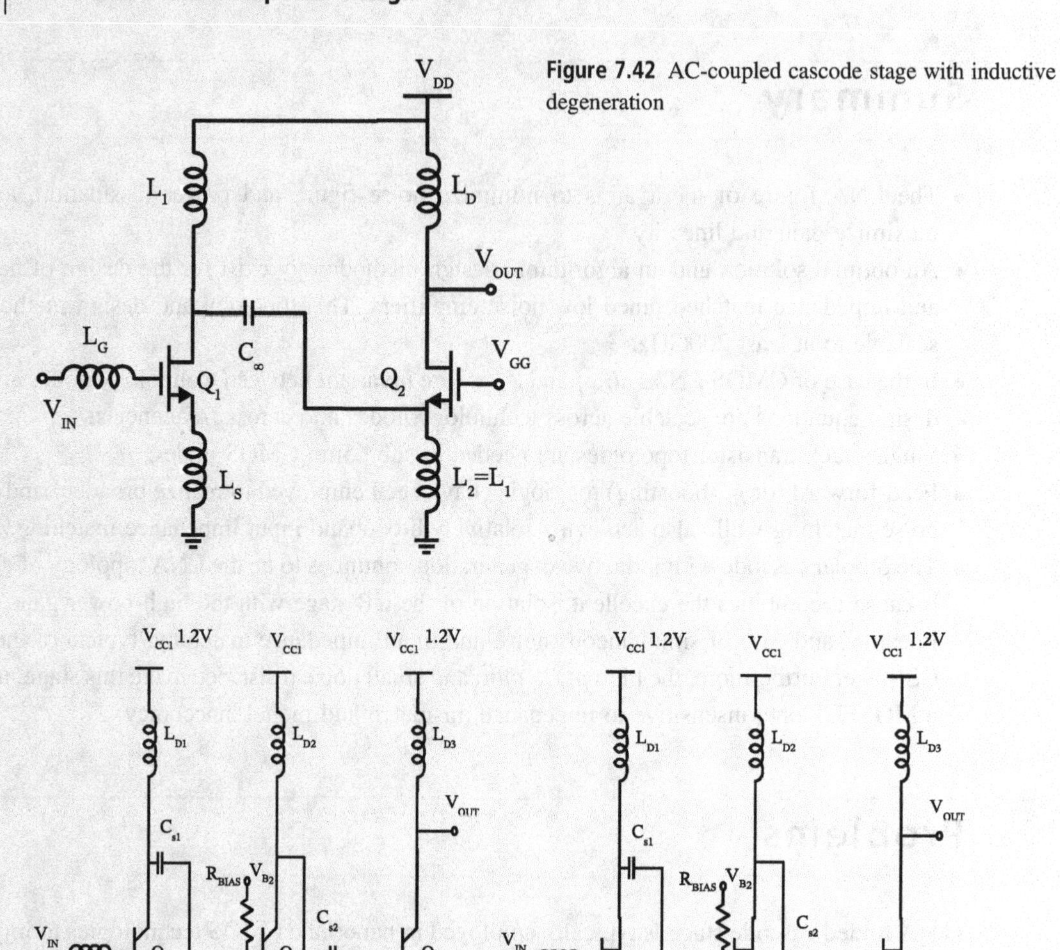

Figure 7.42 AC-coupled cascode stage with inductive degeneration

Figure 7.43 (a) 130nm SiGe HBT LNA schematic (b) 65nm CMOS LNA schematic

(d) Calculate L_G such that the imaginary part of the input impedance is canceled at 60GHz.

(e) Calculate L_D such that it resonates with the output capacitance at 60GHz.

(f) Calculate the maximum power gain of the ac-folded cascode stage when the output is terminated by a matched load.

(g) Calculate the minimum noise figure of the amplifier knowing that $k_1 = 0.75$. Account for the noise from the gate and source resistance of Q_1 and from the inductors L_S and L_G.

(2) Two 62GHz LNAs, biased with $V_{DD} = 1.2$V and implemented with SiGe HBTs and with 65nm MOSFETs, respectively, are shown in Figure 7.43. The transistors and load inductors in the first two stages are identical and biased at the optimal noise current density (8mA/μm^2 for the SiGe HBT). Both the SiGe HBTs and the 65nm MOSFETs are sized for optimal noise matching to 50Ω at 62GHz. The third stage is biased at the maximum linearity bias and its current is twice as large as that of the first two stages.

(a) Calculate the emitter length and bias current of the SiGe HBTs in all stages. Assume that $k_1 = 0.5$ and $f_T = 160$GHz.

(b) Calculate the gate width, number of fingers and bias current of the n-MOSFETs in all stages. Assume that $k_1 = 0.5$ and $f_T = 90$GHz.

(c) Calculate the L_{E1} and L_{S1} such that the real part of the input impedance is 50Ω in both amplifiers.

(d) Calculate the L_{D1} for both circuits such that the output of the first stage resonates at 62GHz. Assuming that inductor *Q*s of 5 are realizable, which one of the two amplifiers will have the higher gain in the first stage?

(e) Why is L_{D2} different from L_{D1}?

(f) Why is there a need for inductive degeneration in the last stage of the bipolar amplifier and not in the CMOS one?

(3) You have access to 130nm SiGe BiCMOS and 90nm CMOS processes.

160GHz SiGe HBT parameters for emitter width of 0.17μm:

Ic@peak $f_T = 8$mA/μm²; $w_E = 0.17$μm; $\beta = 500$; $V_{CESAT} = 0.5$V, $V_{BE} = 0.9$V

tau = 0.6ps; $C_{je} = 20$fF/μm²; $C_\mu = 11$fF/μm²; $C_{cs} = 1.2$fF/μm;

$R'_b = 70$Ωμm, $R_E = 5$Ωμm², $R'_c = 20$Ωμm.

Table 7.6 **Noise parameters for 90nm 20×1μm n-MOSFETs at 0.15mA/μm, $f_T = 120$GHz.**

F(GHz)	*Rn* (Ω)	R_{SOPT} (Ω)	X_{SOPT} (Ω)	X_{IN} (Ω)	F_{MIN}	NF_{MIN} (dB)
5	130.6	275	1237	−979.6	1.05	0.2
30	46.7	79.5	206	−164	1.4	1.46
60	38.26	45	103.5	−83.5	1.75	2.43

Table 7.7 **Noise parameters for 90nm 20/(40) × 1μm CMOS inverter at 0.15 (0.075) mA/μm $f_T = 80$GHz.**

f(GHz)	R_n (Ω)	R_{SOPT} (Ω)	X_{SOPT} (Ω)	X_{IN} (Ω)	F_{MIN}	NF_{MIN} (dB)
5	32.99	232.1	449.45	−349.6	1.06	0.26
30	30.76	40.9	74.75	−59.15	1.55	1.90
60	28.96	22.25	37.38	−31	2.1	3.22

Table 7.8 **Noise parameters for 170nm×2.5μm SiGe HBT at 4mA/μm², $f_T = 150$GHz.**

f(GHz)	R_n (Ω)	R_{SOPT} (Ω)	X_{SOPT} (Ω)	X_{IN} (Ω)	F_{MIN}	NF_{MIN} (dB)
65	56	65.26	80	−66	2	3

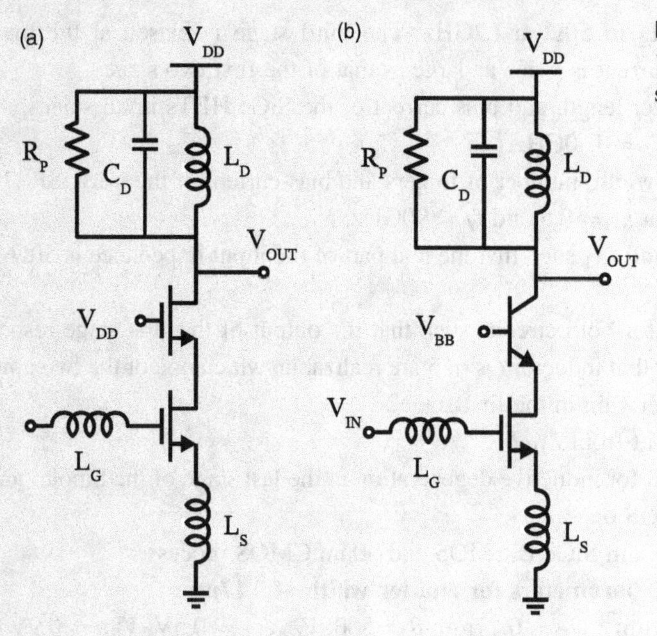

Figure 7.44 (a) 90nm MOSFET, (b) 130nm SiGe BiCMOS cascode LNA schematics

Figure 7.44 shows the schematics of single-stage cascode LNAs implemented in 90nm CMOS and 130nm SiGe BiCMOS technologies.

(a) Using the definition of the current gain as y_{21}/y_{11} and ignoring the series resistances, calculate the f_T of the cascode stages in Figure 7.44(a) and Figure 7.44(b) when all MOSFETs are biased at the optimum noise figure current density and the HBT is biased at the peak f_T current density.

(b) Calculate the transistor size and the bias current such that the real part of the optimum noise impedance is 50Ω at 30GHz.

(c) The following equation holds for the noise factor of the cascode stage as well as for the CS stage

$$F_{MIN} - 1 = k\frac{f}{f_T}.$$

where, in a given, node k has the same value for the cascode and for the CS stages. Depending on the topology, f_T is that of the cascode or of the CS transistor.

To determine the noise parameters of the 130nm n-MOSFET use Table 1 and assume that F_{MIN}-1 scales as $S^{0.5}$ where S is the technology scaling factor (i.e. S = 130nm/ 90nm).

Which one of the LNAs has the lowest bias current? Which one has the lowest NF_{MIN} at 30GHz?

(4) The SiGe HBT LNA is implemented with several single-ended variable gain cascode stages with inductive degeneration, as shown in Figure 7.45. The control voltage can steer the current from one side to the other of the common-base differential pair. All transistors are identical in size. Calculate the emitter length and collector current of all transistors when the

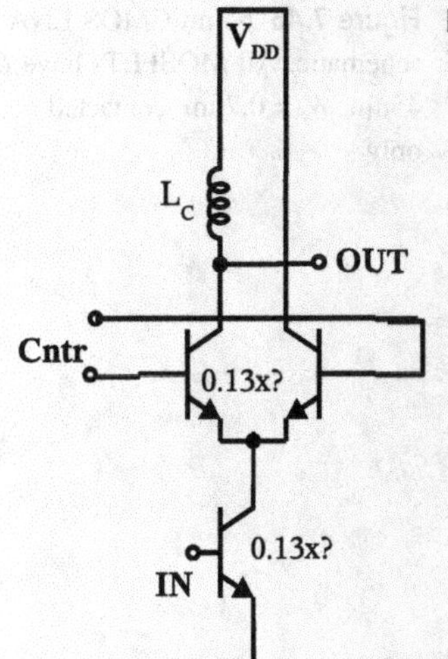

Figure 7.45 Variable gain LNA stage

LNA gain is maximized and the LNA is noise-matched to 50Ω at 80GHz. The optimum noise current density of the SiGe HBT at 80GHz is 8mA/μm^2. Assume that the transistor f_T at J_{OPT} is 200GHz. The common-mode control voltage is 2.2V and V_{DD} is 2.5V.

(d) Calculate the value of the degeneration inductance L_E if the real part of the input impedance of the LNA is 50Ω? Assume that L_E has infinite Q.

(e) The LNA is conjugately matched at the output through a transformer (not shown) at 80GHz to the 150Ω differential input impedance of the mixer. Calculate the maximum power gain of the LNA. What is the LNA power gain when the differential control voltage is 0V? The common-mode control voltage is 2.2V and V_{DD} is 2.5V.

(5) The schematic of a dual-band (8GHz and 12GHz) digitally switched LNA realized in a 65nm CMOS process is shown in Figure 7.46.

(a) Size (N_{f1}, I_{D1}, L_{S1}, L_{G1}) the first branch such that the circuit is noise and input impedance-matched to 50Ω at 8GHz. Account for gate and source resistance. Assume a Q of 20 for all inductors and that $W_3 = W_1$.

(b) Size the second branch (N_{f2}, I_{D2}, L_{S2}, L_{G2}) such that the circuit is noise and input impedance-matched to 50Ω at 12GHz. You may use frequency scaling theory to save time. $W_3 = W_1$.

(c) Find the value of L_{D1} for the 8GHz output branch and L_{D2} for the 12GHz output branch if C_{D1} and C_{D2} are 40fF and 20fF, respectively. (*Hint:* Account for the output capacitance of the transistor.)

(d) Calculate the maximum possible power gain of the LNA at 8GHz and at 12GHz.

(e) Draw a differential schematic version for this LNA and indicate the component values. Use a common-mode inductor (no value required) to improve the common-mode rejection ratio at high frequency. Use varactors for the selectable output capacitance and n-MOSFET switches to select the appropriate output inductor.

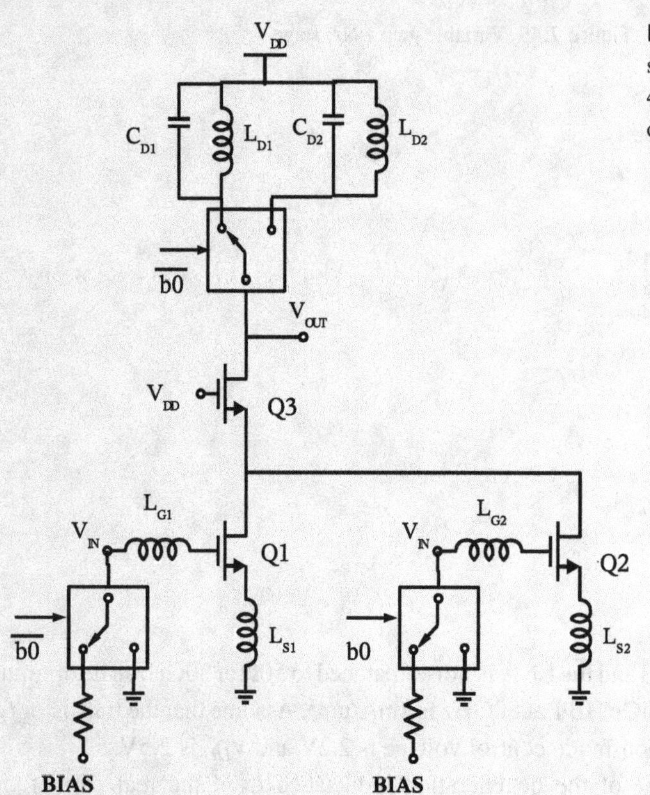

Figure 7.46 65nm CMOS LNA schematic. All MOSFETs have L = 45nm, W_f = 0.7μm, contacted on one side only

(6) Figure 7.47 shows the schematic of a 2.5GHz, differential CMOS cascode LNA fabricated in a 32nm high-k, metal-gate CMOS process. All transistors have minimum gate length (22nm physical) and finger widths of 1μm. The circuit draws 12mA and $V_{DD} = 1.8$V. The differential output load impedance is 100Ω. The measured noise figure is 3.5dB. V_{bgn} and V_{bgp} are DC voltages appropriately biasing the gates of M_{3p}, M_{3n} and of M_{4p} and M_{4n}, respectively. V_{bp}, V_{bn} are bias voltages which can be used for gain control. Assume that C_{cn}, C_{cp} are short circuits at 2.5GHz.

(a) Calculate the effective transconductance, g_{meff}, of the n-MOSFET cascode, of the p-MOSFET cascode and of the CMOS cascode knowing that all transistors are biased at 0.2mA/μm. Account for the impact of R_s. Use the given technology data rather than the plot in Figure 7.48. Note that the differential-mode half circuit consists of an n-MOSFET cascode with inductive degeneration connected in parallel with a p-MOSFET cascode, as illustrated in Figure 7.48.

(b) Calculate the effective f_{Teff} of the CMOS cascode.

(c) Calculate the values of L_{s1n}, L_{s1p} such that the real part of the input impedance of the differential-mode half circuit is 50Ω. Ignore the coupling between L_{gp} and L_{s1p}, L_{s2p}, and between L_{gn} and L_{s1n}, L_{s2n}.

(d) Find the real part of the optimum noise impedance per side. Use $k_1 = 0.5$ in the R_{SOPT} expression or calculate it from P, R, and C. What should be the size and bias current of all transistors such that the optimum noise impedance would be equal to 50Ω per side?

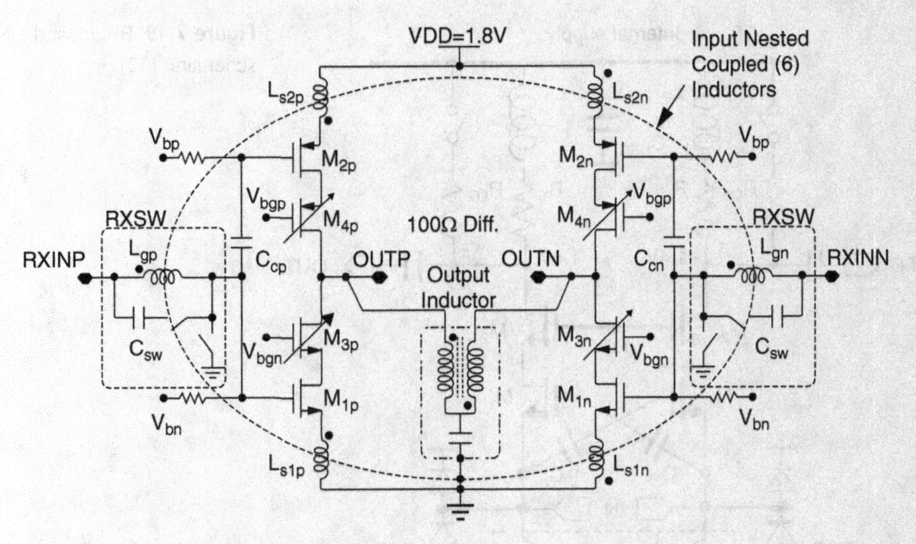

Figure 7.47 Differential LNA schematic [31]

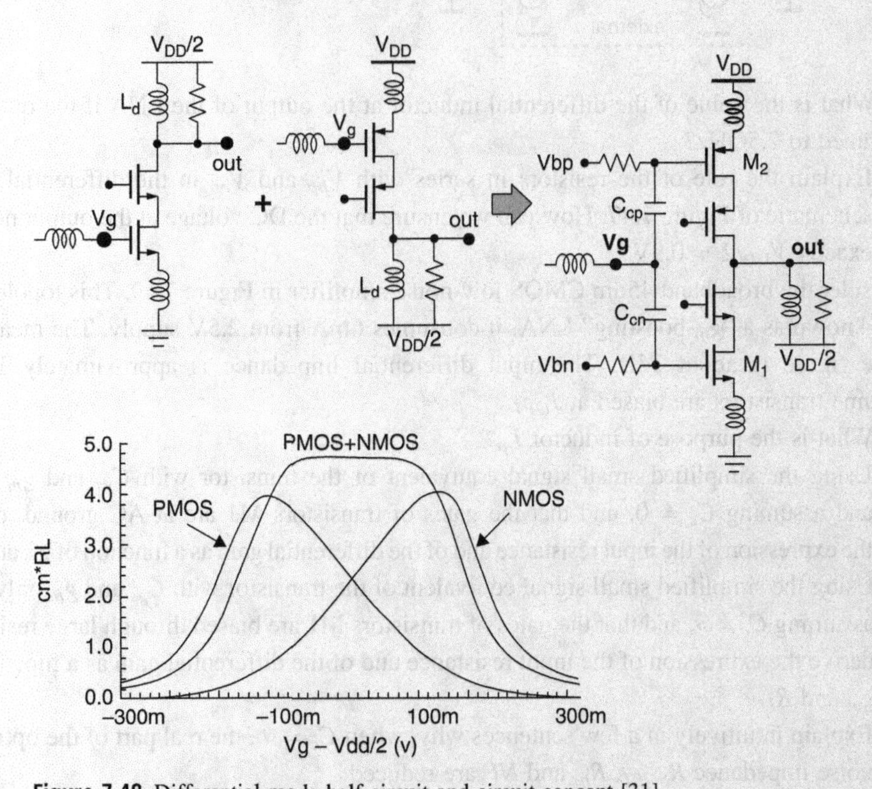

Figure 7.48 Differential mode half-circuit and circuit concept [31]

(e) Find the value of L_{gp}, L_{gpn} such that the imaginary part of the differential input impedance in the differential LNA version in Figure 7.47 is 0Ω at 2.5GHz. Ignore the impact of the input switch.

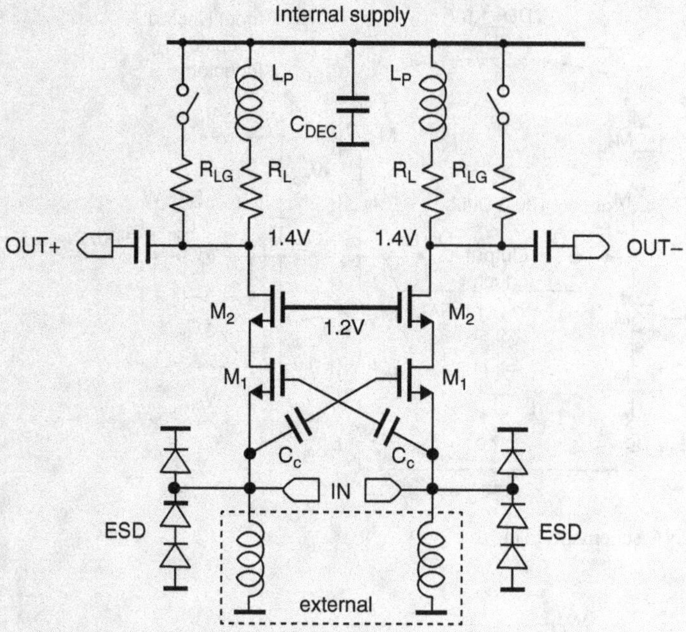

Figure 7.49 Broadband LNA schematic [32]

(f) What is the value of the differential inductor at the output of the LNA if the output is tuned to 2.5GHz?

(g) Explain the role of the resistors in series with V_{bp} and V_{bn} in the differential LNA schematic of Figure 7.47. How can we ensure that the DC voltage at the output node is exactly $V_{DD}/2 = 0.9$V?

(7) Consider the broadband 45nm CMOS low-noise amplifier in Figure 7.49. This topology is also known as a "g_m-boosting" LNA. It consumes 6mA from 2.5V supply. The measured noise figure is about 3dB. The input differential impedance is approximately 100Ω. Assume transistors are biased at J_{OPT}.

(a) What is the purpose of inductor L_p?

(b) Using the simplified small signal equivalent of the transistor with C_{gs} and g_m only, and assuming $C_c = 0$, and that the gates of transistors M1 are at AC ground, derive the expression of the input resistance and of the differential gain as a function of g_m and R_L.

(c) Using the simplified small signal equivalent of the transistor with C_{gs} and g_m only, and assuming $C_c = \infty$ and that the gates of transistors M1 are biased through large resistors, derive the expression of the input resistance and of the differential gain as a function of g_m and R_L.

(d) Explain intuitively in a few sentences why, when $C_c = \infty$, the real part of the optimum noise impedance R_{SOPT}, R_n, and NF are reduced.

(8) Port the 12GHz design of example 7.3 from 90nm to the 32nm CMOS process with high-k metal gates knowing that J_{OPT} changes from 0.15 to 0.3mA/μm.

(9) Draw the schematics of the differential version of the transformer-feedback LNA in Figure 7.33. Elaborate on the number of ports the differential transformer will have. How would you bias the input differential pair using the transformer?

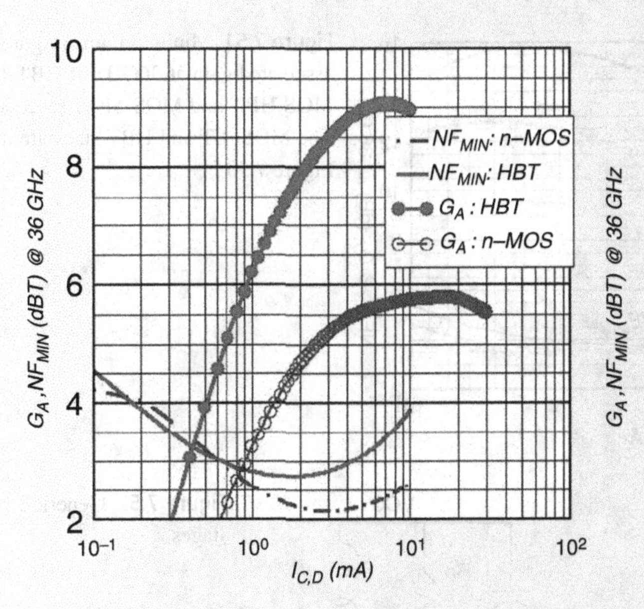

Figure 7.50 Minimum noise figure and associated gain at 36GHz of a 0.2 × 5µm SiGe HBT and of a 12 × 2µm × 0.13µm n-MOSFET

(10) Draw the schematic of the AC-coupled LNA stage in Figure 7.42 where the inductive degeneration is replaced by a shunt–series transformer feedback. Draw a differential schematic of the new circuit.

(11) Redesign the input stage feedback network for the 60GHz SiGe HBT LNA in Figure 7.40 using a shunt–series transformer feedback.

(12) Using the single-stage telescopic cascode CMOS LNA schematic with inductive degeneration, design a scaled LNA stage operating at 500GHz using a 30nm InP HEMT technology with $g'_{meff} = 3\text{mS}/\mu\text{m}$, $g'_{oeff} = 1\text{mS}/\mu\text{m}$, 1µm finger widths, $f_T = 700\text{GHz}$, $C'_{gs} = 0.5\text{fF}/\mu\text{m}$, $C'_{gd} = 0.07\text{fF}/\mu\text{m}$, $R_g = 100\Omega$ per 1µm finger width contacted on one side.

(13) Draw the schematic of a common-gate stage with shunt–series transformer feedback starting from the cascode version in Figure 7.33. Derive the expressions of the input admittance as a function of the MOSFET transconductance and of the transformer parameters: L_P, L_S, M, and k.

(14) Using the information presented in Figure 7.50 for a 130nm SiGe BiCMOS process determine:

(a) The optimum noise current density per unit emitter length and the corresponding associated gain for the SiGe HBT at 36GHz. Will the optimum noise current density be larger or smaller at 20GHz? How about the minimum noise figure?

(b) The optimum noise current density per unit gate width and the corresponding associated gain for the n-MOSFET at 36GHz. Will the optimum noise current density be larger or smaller at 20GHz? How about the minimum noise figure?

(c) You are required to design a 36GHz LNA consisting of a cascade of stages realized using any of the three cascode stages whose characteristics are shown in Figure 7.50:

- Find the combination of stages that results in the lowest overall minimum noise figure and an overall LNA gain larger than 20dB.

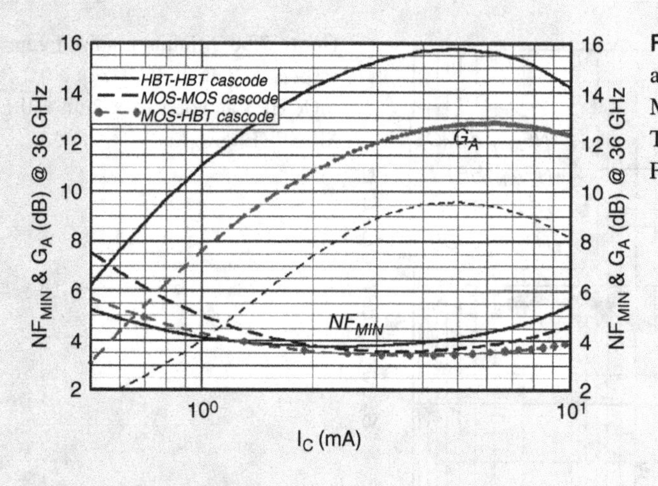

Figure 7.51 Minimum noise figure and associated gain at 36GHz of HBT-HBT, MOS-HBT and MOS-MOS cascodes. The MOSFET and HBT sizes are as in Figure 7.50

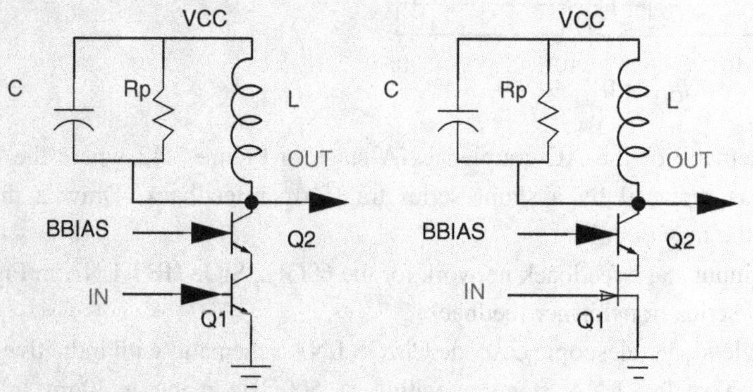

Figure 7.52 Generic LNA stages

- Is this LNA topology also the least sensitive to the signal source impedance mismatch? Support your answer analytically.

(d) Based on Figures 7.51 and 7.52:

- Find the maximum HBT(MOSFET) collector(drain) current swing in the HBT-HBT and MOS-HBT cascodes over which the gain variation remains within 1dB of the peak value? Assume that both cascodes are biased for maximum gain.
- Repeat the calculation for the case when the cascodes are biased at the minimum noise current density.
- For the second case, load the output of each cascode with a *LC*-Rp parallel circuit that resonates at 36GHz. Draw the optimum load line that maintains this linear range and determine the corresponding maximum output voltage swing at the 1dB compression point if $VCC = 2.5$V and $BBIAS = 2$V.
- Which one of the two cascodes has the larger linear input voltage range (defined as the ratio of the collector–drain current swing and the small signal transconductance)?

(e) Draw the location of the input impedance of the HBT-HBT cascode and that of the MOS-HBT cascode from Figure 7.52 on the Impedance Smith Chart at 36GHz. Assume that the cascodes are biased at 6mA. Account for r_π, and R_b, $C\mu$, C_π (in HBTs) and for R_g (gate contacted on one side only with one poly contact per finger), C_{gd}, and C_{gs}. Ignore the emitter resistance and the source resistance:

- What is the quality factor of the input impedance in each case?
- Indicate on the Smith Chart the location of the differential input impedance of a differential circuit based on the HBT-HBT cascode.
- How does the input impedance of the two cascode stages change if the bias current is reduced to 3mA?

REFERENCES

[1] H. T. Friis, "Noise figure of radio receivers," *Proc. IRE*, **32**: 419–422, July 1944.

[2] 2011 ITRS Roadmap. www.itrs.net

[3] S. P. Voinigescu *et al.*, "A scalable high-frequency noise model for bipolar transistors with application to optimal transistor sizing for low-noise amplifier design," *IEEE JSSC*, **32**(9): 1430–1439, September 1997 and S. P. Voinigescu and M. C. Maliepaard on "High-frequency noise and impedance matched integrated circuits" US Patent no. 5789799.

[4] S. P. Voinigescu, T. Chalvatzis, K. H. K. Yau, A. Hazneci, A. Garg, S. Shahramian, T. Yao, M. Gordon, T. O. Dickson, E. Laskin, S. T. Nicolson, and A. C. Carusone, L. Tchoketch-Kebir, O. Yuryevich, G. Ng, B. Lai, and P. Liu, "SiGe BiCMOS for analog, high-speed digital and millimetre-wave applications beyond 50GHz," *IEEE BCTM Digest*, pp. 223–230, October 2006.

[5] T. Yao, M. Gordon, K. Yau, M. T. Yang, and S. P. Voinigescu "60GHz PA and LNA in 90nm RF-CMOS," *IEEE RFIC Symposium Digest*, pp. 147–150, June 2006.

[6] K. H. K. Yau, K. K. W. Tang, P. Schvan, P. Chevalier, B. Sautreuil, and S. P. Voinigescu, "The invariance of the noise impedance in n-MOSFETs across technology nodes and its application to the algorithmic design of tuned low noise amplifiers," *IEEE Si Monolithic Integrated Circuits in RF Systems*, pp. 245–248, Jan. 2007.

[7] C.-H. Jan, M. Agostinelli, H. Deshpande, M. A. El-Tanani, W. Hafez, U. Jalan, L. Janbay, M. Kang, H. Lakdawala, J. Lin, Y.-L. Lu, S. Mudanai, J. Park, A. Rahman, J. Rizk, W.-K. Shin, K. Soumyanath, H. Tashiro, C. Tsai, P. VanDerVoorn, J.-Y. Yeh, and P. Bai, "RF CMOS technology scaling in high-k/metal gate era for RF SoC (system-on-chip) applications," *IEEE IEDM Digest*, pp. 604–607, December 2010.

[8] D. K. Shaffer and T. H. Lee, "A 1.5V 1.5GHz CMOS low noise amplifier," *IEEE JSSC*, **32**(5): 745–759, 1997.

[9] S. T. Nicolson and S. P. Voinigescu, "Methodology for simultaneous noise and impedance matching in W-Band LNAs," *IEEE CSICS Digest*, pp. 279–282, November 2006.

[10] M. Gordon, T. Yao, and S. P. Voinigescu, "65GHz receiver in SiGe BiCMOS using monolithic inductors and transformers," *6th IEEE Topical Meeting on Silicon Monolithic Integrated Circuits in RF Systems Technical Digest*, pp. 265–268, Jan. 2006.

[11] B. Razavi, "A 60GHz direct-conversion CMOS receiver," *IEEE ISSCC Digest*, pp. 400–401, February 2005.

[12] S. T. Nicolson, K. H. K. Yau, S. Pruvost, V. Danelon, P. Chevalier, P. Garcia, A. Chantre, B. Sautreuil, and S. P. Voinigescu, "A low-voltage SiGe BiCMOS 77GHz automotive radar chipset," *IEEE MTT*, **56**: 1092–1104, May 2008.

[13] K. H. K. Yau, P. Chevalier, A. Chantre, and S. P. Voinigescu, "Characterization of the noise parameters of SiGe HBTs in the 70–170 GHz Range," *IEEE MTT*, in print. 2011.

[14] M. Khanpour, K. W. Tang, P. Garcia, and S. P. Voinigescu, "A wideband W-band receiver front-end in 65nm CMOS," *IEEE JSSC*, **43**(8): 1717–1730, August 2008.

[15] A. N. Karanicolas, "A 2–7-V 900MHz CMOS LNA and mixer," *IEEE JSSC*, **31**: 1939–1944, December 1996.

[16] S. B. T. Wang, A. M. Niknejad, and R. W. Brodersen, "A Sub-mW 960MHz ultra-wideband CMOS LNA," *IEEE RFIC Symposium Digest*, pp. 35–38, June 2005.

[17] D. Li and K. Yu, "60GHz LNA radio receiver ICs in 65nm CMOS," 4th year undergraduate project, ECE Dept. University of Toronto, April 2008.

[18] S. T. Nicolson, A. Tomkins, K. W. Tang, A. Cathelin, D. Belot, and S. P. Voinigescu, "A 1.2V, 140GHz receiver with on-die antenna in 65nm CMOS," *IEEE RFIC Symposium Digest*, pp. 239–242, June 2008.

[19] B. Floyd *et al.*, " SiGe bipolar transceiver circuits operating at 60GHz," *IEEE JSSC*, **40**: 156–157, January 2005.

[20] D. J. Allstot, X. Li, and S. Shekhar, "Design considerations for CMOS low-noise amplifiers," *IEEE RFIC Symposium Digest*, pp. 97–100, June 2004.

[21] F. Bruccoleri, E. A. M. Klumperink, and B. Nauta, "Wide-band CMOS low-noise amplifier exploting thermal noise canceling," *IEEE JSSC*, **39**: 275–282, February 2004.

[22] J. R. Long and M. A. Copeland, "A 1.9GHz low-voltage silicon bipolar receiver front-end for wireless personal communication systems," *IEEE JSSC*, **30**: 1438–1448, December 1995.

[23] D. Cassan and J. R. Long, "A 1-V transformer-feedback low-noise amplifier for 5GHz wireless LAN in 0.18-μm CMOS," *IEEE JSSC*, **38**: 427–435, March 2003.

[24] M. T. Reiha, J. R. Long, and J. J. Pekarik, "A 1.2V Reactive-feedback 3.1–10.6 GHz ultrawide-band low-noise amplifier in 0.13μm CMOS," *IEEE RFIC Symposium Digest*, pp. 55–58, 2006.

[25] A. Tomkins, P. Garcia, and S. P. Voinigescu, "A passive W-band imager in 65nm bulk CMOS," *IEEE JSSC*, **45**(10): 1981–1991, October 2010.

[26] A. Tomkins, R. A. Aroca, T. Yamamoto, S. T. Nicolson, Y. Doi, and S. P. Voinigescu, "A zero-IF 60GHz 65nm CMOS transceiver with direct BPSK modulation demonstrating up to 6Gb/s data rates over a 2m wireless link," *IEEE JSSC*, **44**(8): 2085–2099, August 2009.

[27] E. Laskin, A. Tomkins, A. Balteanu, I. Sarkas, and S. P. Voinigescu, "A 60GHz RF IQ DAC transceiver with on-die at-speed loopback," *IEEE RFIC Symposium Digest*, Baltimore, June 2011.

[28] S. P. Voinigescu and M. C. Maliepaard, "5.8GHz and 12.6GHz Si Bipolar MMICs," *IEEE ISSCC Digest*, pp. 372–373, 1997.

[29] M. Gordon and S. P. Voinigescu, "An inductor-based 52GHz 0.18μm SiGe HBT cascode LNA with 22dB gain," *European Solid-State Circuits Conference*, pp. 287–290, September 2004.

[30] V. Subramanian, A. Mercha, B. Parvias, M. Dehan, G. Groeseneken, W. Sansen, and S. Decoutere, "Identifying the bottlenecks to the RF performance of FinFETs," 23rd *International Conference on VLSI Design*, 2010.

[31] C-T Fu, H. Lakdawala, S. S. Taylor, and K. Soumyanath, "A 2.5GHz 32nm 0.35mm^2 3.5dB NF −5dBm P1dB fully differential CMOS push-pull LNA with integrated T/R switch and ESD protection," *IEEE ISSCC Digest*, pp. 56–57, February 2011.

[32] J. Borremans, G. Mandal, V. Giannini, T. Sano, M. Ingels, B. Verbruggen, and J. Craninck, "A 40nm CMOS highly linear 0.4-to-6 GHz receiver resilient to 0dBm out-of-band blockers," *IEEE ISSCC Digest*, pp. 62–63, February 2011.

8 Broadband low-noise and transimpedance amplifiers

In the previous chapter, we have examined the design of front-end low-noise amplifiers for narrowband wireless receivers. The purpose of these LNAs is to amplify a small input signal level while adding as little noise as possible such that the receiver sensitivity is as high as possible. But what about broadband applications – are low-noise front-ends required in high-speed digital receivers? In this chapter, we will build on the small signal properties of broadband amplifiers introduced in Chapter 5 and on the noise analysis and low-noise design techniques developed in Chapters 3 and 7, respectively, to learn how to analyze and design very broadband, DC-coupled, low-noise amplifiers. Along with high-speed logic, discussed in Chapter 11, and broadband large swing output drivers, covered in Chapter 12, these amplifiers form the back-bone of all fiber-optics, backplane, and other wireline communications systems. We will discuss first the specifications and requirements of low-noise broadband amplifiers in high-speed digital receivers. Next we will explore the design of low-noise transimpedance amplifiers (TIAs) commonly found in optical receivers. Finally, design methodologies for other LNA topologies used in broadband electrical links will also be discussed. In all cases, representative design examples in InP HBT, SiGe BiCMOS and nanoscale bulk and SOI CMOS technologies will be provided.

8.1 LOW-NOISE BROADBAND HIGH-SPEED DIGITAL RECEIVERS

Consider the simplified receiver of Figure 8.1, where a noisy input amplifier is directly followed by a decision circuit (a D-type flip-flop) that determines if the received bit is a logical 1 or 0. In situations where small input amplitudes are received, noise added by the input amplifier could cause the decision circuit to make an incorrect decision. Ultimately, this leads to a poor bit error rate (BER) for the broadband receiver. This simple example illustrates the importance of low-noise designs even in high-speed digital receivers.

The main applications for broadband low-noise amplifiers are in:

(a) **optical receivers**, where the signal source is a photodiode and where the design goal is to amplify the electrical current produced by the photodiode and to convert it to a voltage level suitable for a decision circuit (hence the name transimpedance amplifier or TIA) while minimizing the noise of the TIA-photodiode ensemble, maximizing the bandwidth, and minimizing the power consumption; and

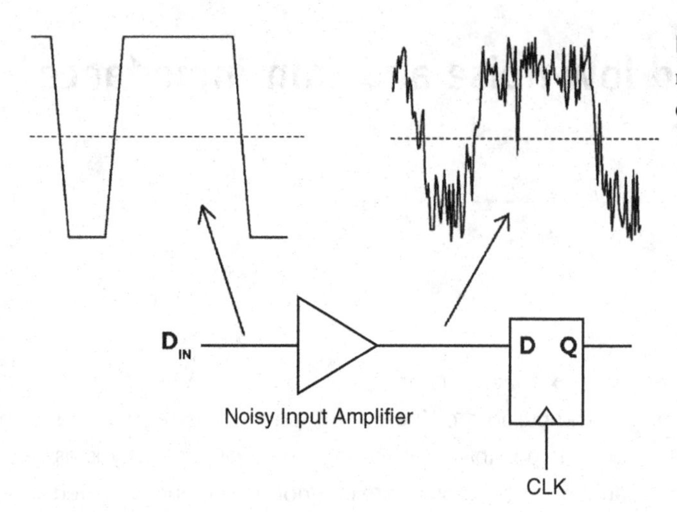

Figure 8.1 Impact of input amplifier noise on bit error rate in a broadband digital receiver

(b) low-noise broadband input comparators in **backplane communications** and **high-speed ADCs**.

In the first application, a suitable figure of merit that captures most of the design goals for the broadband low-noise amplifier is given by

$$FoM = \frac{Z_T \times I_{MAX} \times BW_{3\mathrm{dB}}}{i_{neq}^{rms} \times P_{DC}} \tag{8.1}$$

In the second, the signal source typically has 50Ω impedance and the goal is to amplify the input signal (typically a voltage) while minimizing the input integrated noise voltage or noise current of the amplifier over the bandwidth of the system. A suitable FoM that groups these design goals is

$$FoM = \frac{A_V \times V_{iMAX} \times BW_{3\mathrm{dB}}}{v_{neq}^{rms} \times P_{DC}} \tag{8.2}$$

where Z_T is the transimpedance gain of the TIA following the photodiode, I_{MAX} is the maximum allowed input current of the TIA, A_V is the voltage gain of the receive amplifier, V_{iMAX} is the maximum linear voltage swing at the input of the backplane receiver or ADC, $BW_{3\mathrm{dB}}$ is the 3dB bandwidth of the receiver, $i_{neq}{}^{rms}$ and $v_{neq}{}^{rms}$ are the input equivalent integrated noise current and voltage, respectively, and P_{DC} is the power consumption of the receive amplifier.

As discussed in Chapter 7, the biggest challenge is how to minimize the noise contributed by the amplifier over an extreme broad bandwidth, e.g. DC to 36GHz, as required for a 43Gb/s fiber-optic communication system that forms the backbone of the internet.

Problem: The real parts of the input impedance and of the optimum noise impedance of a transistor are different and have different frequency dependencies.

Consequence: We cannot match both the input and the noise impedance simultaneously over a wide bandwidth with purely reactive elements and, therefore, the noise figure of the

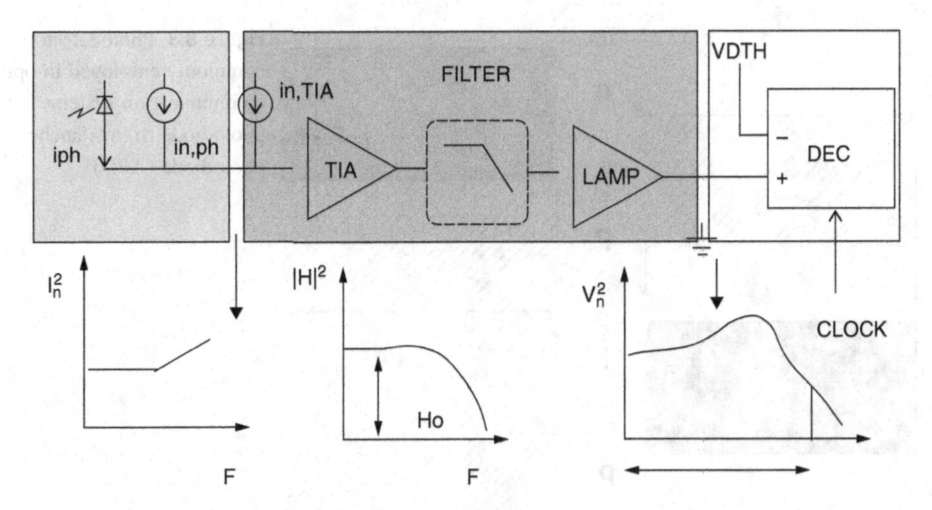

Figure 8.2 Optical receiver architecture and model [1]

amplifier will be degraded compared to the minimum noise figure of the transistor, at least in some range of frequencies.

Solution

Use lossy feedback to accomplish both input and noise impedance matching with minimal degradation of receiver sensitivity. As discussed in Chapter 3, of all the feedback topologies, those with shunt connection at the input are best because they minimize the transistor size and power consumption needed for a certain optimum noise impedance value. In turn, smaller transistor sizes imply smaller capacitance at the input node and, hence, broader bandwidth.

Broadband LNAs may have similar design criteria to their narrowband counterparts, but the circuit topologies are certainly different. Tuned input or output networks obviously cannot be employed in a broadband amplifier, which points to the need for resistive or active loads. This becomes problematic for LNA design because, while ideal inductors and capacitors do not generate noise, resistors and transistors most certainly do! Designing a broadband LNA thus becomes an exercise in minimizing the overall noise from several noise contributors. At the same time, other considerations such as dynamic range or power dissipation must also be kept in mind.

As illustrated in Figure 8.2. in optical communication systems, a photodetector is required at the interface between the optical fiber and the receiver. Typically, a photodiode is employed to generate electrical current from incident photons. Recall that if a reverse-biased diode is illuminated with light of an appropriate wavelength, the photons absorbed in the depletion region generate electron–hole pairs which result in a current flow through the diode. As described in Chapter 3, the electrical current generated by the photodiode, I_{ph}, is related to the incident optical power, P, through the photodetector responsivity, R, as

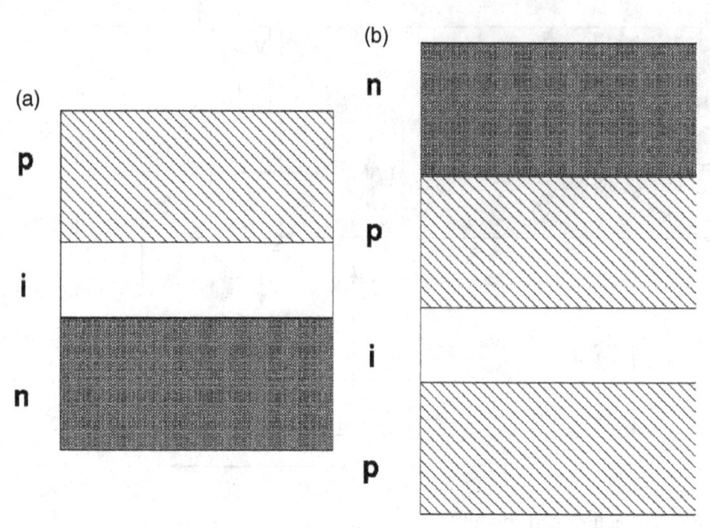

Figure 8.3 Photodetectors commonly employed in optical communication systems (a) p-i-n photodiode (b) avalanche photodiode (APD)

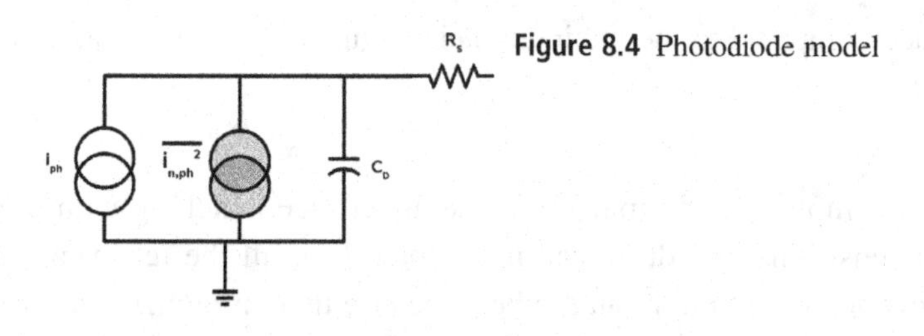

Figure 8.4 Photodiode model

$$I_{ph} = RP. \tag{8.3}$$

Note that the current rises linearly with the incident optical power.

The most common photodetector encountered in fiber-optics receivers is the p-i-n photodiode. As shown in Figure 8.3(a), a thin intrinsic semiconductor layer is inserted between the p and n regions of the diode. This intrinsic layer effectively increases the depletion width of the diode, where most of the electron–hole pairs are generated. The wider the intrinsic region, the greater the number of electron–hole pairs generated, and the greater the responsivity of the detector. However, a wide intrinsic region also increases the travel time of the optically generated carriers to the terminals of the diode, slowing down its frequency response, and reducing its bandwidth. A a result, a responsivity-bandwidth tradeoff exists in all photodiodes.

Higher responsivity can be achieved with the avalanche photodiode (APD) of Figure 8.3(b). A high electric field in the added *p* region of the APD results in avalanche multiplication of the carriers generated in the depletion region. Both photodiodes can be modeled by the small signal electrical circuit shown in Figure 8.4. This simple model consists of the photodiode current, depletion capacitance C_D of the reversed-biased diode, and a series resistance R_S. For a 56Gb/s photodiode, the typical values are 30fF and 10Ω, for C_D and R_S, respectively.

Additionally, the noise generated by the diode must also be considered. This is dominated by the diode shot noise, given by (3.67), which, for a p-i-n photodiode, ignoring the dark current, simplifies to

$$\overline{i_{n,PIN}^2} = 2q \cdot I_{ph} \cdot \Delta f. \tag{8.4}$$

The shot noise is proportional to the receiver bandwidth. Shot noise is appreciably higher in APDs due to the avalanche multiplication factor, M

$$\overline{i_{n,APD}^2} = 2q \cdot I_{ph} \cdot \Delta f \cdot F \cdot M^2. \tag{8.5}$$

Here, F is the excess noise factor of the APD. Notice that the shot noise grows as the square of the avalanche multiplication factor, while the photodiode current only increases proportionally with M.

Now that the optical signal has been converted into a current, a transimpedance amplifier (or TIA) is needed to generate a voltage from this current. Before examining transistor-level TIA implementations, we shall first introduce some common TIA design specifications.

8.2 TRANSIMPEDANCE AMPLIFIER SPECIFICATION

8.2.1 Transimpedance gain

The TIA produces an output voltage in response to a change in the input current generated by the photodiode. The small signal transimpedance gain, Z_T, relates these two quantities

$$Z_T = \frac{v_o}{i_{ph}} \tag{8.6}$$

and has units of Ω. However, it is common to express the gain in decibels. In this case, the gain is normalized with respect to 1Ω and has units of dBΩ $= 20\log(|Z_T|/1\Omega)$.

8.2.2 Bandwidth and group delay

As with any broadband amplifier, the TIA exhibits a low-pass response. Therefore, the TIA must have adequate bandwidth to operate at a given data rate. At a minimum, the bandwidth must be at least equal to the Nyquist bandwidth of the incoming data. However, this value is often too low in an optical receiver. Typically, there are additional amplification stages between the TIA and the decision circuit. The bandwidth of interest for the receiver is that of the entire chain leading to the decision circuit. To prevent limiting the overall receiver bandwidth and causing intersymbol interference, ISI, the TIA bandwidth could be 60–100% of the bit rate frequency depending on the receiver architecture. This translates to a bandwidth of between 25.8GHz and 43GHz for a 43Gb/s OC-768 optical receiver, or 33.6GHz to 56GHz for a

112Gb/s Etherenet system based on serial 56Gb/s data streams. The 3dB bandwidth of the TIA is specified as the frequency at which the magnitude of $Z_T(j\omega)$ becomes equal to $\sqrt{2}/_2$ of its DC value.

To avoid data-dependent jitter at the output of the TIA, the variation in the group delay should be minimized. The group delay, τ, of the TIA is defined as

$$\tau = -\frac{\partial \Phi}{\partial \omega} \tag{8.7}$$

where Φ is the phase of $Z_T(\omega)$. To reduce deterministic jitter, it is desirable that the group delay not vary by more than 10% of a unit interval, corresponding to about 2.3ps and 1.78ps, respectively, for 43Gb/s and 56Gb/s TIAs.

8.2.3 Noise

As with a tuned low-noise amplifier for radio applications, the noise of a TIA is the most critical design parameter. This parameter has important system level ramifications, particularly in the link budget. Lower noise translates into better receiver sensitivity, which allows for a reduction in transmit power. Alternatively for a given transmit power, improved sensitivity means that data can be transmitted over longer distances. Usually the noise of a TIA is specified as an input-referred noise current spectrum, $i_{n,TIA}(f)$ in pA/√Hz or as $i^2_{n,TIA}(f)$ in pA^2/Hz. Since this noise current is frequency dependent, the total noise is often reported as a single, input-referred rms value that can be obtained by integrating the noise spectrum at the output of the TIA and dividing by the midband transimpedance gain

$$i^{rms}_{n,TIA} = \frac{1}{|R_T|}\sqrt{\int_0^{\infty} |Z_T(f)|^2 \cdot i^2_{n,TIA}(f) df} \tag{8.8}$$

where R_T is the midband transimpedance gain [1].

Since at very high frequencies the gain decreases to 0, the noise is typically integrated from DC up to twice the TIA bandwidth. It is important to note that too much bandwidth can be detrimental for the receiver sensitivity.

How does the noise current relate to the TIA sensitivity? Recall from Chapter 2 that the bit error rate is a function of the eye diagram quality factor. For example, a bit error rate of 10^{-12} corresponds to an eye Q of approximately 7. At the output of the TIA, the ratio of the eye amplitude and the total rms noise on the logic high and logic low levels gives the eye Q. Thus the minimum peak-to-peak output voltage (v_o^{pp}) required to achieve a particular Q or bit error rate is

$$v_o^{pp} = 2Qv^{rms}_{n,TIA}. \tag{8.9}$$

Dividing both sides by the transimpedance gain relates the input referred rms noise current to the minimum detectable peak-to-peak photodetector current

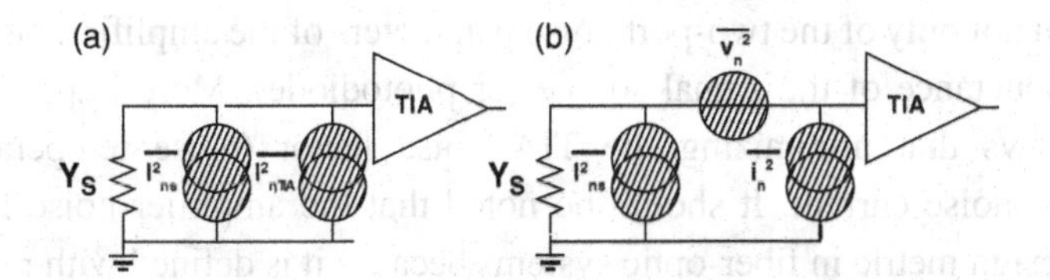

Figure 8.5 TIA noise models: (a) single input-referred noise current and (b) conventional two-port representation with correlated input referred noise voltage and current. The signal source noise current is described by its equivalent noise conductance, G_{seq}, which may be different than $Re(Y_S) = G_S$

$$i_{ph}^{pp} = 2Qi_{n,\,TIA}^{rms}. \tag{8.10}$$

Typical sensitivities are 20μA^{pp} for a 10Gb/s TIA, or 80μA^{pp} for a 40Gb/s TIA.

The input-referred noise current is a convenient quantity for estimating the TIA sensitivity. As discussed in Chapter 3, the noise of a two-port network is fully captured through two correlated noise sources – a noise current $\overline{i_n^2}$ and a noise voltage $\overline{v_n^2}$. It is important to realize that $\overline{i_{n,TIA}^2}$ *is not equal to* $\overline{i_n^2}$.

Consider the two representations of the "noisy" TIA in Figure 8.5. By equating the input short-circuit currents in the two circuits, it can be shown that the input referred noise current must satisfy

$$i_{n,\,TIA}(f) = v_n Y_S + i_n. \tag{8.11}$$

If the two-port noise voltage and noise current sources are correlated as $i_n = i_u + v_n Y_{COR}$, then we can recast (8.11) as

$$\overline{i_{n,\,TIA}^2}(f) = \overline{v_n^2}|Y_S + Y_{COR}|^2 + \overline{i_u^2}. \tag{8.12}$$

From the definition of the noise factor of the two-port network

$$F = \frac{\overline{i_{n,\,TIA}^2} + \overline{i_{ns}^2}}{\overline{i_{ns}^2}} = 1 + \frac{\overline{i_{n,\,TIA}^2}}{\overline{i_{ns}^2}} \tag{8.13}$$

which allows us to link $i_{n,TIA}$ to the noise factor of the TIA

$$\frac{\overline{i_{n,\,TIA}^2}(f)}{4kT\Delta f} = (F-1)\cdot G_{seq} \tag{8.14}$$

where $\overline{i_{ns}}^2 = 4kT\Delta f G_{seq}$ and G_{seq} is the noise conductive of the signal source.

The expression of the noise factor can be further rearranged to highlight its dependence on the noise parameters of the TIA and on the signal source admittance and noise conductance

$$F = 1 + \frac{\overline{i_{n,\,TIA}^2}}{\overline{i_{ns}^2}} = \frac{\overline{v_n}^2}{\overline{i_{ns}^2}}|Y_S + Y_{COR}|^2 + \frac{\overline{i_u}^2}{\overline{i_{ns}^2}} = \frac{R_n}{G_{seq}}|Y_S + Y_{COR}|^2 + \frac{G_u}{G_{seq}}. \tag{8.15}$$

Hence, $i^2_{n,TIA}(f)$ is a function not only of the two-port noise parameters of the amplifier, but also of the equivalent noise conductance of the signal source (or photodiode). More importantly, equation (8.14) clearly shows that minimizing the TIA noise factor (noise temperature) minimizes the input-referred noise current. It should be noted that the amplifier noise factor is not often used as a TIA design metric in fiber-optic systems because it is defined with respect to a non-standard photodiode admittance and noise equivalent conductance, and not with respect to a standard 50Ω signal source. Moreover, the photodiode noise conductance is not equal to its small signal conductance. Nevertheless, as we will see later in the chapter, minimizing this noise factor at a particular frequency with respect to the transistor bias current density and size is equivalent to minimizing the equivalent input noise current of the TIA.

8.2.4 Linearity and overload

Lowering the TIA noise ensures that the receiver can properly detect small input levels, as is the case if the input signal comes from a long stretch of fiber or coaxial cable. But what if the TIA is instead connected to the transmitter via a short cable? In such a case, the input current could be large, and the TIA must be able to operate under these conditions as well. The ability to handle large input signals is described by one of two quantities. The first is the maximum linear input current, which is analogous to the 1dB compression point often defined for wireless components. The maximum linear input current is defined as the input current for which the transimpedance gain is reduced by 1dB from its linear value. This value is particularly important when the TIA is followed by an equalizer whose input must be linear.

Linear operation may not be required for some applications. This is true when the TIA is directly followed by a limiting amplifier. In these cases, the input overload current, which is larger than the maximum linear input current, is a more relevant measure of the maximum input signal. The overload current is defined as the maximum current beyond which the TIA produces serious duty cycle distortion or jitter, causing a degradation of the bit error rate BER from the specified nominal value.

8.3 TRANSIMPEDANCE AMPLIFIER DESIGN

8.3.1 TIA topologies

The design of a broadband low-noise transimpedance amplifier relies on the open-circuit-time-constant technique, OCTC, to determine the 3dB bandwidth [2], and on the linear analysis of noisy networks introduced in Chapter 3 to identify the optimal bias conditions and transistor size that minimize the overall noise added by the amplifier. Before developing a design methodology, we will first analyze several commonly encountered TIA topologies.

As a starting point, we can implement a transimpedance "amplifier" by placing a resistor in series with the photodiode as seen in Figure 8.6. The output voltage is equal to the photodiode current times the resistance R.

$$v_o = i_{\text{ph}} \times R. \tag{8.16}$$

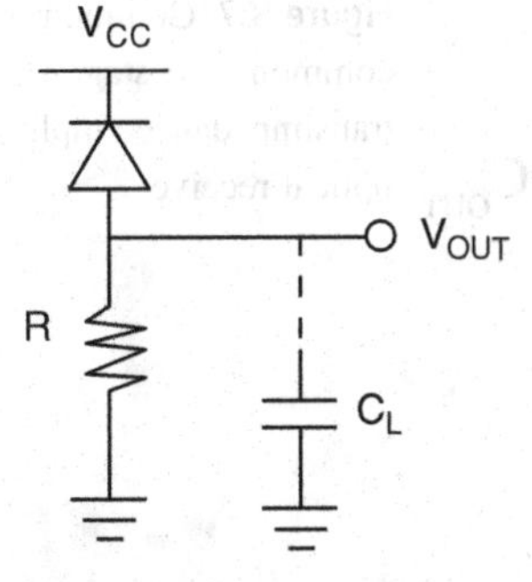

Figure 8.6 Simple resistive front-end used as a transimpedance amplifier

In this simple circuit, the transimpedance gain is equal to the resistance. If we want to have high gain, it appears that using a large resistor is a good idea. But what about the thermal noise from that resistor? The noise current from the resistor over a given bandwidth, BW, is

$$\overline{i_{n,R}^2} = \frac{4kT \times BW}{R}. \tag{8.17}$$

Using a large resistor as the front-end for our optical receiver looks promising since it reduces the noise current. Moreover, the linearity of this front-end is determined only by the supply voltage. This seems too easy! Of course there has to be some tradeoff, and unfortunately it is the bandwidth of the front-end. If we consider that the photodiode/resistor front-ended is loaded by the input capacitance of the subsequent stage, then the 3dB bandwidth of our simple front-end becomes

$$BW_{3\text{dB}} = \frac{1}{2\pi R(C_D + C_L)}. \tag{8.18}$$

We quickly see why using only a resistor becomes a problem. In fact, there is little we can do in the design of such a front-end. If C_D and C_L are given, then R must be chosen to meet the bandwidth specification and the designer must therefore accept whatever gain and noise the front-end generates. Surely there must be a better solution.

What if we buffered the current with a low-impedance current buffer before we collect it on the load resistor R? This allows us to separate the large load resistance and the load capacitance from the input capacitance, and to reduce the input node time constant because of the low impedance of the current buffer. Because of its extremely low input impedance, the common-base stage is an ideal current buffer. Another possibility is a common-gate stage, which has a somewhat higher input impedance. Both circuits are illustrated in Figure 8.7 and have already been analyzed in Chapter 7 for narrow band LNA applications. In either case, the transimpedance gain is R, the same as in Figure 8.6, and the 3dB bandwidth is now reduced to

$$BW_{3\text{dB}} \approx \frac{1}{2\pi\left(\frac{C_D+C_{\text{IN}}}{g_m} + R{\cdot}C_L + R{\cdot}C_{OUT}\right)} \tag{8.19}$$

which is larger than that of (8.18) because $1/g_m$ is significantly smaller than R, and C_{IN} and C_{OUT} can be designed to be only a small fraction of C_D and C_L, respectively.

In addition to the common-gate/common-base transistor and the load resistor, for broadband applications, a current source is needed to provide the bias current, I_B. Since the current source is connected directly at the input, it will contribute to the input capacitance and, more

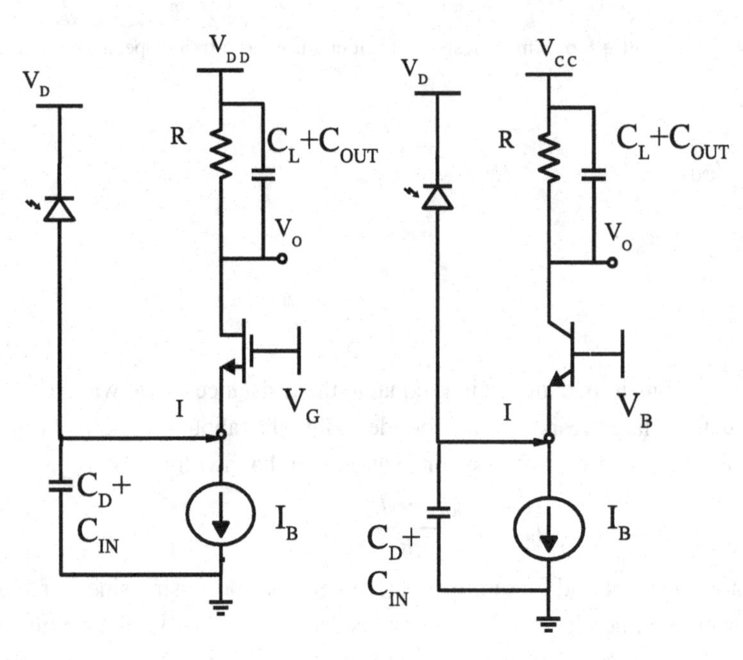

Figure 8.7 Common-gate and common-base stages as transimpedance amplifiers in optical receivers

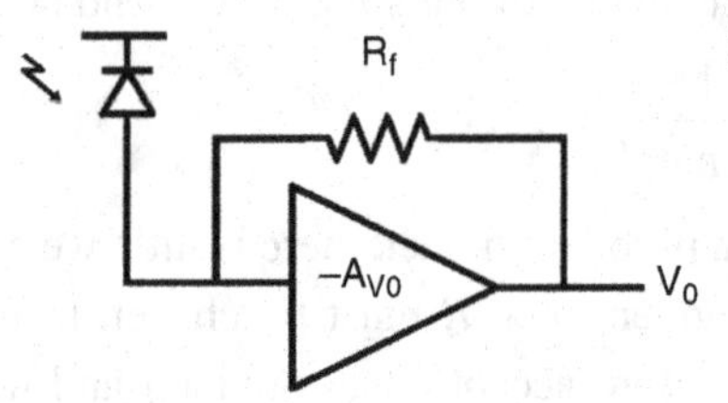

Figure 8.8 Shunt–shunt feedback TIA configuration

importantly, it adds noise. This is unfortunate, because, as we have already seen in Chapter 7, the noise performance of the common-base/common-gate stage is not as good as that of the common-emitter or common-source stages. Nevertheless, common-base and common-source amplifiers have been successfully developed even for 40Gb/s fiber-optic receivers [3].

Instead, let us consider the use of a shunt–shunt feedback amplifier as our TIA. As seen in Figure 8.8, this configuration consists of a feedback resistor R_f placed around an amplifier with a voltage gain of $-A_V$. We have already had a brief encounter with this circuit as part of the Cherry–Hooper stage analyzed in Chapter 5. The low-frequency transimpedance gain of the amplifier can be written as

$$Z_T \stackrel{def}{=} \frac{v_0}{i_{ph}} = -R_f \left[\frac{A_{V0}}{A_{V0}+1} \right]. \tag{8.20}$$

If A_V is much larger than unity, the transimpedance gain of the shunt–shunt feedback amplifier is set by the feedback resistor. To calculate the bandwidth of the amplifier, we start with the simplifying assumption that the forward amplifier has infinite bandwidth. In this hypothetical scenario, the frequency response is determined by the input time constant set by the input resistance R_{IN} of the TIA along with the total capacitance at this node. R_{IN} can be expressed in terms of the voltage gain and the feedback resistor

$$R_{IN} = \frac{R_f}{A_{VO}+1}. \tag{8.21}$$

The capacitance at the input node is the sum of the photodiode capacitance and of the input capacitance of the TIA. For completeness, any bond-pad capacitance would also contribute to this sum. The 3dB bandwidth of the TIA with ideal amplifier is thus given by

$$BW = \frac{1}{2\pi R_{IN}(C_D + C_{IN})} = \frac{A_{VO}+1}{2\pi R_f(C_D + C_{IN})}. \tag{8.22}$$

Now, compare this result to that of the simple resistor front-end from Figure 8.6. To achieve the same transimpedance gain, R_f must equal R times $(A_V+1)/A_V$. In this case, we see that the bandwidth of the transimpedance amplifier is higher by a factor equal to the voltage gain A_V. This is an important benefit of using a TIA in an optical receiver. Note that, if $\frac{A_{VO}+1}{R_f} > g_m$, the bandwidth is also comparable or better than that of the common-base or common-gate amplifiers in Figure 8.7.

Now let us revisit an assumption we made to simplify the analysis regarding the bandwidth of the voltage amplifier. Surely it is not plausible for the amplifier to have infinite bandwidth! One might think that this simplification could be responsible for our assertion that the TIA has higher bandwidth than the resistive front-end. A more realistic scenario is as follows. The TIA feedback resistor and the capacitance at the input node form a pole ω_1

$$\omega_1 = \frac{1}{R_f(C_D + C_{IN})}. \tag{8.23}$$

Additionally, the voltage amplifier has a single dominant pole at ω_2, such that its open-loop response can be expressed as

$$A_V(s) = \frac{-A_{VO}}{1+\frac{s}{\omega_2}}. \tag{8.24}$$

The closed-loop response of this TIA now becomes a second-order system with complex poles. The transfer function of the TIA becomes

$$Z_T(s) = \frac{-R_f\left(\frac{A_{VO}}{1+A_{VO}}\right)}{\frac{s^2}{\omega_1\omega_2(1+A_{VO})} + s\left(\frac{\omega_1+\omega_2}{\omega_1\omega_2(1+A_{VO})}\right) + 1}. \tag{8.25}$$

To better interpret this expression, let us rewrite this transfer function as that of a standard second-order system

$$Z_T(s) = \frac{Z_{TO}}{\frac{s^2}{\omega_0^2} + \frac{2\zeta}{\omega_0}s + 1} = \frac{Z_{TO}}{\frac{s^2}{\omega_0^2} + \frac{s}{Q\omega_0} + 1}. \tag{8.26}$$

Z_{T0} is the DC transimpedance gain, ω_0 is the resonant frequency of the system, and $\zeta = \frac{1}{2Q}$ is the damping factor. Each of these factors can be found by equating the terms in equation (8.25) with those in (8.26)

$$Z_{T0} = -R_f \left[\frac{A_{V0}}{A_{V0}+1} \right]. \tag{8.27}$$

$$\omega_0 = \sqrt{(A_{V0}+1)\omega_1\omega_2} = \sqrt{\frac{(A_{V0}+1)\omega_2}{R_f(C_D+C_{IN})}}. \tag{8.28}$$

$$\zeta = \frac{\omega_1+\omega_2}{2\sqrt{\omega_1\omega_2(1+A_{V0})}} \quad Q = \frac{\sqrt{\omega_1\omega_2(1+A_{V0})}}{\omega_1+\omega_2}. \tag{8.29}$$

The damping factor is a critical parameter governing the response of the second-order system. From a designer's perspective, ζ dictates the spacing of the two open-loop poles ω_1 and ω_2. In general, the system should not be underdamped ($\zeta < \sqrt{2}/2$) in order to avoid excessive ringing in the step response. Additionally, there are two interesting responses which should be considered. The first is the Butterworth response where the frequency response exhibits no peaking, and is met for $\zeta = \sqrt{2}/2$. This response is sometimes called the "maximally flat" frequency response. However, designing for a Butterworth response does result in slight ringing in the step response. Instead, we can choose a Bessel response ($\zeta = \sqrt{3}/2$) which minimizes the variation in the group delay to avoid such ringing. In either case, the 3dB bandwidth can be found by setting the magnitude of equation (8.25) equal to $\frac{\sqrt{2}}{2} Z_{T0}$.

Table 8.1 summarizes the relationship between the open-loop poles, as well as the resulting 3dB bandwidth for the Butterworth and Bessel responses.

Surprisingly, the resulting 3dB bandwidth in both cases is higher than in the situation where the bandwidth of the voltage amplifier was neglected.

EXAMPLE 8.1

As an IC design engineer for a company developing OC-768 optical receivers, you have been asked to design a TIA which operates at data rates up to 43Gb/s. The 3dB bandwidth of the amplifier should be 80% of the maximum bit rate, corresponding to 34.4GHz for the 43Gb/s TIA. It is known that the photodiode suitable for this data rate has a capacitance of approximately 60fF, while the bond pad contributes an additional 30–50fF. You have been asked for an estimate of the performance one can achieve from such a TIA in InP or SiGe bipolar technology, as well as in a CMOS technology. It is known that the bipolar technology can yield a forward amplifier with a voltage gain of $|A_{V0}| \approx 6$, while the gain in a CMOS technology is expected to be only $|A_{V0}| \approx 2-3$. If a Bessel response with minimum group delay variation is desired, determine the maximum transimpedance gain one can expect from such an amplifier, as well as the required 3dB bandwidth of the forward amplifier. Assume that the photodiode and bond pad capacitance dominate the total capacitance at the input of the TIA. (*Note*: in reality the HBT/FET input capacitance is not negligible and should be considered.) How do these specifications change if a Butterworth response is desired?

Table 8.1 **Characteristics of second-order TIAs.**

Response	Butterworth	Bessel
ω_2/ω_1	$2\|A_{VO}\|$	$3\|A_{VO}\|$
3dB Bandwidth	$\sqrt{2}\|A_{VO}\|\omega_1$	$1.36\|A_{VO}\|\omega_1$

Solution

Let us begin with the bipolar TIA. The 3dB bandwidth of the TIA is a function of the feedback resistor, input capacitance, and the voltage gain of the forward amplifier. From Table 8.1, this bandwidth becomes

$$\omega_{3dB} = 2\pi \times 34.4\text{GHz} = \frac{1.36|A_{VO}|}{R_f C} = \frac{1.36 \times 6}{R_f(50\text{fF} + 60\text{fF})}$$

which results in a feedback resistance of 343Ω. The corresponding transimpedance gain is

$$|Z_{TO}| = R_f \left[\frac{|A_{VO}|}{|A_{VO}| + 1}\right] = 343\Omega \times \frac{6}{7} = 294\Omega.$$

To achieve the Bessel response, the forward amplifier must have a 3dB bandwidth ($\omega_2 = 2\pi f_2$) of

$$f_2 \approx 3A_{VO} f_1 = \frac{3|A_{VO}|}{2\pi R_f C} = \frac{3 \times 6}{2\pi \times 343\Omega \times 110\text{fF}} = 75\text{GHz}$$

which implies a gain-bandwidth product $GBW = 450\text{GHz}$! Such a high bandwidth will be difficult to achieve with 150GHz f_T. However, it is acceptable for the advanced InP HBT process described in Chapter 4.3. If we relax our group delay variation requirement and instead opt for a Butterworth response, the feedback resistance needed to satisfy the bandwidth specification becomes

$$R_f = \frac{\sqrt{2}|A_{VO}|}{\omega_{3dB} C} = \frac{\sqrt{2} \times 6}{2\pi \times 34.4\text{GHz} \times (50\text{fF} + 60\text{fF})} = 357\Omega.$$

The transimpedance gain in this case is 306Ω, and the bandwidth of the forward amplifier must be

$$f_2 \approx \frac{2|A_{VO}|}{2\pi R_f C} = \frac{2 \times 6}{2\pi \times 357\Omega \times 110\text{fF}} = 48.6\text{GHz}.$$

This bandwidth is more realistic in a 150GHz f_T SiGe bipolar technology. However, even in this case, the gain-bandwidth product is 291GHz and may require a two-stage amplifier. Alternatively, we can relax the open loop gain from 6 to 4 or 3.

The same analysis can be applied to the CMOS technology with lower voltage gain. However, this lower voltage gain translates into appreciably smaller transimpedance gain. The design requirements for the bipolar and nMOS TIAs with Bessel and Butterworth

Table 8.2 **TIA design parameters in a SiGe bipolar technology with $|A_{VO}| = 6$ and a 65nm CMOS technology with $|A_{VO}| = 2$.**

Technology	HBT		CMOS	
Response	Bessel	Butterworth	Bessel	Butterworth
Feedback resistor (R_f)	343Ω	357Ω	114Ω	119Ω
Transimpedance gain (Z_{TO})	294Ω	306Ω	76Ω	79Ω
Forward amplifier bandwidth (f_2)	75GHz	48.6GHz	75GHz	48.6GHz

responses are summarized in Table 8.2. Note that the forward amplifier bandwidth requirements are the same in SiGe or CMOS for a given response, but the resulting feedback resistors differ by the ratio of the forward amplifier voltage gains.

8.3.2 HBT transimpedance feedback amplifier design

Until now, we have focused on a generic TIA topology without delving into the details of the forward amplifier. However, to analyze the noise performance of a TIA, the actual implementation must be considered. As a starting point, let us examine a common TIA implementation where the voltage amplifier is a resistively loaded HBT common-emitter amplifier. An emitter-follower stage is used to de-couple the transimpedance feedback resistor, R_F, and the load. This topology is illustrated in Figure 8.9. Ignoring the noise contribution from emitter-follower, Q_2, we can identify by inspection the two dominant sources of noise in this amplifier – the feedback resistor and transistor Q_1. The noise contributed by the collector load resistance R_C ($4kT\Delta f/R_C$) is typically A_{VO} times lower than the collector shot noise current of Q_1 ($2q \times \Delta f \times I_C$) and can be neglected in a first-order analysis. As sketched in Figure 8.9, the former has a power spectral density that is constant with frequency and depends only on the resistor value. The transistor noise, on the other hand, is dependent on the device geometry and bias point and increases with frequency.

It has been established in Chapter 4.3 that the minimum noise figure of a HBT is minimized at a particular collector current density J_{OPT}. Once this bias condition is determined, the noise parameters of the transistor scale with device geometry and, some of them, with frequency. These relationships are given by equations (7.4)–(7.7) in Chapter 7. The key to the TIA design philosophy discussed here is to optimize the transistor bias current density and dimension (in this case, the emitter length) in order to minimize the amplifier's noise figure over the entire bandwidth from DC to the 3dB frequency. The noise figure has to be minimized with respect to the signal source impedance and its noise equivalent current/voltage.

In optical applications, the signal source is a photodiode described by its small signal impedance, often approximated by $1/j\omega C_D$, and by its optical-signal-dependent noise equivalent conductance, G_{seq}, given by (3.68). Very often, commercial photodiodes are internally matched to 50Ω, especially those operating at 40Gb/s or higher data rates. In such situations,

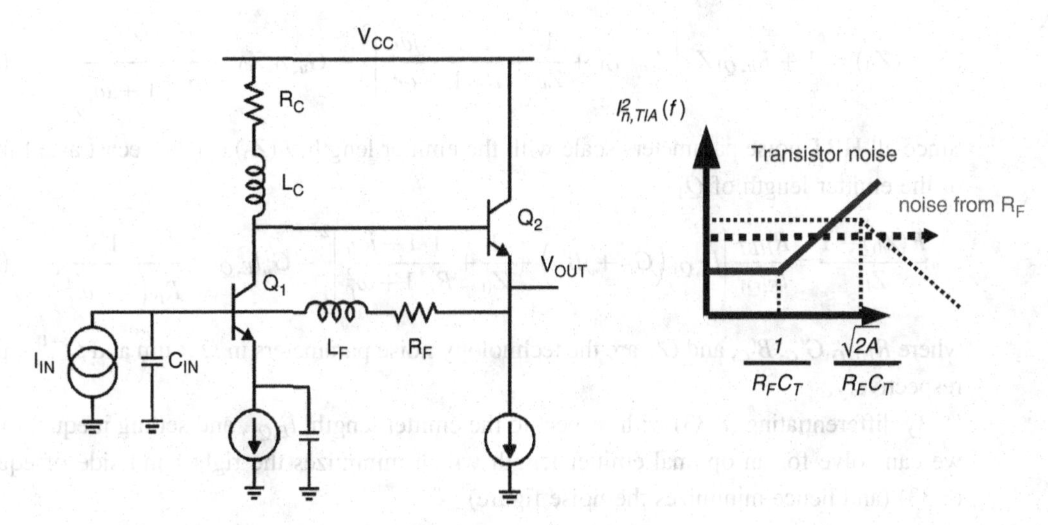

Figure 8.9 SiGe HBT common-emitter TIA [1]

and in cases where the TIA is used as a low-noise broadband amplifier for wireline or ADC applications, the 50Ω noise figure of the TIA must be minimized over the entire frequency bandwidth of interest. As derived in Chapter 3, at low frequencies, the noise figure and optimal noise impedance of the TIA will be dominated by R_F, whereas at higher frequencies, close to the 3dB frequency, the contribution from the transistor itself will dominate. As a result, and since R_F is determined from the input-impedance and/or 3dB bandwidth specification, it is sufficient to focus on minimizing the noise figure of the TIA at the 3dB frequency. Ideally, R_F and, therefore, A_V should be as large as possible to minimize its noise contribution. However, as we have already seen, large A_V may impose unrealistically high bandwidth requirements for the amplifier. First, we will examine the case for a signal source impedance Z_0 and then we will provide the solution when the signal source is a photodiode die, not matched to 50Ω.

50Ω matched broadband amplifier

Treating the TIA as a shunt–shunt feedback amplifier and using the noise-parameter equations (3.168)–(3.169) developed in Chapter 3 for two two-port networks connected in parallel, the noise factor equation (8.15) can be expressed in terms of Z_0 ($Y_S = 1/Z_0$, $G_{seq} = 1/Z_0$), the transistor noise parameters ($R_{n,Q1}$, $G_{u,Q1}$, and $Y_{cor,Q1}$), the feedback resistor R_F and feedback inductor L_F. Starting with the noise parameters of the feedback network

$$R_{nf} = R_f;\ G_{uf} = 0; Y_{cor,f} = 0, y_{11f} = \frac{1}{R_f + j\omega L_f} = \frac{1}{R_f(1 + j\omega_f)} \text{ and } \omega_f = \frac{\omega L_F}{R_F} \tag{8.30}$$

the noise parameters of the TIA become

$$R_n \approx R_{n,Q1}; G_u = G_{u,Q1} + \frac{1}{R_F(1 + \omega_f{}^2)} \text{ and } Y_{cor} = Y_{cor,Q1} + \frac{1}{R_F}\frac{1 - j\omega_f}{1 + \omega_f{}^2}. \tag{8.31}$$

By plugging (8.31) into (8.15), we obtain the expression of the noise factor of the TIA driven by a signal source of impedance Z_0

$$F(Z_0) = 1 + R_{n,Q1}Z_0 \left| Y_{cor,Q1} + \frac{1}{Z_0} + \frac{1}{R_F}\frac{1 - j\omega_f}{1 + \omega_f^2} \right|^2 + G_{u,Q1}Z_0 + \frac{Z_0}{R_F\left(1 + \omega_f^2\right)}. \quad (8.32)$$

Since all HBT noise parameters scale with the emitter length, $F(Z_0)$ can be recast as a function of the emitter length of Q_1

$$\frac{F(Z_0) - 1}{Z_0} = \frac{R_{HBT}}{l_{E,Q1}} \left| l_{E,Q1}\left(G_C' + jB_C'\right) + \frac{1}{Z_0} + \frac{1}{R_F}\frac{1 - j\omega_f}{1 + \omega_f^2} \right|^2 + G_u' l_{E,Q1} + \frac{1}{R_F\left(1 + \omega_f^2\right)} \quad (8.33)$$

where R_{HBT}, G'_C, B'_C, and G'_u are the technology noise parameters in $\Omega \times \mu m$ and $\Omega^{-1} \times \mu m^{-1}$, respectively.

By differentiating (8.33) with respect to the emitter length, $l_{E,Q1}$, and setting it equal to zero, we can solve for an optimal emitter length which minimizes the right-hand side of equation (8.33) (and hence minimizes the noise figure)

$$l_{EOPT} = \sqrt{\left(\frac{1}{Z_0} + \frac{1}{R_F}\frac{1}{1 + \omega_f^2}\right)^2 + \left(\frac{1}{R_F}\frac{\omega_f}{1 + \omega_f^2}\right)^2}\sqrt{\frac{1}{\frac{G_u'}{R_{HBT}} + G_C'^2 + B_C'^2}}$$
$$= \frac{1}{\omega}\left(\frac{1}{Z_0} + \frac{1}{R_F}\frac{1}{1 + \omega_f^2}\right)\sqrt{\frac{1}{\frac{G_{HBT}}{R_{HBT}} + G_{C,HBT}^2 + B_{HBT}^2}}. \quad (8.34)$$

In the last version of (8.34), we have taken advantage of the frequency dependence of the technology noise parameters approximated by (7.4)–(7.7) to illustrate that the optimal emitter length is inversely proportional to frequency.

Optical receiver

In optical applications, assuming that the parasitic photodiode resistance R_S is negligible, the generator impedance is set by the photodiode capacitance

$$Z_0 = \frac{1}{j\omega C_D} \quad (8.35)$$

while the noise of the signal source is described by the noise conductance of the photodiode, G_{seq}. We can define the noise figure of the photodiode-TIA ensemble as

$$F = 1 + \frac{G_{n,TIA} + R_{n,TIA}\left|j\omega C_D + G_{cor,TIA} + jB_{cor,TIA}\right|^2}{G_{seq}}. \quad (8.36)$$

Again, by inserting the expressions of the noise parameters of the TIA as a function of the noise parameters of Q_1 and of the $L_F - R_F$ feedback network, (8.36) can be rearranged as

$$(F-1)G_{seq} = \frac{R_{HBT}}{l_{E,Q1}} \left| l_{E,Q1}\left(G_C' + jB_C'\right) + j\omega C_D + \frac{1}{R_F}\frac{1 - j\omega_f}{1 + \omega_f^2} \right|^2 + G_u' l_{E,Q1} + \frac{1}{R_F\left(1 + \omega_f^2\right)} \quad (8.37)$$

After differentiating (8.37) with respect to $l_{E,Q1}$ and equating the result to 0, the optimal emitter length can be expressed in terms of the photodiode capacitance, the feedback resistance, and the technology-specific noise parameters

$$l_{EOPT} = \frac{1}{\omega_{3dB}} \sqrt{\frac{1}{R_F^2\left(1+\omega_f^2\right)^2} + \left(\omega_{3dB} C_D - \frac{\omega_f}{R_F\left(1+\omega_f^2\right)}\right)^2} \sqrt{\frac{1}{\frac{G_{HBT}}{R_{HBT}} + G_{C,HBT}^2 + B_{HBT}^2}}. \quad (8.38)$$

Again, for completeness, C_D should include the bond pad capacitance. Interestingly, when one considers typical photodiode capacitances of 200–300fF for 10Gb/s applications or 60fF for 40Gb/s receivers, the magnitude of Z_0 (including bond pad parasitics) is close to 50Ω [4].

Equations (8.34) and (8.38) clearly show that, as the 3dB bandwidth of the TIA increases, the optimal transistor size must decrease. Additionally, if driven by a photodiode, the optimal transistor size is proportional to the photodiode capacitance. The feedback inductor helps to compensate for at least part of the photodiode capacitance and to reduce the transistor size and the power consumption.

The above analysis gives rise to an algorithmic methodology for designing a low-noise transimpedance amplifier for optical receivers:

1. Determine the optimal noise figure current density, J_{OPT}, of the TIA stage at the desired 3dB frequency. Fix the current density through Q_1 at J_{OPT} and determine the technology constants R_{HBT}, G_{HBT}, $G_{C,HBT}$, and B_{HBT} for this bias condition and frequency (since the G'_C and G'_u dependencies on frequency given in Chapter 7 are only approximations).
2. Choose the voltage drop across the collector resistor, R_C, that satisfies the maximum output swing (linearity) requirements. The maximum linear peak-to-peak voltage swing is $\Delta V = 2I_C R_C$. This sets the voltage gain of the forward amplifier A_{V0}

$$|A_{V0}| = g_{meff} R_C = \frac{I_C}{V_T + I_C r_E} \cdot \frac{\Delta V}{2I_C} = \frac{\Delta V}{2(V_T + I_C r_E)} \approx \frac{\Delta V}{4V_T}. \quad (8.39)$$

Note that the impact of the emitter parasitic resistance r_E has been accounted for. In advanced SiGe and InP HBT processes, $I_C r_E \approx V_T$. Typical values for ΔV in 10–40 Gb/s TIAs are between 400 and 1000mV$_{pp}$, which sets the magnitude of the voltage gain to between 4 and 10. Note also that in HBT TIAs the linearity may also be limited by the base–emitter diode if no emitter degeneration is employed.

3. From the 3dB bandwidth requirement and the desired TIA response (i.e., Butterworth or Bessel response), the value of the feedback resistance can be estimated. For a Butterworth response, this becomes

$$R_F = \frac{\sqrt{2}|A_{V0}|}{2\pi f_{3dB} C_T} = \frac{\sqrt{2}\Delta V}{8\pi f_{3dB} V_T C_T}. \quad (8.40)$$

Here, C_T is the total capacitance at the input node of the TIA and includes the photodiode and bond pad capacitances (C_D and C_{BP}, respectively), as well as the input capacitance of Q_1

$$C = C_D + C_{BP} + C_{in,Q1} \tag{8.41}$$

where

$$C_{in,Q1} = C_{beeff}(Q_1) + (1 - A_{Vo})C_{bc}(Q_1). \tag{8.42}$$

However, since the emitter length is not known at this time, $C_{in,Q1}$ cannot be determined exactly and some amount of iteration will be required. An initial estimate of l_E can be obtained by ignoring the terms in R_F in (8.34) and (8.38). This is typically not more than 20–25% off the final value and allows us to get initial values for $C_{in,Q1}$ and R_F.

The diffusion capacitance is given by

$$C_{beeff,Q1} = g_{meff}\tau_F \approx \frac{I_C}{2V_T}\tau_F = \frac{J_{OPT}w_E l_{EOPT}}{2V_T}\tau_F. \tag{8.43}$$

For a 150GHz SiGe HBT technology with $\tau_F = 0.6$ps, $w_E = 0.2\mu$m, and $J_{OPT} = 1.8$mA/μm^2 at 34GHz, this yields a C_{be} per unit emitter length of about 4.15fF/μm. In a 43Gb/s TIA design with a photodiode capacitance of 60fF and a bond pad capacitance of 40fF, the device diffusion capacitance is one-fourth to one-third of the total input capacitance for an emitter length of 6μm.

4. Size the emitter length using equation (8.38). This fixes the bias current, as the current density was set in step 1. Additionally, the collector resistance is now known based on the voltage gain requirement.

It becomes important at this step to verify that the forward amplifier has sufficient bandwidth such that the transfer function of the TIA has a second-order Butterworth (or Bessel) response. To first-order analysis, the bandwidth (f_2) of the open-loop common-emitter amplifier is determined by its output time constant

$$f_2 \approx \frac{1}{2\pi R_C[C_{bc} + C_{cs}]} = \frac{1}{2\pi R_C\left[C'_{bc}\cdot w_E + C'_{cs}\right]l_{EOPT}}. \tag{8.44}$$

A more detailed analysis based on the combination of OCTC and short-circuit time constant (SCTC) techniques can be applied to calculate the second pole of the frequency response [2].

Recalling that the collector resistance and voltage gain A_{V0} are related by equation (8.39), the open-loop amplifier bandwidth can be rewritten as

$$f_2 \approx \frac{2I_C}{2\cdot\pi\cdot\Delta V\left(C'_{bc}w_E + C'_{cs}\right)l_{EOPT}} = \frac{I_C}{2\cdot V_T\cdot|A_{V0}|\left(C'_{bc}w_E + C'_{cs}\right)l_{EOPT}} = \frac{J_{OPT}w_E}{4\cdot\pi\cdot V_T\cdot|A_{V0}|\left(C'_{bc}w_E + C'_{cs}\right)}. \tag{8.45}$$

At a fixed bias current density, this bandwidth depends only on the voltage gain since the emitter width is always chosen to be the minimum value allowed by the technology. Consequently, if the bandwidth requirement is not met, there are few options left. The first is to trade off voltage gain for bandwidth, but this comes at the expense of reduced maximum output swing. Alternatively, we can proceed to step 5.

5. Add peaking inductors to extend the bandwidth. The inductors are illustrated in the schematic of Figure 8.9, but have for the most part been ignored until now. Inductive peaking has been covered in Chapter 5, where it was shown that the bandwidth can be extended by up to 60% with minimum group delay variation across the entire bandwidth. At the same time, the peaking inductors provide the additional benefit of filtering thermal noise from the collector and feedback resistor at high frequencies.

Note that in the above methodology we have ignored R_b.

Second, it is obvious that even with inductive peaking, the bandwidth of the forward amplifier (f_2) is dictated solely by the technology parameters (C'_{bc}, C'_{cs}, C'_{beff}), voltage gain, circuit topology, and fanout. This means that, even at Step 2, we can estimate quickly if the chosen voltage gain can be achieved with the desired 3dB bandwidth from (8.45).

A third observation is that this design methodology can also be carried out using computer simulation, especially to identify the optimal noise current density, which depends on the TIA topology, and the noise parameters corresponding to that current density and 3dB frequency.

EXAMPLE 8.2 SiGe HBT 110Gb/s TIA stage

Find the optimal transistor size, bias current, feedback, and collector resistor values needed for a 110Gb/s TIA:

(a) for wireline coaxial cable applications with 50Ω signal source impedance and 20fF pad capacitance;

(b) for a fiber-optic receiver with 15fF diode capacitance and 20fF of pad and device capacitance at the input.

The following technology data are available. 280GHz SiGe HBT parameters for emitter width of 0.1μm. Technology data are provided per emitter length l_E or per emitter area $w_E l_E$:
$J_{OPT} = 10\text{mA}/\mu\text{m}^2$; $w_E = 0.1\mu\text{m}$; $\beta = 500$; $V_{CESAT} = 0.3\text{V}$, $V_{BE} = 0.9\text{V}$, peak $f_T = 280\text{GHz}$.
$\tau_F = 0.25\text{ps}$; $C_{je} = 17\text{fF}/\mu\text{m}^2$; $C'_{bc} = 20\text{fF}/\mu\text{m}^2$; $C_{cs} = 1.2\text{fF}/\mu\text{m}$;
$R'_b = 75\Omega \times \mu\text{m}$, $r'_E = 2\Omega \times \mu\text{m}^2$, $R'_c = 20\Omega \times \mu\text{m}$

At 88GHz: $\sqrt{\dfrac{1}{\frac{G_{HBT}}{R_{HBT}} + G^2_{C,HBT} + B^2_{HBT}}} = 5.98 \cdot 10^{13}\mu\text{m}.$

Solution

Step 1 J_{OPT} has been set at $10\text{mA}/\mu\text{m}^2$ for both cases.

(a) 50Ω signal source.

Step 2 We choose $\Delta V = 400\text{mV}$ which give $SA_{V0} = -\Delta V/(4V_T) = -4$. We can quickly verify that $f_2 > \sqrt{2} \cdot BW_{3\text{dB}} = 1.414 \cdot 88 \cdot 10^9 = 124.4\text{GHz}$.

We are now in a position to calculate the frequency of the output pole from (8.44)

$$f_2 = \frac{J_{OPT} w_E}{2 \cdot \pi \cdot 2 \cdot V_T \cdot |A_{V0}| \left(C'_{bc} w_E + C'_{cs}\right)} = \frac{10 \frac{\text{mA}}{\mu\text{m}^2} \cdot 0.1\mu m}{6.28 \cdot 2 \cdot 0.02587 \cdot 4 \cdot 3.2 \frac{\text{fF}}{\mu\text{m}}} = 240.4\text{GHz}.$$

This pole is much higher than 124.4GHz which means that in theory we could afford higher gain. However, we have ignored any load capacitance!

Step 3 From (8.34), we get an initial value for the optimal emitter width by ignoring the terms in R_F. We obtain l_{EOPT}= 2.166μm. By assuming a 25% larger value in anticipation of the contribution of the terms in R_F, l_{EOPT}= 2.7μm. As a result, I_C = 2.7mA. $R_C = \Delta V/(2I_C) = 74\Omega$; $g_m = \frac{2.7\text{mA}}{25.867\text{mV}} = 104.4\text{mS}$; $r_E = 7.4\Omega$, g_{meff}= 58.9mS, C_{bediff}= 14.725fF, C_{jbeeff}= 2.59fF, $C_{beff} = C_{bediff} + C_{jbeeff}$ = 17.31fF, C_{bc} = 5.4fF and $C_T = C_{BP} + C_{in}$ = 20fF + 17.31fF + 27fF = 64.31fF.

We can now calculate the feedback resistance

$$R_F = \frac{\sqrt{2}\cdot|A_{VO}|}{\omega C_T} = \frac{\sqrt{2}\cdot 4}{6.28\cdot 88\cdot 10^9\cdot 64.3\cdot 10^{-15}} = 159.2\Omega.$$

Step 4 Let us recalculate l_{EOPT} from (8.34) now that we have a value for R_F. For now, we will ignore L_F and, therefore, $\omega_f = 0$. We obtain l_{EOPT} = 2.84μm, which is less than 5% larger than the initial estimate, so we will not bother changing it.

(b) In this case, **Step 2** is identical to case (a).

Step 3 Again, we can get an initial estimate of l_{EOPT} from (8.38) ignoring the terms in R_F which will be much smaller this time because they appear as squared. We obtain

$$l_{\text{EOPT}} = (C_D + C_{BP})\sqrt{\frac{1}{\frac{G_{\text{HBT}}}{R_{\text{HBT}}} + G^2_{C,\text{HBT}} + B^2_{\text{HBT}}}} = 35\text{fF}\cdot 5.98\times 10^{13} = 2.1\mu\text{m}.$$

In reality, if we account for the feedback inductance L_F, an even smaller transistor size and input capacitance is possible. As a result I_C = 2.1mA. $R_C = \Delta V/(2I_C) = 92.2\Omega$ $g_m = \frac{2.1\text{mA}}{25.867\text{mV}} = 81.17\text{mS}$; $r_E = 9.5\Omega$, g_{meff} = 45.82mS, $C_{beff} = C_{bediff} + C_{bejeff}$ = 13.46fF, C_{bc} = 4.2fF and $C_T = C_D + C_{BP} + C_{in}$ = 15fF + 20fF + 13.46fF + 21fF = 69.46fF. We can now calculate the feedback resistance $R_F = \frac{\sqrt{2}\cdot(A_{VO})}{\omega C_T}$ $= \frac{\sqrt{2}\cdot 4}{6.28\cdot 88\cdot 10^9\cdot 69.46\cdot 10^{-15}} = 147.36\Omega$ and the frequency of the input pole

$$f_1 = \frac{1}{2\pi\cdot R_F\cdot C_T} = \frac{1}{6.28\times 147.36\times 69.46\times 10^{-15}} = 15.55\text{GHz}.$$

Step 4 Let us recalculate l_{EOPT} from (8.38) now that we have a value for R_F. For now, we will ignore L_F and, therefore, $\omega_f = 0$. We obtain l_{EOPT}= 2.22μm, which, again, is less than 5% larger than the initial estimate, so we will not bother changing it.

We are now in a position to calculate the frequency of the output pole from (8.44)

$$f_2 = \frac{J_{OPT}w_E}{2\cdot\pi\cdot 2\cdot V_T\cdot|A_{VO}|\left(C'_{bc}w_E + C'_{cs}\right)} = \frac{10\frac{\text{mA}}{\mu\text{m}^2}\cdot 0.1\mu\text{m}}{6.28\cdot 2\cdot 0.02587\cdot 4\cdot(2+1.2)\frac{\text{fF}}{\mu\text{m}}} = 240.4\text{GHz}.$$

This pole is much higher than f_1 but we have ignored any load capacitance. What if the TIA has a fanout of 1, that is $C_L = C_{IN} = 34.46\text{fF}$? In that case, f_2 becomes

$$f_2 = \frac{J_{OPT}w_E}{2\cdot\pi\cdot 2\cdot V_T\cdot|A_{V0}|\left(6\cdot C'_{bc}w_E + C'_{cs} + C'_{beff}W_E\right)}$$

$$= \frac{10\dfrac{\text{mA}}{\mu\text{m}^2}\cdot 0.1\mu\text{m}}{6.28\cdot 2\cdot 0.02587\cdot 4\cdot(12+1.2+6.4)\dfrac{\text{fF}}{\mu\text{m}}} = 39.3\text{GHz}.$$

This is definitely so low, that even with inductive peaking we cannot push f_2 high enough for a Butterworth response which requires that $f_2 > 2\cdot|A_V|\cdot f_1 = 124.4\text{GHz}$. It is obvious that only an emitter-follower stage can be used as load to minimize the load capacitance. Assuming an equal size emitter-follower as load (similar to the topology in Figure 8.9), the output pole frequency becomes

$$f_2 = \frac{J_{OPT}w_E}{2\cdot\pi\cdot 2\cdot V_T\cdot A_{V0}\left(2\cdot C'_{bc}w_E + C'_{cs} + C'_{beff}W_E\right)}$$

$$= \frac{10\dfrac{\text{mA}}{\mu\text{m}^2}\cdot 0.1\mu\text{m}}{6.28\cdot 2\cdot 0.02587\cdot 4\cdot(4+1.2+6.4)\dfrac{\text{fF}}{\mu\text{m}}} = 66.32\text{GHz}.$$

This falls short of the 124.4GHz target, so:

Step 5 If we apply inductive peaking at the output node, then we can push f_2 to at least 106GHz. We need the feedback inductor as well to further push the bandwidth. Alternatively, we may have to accept a 70% bandwidth rather than 80% bandwidth with respect to data rate, which will result in some ISI and larger rise and fall times.

Note from (a) and (b) that, as mentioned, the output pole does not depend on the size of the transistor, only on the chosen topology, its gain, and its fanout. Therefore, we should check the output pole first to determine what the maximum possible gain should be.

Resistively degenerated, self-biased TIA stage

A simpler, self-biased TIA circuit with emitter degeneration is shown in Figure 8.10. It employs only one transistor and can operate with a lower supply voltage. R_E and R_C set the open loop voltage gain, while R_F determines the bandwidth and the input and output impedances. Among its disadvantages is higher noise due to R_E. However, degeneration may be needed in any case in order to satisfy input linearity requirements.

The open loop voltage gain depends primarily on the collector, emitter, and feedback resistors but can also be affected by the collector current and parasitic emitter resistance if the degree of emitter degeneration is low. It can be shown that the low-frequency voltage gain is given by

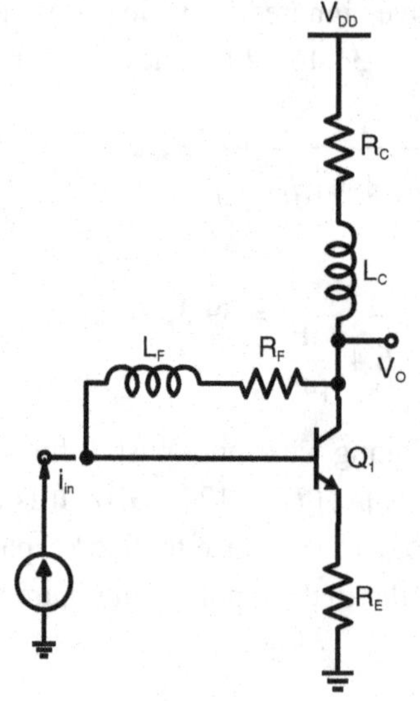

Figure 8.10 Common-emitter TIA topology with emitter degeneration

$$A_V = \frac{-R_C}{R_E + r_E + \frac{1}{g_m}} \left[\frac{R_F - R_E - r_E}{R_F + R_C} \right]. \tag{8.46}$$

When used as a standalone amplifier matched to the same impedance, Z_0, at the input and the output, we want $R_S = R_C = Z_0$, which leads to

$$Z_{in} = Z_{out} = \frac{R_F}{1 + |A_V|} = R_C. \tag{8.47}$$

In order to simplify the noise analysis of this circuit, R_E can be lumped together with r_E in the transistor noise parameter equations. However, since the optimal noise current density of a common-emitter stage with degeneration is affected by R_E, which increases J_{OPT}, the latter must be determined every time the amount of emitter degeneration is modified.

EXAMPLE 8.3

For the 43Gb/s differential InP HBT TIA in Figure 8.10 employed in a fiber-optic receiver, calculate the open loop gain and the transistor sizes knowing that the BW_{3dB} is 36GHz, the optimum noise figure current density is 0.6mA/μm^2 and the total input capacitance is $C_{IN} + C_D + C_{BP} = 80$fF. All transistors have an emitter width of 1μm, $\tau_F = 0.6$ps, $C'_{bc} = 1$fF/μm^2, $r_E = 2\Omega \times \mu m^2$, and $\sqrt{\frac{1}{\frac{G_{HBT}}{R_{HBT}} + G^2_{C,HBT} + B^2_{HBT}}} = 6.25{\cdot}10^{13}$μm. Ignore the base resistance and the collector–substrate capacitance.

Solution

From the information in the schematic, and Chapter 5 we can calculate

$$\omega_f = \omega \frac{L_F}{R_F} = \frac{2{\cdot}3.14{\cdot}36{\cdot}10^9{\cdot}250{\cdot}10^{-12}}{280} = 0.20186.$$

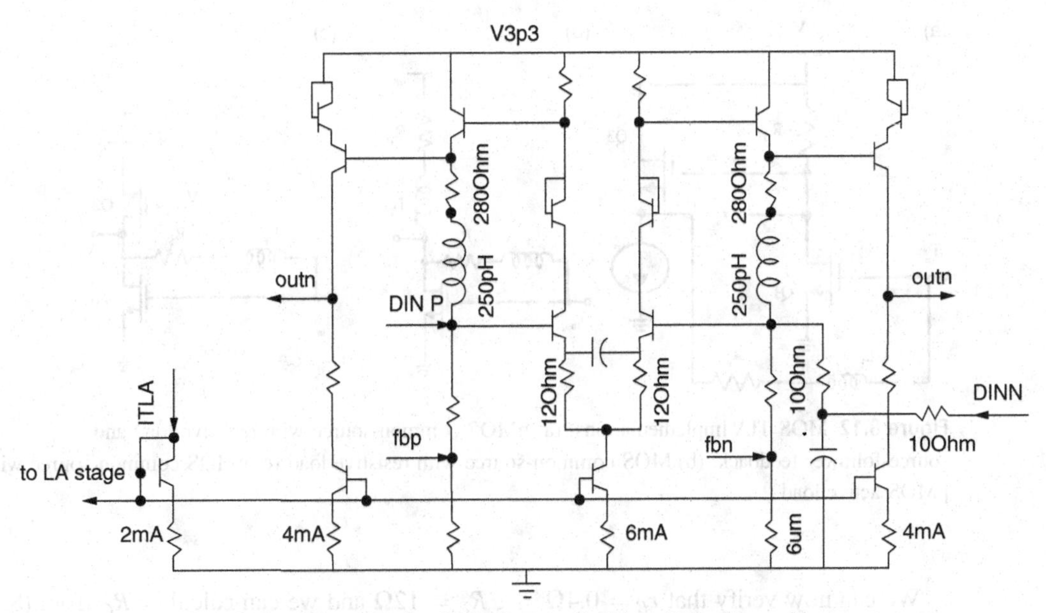

Figure 8.11 InP Differential TIA schematics with emitter degeneration [4]

Since the required bandwidth is 36GHz

$$\frac{R_F}{1+|A_V|} = \frac{1}{2\pi BW_{3dB}(C_D + C_{BP} + C_{IN})} = 55.3\Omega.$$

Therefore

$$|A_V| = \frac{R_F}{55.3} - 1 = \frac{280}{55.3} - 1 = 4.06 \text{ and } A_V = -4.06.$$

The optimal emitter length of the input transistors can be determined from (8.38)

$$l_{EOPT} = \frac{1}{\omega_{3\text{dB}}}\sqrt{\frac{1}{R_F^2\left(1+\omega_f^2\right)^2} + \left(\omega_{3\text{dB}}C_D - \frac{\omega_f}{R_F\left(1+\omega_f^2\right)}\right)^2}\sqrt{\frac{1}{\frac{G_{HBT}}{R_{HBT}} + G_{C,HBT}^2 + B_{HBT}^2}}$$
$$= \frac{1}{2.260\cdot 10^{11}}\sqrt{1.179\cdot 10^{-5} + 3.027\cdot 10^{-4}}\cdot 6.25{,}10^{13}\mu\text{m} = 4.9\mu\text{m}.$$

Since this is a 1μm process, we have to round off the emitter length to 5μm.

This implies that the current through each of the input transistors should be $I_C = J_{OPT} w_E l_E = 0.6\frac{\text{mA}}{\mu\text{m}^2}\cdot 1\mu\text{m}\cdot 5\mu\text{m} = 3\text{mA}$ which results in $I_T = 6\text{mA}$, $I(Q_{1,2}) = 3\text{mA}$, $Q_{1,2,3,4} = 1\mu\text{m}\times 5\mu\text{m}$.

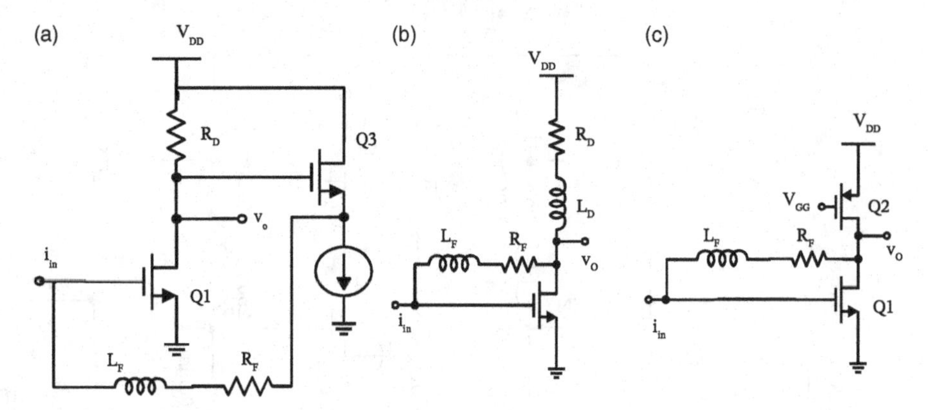

Figure 8.12 MOS TIA implementations: (a) nMOS common-source with resistive load and source-follower feedback, (b) MOS common-source with resistive load (c) nMOS common-source with pMOS active load

We can now verify that $r_E = 0.4\Omega << R_E = 12\Omega$ and we can calculate R_C from (8.46)

$$A_V \cdot \left(R_E + r_E + \frac{1}{g_m} \right) (R_F + R_C) = -R_C[R_F - R_E], \; -85.96(280 + \mathrm{R_C}) = -268 \cdot \mathrm{R_C}.$$

We obtain $R_C = 132.2\Omega$

$$f_1 = \frac{1}{2\pi \cdot R_F(C_D + C_{BP} + C_{IN})} = \frac{1}{6.28 \cdot 280 \cdot 80 \cdot 10^{-15}} = 7.108\text{GHz}.$$

The second pole will be dominated by the output time constant $f_2 = \frac{1}{2\pi R_C\left(2\frac{C_{be}}{3} + C_{bc}\right)} = \frac{1}{6.28 \cdot 132(22.8\text{fF} + 10\text{fF})} = 34\text{GHz}$. Here we have assumed that the two cascoded emitter-followers that load the output node have an effective capacitance of $(2/3)C_{be}$.

These poles have been calculated without accounting for the zero introduced by L_F in the feedback network and also by the emitter degeneration capacitor. Obviously both are needed to increase ω_2 to $2A_{v0}\omega_1$.

8.3.3 FET TIAs

A similar design methodology can be applied to the FET common-source amplifier shown in Figure 8.12(a). This transimpedance amplifier topology, suitable for GaAs MESFETs, HEMTs, and MOSFETs, was first introduced in the 1970s and is based on the Van Tuyl–Hornbuckle gain cell with feedback [5]. For a multi-finger MOSFET layout, the transistor noise parameters (7.8)–(7.11) scale as a function of the total gate width, $W = N \times N_f \times W_f$, where N_f represents the number of gate fingers, each with finger width, W_f, and N describes the number of active areas or fins (for FiNFETs) connected in parallel.

The expression for the TIA noise figure with respect to a signal source of impedance Z_0 becomes identical to that of equation (8.34) except for replacing the HBT subscripts with FET subscripts. Again, we can find an optimal gate width which minimizes the TIA noise figure by following the same procedure as was used for the bipolar common-emitter TIA. The optimal gate width as a function of the signal source impedance, Z_0, becomes

$$W_{OPT} = \frac{1}{\omega}\sqrt{\left(\frac{1}{Z_0} + \frac{1}{R_F}\frac{1}{1+\omega_f^2}\right)^2 + \left(\frac{1}{R_F}\frac{\omega_f}{1+\omega_f^2}\right)^2}\sqrt{\frac{1}{\frac{G_{FET}}{R_{FET}} + G_{C,FET}^2 + B_{FET}^2}}$$
$$= \frac{1}{\omega}\left(\frac{1}{Z_0} + \frac{1}{R_F}\frac{1}{1+\omega_f^2}\right)\sqrt{\frac{1}{\frac{G_{FET}}{R_{FET}} + G_{C,FET}^2 + B_{FET}^2}}. \quad (8.48)$$

In the case when the TIA is driven by a photodiode die which is not internally matched to 50Ω, we can express the optimal gate width as

$$W_{OPT} = \frac{1}{\omega_{3\text{dB}}}\sqrt{\frac{1}{R_F^2\left(1+\omega_f^2\right)^2} + \left(\omega_{3\text{dB}}C_D - \frac{\omega_f}{R_F\left(1+\omega_f^2\right)}\right)^2}\sqrt{\frac{1}{\frac{G_{FET}}{R_{FET}} + G_{C,FET}^2 + B_{FET}^2}}. \quad (8.49)$$

The same design procedure can then be followed as that described for the common-emitter bipolar TIA. However, for a resistively loaded MOS common-source amplifier, like the ones in Figure 8.12(a) and (b), even a voltage gain of 2 is difficult to realize because it requires a relatively large voltage drop on the drain resistance, R_D. The latter cannot be accommodated from a 1.2V or lower supply, typical for nanoscale CMOS technologies. Additionally, because of the small loop gain, the noise contributed by the drain resistance will be only two times smaller than the drain current noise of the MOSFET and can no longer be ignored.

One alternative [6] that has been employed in 10Gb/s TIAs [7], and which allows operation at low supplies with reasonable loop gain, is to replace R_D with an active p-MOS load. The TIA topology is illustrated in Figure 8.12(c). Unfortunately, the p-MOSFET adds even more noise than R_D and also leads to at least a doubling of the output node capacitance. Nevertheless, by inserting series peaking inductors in the feedback loop, at the input [7] and, often, at the output, the output node capacitance can be compensated and bandwidths and loop gains of over 26 GHz and 7dB, respectively, have been reported in 90nm CMOS with 50Ω noise figure values below 7.2dB up to 26GHz [8],[9]. It should be noted that the original Van Tuyl–Hornbuckle cell also used an active load implemented with a depletion-mode n-type GaAs MESFET.

A better solution is illustrated in Figure 8.13. CMOS technologies give rise to the possibility of using p-MOSFETs simultaneously as active load and gain stage in a CMOS inverter, thereby benefiting from the added transconductance of the p-MOS transistor. If the width of the p-MOS device is I_{ON}(nMOS)/I_{ON}(pMOS) times that of the n-channel MOSFET, the p-MOSFET and the n-MOSFET will have equal transconductances

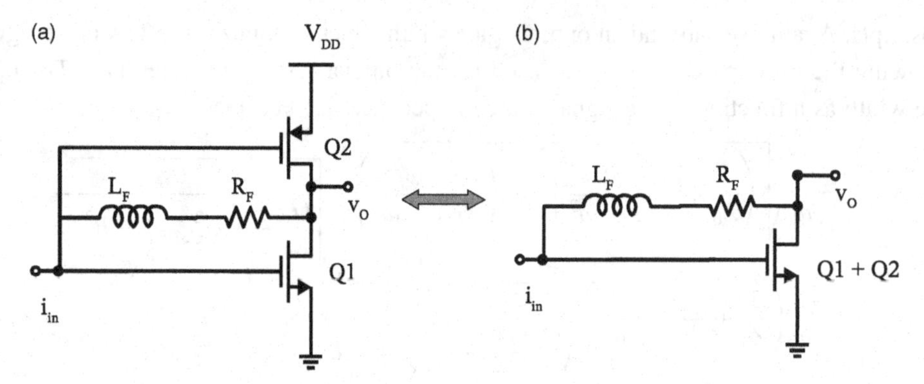

Figure 8.13 (a) CMOS inverter TIA implementation and (b) equivalent circuit

$$W_p = \frac{I_{ON}(n - MOS)}{I_{ON}(p - MOS)} \cdot W_n. \tag{8.50}$$

As discussed in Chapter 3, this CMOS inverter can be treated as a single transistor, where the "equivalent" transistor has twice the transconductance, $1 + \frac{W_p}{W_n}$ times the capacitance, and $\frac{2}{1+\frac{W_p}{W_n}}$ times the f_T and f_{MAX} of a n-channel MOSFET. With this topology, we can achieve comparable gain to that of the TIA of Figure 8.10(b), but with lower power consumption and greater bandwidth. In 45nm CMOS and beyond, where p-MOSFETs are almost as fast as the n-MOSFETs, the noise figure, too, becomes comparable or lower in the CMOS inverter TIA than in the n-MOSFET with active load version shown in Figure 8.12(b). For the optimal sizing equation to be applied to the CMOS inverter TIA, the FET subscripts in the technology noise parameters of (8.48) and (8.49) should be replaced by those of the CMOS inverter. The resulting optimal width corresponds to that of the n-channel device in the CMOS inverter.

The main problem of the self-biased CMOS inverter TIA topology is that the bias current and, to a large extent, the gain and bandwidth of the TIA, are sensitive to threshold voltage and power supply variation. A potential solution is to employ the circuit in Figure 8.14 where a thick-oxide p-MOSFET current mirror provides the tail current as well as common-mode and supply rejection to the differential CMOS inverter TIA. This scheme maximizes the voltage headroom available for the fast, thin-oxide devices in the CMOS inverter by allowing for a larger supply voltage to be applied safely. The latter can be reliably sustained by the thick oxide devices while the common-source node of the thin-oxide p-MOSFETs in the differential CMOS inverter remains below the maximum allowed safe operating voltage, e.g. 1.1V in a 45nm CMOS process or 0.9V in 32nm CMOS.

Finally, as illustrated in Figure 8.15, a method to further increase the loop gain (at the expense of linearity) is to employ cascode topologies, including cascode CMOS inverters, as the open-loop amplifier in the TIA.

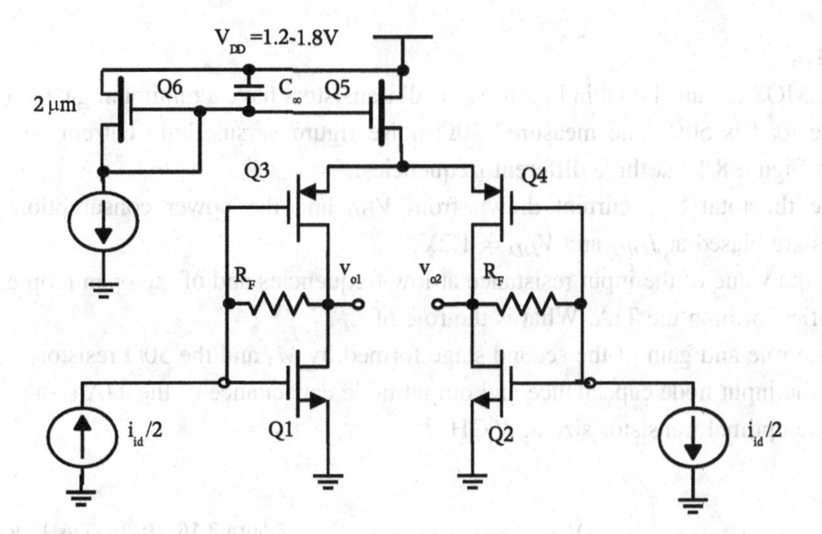

Figure 8.14 Biasing scheme for differential CMOS inverter TIAs using thick-oxide p-MOS (cascode) current source to allow for $V_{DD} > 1.2$V in nanoscale CMOS technologies while the DC voltage at the source of Q_3,Q_4 remains below 1.2V

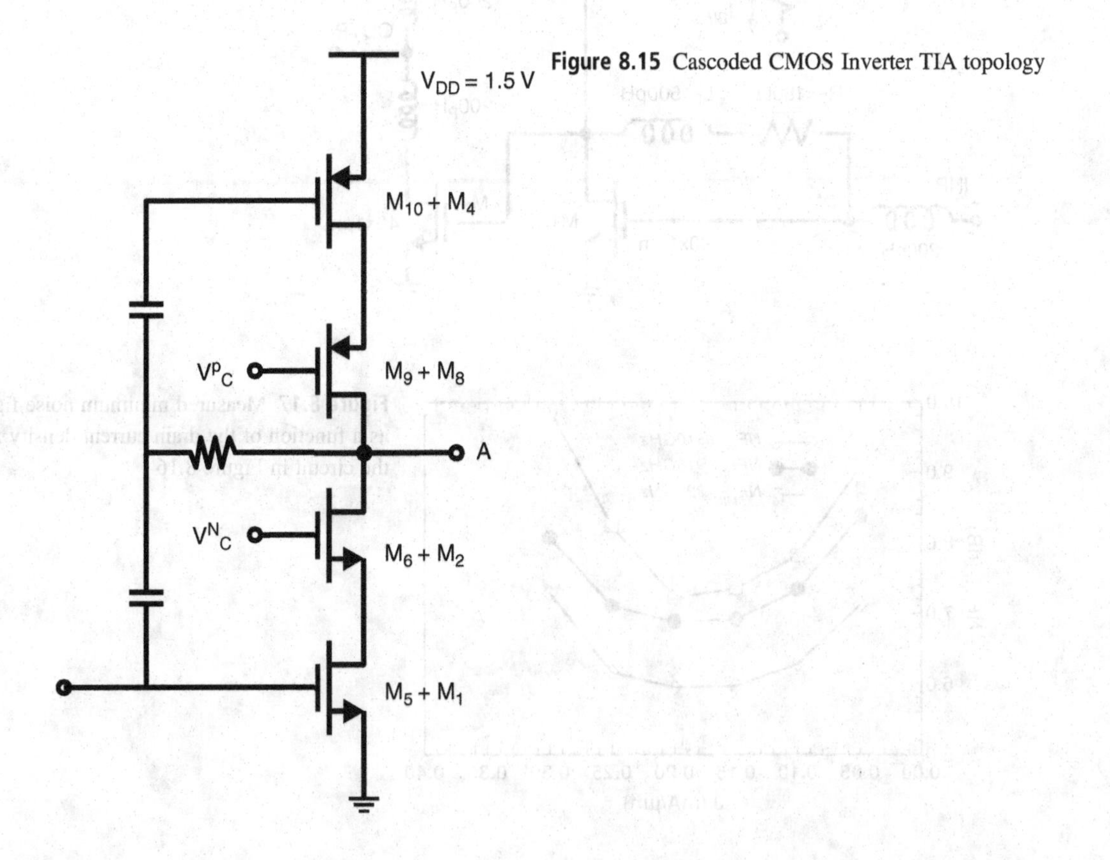

Figure 8.15 Cascoded CMOS Inverter TIA topology

EXAMPLE 8.4

In the 90nm CMOS circuit shown in Figure 8.16, all transistors have a minimum gate length of 90nm and the load is 50Ω. The measured 50Ω noise figure versus drain current density is reproduced in Figure 8.17 at three different frequencies:

(a) Determine the total bias current drawn from V_{DD} and the power consumption if all transistors are biased at J_{OPT} and $V_{DD} = 1.2$V.

(b) Calculate the value of the input resistance at low frequencies and of the open loop gain of the amplifier forming the TIA. What is the role of L_F?

(c) What is the role and gain of the second stage formed by M_4 and the 50Ω resistor?

(d) Calculate the input node capacitance and output node capacitance of the TIA stage.

(e) What is the optimal transistor size at 36GHz?

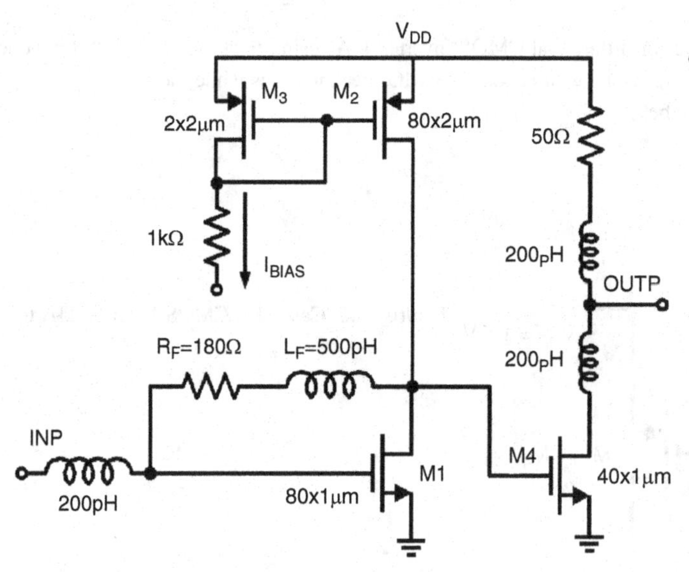

Figure 8.16 10–38 Gb/s 120nm and 90nm n-MOSFET TIAs [8]

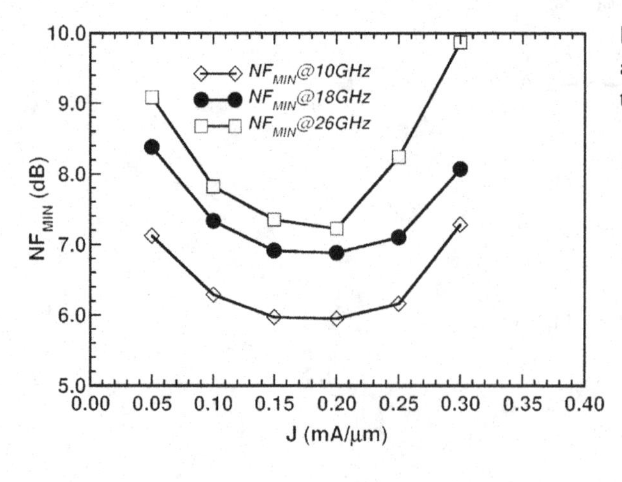

Figure 8.17 Measured minimum noise figure as a function of the drain current density for the circuit in Figure 8.16

90nm n-(p)MOSFET parameters:

$|VT| = 0.3\text{V}$; $g'_{mn} = 2g'_{mp} = 1.1\text{mS}/\mu\text{m}$; $g'_{dsn} = g_{dsp} = 0.15\text{mS}/\mu\text{m}$; the peak f_T of 135GHz occurs at 0.3mA/μm for $0.3\text{V} < V_{DS} < 1V$.

For both n-channel and p-channel devices: $C_{gs} = 1.0\text{fF}/\mu\text{m}$; $C_{gd} = 0.4\text{fF}/\mu\text{m}$; $C_{sb} = C_{db} = 1.1\text{fF}/\mu\text{m}$;

R_s (n-MOS) $= R_d = 200\Omega \times \mu\text{m}$; R_s (p-MOS) $= R_d = 400\Omega \times \mu\text{m}$;

R_{shg}(n/p-MOS) $= 15\Omega$/sq; poly contact resistance $R_{CON} = 10\Omega$. Total R_G/1μm finger $= 150\Omega$.

Noise parameters (n/p-MOS): $P = 1$, $C = 0.4$, $R = 0.25$.

$$R = \frac{P}{g'_m} + R'_s + R'_g(W_f = 1) = 1260\Omega \cdot \mu\text{m};$$

$$G = \frac{PR'_s(C'_{gs} + C'_{gd})^2}{P + \left(R'_s + R'_g(W_f = 1)\right)g'_m} = 2.83 \cdot 10^{-28}\frac{\text{F}^2}{\mu\text{m}^2}; G_C \approx 0$$

$$B = \frac{P(C'_{gs} + C'_{gd})}{P + \left(R'_s + R'_g(W_f = 1)\right)g'_m} = 1.01 \cdot 10^{-15}\frac{F}{\mu\text{m}}$$

$$\text{and} \quad \sqrt{\frac{1}{\frac{G_{FET}}{R_{FET}} + G^2_{C,FET} + B^2_{FET}}} = 5.38 \cdot 10^{14}\mu\text{m}.$$

Solution

(a) From Figure 8.17, we determine that, irrespective of frequency, the optimal bias current density is 0.15–0.2mA/μm. Therefore, to save power we select $J_{OPT} = 0.15\text{mA}/\mu\text{m}$. The drain currents of each transistor become $I_{D1} = I_{D2} = 12\text{mA}$, $I_{D4} = I_{D1}/2 = 6\text{mA}$ and $I_{D3} = 0.6\text{mA}$. The total power consumption of the TIA circuit with output buffer is $18.6\text{mA} \times 1.2\text{V} = 22.32\text{mA}$.

(b) The role of the feedback inductor L_F is to resonate out the capacitance at the TIA node and to improve the bandwidth. It also tunes out the imaginary part of the noise impedance at high frequencies around ω_f. For operation in a 40Gb/s circuit, L_F must be designed for SRF > 80GHz. To determine the open loop gain and input impedance, we must first calculate g_{meff} and g_{oeff} for each transistor

$$g'_{meff} = \frac{g'_m}{1 + R'_s g'_m} = 0.9\frac{\text{mS}}{\mu\text{m}} \text{ and } g'_{oeff} = \frac{g'_o}{1 + R'_s g'_m} = 0.122\frac{\text{mS}}{\mu\text{m}}.$$

$$A = \frac{-g_{m1}}{g_{o1} + g_{o2} + \frac{1}{R_F}} = \frac{-72\text{mS}}{9.76\text{mS} + 19.52\text{mS} + 5.55\text{mS}} = -2.07.$$

$$Z_{IN} = \frac{R_F}{1 - A} = \frac{180}{3.07} = 58.6\Omega.$$

(c) The role of the second stage is to provide matching to the 50Ω external load without allowing the load to affect the performance of the TIA stage.

The voltage gain of this stage is $A_{V2} = \frac{-g_{meff4}}{g_{oeff4}+\frac{2}{Z_0}} = \frac{-36\text{mS}}{4.88\text{mS}+40\text{mS}} = -0.8$. Here we have accounted for the on-chip Z_0 load, as well as for the off-chip one.

(d)

$$C_{IN} = C_{gs1} + (1-A)C_{gd1} = (1 + 3.07 \cdot 0.4)\frac{\text{fF}}{\mu\text{m}}0 \cdot 80\mu\text{m} = 178\text{fF}$$

$$C_{out} = C_{db1} + C_{gd1} + C_{gd2} + C_{db2} + C_{gs4} + (1 - A_{v2})C_{gd4}$$
$$= 88 + 32 + 64 + 176 + 40 + 28.8 = 429\text{fF}.$$

(e) At 36GHz, $\omega_f = 0.628$.

From (8.48), we obtain

$$W_{OPT} = \frac{1}{\omega}\sqrt{\left(\frac{1}{Z_0} + \frac{1}{R_F}\frac{1}{1+\omega_f^2}\right)^2 + \left(\frac{1}{R_F}\frac{\omega_f}{1+\omega_f^2}\right)^2}\sqrt{\frac{1}{\frac{G_{FET}}{R_{FET}} + G_{C,FET}^2 + B_{FET}^2}}$$

$$= \frac{0.0241 \cdot 5.38 \cdot 10^{14}}{2.26 \cdot 10^{11}} = 57.4\mu\text{m}$$

which is smaller than the gate width of the input MOSFETs in the schematic of Figure 8.16.

The 80μm width of the input device is optimal for a 3dB bandwidth of 25.8GHz, reflecting the measurements in [8].

EXAMPLE 8.5 Porting CMOS TIAs across technology nodes

Consider the CMOS TIAs shown in Figure 8.18 used in a wireline 43Gb/s receiver. The load resistor in the output buffer is 60Ω. This circuit was first designed in a 90nm CMOS technology with the same technology parameters as those in Example 8.4. It was then ported into a 65nm CMOS process without any modifications. Computer simulations of the TIA noise figure at 36GHz for a signal source impedance Z_0 = 50Ω are illustrated in Figure 8.19. The 65nm CMOS technology parameters are listed below:

$L = 45\text{nm}$, $|VT| = 0.3\text{V}$; Idsn at minimum noise = 0.15mA/μm; $g'_{mn} = 2g'_{mp} = 1.3\text{mS}/\mu\text{m}$; $g'_{dsn} = g'_{dsp} = 0.18\text{mS}/\mu\text{m}$; the peak f_T of 170GHz occurs at 0.3mA/μm for $0.3\text{V} < V_{DS} < 1\text{V}$. For both n-channel and p-channel devices: $C'_{gs} = 0.8\text{fF}/\mu\text{m}$; $C'_{gd} = 0.4\text{fF}/\mu\text{m}$; $C'_{sb} = C'_{db} = 0.7\text{fF}/\mu\text{m}$;

R'_s (n-MOS) $= R'_d = 160\Omega \times \mu\text{m}$; R'_s (p-MOS) $= R'_d = 320\Omega \times \mu\text{m}$;

R_{shg} (n/p-MOS) = 15Ω/sq; poly contact resistance $R_{CON} = 40\Omega$.

Total R_g/0.7μm finger contacted on one side is 158Ω. Noise parameters: $P = 1$, $C = 0.4$, $R = 0.25$.

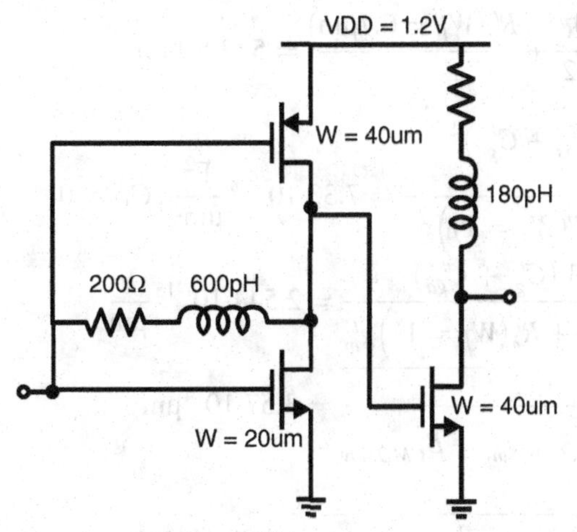

Figure 8.18 Scaling of CMOS inverter TIAs [8]

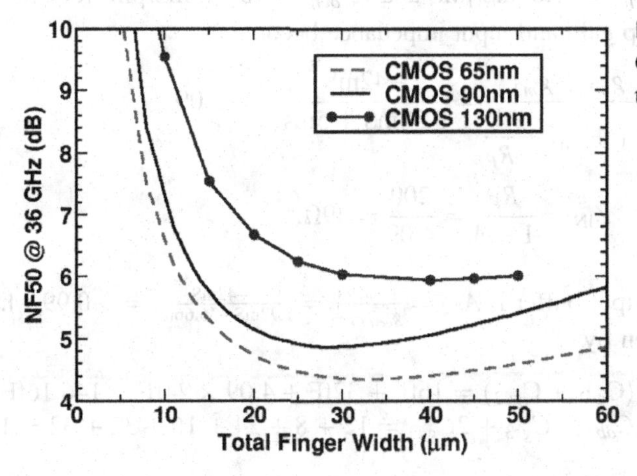

Figure 8.19 Simulated noise figure of the CMOS inverter TIA in three different technology nodes

(a) Calculate the technology parameters of the CMOS inverter in both processes.
(b) Determine the open loop voltage gain, the input impedance and the input and output capacitance of the TIA stage in the 65nm CMOS process.
(c) Calculate the optimal n-MOSFET gate width in both technologies for a TIA bandwidth of 36GHz assuming 0.7μm gate fingers in 65nm CMOS and 1μm gate fingers in 90nm CMOS.

Solution

(a) Given that $g'_{mn}/g'_{mp} = 2$, the technology noise parameters for the CMOS inverter in the 65nm CMOS process become

$$R(CMOS_{inv}) = \frac{P}{2g'_m} + \frac{R'_s}{2} + \frac{R'_g(W_f = 0.7\mu m)}{2} = 543.5\Omega \cdot \mu m;$$

$$G(CMOS_{inv}) = \frac{P\frac{R'_s}{2} \cdot 9(C'_{gs} + C'_{gd})^2}{P + \left(R'_s + R'_g(W_f = 1)\right)g'_m} = 7.33 \cdot 10^{-28}\frac{F^2}{\mu m^2}; G_C \approx 0,$$

$$B(CMOS_{inv}) = \frac{P \cdot 3 \cdot (C'_{gs} + C'_{gd})}{P + \left(R'_s + R'_g(W_f = 1)\right)g'_m} = 2.544 \cdot 10^{-15}\frac{F}{\mu m}$$

and $\sqrt{\frac{1}{\frac{G_{CMOSinv}}{R_{CMOSinv}} + G^2_{C,CMOSinv} + B^2_{CMOSinv}}} = 3.57 \cdot 10^{14} \mu m$

Similarly, in the 90nm process $\sqrt{\frac{1}{\frac{G_{CMOSinv}}{R_{CMOSinv}} + G^2_{C,CMOSinv} + B^2_{CMOSinv}}} = 2.99 \cdot 10^{14} \mu m.$

(b) First, we determine $g'_{meff} = 1.05 mS/\mu m$ and $g'_{oeff} = 0.143 mS/\mu m$ for the 65nm n-MOSFET. The open loop gain and input impedance become

$$A = -\frac{g_{m1} + g_{m2}}{g_{o1} + g_{o2} + \frac{1}{R_F}} = \frac{-42mS}{8.58mS + 5ms} = -3.09$$

$$Z_{IN} = \frac{R_F}{1 - A} = \frac{200}{4.09} = 49\Omega.$$

The voltage gain of the output buffer is $A_{V2} = \frac{-g_{meff3}}{g_{oeff3} + \frac{1}{25}} = \frac{-42mS}{5.72mS + 36.66mS} = -0.99$ while the node capacitances are given by

$$C_{IN} = C_{gs1} + C_{gs2} + (1 - A)(C_{gd1} + C_{gd2}) = 16fF + 32fF + 4.09 \times 24fF = 146.16fF$$
$$C_{out} = C_{db1} + C_{gd1} + C_{gd2} + C_{db2} + C_{gs3} + 2C_{gd3} = 12 + 8 + 24 + 16 + 32 + 32 = 124fF.$$

Ignoring L_F, f_1 and f_2 are given by

$$f_1 = \frac{1}{2\pi C_{IN} R_F} = 5.43 GHz \text{ and}$$

$$f_2 = \frac{g_{out}}{2\pi C_{out}} = \frac{g_{o1} + g_{o2} + \frac{1}{R_F}}{2\pi C_{out}} = 17.44 GHz.$$

(c) In both circuits, at 36GHz, $\omega_f = 0.678$.

For the 90nm TIA

$$W_{OPT} = \frac{1}{\omega}\sqrt{\left(\frac{1}{Z_0} + \frac{1}{R_F}\frac{1}{1 + \omega_f^2}\right)^2 + \left(\frac{1}{R_F}\frac{\omega_f}{1 + \omega_f^2}\right)^2}\sqrt{\frac{1}{\frac{G_{FET}}{R_{FET}} + G^2_{C,FET} + B^2_{FET}}}$$

Table 8.3 **Comparison of measured performance of 90nm CMOS and n-MOS TIA stages [8],[9].**

Parameter	CMOS	n-MOS
Power	3.9mW	14.4mW
Bandwidth	28GHz	21GHz.

$$= \frac{0.02342 \cdot 2.99 \cdot 10^{14}}{2.26 \cdot 10^{11}} = 31\,\mu\text{m}.$$

For the 65nm TIA

$$W_{OPT} = \frac{1}{\omega}\sqrt{\left(\frac{1}{Z_0} + \frac{1}{R_F}\frac{1}{1+\omega_f^2}\right)^2 + \left(\frac{1}{R_F}\frac{\omega_f}{1+\omega_f^2}\right)^2}\sqrt{\frac{1}{\frac{G_{FET}}{R_{FET}} + G_{C,FET}^2 + B_{FET}^2}}$$

$$= \frac{0.02342 \cdot 3.57 \cdot 10^{14}}{2.26 \cdot 10^{11}} = 37\,\mu\text{m}.$$

Both values calculated using analytical expressions are very close to the simulated values in Figure 8.19. Both are larger than the values used in the schematic. Note that the optimal size increases in more advanced nodes due to the fact that the gate–source and drain–bulk capacitances have decreased between the 90nm and 65nm nodes. Constant-field scaling no longer applies between the 90nm and 65nm nodes. Nevertheless, because g'_m increases and C'_{gs} and C'_{db} decrease, the noise and bandwidth of the TIA improve from the 90nm to the 65nm nodes.

Note also that the transistor size and the node capacitances have decreased between the n-MOS – only and the CMOS inverter TIA. The measured data for 90nm implementations of the n-MOS and CMOS TIA stages are summarized in Table 8.3.

8.4 OTHER BROADBAND LOW-NOISE AMPLIFIER TOPOLOGIES

Until now, we have considered the design of low-noise transimpedance amplifiers employed in conjunction with a photodetector in an optical communication receiver. In electrical links such as those used for chip-to-chip communications over backplanes, low-noise design still plays a critical role in improving receiver sensitivity. This is particularly true when transmitting at very high data rates where the receiver noise may dominate over noise from the backplane, cables, and connectors. A TIA can also be employed as a low-noise broadband amplifier for wireline applications. However, in this section, we will develop a low-noise design methodology for other input stages commonly found in electrical receivers. These include bipolar or FET differential pairs with and without emitter/source degeneration and EF stages followed by differential pairs.

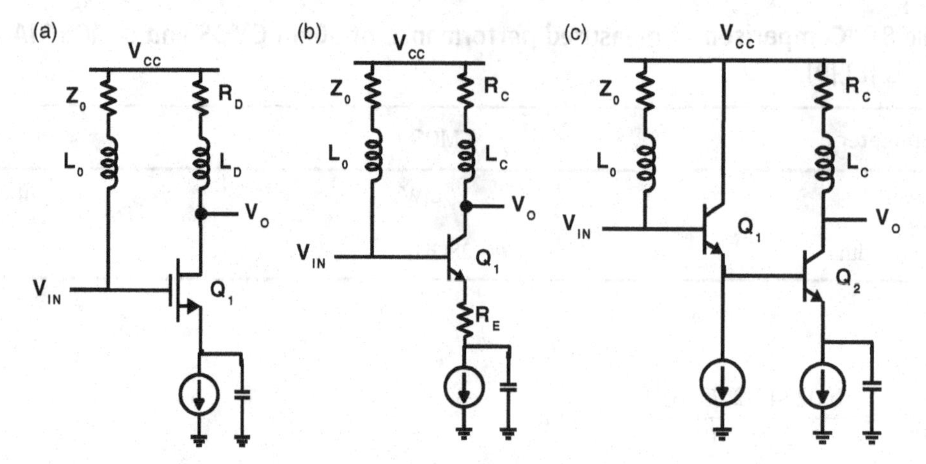

Figure 8.20 (a) FET inverter, (b) HBT inverter and (c) EF-inverter resistively matched differential half circuits

Before examining the topologies, let us consider the design criteria for these electrical input stages. The paramount design consideration in any low-noise amplifier is the noise figure. As with the TIA discussed earlier in this chapter, a broadband LNA design methodology involves minimizing the amplifier noise figure at its 3dB bandwidth, knowing that the noise figure at any frequency below the 3dB frequency will be lower. We would like our input stage to have high gain, but linearity must not be compromised if an equalizer follows the input stage. Also, impedance matching is required to avoid reflections. This means matching the input to 50Ω single-ended, 100Ω differentially, and, in most cases, both.

8.4.1 Low-noise impedance matched differential pair inverters

The half circuit of MOSFET and HBT differential pair input stages are shown in Figure 8.20. The HBT circuit consists of a common-emitter stage with optional resistive degeneration, R_E, and a matching resistor, Z_0, at the input. Inductive peaking may be employed at the input, L_O, and output nodes (L_C, L_D) to extend bandwidth as well as to improve the input match at high frequencies. Additionally, these inductors serve to filter high-frequency noise from the matching and load resistors. The noise factor of the amplifier can be expressed as a function of the generator impedance Z_0, the noise parameters (R_n, G_u, and Y_{COR}) of transistor Q_1, and of the noise parameters of the input matching network formed by Z_0 and L_0

$$F(Z_0) = 1 + \frac{1}{1 + \left(\frac{\omega L_0}{Z_0}\right)^2} + R_n Z_0 \left| Y_{COR} + \frac{2}{Z_0} \right|^2 + G_u Z_0. \tag{8.51}$$

Again, as in the TIA noise analysis, we have ignored the noise generated by R_C. By inserting the expressions of the HBT noise parameters in (8.51), the expression for the noise figure with respect to the signal source impedance Z_0 can be recast as a function of the emitter length

$$\frac{F(Z_0)-1}{Z_0}=\frac{1}{Z_0\left\{1+\left(\frac{\omega L_0}{Z_0}\right)^2\right\}}+\frac{R_{HBT}}{l_E}\left|\omega l_E\left(G_{C,HBT}+jB_{HBT}\right)+\frac{2}{Z_0}\right|^2+G_{HBT}\omega^2 l_E \quad (8.52)$$

An optimal emitter length which minimizes the noise factor can be found by differentiating equation (8.55) with respect to the emitter length l_E. The resulting optimal emitter length l_{EOPT} becomes

$$l_{EOPT}=\frac{1}{\omega}\frac{2}{Z_0}\sqrt{\frac{1}{\frac{G_{HBT}}{R_{HBT}}+G_{C,HBT}^2+B_{HBT}^2}}. \quad (8.53)$$

By setting ω equal to the desired amplifier bandwidth, this emitter length minimizes the noise factor over the entire operating frequency range. Note that, for $Z_0 = 50\Omega$, this is equivalent to noise matching the transistor to a 25Ω generator impedance due to the fact that the device sees a 50Ω on-chip matching resistor in parallel with the 50Ω off-chip signal source.

The same analysis can be applied to the MOS differential inverter input stage in Figure 8.21(a). A similar expression is obtained for the optimal gate width which minimizes the 50Ω noise figure

$$W_{OPT}=\frac{1}{\omega}\frac{2}{Z_0}\sqrt{\frac{1}{\frac{G_{MOS}}{R_{MOS}}+G_{C,MOS}^2+B_{MOS}^2}}. \quad (8.54)$$

It should be pointed out that, at the same frequency, the optimal gate width of a silicon n-channel MOSFET is often significantly larger than the optimal emitter length of a SiGe HBT due to its larger noise resistance, R_n, and larger optimal noise resistance, R_{SOPT}.

At this point, an algorithmic design methodology for designing a low-noise HBT or FET inverter input stage becomes apparent:

1. Find the optimal noise current density, J_{OPT}, of the stage, either through simulations or from measured data, at the desired 3dB bandwidth of the amplifier. For n-channel MOSFETs, this value is approximately 0.15–0.2mA/μm, irrespective of frequency or technology. For HBT circuits, J_{OPT} is frequency and technology dependent and increases if emitter degeneration is employed.
2. Determine the technology constants R, G, G_c, and B for the transistor when biased at J_{OPT}, either by hand calculations, by computer simulation using the design kit model, or from measurements if available.
3. Bias the transistor Q_1 at J_{OPT} and size its emitter length or gate width using equation (8.53) or (8.54). In a bipolar input stage, this fixes the collector bias current $I_C = J_{OPT} \times w_E \times l_{EOPT}$, where w_E is the emitter width. For a MOSFET stage, the drain current is simply $I_D = J_{OPT} \times W$, where J_{OPT} has units of mA/μm. In a differential implementation, the tail current in the differential pair becomes $I_T = 2I_C$ or $I_T = 2I_D$.
4. Choose the value of the load resistor based on the gain or maximum output swing (ΔV) requirements. $R_C = \Delta V/I_T$.
5. Add peaking inductors to extend the amplifier bandwidth by up to 60% while maintaining constant group delay.

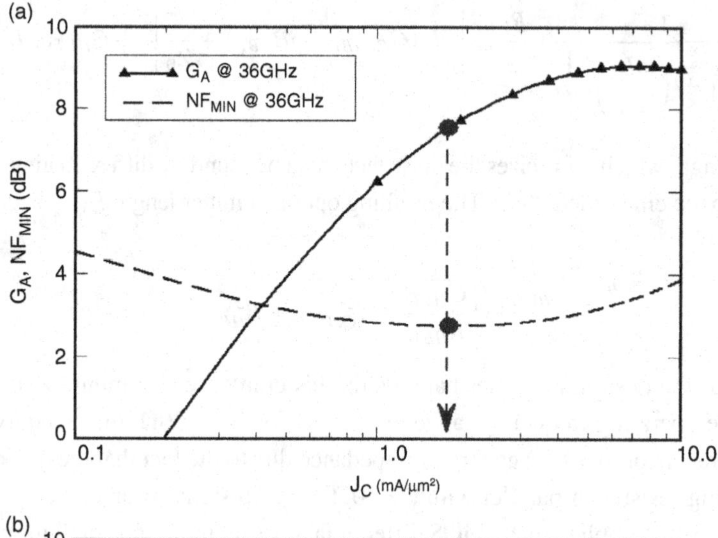

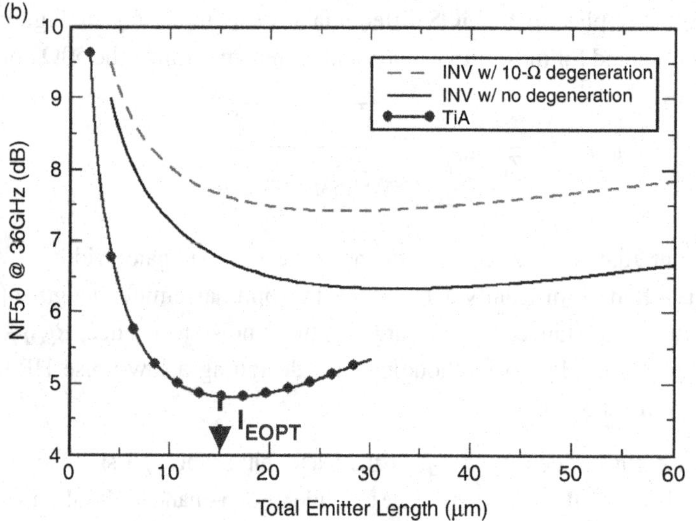

Figure 8.21 (a) Simulated noise figure and power gain as a function of collector current density at 36GHz for a 0.2μm emitter SiGe HBT (b) simulated 50Ω noise figure as a function of emitter length for several SiGe HBT broadband amplifier stages

As in the case of tuned LNAs, the linearity of the amplifier is determined by the choice we have already made for I_C and I_D, and by the emitter/source degeneration resistance.

This design methodology ensures the lowest noise figure (close to the transistor NF_{MIN}) over the band of interest. However since the transistor is noise-matched to such a small (25Ω) generator impedance, the resulting device size and power consumption can be large. This large transistor size ultimately limits the amplifier bandwidth. Neglecting the inductors and the base resistance, the 3dB bandwidth of a SiGe HBT input stage can be approximated using the sum of the open-circuit time constants

$$\frac{1}{2\pi f_{3dB,\,HBT-INV}} \approx \frac{Z_0}{2}\left[C_{beeff} + (1 - A_V)C_{bc}\right] + \frac{\Delta V}{2I_C}\left[C_{bc} + C_{cs}\right]. \tag{8.55}$$

The first term represents the input time constant, which is determined by the generator and matching resistor as well as by the input capacitance. The output time constant is the product of the load resistance (which is a function of the tail current and logic swing), the load capacitance, ignored in (8.55), and the output capacitance of the transistor. Recall that, at a given bias current density, the transistor capacitances scale as a function of the emitter length. Hence, C_{beeff}, C_{bc}, and C_{cs} can be expressed in terms of l_{EOPT}

$$C_{beeff} = C'_{beeff} w_E l_{EOPT},\ C_{bc} = C'_{bc} w_E l_{EOPT} \text{ and } C_{cs} = C'_{cs} l_{EOPT} \tag{8.56}$$

where C'_{beff}, C'_{bc}, and C'_{cs} are technology constants.

Additionally, as derived earlier, the small signal gain of the bipolar amplifier is

$$A_V = -g_{meff} R_C = -\left(\frac{I_C}{V_T + I_C r_E}\right) \cdot \left(\frac{\Delta V}{2I_C}\right) \approx \frac{-\Delta V}{4V_T}. \tag{8.57}$$

Thus, the amplifier bandwidth can be recast as a function of the optimal emitter length

$$\frac{1}{2\pi f_{3dB,\,HBT-INV}} \approx l_{EOPT} w_E \frac{Z_0}{2}\left[\frac{C'_{beeff}}{2} + \left(1 + \frac{\Delta V}{4V_T}\right) C'_{bc}\right] + \frac{\Delta V}{2J_{OPT} w_E}\left[w_E C'_{bc} + C'_{cs}\right]. \tag{8.58}$$

The second term is, to a first-order analysis, independent of the emitter length due to the fact that the load resistance is reduced as the bias current and device size are increased in the design methodology. This time constant can be improved by reducing the maximum single-ended output swing, although the latter is usually set between 250 and 300mV to ensure complete switching of a bipolar CML logic gate over all process and temperature corners. If the logic swing is set, the output time constant becomes a technology constant and can only be improved by scaling. At the input, the time constant increases linearly with the emitter length. The resistance at this node is fixed by matching requirements, which, when combined with the high input capacitance from the large transistor size, leads to low bandwidth.

Similarly, the bandwidth of the MOS inverter input stage can be determined as

$$\frac{1}{2\pi f_{3dB,\,MOS-INV}} \approx \frac{Z_0}{2}\left[C_{gseff} + (1 - A_V) C_{gd}\right] + \frac{\Delta V}{2I_D}\left[C_{gd} + C_{db}\right]. \tag{8.59}$$

Since the transistor capacitances scale with the gate width as

$$C_{gseff} = C'_{gseff} W_{OPT},\ C_{gd} = C'_{gd} W_{OPT} \text{ and } C_{db} = C'_{db} W_{OPT} \tag{8.60}$$

and the gain of the MOS inverter is given by

$$A_V = -g_{meff} R_L = -g'_{meff} W_{OPT}\left(\frac{\Delta V}{2I_D}\right) = -\frac{g_{meff}\Delta V}{2J_{OPT}} \tag{8.61}$$

the bandwidth of the impedance-matched MOS inverter can be expressed in terms of the optimal gate width

$$\frac{1}{2\pi f_{3dB,MOS-INV}} \approx W_{OPT}\frac{Z_0}{2}\left[C'_{gseff} + \left(1+\frac{g'_{meff}\Delta V}{2J_{OPT}}\right)C'_{gd}\right] + \frac{\Delta V}{2J_{OPT}}[C'_{gd} + C'_{db}]. \quad (8.62)$$

As with the HBT inverter, the output time constant becomes a technology constant and the input time constant grows with the total gate width. Consequently, the actual 3dB bandwidth may be lower than the desired value. One could improve bandwidth by trading off gain and using resistive emitter/source degeneration. This improvement comes at the expense of higher noise–series–series feedback increases the optimal source impedance as well as the minimum noise figure, resulting in even larger device sizes to noise match the amplifier.

Since the input time constant is the limiting factor in both the matched HBT inverter and MOS inverter, another option for bandwidth improvement is to add an emitter-follower or source-follower buffer at the input. This arrangement also improves the input linear range of the amplifier. The penalty for using the emitter-follower is higher noise and higher power consumption.

Figure 8.21(a) shows simulation of the optimum current density at 36GHz in a SiGe HBT with 160GHz f_T, while Figure 8.21(b) compiles the simulated 50Ω noise figure at 36GHz as a function of emitter length for optimally noise-biased SiGe HBT common-emitter (INV), common-emitter with resistive degeneration (INV with 10Ω degeneration) and TIA stages. These simulations clearly show that the best noise figure and lowest power consumption with minimum transistor size, and therefore, largest bandwidth, are obtained with a shunt–shunt feedback TIA topology. The simulation results are corroborated by the measurements of 50Ω noise figure for various SiGe HBT and CMOS TIA stages, as shown in Figure 8.22. It is noteworthy that the noise figures of the 160GHz SiGe HBT TIA and 90nm n-MOS TIA are comparable. Similarly the noise figure of the SiGe HBT and SiGe CE stages are comparable but the power consumption and bandwidth are better for the TIA topology. The emitter-follower + differential pair input stage has the worst noise figure.

8.5 DC OFFSET COMPENSATION AND VGA-TIA TOPOLOGIES

In many low-cost data communication applications, the TIA stage is integrated together with a differential limiting amplifier (LA). This high-gain circuit is known as a TIALA [4],[7],[10]. In such situations, the wide range of input photodiode currents can cause a significant DC offset to appear at the differential output of the TIALA which, in turn, can lead to false decisions by the decision circuit. The inherent offset is typically canceled with a DC offset compensation loop whose role is to suppress any gain on the signal path at low frequencies (typically 10s of kHz to 1MHz) but to allow higher-frequency signals to pass unaffected [1].

The frequency response analysis of the TIALA with a DC offset compensation loop can be conducted using the block diagram in Figure 8.23 where:

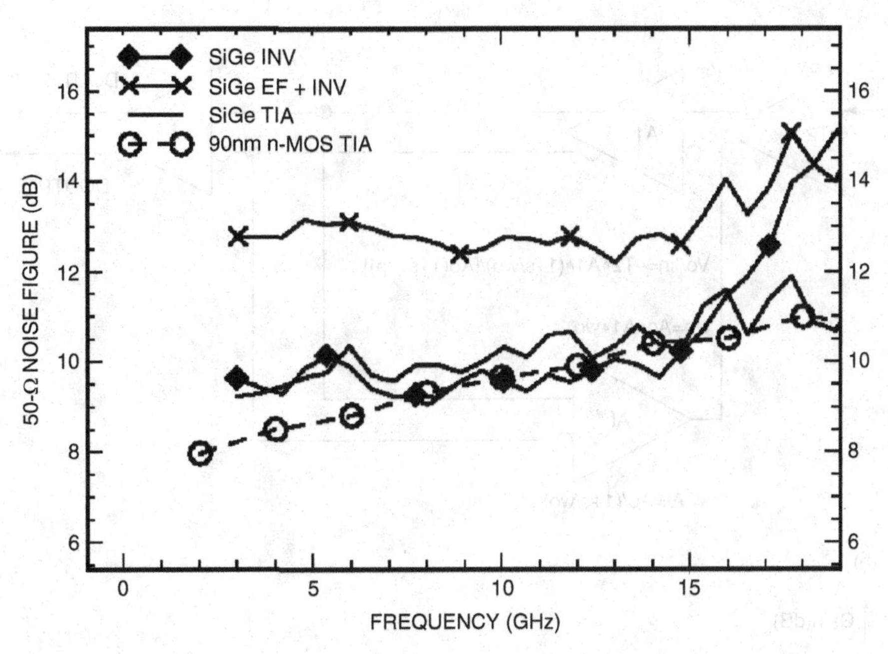

Figure 8.22 Measured 50Ω noise figure for SiGe HBT and CMOS TIAs, differential pairs and EF-INV input stages

- T_z is the frequency-dependent transimpedance gain of the single-ended TIA stage
- A_1 is the frequency-dependent gain of the limiting amplifier,
- $A_f(s) = \frac{A_0}{1+\frac{s}{\omega_0}}$ is the gain of the DC offset feedback amplifier

with low-frequency voltage gain A_0 and 3dB bandwidth f_0.

The transfer characteristic of the TIALA is given by

$$Z_T(s) \stackrel{\text{def}}{=} \frac{v_o}{i_{in}} = \frac{T_Z A_1}{1 + A_f A_1} = \frac{T_z(s) \cdot A_1(s) \cdot \left(1 + \frac{s}{\omega_0}\right)}{[1 + A_0 A_1(s)]\left(1 + \frac{s}{\omega_p}\right)} \tag{8.63}$$

where

$$\omega_p = [A_0 \cdot A_1(0) + 1] \cdot \omega_0 \tag{8.64}$$

is the low-frequency pole of the TIALA and the DC gain is given by

$$Z_T(0) = \frac{T_Z(0)}{A_0}. \tag{8.65}$$

At high frequencies ($f >> f_p$), the transimpedance gain of the TIALA becomes

$$Z_T(s) = T_z(s) \cdot A_1(s) \tag{8.66}$$

and the upper 3dB frequency is determined by the poles of $T_z(s)$ and $A_1(s)$.

Equations (8.64)–(8.66) illustrate the compromise that must be made when selecting the values for A_0 and f_0 such that the DC gain be made smaller than 1, while ensuring that f_p is

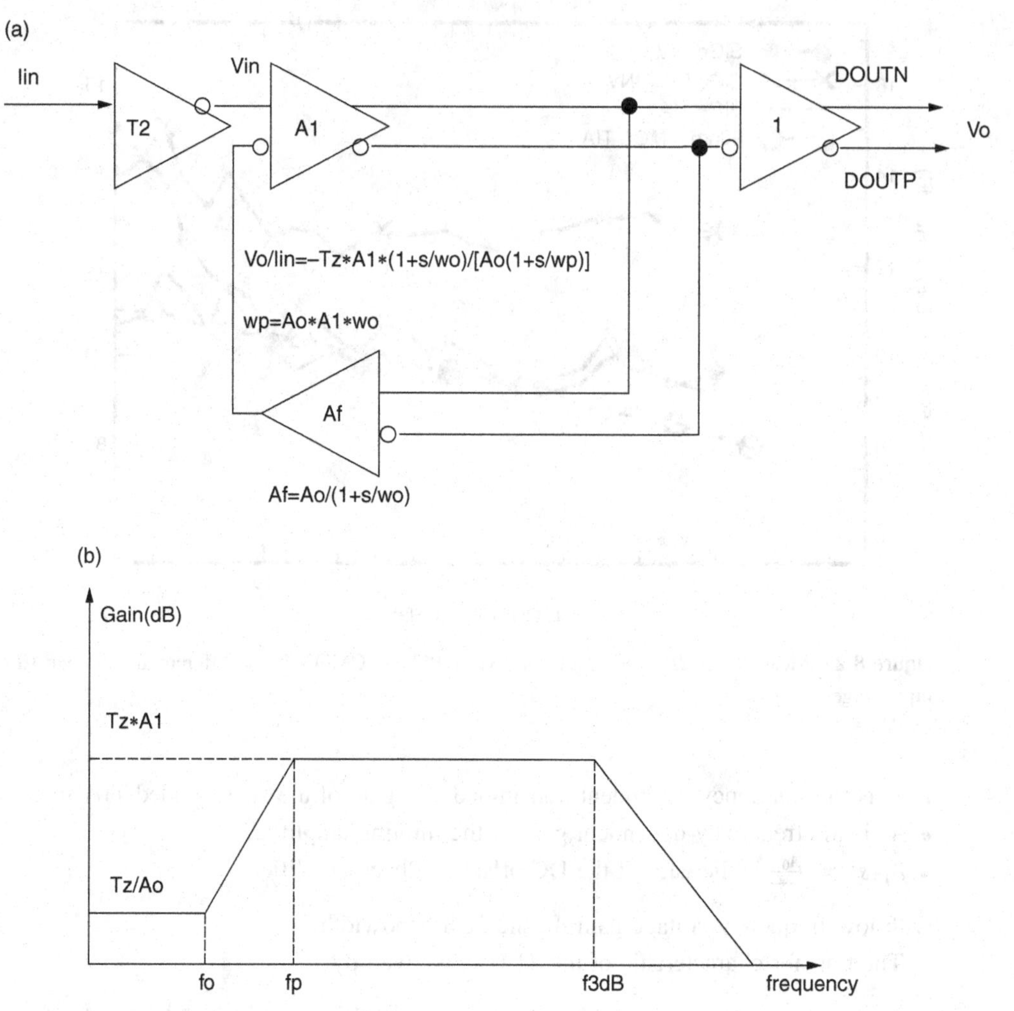

Figure 8.23 (a) TIALA bock diagram and DC-offset compensation loop bandwidth and gain analysis (b) TIALA frequency response

lower than a few hundred KHz. Typical values for the in-band gain T_zA_1 are in the 1kΩ to 10-kΩ range.

Figure 8.24 shows the block diagram a differential 10Gb/s receiver channel for high-density optical interconnects [10]. It consists of a pseudo-differential TIA and a multi-stage differential limiting amplifier. Each single-ended TIA features a DC offset compensation feedback loop. One TIA amplifies the photodiode current while the other is just a replica to maintain symmetry and to reject supply noise. The DC offset compensation feedback loop consists of a low-frequency operational transconductance amplifier (OTA) and a compensation capacitor, as shown in Figure 8.25. The TIA employs a modified regulated-cascode topology, similar to that discussed in Chapter 7. The common-gate input transistor allows for a low-input impedance of approximately 42Ω to be achieved, along with a 3dB bandwidth of 7GHz while consuming only 1.3mA from a 1.1V supply. The overall transimpedance gain is 1.13kΩ.

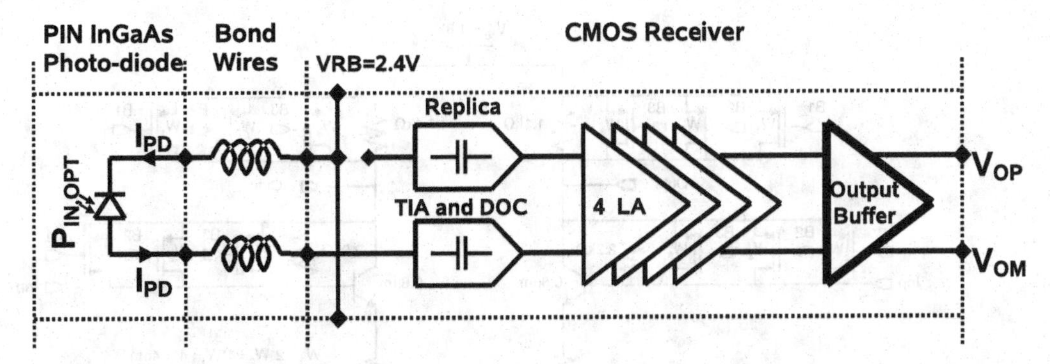

Figure 8.24 Block diagram of a low-power fiber-optic receiver for chip–chip communication [10]

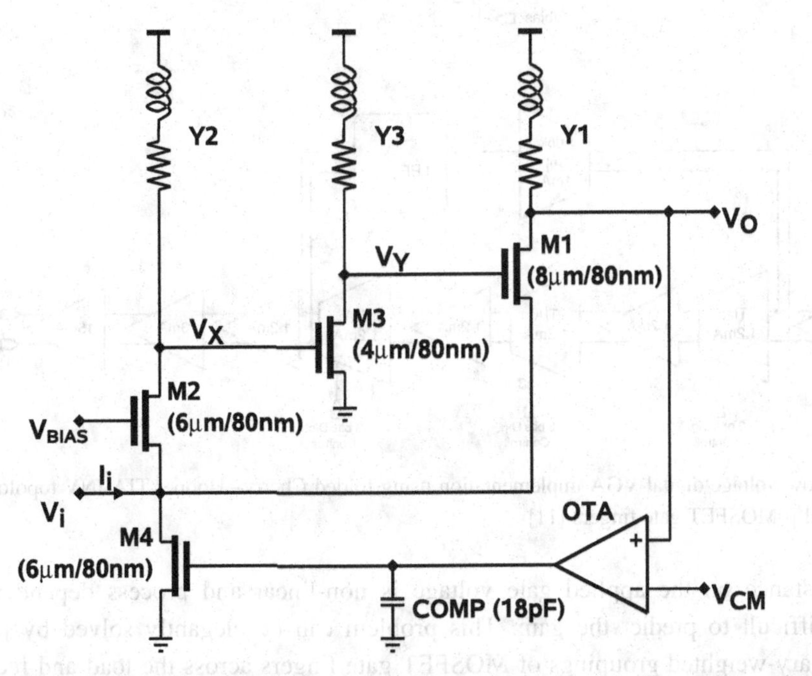

Figure 8.25 Modified regulated cascode TIA topology with DC offset compensation [10]

In some applications where the received signal level varies over a large range, it is necessary to adjust the gain of the transimpedance amplifier. Since the transimpedance gain depends primarily on the feedback resistance value, R_F, the first idea that comes to mind is to replace R_F with a variable resistor. However, according to (8.22), if R_F changes the bandwidth of the amplifier also changes and may cause peaking and group delay variation in the frequency response. A solution to this problem is to vary the open loop amplifier gain, A_V, in tandem with the feedback resistance value. This can be done by changing the load resistance value R_C, as illustrated in Figure 8.26.

The variable resistance has been typically implemented by applying a control voltage to the gate of a FET operated in the triode region. However, the dependence of the FET

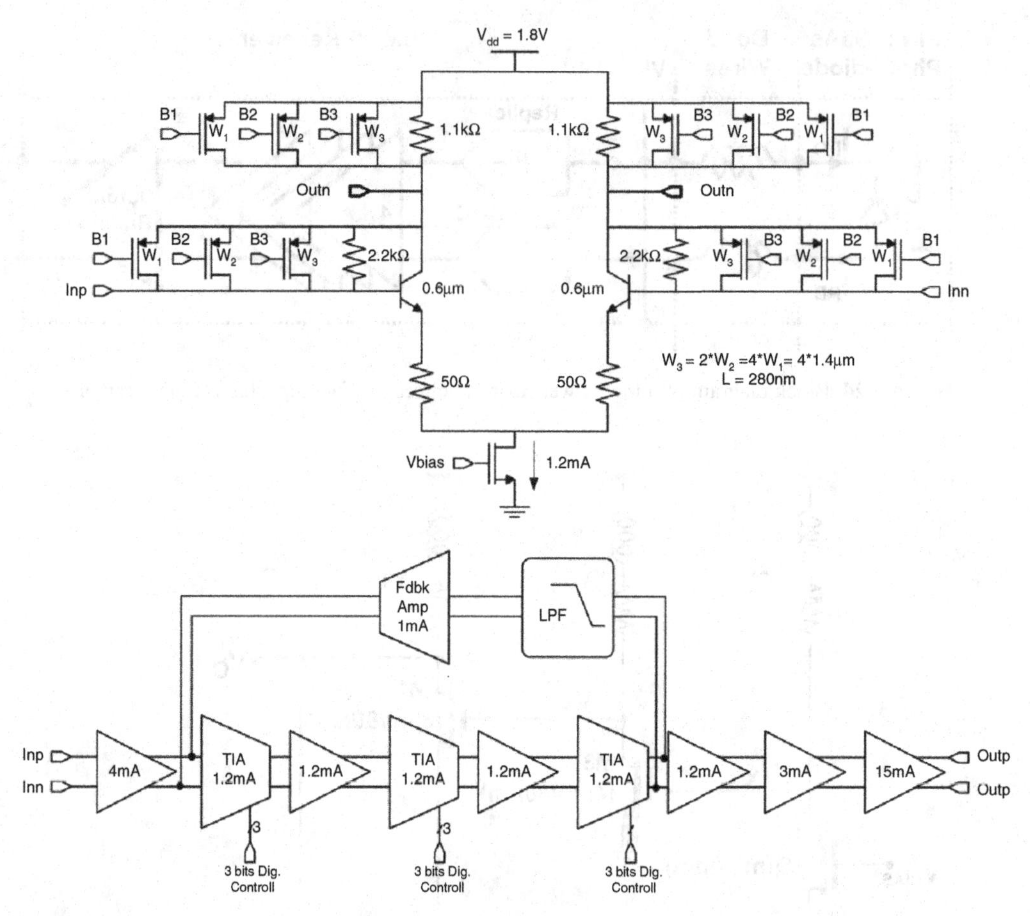

Figure 8.26 Low-voltage digital VGA implementation using folded Cherry–Hooper TIA-INV topology with segmented p-MOSFET gate fingers [11]

channel resistance on the applied gate voltage is non-linear and process dependent [6], making it difficult to predict the gain. This problem can be elegantly solved by placing banks of binary-weighted groupings of MOSFET gate fingers across the load and feedback resistors and by simply turning those MOSFET channels fully on or fully off. A multi-bit DAC is thus obtained without changing the bias current density and bandwidth of the TIA stage.

A circuit based on this concept, shown in Figure 8.26(a), features three control bits realized with thick oxide p-MOSFETs as digitally controlled resistors. The least significant bit controls a single gate finger, 1.4µm wide. The second bit controls two gate fingers, each 1.4µm wide, while the third bit controls four gate fingers. By cascading three such digital variable gain TIA stages (DVG_TIA), an amplifier with 9 control bits can be implemented. To maximize the bandwidth while minimizing power consumption and ensuring that the digital control in one stage does not affect the bias point in the next TIA stage, differential inverter pairs are placed between the TIA stages, as shown in Figure 8.26(b). This INV-TIA arrangement forms a folded Cherry–Hooper stage, known for its excellent bandwidth. Note that the MOSFETs contribute additional capacitance at the input and output nodes and will lead to some degradation of the TIA bandwidth compared to a TIA stage without gain control. However, in this implementation the total gate width of the MOSFETs forming the three bits of control is only $7 \times 1.4 = 9.8\mu m$, which contributes only about 15fF of capacitance between the drain and bulk and the source and bulk nodes.

Other variable-gain and digital-variable gain amplifier topologies based on the Gilbert cell topology will be discussed in Chapter 9.

Summary

- TIAs feature the broadest bandwidth with the lowest noise and lowest power dissipation amongst lumped broadband amplifier topologies.
- TIAs offer the best solution for broadband input stages in fiber-optics, backplane, and UWB receivers.
- An optimal input transistor size exists for a given photodiode capacitance and 3dB bandwidth of the TIA.
- In all broadband amplifier topologies, the transistors must be biased at J_{OPT} to minimize noise.
- The CMOS inverter TIA is the lowest power, highest bandwidth, and lowest noise topology for broadband low-noise amplifiers fabricated in nanoscale CMOS technologies.
- Digital gain control can be incorporated into the shunt–shunt feedback TIA by simultaneously changing the feedback resistance value and the gain of the forward amplifier.

Problems

(1) Repeat Example 8.1 for a system operating at a serial data stream of 56Gb/s with a photodiode capacitance of 30fF and a pad capacitance of 30fF. Design for a voltage gain of 3.

(2) Describe a technique to obtain the four technology noise parameters R_{FET}, G_{FET}, G_{CFET}, and B_{FET} using computer simulation with the model provided in the design kit.

(3) Derive the expressions of the four technology noise parameters for a SiGe HBT process using the intrinsic transistor noise parameter expressions presented in Chapter 4.1.

(4) Develop a step-by-step computer design methodology for the HBT TIA in Figure 8.9.

(5) Develop a step-by-step computer design methodology for the HBT TIA stage in Figure 8.10.

(6) Develop a step-by-step computer design methodology for the CMOS TIA in Figure 8.14.

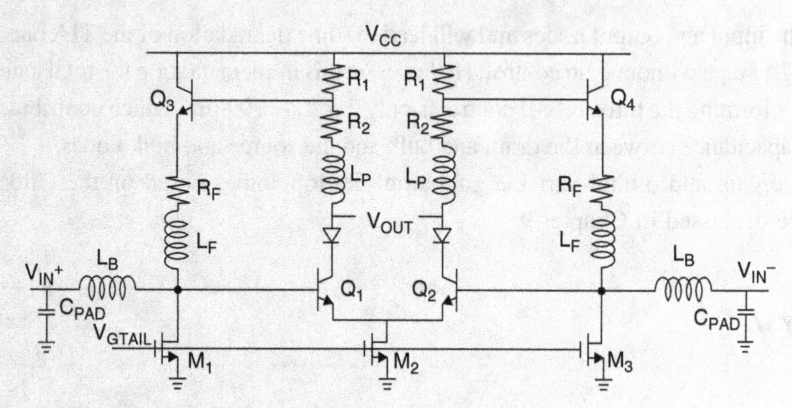

Figure 8.27 45Gb/s SiGe BiCMOS differential TIA [8]

(7) For the 43Gb/s, 130nm SiGe BiCMOS TIA with series and shunt peaking in Figure 8.27, calculate the value of R_F for 50Ω matching and determine the value of L_P which maximizes the bandwidth with a maximally flat group delay. The optimal noise current density at 36GHz is $J_{OPT} = 1.5\text{mA}/\mu\text{m}^2$ $I_T = 8\text{mA}$, $I(Q_{1,2}) = 2 \times I(Q_{3,4}) = 4\text{mA}$, $Q_{1,2,3,4} = 2 \times 0.17\mu\text{m} \times 8\mu\text{m}$, $R_1 = R_2 = 27\Omega\cdot$ Other technology data are: $Ic@peak$ $f_T = 8\text{mA}/\mu\text{m}^2$; $w_E = 0.17\mu\text{m}$; $\beta = 100$; $V_{CESAT} = 0.4\text{V}$, $V_{BE} = 0.85\text{V}$
$\tau_F = 0.6\text{ps}$; $C_{je} = 10\text{fF}/\mu\text{m}^2$; $C_\mu = 11\text{fF}/\mu\text{m}^2$; $C_{cs} = 1.1\text{fF}/\mu\text{m}$;
$R_b = 150\Omega\mu\text{m}$, $r_E = 4\Omega\mu\text{m}^2$, $R_c = 20\Omega\mu\text{m}$.

(8) Figure 8.28 shows an ultra-wideband (UWB) radio low-noise amplifier operating in the 2–10GHz band implemented as a two-stage transimpedance feedback amplifier [12].

(a) Draw the schematic of the differential-mode half circuit and identify the main amplifier network and the feedback network. Indicate the type of feedback (shunt–shunt, shunt–series, etc.) Assume that C_X is a short circuit at frequencies above 2GHz.

(b) Write the expression of the Y-parameter matrix and of the noise admittance correlation matrix C_Y of the feedback network.

(c) Using the noise parameter admittance formalism $\{R_n$, G_u, Y_{SOPT}, and $F_{MIN}\}$ and assuming that the noise parameters of the amplifier network are identical to those of the input n-MOSFET, derive the condition (as a function of the transistor g_m and noise params: R, P, and C) that must be satisfied by R_F such that its noise contribution to the overall amplifier noise figure is at least ten times smaller than that of the transistor. Assume that the following noise parameter equations apply for a MOSFET;

$$\langle i_{n1}, i_{n1}^* \rangle = 4KT\Delta fRg_m\frac{f^2}{f_T^2} \quad \langle i_{n2}, i_{n2}^* \rangle = 4KT\Delta fPg_m$$

$$\langle i_{n1}, i_{n2}^* \rangle = 4KT\Delta fjC\sqrt{PR}g_m\frac{f}{f_T}.$$

(9) The circuit in Figure 8.29 represents the differential-mode half circuit of a broadband differential CMOS cascode TIA amplifier with variable gain. It forms the low-noise

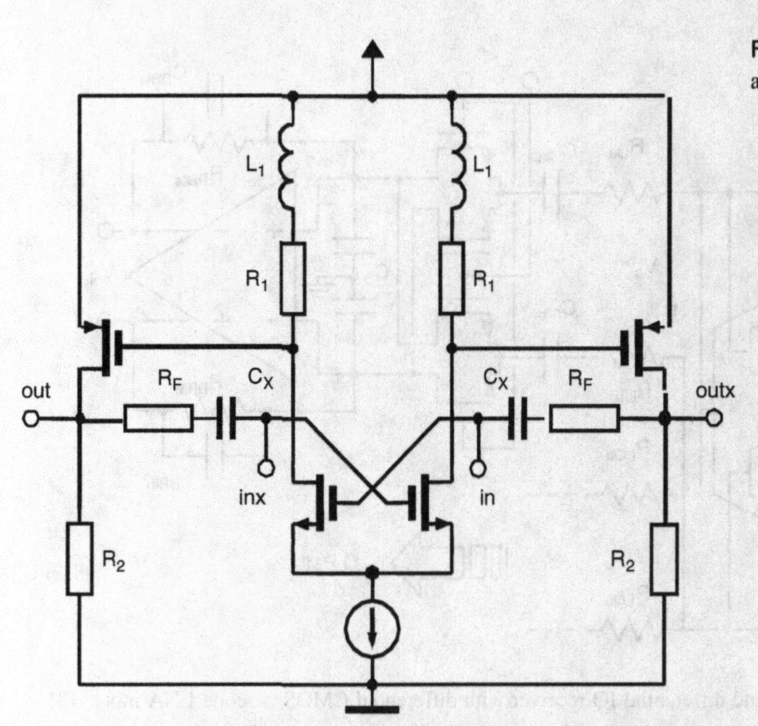

Figure 8.28 UWB low-noise amplifier schematics [12]

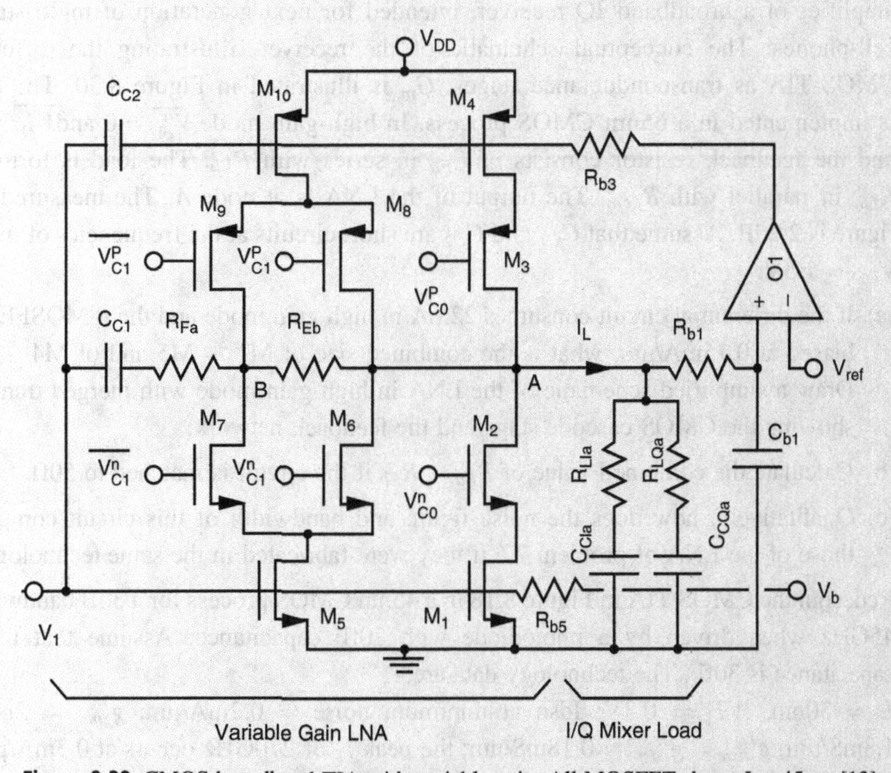

Figure 8.29 CMOS broadband TIA with variable gain. All MOSFETs have L = 45nm [13]

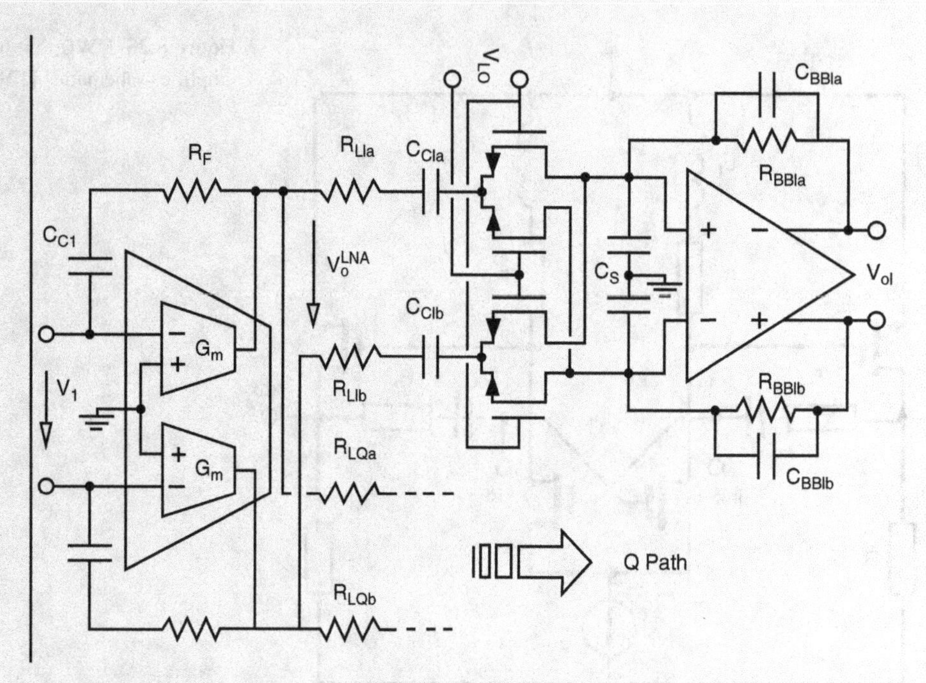

Figure 8.30 Broadband differential IQ receiver with differential CMOS cascode LNA input [13]

amplifier of a broadband IQ receiver, intended for next generation of multi-standard cell-phones. The conceptual schematic of the receiver, illustrating the differential CMOS-TIA as transconductance stages, G_m is illustrated in Figure 8.30. The circuit is implemented in a 65nm CMOS process. In high-gain mode, $\overline{V_{c1}^{n}} = 0$ and $\overline{V_{c1}^{p}} = V_{DD}$ and the feedback resistor consists of R_{Fa} in series with R_{Fb}. The load is formed by R_{LIa} in parallel with R_{LQa}. The output of the LNA is at node A. The measured noise figure is 2.2dB. Assume that C_{C1} and C_{C2} are short circuits at the frequencies of interest. $V_{DD} = 1.5$V.

(a) If the differential circuit consumes 22mA in high gain mode and the n-MOSFETs are biased at 0.15mA/µm, what is the combined size of M1 + M5 and of M4 + M10. Draw a simplified schematic of the LNA in high gain mode with merged transistors showing the CMOS cascode stage and the feedback network.

(b) Calculate the combined value of $R_{Fa} + R_{Fb}$ if the circuit is matched to 50Ω.

(c) Qualitatively, how does the noise figure and bandwidth of this circuit compare to those of the LNA of problem 7.6 if they were fabricated in the same technology?

(10) Redesign the CMOS TIA in Figure 8.18 in a 45nm CMOS process for a 3dB bandwidth of 45GHz when driven by a photodiode with 20fF capacitance. Assume that the pad capacitance is 30fF. The technology data are:
L = 30nm, $|VT|$ = 0.3V; Idsn at minimum noise = 0.2mA/µm; $g'_{mn} = 2g'_{mp}$ = 1.8mS/µm; $g'_{dsn} = g'_{dsp}$ = 0.18mS/µm; the peak f_T of 270GHz occurs at 0.3mA/µm for 0.3V < V_{DS} < 1V.

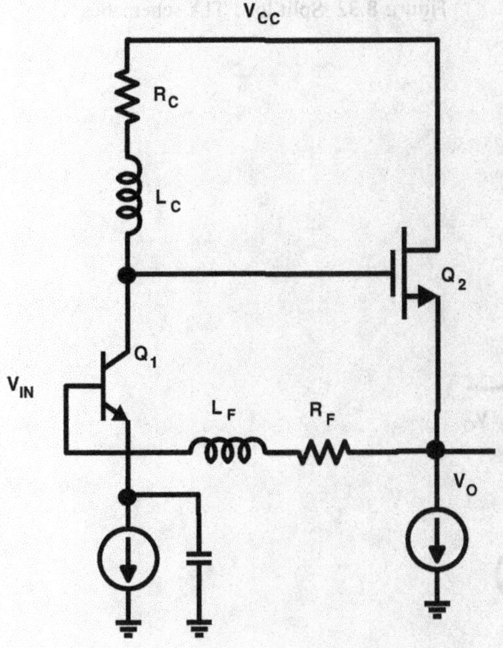

Figure 8.31 SiGe BiCMOS TIA schematic

For both n-channel and p-channel devices: $C'_{gs} = 0.7\text{fF/}\mu\text{m}$; $C'_{gd} = 0.3\text{fF/}\mu\text{m}$; $C'_{sb} = C'_{db} = 0.4\text{fF/}\mu\text{m}$;

R'_s (n-MOS) $= R'_d = 200\Omega \times \mu\text{m}$; R'_s (p-MOS) $= R'_d = 400\Omega \times \mu\text{m}$;

R_{shg}(n/p-MOS) $= 20\Omega/\text{sq}$; poly contact resistance $R_{CON} = 70\Omega$.

Total R_g/0.7μm finger contacted on one side $= 200\Omega$. Noise params: $P = 1$, $C = 0.4$, $R = 0.25$.

(11) For the circuit in Figure 8.26(a), calculate the transimpedance gain and the 3dB bandwidth when all the control bits are "0" and all the control bits are "1." Assume that the off resistance of the MOSFET is infinite and that the on resistance is $1000\Omega \times \mu\text{m}$ and that, in the triode region $C'_{gs} = C'_{gd} = 0.7\text{fF/}\mu\text{m}$, $C'_{sb} = C'_{db} = 1.5\text{fF/}\mu\text{m}$. The 130nm SiGe HBT parameters are $Ic@peak\, f_T = 14\text{mA/}\mu\text{m}^2$, $w_E = 0.13\mu\text{m}$, $\beta = 500$; $V_{CESAT} = 0.3\text{V}$, $V_{BE} = 0.9\text{V}$, peak $f_T = 230\text{GHz}$, $\tau_F = 0.3\text{ps}$, $C_{je} = 17\text{fF/}\mu\text{m}^2$, $C'_{bc} = 25\text{fF/}\mu\text{m}^2$, $C'_{cs} = 1.2\text{fF/}\mu\text{m}$, $R'_b = 75\Omega \times \mu\text{m}$, $r'_E = 2\Omega \times \mu\text{m}^2$, $R'_c = 20\Omega\times\mu\text{m}$.

(12) Design a 65nm SiGe BiCMOS version of the circuit in Figure 8.9 for a 110Gb/s coaxial cable receiver. The input should be matched to 50Ω. Assume a 20fF pad capacitance and replace transistor Q_2 with a low-V_T 65nm n-MOSFET with 10μm gate width, as illustrated in Figure 8.31.

Assume the following technology data:

65nm n-(p)MOSFET parameters:

$L = 45\text{nm}$, $|VT| = 0.3\text{V}$; Idsn at minimum noise $= 0.15\text{mA/}\mu\text{m}$; $g'_{mn} = 2g'_{mp} = 1.3\text{mS/}\mu\text{m}$; $g'_{dsn} = g'_{dsp} = 0.18\text{mS/}\mu\text{m}$; the peak f_T of 170GHz occurs at 0.3mA/μm for $0.3\ V < V_{DS} < 1\text{V}$.

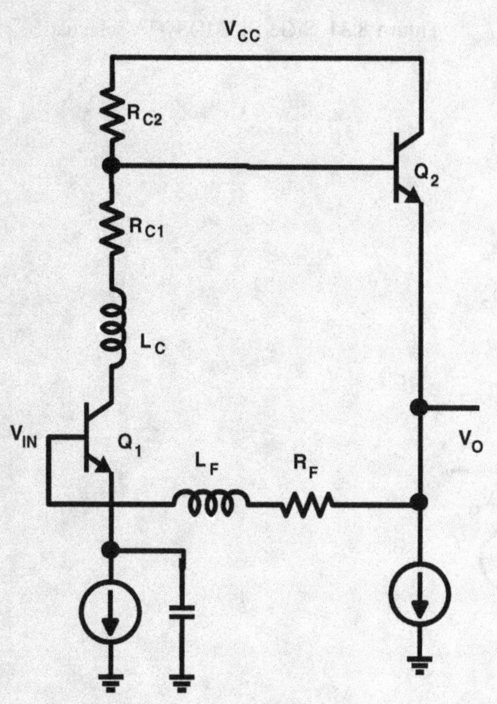

Figure 8.32 Split load TIA schematics

For both n-channel and p-channel devices: $C'_{gs} = 0.8$fF/μm; $C'_{gd} = 0.4$fF/μm; $C'_{sb} = C'_{db} = 0.7$fF/μm;
R'_s (n-MOS) $= R'_d = 200\Omega \times$ μm; R'_s (p-MOS) $= R'_d = 400\Omega \times$ μm;
R_{shg}(n/p-MOS) $= 15\Omega$/sq; poly contact resistance $R_{CON} = 40\Omega$.
Total R_g/0.7μm finger contacted on one side is 158Ω. Noise params: $P = 1$, $C = 0.4$, $R = 0.25$.

280GHz SiGe HBT parameters for emitter width of 0.1μm. Technology data are provided per emitter length l_E or per emitter area $w_E \times l_E$:
Ic@peak $f_T = 14$mA/μm^2; $w_E = 0.1$μm; $\beta = 500$; $V_{CESAT} = 0.3$V, $V_{BE} = 0.9$V, peak $f_T =$ 280GHz.

$\tau_F = 0.25$ps; $C_{je} = 17$fF/μm^2; $C'_{bc} = 25$fF/μm^2; $C_{cs} = 1.2$fF/μm;
$R'_b = 75\Omega \times$ μm, $r'_E = 2\Omega \times$ μm^2, $R'_c = 20\Omega \times$ μm

(13) Repeat problem 8.12 in the same technology but using a SiGe HBT for Q_2. Also, as shown in Figure 8.32, to avoid the V_{CE} of Q_1 being larger than 1.6V, use a split resistive load ($R_{C1} + R_{C2}$) for R_C, with the base of Q_2 being connected at the node between the two resistors R_{C1} and R_{C2}. Derive the expression of the low-frequency gain and input impedance when $R_{C2}/R_{C1} = 2$ with R_{C2} being connected to the supply and R_{C1} to the collector of Q_1.

(14) Consider the TIAs in Figure 8.33 based on the f_T doubler concept introduced in Chapter 5. Assume the same technology parameters as in the previous two problems and assume that Q_1 is half the size of Q_2.

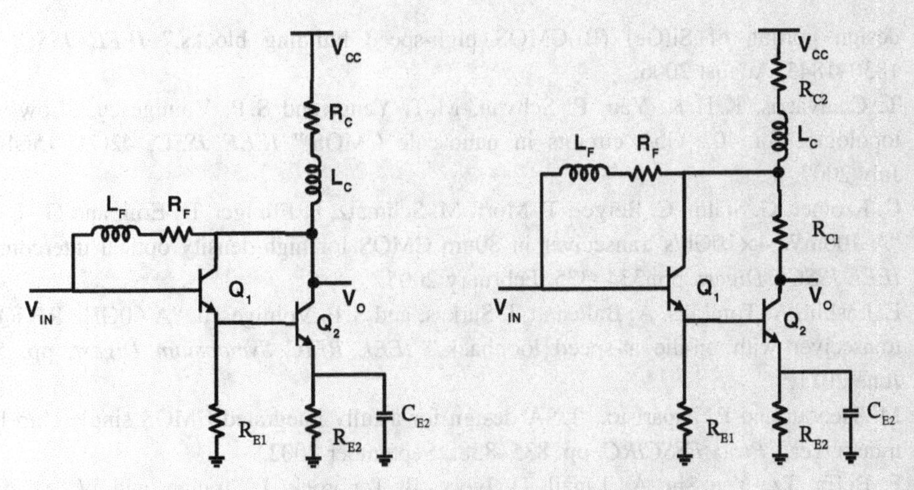

Figure 8.33 TIA schematics based on the Darlington f_T doubler concept

(a) Explain how the bandwidth is improved compared to the TIAs in Figures 8.31 and 8.32. What are the disadvantages?

(b) Design the value of the load and feedback resistors such that a Bessel response is obtained with a 3dB frequency of 88GHz. What should be the value of R_{E2} if Q_1 and Q_2 are biased at J_{OPT} and the circuit should remain linear for input voltages as high as 800mV$_{pp}$? Assume that the emitter lengths of Q_1 and Q_2 are 1μm and 2μm, respectively.

REFERENCES

[1] E. Säckinger, *Broadband Circuits for Optical Fiber Communication*, John Wiley & Sons, 2005, Chapter 5.

[2] S. Long, "High-speed circuit design principle," in *VLSI Handbook*, edited by W.-K. Chen, CRC Press and IEEE Press, 2nd Edition, 2006.

[3] K.W. Kobayashi, "An InP HBT common-base amplifier with tuneable transimpedance for 40Gb/s applications," *IEEE GaAs IC Symposium Tech. Digest*, pp. 155–158, November 2002.

[4] H. Tran, F. Pera, D.S. McPherson, D. Viorel, and S.P Voinigescu. "6-kΩ 43GB/s differential transimpedance-limiting amplifier with auto-zero feedback and high dynamic range," *IEEE JSSC*, **39**: 1680–1689, October 2004.

[5] D. Hornbuckle and R. Van Tuyl, "Monolithic GaAs direct-coupled amplifiers," *IEEE ED*, pp. 166–170, February 1981.

[6] Behzad Razavi, *Design of Integrated Circuits for Optical Communications*, McGraw-Hill, 2003, Chapter 4.

[7] F. Pera and S.P. Voinigescu, "An SOI CMOS, high gain and low noise transimpedance-limiting amplifier for 10Gb/s applications," *IEEE RFIC Symposium Digest*, pp. 401–404, June 2006.

[8] T.O. Dickson, K.H.K. Yau, T. Chalvatzis, A. Mangan, R. Beerkens, P. Westergaard, M. Tazlauanu, M.T. Yang, and S.P. Voinigescu, "The invariance of characteristic current densities in nanoscale MOSFETs and its impact on algorithmic design methodologies and

design porting of Si(Ge) (Bi)CMOS high-speed building blocks," *IEEE JSSC*, **41**(8): 1830–1845, August 2006.

[9] T. Chalvatzis, K.H.K. Yau, P. Schvan, M.T. Yang, and S.P. Voinigescu, "Low-voltage topologies for 40 + Gb/s circuits in nanoscale CMOS," *IEEE JSSC*, **42**(7): 1564–1573, July 2007.

[10] C. Kromer, G. Sialm, C. Berger, T. Morf, M. Schmatz, F. Ellinger, D. Erni, and G.-L. Bona, "A 100mW 4×10Gb/s transceiver in 80nm CMOS for high-density optical interconnects," *IEEE ISSCC Digest*, pp. 334–335, February 2005.

[11] E. Laskin, A. Tomkins, A. Balteanu, I. Sarkas, and S.P. Voinigescu, "A 60GHz RF IQ DAC transceiver with on-die at-speed loopback," *IEEE RFIC Symposium Digest*, pp. 57–60, June 2011.

[12] M. Tiebout and E. Paparisto, "LNA design for a fully integrated CMOS single chip UMTS transceiver," *Proc. ESSCIRC*, pp. 835–838., September 2002.

[13] F. Beffa, Tze Yee Sin, A. Tanzil, D. Ivory, B. Tenbroek, J. Strange, and W. Ali-Ahmad, "A receiver for WCDMA/EDGE mobile phones with inductorless front-end in 65nm CMOS," *IEEE ISSCC Digest*, pp. 370–371, February 2011.

9 Mixers, switches, modulators, and other control circuits

WHAT IS A MIXER?

A mixer is a three-port circuit that employs a non-linear or time-varying device in order to perform the critical frequency translation function in wireless communication systems. The non-linear or time-varying parameter can be either a conductance/resistance or a transconductance. If the time-varying element is a resistance or conductance, the mixer is called *resistive*. Mixers that rely on a time-varying transconductance are known as *active* mixers.

When used in a transmitter, the mixer acts as an upconverter by shifting the data signal from a low frequency to the carrier frequency, making it suitable for transmission by the antenna. In the receiver, it serves as a downconverter by separating the data signal from the carrier and shifting it to a low frequency, where it can be demodulated and processed in a cost-effective manner. Ideally, in both cases, the signal at the output is a replica of the signal at one of the mixer inputs, translated to a lower or higher frequency, with no loss of information and no added distortion.

Most IC mixers are implemented with switches. In addition, image-reject mixers also require 90 degree phase shifters and in-phase power combiners or splitters. Finally, mixers can be employed to realize digital modulators. The final part of this chapter will review switches, phase shifters, and M-ary phase and QAM modulators based on phase shifters and Gilbert cell mixer topologies.

9.1 MIXER FUNDAMENTALS

The ideal mixer is a multiplier which produces an output consisting of the sum and difference frequencies of the two input signals. Its symbol, shown in Figure 9.1, is a multiplication sign placed within a circle, featuring two inputs, at frequencies ω_1 and ω_2, and one output at frequency ω_3. The second input is the local oscillator (LO) signal, and its amplitude is usually (but not always) larger than that of the first input.

In an upconversion mixer, the first input is a baseband or intermediate frequency (IF) signal with the output port providing the RF signal. In a downconversion mixer, the first input is the RF signal while the output port provides the intermediate frequency or baseband signal.

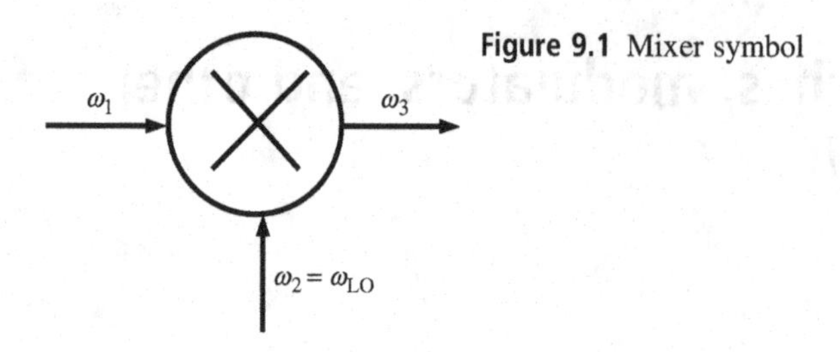

Figure 9.1 Mixer symbol

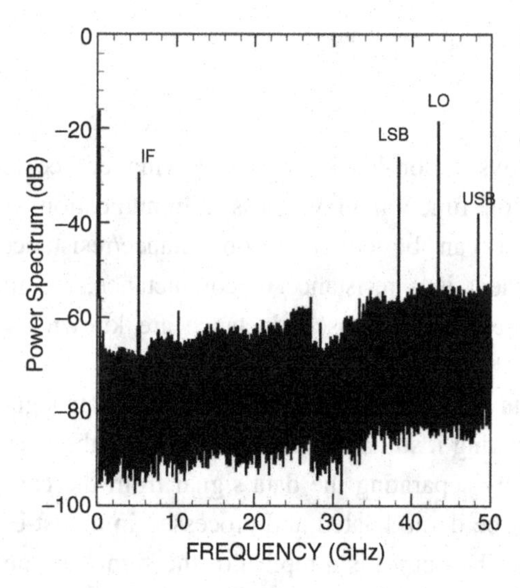

Figure 9.2 Measured spectrum at the RF output of an upconversion mixer showing the IF signal at 5GHz, the LO at 43GHz and the USB and LSB signals at 38GHz and 48GHz, respectively [1]

9.1.1 Frequency conversion principle

Equation (9.1) describes the ideal behavior of the mixer as a multiplier

$$A\cos(\omega_1 t) \times B\cos(\omega_2 t) = \frac{AB}{2}[\cos(\omega_1 - \omega_2)t + \cos(\omega_1 + \omega_2)t] \tag{9.1}$$

and illustrates that signals are produced at the output at the sum and difference frequencies of the two inputs. Usually, only one of the signals at the output of the ideal mixer is retained. The other one is discarded, half of the output power being lost.

In practice, the multiplier is implemented with a non-linear element such as a diode or transistor, or with a time-varying element. In either case, a large number of harmonics, as well as undesired products of the input signals, appear at the output. To eliminate the unwanted harmonics, filters are placed at the RF, LO, and IF ports of the mixer to select only the signal frequency of interest.

9.1.2 Upconverter mixer operation

In this situation, encountered in transmitters, the mixer input signal is applied at the IF port $\omega_1 = \omega_{IF}$, $\omega_2 = \omega_{LO}$ and both the $\omega_{LO} - \omega_{IF} = \omega_{RF}$ and the $\omega_{LO} + \omega_{IF} = \omega_{RF}$ sidebands are generated at the RF (output) port. The sum frequency $f_{RF} = f_{LO} + f_{IF}$ is known as the *upper sideband* (*USB*). The mixer also produces a *lower sideband* (*LSB*) at $f_{LO} - f_{IF}$, as illustrated in Figure 9.2.

A bandpass filter, placed at the RF port, selects the desired signal and rejects any leakage from the IF and LO ports. In other words, the output filter must suppress signals at f_{IF} and f_{LO} in addition to those at the unwanted sideband.

9.1.3 Downconverter mixer operation

In a receiver, the input signal is applied to the RF port and $\omega_1 = \omega_{RF}$, where $\omega_{RF} = \omega_{LO} + \omega_{IF}$ or $\omega_{RF} = \omega_{LO} - \omega_{IF}$. The LO signal is applied at port 2, as in an upconverter, $\omega_2 = \omega_{LO}$, and signals both at $\omega_{RF} - \omega_{LO} = \omega_{IF}$ and $\omega_{LO} - \omega_{RF} = \omega_{IF}$ are collected at the IF port.

9.1.4 Image frequency

As we have noticed, in the receiver, the mixer downconverts signals present both at the upper sideband, $\omega_{LO} + \omega_{IF}$, and at the lower sideband, $\omega_{LO} - \omega_{IF}$, to the same IF frequency. Typically, only one of the two sidebands around the LO signal contains the desired RF signal. The other sideband is called the *image frequency* (*IM*) or, in short, *image*. Once downconverted to the IF frequency, the RF and the IM signals are indistinguishable. In most applications, the image signal is undesirable and therefore must be filtered out using a bandstop (also called band reject) filter centered at the image frequency.

The choice of upper or lower sideband for the RF frequency is dictated by system design considerations. We note that the desired RF signal and the image signal are always separated in frequency by $2\omega_{IF}$.

Another important observation is that there are always two possible LO frequencies for a given RF and IF frequency combination

$$\omega_{LO} = \omega_{RF} \pm \omega_{IF}. \tag{9.2}$$

The best choice of LO frequency is determined by practical implementation constraints. For example, an upper-sideband LO is often preferred because it requires a VCO with a smaller tuning ratio for a given range of RF signal frequencies. Low-noise, voltage-controlled oscillators with wide tuning ratios are difficult to implement. However, if the VCO operates at mm-wave frequencies and is part of a phase locked-loop (PLL), a higher LO frequency may be beyond the reliable input range of the first frequency divider stage of the PLL. In this situation, a lower sideband LO can be an acceptable solution.

EXAMPLE 9.1

In a 5.2GHz WLAN, the RF signal is at 5.2GHz, the IF = 1GHz, while the LO could be at 4.2GHz, resulting in an IM at 3.2GHz. If an LO at 6.2GHz is selected and the IF remains at 1GHz, the new IM will be at 7.2GHz.

EXAMPLE 9.2

In a 60GHz wireless video area network (WVAN) radio, the RF signal occupies a 2GHz channel in the 57–66GHz band and the LO is tunable from 66 to 73GHz. This results in a 2GHz wide IF centered at 8GHz and an IM band from 73 to 82GHz, as illustrated in Figure 9.3. When the LO is at 66GHz, the receiver downconverts the 57–59GHz channel to 7–9GHz. The image channel occupies the 73–75GHz frequency slot. This frequency allocation scheme takes advantage of the naturally decreasing gain with increasing frequency of the low-noise amplifier which provides cost free image rejection. Alternatively, with the RF signal at 57–59GHz we could select an LO frequency of 50GHz which will also produce an IF at 7–9GHz. However, the image channel is now in the 41–43GHz range and may be more easily amplified by gain stages in the receiver.

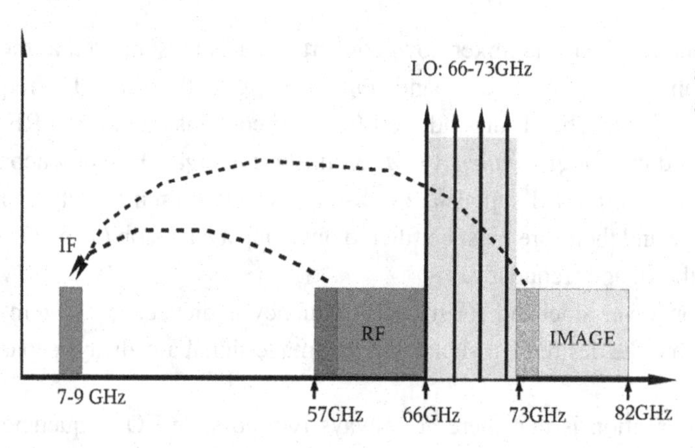

Figure 9.3 IF, RF, LO, and IM band allocation in 60GHz wireless video area network

9.1.5 How does one make a multiplier?

As illustrated in Figure 9.4, this can be accomplished using (i) a non-linear device, as already discussed in Chapter 5, or (ii) a time-varying device. Examples of *non-linear devices* that have been employed as multipliers in mixers are:

- diodes, exploiting the exponential I-V characteristics of the pn junction or Schottky barrier,
- bipolar transistors such as BJTs and HBTs, also relying on the exponential I-V characteristics of the pn junction, and
- MOSFETs or HEMTs which exhibit weaker non-linearity with square-law behavior at low effective gate voltages, but which behave linearly in nanoscale nodes for most of their bias range.

Although MOSFETs operated in the subthreshold region exhibit exponential I-V characteristics similar to those of a bipolar transistor, this regime of operation results in poor high-frequency performance and is rarely used. Instead, in mixers, FETs are almost always operated as a switch, swinging from the cutoff to the active region.

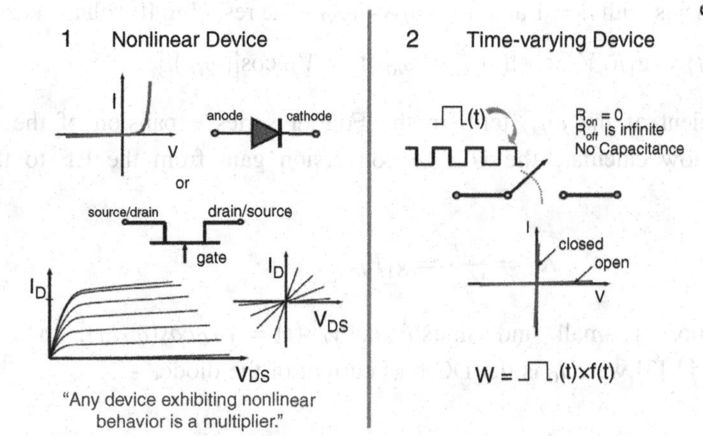

Figure 9.4 Devices and techniques employed to build a multiplier [2]

Diode multiplier

One of the earliest resistive mixers was implemented with Schottky diodes, whose I-V characteristics are described by

$$I(V) = I_S\left[\exp\left(\frac{qV}{nKT}\right) - 1\right] \tag{9.3}$$

When a large time-varying LO signal, $v_{LO}(t)$, is applied to the junction, the diode conductance also becomes time variant

$$g(t) = \frac{dI}{dV} = \frac{qI_S}{nKT}\exp\left(\frac{qv_{LO}(t)}{nKT}\right) \approx \frac{qI(v_{LO})}{nKT}. \tag{9.4}$$

In the general case, $v_{LO}(t)$ is not perfectly sinusoidal and $g(t)$ is strongly distorted by the non-linear I-V characteristics of the diode. If a small signal RF perturbation, $v_{RF}(t)$, is added to the diode voltage (dominated by the LO voltage), mixing will occur resulting in a time-varying diode current [3]

$$i(v(t)) = g(t)v_{RF}(t) \approx \frac{qI(v_{LO}(t))}{nKT}v_{RF}(t) = \left[g_0 + \sum_{n=1}^{\infty} 2g_n\cos(n \times \omega_{LO}t)\right]v_{RF}(t) \tag{9.5}$$

where the time-varying diode conductance has been expanded in a Fourier series to illustrate its rich harmonic content. Coefficients, g_n, depend on the average DC current through the diode, $I(V_{LO})$ [3]. Here, V_{LO} represents the amplitude of the first harmonic of the LO. The first term in (9.5), $g_0 = I(V_{LO})/V_T$, has the dimensions of a conductance and is responsible for the diode mixer also being called a "resistive" mixer. If the RF perturbation is sinusoidal, $v_{RF}(t) = V_{RF}\cos(\omega_{RF}t)$, the diode current expressed in (9.5) consists of RF, LO, and IF components, as well as of a large number of mixing frequencies given by $n \times \omega_{LO} \pm \omega_{IF}$, where n is an integer.

This current can be converted to voltage by passing it through a load resistor R_L at the IF port and filtering only the components with $n = 1$ at $\omega_{IF} = \omega_{LO} - \omega_{RF}$. The resulting IF voltage becomes

$$v_{IF}(t) = g_1 R_L V_{RF} \cos[(\omega_{LO} - \omega_{RF})t] = V_{IF} \cos[(\omega_{IF})t] \tag{9.6}$$

where g_1 is the coefficient of the ω_{LO} term in the Fourier series expansion of the diode conductance. We can now calculate the voltage conversion gain from the RF to the IF frequencies as

$$A_C = \frac{V_{IF}}{V_{RF}} = g_1 R_L. \tag{9.7}$$

If the LO voltage, too, is small and sinusoidal, $v_{LO}(t) = V_{LO} cos(\omega_{LO} t)$, then $g_1 = g'_o V_{LO}/2 = V_{LO} I_0/(2V_T^2)$ [3] were I_0 is the DC bias current of the diode.

FET multiplier

A classical example of a second-order non-linearity is that encountered in the I-V characteristics of a long-channel MOSFET operating in the saturation region

$$i_{DS} = \frac{\beta}{2}[v_{GS}(t) - V_T]^2. \tag{9.8}$$

If both a LO signal $v_{LO}(t) = V_{LO} cos(\omega_{LO} t)$ and a RF signal $v_{RF}(t) = V_{RF} cos(\omega_{RF} t)$ are applied to the gate terminal, the drain current becomes

$$\begin{aligned} i_{DS}(t) = &\frac{\beta}{2}[V_{LO}\cos(\omega_{LO}t) + V_{RF}\cos(\omega_{RF}t) - V_T]^2 = \beta V_{LO} V_{RF} \cos(\omega_{LO}t)\cos(\omega_{RF}t) \\ &+ \frac{\beta}{2} V_{LO}^2 \cos^2(\omega_{LO}t) + \frac{\beta}{2} V_{RF}^2 \cos^2(\omega_{RF}t) + \frac{\beta}{2} V_T^2 - \beta V_T V_{LO} \cos(\omega_{LO}t) - \beta V_T V_{RF} \cos(\omega_{RF}t). \end{aligned} \tag{9.9}$$

We notice that the first term contains the product of the LO and RF signals, whereas the rest of the terms are either scaled versions of the LO and RF signals, or represent the DC and second harmonics of the LO and RF signals, and can be easily filtered out. After filtering, the drain current becomes a linear function of the LO and RF signal amplitudes at the sum and difference frequencies, as described by (9.1) for an ideal multiplier

$$i_{DS}(t) = \frac{\beta}{2} V_{LO} V_{RF} \cos[(\omega_{LO} - \omega_{RF})t] + \frac{\beta}{2} V_{LO} V_{RF} \cos[(\omega_{LO} + \omega_{RF})t]. \tag{9.10}$$

This current can be picked up by an IF resistor, R_L, to produce an IF voltage of amplitude

$$V_{IF} = \frac{\beta}{2} \cdot R_L \cdot V_{LO} \cdot V_{RF}. \tag{9.11}$$

The voltage transfer function from RF to IF is called *voltage conversion gain*, A_C. It is equal to the ratio of the amplitude of the IF signal at the IF port and the amplitude of the RF signal at the RF port, and depends solely on the second-order derivative of the MOSFET I-V characteristics, β, which is a constant, and on the amplitude of the LO signal. As long as the long-channel MOSFET remains in the saturation region, A_C does not depend on the amplitude of the RF signal

$$A_C = \frac{V_{IF}}{V_{RF}} = \frac{\beta}{2} \times R_L \times V_{LO}. \qquad (9.12)$$

It is instructive to compare (9.12) with the voltage gain of a common-source amplifier with resistive load, R_L. In the latter, $A_V = -g_m R_L = \beta(V_{GS} - V_T)R_L$, indicating that we should typically expect lower gain in the mixer than in the amplifier.

We conclude by noting that a second-order non-linearity produces the ideal mixer response. Stronger non-linearity, as encountered in bipolar transistors, can be employed in mixers but it will produce many more multiplication products. Filtering them becomes more challenging (but not impossible) and therefore more expensive.

EXAMPLE 9.3

Let us assume a single-FET mixer realized in a 90nm n-MOSFET with 65nm physical gate length, $W/L = 200$, $\beta = 50\text{mA/V}^2$, biased at $V_{GS} - V_T = 0.25\text{V}$ and $I_{DS} = 1.56\text{mA}$. Note that the n-MOSFET operates at a bias current density of 0.12mA/μm (i.e. is lower than 0.15mA/μm and, therefore, in the square-law region). The IF load resistance R_L is 200Ω. Based on these bias data, we can calculate $g_m = \beta(V_{GS} - V_T) = 0.050 \times 0.025 = 12.5\text{mS}$. If a 24GHz LO signal with an amplitude of 100mV is applied at its gate, then, according to (9.12), the voltage conversion gain of the mixer becomes

$$A_C = \frac{\beta}{2} \times R_L \times V_{LO} = \frac{0.05}{2} 200 \times 0.1 = 0.5.$$

For comparison, the voltage gain of the corresponding amplifier is $g_m R_L = 0.0125 \times 200 = 2.5$.

It is instructive to estimate the conversion gain of a diode mixer biased at the same current as the FET mixer. The g_o of the diode is $I_0/V_T = 0.00156/0.02586 = 6mS$ and $g_1 = g_o \times V_{LO}/(2V_T) = 12\text{mS}$. According to (9.7), the voltage conversion gain of the diode mixer terminated on the same load resistor at IF is $g_1 R_L = 0.012 \times 200 = 2$.

Note: Above, we have assumed that an LO signal with 100mV amplitude can still be considered "small signal," a rather crude approximation. Also, it is important to note that the maximum power conversion gain ($<= 1$) of the diode mixer is achieved when the RF and IF ports are matched and the image port is terminated in a short circuit.

Even an almost ideally linear device, such as a nanoscale MOSFET, can be transformed into a *time-varying element* when operated as a switch. The control voltage, consisting of a periodic train of pulses (Figure 9.4(b)), is applied to the gate. The ideal switch, described in detail in Section 9.9.1, is characterized by zero resistance when on and by an infinite resistance when turned off. Its capacitance should be zero. Historically, high-frequency switches have been realized using Schottky diodes and FETs, with the latter becoming ubiquitous in the last two decades with the emergence of III-V and, more recently, silicon CMOS microwave monolithic ICs. PIN diodes and MEM switches are also employed in some applications where the switching speed is not very high and monolithic integration is not possible or not economical.

9.1.6 How does a switch-based commutating mixer operate?

The simplified equivalent circuit of the mixer can be approximated as illustrated in Figure 9.5. The current in the switch and through the IF load resistance R_L can be calculated by multiplying the switch conductance with the RF input voltage $v_{RF}(t)$. Because of the periodic train of pulses that acts as the LO signal controlling the switch, the combination of the normalized conductance of the switch in series with the load resistance R_L can be described by a square wave, between 0 and 1, whose Fourier series transform is given by [3]

$$g(t) = \frac{1}{2R_L} + \sum_{n=1}^{\infty} \frac{2}{n\pi R_L} \sin\left(\frac{n\pi}{2}\right)\cos(n\omega_{LO}t). \tag{9.13}$$

In (9.13), we have assumed that the switch conductance is 0 when the switch (FET or diode) is off and infinite when the switch is on.

If a sinusoidal voltage $V_{RF}\cos(\omega_{RF}t)$ is applied at the RF port, the current through the load resistance RL becomes

$$\begin{aligned} i(t) = g(t)v_{RF}(t) &= \frac{V_{RF}}{2R_L}\cos(\omega_{RF}t) + \frac{V_{RF}}{R_L}\sum_{n=1}^{\infty}\frac{2}{n\pi}\sin\left(\frac{n\pi}{2}\right)\cos(n\omega_{LO}t)\cos(\omega_{RF}t) \\ &= \frac{V_{RF}}{2R_L}\cos(\omega_{RF}t) + \frac{V_{RF}}{R_L}\sum_{n=1}^{\infty}\frac{1}{n\pi}\sin\left(\frac{n\pi}{2}\right)[\cos((n\omega_{LO}+\omega_{RF}))t + \cos((n\omega_{LO}-\omega_{RF})t)] \end{aligned} \tag{9.14}$$

After filtering the IF component corresponding to $n = 1$ ($\omega_{IF} = \omega_{LO} - \omega_{RF}$), we obtain:

$$i_{IF}(t) = \frac{V_{RF}}{\pi R_L}\cos(\omega_{IF}t). \tag{9.15}$$

From (9.15), the voltage conversion gain from the RF port to the IF port is derived as

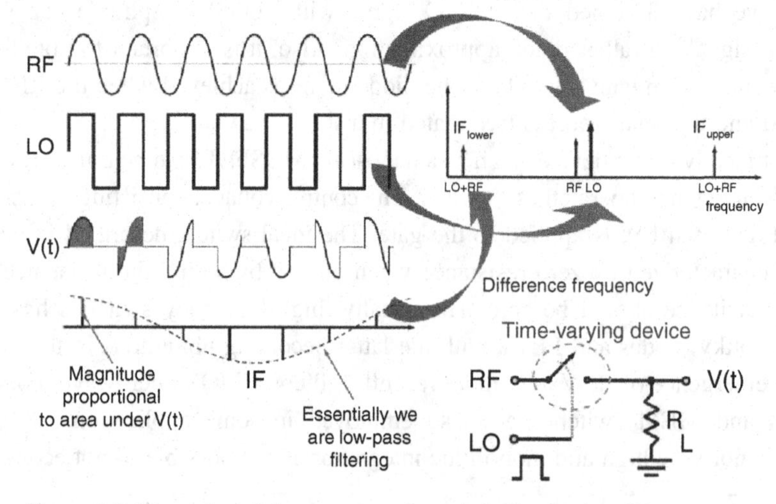

Figure 9.5 Illustration of the operation of a switch-based mixer [2]

$$A_C = \frac{V_{IF}}{V_{RF}} = I_{IF}R_L = \frac{\frac{V_{RF}R_L}{\pi R_L}}{V_{RF}} = \frac{1}{\pi}. \tag{9.16}$$

Equations (9.16) indicates that the maximum possible voltage conversion gain of a single-switch mixer is $1/\pi$.

In the previous derivations, it has been assumed that the switch is ideal and that its "on" resistance is zero. In reality, the non-zero switch resistance, r_{sw}, will form a voltage divider with the load resistor and the maximum voltage conversion gain is limited to

$$A_C = \frac{1}{\pi}\frac{R_L}{R_L + r_{sw}}.$$

EXAMPLE 9.4

In a MOSFET switch, the lowest r_{sw} value is limited by the parasitic source and drain resistance to at least 300Ω × μm, for a MOSFET switch with a unit gate width of 1μm. For example, if R_L is 150Ω, at least a 20μm wide FET is needed to ensure that the voltage gain degradation due to the "on" resistance of the switch is less than 10%. This design aspect is often ignored in MOSFET-based Gilbert cell mixers where reducing the gate width of the MOSFET switch is often pursued in order to minimize the capacitive loading on the LO path.

9.1.7 Modeling the mixer as a three-port device

It is sometimes useful to treat the large LO input signal as a DC bias voltage and thus simplify the mixer model to that of a two-port with a physical input at RF and a physical output at IF, or vice-versa. However, as we have seen already, this is a rather crude approximation because it ignores many other responses besides the RF response, all of which are downconverted to the IF port. At the very least, for mixer modeling purposes, we must account for the image response by adding a third, fictitious, port corresponding to the image frequency IM, as shown in Figure 9.6 [3],[4]. In the real world, the IM and the RF signals are both applied at the physical RF port of the device. We note that the IF, RF, and IM responses are all the result of mixing

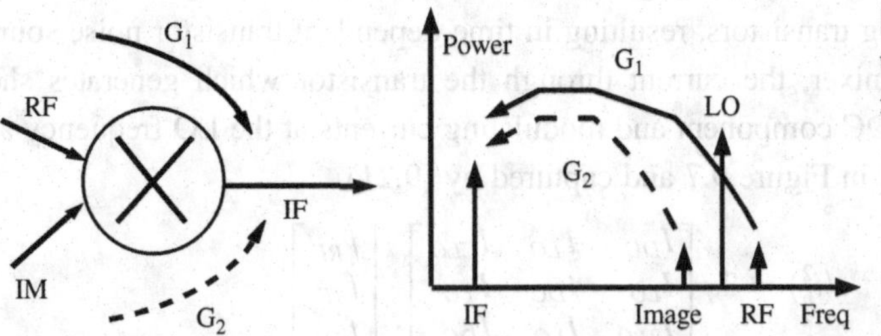

Figure 9.6 Mixer model as a three-port device with RF and IM inputs

products with the first three terms, at DC, at ω_{LO}, and at $2\omega_{LO}$, in the Fourier series expansion of the diode (9.5) or switch (9.13) conductance

$$g(t) = g_0 + 2g_1\cos(\omega_{LO}t) + 2g_2\cos(2\omega_{LO}t). \tag{9.17}$$

For example, the IM $= 2\omega_{LO} - \omega_{RF}$ is obtained from the $2g_2\cos(2\omega_{LO}) \times \cos(\omega_{RF})$ product.

The mixer can now be modeled by its three-port Y-, S-, or Z-matrix, noting that the entries of these symmetrical matrices are at three different frequencies:

$$\begin{bmatrix} I_{RF} \\ I_{IF} \\ I_{IM} \end{bmatrix} = \begin{bmatrix} Y_{11} & Y_{12} & Y_{13} \\ Y_{21} & Y_{22} & Y_{23} \\ Y_{31} & Y_{32} & Y_{33} \end{bmatrix} \begin{bmatrix} V_{RF} \\ V_{IF} \\ V_{IM} \end{bmatrix} = \begin{bmatrix} g_0 & g_1 & g_2 \\ g_1 & g_0 & g_1 \\ g_2 & g_1 & g_0 \end{bmatrix} \begin{bmatrix} V_{RF} \\ V_{IF} \\ V_{IM} \end{bmatrix} \tag{9.18}$$

$$\begin{bmatrix} b_{RF} \\ b_{IF} \\ b_{IM} \end{bmatrix} = \begin{bmatrix} S_{11} & S_{12} & S_{13} \\ S_{12} & S_{22} & S_{23} \\ S_{13} & S_{23} & S_{33} \end{bmatrix} \begin{bmatrix} a_{RF} \\ a_{IF} \\ a_{IM} \end{bmatrix}. \tag{9.19}$$

If the mixer is matched at all ports, then:

$$\begin{bmatrix} b_{RF} \\ b_{IF} \\ b_{IM} \end{bmatrix} = \begin{bmatrix} 0 & S_{12} & S_{13} \\ S_{12} & 0 & S_{23} \\ S_{13} & S_{23} & 0 \end{bmatrix} \begin{bmatrix} a_{RF} \\ a_{IF} \\ a_{IM} \end{bmatrix} \tag{9.20}$$

where, in an upconverter, $G_1 = |S_{12}|^2$ and $G_2 = |S_{23}|^2$ are the power conversion gains from IF to RF and from IF to IM, respectively. Ideally, there should be no conversion from IM to RF and, therefore, $S_{13} = 0$.

9.1.8 Mixer noise

Noise generated inside a mixer is normally a cause for concern only in downconverters. Noise performance, in terms of noise temperature or noise figure, is not specified for upconversion mixers.

The concepts of mixer noise temperature and mixer noise figure can be more difficult to understand than those defined for low-noise amplifiers. Some confusion could arise from the fact that, theoretically, mixers have an infinite number of responses from $n \times f_{LO} + / - m \times f_{RF}$ to f_{IF}. A process of frequency translation of noise sources occurs from small signals on either side of the LO harmonics (Figure 9.7) $nf_{LO} + / - f_{IF}$ to IF. Practically, most of these responses are filtered out. However, in some broadband mixers, more than two responses contribute noise at IF.

To further complicate matters, large signals at harmonics of the LO (Figure 9.8) modulate the DC current of the mixing transistors, resulting in time-dependent transistor noise sources. For example, in a bipolar mixer, the current through the transistor which generates shot noise consists of the average DC component and modulating currents at the LO frequency and at its harmonics, as illustrated in Figure 9.7 and captured by (9.21)

$$\langle i_n^2 \rangle = 2q \begin{bmatrix} I_{DC} & I_{LO} & I_{2LO} \\ I_{LO} & I_{DC} & I_{LO} \\ I_{2LO} & I_{LO} & I_{DC} \end{bmatrix} \cdot \begin{bmatrix} f_{RF} \\ f_{IF} \\ f_{IM} \end{bmatrix}. \tag{9.21}$$

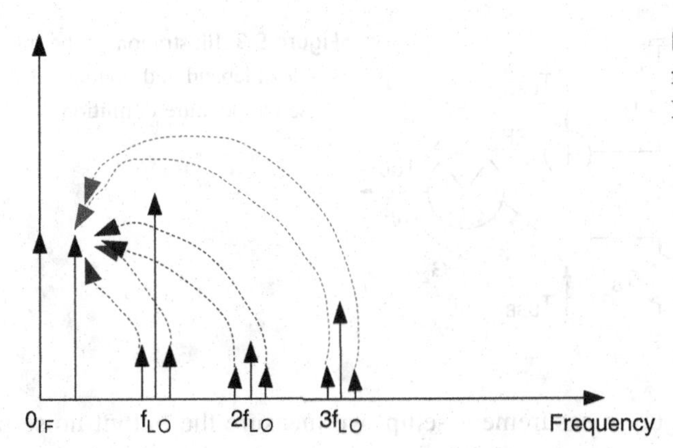

Figure 9.7 Noise downconversion paths from sidebands around the LO harmonics to IF

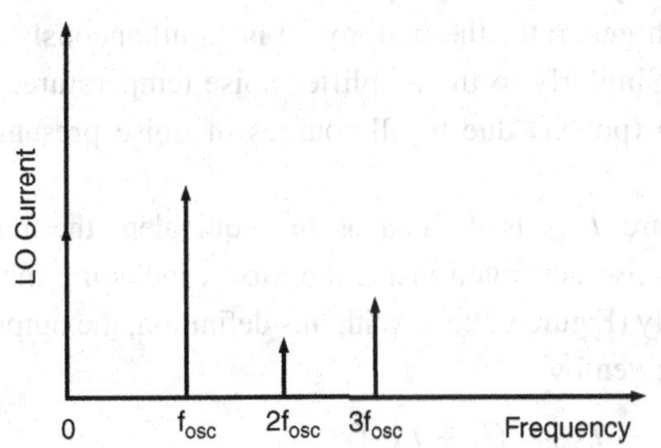

Figure 9.8 LO current harmonics which modulate the bias current of the mixing bipolar transistor and cause shot noise

In general, for broadband mixers which have many responses, the output noise consists of noise generated within the mixer, as well as of the noise of the terminations at each response, downconverterd to the IF frequency [4]

$$T_{out} = T_{m,Ln} + \sum_{n=1}^{\infty} G_{Cn} T_S \tag{9.22}$$

where $T_{m,Ln}$ is the output noise temperature of the mixer due only to noise generated within the mixer, T_S is the termination noise temperature (Z_0 is typically equal to 50Ω), and G_{Cn} is the conversion gain at the nth response.

At the very least, as shown in Figure 9.6, two of these responses, from the RF and from the IM to IF, remain. Dealing with the termination noise at IM is the fundamental problem in defining the noise temperature and the noise figure of a mixer [4].

In order to define the mixer noise temperature and to carry out the analysis, we employ the three-port mixer model described earlier, with two inputs (at RF and at IM) and one IF output, as illustrated in Figure 9.9 [4]. First, we denote the conversion power gain from the RF frequency to IF as $G_1 = |S_{12}|^2$ and the conversion power gain from IM to IF as $G_2 = |S_{23}|^2$. For simplicity, we assume that the LO port does not contribute noise at IF. This is not valid in general but constitutes a reasonable assumption considering that mixers are designed in such a way that the LO signal is filtered at the IF port.

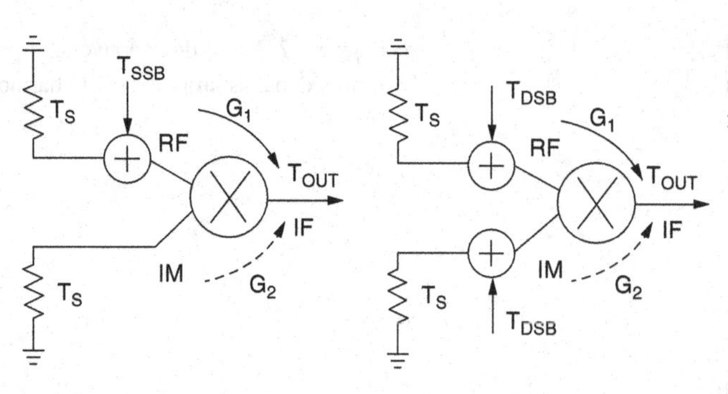

Figure 9.9 Illustration of the mixer single-sideband and double-sideband noise temperature definition

In a typical mixer noise figure measurement setup, we measure the output noise power, characterized by its equivalent thermal noise temperature T_{OUT}, as a function of the noise power of the signal source, described by its noise temperature T_S. The noise source (in most cases, a simple termination which generates thermal noise) is simultaneously applied to the RF and IM ports of the mixer. Similarly to the amplifier noise temperature, we can calculate the output noise temperature (power) due to all sources of noise present in the circuit.

The single-sideband noise temperature T_{SSB} is defined as the equivalent thermal noise temperature of the mixer when all the noise generated inside the mixer, including that of the termination, is referred to the RF port only (Figure 9.9(a)). With this definition, the output noise temperature due to all noise sources is given by

$$T_{out} = (T_S + T_{SSB}) \cdot G_1 + T_S \cdot G_2. \tag{9.23}$$

From it, we can derive the expression of T_{SSB} as

$$T_{SSB} = \frac{T_{OUT} - T_S(G_1 + G_2)}{G_1} = \frac{T_{OUT}}{G_1} - T_S\left(1 + \frac{G_2}{G_1}\right). \tag{9.24}$$

For the definition of the double-sideband noise temperature, T_{DSB}, the equivalent noise temperature of the mixer is equally attributed to the RF and to the image ports (Figure 9.9(b))

$$T_{out} = (T_S + T_{DSB}) \cdot G_1 + (T_S + T_{DSB}) \cdot G_2 \tag{9.25}$$

and

$$T_{DSB} = \frac{T_{OUT} - T_S(G_1 + G_2)}{G_1 + G_2} = \frac{T_{OUT}}{G_1 + G_2} - T_S. \tag{9.26}$$

T_{DSB} is important in double-sideband receivers in radiometers and automotive radar where information from both bands needs to be processed. We notice that if $G_1 = G_2$, $T_{SSB} = 2\,T_{DSB}$. When $G_2 = 0$, as is the case of an image-reject mixer, $T_{SSB} = T_{DSB}$.

Equations (9.24) and (9.26) indicate how the mixer double-sideband and single-sideband noise temperatures and small signal gain can be obtained experimentally by measuring the output noise temperature (power) at the IF port for at least two different temperatures of the termination at the RF port. This method is identical to the Y-factor method employed for LNA and receiver noise measurements.

EXAMPLE 9.5

Figures 9.10 and 9.11 illustrate the relevant noise temperatures and gains in receivers consisting of an LNA followed by a mixer without, and with image-reject filter, respectively. In the first case, in order to be able to determine the correct noise temperature of the receiver, it is critical that the LNA gain, G_{A2}, and output noise temperature, T_{AL}, be measured at the image frequency. Derive the expressions of the equivalent single-sideband noise temperatures at the input of each receiver.

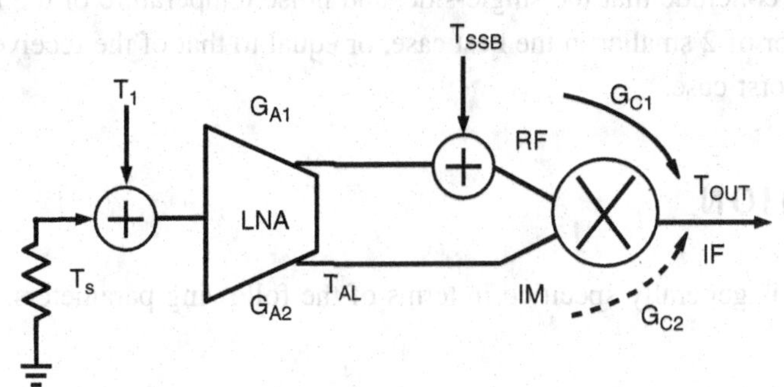

Figure 9.10 Receiver chain in which the mixer has gain at the image frequency

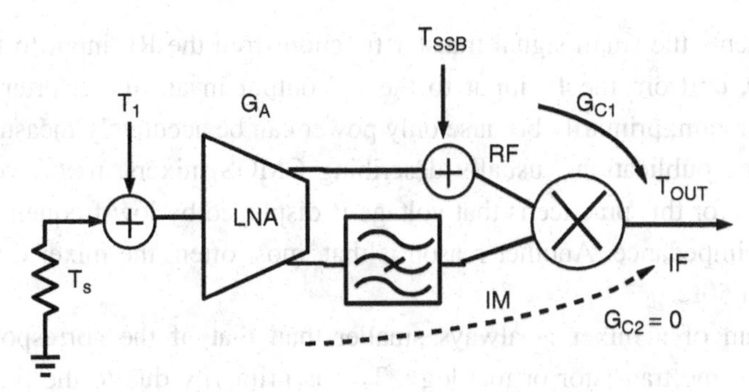

Figure 9.11 Receiver chain in which the mixer has no gain at the image frequency

Solution

For the circuit in Figure 9.10, assuming matching between the mixer and the LNA, the output noise temperature at the IF frequency is calculated as

$$T_{OUT} = (T_S + T_1)G_{A1}G_{C1} + T_{SSB}G_{C1} + T_{AL}G_{C2} \tag{9.27}$$

where T_{AL} is the noise temperature at the output of the LNA at the image frequency and T_1 is the noise temperature at the input of the LNA at the RF frequency. Note that the difficulty arises from the fact that the noise temperature at the input of the LNA at IM is T_{AL}/G_{A2} and is usually not equal to T_1, because the amount of noise added by the mixer is frequency dependent.

The equivalent single-sideband noise temperature $T_{SSB(R)}$ at the receiver input is given by

$$T_{SSB(R)} = \frac{T_{OUT}}{G_{A1}G_{C1}} - T_S = T_1 + \frac{T_{SSB}}{G_{A1}} + \frac{T_{AL}G_{C2}}{G_{A1}G_{C1}}. \tag{9.28}$$

If $G_{C1} = G_{C2}$ and $G_{A1} = G_{A2}$ then

$$T_{SSB(R)} = \frac{T_{OUT}}{G_{A1}G_{C1}} - T_S = 2T_1 + \frac{T_{SSB}}{G_{A1}}. \tag{9.29}$$

For the circuit in Figure 9.11, $T_{SSB(R)}$ is obtained by making $G_{C2} = 0$ in (9.28)

$$T_{SSB(R)} = T_1 + \frac{T_{SSB}}{G_{A1}}. \tag{9.30}$$

From (9.29) and (9.30), we conclude that the single-sideband noise temperature of the image-reject receiver can be a factor of 2 smaller in the best case, or equal to that of the receiver with no image rejection in the worst case.

9.2 MIXER SPECIFICATION

The performance of mixers is generally specified in terms of the following parameters.

9.2.1 Conversion gain

The conversion gain represents the small signal transfer function from the RF input to the IF output in a downconverter, or from the IF input to the RF output in an upconverter. It is normally defined as a power gain, primarily because only power can be accurately measured at the RF and LO ports. Some publications, usually describing CMOS mixers, report voltage conversion gain. One reason for this practice is that voltage is displayed by low-frequency test equipment with high input impedance. Another reason is that, most often, the mixer drives a high impedance, rather than 50Ω.

The downconversion gain of a mixer is always smaller than that of the corresponding amplifier realized with the same transistor or topology. This is primarily due to the fact that one of the signals produced by the mixer (at $\omega_{LO} + \omega_{RF}$) is almost invariably discarded, resulting in a 3dB power loss, that is a factor of 2 decrease in power gain. Secondly, the mixer transconductance is modulated by the large LO signal, resulting in an average transconductance value in the mixer that is smaller than the peak transconductance of an amplifier stage.

As we have seen in the preceding sections, conversion gain depends on the LO power and on the mixer topology. To maximize conversion gain and to minimize noise figure, the RF, LO, and IF ports should be terminated by conjugately matched impedances at the RF, LO, and IF frequencies, respectively.

Resistive mixers implemented with diodes and FETs exhibit conversion loss, whereas active mixers implemented with transistors can exhibit conversion gain.

9.2.2 Linearity

The linearity of the downconversion mixer is defined in much the same way as that of a low-noise amplifier or receiver, using the 1dB *compression point* P_{1dB} and the *third-order intercept point at the input*, IIP3. Equation (9.31) describes the linearity of a chain of power-matched stages. It indicates that the mixer, which is normally the second block in the receiver chain, most often limits the linearity and dynamic range of the entire receiver

$$\frac{1}{IIP3} = \frac{1}{IIP3_1} + \frac{G_1}{IIP3_2} + \frac{G_1 \times G_2}{IIP3_3} + \ldots + \frac{G_1 \times G_2 \times \ldots G_{n-1}}{IIP3_n}. \tag{9.31}$$

In zero-IF or low-IF receiver architectures, the mixer is also vulnerable to the second-order intermodulation products which are downconverted to DC. For these mixers, in addition to the IIP3, the second-order non-linearity is specified in terms of the input second-order intercept point, IIP2.

In upconversion mixers, the signal levels are usually higher than in a downconverter and the linearity is often specified as in a power amplifier, in terms of the input and output 1dB compression point.

In general, zero-bias (no bias current required) passive resistive mixers implemented with diodes or with FETs in balanced topologies, exhibit the best linearity. Unfortunately, they require larger LO signals and suffer from high conversion loss and high noise figure.

9.2.3 Noise figure

The noise figure (factor), rather than the noise temperature, is typically specified for mixers. As in the case of LNAs, there is a direct relationship between the noise factor at temperature T and the noise temperature T_a of a mixer. Therefore, we can employ the classical definition of the noise figure as the SNR ratio between the RF input and the IF output, or we can calculate it from the noise temperature. Either way, we end up with double-sideband and single-sideband definitions of the mixer noise figure. As an example, we will derive the noise figure expression using the same approach as for LNAs. The total noise at the output is given by

$$N_o = kT\Delta f(G_1 + G_2) + N_a \tag{9.32}$$

where:

- G_1 and G_2 are the conversion gains to IF from the RF and IM frequencies, respectively,
- $N_i = kT\Delta f$ is the input noise power at both the IM and RF bands, and
- N_a is the total noise (from all responses) added by the mixer at the IF output.

If we assume that the input signals at the RF and IM frequencies have the same amplitude, V_{RF}, then

$$V_{iDSB} = V_{RF}\cos[(\omega_{LO} - \omega_{IF})t] + V_{RF}\cos[(\omega_{LO} + \omega_{IF})t]. \tag{9.33}$$

After downconversion of both the RF and IM bands to IF, the output power at the IF port is

$$S_{oIF} = (G_1 + G_2)\left(V_{RF}^2/2\right). \tag{9.34}$$

Similarly, if the input signal is single-sideband, then

$$S_{oIF} = G_1\left(V_{RF}^2/2\right). \tag{9.35}$$

Using the signal-to-noise ratio definition, we can now derive the expressions the DSB and SSB noise factors

$$F_{DSB} = \frac{S_{iDSB} N_o}{S_o N_i} = \frac{V_{RF}^2[(G_1 + G_2)kT\Delta f + N_a]}{(G_1 + G_2)V_{RF}^2 kT\Delta f} = \left[1 + \frac{N_a}{kT\Delta f(G_1 + G_2)}\right]. \quad (9.36)$$

$$F_{SSB} = \frac{S_{iSSB} N_o}{S_o N_i} = \frac{V_{RF}^2[(G_1 + G_2)kT\Delta f + N_a]}{G_1 V_{RF}^2 kT\Delta f} = \left(1 + \frac{G_2}{G_1}\right)\left[1 + \frac{N_a}{kT\Delta f(G_1 + G_2)}\right]. \quad (9.37)$$

If the mixer does not have an image rejection filter or an image-reject topology, and if $G_1 = G_2$, then $NF_{SSB} = NF_{DSB} + 3\text{dB}$. In an image-reject mixer, $G_2 = 0$ and $F_{DSB} = F_{SSB}$. More importantly, (9.37) shows that if the mixer is ideal and does not contribute noise of its own, that is $N_a = 0$, the single sideband noise figure is 3dB, and equals the mixer conversion loss. This is an expected result given that half of the mixer power at the IF port, corresponding to $\omega_{LO} + \omega_{RF}$, is filtered out.

The noise figure of passive mixers is close to, but slightly higher than, their conversion loss. Typical NF_{SSB} values for Schottky diode mixers are 5–6dB [4].

EXAMPLE 9.6

The measured DSB noise figure and DSB downconversion gain of a 77GHz receiver are 3.8dB and 41dB, respectively, at an IF frequency of 1GHz. The noise figure and gain of the LNA were measured separately as 3.7dB and 30dB. What is the SSB noise figure of the receiver at 1GHz IF if an image-reject filter is placed in front of the mixer, after the LNA? Assume that the gain and noise figure of the LNA and of the mixer remain constant from 76GHz to 78GHz.

Solution

According to (9.37), when an image-reject filter is placed at the LNA output $G_2 = 0$ and the single-sideband noise figure and noise temperature of the receiver is going to be the same as the double-sideband one.

9.2.4 Isolation

It is always desirable to minimize the interaction between the LO, RF, and IF ports. This is specified in terms of the isolation between the various ports and is measured in dB:

- LO to IF, $IS_{LO\text{-}IF}$
- LO to RF, $IS_{LO\text{-}RF}$
- IF to LO, $IS_{IF\text{-}LO}$.

Higher isolation is always the goal and can be achieved by the appropriate choice of mixer topology and/or filtering. Ideally, the output port (IF or RF) impedance should be a short circuit at the LO frequency and at all its harmonics. This will prevent the LO signal from leaking at the RF and IF ports. In a downconverter, the IF port should also present a short circuit for the RF and IM frequencies. Similarly, in an upconverter, the RF port should behave as a short circuit for signals at the IF and LO frequencies. Isolation higher than 40dB is possible with double-balanced mixer topologies, as will be discussed later in this chapter.

9.2.5 Figure of merit

It is possible and sometimes useful to define a figure of merit for mixers, in much the same way as we did for LNAs and PAs, by combining the relevant parameters that describe the performance of the mixer.

In downconverters, at the very least, we want to maximize the linearity (IIP3 or IIP2), increase the conversion gain, and decrease the noise figure and power dissipation while accounting for the fact that mixers with good performance at high RF frequencies are more difficult to realize. Based on these goals, not surprisingly, we end up with the same figure of merit as that of LNAs, to which we have added the LO–RF isolation and the LO power

$$FoM\ (downconverter) = \frac{IIP_3 G_C IS_{LO-RF} f}{(F_{SSB} - 1) P_{DC} P_{LO}}. \tag{9.38}$$

Similarly, for an upconverter, we build on the PA FoM and add incentives for isolation and LO power

$$FoM\ (upconverter) = \frac{G_C OP_{1dB} IS_{LO-RF} f^2}{P_{DC} P_{LO}} \tag{9.39}$$

where $IS_{LO\text{-}RF}$ is the isolation between the LO and RF ports and P_{LO} is the power needed at the local oscillator port.

9.3 MIXER TOPOLOGIES

9.3.1 Classification of mixer topologies

As we have seen earlier, mixers can be classified as:

- *passive*, if they are realized with diodes or FETs operated as resistors, or
- *active*.

Active mixers are implemented with transistors.

Another classification is based on the number of terminals of the non-linear device employed to realize the mixing function. According to this criterion, the mixer can have:

- a one-port device (usually a diode) and filters, as in Figure 9.12,
- a two-port device (such as a transistor) and filters, as in Figure 9.13, or

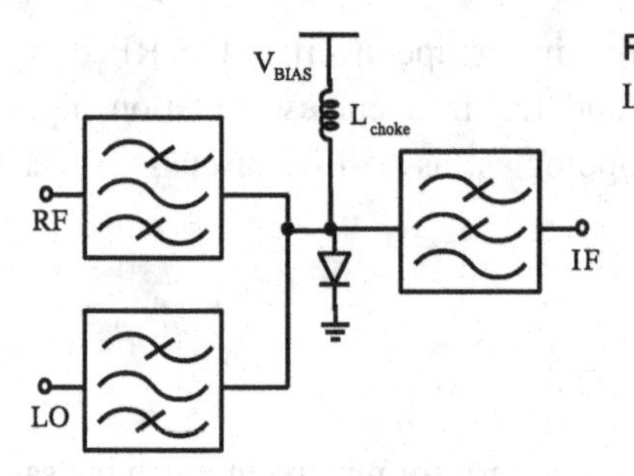

Figure 9.12 Single diode mixer schematics with filters at the RF, LO, and IF ports

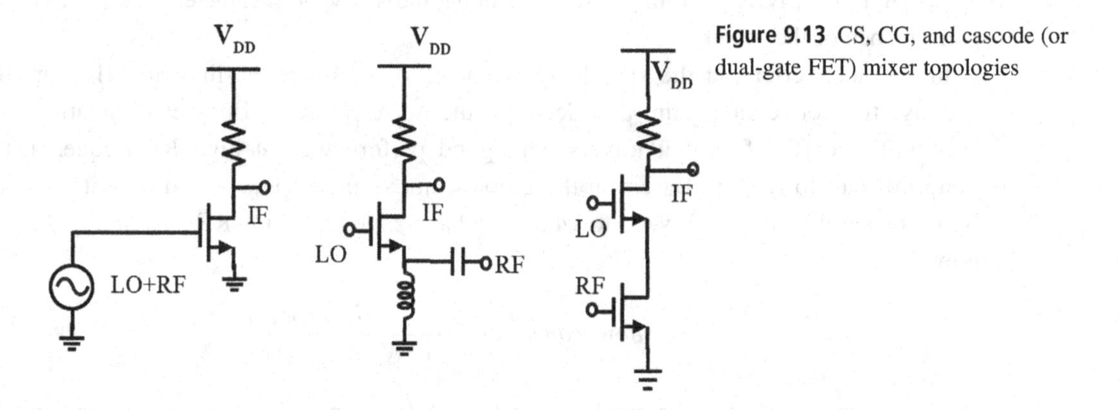

Figure 9.13 CS, CG, and cascode (or dual-gate FET) mixer topologies

- three-port devices (such as transistors, dual-gate FETs, or cascode stages) using different device terminals (source, gate, drain, or bulk) for each mixer port.

Although mixers require three physical ports to separate the LO, IF, and RF signals, historically it has been possible to realize mixers using filters and non-linear devices with fewer than three ports. Figure 9.12 shows the block diagram of a single-diode passive mixer, still used today, especially at frequencies above 100GHz. Filters or duplexers and an inductive choke (large inductor) are employed to provide isolation between the port signals and between the port signals and the power supply. In 0-bias mixers, the LO signal must be large enough to self-bias the diode.

Transistor and cascode-based active mixers are illustrated in Figure 9.13. We note that the isolation between the drain, gate, and source terminal of the transistor offers inherent filtering between some of the mixer ports, resulting in smaller size and lower cost since an external filter is no longer needed.

In mm-wave amplifiers, it is customary to bias the FET at large effective gate voltage where the transconductance is practically constant, thus guarantying good linearity, as illustrated in Figure 9.14. In the mixer topologies shown in Figure 9.13, the FET to which the LO signal is applied is normally biased in the square-law region, close to V_T, where g_m and C_{gs} are highly non-linear and change rapidly with V_{GS} and I_{DS}. This biasing scheme ensures that mixing of the LO and RF signals occurs by exploiting the non-linearity of the transistor while minimizing the noise generated inside the mixer.

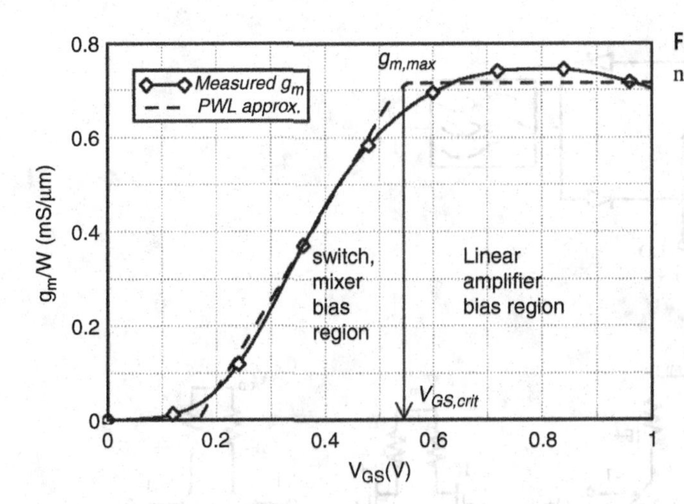

Figure 9.14 Transconductance of a nanoscale MOSFET as a function of V_{GS}

EXAMPLE 9.7

A 60GHz cascode mixer with the same schematic as in Figure 9.13 is implemented in a 65nm CMOS process. The bottom transistor has 25 gate fingers, each 1µm wide and 45nm long and is biased at the peak f_T current density of 0.3mA/µm to ensure good linearity and high transconductance of 1.1mS/µm. The common-gate transistor also has minimum gate length and is biased in the square region, close to V_T, at 0.1mA/µm to ensure that its transconductance is a strong function of V_{GS} and therefore highly non-linear. Calculate the bias current through the two transistors. What is the gate width of the common-gate transistor? Estimate the voltage gain of the mixer if the load resistance is 100Ω.

Solution

If the common-source transistor is biased at 0.3mA/µm and its total gate width is 25µm, the corresponding drain current becomes 0.3 × 25 = 7.5mA. Since the top, common-gate transistor is designed to operate at 0.1mA/µm, its total gate width must be set to W2 = 7.5mA/(0.1mA/µm) = 75µm.

Since the common-gate transistor is turned on and off by the large LO signal, it can be viewed as an ideal switch with $R_{SW} = 0$ and infinite impedance when turned off. The small signal RF current $g_{m1} \times V_{RF}$ provided by the common-source transconductor with transconductance g_{m1} is thus multiplied by a square wave and according to (9.16), at the fundamental LO frequency, can be approximated by

$$A_C = \frac{-1}{\pi} g_{m1} R_L = 0.318 \cdot 1.1(\text{mS}/\mu\text{m}) \cdot 25\mu\text{m} \cdot 100\Omega = -0.875.$$

Finally, we can use the mixer topology to classify mixers in

- single-balanced (Figure 9.15), or
- double-balanced topologies (Figure 9.16).

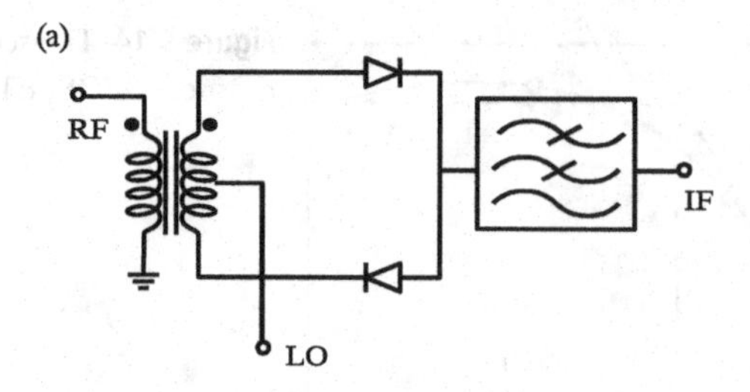

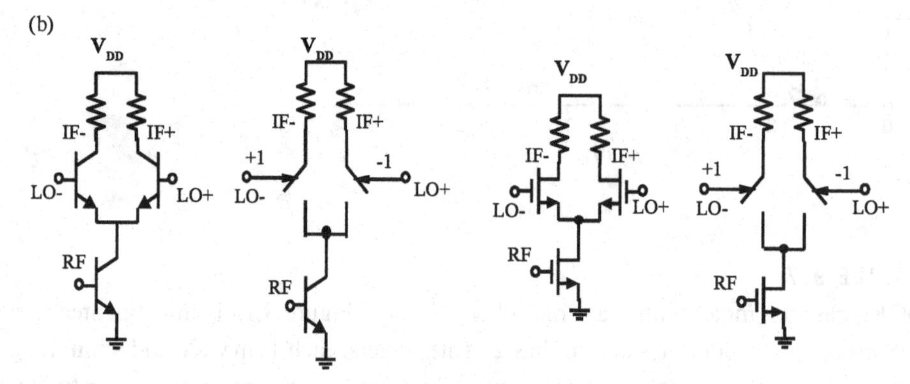

Figure 9.15 (a) Single-balanced diode and (b) BJT/HBT and MOSFET mixer implementations and their conceptual equivalent circuit with anti-phase switches controlled by the LO signal

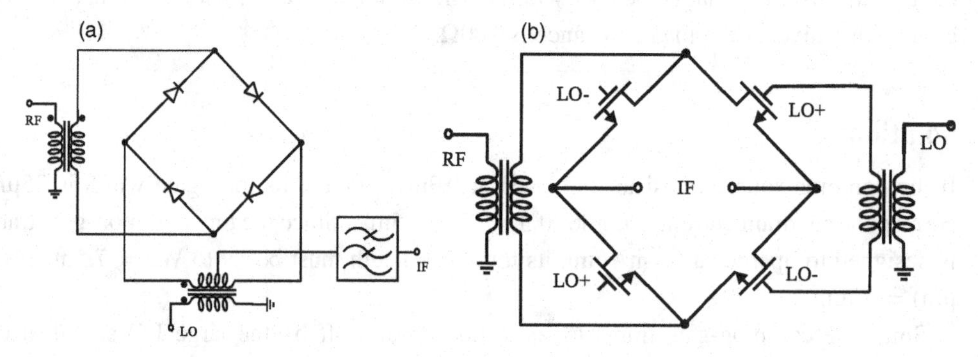

Figure 9.16 Double-balanced (a) diode and (b) resistive FET mixer topologies

Isolation of the mixer ports without filtering can be accomplished by employing single-balanced topologies implemented either with diodes or with transistors, as shown in Figure 9.15. Although in single-balanced mixers the LO and RF ports are isolated from each other, the IF port still requires external filtering.

Building on the signal balancing concept, the passive double-balanced mixer topologies from Figure 9.16, employ either a diode or a FET bridge and, because of their symmetry, exhibit excellent LO-RF and RF-LO isolation (40dB or higher) while reducing the conversion loss in half when compared to the single-balanced diode mixer. This factor of 2 improvement in conversion gain can be explained by the fact that the RF current passes through one of the two switches which are 180 degrees out of phase, doubling the voltage on the IF load resistor (Figure 9.15). Therefore, the IF output voltage of the single-balanced mixers in Figure 9.15(b) is equal to the RF current, multiplied by the load resistance, R_L, and by $2/\pi$ (which comes from the first harmonic term in the Fourier series expansion of the square wave). The voltage conversion gain of the mixer becomes

$$A_C = \frac{-2}{\pi} g_m R_L. \tag{9.40}$$

It can be demonstrated, using the same Fourier series expansion of the switch conductance and accounting for the two 180 degrees out of phase switches, that the voltage gain of the double-balanced FET passive mixer in Figure 9.16(b) is $2/\pi$.

Although single-balanced and double-balanced mixers were first implemented with diodes, the most common double-balanced topology today is the Gilbert cell [5] illustrated in Figure 9.17 in both bipolar and FET versions.

9.3.2 The double-balanced Gilbert cell mixer topology

The Gilbert cell topology, shown in Figure 9.17, consists of a switching quad formed by two, cross-coupled differential transistor pairs preceded by a differential voltage-to-current converter (or transconductor) which also acts as gain stage.

In the bipolar implementation, the AC difference current ΔI between Q_2 and Q_3 and flowing through the combined IF load resistors, R_L, can be calculated by taking into consideration the exponential dependence of the collector current on the base–emitter voltage

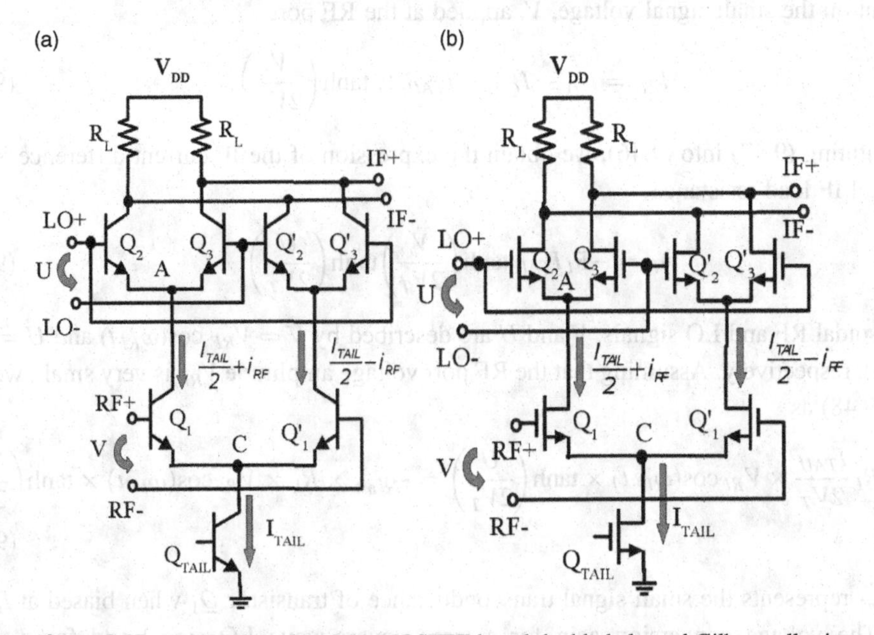

Figure 9.17 (a) HBT-based and (b) MOSFET-based double-balanced Gilbert cell mixers

$$I_C = I_S \exp\left(\frac{V_{BE}}{V_T}\right) \text{ and } V_{BE} = V_T \ln\left(\frac{I_C}{I_S}\right). \tag{9.41}$$

Assuming that a small signal voltage difference U is applied between the bases of Q_2 and Q_3, and knowing that $I_{Q2} + I_{Q3} = I_{Q1}$, we can derive the expression of the difference and sum currents

$$\Delta I = I_{Q2} - I_{Q3} = I_S\left[\exp\left(\frac{V_{BE2}}{V_T}\right) - \exp\left(\frac{V_{BE3}}{V_T}\right)\right] = I_S \exp\left(\frac{V_{BE3}}{V_T}\right)\left[\exp\left(\frac{U}{V_T}\right) - 1\right] \tag{9.42}$$

$$I_{Q1} = I_{Q2} + I_{Q3} = I_S\left[\exp\left(\frac{V_{BE2}}{V_T}\right) + \exp\left(\frac{V_{BE3}}{V_T}\right)\right] = I_S \exp\left(\frac{V_{BE3}}{V_T}\right)\left[\exp\left(\frac{U}{V_T}\right) + 1\right]. \tag{9.43}$$

From (9.42) and (9.43), we obtain

$$\Delta I = I_{Q2} - I_{Q3} = I_{Q1} \tanh\left(\frac{U}{2V_T}\right). \tag{9.44}$$

Similarly

$$\Delta I = I_{Q'3} - I_{Q'2} = I_{Q'1} \tanh\left(\frac{U}{2V_T}\right) \tag{9.45}$$

and

$$\Delta I_{IF} = I_{Q2} - I_{Q3} - I_{Q'2} + I_{Q'3} = \left(I_{Q1} - I_{Q'1}\right) \times \tanh\left(\frac{U}{2V_T}\right). \tag{9.46}$$

Following a similar derivation, the currents through Q_1 and its transconductor pair Q'_1 are dependent on the small signal voltage, V, applied at the RF port

$$\Delta I_{RF} = I_{Q1} - I_{Q'1} = I_{TAIL} \times \tanh\left(\frac{V}{2V_T}\right). \tag{9.47}$$

By substituting (9.47) into (9.46), we obtain the expression of the IF current difference in the differential IF load resistance

$$v_{IF} = -R_L I_{TAIL} \tanh\left(\frac{V}{2V_T}\right) \tanh\left(\frac{U}{2V_T}\right). \tag{9.48}$$

For sinusoidal RF and LO signals, V and U are described by $V = V_{RF}\cos(\omega_{RF}t)$ and $U = V_{LO}\cos(\omega_{LO}t)$, respectively. Assuming that the RF port voltage amplitude V_{RF} is very small, we can rewrite (9.48) as

$$v_{IF} = -R_L \frac{I_{TAIL}}{2V_T} \times V_{RF}\cos(\omega_{LF}t) \times \tanh\left(\frac{U}{2V_T}\right) = -g_{m1} \times R_L \times V_{RF}\cos(\omega_{LF}t) \times \tanh\left(\frac{U}{2V_T}\right) \tag{9.49}$$

where g_{m1} represents the small signal transconductance of transistor Q_1 when biased at $I_{C1} = I_{TAIL}/2$. The voltage conversion gain for a large square-wave LO signal and for a small sinusoidal RF input signal is obtained by recognizing that for large V_{LO}, $\tanh(U/2V_T)$ becomes a square wave with + 1, −1 values, which can be expanded as the Fourier series (9.50)

$$\tanh(V_{LO}\cos(\omega_{LO}t)) = \frac{4}{\pi}\cos(\omega_{LO}t) + \frac{4}{3\pi}\cos(3\omega_{LO}t) + \frac{4}{5\pi}\cos(5\omega_{LO}t) + \ldots \quad (9.50)$$

9.3.3 Gilbert cell mixer conversion gain

By substituting (9.50) in (9.49) and filtering the third- and higher-order harmonics of the LO signal, we obtain the voltage conversion gain, A_{VC}, for small RF input signals V_{RF}

$$A_{VC} = \frac{v_{IF}}{v_{RF}} = \frac{-2}{\pi} g_{m1} R_L \quad (9.51)$$

where $g_{m1} = I_{TAIL}/(2V_T)$ and the "−" sign indicates that the mixer is inverting.

Equation (9.51) applies equally well to HBT and FET-based Gilbert cell mixers.

We note that in a downconverter, R_L represents the equivalent resistance at the IF output of the mixer, including the output resistance of the cascode stage. In the case of an upconverter, R_L describes the equivalent resistance at the *RF* output, and includes the resonant tank losses in parallel with the output resistance of the mixing quad transistors.

Equation (9.51) shows that the voltage gain of the Gilbert cell mixer is smaller than that of the corresponding cascode amplifier by a factor of $\pi/2$. If resistive or inductive degeneration is employed, as shown in Figure 9.18, g_{m1} must be replaced in the voltage gain expression by $g_{m1}/(1 + g_{m1}R_E)$ or $g_{m1}/(1 + j\omega_{RF}L_E g_{m1})$, respectively.

Similarly to the matched LNA power gain, we can derive the expression of the maximum downconversion power gain of a mixer matched at the RF port to Z_{IN} and at the IF port to R_L as

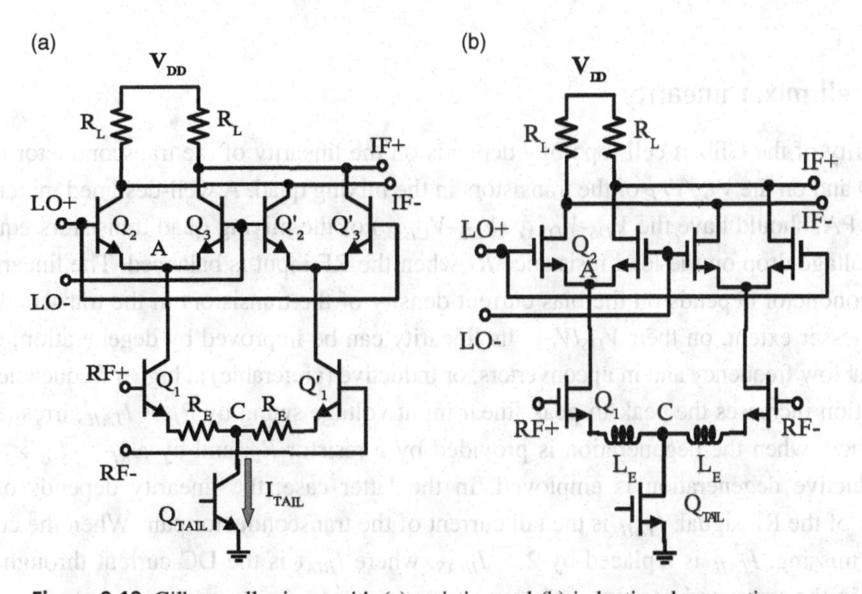

Figure 9.18 Gilbert cell mixers with (a) resistive and (b) inductive degeneration

$$G_{Cmax} = \frac{1}{2\pi}\frac{f_{Teff}^2}{f_{RF}^2}\frac{R_L}{Z_{IN}} \tag{9.52}$$

In (9.52), f_{RF} is the frequency of the RF signal, f_{Teff} is the cutoff frequency of the cascode, and inductive degeneration (through L_E as in Figure 9.18(b)) was assumed to achieve input matching $Z_{IN} = 2\pi f_{Teff} L_E$.

Finally, we note that (9.51) is equal to the conversion gain of the double-balanced FET passive mixer in Figure 9.16, multiplied by $g_m R_L$.

9.3.4 Isolation of the Gilbert cell topology

By differentiating (9.46) with respect to U, we can define the small signal transconductance from the LO port to the IF port

$$g_{mLO-IF} = \frac{(I_{Q1} - I_{Q'1})}{2V_T} \cdot sech^2\left(\frac{U}{2V_T}\right). \tag{9.53}$$

Equation (9.53) indicates that, if the RF differential pair is balanced ($I_{Q1} = I_{Q'1}$), the LO signal does not leak at the IF port. Similarly, the LO signal does not leak to the RF port because the path from the LO to the RF port appears in common-mode to the differential LO signal. This self-filtering is an important and desirable property of the double-balanced Gilbert cell mixer topology and, in the ideal case when all transistors and resistors are perfectly matched, results in infinite LO–IF isolation. However, in practical implementations, any *device mismatch* at the *transistor* or *resistor* level will create DC offsets, degrading isolation over a broad spectrum of frequencies. Furthermore, *capacitor* or *inductor mismatch* and asymmetries introduced by *capacitive layout parasitics* degrade the port-to-port isolation at higher frequencies.

9.3.5 Gilbert cell mixer linearity

The linearity of the Gilbert cell topology depends on the linearity of the transconductor (input amplifier) and on the V_{DS}/V_{CE} of the transistors in the mixing quad. A well-designed mixer, like a class A PA, should have the V_{DS}-V_{DSAT} (V_{CE}-V_{CEsat}) of the mixing quad transistors equal to the DC voltage drop on the load resistance R_L when the RF input is balanced. The linearity of the transconductor depends on the bias current density of the transistors in the transconductor and, to a lesser extent, on their V_{DS}/V_{CE}. Its linearity can be improved by degeneration, either resistive, at low frequency and in upconverters, or inductive (preferable) at higher frequencies [6]. Degeneration increases the peak-to-peak linear input voltage swing by $R_E \times I_{TAIL}$, irrespective of frequency, when the degeneration is provided by a resistor R_E, and by $\omega_{RF} \times L_E \times I_{TAIL}$, when inductive degeneration is employed. In the latter case, the linearity depends on the frequency of the RF signal. I_{TAIL} is the tail current of the transconductor pair. When the current source is missing, I_{TAIL} is replaced by $2 \times I_{BIAS}$, where I_{BIAS} is the DC current through each transistor in the transconductor.

It now becomes straightforward to assess which part of the mixer, the mixing quad output or the transconductor, limits its linearity by comparing $V_{OMAX} = min\{R_L \times I_{TAIL}, V_{CE}(Q_2) - V_{CESAT}\}$ and $\omega_{RF} \times L_E \times I_{TAIL} \times A_{VC}$.

It should be noted that the design and topology of the transconductor in a downconvert mixer are practically identical to those of an LNA stage. The transconductor can be either CE/CS or CB/CG. The CB transconductor with a Gilbert cell quad topology has been often employed in automotive radar mixers for improved linearity. However, as discussed in Chapter 7, the common-base or common-gate stages require much lower transconductance for matching to 50Ω. This typically results in worse linearity than that of a CE or CS stage matched to 50Ω. Only a large bias current (and therefore much smaller input impedance) increases the linearity of the CB or CG transconductor. As we will see in Chapter 14, a large bias current and a large supply voltage are, indeed, the main features of high linearity common-base transconductors employed in automotive radar mixers.

EXAMPLE 9.8

Let us compare the performance of 6GHz 65nm CMOS and SiGe HBT mixers whose schematics are shown in Figure 9.17, and which are biased from 1.2V and 2.5V, supplies, respectively, each operating with a tail current of 2mA. We can allocate 0.3V for the HBT tail current source, 0.2V for the MOSFET tail current source, 1V for the V_{CE} of each HBT, and 0.4V for the V_{DS} of each MOSFET, leaving the same DC voltage drop $V_{RL} = I_{TAIL} \times R_L/2 = 0.2V$ on the IF load resistor. The linear voltage swing at the output of both mixers is limited then to $2V_{RL} = 0.4V_{pp}$.

In mixers fabricated in nanoscale CMOS technologies, the MOSFETs in the transconductor should be biased at the peak f_{MAX} or peak f_T current density (as in GaAs MESFET or p-HEMT mixers) where g_m, C_{gs}, and C_{gd} are practically constant with respect to V_{GS} (Figure 9.14) and I_{DS}. In this bias region, the noise figure is only slightly higher than at the optimum noise bias. The main benefit of this bias scheme is that the transconductor remains linear for input voltage swings as high as $0.4V_{pp}$, that is about one third of the supply voltage (typically 1V to 1.2V in sub-130 nm CMOS nodes), without having to resort to inductive or resistive source degeneration which reduce gain. Furthermore, the transistor transconductance, and therefore the mixer gain, are maximized. Assuming that Q_1 and Q'_1 are biased at 0.3mA/μm, the total device width is set to 1mA/(0.3mA/μm) = 3.33μm. At this bias, the typical transconductance for a 65nm GP MOSFET is 1.1mS/μm, resulting in a g_m of 3.67mS and a downconversion gain of the mixer of $(2/\pi)(g_m R_L) = 0.4$.

To achieve a similar linear input voltage swing of $0.4V_{pp}$, an-HBT based Gilbert cell mixer will require inductive degeneration $L_E = 0.4V_{pp}/(\omega I_{TAIL})$. For a 6GHz mixer with $I_{TAIL} = 2$mA, $L_E = 5.3$nH. This is an extremely large inductor that will occupy significant area and will limit the downconversion voltage gain of the mixer to $(2/\pi)(V_{RL}/0.2V) = 0.63$. If the degeneration is implemented with a resistor, the downconversion gain of the HBT mixer remains 0.63 but its noise is significantly degraded. We note that the bipolar mixer has 4dB higher gain while its DC supply voltage and power consumption are double those of the MOSFET mixer for identical linearity.

We conclude this example by noticing that it is possible to further increase the gain, especially of the HBT mixer, by replacing R_L with an active load. However, the noise will increase.

9.3.6 Examples of 5GHz upconversion and downconversion mixers

A typical **upconverter** mixer employed in radio transmitters features a double-balanced Gilbert cell topology consisting of:

- *a linear* transconductor at the IF input,
- a mixing quad, and
- a tuned bandpass filter (BPF) at the RF output.

Both the transconductor and the mixing quad are realized with MOSFETs or HBTs or with a combination of them. The IF signal is applied to the bottom (transconductor) pair while the LO signal drives the gates of the mixing quad transistors directly. If needed, an image-reject filter must be placed after the mixing quad or built into the topology.

An example of a SiGe HBT double-balanced Gilbert cell upconverter mixer integrated in a 5GHz radio transmitter is shown in Figure 9.19. The IF input is at 1GHz, the LO is at 4GHz, and the RF and IM are at 5GHz and 3GHz, respectively. A differential image-reject filter, not shown, is connected to the driver amplifier stage which follows the mixer [7] at the nodes indicated with dashed lines in Figure 9.19. To save voltage headroom and to maximize the upconversion gain, resistors are employed in lieu of a current source, the RF BPF filter is realized with parallel LC tanks, and the transconductor linearity is increased by employing differential resistive emitter degeneration.

A typical **downconverter** mixer features a double-balanced Gilbert cell topology with:

- a RF (low-noise) *linear* input amplifier,
- a mixing quad,
- a low-pass filter at the IF output, and
- an LO and/or RF trap (i.e. a band-reject filter centered on the RF frequency) at the IF output [6].

Again, both FET and BJT devices, or a combination of both, can be employed in the transconductor and in the mixing quad. The RF signal is applied to the bottom transconductor pair while the LO signal drives the mixing quad. An image-reject filter, if needed, must be placed before the signal reaches the mixing quad or built into the topology.

The transistor-level schematic of a 5GHz image-reject receiver, implemented in the same technology as the transmitter discussed earlier, is shown in Figure 9.20 [7]. The frequency plan is identical to that in the transmitter. The receiver includes simultaneously a noise and input impedance matched differential cascode LNA with inductive degeneration and a Gilbert cell mixer. The transconductor in the mixer is designed as an LNA stage with inductive degeneration for improved linearity and low-noise. Both the LNA and the mixer feature a parallel LC resonant circuit, tuned at $2\times f_{RF}$, as the AC current source. This scheme ensures good even-mode harmonic rejection and, therefore, good linearity. As in the transmitter, the mixer IF output features a parallel LC tank, which, in this case, is tuned to the 1GHz IF. A 3GHz image-reject filter, not shown, is connected at the collectors of the mixer transconductor transistors Q_1 and Q_2. We note that the entire 3.3V supply voltage drops on only two vertically stacked SiGe

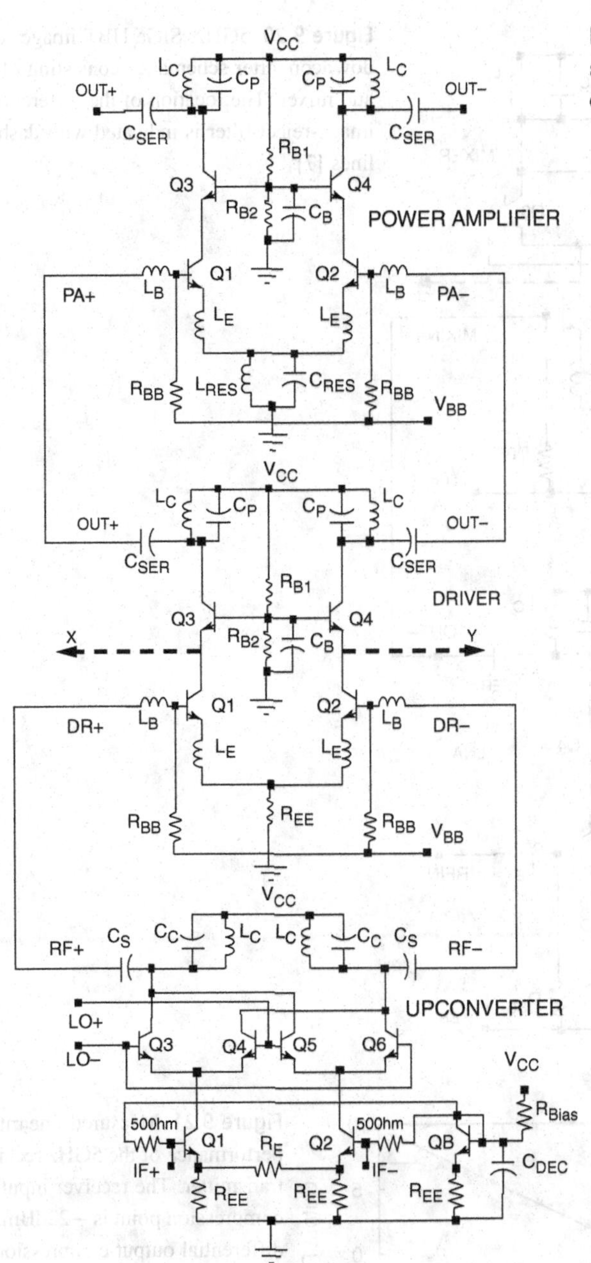

Figure 9.19 5GHz, SiGe HBT differential single-sideband upconverter using a double-balanced Gilbert cell mixer [7]

HBTs, in both the LNA and the mixer. This maximizes the voltage swing and the linearity of the receiver. Figure 9.21 reproduces the measured linearity of the transmitter and receiver at 5GHz.

As a result of its symmetry, compact size, built-in filtering and isolation, good linearity and gain, the Gilbert cell has become the *de-facto* standard mixer topology employed in RF, microwave, and millimetre-wave monolithic integrated circuits. Because its 4-quadrant

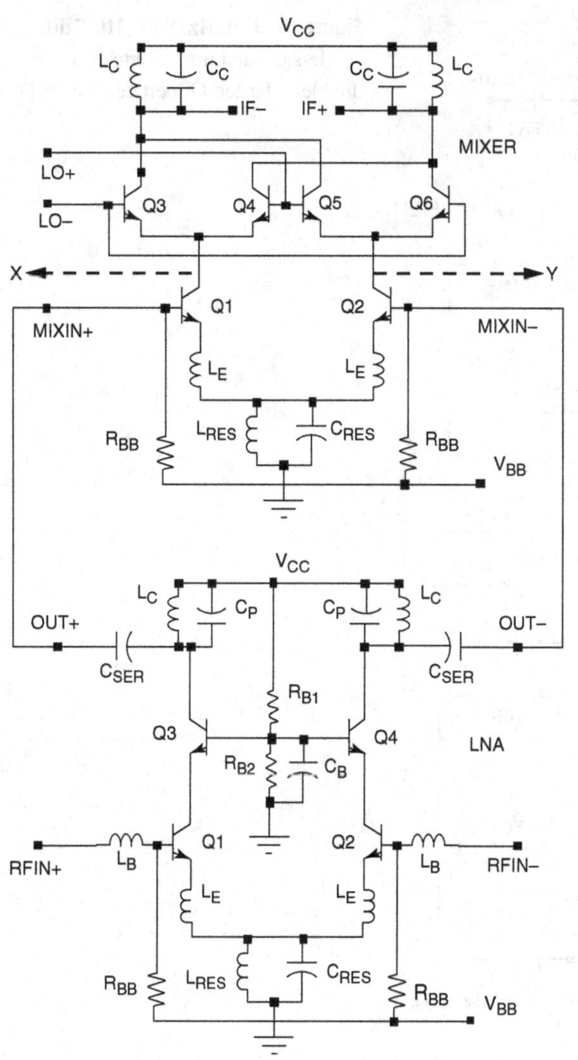

Figure 9.20 5GHz, SiGe HBT image-reject downconverter schematics consisting of LNA and mixer. The location of the differential image-reject filter is indicated with dashed lines [7]

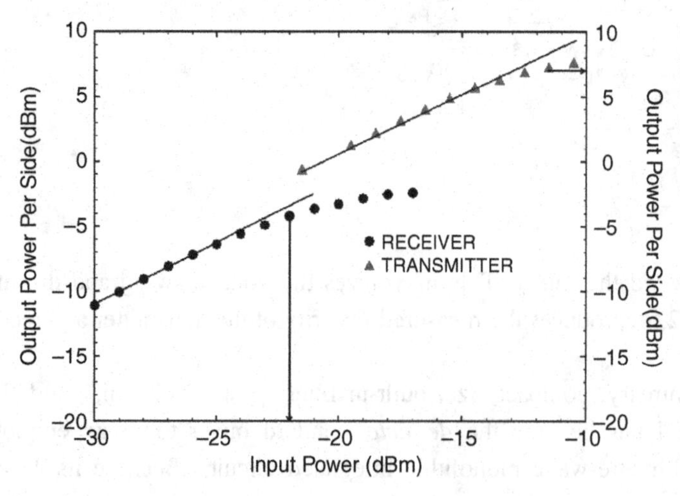

Figure 9.21 Measured linearity performance of the 5GHz receiver and transmitter. The receiver input compression point is −22dBm and the differential output compression point of the transmitter is +10dBm [7]

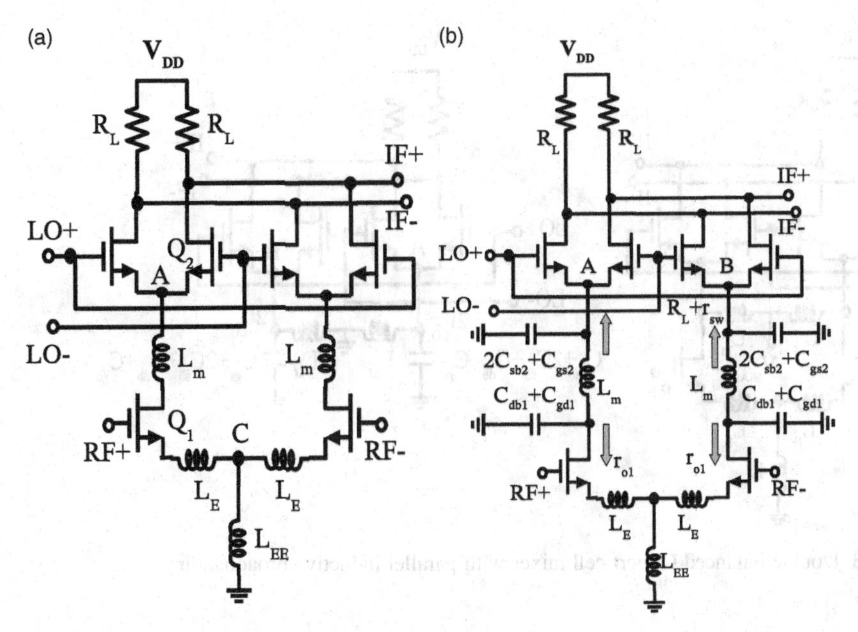

Figure 9.22 (a) Double-balanced Gilbert cell mixer with series inductive degeneration and broadbanding (b) Illustration of the critical node capacitances

multiplication properties remain valid up to the cutoff frequency of the transistor [5], the double-balanced Gilbert cell has been successfully employed at frequencies as high as 180GHz in SiGe HBT [8], and up to 140GHz in 65nm CMOS technologies [9]. Undoubtedly, as transistor performance improves with every new SiGe BiCMOS and CMOS technology node, this trend towards higher frequencies is expected to continue for many years.

9.3.7 Second-order non-linearity and stability problems

To suppress the even-mode harmonics of the RF and LO signals, good common-mode rejection must be built into the topology. We note that this is ensured at DC and low-frequency because the common-mode gain from the RF and from the LO ports to the IF port is approximately $-R_L/(2r_o(\text{TAIL}))$ and $-R_L/(2g_{m1}r_{o1}{}^2)$, respectively. However, as discussed in Chapter 5, at higher frequencies the common-mode impedance of the transconductance pair becomes capacitive. With the help of Figure 9.22, we notice that if the degeneration inductors are ignored, the total capacitance at the common-source node A of the mixing quad is $C_A = C_{gd1} + C_{db1} + 2C_{sb2} + C_{gs2}$. Here we have assumed that only one transistor is on in the mixing quad pair connected to node A. As the frequency increases, and especially at the second harmonic of the LO and RF signals, the common-mode impedances at nodes A and C (if a current tail instead of a simple inductor is used) become so small that second-order LO and RF harmonics are not suppressed.

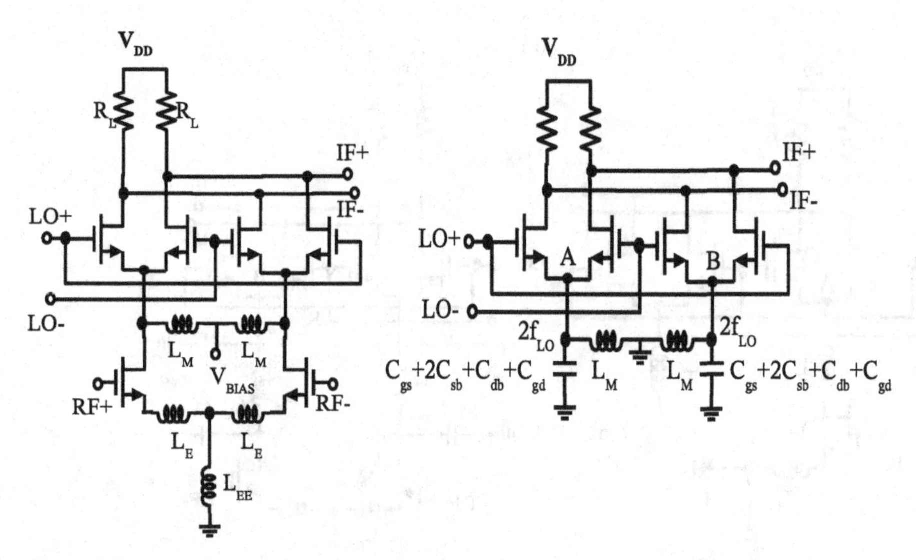

Figure 9.23 Double-balanced Gilbert cell mixer with parallel inductive broadbanding

Two solutions to this problem have been proposed. The first one is illustrated in Figure 9.22 [10] and employs series common-mode inductors L_m on the LO path, and common-mode inductor L_{EE} on the RF path. These provide a relatively large common-mode impedance which suppresses the even mode harmonics of the LO and RF signals, improving IIP2 and simultaneously increasing the gain and bandwidth and reducing the noise figure of the mixer. In this topology, the inductors L_m form an artificial transmission line with the parasitic capacitance at node A and at the drain of Q_1, thus providing a relatively large common-mode resistance (approximately equal to r_{o1}) to the LO signal over a very broadbandwidth.

The second solution [11], shown in Figure 9.23, also relies on inductors, L_M, and places them in parallel with the capacitance at node A, to form a parallel LC tank which ideally resonates at $2 \times f_{LO}$, providing a large impedance to the second harmonic of the LO signal. The latter, too, is approximately equal to r_{o1}.

In addition to compromising common-mode rejection at high frequencies, the parasitic capacitances at nodes A and B (Figure 9.23) can generate a negative resistance at the LO port, rendering the mixer vulnerable to oscillations either in common-mode or in differential-mode. Instability is a particularly common occurrence in mixers realized with bipolar transistors which, for a given bias current, have higher transconductance and produce a larger negative resistance than FETs.

9.3.8 Low-voltage mixer topologies

Despite its many advantages, the conventional double-balanced Gilbert cell mixer has some shortcomings. The most important one is the vertical stacking of at least two high-frequency path transistors between the supply voltage and ground. This makes operation with 1.2V or

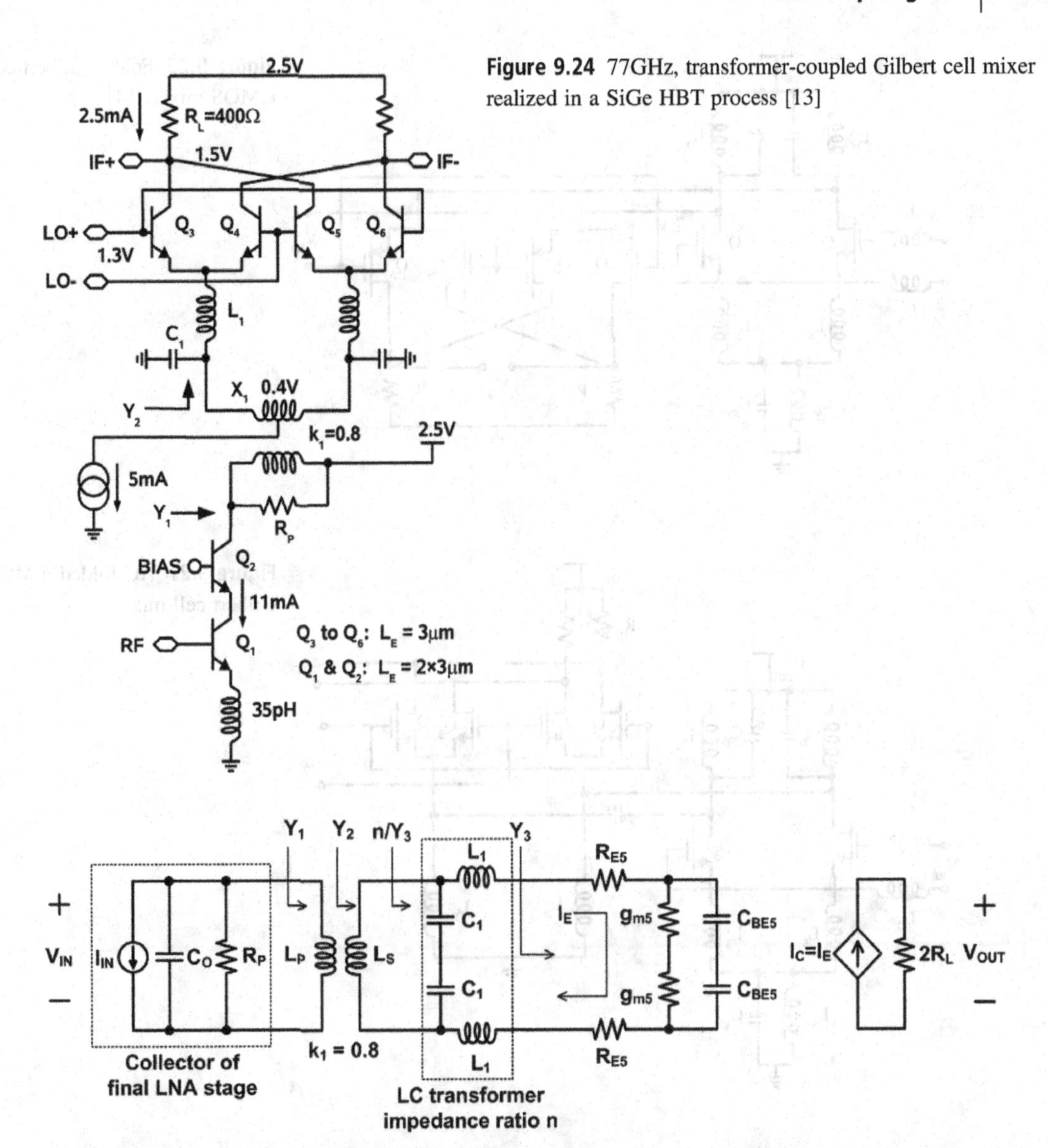

Figure 9.24 77GHz, transformer-coupled Gilbert cell mixer realized in a SiGe HBT process [13]

Figure 9.25 Small signal equivalent circuit describing the impedance matching between the LNA and the mixing quad for the circuit in Figure 9.24 [13]

lower supplies and high linearity an extremely challenging goal. Over the years, several solutions have been proposed:

- Gilbert cell mixing quad with transformer coupling of the RF signal (Figures 9.24 and 9.25), which results in lower supply voltage for a given linearity performance but also in higher noise figure and conversion loss [12],[13];
- folded Gilbert cell implemented with either p-MOSFET, as in Figure 9.26 [14] or n-MOSFET switching quad, as in Figure 9.27;
- passive (resistive) FET mixer (Figure 9.16(b) and 9.28) where the mixing quad consists of zero-biased FET switches [15] and which does not consume DC power, has high IIP3 but which suffers from high conversion loss, high LO power requirements, and high noise figure.

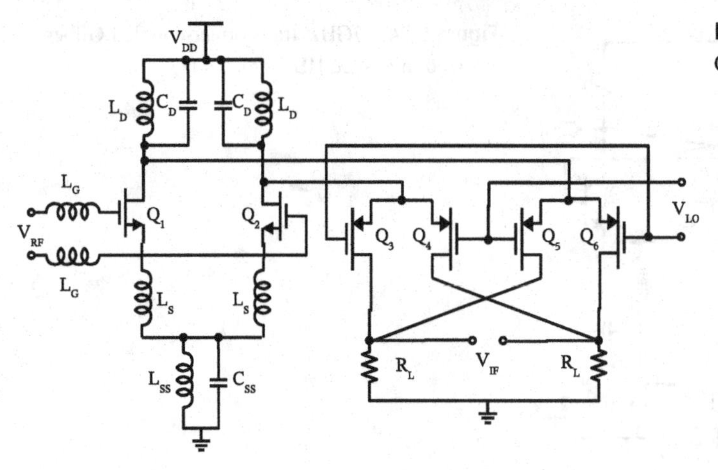

Figure 9.26 Folded Gilbert cell CMOS mixer [14]

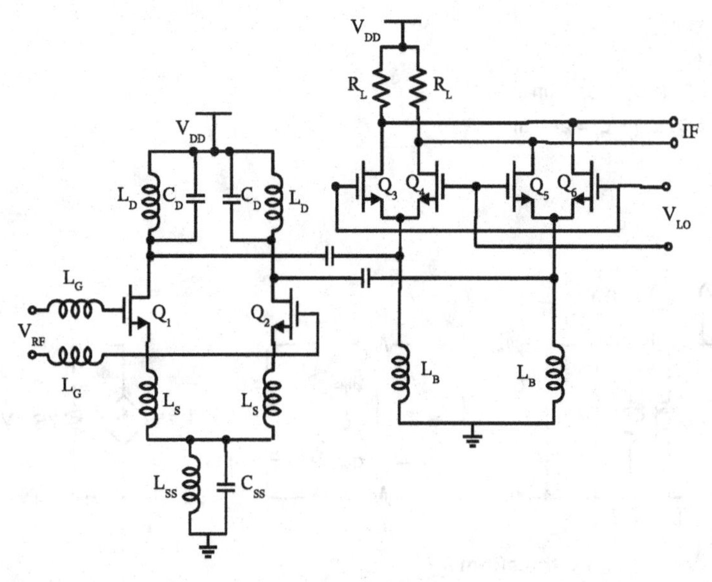

Figure 9.27 AC-folded n-MOS Gilbert cell mixer

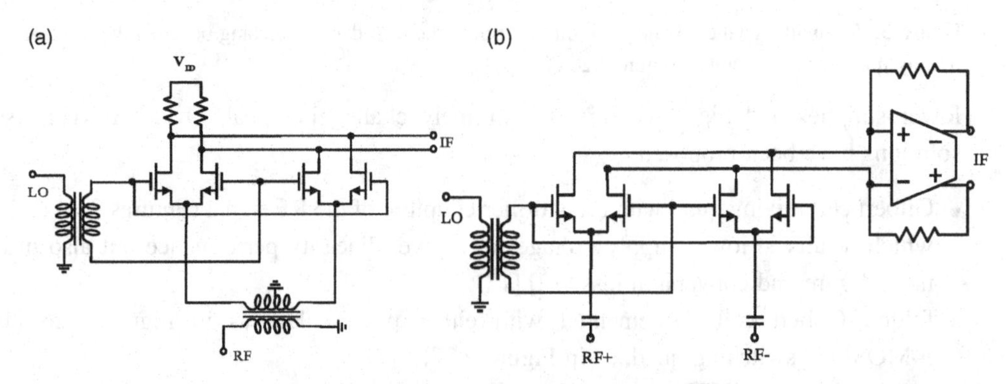

Figure 9.28 (a) Transformer-coupled, voltage-controlled passive MOSFET mixer (b) ac-coupled passive FET mixer with baseband transimpedance amplifier. The locations of the LO and RF ports can also be reversed [15]

EXAMPLE 9.9

Let us analyze the input impedance matching of the transformer-coupled mixer whose schematic is shown in Figure 9.24. Using the small signal equivalent circuit from Figure 9.25, the differential input impedance looking into the emitter nodes of the mixing quad is

$$Z_{indiff} = \frac{1}{Y_3} = 2R_E + \frac{2}{g_m + j\omega C_{be}} = 2R_E + 2\frac{g_m}{g_m^2 + \omega^2 C_{be}^2} - 2j\frac{\omega C_{be}}{g_m^2 + \omega^2 C_{be}^2} \quad (9.54)$$

where $g_{m3,6} = g_m$ and $C_{be3,6} = C_{be}$, and we have made allowance for the fact that only two transistors in the switching quad are on at any given time. The other two transistors are in cutoff and their C_{be} is negligible.

One matching approach is to employ series inductors L_1 to tune out the imaginary part of the input impedance looking into the mixing quad, and capacitors C_1 to cancel the inductance of the transformer secondary L_S. Alternatively, the transformer can be designed such that L_S resonates with the capacitance of the mixing quad at the desired frequency.

Next, the exact expression of the admittance looking into the primary of the transformer whose secondary is terminated on a real impedance R_L is derived as

$$Y_1 = \frac{R_L + j\omega L_S}{-\omega^2 (L_S L_P - M^2) + j\omega L_P R_L} = \frac{R_L + j\omega L_S}{-\omega^2 L_S L_P (1 - k_1^2) + j\omega L_P R_L} \quad (9.55)$$

$$Y_1 = \frac{\omega^2 R_L L_S L_P k_1^2 - j\omega L_P \left[R_L^2 + \omega^2 L_S^2 (1 - k_1^2)\right]}{\omega^2 L_P^2 \left[R_L^2 + \omega^2 L_S^2 (1 - k_1^2)^2\right]} \quad (9.56)$$

and

$$Z_1 = \frac{\omega^2 k_1^2 L_S L_P R_L + j\omega L_P \left[R_L^2 + \omega^2 L_S^2 (1 - k_1^2)\right]}{R_L^2 + \omega^2 L_S^2}. \quad (9.57)$$

The output impedance of the LNA is represented by the parallel connection of C_o and R_P. The latter can be rearranged as a series R-C circuit which must be conjugately matched to Z_1. The series representation of the LNA output impedance is derived as

$$Z_S = \frac{R_P - j\omega C_O R_P^2}{1 + \omega^2 C_O^2 R_P^2} \quad (9.58)$$

and must be equal to Z^*_1.

For this mixer, $L_P = L_S = 120\text{pH}$, $k_1 = 0.8$, $R_P = 350\Omega$, $C_O = 24\text{fF}$, $C_{be} = 52\text{fF}$, $g_m = 0.106\text{S}$, $R_E = 6.66\Omega$, making $R_L = R_{indiff} = 32\Omega$. As a result, at 80GHz, $Z_S = 18.6\Omega - j\,78.56\Omega$, and $Z_1 = 16.14\Omega + j\,30.5\Omega$, where

$$\Re(Y_3) = \frac{1}{2}\frac{g_m(1 + R_E g_m) + \omega^2 C_{be}^2 R_E}{(1 + g_m R_E)^2 + \omega^2 C_{be}^2 R_E^2} \approx 0.3 g_m = 31\text{mS at 80GHz}$$

$$\Im(Y_3) = \frac{1}{2}\frac{j\omega C_{be}}{(1 + g_m R_E)^2 + \omega^2 C_{be}^2 R_E^2} \approx \frac{j\omega C_{be}}{5.55} = j5\text{mS at 80GHz.}$$

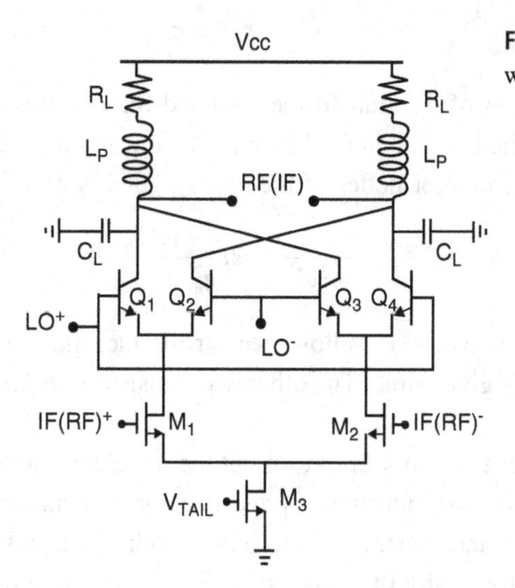

Figure 9.29 SiGe BiCMOS broadband up/down converter with n-MOS transconductors and HBT mixing quad [1]

The folded cascode CMOS mixer in Figure 9.26 is designed for operation with low supply voltage. In order to ensure that most of the RF current from Q_1 and Q_2 flows into the mixing quad formed by p-MOSFETs Q_3–Q_6, the parallel L_D–C_D tank at the output of the transconductor pair Q_1–Q_2 is designed to resonate at the RF frequency. The ac-folded n-MOSFET mixer in Figure 9.27 builds on the same idea but employs a n-MOSFET mixing quad to further improve switching speed and, therefore, the frequency range allowed at the LO port. If the LC tanks have high Q, the gain of these mixers is the same as that of the conventional Gilbert cell mixer. We note that, like the transformer-coupled mixer in Figure 9.24, the two folded cascode mixers require two times larger bias current than the conventional vertically stacked Gilbert cell mixer, therefore consuming more power. Nevertheless, sometimes situations occur when the available supply voltage is too small for two vertically stacked FETs to be deployed and to maintain good linearity.

Although it features two transistors stacked vertically on the RF path, the BiCMOS Gilbert cell with MOS input and HBT mixing quad, shown in Figure 9.29, exhibits improved transconductor linearity compared to an HBT-only transconductor, and requires lower LO swing compared to a MOSFET-only implementation [1]. In addition, because at the peak f_T bias, the V_{GS} of the low-V_T MOSFET is lower than the V_{BE} of the HBT, the circuit can operate from a lower supply when compared to an all-HBT mixer. As with MOS-only or HBT-only topologies, the supply can be further reduced by employing transformer coupling between the MOS transconductor and the HBT mixing quad.

9.4 DESIGN METHODOLOGY FOR DOWNCONVERTERS

This procedure closely follows the LNA design methodology described in Chapter 7. For low-noise mixers, the input and noise impedances of the mixer are designed to be conjugately matched to the output impedance of the LNA stage [6]. However, the emphasis in the mixer design is placed primarily on linearity. As the second block in the receiver chain, the mixer usually limits the overall linearity of the receiver. The step-by-step methodology is described below:

1. If the mixer has conversion gain, set the DC voltage drop V_{RL} on the IF load resistors R_L (if present) and the V_{DS}/V_{CE} of the mixing quad transistors to satisfy the desired peak-to-peak output linear voltage swing V_{OMAX}

$$V_{RL} = \frac{V_{OMAX}}{2} = V_{DS} - V_{DSAT} = \frac{I_{TAIL}R_L}{2} \tag{9.59}$$

2. Set the bias current density of the transistors in the transconductor pair to the minimum NF current density J_{OPT} (0.15–0.2mA/μm for MOSFET mixers). The following condition must therefore be satisfied

$$W_1(A_{E1}) = \frac{I_{TAIL}}{2J_{OPT}} \tag{9.60}$$

3. Set the bias current density of the mixing quad transistors for maximum switching speed to $J_{pfT}/2$ (typically 0.15–0.2mA/μm) for MOSFETs, and at $J_{pfT}/1.5$ for bipolar mixing quads.

$$W_2 = \frac{I_{TAIL}}{2J_{pfT}} \text{ for MOSFETs and } A_{E2} = \frac{I_{TAIL}}{3J_{pfT}} \text{ for HBTs.} \tag{9.61}$$

This step fixes the size ratio of the transconductor (W_1/A_{E1}) and mixing quad transistors (W_2/A_{E2}) to

$$\frac{W_2}{W_1} = \frac{J_{OPT}}{J_{pfT}} \approx 0.5 \text{ for MOSFETs and} \frac{A_{E2}}{A_{E1}} = \frac{J_{OPT}}{1.5J_{pfT}} \text{ for HBTs.} \tag{9.62}$$

4. Size (i.e. find W_1, A_{E1}) the transistors in the RF transconductor for the desired R_{SOPT} at the RF frequency f_{RF}. The "source" impedance is the LNA output impedance. In the case of a differential mixer

$$R_{out}(LNA) = R_{SOPT}(\text{mixer}) \approx \frac{2 \cdot f_{Teff}}{f \cdot g'_{meff} W_1} \sqrt{\frac{g'_m \left[R'_s + W_f R'_g (W_f)\right]}{k_1}} \tag{9.63}$$

where f_{Teff} and g'_{eff} are the f_T of the cascode stage and transconductance per gate width (per emitter area) of the transconductor, respectively, when the bottom transistor is biased at J_{OPT}., k_1 is the transistor noise parameter that was introduced in Chapter 4.

At this stage in the design methodology, the size of all transistors in the transconductor and mixing quad, and the tail current are fixed from (9.63) and (9.61), and (9.60), respectively

$$W_1 = N_{f1} \cdot W_f = \frac{2 \cdot f_{Teff}}{f_{RF}\, g'_{meff} R_{out}(LNA)} \sqrt{\frac{g'_m \left[R'_s + W_f R'_g (W_f)\right]}{k_1}}$$

for MOSFETs and

$$N \cdot l_E = \frac{2 \cdot f_{Teff}}{f_{RF} \cdot g'_{meff} \cdot R_{out}(LNA)} \sqrt{\frac{g'_m}{2}\left(r'_E + R'_b\right)} \text{ for HBTs.} \quad (9.64)$$

5. Add inductive emitter/source degeneration L_E to satisfy the linearity target (more important than noise and conversion gain). The MOSFET input is more linear than an HBT input without degeneration. If the mixer is designed for noise matching, then linearity is given by

$$IIP_3 \propto \frac{f_{RF} g_m R_{out}(LNA)}{f_{Teff}} \text{ where } L_E = \frac{R_{out}(LNA)}{2\pi f_{Teff}}. \quad (9.65)$$

6. Add inductor L_G in series with the gate/base of Q_1 to tune out the imaginary part of the input impedance.
7. The LO swing must be large enough to fully switch the mixing quad, yet not too large to cause the transistors in the quad to exit the active region. From the perspective of the LO swing requirement to drive the mixing quad, the MOSFET quad is inferior to the HBT implementation. Typical LO swing values are 300mV$_{pp}$ per side for HBT mixing quads and 400–500mV$_{pp}$ per side in 65nm and 90nm CMOS mixing quads.

It is important to note that the design of I_{TAIL} can be de-coupled from the output swing and gain conditions if part of I_{TAIL} is redirected to V_{BIAS} through inductors or resistors, as illustrated in Figure 9.23(a). This allows the designer to increase the size, transconductance, gain, and linearity of the transconductor pair without having to increase the size of the mixing quad transistors.

9.5 UPCONVERTER MIXER DESIGN METHODOLOGY

The design follows that of class A power amplifiers. The transistors in the transconductor are biased at the optimum linearity current density which, in both FETs and HBTs, roughly coincides with the peak f_T current density J_{pfT}.

1. If the mixer has conversion gain, set the swing on the resonant load resistors R_P (which includes the transistor output resistance) at the RF output and the V_{DS}/V_{CE} of the mixing quad transistors to satisfy the peak-to-peak linear output voltage swing condition

$$V_{OMAX} = 2(V_{DS} - V_{DSAT}) = I_{TAIL} \times R_P. \quad (9.66)$$

I_{TAIL} is set by the power budget ($V_{DD} \times I_{TAIL}$) allocated to the upconverter but can be de-coupled from the output swing and gain condition as was discussed earlier in the downconverter mixer design section. Alternatively, when designing for the minimum power consumption, the smallest possible I_{TAIL} results from (9.66) with V_{OMAX} given as a design specification and the maximum realizable R_P at f_{RF} being a technology constant.

2. Set the bias current density of the transistors in the transconductor pair to the peak f_T current density J_{pfT} (0.3–0.4mA/μm for MOSFET mixers). That fixes the transistor size to

$$W_1(A_{E1}) = \frac{I_{TAIL}}{2J_{pfT}}. \tag{9.67}$$

3. Set the bias current density of the mixing quad transistors for maximum switching speed to $J_{pfT}/2$ (typically 0.15–0.2mA/μm) for MOSFETs, and at $J_{pfT}/1.5$ for bipolar mixing quads. The quad transistor size becomes

$$W_2 = \frac{I_{TAIL}}{2J_{pfT}} = W_1 \text{ for MOSFETs and } A_{E2} = \frac{I_{TAIL}}{3J_{pfT}} = \frac{A_{E1}}{1.5} \text{ for HBTs.} \tag{9.68}$$

4. Add resistive emitter/source degeneration R_E to meet the linearity target $IIP_3 \propto R_E I_{TAIL}$.

EXAMPLE 9.10

Size the transistors in the mixer in Figure 9.29 for a broadband upconverter with maximum linearity and a tail current of 6mA. Assume a 130nm SiGe BiCMOS process where the peak f_T current density is 0.3mA/μm in the MOSFETs and 12mA/μm^2 in the HBTs. The MOSFETs have g'_m = 0.8mS/μm, C'_{gs} = 1.25fF/μm, C'_{gd} = 0.5fF/μm, and C'_{db} = 1.5fF/μm. All HBTs have w_E = 0.13μm, τ_F = 0.32ps, C_{bc}/l_E = 2.5fF/μm, C'_{CS}/l_E = 1.5fF/μm, and C'_{je}/l_E = 5fF/μm. When biased at peak f_T, the V_{BE} of the HBT is 0.9V and V_{CEsat} = 0.4V. Assume that the power supply is 2.5V.

Redesign the mixer to increase the voltage conversion gain by 3dB without changing the size of the HBTs.

Solution

We allocate 0.3V for the MOSFET tail current source. That leaves 2.2V for the MOSFET, HBT, and load resistors R_L. To ensure maximum speed, the V_{DS} of the MOSFETs should be larger than 0.8V but cannot exceed 1.2V. We allocate 0.8V to the V_{DS} of the transconductor FETs, 0.5V to R_L, and 0.9V to the V_{CE} of the HBTs, resulting in the largest possible linear swing at the output of 1V$_{pp}$ and R_L = 167Ω. The MOSFETs must be biased at peak f_T for maximum g_m and linearity. Hence $W_1 = W_2$ = (3mA)/(0.3mA/μm) = 10μm and $g_{m1} = g_{m2}$ = 8mS. The HBTs must be biased at 0.75 × J_{PfT} = 8–9mA/μm^2, resulting in an emitter length of l_E = (1.5mA)/(0.13 × 9mA) = 1.28μm. The upconversion gain is 0.816.

To increase the voltage gain by 3dB, we must increase the size and bias current of the MOSFETs in the transconductor by 42%. The extra current can be steered through shunt inductors and a common resistor to V_{DD}, as illustrated in Figure 9.23.

9.6 EXAMPLES OF MM-WAVE GILBERT CELL MIXERS

9.6.1 60GHz downconverter in 160GHz 0.18μm SiGe BiCMOS technology [16]

Figure 9.30 illustrates the schematics of a 65GHz double-balanced Gilbert cell mixer which operates from a 3.3V supply and is realized in a 180nm SiGe BiCMOS process. The measured peak f_T and peak f_{MAX} of the SiGe HBTs is 160GHz at a current density of 6mA/μm^2. In this

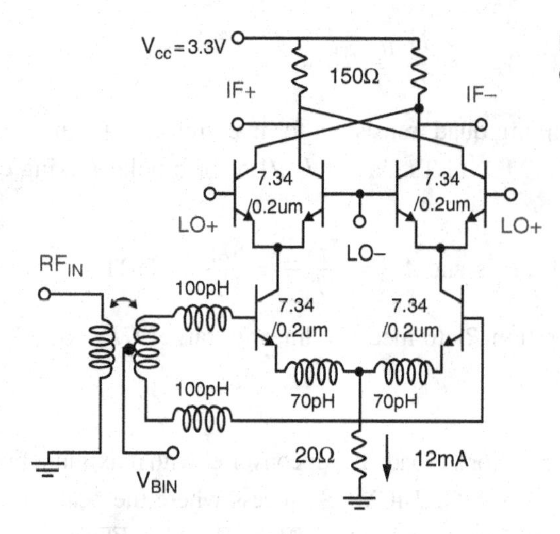

Figure 9.30 Schematics of a 65GHz, double-balanced Gilbert cell mixer [16]

circuit, the HBTs are biased at a current density of 4mA/μm^2, which corresponds to the optimal noise bias at 60GHz. The 20Ω common-mode resistor provides some frequency-independent rejection of both DC and mm-wave common-mode signals, compensating for any common-mode leakage in the single-ended to differential transformer placed at the RF port. The 70pH inductors ensure good linearity and, along with the 100pH inductors, provide simultaneous noise and input impedance matching to the output impedance of the LNA. The latter is approximately 150Ω. The transformer at the RF port has a 1:1 turns ratio. The mixer has a conversion gain of 7dB, an input compression point of +1dBm, and the IF bandwidth extends from DC beyond 5GHz.

9.6.2 60GHz downconverter in 90nm GP CMOS [10]

The same mixer topology, with minor modifications, was also implemented at 60GHz in a 90nm GP CMOS process, as illustrated in Figure 9.31 [10]. Since this circuit operates from a lower supply of only 1.2V, the DC voltage headroom is very tight and, therefore, the common-mode resistor was replaced by a 70pH inductor. In addition, series 140pH inductors were placed between the transconductor pair and the mixing quad to increase gain and reduce noise figure. The mixer draws 6mA and the transconductor MOSFETs have the minimum physical gate length of 65nm and are biased at a drain current density of 0.18mA/μm. The mixing quad MOSFETs feature 32 gate fingers, each 1μm wide and 65nm long. The IF output consists of a low-*Q* resonant tank, with a 3dB bandwidth of 3.5GHz to 5GHz, realized with 2.8nH inductors and 250fF MIM capacitors. The mixer downconversion gain is 2–3dB.

9.6.3 60GHz upconverter in 90nm GP CMOS

Using the same 90nm GP CMOS technology and Gilbert cell topology as in the previous example, a 60GHz upconverter was designed and fabricated. Its schematic is shown in Figure 9.32. Since large output power and linearity are the main concerns, all transistors

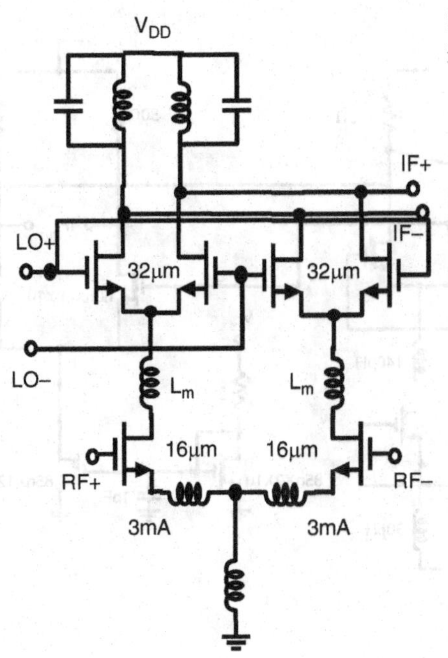

Figure 9.31 Schematics of double-balanced Gilbert cell mixer in 90nm GP CMOS technology [10]

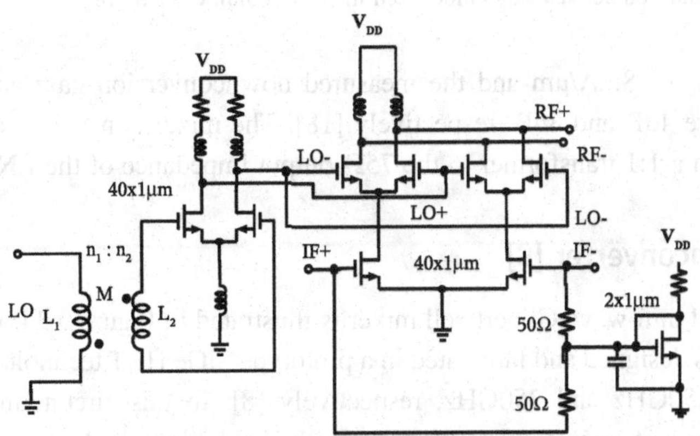

Figure 9.32 Schematics of a 60GHz upconversion mixer in 90nm CMOS [17]

have a total gate periphery of 40μm and are biased at a current density 0.3mA/μm^2 from a 1.2V supply. The IF input is broadband, from DC to 6GHz and is driven directly from an off-chip 50Ω signal source. The upconversion gain is −6dB, half the input power being lost on the 50Ω resistors. IF-input matching resistors, and the isolation is better than 40dB. Unlike in the downconverter, inductive broadbanding is not employed since its efficacy is greatly diminished because of its location on the IF, rather than on the RF path.

9.6.4 70–100GHz downconvert mixers in 65nm and 45nm CMOS

The downconverter topology from Figure 9.31 was scaled in frequency from 60GHz to 90GHz and ported to a 65nm GP CMOS technology, demonstrating that entire mm-wave circuits are scalable across frequency and technology nodes. The schematics are shown in Figure 9.33. The

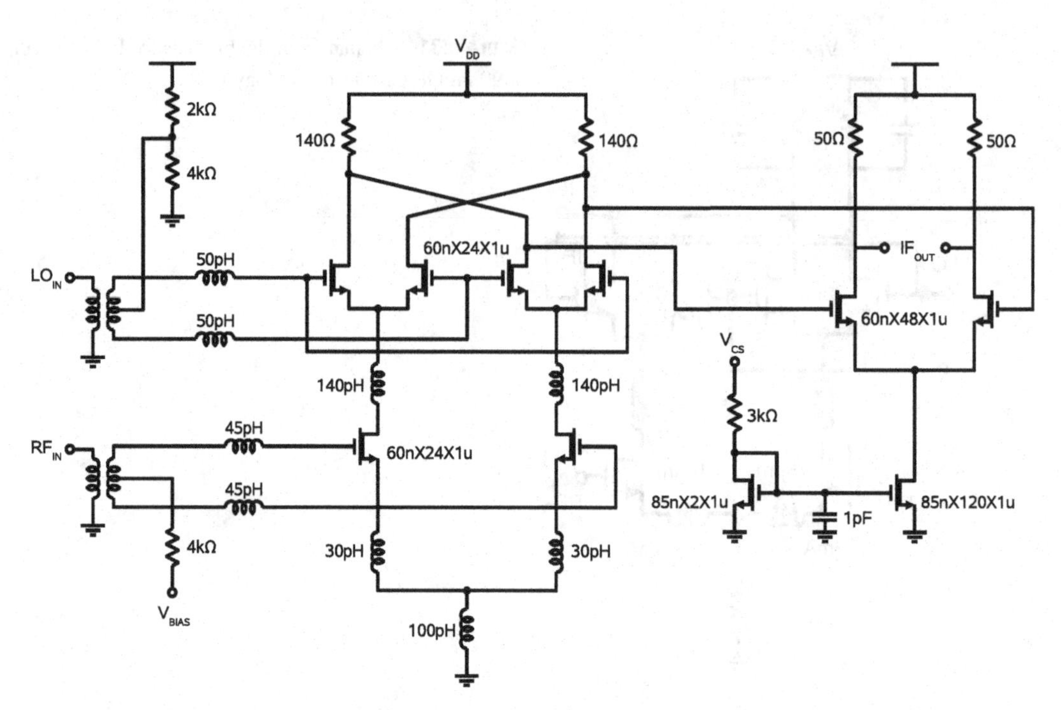

Figure 9.33 Schematic of W-Band double-balanced Gilbert cell mixer in 65nm CMOS [18]

bias current density remains 0.18mA/μm and the measured downconversion gain and DSB noise figure of the mixer are 4dB and 8dB, respectively [18]. The mixer is noise- and input-impedance-matched through a 1:1 transformer to the 75Ω output impedance of the LNA.

9.6.5 180GHz SiGe HBT downconverter [8]

Another example of a scaled mm-wave Gilbert cell mixer is illustrated in Figure 9.34, where a 180GHz downconverter was designed and fabricated in a prototype SiGe HBT technology with transistor f_T and f_{MAX} of 250GHz and 350GHz, respectively [8]. In this circuit, inductive degeneration using 5.8pH inductors is employed to match the switching quad impedance to that of the LO buffer. Transformers with 1:1 turn ratio are placed at the RF and LO ports of the mixer for single-ended to differential conversion.

9.6.6 140GHz transformer-coupled downconverter in 65nm CMOS technology

Finally, the last example of a scaled mm-wave Gilbert cell mixer is illustrated in Figure 9.35, where a 140GHz transformer-coupled downconverter was designed and fabricated in a 65nm GP CMOS technology [19]. In this circuit, which operates with 1.2V supply, inductive degeneration is implemented with 21pH inductors and a 100pH inductor is employed in common-mode on the RF path. Transformers with 1:1 turn ratio are placed at the RF and LO ports of the mixer for single-ended to differential conversion and between the transconductor pair and the mixing quad. Both the mixing quad and the transconductor pair draw 6mA. The transconductor pair is biased at 0.18mA/μm for lowest noise figure [19].

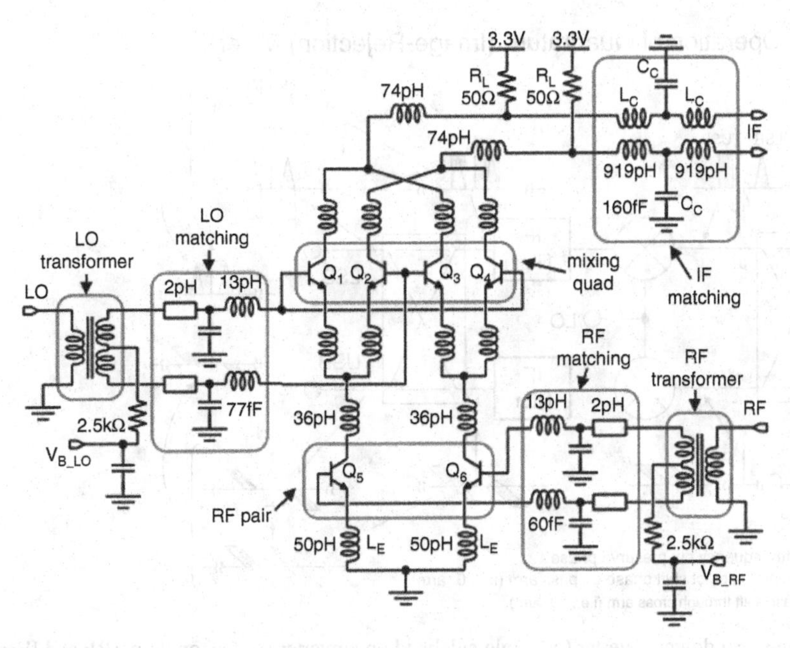

Figure 9.34 Schematics of 180GHz SiGe HBT mixer [8]

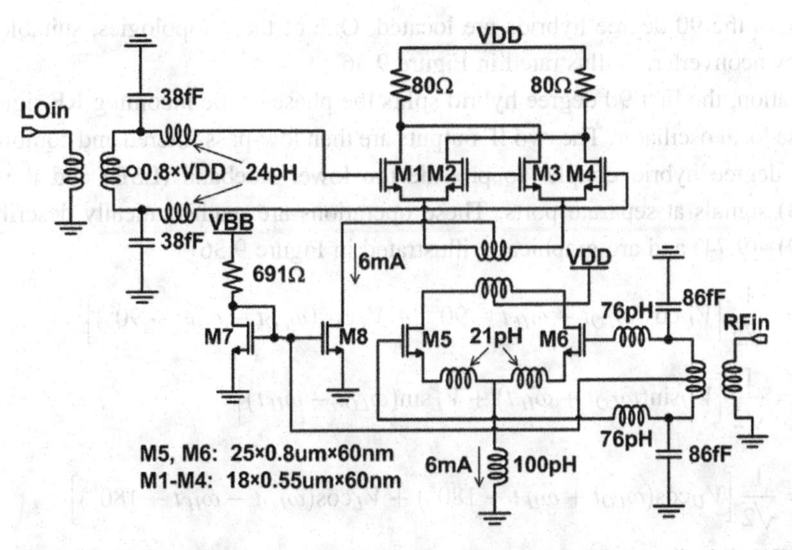

Figure 9.35 Schematics of a 140GHz transformer-coupled Gilbert cell mixer [19]

9.7 IMAGE-REJECT AND SINGLE-SIDEBAND MIXER TOPOLOGIES

We have seen earlier that the physical size of the mixer can be considerably reduced if a topology with built-in filtering is employed, as in the case of the Gilbert cell. Following the same idea, mixer topologies that reject the IM signals, without the need for an image-reject filter, have also been developed. They employ two 90 degree hybrid couplers and an in-phase power splitter or adder (combiner). Three such topologies can be envisioned, depending at

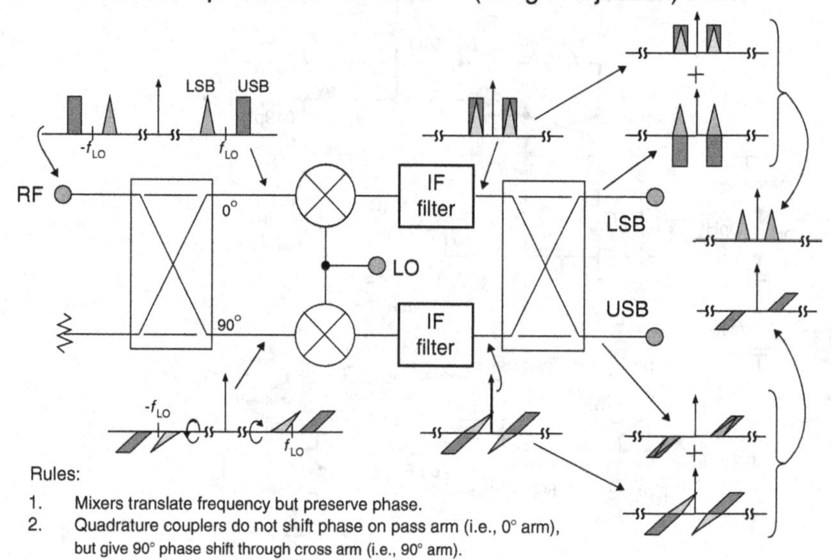

Figure 9.36 Image-reject downconverter (or single-sideband upconverter mixer) topology (RF and IF ports are interchanged) [2]

which mixer ports the 90 degree hybrids are located. One of these topologies, suitable as an image-reject downconverter, is illustrated in Figure 9.36.

In this realization, the first 90 degree hybrid splits the phase of the incoming RF signal and mixes it with the local oscillator. The two IF outputs are then low-pass filtered and combined by the second 90 degree hybrid coupler to provide the lower sideband (LSB) and the upper sideband (USB) signals at separate ports. These operations are mathematically described by equations (9.69)–(9.74) and are graphically illustrated in Figure 9.36

$$\begin{aligned} v_{RFI}(t) &= \frac{1}{\sqrt{2}}\left[V_U\cos(\omega_{LO}t + \omega_{IF}t - 90^{\circ}) + V_L\cos(\omega_{LO}t - \omega_{IF}t - 90^{\circ})\right] \\ &= \frac{1}{\sqrt{2}}\left[V_U\sin(\omega_{LO}t + \omega_{IF}t) + V_L\sin(\omega_{LO}t - \omega_{IF}t)\right] \end{aligned} \tag{9.69}$$

$$\begin{aligned} v_{RFQ}(t) &= \frac{1}{\sqrt{2}}\left[V_U\cos(\omega_{LO}t + \omega_{IF}t - 180^{\circ}) + V_L\cos(\omega_{LO}t - \omega_{IF}t - 180^{\circ})\right] \\ &= -\frac{1}{\sqrt{2}}\left[V_U\cos(\omega_{LO}t + \omega_{IF}t) + V_L\cos(\omega_{LO}t - \omega_{IF}t)\right]. \end{aligned} \tag{9.70}$$

Here subscripts U and L describe the upper and lower sidebands, respectively. After mixing and low-pass filtering at IF, we obtain

$$v_{IFI}(t) = \frac{A_C V_{LO}}{2\sqrt{2}}[V_U\sin(\omega_{IF}t) - V_L\sin(\omega_{IF}t)] \tag{9.71}$$

$$v_{IFQ}(t) = \frac{-A_C V_{LO}}{2\sqrt{2}}[V_U\cos(\omega_{IF}t) + V_L\cos(\omega_{IF}t)] \tag{9.72}$$

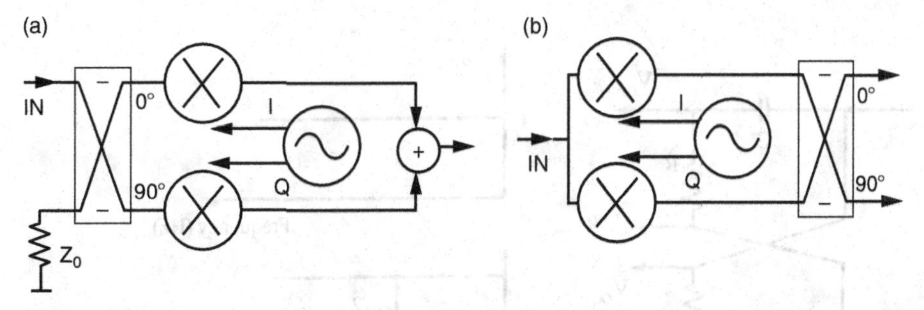

Figure 9.37 (a) Single-sideband modulator and (b) image-reject mixer topologies based on quadrature VCOs

where A_C is the voltage conversion gain and which is assumed here to be equal for the RF and IM bands. Following the 90 degree hybrid at the IF output, we arrive at the expression of the LSB and USB signal voltages at the IF ports of the image-reject mixer:

$$v_{LSB}(t) = \frac{A_C V_{LO}}{4}\left[(V_U - V_L)\sin(\omega_{IF}t - 90°) - (V_L + V_U)\cos(\omega_{IF}t - 180°)\right]$$

$$= \frac{A_C V_{LO} V_L}{2}\cos(\omega_{IF}t) \tag{9.73}$$

and

$$v_{USB}(t) = \frac{A_C V_{LO}}{4}\left[(V_U - V_L)\sin(\omega_{IF}t - 180°) - (V_L + V_U)\cos(\omega_{IF}t - 90°)\right]$$

$$= \frac{-A_C V_{LO} V_U}{2}\sin(\omega_{IF}t) \tag{9.74}$$

Two other image-reject and single-sideband mixer topologies are possible, if a quadrature VCO is employed instead of a second quadrature hybrid, as illustrated in Figure 9.37. The choice between the three image-reject topologies depends on the economics and feasibility of implementing linear, low-loss, low-power 90 degree hybrids or precisely phase-controlled quadrature VCOs. A quadrature LO signal can be realized either with a VCO operating at two times the desired frequency followed by a static frequency divider (the preferred method at GHz-range frequencies) or using another 90 degree hybrid.

The image-reject mixer topologies in Figures 9.36 and 9.37 can best be analyzed using multi-port *S*-parameter conversion matrices with RF, IM, and USB and LSB IF ports. In all cases, in-phase, 180 degree, and quadrature hybrids are required. These hybrids must have a relatively broad bandwidth to cover the entire RF and image frequency bands. In a radio transceiver with high IF frequency, as in 60GHz radio implementations, designing a sufficiently broadband quadrature hybrid for the RF path is not trivial. However, the bandwidth requirements are significantly relaxed for a hybrid placed on the LO path. These considerations will influence the choice of image-reject mixer topology.

180 degree couplers can be realized with:

- transformers,
- rat-race couplers,
- Marchand baluns.

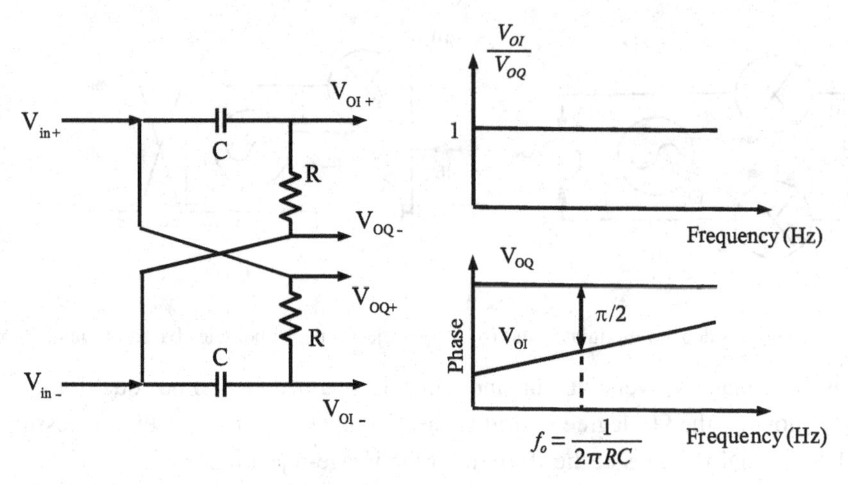

Figure 9.38 RC polyphase filter [19]

90 degree couplers are typically implemented as:

- Lange couplers,
- branchline couplers,
- RC phase shifters [19],
- RLC phase shifters or [20],
- transformer-C phase shifters [21].

An ideal 90 degree hybrid is modeled by the following *S*-parameter matrix

$$[S] = \frac{1}{\sqrt{2}} \begin{bmatrix} 0 & 1 & j & 0 \\ 1 & 0 & 0 & j \\ j & 0 & 0 & 1 \\ 0 & j & 1 & 0 \end{bmatrix}. \tag{9.75}$$

An example of a 90 degree RC phase shifter, also known as poly phase filter, is shown in Figure 9.38.

Versions of the RC-polyphase filter with constant phase difference versus frequency or with constant amplitude versus frequency response are possible, but not with both. To increase the polyphase filter bandwidth, several cascaded filter sections with staggered center frequencies are typically employed.

An improved quadrature all-pass filter [20] employs inductors to extend the bandwidth over which the 90 degree phase shift is maintained with little amplitude imbalance, Figure 9.39. The transfer function of the differential version is described by

$$\begin{bmatrix} V_{OI+,-} \\ V_{OQ+,-} \end{bmatrix} = V_{in} \begin{bmatrix} \dfrac{\pm s^2 + 2\dfrac{\omega_0}{Q}s - \omega_0^2}{s^2 + 2\dfrac{\omega_0}{Q}s + \omega_0^2} \\ \\ \dfrac{\mp s^2 - 2\dfrac{\omega_0}{Q}s - \omega_0^2}{s^2 + 2\dfrac{\omega_0}{Q}s + \omega_0^2} \end{bmatrix} \tag{9.76}$$

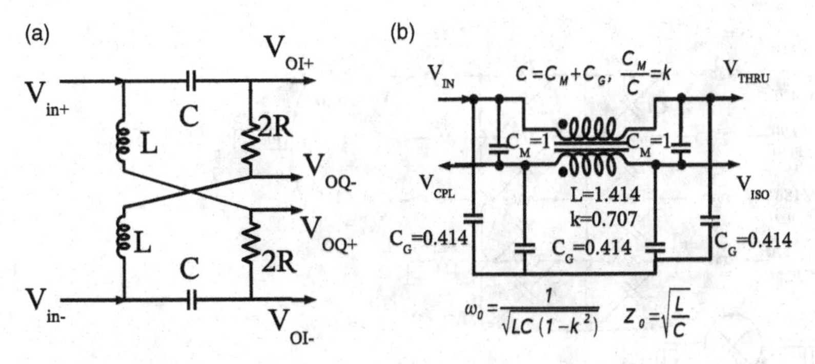

Figure 9.39 (a) LRC polyphase filter [20] and (b) lumped 90 degree hybrid coupler [21]

where

$$Q = \frac{\sqrt{\frac{L}{C}}}{R} \text{ and } \omega_0 = \frac{1}{\sqrt{LC}}. \tag{9.77}$$

This network shows that the I and Q outputs have the same magnitude at all frequencies (hence the "all-pass" term) with orthogonal phase splitting at $\omega = \omega_0$. If $L = 0$, the dual version of the RC polyphase filter from Figure 9.36 is obtained. Its phase transfer function exhibits constant 90 degree phase shift with frequency, but the amplitude response is balanced only at a single frequency

$$f_0 = \frac{1}{4\pi RC}. \tag{9.78}$$

Furthermore, especially at mm-wave frequencies, the capacitance of the following stage must be accounted for in the design process.

EXAMPLE 9.11 60GHz single-sideband upconverter with polyphase filters

The two-stage, stagger-tuned polyphase filter was implemented as part of a 60GHz radio transmitter based on a single-sideband (SSB) upconverter mixer topology. The schematics of the filter and transmitter are shown in Figure 9.40 [22]. The filter component values were designed based on the requirements of phase and amplitude match between the quadrature pairs, as given by

$$\text{Phase}\left(\frac{V_{p3}}{V_{p2}}\right) = \text{Phase}\,(j\omega RC) = 90 \text{ degree} \tag{9.79}$$

$$\left|\frac{\frac{V_{p3}}{V_{p1}}(j\omega_o)}{\frac{V_{p2}}{V_{p1}}(j\omega_o)}\right| = 1 \rightarrow |j\omega_o CR| = 1 \rightarrow \omega_o CR = 1 \tag{9.80}$$

where $\omega_0 = 65\text{GHz}$.

To achieve an input/output impedance of 100Ω, as needed in a differential, 50Ω based system with interstage matching, the real part of Z_{in} (R) should be sized to be 100Ω, while the capacitance C is determined by the desired center operating frequency ω_o of the stage. Inductors are used for matching by canceling the imaginary part of the input/output impedance at mm-waves

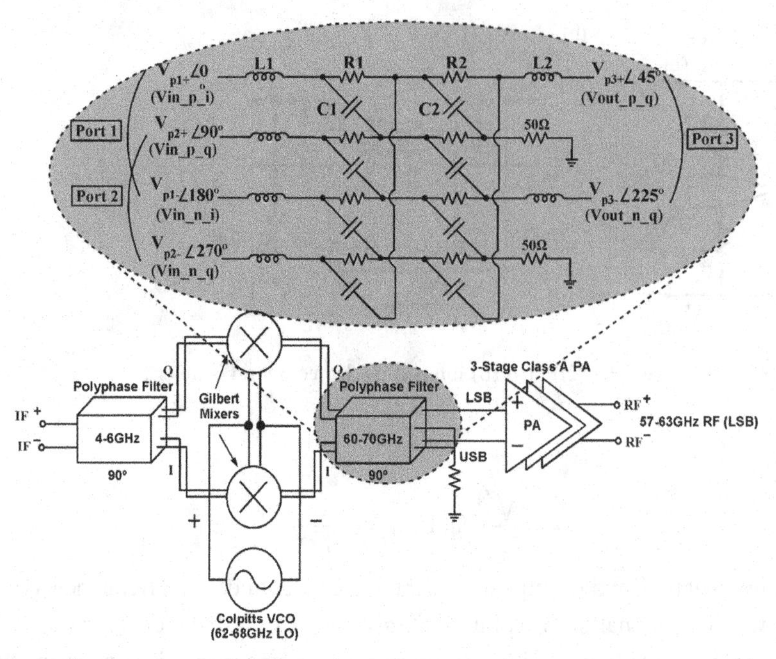

Figure 9.40 Schematic of 60–70GHz 2-stage RC polyphase filter as part of a single-sideband 60GHz WLAN SiGe BiCMOS transmitter [22]

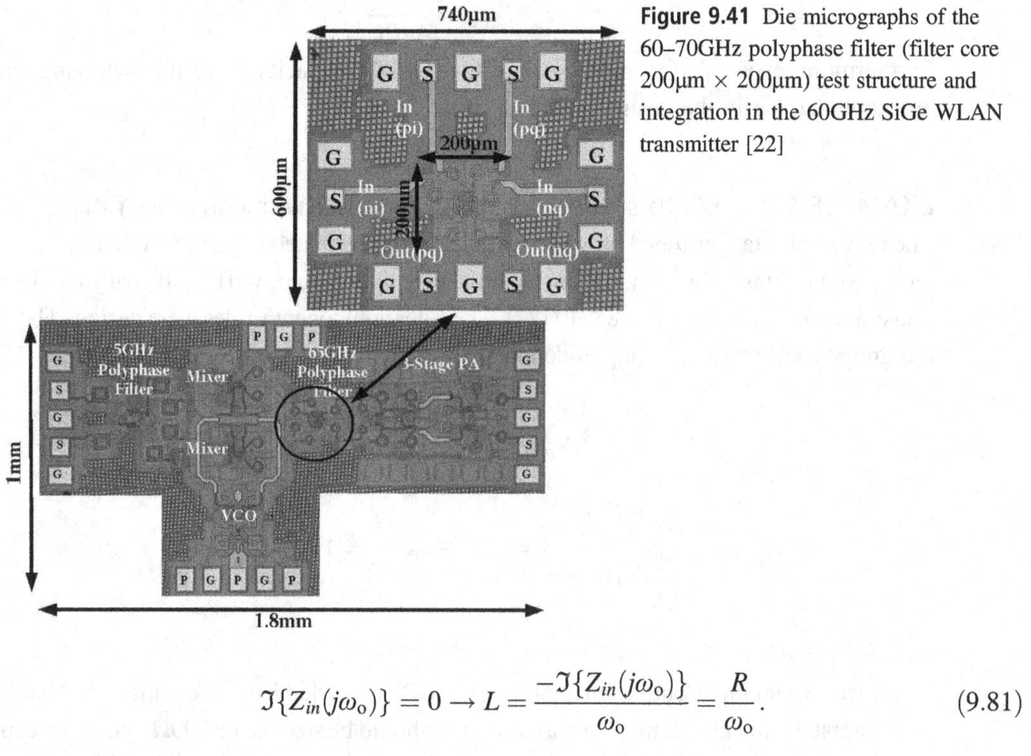

Figure 9.41 Die micrographs of the 60–70GHz polyphase filter (filter core 200μm × 200μm) test structure and integration in the 60GHz SiGe WLAN transmitter [22]

$$\Im\{Z_{in}(j\omega_o)\} = 0 \rightarrow L = \frac{-\Im\{Z_{in}(j\omega_o)\}}{\omega_o} = \frac{R}{\omega_o}. \tag{9.81}$$

Figure 9.41 shows the die micrographs of the filter test structure and integrated transmitter. The six inductors are implemented as planar octagonal spirals for reduced on-chip radiation

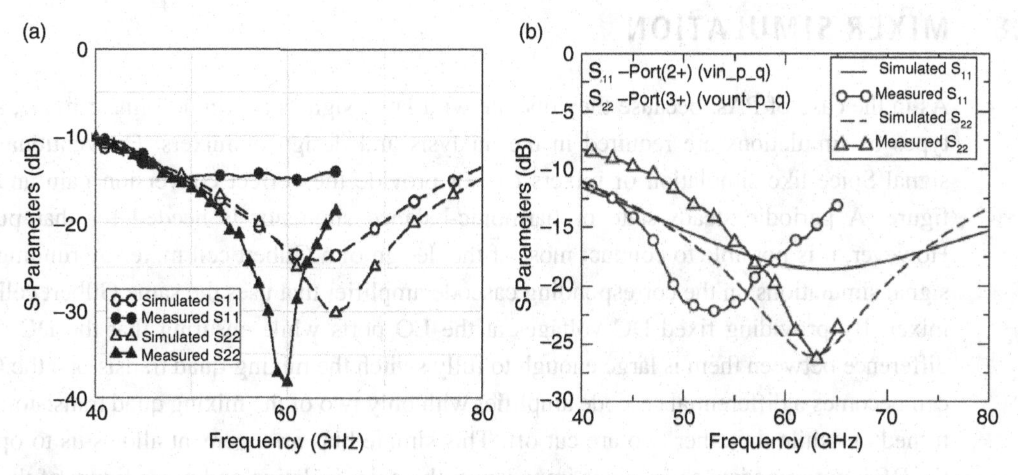

Figure 9.42 (a) 65GHz polyphase filter matching at ports 1+ and 3− (b) 65GHz polyphase filter matching at ports 2+ and 3+ [22]

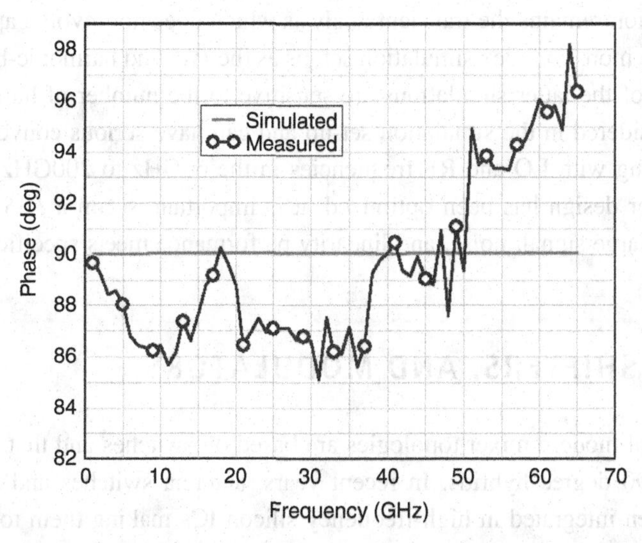

Figure 9.43 65GHz polyphase filter measured vs. simulated phase variation [22]

effects. Additional interconnect inductances are also accounted for at 65GHz. The filter structure, including all pads, measures 740μm × 600μm, while the filter core occupies 200μm × 200μm. The accurate modeling of inductors and minimization of layout parasitics lead to better than 25dB return loss at the output port, while at the input port, the use of 15μm wide transmission lines which connect the octagonal inductors to the pads results in a broadband match (return loss >10dB) across the 40–65GHz range (Figures 9.42(a) and (b)). There is also good agreement between the measured and simulated phase response for the 65GHz filter (Figure 9.43).

9.8 MIXER SIMULATION

As in the case of PAs, because they operate with large signals and in non-linear mode, several types of simulations are required in the analysis and design of mixers. Conventional small signal Spice-like simulation of mixers cannot provide the correct conversion gain and noise figure. A periodic steady-state or harmonic-balance simulator is needed for that purpose. However, it is possible to conduct most of the design of a Gilbert cell mixer by running small signal simulations on the corresponding cascode amplifier that uses the same Gilbert cell as the mixer. By providing fixed DC voltages at the LO ports while ensuring that the DC voltage difference between them is large enough to fully switch the mixing-quad transistors, the Gilbert cell becomes a differential cascode amplifier with only two of the mixing quad transistors being turned on while the other two are cut off. This simple bias arrangement allows us to optimize the RF input impedance, the noise impedance, the gain, isolation and noise figure of the mixer using only small signal and noise analysis [7]. The very same approach can be employed in the lab to measure the *S*-parameters of the mixer or receiver when the mixer is biased as an amplifier [7].

The most reliable simulation remains the transient analysis which does not involve approximations and does not require more complex simulation setups as the *PSS* and harmonic-balance simulations do. The results of the latter simulations are sensitive to the number of harmonics and the frequency plan considered in the simulation set-up and can have serious convergence problems for mixers operating with LO and RF frequencies in the 60GHz to 200GHz range. Nevertheless, once the mixer design has been optimized, it is important to run a *PSS* or *HB* simulation to verify that its large signal, noise, and linearity performance meets specifications.

9.9 SWITCHES, PHASE SHIFTERS, AND MODULATORS

We have already seen that all modern mixer topologies are based on switches and that image-reject mixers also employ 90 degree hybrids. In recent years, antenna switches and phased arrays have increasingly been integrated in high-frequency silicon ICs making them topics of intense research. QPSK and 16QAM modulators, like mixers, rely on switches and 90 degree phase shifters to produce digitally modulated carriers. For these reasons, the most common FET-based switches and phase shifters will be reviewed next along with their application in BPSK, QPSK, and 16QAM modulators.

9.9.1 Switches

Basic shunt and series diode and FET switches are illustrated in Figure 9.44. Both p-i-n (PIN) and Schottky diodes have been employed as switches. The diodes operate as an electronic on-off switch when switched between forward and reverse bias conditions. PIN diodes provide the best switch performance, with a very low on-resistance, when forward biased, and a very small

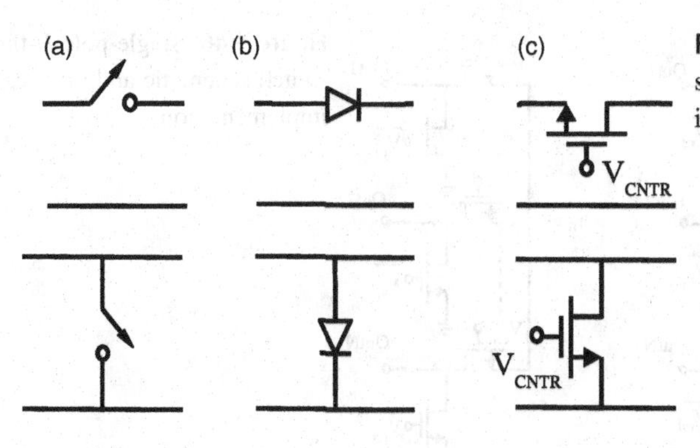

Figure 9.44 Schematics of (a) basic series and shunt switches and (b) p-i-n diode and (c) FET implementations

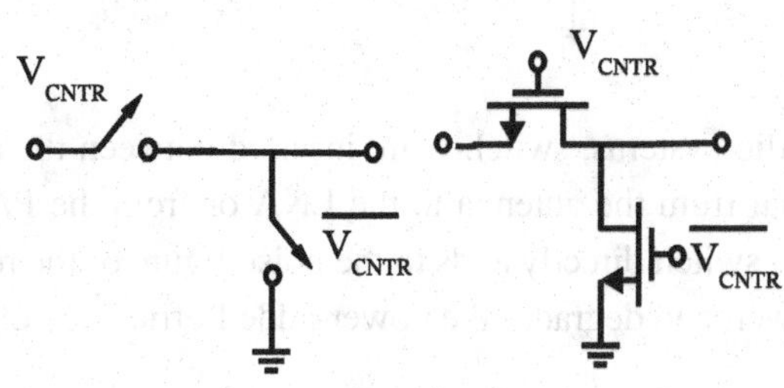

Figure 9.45 Single-pole single-throw switch schematic and practical implementation

capacitance when reverse biased. However, on the negative side, the bias supply voltage must be applied directly to the RF signal port through a low-pass filter, and integration with transistor technology is expensive, though possible. FETs, on the other hand, have the advantage of a third terminal, allowing for the control voltage to be isolated from the signal path without the need for a filter. A FET can be used as a switch between its source and drain terminals, while the control voltage is applied to the gate terminal. The most important performance parameters for a switch are the insertion loss, *IL*, and the isolation, *I*.

The insertion loss and isolation of a series switch with on-resistance R_{SW} and reverse-bias capacitance C_{SW}, embedded into a system of Z_0 characteristic impedance can be determined from the expressions:

$$IL = 20 \cdot \log_{10}\left(1 + \frac{R_{SW}}{2 \cdot Z_0}\right) \text{and } I = 10 \cdot \log_{10}\left[1 + (4 \cdot \pi \cdot f \cdot C_{SW} \cdot Z_0)^{-2}\right]. \quad (9.82)$$

Similarly, the insertion loss and isolation of a shunt switch is obtained from

$$IL = 10 \cdot \log_{10}\left[1 + (\pi \cdot f \cdot C_{SW} \cdot Z_0)^2\right] \text{and } I = 20 \cdot \log_{10}\left(1 + \frac{Z_0}{2 \cdot R_{SW}}\right). \quad (9.83)$$

It becomes immediately apparent that the switch performance is dictated by the on-resistance (as low as possible) and the off-capacitance (as low as possible) of the switch. For d 65nm GP n-MOSFET, R_{SW} is about 370Ω × μm and C_{SW} is 0.8fF/μm with the switch FoM = $R_{on} \times C_{off}$ = 300fs.

Figure 9.45 shows the symbolic representation and a FET-based implementation of a series–shunt switch, also known as a single-pole single-throw (SPST) switch. The concept is expanded in Figure 9.46 to single-pole double-throw and single-pole *N*-throw switches, typically employed as antenna switches in radio and phased array radar systems.

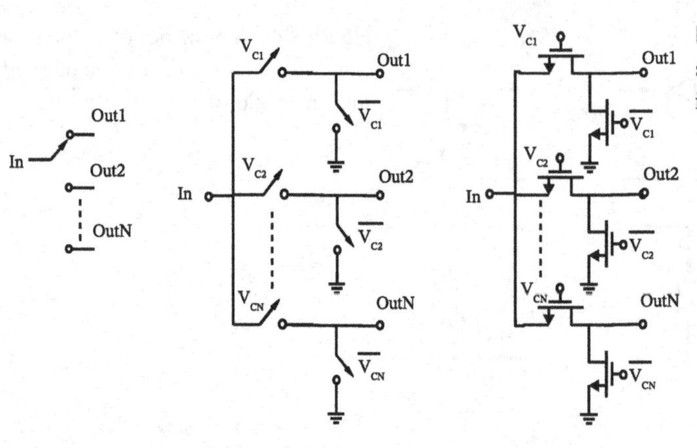

Figure 9.46 Single-pole *N*-throw switch schematic and practical implementation

In time-division-multiplex (TDM) radio systems, switches are inserted between the antenna and the transceiver, redirecting the signal from the antenna to the LNA or from the PA to the antenna. The insertion loss of an antenna switch directly adds to the noise figure of the receiver, subtracts from the transmitter output power, and degrades the power-added-efficiency of the PA according to the equation

$$PAE' = PAE \cdot 10^{\frac{-IL}{10}}. \tag{9.84}$$

Minimizing the insertion loss of the antenna switch is thus a critical design goal. Since the large transmitter signal passes through the switch, the linearity, characterized by the P_{1dB} or IIP3, is also an important design parameter for the switch.

When off, the output impedance of the FET switch is formed by a small capacitance, $C_{sw} = C_{db} + C_{ds} + C_{gd}$, in parallel with a finite, yet very large, resistance. To maximize the off-impedance at high frequencies, an inductor is typically placed in parallel with the FET, to resonate out its output capacitance, as illustrated in the shunt–series antenna switch of Figure 9.47. When fabricated in GaAs or InP HEMT technologies, where the C_{db} term is negligible, this circuit has been traditionally employed as an antenna switch. It is now possible to integrate it in nanoscale CMOS technology. In fact, all the FET-based switch circuits developed in the 1980s and 1990s in III-V technologies can now be transferred to nanoscale CMOS technology. Even more exciting is the fact that new, improved performance, symmetric switch topologies may emerge, which take advantage of the strain-engineering improved p-MOSFETs whose transconductance and cutoff frequency are now within 20% of the n-MOSFET values.

We note that, unlike diode switches, which consume bias current when on, a FET switch draws no current in either the "on" or the "off" state. However, a DC path must be provided to the drain and source of the FET used in the switch for proper operation.

If fast switching is not a requirement, a resistance of a few kΩ is typically added in series with the gate, as illustrated in Figure 9.47, in order to reduce the capacitance seen between the drain and ground. This scheme makes the gate–drain and gate–source capacitance appear in series, thus reducing the gate–drain component of the output capacitance and improving isolation and insertion loss.

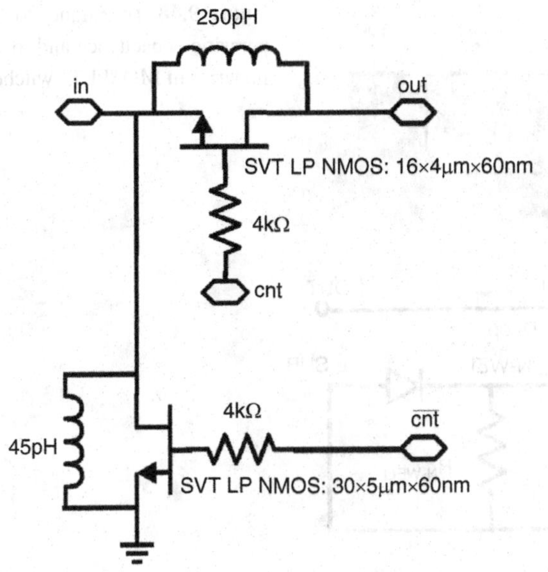

Figure 9.47 Series–shunt 60GHz FET switch [24]

Since, unlike SOI, silicon-on-sapphire (SOS) and III-V FETs, silicon bulk MOSFETs suffer from large drain–bulk, and source–bulk capacitances, placing the switch inside an isolated deep n-well and adding a large resistor of a few kΩ in series with the substrate can further reduce loss and improve isolation. This is shown in the top part of Figure 9.48. In all FET switches, linearity can be increased by vertically stacking transistors, as illustrated in Figure 9.48, at the expense of slightly increased loss.

From the isolation equation of a shunt switch, (9.83), we can observe that a variable attenuator can be realized if the switch on-resistance, R_{SW}, can be varied. Thus, a set of shunt switches can be used to realize a digitally controlled attenuator. The schematic and layout of a n-bit attenuator in 65nm CMOS is illustrated in Figure 9.49 [23]. Each transistor has a drawn gate length of 60nm and a gate finger width of 1μm. They share the same well and source and drain terminals. The expression of the digitally controlled on-resistance of the shunt switch is given by

$$R_{SW} = \frac{R_{unit}}{(b_0 + b_1 2 + b_2 2^2 + \ldots + b_{n-1} 2^{n-1})} \tag{9.85}$$

which leads to the following expression for the attenuation (from the isolation formula)

$$I = 20 \cdot \log_{10}\left(1 + \frac{Z_0}{2 \cdot R_{SW}}\right) = 20 \cdot \log_{10}\left[1 + \frac{25}{R_{unit}}(b_0 + b_1 2 + b_2 2^2 + .. + b_{n-1} 2^{n-1})\right] \tag{9.86}$$

where R_{unit} is the on resistance of the unit gate finger. Note that the total output capacitance remains the same, irrespective of the control word.

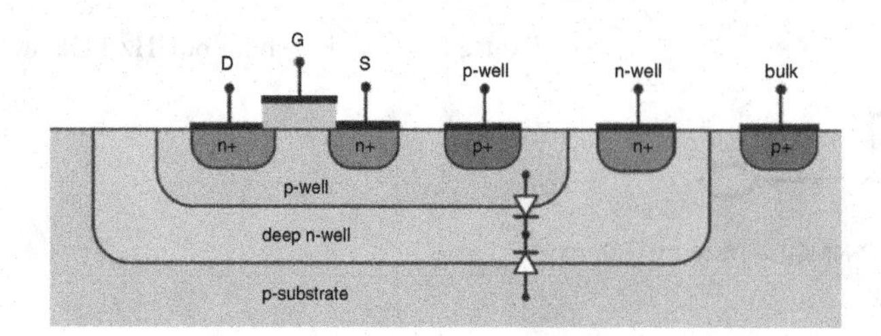

Figure 9.48 Techniques to reduce parasitic capacitance and to increase linearity in MOSFET switches

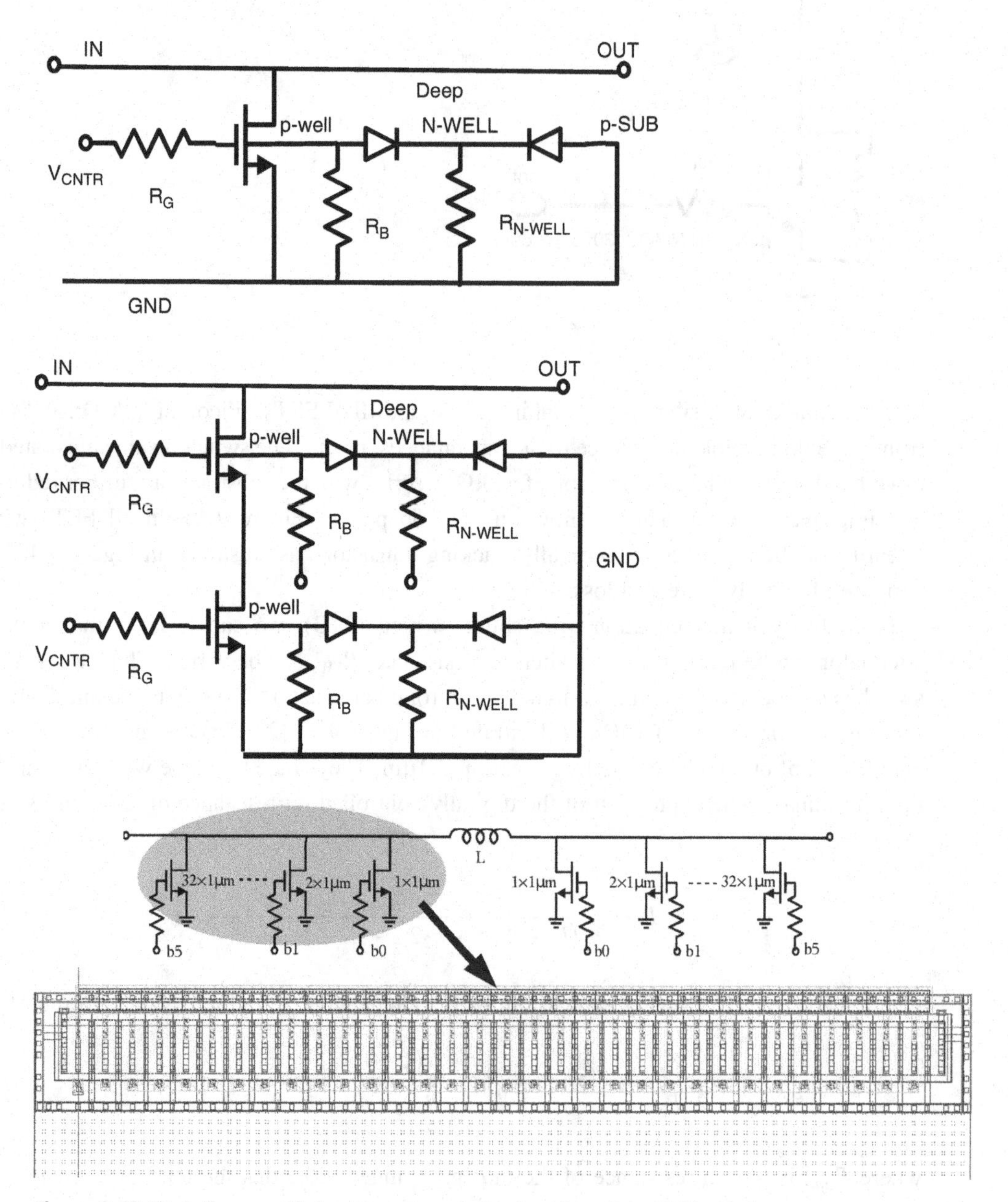

Figure 9.49 Digitally controlled mm-wave attenuator based on shunt FET switches

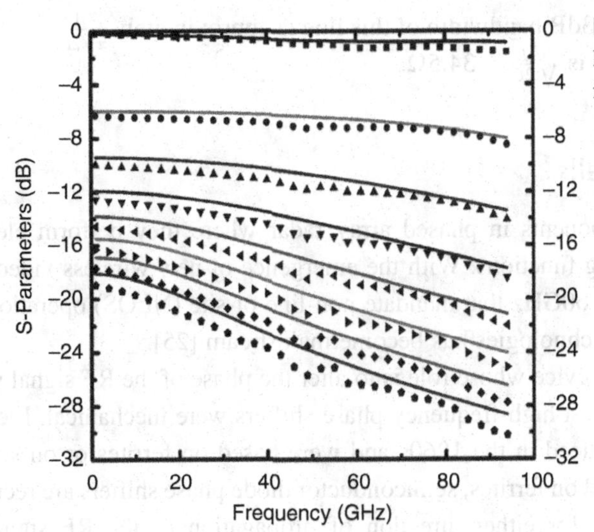

Figure 9.50 Measured vs. simulated insertion loss of a 3-bit variable attenuator manufactured in 65nm CMOS [23]

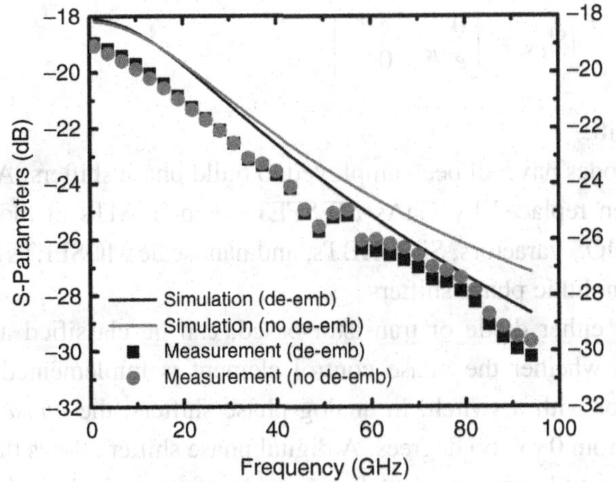

Figure 9.51 Measured vs. simulated isolation before and after de-embedding pad parasitics [23]

Figure 9.50 reproduces the measured attenuation of a 3-bit attenuator fabricated in a 65nm CMOS process and based on the concept illustrated in Figure 9.49. The least significant bit controls the smallest shunt FET which has six parallel gate fingers, each 1μm wide, while the most significant bit is applied to the gate of the largest FET with 24 parallel fingers, each 1μm wide. When all the control bits are "0," all the MOSFETs are off and the insertion loss of the attenuator is almost 0, determined by the output capacitance. In practice, 1–2dB of loss is typically realizable up to 94GHz. The largest attenuation, approximately 30dB, is obtained when all bits are set to "1," as illustrated in the measured and simulated characteristics of Figure 9.51. This attenuator is designed to operate in the 60–110GHz range and employs two shunt switches separated by a series inductor which forms an artificial transmission line with the output capacitance ($C = (N_{f1} + N_{f2} + N_{f3}) \times W_f \times (C'_{db} + C'_{gd}/2) = 34\text{fF}$) of the two switches for bandwidth extension. Ignoring the parasitic capacitance of the inductor, the 3dB bandwidth of this line is approximately $\frac{1}{\pi\sqrt{2LC}} = 136\text{GHz}$ and its characteristic impedance is $\sqrt{\frac{L}{2C}} = 34.5\Omega$.

9.9.2 Phase shifters and delay cells

Phase shifters are critical components in phased array radar where they perform electronic beam-steering and beam-forming functions. With the emergence of new wireless video standards for consumer electronics at 60GHz that mandate non-line-of-site (NLOS) operation, their integration in low-cost silicon technologies has become mainstream [25].

A phase shifter is a two-port device whose role is to alter the phase of the RF signal without (ideally) attenuating it. The earliest high-frequency phase shifters were mechanical. Electronic phase shifters where first introduced in the 1960s and were based on ferrites or on semiconductor diodes. Unlike those based on ferrites, semiconductor diode phase shifters are reciprocal, imparting the same phase shift for either direction of propagation of the RF signal. The scattering matrix of a phase shifter is represented as

$$[S]_{PS} = \begin{bmatrix} 0 & e^{-j\phi} \\ e^{-j\phi} & 0 \end{bmatrix} \tag{9.87}$$

where ϕ is the insertion phase shift.

PIN, Schottky, and varactor diodes have all been employed to build phase shifters. As in the case of mixers, diodes have been replaced by GaAs MESFETs or p-HEMTs in monolithic phase shifters. More recently, AMOS varactors, SiGe HBTs, and nanoscale MOSFETs have all been demonstrated in silicon monolithic phase shifters.

Semiconductor phase shifters, either diode or transistor-based, can be classified as either *analog* or *digital*, depending on whether the phase control element is implemented with a continuously variable reactance or with a switch. In analog phase shifters, the *phase shift* is continuously changed, typically from 0 to 360 degrees. A digital phase shifter allows the phase to be controlled in discrete steps and may consist of a cascade of several phase bits. For example, in the block diagram of the 4-bit phase shifter ($n = 4$), shown in Figure 9.52, the smallest phase step is $360/2^4 = 22.5$ degrees.

Another classification identifies phase shifters as *passive* or *active*, depending on the type of semiconductor device employed to produce the phase shift. Passive phase shifters use diodes (varactors) or cold (no bias current) FET switches, while active ones are transistor-based and can exhibit gain, in much the same way as active mixers do. Examples of semiconductor

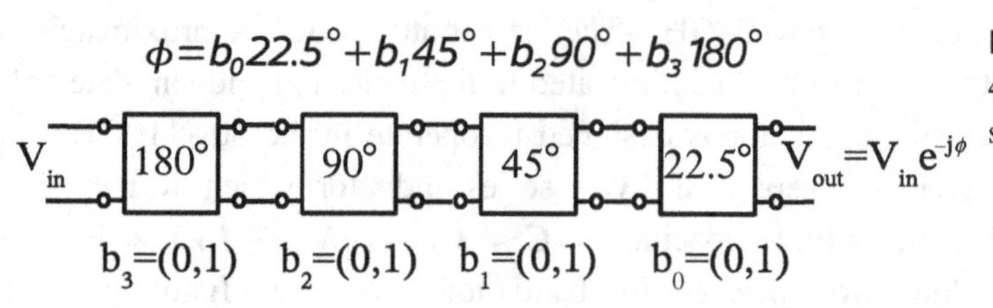

Figure 9.52 Block diagram of a 4-bit digitally controlled phase shifter

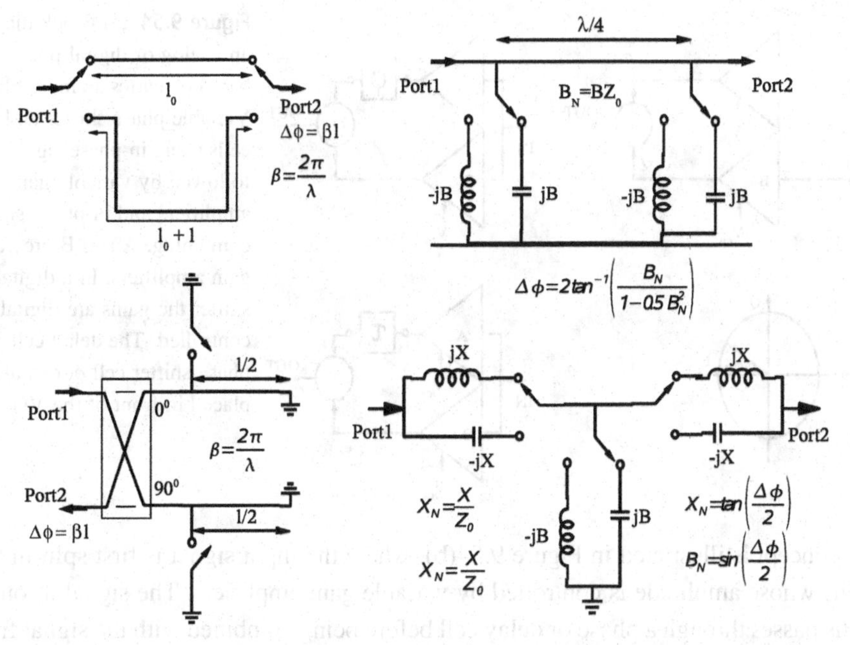

Figure 9.53 Block diagrams of switched line, hybrid-coupled, loaded line, and high-pass low-pass

devices that can be employed as switches have been discussed earlier. Passive analog phase shifters are implemented with FETs and varactor diodes. Depending on the circuit topology, passive phase shifters can be classified in [26]:

- switched line,
- hybrid coupled,
- loaded line,
- high-pass low-pass.

All these topologies are illustrated in Figure 9.53 at the block level and can be realized with either diodes or FETs.

Active phase shifters employ the interpolation concept based on a fixed phase or delay cell and variable gain amplifiers. The cartesian interpolation topology shown in Figure 9.54(a) is based on a 90 degree hybrid which feeds the RF signal in quadrature to two parallel paths, each with its own variable gain amplifier. It has become the most popular architecture for both analog and digital phase shifters [26]. Analog phase shifters differ from digital ones in that the VGA control is provided by an analog, as opposed to a digital signal. In either case, the output voltage can be expressed as a function of the amplitude of the in-phase I, and quadrature Q, signals, which can be controlled separately.

$$OUT = I + jQ = |OUT|e^{j\phi} \tag{9.88}$$

where

$$|OUT| = \sqrt{I^2 + Q^2} \text{ and } \phi = \tan^{-1}\left(\frac{Q}{I}\right). \tag{9.89}$$

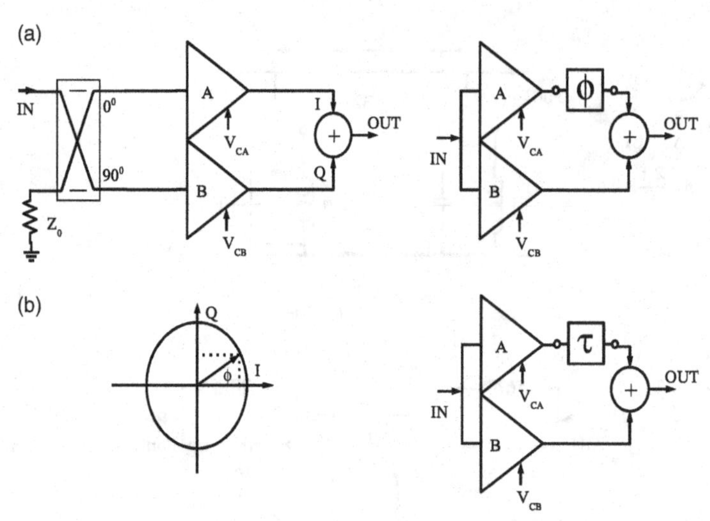

Figure 9.54 (a) Block diagram of an analog or digital phase shifter based on cartesian interpolation (b) Variable phase shifter and delay cells using in-phase signal splitting followed by variable gain amplifiers and in-phase signal combining. A and B are variable gain amplifiers. In a digital phase shifter the gains are digitally controlled. The delay cell τ or the phase-shifter cell ϕ, can also be placed in front of the VGA

A similar concept is illustrated in Figure 9.54(b), where the input signal is first split in two in-phase paths whose amplitude is controlled by variable gain amplifiers. The signal through one of the paths passes through a phase or delay cell before being combined with the signal from the second path.

9.9.3 BPSK, QPSK, and QAM modulators

As discussed in Chapter 2, all M-ary PSK and QAM modulators with $M >= 4$ can be synthesized with a circuit that consists of a 90 degree hybrid to generate the I and Q replicas of the carrier, and two digital amplitude modulators – one for each of the quadrature versions of the carrier. The amplitude modulator can be a switch, a mixer, or a DAC.

A direct modulator is a circuit in which the local oscillator input signal is directly modulated by the digital data signals to produce a modulated carrier at its output, without resorting to linear upconversion. The main advantage is that the circuit can be operated in saturated power mode, presumably with improved efficiency. The main disadvantage is that the output spectrum has multiple lobes which must be filtered either with a physical filter placed at the output of the modulator, or by employing oversampling of the data signals.

A BPSK modulator is typically implemented using a balanced (differential) switch. Historically the first BPSK modulators were realized with diodes and diode bridges or with isolators or 90 degree hybrids and phase shifters. In FET or HBT technologies, the Gilbert cell, the equivalent of a diode bridge, is the preferred topology for a differential BPSK modulator. The modulating signal is applied at the LO port while the carrier signal is applied at the input of the differential transconductor of the Gilbert cell mixer. The modulator output is collected differentially at the mixer IF port. BPSK modulators based on this topology (Figure 9.55) have been realized at frequencies as high as 65GHz [27] and 77GHz [28] in SiGe HBT technology. A 60GHz CMOS version [24] is illustrated in Figure 9.56.

QPSK modulators have been realized in series or parallel configurations using 90 degree hybrids (or isolators) and 180 degree phase shifters, as shown in Figure 9.57. In transistor

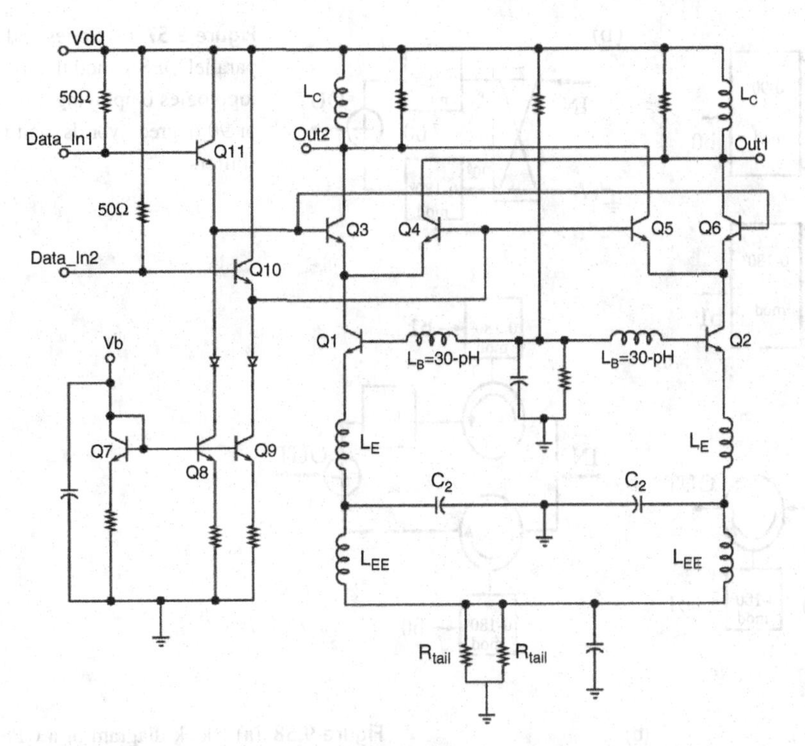

Figure 9.55 65GHz oscillator with built-in BPSK modulator [27]

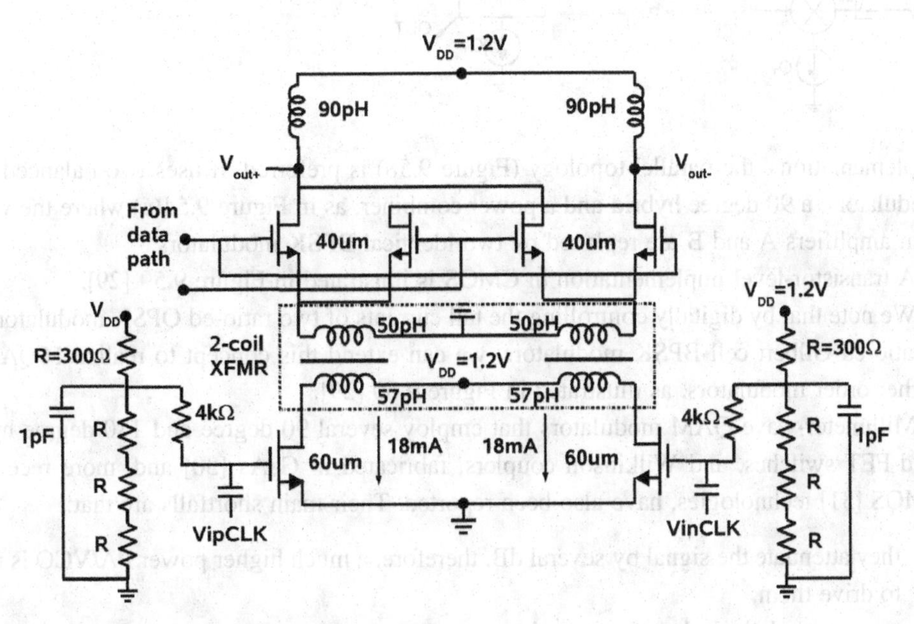

Figure 9.56 60GHz, BPSK modulator [24]

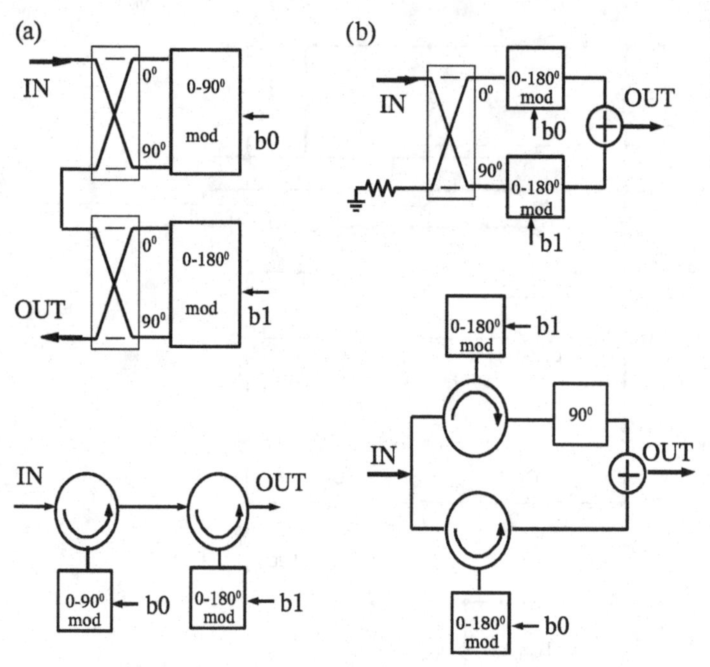

Figure 9.57 (a) Series and (b) parallel QPSK modulator topologies employing circulators or 90 degree hybrids, and phase shifters

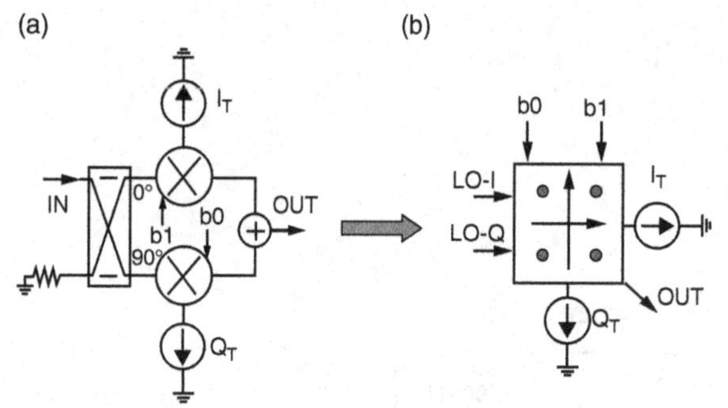

Figure 9.58 (a) Block diagram of a QPSK modulator using two Gilbert cell BPSK modulators, a 90 degree hybrid and a power combiner (b) Equivalent symbolic representation

implementations, the parallel topology (Figure 9.58) is preferred. It uses two balanced BPSK modulators, a 90 degree hybrid and a power combiner, as in Figure 9.54(a) where the variable gain amplifiers A and B are replaced by two identical BPSK modulators.

A transistor-level implementation in CMOS is illustrated in Figure 9.59 [29].

We note that by digitally controlling the tail currents of two ratio-ed QPSK modulators or of 4 ratio-ed Gilbert cell BPSK modulators, we can extend this concept to realize 16QAM and higher-order modulators, as illustrated in Figure 9.60 [29].

Millimeter-wave QAM modulators that employ several 90 degree and 180 degree hybrids, cold-FET switches, and Wilkinson couplers, fabricated in GaAs [30] and, more recently, in CMOS [31] technologies, have also been reported. Their main shortfalls are that:

(i) they attenuate the signal by several dB, therefore, a much higher power PA/VCO is needed to drive them,
(ii) occupy a relatively large area, and

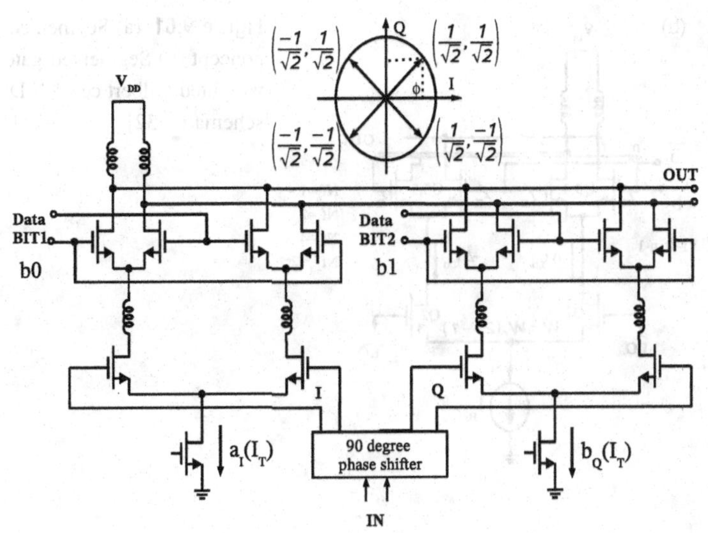

Figure 9.59 CMOS schematic of a mm-wave QPSK modulator based on quadrature generation and Gilbert cells [29]

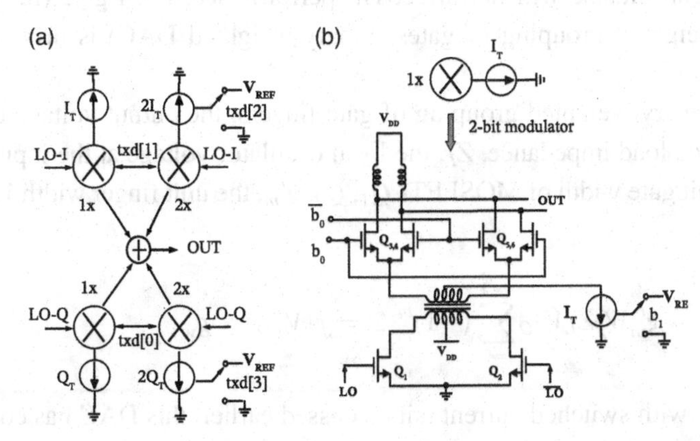

Figure 9.60 (a)16QAM modulator topology based on binary-weighted, tail-current switched BPSK modulator blocks (b) Two-bit amplitude and phase modulated unit modulator cell [29]

(iii) that, despite their passive nature, the linearity of the switch resistance limits the maximum output power.

The Gilbert cell based 16QAM modulator shown in Figure 9.60 overcomes the first two of these problems. Its main drawbacks are that (i) as the tail currents are turned on and off, the output impedance changes and (ii) that the finite isolation between data paths and the LO leakage will ultimately limit the dynamic range and effective resolution of this direct IQ modulator. Both these weaknesses can be alleviated by adding a common-base or a common-gate transistor pair at the output of the modulator.

By connecting unit Gilbert cells in parallel, each controlled by a separate bit, the direct Gilbert cell BPSK modulator concept can be extended to a multi-bit DAC. As illustrated in Figure 9.61, a multi-bit RF DAC can be realized using a segmented-gate Gilbert cell. The high-frequency LO signal is provided to the source, modulated by the gate, and collected at drain nodes. The source and drain nodes are common to all gates and thus have minimal layout parasitics. Control signals (bits) are applied at the individual gates, which are low-speed nodes

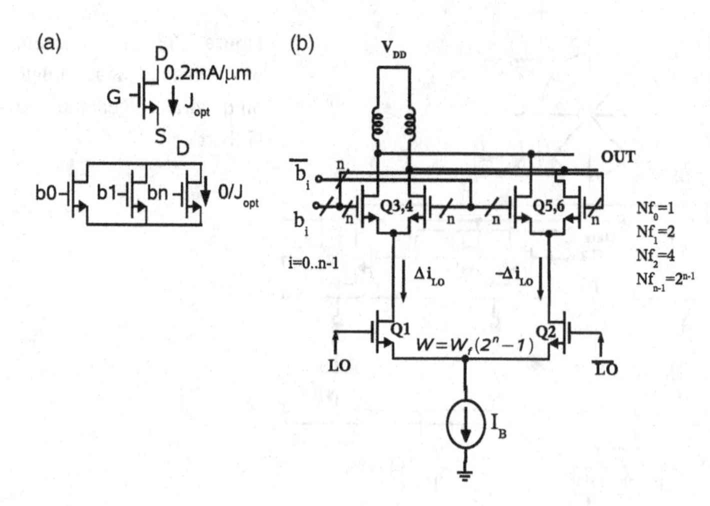

Figure 9.61 (a) Segmented-gate concept (b) Segmented-gate, binary-weighted Gilbert cell RF DAC schematic [32]

where additional parasitic capacitance will not affect HF performance. Each gate (fully segmented DAC) or binary-weighted grouping of gates (binary-weighted DAC) is driven by a CMOS inverter [32].

In the RF-DAC with a binary-weighted grouping of gate fingers, the output voltage can be described as a function of the load impedance, Z_L, the local oscillator voltage at the input V_{LO}, the transconductance per unit gate width of MOSFETs Q_1, Q_2, g'_m, the unit finger width W_f, and the digital bit word, g_w

$$V_{OUT} = g'_m W_f Z_L V_{LO} \sum_{i=0}^{n-1} (-1)^{b_i} 2^i = f(V_{LO}) \times g_w \tag{9.90}$$

Unlike the QAM modulator with switched current tail discussed earlier, this DAC has constant tail current and the output impedance is constant for all bit settings. However, it still suffers from resolution degradation due to finite isolation, especially at mm-waves.

By combining two such RF DACs driven in quadrature, a multi-bit direct RF IQ DAC is obtained which can be employed to generate any QAM and phase modulated carriers. This is illustrated in Figure 9.62. The corresponding equation that describes the output voltage is given by

$$V_{OUT} = g'_m W_f Z_L V_{IN} \left[\cos(\omega t) \sum_{i=0}^{n-1} (-1)^{a_i} 2^i + j\sin(\omega t) \sum_{i=0}^{n-1} (-1)^{b_i} 2^i \right]. \tag{9.91}$$

In all the Gilbert cell based modulators, the design of the Gilbert cell is similar to that of an upconverter mixer, or of a switch, because both the LO and the data signals are typically large. Moreover, as (9.90) and (9.91) indicate, these cells can be operated in saturated output power mode, to maximize their PAE, as in a PA. Although the large LO signal at the input will distort g'_m, it will do so equally for all combinations of bits and, therefore, the

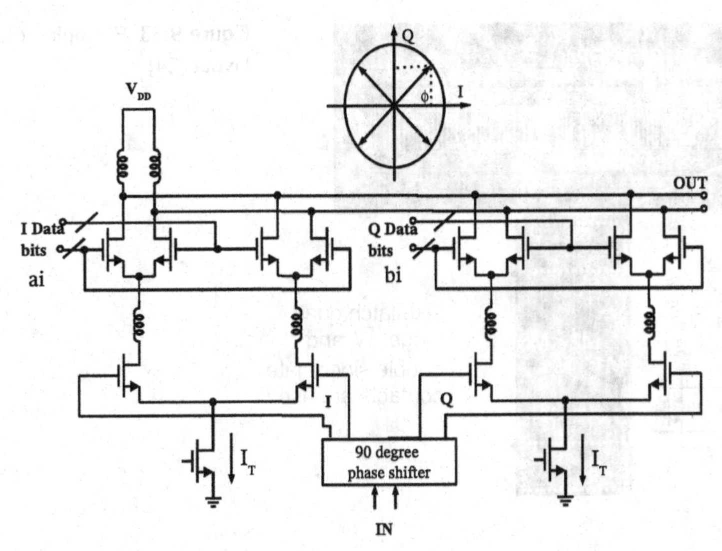

Figure 9.62 Schematics of binary-weighted phase rotator or IQ DAC

relative amplitudes of a_k and b_k remain unaffected, that is the terms in the square bracket of (9.96) do not depend on g'_m.

Finally, it should be noted that it is possible to build high PAE IQ power DACs out of saturated power amplifiers driven in quadrature, each with its own antenna, and employing free-space power combining. Such a proof-of-concept architecture has recently been demonstrated at frequencies as high as 45GHz with 24dBm output power and over 20% PAE per PA cell, in 45nm SOI CMOS technology. In this arcchitecture, amplitude modulation is implemented with a current source that turns on and off the power-amplifier cell [34], and binary phase modulation is implemented in the stages preceding the PA cell.

9.10 GILBERT CELL LAYOUT

The symmetry of the Gilbert cell layout is critical in achieving high isolation and high image rejection in image-reject mixer topologies. Both DC (resistors, transistors, via, and metal resistance) and AC (parasitic capacitance and inductive coupling) layout symmetry is important. Furthermore, as with all HF circuits, especially in nanoscale CMOS technologies, minimizing the layout parasitic capacitance is extremely important for achieving high conversion gain and low-noise figure. Some basic layout rules are listed here:

- minimize cell footprint since the parasitic capacitance to the substrate is roughly proportional to the cell area,
- merge all mixing quad transistors in a single active region, using an interdigitated, common centroid layout,
- merge transconductor and current source transistors in a single active region using an interdigitated, common-centroid layout,

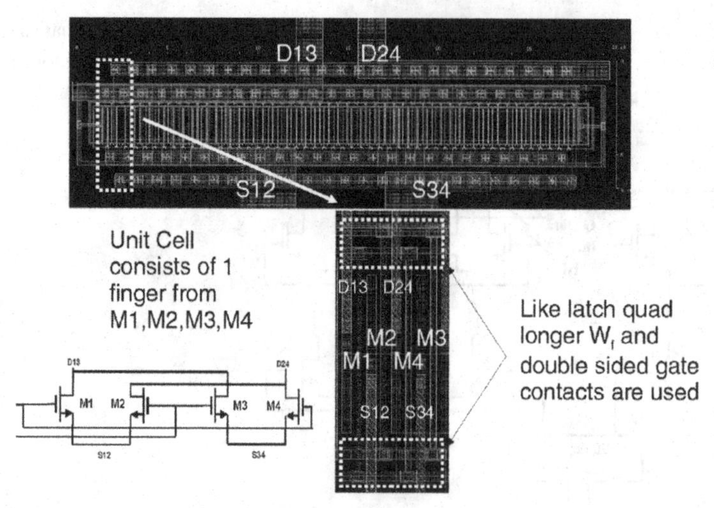

Figure 9.63 Example of latch layout [24]

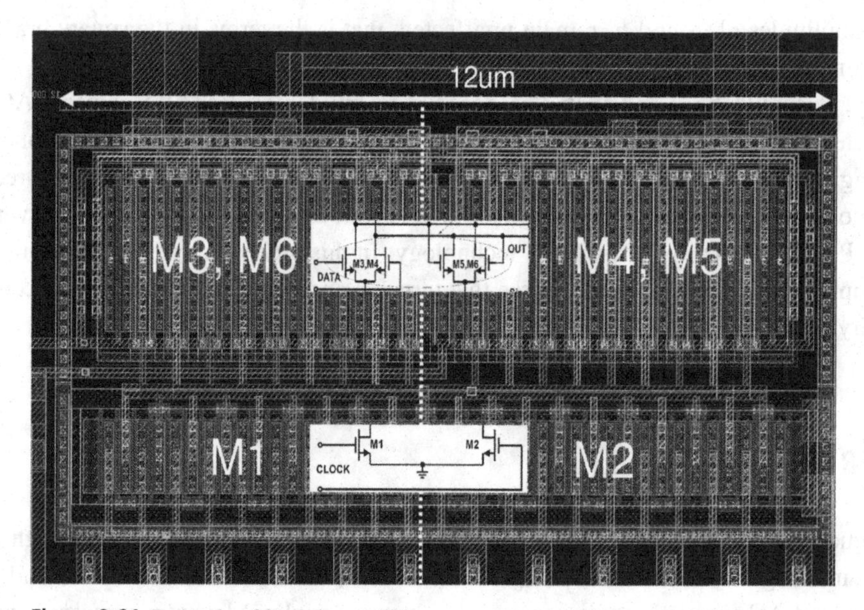

Figure 9.64 Example of latch layout [33]

- employ lower-level metal for interconnect between transconductor and mixing quad to reduce the overall size of the Gilbert cell,
- bring out the inputs and outputs of the Gilbert cell in the top two metal layers to facilitate global interconnect,
- minimize the capacitance at the output node in upconverters and direct modulators at the expense of the capacitance at the middle node.

Examples of these layout design principles being applied to the BPSK modulator schematic in Figure 9.56 are illustrated in Figures 9.63–9.65.

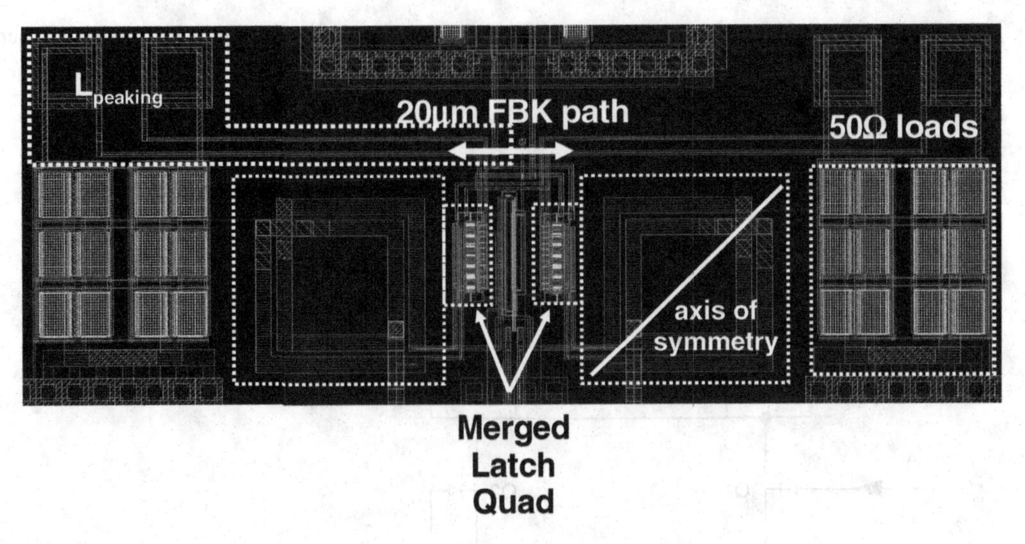

Figure 9.65 Example of transformer-coupled latch layout [24]

Problems

(1) Derive the expressions of the single-sideband and double-sideband mixer noise factor starting from the definition of the noise factor as the degradation of the signal-to-noise ratio.

(2) The single-balanced and double-balanced mixers shown in Figures 9.15 and 9.17 have the same conversion gain. Explain what are the benefits (if any) of the double-balanced topology over those of the single-balanced mixers.

(3) The selector schematic in Figure 9.66 can be used to create a 36GHz low-noise SiGe BiCMOS Gilbert cell mixer. The MOS-HBT cascode characteristics are those reproduced in Figure 9.67.

(a) Indicate how the differential local oscillator signal should be applied to the CC′ and BB' nodes if the circuit is to be used as an upconverter mixer with the output at 36GHz and the LO signal at 40GHz. Determine the value of the intermediate frequency applied at AA'. Replace R_L with an inductor.

(b) What is the current density at which the transistors (HBTs and MOSFETs) in the upconvert mixer should be biased for maximum gain?

(c) Knowing that the largest inductor value that resonates with the output capacitance of the mixing quad at 36GHz is 400pH and that its desired Q is 4, size the MOSFETs and HBTs in the mixer such that the mixer upconversion gain is equal to 4 (i.e. 12dB).

(4) A 24GHz mixer employed in an automotive collision avoidance radar uses the topology in Figure 9.23 Assume 90nm CMOS technology with n-MOS f_T of 120GHz at V_{DS} = 0.5V and a power supply of 1.2V.

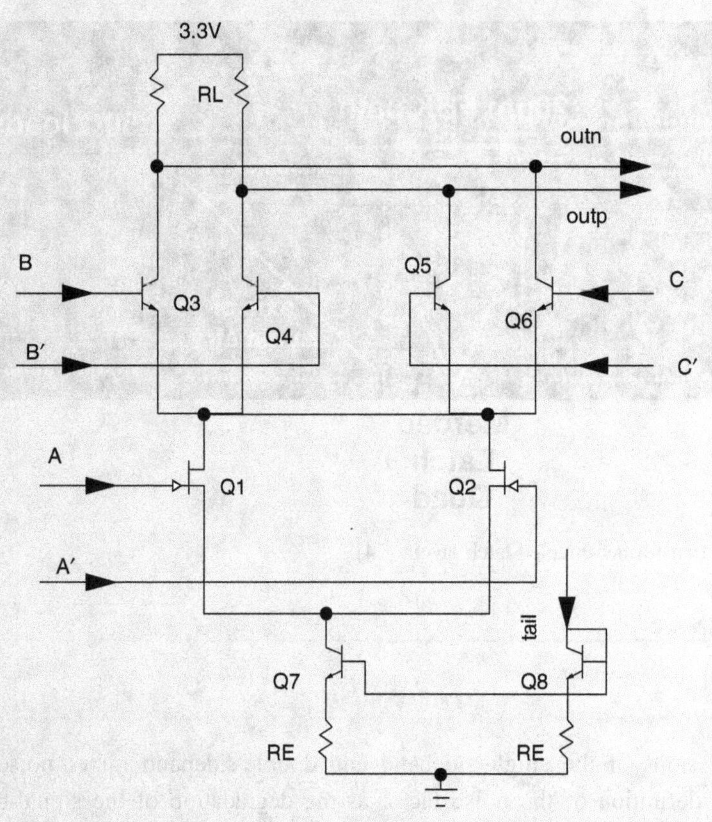

Figure 9.66 Selector schematics

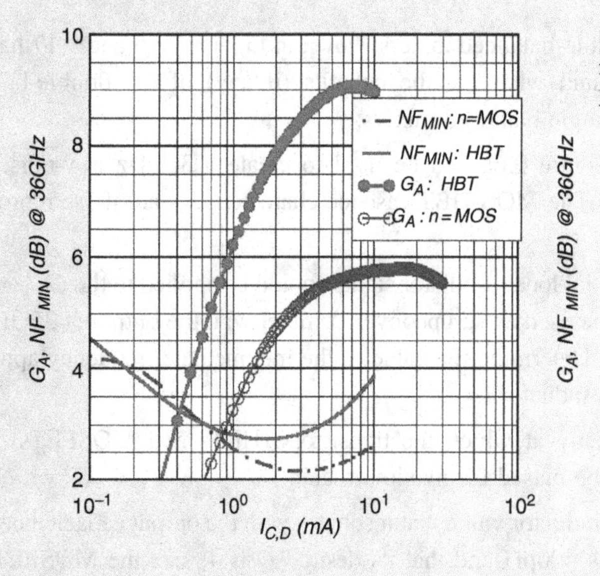

Figure 9.67 Associated gain and minimum noise figure vs. I_C/I_D characteristics for the SiGe HBT and the 130nm n-MOSFET

(a) Find the value of L_E, the bias current and the transistor sizes if the mixer has a conversion gain of 0dB and load resistors of 200Ω. Assume that the transistors in the transconductor are biased at the peak f_T current density of 0.3mA/μm and that the differential input impedance of the mixer is 100Ω. Assume that the total gate width of the transistors in the mixing quad is 10μm.

(b) What is the bias current flowing through L_M?

(c) Calculate the value of L_M for maximum second LO harmonic rejection.

(5) Design a 0.9V, 94GHz downconvert mixer using the topology in Figure 9.26 in a 45nm CMOS process with the following technology parameters $L = 40$nm, $|VT| = 0.3$V; $Idsn$ at minimum noise = 0.15mA/μm; $g'_{mn} = 1.25 \times g'_{mp} = 1.8$mS/μm; $g'_{dsn} = g'_{dsp} = 0.18$mS/μm; the peak f_T of 250/180GHz for the n-MOSFETs and p-MOSFETs, respectively, occurs at 0.3mA/μm for $0.3\text{V} < V_{DS} < 1\text{V}$.

For both n-channel and p-channel devices, $C'_{gs} = 0.7$fF/μm; $C'_{gd} = 0.3$fF/μm; $C'_{sb} = C'_{db} = 0.4$fF/μm;

$$R'_s(n-\text{MOS}) = R'_d = 200\Omega \times \mu m;\ R'_s(p-\text{MOS}) = R'_d = 300\Omega \times \mu m;$$
$$R_{shg}(n/p-\text{MOS}) = 20\Omega/\text{sq; poly contact resistance } R_{CON} = 70\Omega.$$

Total R_g/0.7μm finger contacted on one side = 200Ω. Noise params: $P = 1$, $C = 0.4$, $R = 0.25$.

(a) Determine the size and bias current of the transconductor stage MOSFETs Q_1 and Q_2 if the transconductor is designed as a minimum noise figure LNA stage, noise matched to 50Ω per side.

(b) If Q_3–Q_6 have 10μm gate width and are biased to switch between 0 and 0.3mA/μm, what should be the value of the load resistance R_L if the linear voltage swing at the output of the mixer is maximized (assume that $V_{DSAT} = 0.1$V).

(c) If L_D, C_D are designed to resonate with the input capacitance of the mixing quad at 94GHz, derive the expression of the downconversion gain and calculate its value.

(6) What is the main difference in the bias schemes employed in the mixers in Figure 9.31 and 9.33? In each case, the mixing quad MOSFETs are switching between 0 and 0.3mA/μm.

(7) For the mixer in Figure 9.23, derive the expression of the common-mode impedance seen by the LO signal at $2 \times f_{LO}$ as a function of L_M and the capacitances in parallel with it. If all MOSFETs have 65nm×10×1μm gates, calculate the optimal value of L_M if the LO signal is at 20GHz. Assume the same 65nm CMOS technology parameters as in Appendix 9.

(8) Redesign the 140GHz downconvert mixer in Figure 9.35 to operate as a 1V, 140GHz direct BPSK modulator with a tuned output load matched to 50Ω and +10dBm output power. Use the same 45nm CMOS process as in problem 5. Where will the data bit be applied? If the peak-to-peak output voltage swing is 1.4Vpp, what should be the total gate width of M1–M4?

(9) Design a single-stage RC-polyphase filter centered at 20GHz, 60GHz, and 80GHz, respectively. Discuss the implementation issues.

(10) Design a single-stage LRC-polyphase filter centered at 60GHz. Calculate the bandwidth over which the phase imbalance remains smaller than +/−5 degrees.

(11) Develop a mathematical analysis of the two IR mixer topologies in Figure 9.37 as was conducted for the IR mixer topology in Figure 9.36.

(12) Demonstrate that the circuit in Figure 9.37(a) can also be used as a doubler if the LO signal is also applied at the IN port.

(13) What is the bandwidth required for the 90 degree hybrid on the RF path if the IR mixer topology from Figure 9.37(a) is employed in a 60GHz wireless Video Area Network receiver whose frequency plan was discussed in Example 9.2? What is the minimum bandwidth of the 90° hybrid that could be used to generate the quadrature LO signal?

(14) Design the 65nm MOSFET switch in Figure 9.47 to be centered at 62GHz. Do we still need inductors if the switch is employed in the DC to 10GHz range? In the latter case, calculate the insertion loss and the isolation of the switch at 5GHz.

(15) Using equations (9.88) and (9.89) calculate the normalized gain values (on a 0 to 1 scale) of amplifiers A and B in Figure 9.54(a) such that the amplitude of the output signal remains constant when the phase of the output signal is 0, 45, 90, 135, 180, 225, 270, and 315 degrees.

(16) Derive equations (9.90) and (9.91). Using (9.88), (9.89) and (9.91) determine how many bits of resolution are required in each DAC to in order to generate a 64 QAM constellation with less than 2 degree phase error.

(17) Redesign the BPSK modulator in problem 8 as an 8-bit direct mm-wave DAC with +10dBm output power at 140GHz. What are the finger width and number of fingers of the least significant bit?

(18) Design a 60GHz 16 QAM modulator based on the topology in Figure 9.59. *Hint*: use 1-bit current tail sources on the I and Q paths. Each current source must switch current in a 1/3 ratio. This will work well if the bottom differential pairs on the I and Q paths are implemented with bipolar transistors. In CMOS implementations, the differential pair size must also be switched in a 3:1 ratio, simultaneously with the current source.

REFERENCES

[1] T. O. Dickson, M.-A. LaCroix, S. Boret, D. Gloria, R. Beerkens, and S. P. Voinigescu, "30–100GHz inductors and transformers for millimeter-wave (Bi)CMOS integrated circuits," *IEEE MTT*, **53**(1): 123–133, 2005.

[2] S. L. Long and D. B. Estreich, "Basics of compound semiconductor ICs," Primer Course, IEEE CSICS, San Antonio, Texas, November 2006.

[3] D. M. Pozar, *Microwave and RF Wireless Systems*, John Wiley & Sons, 2001, Chapter 7.

[4] S. A. Mass, *Noise in Linear and Nonlinear Circuits*, Artech House, 2005.

[5] B. Gilbert, "A precise four-quadrant multiplier with subnanosecond response," *IEEE JSSC*, **3**(4): 365–373, 1968.

[6] S. P. Voinigescu and M. C. Maliepaard, "5.8GHz and 12.6GHz Si bipolar MMICs," *IEEE ISSCC Digest*, pp. 372–373, 1997.
[7] M. A. Copeland, S. P. Voinigescu, D. Marchesan, P. Popescu, and M. C. Maliepaard, "5GHz SiGe HBT monolithic radio transceiver with tunable filtering," *IEEE MTT*, pp. 170–181, February 2000.
[8] E. Laskin, P. Chevalier, A. Chantre, B. Sautreuil, and S. P. Voinigescu, "80GHz /160GHz transceiver in SiGe HBT Technology," *IEEE RFIC Symposium Digest*, pp. 153–156, Honolulu, HI, June 2007.
[9] S. T. Nicolson, A. Tomkins, K. W. Tang, A. Cathelin, D. Belot, and S. P. Voinigescu, "A 1.2V, 140GHz receiver with on-die antenna in 65nm CMOS," *IEEE RFIC Symposium Digest*, pp. 229–232, 2008.
[10] D. Alldred, B. Cousins and S. P. Voinigescu, "A 1.2V, 60GHz radio receiver with on-chip transformers and inductors in 90nm CMOS," *Proc. IEEE CSICS Digest*, pp. 51–54, November 2006.
[11] M. Brandolini, P. Rossi, D. Sanzogni, and F. Svelto, "A CMOS direct downconversion mixer with +78dBm minimum IIP2 for 3G Cell-Phones," *IEEE ISSCC Digest*, pp. 320–321, February 2005.
[12] J. R. Long and M. A. Copeland, "A 1.9GHz low-voltage silicon bipolar receiver front-end for wireless personal communication systems," *IEEE JSSC*, **30**: 1438–1448, December 1995.
[13] S. T. Nicolson, K. A. Tang, K. H. K. Yau, P. Chevalier, B. Sautreuil, and S. P. Voinigescu, "A low-voltage 77GHz automotive radar chipset," *IEEE IMS Digest*, pp. 487–490, June 2007.
[14] E. Abou-Allam, J. J. Nisbet, and M. C. Maliepaard, "A 1.9V front-end receiver in 0.5μm CMOS technology," *IEEE JSSC*, **36**: 1434–1443, October 2001.
[15] J. Crols and M. S. J. Steyaert, "A single-chip 900MHz CMOS receiver front-end with high performance low-IF topology," *IEEE JSSC*, **30**: 1483–1492, December 1995.
[16] M. Q. Gordon, T. Yao, and S. P. Voinigescu, "65GHz receiver in SiGe BiCMOS using monolithic inductors and transformers," *IEEE SiRF Digest*, pp. 265–268, January 2006.
[17] S. P. Voinigescu, S. T. Nicolson, M. Khanpour, K. K. W. Tang, K. H. K. Yau, N. Seyedfathi, A. Timonov, A. Nachman, G. Eleftheriades, P. Schvan, and M. T. Yang, "CMOS SOCs at 100GHz: system architectures, device characterization, and IC design examples," *IEEE ISCAS*, New Orleans, pp. 1971–1974, May 2007.
[18] K. W. Tang, M. Khanpour, P. Garcia, C. Garnier, and S. P. Voinigescu, "65nm CMOS, W-band receivers for imaging applications," *IEEE CICC Digest*, pp. 749–752, September 2007.
[19] M. J. Gingell, "Single sideband modulation using sequence asymmetric polyphase networks," *Elect. Commn*, **48**: 21–25, 1973.
[20] K.-J. Koh and G. M. Rebeiz, "0.13μm CMOS phase shifters for X-, Ku-, and K-band phased arrays," *IEEE JSSC*, **42**(11): 2535–2546, November 2007.
[21] R. C. Frye, S. Kapur, and R. C. Melville, "A 2GHz quadrature hybrid implemented in CMOS technology," *IEEE JSSC*, **38**(3): 550–555, March 2003.
[22] T. Yao, "Transmitter front end ICs for 60GHz radio," M.Sc. thesis, University of Toronto, 2006.
[23] A. Tomkins, P. Garcia, and S. P. Voinigescu, "A 94GHz SPST switch in 65nm bulk CMOS," *IEEE CSICS*, Monterey, CA, October 2008.
[24] A. Tomkins, R. A. Aroca, T. Yamamoto, S. T. Nicolson, Y. Doi, and S. P. Voinigescu, "A zero-IF 60GHz transceiver in 65nm CMOS with > 3.5Gb/s links," *IEEE CICC*, San Jose, CA, September 2008.

[25] A. M. Niknejad and H. Hashemi (eds.), *mm-Wave Silicon Technology*, Springer 2008.
[26] S. K. Koul and B. Bhat, *Microwave and Millimeter Wave Phase Shifters*, Artech House, 1991.
[27] C. Lee, T. Yao, A. Mangan, K. Yau, M. A. Copeland, and S. P. Voinigescu, "SiGe BiCMOS 65GHz BPSK transmitter and 30 to 122GHz LC-varactor VCOs with up to 21% tuning range," *IEEE CSICS Digest*, Monterey, CA, pp. 179–182, October 2004.
[28] S. Trotta, H. Knapp, D. Dibra, K. Aufinger, T.F. Meister, J. Bock, W. Simburger, and A. Scholtz, "A 79GHz SiGe-bipolar spread-spectrum TX for automotive radar," *IEEE ISSCC Digest*, pp. 430–431, February 2007.
[29] S.P. Voinigescu, A. Tomkins, M.O. Wiklund, and W.W. Walker, "Direct m-ary quadrature amplitude modulation (QAM) modulator operating in saturated power mode," US Patent Application 20110013726A1, January 20, 2011.
[30] D. S. McPherson and S. Lucyszyn, "Vector modulator for W-band software radar techniques," *IEEE MTT*, **49**(8): 1451–1461, 2001.
[31] C.-H. Wang, H.-Y. Chang, P.-S. Wu, K.-Y. Lin, T.-W. Huang, H. Wang, and C. H. Chen, "A 60GHz low-power six-port transceiver for gigabit software-defined transceiver applications," *IEEE ISSC*, pp. 192–193, February 2007.
[32] E. Laskin, M. Khanpour, S. T. Nicolson, A. Tomkins, P. Garcia, A. Cathelin, D. Belot, and S. P. Voinigescu, "Nanoscale CMOS transceiver design in the 90–170 GHz range," *IEEE MTT*, **57**: 3477–3490, December 2009.
[33] S. Shahramian, A. C. Carusone, P. Schvan, and S. P. Voinigescu, "An 81GB/s, 1.2V TIALA-retimer in standard 65nm CMOS," *IEEE CSICS Digest*, pp. 215–218, October 2008.
[34] I. Sarkas, A. Balteanu, E. Dacquay, A. Tomkins, and S. P. Voinigescu, "A 45nm SOI CMOS class D mm-wave PA with >$10V_{pp}$ differential swing," *IEEE ISSCC Digest*, pp. 24–25, February 2012.

10 Design of voltage-controlled oscillators

WHAT IS AN OSCILLATOR?

Oscillators are critical building blocks in both wireline and wireless systems and, along with synthesizers and clock and data recovery circuits (CDR), ensure that the synchronization of all receive and transmit functions is correctly carried out.

In most practical applications, the frequency of the signal generated by the oscillator must be controllable over some range. This is typically realized with a voltage-controlled reactance element in the oscillator circuit, hence the name voltage-controlled-oscillator or, in short, VCO.

The most important design considerations for VCOs refer to their:

- oscillation frequency,
- frequency tuning range and VCO gain, specified in GHz/V or MHz/V,
- output power, specified in mW or dBm,
- phase noise, measured in dBc/Hz and specified usually at 100KHz or 1MHz offset from the carrier frequency,
- frequency stability over temperature (in ppm/°C).

10.1 VCO FUNDAMENTALS

An oscillator can generally be described as a non-linear circuit that converts DC power to an AC waveform. In most RF, microwave, mm-wave, and fiber-optic communication systems, the oscillator must provide a purely sinusoidal waveform. Therefore, the design effort is focussed on minimizing the undesired harmonics and the phase noise, and on ensuring the long-term and short-term stability of the oscillation frequency and amplitude.

10.1.1 Oscillator model

As with many non-linear circuits, the basic concept can be understood by resorting to a simplified, linear model. However, it should be noted that the oscillator is a strongly non-linear circuit and large signal analysis is ultimately required to accurately describe and design the behavior of the oscillator. In the linear approximation, oscillators have been traditionally described as:

- an amplifier with positive and selective feedback, as in Figure 10.1(a), or
- as a negative resistance single-port connected in parallel with a resonant tank, Figure 10.1(b).

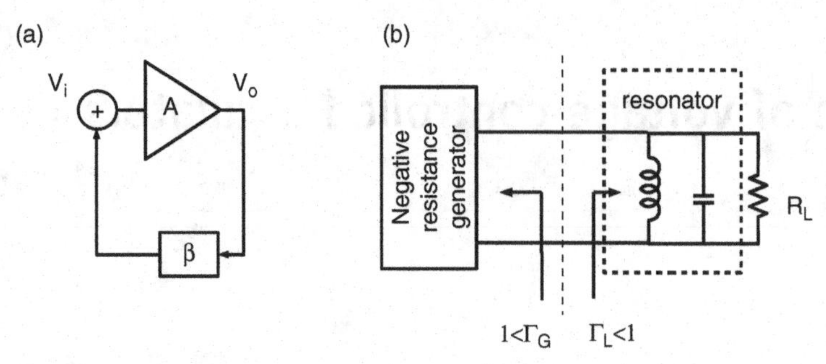

Figure 10.1 Oscillator model as (a) linear feedback system and (b) as a linear negative resistance generator in parallel with a resonant tank load

The negative resistance model can be used for both transistor and diode oscillators while the amplifier-with-feedback model is applicable only to oscillators built with transistors.

10.1.2 Conditions for oscillation

Linear feedback model

In the linearized amplifier-with-feedback model, we start by assuming that the gain of the amplifier, A, is a function of the angular frequency, ω. Its dependence on signal amplitude is, for now, ignored. The transfer function of the feedback network varies with frequency and is described by $\beta(\omega)$. By inspecting Figure 10.1(a), we can quickly derive the expression of the output voltage V_o as a function of the input voltage V_i as

$$V_o = \frac{A(\omega)}{1-\beta A(\omega)A(\omega)} V_i \tag{10.1}$$

When the denominator in (10.1) becomes equal to 0 for some frequency f_{OSC}, the gain from the input to the output of the loop reaches infinity and it is possible for the output voltage to develop simply out of nothing. In reality, when the circuit is biased, the broadband noise present at the input of the oscillator, or the broadband noise generated by the active and passive devices in the oscillator, is responsible for initiating the oscillation. Only noise at and around the oscillation frequency will be amplified by the selective, positive feedback loop and develop into an oscillation. Noise outside the 3dB bandwidth of the feedback network will be attenuated.

We can safely say that, at the onset of the oscillation, the oscillation amplitude is practically zero and hence the linear model quite accurately captures that initial phase in the operation of the oscillator. As the voltage at the output of the loop, V_o, increases, so does the input voltage to the amplifier. The amplifier gain, and in some cases the transfer function of the selective feedback network, become a function of the oscillation amplitude. In general, the amplifier gain decreases as the amplitude of the oscillation increases, ultimately limiting it to $V_o = V_{OSC}$. In this situation, (10.1) becomes non-linear

$$V_{OSC} = \frac{A(V_{OSC},\omega)}{1-\beta(V_{OSC},\omega)A(V_{OSC},\omega)} V_i \tag{10.2}$$

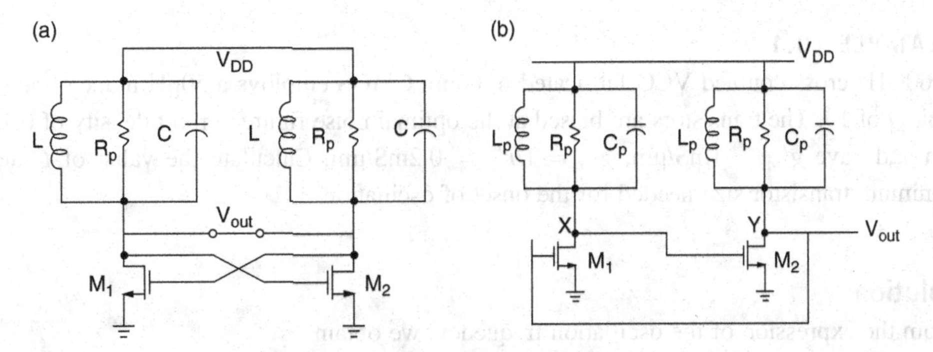

Figure 10.2 (a) Cross-coupled LC VCO topology (b) Amplifier-with-feedback model of the cross-coupled VCO topology

and the oscillation condition can be mathematically described by (10.3). The latter consists of an inequality that must be satisfied by the magnitude of the loop gain at the onset of oscillation, and of two equations, one for the magnitude and one for phase of the loop-gain under stable oscillation regime when the oscillation amplitude becomes equal to V_{OSC}

$$|\beta(\omega_{OSC})A(0)| > 1; |\beta(\omega_{OSC})A(V_{OSC})| = 1; \text{ PHASE } [\beta(\omega_{OSC})A] = 360^{\circ}. \quad (10.3)$$

In practice, the amplitude inequality is made more stringent in order to ensure that the circuit oscillates over process, temperature, and power supply corners.

There are many examples of RF and microwave oscillators that can be described by the amplifier-with-feedback model. The most common CMOS one is the cross-coupled VCO topology [1], discussed in more detail in Section 10.2.5. It consists of a matched MOS (or bipolar) differential pair with parallel LC tank loads, as illustrated in Figure 10.2(a). The circuit is redrawn in Figure 10.2(b) to emphasize the amplifier-with-feedback topology, where the amplifier gain depends both on frequency and on voltage amplitude, whereas the feedback factor is constant and equal to 1.

For the sake of simplicity, we assume that C includes the transistor capacitances and the parasitic capacitance of the inductor and of the interconnect, and that the loss resistance of the inductor, of the capacitor, and the output resistance of the transistor, r_o, are absorbed in R_p. The voltage gain of each common-source stage is $-g_m Z_L$, where Z_L is the impedance of the R_pLC tank whose magnitude reaches a peak value equal to R_p at

$$\omega_{OSC} = \frac{1}{\sqrt{LC}}. \quad (10.4)$$

For this circuit, at ω_{OSC}, $\beta = 1$ and $A(\omega_{OSC}) = (-g_m R_P)(-g_m R_P)$. The oscillation condition becomes

$$g_m R_P > 1. \quad (10.5)$$

EXAMPLE 10.1

A 60GHz cross-coupled VCO fabricated in 65nm CMOS employs a 50pH inductor and has a tank Q of 10. The transistors are biased at the optimal noise figure current density of 0.15mA/μm and have $g'_m = 1\text{mS}/\mu\text{m}$, $g'_o = 1/r'_o = 0.2\text{mS}/\mu\text{m}$. Calculate the value of C and the minimum transistor size needed for the onset of oscillation.

Solution

From the expression of the oscillation frequency, we obtain

$$C = \frac{1}{L(2\pi f_{OSC})^2} = \frac{1}{5\times10^{-11}(6.28\times6\times10^{10})^2} = 141\text{fF}.$$

For the VCO design to be realizable, this capacitance must be larger than the total transistor and parasitic capacitance present at the drain nodes of M1 or M2.

To determine the transconductance and the transistor size, we must first calculate the expression of R_P as a function of the inductor Q.

The oscillation condition can be written as a function of the transistor gate width

$$1 < g_m R_P = \frac{g_m}{g_o + \frac{1}{Q\omega_{OSC}L}} = \frac{g'_m W}{g'_o W + \frac{1}{Q\omega_{OSC}L}} = \frac{g'_m}{g'_o + \frac{1}{WQ\omega_{OSC}L}}$$

which can be recast as

$$W > \frac{1}{(g'_m - g'_o)Q\omega_{OSC}L} = \frac{1}{0.0008\times10\times6.28\times6\times10^{10}\times5\times10^{-11}} = 6.63\mu\text{m}. \tag{10.6}$$

This result indicates that for a given inductor Q a minimum transistor size exists below which the circuit will not oscillate. The corresponding minimum bias current through each transistor becomes 6.62μm × 0.15mA/μm = 1mA. Thus, the minimum power consumption of the VCO is 2.4mW from a 1.2V power supply. In practice, the transistor size will be chosen two to three times larger than the minimum value to account for process and temperature variation, and for inductor Q degradation over temperature.

We notice that the transistor size and, therefore, bias current are inversely proportional to the inductor value, the Q of the inductor, and the oscillation frequency. It is also instructive to observe that the technology constant $g'_m - g'_o$ in (10.6) scales (increases) with the technology scaling factor S. We conclude from here that it becomes increasingly easier to satisfy the oscillation condition in advanced CMOS technology nodes. Transistor technology scaling helps in building higher- and higher-frequency oscillators.

Negative resistance model

Figure 10.3 illustrates the alternate negative resistance/conductance model of the VCO. The oscillation condition, which corresponds to the overall single-port impedance or admittance being equal to zero, is described by

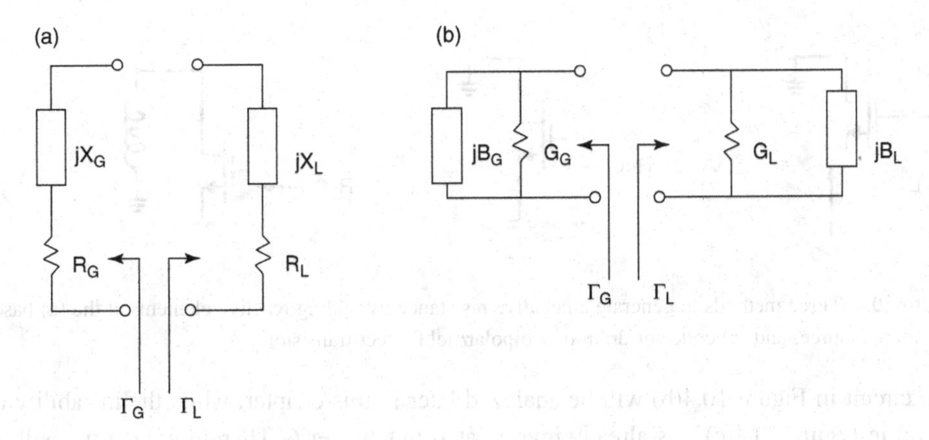

Figure 10.3 Oscillator model as (a) a negative resistance and (b) as negative conductance

$$|R_G(0)| > R_L(\omega_{OSC}); R_G(V_{OSC}) + R_L(\omega_{OSC}) = 0; X_G(V_{OSC}, \omega_{OSC}) + X_L(\omega_{OSC}) = 0 \quad (10.7)$$

in the impedance formalism (Figure 10.3(a)) or by

$$|G_G(0)| > G_L(\omega_{OSC}); G_G(V_{OSC}) + G_L(\omega_{OSC}) = 0; \ B_G(V_{OSC}, \omega_{OSC}) + B_L(\omega_{OSC}) = 0 \quad (10.8)$$

in the admittance formalism, Figure 10.3(b).

The negative resistance generator ($Z_G = R_G + jX_G$ or $Y_G = G_G + jB_G$) can be either a transistor or a diode. Historically, tunnel, IMPATT, Gun, BARITT, and TRAPATT diodes [2] have been employed in microwave and mm-wave oscillators. Today, the overwhelming majority of high-frequency oscillators are realized with transistors.

As in the amplifier-with-feedback model, the amplitude is stabilized by the non-linearity of the negative resistance device. A factor of 3 is typically chosen in the inequalities (10.7) and (10.8) based on the empirical observations collected over many decades by various designers that, for most semiconductor diodes and bipolar transistors, under large signal conditions, the absolute value of the negative resistance/conductance decreases approximately three times from the small signal value.

The requirement that the oscillation condition be satisfied in the context of process, temperature, and supply voltage variation also dictates that the negative resistance be much larger than the minimum value determined from the oscillation condition.

How can a negative resistance or conductance be developed at the terminals of a transistor? The obvious answer is provided by Figure 10.1(a): by building a positive feedback loop around the transistor. As illustrated in Figure 10.4, this can be accomplished with a reactive element in three ways:

(i) adding an inductor in the base of a bipolar transistor or in the gate of a field-effect transistor causes a negative resistance to develop at its emitter/source;
(ii) adding a capacitor in the emitter of a bipolar transistor or in the source of a field-effect transistor causes a negative resistance to develop at its base/gate;
(iii) adding an inductor in the collector of a bipolar transistor or in the drain of a field-effect transistor may cause a negative resistance to develop at its base/gate.

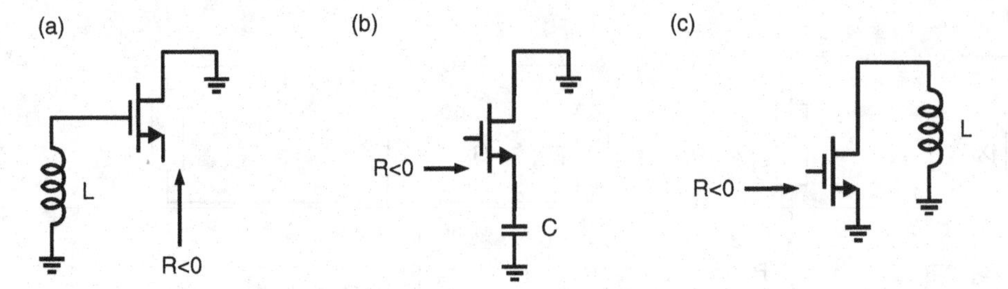

Figure 10.4 Three methods to generate a negative resistance by adding reactive elements at the (a) base/gate, (b) emitter/source, and (c) collector/drain of a bipolar/field effect transistor

The circuit in Figure 10.4(b) will be analyzed later in this chapter, while the instability of the circuit in Figure 10.4(c) was already investigated in Chapter 6. Therefore, next we will derive the expression of the negative resistance appearing at the source of the circuit in Figure 10.4(a). For brevity, we will ignore the source resistance and the output conductance of the small signal equivalent circuit of the FET, but we will account for g_m, C_{gs}, C_{gd}, and R_g. The input impedance can be expressed as

$$Z_{in} = \frac{Z_G + Z_{gs}}{1 + g_m Z_{gs}} \tag{10.9}$$

where

$$Z_G = \frac{R_g + j\omega L}{1-\omega^2 C_{gd} L + j\omega R_g C_{gd}} \text{ and } Z_{gs} = \frac{1}{j\omega C_{gs}}. \tag{10.10}$$

After substituting (10.10) in (10.9), we obtain

$$Z_{in} = \frac{1 + j\omega C_{gs} Z_G}{g_m + j\omega C_{gs}} = \frac{(1 + j\omega C_{gs} Z_G)(g_m - j\omega C_{gs})}{g_m^2 + \omega^2 C_{gs}^2}. \tag{10.11}$$

If we ignore the terms in R_g and assume $(f_T/f)^2 >> 1$, then the real part of the input impedance in (10.11) simplifies to

$$\Re(Z_{in}) \approx \frac{1}{g_m} - \frac{\omega^2 C_{gs} L}{g_m (1-\omega^2 C_{gd} L)}. \tag{10.12}$$

The resistance becomes negative if L satisfies the condition

$$L > \frac{1}{\omega^2 (C_{gd} + C_{gs})}. \tag{10.13}$$

EXAMPLE 10.2

A 65nm n-MOSFET with a total gate width of 50μm is chosen to realize a 60GHz oscillator. The transistor is biased for low-phase noise at 0.15mA/μm resulting in $g_m = 50\text{mS}$, $C_{gs} = 30\text{fF}$, and $C_{gd} = 15\text{fF}$. Calculate the minimum value of the inductor which must be connected between the gate and ground to produce a negative resistance at 60GHz at the source terminal of the MOSFET.

Solution

Applying (10.13), we get

$$L > \frac{1}{\left(6.28 \times 6 \times 10^{10}\right)^2 (15 + 30) \times 10^{-15}} = 156.51\text{pH}.$$

In practice, the inductor has to be larger to compensate for the parasitic gate and source resistance. If we pick $L = 0.3$nH, then the negative resistance calculated from (10.12) at 60GHz becomes -50.77Ω.

10.1.3 Resonators

Unlike ring and relaxation oscillators used at low frequencies, high-frequency oscillators feature a highly selective resonator which is employed to realize a bandpass or bandstop filter function. This has several advantages:

(i) it removes higher-order harmonics making the waveform appear as an almost ideal sinusoid,

(ii) it prevents noise from outside the filter bandwidth from degrading the VCO phase noise, and

(iii) it provides a highly stable (over process, temperature and time) oscillation frequency.

The quality factor of the resonator is critical in dictating the overall performance of the oscillator. A resonator with high Q is desirable to capitalize on these benefits.

VCO resonators can be classified in high Q and low Q:

- The high-Q resonators have quality factors in excess of 1000. They have yet to be integrated into ICs. Several types of resonators have been employed in VCOs:
  - quartz crystals (usually in the 1MHz to 600MHz range),
  - surface (SAW) and bulk (BAW) acoustic wave resonators,
  - dielectric puck (based on dielectric materials with high permittivity),
  - ferroelectric (interesting for tunable resonators),
  - magnetic (the most widely tunable, typically over octave bandwidths):
    ferrite YIG (yttrium-iron-garnet) sphere
    MSW (magnetostatic wave) thin film.
- The low-Q resonators exhibit quality factors generally below 100 and can be integrated in ICs. They are:
  - lumped LC: tunable if the capacitor element, C, is a varactor,
  - distributed transmission line with varactor loading for tunability.

Resonant tank model

LC resonators are the most common type of resonator encountered in silicon integrated circuits. Their small signal equivalent circuit is represented either as a parallel or as a series RLC circuit, where R describes the loss resistance of the resonator. The two representations are

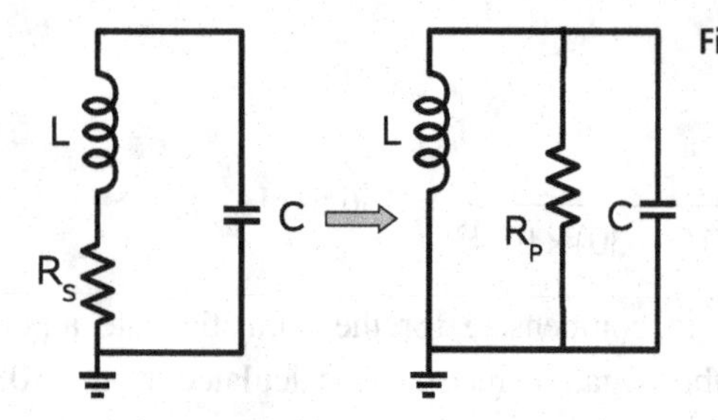

Figure 10.5 Series to parallel resonant circuit conversion

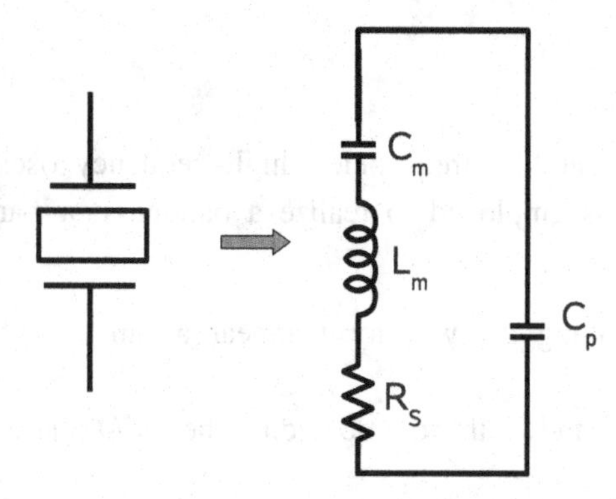

Figure 10.6 Equivalent small signal circuit of a quartz crystal resonator

equivalent and, as shown in Figure 10.5, conversion from one equivalent circuit to another is straightforward. For example, the series, R_S, or parallel loss resistance, R_P, of an inductor can be readily expressed as a function of each other, of the inductor Q, and of the inductance value: $R_p = \omega_1 LQ$, $R_s = R_p/Q^2$, $R_s = \omega L/Q$.

Quartz crystals, which rely on their piezoelectric properties to convert electrical charge into mechanical strain and vice versa, are the most common non-RLC resonator and have the highest Q, typically 10^4–10^6. Although limited to frequencies below 1GHz due to their unmatched mechanical and electrical (frequency) stability, they are used in highly stable low-frequency reference oscillators for microwave and mm-wave PLLs. Most quartz crystals are manufactured to have fundamental frequencies between 2MHz and 60MHz. The equivalent circuit of the quartz crystal, shown in Figure 10.6, consists of a highly selective series resonant circuit formed by L_m and C_m which describe the mechanical oscillation mode in the crystal. Typical values for L_m and C_m are in the mH to tens of mH, and fF range, respectively. R_s, the equivalent series resistance (known as ESR), describes the loss of the resonator, with values between 40Ω and 80Ω, while C_p describes the parasitic capacitance associated with the contacts and lead wires, with values in the pF range. The frequency stability of ə quartz crystal is better than 50ppm (5×10^{-5}).

Surface and bulk acoustic wave resonators

These are extensions of the crystal resonator concept at frequencies from a few hundred MHz to a few GHz. As their name indicates, surface acoustic wave (SAW) resonators take advantage of the interaction and excitation of surface waves with the high-frequency electric field applied to

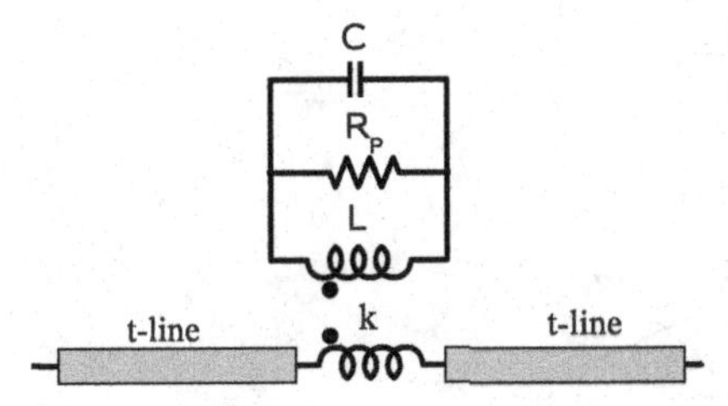

Figure 10.7 Equivalent small signal circuit of a dielectric resonator coupled to a microstrip line

their electrodes. The most common material for the fabrication of SAW resonators and filters is lithium niobate, which, until now, has proven to be incompatible with integration in silicon or III-V semiconductor technologies. Bulk acoustic wave (BAW) resonators have emerged in the last few years as a potential candidate for above-IC integration of high-Q resonators in RF ICs.

Dielectric resonators

A dielectric resonator is realized from low-loss dielectric materials with high relative permittivity (40–80). They are normally cut in a cylindrical shape. The resonant frequency is determined by the dimensions of the resonator and by its permittivity. The unloaded quality factor is of the order of several thousand, making them ideal resonators for very low-phase noise, highly stable microwave and mm-wave oscillators. When used in an oscillator, the resonator is placed in close proximity to an unshielded microstrip line. The corresponding equivalent circuit is that of a low-loss parallel RLC circuit, magnetically coupled to the microstrip line, as shown in Figure 10.7.

Below the resonance frequency, the resonator behaves inductively and presents a very low impedance to the line. Above the resonance frequency, it acts capacitively and also presents a small impedance in series with the microstrip line, without perturbing its magnetic field. It is only at resonance that the resonator exchanges energy with the line and produces a very high reflection coefficient along its length. Stronger coupling results in a larger energy being transferred to the line but also lowers the loaded Q of the resonator. The physical dimensions of the resonator at microwave and mm-wave frequencies have rendered it, up to now, impractical for integration in monolithic ICs. As a result, all present dielectric resonator oscillators (DROs) are realized as hybrid integrated circuits. Another disadvantage of dielectric resonators is the fact that they are not tunable, although, unlike quartz crystals, mechanical tuning over bandwidths smaller than 1% is possible. Dielectric resonators are typically realized for frequencies ranging from a few GHz to 100GHz.

10.1.4 Phase noise

In tuned and broadband communication systems, minimizing the phase noise of the oscillator is extremely important because it degrades the sensitivity of the receiver. In a wireless system, oscillator phase noise can result in the downconversion of undesired signals from

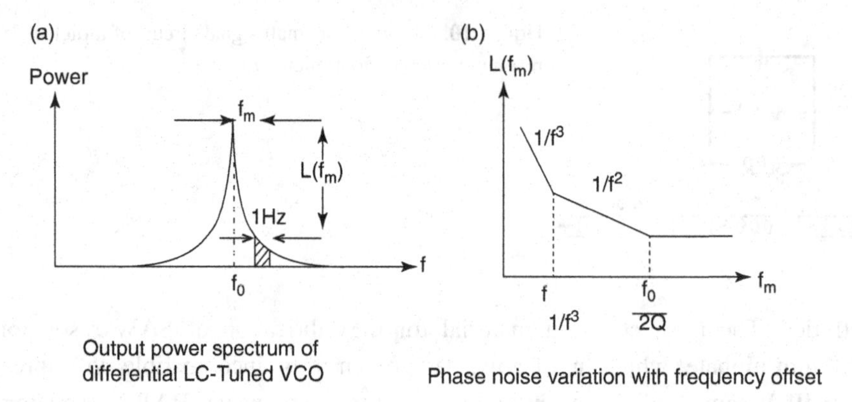

Figure 10.8 (a) Noisy spectrum of a VCO output signal and (b) the downconverted phase noise of the VCO

adjacent communication channels, causing receiver sensitivity degradation and limiting how closely channels can be spaced in frequency. In an automotive Doppler radar, local oscillator phase noise can completely drown the reflected signal from a slow-moving vehicle or pedestrian, limiting the minimum vehicle speed that the radar can resolve. Finally, in a broadband fiber-obtic receiver or in a direct-conversion (zero-IF) radio receiver, VCO phase noise interferes with the baseband digital signal, limiting receiver sensitivity.

As illustrated in Figure 10.8(a), phase noise manifests itself through the broadening of the oscillator spectral signature. Hypothetically, if the noisy oscillator signal is mixed with that of an ideal, noise-free oscillator operating at the same frequency f_{OSC}, then the noise power spectrum shown in Figure 10.8(b) is obtained. The *X*-axis represents the frequency offset, f_m, from the oscillation frequency f_{OSC}. The plot in Figure 10.8(b) is the typical way of representing and characterizing the phase noise of a VCO. The noise power is largest near the oscillation frequency and decreases exponentially away from it. The shape of the phase noise versus frequency characteristics of the VCO is a manifestation of the fact that the noise in the passband of the resonator is amplified by the oscillation process.

Definition: Phase noise is a measure of oscillator stability and refers to short-term random fluctuations in frequency *f* or phase, ϕ. It is is defined as the ratio of the noise power integrated in a 1-Hz sideband at a frequency offset f_m from the carrier frequency f_{OSC}, and the power of the carrier, P_{AVS}.

Phase noise as frequency modulation. The output voltage of the oscillator can be expressed as

$$v_o(t) = V_{OSC}\cos[\omega_{OSC}t + \theta(t)] \tag{10.14}$$

where $\theta(t)$ represents the random phase fluctuations.

In turn, the small phase fluctuations can be modeled as sinusoidal signals

$$\theta(t) = \frac{\Delta f}{f_m}\sin(\omega_m t) = \theta_p\sin(\omega_m t). \tag{10.15}$$

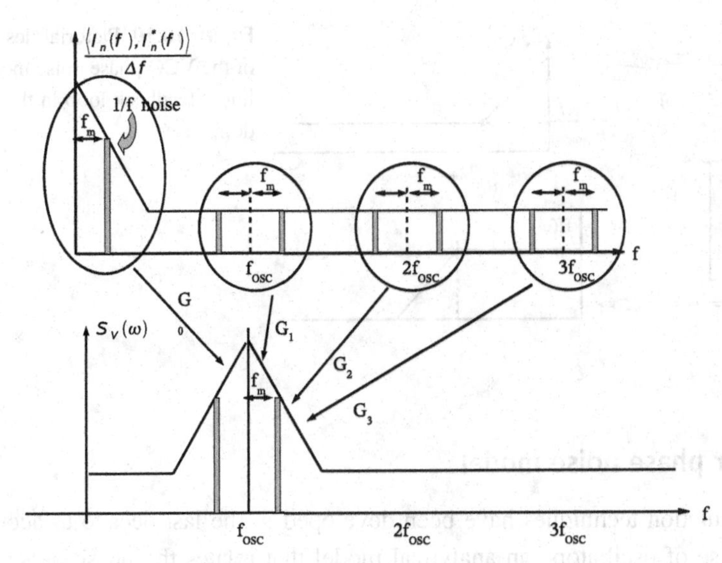

Figure 10.9 Pictorial illustration of the noise mixing process from DC and harmonics of the oscillation frequency. G_0, G_1, G_2, and G_3 represent the conversion power gains from sidebands at DC, fundamental and harmonic frequencies to the fundamental frequency

By inserting (10.15) into (10.14), and applying well-known trigonometric identities, we obtain

$$v_o(t) = V_{OSC}\left[\cos(\omega_{OSC}t)\cos(\theta_p\sin(\omega_{OSC}t)) - \sin(\omega_{OSC}t)\sin(\theta_p\sin(\omega_{OSC}t))\right] \quad (10.16)$$

which can be recast as

$$v_o(t) \approx V_{OSC}\left[\cos(\omega_{OSC}t) - \frac{\theta_p}{2}\left[\cos((\omega_{OSC}+\omega_m)t) - \cos((\omega_{OSC}-\omega_m)t)\right]\right]. \quad (10.17)$$

Equation (10.17) shows that the oscillator signal has three frequency components, one at ω_{OSC}, representing the power of the oscillator fundamental signal (or carrier power) and two "sidebands" of equal amplitude, at $\omega_{OSC} + \omega_m$ and at $\omega_{OSC} - \omega_m$, symmetrically located around the oscillator frequency and which describe the phase noise of the oscillator. Understanding this aspect, and the fact that the two noise sidebands are partially correlated is important. Appendix 10 expands on this topic, explaining the physical origins of the two noise sidebands and their partial correlation based on a non-linear power series analytic model of oscillator phase noise.

Phase noise due to noise mixing from harmonics of the LO

Although this simple model quite accurately captures the phase noise of an oscillator with a high-Q resonant tank, noise that mixes down or up to the oscillation frequency from higher harmonics of the oscillation frequency, or especially from DC, is not accounted for. This noise upconversion and downconversion process, similar to that which occurs in mixers, and caused by the non-linearity of the negative resistance device in the oscillator, is illustrated in Figure 10.9, and can be captured by non-linear large signal noise simulation techniques, as will be discussed in Section 10.3.

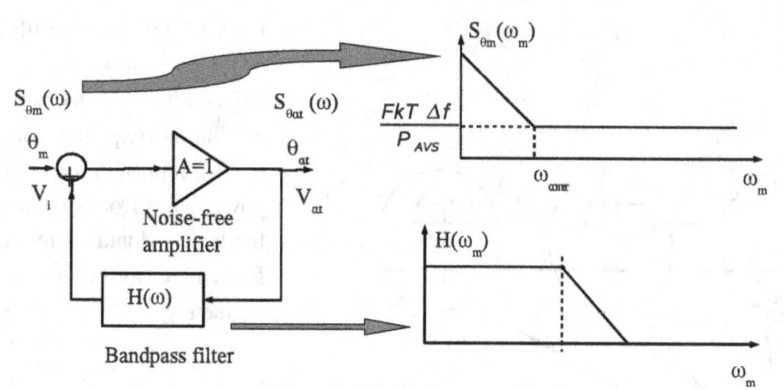

Figure 10.10 Pictorial description of the VCO phase noise model as a linear feedback loop in the phase domain

10.1.5 Leeson's oscillator phase noise model

Although several simulation techniques have been developed in the last decade to accurately capture the phase noise of oscillators, an analytical model that relates the phase noise of the VCO to the transistor bias current, transistor size, and to the values of the resonator components is needed for design purposes. Accurately describing the phase noise of an oscillator analytically is one of the most challenging problems in electronic circuit analysis and design. The difficulty arises from the fact that noise is a small signal perturbation of the large signals present at the transistor terminals. Small signal approximations therefore do not work well in this case. Furthermore, accurate large signal transistor models have become too complex for simple analytical formulations, and their parameters are difficult to measure and extract with precision in nanoscale CMOS technologies.

Leeson was the first to develop a simple and intuitive VCO phase noise model [3] which continues to be popular 50 years later. His approach was to employ the amplifier-with-feedback model of the VCO, as illustrated in Figure 10.10, where the amplifier and the bandpass feedback resonator are described in the phase (rather than voltage) domain. This model closely represents selective feedback oscillators with moderate and high resonator Qs.

The transfer function $H(f_m)$ of the bandpass resonator around f_{OSC} is modeled as that of a low-pass filter

$$H(jf) = \frac{1}{1 + jQ\left(\frac{f}{f_{OSC}} - \frac{f_{OSC}}{f}\right)} \text{ hence } H(jf_m) \approx \frac{1}{1 + j\frac{2f_m Q_L}{f_{OSC}}} \tag{10.18}$$

where $f_m = f - f_{OSC}$ and it is assumed that $f_m << f_{OSC}$.

Expression (10.18) is in accordance with the phase noise representation in Figure 10.8(b) where, after mixing with an ideal oscillator at f_{OSC}, the bandpass filter centered at f_{OSC} becomes a low-pass filter with a 3dB bandwidth equal to $f_{OSC}/2Q_L$. Here, Q_L describes the loaded Q of the resonator.

The amplifier is modeled as a noise-free amplifier preceded by a phase modulator, as illustrated in the inset of Figure 10.10. The modulating phase signal represents the noise power spectral density of the amplifier as a function of frequency. It features a $1/f$ spectral density region from DC up to the 1/*f* noise corner frequency, f_{corner}, and a white noise region, with a constant power spectral density *kTF*, above the corner frequency.

Using Figure 10.10, we are now in a position to derive the expression of the phase at the output of the loop as

$$\theta_{out}(f_m)\times H(f_m)+\theta_m=\theta_{out}(f_m) \text{ and}$$

$$\theta_{out}(f_m)=\frac{\theta_m}{1-H(f_m)}=\theta_m\frac{1+j\dfrac{2f_mQ_L}{f_{OSC}}}{j\dfrac{2f_mQ_L}{f_{OSC}}}=\theta_m\left(1-j\frac{f_{OSC}}{2f_mQ_L}\right) \quad (10.19)$$

The output noise power P_{noise} becomes

$$P_{noise}=\langle\theta_{out},\theta_{out}\rangle=\langle\theta_m,\theta_m\rangle\left[1+\left(\frac{f_{OSC}}{2f_mQ_L}\right)^2\right] \quad (10.20)$$

where

$$\langle\theta_m,\theta_m\rangle=FkT\Delta f\left(1+\frac{f_{corner}}{f_m}\right). \quad (10.21)$$

By inserting (10.21) into (10.20), and using (10.17) and the phase noise definition, we obtain the expression of the VCO phase noise

$$L(f_m)=\frac{P_{noise}}{P_{AVS}}=\frac{\dfrac{1}{2}\left(\dfrac{V_{OSC}\theta_{outP}}{2}\right)^2}{\dfrac{1}{2}V_{OSC}^2}=\frac{\theta_{outP}^2}{4}=\frac{\theta_{out}^2}{2}=\frac{FkT\Delta f}{2P_{avs}}\left[1+\left(\frac{f_{OSC}}{2Q_Lf_m}\right)^2\right]\left(1+\frac{f_{corner}}{f_m}\right) \quad (10.22)$$

where:

- P_{noise} is the noise power in a single sideband of 1Hz,
- *F* is the transistor (amplifier) noise factor with respect to resonator impedance at resonance, and
- Δf is the noise bandwidth, typically 1Hz.

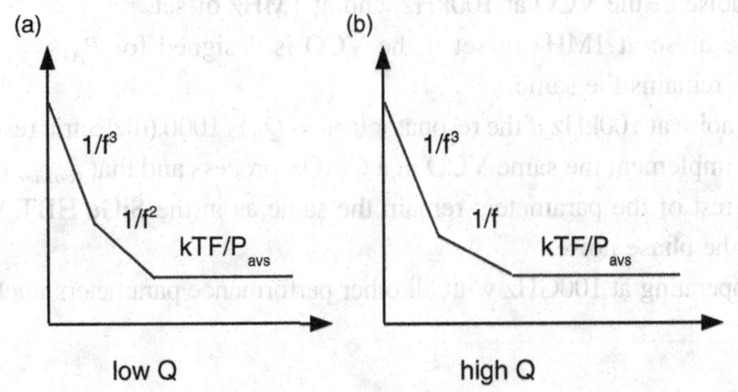

Figure 10.11 Typical phase noise spectra for (a) low *Q* and (b) high *Q* VCOs

Formula (10.22) indicates that there can be up to four distinct regions in the $L(f)$ characteristics, described by

- $1/f$,
- $1/f^2$,
- $1/f^3$, and
- f^0

dependencies.

The main parameters that determine the phase noise of the VCO are summarized below in order of their significance:

- The resonator Q. A higher Q always results in lower phase noise.
- The oscillation amplitude. The higher the VCO output power or voltage swing, the lower the phase noise. Optimally designed VCOs deliver the maximum voltage swing allowed for the safe operation of the transistor.
- Transistor noise. This is described by the noise factor F in (10.22). Lower transistor noise always results in lower VCO phase noise. However, it should be noted that it is not entirely clear how F should be calculated. This has been a topic of much confusion and research over the last 40 years.
- The VCO amplitude limitation mechanism. In general, it is desirable to avoid driving HBTs/ BJTs in the saturation region which degrades phase noise.
- Noise leaking from the bias supply and tuning control voltage nodes, or noise generated by the current source/tail. A differential control voltage and a differential VCO topology alleviate the insertion of noise from the power supply or control voltage. However, common-mode/even-mode components still contribute noise because a differential oscillator operates under large signal conditions, with one of its half circuits being practically cut off.
- Buffer amplifier or load noise. Loading the tank of the VCO directly by the buffer increases its phase noise and also reduces its tuning range.

EXAMPLE 10.3

Consider a cross-coupled SiGe HBT VCO with $Q_L = 10, f_{OSC} = 10\text{GHz}, f_{corner} = 1\text{kHz}$, $F = 4$, $P_{avs} = 1\text{mW}$, and assume that F describes the noise factor of the common-emitter amplifier stage when driven by a signal source of impedance R_P, equal to the VCO tank impedance.

(a) Calculate the phase noise of the VCO at 100kHz, and at 1MHz offset.

(b) Recalculate the phase noise at 1MHz offset if the VCO is designed for $P_{AVS} = 10\text{mW}$, while everything else remains the same.

(c) Recalculate the phase noise at 100kHz if the resonator loaded Q_L is 1000 (dielectric resonator).

(d) Assume that we now implement the same VCO in a CMOS process and that f_{corner} changes to 10MHz while the rest of the parameters remain the same as in the SiGe HBT VCO in part (a). Recalculate the phase noise.

(e) Redo (a) for a VCO operating at 100GHz with all other performance parameters unchanged.

Solution

(a)

$$L(100\text{kHz}) = -174\text{dBm} + 10\log\left(\frac{4}{2}\right) - 0\text{dBm} + 10\log\left[1 + \left(\frac{10^{10}}{20\times 10^{5}}\right)^{2}\right] + 10\log\left(1 + \frac{10^{3}}{10^{5}}\right) = -97\text{dBc/Hz}$$

$$L(1\text{MHz}) = -170.99\text{dBm} + 10\log\left[1 + \left(\frac{10^{10}}{20\times 10^{6}}\right)^{2}\right] + 10\log\left(1 + \frac{10^{3}}{10^{6}}\right) = -117\text{dBc/Hz}$$

(b)

$$L(1\text{MHz}) = -174\text{dBm} + 10\log(2) - 10\text{dBm} + 10\log\left[1 + \left(\frac{10^{10}}{20\times 10^{6}}\right)^{2}\right] + 10\log\left(1 + \frac{10^{3}}{10^{6}}\right) = -107\text{dBc/Hz}$$

(c)

$$L(100\text{kHz}) = -174\text{dBm} + 10\log(2) - 0\text{dBm} + 10\log\left[1 + \left(\frac{10^{10}}{10^{3}\times 10^{5}}\right)^{2}\right] + 10\log\left(1 + \frac{10^{3}}{10^{6}}\right) = -131\text{dBc/Hz}$$

(d)

$$L(100\text{kHz}) = -174\text{dBm} + 10\log(2) - 0\text{dBm} + 10\log\left[1 + \left(\frac{10^{10}}{20\times 10^{5}}\right)^{2}\right] + 10\log\left(1 + \frac{10^{7}}{10^{5}}\right) = -76.56\text{dBc/Hz}$$

$$L(1\text{MHz}) = -170.99\text{dBm} + 10\log\left[1 + \left(\frac{10^{10}}{20\times 10^{6}}\right)^{2}\right] + 10\log\left(1 + \frac{10^{7}}{10^{6}}\right) = -106.58\text{dBc/Hz}$$

(e)

$$L(100\text{kHz}) = -174\text{dBm} + 10\log(2) - 0\text{dBm} + 10\log\left[1 + \left(\frac{10^{11}}{20\times10^5}\right)^2\right]$$
$$+ 10\log\left(1 + \frac{10^3}{10^5}\right) = -77\text{dBc/Hz}$$

$$L(1\text{MHz}) = -170.99\text{dBm} + 10\log\left[1 + \left(\frac{10^{11}}{20\times10^6}\right)^2\right] + 10\log\left(1 + \frac{10^3}{10^6}\right)$$
$$= -97\text{dBc/Hz}.$$

10.1.6 Frequency tuning

Most oscillators are used in a phase lock loop and, therefore, their frequency needs to be electronically tunable and controlled, typically by a voltage. Ideally, the VCO (angular) frequency should be a linear function of the control voltage, as described by

$$\omega_{OSC} = \omega_0 + K_{VCO}V_{cont} \tag{10.23}$$

where:

- $\omega_2 - \omega_1$ is the tuning range,
- $V_2 - V_1$ is the control voltage range, and
- K_{VCO} is the VCO gain, sometimes also called VCO sensitivity.

A typical tuning curve is illustrated in Figure 10.12. It suggests that the frequency tuning curve is only linear over a limited range of frequencies and that the VCO gain, defined at the center of the tuning range, satisfies

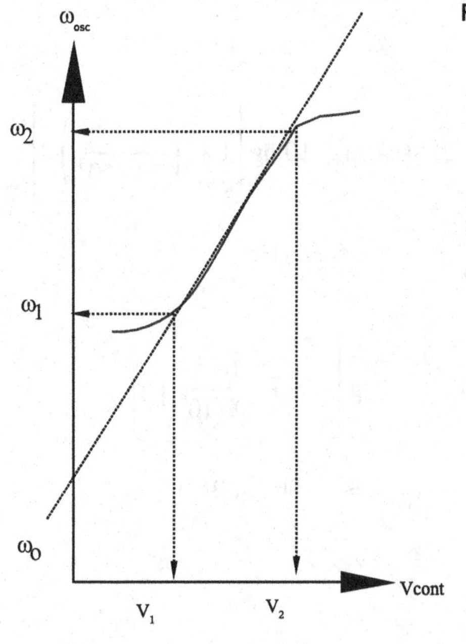

Figure 10.12 Typical VCO tuning characteristics

$$K_{VCO} > \frac{\omega_2 - \omega_1}{V_2 - V_1} \tag{10.24}$$

10.1.7 VCO specification and figure of merit

The most important performance parameters specified in the design of VCOs are:

(a) the oscillation frequency, f_{OSC},
(b) the tuning range, $(f_2 - f_1)/f_{OSC}$,
(c) the gain and tuning linearity,
(d) the phase noise,
(e) the output power, P_{AVS}, or output voltage amplitude, V_{OSC},
(f) the DC power consumption, P_{DC},
(g) power supply noise rejection, also known as **pushing,** and defined as the change in frequency as a function of supply voltage: $\frac{\Delta f_{OSC}}{\Delta V_{supply}}$, and
(h) load mismatch rejection or **pulling**, defined as the change in frequency as a function of the load impedance: $\frac{\Delta f_{OSC}}{\Delta \Gamma_L}$.

In general, we want to maximize the tuning range of integrated circuit VCOs in order to cover the oscillation frequency variation over process and temperature. The large tuning range must be balanced against the inevitably higher VCO gain and higher sensitivity to noise leaking through the control voltage node. The latter tends to increase the VCO phase noise. To alleviate the degradation of phase noise due to a high K_{VCO}, in practice, most VCOs are designed with coarse and fine voltage controls. The coarse control, characterized by a large VCO gain, is employed in frequency acquisition or post fabrication setting, while the fine tuning control, featuring small VCO gain, is used in phase acquisition in PLLs.

The output power is normally dictated by the circuits that need to be driven by the VCO signal. These include the upconvert and downconvert mixer, power amplifier, and divider in wireless systems, retimers, multiplexers, dividers, and demultiplexers in optical fiber systems. While there is a tradeoff between output power and DC power consumption, larger output power always leads to lower phase noise. Another important design aspect is to ensure that the output power remains constant, or at least above the minimum required value, over the entire tuning range of the VCO.

One of the conflicting requirements of VCO design refers to *pushing*. While it is desirable to be able to change the VCO frequency through the control voltage, the oscillation frequency must be insensitive to supply voltage variation. Therefore, a lower pushing number is always sought for. For this reason, differential VCOs and differential control voltage schemes are very common in ICs because of their built-in common-mode supply rejection and control voltage noise rejection.

As with pushing, VCO pulling must be minimized, especially in direct-conversion wireless transceivers. This is typically accomplished by employing VCO topologies that have built-in isolation between the VCO tank and the load, and/or by adding high reverse-isolation buffers between the VCO and its load.

The VCO figure of merit

As with LNAs and PAs, it should be possible to combine all the specification parameters of a VCO into a figure of merit that allows for a more realistic comparison of various VCO designs operating at different frequencies. However, not all oscillators are specified for all the parameters listed earlier. For example, some have a fixed oscillation frequency and, therefore, tuning range is not applicable. Pushing and pulling, though important, are rarely reported in publications, while VCO output power and voltage swing are not easily interchangeable if the VCO load impedance is not specified. The wireless committee of the ITRS [4] defines several figures of merit for VCOs. The most common includes the phase noise, the oscillation frequency, and the DC power consumption

$$FoM_1 = \left(\frac{f_{OSC}}{f_m}\right)^2 \frac{1}{L(\Delta f)P_{DC}}. \tag{10.25}$$

A second FoM, that reflects the usefulness of the VCO in a wireless or fiber-optic system, also includes its output power, P_{AVS}, and avoids the ambiguity of specifying voltage swing without specifying the load impedance

$$FoM_2 = \left(\frac{f_{OSC}}{f_m}\right)^2 \frac{P_{AVS}}{L(\Delta f)P_{DC}}. \tag{10.26}$$

A third figure of merit includes the normalized tuning range of the VCO

$$FoM_{VCO} = \left(\frac{f_{OSC}}{\Delta f}\right)^2 \frac{P_{AVS} \cdot FTR}{L(\Delta f) \cdot P_{DC}} \tag{10.27}$$

where P_{AVS}[W] is the VCO output power, L is the phase noise in a 1Hz band at offset frequency Δf, and

$$FTR = 200(F_{MAX} - F_{MIN})/(F_{MAX} + F_{MIN}) \tag{10.28}$$

is the normalized frequency tuning range of the VCO.

10.2 LOW-NOISE VCO TOPOLOGIES

Due to the significant speed improvement brought about by the scaling of MOSFETs to nanoscale gate lengths, ring oscillators can now be used at microwave and mm-wave frequencies. However, their phase noise remains inadequate for most wireless and fiber-optic communication systems. Therefore, in this chapter, we will focus our attention only on low-noise VCO topologies. The latter can be classified according to the type of resonator as:

- LC-tank VCO topologies, and
- high-Q resonator topologies.

10.2.1 LC-tank VCO topologies

LC VCOs are divided in two large classes: selective feedback and negative g_m (or cross-coupled.) The latter consists of a tuned amplifier with unity feedback.

10.2.2 Selective feedback oscillators

The selective feedback topologies are realized using transformers, inductors, and capacitors, and include the following ocillators:

- Armstrong, where the resonator is formed from one transformer and one capacitor,
- Hartley, which features two inductors and one capacitor,
- Colpitts, with two capacitors and one inductor,
- Clapp, with two capacitors and a series (or parallel) inductor-capacitor circuit, and
- Pierce, which comprises two capacitors and a quartz crystal resonator,

These are the earliest RF oscillators and employ essentially single transistor, negative resistance topologies, which can be described by the general three-node small signal equivalent circuit shown in Figure 10.13. The first oscillator of this family was invented by Armstrong in 1914 as part of the regenerative radio receiver. An improved version was perfected by Ralph Hartley of Western Electric (the precursor of Bell Labs) in 1915 and it bears his name. The most common topology is the one invented by Canadian-born Edwin Colpitts, also of Western Electric, and is illustrated in Figure 10.14(a). By replacing the inductor with a quartz crystal, in 1923, Pierce developed yet another oscillator that is in heavy use today and bears his name.

One could argue that the Clapp and Pierce topologies are essentially variants of the original Colpitts topology, while the Hartley VCO can be regarded as a particular case of the Armstrong topology.

In all cases, the oscillation condition can be rigorously derived by writing Kirchoff's law for the three nodes of the circuit in the absence of any external excitation. This results in a system of three equations with the three node voltages as unknowns.

For simplicity and without losing generality, we can assume that the input and output admittances of the transistor have been absorbed in Y_1 and Y_2, respectively, and that the

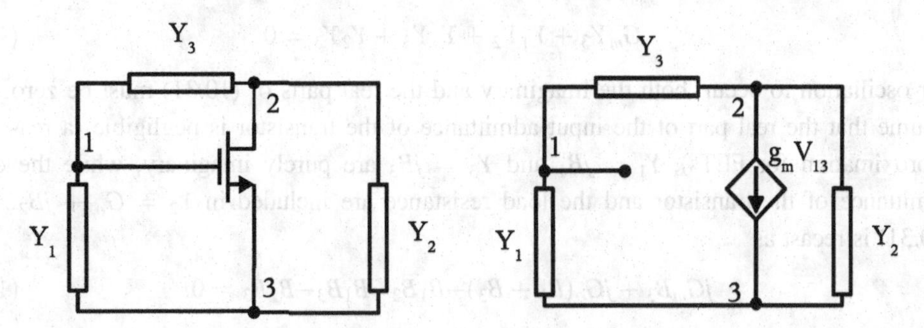

Figure 10.13 Generic representation of a selective feedback oscillator. Both FETs and bipolar transistors can be used

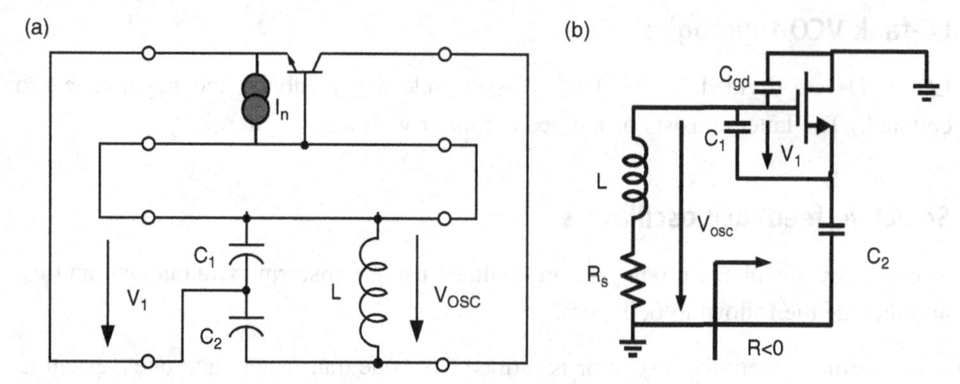

Figure 10.14 (a) General topology of a Colpitts oscillator showing the equivalent noise current source at the input of the transistor and (b) simplified small signal circuit for a FET-based design

feedback admittance of the transistor, Y_{bc} or Y_{gd}, is included in Y_3. The equivalent circuit now reduces to that in Figure 10.13(b) and the matrix equation becomes

$$\begin{bmatrix} Y_1 + Y_3 & -Y_3 & -Y_1 \\ G_m - Y_3 & Y_2 + Y_3 & -G_m - Y_2 \\ -G_m - Y_1 & -Y_2 & G_m + Y_1 + Y_2 \end{bmatrix} \begin{bmatrix} V_1 \\ V_2 \\ V_3 \end{bmatrix} = \begin{bmatrix} 0 \\ 0 \\ 0 \end{bmatrix}. \tag{10.29}$$

The large signal transconductance G_m is employed in (10.29) to account for the large signal nature of the oscillation condition. G_m is equal to g_m at the onset of oscillation and decreases to a fraction of g_m as the oscillation amplitude stabilizes. Depending on which of the three nodes is connected to ground, common-collector (drain), common-base (gate), and common-emitter (source) versions of this oscillator topology are possible. For the common-collector–drain circuit, by making $V_2 = 0$ in (10.29) and by substituting in the other two equations of (10.29), we end up with a system of two equations with two unknowns: V_1 and V_3. If this circuit is to operate as an oscillator, non-zero solutions for V_1 and V_3 must exist, and therefore the determinant of the system matrix must be equal to zero

$$\begin{bmatrix} G_m - Y_3 & -G_m - Y_2 \\ -G_m - Y_1 & G_m + Y_1 + Y_2 \end{bmatrix} = 0 \tag{10.30}$$

which is equivalent to

$$G_m Y_3 + Y_1 Y_2 + Y_1 Y_3 + Y_2 Y_3 = 0. \tag{10.31}$$

For oscillation to occur, both the imaginary and the real parts of (10.31) must be zero. If we assume that the real part of the input admittance of the transistor is negligible (a reasonable approximation for FETs), $Y_1 = jB_1$ and $Y_3 = jB_3$ are purely imaginary, while the output admittance of the transistor and the load resistance are included in $Y_2 = G_o + jB_2$. Then, (10.31) is recast as

$$jG_m B_3 + jG_o(B_1 + B_3) - B_1 B_2 - B_1 B_3 - B_2 B_3 = 0. \tag{10.32}$$

By breaking it down in its real and imaginary parts, the conditions to satisfy the oscillation become

$$\frac{G_m}{G_o} = -\left(1 + \frac{B_1}{B_3}\right) = \frac{B_1}{B_2} \tag{10.33}$$

and

$$\frac{1}{B_1} + \frac{1}{B_2} + \frac{1}{B_3} = 0. \tag{10.34}$$

The last equation, which gives the oscillation frequency, clearly indicates that one of B_1, B_2, or B_3 must be of opposite type with respect to the other two. For example, they cannot all be inductors or all capacitors. Since both G_m and G_o are positive, for (10.33) to be satisfied B_1 and B_3 must have opposite sign and $|B_1/B_3|>1$. Depending on whether B_1 and B_2 are inductive or capacitive, the circuit in Figure 10.13 can become a Hartley (Armstrong), or a Colpitts oscillator. For example, in a Hartley oscillator, $B_1 = -1/\omega L_1$, $B_2 = -1/\omega L_2$, and $B_3 = \omega C_3$. From (10.34), the oscillation frequency is derived as

$$f_{OSC} = \frac{1}{2\pi\sqrt{C_3(L_1 + L_2)}}. \tag{10.35}$$

According to (10.33), in order for the circuit to start oscillating, the large voltage gain of the transistor G_m/G_o must be larger than L_2/L_1. This represents a very attractive feature of the Hartley topology because it allows for frequency tuning, by making C_3 a variable capacitance, while the condition for the onset of oscillation (and the output power) remains, to first order, independent of the capacitance value. Nevertheless, perhaps because of the area occupied by inductors, so far the Hartley topology has been rarely integrated in silicon [5]. However, as discussed in [5], due to the favorable scaling of the size and Q of monolithic inductors at mm-wave frequencies, this topology is attractive for the fabrication of very low-phase noise, low-voltage, mm-wave oscillators in nanoscale CMOS technologies.

10.2.3 Simplified analysis and design equations for the Colpitts topology

The selective feedback oscillator can also be analyzed (and designed) as a negative resistance circuit. For example, using the simplified equivalent circuit of a Colpitts oscillator shown in Figure 10.14(b), and a simplified small signal model of the MOSFET which only considers its transconductance, the expression of the small signal negative resistance seen at the gate of the transistor is derived as

$$R = \frac{-g_m}{\omega^2 C_1 C_2} \tag{10.36}$$

which indicates that a large transconductance and small capacitors will produce a larger negative resistance and stronger oscillation.

How small can C_1 and C_2 be? The ultimate values are limited by the junction capacitances of the transistor with which they appear in parallel.

The negative resistance R from (10.36) must be larger in absolute value than the equivalent series loss resistance of the tank which, in most cases, is dominated by the inductor series resistance R_S. The latter can be approximated as

$$R_S = \frac{\omega_{OSC} L}{Q}. \tag{10.37}$$

Using the equivalent circuit in Figure 10.14(b), the condition for the tank inductance to cancel out the imaginary part of the impedance looking into the gate of the transistor is satisfied when the series combination of C_1 (which includes C_{gs}) and C_2 (which includes C_{sb}), in parallel with C_{gd}, resonate with L. This occurs at the oscillation frequency

$$\frac{1}{f_{OSC}} = 2\pi\sqrt{L\left[C_{gd} + \frac{C_1 C_2}{C_1 + C_2}\right]} \approx 2\pi\sqrt{\frac{LC_1C_2}{C_1 + C_2}}. \tag{10.38}$$

The oscillation condition can be derived from (10.36)–(10.38) by replacing the small signal transconductance g_m with the large signal transconductance G_m in (10.36) and (10.37), and substituting for the oscillation frequency from (10.38)

$$\frac{G_m}{\omega_{OSC}^2 C_1 C_2} = R_S \tag{10.39}$$

and

$$\frac{G_m L C_1 C_2}{(C_1 + C_2) C_1 C_2} = R_S = \frac{\omega_{OSC} L}{Q} => \frac{G_m}{\omega_{OSC}(C_1 + C_2)} = \frac{1}{Q}. \tag{10.40}$$

Equation (10.39) can be recast in a form similar to (10.5) if the parallel representation of the inductor loss is employed $R_P = Q^2 R_S$, and if the expression of the oscillation frequency from (10.38) is substituted into (10.39)

$$\frac{G_m}{\omega_{OSC}^2 C_1 C_2} = \frac{R_P}{Q^2} => G_m R_P = \frac{(C_1 + C_2)^2}{C_1 C_2} \geq 4. \tag{10.41}$$

For a given equivalent resonator capacitance,

$$C_{eq} = \frac{C_1 C_2}{C_1 + C_2}$$

equation (10.41) is easiest to satisfy when $C_1 = C_2$. The latter condition corresponds to $G_m R_P = 4$ and implies that a four times larger transconductance is needed in the Colpitts topology than in the cross-coupled VCO. This explains why the cross-coupled VCO has been preferred in CMOS implementations where a large transconductance is difficult to achieve with low bias currents.

A reasonably accurate approximation for G_m is the ratio of the large signal drain current amplitude of the transistor at the fundamental frequency and the amplitude of the large signal gate–source voltage, V_1. Given that the tank voltage is very large, we can assume that the transistor acts like a switch. Following a similar Fourier series approach as in the class B or class F PA analyses in Chapter 6, or as in the balanced mixer analysis of Chapter 9, the drain–collector current can be (approximately) described by a periodic train of pulses with peak value I_{MAX}, from which the average DC current I_{BIAS} and I_1 can be derived. For example, if we approximate the drain current pulses as a periodic train of rectified (half) sinusoids, we obtain that $I_{BIAS} = I_{DC} = I_{MAX}/\pi$ and $I_1 = I_{MAX}/2$. Therefore, in this case $I_1 = (\pi/2)I_{BIAS}$. Another possible approximation, corresponding to very narrow current pulses, is that $I_1 = 2I_{BIAS}$ [6]. While both these approximations are crude because the actual shape of the current pulses depends on the class of operation (bias condition) of the transistor in the VCO and how close the oscillation frequency is to the transistor f_T and f_{MAX}, they are useful in providing an intuitive link between the bias current and the oscillation amplitude. They will prove valuable in deriving a hand design methodology for low-noise oscillators which can then be refined using computer simulation. Staying with the second approximation, we obtain

$$G_m \approx \frac{2I_{BIAS}}{V_1} = \frac{2I_{BIAS}\left(1 + \frac{C_1}{C_2}\right)}{V_{OSC}} = 2\frac{I_{BIAS}}{V_{OSC}} C_1 L \omega_{OSC}^2 \tag{10.42}$$

where:

- V_1 is the amplitude of the voltage across C_1,
- I_{BIAS} is the transistor bias current.

By combining (10.42) with (10.40), we obtain a large signal design equation that links the oscillation amplitude on the tank, V_{OSC}, with the transistor bias current

$$2\frac{I_{BIAS}}{V_{OSC}} C_1 L \omega_{OSC}^2 = G_m = \frac{\omega_{OSC}(C_1 + C_2)}{Q} => V_{OSC} = \frac{2I_{BIAS}Q}{C_2 \omega_{OSC}}. \tag{10.43}$$

Equation (10.43) shows the tradeoff between bias current and tank Q. It also becomes apparent that a larger bias current and DC power are needed to maintain a given oscillation amplitude at higher frequencies.

To reduce pushing, C_1 must be chosen such that it is much larger than the transistor capacitance $C_1 \gg C_\pi(C_{gs})$. The latter varies with the bias voltage or current and would otherwise cause the oscillation frequency to change with the bias.

Finally, it has been demonstrated [8] that the VCO phase noise is a function of the oscillation amplitude, C_1, the C_1/C_2 ratio, and of I_n, the total equivalent input noise current of the transistor

$$L(f_m) = \frac{P_{noise}}{P_{AVS}} = \frac{|I_n^2|}{V_{OSC}^2 f_m^2} \frac{1}{C_1^2\left(\frac{C_1}{C_2} + 1\right)^2}. \tag{10.44}$$

Equation (10.44) can also be recast as a function of I_{BIAS} using (10.43)

$$L(f_m) = \frac{|I_n|^2 \omega_{OSC}^2}{I_{BIAS}^2 f_m^2 4Q^2} \frac{C_2^2}{C_1^2} \times \frac{1}{\left(\frac{C_1}{C_2} + 1\right)^2}. \tag{10.45}$$

From (10.44) and (10.45), it follows that, to reduce phase noise, C_1 and the C_1/C_2 ratio must be maximized along with the tank voltage and bias current, while the equivalent noise current I_n at the input of the transistor must be minimized.

Equations (10.38)–(10.45) form the basis of the Colpitts oscillator design methodology.

EXAMPLE 10.4

Find the minimum bias current, the value of the external capacitors C_1, C_2, and the negative resistance of a 40GHz Colpitts VCO fabricated in 65nm CMOS which uses a tank inductor of 80pH and a tank Q (dominated by the inductor) of 5. Assume that $I_{DS} = 0.15$mA/μm, $g'_m =$ 0.8mS/μm, $C'_{gs} = 0.7$fF/μm, $C'_{gd} = 0.35$fF/μm, and $C'_{sb} = 0.6$fF/μm and that the tank voltage amplitude V_{OSC} is 0.6V.

Solution

According to (10.41), the minimum transconductance and bias current needed for oscillation are obtained when $C_1 = C_2$ and $G_m R_P = 4$ where $R_P = Q\omega_{OSC}L = 100.5\Omega$. This implies that $G_m = 40$mS and

$$I_{BIAS} = G_m \frac{V_{OSC}}{2} = \frac{0.04 \times 0.6}{2} = 12\text{mA}.$$

The required MOSFET size is $W = I_{BIAS}/(0.15\text{mA}/\mu\text{m}) = 80\mu$m and $g_m = 64$mS, $C_{gs} = 56$fF, $C_{gd} = 28$fF, $C_{sb} = 48$fF.

The equivalent capacitance needed for oscillation at 40GHz is calculated next

$$C_{eq} = \frac{1}{\omega_{OSC}^2 L} = \frac{1}{(6.28 \times 4 \times 10^{10})^2 \times 80 \times 10^{-12}} = 198 fF.$$

From (10.36) and accounting for C_{gd}, C_{gs}, and C_{sb}

$$C_{eq} - C_{gd} = 198\text{fF} - 28\text{fF} = 170\text{fF} = \frac{C'_1 + C_{gs}}{2} = \frac{C'_2 + C_{sb}}{2};$$

$$C'_1 = 340\text{fF} - C_{gs} = 284\text{fF}\ ;\ C'_2 = 340\text{fF} - C_{sb} = 292\text{fF}$$

where $C_1 = C'_1 + C_{gs}$ and $C_2 = C'_2 + C_{sb}$.

The small signal negative resistance is

$$R = \frac{-g_m}{\omega^2 C_1 C_2} = \frac{0.064}{(6.28 \times 4 \times 10^{10} \times 340 \times 10^{-15})^2} = -7\Omega.$$

In a practical implementation, MOSFET, varactor, inductor, and capacitor layout parasitic capacitance will lead to smaller values for C'_1 and C'_2.

We conclude this example by noting that g_m/G_m is only 1.6, and not 2–3, as discussed in Section 10.1.2. When biased at relatively high current densities, MOSFETs exhibit a flat g_m–V_{GS} characteristics which leads to only a small degradation of the transconductance under large signal operation. In CMOS Colpitts VCOs, it is thus possible to use a smaller g_m/G_m ratio.

Frequency tuning

In theory, either C_1 or C_2 could be realized as varactors to ensure the tunability of the oscillation frequency. In practice, C_1 appears in parallel with the large gate–source or base–emitter capacitance of the transistor, seriously limiting the tuning range. For this reason, the preferred

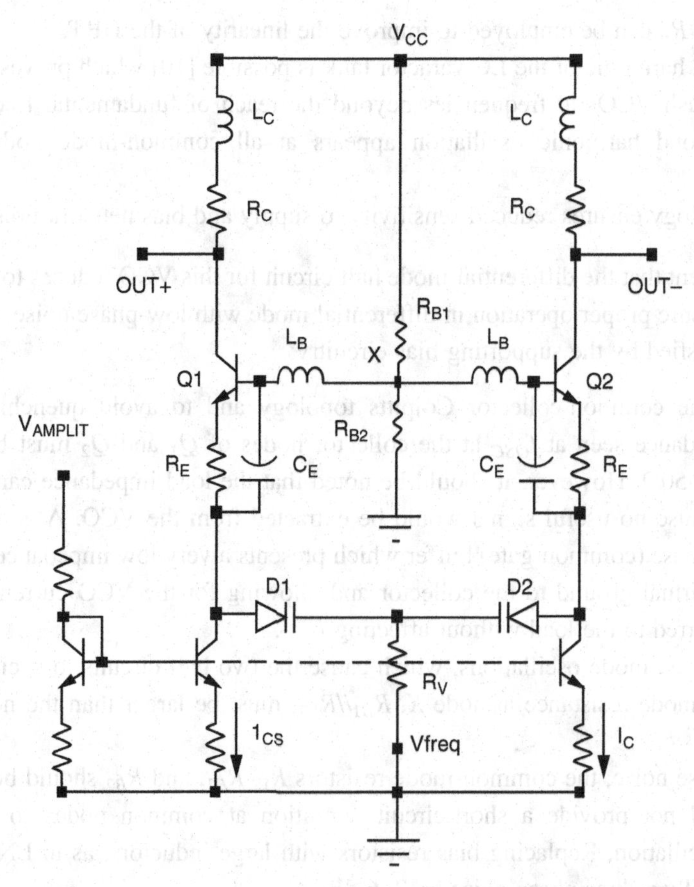

Figure 10.15 Differential Colpitts oscillator topology [9],[11],[12]

method to introduce frequency tuning is by implementing C_2 as a varactor diode. This has the added benefit that, if $C_1 >> C_2$, both the phase noise and the tuning range are improved. From (10.39), we note that, if a varactor is employed as C_2, the condition for the onset of oscillation and the output power of the VCO will change across the tuning range. Finally, if L, C_1, and C_2 are scaled by the same factor k, and the C-V ratio of C_2 remains unchanged, the oscillation frequency, too, scales by $1/k$, while the normalized tuning range remains the same.

The differential Colpitts topology

In the late 1990s, with the emergence of silicon technology as a dominant candidate for RF ICs, a differential version of the Colpitts oscillator was developed [9]. Although most frequently realized with bipolar transistors, as in Figure 10.15, this topology has also become popular in CMOS, especially at mm-wave frequencies. Some of the features that make it very attractive at mm-wave frequencies are listed below:

- The negative resistance transistors $Q_{1,2}$ also act as buffers, thus improving the phase noise and tuning range since no additional transistors are needed to isolate the VCO from its load.
- Q_1 and Q_2 can be independently sized and biased for optimal noise and maximum oscillation amplitude.

- Emitter degeneration R_E can be employed to improve the linearity of the HBT.
- Operation on the 2nd harmonic of the LC varactor tank is possible [10] which proves useful in realizing push–push VCOs at frequencies beyond the reach of fundamental frequency topologies. The second harmonic oscillation appears at all common-mode nodes, for example at node X.
- The differential topology ensures reduced sensitivity to supply and bias network noise.

It is immediately apparent that the differential-mode half circuit for this VCO reduces to that of Figure 10.14(b). To ensure proper operation in differential mode with low-phase noise, certain conditions must be satisfied by the supporting bias circuitry:

- First, to establish the common-collector Colpitts topology and to avoid quenching the oscillation, the impedance seen at f_{OSC} at the collector nodes of Q_1 and Q_2 must be kept low, typically below 50Ω. However, it should be noted that the load impedance cannot be reduced to zero because no useful signal would be extracted from the VCO. A good load circuit is a common-base (common-gate) buffer which presents a very low impedance to the VCO, providing a virtual ground to the collector and allowing for the VCO current to be extracted and transferred to the load without affecting f_{OSC}.
- Second, to prevent even mode oscillations, which cause the two half circuits to oscillate in phase, the common-mode resistance at node X, $R_{B1}//R_{B2}$, must be larger than the negative resistance $\frac{g_m}{\omega_{OSC}^2 C_1 C_2}$.
- Third, to reduce phase noise, the common-mode resistors R_V, R_{B1}, and R_{B2} should be small ($<2K\Omega$), but should not provide a short-circuit condition at common nodes to ensure differential-mode oscillation. Replacing bias resistors with large inductors, as in LNA bias networks, will typically reduce phase noise by 3–6 dB.
- Fourth, the DC voltage at the emitter nodes is chosen to be 0.6V, or higher, to allow for the proper voltage excursion to be exercised at the control terminal of the varactor diodes and thus maximize the tuning range.
- Finally, the output impedance of the two current sources must be significantly larger than the impedance of C_2 at the oscillation frequency. The latter condition is more difficult to satisfy at mm-wave frequencies.

Numerous improvements to the differential Colpitts topology, captured in Figures 10.16–10.18, have been reported in the past ten years. These include:

- inductive peaking, L_C, to improve gain at high frequencies [12],
- replacing the current source tails with resistor R_B (R_{SS}), capacitor C_B (C_{SS}), and inductor L_{SS} to filter out the noise introduced by the bias circuitry [13],[14], Figure 10.16 and Figure 10.17,
- replacing resistive degeneration with inductive emitter degeneration L_{E1} and adding L_{E2} to reduce noise at f_{OSC} and $2f_{OSC}$ respectively, such that [15]

$$\frac{1}{2\pi\sqrt{L_{E1}C_{var}}} > f_{OSC} \text{ and } \frac{1}{2\pi\sqrt{L_{E2}C_{var}}} < f_{OSC},$$

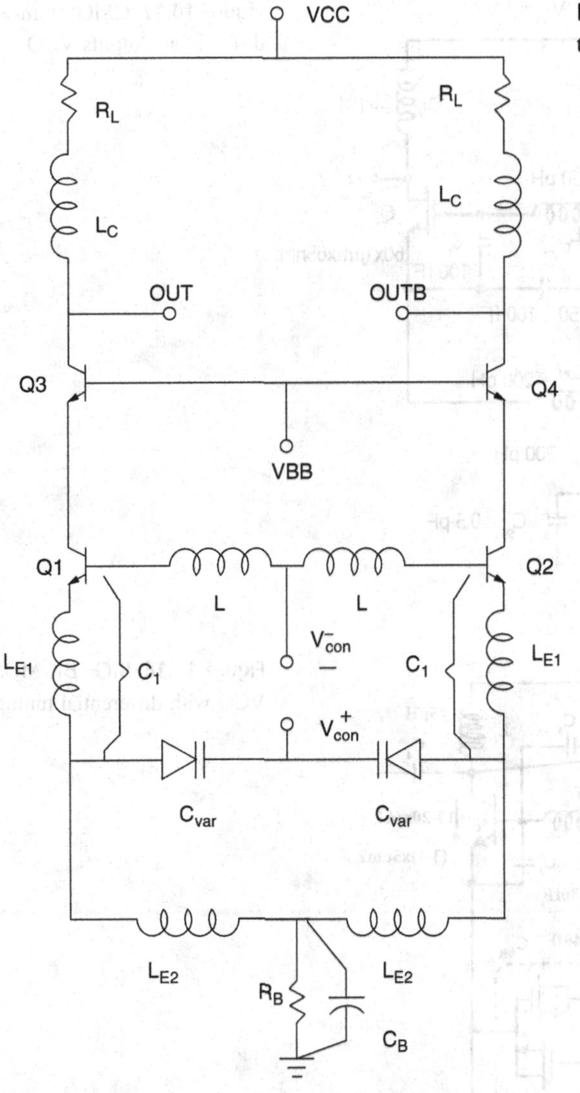

Figure 10.16 mm-wave differential Colpitts topology with inductive degeneration

- applying the control voltage differentially (Figure 10.18), and
- adding C_3 (Figure 10.18) to compensate for the Miller capacitance and further increase f_{OSC}.

10.2.4 Other selective feedback topologies

Clapp

This topology, illustrated in Figure 10.19, is named after James K. Clapp who invented it in 1948. The inductor from the Colpitts topology has been replaced by a series (and sometimes parallel) [16],[17] LC circuit. Frequency control can be introduced if the third

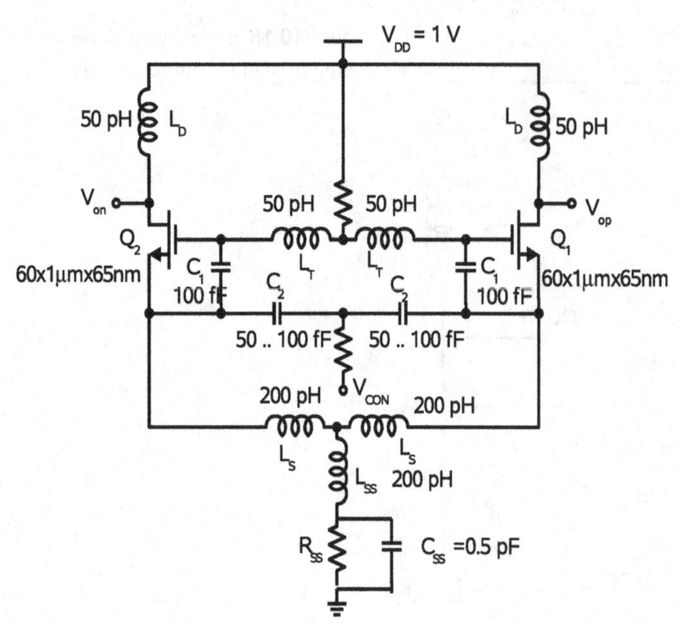

Figure 10.17 CMOS mm-wave differential Colpitts VCO

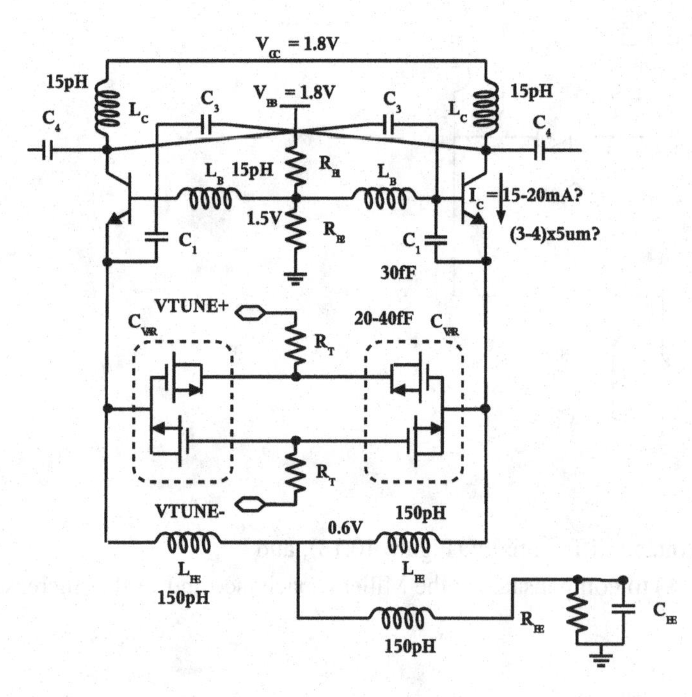

Figure 10.18 SiGe BiCMOS Colpitts VCO with differential tuning

capacitor, C_{VAR} is a varactor diode. The oscillation frequency is given by the series resonance of the three capacitors and inductor

$$f_{OSC} = \frac{1}{2\pi}\sqrt{\frac{1}{L}\left(\frac{1}{C_{var}} + \frac{1}{C_1} + \frac{1}{C_2}\right)}. \tag{10.46}$$

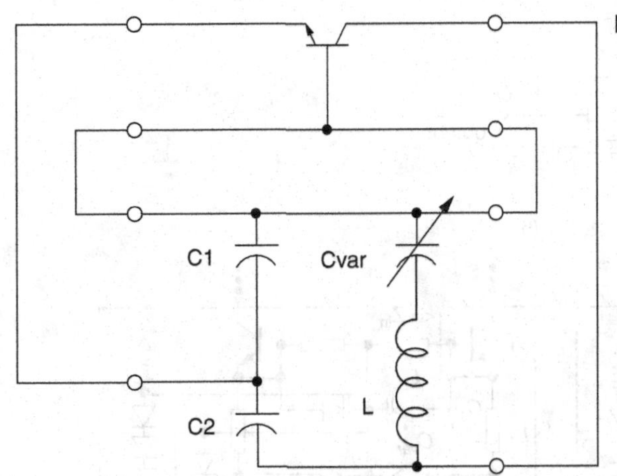

Figure 10.19 Generic Clapp oscillator topology

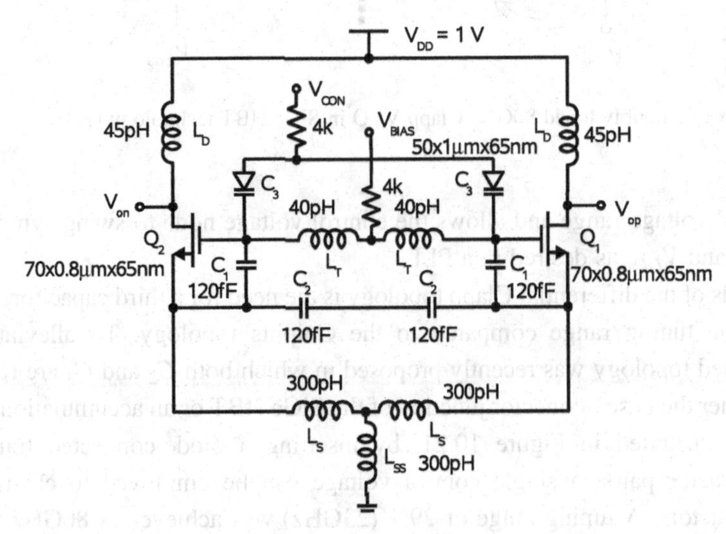

Figure 10.20 Schematics of a 60GHz Clapp VCO

In discrete, single-ended implementations of a VCO, the Clapp oscillator is preferred over the Colpitts topology. In the latter, when C_2 is varied to change the frequency, the voltage division ratio and negative resistance also change, making it sometimes difficult to satisfy the oscillation condition over the entire tuning range, or results in a large variation of the output power over the tuning range. This problem is avoided in the Clapp oscillator since both C_1 and C_2 are kept constant and only C_3 is varied. In a differential Clapp oscillator, the varactor is usually placed in parallel, rather than in series, with the inductor, as illustrated in the 60GHz VCO schematic of Figure 10.20. This arrangement ensures that the size of the varactor does not become prohibitively large, thus minimizing the undesired impact of parasitic capacitances. Another advantage of the Clapp topology, which becomes increasingly important in nanoscale CMOS technologies, is that one terminal of the varactor diode is connected to the gate of the MOSFET. This

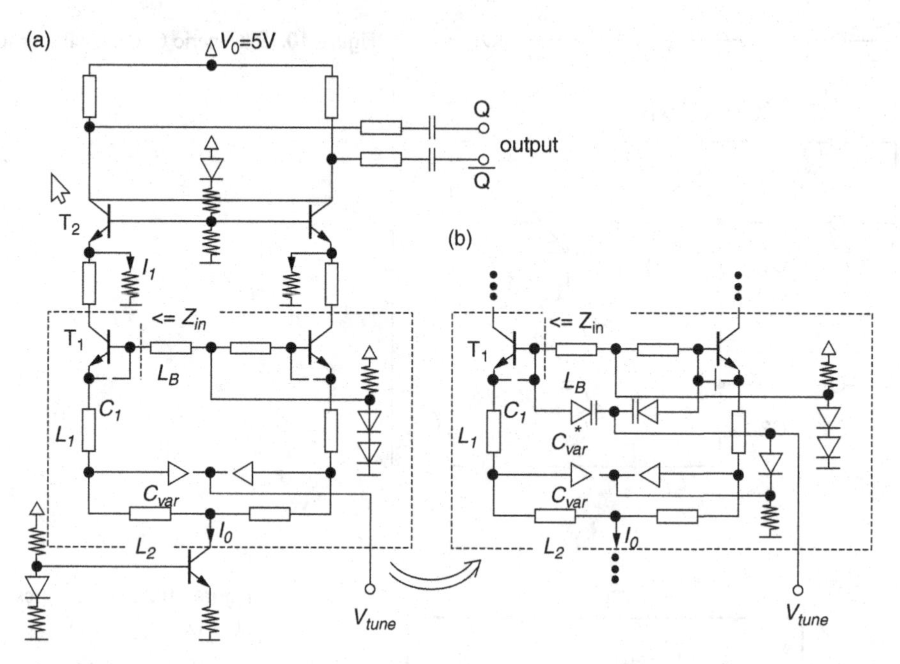

Figure 10.21 Schematics of a doubly tuned 80GHz Clapp VCO in SiGe HBT technology [17]

maximizes the control voltage range and allows the control voltage node to swing symmetrically between ground and V_{DD}, as desired in a PLL.

One of the shortfalls of the differential Clapp topology is the need for a third capacitor, which inevitably reduces the tuning range compared to the Colpitts topology. To alleviate this problem a doubly tuned topology was recently proposed in which both C_2 and C_3 are realized as varactors using either the base–collector junction of the SiGe HBT or an accumulation mode MOS capacitor. As illustrated in Figure 10.21, by inserting a diode-connected transistor between the two varactor pairs, a single control voltage can be employed to change the capacitance of all varactors. A tuning range of 29% (23GHz) was achieved at 80GHz with a phase noise of −95dBc/Hz at 1MHz offset [17].

Armstrong

Since only one external capacitor is employed, like the Hartley, this topology, shown in Figure 10.22, holds the prospect of a wide tuning range if the capacitor is realized with a varactor diode.

By simply changing the sign of the feedback in the shunt–series transformer-feedback LNA topology discussed in Chapter 7, we obtain yet another transformer-based differential VCO topology that differs from the traditional Armstrong topology in Figure 10.22(b). Possible differential IC implementations of fundamental frequency modified Armstrong VCOs at 60GHz and 120GHz are illustrated in Figures 10.23 and 10.24, respectively.

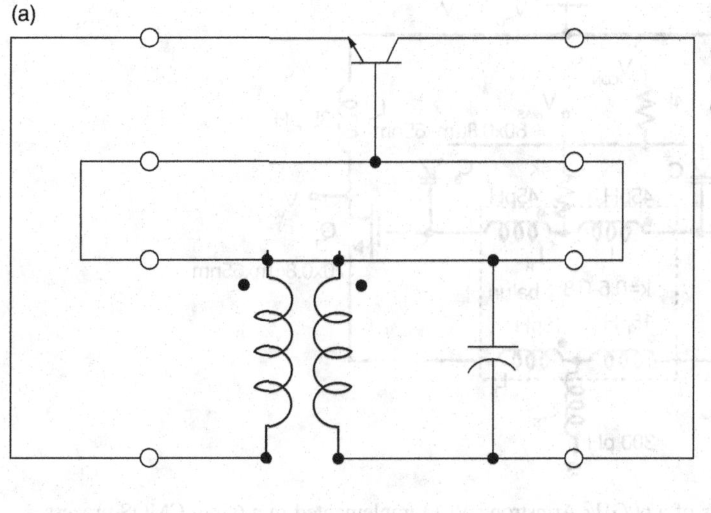

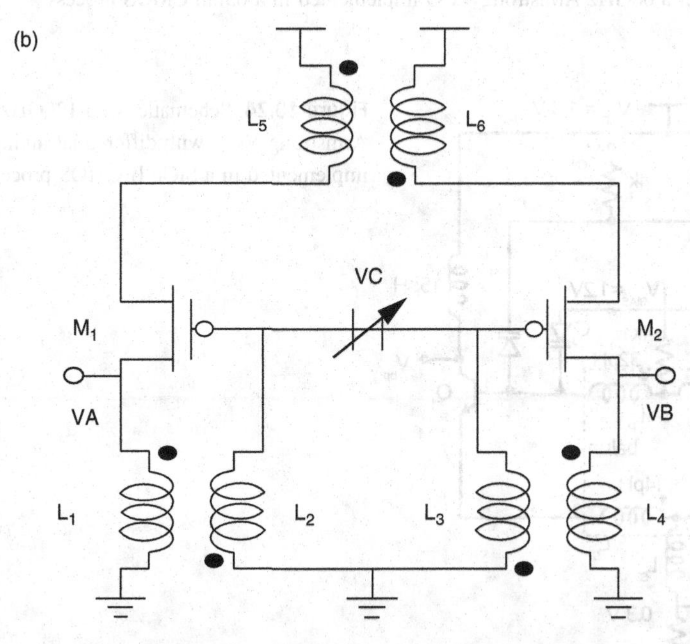

Figure 10.22 (a) Generic Armstrong oscillator topology and (b) differential implementation in CMOS using p-MOSFETs [18]

The oscillation frequency for these transformer-based topologies can be derived using the three-node network analysis described earlier, where the transistor is modeled by its large signal transconductance G_m, output conductance G_o, and its parasitic capacitances C_{gs} (C_{be}), C_{gd} (C_{bc}), and C_{sb}. The latter has no equivalent in BJTs/HBTs. The transformer is described by its corresponding *y*-matrix while the load conductance G_L includes the effect of the output conductance of the transistor, transformer loss resistance, and the load resistance connected at the source/emitter node of the transistor. Following straightforward but lengthy algebraic manipulations, the expression of the oscillation frequency is obtained from the condition that the real part of the oscillation equation is equal to zero

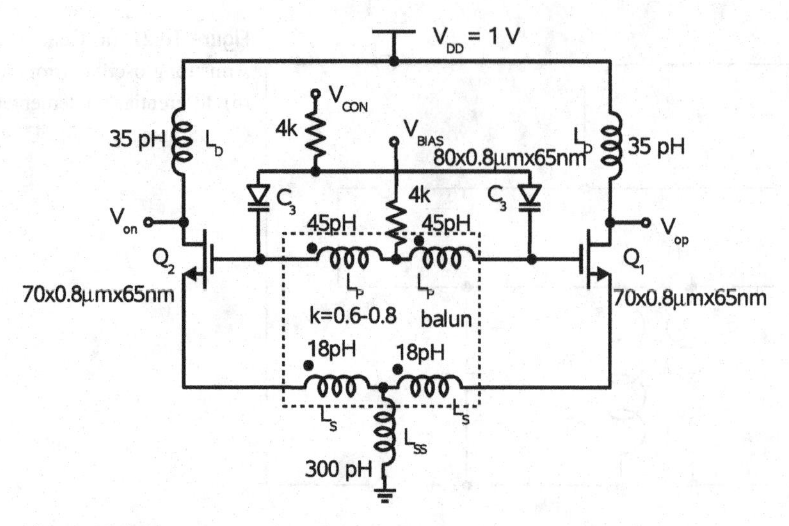

Figure 10.23 Schematics of a 60GHz Armstrong VCO implemented in a 65nm CMOS process

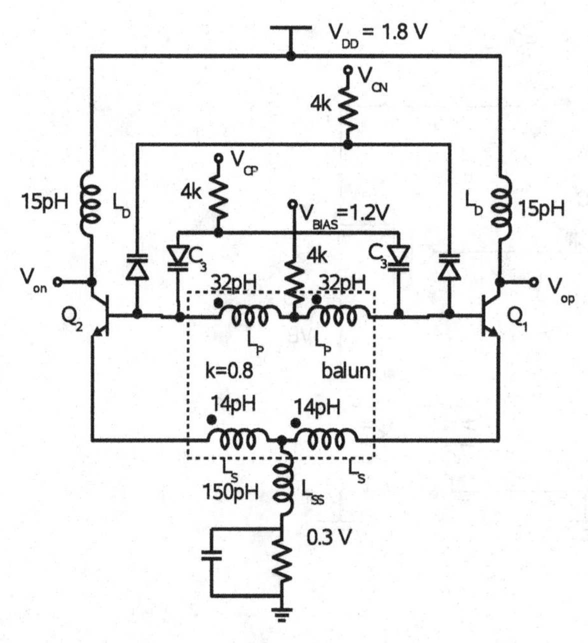

Figure 10.24 Schematics of a 120GHz Armstrong VCO with differential tuning implemented in a SiGe BiCMOS process

$$\omega^2\left(C_{var}C_{gs} + C_{Var}C_{sb} + C_{sb}C_{gs}\right) + \frac{1}{\omega^2 L_P L_S\left(1-k^2\right)}$$
$$= \frac{C_{var}L_P + C_{sb}L_S + C_{gs}(L_P + L_S - 2M)}{L_P L_S\left(1-k^2\right)}.$$

In a first-order approximation, the first term can be ignored, and the oscillation frequency is determined by

$$f_{OSC} = \frac{1}{2\pi\sqrt{C_{var}L_P + C_{sb}L_S + C_{gs}(L_P + L_S - 2M)}} \tag{10.47}$$

in which the dominant contribution comes from the $C_{var} \times L_P$ term since $L_S < L_P/2$. The latter design choice also maximizes the tuning range of the oscillator.

From the imaginary part of the oscillation condition, the transformer and the large signal voltage gain of the circuit must satisfy the inequality

$$\frac{G_m}{G_L} \geq \frac{\omega\left(C_{gs} + C_{var}\right) - \dfrac{1}{\omega L_P\left(1 - k^2\right)}}{\omega C_{var} - \dfrac{L_S - M}{\omega L_P L_S\left(1 - k^2\right)}}. \tag{10.48}$$

This VCO configuration with positive transformer feedback at the gate(base) and source(emitter) nodes of the transistor has two important advantages over the conventional Armstrong topology of Figure 10.22. First, as in the Hartley topology, the varactor is connected in parallel with the smallest parasitic capacitance of the transistor: $C_{gd}(C_{bc})$. Second, as in the Clapp VCO of Figure 10.23, the varactor control voltage excursion is maximized, allowing it to swing between ground and V_{DD}, practically symmetrically below and above the DC voltage at the gate/base node. Both these features maximize the tuning range of the VCO and allow this topology to operate with the lowest possible supply voltage, making it ideally suited for scaling in the 32nm and 22nm CMOS technology nodes.

10.2.5 Cross-coupled VCO topologies

This topology was already introduced in a simplified and intuitive manner in paragraph 10.1.2. Its origins can be traced back to 1934 and a paper by Millar, who investigated magnetically cross-coupled oscillators [19]. The earliest IC implementation appears to be, perhaps surprisingly, as late as 1995, and in CMOS [1]. It works well with both FETs and HBTs and it is by far the most popular VCO topology encountered in CMOS ICs. A schematic with bipolar transistors is illustrated in Figure 10.25, along with a simplified small signal equivalent circuit that

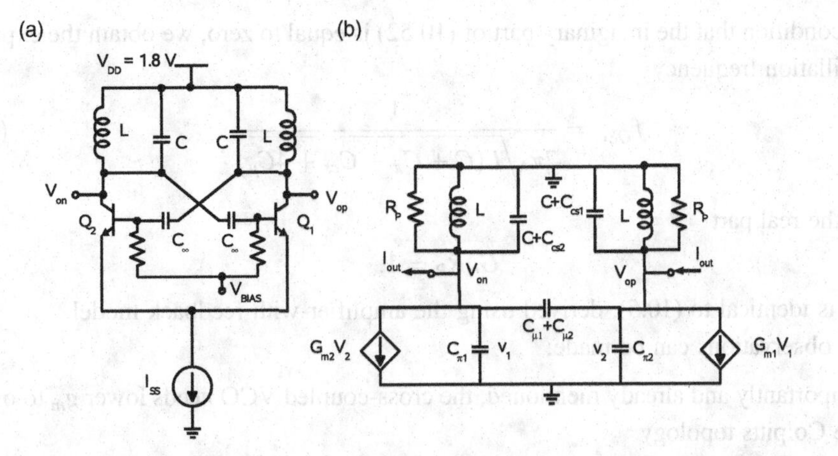

Figure 10.25 (a) Cross-coupled VCO with bipolar transistors, (b) small-signal equivalent circuit

allows us to analyze it as a negative resistance oscillator. Capacitor C is normally realized as a varactor, to introduce tunability. As usual, we can lump together the output resistance of the transistor, the loss resistance of the varactor, and of the tank inductor L as a resistor R_P in parallel with the LC tank.

The output admittance of the cross-coupled differential pair can be derived as a function of the small signal equivalent circuit illustrated in Figure 10.25(b) using the definition

$$Y_{OUT} = \frac{I_{OUT}}{V_{OUT}} = \frac{I_{OUT}}{V_{OP} - V_{ON}}$$

$$\frac{G_{m1}V_1 + \left[j\omega(C_{\pi2} + C_{cs1} + C_{\mu1} + C_{\mu2} + C) + \frac{1}{R_P} + \frac{1}{j\omega L}\right]V_2 - j\omega(C_{\mu1} + C_{\mu2})V_1}{V_2 - V_1}. \quad (10.49)$$

Since the current entering the circuit at node V_{OP} must be equal to the one exiting at node V_{ON}, the following condition must be satisfied

$$\begin{aligned} &G_{m1}V_1 + \left[\frac{1}{R_P} + j\omega(C_{\pi2} + C_{\mu1} + C_{\mu2} + C_{cs1} + C) + \frac{1}{j\omega L}\right]V_2 - j\omega(C_{\mu1} + C_{\mu2})V_1 \\ &= -G_{m2}V_2 \\ &\quad - \left[\frac{1}{R_P} + j\omega(C_{\pi1} + C_{\mu1} + C_{\mu2} + C_{cs2} + C) + \frac{1}{j\omega L}\right]V_1 + j\omega(C_{\mu1} + C_{\mu2})V_2. \end{aligned} \quad (10.50)$$

When accounting for the symmetry of the circuit, (10.50) simplifies to

$$\left[G_m + \frac{1}{R_P} + j\omega(C_\pi + C_{cs} + C) + \frac{1}{j\omega L}\right](V_1 + V_2) = 0 \quad (10.51)$$

which holds only if V_1 is equal to $-V_2$. By substituting into (10.49), we obtain

$$Y_{out} = \frac{-G_m}{2} + \frac{1}{2R_P} + j\omega\left(\frac{C_\pi}{2} + \frac{C_{cs}}{2} + 2C_\mu\right) + \frac{1}{j\omega L}. \quad (10.52)$$

From the condition that the imaginary part of (10.52) is equal to zero, we obtain the expression of the oscillation frequency

$$f_{OSC} = \frac{1}{2\pi\sqrt{L(C + C_\pi + C_{cs} + 4C_\mu)}} \quad (10.53)$$

and from the real part

$$G_m R_P = 1.$$

The latter is identical to (10.5), derived using the amplifier-with-feedback model.

Several observations can be made:

- Most importantly and already mentioned, the cross-coupled VCO needs lower g_m to oscillate than the Colpitts topology.
- Second, the negative conductance g_m increases with the bias current.

- Third, assuming the same current waveforms as in the case of the Colpitts VCO, the oscillation voltage (per side) can be described as a function of the bias current and of the tank impedance at resonance: $V_{OSC} = I_{SS}R_p$, or $(\pi/4)\ I_{SS}R_p$, depending on the approximation used.
- Fourth, as in all other topologies, by biasing transistors at J_{OPT}, the phase noise is minimized.
- Fifth, in the HBT case, de-coupling capacitors must be used for separately biasing the bases of Q_1 and Q_2.
- Sixth, unlike the Colpitts oscillator, there is no built-in buffering of the VCO load from the tank. The buffer amplifier directly loads the tank and affects both its phase noise and the oscillation frequency.
- Finally, a shortfall of this topology is that MOSFETs enter the triode region at least for some portion of the oscillation period, which degrades their noise and high-frequency performance.

Similarly to the Colpitts VCO, the large signal transconductance can be defined as a function of the bias tail current and of the oscillation amplitude

$$G_m = \frac{I_{SS}}{V_{OSC}} \tag{10.54}$$

Figure 10.26 shows a possible implementation of a cross-coupled VCO with differential frequency control. The role of the tail current source is to set the operating point of the oscillator transistors and also to lower the DC level at nodes X and Y, ideally to $V_{DD}/2$, such that the voltage across the varactor diodes varies between $-V_{DD}/2$ and $V_{DD}/2$, with the ideal DC node voltage at X, and Y being equal to $V_{DD}/2$. As in the case of the differential Colpitts VCO, the high-frequency portion of the noise generated by the tail current source can be

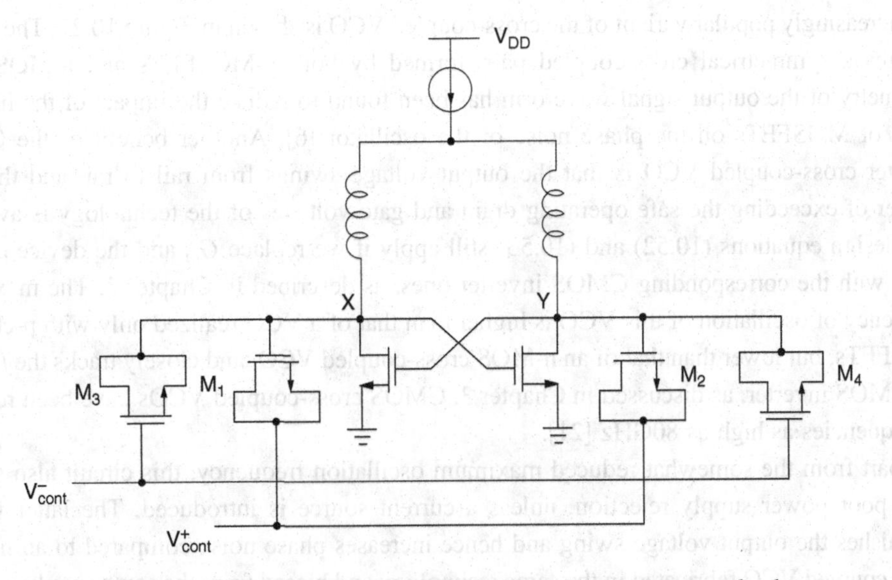

Figure 10.26 Cross-coupled CMOS VCOP with differential varactor control voltage

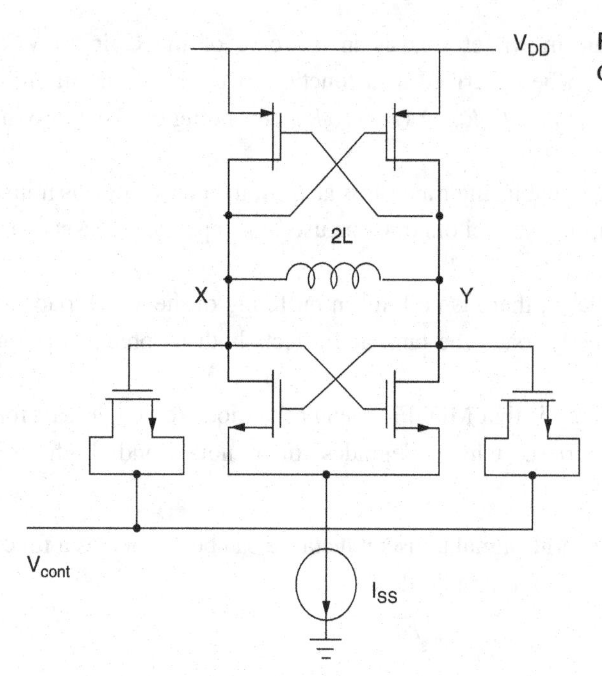

Figure 10.27 Schematics of a varactor-tuned CMOS cross-coupled VCO

filtered out by placing a de-coupling capacitor of a few pF across the current source. A common-mode resistor, or even an inductor [5], can be used instead of the current source to reduce the noise and the supply voltage at the expense of limited control voltage range for the varactor.

The cross-coupled oscillator with the highest oscillation frequency to date has been reported in 65nm CMOS reaching 300GHz, while push–push versions in 45nm CMOS exceed 400GHz [20].

Symmetrical cross-coupled VCO with CMOS inverter

An increasingly popular variant of the cross-coupled VCO is shown in Figure 10.27. The circuit features a symmetrical cross-coupled pair, formed by both p-MOSFETs and n-MOSFETs. Symmetry of the output signal waveform has been found to reduce the impact of the high $1/f$ noise of MOSFETs on the phase noise of the oscillator [6]. Another benefit of the CMOS inverter cross-coupled VCO is that the output voltage swings from rail to rail and thus the danger of exceeding the safe operating drain and gate voltages of the technology is avoided. The design equations (10.52) and (10.53) still apply if we replace G_m and the device capacitance with the corresponding CMOS inverter ones, as described in Chapter 3. The maximum frequency of oscillation of this VCO is higher than that of a VCO realized only with p-channel MOSFETs, but lower than that of an n-MOS cross-coupled VCO, and closely tracks the f_{MAX} of the CMOS inverter, as discussed in Chapter 3. CMOS cross-coupled VCOs have been realized at frequencies as high as 80GHz [21].

Apart from the somewhat reduced maximum oscillation frequency, this circuit also suffers from poor power supply rejection, unless a current source is introduced. The latter further diminishes the output voltage swing and hence increases phase noise compared to an n-MOS cross-coupled VCO fabricated in the same technology and biased from the same supply voltage.

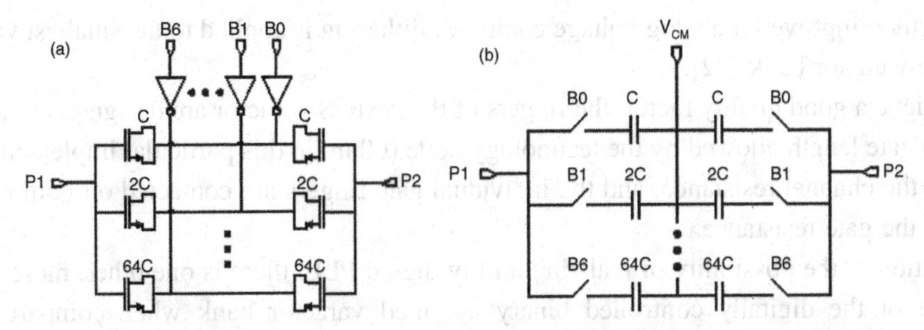

Figure 10.28 Possible implementations of binary-weighted, digitally-controlled varactor cells (a) using AMOS varactor fingers and (b) using fixed MIM or MOM capacitors and MOSFET switches

10.2.6 The digitally controlled oscillator (DCO)

Digitally controlled oscillators (DCO), realized using binary-weighted, AMOS varactors, have recently been demonstrated for mobile phones and multi GHz wireless applications [22]. This topology yields a linear tuning curve, with the oscillation frequency being determined by the digital word. With the implementation of a DCO, we can take advantage of the advanced digital CMOS technologies to realize a fully digital synthesizer [22].

Figure 10.28(a) shows the schematic of a differential bank of binary-weighted n-type AMOS varactors with *n*-bit digital control. The gate represents one port of the varactor, while the source and drain (shorted together) represent the second port, where the digital control is applied through CMOS inverters. The substrate, not shown, forms the third terminal of the varactor. The unit cell has a maximum capacitance value C which can be reduced by at least a factor of two, to C_{MIN}, as the binary control voltage drives the structure from the accumulation to the depletion regions. The overall capacitance of the varactor bank can thus be controlled over at least a 2:1 range with *n*-bit resolution, by the set of *n* conventional CMOS inverters $B_0 \ldots B_{n-1}$.

$$C_T = (2^n - 1)C_{MIN} + (C - C_{MIN})\left(b_0 + b_1 2^1 + b_2 2^2 + \ldots + b_{n-1} 2^{n-1}\right) \qquad (10.55)$$

where $b_0 \ldots b_{n-1}$ are either 0 or 1.

Each CMOS inverter sets the control voltage of the corresponding binary-weighted varactor cell either to V_{HIGH} (depletion) or to V_{LOW} (accumulation). V_{HIGH} is a low-noise voltage reference whose value is usually equal to V_{DD}, while V_{LOW} is a low-noise supply voltage whose value is usually equal to 0V. We note that the parasitic capacitance and resistance of the CMOS inverters do not affect the quality factor and the capacitance of the differential varactor bank because they appear only in common-node.

The minimum realizable unit cell capacitance corresponds roughly to the total gate capacitance of the smallest feature size MOSFET, and therefore decreases from one technology node to another. For example, in the 65nm CMOS node the minimum size n-MOSFET can be as small as 60nm × 200nm with a corresponding accumulation-mode capacitance of 0.3fF. After accounting for a 2:1 reduction in capacitance as a function of the control voltage in the depletion region, capacitance control with 0.13fF resolution is thus possible. This resolution can be further improved if analog voltage control or dithering is applied to the smallest varactor cell in the varactor bank [22].

To achieve a good quality factor, the fingers of the AMOS varactor are designed to have the minimum gate length allowed by the technology node (60nm in this particular implementation) to reduce the channel resistance, and the individual gate fingers are contacted on both sides to minimize the gate resistance.

In addition to the possibility of realizing a fully digital PLL, there is one other, more subtle, advantage of the digitally controlled binary-weighted varactor bank when compared to a conventional AMOS varactor with analog control. Because the control voltage drives the varactor either in accumulation or in depletion, where its capacitance is practically constant, noise present on V_{LOW} and on V_{HIGH} will have an attenuated impact on the instantaneous value of the varactor capacitance and therefore will not produce significant phase noise.

Alternatively, a digitally controlled varactor bank could be realized with binary-weighted MIM or MOM capacitors, controlled by MOSFET switches (Figure 10.28(b)). The main shortcomings of the latter topology are that:

(i) the series resistance (on the order of at least a few Ohms) and the parasitic capacitance of the switch, would (seriously) degrade the quality factor and the capacitance tuning range of the synthesized varactor, and

(ii) the minimum capacitance value that could be realized with MIM or MOM caps is larger than that of a minimum size AMOS varactor gate finger.

10.2.7 VCO banks

There are many practical situations when the VCO tuning range is not large enough to cover the bandwidth required by the application over process, supply, and temperature (PVT) corners. Even if it does, the VCO may not meet the phase noise and output power specification over the entire tuning range. Examples of such applications are the 60GHz wireless PAN standard where a 9GHz (15%) bandwidth must be covered worldwide. Similarly, 10–11.1Gb/s, 40–44.4Gb/s, and 100–110Gb/s SONET and Ethernet standards specify data rates with or without forward error correction (FEC) that typically require 10% tuning range over all PVT corners. In all these cases, it is still possible to satisfy all specifications if the VCO is replaced by a bank of VCOs, each with smaller, yet overlapping tuning ranges. A multiplexer is then used to select the desired VCO for each frequency subband while the other VCOs are turned off to save power and to avoid crosstalk and intermodulation problems, as shown in Figure 10.29.

Although it is tempting to change only one of the tank components (inductor, varactor, or capacitor sizes) between the different VCOs in the bank, in order to minimize phase noise and power dissipation, and to maximize output power, each VCO design should be separately optimized. This implies that all inductors, varactors, capacitors, transistor sizes, and bias currents should be scaled according to the VCO center frequency. This aspect will be revisited later in this chapter.

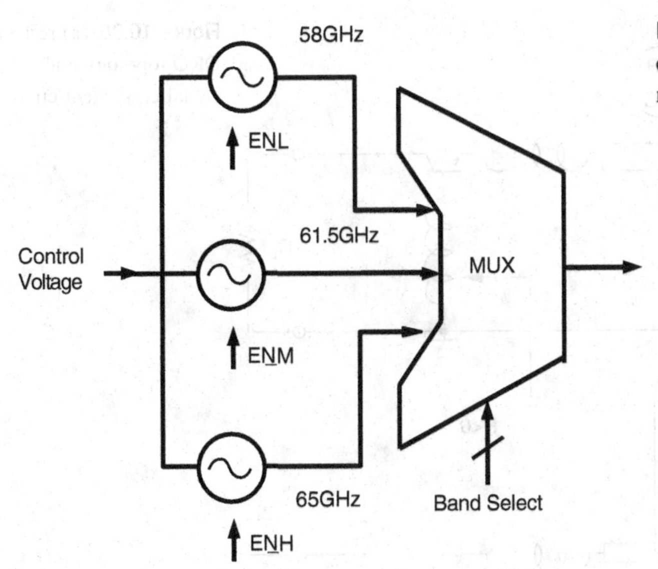

Figure 10.29 60GHz VCO bank consisting of three oscillators and a multiplexer

10.2.8 High-*Q* resonator oscillator topologies

These topologies are typically implemented in hybrid integrated circuits [23] where the resonator is mounted on a ceramic substrate, in close proximity of the oscillator IC. However, there is compelling evidence from recent research work that BAW, dielectric, and ferroelectric resonators may soon be integrated on top of the silicon oscillator IC, using a so-called "above-IC" process. Three types of oscillator topologies are employed to realize quartz crystal, DR, YIG, or magnetostatic-wave device oscillators: the reflection line, the feedback (or Pierce) and the Clapp topology. The similarity to the negative resistance and amplifier-with-feedback oscillator models is immediately apparent.

Reflection line

A typical reflection line oscillator topology that works with both dielectric and magnetic resonators is illustrated in Figure 10.30, where the negative resistance, as in (10.12), is generated by a common-base or common-gate transistor and the resonator is magnetically coupled to a transmission line. The transmission line is terminated at the far end on its characteristic impedance.

The physical length of transmission line between the dielectric resonator (or YIG sphere) and the transistor determines the load resistance seen by the transistor. At the onset of oscillation, the latter must be smaller than the negative resistance at the emitter (source) node

$$R \approx r_E + \frac{1}{g_m} - \frac{\omega^2 L_B}{\omega_T}. \tag{10.56}$$

The length of the microstrip line can be adjusted to ensure that the oscillation condition is satisfied.

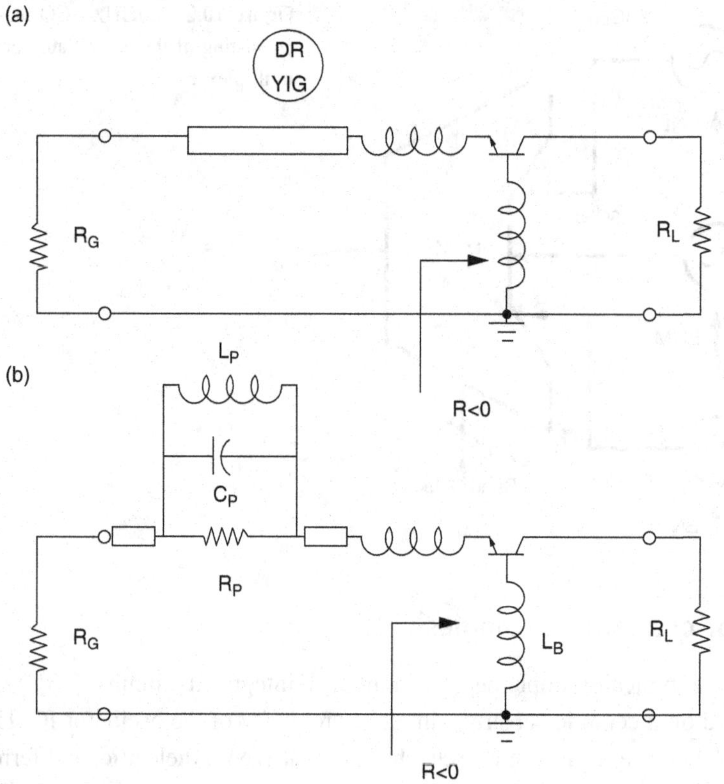

Figure 10.30 (a) reflection line DRO topology and (b) small-signal equivalent circuit

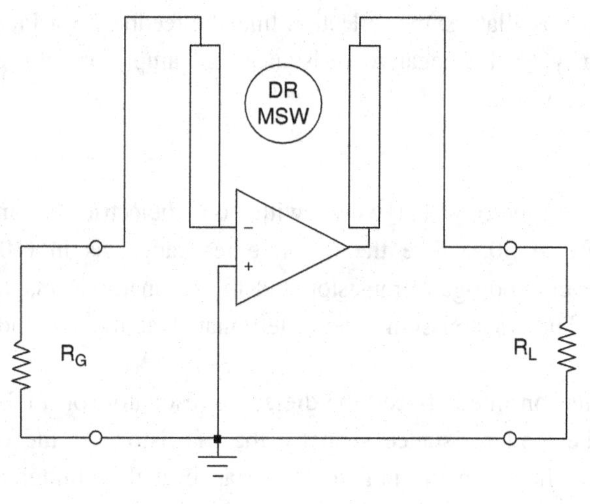

Figure 10.31 A high Q resonator oscillator where the resonator is in the feedback network

Feedback

An oscillator employing the amplifier-with-feedback topology is illustrated in Figure 10.31. The amplifier can be a CS or CE transistor stage, a CMOS inverter, or even a distributed broadband amplifier. The resonator can be either a DR, a quartz crystal (as in the Pierce

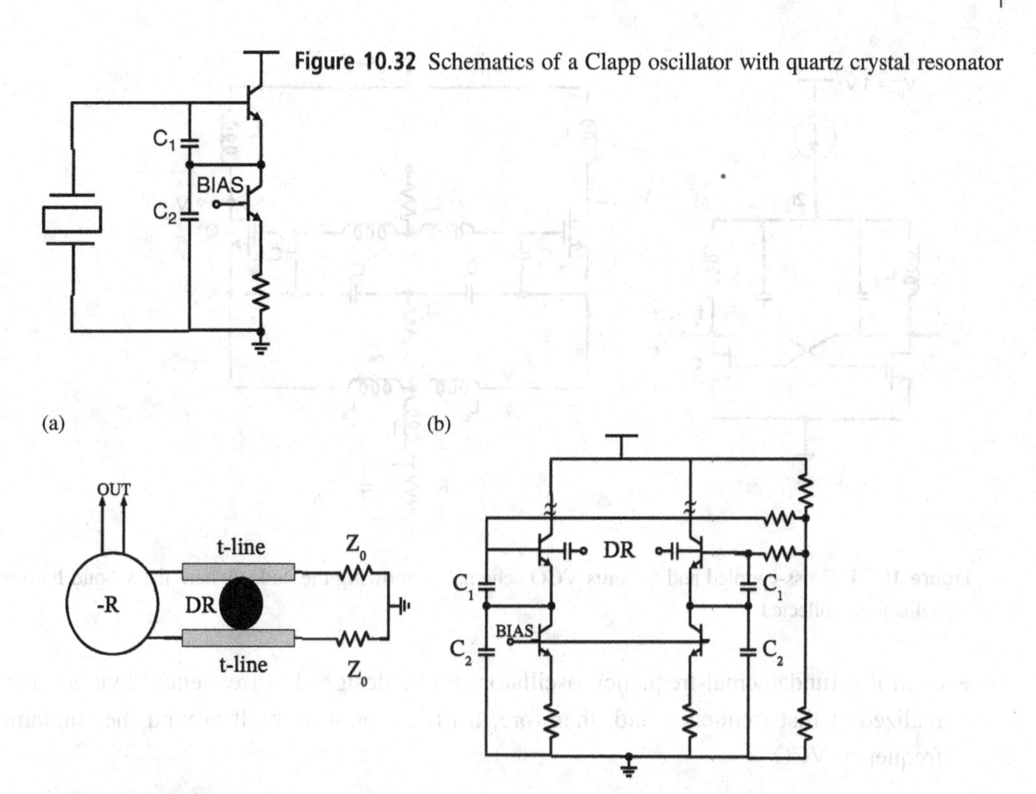

Figure 10.32 Schematics of a Clapp oscillator with quartz crystal resonator

Figure 10.33 (a) Block diagram and (b) negative resistance generator circuit of a 19GHz SiGe HBT DRO [24]

oscillator), or a magnetostatic wave resonator. In the latter case, because of its wide tunability, the amplifier is realized as a distributed amplifier with octave bandwidth. This circuit can be analyzed and designed as an amplifier with positive, selective feedback.

Clapp topology

This is a variation on the Clapp topology where the series LC resonator is replaced by a quartz crystal, Figure 10.32. More recently [24] a differential dielectric resonator version of this circuit, shown in Figure 10.33, was realized in a production SiGe HBT technology. The circuit oscillates at 19GHz and is employed in an offset PLL as a fixed, low-phase noise second local oscillator for a 77GHz automotive radar transceiver.

10.2.9 Push–push topologies

There are some very high-frequency system applications where a fundamental-frequency oscillator cannot be employed. The reasons can be diverse:

- a fundamental-frequency oscillator cannot be realized at this frequency in the given technology,
- the application frequency is too high for a sufficiently high-Q resonator to be fabricated on chip,

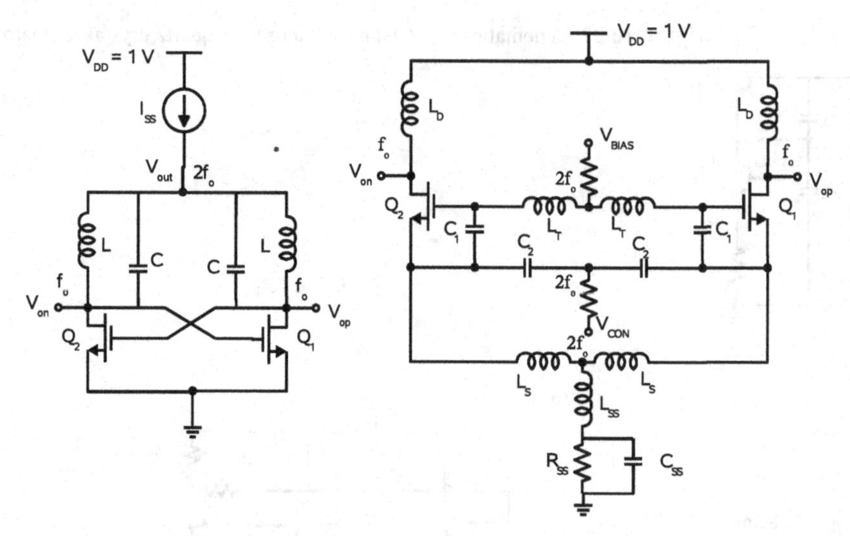

Figure 10.34 Cross-coupled and Colpitts VCO schematics showing the nodes where the second harmonic signal can be collected

- even if a fundamental-frequency oscillator can be designed, a frequency divider cannot be realized at that frequency and, therefore, a PLL cannot be built around the fundamental-frequency VCO.

In such situations, we can employ either a lower-frequency oscillator followed by a multiplier, or an oscillator that generates signals at the fundamental frequency and at higher-order harmonics of the fundamental. In general, the latter solution results in a smaller die area and power consumption.

In any given transistor technology, it is possible to generate signals at higher-order harmonics of the resonant frequency of the oscillator tank circuit by taking advantage of the strong non-linearity of the oscillating transistor(s). For example, in the differential fundamental oscillator topologies discussed earlier, the second harmonic signal is available at all common-mode nodes, as illustrated in Figure 10.34. By properly designing the impedance seen at the common nodes at the second harmonic of the resonator, we can maximize the power of the second harmonic. Such oscillator topologies, consisting of two identical half circuits, are known as push–push when operating on the second harmonic of the tank resonance frequency. *N*-push topologies, consisting of *n* identical suboscillators, are also possible. They oscillate on the *n*th harmonic of the resonator frequency.

One of the first reported integrated push–push oscillators employed a Colpitts topology and was fabricated in InP HBT technology [10]. More recently, several push–push VCOs based on cross-coupled [20] and Clapp topologies [16],[25] have also been demonstrated. The schematic of a push–push Clapp VCO used in a 60GHz CMOS radio transceiver is illustrated in Figure 10.35 [25].

10.2.10 Quadrature and multi-phase VCO topologies

Precise quadrature signals are required in many radio and fiber-optic communication transceivers for carrier and data recovery, as well as for single-sideband upconversion. They can be generated using:

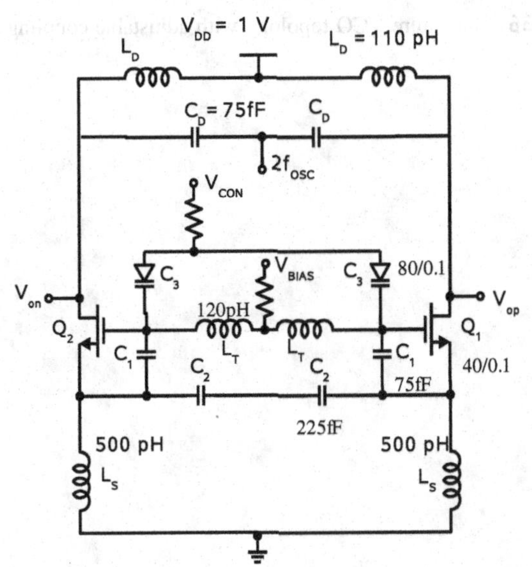

Figure 10.35 30GHz/60GHz push-push Clapp VCO [25]

(i) an oscillator operating at twice the desired frequency followed by a static frequency divider,
(ii) a 90 degree hybrid coupler, phase shifter, or polyphase filter, or
(iii) a quadrature oscillator [26].

The first solution, the more costly of the three, guarantees that the signals are in perfect quadrature over process, power supply, and temperature variations. As we have seen in Chapter 9, quadrature hybrids or polyphase filters have limited bandwidth and attenuate the signal. It comes as no surprise that a significant body of theoretical and experimental research has been dedicated to developing quadrature VCOs. One of the most common topologies, shown in Figure 10.36, relies on two coupled, negative-transconductance VCOs. Coupling must be at least 25% for the two oscillators to be 90° out of phase. Both VCO tanks operate slightly off resonance, leading to higher noise than in the corresponding single-tank VCO. Furthermore, asymmetries, component mismatch, and inductor coupling cause phase errors [27]. Although topologies with even higher number of phases have been investigated [28], quadrature VCOs remain a risky solution for the reasons mentioned above.

10.2.11 Intuitive model of differential HF VCOs as commutating switches

As in the case of the Gilbert cell mixers, a qualitative large signal model of differential HF VCOs, irrespective of their topology, can be developed based on the observation that the large tank voltage applied to the gates/bases of the transistors makes them operate almost as ideal commutating switches. The tank voltage itself develops as the tail current, I_{SS}, flows in and out of the resonant tank, Figure 10.37. Although the tank current waveform closely approaches that of a periodic train of rectangular pulses varying between $+/-I_{SS}$, the high-Q tank acts as a high

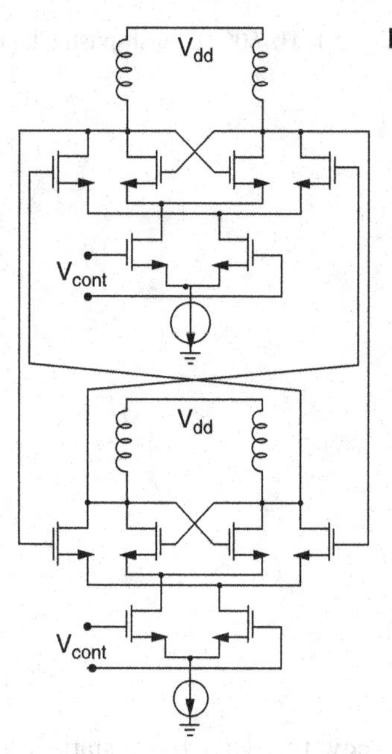

Figure 10.36 Quadrature VCO topology with adjustable coupling

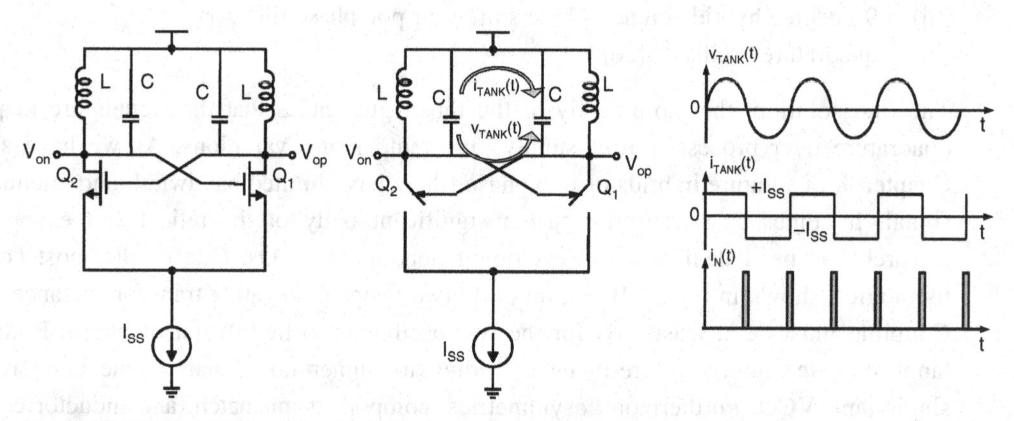

Figure 10.37 Pictorial large signal model of a cross-coupled VCO as commutating switch with periodic noise injection in the resonant tank

selectivity bandpass filter and only the fundamental component of the current, $\frac{4}{\pi}I_{SS}$, contributes to the sinusoidal tank voltage whose amplitude becomes $V_{OSC} = \frac{4}{\pi}I_{SS}R_P$, where R_P is the tank (parallel) resistance at resonance and V_{OSC} is the oscillation amplitude across the tank impedance at resonance, R_P.

Noise from the commutating transistors is periodically injected into the tank as short current pulses only when the tank voltage goes through a zero-crossing. At this instance in time, the VCO transistors are at their quiescent bias point and the VCO can be described as a small signal amplifier driven by a signal source whose impedance is equal to the tank impedance at resonance. Therefore, it makes sense to assume that the amplitude of the "noise current" pulses is proportional to the noise factor of the "VCO amplifier" at the quiescent operating bias point. This is essentially the "Leeson" phase noise model. As already discussed, the latter ignores the spectral folding of the noise from sidebands around the harmonics of the VCO frequency to sidebands around f_{OSC}, due to the mixing action of the commutating transistors. Spectral folding results in a higher phase noise than the Leeson model would predict if F were strictly equal to the noise factor of the amplifier. Naturally, the conversion gain of the noise folding process, and ultimately the calculated or simulated phase noise, depend on the symmetry of the VCO topology, on the physical non-linearity of the transistor, and on the accuracy of the non-linear compact model employed in simulation.

In the case of MOSFETs, only surface-potential based models, such as PSP, are able to accurately capture transistor non-linearity and symmetry. The reader should be warned against putting too much faith in analytical estimates of phase noise or in simulation results based on the various versions of the popular BSIM model.

Irrespective of the transistor model, MOSFETs, and nanoscale MOSFETs in particular, exhibit very high $1/f$ noise corner frequencies in the tens or hundreds of MHz range, where the phase noise of most VCOs is specified. Furthermore, MOSFET $1/f$ noise is a strong function of the quality of the channel-to-gate stack interface which varies from foundry to foundry, making it difficult to predict. On the other hand, HBTs exhibit $1/f$ noise corner frequencies in the 100Hz to 1kHz range, and their non-linearities, and therefore phase noise, can be reliably simulated using the MEXTRAM or HICUM compact models.

10.3 VCO SIMULATION TECHNIQUES

10.3.1 Survey of computer simulation of oscillators

Up to now, the emphasis has been on simplified design and analysis equations for various oscillator topologies. These are important in the first decision of what circuit topology to choose. However before committing to fabrication, a more accurate expectation of how the fabricated circuit will perform is desired, so as to reduce the need for iterations in the fabrication. This is the role of proprietary versions of computer circuit simulators such as Spice, ADS, Spectre, etc. Even though the initial source of these simulators may be in the open literature, e.g. for Spice, proprietary versions have arisen because support is needed to keep the transistor and other device models in the simulator as faithful as possible to the final circuit that will be fabricated. For example, there will be specific values of the circuit elements within the model that are tied to the fabrication process, as well as possible added circuit elements. Figure 10.38 shows a typical lumped parameter transistor model (in this case bipolar) which will be included along with a given proprietary simulator. The inter-connections between devices in integrated circuits must deal with distributed parameter

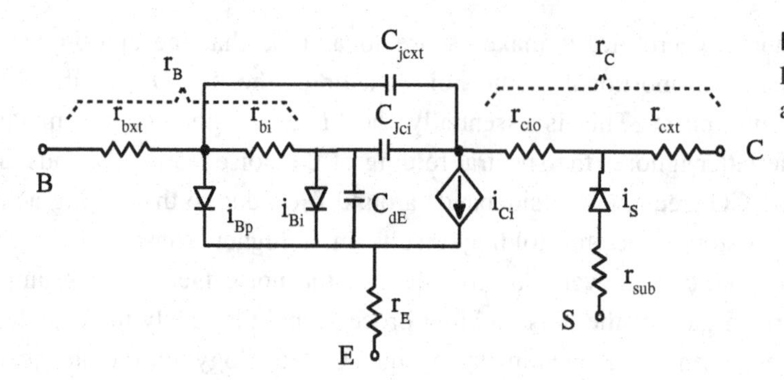

Figure 10.38 Simplified non-linear lumped-equivalent circuit of a bipolar transistor

effects as frequencies of operation have increased. The use of lumped models such as in Figure 10.38 however remains valid due to the continual reduction of transistor and other device sizes as the technology progresses. Certain elements within the lumped model are non-linear. For example, in the bipolar case, a dominant non-linearity is between the base and the emitter, which is diode-like. In FETs, non-linear behavior occurs in the gate–source and drain–bulk "junctions" impacting the transconductance, output resistance, as well as the gate–source and drain–bulk capacitances. The large signal behavior of the oscillator will exercise such non-linearities.

An oscillator is usually designed with an arbitrary choice of parameters which are arrived at by simplified design and analysis equations, as has been covered in previous sections. These parameters are then used within a circuit simulator such that the desired result is finally obtained. Since the oscillator is free running, the frequency and amplitude cannot generally be specified up front in simulations, but are an outcome of the various circuit parameters chosen. This will require iteration of the circuit values in the simulations. Thus the accuracy of the circuit simulator and its models is very important. Finally, testing of the behavior of the actual fabricated oscillator in comparison with the design goals will establish how well the simulator mimics the actual process. There are a number of possible computer simulation modes for the oscillator prototype:

- small signal simulation,
- transient simulation,
- harmonic balance and other periodic steady-state simulation techniques.

In the following discussion, the DC biases of the circuit are applied but large signal non-linear excursions are not assumed (although the non-linear models are present in the simulator).

Small signal simulation

Although the oscillator is a strongly non-linear circuit that operates under large signal conditions, a very efficient way to conduct the initial design is to start off with a small signal (*S*-parameter) analysis of its half circuit (Figure 10.39), based on the component values arrived at by hand analysis. Compared to transient, periodic steady state (pss) or

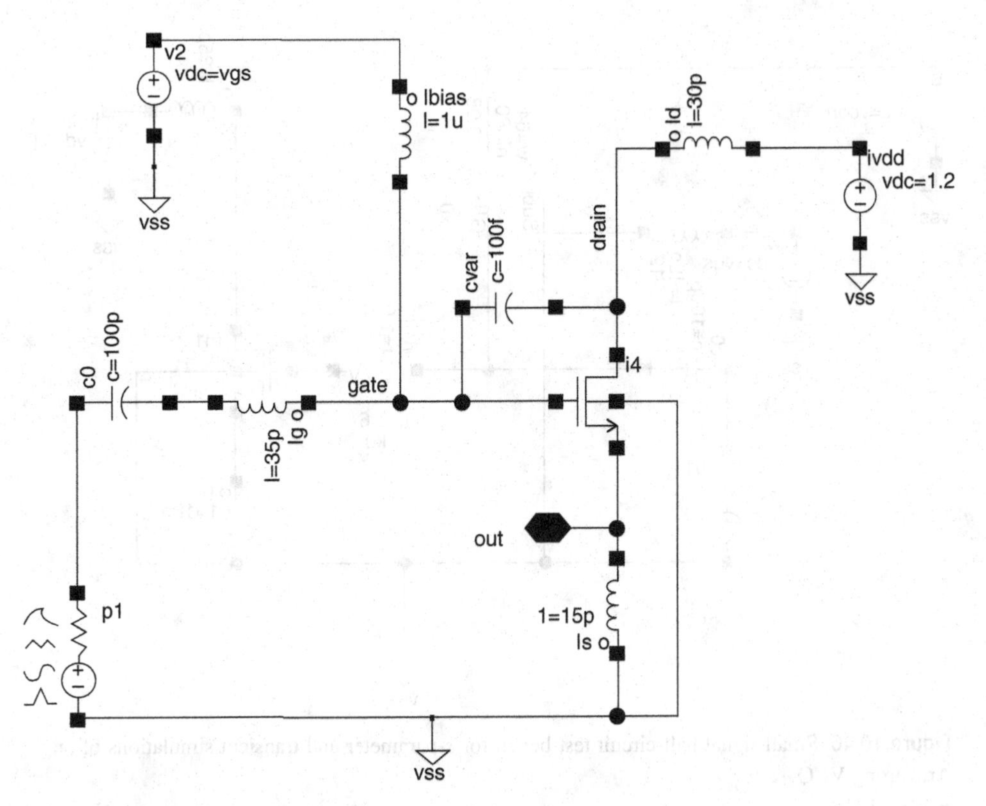

Figure 10.39 Small signal half-circuit testbench for *s*-parameter and transient simulations of a Hartley VCO

harmonic balance (HB) simulations, small signal simulations are fast, do not suffer from convergence problems and do not require initial conditions, allowing us to optimize the VCO design in a very short time.

By plotting the real and imaginary parts of the impedance looking into the gate of the MOSFET, we can quickly assess if the circuit produces an adequate negative conductance (Figure 10.40) or resistance (Figures 10.39 and 10.41) at the desired frequency, and we can determine the optimal values of L, C_1, and C_2 that cancel out the imaginary part of the input impedance, as well as the tuning range of the oscillator. We note that the *S*-parameter port impedance must be set either very high, when exploring parallel resonant tanks as in the case of the Armstrong topology in Figure 10.40, or very small, when investigating the negative resistance of a Hartley (Figure 10.39) or Colpitts VCO half circuit, Figure 10.41.

Figures 10.42 and 10.43 illustrate how the half-circuit *S*-parameter simulation schematic in Figure 10.41 can be employed to optimize the negative resistance and the tuning range, respectively, as a function of the transistor size for an 80GHz Colpitts DCO [29]. In both plots, as the transistor size is changed, so is the bias current such that the transistor is always biased at the optimal noise figure current density. The value of the gate inductor is held constant and the

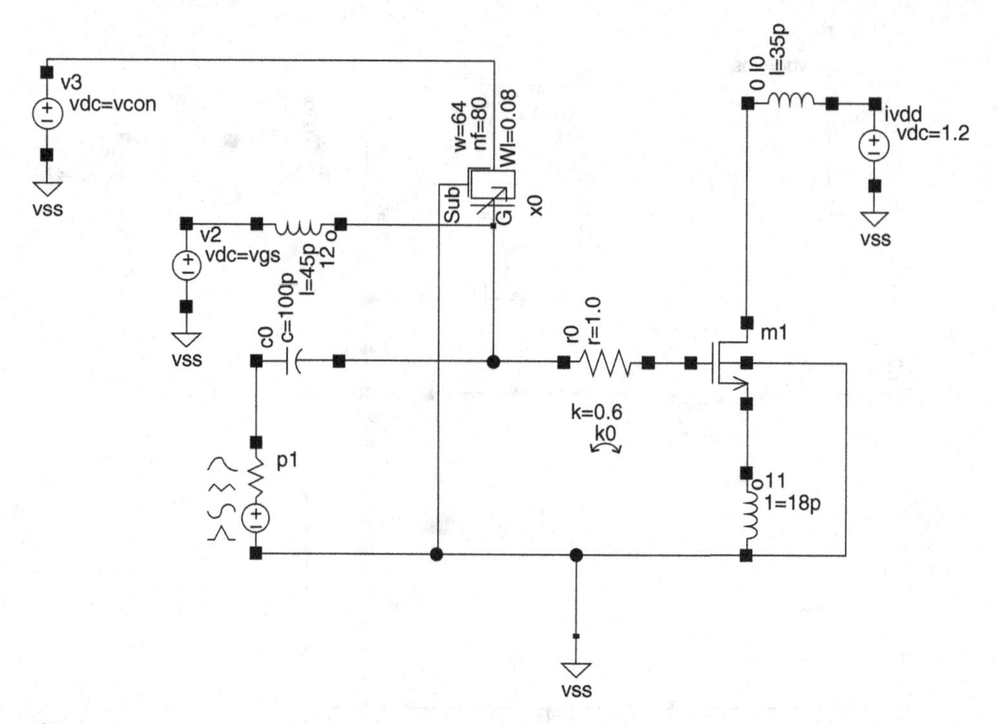

Figure 10.40 Small signal half-circuit test-bench for *s*-parameter and transient simulations of an Armstrong VCO

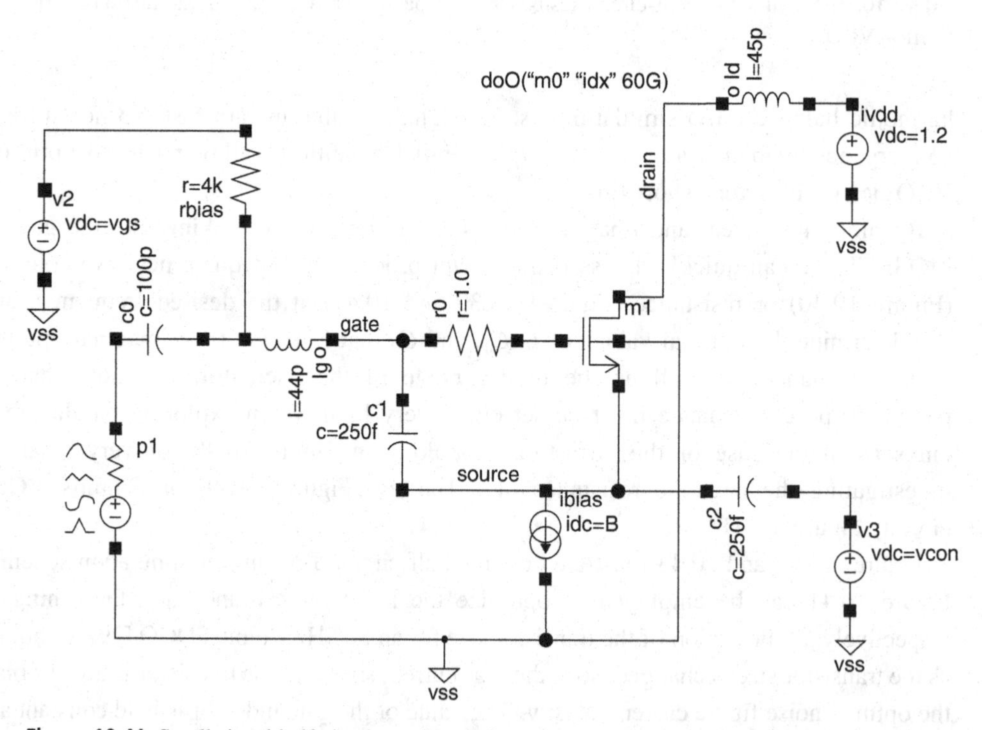

Figure 10.41 Small signal half-circuit test-bench for s-parameter and transient simulations of a Colpitts VCO

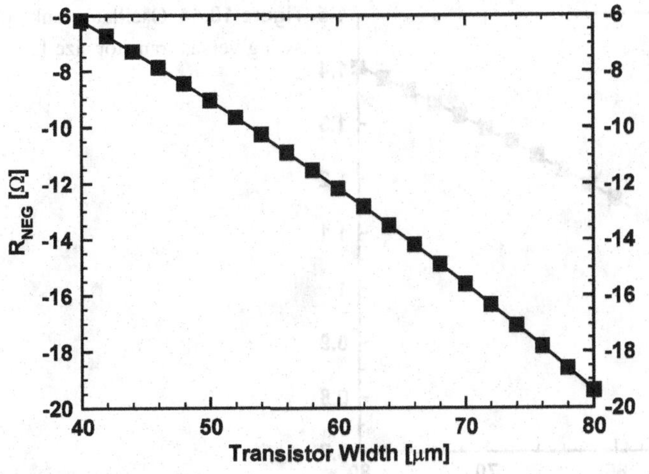

Figure 10.42 Oscillator negative resistance versus transistor size [29]

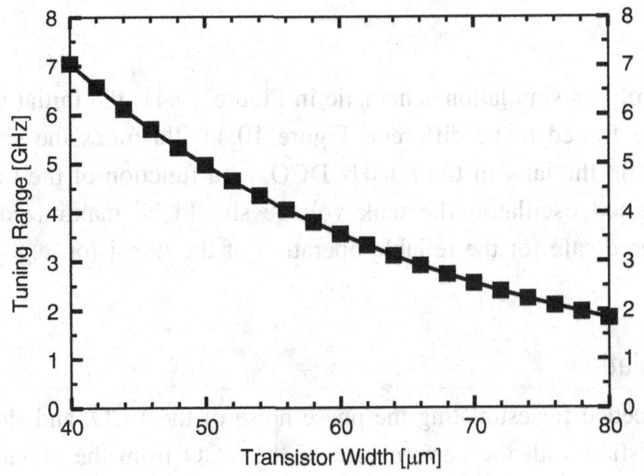

Figure 10.43 Oscillator tuning range versus transistor size [29]

size of the varactor diode is modified concomitantly with that of the transistor to maintain a fixed center frequency of 80GHz.

Transient simulation

Once the oscillation condition and tuning range have been validated by small signal simulation, we can proceed with the time domain simulation of the oscillator half circuit or of the fully differential oscillator circuit. Transient simulations can provide an accurate estimate of the oscillation amplitude and frequency, as well as of the oscillation start-up, providing that the proprietary non-linear device models in the simulator are correct. This correctness must apply up to a large number of harmonics of the fundamental oscillator frequency since device non-linearities will generate time domain harmonics. This means that the time steps of the simulation must be small enough to allow these harmonics to be realized adequately. In most cases, initial conditions must be set at appropriate nodes to speed up the oscillation start-up and reduce simulation time. For example, in the

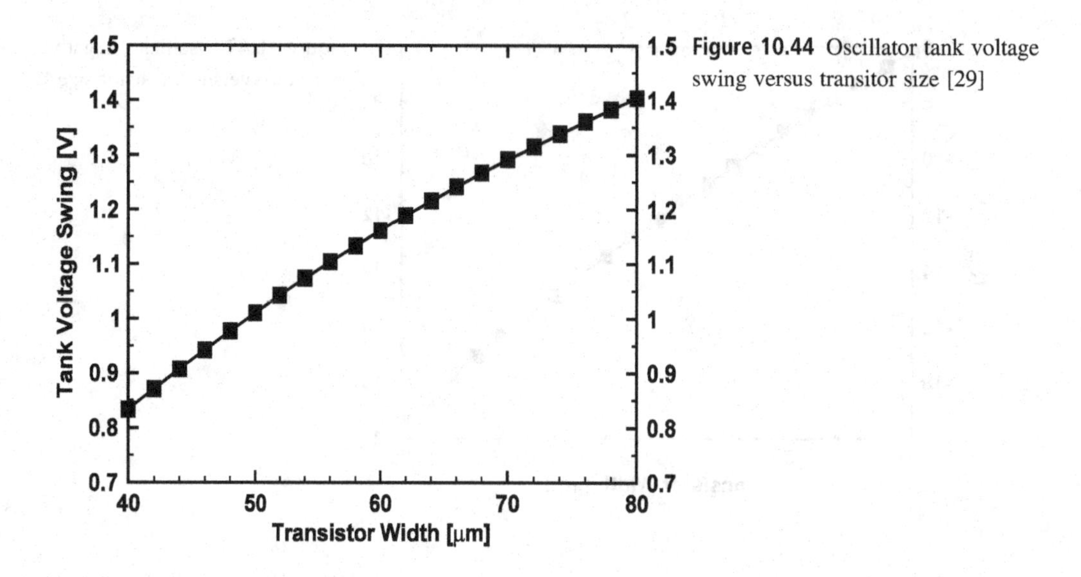

Figure 10.44 Oscillator tank voltage swing versus transitor size [29]

differential version of the Colpitts simulation schematic in Figure 10.41, the initial voltages at the two source nodes are forced to be different. Figure 10.44 illustrates the simulated peak-to-peak voltage swing on the tank of the 80GHz DCO, as a function of the transistor gate width. In a well designed oscillator, the tank voltage should be maximized to the largest value that is considered safe for the reliable operation of the transistor, e.g. 1.4–1.5 V_{pp} in 65nm CMOS.

PSS, *HB*, and *p*-noise simulation

This type of simulation is needed for estimating the phase noise of the VCO, and should be performed when we are satisfied with the performance of the VCO from the previous two simulations. Unlike the previous two, the success of this simulation is highly dependent on the simulation set-up, including the number of harmonics.

The main problem with pss or *HB* simulations for circuits operating in the 60–300GHz range lies with the fact that at least 3 to 5 harmonics of the fundamental must be present for the simulation to be accurate. This applies whether the harmonics are in the time domain for transient simulation or in the frequency domain for *HB* simulation. While this poses no problems when simulating a 6GHz radio, the third harmonic of the 60GHz fundamental is at 180GHz, which typically far exceeds the measurement and modeling range of all the BJTs, MOSFETs, varactors, resistors, inductors, transmission lines, and transformers employed in the circuit. This makes such high-frequency simulations problematic. Nevertheless, it is the only way to determine how each circuit component, especially bias resistors and transistors, qualitatively contributes to the overall phase noise of the VCO.

We end this section by noting that the half-circuit test benches in Figures 10.40 and 10.41 can be employed to run small signal, transient, as well as phase noise simulations.

10.4 VCO DESIGN METHODOLOGY

10.4.1 Design philosophy

A common design philosophy can be applied to all LC oscillators, irrespective of their topologies.

First, since the phase noise is inversely proportional to the square of the tank Q and the square of the oscillation amplitude V_{OSC}, both must be maximized. For a given technology, the tank Q is determined by the back-end (assuming optimal design of the varactor and inductor). The peak-to-peak voltage on any transistor junction cannot exceed the breakdown voltage or the maximum safe voltage specified by the foundry.

Second, based on ample experimental evidence and on modeling the transistors in the oscillator as ideal switches, the phase noise is minimized (i.e. I_n is reduced) by biasing the transistors in the VCO at the optimal noise figure current density. As discussed in Chapter 4, the latter is a function of frequency $J_{OPT}(\omega_{OSC})$ and of V_{DS}/V_{CE}. When transistors act as switches, an approximation that applies to all differential selective feedback and cross-coupled topologies, they only contribute noise for a short fraction of the oscillation period when both transistors in the differential oscillator operate in the active region and conduct an equal amount of current, that is $I_{SS}/2$. Once known, $I_{SS}/2$ and J_{OPT} determine the size of the negative resistance transistors in the VCO.

The first decision that the circuit designer has to make involves the value of the tank inductor L which affects whether the VCO will achieve:

- the lowest power, demanding a large L, or
- the lowest phase noise, which calls for the lowest L and largest bias current,

as can be concluded from the phase noise design equations (10.44) and (10.45).

10.4.2 Low power VCO design methodology

In the case of low-power design, in addition to V_{OSC} (set to the maximum allowed transistor voltage), the power consumption and supply voltage are given, therefore I_{BIAS} is known.

1. To minimize phase noise, the transistors are biased at J_{OPT}. This also sets the total MOSFET gate width or HBT emitter area A_E as $W = I_{BIAS}/2J_{OPT}$ or $A_E = I_{BIAS}/2J_{OPT}$ where we have accounted for the two transistors in the differential VCO.
2. Knowing the transistor bias current and size, we also know its transconductance and junction capacitances. We are now in a position to calculate the tank resistance at resonance R_{PMIN} required for oscillation at f_{OSC} $R_{PMIN} = \frac{V_{OSC}}{2I_{BIAS}}$ (for a cross-coupled VCO) and the minimum inductor value that can be used based on the Q of the technology back-end $L_{MIN} = \frac{R_{PMIN}}{Q\omega_{OSC}} = \frac{V_{OSC}}{2I_{BIAS}Q\omega_{OSC}}$.

By choosing the lowest inductor value for the tank, we ensure that the VCO will achieve the lowest possible phase noise for the given power consumption. However, we must make sure that the condition for the onset of oscillation is satisfied: $g_m R_{PMIN} > 1$ (cross-coupled VCO) or $g_m R_{PMIN} > 4$ (Colpitts or Clapp VCOs when $C_1 = C_2$)

3. Calculate the total effective capacitance that must be added to the transistor (C_{VAR} for cross-coupled VCOs, C'_1, C'_2, and C'_3 for Colpitts and Clapp VCOs) after subtracting the contribution of the transistor capacitances $C_{eq} = \frac{1}{\omega_{OSC}^2 L}$.

EXAMPLE 10.5

Design a 60GHz low-power cross-coupled VCO that consumes less than 4mW from a 1.2V supply using a low-power 45nm logic CMOS process with a tank Q of 5 (dominated by the inductor) or a 130nm SiGe BiCMOS process with 230GHz/300GHz f_T/f_{MAX} HBT and a tank Q of 5 (dominated by the varactor diode)

Assume that:

(a) for the 45nm n-MOSFET at $V_{DS} = 0.6$V, and $I'_D = 0.15$mA/μm, $C'_{gs} = 0.65$fF/μm, $C'_{db} = 0.3$fF/μm, $C'_{gd} = 0.38$fF/μm, $g'_{mn} = 2g'_{mn} = 0.8$mS/μm, $g'_{o\,n} = 2g'_{o\,p} = 0.125$mS/μm. $V_{OSC} = 0.6$V. These values include the impact of the source resistance.

(b) for the 130nm HBT: J_{OPT} (@60GHz) = 6mA/μm², $w_E = 0.13$μm; $\beta = 500$; $V_{CESAT} = 0.3$V, $V_{BE} = 0.85$V, $\tau_F = 0.3$ps; $C_{je} = 17$fF/μm²; $C_\mu = 15$fF/μm²; $C_{cs} = 1.2$fF/μm; $R'_b = 75\Omega \times$ μm, $R'_E = 2\Omega \times$ μm², $R'_c = 20\Omega \times$ μm, $V_{OSC} = 0.6$V.

Solution

(a) CMOS version

The maximum allowed tail current is 4mW/1.2V = 3.33mA.

1. The maximum n-MOS transistor W = 3.33mA × μm /(2 × 0.15mA) = 11.1μm. We can select W = 11μm and I_{BIAS} = 11μm × 0.15mA = 1.65mA. The small signal transconductance is $g_m = 0.8\text{mS} \times 11 = 8.8\text{mS}$
   for an n-MOS cross-coupled VCO, and $g_{mn} + g_{mp} = 17.6$mS for a CMOS cross-coupled VCO with the same bias current.
2. The minimum tank impedance at resonance is $R_{PMIN} = \frac{V_{OSC}}{2I_{BIAS}} = \frac{0.6}{3.3\text{mA}} = 182\Omega$
   and $L_{MIN} = \frac{R_{PMIN}}{Q\omega_{OSC}} = \frac{182}{5\times 6.28\times 6\times 10^{10}} = 96.6\text{pH}$.

With this inductance value, the n-MOS cross-coupled VCO meets the oscillation condition with small margin, but a CMOS one, where $W_p = 2W_n$, has more margin.

$$g_{mn}R_{PMIN} = 0.0088\times 182 = 1.6 \text{ and } (g_{mn} + g_{mp})R_{PMIN} = 0.0176\times 182 = 3.2.$$

For the n-MOS VCO with 100pH inductance per side, we can calculate C_{VAR} from the expression of the oscillation frequency

$$f_{OSC} = \frac{1}{2\pi\sqrt{L(C_{var} + C_{gs} + C_{db} + 4C_{gd})}} \text{ where } C_{gs} = C_{gsn} +$$
$$C_{gsp}, C_{gd} = C_{gdn} + C_{gsp}, C_{gs} = C_{gdn} + C_{gdp}, C_{db} = C_{dbn} + C_{dbp},$$

$$C_{VAR} = \frac{1}{L\omega_{OSC}^2} - C_{gs} - C_{db} - 4C_{gd}$$

$$C_{VAR} = \frac{1}{10^{-10}\times(6.28\times6\times10^{10})^2} - (0.65 + 0.3 + 4\times0.38)\mathrm{W}\times10^{-15} = 43.26\mathrm{fF}.$$

(b) HBT solution

1. The maximum HBT emitter area is calculated as $A_E = 3.33\mathrm{mA}/(2\times6\mathrm{mA}/\mu\mathrm{m}^2) = 0.2775\mu\mathrm{m}^2$. The corresponding emitter length l_E is 2.13μm. We can select $l_E = 2.0\mu\mathrm{m}$. $A_E = 0.26\mu\mathrm{m}^2$ and $I_{BIAS} = 1.56\mathrm{mA}$. For this HBT size, we can calculate the small signal parameters $C_\pi = 17\mathrm{fF} \times 0.26 + 0.3\mathrm{ps} \times 38 \times 1.56\mathrm{mA} = 22.2\mathrm{fF}$; $C_\mu = 3.9\mathrm{fF}$; $C_{cs} = 2.4\mathrm{fF}$, $R_E = 7.7\Omega$, $R_b = 37.5\Omega$.
   The small signal transconductance is $g_m = \frac{38I_C}{(1+38I_C R_E)} = 41.5\mathrm{mS}$.
2. The minimum tank impedance at resonance is obtained as

$$R_{PMIN} = \frac{V_{OSC}}{2I_{BIAS}} = \frac{0.6}{3.12\mathrm{mA}} = 192\Omega \text{ and } L_{MIN} = \frac{R_{PMIN}}{Q\omega_{OSC}} = \frac{192}{5\times6.28\times6\times10^{10}} = 102\mathrm{pH}.$$

With this value, the HBT cross-coupled VCO meets the oscillation condition by a large margin

$$g_{mn}R_{PMIN} = 0.0415\times192 = 7.96 \gg 1.$$

We can calculate C_{VAR} from the expression of the oscillation frequency

$$C_{VAR} = \frac{1}{L\omega_{OSC}^2} - C_\pi - 4C_\mu = \frac{1}{1.02\times10^{-10}\times(6.28\times6\times10^{10})^2} - 22.2\mathrm{fF} - 15.6\mathrm{fF} = 31.25\mathrm{fF}$$

which is about 20% smaller than the capacitance of the varactor in the CMOS cross-coupled VCO. As a result, the latter will have higher tuning range.

10.4.3 Low-phase noise cross-coupled VCO design methodology

1. Select L as small as possible.
2. Once L is selected, C_{eq} is known and, if we assume that the tank Q is dictated by the back-end, R_p is also known.
3. Since R_P and V_{OSC} are known, the minimum value of G_m can be estimated from (10.5). An initial guess on I_{BIAS} can be made from (10.54), assuming $C_1 = C_2 = 2C_{eq}$.
4. To minimize phase noise, the transistors are biased at J_{OPT}. This also sets an initial minimum value for the total MOSFET gate width or the HBT emitter area A_E.
5. Calculate C_{VAR} from the operating frequency and C_{eq}.

10.4.4 Low-phase noise Colpitts VCO design methodology

1. Select L as small as possible.
2. Once L is selected, C_{eq} is known. Again, assuming that the tank Q is determined by the back-end, R_p is also known.

3. Since R_P and V_{OSC} are known, the minimum value for G_m can be estimated from (10.39) and (10.41). An initial guess on I_{BIAS} can be made from (10.42), assuming $C_1 = C_2 = 2C_{eq}$.
4. To minimize phase noise, the MOSFETs or HBTs are biased at J_{OPT}. This also sets an initial minimum value for the total MOSFET gate width.
5. Select $C_1 = C'_1 + C_{gs} >> C_{gs}$ and C_{var} for the target f_{OSC} and maximum tuning range. $C_1 = C_{var}$ gives the largest f_{OSC} and the lowest power consumption but not the best phase noise and tuning range. Here, C'_1 represents the value of the MiM capacitor that is added in parallel with C_{gs}. The exact expression for C_{eq}, accounting for the transistor parasitic capacitances, is

$$C_{eq} = C'_{gd}W + \frac{(C'_1 + C'_{gs}W)(C'_2 + C'_{sb}W)}{C'_1 + C'_2 + (C'_{gs} + C'_{sb})W} \quad (10.57)$$

for a FET VCO, or

$$C_{eq} = C_{bc} + \frac{(C'_1 + C_{be})C'_2}{C'_1 + C'_2 + C_{be}} \quad (10.58)$$

for an HBT VCO.

6. Since (10.42) is approximate, and since L, C'_1, and C_{var} introduce parasitic capacitances, we must iterate (3)–(5) by changing $W(A_E)$, C'_1, and C_{var} for best phase noise and VCO tuning range. This is best carried out with a simulator.

Note that an HBT has fewer parasitic capacitances than a MOSFET (no C_{sb} equivalent) leading to a wider tuning range.

EXAMPLE 10.6

Design a low-noise 60GHz differential Colpitts VCO in the 45nm CMOS technology from Example 10.5. Calculate its frequency tuning range assuming a tank Q of 5 and a varactor capacitance ratio of 2:1.

Solution

We start by setting V_{OSC} to 0.6V, the maximum value allowed for safe operation, such that $V_{DD} <= 1.2$V:

1. We select $L = 44$pH as a reasonably small value. We note that an uncertainty of +/−2pH in interconnect inductance translates into a +/−4.5% error in tank inductance.
2. A C_{eq} value of 160fF is calculated from $f_{OSC} = 60$GHz and $L = 44$pH. If we assume a tank Q of 5, an R_p of 83Ω is calculated at 60GHz.
3. Since R_P and V_{OSC} are known, a minimum value for $G_m = 48.2$mS is estimated from (10.41). An initial guess on I_{BIAS} >7.23mA is made from (10.42), assuming $C_1 = C_2 = 2C_{eq}$.
4. To minimize phase noise, the MOSFETs are biased at $J_{OPT} = 0.15$mA/μm at $V_{DS} = V_{DD}/2 = 0.6$V. This sets the minimum MOSFET gate width to 48.2μm. We select $W = 72$μm, about 1.5 times larger, which results in a total bias current of 2 × 72μm × 0.15mA/μm = 21.6mA.

5. Knowing W and $C_1 = C_2 = 2\,(C_{eq} - C_{gd}) = 2(160\text{fF} - 27.36\text{fF}) = 265.28\text{fF}$ we calculate $C'_1 = 265.28\text{fF} - (0.65\text{fF}) \times 72\mu\text{m} = 218.48\text{fF}$ and $C'_2 = 265.28\text{fF} - (0.3\text{fF}) \times 72\mu\text{m} = 243.68\text{fF}$.

We can now calculate the high-frequency limit of the tuning range knowing that the minimum value of the varactor capacitance C'_2 is $243.68/2\text{fF} = 121.89\text{fF}$. This makes $C_2 = 27.36\text{fF} + 121.89\text{fF} = 149.25\text{fF}$ and

$$C_{eq} = 27.36\text{fF} + \frac{(265.38\text{fF})(149.25\text{fF})}{265.38\text{fF} + 149.25\text{fF}} = 122.87\text{fF}.$$

which leads to $f_{OSC2} = 68.48\text{GHz}$. The resulting VCO tuning range extends from 60GHz to 68.48GHz.

In reality the tuning range will be smaller and will shift to a lower frequency because we have ignored the parasitic capacitance associated with the inductor and the interconnect.

10.5 FREQUENCY SCALING AND TECHNOLOGY PORTING OF CMOS VCOs

We learnt in Chapter 4 that the self-resonance frequency and the peak Q frequency of inductors and transformers are inversely proportional to the component size. This allows inductors and transformers to be scaled for operation at higher frequency while their peak Q value remains approximately the same. If we ignore the resistive parasitics (gate resistance and source–drain series resistance), the size of varactors and MOSFETs and their Q also scale with frequency. However, it should be noted that in practice, due to the source–drain resistance which increases as the varactor size is reduced, the AMOS varactor Q actually decreases with increasing application frequency. We have also seen in this chapter that the Hartley, Colpitts, Clapp, and cross-coupled VCOs have simple topologies consisting of a single transistor half circuit, and an LC varactor resonant tank. As long as the resonator tank Q is dominated by the Q of the inductor/transformer, it should be straightforward to scale an existing VCO design in frequency. If the oscillation frequency scales by k, then L, C, transistor gate width W and I_{BIAS} should also be scaled by k, as indicated in (10.60).

$$L' \rightarrow \frac{L}{k}, C' \rightarrow \frac{C}{k}, W \rightarrow \frac{W}{k}, f'_{OSC} = \frac{1}{2\pi\sqrt{L'C'}} \rightarrow \frac{1}{2\pi\sqrt{\frac{L}{k}\frac{C}{k}}} = kf_{OSC} \tag{10.59}$$

where L and C represent the total equivalent inductance (in the case of a Hartley VCO) and capacitance (for Colpitts, Clapp, and cross-coupled VCOs) of the resonator tank, including the parasitic capacitances associated with the transistor, inductor, varactors, and MiM capacitors.

The oscillation condition, too, must be satisfied at the new frequency even after accounting for the transistor gate width scaling

$$g_m R_P = g'_m W R_P \rightarrow g'_m \frac{W}{k} R_P > 1 \text{or} \frac{g_m}{\omega_{OSC}^2 C_1 C_2 R_S} = \frac{g'_m W}{\omega_{OSC}^2 C_1 C_2 R_S} \rightarrow \frac{g'_m \frac{W}{k}}{k^2 \omega_{OSC}^2 \frac{C_1}{k}\frac{C_2}{k} R_S} > 1 \tag{10.60}$$

Table 10.1 **Measured phase noise and output power of fabricated CMOS Colpitts VCO test structures as a function of frequency, technology node and MOSFET finger width.**

Colpitts VCO	90nm 10GHz	90nm 80GHz	180nm 25GHz	180nm 50GHz	90nm 50GHz	90nm 80GHz	65nm 80GHz
L_{TANK} [pH]	435	50	200	100	100	60	40
C'_1 [fF]	800	100	100	50	50	35	80
C_{VAR} [fF]	800	100	100	50	50	35	80
W_f [μm]	1	1	2	2	2	2	0.8
N_f [μm]	100	60	40	20	20	16	76
f_{OSC} [GHz]	10–12	74–80	23–24.5	49–50.5	49–54	80–85	79–84
P_{DC} [mW]	36	37.5	86.4	57.6	50	37.5	37
P_{OUT} [dBm]	4	−13.6	−1	−9	−12	−17	−3
L[1MHz]	−117.5	−100.3	−98.8	−92.6	−76	−80	−95.7

knowing that the equivalent (parallel R_P or series R_S) tank loss resistance remains unchanged. This often implies that, instead of reducing the transistor W, g_m, and I_{BIAS} as the VCO is scaled in frequency, they should remain unchanged from the original lower-frequency VCO design. The size of the other (MIM or varactor) capacitors in the VCO can be reduced further to compensate for the lack of size reduction in the transistor. Obviously, this frequency scaling exercise is ultimately going to be limited by the transistor f_{MAX} when C_1, C_2, and/or C_3 become vanishingly small. In fact, in most mm-wave Colpitts and Clapp VCOs, C_1 typically reduces to the transistor capacitance [15]–[17]. When pushing the oscillation frequency even higher, as the transistor size is reduced, the impact of series parasitics due to R_g/R_b and R_S/r_E also increases, further eroding the tank Q and making it difficult to satisfy the oscillation condition. In the case of the Colpitts and Clapp VCOs, accounting for the series transistor parasitics is as simple as adding them to the R_S term in (10.61).

Equations (10.59) and (10.60) can also be employed to illustrate that an existing VCO design at one frequency can be ported to a more advanced technology node without redesign, while the VCO performance is improved. Alternatively, the transistor size and bias current can be reduced S times in the new technology node while still satisfying (10.60) at the same oscillation frequency. In other words, we can obtain the same performance with reduced power dissipation. This is possible because, due to advances in the gate dielectric stack, the maximum allowed transistor voltage swing has remained practically constant (1.2V to 1.1V) between the 130nm and the 45nm nodes, while the transconductance per unit gate width has continuously increased.

An experimental study was recently conducted in 180nm, 90nm [21] and 65nm CMOS [29] technologies to validate these simple theoretical frequency scaling and technology porting predictions. The description of the test structures and measurement results for CMOS Colpitts and cross-coupled VCOs are summarized in Tables 10.1 and 10.2, respectively. The

Table 10.2 **Measured phase noise and output power of fabricated cross-coupled CMOS VCO test structures as a function of frequency, technology node, and MOSFET finger width.**

Cross-coupled VCO Component value	90nm 10GHz	90nm 12GHz	180nm 17GHz
L_{TANK} [pH]	435	273	70
C_{VAR} [fF]	260	260	70
W_f [μm]	1	1	2
N_f	24	24	40
P_{OUT} [Bm]	−2.2	−10	−8.6
PN [dBc/Hz]@1MHz	−109	−105	−106

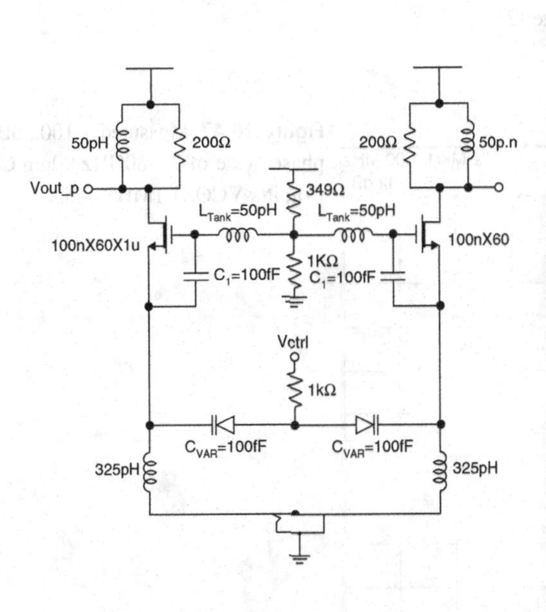

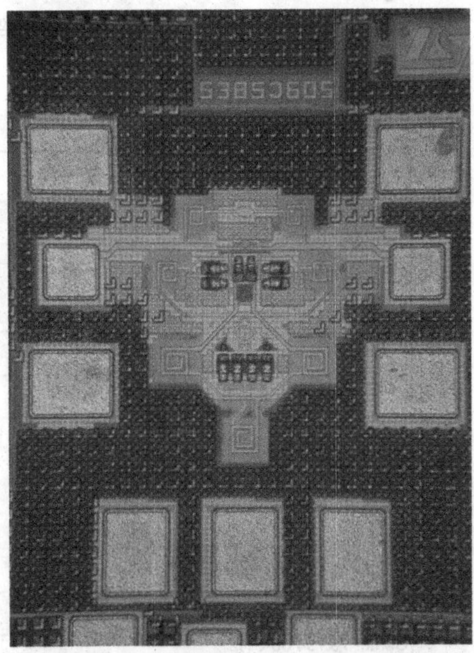

Figure 10.45 Schematics and die photo of 74–80GHz 90nm CMOS Colpitts VCO. The die area is 0.4mm × 0.42mm.

schematic and layout of the 74–80GHz Colpitts VCO are shown in Figure 10.45, while the measured phase noise, tuning range and output power as a function of the control voltage are reproduced in Figure 10.46. Figure 10.47 shows the measured phase noise spectrum at 79GHz.

By examining the measurement results in Table 10.1, we note that there is almost perfect phase noise scaling of 17dB between the 10GHz and 80GHz 90nm CMOS VCOs in columns 1

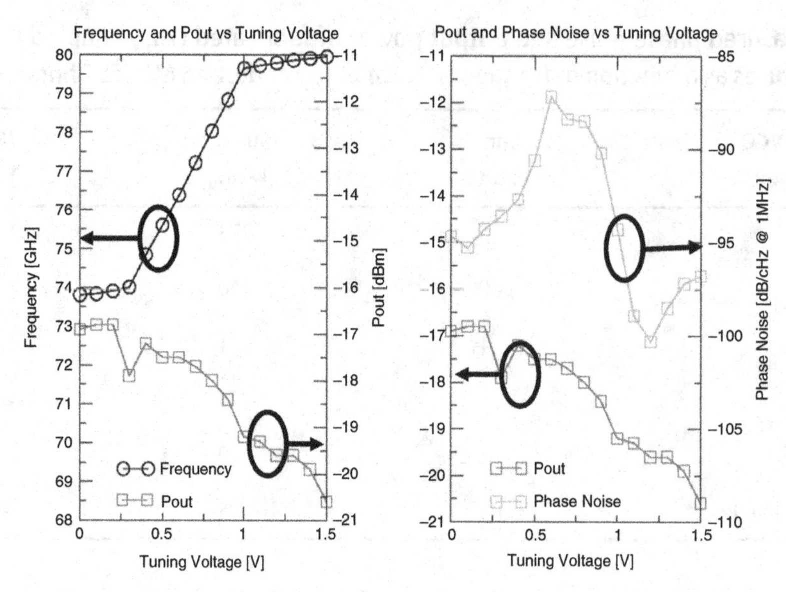

Figure 10.46 Measured oscillation frequency, phase noise, and output power of 73–80GHz 90nm CMOS Colpitts VCO as a function of control voltage [21]

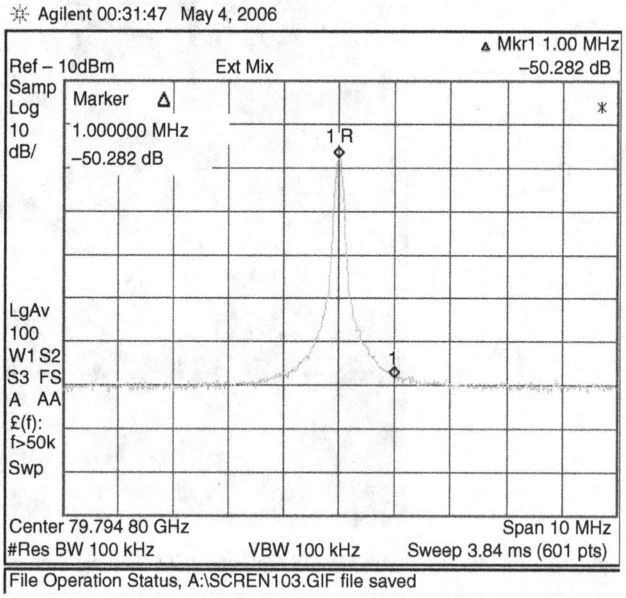

Figure 10.47 Measured −100.3dBc/Hz phase noise of 73–80GHz 90nm CMOS Colpitts VCO at 1MHz offset

and 2, and between the 25GHz and 50GHz 180nm CMOS VCOs in columns 3 and 4. From the VCOs in columns 4 and 5, we may conclude that VCOs can be ported from one technology node to another without redesign at the same frequency. However, unless the finger width is scaled with the technology node (while keeping the total gate width constant), the phase noise is considerably degraded. Finally, from the various VCO designs at 80GHz in 90nm and 65nm

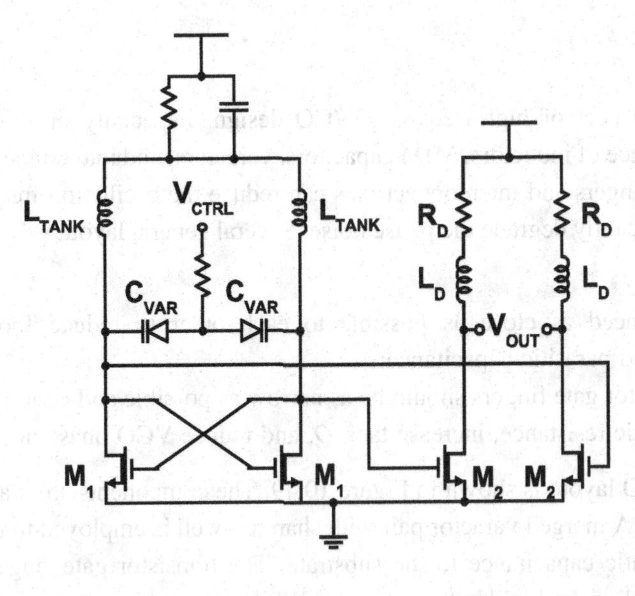

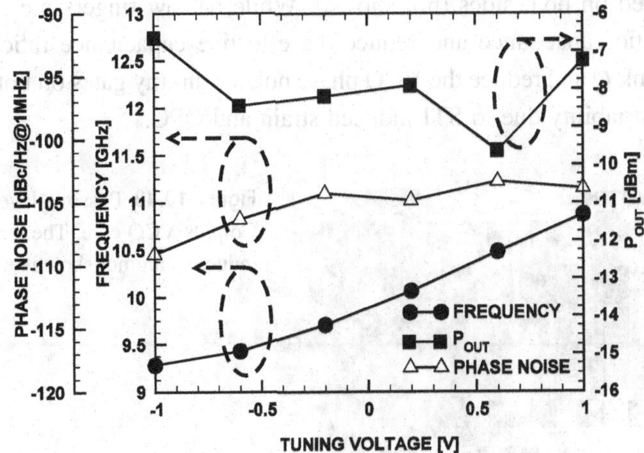

Figure 10.48 Schematics and measured oscillation frequency, output power and phase noise as a function of tuning voltage of the 10GHz 90nm n-MOS cross-coupled VCO [21]

CMOS, we can verify that the phase noise is improved by scaling only the tank inductor and capacitors, and by retaining a large transistor gate width as the oscillation frequency is increased.

Figure 10.48 shows the schematics and measured performance of the cross-coupled CMOS VCOs employed in the experiments summarized in Table 10.2. The same technologies and fabrication runs were employed to realize the Colpitts and cross-coupled VCOs. The 90nm and 65nm technologies have a digital back-end, whereas the 180nm CMOS process features an RF thick metal, copper back-end with higher inductor Q. To reduce the impact of the bias network on phase noise, no transistor current sources are employed and an RC filter is used on the power supply node. The 10GHz 90nm CMOS Colpitts VCOs show lower phase noise, but consume higher power, than the cross-coupled VCOs fabricated on the same wafer and operating at the same frequency. However, their FoM is practically the same.

10.6 VCO LAYOUT

Layout is the most critical aspect of high-frequency VCO design. Especially in nanoscale CMOS, the parasitic capacitance of inductors, MIM capacitors, varactors, and interconnect, and the series resistance of gate fingers and interconnect vias can reduce the oscillation frequency by more than 30% and significantly degrade the phase noise. Several general layout rules apply to all high-frequency VCOs:

- components should be placed as close as possible to each other to reduce footprint, interconnect inductance, and parasitic capacitance;
- transistor and AMOS varactor gate fingers should be as narrow as possible and contacted on both sides to reduce parasitic resistance, increase tank Q, and reduce VCO phase noise.

An example of a Colpitts VCO layout is shown in Figure 10.49. The components are placed in close proximity to each other. A merged varactor pair with shared n-well is employed to reduce varactor mismatch and parasitic capacitance to the substrate. The transistor gate fingers are 0.8μm wide and are contacted on both sides (not shown). While narrow fingers increase the preponderance of the parasitic capacitance and reduce the effective capacitance ratio of the varactor, they improve the tank Q and reduce the VCO phase noise. Dummy gates on both sides are employed to minimize variability due to STI-induced strain and OPC.

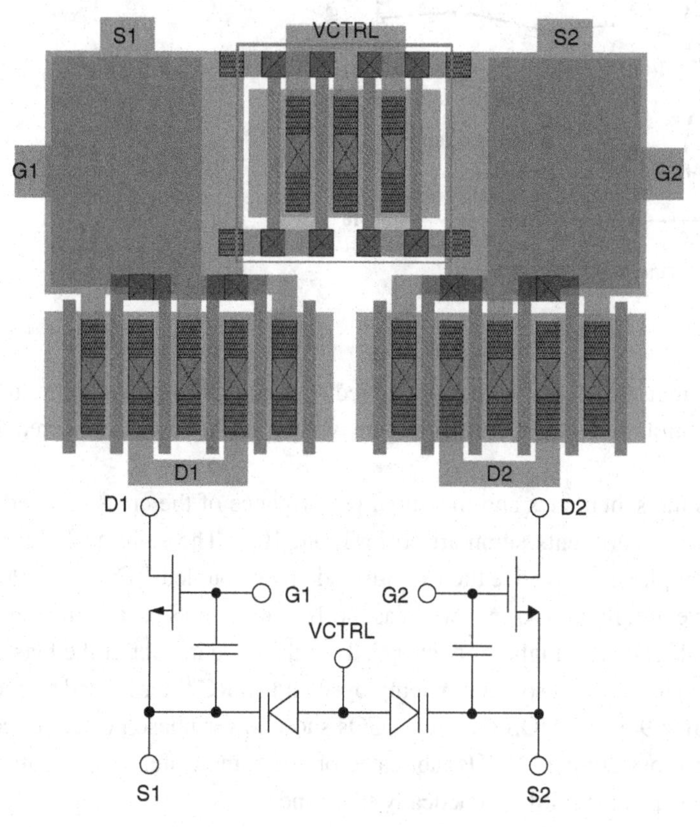

Figure 10.49 Possible layout of a Colpitts VCO core. The tank inductors are not shown

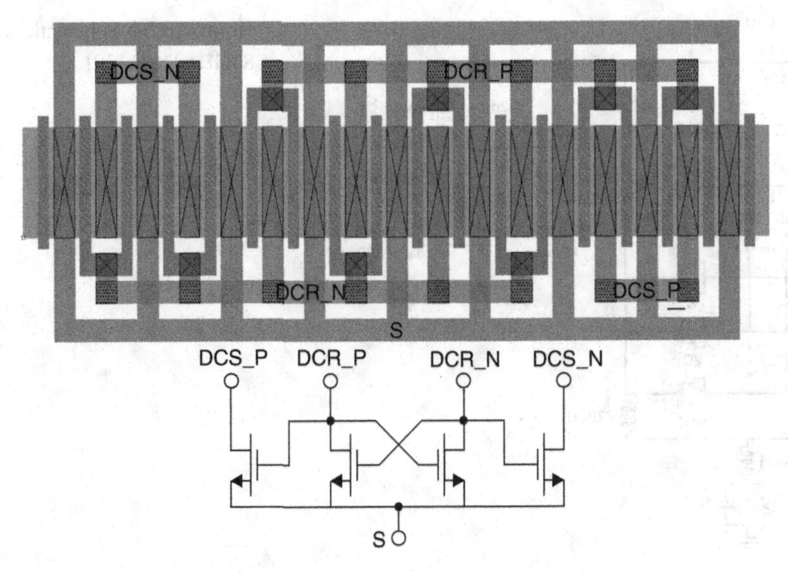

Figure 10.50 Possible layout of the transistors in a cross-coupled CMOS VCO, including the VCO buffer transistors. The tank inductor and varactors are not shown

Figure 10.50 reproduces the core layout of a cross-coupled CMOS VCO with output buffer. Merged cross-coupled and buffer transistor pairs are employed to minimize interconnect capacitance and to reduce mismatch.

10.7 MM-WAVE VCO EXAMPLES

In this section, we discuss in more detail two CMOS and BiCMOS VCOs designed for operation in the W-Band.

10.7.1 80GHz 65nm CMOS Colpitts DCO

Figure 10.51 shows the schematic of an 80GHz DCO [29] with a binary-weighted 7-bit varactor bank. All transistors and varactors have minimum gate length (60nm drawn, 45nm physical) and 0.8μm wide fingers which are contacted on both sides with two rows of contacts (see Figure 10.52) in order to minimize the phase noise of the oscillator. The unit varactor size is formed by a single 60nm × 0.8μm finger. The MOSFET gates are biased at $V_{DD} = 1.2$V through a series combination of 220Ω and 880Ω resistors. The 0.8 pF capacitor filters high-frequency noise that might be injected at the gates from V_{DD}. The role of the 250pH common-mode inductor and of the 1 pF capacitor across the 10Ω biasing resistor is to filter the resistor noise. The 250pH inductor is designed to be as large as possible, while still maintaining a high-Q and a self-resonance frequency above 2 × f_{OSC} (i.e. 170GHz). Ideally, the DC voltage drop on the 10Ω bias resistor should be 0.6 V, leaving another 0.6V for the gate–source voltage of the MOSFETs. In a low-V_T GP n-MOSFET, the latter happens to correspond to the optimum noise figure current density of

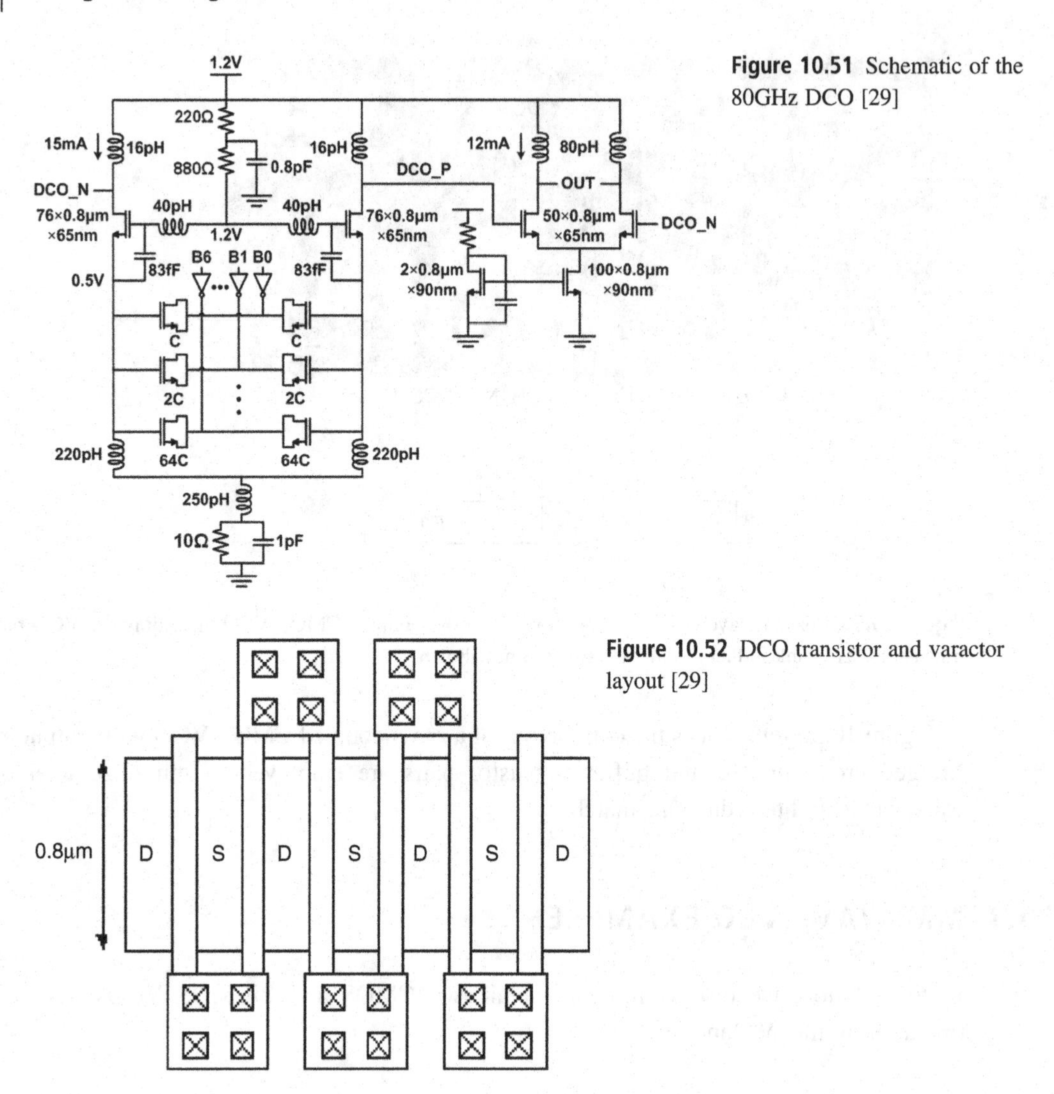

Figure 10.51 Schematic of the 80GHz DCO [29]

Figure 10.52 DCO transistor and varactor layout [29]

0.15 mA/μm for a V_{DS} of 0.6V. With this bias scheme, the gate–source, drain–source, and gate–drain junctions of the MOSFET swing between 0 and 1.2 V, the varactor gate voltage is pinned at 0.6V, and its C-V characteristic is fully exercised between the accumulation and depletion regions as the outputs of the CMOS inverters switch between 0V and 1.2V. Unfortunately, in this design the transistors are biased at a higher current density of 0.25mA/μm, resulting in somewhat higher phase noise and in a reduced tuning range than the optimum. This can be easily fixed by increasing the common-mode bias resistor from 10Ω to 33Ω.

The differential VCO buffer is DC-coupled to the oscillator and is tuned to 80GHz with 80pH inductive loads. Its role is to minimize load pulling and to provide a large power to the power amplifier and/or to the receive mixer in an automotive collision avoidance radar or in an active imager.

Figure 10.53 reproduces the simulated oscillation frequency as a function of the three most significant bits. The simulated tuning range is 79GHz to 84GHz, in good agreement with the measured 79–83.5GHz range. The simulated transient waveforms of the tank voltage, as well as

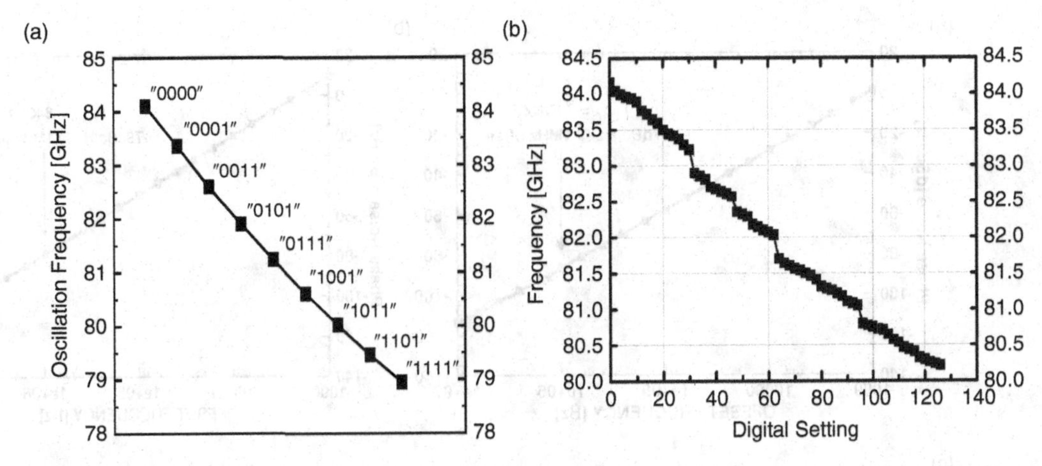

Figure 10.53 (a) Simulated DCO oscillation frequency as a function of the four most significant bits (b) Measured oscillation frequency as a function of the digital control word [29]

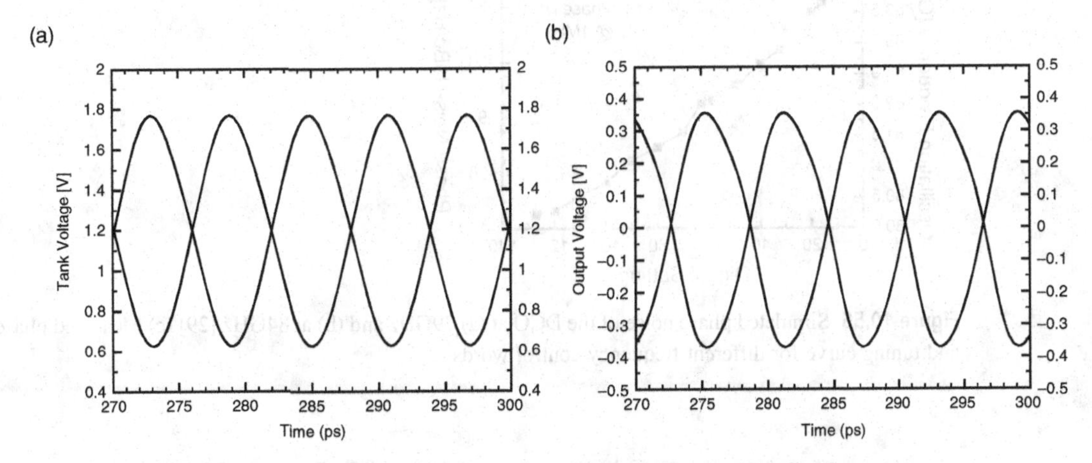

Figure 10.54 Simulated tank and output voltage of the DCO oscillating at 84GHz [29]

of the DCO output voltage at 84GHz, are illustrated in Figure 10.54. The voltage swing on the tank varies between $1V_{pp}$ and $1.2V_{pp}$. The measured phase noise varied over the tuning range between −79dBc/Hz and −94dBc/Hz at 1MHz offset, in close agreement with the phase noise simulations, Figure 10.55.

10.7.2 SiGe bipolar VCOs

The schematic in Figure 10.18 was employed to design SiGe BiCMOS VCOs operating at 77GHz, 104GHz [30] and 120–150GHz [31]. A MiM capacitor (C_1) in parallel with C_{be} increases the negative resistance and reduces phase noise by shunting the base resistance. Optionally, negative Miller capacitors (C_3) are placed at the BC junctions to cancel the effect of C_{bc}. Finally, differential tuning with AMOS varactors is used for better supply noise rejection. This also reduces the modulation of the varactor capacitance (C_{VAR}) by the tank

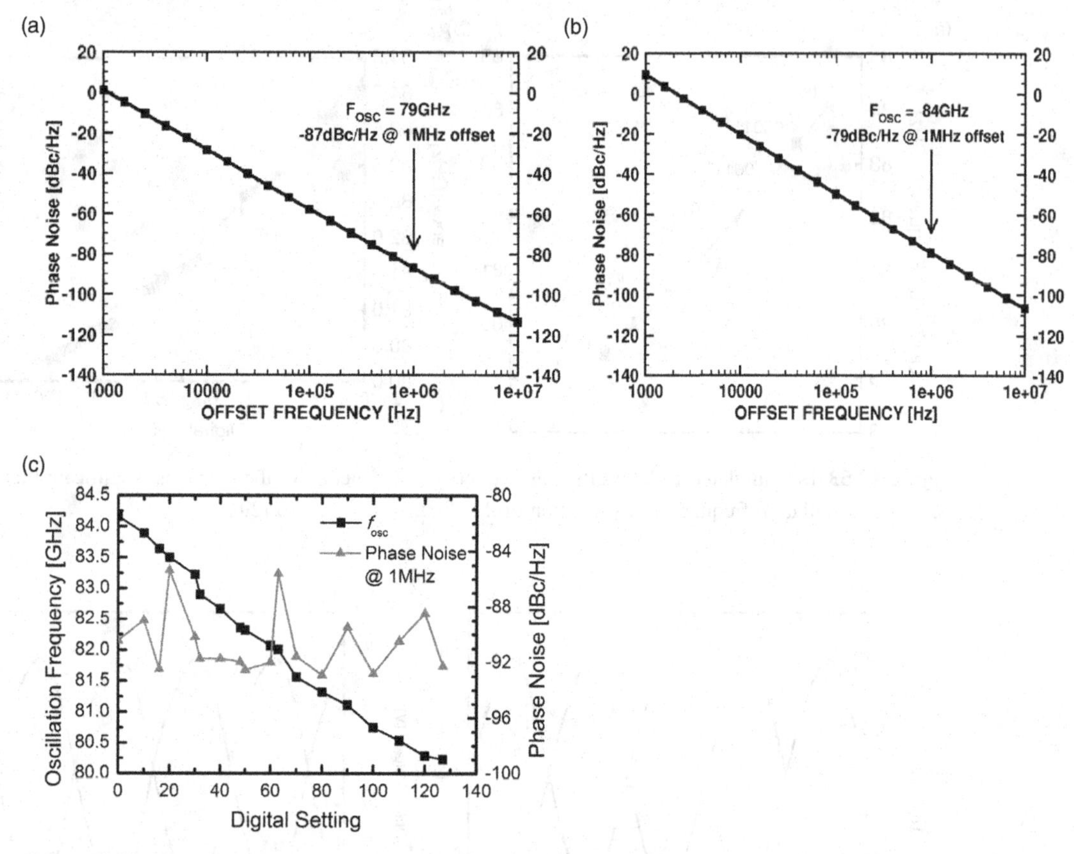

Figure 10.55 Simulated phase noise of the DCO at (a) 79GHz and (b) at 84GHz [29] (c) Measured phase noise and tuning curve for different frequency control words

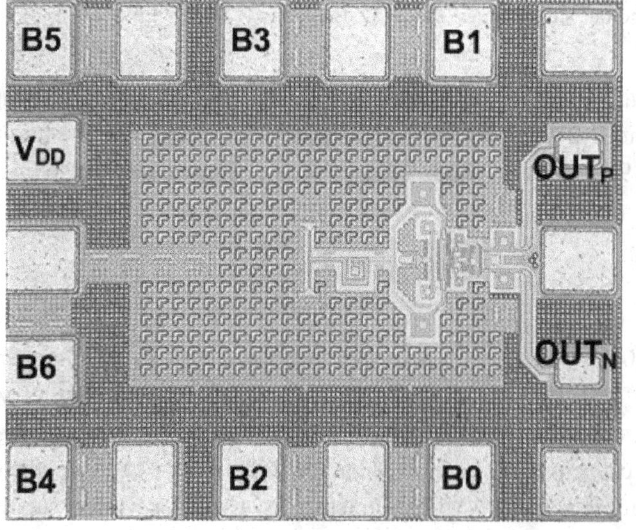

Figure 10.56 Die photograph of 80GHz DCO [29]

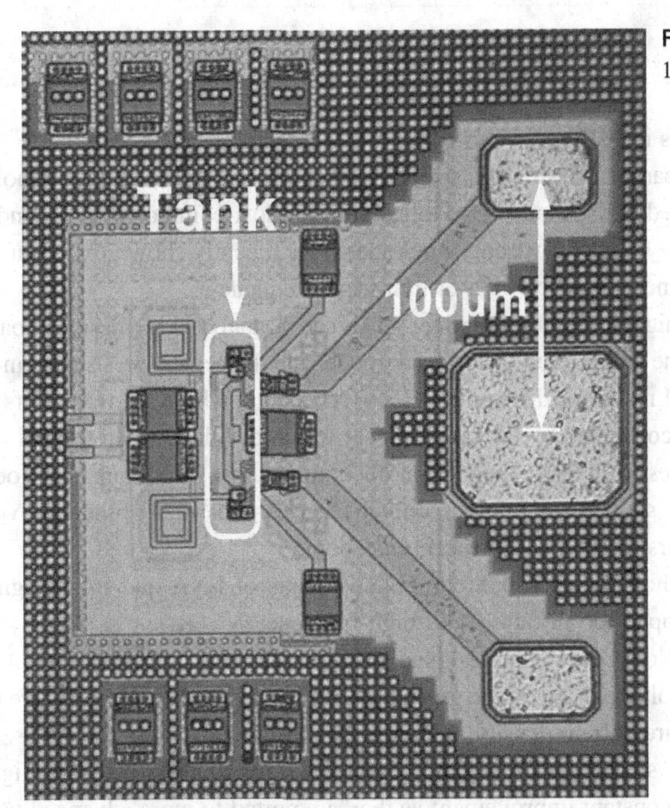

Figure 10.57 Die photograph of a 120GHz Colpitts VCO [31]

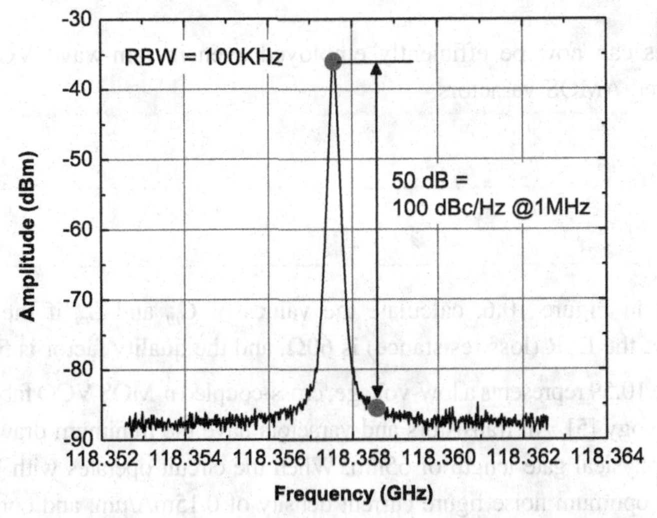

Figure 10.58 Measured phase noise of a 120GHz SiGe HBT Colpitts VCO [31]

voltage, which helps to suppress the maximum in phase noise at the center of the tuning range seen in other VCOs [21]. The varactor layouts are optimized for high-Q, as described earlier. Wherever possible, spiral inductors are used in place of transmission lines to achieve a compact layout, as shown in the die photograph of Figure 10.57. The circuit consumes 72mW from 1.8V supply and has a core size of 0.1mm × 0.1mm [31]. The measured phase noise spectrum is shown in Figure 10.58 with a value of −101.3dBc/Hz at 1MHz offset.

Summary

- VCOs are critical blocks in both radio and optical fiber systems.
- High-frequency VCOs can be classified in selective feedback and cross-coupled topologies.
- Differential cross-coupled and Colpitts topologies are the most common CMOS and HBT VCOs, but Clapp, Hartley, and Armstrong topologies have recently shown excellent potential for integration in nanoscale silicon ICs at mm-wave frequencies.
- Although operating in a highly non-linear regime, VCOs can be analyzed and designed based on two linearized models: the amplifier-with-feedback model and the negative resistance model.
- An intuitive large signal model of differential HF VCOs is that where the transistors act as commutating switches, controlled by the tank voltage which develops when the tail current flows in and out of the resonant tank. Noise from the commutating transistors is periodically injected into the tank, as short current pulses, only at the zero-crossings of the tank voltage, when the VCO transistors are at their quiescent bias point.
- Algorithmic design methodologies which combine elements of LNA and PA design techniques have been developed for all families of high-frequency VCOs.
- Irrespective of their topology, optimally designed VCOs employ the highest resonator Q and the maximum allowed voltage swing in the given technology, while the transistors are biased at the optimal noise figure current density at the oscillation frequency.
- All VCO topologies are scalable in frequency beyond at least 200GHz and their figure of merit has demonstrated constant improvement as they are ported to more advanced technology nodes.
- Digital tuning techniques can now be efficiently employed even in mm-wave VCOs by relying on binary-weighted AMOS varactors.

Problems

(1) For the quartz crystal in Figure 10.6, calculate the values of C_m and L_m if the series resonance is at 57MHz, the ESR (loss resistance) is 60Ω, and the quality factor is 50,000.

(2) The schematic in Figure 10.59 represents a low-voltage, cross-coupled n-MOS VCO fabricated in 90nm CMOS technology [5]. All transistors and varactors have the minimum drawn gate length of 90nm and a physical gate length of 65nm. When the circuit operates with $V_{DD} =$ 0.6V it is biased at the optimum noise figure current density of 0.15mA/μm, and consumes 9mA, less power than a CMOS inverter cross-coupled version while still allowing for a tank swing of 1.2V_{pp}. Knowing that $C'_{gs} = C'_{db} =$ 1fF/μm, $C'_{gd} =$ 0.35fF/μm, $g'_m =$ 0.75mS/μm, $g'_o =$ 0.15mS/μm, $L_D =$ 130pH and has a Q of 20, calculate the transistor width and the varactor width for the circuit to oscillate at a minimum frequency of 10.5GHz. What is the maximum oscillation frequency if the varactor capacitance ratio is 2:1? Assume that $C_{VAR} =$ 0.75fF/μm to 1.5fF/μm of gate finger width. Scale this 12GHz design to 24GHz.

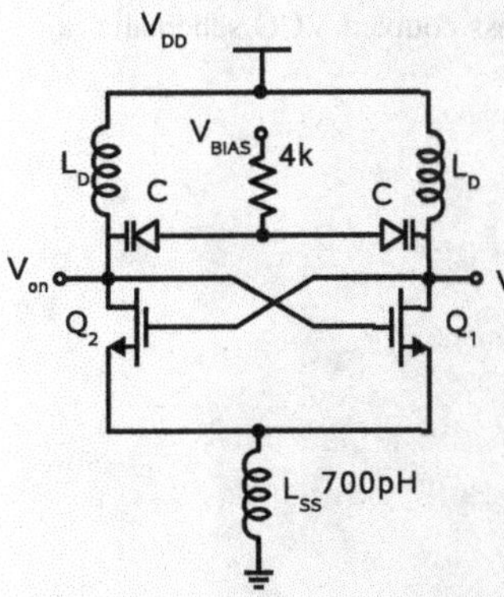

Figure 10.59 12GHz, low-voltage cross-coupled oscillator schematic

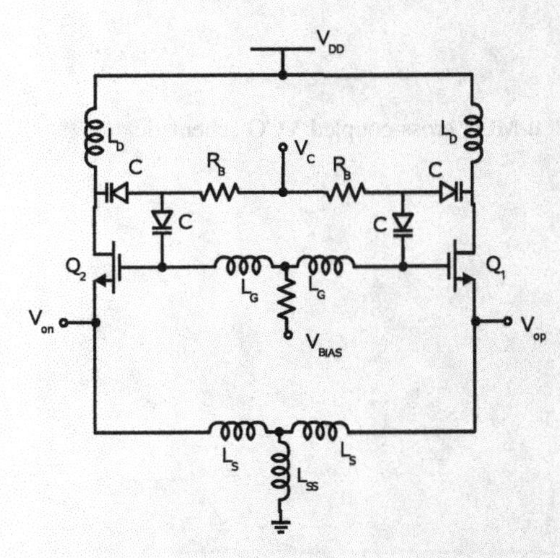

Figure 10.60 Differential Hartley oscillator schematics

(3) The same 90nm CMOS technology is employed to fabricate the differential Hartley oscillator shown in in Figure 10.60. Transistors Q_1 and Q_2 and varactors C, all have 80μm total gate width and minimum gate length.

Find the expression of the oscillation frequency as a function of L_D, L_G, and W. If $f_{OSCMIN} = 17.5$GHz, calculate the value of $L_D = L_G$ and f_{OSCMAX}. Ignore the impact of R_B (virtual open) and $L_S = 20$pH (virtual short) on the oscillation frequency. Comment on the impact of the varactor n-well to substrate capacitance. Scale the VCO from 18GHz to 60GHz.

(4) Design a 65nm CMOS crystal oscillator which operates as a 45MHz stable reference and consumes 1mA from 1V. Assume that the crystal series loss resistance (ESR) is 60Ω and that $g'_m = 0.7$mS/μm. Ignore the parasitic capacitances of the transistor.

(5) For the fundamental frequency Clapp VCO shown in Figure 10.20, draw the differential-mode half circuit and the common-mode half circuit. Calculate the oscillation frequency and verify the oscillation condition in differential mode and in common mode. $C'_{gs} = C'_{db} = 0.7$fF/μm, $C'_{gd} = 0.35$fF/μm, $g'_m = 0.8$mS/μm, $g'_o = 0.125$mS/μm, $L_T = 40$pH. $C_{VAR} = 0.75$fF/μm to 1.5fF/μm of gate finger width. Assume that the transistors are biased at 0.15mA/μm.

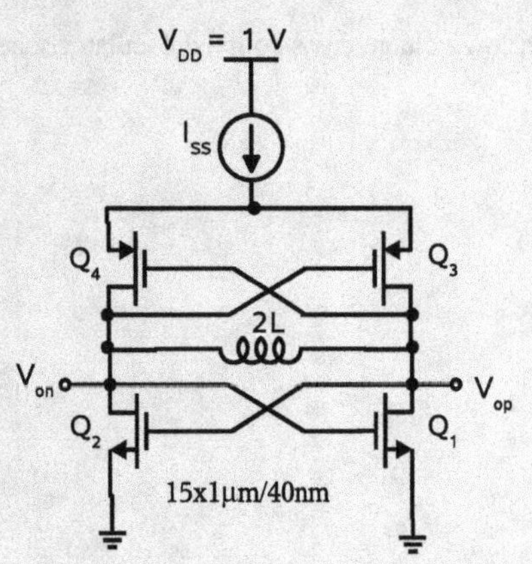

Figure 10.61 CMOS cross-coupled VCO schematic

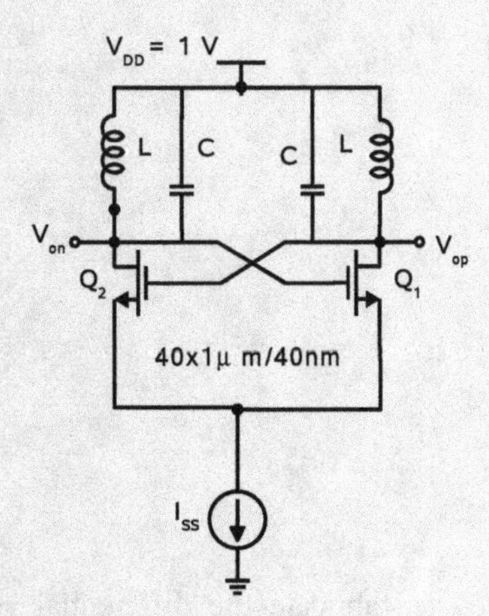

Figure 10.62 n-MOS cross-coupled VCO schematic

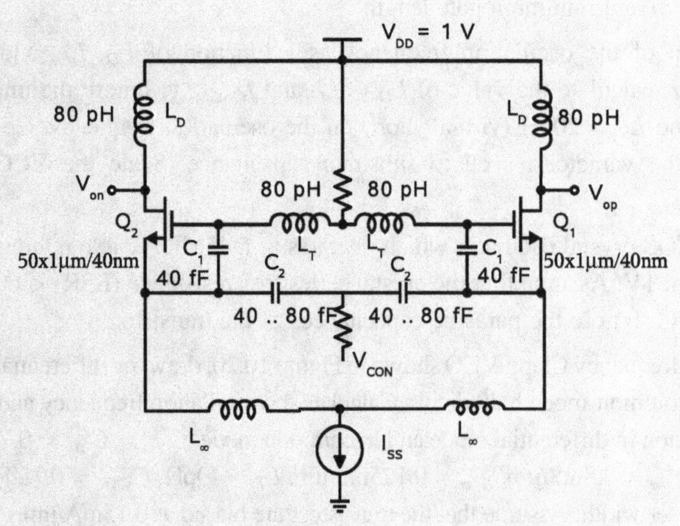

Figure 10.63 n-MOS Colpitts VCO schematic

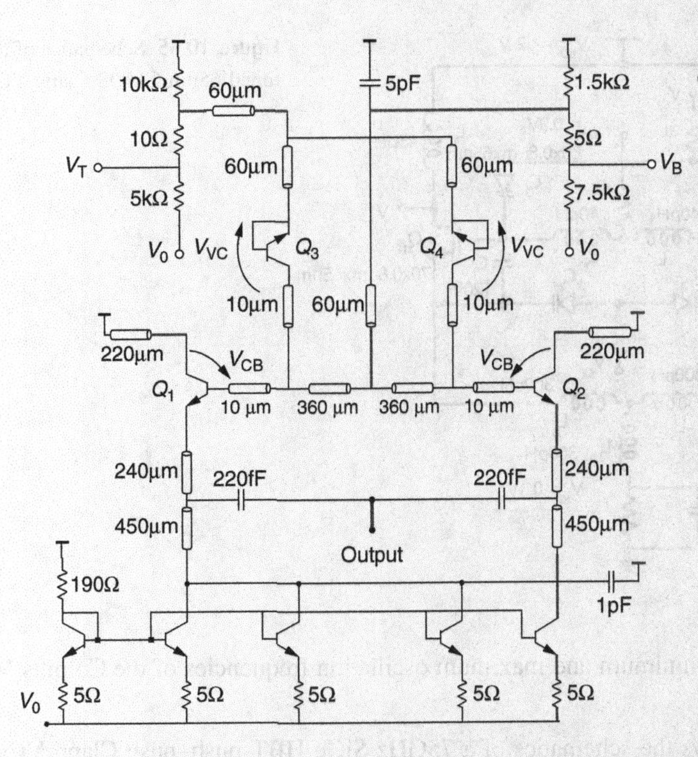

Figure 10.64 Push–push SiGe HBT Clapp VCO schematics [16]

(6) You have been assigned to design 80GHz PLL building blocks in a low-power 45nm CMOS logic process. The first block is a VCO which could be implemented in any of the 3 topologies shown in Figures 10.61, 10.62, and 10.63. In all cases, the supply voltage is 1V and I_{SS} is 3mA, 4.5mA, and 16mA, respectively. The p-MOSFETs are sized to be twice as large as the n-MOSFETs and the n-MOSFETs are biased at the optimum noise current density of 0.15mA/μm when I_{SS} is equally split between the two halves of the circuit. Assume that at $V_{DS} = 0.6$V, $C'_{gs} = 0.65$fF/μm, $C'_{db} = 0.3$fF/μm, $C'_{gd} = 0.38$fF/μm, $g'_m = 0.8$mS/μm, $g'_o = 0.125$mS/μm.

(a) For the topology in Figure 10.61, find the expression and the value of the negative conductance between nodes V_{op} and V_{on}.

(b) For the VCOs in Figures 10.61, 10.62, and 10.63:

- Derive the expression of the maximum possible oscillation frequency as a function of inductor Q and of the small signal parameters (g'_m, C'_{gs}, C'_{gd}, C'_{sb}, and C'_{db}) of the n-MOSFET if all transistors are biased at the minimum noise current density and if the external capacitors are removed. Assume that the Q of the inductor is constant (e.g. $Q = 10$), irrespective of its size, and ignore the gate resistance and output conductance of transistors. Assume that $g_{mp} = g_{mn}/2$.
- Which one of the three topologies will result in the highest frequency of oscillation?
- What will ultimately limit the oscillation frequency if the gate resistance and output conductance of the transistors are accounted for?

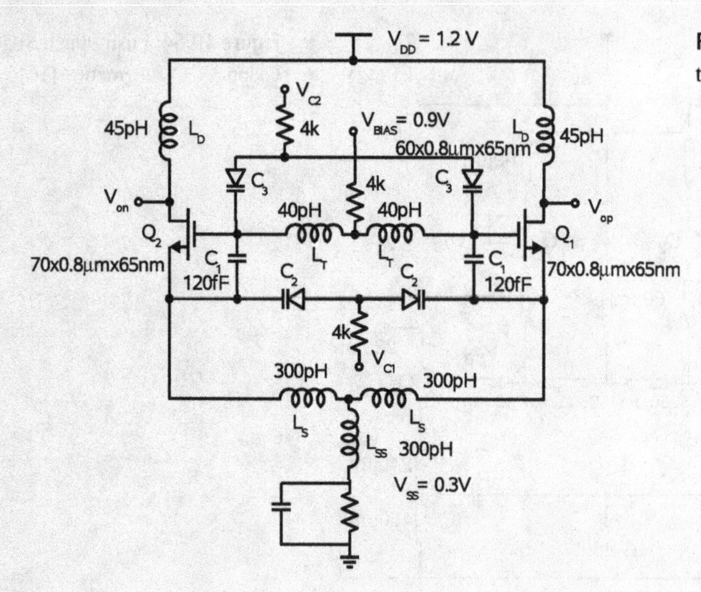

Figure 10.65 Schematics of a doubly tuned 65nm CMOS Clapp VCO

(c) Calculate the minimum and maximum oscillation frequencies of the Colpitts VCO in Figure 10.63.

(7) Figure 10.64 shows the schematics of a 75GHz SiGe HBT push–push Clapp VCO [16]. Draw the equivalent differential and common-mode half circuits. Calculate f_{OSC}. Verify the oscillation condition at 37.5GHz. $I_{C1} = I_{C2} =$ 20mA, $\tau_F =$ 0.3ps, $C_{bc} =$ 50fF, $C_{cs} =$ 25fF, $C_{be} =$ 35fF, $R_B = 3.5\Omega$, $R_E = 1.5\Omega$. $A_{E1} = A_{E2} = 2 \times 0.14\mu m \times 10\mu m$. Assume that the inductance of each t-line is 0.5pH/μm of length.

(8) Figure 10.65 shows a doubly tuned version of the 60GHz Clapp CMOS VCO from Problem 5. Calculate the new frequency tuning range. Assume that varactors C_2 and C_3 are identical.

(9) Design the AMOS varactor bank for the DCO in Figure 10.51 with 8-bit frequency control. Calculate the minimum and maximum oscillation frequencies and the minimum frequency step.

(10) For the 90nm CMOS push–push Clapp VCO in Figure 10.35, assume that the same technology parameters as in Problems 2 and 3 apply. Calculate the tuning range at the second harmonic output. Calculate the negative resistance seen at the gates of the MOSFETs at the fundamental frequency and estimate V_{OSC} if the tank $Q = 10$. Replace the varactors with a 7-bit binary-weighted AMOS varactor bank. What is the size of the least significant bit varactor? What is the frequency tuning resolution?

(11) Design a push–push, 36GHz version of the Hartley VCO in Figure 10.60. Where can the second harmonic signal be collected? Redesign the circuit as a 8-bit DCO.

(12) Consider 65nm n-MOS cross-coupled (Figure 10.61), Clapp (Figure 10.20), Hartley (Figure 10.60) and Armstrong (Figure 10.23) VCOs designed for operation at 60GHz from 1V supply. Which one requires the lowest supply voltage? Which VCO has the largest

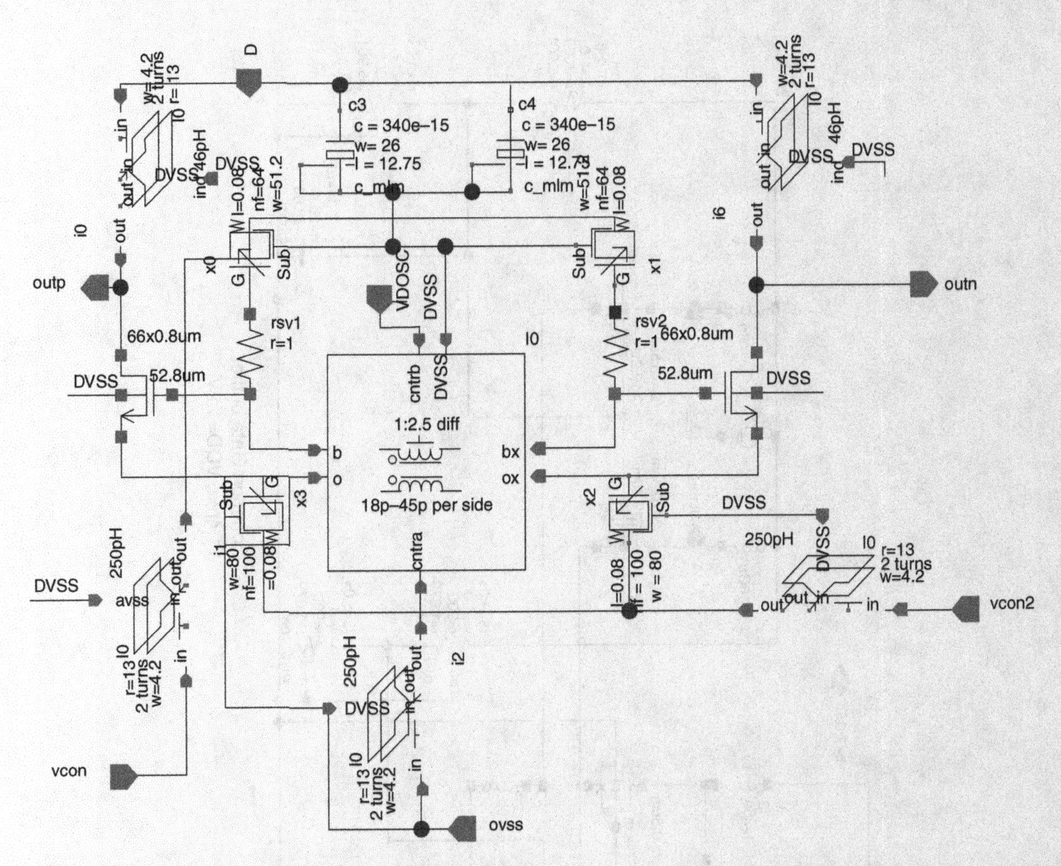

Figure 10.66 Armstrong VCO with dual varactor tuning

tuning range? Assuming classical constant field scaling between the 65nm and 32nm nodes, redesign all four VCOs for operation at 120GHz in 32nm CMOS from 0.8 V supply. Are their any issues with the control voltage excursion? Can the same VCO topologies be designed with bipolar transistors and with 1V supply? What about 0.8V supply?

(13) For the Armstrong VCO in Figure 10.22(b), draw the equivalent small signal half circuit and derive the expression of the oscillation frequency and of the oscillation condition as a function of the transistor transconductance, tank loss resistance, total equivalent tank capacitance C, transformer inductance L, and transformer coupling coefficient k. Assume that the primary and secondary inductances are identical.

(14) Figure 10.66 shows the schematics of a 65nm CMOS Armstrong VCO with dual varactor tuning. Explain the role of each component and estimate the tuning range assuming that the varactor capacitance varies between 0.8fF/μm and 1.6fF/μm.

(15) Figures 10.68 and 10.69 show the schematics of 60GHz Armstrong and Hartley 8-bit DCOs, respectively. The digital varactor bank schematic and the unit varactor pair are reproduced in Figures 10.70 and 10.71, respectively. Estimate the tuning range and frequency resolution in each case.

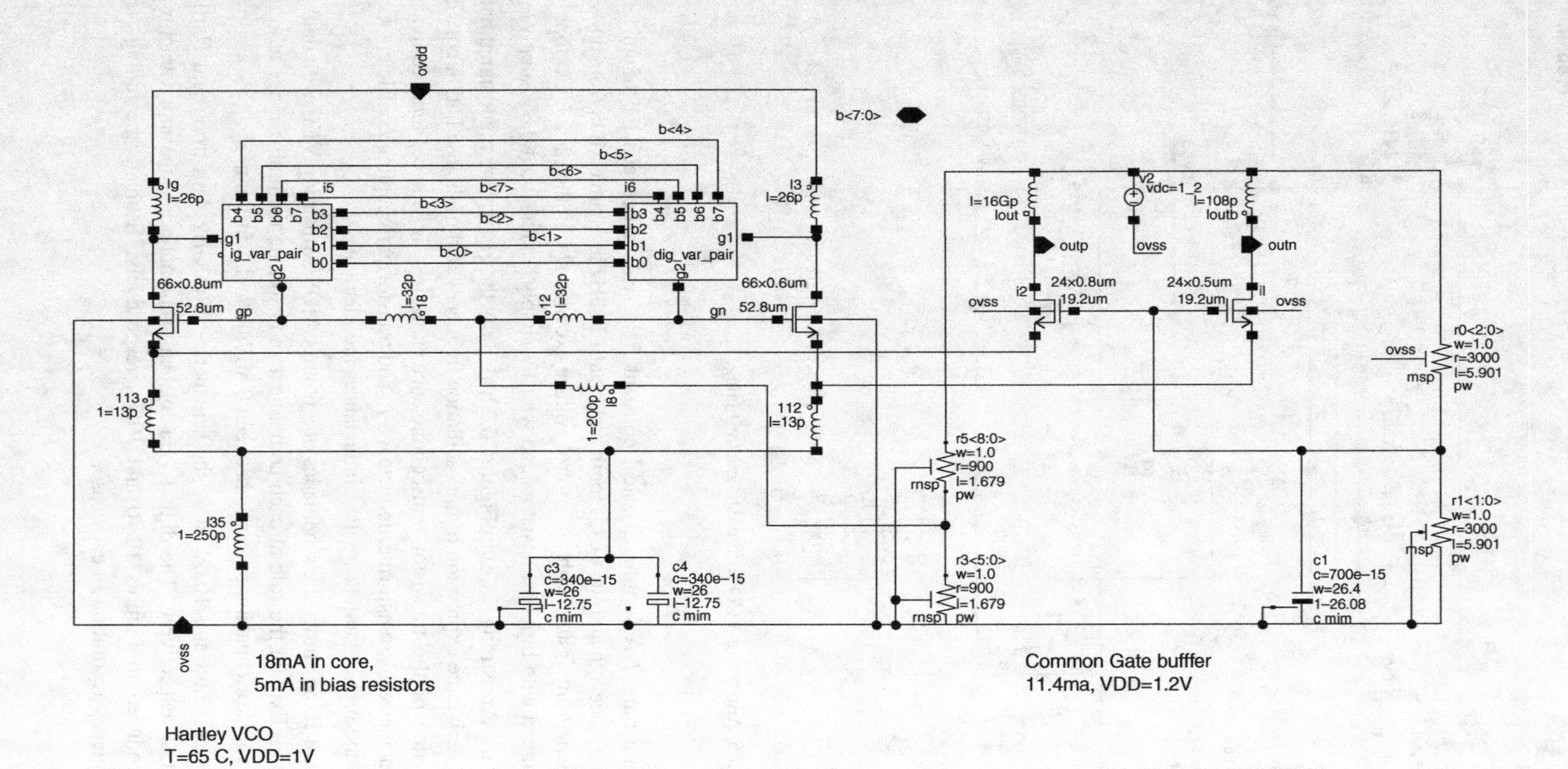

ovdd
b<7:0>
b<4>
b<5>
b<6>
b<7>
b<3>
b<2>
b<1>
b<0>
i5
i6
lg
l=26p
l3
l=26p
ig_var_pair
dig_var_pair
66×0.8um
52.8um
66×0.6um
52.8um
gp
gn
l=32p
l8
l2
l=32p
l13
1=13p
l12
l=13p
1=200p
l8
l35
1=250p
c3
c=340e−15
w=26
l−12.75
c mim
c4
c=340e−15
w=26
l−12.75
c mim
ovss
18mA in core,
5mA in bias resistors
Hartley VCO
T=65 C, VDD=1V
l=16Gp
lout
v2
vdc=1_2
l=108p
loutb
outp
outn
ovss
i2
24×0.8um
19.2um
24×0.5um
19.2um
il
r0<2:0>
w=1.0
r=3000
l=5.901
pw
msp
r5<8:0>
w=1.0
r=900
l=1.679
pw
rnsp
r3<5:0>
w=1.0
r=900
l=1.679
pw
r1<1:0>
w=1.0
r=3000
l=5.901
pw
c1
c=700e−15
w=26.4
1−26.08
c mim
Common Gate bufffer
11.4ma, VDD=1.2V

Figure 10.67

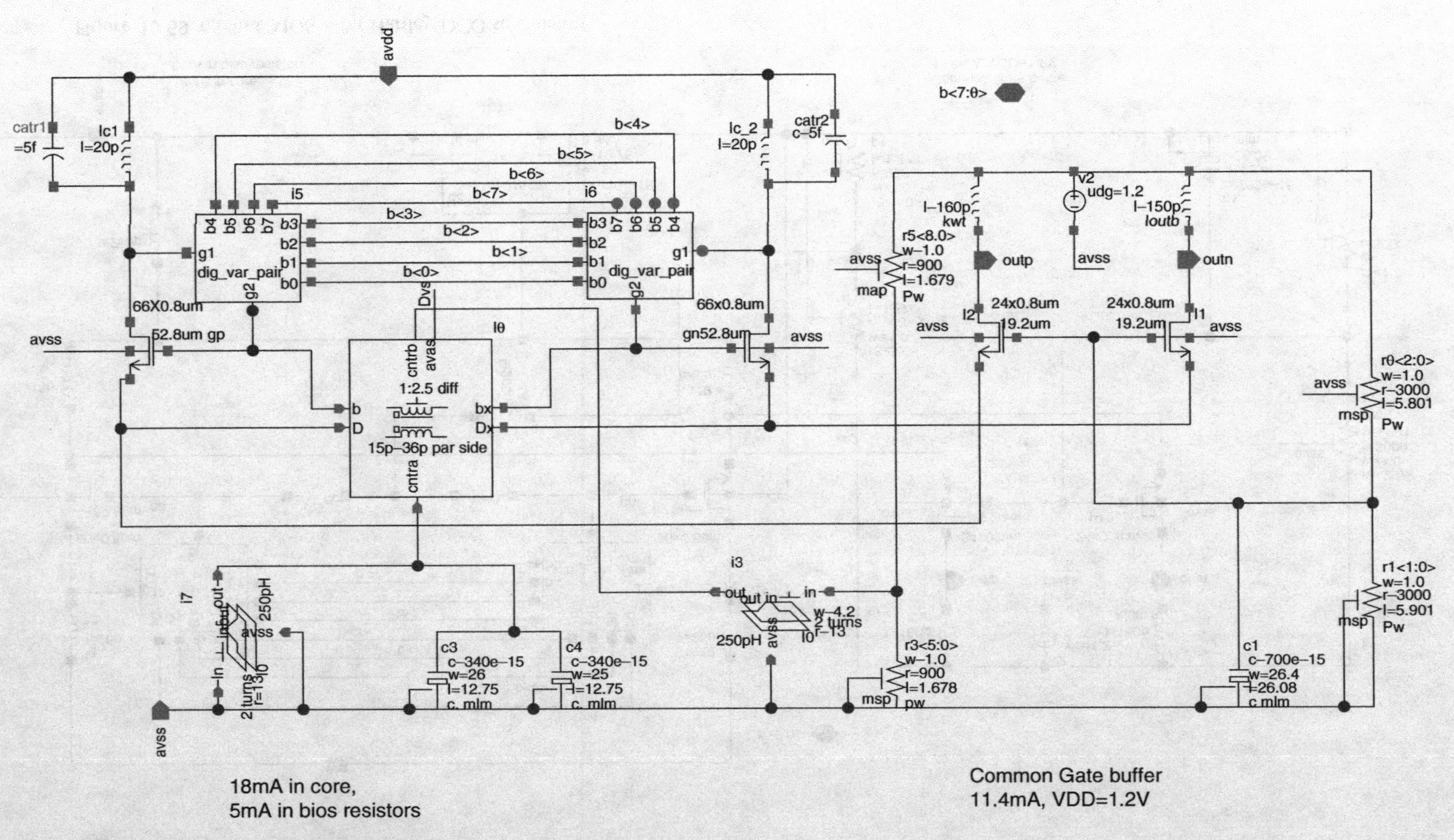

Figure 10.68 65nm CMOS 8-bit Armstrong DCO schematics

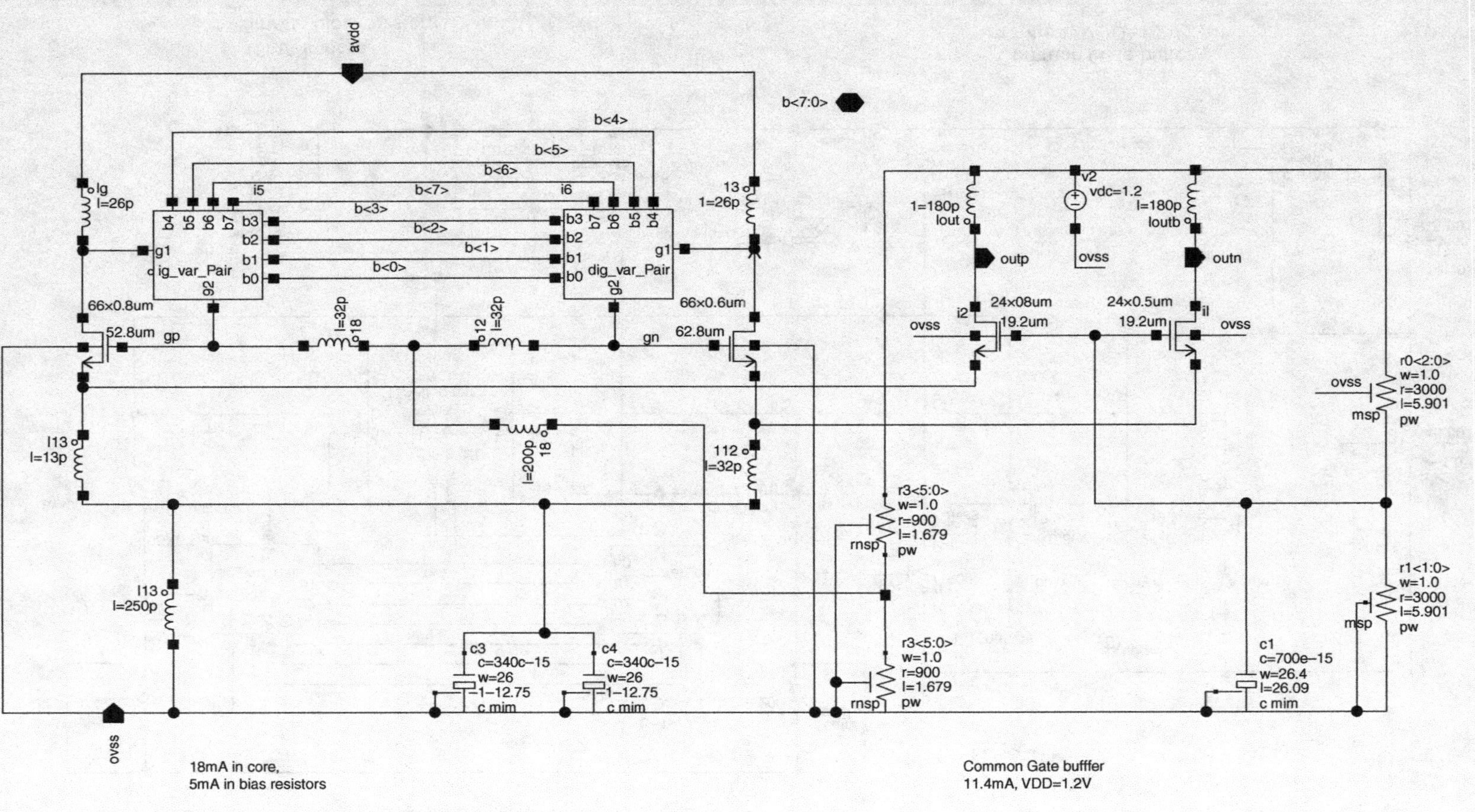

Figure 10.69 65nm CMOS 8-bit Hartley DCO schematics

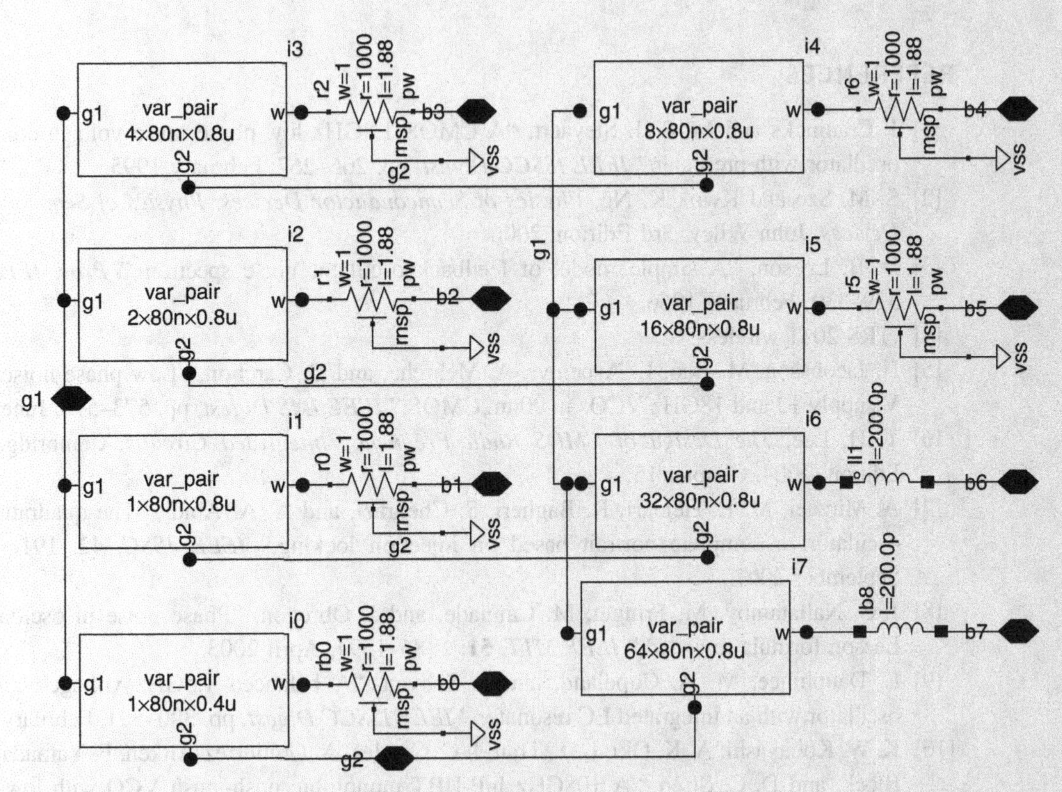

Figure 10.70 65nm CMOS 8-bit digital varactor bank

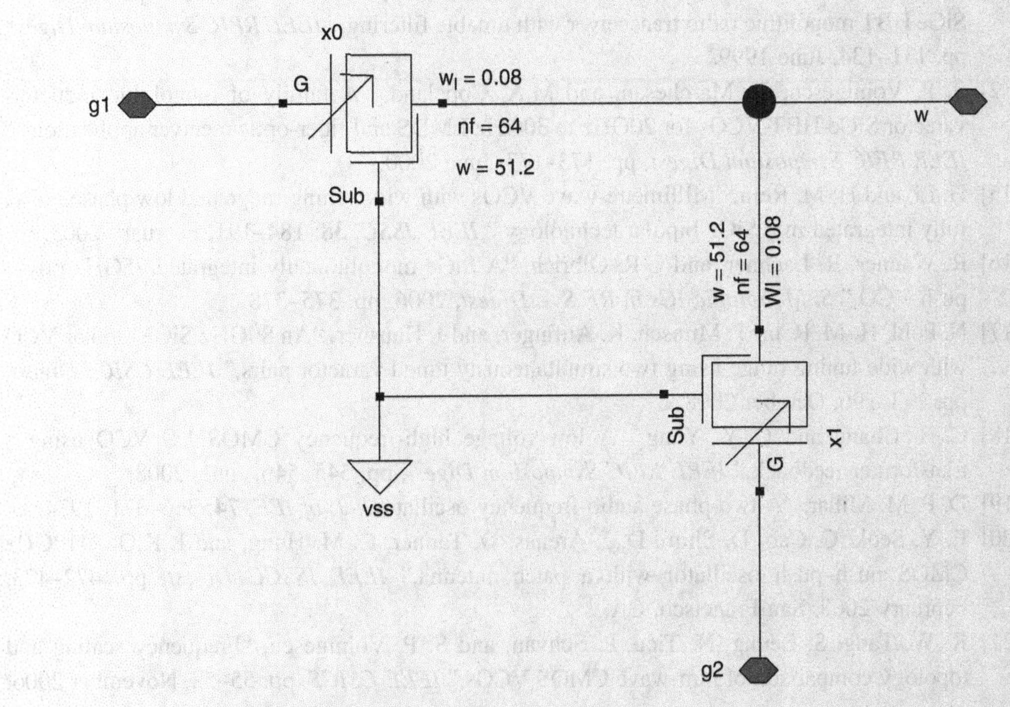

Figure 10.71 65nm CMOS digital varactor pair

REFERENCES

[1] J. Craninckx and M. S. J. Steyaert, "A CMOS 1.8GHz low-phase noise voltage controlled oscillator with prescaler," *IEEE ISSCC Digest*, pp. 266–267, February 1995.

[2] S. M. Sze and Kwok K. Ng, *Physics of Semiconductor Devices, Physics of Semiconductor Devices*, John Wiley, 3rd Edition, 2006.

[3] D. B. Leeson, "A simple model of feedback oscillator noise spectrum," *Proc. IEEE*, **54**: 329–330, February 1966.

[4] ITRS 2011 wireless.

[5] H. Jacobsson, M. Bao, L. Aspemyr, A. Mehrcha, and G. Carchon, "Low phase noise sub-1 V supply 12 and 18GHz VCOs in 90nm CMOS," *IEEE IMS Digest*, pp. 573–576, June 2006.

[6] T. H. Lee, *The Design of CMOS Radio-Frequency Integrated Circuits*, Cambridge, 2nd Edition, 2004, Chapter 15.

[7] A. Mirzaei, M. E. Heidari, R. Bagheri, S. Chehrazi, and A. A. Abidi, "The quadrature LC oscillator: a complete portrait based on injection locking," *IEEE JSSC*, **42**: 1916–1932, September 2007.

[8] J.-C. Nallatamby, M. Pringet, M. Camiade, and J. Obregon, "Phase noise in oscillators – Leeson formula revisited," *IEEE MTT*, **51**: 1386–1394, April 2003.

[9] L. Dauphinee, M. A. Copeland, and P. Schvan, "A balanced 1.5GHz voltage-controlled oscillator with an integrated LC resonator," *IEEE ISSCC Digest*, pp. 390–391, February 1997.

[10] K. W. Kobayashi, A. K. Oki, L. T. Tran, J. C. Cowles, A. Guitierrez-Aitken, F. Yamada, T. R. Block, and D. C. Streit, "A 108GHz InP-HBT monolithic push–push VCO with low phase noise and wide tuning bandwidth," *IEEE JSSC*, **34**(9): 1225–1232, September 1999.

[11] S. P. Voinigescu, M. A. Copeland, D. Marchesan, P. Popescu, and M. C. Maliepaard, "5GHz SiGe HBT monolithic radio transceiver with tunable filtering," *IEEE RFIC Symposium Digest*, pp. 131–134, June 1999.

[12] S. P. Voinigescu, D. Marchesan, and M.A. Copeland, "A family of monolithic inductor-varactor SiGe-HBT VCOs for 20GHz to 30GHz LMDS and fiber-optic receiver applications," *IEEE RFIC Symposium Digest*, pp. 173–177, June 2000.

[15] H. Li, and H.-M. Rein, "Millimetre-wave VCOs with wide tuning range and low phase noise, fully integrated in a SiGe bipolar technology," *IEEE JSSC*, **38**: 184–191, February 2003.

[16] R. Wanner, R. Lachner, and G.R. Olbrich, "A SiGe monolithically integrated 75GHz push–push VCO," *Si Monolithic ICs in RF Sys. Digest*, 2006, pp. 375–378.

[17] N. Pohl, H.-M. Rein, T. Munsch, K. Aufinger, and J. Hausner, "An 80GHz SiGe bipolar VCO with wide tuning range using two simultaneously tuned varactor pairs," *IEEE CSICS Digest*, pp. 193–196, October 2008.

[18] C.-H. Chang and C.-Y. Yang, "A low-voltage high-frequency CMOS LC VCO using a transformer feedback," *IEEE RFIC Symposium Digest*, pp. 545–546, June. 2008.

[19] D. P. M. Millar, "A two-phase audio-frequency oscillator," *J. of IEE*, **74**: 366–371, 1934.

[20] E. Y. Seok, C. Cao, D. Shim, D. J. Arenas, D. Tanner, C.-M. Hung, and K.K.O, "410GHz CMOS push–push oscillator with a patch antenna," *IEEE ISSCC Digest*, pp. 472–473, February 2008, San Francisco, CA.

[21] K. W. Tang, S. Leung, N. Tieu, P. Schvan, and S. P. Voinigescu, "Frequency scaling and topology comparison of mm-wave CMOS VCOs," *IEEE CSICS*, pp. 55–58, November 2006.

[22] R. B. Staszewski and P. T. Balsara, *All-digital Frequency Synthesizer in Deep-Submicron CMOS*, Chapter 4, John Wiley, 2006.

[23] D.M Pozar, *Microwave and RF Wireless Systems*, Chapter 7, John Wiley & Sons, 2001.

[24] H.P. Forstner, H.D. Wohlmuth, H. Knapp, C. Gamsjäger, J. Böck, T. Meister, and K. Aufinger, "A 19GHz DRO downconverter MMIC for 77GHz automotive radar frontends in a SiGe bipolar production technology," *IEEE BCTM Proc.*, pp. 117–120, October 2008.

[25] C. Marcu, D. Chowdhury, C. Thakkar, L. Kong, M. Tabesh, J. Park, Y. Wang, B. Afshar, A. Gupta, A. Arbabian, S. Gambini, R. Zamani, Ali Niknejad, and E. Alon, "A 90nm CMOS low-power 60GHz transceiver with integrated baseband circuitry," *IEEE ISSCC Digest*, pp. xxx–xx + 1, February 2009.

[26] A. Rofourgaran, J. Rael, M. Rofougaran, and A. Abidi, "A 900MHz CMOS LC-oscillator with quadrature outputs," *IEEE ISSCC Digest*, pp. 392–393, February 1996.

[27] A. Mazzanti, F. Svelto, and P. Andreani, "On the amplitude and phase errors of quadrature LC-tank CMOS oscillators," *IEEE JSSC*, **38**: 1305–1313, June . 2006.

[28] J. Lee and B. Razavi, "40Gb/s clock and data recovery circuit in 0.18μm CMOS technology," *IEEE ISSCC Digest*, pp. 392–393, February 2003.

[29] M. Khanpour, M.Sc. Thesis, University of Toronto, 2008.

[30] S. T. Nicolson, K. H. K Yau, P. Chevalier, A. Chantre, B. Sautreuil, K. A. Tang, and S. P. Voinigescu, "Design and scaling of W-band SiGe BiCMOS VCOs," *IEEE JSSC*, **42**(9): 1821–1833, September 2007.

[31] I. Sarkas, J. Hasch, A. Balteanu, and S. Voinigescu, "A fundamental frequency, single-chip 120GHz SiGe BiCMOS precision distance sensor with above IC antenna operating over several meters," *IEEE MTT*, **60**(3): 795–812, March 2012, in print.

11 High-speed digital logic

Broadband serial "wireline" links today routinely reach data rates of at least 10Gb/s with the highest capacity networks, based on the OC-768 standard, running at serial data rates of 43Gb/s. At the time of writing, summer 2011, 110Gb/s links, typically achieved by aggregating four 28Gb/s serial data streams onto a single fiber, are being introduced. Even in microprocessors or ASICs, I/Os with data rates of 10Gb/s and (soon) 25Gb/s are becoming common. Despite the continued progress in CMOS I/Os, these output data rates exceed the speed capabilities of conventional static CMOS logic I/Os even in the advanced 45nm and 32nm nodes, raising the question of how such high speeds can be realized in digital networks. Instead, current-mode logic (CML), or variations thereof, is most often used in the design of very high-speed serial transmitters and receivers.

So, if CML logic is so fast, why is it only used in these high-speed applications? Why isn't this logic family used in microprocessor designs? In fact at one point it was! However, there are two main drawbacks of CML which make it unsuitable for large-scale integration. First, a CML logic gate occupies more area than its CMOS counterpart, due to the use of resistors as well as non-minimum-size transistors. The second, and more costly, drawback is the static power consumption. Even if each logic gate had a tail current of 100μA (a fairly low value for a very high-speed gate), a microprocessor with 10million gates would draw more than 1000 A. In a CMOS technology with a 1V supply, the power could exceed 1000W! Since static CMOS does not consume static current, it is much better suited for highly integrated digital ICs. Instead, CML finds use only in high-speed applications where conventional CMOS logic cannot meet the performance requirements.

This chapter first briefly introduces typical communication transceivers that employ high-speed CML logic. The basic operation of the CML logic families in CMOS and bipolar implementations are discussed in detail next, along with a design methodology, design examples and techniques for switching speed enhancement such as inductive peaking and inductive broadbanding. Practical implementations of the most common logic gates such as inverters (INVs), selectors, ANDs, XORs, latches, flip-flops, multiplexers and demultiplexers manufactured in CMOS, SiGe HBT and SiGe BiCMOS technologies are described next. The chapter ends with the analysis and design of dividers and prescalers. Examples of GaAs p-HEMT SCFL gates will be provided in Chapter 12.

11.1 SYSTEMS USING HIGH-SPEED LOGIC

The high-speed logic building blocks discussed in this chapter are primarily found in high-speed serial links, such as fiber-optic networks, over the backplane, or even short-range chip-to-chip communication systems. Let us review the block diagram of the 10Gb/s SONET transceiver or serializer-deserializer (SerDes) [1] shown in Figure 11.1, to understand what high-speed digital building blocks are required. In this SerDes, the transmitter, shown in the upper half, receives 16 parallel NRZ data inputs, each at 622Mb/s, synchronizes them in the phase alignment block, and then time-division multiplexes them into a single NRZ output stream with a data rate 16 times higher than that of the individual inputs. The serialized 10Gb/s data are then retimed and sent by the 50Ω driver to modulate the current of a VCSEL or to a large-current laser driver.

The timing of all the signals in the transmitter is controlled by the transmitter PLL block, sometimes known as the clock multiplication unit (CMU). In the receiver, or de-serializer, located in the bottom half of the diagram, the inverse operations occur. The incoming 10Gb/s NRZ serial data stream is first amplified by the input comparator and then passed on to the clock-and-data recovery block, CDR, which extracts the clock signal from the incoming data,

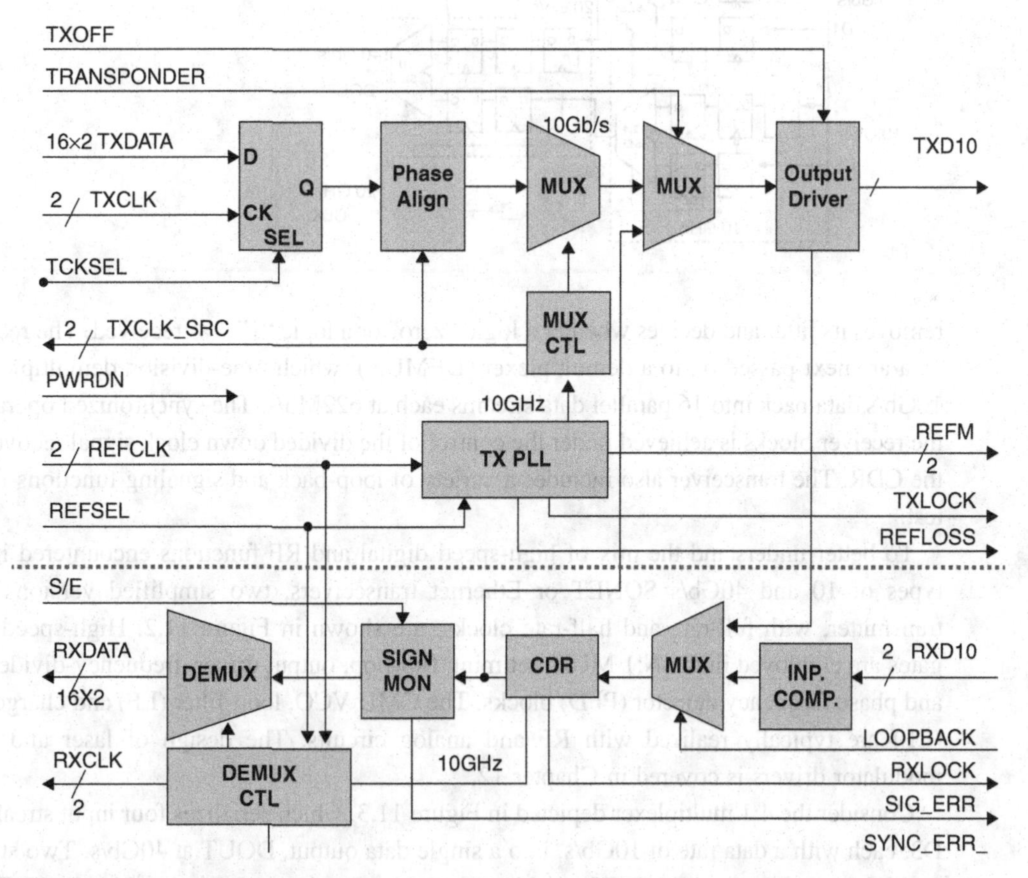

Figure 11.1 Block diagram of a commercial 10 Gbs/ SONET single-chip transceiver [1]

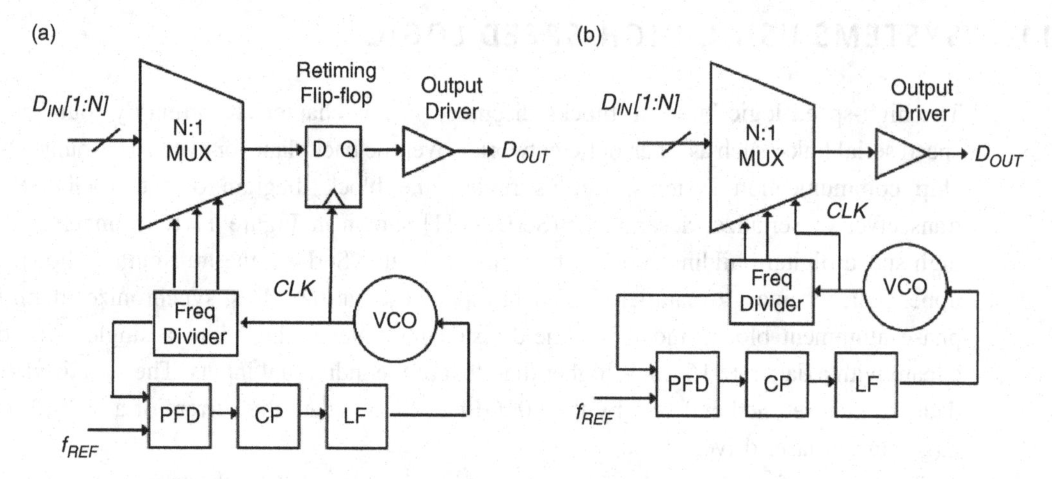

Figure 11.2 Serial transmitter with (a) full-rate and (b) half-rate retiming

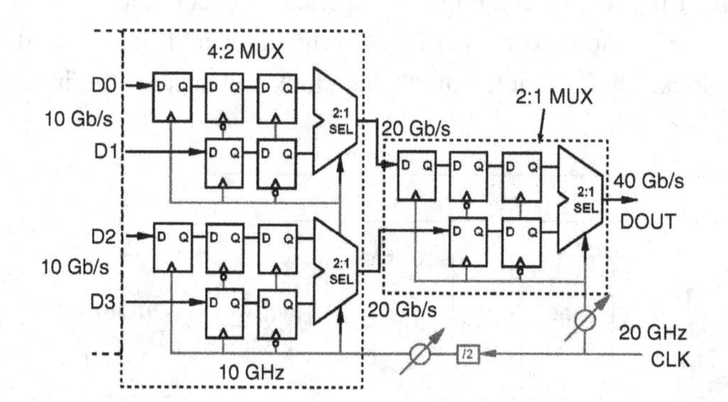

Figure 11.3 4:1 MUX schematic

removes its jitter and decides whether a logic "zero" or a logic "1" was received. The recovered data are next passed on to a demultiplexer (DEMUX), which time-division demultiplexes the 10Gb/s data back into 16 parallel data streams each at 622Mb/s. The synchronized operation of the receiver blocks is achieved under the control of the divided down clock signal recovered by the CDR. The transceiver also includes a variety of loop-back and signaling functions for self-test.

To better understand the mix of high-speed digital and RF functions encountered in these types of 10 and 40Gb/s SONET or Ethernet transceivers, two simplified versions of the transmitter, with full-rate and half-rate clocks, are shown in Figure 11.2. High-speed digital gates are employed in the N:1 MUX, retiming flip-flop, output driver, frequency-divider chain and phase frequency detector (PFD) blocks. The CMU VCO, loop-filter (LF) and charge-pump (CP) are typically realized with RF and analog circuits. The design of laser and optical modulator drivers is covered in Chapter 12.

Consider the 4:1 multiplexer depicted in Figure 11.3, which serializes four input streams D0-D3, each with a data rate of 10Gb/s, into a single data output, DOUT at 40Gb/s. Two stages of multiplexing are required to perform this task. In the first stage, D0 is combined with D1, and

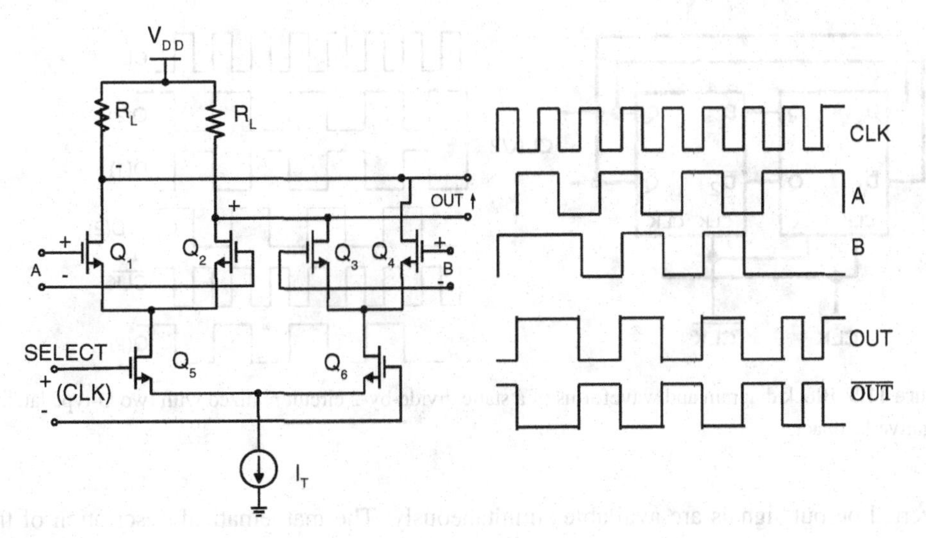

Figure 11.4 MOS-CML 2:1 selector schematics

D2 is combined with D3. Hence, this stage can be referred to as the "4:2 MUX" as it generates two outputs, each at 20Gb/s, from the four 10Gb/s inputs. These two outputs then serve as the inputs for the final 2:1 MUX which produces DOUT at 40Gb/s. Note that the inputs to the 2:1 MUX are delayed by half a clock cycle with respect to each other. This prevents glitches at the output of the 2:1 selector, if the two data inputs were to switch simultaneously [2].

At the heart of each 2:1 multiplexer (MUX) is a selector which chooses one of the two inputs depending on the clock signal. Additionally, the input data must be properly aligned with respect to the clock. Phase alignment between clock and data is accomplished through a series of latches. Two latches are placed along the path of the first input to form a master–slave flip-flop. A third latch is inserted in the path of the second input, creating a master–slave–master flip-flop. This ensures that the selector always samples an input while the data are latched.

Let us examine the operation of the 2:1 selector in more detail. This CML gate selects one of its two inputs A, or B depending on the clock signal (CLK). The latter is a periodic 1010 sequence at twice the data rate of the input data streams. A MOS CML implementation of this selector is depicted in Figure 11.4, along with an idealized timing diagram. The details of its operation and design will be addressed later; here we will provide just an intuitive explanation. For simplicity, we will assume that each MOSFET behaves as an ideal switch, with zero on-resistance and infinite off-impedance.

When the clock signal, CLK, is high, transistor Q_5 is turned on and Q_6 is turned off. As a result, the entire tail current, I_T, is steered through the differential pair Q_1 and Q_2 while the differential pair formed by Q_3 and Q_4 is cut off because no current flows through it. Therefore, only the signal applied at the input of the $Q_1 - Q_2$ pair (A) will be passed to the output.

When *CLK* is low, Q_5 is turned off and Q_6 is turned on, with the entire tail current I_T flowing through Q_6 to the differential pair formed by Q_3 and Q_4, while Q_1 and Q_2 are cut off. As a result, only the data input of the Q_3-Q_4 differential pair, signal B, is transmitted to the output. Note that, depending on how the signal is collected at the output, both the positive and the

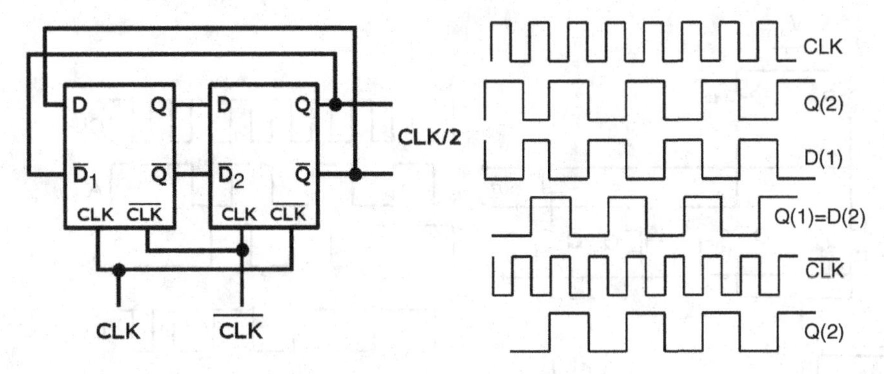

Figure 11.5 Block diagram and waveforms of a static divide-by-2 circuit realized with two D-type latches with negative feedback

inverted output signals are available simultaneously. The mathematical description of the 2:1 selector operation is given by

$$OUT = A{\cdot}CLK + B{\cdot}\overline{CLK}. \tag{11.1}$$

Higher levels of multiplexing can be obtained by connecting multiple 2:1 MUXs in a tree-like network. For example, we can extend the network in Figure 11.3 further to multiplex eight input streams by placing an 8:4 MUX before the 4:1 MUX. In general, 2^N inputs can be time-division multiplexed into a single serial stream using a binary multiplexing tree with N stages.

In the previous example, the 4:2 MUX stage needs a clock at half the frequency of the 2:1 MUX. In broadband transmitters, the CMU generates all the clock signals required in the transmitter. The CMU phase locks a voltage-controlled oscillator to an external reference, as seen in Figure 11.2. In this PLL, the VCO output is divided down to the same frequency as the reference, f_{REF}, through a series of cascaded divide-by-2 frequency dividers. In the process, clock signals are generated at half the VCO frequency, ¼, 1/8, 1/16, and so forth. A single divide-by-2 stage is implemented using a D-type flip-flop with negative feedback, as depicted in Figure 11.5. To understand its principle of operation and waveforms, we must first examine the operation of the D-type latch.

A MOS CML latch and the corresponding timing waveforms are shown in Figure 11.6. Its topology resembles that of the selector, except that the inputs of the second differential pair at the upper level are connected to each other's outputs to form a latching pair. The latch has only one differential data input, D, corresponding to the A input of the 2:1 selector gate. The differential clock input is identical to that of the selector. Note that the output capacitance of the latch is larger than that of the selector since two drains and one gate, rather than just two drains, are connected at each output. This already gives us a hint that, in a given technology, the maximum data rate achieved by the latch is going to be lower than that of the selector.

Again, the operation of the latch can be understood by treating each MOSFET as an ideal switch. However, unlike the selector, the inputs to the second differential pair formed by Q_3 and Q_4 are connected to each other's inverting outputs. When the clock signal is high, Q_6, Q_3, and Q_4 are off and the output will simply follow the input signal with a small delay, τ. When the clock signal is

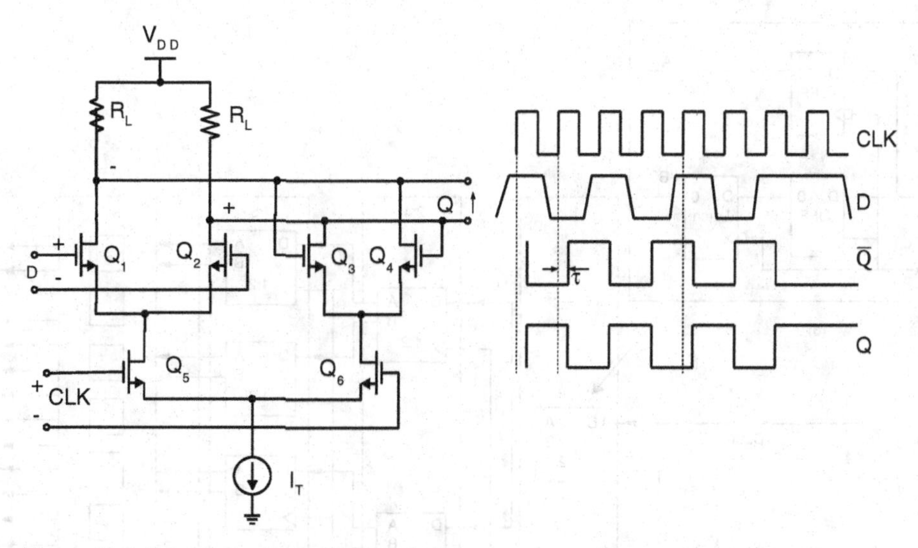

Figure 11.6 Schematic of a MOS-CML latch and corresponding waveforms

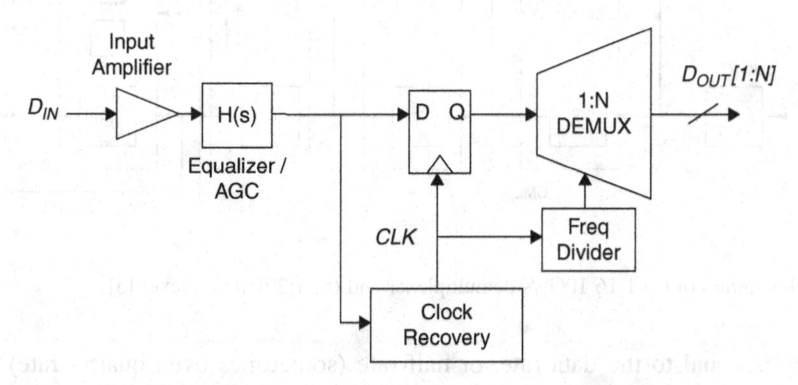

Figure 11.7 Block diagram of a serial receiver with demultiplexer

low, Q_5 and the Q_1–Q_2 differential pair are turned off, and the entire tail current I_T flows through the latching pair to the output side that is already at a low logic level, thus re-enforcing it. Until the next positive clock pulse, the output no longer responds to changes in the data input. In other words, the output remains latched to its previous state when the CLK signal was high.

One important function of the latch that is not be obvious from the idealized waveforms is that, besides aligning the data with the clock, the typically low-jitter (i.e. low-phase noise) clock signal also removes deterministic and random jitter from the data stream and speeds up the rising and falling edges of the data. This is the main reason why in high-performance transmitters for long-haul fiber-optic communications, the data are always re-timed with a full-rate flip-flop before being sent to the output driver. The full-rate latch is the fastest logic block in the transceiver and most challenging to design.

The previous discussion gave us a sneak preview of two critical high-speed digital gates – the selector and the latch. In fact, the latter is widely used in the receiver as well, as shown in Figure 11.7. As in the transmit PLL, the CDR can be full-rate, when the

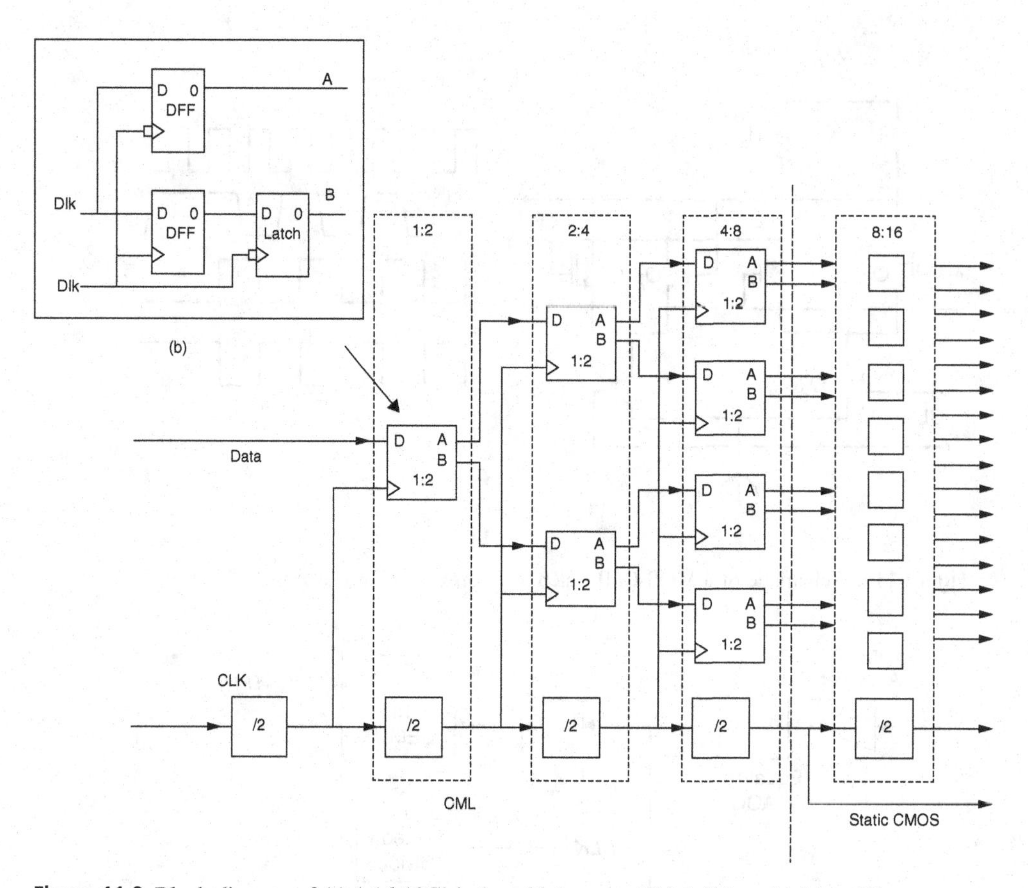

Figure 11.8 Block diagram of (a) 1:16 10Gb/s demultiplexer and (b) 1:2 demultiplexer [3]

clock frequency is equal to the data rate, or half-rate (sometimes even quarter-rate) when the recovered clock frequency is half that of the input data rate. In all cases the CDR includes a VCO.

Figure 11.8 illustrates a 1:16 demultiplexer (DEMUX) that was part of the first 10Gb/s CMOS SerDes [3]. The 1:8 DEMUX is implemented with MOS CML gates, whereas the lower speed 8:16 DEMUX uses standard static CMOS logic gates. The inset shows the schematic of a 1:2 DEMUX stage which, somewhat similar to the 2:1 MUX, consists of two (output) paths driven in parallel by the data input and in anti-phase by the clock input. One output path has a D-type flip-flop while the other consists of a D-type flip-flop followed by a D-type latch. This arrangement ensures that the A and B outputs are delayed by half a clock cycle with respect to each other. The input data are sampled using both edges of a half-rate clock (e,g. a 5GHz clock in a 10Gb/s system).

In this section, we have only taken a quick glance at the more traditional NRZ SerDes family. However, even the latest DAC- and ADC-based 110Gb/s transmitter and receiver system architecture introduced in Chapter 2 for fiber-optic transceivers with 16QAM and OFDM modulation schemes, high-speed latches and inverters are required at least in the transmitter. These are implemented using CML and/or ECL gates.

11.2 HIGH-SPEED DIGITAL LOGIC FAMILIES

As you may have already noticed, CML and all high-speed logic families derived from it, such as emitter-coupled-logic (ECL) and source-coupled-FET-logic (SCFL), use current-steering differential amplifier topologies in which the transistors switch between the active region and cut off. They consist of differential-pair inverters, differential-cascode inverter, and/or differential emitter-follower or source-follower topologies. Unlike standard CMOS, to prevent degradation of the switching speed, operation in the triode (for FETs) or the saturation (for HBTs) regions is avoided.

To illustrate the basic operation, consider the CML inverters in Figures 11.9(a) and (b) and the associated transfer characteristics in Figure 11.10. If a sufficiently large voltage is applied between the differential inputs, the tail current, I_T, is fully switched to one side of the differential pair, turning one transistor fully on while the other is cut off. For example, if IN_P is high and IN_N is low, Q2 (or M2 in the MOS implementation) conducts all of the tail current, creating a voltage drop across its load resistor R_L. As a result, the voltage at OUT_N will be $V_{CC} - R_L I_T$.

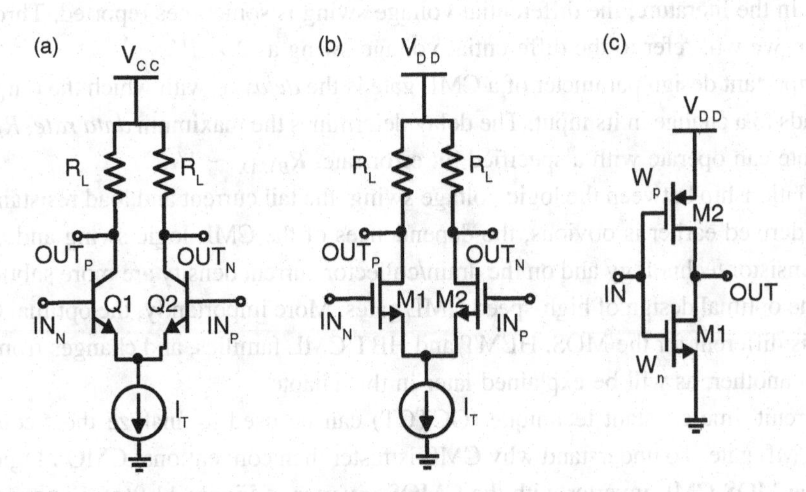

Figure 11.9 Logic inverters in (a) bipolar CML, (b) MOS CML, and (c) static CMOS

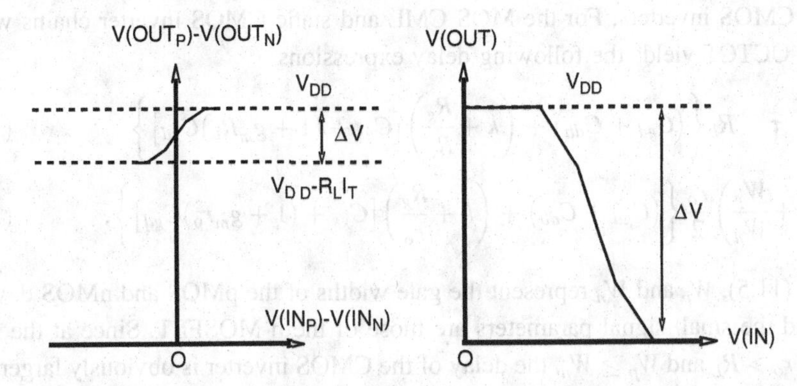

Figure 11.10 DC transfer characteristics $V_{OUT}(V_{IN})$ for (a) CML and (b) static CMOS inverters

Since no current flows through $Q1$(M1), the voltage drop across its load resistance is zero and the voltage at OUT_P equals V_{CC}. When IN_P is low and IN_N is high, no current flows through $Q2$ (M2), and the output voltage OUT_N is equal to the supply voltage while $OUT_N = V_{CC} - R_L I_T$. Hence, the voltages corresponding to the two logic levels are $V_{OH} = V_{DD}$ or V_{CC} (logic level "1") and $V_{OL} = V_{DD} - I_T R_L$ or $V_{CC} - I_T R_L$ (logic level "0"). We can define the *logic swing* (or *voltage swing*) ΔV as

$$\Delta V = I_T \cdot R_L \tag{11.2}$$

and calculate the power consumption of the logic gate as

$$P_D = I_T \cdot V_{DD} \text{ or } P_D = I_T \cdot V_{CC}. \tag{11.3}$$

The voltage swing, ΔV, is an important design parameter in a CML gate. It has to be equal or larger than the minimum voltage required to fully switch the tail current to one side of the differential pair. However, in all CML families, it is typically smaller than $V_{DD}/2$. It is important to point out that CML inverters are fully differential, which makes them more robust in the presence of common-mode perturbations such as supply noise. Note that ΔV is the single-ended voltage swing. In the literature, the differential voltage swing is sometimes reported. Throughout this chapter, we will refer to the differential voltage swing as $2 \times \Delta V$.

The other important design parameter of a CML gate is the *delay*, τ, with which the output of the gate responds to a change in its input. The delay determines the maximum *data rate*, R_{BMAX}, at which the gate can operate with a specified bit error rate: $R_{BMAX} \sim 1/\tau$.

While the relationship between the logic voltage swing, the tail current and load resistance of the CML gate derived earlier is obvious, the dependences of the CML logic swing and of the delay on the transistor technology and on the drain/collector current density are more subtle and are central to the optimal design of high-speed CML gates. More importantly, the optimal CML voltage swing is different for the MOS, HEMT and HBT CML families, and changes from one CMOS node to another, as will be explained later in this chapter.

The open-circuit-time-constant technique (OCTCT) can be used to analyze the frequency response of a CML gate. To understand why CML is faster than conventional CMOS logic, we can compare the MOS CML inverter with the CMOS inverter of Figure 11.9(c). For example, the inverter delay can be estimated from the 3dB frequency of a chain of MOS CML, bipolar CML, or static CMOS inverters. For the MOS CML and static CMOS inverter chains with a fanout of k, the OCTCT yields the following delay expressions

$$\tau = R_L \left\{ (C_{gd} + C_{db}) + \left(k + \frac{R_g}{R_L}\right) [C_{gs} + (1 + g_m R_L) C_{gd}] \right\} \tag{11.4}$$

$$\tau = \left(1 + \frac{W_p}{W_n}\right) \frac{r_o}{2} \left\{ (C_{gd} + C_{db}) + \left(k + \frac{R_g}{r_o}\right) [C_{gs} + (1 + g_m r_o) C_{gd}] \right\}, \tag{11.5}$$

respectively. In (11.5), W_p and W_n represent the gate widths of the pMOS and nMOS devices, respectively, and the small signal parameters are those of the n-MOSFET. Since at the same average current $r_o > R_L$ and $W_p \geq W_n$, the delay of the CMOS inverter is obviously larger than that of the CML inverter.

An intuitive explanation is that for the same peak switching current, the CMOS inverter must swing a larger voltage across a larger load capacitor, hence limiting its speed. Nevertheless, as discussed in Chapters 8 and 12, the switching speed of a nanoscale CMOS inverter can be pushed beyond 40Gb/s by adding transimpedance feedback and/or series inductive peaking. In a given technology node though, it remains lower than the speed of the MOS CML inverter with inductive peaking.

Historically, current-mode logic (CML) and emitter-coupled logic (ECL) gates were first introduced in the 1950s and were implemented using silicon bipolar transistors. Most mainframe computers employed bipolar ECL circuits well into the 1980s and early 1990s. The same topologies were naturally extended in the 1980s to the newly developed III-V and SiGe HBTs. In the 1970s, FET-based CML logic families, most often in the form Source-Coupled-FET Logic, SCFL, evolved from ECL and were extensively used throughout the 1980s with GaAs MESFETs and III-V HEMTs in the Cray family of supercomputers, the fastest of their time. Most recently, since the early 2000s, CMOS CML families have appeared in commercial 10Gb/s fiber-optic transceivers [3].

Irrespective of the transistor technology, all high-speed logic families share several common features:

- They use current-steering differential pair (INV) topologies which switch between the cutoff and the active region of the transistor.
- The power consumption of a CML gate is independent of the operation speed, which unfortunately implies that the gate consumes the same power even when not switching.
- To maximize switching speed, operation in the triode (for FETs) or saturation (for HBTs) regions is avoided.
- The switching speed is determined by the small signal bandwidth of the corresponding differential pair.
- Since switching speed rather than power consumption is the ultimate goal, the current-steering differential pair can be replaced by current-steering topologies with higher bandwidth such as differential cascode, EF/SF-INV, and/or EF/SF-cascode inverter topologies.
- Inductive peaking can be employed to further improve switching speed.
- Because of the use of differential topologies and small logic swings, they have the lowest noise of all logic families, facilitating their integration along sensitive analog, RF, and mm-wave circuits.

The three CML gates discussed so far: the selector, the latch, and the inverter, have a single tail current and their power consumption is simply the product between the tail current and the supply voltage. When emitter-followers or source-followers are employed, the number of tail currents and therefore the power consumption of the logic gate increases. The latter logic families are known as ECL or SCFL, respectively, or as E^2CL when two cascaded emitter-followers are employed.

We can wrap up this introduction on the common features of the CML/ECL logic families by defining a figure of merit to assess the performance and the quality of the design of the basic logic gate. This figure of merit should aim to maximize switching speed first and foremost, otherwise we should use static CMOS, while minimizing power consumption

$$FoM_{gate} = \frac{R_{BMAX}}{P_D}. \quad (11.6)$$

In the next two sections, we will focus on the technology-specific $\Delta V(J_C/J_{DS})$ and $\tau(J_C/J_{DS})$ characteristics of FET and HBT CML gates in order to identify the optimal bias current and component sizes required to maximize the logic gate FoM.

11.2.1 Bipolar current mode and emitter-coupled logic

A large signal analysis of the DC transfer characteristics of a bipolar differential pair with resistive loads, reveals that a differential input voltage swing of four times the thermal voltage, V_T, will completely steer the current from one side to another of the differential pair. This suggests that the CML logic swing could be as low as 100mV. Since V_T is a fundamental physical constant, the bipolar CML swing does not depend on the material system: III-V, silicon, or SiGe. However, this simplified analysis included in most introductory textbooks on analog electronics, ignores the voltage drop on the parasitic emitter resistance, r_E. In new HBT technology nodes, the peak f_T current density, J_{pfT}, increases and the emitter width decreases, thus increasing the parasitic emitter resistance. Together, the higher r_E and higher J_{pfT} raise the minimum ΔV. Hence, for complete switching of the tail current, I_T, the logic swing of bipolar current-steering pairs must be corrected to

$$\Delta V \geq 4V_T + I_T \cdot r_E \quad (11.7)$$

Even larger swing is required at higher temperatures due to the increased thermal voltage and emitter resistance. For example, in a 130nm SiGe HBT or 130nm InP HBT process with an emitter resistance times emitter area of $3\Omega \times \mu m^2$ and J_{pfT} of $12mA/\mu m^2$, the minimum required CML voltage swing at 125°C becomes 186mV when the maximum current flowing through the transistor reaches 1.5 J_{pfT}.

Typically, in bipolar CML gates ΔV is chosen to be between 250 and 300mV. This provides enough margin to account for the impact of process variation on the value of the load resistance. Larger voltage swing should be avoided, as it will either raise the power consumption due to the need for higher tail current, or, according to (11.4), increase the gate delay due to the larger load resistance.

Aside from the logic swing, there are three basic parameters which must be chosen when designing a bipolar CML inverter: the bias (or tail) current I_T, the load resistance R_L, and the emitter length of the bipolar transistor, l_E. In fact, all of these parameters are related to each other. As seen from equation (11.2), the product of I_T and R_L determines the single-ended logic swing ΔV, which is already set to about 300mV for the reasons described above.

Assuming that the tail current is known, R_L follows immediately from $\Delta V/I_T$, but how do we select the transistor emitter length? To maximize speed, the transistor should be biased in the active region and at a collector current density where the gate delay is minimized. Once we identify that optimal current density, $J_{C\tau}$, the emitter length is obtained as $l_E = I_T/J_{C\tau}$.

How do we determine that optimal collector current density?

Let us apply the open-circuit-time-constant technique, to estimate the expression of the inverter delay with a fanout of k as a function of the collector current density. The equation derived in Chapter 5 for the dominant pole of a chain of cascaded common-emitter stages with resistive loads can be recast as a function of the CML voltage swing and tail current by replacing R_L with $\Delta V/I_T$

$$\tau = \Delta V\left[\frac{C_{bc}+C_{cs}+C_{int}}{I_T}+\left(k+\frac{R_b}{R_L}\right)\frac{C_{be}+(1-A_V)C_{bc}}{I_T}\right] \approx \frac{\Delta V}{SL_i}+\frac{\Delta V}{2V_T}\left(k+\frac{R_b}{R_L}\right)\left(\frac{1}{\omega_T}+R_L C_{bc}\right) \tag{11.8}$$

The first term in the square bracket describes the time constant at the output of the driving inverter and is approximately equal to the voltage swing divided by the intrinsic slew rate of the HBT. This slew rate is degraded by any interconnect capacitance, C_{int}, present at the output node. The second term in (11.8) accounts for the input time constant of the inverter and, in the low k limit, is dominated by the R_bC_{bc} component. Although in HBTs the intrinsic and extrinsic components of the base resistance can be reduced by increasing the emitter length or by using multiple emitter stripes connected in parallel, both techniques increase the base–collector capacitance proportionally with the decrease in base resistance. Hence, the R_bC_{bc} product is approximately constant for a given bipolar technology node. The small signal voltage gain, A_V, further increases the input time constant due to the Miller effect. Accounting for the parasitic emitter resistance, the small signal gain can be derived as

$$A_V \approx -g_{meff}R_L = \frac{g_m R_L}{1+g_m r_E} = -\frac{\frac{I_T}{2V_T}}{1+r_E\frac{I_T}{2V_T}}\cdot\frac{\Delta V}{I_T} = -\frac{\Delta V R_L}{2V_T+r_E I_T}. \tag{11.9}$$

In (11.9) the small signal parameters are calculated for the case when the two inputs of the differential pair are balanced, as in an amplifier, and the tail current is equally split between the two HBTs of the differential pair.

If ΔV is in the range of 250mV to 300mV, and if we account for the resistive degeneration due to the parasitic emitter resistance, the small signal gain lies between −2.5 and −3 at room temperature. Equations (11.8) and (11.9) illustrate the importance of minimizing the voltage swing, as was mentioned earlier. If the voltage swing is too large, the gate delay will be degraded. Moreover, a large voltage swing also implies a large small signal gain and hence will exacerbate the Miller effect at the input.

Since R_b and C_{bc} are weak functions of current and ΔV and R_L are fixed, (11.8) indicates that the gate delay dependence on current (density) follows that of $1/f_T$ and $1/SL_i$, both of which have a minimum at J_{pfT}. This suggests that, to minimize the gate delay, we should bias the transistors in the CML gate close to the peak f_T current density. The emitter length can be sized such that the transistor reaches about $1.5J_{pfT}$ when all of the tail current flows through the transistor

$$l_E = \frac{I_T}{1.5 \cdot J_{pfT} \cdot w_E}. \tag{11.10}$$

Sizing according to (11.10) keeps the transistor near the peak $-f_T$ over most of the switching cycle. Sometimes, switching to $2J_{pfT}$ is tolerated [4]. However, larger current densities will actually slow-down the logic gate due to the Kirk effect causing a steep decrease in f_T and f_{MAX} (Chapter 4.3).

Remember that for bipolar transistors, J_{pfT} is specified as current per unit emitter area, accounting for the appearance of the emitter width w_E in the sizing equation. The emitter width, however, is not a design variable and should be kept as small as is supported in the technology to maximize the transistor speed.

By now, maybe you've realized how little work is involved in designing a high-speed bipolar logic gate. It's as simple as selecting a bias current! Once I_T is selected (perhaps it is dictated by the power consumption requirements or by speed), the device geometry and load resistance are set as well. However, the power consumption of the CML gate also depends on the value of the supply voltage, V_{CC}. Usually V_{CC} is given, but how low can it be?

In InP HBT and SiGe HBT CML inverters, the V_{BE} value corresponding to a collector current density of 1.5 J_{pfT} is in the 0.7V to 0.9V range. Thus, the voltage headroom allowed for the current tail becomes V_{CC}-V_{BE}, and should be at least 0.6–0.7V since a V_{CE} of at least 450mV is needed for most SiGe and InP HBTs to remain in the active region, and the current source should have a minimum of 150mV of emitter degeneration. Consequently, even for a CML inverter, which is the simplest logic gate, the supply voltage must be larger than 1.5V.

In a 130nm or finer lithography SiGe BiCMOS process it is possible to reduce V_{CC} to 1.2V if the bipolar current source is replaced by a MOSFET current source. Despite its worse output resistance, the latter can operate reasonably well with a V_{DS} of only 0.25–0.3V. However, as we have already seen, all other CML gates: latches, selectors ANDs, XORs, etc. require at least two transistors stacked up vertically, in addition to the tail current source. As a result, most bipolar CML and ECL families operate with 2.5V or higher supply voltages.

If the input time constant of the gate delay is too large for the intended data rate, the use of a cascode inverter topology can improve the bandwidth by reducing the impact of the Miller capacitance, as discussed in Chapter 5. The schematics of HBT-HBT, FET-HBT and FET-FET cascode INVs are depicted in Figure 11.11. As in the case of the simple INV,

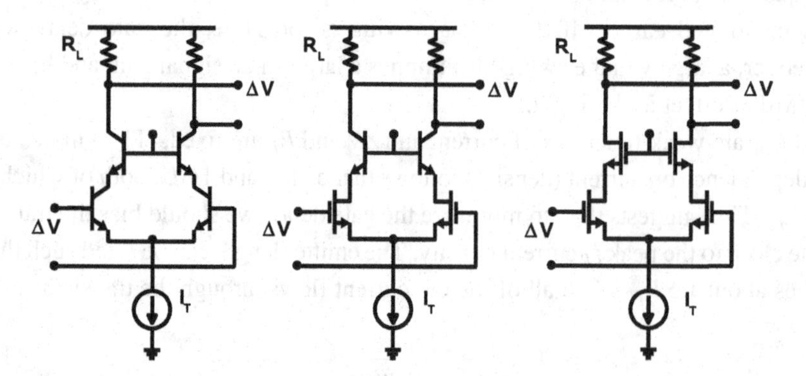

Figure 11.11 Differential cascode inverters available in a SiGe BiCMOS process

we can estimate the bandwidth of a chain of such cascode INVs with a fanout of k by examining the dominant time constant.

For the HBT-HBT cascode, by adding up the time constants at the input, at the output and at the middle node, we obtain an overall delay

$$\tau = \Delta V \frac{C_{be} + C_{cs} + C_{int}}{I_T} + \left(k + \frac{R_b}{R_L}\right) \Delta V \frac{C_{be} + 2C_{bc}}{I_T} + \frac{C_{be} + C_{cs} + C_{bc}}{g_m}. \qquad (11.11)$$

In this amplifier stage, the small signal gain from the base to the collector of the input transistor is −1, which suppresses the Miller effect in the second term. The third term in (11.11) represents the time constant at the emitter of the common-base stage and is approximately equal to $1/\omega_T$ of the HBT, making it negligible in most cases.

This first-order analysis is particularly useful in the most common cases when $k \times R_L$ is much larger than the base resistance R_b of the bipolar transistor.

DESIGN EXAMPLE 11.1

You are designing a selector, like that of Figure 11.12, to be used in the final 2:1 multiplexer of an OC-768 fiber-optic transmitter with a maximum output data rate of 43Gb/s. The selector drives the base of another bipolar transistor biased at 6mA with a single-ended logic swing of 300mV, for example, a CML inverter with a tail current of 12mA. The design is to be implemented in a 160GHz SiGe HBT technology with a peak f_T current density of

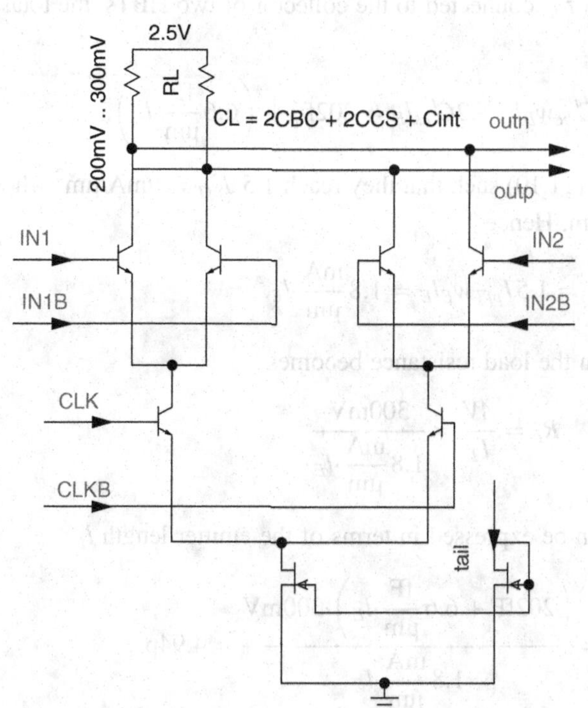

Figure 11.12 Schematic of a bipolar CML selector in a SiGe BiCMOS process

$6\text{mA}/\mu\text{m}^2$. The minimum emitter width is 0.2μm, the base-to-collector capacitance is $11\text{fF}/\mu\text{m}^2$, the collector-to-substrate capacitance is 1.1fF/μm of emitter length, and the forward transit time τ_F is 0.6ps. What is the minimum tail current needed to achieve the required bandwidth? Determine the appropriate emitter lengths for the SiGe HBTs, as well as the load resistance for the selector. Ignore the interconnect capacitance, C_{int}, and the base resistance.

Solution

For reliable operation at 43Gb/s, the minimum bandwidth should be 75% of the bit rate, i.e. 32.25GHz, and hence the output time constant must be lower than

$$\tau = R_L C_{TOT} = \frac{1}{2 \cdot \pi \cdot 32.25\text{GHz}} = 4.94ps.$$

C_{TOT} represents the total capacitance per side at the output node. The load capacitance from the next bipolar transistor can be approximated by $C_{be} + (1 - A_V)\, C_{bc}$. As we have seen, $A_V = -2.5$ to -3. Let us assume $A_V = -3$.

In a switching differential pair sized according to (11.10), we can calculate that, for a tail current of 12mA, the emitter length should be about 6.6μm. If the base-to-collector capacitance per unit emitter area is $11\text{fF}/\mu\text{m}^2$, this results in a C_{bc} of 14.5fF.

$$C_L = C_{be} + 4C_{bc} \approx g_m \tau_F + 4 \cdot 11 \frac{\text{fF}}{\mu\text{m}^2} \cdot 0.2\mu\text{m} \cdot 6.6\mu\text{m} = \frac{6\text{mA}}{25\text{mV}} \cdot 0.6\text{ps} + 58\text{fF} = 202\text{fF}.$$

Note that the Miller capacitance is still less than one third of the total input node capacitance.

Since each output of the selector is connected to the collector of two HBTs, the total output node capacitance becomes

$$C_{TOT} = C_L + 2C'_{bc} w_E l_E + 2C'_{cs} l_E = 202\text{fF} + \left(6.6 \frac{\text{fF}}{\mu\text{m}} \cdot l_E\right).$$

The HBTs are sized according to (11.10) such that they reach $1.5\, J_{pfT} = 9\text{mA}/\mu\text{m}^2$ when all of the tail current flows through them. Hence

$$I_T = 1.5 J_{pfT} w_E l_E = 1.8 \frac{\text{mA}}{\mu\text{m}} \cdot l_E.$$

If the logic swing is 300mV, then the load resistance becomes

$$R_L = \frac{\Delta V}{I_T} = \frac{300\text{mV}}{1.8 \frac{\text{mA}}{\mu\text{m}} \cdot l_E}$$

The output time constant can then be expressed in terms of the emitter length l_E

$$\tau = R_L C_{TOT} = \frac{\left(202\text{fF} + 6.6 \frac{\text{fF}}{\mu\text{m}} \cdot l_E\right) \cdot 300\text{mV}}{1.8 \frac{\text{mA}}{\mu\text{m}} \cdot l_E} = 4.94\text{ps}.$$

This results in an emitter length of 8.7μm. The corresponding tail current is 15.64mA, which then requires a load resistance of 19.2Ω to generate a single-ended voltage swing of 300mV. The implications are that, to drive another gate with 12mA tail current, a selector with 15.64mA tail current is needed. Therefore, to reach speeds of 43Gb/s in this technology using CML gates, only a fanout of 0.75 is allowed.

As we shall see later in this chapter, inductive peaking can be applied to obtain the same bandwidth with lower current consumption and thus increase the fanout.

Another solution to reduce the power consumption is to use a faster SiGe BiCMOS process (if available). For example, in a technology with $\tau_F = 0.3$ps, $J_{pfT} = 12$mA/μm^2, $w_E = 0.11$μm, $C'_{bc} = 12$fF/μm^2 and $C'_{cs} = 2$fF/μm, the calculated transistor size and tail current become $l_E =$ 4μm and 7.92mA, respectively. Thus, by reducing the transit time of the HBT without significantly increasing C_{bc} and C_{cs}, the power consumption is reduced in half at the same data rate, R_B.

Emitter-followers and ECL

As already mentioned, in high-speed logic gates with bipolar transistors, operating the transistor in the saturation region should be avoided because it slows down the switching response. Let us consider the case of the bipolar selector shown in Figure 11.12 and let us assume that the clock and data inputs are driven by bipolar CML inverters. Since the logic high level is equal to the supply voltage, V_{CC}, it is guaranteed that one of the inputs to each differential pair in the selector will always be at V_{CC}. As a result, the V_{CE} of the clock switching pair will equal 0V and hence these bipolar transistors will operate in saturation. This scenario points to the need for inserting emitter-followers to provide DC level shifting to ensure that all bipolar transistors remain biased in the active region. As discussed in Chapter 5, EFs have the added benefit of acting as buffers between cascaded logic cells, improving the bandwidth and the switching speed of the gate. The role of the emitter-followers is to increase the input impedance of the gate and thus reduce the loading on the driving gate. When EFs are used, the logic family is referred to as ECL (emitter-coupled-logic) or E^{2}CL when two cascaded EF stages are employed, as illustrated in Figure 11.13.

As we have seen, bipolar CML gates can be implemented with a 1.2-to-1.8 V supply. However, for ECL and E^{2}CL gates supply voltages of at least 2.5V and 3.3V, respectively, are needed. Figure 11.14 shows detailed schematics of ECL and E^{2}CL latches indicating the DC levels at various nodes.

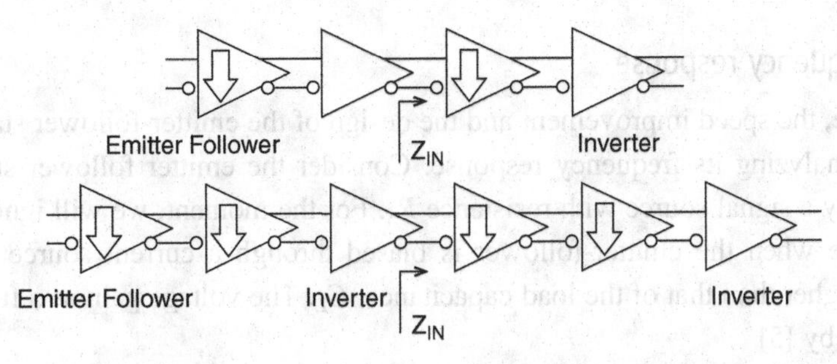

Figure 11.13 Block diagrams illustrating chains of ECL and E^{2}CL inverters

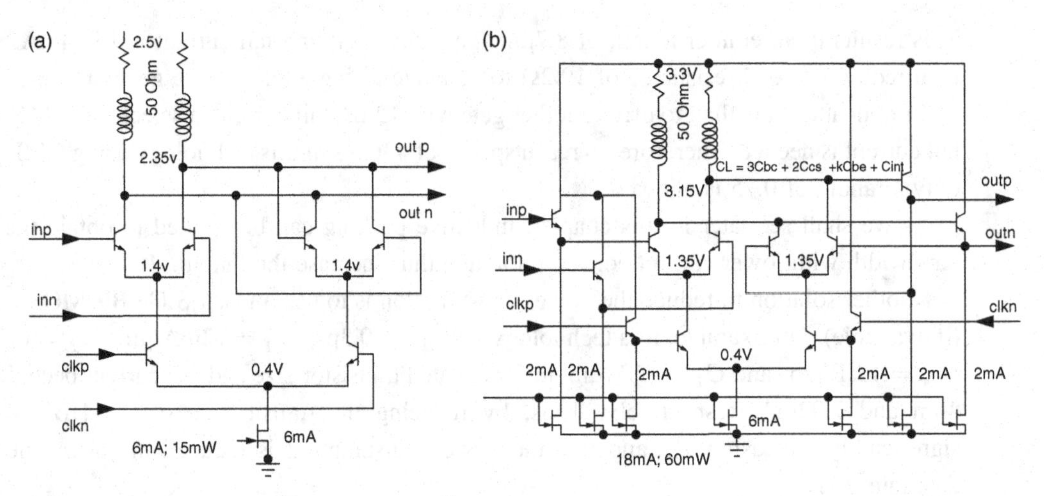

Figure 11.14 Example of (a) ECL and (b) E^2CL latches

In a 130nm or finer lithography SiGe BiCMOS process, level shifting can also be realized with a source-follower stage. If low V_T MOSFETs are available, the latter approach has the added benefit of operation with lower supply voltage, as we shall see later.

Let us now turn our attention to the role that V_{CE} plays in the design and operation of bipolar CML, ECL and E^{2}CL gates. We already know from Chapter 4 that the HBT f_T and f_{MAX} increase with increasing V_{CE} and that they become relatively constant when $V_{CE} >$ 1.2V. Therefore, to maximize switching speed V_{CE} should be set larger than 1.2V, but low enough to avoid breakdown. Alas, this is not possible in bipolar CML logic. As we have seen, one of the problems with bipolar CML is that, in a chain of CML inverters, the V_{CE} of the differential pair transistors varies between $V_{BE} - \Delta V$ and V_{BE}. In other words, the base–collector junction is weakly forward biased at $\Delta V = 300$mV. It is yet another reason why increasing the voltage swing in CML logic beyond 300mV is not advisable. Since the HBT is on the verge of quasi-saturation, its full speed potential is not exploited. This problem is avoided in ECL where emitter-followers are used for level shifting and to increase the V_{CE} of the HBTs in the differential pair.

In the MOS and bipolar CML/ECL latches and selectors discussed so far, the clock signal, which is always faster than the input data signal, is applied at the lower-level differential pair. This scheme is important in maximizing R_B because the path from the clock input to the latch output sees a cascode which has a higher bandwidth than the differential pair on the data path.

Emitter-follower frequency response

As in the inverter case, the speed improvement and the design of the emitter-follower stage are best understood by analyzing its frequency response. Consider the emitter follower stage in Figure 11.15 driven by a signal source with resistance R_S. For the moment, we will ignore R_L. This is often the case when the emitter-follower is biased through a current source whose impedance is much higher than that of the load capacitance, C_L. The voltage gain as a function of frequency is given by [5]

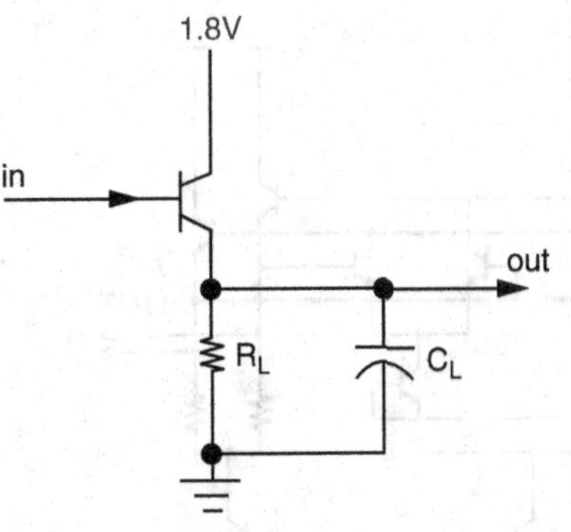

Figure 11.15 Emitter follower schematic used in frequency response analysis

$$A_V(s) = \frac{g_m + sC_{be}}{R_S(C_{be}C_L + C_{be}C_{bc} + C_{bc}C_L)s^2 + (g_m R_S C_{be} + C_L + C_{be})s + g_m} \tag{11.12}$$

Equation (11.12) has two poles, the first being approximated by

$$\tau_1 \approx R_S C_{bc} + \frac{C_{be} + C_L}{g_m}. \tag{11.13}$$

Since its frequency response function has two poles, the EF stage can become potentially unstable. The first symptom of instability in a high-speed digital gate is "ringing" in the large signal response. The origins of this instability can be identified easily by examining the expression of the input impedance looking into the base

$$Z_{in} = R_b + Z_{bc} II [Z_{be} + (1 + g_m Z_{be}) Z_L] \tag{11.14}$$

$$Z_{in} = R_b + \frac{1}{j\omega C_{bc}} II \left[\frac{1}{j\omega C_{be}} + \frac{1}{j\omega C_L} - \frac{g_m}{C_{be}C_L\omega^2}\right]. \tag{11.15}$$

Note the last term, which represents a negative real part of the input impedance. When the EF is connected to the output of another logic stage, one must ensure that the total resistance at this node remains positive such that oscillations do not occur.

The expression of the output impedance is approximated by

$$Z_{out} \approx \frac{R_S C_{be} s + 1}{g_m + C_{be} s} \tag{11.16}$$

which is inductive at frequencies extending from DC up to the f_T of the device.

A technique that has been found to be effective in stabilizing the emitter follower stage is to bias it at collector current densities of 1/3 to ½ J_{pfT}. Obviously this also reduces the bandwidth but is the lesser of two evils. Note that when biasing at $J_{pfT}/2$, the f_T of the HBT is degraded by less than 20% from its peak value.

Another solution which can be used separately or together with the first is to place a resistor R_L in series with the current source or to replace the current source, as in Figure 11.15. The

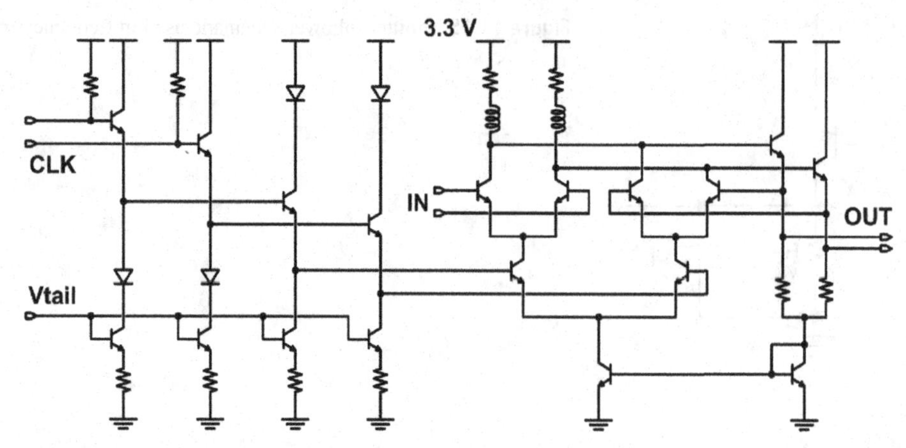

Figure 11.16 E^2CL latch schematic illustrating the use of resistive padding to stabilize emitter-follower stages and of diodes to avoid transistor breakdown [6]

impact of R_L can best be understood by re-examining the expression of the input resistance (11.13), this time including R_L and continuing to ignore C_{bc}

$$R_{in} = R_b + \frac{R_L}{1 + \omega^2 C_L R_L} - \frac{g_m R_L^2 C_L}{C_{be}\left(1 + \omega^2 C_L^2 R_L^2\right)}. \tag{11.17}$$

Equation (11.17) indicates that the presence of R_L adds a positive term and reduces the negative term compared to (11.15).

A word of caution is important at this stage. In ECL and E^2CL gates, care must be taken to avoid transistor breakdown by placing diode-connected transistors at appropriate locations in the schematic, as illustrated in Figure 11.16. Even worse, in advanced SiGe HBT processes with $f_T >$ 300GHz, BV_{CEO} is smaller than 1.6V, and a topology like the latch in Figure 11.14(b), where the collector–emitter voltage of the latching pair is equal to $2V_{BE}$, is no longer feasible.

11.2.2 MOS CML

Understanding the design of MOS CML begins by examining the basic INV gate. As with the emitter width in bipolar CML/ECL, to maximize speed, the gate length of all switching MOSFETs is set to the minimum allowed by the technology. Again, there are three variables at the discretion of the circuit designer: the tail current I_T, the load resistance R_L, and, in this case, the MOSFET gate width, W. Similar to the situation with bipolar designs, the product of the tail current and the load resistance sets the logic swing ΔV. However, unlike bipolar CML where the voltage swing is fixed at 250–300 mV and does not depend on the size of the HBT or the collector current density, in MOS CML the logic swing is a function of the current density switching through the MOSFET differential pair.

A similar question arises as in the bipolar CML design: what is the optimal current density at which the switching speed is maximized? To answer this question, we must first derive the relationship between the gate delay, the voltage swing, and the tail current.

The gate delay of a MOS CML inverter with a fanout of k is given by (11.4), where $A_V = -g_m R_L$ is the small signal gain of the current-steering differential pair and $R_L = \Delta V/I_T$. If $I_T/(2W) <= 0.15$mA/µm, $g_m \approx I_T/V_{EFF}$ and $A_V \approx -\Delta V/V_{EFF}$ where $V_{EFF} = V_{GS} - V_T$ and V_{GS} is the gate–source voltage of the MOSFET in the differential pair when biased at $I_T/2$. Equation (11.4) can be recast as a function of the MOS CML logic swing

$$\tau = \frac{\Delta V}{I_T}\left[C_{gd} + C_{db} + C_{int} + \left(k + \frac{R_g}{R_L}\right)\left[C_{gs} + (1 - A_V)C_{gd}\right]\right]. \quad (11.18)$$

For a given total gate width, the gate resistance can be reduced to less than $0.1R_L$ by increasing the number of gate fingers, N_f. This further simplifies (11.18) to

$$\tau = \frac{\Delta V}{\frac{I_T}{W}}\left[C'_{gd} + C'_{db} + \frac{C_{int}}{W} + k\left[C'_{gs} + (1 - A_V)C'_{gd}\right]\right] \quad (11.19)$$

where ΔV, A_V and the MOSFET capacitances are all functions of the peak drain current density in the gate: I_T/W.

At first glance, as in the case of the bipolar CML inverter, minimizing the voltage swing improves the delay. This voltage swing is determined by the minimum voltage required to fully switch the tail current in a differential pair. In MOS CML designs, the required voltage swing is strongly dependent on the bias current density of the MOSFETs in the differential pair but does not depend on threshold voltage.

In theory, given the tail current, the expression of the minimum voltage swing can be derived from the I-V characteristics of the MOSFET. In practice this is more complicated since the generalized I-V characteristics derived for a nanoscale MOSFET in Chapter 4.2 are not that easy to manipulate. They transition from a square-law at low $V_{GS} - V_T$ and $I_{DS}/W < 0.15$mA/µm, to a linear-law at $I_{DS}/W > 0.3$mA/µm.

Without having to restrict ourselves to a particular I_{DS}–V_{GS} and $g_m - V_{GS}$ model equation, the minimum voltage swing required to fully steer the tail current from one side of the MOS differential pair to another can be defined as

$$\Delta V_{MIN} = 2 \cdot \left[V_{GS}\left(\frac{I_T}{W}\right) - V_{GS}\left(\frac{I_T}{2W}\right)\right]. \quad (11.20)$$

Using (11.20), ΔV_{MIN} (I_T/W) can be determined from the measured transfer characteristics (I_{DS} versus V_{GS}) and plugged into (11.19) along with the measured dependence of A_V, C'_{gd}, C'_{gs} and C'_{db} on I_T/W to determine the gate delay as a function of I_T/W [7].

The experimental dependence of the delay of a nMOS CML inverter with $k = 1$ on I_T/W in three CMOS nodes is reproduced in Figure 11.17. It indicates that there is little speed improvement when I_T/W increases beyond 0.3–0.4 mA/µm, which corresponds to the peak f_T current density in a n-MOSFET, irrespective of the technology node. This result is similar to that obtained for bipolar CML and should not be surprising. In practice, to ensure that the CML gate has a gain $|A_V|$ of at least 1.5, the logic swing is typically set at

$$\Delta V = 1.5\Delta V_{MIN}. \quad (11.21)$$

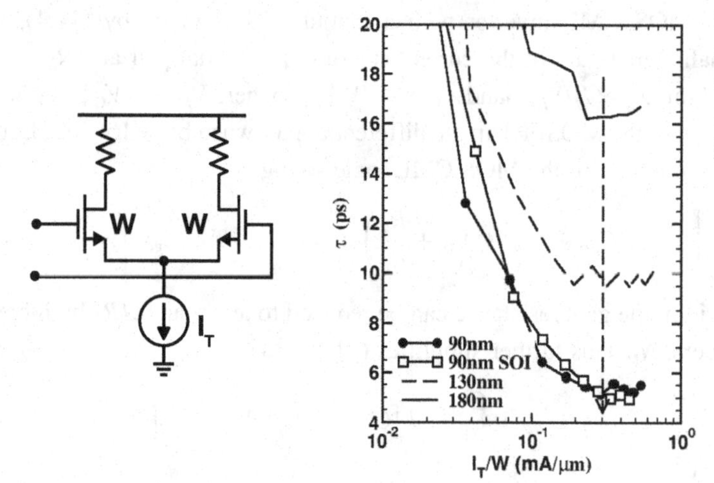

Figure 11.17 Measured CML gate delay as a function of the peak current density in the MOSFET for 180nm, 130nm and 90nm bulk and SOI n-MOSFETs [7]

Consequently, a current-density-centric design philosophy, similar to that which is commonly employed in bipolar CML, is also applicable to MOS CML logic gates. In a current-centric CML design scenario, the gate width, W, of the MOSFET is sized relative to the tail current of the differential pair such that

$$W = \frac{I_T}{J_{pfTMOS}}. \tag{11.22}$$

Increasing ΔV and the current density beyond the values given by (11.21) and (11.22) will reduce the already small A_V without improving the gate delay.

The experiments in Figure 11.17, in conjunction with (11.19), highlight the need to minimize voltage swing in order to improve digital circuit speed. It should be noted that, unlike that of HBT CML, the minimum MOS CML voltage swing given by (11.21) scales with the technology node. For example, in 130nm CMOS, ΔV is approximately 600mV and decreases to about 450mV and 320mV, respectively, in the 90nm and 65nm GP CMOS nodes. However, since constant field scaling has essentially run out of steam at the 45nm node and beyond, it is unlikely that the MOS CML logic swing will decrease below 300mV in the 32nm and 22nm CMOS nodes. While scaling will be advantageous for future CMOS generations because of the continued reduction in C_{gs}/W and C_{db}/W, overcoming the intrinsic slew rate and C_{gd} limitations remains a challenge for MOS CML design. Still, with proper attention, MOS CML can be made to profit from its low input time constant. The contribution of the gate resistance can be diminished through layout techniques. Note however that the gain must still be greater than unity to maintain the regenerative properties of the logic family.

Although the discussion so far has applied to a n-MOSFET CML inverter, a p-MOSFET implementation is also possible. While in the past it was avoided because of the slower carrier transport in p-channel devices, the situation has changed in 65nm, 45nm and 32nm nodes with the p-MOSFET becoming almost as fast as the n-MOSFET. In fact, a ground-referenced MOS

CML logic family based on p-MOSFET current-steering differential pairs with the current tail at the supply has several advantages. These include:

- better power supply rejection,
- ease of DC-coupling to other circuits on a PCB, and
- the possibility of using higher supply voltage and cascode current sources with long gates to generate the tail current, without danger of breaking down the high-speed devices in the CML gate.

Examples of a 56Gb/s p-MOS CML inverter will be discussed in Chapter 12. See also Problem 11.4.

DESIGN EXAMPLE 11.2

Let us compare the 43Gb/s SiGe HBT selector designed earlier to 130nm, 90nm and 65nm CMOS implementations. Again, we will assume that the load is an inverter with 12mA tail current, which could serve as a 50Ω output driver, similar to the ones discussed in Chapter 12. We know from Chapter 4.2 that an n-channel MOSFET has a peak f_T current density of approximately 0.3mA/μm, irrespective of the technology node. Although the transistor capacitances are somewhat smaller in a 65nm GP CMOS process than in a 130nm CMOS process, for simplicity, we will assume average device capacitances per unit gate width for all three technologies of approximately $C_{gs}' = 1.0$fF/μm, $C_{gd}' = 0.5$fF/μm, and $C_{db}' = 1.0$fF/μm. At what technology node can such a selector be realized using MOS CML? Find the minimum tail current to achieve the necessary bandwidth along with the corresponding gate width and load resistance.

Solution

All MOSFETs in the CML gate are sized to reach their peak f_T when all of the tail current is switched through the device. Assuming that the loading inverter was designed for a small signal gain $A_V = -2$, then its MOSFET gate width is

$$\frac{I_T}{J_{pfT}} = W = \frac{12\text{mA}}{0.3\dfrac{\text{mA}}{\mu\text{m}}} = 40\mu\text{m}$$

and the load capacitance becomes

$$C_L = C_{gs} + (1 - A_V)C_{gd} = \frac{12\text{mA}}{0.3\dfrac{\text{mA}}{\mu\text{m}}}\left[1\frac{\text{fF}}{\mu\text{m}} + \left(3\cdot 0.5\frac{\text{fF}}{\mu\text{m}}\right)\right] = 100\text{fF}.$$

When compared with the load capacitance presented by the bipolar transistor from Example 11.1, we note that, for the same current, the MOS input capacitance is half that of the 0.2μm HBT, but the same as that of the HBT in the 110nm process.

An important note here is that the Miller capacitance forms a larger portion of the input capacitance in a MOS CML gate than in a bipolar CML gate.

The total output capacitance C_{TOT} per side is a function of the gate width W of the devices in the selector.

$$C_{TOT} = C_L + 2C'_{gd}W + 2C'_{db}W = 100\text{fF} + \left(3\frac{\text{fF}}{\mu\text{m}}\cdot W\right)$$

Although at this stage neither the tail current nor the size of the MOSFETs are known, the MOSFETs are sized to reach their peak f_T when all of the tail current is switched through the device

$$I_T = J_{pfT}W = 0.3\frac{\text{mA}}{\mu\text{m}}\cdot W.$$

This equation provides us with a relationship between W and I_T that we can use to reduce the number of unknowns in the design process.

The logic swing required for complete switching of the differential pair depends on the technology node. The load resistance is a function of the logic swing.

$$R_L = \frac{\Delta V}{I_T} = \frac{\Delta V}{0.3\frac{\text{mA}}{\mu\text{m}}\cdot W}$$

To achieve the 4.9ps time constant at the output node for 32.25GHz bandwidth, the following condition must be satisfied

$$\tau = R_L C_{TOT} = \frac{\left(100\text{fF} + 3\frac{\text{fF}}{\mu\text{m}}\cdot W\right)\cdot\Delta V}{3\frac{\text{fF}}{\mu\text{m}}\cdot W} = 4.94\text{ps}$$

Rearranging the terms yields an expression for the gate width as a function of the logic swing

$$W = \frac{100\text{fF}\cdot\Delta V}{\left(4.94\text{ps}\cdot 0.3\frac{\text{mA}}{\mu\text{m}}\right) - \left(3\frac{\text{fF}}{\mu\text{m}}\cdot\Delta V\right)}$$

In order for the denominator to remain positive (otherwise a physical solution is not possible), the single-ended logic swing can be at most 500mV. We have already established that in an optimally designed CML gate, 130nm n-channel MOSFETs require a voltage swing of about 600mV, meaning that at least a 90nm technology will be necessary for this design. However if we design for a 450mV swing, we quickly find that the required gate width is about 341μm, and the corresponding tail current becomes 102mA! This high-power consumption can be reduced if we move to a 65nm technology with a logic swing of 320mV. In this case, the gate width is reduced considerably to 62μm, and the current consumption is still large but a more reasonable 18.6mA. The load resistance for the 65nm design would be about 17.2Ω. Note that the fanout is 0.64. In other words, without inductive peaking, even in 65nm CMOS we cannot build a MOS CML selector that can drive itself at 43Gb/s.

If the bipolar designs from the previous example operate from a 2.5-V power supply, the power consumption of the selector with a 5.64mA tail current would be 39mW for the technology with 0.2μm emitter and 19.8mW in the technology with 0.11μm emitter. If the 65nm CMOS design is operated from a 1.1-V supply, its power consumption is 20.46mW, but only the 110nm HBT CML gate can drive itself without inductive peaking at 43Gb/s!

These design examples highlight a subtle difference in how MOS and bipolar technology scaling improves the speed of CML gates. In bipolar designs, the logic swing remains approximately constant across technologies. Scaling reduces the transit time and C_{be} and increases the HBT peak f_T current density, and improves the intrinsic slew rate of the device. On the other hand, to first order, the peak f_T current density in MOS technologies remains constant, as does the device slew rate. Instead, higher speed is obtained by reducing the voltage swing required to completely switch the differential pair. This observation is true so long as CMOS foundries follow constant field scaling rules. However, in recently developed 45nm and 32nm low-power CMOS processes, we already see that constant voltage (rather than constant field) scaling is being applied and the MOSFET capacitances per unit gate width do decrease without increasing g'_m. This suggests that, the CML voltage swing would remain approximately constant in these technologies while the peak f_T current density increases only slightly.

11.3 INDUCTIVE PEAKING

We have seen in Chapter 5 how inductors can be added to a resistively loaded common-emitter or common-source amplifier to extend its bandwidth. By placing an inductor in series with the load resistor (known as shunt peaking), the 3dB bandwidth of the amplifier can be extended by up to 60%. In this situation, the shunt peaking inductor is sized such that

$$L_P = \frac{C_L R_L^2}{3.1} \tag{11.23}$$

where C_L is the total capacitance at the output node of the amplifier. Recall that sizing the inductor according to (11.23) ensures no variation in the group delay of the amplifier over its 3dB bandwidth, which is critical in high-speed digital circuits. The expression for the shunt peaking inductance can be recast as a function of the logic swing and tail current

$$L_P = \frac{C_L}{3.1}\frac{\Delta V^2}{I_T^2}. \tag{11.24}$$

To achieve the maximum possible speed in a given technology, one could develop a very simple design methodology – design the logic gate to be as fast as possible without inductors, and then add shunt (or even shunt–series) inductive peaking in order to further increase the data rate. If applied correctly, this approach will almost certainly yield the fastest possible digital gate in a given technology. At the same time, this design methodology has completely ignored power consumption, which will undoubtedly be high to achieve record-breaking bandwidth.

Let us examine inductive peaking from another point of view – how can we apply shunt peaking to a logic gate in order to reduce the power consumption? Recall the basic design equations for a CML gate without inductive peaking. Assuming the output time constant limits the speed, the 3dB bandwidth of the gate can be approximated as

$$BW_{3\text{dB}} = \frac{1}{2\pi R_L \cdot C_L} = \frac{I_T}{2\pi \Delta V \cdot C_L}. \tag{11.25}$$

To a first-order approximation, the bandwidth increases proportionally to the tail current, which explains why record-breaking performance and high-power consumption go hand-in-hand. However, adding inductive peaking yields a 60% increase in bandwidth.

$$BW_{3dB} = \frac{1.6}{2\pi R_L C_L} = \frac{1.6 I_T}{2\pi \Delta V \cdot C_L}. \tag{11.26}$$

This gives the designer the freedom to aim for a lower nominal bandwidth without inductive peaking in order to reduce the current consumption, and then add inductive peaking to reach the bandwidth specification.

In a given technology BEOL, there is a maximum realizable inductor value, L_{Pmax}, that can be manufactured with an adequate self-resonance frequency for use at a particular data rate. A larger inductor value will result in a self-resonance-frequency that is too low for that particular speed, and the shunt inductor will fail to improve the circuit bandwidth because of its parasitic capacitance. Consequently, a minimum tail current (and hence power consumption) exists that is needed to achieve that speed. Its expression can be derived from (11.24) [8]

$$I_{Tmin} = \Delta V \sqrt{\frac{C_L}{3.1 \cdot L_{Pmax}}}. \tag{11.27}$$

This leads us to the conclusion that one could trade off bias current for inductive peaking in order to save power. The area penalty of this approach can be mitigated by using the three-dimensional (or "stacked") spiral inductor structures discussed in Chapter 4 which allow for a larger inductance value to be generated in a given die area. As a result, a larger number of inductors can be integrated on a single chip to implement low-power broadband transceiver circuits at data rates of 10Gb/s or higher. In fact, all CMOS high-speed logic gates operating above 2Gb/s need some form of inductive peaking to achieve that speed. In bipolar CML, inductive peaking is only necessary at data rates of 40Gb/s or higher.

DESIGN EXAMPLE 11.3

Suppose now we wish to apply shunt peaking to our previously designed HBT and MOS 43Gb/s selectors. How much power can we save in each design, and how large of an inductance value is required? Does the choice of CMOS technology node change as a result of our decision to employ inductors in the design?

Solution

Again, assuming that the output time constant sets the overall bandwidth, the use of inductive peaking allows us to increase the RC time constant at the output knowing that the bandwidth with peaking will be 60% higher.

$$\tau = R_L C_{TOT} = \frac{1.6}{2\pi 32.25\text{GHz}} = 7.9\text{ps}$$

For the bipolar design, the emitter length can be determined as

$$\tau = R_L C_{TOT} = \frac{\left(202\text{fF} + 6.6\frac{\text{fF}}{\mu\text{m}} \cdot l_E\right) \cdot 300\text{mV}}{1.8\frac{\text{mA}}{\mu\text{m}} \cdot l_E} = 7.9\text{ps}$$

which yields an emitter length of 4.9μm and a tail current of 8.81mA. In a 2.5-V system, the power consumption is reduced from 39mW to 22mW by applying inductive peaking. For this HBT device size, the output capacitance is

$$C_{TOT} = C_L + 2C'_{bc} w_E l_E + 2C'_{cs} l_E = 202\text{fF} + \left(6.6\frac{\text{fF}}{\mu\text{m}} \cdot l_E\right) = 234\text{fF}$$

and the necessary peaking inductance then becomes

$$L_P = \frac{C_L}{3.1}\frac{\Delta V^2}{I_T^2} = \frac{234\text{fF}}{3.1}\left(\frac{300\text{mV}}{8.81\text{mA}}\right)^2 = 87.5\text{pH}.$$

This value is small enough that an inductance with a SRF higher than 100GHz, suitable for a 40Gb/s logic gate, can be easily realized even in a "digital" back-end.

For the MOS CML selector, accounting for inductive peaking, the gate width is now given by

$$W = \frac{100\text{fF} \cdot \Delta V}{\left(7.9\text{ps} \cdot 0.3\frac{\text{mA}}{\mu\text{m}}\right) - \left(3\frac{\text{fF}}{\mu\text{m}} \cdot \Delta V\right)}.$$

The maximum voltage swing which can be achieved while still meeting the required bandwidth is about 790mV. This suggests that a 130nm CMOS technology with a logic swing of 600mV would be sufficient. Assuming the voltage swing scales by about $\sqrt{2}$ with each generation, a CML gate in 180nm CMOS would need a voltage swing of about 850mV. Hence, the 130nm CMOS is the "slowest" technology we could use for the shunt-peaked selector design. Using a similar methodology to the bipolar design, we can easily determine the gate widths, tail currents, and inductance values for MOS designs in 130nm, 90nm, and 65nm technologies.

The component values for the CMOS selectors are summarized in Table 11.1, and are compared with those of the gates designed without inductive peaking. Clearly we can see the substantial savings in current consumption that can be achieved by adding inductive

Table 11.1 Component values for the CMOS selectors.

		Without inductive peaking				With inductive leaking				
Technology node	ΔV (mV)	W_G (μm)	I_T (mA)	R_L (Ω)	C_T (fF)	W_G (μm)	I_T (mA)	R_L (Ω)	C_T (fF)	L_P (pH)
130nm	600					105.2	31.6	19	416	48.5
90nm	450	341	102	4.3	1123	44	13.2	34.1	232	87
65nm	320	62	18.2	17.2	286	22.7	6.8	47	168	120

peaking. Note however, that the value of the peaking inductor actually increases with technology scaling despite the reduction in capacitance. Recall that the peaking inductance is proportional to the load capacitance, but increases with the square of the load resistance.

Finally, note that the 65nm CMOS selector has similar tail current and peaking inductor value as the SiGe HBT selector fabricated in the 150-GHz, 0.2μm process. However, the power consumption is reduced to half because of the lower 1.1V as opposed to 2.5V supply. At comparable f_T and f_{MAX}, the MOS CML gate has higher FoM than the HBT CML one.

11.4 INDUCTIVE BROADBANDING

As in the case of the low-noise amplifiers and mixers discussed in Chapters 7 and 9, respectively, in logic gates with cascode topologies such as latches, selectors, XOR and AND gates, an inductor can be placed in series between the drain/collector of the CS/CE device and the source/emitter of the CG/CB devices to create an artificial transmission line and maximize the bandwidth. The technique is illustrated in Figure 11.18. It was first applied in an InP HEMT latch [9]. Its impact is significant in FET-based circuits but less so in HBT circuits.

11.5 DESIGN METHODOLOGY FOR MAXIMUM DATA RATE

Based on the previous sections, a general design methodology for highest data-rate CML/ECL gates is summarized below.

- Set ΔV to 250–300 mV in bipolar CML/ECL and to $1.5\times\Delta V_{MIN}$ in MOS CML (600mV in 130nm CMOS, and 350mV in 65nm CMOS or finer lithography)
- I_T is usually given, otherwise set I_T as low as possible to achieve the desired data rate using (11.27) for guidance

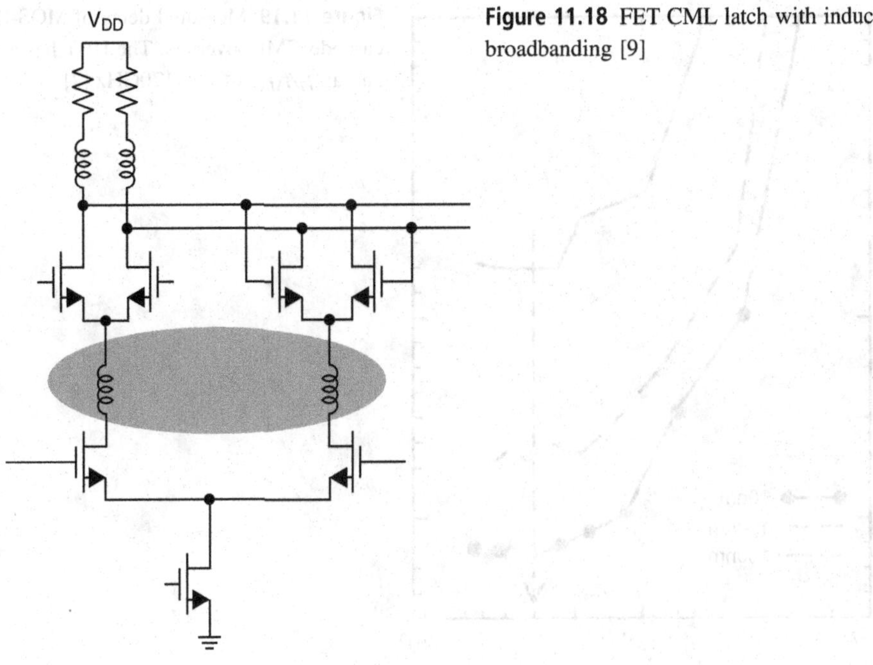

Figure 11.18 FET CML latch with inductive broadbanding [9]

- Size the total gate width, W, of the switching MOSFETs such that $I_T/W = J_{pfT}$ (typically 0.3–0.4 mA/μm in n-MOSFETs). In HBTs, set the emitter length such that $l_E = I_T/(1.5w_E J_{pfT})$
- Start by selecting a differential pair topology. If the topology proves not to be sufficiently fast, add inductive peaking and inductive broadbanding. If the topology is still not fast enough, emitter-followers and cascodes can be employed at the expense of higher power consumption.
- Bias emitter-followers and source-followers at $J_{pfT}/2$ and add resistive padding for stability.
- Optimize the scaling/buffering ratio between the EF/SF stage and the inverter (i.e. the fanout of the emitter-follower stage). A ratio of 1.5–2 is a good start. However, in general, if the gate cannot operate at the desired speed with a fanout of at least 2, consider using a faster transistor technology.

11.6 BiCMOS MOS-HBT LOGIC

Since, at a given tail current, the collector-to-substrate capacitance (C_{cs}) of the HBT is smaller than the corresponding drain–bulk capacitance (C_{db}) of the MOSFET, the output time constant is lower in the SiGe HBT CML inverter for identical tail currents. However, the input time constant is usually smaller in the MOS CML gate [10]. In a 130nm or finer lithography SiGe BiCMOS technology, a MOS-HBT cascode can lead to bandwidth

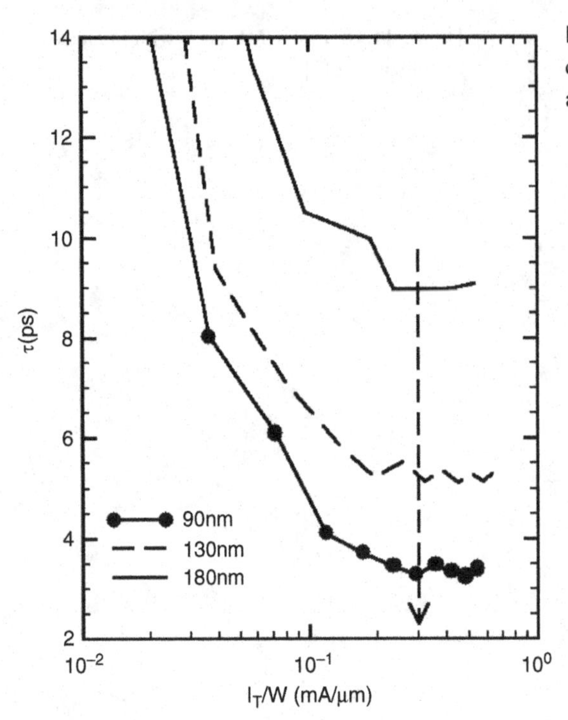

Figure 11.19 Measured delay of MOS-HBT cascode CML inverters. The HBT has a τ_F of 6ps and an f_T/f_{MAX} of 160/170GHz [7]

improvement, supply voltage and power consumption reduction compared to a bipolar-only cascode. The same effect can be obtained with a HEMT-HBT combination in III-V technologies.

The delay of the MOS-HBT cascode CML inverter is given by [10]

$$\tau_{BiCMOS} \approx \Delta V \frac{C_{bc} + C_{cs}}{I_T} + \left(k + \frac{R_g}{R_L}\right) \Delta V \frac{C_{gs} + C_{gd}}{I_T} \tag{11.28}$$

As can be seen in Figure 11.19, adding the 0.2μm emitter SiGe HBT discussed in Example 11.1 to a 180nm, 130nm, or 90nm CMOS process almost doubles the speed of the MOS CML gate [7].

As illustrated in Figure 11.20, the main idea is to use MOSFETs in the lower-level differential pair to reduce the $R_g C_{in}$ time constant, and HBTs in the upper levels of the gate to reduce the output capacitance and improve the slew-rate. With this arrangement, the logic swing at the output of the latch or selector remains that of the HBT CML family: 250 to 300mV, while the required logic swing at the clock inputs is that of a MOS CML gate: 600mV, 450mV or 320mV, respectively, in 130nm, 90nm, and 65nm SiGe BiCMOS technologies. In a 65nm or 45nm SiGe BiCMOS process the HBT and MOS logic swings become almost identical, leading to the most speed and power consumption benefit being obtained by combining MOSFETs with HBTs on the signal path.

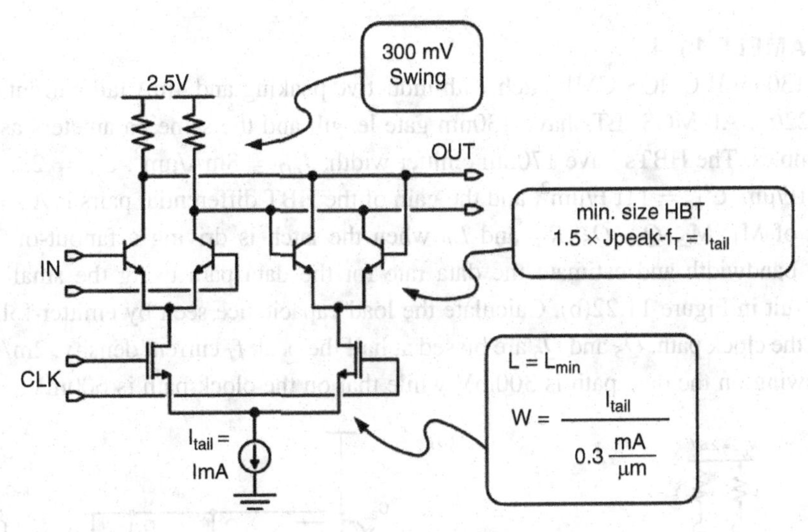

Figure 11.20 BiCMOS-CML latch schematic [11]

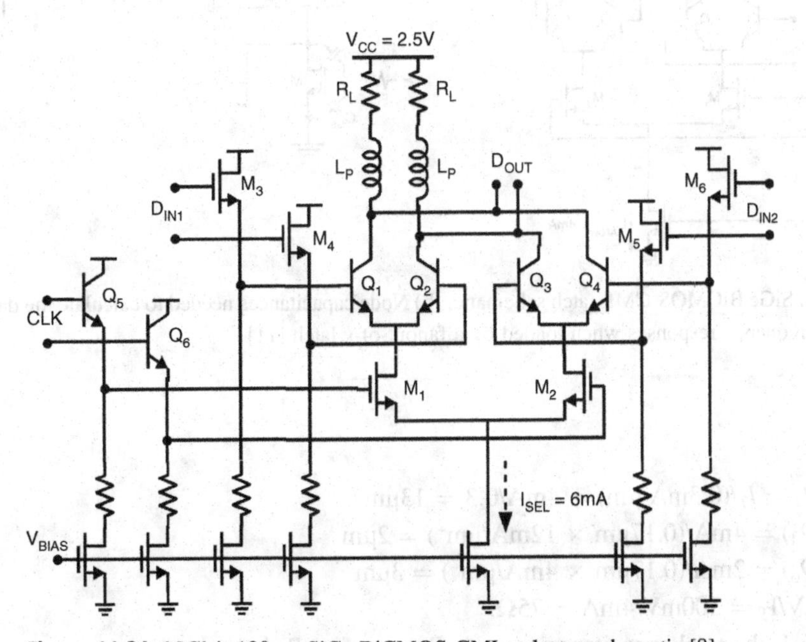

Figure 11.21 90Gb/s 130nm SiGe BiCMOS-CML selector schematic [8]

Based on these observations, a BiCMOS-E^{2}CL library can be developed operating from 2.5V supply at speeds exceeding 80Gb/s, even in a 130nm node [8]. Figure 11.21 illustrates the schematics of a 2.5V selector with SF and EF level shifting operating at up to 90Gb/s [8]. Simulations predict that this logic family can operate at over 110Gb/s in a future 65nm SiGe BiCMOS process with 280/400GHz f_T/f_{MAX} HBTs.

DESIGN EXAMPLE 11.4

Consider the 130nm BiCMOS CML latch with inductive peaking and 4mA tail current shown in Figure 11.22(a). All MOSFETs have 130nm gate length and the same parameters as in the previous examples. The HBTs have 170nm emitter width, $J_{pfT} = 8\text{mA}/\mu\text{m}^2$, $C_{be} = 23.2\ pF \times I_C$, $C'_{cs} = 1.1\text{fF}/\mu\text{m}$, $C'_{bc} = 11\text{f F}/\mu\text{m}^2$, and the gain of the HBT differential pairs is $A_V = 2.85$. Find the size of M_1, M_2, Q_1–Q_6, R_L, and L_P when the latch is driving a fanout-of-2 load. Calculate the bandwidth and estimate the data rate for the data path using the small signal equivalent circuit in Figure 11.22(b). Calculate the load capacitance seen by emitter-followers Q_5 and Q_6 on the clock path. Q_5 and Q_6 are biased at half the peak f_T current density, 2mA each. The voltage swing on the data path is 300mV while that on the clock path is 600mV.

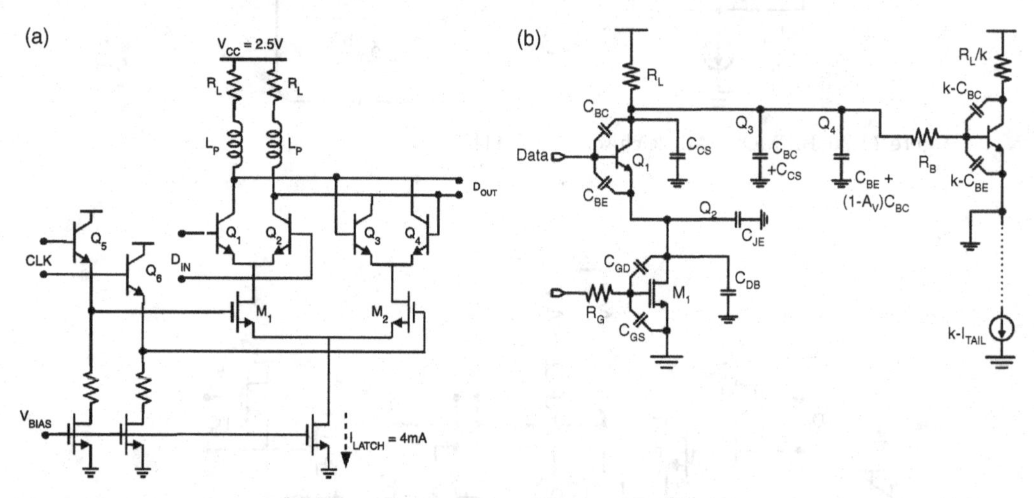

Figure 11.22 (a) SiGe BiCMOS-CML latch schematic. (b) Node capacitances needed to calculate the data path and clock path frequency responses when loaded by a fanout-of-*k* latch [11]

Solution

(a) $W(\text{M1,M2}) = I_T/(0.3\text{mA}/\mu\text{m}) = 4\text{mA}/0.3 = 13\mu\text{m}$

$l_E(Q_1\text{-}Q_4) = 4\text{mA}/(0.17\mu\text{m} \times 12\text{mA}/\mu\text{m}^2) = 2\mu\text{m}$

$l_E(Q_5\text{-}Q_6) = 2\text{mA}/(0.17\mu\text{m} \times 4\text{mA}/\mu\text{m}^2) = 3\mu\text{m}$

$R_L = \Delta V/I_T = 300\text{mV}/4\text{mA} = 75\Omega$.

For $k = 1$, the total capacitance seen by R_L, ignoring R_b is

$$C_T = C_{cs1} + C_{bc1} + C_{bc3} + C_{cs3} + C_{be4} + (1 - A_V)C_{bc4} + k(C_{be} + (1 - A_V)C_{bc}) = 72\text{fF}.$$

With these values for R_L and C_T, we obtain $L = 130\text{pH}$ and a bandwidth $BW_{3\text{dB}} = 1.6 \times \frac{1}{2\pi R_L C_T} = 47\text{GHz}$ which is fast enough for operation at 62Gb/s.

For $k = 2$

$$C_T = C_{cs1} + C_{bc1} + C_{bc3} + C_{cs3} + C_{be4} + (1 - A_V)C_{bc4} + k(C_{be} + (1 - A_V)C_{bc}) = 100.5\text{fF}$$

$L_P = 184\text{pH}$ and $BW_{3\text{dB}} = 33.8\text{GHz}$ which is good enough for operation at 45Gb/s.

(b) The load capacitance seen by each emitter-follower is C_{gs}(M1) + C_{gd} (M1) = 13fF + 6.5fF = 19.5fF, much smaller than the load capacitance, and the resistance at the clock input is also much smaller than R_L.

Note that the HBT has no capacitance between the emitter and the substrate (or ground). The only other capacitance at that node comes from the padding resistance. The small load capacitance and the high f_T of the EF stage explains why the clock path is sufficiently fast to pass a 45GHz clock signal even when the emitter-follower is biased at $J_{pfT}/2$, where the HBT f_T is still larger than 100GHz.

11.7 PSEUDO-CML LOGIC

What can be done to further increase the switching speed of MOS CML gates? The question is particularly important in sub 65nm CMOS technologies where the supply voltage is so low that stacking three transistors between V_{DD} and ground severely reduces their V_{DS} and, therefore, f_T and f_{MAX}. A solution, illustrated in Figure 11.23 for a MOSFET latch [12] and in Figure 11.24 for a BiCMOS latch [11], is to remove the current tail and to find a method to self-bias the clock path transistors [12],[13]. We thus gain 0.2–0.4V of voltage headroom which can be used for operation at higher data rates, or for operation at the same speed with a lower supply voltage.

Ideally, the clock path MOSFETs should be biased at I_{DS}/W = 0.15mA/μm and their drain current should swing up to 0.3mA/μm when the clock input is high, and down to 0mA/μm when the clock input is low. To maximize the data rate, the clock path MOSFETs must have a high V_T while those on the data path must have low V_T. In MOSFET-only implementations, this arrangement allows the data and the clock paths to be driven from the same logic level without pushing the clock path MOSFETs in the triode region and without requiring level-shifters. The high V_T on the clock path ensures that the clock path MOSFETs can be turned off even when V_{OL} is 0.6V.

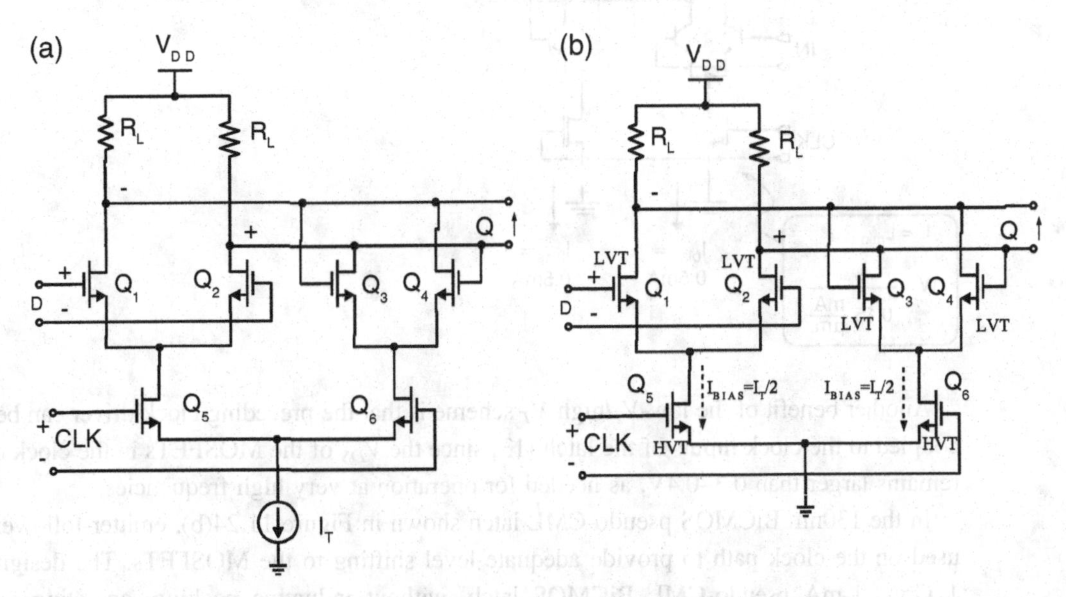

Figure 11.23 CMOS (a) CML and (b) pseudo-CML latch schematic [12],[13]

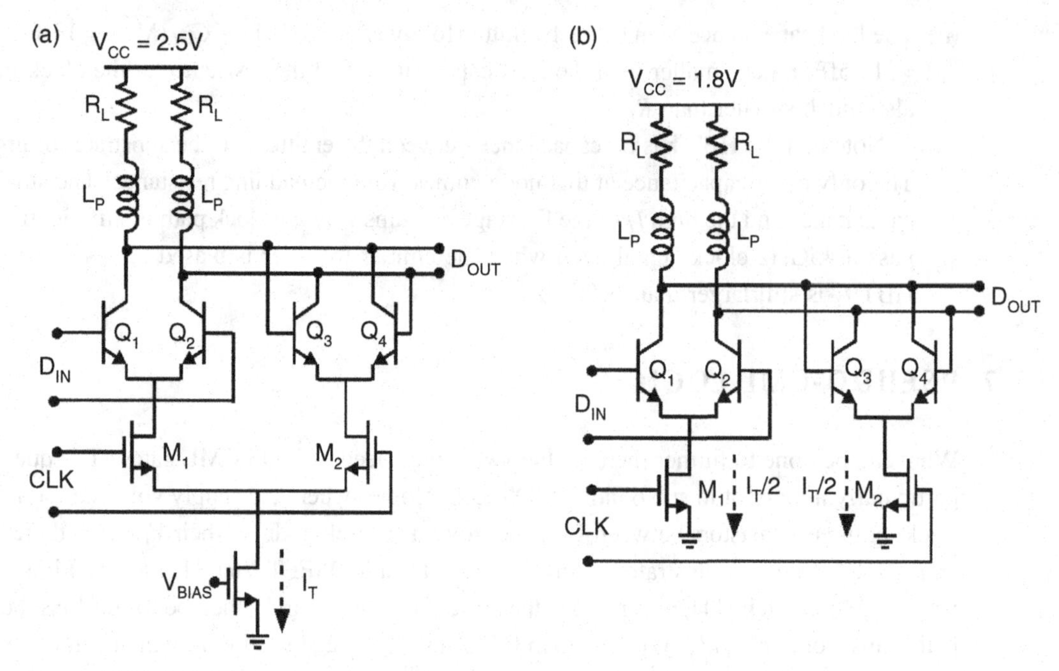

Figure 11.24 SiGe BiCMOS (a) CML and (b) pseudo-CML latch schematic [11]

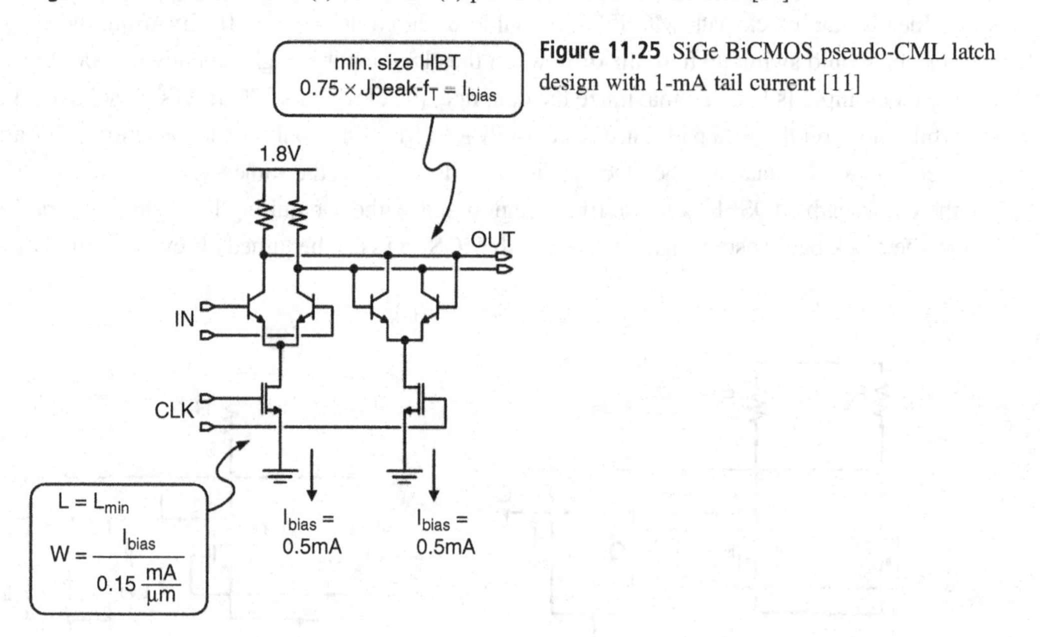

Figure 11.25 SiGe BiCMOS pseudo-CML latch design with 1-mA tail current [11]

Another benefit of the low-V_T/high-V_T scheme is that the preceding clock driver can be DC-coupled to the clock inputs of the latch [13] since the V_{DS} of the MOSFETs in the clock driver remains larger than 0.3–0.4V, as needed for operation at very high frequencies.

In the 130nm BiCMOS pseudo-CML latch shown in Figure 11.24(b), emitter-followers are used on the clock path to provide adequate level shifting to the MOSFETs. The design of a 12Gb/s, 1-mA pseudo-CML BiCMOS latch, without inductive peaking, operating from a 1.8V supply and consuming 1.8mW is illustrated in Figure 11.25 [11].

DESIGN EXAMPLE 11.5

Using the latch schematic in Figure 11.25 and a 130nm SiGe BiCMOS technology with HBT $w_E = 130\text{nm}, f_T = 230\text{GHz}$ at $J_{pfT} = 14\text{mA}/\mu\text{m}^2$, find the gate width and emitter lengths of all transistors and the value of the load resistor for an output logic swing of 600mV.

Solution

Since the voltage swing is given by

$\Delta V = 2{\cdot}R_L I_{bias}$, the value of the load resistance becomes

$$R_L = \frac{\Delta V}{2{\cdot}I_{bias}} = \frac{600\text{mV}}{2{\cdot}0.5\text{mA}} = 600\Omega.$$

According to the design equations, the gate width of the MOSFETs is given by $W = \frac{2I_{bias}}{J_{pfTMOS}} = \frac{2{\cdot}0.5\text{mA}}{0.3\frac{\text{mA}}{\mu\text{m}}} = 3.3\mu\text{m}$. The MOSFET layout can be realized with two gate fingers, each 1.67μm wide. The fingers of the two MOSFETs can be interdigitated and placed in the same active area.

Finally, the emitter length of the HBT is calculated from

$$l_E = \frac{I_{bias}}{w_E{\cdot}0.75{\cdot}J_{pfT}} = \frac{0.5\text{mA}}{0.13\mu\text{m}{\cdot}0.75{\cdot}14\frac{\text{mA}}{\mu\text{m}^2}} = 0.37\mu\text{m}.$$

This value is actually smaller than the minimum emitter length allowed in this technology, which is 0.6μm. We will have to use 130nm × 0.6μm HBTs and operate them below the peak f_T current density.

11.8 OTHER BIPOLAR, MOS AND BiCMOS CML, AND ECL GATES

In addition to inverters, selectors, and latches, AND and XOR gates are needed to realize the various logic functions encountered in high-speed digital transceivers, PLLs, synthesizers, and CDRs. Two slightly different implementations of a bipolar AND gate are depicted in Figure 11.26. A MOS CML version can be obtained simply by replacing the HBTs with MOSFETs. The AND CML gate is inherently asymmetrical. In both circuits, to ensure that all transistors have approximately the same V_{CE}/V_{DS}, a common-base/gate transistor is added in the right branch of the low-level input. The circuit on the left has two collectors connected to *outn* and only one collector connected to *outp*, which results in different capacitances at the *p* and *n* outputs. In the AND gate in Figure 11.26(b), a dummy cascode is added to equalize the capacitances at the two outputs at the cost of introducing a DC offset at the output.

A BiCMOS CML XOR gate is shown in Figure 11.27 [11]. This circuit was manufactured in a 130nm SiGe BiCMOS process and requires a 600mV logic swing at the IN0 input. The logic swing at the IN1 input and at the output can be either 250–300 mV or, for

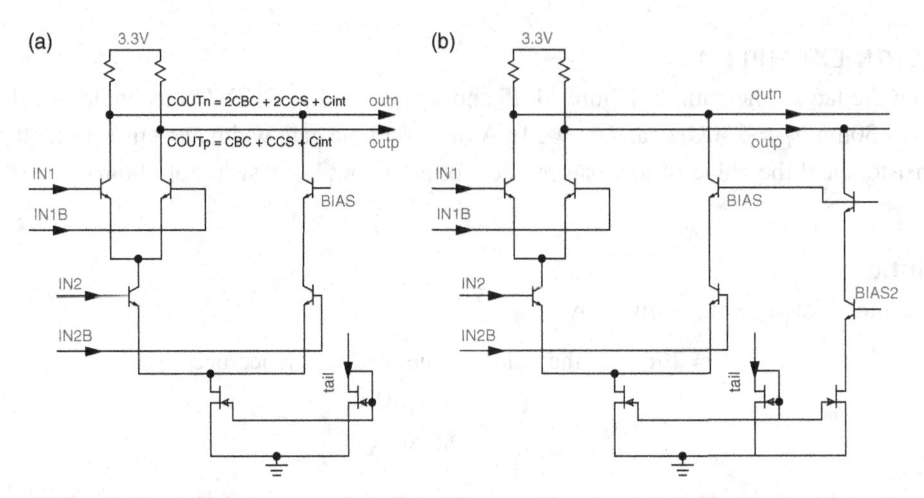

Figure 11.26 Bipolar CML AND gates: (a) balanced current and (b) balanced output capacitance

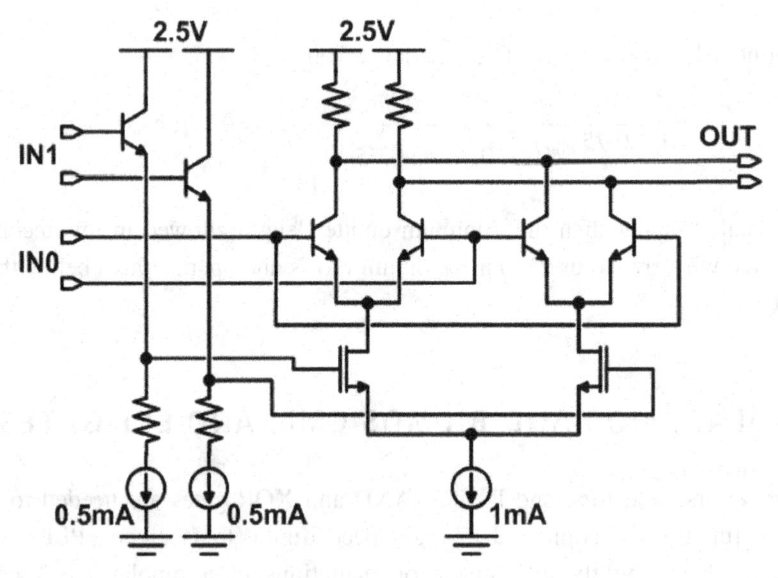

Figure 11.27 BiCMOS-CML XOR gate schematic [11]

symmetry reasons, 600mV. In a future 65nm SiGe BiCMOS technology, both inputs could operate at a logic swing of 320mV. MOS CML and bipolar CML versions of this gate can be obtained by replacing the HBTs with MOSFETs and vice-versa, respectively, while a 1.8V supply, pseudo-CML XOR gate can be derived from it by removing at least the 1mA tail current source.

A technique to further reduce the number of vertically stacked transistors, which is applicable to all MOS, bipolar and BiCMOS CML, and pseudo-CML latches and selectors, is to

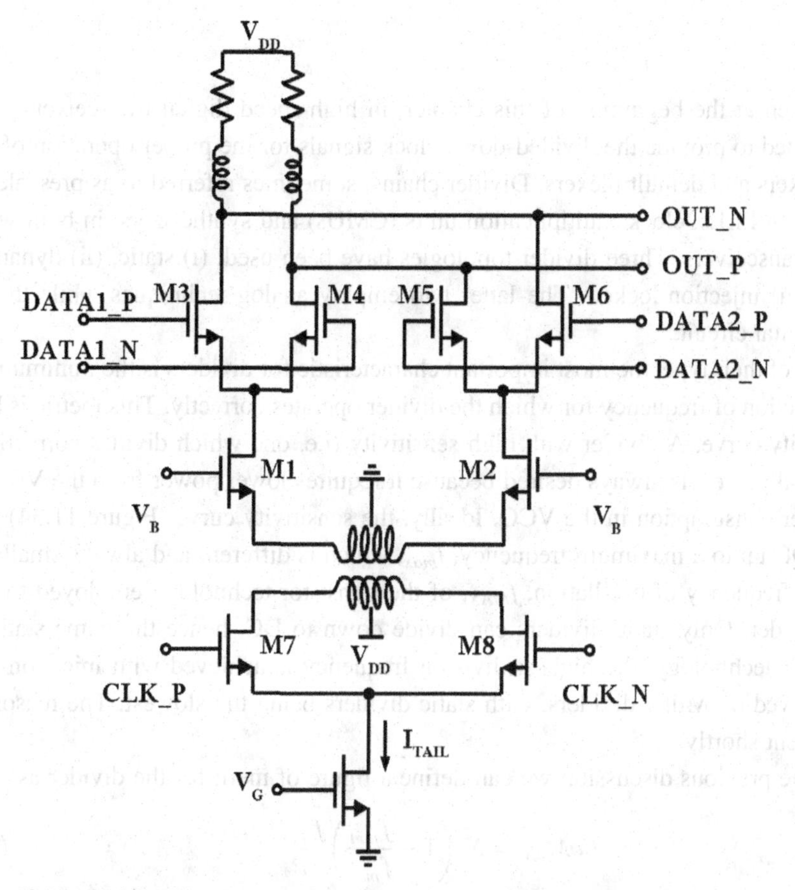

Figure 11.28 MOS-CML selector with transformer-coupled clock path [14]

employ AC-coupling between the clocking pair and the data path differential pairs. AC-coupling can be implemented either with transformers or with capacitors. It helps in pushing the maximum clock frequency higher than in all other topologies by allocating all, or at least a greater portion, of the supply voltage to each switching transistor in the gate, thus increasing their f_T and f_{MAX}.

The schematics of a MOS CML 2:1 selector with tuned clock path based on transformer coupling is depicted in Figure 11.28. This circuit was manufactured in a 90nm standard CMOS process and operates at up to 60-Gbs with a 30GHz clock signal [14]. By using transformer coupling, the power-supply voltage can be reduced, in this case to 0.7V, at the expense of increasing the total current. Nevertheless, unlike the telescopic-cascode selector or telescopic-cascode latch, this topology scales well in low-voltage 32nm and 22nm CMOS technologies. The same topology can be used in latches. Its main disadvantage is that it does not operate at low clock frequencies. However, the latter is not normally a problem in most applications requiring the highest switching speed as long as the data path inputs and outputs are DC-coupled.

11.9 DIVIDERS

As we have seen at the beginning of this chapter, in high-speed digital transceivers, divider chains are needed to provide the divided-down clock signals for the proper operation of multi-stage multiplexers and demultiplexers. Divider chains, sometimes referred to as prescalers, are also employed in PLLs, clock multiplication units (CMUs) and synthesizers, in both wireline and wireless transceivers. Three divider topologies have been used: (i) static, (ii) dynamic (or Miller), and (iii) injection-locked. The latter two employ analog techniques while the static divider is a digital circuit.

Irrespective of their type, the most important characteristic for dividers is the minimum input power as a function of frequency for which the divider operates correctly. This metric is known as the sensitivity curve. A divider with high sensitivity (i.e. one which divides correctly with low input signal power) is always desired because it requires lower power from the VCO, thus reducing power consumption in the VCO. Ideally, the sensitivity curve (Figure 11.34) should extend from DC up to a maximum frequency, f_{max}, which is different and always smaller than the maximum frequency of oscillation, f_{MAX}, of the transistor technology employed to manufacture the divider. Only static dividers can divide down to DC, hence the name static. In a given transistor technology, the highest division frequency is achieved with injection-locked dividers, followed by Miller dividers, with static dividers being the slowest. The reasons will become apparent shortly.

Based on the previous discussion we can define a figure of merit for the divider as

$$FoM_{div} = N \cdot \left(1 - \frac{f_{min}}{f_{max}}\right) \frac{f_{max}}{P} \qquad (11.29)$$

where f_{min} and f_{max} are the maximum and minimum input frequencies at which the divider operates correctly, N is the division ratio, and P is the power dissipation of the divider. In fiber-optic transceivers N is typically a power of 2. Ideally, in static dividers, f_{min} should be 0. However, in some "static" dividers that use inductive peaking and in those that employ transformer-coupling, this is not true. A typical measured sensitivity curve for a static divider is shown in Figure 11.29 [15].

In general, dividers contribute negligible noise to the output signal, except at the edges of the sensitivity curve. In fact, if the output signal becomes noisy, this is an indication that the divider is no longer operating correctly. Low-noise dividers are particularly important in PLLs where the phase noise added by the divider chain should be negligible. In a well-designed divider, the phase noise of the output signal should be $20\log_{10}N$ dB lower than the phase noise of the input signal.

11.9.1 Static dividers

Static dividers are truly digital circuits. The most common static divider topology is a divide-by-2 stage realized with two cascaded D-type latches with negative feedback from the data outputs of second latch to the data inputs of the first latch. The signal to be divided is applied to the clock-inputs of the two latches in parallel, but 180 degrees out of phase, as illustrated in

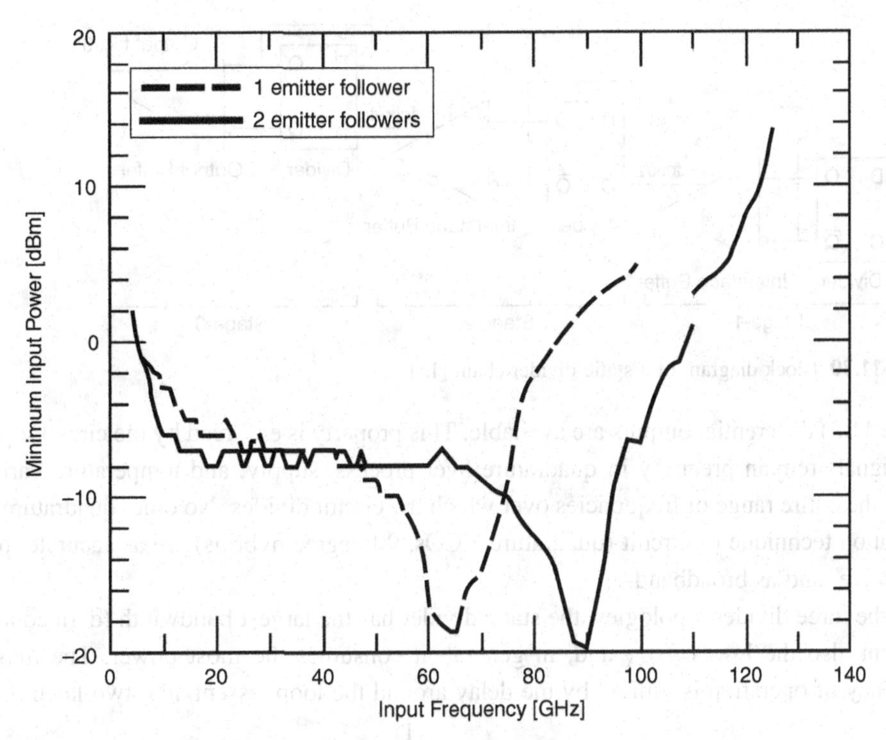

Figure 11.29 Measured sensitivity of ECL and E²CL SiGe HBT static dividers [15]

Figure 11.5. Dividers with other integer ratios are realized using a combination of latches, XOR, and/or AND gates.

Since CML latches are differential, the divide-by-2 stage can also be viewed as a two-stage differential ring oscillator. The oscillation frequency of the static divide-by-2 circuit is known as the self-oscillation frequency (*SOF*). Indeed, if no signal is applied at the clock input, the circuit oscillates and a signal at *SOF*/2 appears at its output. Obviously, if the input signal frequency is close to the *SOF*, only a very low input power is required for division. This explains the sharp dip observed in the sensitivity curves of the two dividers in Figure 11.29 at 65GHz and 90GHz.

For static dividers only, we can define a simpler figure of merit that is much easier to measure and map across wafer runs and technologies

$$FoM_{statdiv} = \frac{SOF}{P}. \tag{11.30}$$

Because the SOF is easy to measure (compared to the sensitivity curve), and because it has been shown to scale with f_{MAX} [6], static dividers and their SOF are often used as process monitors to asses the speed of bipolar [6],[15],[16] and, less often, CMOS technologies [17],[18]. The SOF is determined by the time constant at the load-resistor node of the second latch.

A very useful property of any differential 2-stage ring oscillator is that both the in-phase (at the data outputs) and the in-quadrature (at the middle nodes between the two latches in

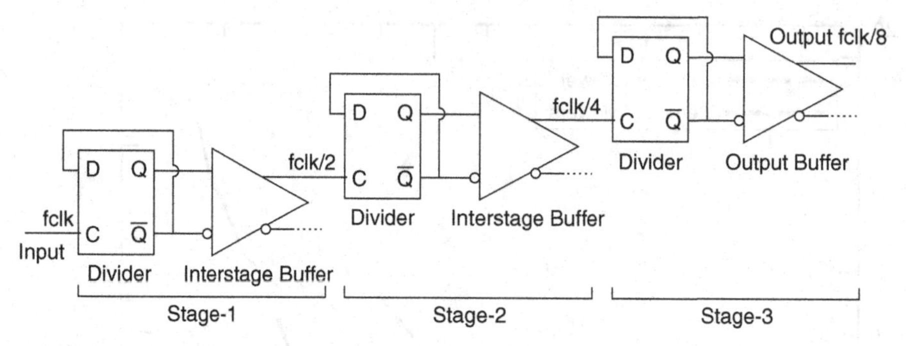

Figure 11.30 Block diagram of a static divider chain [15]

Figure 11.5) differential outputs are available. This property is enforced by the circuit topology. The signals remain precisely in quadrature over process, supply, and temperature variations, and in the entire range of frequencies over which the circuit divides. No other quadrature signal generation technique or circuit (quadrature VCOs, 90 degree hybrids) are as accurate, process insensitive, and as broadband.

Of the three divider topologies, the static divider has the largest bandwidth (defined as $f_{max}-f_{min}$) but also the lowest f_{MAX} and, in general, it consumes the most power. The maximum frequency of operation is limited by the delay around the loop, essentially, two latch delays

$$f_{max} = \frac{1}{2\tau_{latch}}. \tag{11.31}$$

Operation down to DC requires that the absolute gain of the latching differential pair in each latch, $g_m R_L$, be larger than 1. In latches with inductive peaking, the gain at DC, and therefore operation at low frequency, is sometimes sacrificed in favor of higher peaking to push the divider f_{MAX} to higher frequencies.

A static divider chain with a division ratio $N = 2^P$ is typically realized by cascading divide-by-2 stages. The output time constant and, therefore, f_{max} is determined by the fanout of the second latch in the first D-type flip-flop of the chain, which must drive both the first latch as well as the following divider stage. To reduce the fanout below 2, a small buffer is sometimes placed between the first and second divider stages, as illustrated in Figure 11.30. The buffer also drives other blocks, for example in a multi-stage multiplexer or demultiplexer. Unless the quadrature output is needed, the first latch has only a fanout of 1 and does not limit the maximum speed of the divider stage.

Since in a divider chain the toggling frequency decreases by a factor of 2 at each divider stage, to save power, the tail currents and transistor sizes are scaled down in each lower speed divider stage, usually by a factor of 2. Thus, almost half of the total power consumption of the chain is due to the first stage.

To reduce the capacitive loading at the output node, the fastest divider stages employ ECL or E^2CL latch topologies, as illustrated in Figures 11.31 and 11.32, respectively. The ECL divider has a *SOF* of 77GHz and consumes 122mW from 3.3V while operating up to 100GHz at room temperature, and up to 91GHz at 100°C. The E^2CL divider operates at up to 125GHz with a *SOF* of 90GHz and consuming 232mW from a 5.5V supply.

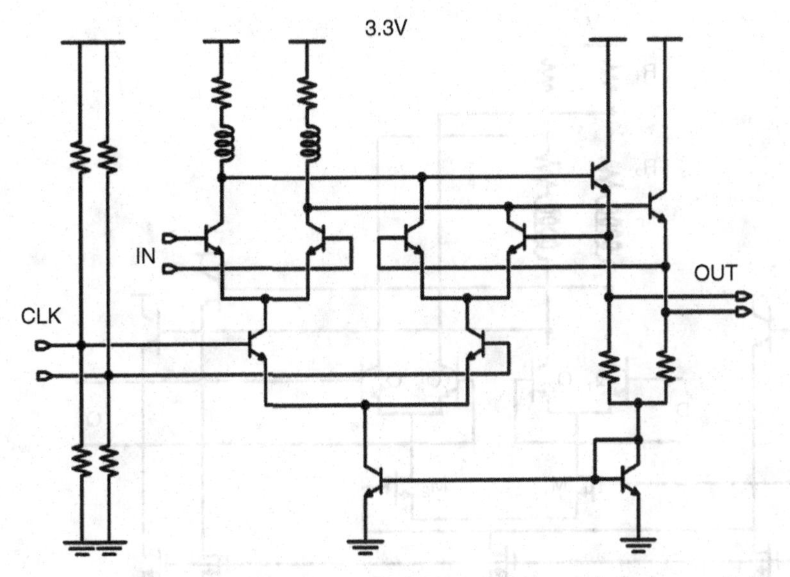

Figure 11.31 ECL latch with inductive peaking used in a static frequency divider [6]

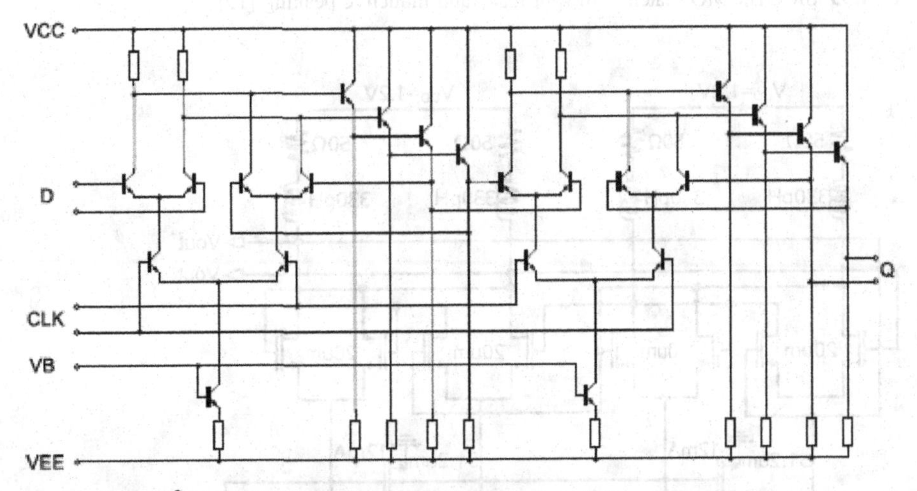

Figure 11.32 E^2CL implementation of a divide-by-2 stage [15]

Techniques that further increase the speed of the latch, by acting to reduce the output capacitances or to compensate it, include:

- inductive peaking as in Figure 11.31,
- split loads, as in Figure 11.33 [19],
- Cherry–Hooper loads [20],
- asymmetrically sized (smaller) latching pair (in CMOS),
- tuned clock path relying on transformers (Figure 11.34), and
- a combination of two or more of the techniques above.

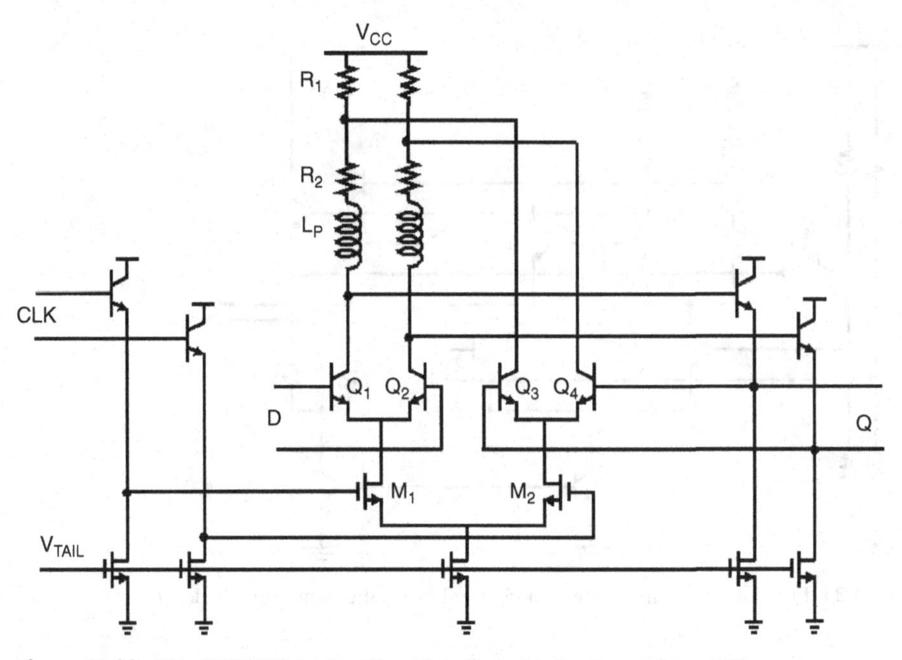

Figure 11.33 SiGe BiCMOS latch with split loads and inductive peaking [19]

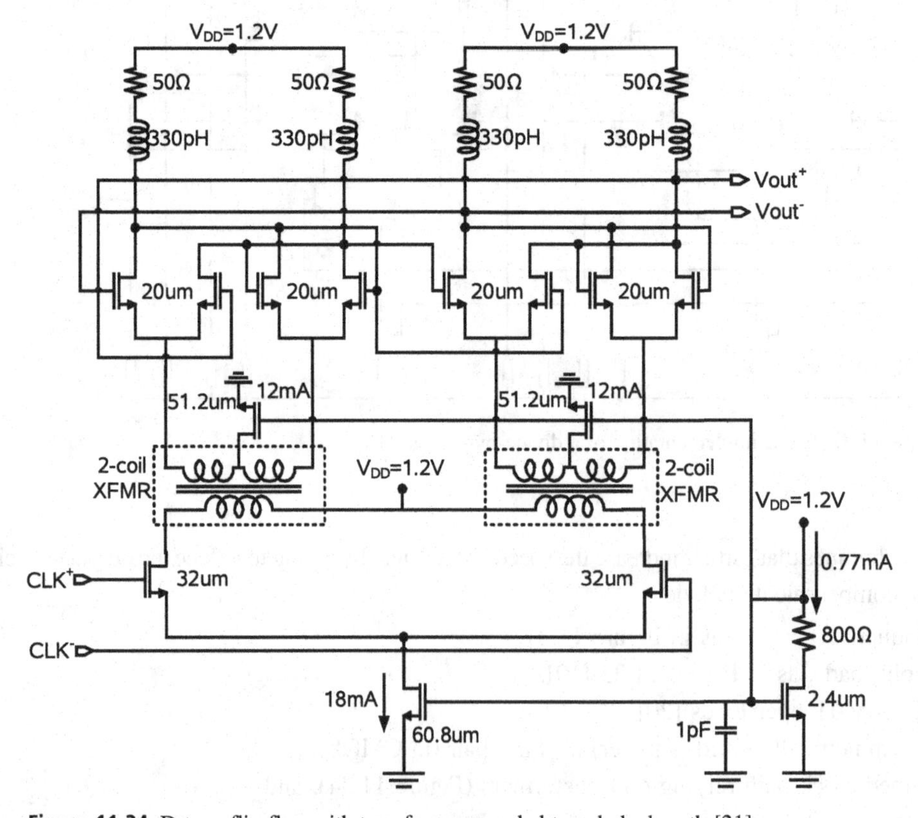

Figure 11.34 D-type flip-flop with transformer-coupled tuned clock path [21]

The highest frequency static dividers reported to date have been fabricated in an InP HBT technology with f_T/f_{MAX} of 400/650GHz and operate up to 205GHz [16]. The fastest static-dividers in standard CMOS reported to date operate at up to 95GHz at room temperature but their bandwidth does not extend down to DC and practically vanishes at 100°C [22].

11.9.2 Dynamic dividers

Dynamic, also known as regenerative or Miller, dividers were first proposed in the 1930s [23]. The basic concept is illustrated in Figure 11.35. The circuit consists of an active mixer with gain followed by a low-pass filter, or of a mixer followed by a LPF and an amplifier. In either case, the output signal is fed back into the mixer where it mixes with the input. In order for the circuit to operate correctly, the loop gain must be larger than 1 (hence the name regenerative) and the phase around the loop must be a multiple of 360 degrees. If the mixer is active, it amplifies the fed-back signal before it mixes it with the input. The input signal is applied to one of the inputs of the mixer (RF or LO) while the amplified fed-back signal goes to the second mixer input. If f_o is the frequency of the output signal and f_{in} is the frequency of the input signal, the following equation must hold

$$nf_{in} \pm mf_o = f_o \text{ and } f_o = \frac{n}{1 \pm m} f_{in} \tag{11.32}$$

where m and n are integers representing the mixing product harmonics. The first-order mixing product, $m = 1$ and $n = 1$, has the largest power and results in

$$f_o = \frac{1}{2} f_{in}. \tag{11.33}$$

The output signal frequency is low enough to be amplified by the mixer and to pass unattenuated through the LPF.

Another possible mixing product at the output occurs when $n = 3$ and $m = 1$, resulting in

$$f_o = \frac{3}{2} f_{in}. \tag{11.34}$$

This explains the role played by the low-pass filter, which must attenuate the 3/2 f_{in} signal.

The regenerative divider finds its way in many practical systems when a static divider either consumes too much power or is not sufficiently fast for operation at the desired frequency. However, because the third harmonic of the output signal must be filtered by the low-pass filter, the minimum frequency of operation of the Miller divider cannot be lower than its $f_{max}/3$. Another limitation of the Miller divider is that it does not provide quadrature signals at the output.

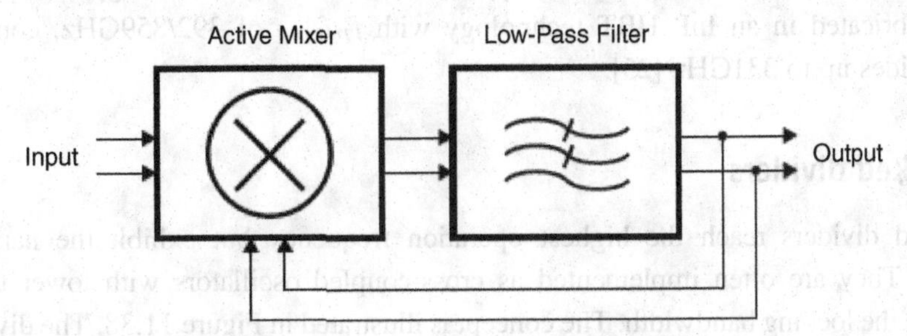

Figure 11.35 Dynamic (Miller or regenerative) divider concept [20]

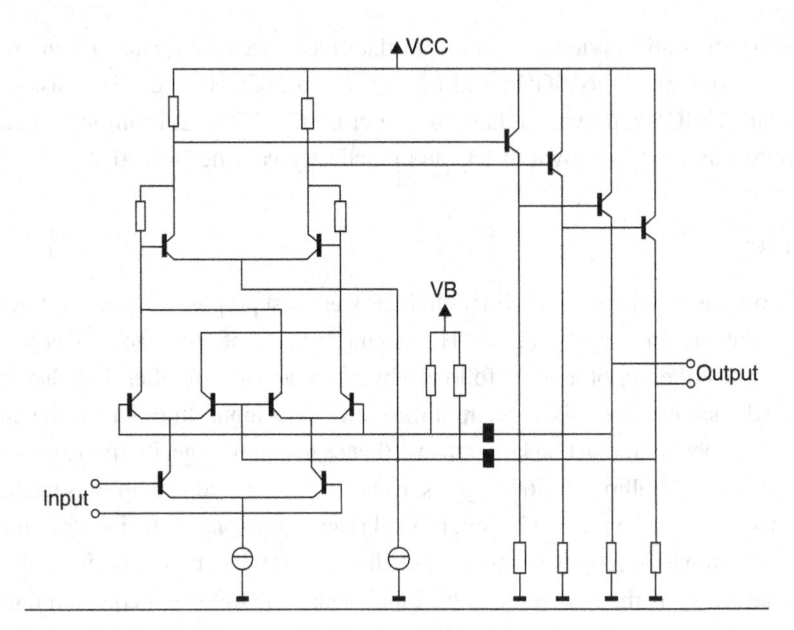

Figure 11.36 Dynamic divider schematic with Cherry–Hooper mixer load [20]

As in the case of the static divider, the time constant at the mixer output node determines the maximum frequency. Figure 11.36 shows the schematic of the fastest (Miller) divider in silicon which operates from 55 to 168GHz [20]. A Cherry–Hooper stage and two cascaded emitter-follower stages are employed at the output of the Gilbert cell mixer to reduce the output time constant. To maximize the overall bandwidth of the divider, the input signal is applied at the traditional, lower capacitance RF port of the Gilbert cell mixer, with the divided-by-2 output fed to the bases of the mixing quad. This circuit consumes 104mW and is manufactured in a SiGe HBT process with f_T/f_{MAX} of 225/320GHz.

A lower power and lower voltage version with transformer coupling in the feedback path is reproduced in Figure 11.37. This circuit divides from 75GHz up to at least 153GHz while consuming 32mW from 1.8V [24]. The input signal is applied directly to the mixing quad and only one emitter-follower stage is employed at the output. Using one instead of two emitter-follower stages helps to reduce the power consumption but also reduces f_{MAX}. At the same time, the transformer limits the divider f_{min}.

As in the case of static dividers, the fastest dynamic divider in any technology at the time of writing was fabricated in an InP HBT technology with f_T/f_{MAX} of 392/859GHz, consumes 85mW and divides up to 331GHz [25].

11.9.3 Injection-locked dividers

Injection-locked dividers reach the highest operation frequency but exhibit the narrowest locking range. They are often implemented as cross-coupled oscillators with lower tank Q [28] to improve the locking bandwidth. The concept is illustrated in Figure 11.38. The divider is normally excited by a signal at approximately two times its oscillation frequency. Although

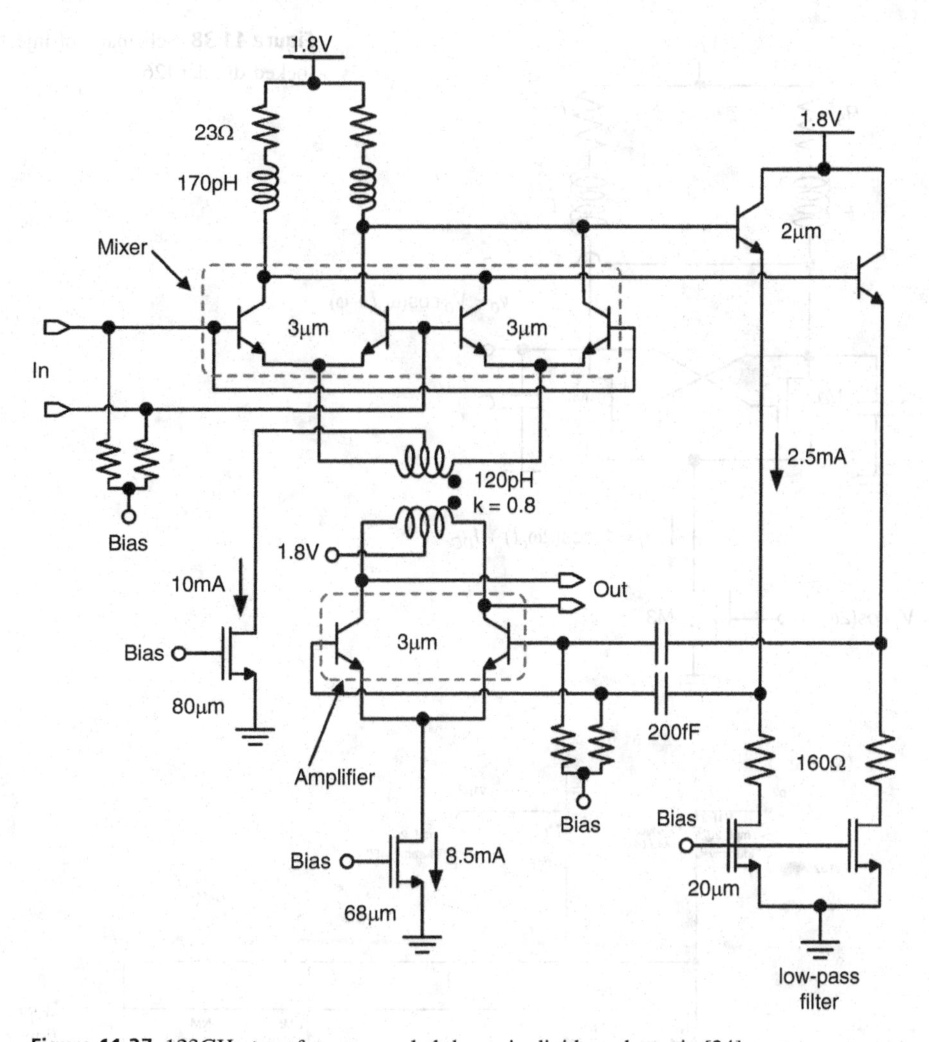

Figure 11.37 123GHz transformer-coupled dynamic divider schematic [24]

numerous low-power CMOS injection-locked dividers have been published, they are rarely used in products due to their poor performance over process and temperature corners. Like the Miller divider, the injection-locked divider does not provide quadrature signals at the output. In general, they should be avoided.

11.10 CML/ECL GATE LAYOUT TECHNIQUES

Just like for conventional static CMOS, the industry practice is to develop a library of standard CML cells that are then used to build more complex high-speed digital circuits. Such high-speed libraries are not usually provided by the semiconductor foundry and they have to be custom-developed. The schematic and template layout of a 130nm CMOS pseudo-CML latch

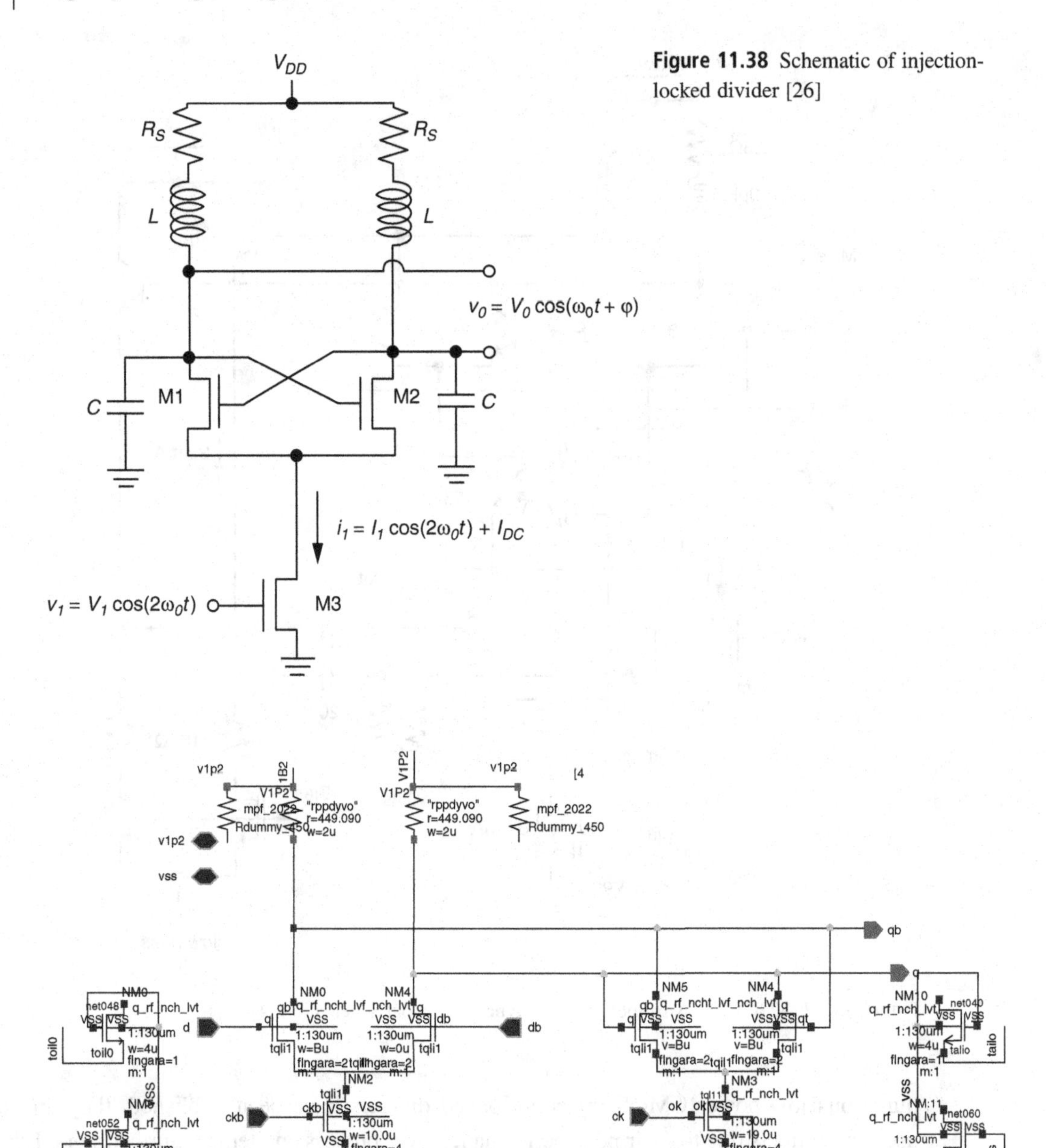

Figure 11.38 Schematic of injection-locked divider [26]

Figure 11.39 Schematics of pseudo-CML 130nm MOSFET latch cell without inductive peaking but with dummy resistors and dummy MOSFETs [12]

cell are reproduced in Figures 11.39. and 11.40, respectively. Given the mixed-signal, differential nature of the CML gate, both analog and digital techniques are employed. For example, dummy poly resistors and dummy gate fingers are placed on each side of the load resistors and of the clock and data path MOSFETs, as in analog blocks. The Metal-2 V_{DD} bar at the top is

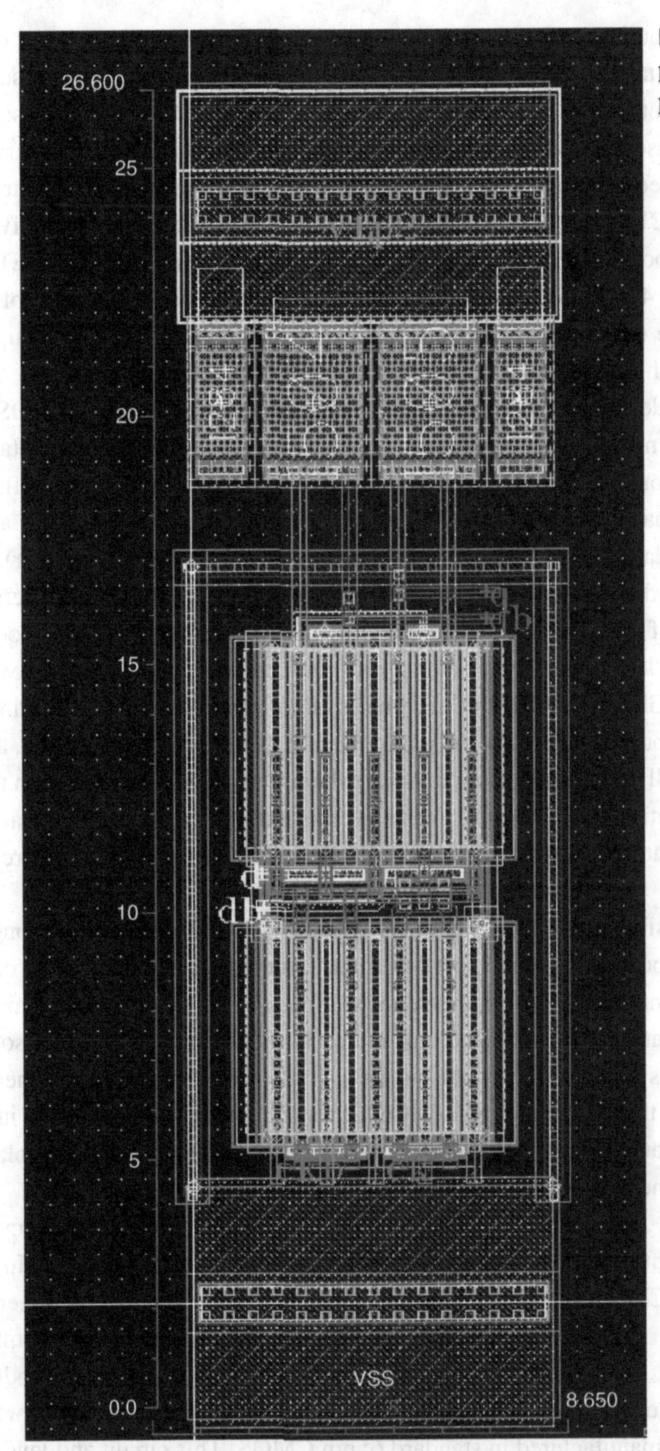

Figure 11.40 Layout of 130nm CMOS pseudo-CML latch cell without inductive peaking

placed above a Metal-1 ground bar for bias decoupling. Below both metal bars lies a deep n-well connected through n-taps to V_{DD}. A V_{SS} (ground) bar in Metal-1 with p-tap substrate contacts is located at the bottom, as in CMOS standard cells. The polysilicon resistors, placed above the same reverse-biased n-well as the V_{DD} bar, form the CML loads in the upper half of the cell. A ring of grounded substrate p-taps surrounds the MOSFET core of the latch. The layouts of the four MOSFETs in the top differential pairs are merged into a single active area. The two MOSFETs at the bottom are also merged in a single active region. All MOSFETs have 130nm gate length and use 4-μm wide gate fingers. Note also the identical placement of all the components, including the dummies, in the schematic and the layout. The cell occupies an area of 8μm × 27μm which is larger than a standard static CMOS latch cell.

A critical aspect in the design of the layout of ECL and E^2CL bipolar and BiCMOS CML and pseudo-CML gates is ensuring that the interconnect between the output of the EF stage and the input of the next stage or gate, is as short as possible. This is important to avoid reflections because of the large mismatch between the low output impedance of the EF stage and the relatively large input impedance at the base/gate of the next gate (EF or differential pair). As an example, the schematics and layout of a 80Gb/s 130nm BiCMOS ECL selector with very short interconnect at the output of the EF stage are depicted in Figure 11.41 [27]. Only the clock path emitter-followers are included in this layout along with the latching quad and the clock path MOSFETs. To achieve 80Gb/s operation, this cell employs both shunt and series peaking. The shunt peaking inductor layouts are placed at the top of the cell, below the load resistors, and are surrounded by a deep n-well connected to V_{DD} to isolate them from each other and from the rest of the circuit. The metallization contacting the deep n-well is interrupted at specific locations to avoid forming a ring resonator. The shunt inductors are provided "for free" by the interconnect to the next gate.

In contrast, as clearly visible in the layout of the D-type flip-flop of Figure 11.42, long lines can be placed from the outputs (drain–collector) of a differential pair to the input (gate or base) of another EF/SF or differential stage in a shunt–series feedback arrangement which maximizes the bandwidth. In this situation, the impedance mismatch is much smaller. Another solution, applied when circuit blocks need to be several hundred microns away from each other, is to place the resistive loads of the differential pair (MOS or bipolar) at the far end, at the input of the next gate [4]. We reiterate the strict layout rule that long interconnect can only be placed at the drain/collector output, never at the source or emitter output.

As mentioned previously, because of its larger output capacitance and high-frequency clock input, the design and layout of the latch is the most challenging of all CML gates. In 65nm CMOS, only the pseudo-CML latch topology appears to meet the speed requirements for operation at 40Gbs with a sub 1.2V supply. Layout techniques that reduce the footprint of the 6-transistor latch layout and the interconnect parasitic capacitance at the output node have large impact on performance. Figure 11.43 reproduces the schematics and core transistor layout detail of a D-type flip-flop manufactured in standard 65nm CMOS. This circuit and layout was employed both in a 96GHz static divider chain [18],[22] and in a 81Gbs retimer [28]. Although all MOSFETs have approximately the same total gate width, the four low-V_T transistors at the top have 1.6μm wide fingers and are merged in a single active area, while the two high-V_T

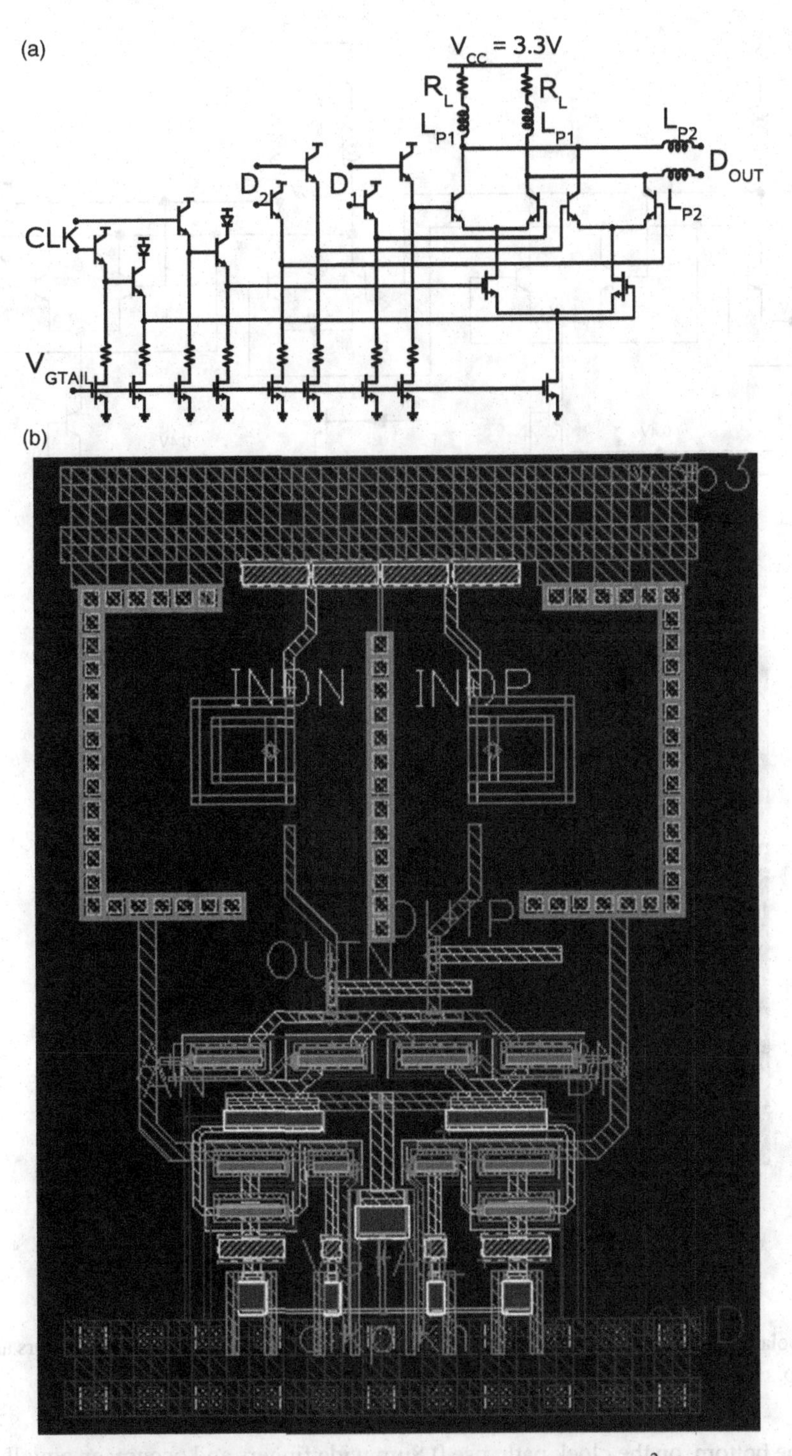

Figure 11.41 (a) Schematics and (b) layout of a 3.3V BiCMOS-E²CL 2:1 selector cell with series–shunt inductive peaking and emitter-followers on the clock path [27]

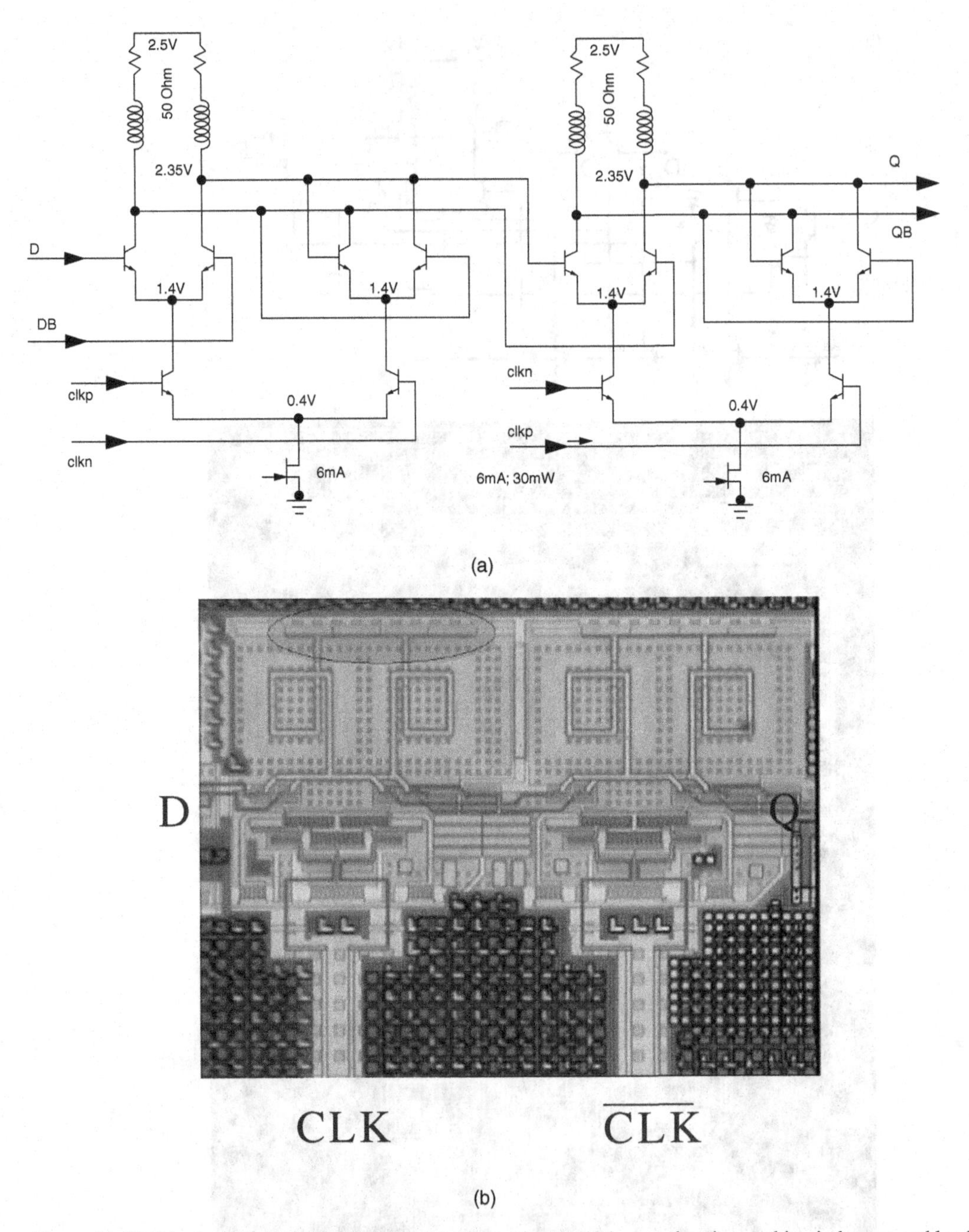

Figure 11.42 Bipolar ECL/BiCMOS flip-flop (a) schematic and (b) layout showing peaking inductors and load resistors at the top, clock inputs at the bottom, data inputs and outputs at left and right

MOSFETs at the bottom, on the clock path, use 0.8μm wide fingers and occupy an equally wide active area. The height of the latching quad active area is two times larger.

This arrangement leads to several speed-improving features. First, it minimizes the overall footprint, thus reducing interconnect capacitance at all circuit nodes. Second, the wider gate fingers

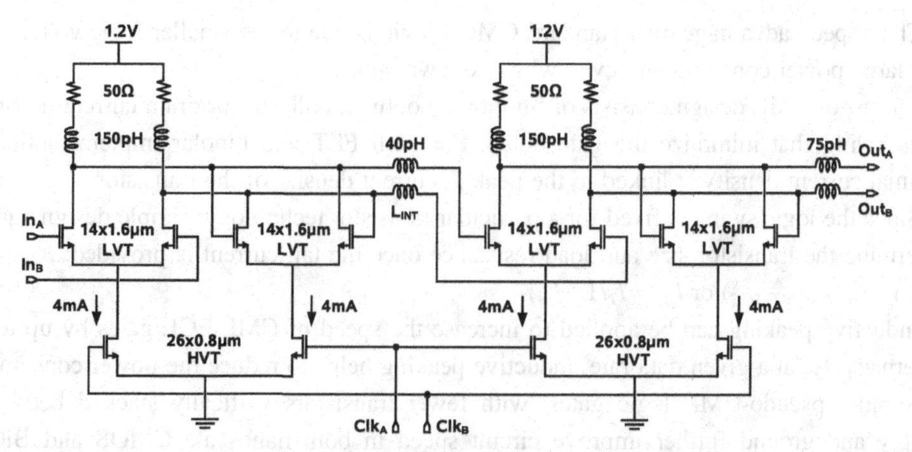

Figure 11.43 Schematic of a 81Gb/s D-type flip-flop in 65nm GP CMOS [28]

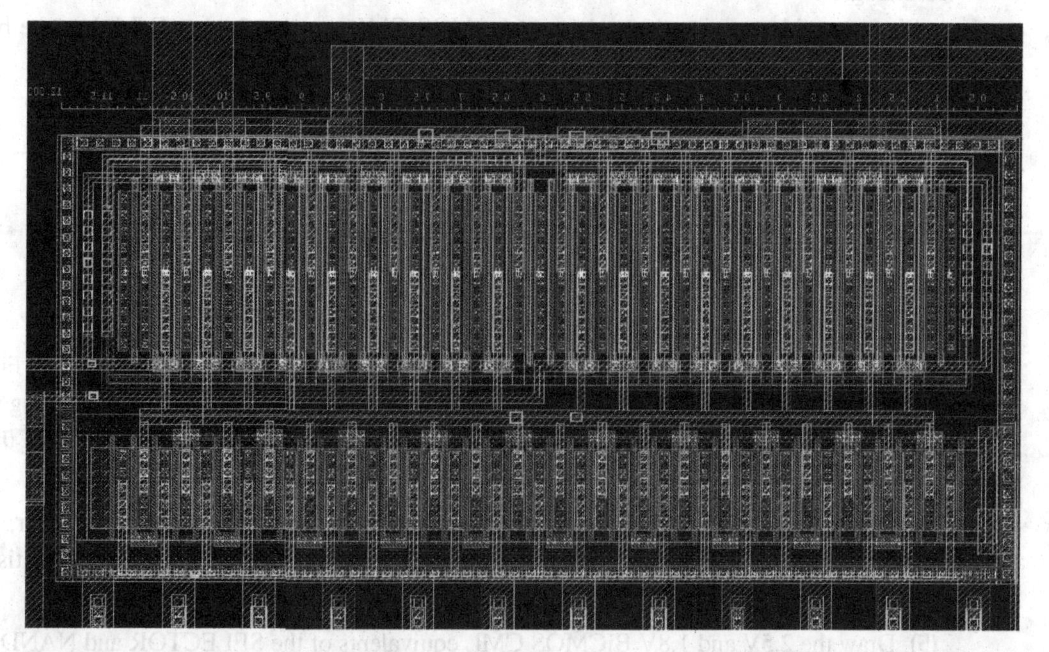

Figure 11.44 Layout detail showing the clock-pair and latching quad of the 80GHz latch in Figure 11.43

at the top reduce the output node capacitance by requiring fewer cross-overs between the gate and the drain interconnects. Third, the narrower fingers of the clock path MOSFETs reduce the gate resistance and the input time constant, further increasing the bandwidth of the critical clock path.

Summary

High-speed communication transceivers rely heavily on CML and ECL gates. These are realized with differential current-steering topologies which direct the tail current I_T from one side to the other of the differential pair with resistive loads.

Their speed advantage over standard CMOS logic is due to the smaller logic voltage swing and large power consumption, even when not switching.

The art of CML design consists of finding the optimal collector or drain current density and logic swing that minimize the gate delay. For both FET and bipolar implementations, the optimal current density is linked to the peak f_T current density of the transistor.

Since the logic swing is fixed for a particular transistor technology, simple design equations determine the transistor size and load resistance once the tail current is provided: $R_L = \Delta V/I_T$ and $W = I_T/(J_{pfTMOS})$ or $l_E = I_T/1.5J_{pfT}$.

Inductive peaking can be applied to increase the speed of CML/ECL gates by up to 60%. Alternatively, at a given data rate, inductive peaking helps to reduce the power consumption.

Simpler pseudo-CML logic gates, with fewer transistors vertically stacked between the supply and ground further improve circuit speed in both nanoscale CMOS and BiCMOS technologies.

At comparable transistor f_T and f_{MAX}, the MOS CML gate has higher FoM than the bipolar CML gate because of the lower supply voltage.

Problems

(1) Draw the timing diagrams of a 40Gb/s 2:1 MUX with 20GHz clock signal.

(2) Draw the timing diagrams of a 40Gb/s 1:2 DEMUX with 20GHz clock signal.

(3) Draw the schematic of a MOS CML selector with shunt peaking in 65nm CMOS. Find the transistor sizes, resistor sizes, and inductor sizes needed to maximize the switching speed. Assume that the tail current is 4mA, $\Delta V = 400$mV, a voltage gain of -1.5 and that the circuit has a fan-out of 2.

(4) Draw the p-MOSFET equivalents of a MOS CML logic family with inverter, XOR, SELECTOR, NAND, and LATCH. Design a 45nm p-MOS CML selector to satisfy the requirements in design Example 11.2.

(5) Draw the 2.5V and 1.8V BiCMOS CML equivalents of the SELECTOR and NAND gates. Design a 1.8V BiCMOS CML selector to satisfy the requirements in design Example 11.2. Use the 130nm SiGe BiCMOS technology data provided with Appendix g.

(6) Using the open-circuit-time-constant technique, find the expression of the dominant pole between the clock input and the data output of the MOS and BiCMOS CML-latch. Derive the expression as a function of the small signal parameters (capacitances, transconductance, load resistor). How can you account for inductive peaking at the output node? Redo the derivation for the pseudo-CML MOS and BiCMOS latches. Is the pseudo-CML latch faster, slower or the same as the corresponding CML latch?

(7) Scale the 1mA 130nm SiGe BiCMOS pseudo-CML latch to the next generation 65nm SiGe BiCMOS process and add inductive peaking. Determine the maximum speed. Assume that the HBT f_T increases from 160GHz to 280GHz, τ_F decreases from 0.6ps

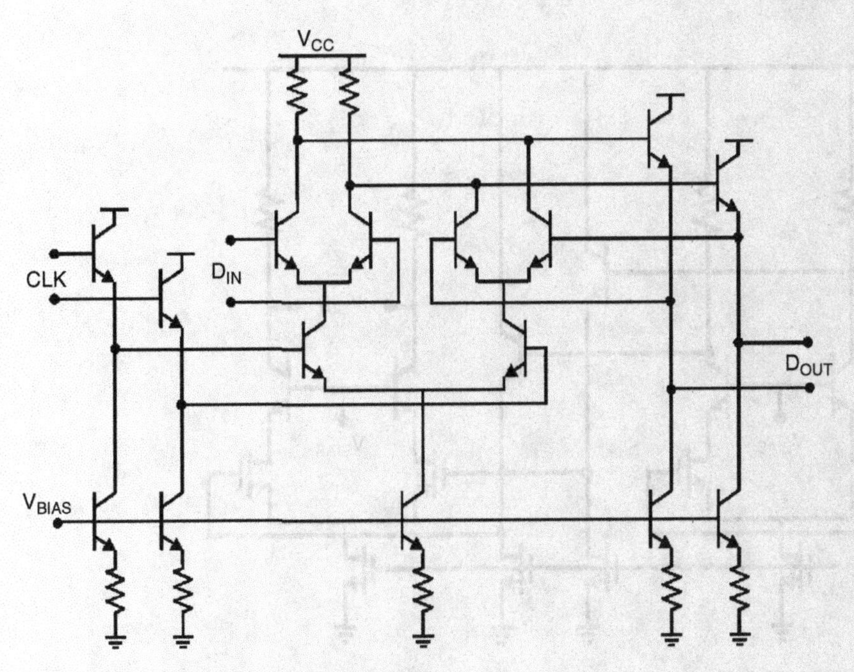

Figure 11.45 Bipolar ECL latch schematic

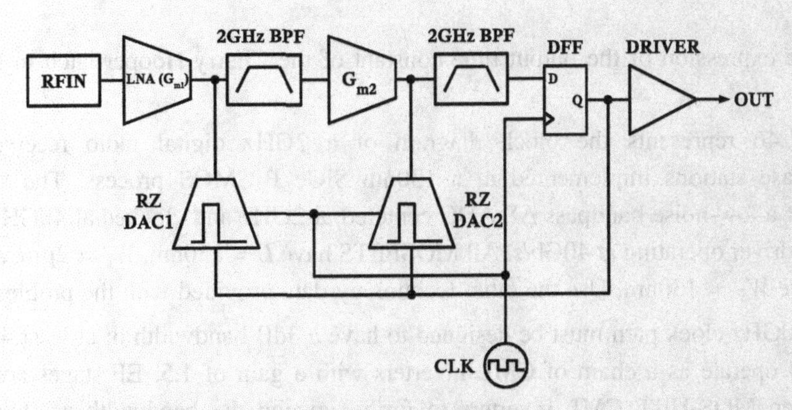

Figure 11.46 Digital radio receiver block diagram

to 0.25ps, w_E decreases from 170nm to 110nm, while C_{cs} remains unchanged and C_{bc} doubles from 11fF/μm² to 22fF/μm². J_{pft} changes from 8mA/μm² to 16mA/μm². For 65nm MOSFETs, use the data provided with the problem set.

(8) For the 65nm pseudo-CML CMOS latch in Figure 11.43, assume $J_{pfT} = 0.4$mA/μm, $A_V = -1.5$, $\Delta V = 350$mV. Calculate the 3dB bandwidth and the maximum data rate on the data path.

(9) Analyze the output time constant in a flip-flop consisting of two cascaded ECL latches like the one in Figure 11.45. Knowing that each degeneration resistor has a voltage drop of 150mV, that $V_{BE} = 0.85$V and that V_{CE} should be larger than 600mV on all HBTs, find the minimum supply voltage needed to operate this circuit.

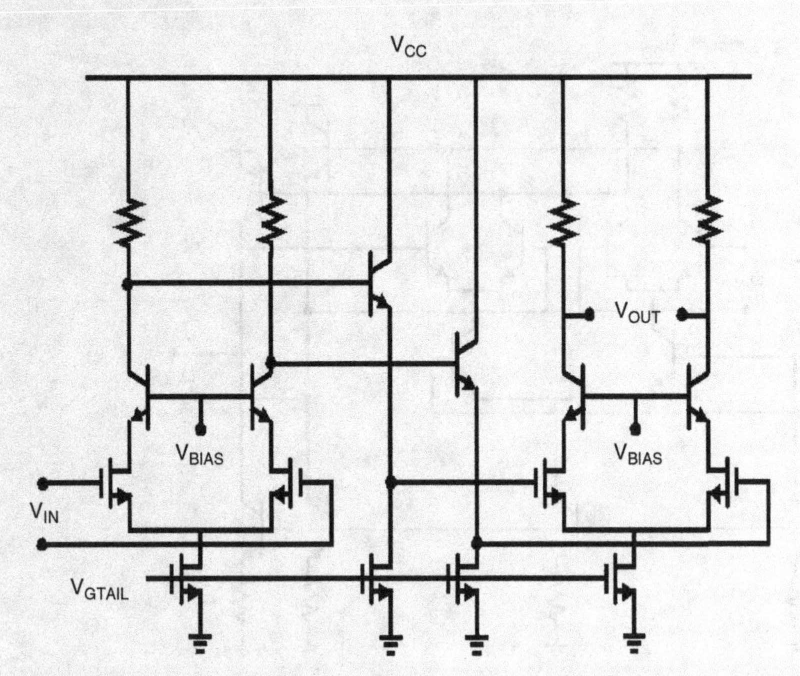

Figure 11.47 EF-CML-inverter chain employed on clock path. Peaking inductors are not shown

(10) Derive the expression of the output time constant of the Cherry-Hooper latch in Figure 11.36.

(11) Figure 11.46 represents the block diagram of a 2GHz digital radio receiver for cellular base stations implemented in a 130nm SiGe BiCMOS process. The system consists of a low-noise bandpass ΔΣ ADC centered at 2GHz and clocked at 40GHz, and an output driver operating at 40Gb/s. All MOSFETS have $L = 130\text{nm}$, $W_F = 2\mu\text{m}$, and all HBTs have $W_E = 130\text{nm}$. Use the other technology data provided with the problem set.

(a) The 40GHz clock path must be designed to have a 3dB bandwidth of at least 40GHz and to operate as a chain of CML inverters with a gain of 1.5. EF stages are used between MOS-HBT CML inverters to further extend the bandwidth as shown in Figures 11.47 and 11.48. Design the optimal size of the MOS-HBT CML inverter components for a tail current of 3mA and shunt peaking, knowing that the minimum voltage swing required to fully switch a 130nm MOS differential pair biased for maximum switching speed is 600mV_{pp}. Ignore the gate resistance. What is the bandwidth of the circuit if the CML inverter drives an EF stage with 2-mA current per side? The EF transistors are biased at ½ peak f_T current density.

(b) What is the lowest total power dissipation of the 40GHz clock path in Figure 11.48 if, to ensure adequate bandwidth, the scaling factor between two consecutive inverters in the chain must be smaller or equal to 2?

(c) The schematic of the master latch in the master–slave–master DFF is shown in Figure 11.49. Calculate the total node capacitance seen by the 35Ω load resistor.

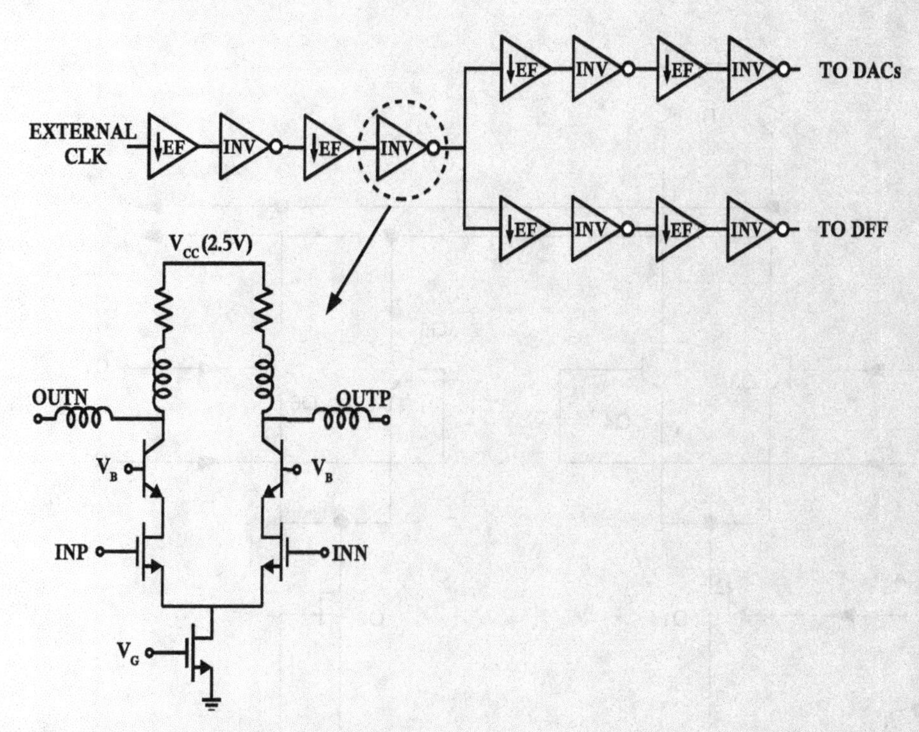

Figure 11.48 40GHz ADC clock path

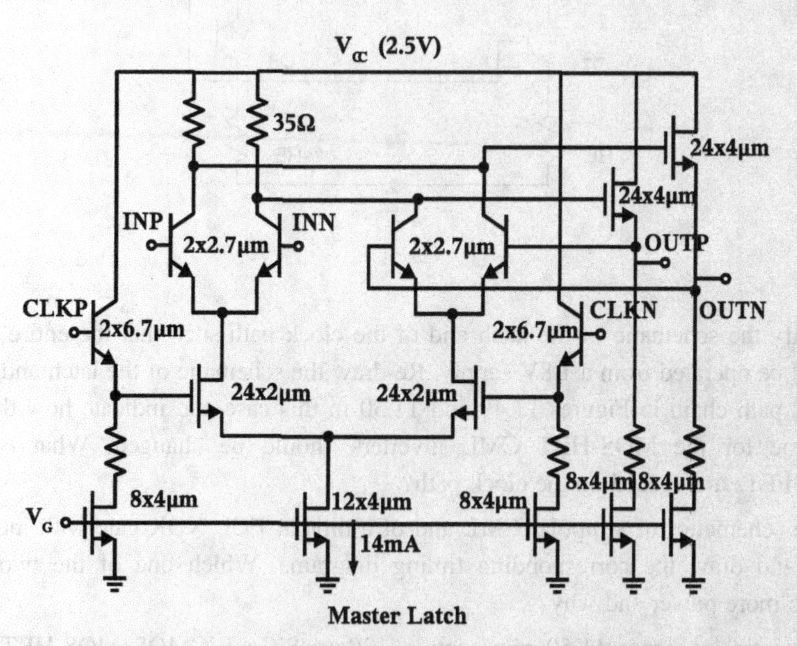

Figure 11.49 130nm SiGe BiCMOS master latch. All MOSFETs have L =130nm and all HBTs have w_E = 130nm

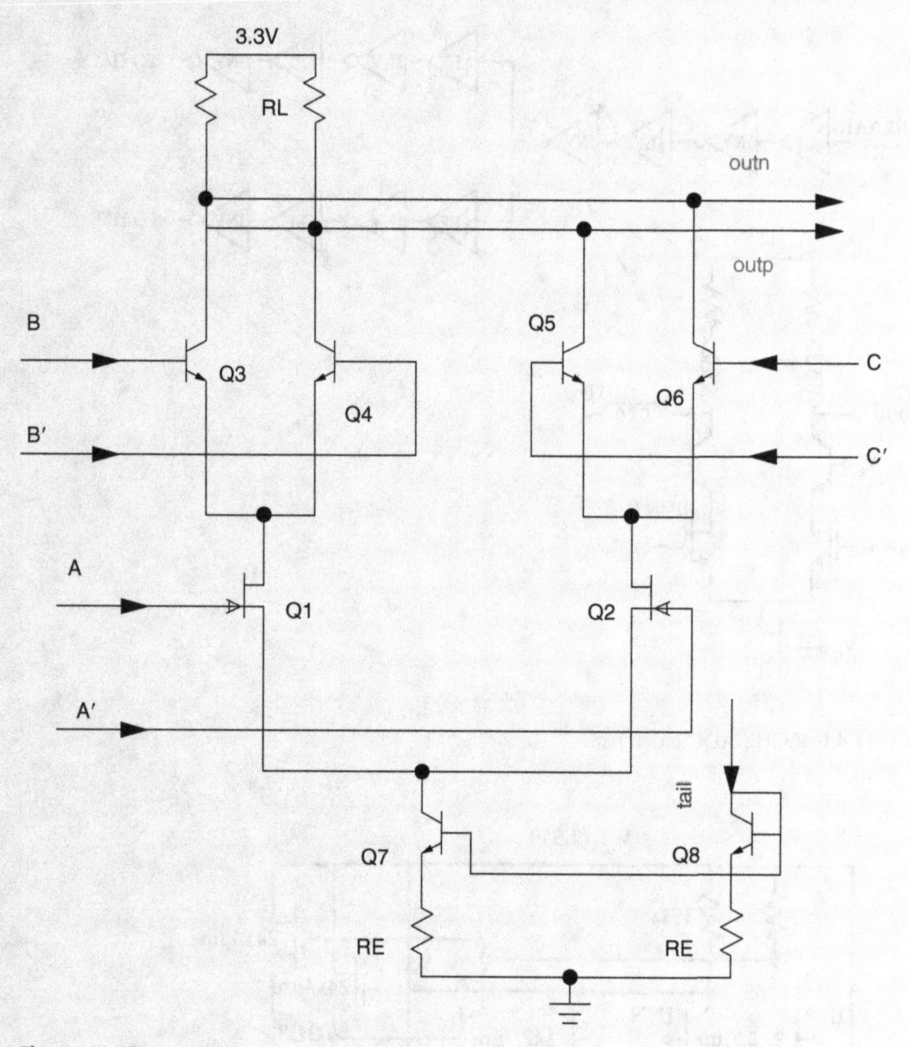

Figure 11.50

(d) Modify the schematic of the latch and of the clock path such that the entire circuit could be operated from a 1.8V supply. Re-draw the schematic of the latch and of the clock path chain in Figures 11.49 and 11.50 in this case and indicate how the bias scheme for the MOS-HBT CML inverters should be changed. What type of MOSFETs are needed on the clock path?

(12) Draw the schematics of a bipolar CML and of a bipolar ECL XOR gate with inductive peaking and draw the corresponding timing diagrams. Which one of the two gates consumes more power and why?

(13) The schematic in Figure 11.50 represents a 130nm SiGe BiCMOS MOS-HBT CML selector cell.

(a) If the circuit is operated as a selector (MUX) cell, write the logic function that describes the differential output signal as a function of the differential inputs AA′, BB′, and CC′.

(b) Determine the minimum tail current and the corresponding size of transistors $Q1$-$Q6$ and value of the load resistance R_L such that the circuit operates with 300 mVp-p logic swing per side, a load capacitance of 200fF at each output, and with a small signal open-circuit time constant of 8ps. You may use inductive peaking to minimize power dissipation but you must ensure that a flat group delay response is maintained. Assume that the inputs BB' and CC' are driven 0-impedance signal sources.

(14) Using the 45nm CMOS technology data, design a static CMOS inverter with minimum gate length. Assume that the n-MOSFET has a total gate width of 5μm and the gate finger width of 0.4μm and that the power supply voltage is 1.1V.

(a) If the p-MOSFET has the same number of gate fingers as the n-MOSFET, what is the optimal finger width of the p-MOSFET to obtain identical transconductance and I_{ON} as the n-MOSFET?

(b) Find the delay of a chain of minimum gate length static CMOS inverters with a fanout of 2. Assume that the first inverter in the chain is the one in part (a) of this problem.

(c) What happens to the bandwidth of the chain and to the delay of the CMOS inverter if series inductors are placed between each CMOS inverter in the chain? How should those inductors be scaled from the first inverter in the chain to the last?

(d) Assume now that you want this chain of inverters (without inductors) to drive an external 50Ω load with a rail-to-rail swing of $1.1V_{pp}$. How many inverters do you need in the chain? What is the power consumption of this inverter chain when it is driven by a 20GHz VCO signal with $1.1V_{pp}$ swing? How does this power consumption compare to that of a MOS CML inverter with 25Ω load resistors and 400mV swing per side operating from 1.1V supply running at the same speed (*Note*: 20GHz means 40Gb/s).

REFERENCES

[1] S.P. Voinigescu, P. Popescu, P. Banens, M. Copeland, G. Fortier, K. Howlett, M. Herod, D. Marchesan, J. Showell, S. Szilagyi, H. Tran, and J. Weng, "Circuits and technologies for highly integrated optical networking ICs at 10Gb/s to 40Gb/s," *Proc. of the IEEE CICC*, pp. 331–338, May 2001.

[2] Behzad Razavi, *Design of Integrated Circuits for Optical Communications*, Chapter 11, McGraw-Hill, 2003.

[3] J. Cao, M. Green, A. Momtaz, K. Vakilian, D. Chung, K-C. Jen, M. Caresosa, X. Wang, W-G. Tan, Y. Cai, I. Fujimori, and A. Hairapetian, "OC-192 transmitter and receiver in standard 0.18μm CMOS," *IEEE JSSC*, **37**(12):1768–1780, December 2002.

[4] H.-M. Rhein "Si and SiGe Bipolar ICs for 10 and 40Gb/s optical-fiber TDM links," in *Int.J. Of High-Speed Electron.Syst.*, **9**):1–37, 1998.

[5] D. Johns and K. Martin, *Analog Integrated Circuit Design*, *John* Wiley & Sons, 1997.

[6] E. Laskin, S. T. Nicolson, P. Chevalier, A. Chantre, B. Sautreuil, and S. P. Voinigescu. "Low-power, low-phase noise SiGe HBT static frequency divider topologies up to 100GHz," *IEEE BCTM Digest*, pp. 235–238, October 2006.

[7] T. O. Dickson, K. H. K. Yau, T. Chalvatzis, A. Mangan, R. Beerkens, P. Westergaard, M. Tazlauanu, M. T. Yang, and S. P. Voinigescu, "The invariance of characteristic current densities in nanoscale MOSFETs and its impact on algorithmic design methodologies and design porting of Si(Ge) (Bi)CMOS high-speed building blocks," *IEEE JSSC*, **41**(8): 1830–1845, August 2006.

[8] T. O. Dickson and S. P. Voinigescu, "Low-power circuits for a 10.7-to-86 Gb/s serial transmitter in 130nm SiGe BiCMOS," *IEEE JSSC*, **42**(10): 2077–2085, October 2007.

[9] T. Suzuki, Y. Nakasha, T. Takahashi, K. Makiyama, K. Imanishi, T. Hirose, and Y. Watanabe, "A 90Gb/s 2:1 Multiplexer IC in InP-based HEMT technology," *IEEE ISSCC Digest*, pp. 192–193, February 2002.

[10] T. O. Dickson, R. Beerkens, and S.P. Voinigescu, "A 2.5-V, 45Gb/s decision circuit using SiGe BiCMOS logic," *IEEE JSSC*, **40**(4): 994–1003, 2005.

[11] E. Laskin and S.P. Voinigescu, "A 60mW per Lane, 4 × 23Gb/s 2^7–1 PRBS generator," *IEEE JSSC*, **41**(10): 2198–2208, October 2006.

[12] P. Popescu, D. Dobson, G. Fortier, and J. Weng, "High speed logic circuits," US Patent 6,774,721 B1 granted August *10, 2004*.

[13] T. Chalvatzis, K.H.K. Yau, P. Schvan, M.T. Yang, and S. P. Voinigescu, "Low-voltage topologies for 40+ Gb/s circuits in nanoscale CMOS," *IEEE JSSC*, **42**(7): 1564–1573, July 2007.

[14] D. Kehrer and H.-D. Wohmuth, "A 60Gb/s 0.7V 10-mW monolithic transformer-coupled 2:1multiplexer in 90nm CMOS," in *IEEE CSICS Digest*, pp. 105–108, October 2004.

[15] H. Knapp *et al.*, "Static frequency dividers up to 125GHz in SiGe:C bipolar technology," *IEEE BCTM Digest* October 2010.

[16] Z. Griffith, M. Urteaga, R. Pierson, P. Rowell, M. Rodwell, and B. Brar, "A 204.8GHz static divide-by-8 frequency divider in 250nm InP HBT," *IEEE CSICS Digest*, pp. 53–56, October 2010.

[17] J. Kim, and J.O. Plouchart, "Wideband mm Wave CML static divider in 65nm S01 CMOS technology," *ISSCC Digest*, February 2008.

[18] S.P. Voinigescu, R. Aroca, T.O. Dickson, S.T. Nicolson, T. Chalvatzis, P. Chevalier, P. Garcia, C. Garnier, and B. Sautreuil, "Towards a sub-2.5V, 100Gb/s serial transceiver," IEEE CICC, pp. 471–478, *San Jose, CA*, Sept. 2007.

[19] T.O. Dickson and S.P. Voinigescu, "SiGe BiCMOS topologies for low-voltage millimeter-wave voltage-controlled oscillators and frequency dividers," *Si Monolithic Integrated Circuits in RF Systems, Technical Digest*, pp. 273–276, Jan. 2006.

[20] H. Knapp, T. F. Meister, W. Liebl, K. Aufinger, H. Schäfer, Josef Böck, S. Boguth, and R. Lachner, "168GHz dynamic frequency divider in SiGe:C bipolar technology," *IEEE BCTM Digest*, pp 190–193, October 2008.

[21] A. Tomkins, R.A. Aroca, T. Yamamoto, S.T. Nicolson, Y. Doi, and S.P. Voinigescu, "A Zero-IF 60GHz 65nm CMOS Transceiver with direct BPSK modulation demonstrating up to 6Gb/s data rates over a 2m wireless link," *IEEE JSSC*, **44**(8): 2085–2099, August 2009.

[22] E. Laskin, M. Khanpour, R. Aroca, K.W. Tang, P. Garcia, S.P. Voinigescu, "95GHz receiver with fundamental frequency VCO and static frequency divider in 65nm digital CMOS," *IEEE ISSCC Digest*, pp. 180–181, February 2008.

[23] R.L. Miller, "Fractional-frequency generators utilizing regenerative modulation," *Proc. IRE*, **27**: 446–457, July 1939.

[24] I. Sarkas, J. Hasch, A. Balteanu, and S. Voinigescu, "A fundamental frequency, single-chip 120GHz SiGe BiCMOS precision distance sensor with above IC antenna operating over

several meters," submitted to the special mm-wave circuits and systems issue of *IEEE MTT*, **60**(3): 735–812, March 2012.

[25] M. Seo, M. Urteaga, A. Young, and M. Rodwell, "A 305–330+ GHz 2:1 dynamic frequency divider using InP HBTs," *IEEE Microwave and Wireless Component Letters*, pp. 468–470, June 2010.

[26] Shwetabh Verma, Hamid R. Rategh, and Thomas H. Lee, "A unified model for injection-locked frequency dividers," *IEEE JSSC*, **36**(6):1015–1027, June 2003.

[27] T. O. Dickson, E. Laskin, I. Khalid, R. Beerkens, J. Xie, B. Karajica, and S.P. Voinigescu, "An 80Gb/s 2^{31}–1 Pseudo-random binary sequence generator in SiGe BiCMOS technology," *IEEE JSSC*, **40**(12): 2735–2745, December 2005.

[28] S. Shahramian, A. C. Carusone, P. Schvan, and S.P. Voinigescu, "An 81Gb/s, 1.2V TIALA-retimer in Standard 65nm CMOS," *IEEE CSICS Digest*, pp. 215–218, Monetery, CA, October 2008.

12 High-speed digital output drivers with waveshape control

WHAT IS A HIGH-SPEED DIGITAL OUTPUT DRIVER?

High-speed digital output drivers can be regarded as very broadband, DC- or AC-coupled power amplifiers which operate in switching (or limiting) mode and transmit high-speed digital signals off chip to the outside world. In wireline applications, the output signal is sent into a 50Ω transmission line on a board, or to a coaxial cable. In fiber-optic systems, the signal (voltage or current, rather than power) from the output driver modulates the bias current of a laser diode or the bias voltage of an electro-optical modulator. As illustrated in Figure 12.1, the data input is typically differential and operates with signal levels in the 50mV_{pp} to 1.5V_{pp} range, while the data output can be either differential or single-ended. In some standalone laser or modulator drivers, the data are retimed using an external clock input before being amplified and re-shaped.

Like tuned PAs, high-speed digital output drivers have widely varying requirements and typically impose the toughest demands on semiconductor technologies in terms of breakdown voltage and transistor speed. As a result, linear-mode operation is avoided because it is very costly and very difficult to accommodate.

There has been a growing trend towards introducing flexibility in the shape of the output signal. Control over the output waveform is often needed to overcome imperfections in the response of wireline channels or of optical devices such as lasers and modulators. Initially, because of power consumption constraints and bandwidth requirements, all waveshape control functions were realized with analog techniques. Over the years, as semiconductor technologies have become faster and higher-order modulation formats have been introduced, there has been a marked shift towards digital techniques, first at the lower data rates up to

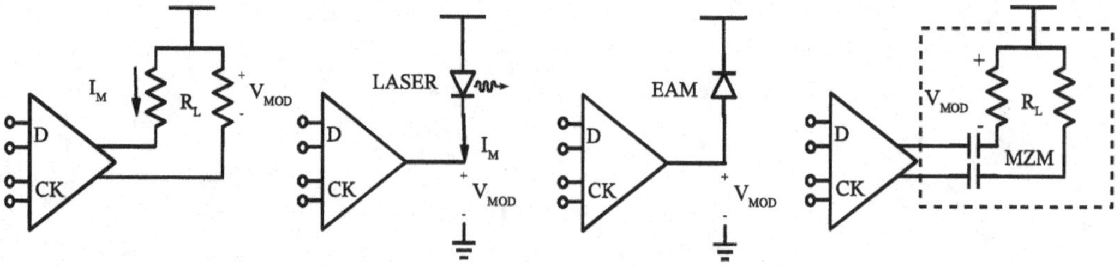

Figure 12.1 General representation of the main types of output drivers

10 Gb/s, and culminating with 6-bit large-swing DAC implementations of output drivers at data rates as high as 60Gb/s [1].

This chapter discusses the main types of high-speed digital drivers with output waveform control, their specifications and typical waveshape control functions. Design methodologies and examples are provided for the most common driver topologies.

12.1 TYPES OF HIGH-SPEED DRIVERS

In general, based on the type of load which determines the signal swing and circuit topology, high-speed output drivers can be classified in:

- backplane or coaxial cable drivers, often referred to as I/Os in most digital chips,
- laser drivers,
- Electro-Absorption Modulator (EAM) drivers, and
- Mach–Zehnder Modulator (MZM) drivers.

12.2 DRIVER SPECIFICATION AND FoMs

In addition to the bit-rate, R_B, the following parameters are specified for on-off keying, high-speed digital output drivers:

- output voltage swing for backplane and coaxial drivers,
- modulation and bias current range for lasers drivers,
- voltage swing and bias voltage for modulator drivers
- power dissipation,
- rise and fall times,
- pulse-width distortion,
- jitter generation,
- eye-diagram mask test, and
- input/output return loss.

Each one of them is discussed in more detail next.

12.2.1 Output voltage swing and DC output level

The peak-to-peak output voltage swing, V_{MOD}, in a 50Ω load is always specified for backplane and cable drivers. Typical values range from 300mV$_{pp}$ per side to over 5V$_{pp}$ per side, depending on the application. For applications where the driver is DC-coupled to the external load, the DC voltage range at the output is also specified. The latter is typically set to $V_{DD} - V_{MOD}/2$, also known as a "CML level" where V_{DD} is the supply voltage of the output stage and V_{MOD} is equivalent to the CML voltage swing of the output stage.

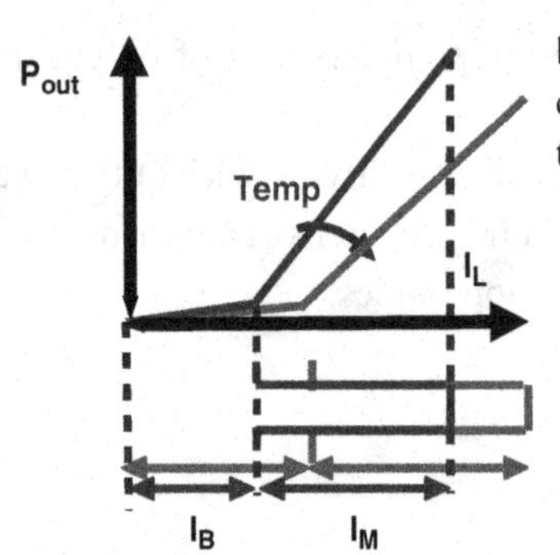

Figure 12.2 Typical optical power, P_{out}, versus current, I_L, characteristics of a laser diode at two different temperatures illustrating the bias current, I_B, and modulating current waveform, I_M

12.2.2 Modulation and bias current range

The generic optical power versus bias current (known as *L-I*) characteristics of a laser diode are illustrated in Figure 12.2. The DC current, I_B, forward biases the laser diode close to its threshold current, I_{TH}. Consequently, the total current flowing through the laser is the sum of the DC current I_B and the high-speed data modulated current which switches between 0 and I_M.

Several types of lasers are used in fiber-optic communications. The most common are the Fabry Perot, the distributed-feedback (DFB), and Vertical-Cavity-Surface-Emitting Lasers or VCSELs. They each have different *L-I* characteristics and therefore impose different requirements on the laser driver. The nominal range of bias currents varies between 20mA and 80mA in DFB and Fabry-Perot lasers, and from 0 to 20mA in VCSELs. Similarly, the required modulation current, I_M, ranges between 5 and 80mA for DFB and Fabry-Perot lasers, and between 5 and 15mA in VCSELs.

The threshold current and slope of the *L-I* characteristics are relatively strong functions of temperature. As a result, the laser driver is required to provide a wide range of programmable bias and modulation current values, with programmable temperature coefficients.

12.2.3 Modulation voltage and bias voltage range

An electro-absorption modulator (EAM) is typically integrated along with a continuous-wave (CW) DFB laser on an InP substrate, as illustrated in Figure 12.3(a) [2]. The modulator is formed by inserting a quantum-well layer between the *p* and *n* regions of a diode. According to the *Franz-Keldysh* and the *quantum-confined Stark* effects, the bandgap of the semiconductor in the quantum-well can be modulated by applying an electric field across it. The quantum-well, QW, region is designed such that, when the voltage across the *pn* junction is 0V, its bandgap is wide enough to be transparent to the optical wavelength emitted by the laser. When a reverse voltage larger than V_{SW} is applied to the diode, the bandgap is reduced and the semiconductor region becomes opaque to the radiated laser light, attenuating it.

The relationship between the optical power at the output of the modulator, P_{OUT}, and the reverse bias voltage applied on the modulator, V_M, shown in Figure 12.3(b), is known as the switching curve of the EAM. The difference between the optical power at 0V and at a

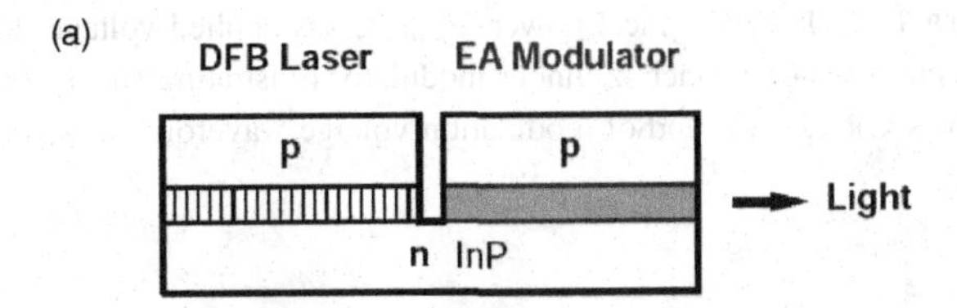

Switching curve and electrical equivalent circuit:

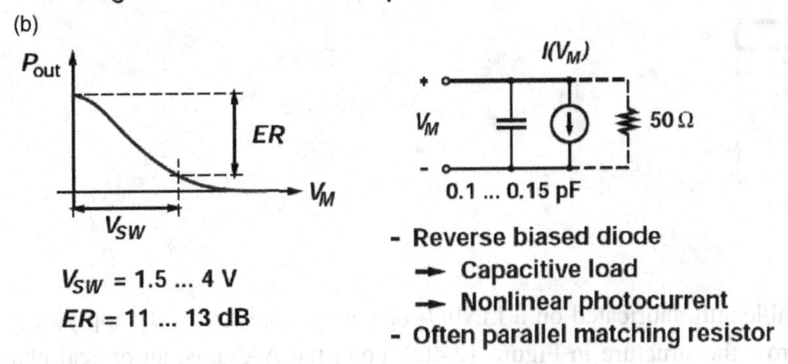

Figure 12.3 (a) Laser + EA modulator cross-section (b) Switching characteristics, and equivalent electrical circuit [2]

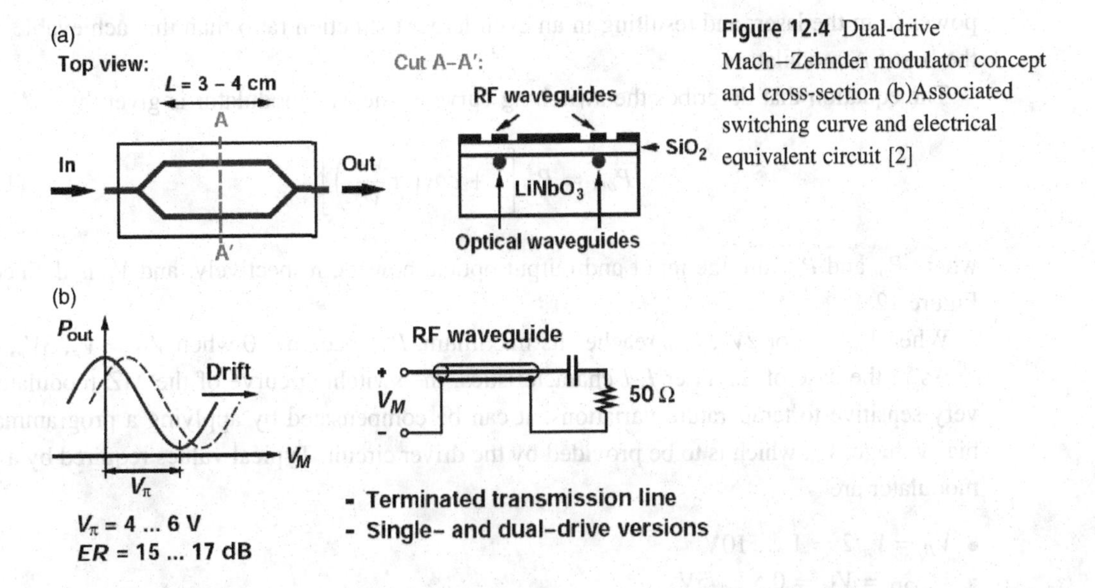

Figure 12.4 Dual-drive Mach–Zehnder modulator concept and cross-section (b)Associated switching curve and electrical equivalent circuit [2]

sufficiently large reverse voltage ($>V_{SW}$) is known as the extinction ratio, *ER*. *ER* is specified in dB and is typically in the 11–13 dB range [2].

The EAM is DC-coupled to the driver and the following bias voltage and modulating voltage values need to be provided by the driver

- $V_B = 0 \ldots 1\text{V}$ (also known as the DC offset)
- $V_{MOD} = 0.5 \ldots 3\text{V}$.

A Mach–Zehnder modulator, illustrated in Figure 12.4, is based on the interferometer principle. It consists of an optical power splitter and power combiner separated by two optical waveguides

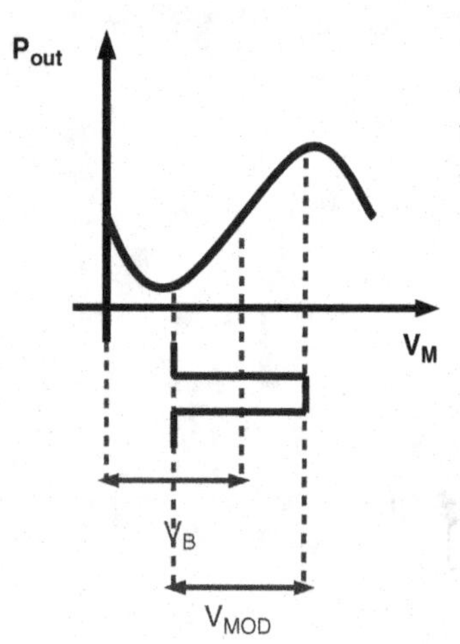

Figure 12.5 Typical optical power, P_{out}, versus applied voltage, V_M, characteristics of a Mach–Zehnder modulator illustrating superposition of the bias voltage, V_B, and of modulation voltage waveform, V_{MOD}

of equal physical length, fabricated on a $LiNbO_3$ or polymer substrate. By applying a vertical electric field across the structure in Figure 12.4(a) along the AA' axis, an optical phase delay develops between the two optical waveguides. By adjusting the phase delay, the signals at the output of the modulator can add constructively or destructively, thus modulating the optical power from the laser, and resulting in an even larger extinction ratio than that achievable with the best EA modulators.

The equation that describes the switching curve of the MZ modulator is given by

$$P_{out} = P_{in}\left[1 + cos\left(\pi\frac{V_M}{V_\pi}\right)\right]. \tag{12.1}$$

where P_{in} and P_{out} are the input and output optical powers, respectively, and V_π is defined in Figure 12.4.

When $V_M = 0$ or $2V_\pi$, P_{out} reaches its maximum, P_{out} becomes 0 when $V_M = V_\pi$, $3V_\pi$, etc.

As in the case of the laser *L-I* characteristics, the switching curve of the MZ modulator is very sensitive to temperature variations. It can be compensated by applying a programmable bias voltage, V_B, which is to be provided by the driver circuit. Typical values required by a MZ modulator are

- $V_B = V_\pi/2 = 1 \ldots 10V$
- $V_{MOD} = V_M = 0.5 \ldots 5V$.

The high-frequency electrical inputs of the MZ modulator are AC-coupled to the driver circuit and, therefore, as illustrated in Figure 12.5, the modulating voltage waveforms swing below and above the bias voltage, V_B.

A considerable research effort has been dedicated in recent years, with some success [3] in realizing electro-optical modulators in silicon SOI technologies. This is particularly important for future high density and high-data-rate optical chip-to-chip I/O links [4]. Ideally, the switching voltage, V_{SW}, should be 2V or lower such that low breakdown voltage nanoscale MOSFETs could be employed as drivers.

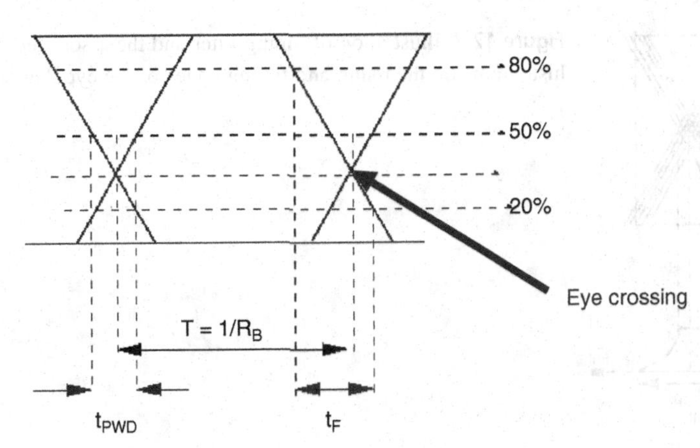

Figure 12.6 Illustration of the definitions of rise and fall times and of pulse-width (or duty-cycle) distortion, t_{PWD}

12.2.4 Power dissipation

Because they have to deliver large voltage or current swings into 25–50 Ω loads, output drivers consume significantly larger power than other high-speed circuits such as TIAs and logic cells. Minimizing the power consumption for a given data rate and voltage swing is always one of the most important goals of high-speed output driver design. The power consumption of the output stage (ignoring the DC power required to bias the laser or optical modulator) is given by

$$P_{DC} = I_M \cdot V_{DD}. \tag{12.2}$$

Since the modulation current and bias current/voltage in most drivers are programmable, when specifying the power consumption of the output driver, the programmed values of these parameters must also be indicated.

Unlike standard I/Os and cable or wireline drivers, a large part of the overall power consumption of a laser driver is dissipated in the laser itself.

Typical power consumption values range from several tens of mW, for wireline and VCSEL drivers, to hundreds of mW in Fabry–Perot laser drivers, and several watts in Mach–Zender modulator drivers.

12.2.5 Rise and fall time

The rise and fall time, τ_R and τ_F, of an output driver are specified with respect to the amplitude of the output signal and are measured under nominal loading conditions at the nominal data rate, R_B. The most common definition, between the 20% and 80% points of the full output signal swing, is illustrated in Figure 12.6. The more stringent 10% to 90% definition, which results in longer rise and fall times, is less often employed. To ensure that the rise and fall times are not underestimated because the output waveform has not reached its full swing, the rise and fall times are measured for a repeating 11110000 bit pattern. Typically, the rise and fall time values specified for the driver should be less than $0.2UI$ (unit interval) relative to the pulse period T.

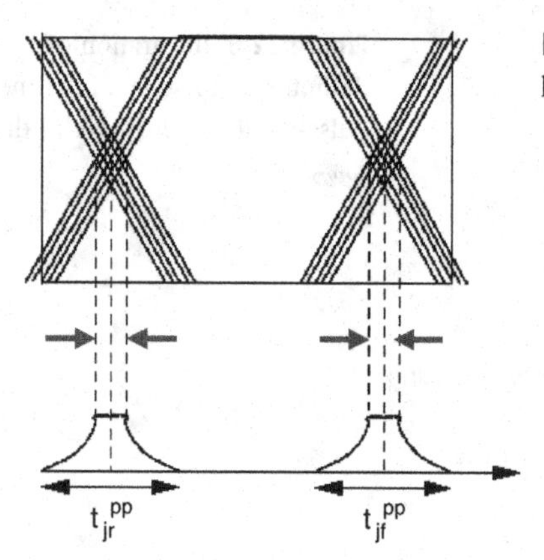

Figure 12.7 Illustration of timing jitter and the associated histograms on the rising and falling edges of the eye diagram

12.2.6 Pulse-width or duty-cycle distortion

The pulse-width distortion, t_{PWD}, is caused by DC offsets in the driver circuit or by non-idealities of the input data signal. It is often also referred to as duty-cycle distortion (DCD) and is defined relative to the ideal 50%-crossing of the eye diagram, as illustrated in in Figure 12.6. If the eye crossing is ideal at the 50% level, $t_{PWD} = 0$. For most drivers, the maximum allowed t_{PWD} is specified to be less than $+/-0.1UI$. In more advanced drivers, a pulse-width adjustment range of $+/-0.2UI$ is specified.

12.2.7 Jitter generation

In addition to the pulse-width distortion, high-speed digital output drivers introduce timing jitter, t_j, due to imperfections in their circuits. This can be:

- deterministic jitter, specified as a peak-to-peak value and caused by insufficient bandwidth of some or all circuits in the driver, and
- random jitter, which is specified in rms values, as illustrated in Figure 12.7, where $t_j^{pp} = 14 \times t_j^{rms}$.

Distinct double edges in the zero-crossing of the eye diagram are a typical manifestation of insufficient bandwidth in the driver. They produce the flat region of the jitter histogram, whereas the random jitter shows up as Gaussian skirts in Figure 12.7.

The jitter generation specification is prescribed by standards. The rms jitter is typically required to be below 0.01 *UI* while the peak-to-peak jitter should be lower than 0.1 *UI*, e.g. over a jitter bandwidth from 50Hz up to 80MHz in the case of the OC192 10Gb/s SONET standard.

Jitter generation is measured with "*jitter-free*" external clock and data signals. In practice, the jitter of the clock and data test signals is never 0, but should be less than 1/3 of the best jitter value expected to be measured.

12.2.8 Eye diagram mask test

The eye digram mask test checks the output eye diagram simultaneously for many imperfections: eye amplitude, rise and fall time, jitter, pulse-width distortion, etc. It is specified for each standard. An example eye-mask test is illustrated in Figure 12.8. The eye diagram is

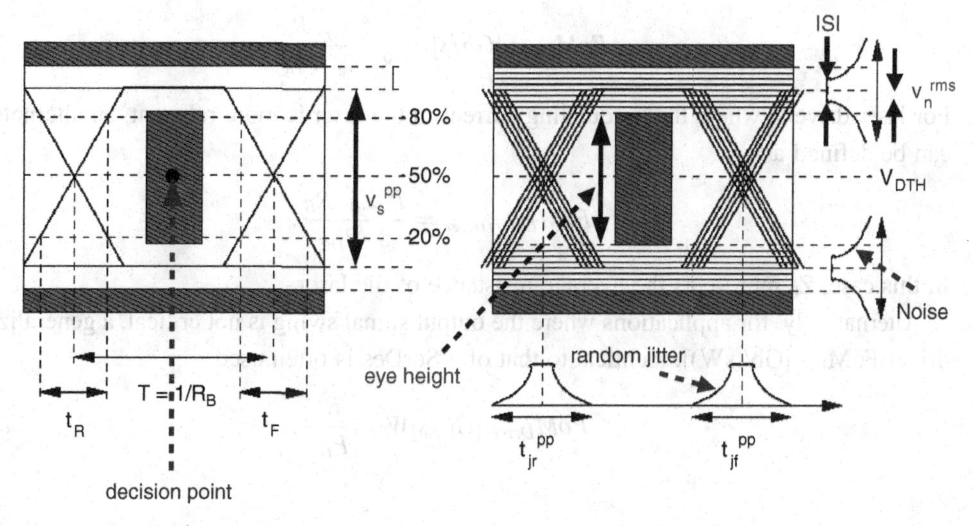

Figure 12.8 Eye-diagram mask test

required to avoid the shaded regions. The shaded rectangle in the center of the eye defines the minimum eye opening and quantifies the timing and voltage margin available for the decision circuit. Related to this is the *eye quality factor*, or *Q*-factor

$$Q = min\left(\frac{1}{2 \cdot R_B \cdot t_j^{rms}}, \frac{v_s^{pp}}{2 \cdot v_n^{rms}}\right) \tag{12.3}$$

where v_s^{PP} is the peak-to-peak eye amplitude in volts and v_n^{rms} is the rms noise voltage on the logic "1" and logic "0" levels, assumed to be equal.

12.2.9 Input and output return loss

For standalone drivers, the single-ended S_{11} and S_{22} must typically satisfy the following specification:

$S_{11}, S_{22} < -15\text{dB for } f < R_B/2$
$S_{11}, S_{22} < -10\text{dB for } f < R_B$
$S_{11}, S_{22} < -5\text{dB for } R_B < f < 2R_B.$

Most standards also specify the differential-mode and common-mode return loss.

12.2.10 Driver figures of merit

Based on the previous discussion of driver specification, a figure of merit can be derived that accounts for the data rate, R_B [Gb/s], DC power consumption, P_{DC} [W], and peak-to-peak output voltage swing, V_{MOD}, in a given load impedance Z_0

$$FoM_{Driver}[Gb/s] = \frac{V_{MOD}^2 \cdot R_B}{8 \cdot Z_0 \cdot P_{DC}} \tag{12.4}$$

For laser drivers, where the modulating current in the laser is more relevant, an alternate *FoM* can be defined as

$$FoM_{LaserDriver} = \frac{I_M^2 Z_0 \cdot R_B}{8 \cdot P_{DC}}. \tag{12.5}$$

In this case, Z_0 represents the average resistance of the laser.

Alternatively, for applications where the output signal swing is not critical, a generalized I/O driver $FoM_{I/O}$ [Gb/(sW)], identical to that of a SerDes is often used

$$FoM_{Driver}[Gb/s]W = \frac{R_B}{P_{DC}}. \tag{12.6}$$

12.3 DRIVER ARCHITECTURE AND BUILDING BLOCKS

A traditional high-speed digital output driver architecture, illustrated in Figure 12.9, incorporates the following building blocks:

- predriver,
- output buffer with waveshape control, and
- laser/modulator bias and control block (optional).

The generic architecture in Figure 12.9 also incorporates a number of signal (or waveshape) conditioning functions such as:

- pulse-width (duty cycle) control,
- output signal amplitude control,

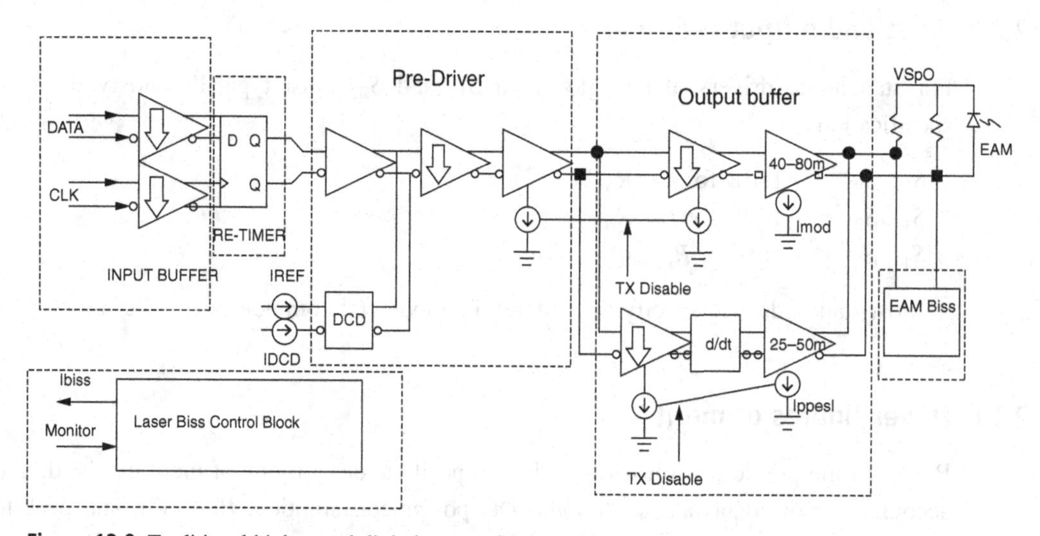

Figure 12.9 Traditional high-speed digital output driver architecture with waveshape control

- output signal pre-emphasis control,
- output DC offset level control.

Except for the pulse-width (or duty-cycle) control function, typically realized in the predriver, the signal conditioning functions are implemented in the output buffer.

Next, we will discuss topologies and design methodologies for each of these blocks and functions. As in the real design, we will start from the output stage and move back step-by-step towards the input.

It should be noted that this, mostly analog, high-speed output driver architecture is increasingly replaced at the lower data rates by large-swing digital-to-analog converters where all the signal-conditioning functions are performed by digital-signal-processing (DSP) techniques. Examples of such high-data-rate DAC-based drivers will be discussed at the end of the chapter.

12.4 OUTPUT BUFFERS

12.4.1 Backplane or coaxial cable drivers

The backplane output buffer is designed to drive differential 50Ω transmission lines on a PCB. It is realized using the differential current-steering topology with on-chip 50Ω loads, (also called back terminations [2]) shown in Figure 12.10. The same circuit can also be used in single-ended fashion to drive 50Ω or (with appropriate on-chip load resistors) 75Ω coaxial cables. Its operation and design are identical to those of a CML gate, where the tail (or modulation) current is steered from one side to another of the differential pair by the data signal applied at the differential input *in*/*inb*. The peak-to-peak output voltage swing per side (or CML logic swing) is given by

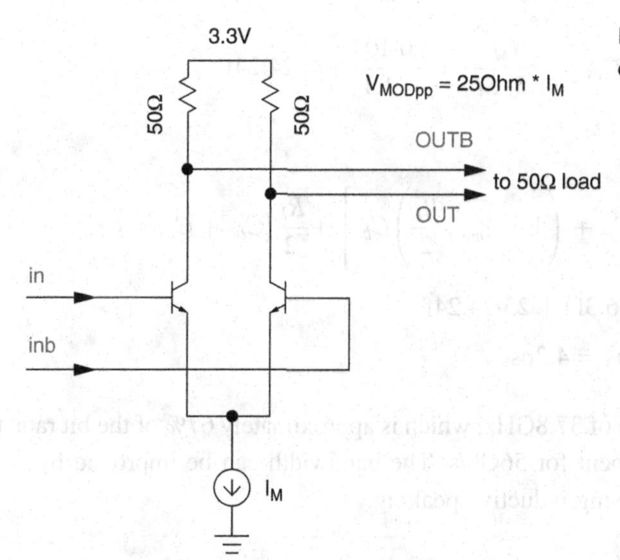

Figure 12.10 Current-steering 50Ω backplane driver circuit

$$V_{MOD} = \Delta V = \frac{I_M \cdot R_L}{2} \tag{12.7}$$

where R_L is the on-chip load resistance whose value is equal to the characteristic impedance of the PCB trace or of the coaxial cable. The DC voltage drop across the current source I_M is set by the DC voltage level at the input of the output stage.

EXAMPLE 12.1 56Gb/s SiGe HBT backplane driver with 0.5V$_{pp}$ swing per side

To obtain a voltage swing $V_{MOD} = 0.5V_{pp}$ per side, the required tail current becomes

$$I_M = \frac{2V_{MOD}}{R_L} = \frac{1V}{50\Omega} = 20mA.$$

As in CML gates, to maximize the switching speed, the emitter length of the SiGe HBTs in the differential pair must be chosen such that, when the entire I_M flows through one transistor, the maximum current density is approximately $1.5J_{pfT}$. Assuming a SiGe BiCMOS process with 0.13μm emitter width, $r_E = 4\Omega \times \mu m^2$, $R_b \times l_E = 100\Omega \times \mu m$, $\tau_F = 0.3ps$, $C_{je}/A_E = 17fF/\mu m^2$; $C_{jc}/A_E = 12fF/\mu m^2$; $C_{cs}/l_E = 1fF/\mu m$, $J_{pfT} = 12mA/\mu m^2$, and $BV_{CEO} = 1.8V$, we obtain the emitter length, l_E

$$l_E = \frac{I_M}{1.5 \cdot J_{pfT} \cdot W_E} = \frac{20}{1.5 \cdot 12 \cdot 0.13} = 8.5\mu m.$$

Let us now estimate the output pole, assuming a pad capacitance, C_{PAD}, of 50fF, and an intrinsic transistor slew rate, SL_i, of 0.55V/ps. For this transistor size and $I_C = 10mA$, we obtain $R_b = 11.8\Omega$, $r_E = 3.6\Omega$, $g_m = 380mS$, $g_{meff} = 160.5mS$, $C_{be} = g_m \tau_F + C_{je} = 114fF + 18.8fF = 132.8fF$.

From the slew-rate formula, we can estimate the total output capacitance of the HBT as

$$C_{bc} + C_{cs} = \frac{I_M}{1.5 \cdot SL_i} = \frac{20 \cdot 10^{-15}}{1.1} = 24.24fF.$$

The dominant pole is given by

$$\tau = R_b \left[g_{meff}\tau_F + C_{je} + \left(1 + g_{meff}\frac{R_L}{2}\right) C_{bc} \right] + \frac{R_L}{2}[C_{bc} + C_{cs} + C_{PAD}]$$

$$= 11.8(132.8f + 66.3f) + 25 \cdot 74.24f$$

$$= 2.35ps + 1.856ps = 4.2ps$$

This results in a 3dB bandwidth of 37.8GHz, which is approximately 67% of the bit rate, falling short of the minimum requirement for 56Gb/s. The bandwidth can be improved by compensating the output capacitance using inductive peaking.

EXAMPLE 12.2 56Gb/s backplane driver in 65nm CMOS

Similarly, for a 65nm CMOS process with J_{pfT} = 0.4mA/μm, g_{meff} = 1mS/μm C'_{gd} = 0.35fF/μm $C'_{gs} = C'_{db}$ = 0.65fF/μm, R'_g = 200Ω per 1μm wide fingers, and V_{MOD} = 0.5V_{pp} per side, we obtain:

I_M = 20mA, $I_M/J_{pfT} = W(Q_{1,2})$ = 20mA/(0.4mA/μm) = 50μm. If W_f = 1μm then N_f = 50. Based on this transistor size, g_{meff} = 50mS and $A_v = -1.25$ are calculated. Finally, the dominant pole is calculated as

$$\tau = R_g(C_{gs} + 2.25C_{gd}) + R_L/2\,(C_{db} + C_{gd} + C_{PAD}) = 4(32.5\text{fF} + 39.375\text{fF}) + 25\,(50\text{fF} + 50\text{fF})$$
$$= 0.283\text{ps} + 2.5\text{ps} => \text{BW}_{3\text{dB}} = 57\text{GHz}.$$

Although there seems to be sufficient margin, in a full CMOS driver implementation, the input node time constant is limited by the predriver stage. Nevertheless, the previous design examples illustrate the significant impact of the Miller capacitance and of the base resistance in bipolar drivers.

12.4.2 Laser driver

Examples of typical laser driver output stages, where the bias current of the laser diode is modulated by the high-speed data signal coming from the driver, are shown in Figure 12.11. The circuits are based on the same current steering concept described earlier. In the simplest version, shown in Figure 12.11(a) [2], the laser diode replaces one of the two load resistors R_L and is DC-coupled to the driver. This topology is known as open-collector or open-drain because there is no back-termination on the driver itself, only at the far end. Since the small signal equivalent circuit of the laser diode is formed by a resistor (R_L = 5–25 Ω) in parallel with a capacitance (C_L = 0.1–0.3 pF), the load in the opposite branch of the current-steering output stage is adjusted to match the impedance of the laser diode and thus ensures that symmetrical waveforms are maintained. The relationship between the modulating current and the output voltage swing of the laser driver remains the same as (12.1) where R_L can vary between 15 and 50Ω, depending on the laser type.

This topology works well at data rates below 10Gb/s, or in monolithic optoelectronic circuits where the laser diode and the driver are integrated on the same (III-V) semiconductor substrate. However, in most practical electro-optical modules, the laser diode and the laser driver are fabricated in different technologies and close proximity of the laser and the driver circuit cannot always be guaranteed. In such cases the driver and laser diode are separately matched to 50Ω and can be connected via 50Ω transmission lines. Matched 50Ω back-terminations are placed on the driver chip itself to avoid degradation of the signal waveform due to reflections along the t-line caused by the variations of the laser diode impedance with bias current and modulation current amplitude. Figure 12.11(b) illustrates such a situation where the laser diode is mounted differentially and is AC-coupled to the driver. The bias current for the laser still needs to be provided by the driver circuit. However, since the modulating current is now split between the laser diode and the on-chip back termination, the driver modulating current, I_M, is twice as large as in the open-collector–drain topology. Although the size and output capacitance of the transistors in the current steering stage doubles, the load resistance is reduced in half and, to

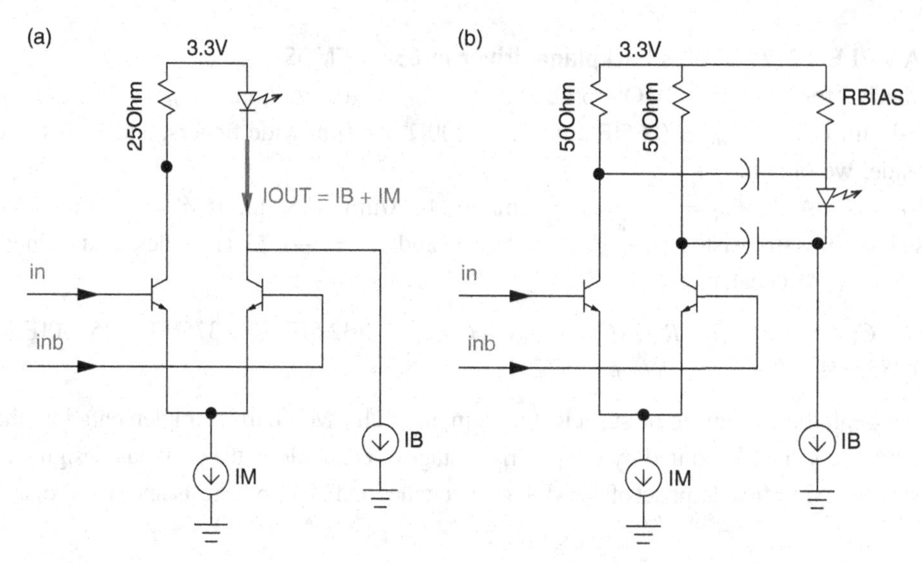

Figure 12.11 Laser driver schematics: (a) DC-coupled, single-ended driven laser and (b) AC-coupled, differentially driven laser

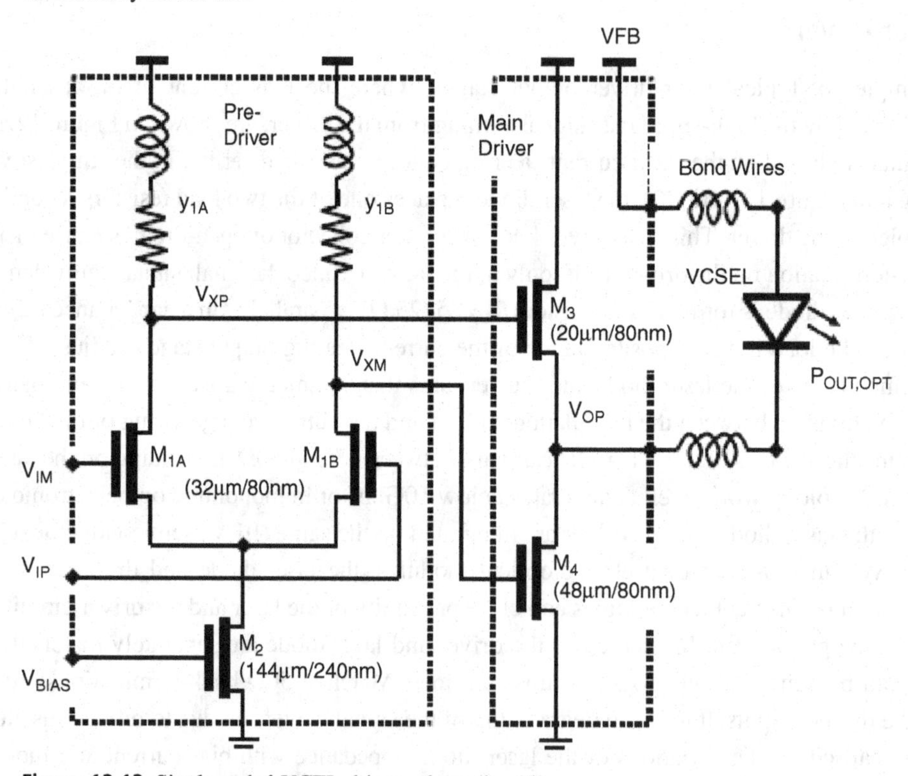

Figure 12.12 Single-ended VCEL driver schematics [5]

first order, the bandwidth should be preserved. However, in practice, because the interconnect losses increase for larger currents, the bandwidth is reduced. Even more importantly, the power consumption is also doubled.

Figure 12.12 shows an example of a 10Gb/s VCSEL driver fabricated in a 90nm CMOS technology where the output stage is single-ended and the VCSEL is DC-coupled to the driver [5]. Transistor Q_4 provides both the bias current and one half of the modulating current to the VCSEL. While the driver itself operates from a low supply voltage of 1.2V, the VCSEL bias voltage is usually larger than 1.8V, in this case 2.4V, since the turn-on voltage of an InP – or GaAs-based diode can be as high as 0.7V and 1.3V, respectively. Assuming a maximum drain current density of 0.4mA/μm in Q_4 at $V_{DS} = 0.6$V and a voltage swing of 0.5V_{pp} at the output of the CML inverter, the maximum instantaneous current through the VCSEL never exceeds 20mA.

EXAMPLE 12.3 80-mA, 10Gb/s Fabry–Perot laser driver output stage design

Let us assume first an HBT implementation and a DC-coupled laser with $R_L = 25\Omega$ and $C_L = 0.3$ pF, a threshold current $I_B = 20$mA and a modulating current of 80mA. Since the only load is the laser diode itself, the resulting output voltage swing becomes $\Delta V = V_{MOD} = I_M \cdot R_L = 0.08 \cdot 25 = 2V_{pp}$.

A DC voltage drop of about 1V is expected on the laser diode and on the load resistor in the opposite branch. Although the output voltage swing is 2V_{pp} and the breakdown voltage of the HBT is only 1.8V, we can still use the same SiGe BiCMOS process as in Example 12.1 as long as the DC collector–emitter voltage does not exceed 1.8V and as long as the output impedance of the predriver is not higher than a few hundred Ohm.

Again, the emitter length of the SiGe HBTs in the differential pair is chosen such that the maximum current density when the entire I_M flows through one transistor is approximately 1.5 J_{pfT}

$$l_E = \frac{I_M}{1.5 \cdot J_{pfT} \cdot W_E} = \frac{80}{1.5 \cdot 12 \cdot 0.13} = 34\mu\text{m}.$$

This is a very large transistor size, four times larger than in Example 12.1 and can be realized by connecting four unit transistors, each with a 8.5μm long emitter stripe, in parallel. The transistor capacitances and transconductance are also four times larger. We obtain $R_b = 3\Omega$, $r_E = 0.8\Omega$, $g_m = 1.52$ S, $g_{meff} = 642$mS, $C_{be} = g_m\, \tau_F + C_{je} = 456\text{fF} + 79.2\text{fF} = 535.2\text{fF}$.

From the slew-rate formula, we can estimate the total output capacitance of the HBT as

$$C_{bc} + C_{cs} = \frac{I_M}{1.5 \cdot SL_i} = \frac{80 \cdot 10^{-15}}{1.1} = 97\text{fF}.$$

The dominant pole becomes

$$\begin{aligned}\tau &= R_b \left[g_{meff}\tau_F + C_{je} + \left(1 + g_{meff}\frac{R_L}{2}\right) C_{bc} \right] + \frac{R_L}{2}[C_{bc} + C_{cs} + C_{laser}] \\ &= 3(535.2\text{f} + 132.6\text{f}) + 12.5 \cdot 397\text{f} \\ &= 2.03\text{ps} + 4.96\text{ps} = 7\text{ps}.\end{aligned}$$

This results into a 3dB bandwidth of 22.74GHz which is adequate even for 28Gb/s operation. It is not possible to implement this DC-coupled driver with a supply voltage of less than 2.5V. Even if a MOSFET current source is employed for I_M (which requires only 0.3–0.4V

DC drop, allowing for $V_{CE} = 1.5$V for the HBTs and 1-V drop on the laser diode, the minimum supply voltage becomes 2.8–2.9V.

Note that the input time constant is no longer dominant. We could not have implemented this output stage in a standard 65nm CMOS technology because of the very large output voltage swing.

Another problem caused by the large output current is the increased series loss and parasitic capacitance associated with the metal interconnect at the output of the driver which must satisfy the electromigration rules. Since the load impedance remains 50Ω, the larger the output current, the larger the required transistor breakdown voltage, transistor size, transistor capacitances and parasitic interconnect capacitances. All these practical physical constraints reduce the speed of the driver at large modulating currents and pose significant challenges on the semiconductor technology and its back-end.

12.4.3 Electro-absorption modulator driver

The electrical equivalent circuit of the EAM is similar to that of a reverse-biased diode, dominated by the *pn* junction capacitance C_L, in the 0.05 to 0.2 pF range. In order to avoid reflections between the driver and the modulator, a 50Ω resistor is placed in parallel with the EAM. The resulting EAM electrical equivalent circuit is shown in Figure 12.3(b).

Figure 12.13 illustrates a conceptual EAM driver schematic based on the same current steering topology employed in the output stage of the laser driver. The data inputs and the tail current, I_M, develop a modulating voltage, V_{MOD}, across the EA modulator. Bias currents I_B are employed to adjust the *DC offset voltage* at the output of the driver and to ensure that the

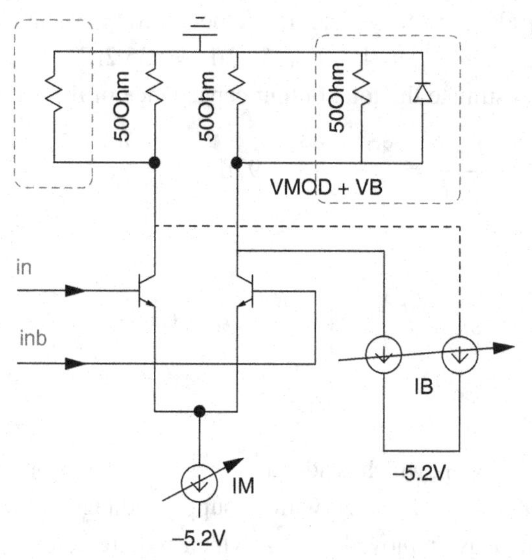

Figure 12.13 EA modulator driver concept

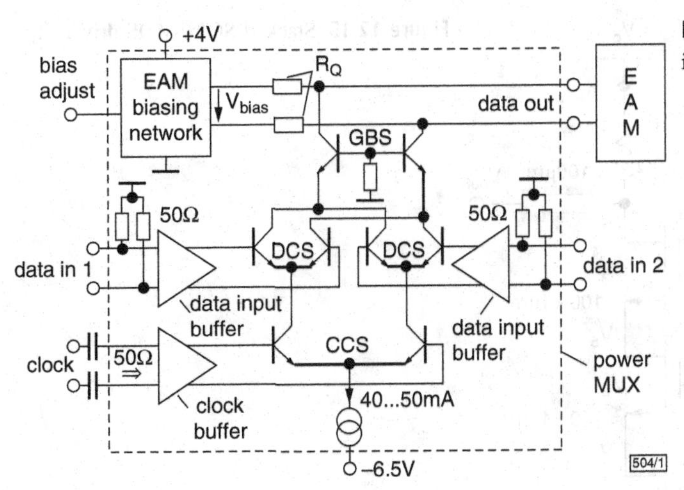

Figure 12.14 EAM driver with integrated 2:1 MUX [7]

extinction ratio is maximized. (Note that the DC offset voltage of the EAM is an entirely different concept than the usual DC offset voltage of a differential pair.) Back terminations are provided on chip to minimize reflections, while the EAM and the matching resistor in the opposite arm, shown in the dashed boxes, are left off-chip. Since the EA modulator is a reverse-biased diode, the DC currents I_B flow mainly through the on-chip 50Ω resistors.

Given that present commercial EA modulators require a V_{MOD} of 3V$_{pp}$, the modulating current (> 120mA) and breakdown voltage demands on the driver are even tougher than in the case of Fabry–Perot or DFB laser drivers. To reduce the impact of the large output node capacitance, distributed output buffer topologies are often used for applications at 40Gb/s and above [6].

Figure 12.14 shows an example of a lumped 40Gb/s EAM driver with integrated 2:1 power MUX which was fabricated in a 72GHz SiGe HBT process [7]. The data and clock inputs operate at 20Gb/s and 20GHz, respectively, which simplifies signal and clock distribution from the SerDes to the driver. To reduce I_M and save power, the EAM is mounted differentially. The differential voltage swing across the modulator varies with I_M from 2V$_{pp}$ to 2.5V$_{pp}$. Apart from the fact that this swing may not be sufficient for commercial modulators, the disadvantage of this arrangement is that the multiplexer operates with a very large tail current. The circuit consumes 2W from −6.5V and + 4V supplies.

As mentioned, modulators requiring lower voltage drive are currently being researched for integration in CMOS technologies. An example of a single-ended driver, implemented in 45nm SOI CMOS and operating at 40Gb/s with 2V$_{pp}$ swing is illustrated in Figure 12.15 [8]. Because of the very large intrinsic f_T and f_{MAX} of over 350GHz for both the n-channel and p-channel transistors in this technology, a purely digital implementation of the predriver as a chain of CMOS inverters is possible. This current-switching, rather than current steering, single-ended topology leads to significant power savings and a large FoM. The output stage is based on the same super-cascode concept employed in tuned PAs and analyzed in Chapter 6. By deploying appropriate capacitive loading at the gates of vertically stacked low-voltage SOI MOSFETs, a large voltage swing is developed at the output while the

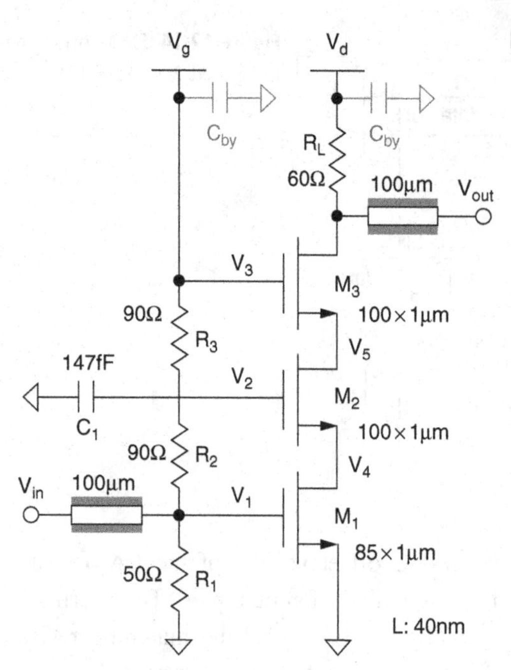

Figure 12.15 Stacked SOI CMOS driver [8]

instantaneous gate–source, gate–drain, and drain–source voltages never exceed the limits allowed by device reliability. The I_M current in the output stage swings between 0 and $I_{ON} \approx 85\text{mA}$. The necessary large supply voltage is only feasible in SOI (as opposed to bulk) CMOS technologies where the body of vertically stacked transistors is isolated from the silicon handle wafer by a relatively thick buried oxide (BOX) whose breakdown voltage is larger than the supply voltage.

12.4.4 Mach–Zehnder modulator driver

From the electrical point of view, a Mach–Zehnder modulator can be represented as a capacitively coupled 50Ω load. It can be driven single-endedly or differentially. Because of the very large modulation voltage, typically exceeding $5V_{pp}$ per side, and the corresponding $I_M > 200\text{mA}$, all commercial MZ modulator drivers to date have been realized with a distributed output stage and only in III-V technologies. The output stage must also provide an adjustable bias voltage for the modulator. Despite the tremendous challenges of designing a suitable large-swing driver, because of their outstanding ER and bandwidth, MZ modulators remain by far the most dominant optical modulation solution for high-performance long-haul fiber-optic links at 10Gb/s and beyond.

One for the first 10Gb/s MZM drivers operating in limiting mode is shown in Figure 12.16. It consists of a lumped predriver and of a large-swing 5-stage differential distributed amplifier. Each distributed amplifier stage consists of two identical cells connected in parallel. The cell itself features a InGaP/GaAs HBT current-steering differential pair with emitter degeneration

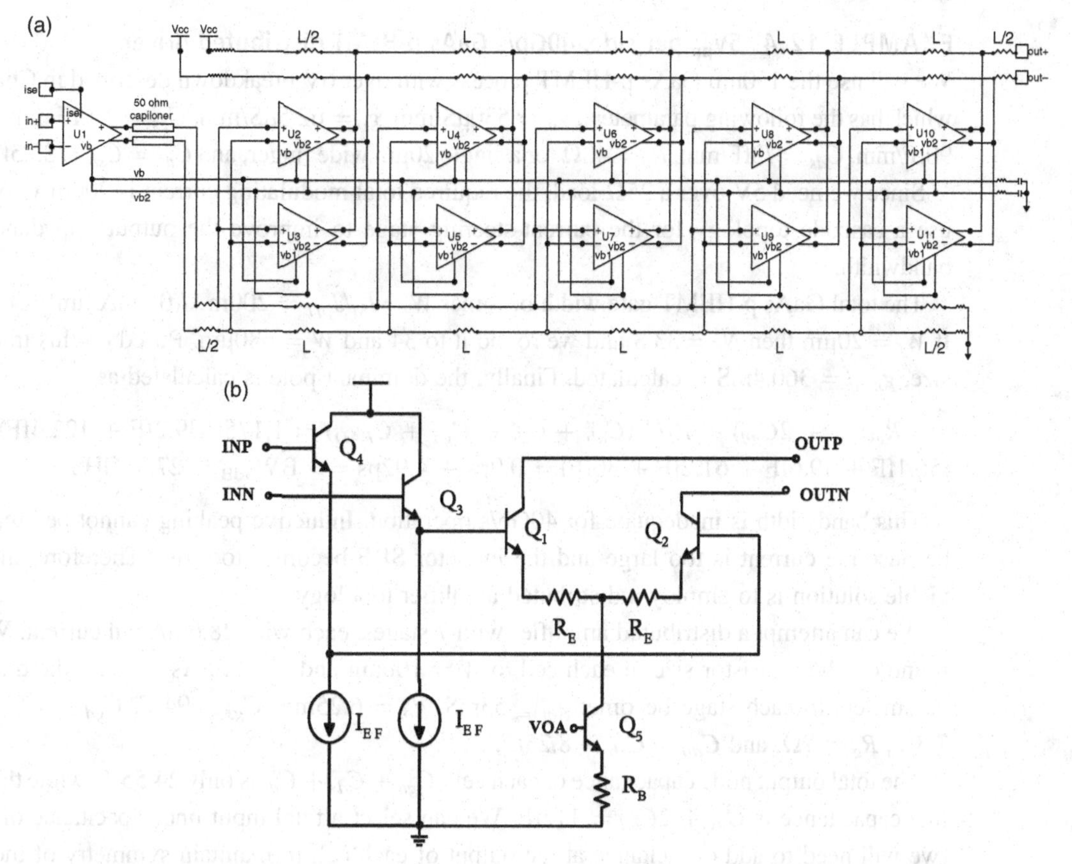

Figure 12.16 (a) 5-stage 10Gb/s differential limiting-mode distributed MZM driver (b) DA cell schematics [9]

for improved bandwidth and increased input impedance. Emitter-followers are employed to further increase the input impedance of the cell and to reduce loading of the input transmission lines. The bias voltage applied at the VOA node determines the total modulation current. Since the desired output voltage swing is $3V_{PP}$ per side into effective 25Ω loads (50Ω on chip and 50Ω on the Mach–Zehnder modulator itself), a total modulating current of 120mA (24mA per cell) is needed. As discussed in Chapter 5, the inductors L are chosen based on the input and output capacitance C of the DA stage such that
$Z_0 = \sqrt{\frac{L}{C}}$. The cutoff frequency of the artificial transmission line $f_c = \frac{1}{\pi\sqrt{LC}}$ must be larger than 7GHz for the circuit to function with adequate margin at 10Gb/s data rates.

The limiting mode DA presents several advantages over a linear DA. Most importantly, the bandwidth requirements are relaxed. Second, the gain flatness and phase linearity are not as critical to the integrity of the eye diagrams as in the case of a linear DA. Finally, power consumption is minimized since the limiting DA does not have to be backed off from the 1dB compression point to maintain linearity.

EXAMPLE 12.4 $5V_{pp}$ per side, 40Gb/s GaAs p-HEMT distributed driver

We will use the 150nm GaAs p-HEMT process with over 6V breakdown described in Chapter 4 which has the following parameters: $g_m = 530\mu S/mm$, $g_o = 62.5\mu S/mm$, $C_{gs} = 940fF/mm$, $C_{gd} = 90fF/mm$, $C_{ds} = 73fF/mm$, $R_g = 40\Omega$ for a single 20μm wide finger, and $C_{db} = C_{sb} = 82.5fF/mm$.

Since we need 5V over a 25Ω load, the required total modulating current is 200mA. We will use a cascode topology for the current-steering stage to improve the output impedance and bandwidth.

The total GaAs p-HEMT gate width becomes $W = I_M/J_{pfT} = 200mA/(0.3mA/\mu m) = 666\mu m$. If $W_f = 20\mu m$ then $N_f = 33.3$ and we round it to 34 and $W = 680\mu m$. Based on this transistor size, $g_{meff} = 360.4mS$ is calculated. Finally, the dominant pole is calculated as

$$\tau = R_g(C_{gs} + 2C_{gd}) + R_L/2\ (C_{db} + C_{ds} + C_{gd} + C_{PAD}) = 1.175(639.2fF + 122.4fF) + 25\ (56.1fF + 49.6fF + 61.2fF + 30fF) = 0.9ps + 4.92ps \Rightarrow BW_{3dB} = 27.36GHz.$$

This bandwidth is inadequate for 40Gb/s operation. Inductive peaking cannot be employed because the current is too large and the inductor SRF becomes too low. Therefore, the only viable solution is to employ a distributed amplifier topology.

We can attempt a distributed amplifier with 7 stages, each with 28.6mA tail current. We can round off the transistor size in each cell to $W = 100\mu m$ and $N_f = 5$. As a result, the transistor parameters in each stage become: $g_m = 53mS$, $g_o = 6.25mS$, $C_{gs} = 94fF$, $C_{gd} = 9fF$, $C_{ds} = 7.3fF$, $R_g = 7\Omega$, and $C_{db} = C_{sb} = 8.25fF$.

The total output node capacitance of each cell, $C_{gd} + C_{ds} + C_d$, is only 24.55fF, while the input line capacitance is $C_{gs} + 2C_{gd} = 112fF$. We can select a total input line capacitance of 120fF (we will need to add capacitance at the output of each cell to maintain symmetry of the input and output lines). The resulting inductance of the input and output transmission lines becomes: $L = CZ_o^2 = 300pH$ and $BW_{3dB} = 1/[\pi(LC)^{1/2}] = 53GHz$. In practice, the bandwidth will be limited by the frequency at which the MAG of the cascode stage $= 18.8dB$ ($A_v = 8.75$) – 3dB = 15.8dB.

12.4.5 Output swing (I_M/V_{MOD}) control

In most high-speed digital output drivers, output swing control is implemented by varying the tail current I_M. To avoid distortion of the output waveform as the tail current changes, the voltage swing provided by the predriver at the input of the output stage must be large enough to fully switch the current-steering pair for the entire range of desired I_M and V_{MOD} values. This ensures that the peak-to-peak output voltage swing V_{MOD} is always given by $I_M \times R_L/2$ as I_M is varied.

As we have learned in Chapter 4, in CMOS or HEMT output drivers, as long as the tail current density varies between 0.15mA/μm and 0.5mA/μm, the device capacitances, transconductance and output conductance remain practically constant. This means that the input and output impedances and the bandwidth of the output stage remain largely constant as the output amplitude changes over a 3:1 range. This in turn, means that the load seen by the predriver does not change and the integrity of the signal waveforms throughout the driver is

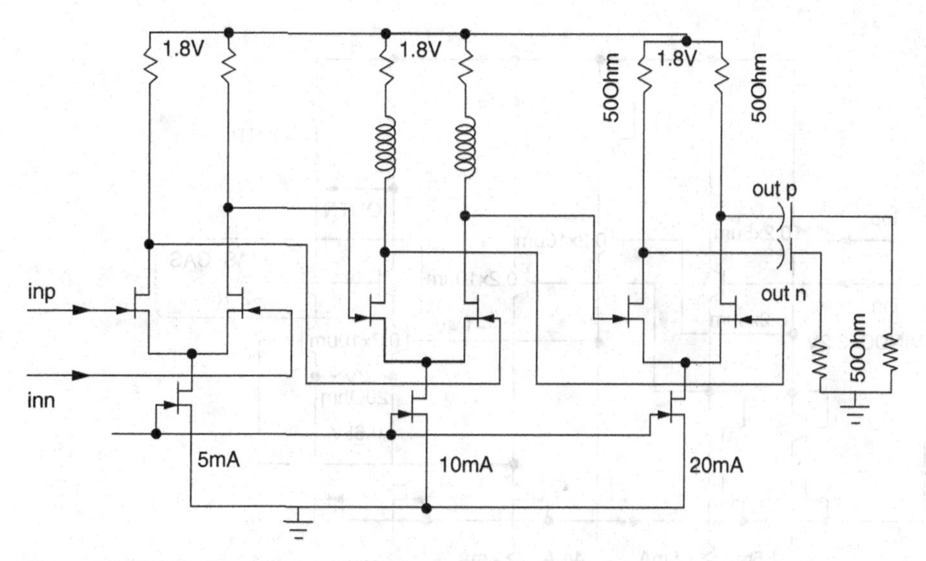

Figure 12.17 Wireline 180nm CMOS output stage and predriver schematics [10]

maintained. An example of a MOS CML 50Ω backplane driver operating at 10Gb/s is illustrated in Figure 12.17 [10]. The output swing can be varied up to a maximum of $500mV_{pp}$ per side by changing the tail current in the last stage. Only the current in the last stage needs to be varied to control the amplitude of the output signal.

In bipolar output stages the input capacitance changes almost linearly with the tail current. As a result, the tail current in the predriver stages must be varied in tandem with I_M in order to avoid overshoot or undershoot in the voltage waveform. This is illustrated in the 40Gb/s bipolar output stage in Figure 12.18 realized with two cascaded (double) emitter-followers and a cascode current-steering pair with resistive degeneration. The resistive degeneration improves stability and, along with the cascode topology, maximizes the bandwidth and desensitizes the output impedance to I_M variation. The role of the emitter-follower stages is to minimize the loading on the predriver. When the control voltage ***VOA*** is varied to modify the amplitude of the output voltage, the tail currents of all three stages change proportionally.

Finally, a single-ended 156 Mb/s Ethernet CMOS driver with digital output amplitude control is illustrated in Figure 12.19 where binary-weighted groupings of MOSFETs (i.e. a binary-weighted current DAC) are turned on and off to control the output current. Rather than employing the current-steering concept, this circuit is based on current switching and, as the stacked SOI MOSFET driver in Figure 12.15, is suitable for burst-mode fiber-optic communications [2]. Although the power efficiency of this purely digital circuit is high, the output impedance changes with the signal amplitude level affecting the integrity of the output waveforms if it was to be used at higher data rates.

12.4.6 Pre-emphasis control

Pre-emphasis is needed in the output driver to compensate for frequency-dependent loss in the transmission medium (coaxial cable or PCB trace) at the output of the driver, or to correct the asymmetrical rise and fall times and signal waveforms of lasers and optical modulators. It

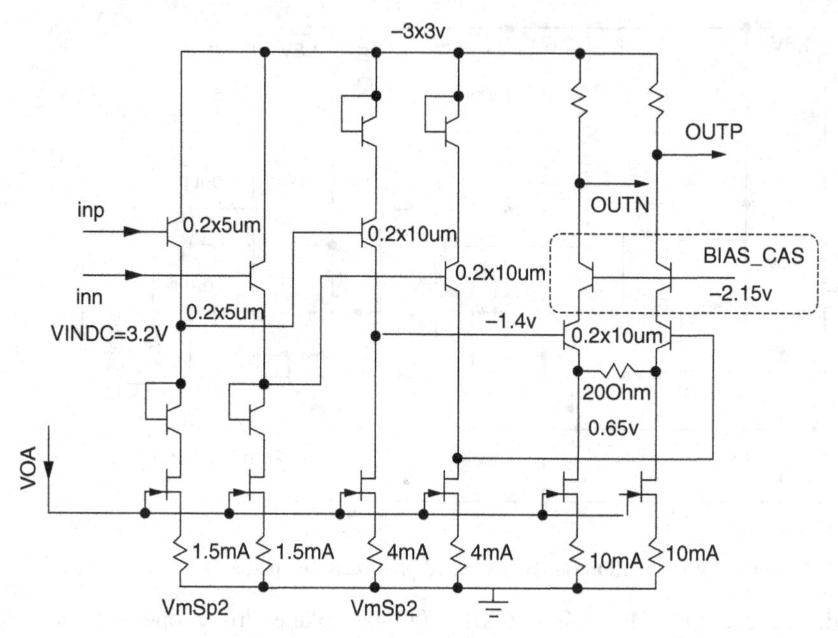

Figure 12.18 Schematics of a SiGe HBT output stage with output swing control

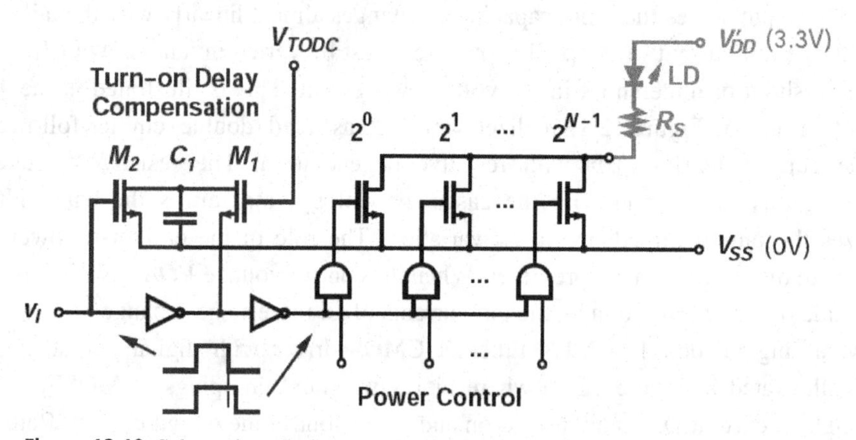

Figure 12.19 Schematics of a burst-mode CMOS laser driver with digital control of the output current [2]

manifests itself as gain peaking in the frequency domain and as overshoot and undershoot in time domain.

Pre-emphasis can be introduced by adding short positive signal pulses at the rising edge, and short negative pulses on the falling edge of the data signal. These pulses can be generated by differentiating or delaying the data signal as illustrated in Figure 12.20. At the circuit level, pre-emphasis has been implemented using analog or digital techniques. Pre-emphasis (v_{emph} in Figure 12.21) is specified as a percentage of the peak-to-peak nominal voltage swing at low frequency, v_o^{pp}.

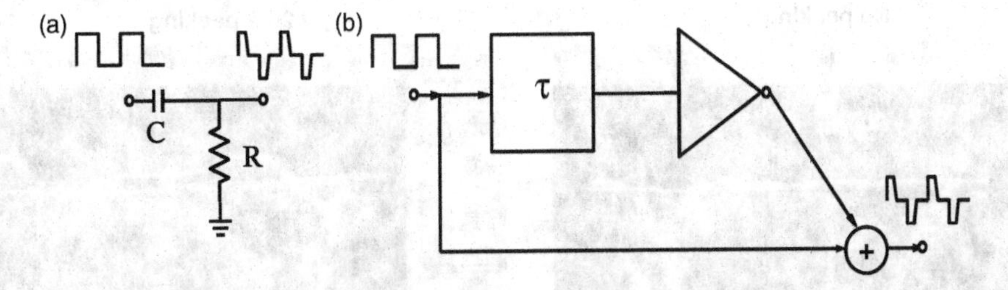

Figure 12.20 (a) Analog and (b) (quasi-)digital differentiator circuits

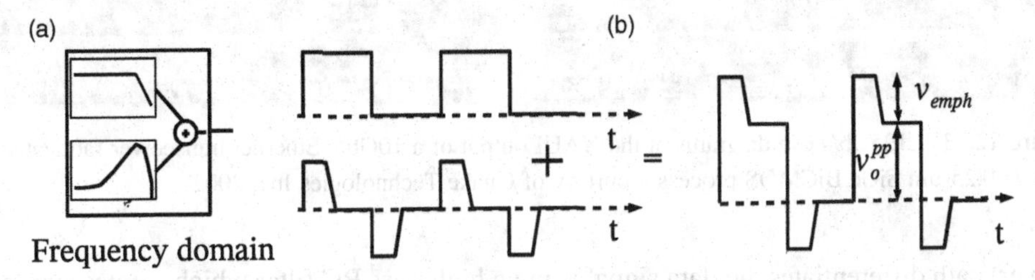

Figure 12.21 Illustration of pre-emphasis and its manifestation in (a) frequency and (b) time domains

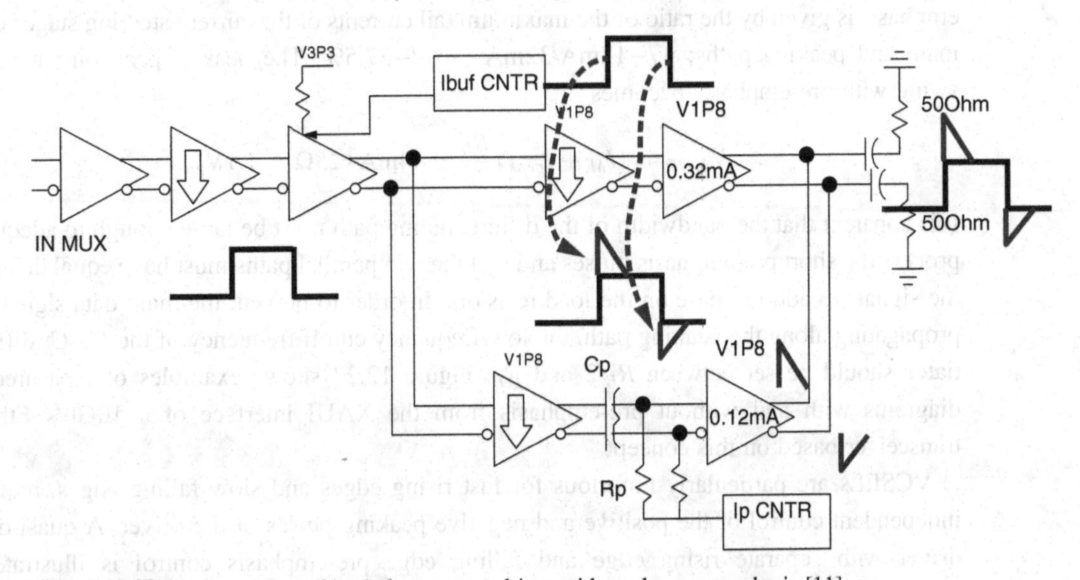

Figure 12.22 Basic concept of backplane output driver with analog pre-emphasis [11]

An example of an analog implementation suitable for backplane and coaxial cable drivers is illustrated conceptually in Figure 12.22 where the triangular symbols with down-pointing arrows represent emitter-follower stages. The data signal is split in two parallel paths at the output of the predriver. The amplitude of the signal on the main (upper) path is controlled by the tail current of the last current-steering stage, I_M, which varies between 0 and 32mA. The second

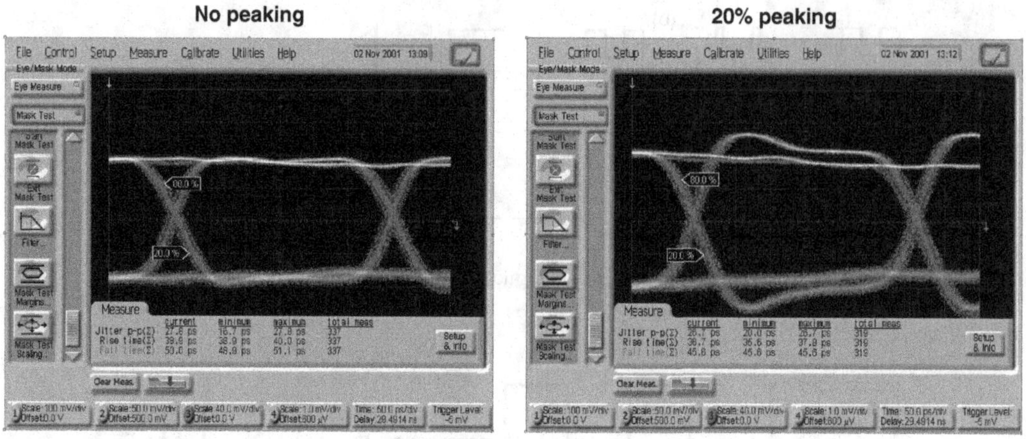

No peaking 20% peaking

Figure 12.23 3.125Gb/s eye diagrams at the XAUI output of a 10Gb/s Ethernet transceiver fabricated in a 70GHz 0.25-μm SiGe BiCMOS process (courtesy of Quake Technologies Inc. 2002)

(lower) path differentiates the data signal using a high-pass RC filter which creates symmetrical positive and negative pre-emphasis pulses. Their amplitude is adjusted by changing the tail current of the current-steering stage, I_{peak}, from 0 to 12mA. The signals from the two paths are then added on the output load resistors which act as current summers. The maximum pre-emphasis is given by the ratio of the maximum tail currents of the current steering stages on the main and peaking paths: $+/-12\text{mA}/32\text{mA} = +/-37.5\%$. The peak-to-peak output voltage swing with pre-emphasis becomes

$$V_{MOD} = \left(I_M + I_{peak}\right)\frac{R_L}{2} = 44\text{mA} \cdot 25\Omega = 1.1\text{V}_{\text{pp}}. \tag{12.8}$$

It is apparent that the bandwidth of the differentiating path must be large enough to adequately process the short pre-emphasis pulses and that the two parallel paths must have equal delays for the signals to add in phase on the load resistors. In order to prevent the main data signal from propagating along the peaking path, the low-frequency cutoff frequency of the R_p–C_p differentiator should be set between $R_B/2$ and R_B. Figure 12.23 shows examples of measured eye diagrams with and without pre-emphasis from the XAUI interface of a 10Gb/s Ethernet transceiver based on this concept.

VCSELs are particularly notorious for fast rising edges and slow falling edges, requiring independent control of the positive and negative peaking pulses of the driver. A quasi-digital driver with separate rising edge and falling edge pre-emphasis control is illustrated in Figure 12.24 [11]. The differentiator circuits are realized using delay cells (in the *ppeak* and *npeak* blocks) and CML AND gates operating at 10Gb/s. The amplitude of the positive and negative electrical pre-emphasis pulses can be independently adjusted from the tail currents I_{ppeak} and I_{npeak}, respectively. In this circuit, I_M is adjustable from 5 to 20mA, while I_{ppeak} and I_{npeak} can be varied between 2.5 and 10mA. The VCSEL is AC-coupled differentially at the output of the driver. The VCSEL bias current is adjustable between 2.5mA and 16mA. The corresponding 10.3Gb/s eye diagrams with and without pre-emphasis and an output amplitude

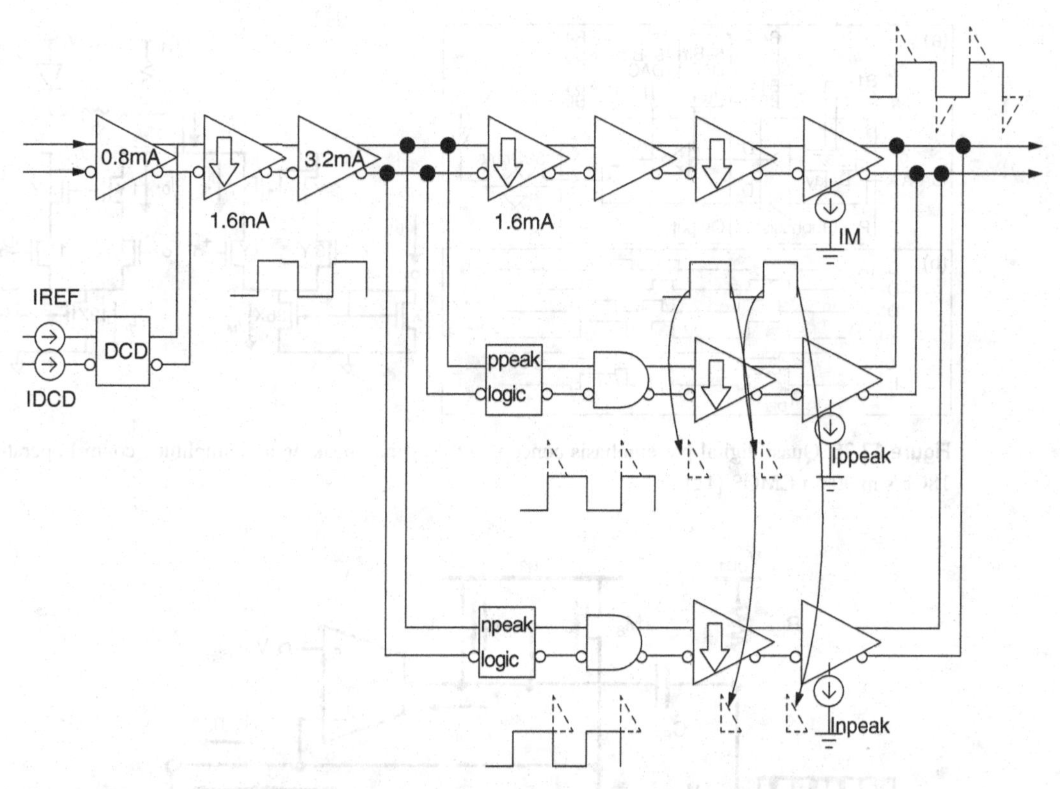

Figure 12.24 Quasi-digital pre-emphasis concept with separate positive and negative peaking amplitude control [11]

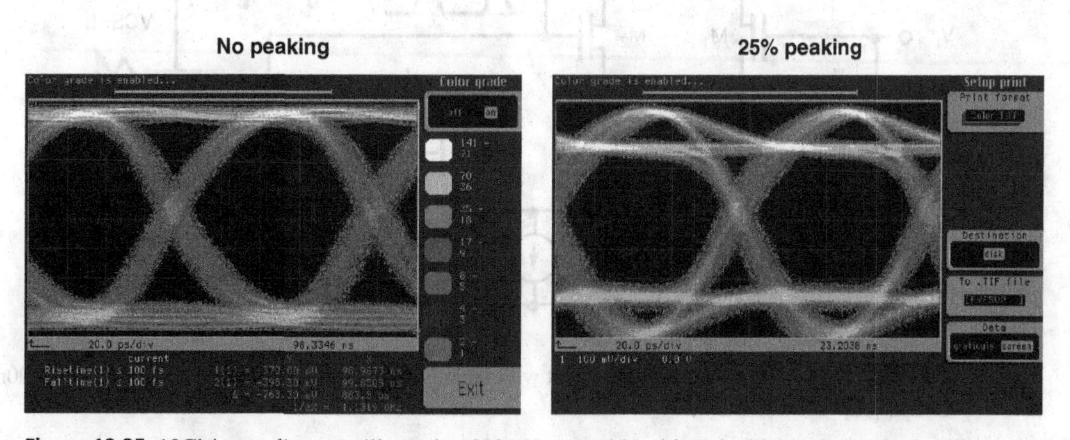

Figure 12.25 10Gb/s eye diagrams illustrating $800mV_{pp}$ per side with and without pre-emphasis [11]

of $800mV_{pp}$ per side are illustrated in Figure 12.25. The circuit was fabricated in a 0.35μm SiGe BiCMOS technology with SiGe HBT f_T and f_{MAX} of 45GHz [11].

A digital approach with symmetrical rise and fall-edge pre-emphasis is implemented in the 18Gb/s, 90nm CMOS VCSEL driver depicted in Figure 12.26 [12]. The pre-emphasis pulses are generated by adding the data signal, *D*, with a delayed, inverted and down-scaled version of

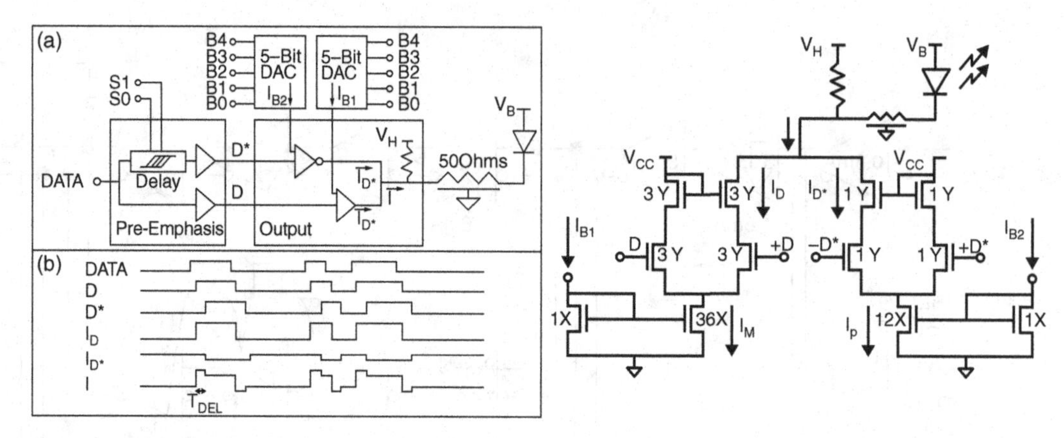

Figure 12.26 Quasi-digital pre-emphasis concept with separate peaking and amplitude control operating at 18Gb/s in 90nm CMOS [12]

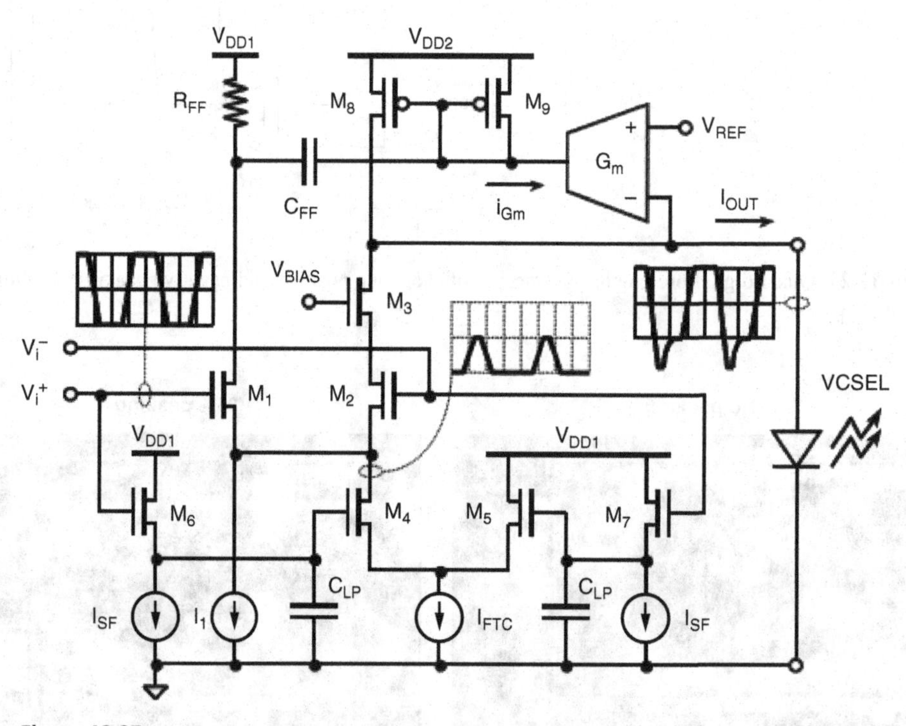

Figure 12.27 Analog pre-emphasis concept with falling edge peaking only operating at 20Gb/s in 90nm CMOS [13]

itself, D^*. The amplitude ratio between the main data and the delayed data signals is 3:1, determined by the ratio of the tail currents I_M and I_P.

In contrast, the 20Gb/s VCSEL driver in Figure 12.27 introduces pre-emphasis (undershoot) only on the falling edge of the data to compensate for the slow turn-off of the optical power in the VCSEL. The data signal is applied to the main current-steering stage formed by M1 and M2 and current source I_1. The second current steering pair, M4 and M5, forms a CML AND gate

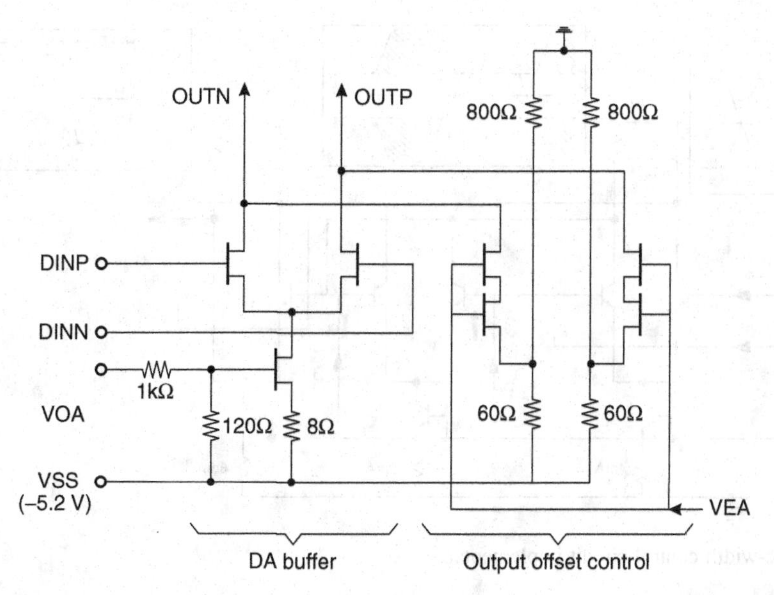

Figure 12.28 Implementation of DC output offset control in a 40Gb/s GaAs p-HEMT EAM driver [6]

with M1 and M2 and injects current pulses only on the negative edge of the input data signals. By adding I_1 and the pre-emphasis current at a low impedance node, and by buffering the CML AND gate from the output node using the common-gate transistor M3, the output node capacitance is minimized and the bandwidth of the driver is maximized.

12.4.7 Output DC offset control for EA modulators

An example of a DC output offset control circuit, connected in parallel with the output stage of a 40Gb/s EA modulator driver and realized in a 150nm GaAs p-HEMT technology is illustrated in Figure 12.28 [6]. By adjusting the voltage on the *VEA* pin, the DC current through the cascoded HEMTs is modified, resulting in a change in the total DC current flowing through the on-die 50Ω loads connected at *OUTN* and *OUTP*. This, in turn, controls the DC offset voltage at the output nodes.

12.5 PREDRIVER

A predriver is placed in front the output stage for several reasons. First, because of the very large output current and voltage swing of a laser or modulator driver, it is difficult for a logic gate to drive the output stage directly without consuming very large power. Second, if the large output stage with large input capacitance were to be driven directly from off chip, the input return loss and bandwidth would be severely degraded. Third, the relatively small input voltage available from off chip or from CML gates is often inadequate to fully switch the modulating current in the output stage. Additional gain is needed. The predriver addresses all these issues.

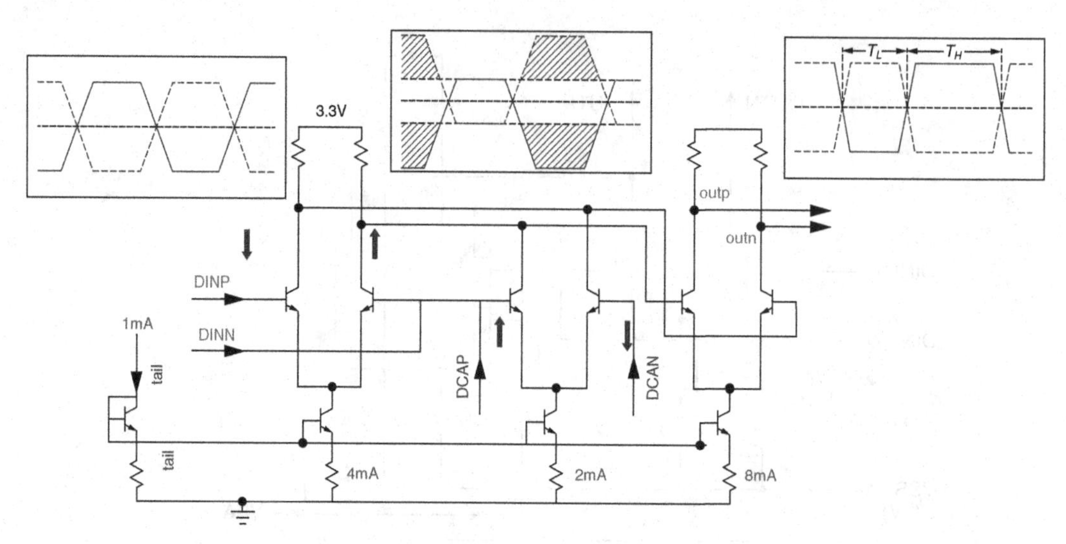

Figure 12.29 Pulse-width control circuit implementation

The input buffer, retiming, and the pulse-width control functions are implemented in the predriver.

The predriver is realized with one or more cascaded current-steering stages with increasing output voltage swing from the input to the output. As discussed in Chapter 5, stage-to-stage size and current scaling can be employed to save power while maintaining the required switching speed. For example, in the 180nm CMOS output driver in Figure 12.17, the predriver consists of two MOS CML differential inverters with a stage-to-stage scaling factor of two.

12.5.1 Pulse-width control block

Pulse-width control is introduced most often with the same circuit used to compensate the DC offset of a differential pair. As illustrated in Figure 12.29, by applying a DC voltage between the **DCAP** and **DCAN** inputs, a DC offset voltage arises at the output of the first current-steering stage. This triggers a shift in the decision threshold of the second current-steering stage, which operates in limiting mode. The hard-limiting action causes the pulse width or duty cycle of the data signal to change. The pulse-width change relative to the pulse period is proportional to the amount of DC offset introduced relative to the CML logic swing of the first current-steering stage. For example, if the first stage in Figure 12.29 has an output logic swing $\Delta V =$ 300mV$_{pp}$ per side, the maximum DC offset at its output becomes

$$V_{OS} = \Delta V \frac{I_{OS}}{I_{OS} + I_T} = 300\text{mV}\frac{2\text{mA}}{6\text{mA}} = 100\text{mV}$$

and the pulse width can be adjusted approximately between 33% and 66%.

One problem with this approach is that the pulse width is somewhat sensitive to the slope of the rising and falling edges of the data signal [2],[6]. In practice, the pulse-width control range is chosen with adequate margin to compensate for this uncertainty.

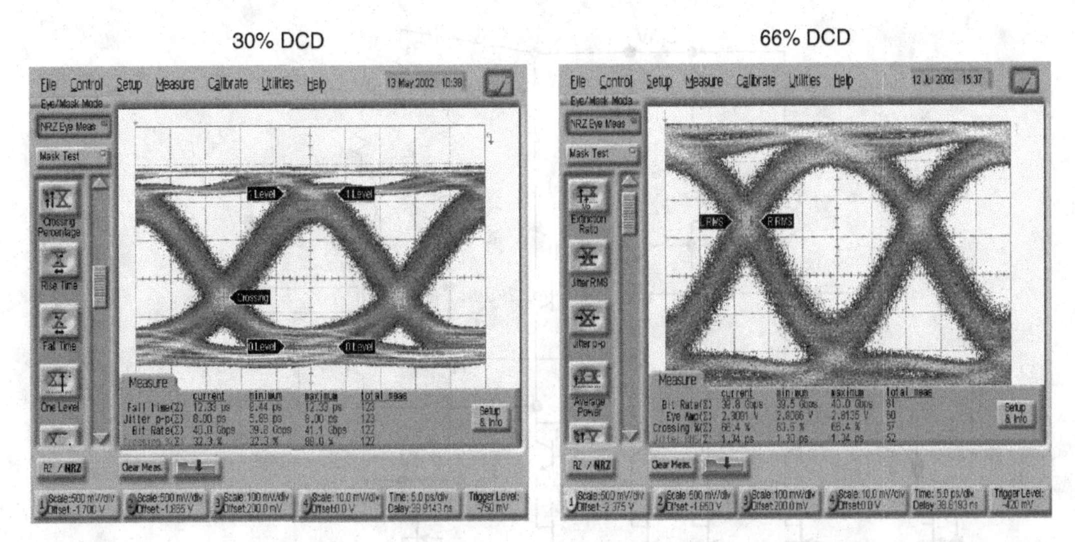

Figure 12.30 Measured 40Gb/s eye diagrams of a GaAs p-HEMT EAM driver with pulse-width control [6]

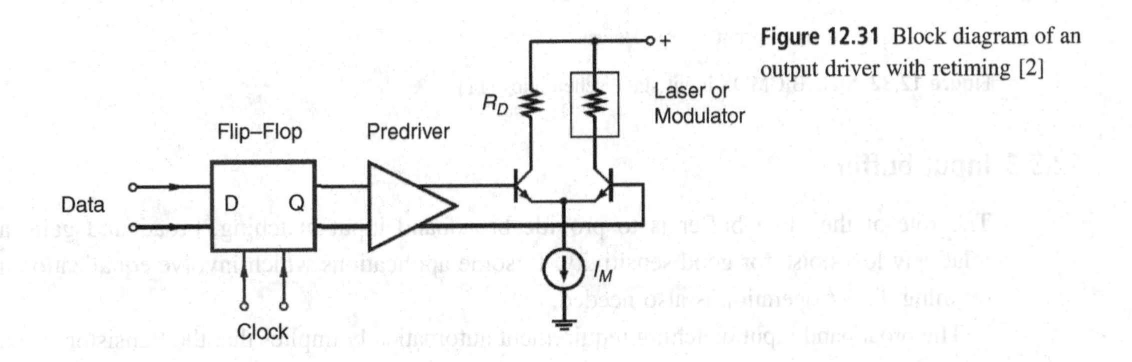

Figure 12.31 Block diagram of an output driver with retiming [2]

Measured eye-diagrams with pulse-width control are shown in Figure 12.30 for a 40Gb/s EAM driver with an output swing of $3V_{pp}$ per side and implemented in a 150nm GaAs p-HEMT process [6].

12.5.2 Optional retimer

In some high-performance standalone laser and modulator drivers, the data signal is retimed by a low-phase noise external clock before it is applied to the output stage. The retiming function is realized with a flip-flop. In addition to increased power consumption, one of the challenges associated with retiming is to maintain good phase alignment between the data and clock signals, both of which are provided by the SerDes chip. This alignment is very difficult to achieve at data rates of 10Gb/s and higher without a PLL or DLL. Furthermore, the external clock must have very low-phase noise (jitter), otherwise the eye diagram at the output of the driver will exhibit significant random jitter.

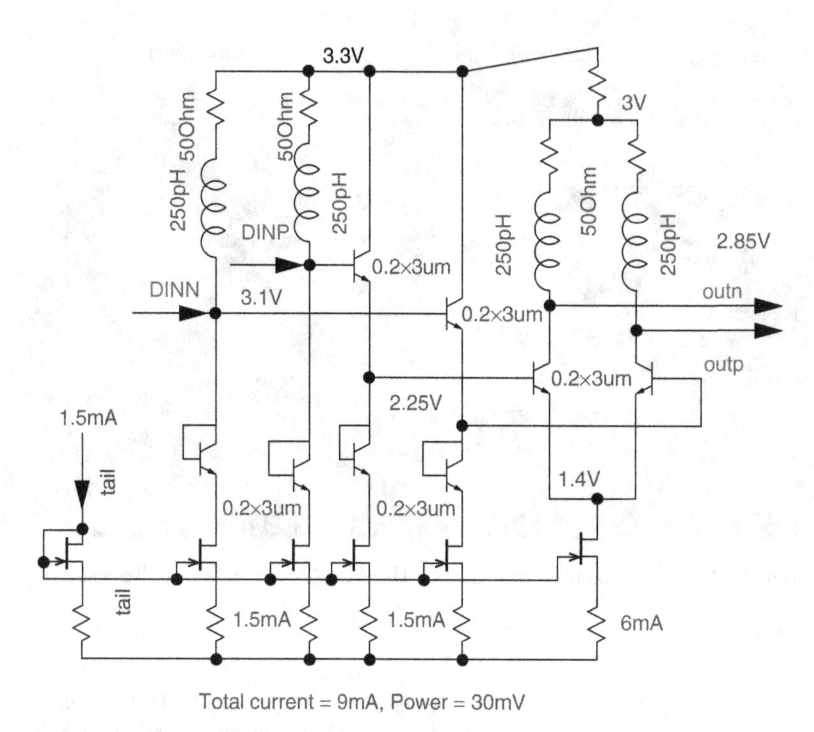

Figure 12.32 SiGe BiCMOS input stage schematics [11]

12.5.3 Input buffer

The role of the input buffer is to provide broadband input matching, broadband gain, and relatively low noise for good sensitivity. In some applications which involve equalization and retiming, linear operation is also needed.

The broadband input matching requirement automatically implies that the transistor size and bias current should be small, and that the input impedance and return loss should be relatively insensitive to data signal level variation and data switching. Since the input impedance of a FET is relatively constant with bias current, FET-based input buffers can be realized with a differential current-steering stage. However, in the case of III-V and SiGe HBT implementations, an emitter-follower stage is typically placed in front of the current-steering stage, as illustrated in Figure 12.32. A higher input sensitivity is obtained if a TIA stage is employed [14].

Linear input buffer with pre-emphasis

In active cables and repeaters, used for example in HDTV signal transmission applications, the input stage must provide linear equalization, before retiming, to compensate for the increasing coaxial cable loss as a function of frequency. Figure 12.33 reproduces the measured loss for several types of coaxial cables of different lengths. In all cases, the cable loss can be described as [15]

$$C(f) = e^{-l\left[k_s(1+j)\sqrt{f}+k_d f\right]} \tag{12.9}$$

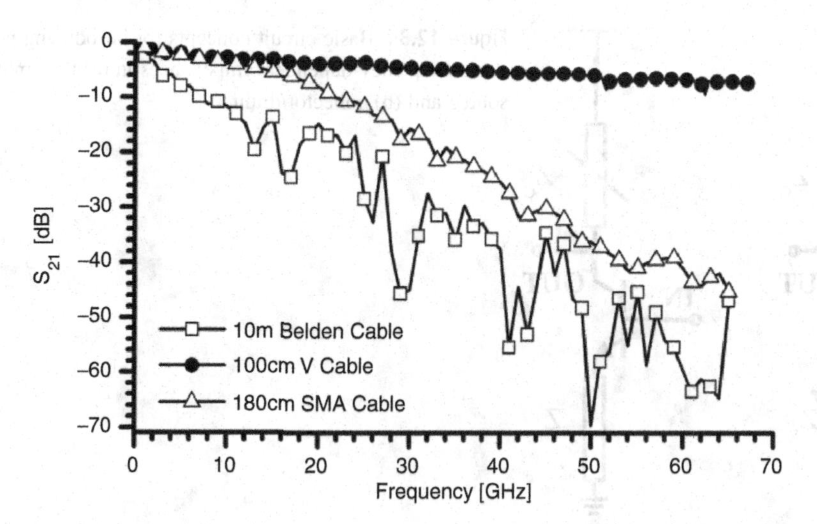

Figure 12.33 Measured attenuation vs. frequency for a variety of coaxial cables [16]

where k_s is the skin effect constant, k_d is the dielectric constant, and l is the cable length.

It becomes immediately apparent that the low-frequency components of the high-speed digital data signals will pass through the cable practically unaffected whereas the high-frequency components will be severely attenuated.

Equalization of the cable loss can be achieved with a circuit whose frequency response is represented by the inverse of (12.9). The following requirements must be satisfied:

- a high-pass transfer function with multiple zeros is needed to compensate for the increasing attenuation of the cable with frequency;
- a low-pass transfer function is needed for short cables with little or no attenuation;
- in-phase summation of the two signal paths;
- the circuit must be linear at all frequencies;
- the circuit must have low-noise figure at high frequency where the signal amplitudes are very low.

As in the pre-emphasis concept of Figure 12.21, this circuit can be described by the superposition of two functions: one that has constant gain over frequency and one whose gain increases exponentially with frequency

$$H(s) = G_{LF} + G_{HF}\frac{(s+\alpha_1)(s+\alpha_2)\ldots(s+\alpha_n)}{(s+\gamma_1)(s+\gamma_2)\ldots(s+\gamma_{n+1})} \tag{12.10}$$

where G_{LF} is the DC gain, G_{HF} is a scaling constant for the frequency-dependent gain term, and α_n and γ_n represent the zeros and poles of the frequency-dependent term, chosen to best match the inverse of the cable response.

The function can be implemented with the basic circuit concept shown in Figure 12.34 where frequency-dependent impedances are employed. Several such stages can be cascaded, with the voltage gain expression of one stage given by

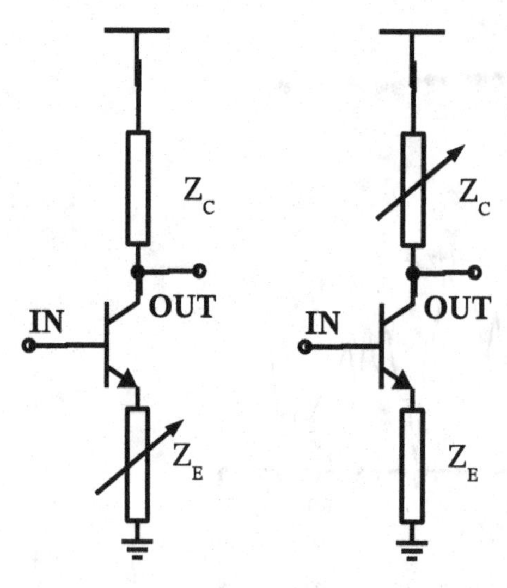

Figure 12.34 Basic circuit concepts for introducing peaking using frequency-dependent impedances in (a) the emitter/source and (b) collector/drain

$$A_v(s) = \frac{-Z_C(s)}{Z_E(s)}. \tag{12.11}$$

The zeros of Z_C(s) and/or the poles of Z_E(s) introduce the necessary zeros in the transfer function. Z_E(s) also provides the necessary emitter degeneration to improve linearity. In fact, by appropriately choosing the poles of $Z_{E(s)}$, the amount of resistive emitter/source degeneration (essentially the amount of feedback in this negative series–series feedback amplifier stage) decreases with increasing frequency. This results in three important beneficial trends: (i) the gain increases with frequency, thus compensation for the cable loss, (ii) the noise contribution of the degeneration impedance to the noise figure of the circuit decreases with frequency, and (iii) linearity is high at low frequency, where the data signal is large, and decreases at higher frequencies where the data signal components are weak.

Figure 12.35 shows a very compact bipolar circuit implementation of this concept [15] which achieved a maximum peaking frequency of 55GHz [16]. It includes the low-pass filter (highlighted in Figure 12.35(a)), high-pass filter (highlighted in Figure 12.35(b)) and the summing functions in a single stage. By simple inspection, and assuming that the emitter-follower stage has a voltage gain of 1, we see that $G_{LF} \approx -1$ and $G_{HF} \approx -Z_C(s)/Z_E(s)$. Apart from the peaking inductors introduced to extend the bandwidth, Z_C(s) is constant. The three zeros, scaled in frequency by a factor of 10 to compensate the cable loss over three decades of frequency, are implemented with series RC circuits as part of Z_E(s).

$$Z_E(s) = \frac{1}{G_E + G_1(s) + G_2(s) + G_3(s)} \tag{12.12}$$

where

$G_E = 1/100\Omega^{-1}$, $G_i(s) = \frac{sC}{1+sCR_i} = \frac{sC}{1+\frac{s}{\omega_{zi}}}$, $C = 2\times27\text{fF}$, $R_i = 10^{i+1}\,\Omega\ (i = 1 \cdots 3)$, and the bias current $I_E = 2\text{mA}$.

The degree of gain peaking in the transfer function can be adjusted to suit variable cable lengths by applying a control voltage between the nodes Main and Tune. The two Gilbert cells

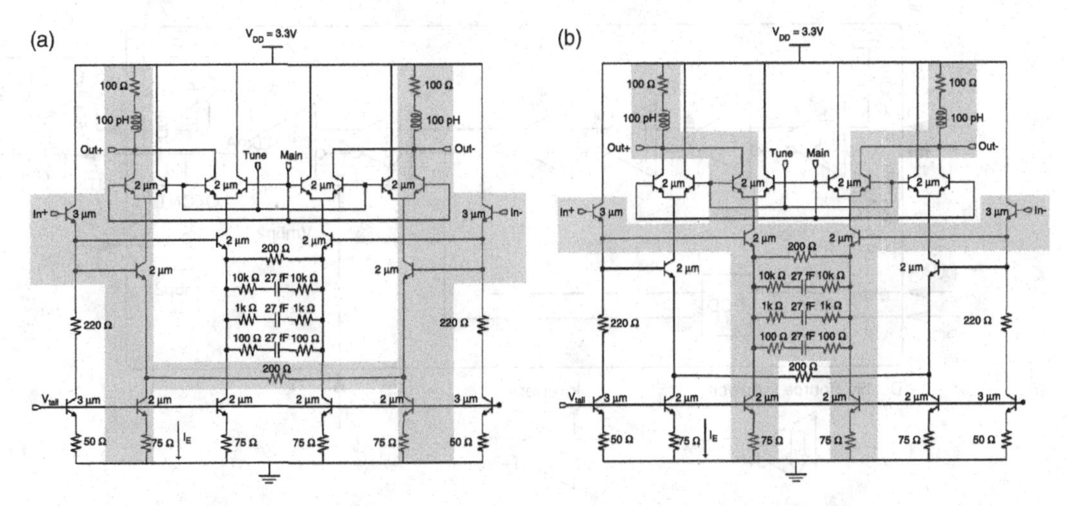

Figure 12.35 Illustration of the (a) low-pass and (b) high-pass signal transfer paths [16]

perform the weighted summation of the low-pass and high-pass transfer functions and ensure that the DC current and voltage on the 100Ω load resistors, the output impedance, and the bandwidth of the stage remain constant as the shape of the frequency response is changed. The circuit is linear for input voltages of up to 400mV$_{pp}$ per side and consumes 14mA from 3.3V supply.

Finally, it should be noted that the circuit in Figure 12.35 is yet another form of analog differentiator and can be used both as a receive equalizer or as an output stage with pre-emphasis. The highest peaking frequency is limited by the speed of the transistor technology.

12.6 EXAMPLES OF DISTRIBUTED OUTPUT DRIVERS OPERATING AT 40Gb/S AND BEYOND

12.6.1 6V$_{pp}$, 40Gb/s EAM driver in 150nm GaAs p-HEMT technology

We start with an EAM driver [6] fabricated in a 150nm GaAs p-HEMT process with f_T of 110GHz, f_{MAX} of 180GHz, $J_{pfT} = 0.3$mA/μm, $J_{pfMAX} = 0.2$mA/μm, $V_T = -0.9$V, and a gate–drain breakdown voltage larger than 6V. The block diagram is shown in Figure 12.36. It features a lumped predriver with duty-cycle distortion (DCD) control, as illustrated in Figure 12.37, and a distributed output stage with output swing (VOA pad) and DC offset control (VEA pad). The triangles with down-pointing arrow indicate differential double-source-follower stages, biased at J_{pfT}, while the open triangle describes a current-steering differential HEMT pair with resistive loads [6]. The p-HEMTs in the current-steering stages operate in limiting mode, are biased at $J_{pfT}/2$, and swing up to J_{pFT}.

The distributed limiting output stage consists of 5 identical stages, each drawing a total of 63mA from a single −5.2V supply. The output is referenced to ground and allows for a DC-coupled connection of the EAM modulator whose cathode is grounded to minimize parasitic capacitance and thus improve the bandwidth. By including the output amplitude and

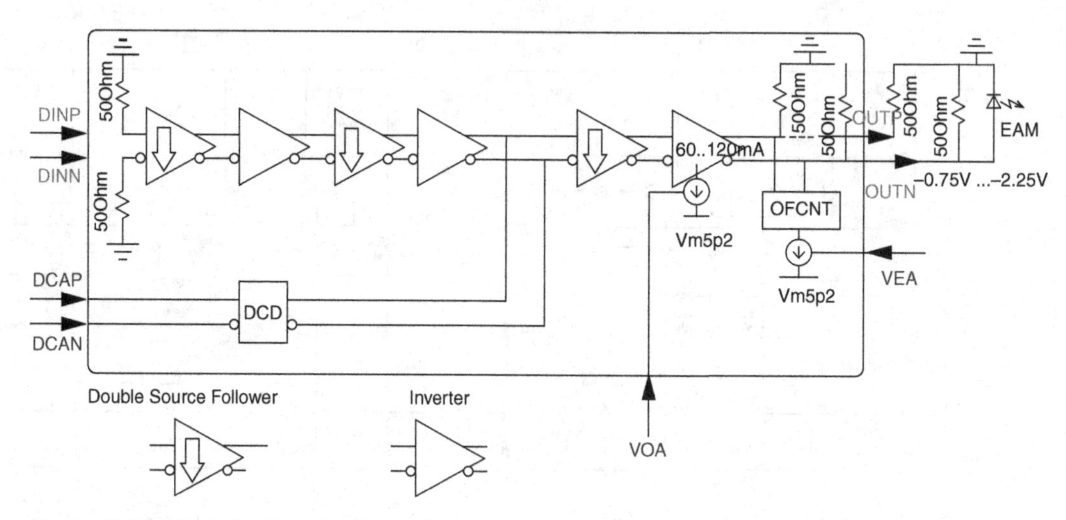

Figure 12.36 Block diagram of a 40Gb/s EAM driver [6]

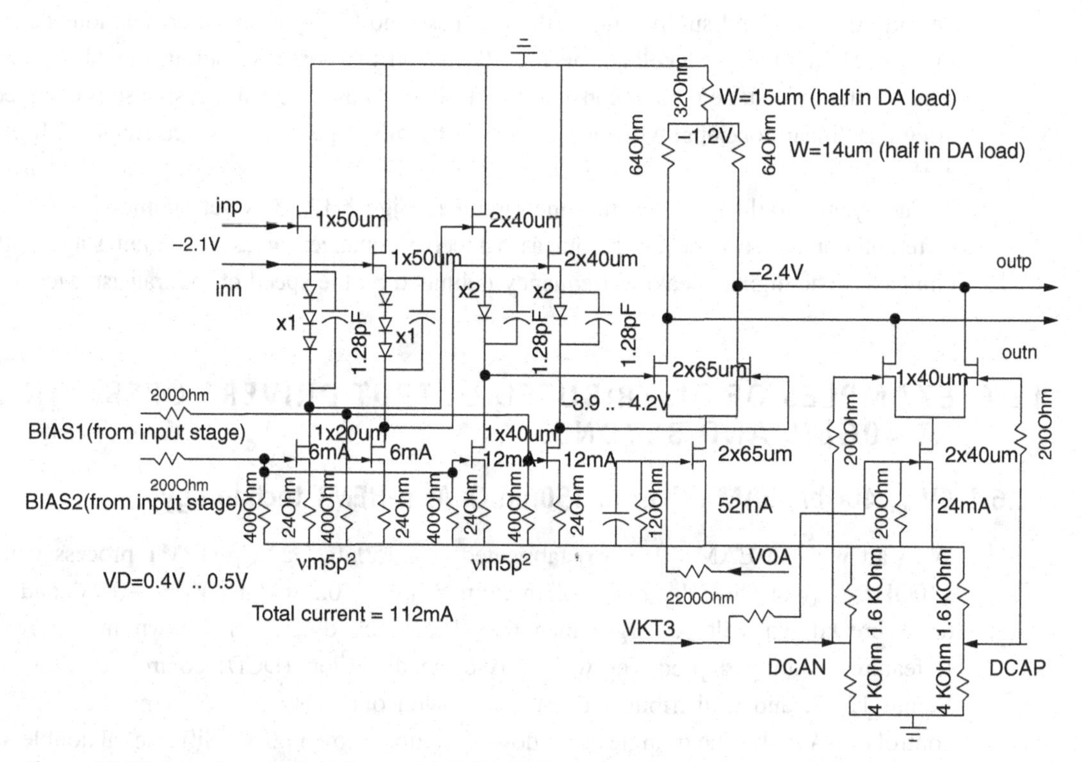

Figure 12.37 Second stage of predriver with PWC (DCD) control circuit

DC offset control in each DA cell, the extra capacitance at the output node associated with these functions is distributed along the output transmission lines, thus mitigating its impact on bandwidth. The simplified block diagram of the distributed output stage is illustrated in Figure 12.38, while the schematic of a DA cell is reproduced in Figure 12.39. To provide a voltage swing of $3V_{pp}$ per side into 50Ω differential loads, a total modulating current of

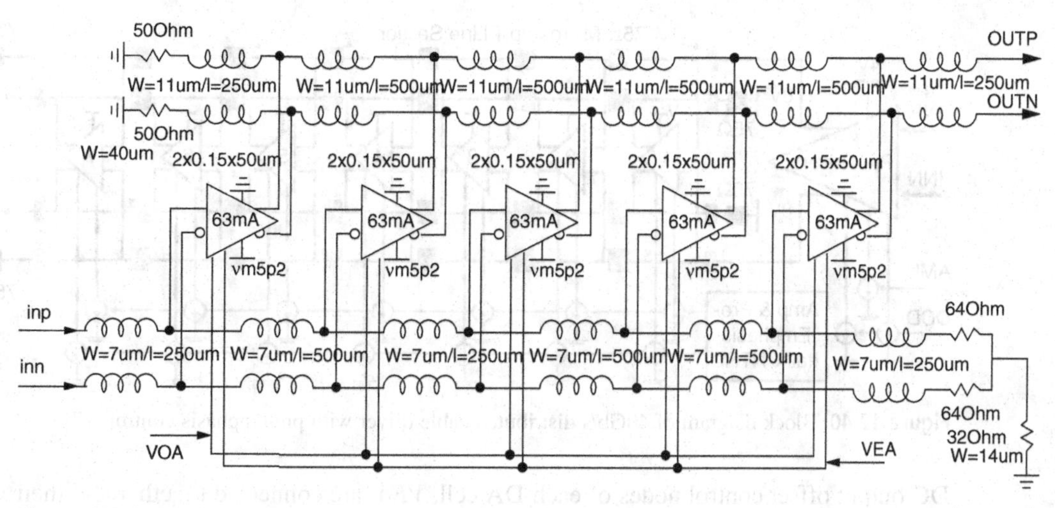

Figure 12.38 Simplified block diagram of distributed output stage [6]

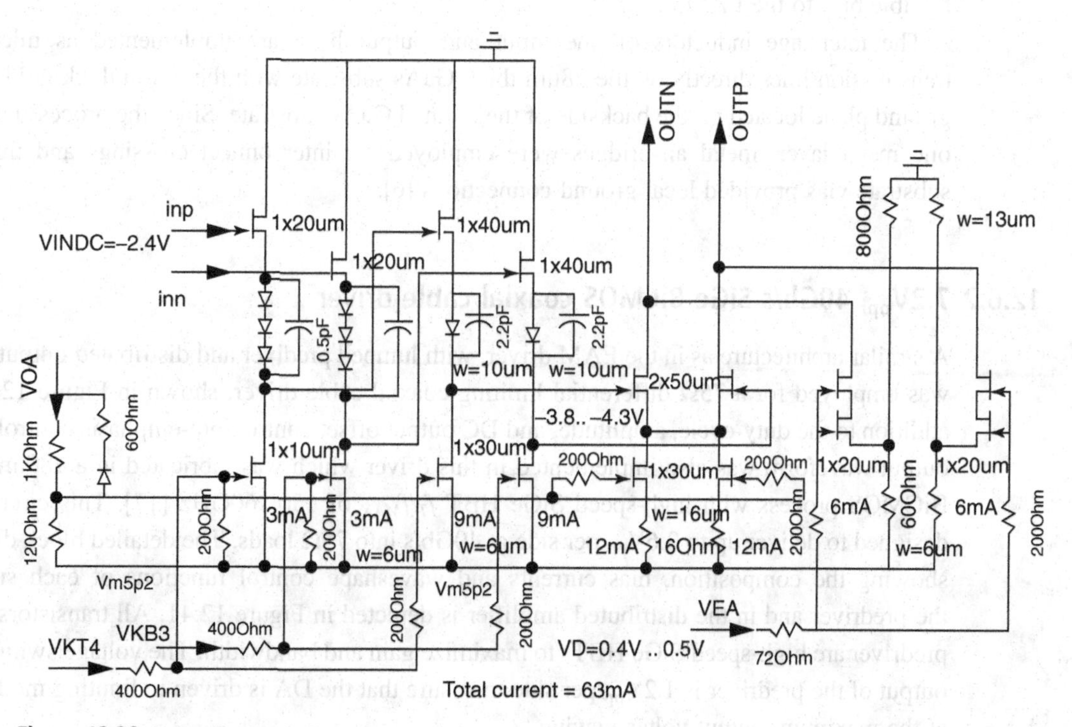

Figure 12.39 DA cell schematics

120mA, i.e. 24mA per stage is needed. The rest of the 39mA per stage is consumed by the two source-followers and by the DC output offset control circuit. As seen in Figure 12.39, the gates of the tail current sources in the current-steering stage and in the source-follower stages are tied together to the *VOA* pad. By changing the voltage on this pad, the total output current and output voltage swing can be adjusted from $1.5V_{pp}$ per side to $3V_{pp}$ per side without degrading the quality of the eye diagram. The *VOA* nodes of all 5 DA cells are tied together. Similarly, the

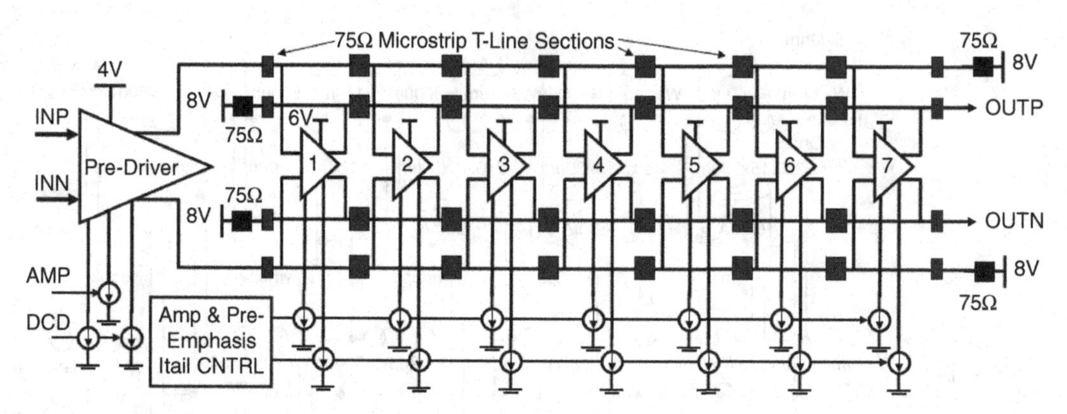

Figure 12.40 Block diagram of 40Gb/s distributed cable driver with pre-emphasis control [17]

DC output offset control nodes of each DA cell, *VEA*, are connected together and their voltage is adjusted to modify the DC output offset voltage between −0.75V and −2.25V, providing flexible bias to the EAM.

The interstage inductors on the input and output lines are implemented as microstrip transmission lines directly on the 28μm thick GaAs substrate with the 32μm thick gold-plated ground plane located on the backside of the thinned GaAs substrate. Since the process has only one metal layer, metal air-bridges were employed for interconnect crossings and through-substrate vias provided local ground connections [6].

12.6.2 7.2V_{pp}, 40Gb/s SiGe BiCMOS coaxial cable driver

A similar architecture as in the EAM driver, with lumped prediver and distributed output stage, was employed for a 75Ω differential limiting coaxial cable driver, shown in Figure 12.40. In addition to the duty-cycle, amplitude, and DC output offset control, pre-emphasis control of the output waveform was also implemented in this driver which was fabricated in a 180nm SiGe BiCMOS process with high-speed SiGe HBT f_T/f_{MAX} of 150/160GHz [17]. This circuit was designed to deliver up to 3.6V_{pp} per side at 40Gb/s into 75Ω loads. The detailed block diagram showing the composition, bias currents and waveshape control functions of each stage in the predriver and in the distributed amplifier is depicted in Figure 12.41. All transistors in the predriver are high-speed SiGe HBTs to maximize gain and bandwidth. The voltage swing at the output of the predriver is 1.2V_{pp} per side to ensure that the DA is driven in limiting mode even at the maximum output voltage swing.

The DA consists of 7 stages, each featuring a source-follower/emitter-follower/cascode current-steering topology. Despite having a low f_T of only 60GHz, the 180nm MOSFETs employed in the source-follower stage are useful in reducing signal loss along the input transmission line by providing a small capacitive load. As illustrated in Figure 12.42, analog pre-emphasis is implemented by splitting the signal into low-pass (LPF) and high-pass filter (HPF) paths after the emitter-follower stage. The HPF path consists of an analog R-C differentiator (R_p − C_p) followed by a high-speed HBT current-steering pair with emitter

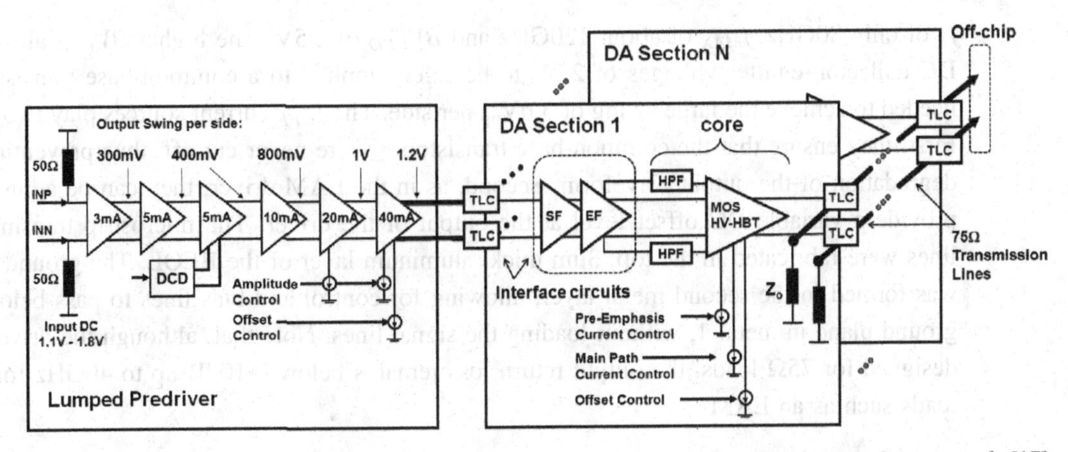

Figure 12.41 Detailed block diagram of the lumped predriver and DA cells showing waveshape controls [17]

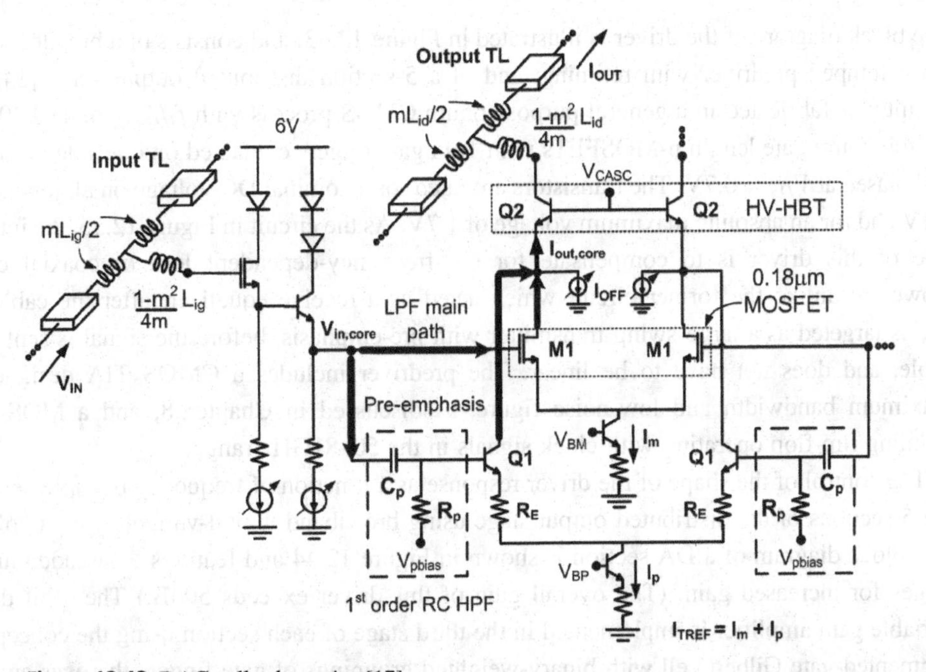

Figure 12.42 DAC cell schematics [17]

degeneration ($Q1$) whose bandwidth extends far beyond 40GHz. The amplitude of the symmetrical pre-emphasis pulses on the rising and falling edges of the data signal is controlled by the tail current source I_p. The low-pass path is formed by the 180nm n-MOSFET differential pair (M1) whose tail current, I_m, controls the output voltage amplitude. The signals from the two paths are then recombined at the low-impedance nodes formed by the emitters of the common-base transistors Q_2. This scheme avoids adding significant capacitance at the higher impedance output nodes of the DA cell. Since the BV_{CEO} of the high-speed HBTs in this process is only 1.8V, in order to ensure that the circuit meets the long-term reliability requirement, transistors Q_2 are implemented as high-voltage HBTs with

f_T of only 80GHz, f_{MAX} of about 120GHz and BV_{CEO} of 2.5V. The higher BV_{CEO} allows for DC collector–emitter voltages of 2.2V to be safely applied to a common-base transistor, as needed to achieve the large swing of 3.6V_{pp} per side. The I_{OFF} current sources play two roles. First, they ensure that the common-base transistors Q_2 are never cut off, thus preventing the degradation of the output waveform. Second, as in the EAM driver, they can be adjusted to provide a variable DC offset level at the output of the driver. The microstrip transmission lines were fabricated in the top, 3μm thick, aluminum layer of the BEOL. The ground plane was formed in the second metal layer, allowing for control and bias lines to pass below the ground plane in metal 1, without loading the signal lines. Note that, although the driver was designed for 75Ω loads, the output return loss remains below −10dB up to 40GHz for 50Ω loads such as an EAM.

12.6.3 2.4V_{pp}, 60Gb/s coaxial cable driver with retiming in 65nm CMOS

The block diagram of the driver is illustrated in Figure 12.43. and consists of a broadband low-noise lumped predriver with retiming, and of a 5-section distributed output stage [14]. The circuit was fabricated in a general purpose 65nm CMOS process with f_T/f_{MAX} of 180/200GHz for minimum gate length n-MOSFETs with 1μm gate fingers contacted on one side of the gate and biased at $V_{DS} = 0.7$V. The transistors are rated for a nominal DC voltage on all junctions of 1.1V and for an absolute maximum voltage of 1.7V. As the circuit in Figure 12.35, the intended role of this driver is to compensate for the frequency-dependent loss of coaxial cables. However, unlike the former circuit which acted as a receive equalizer, after the cable, this one is targeted as a large-swing transmitter with pre-emphasis, before the signal is sent to the cable, and does not have to be linear. The predriver includes a CMOS TIA designed for maximum bandwidth and low-noise figure, as discussed in Chapter 8, and a MOS CML retiming flip-flop operating with clock signals in the 50–81GHz range.

The control of the shape of the driver response as a function of frequency is implemented in the 5 sections of the distributed output stage using broadband digital-variable gain amplifiers. The block diagram of a DA section is shown in Figure 12.44 and features 5 cascaded lumped stages for increased gain. (The overall gain of this driver exceeds 50dB.) The 4-bit digital-variable gain amplifier is implemented in the third stage of each section using the concept of a segmented-gate Gilbert cell with binary-weighted groupings of gate fingers that was analyzed in Chapter 9. The transfer function of the first two DA sections has a low-pass filter (LPF) characteristic and their outputs are DC-coupled to the output transmission line. The last three DA sections have a bandpass filter (BPF) response tuned at three different fixed frequencies: 25GHz, 35GHz and 45GHz, and are AC-coupled through transformers to the output line. Because the transformer primary coils act as loads at the output of the BPF DA sections, the tuning frequency of each of these sections is set by the number of series-connected transformers between a particular section and the output line.

The transfer function of the distributed output stage is synthesized as the weighted sum of the transfer functions of each DA section, with the output transmission line acting as the summer and the digital variable gain amplifiers in each DA section providing the weighting functions.

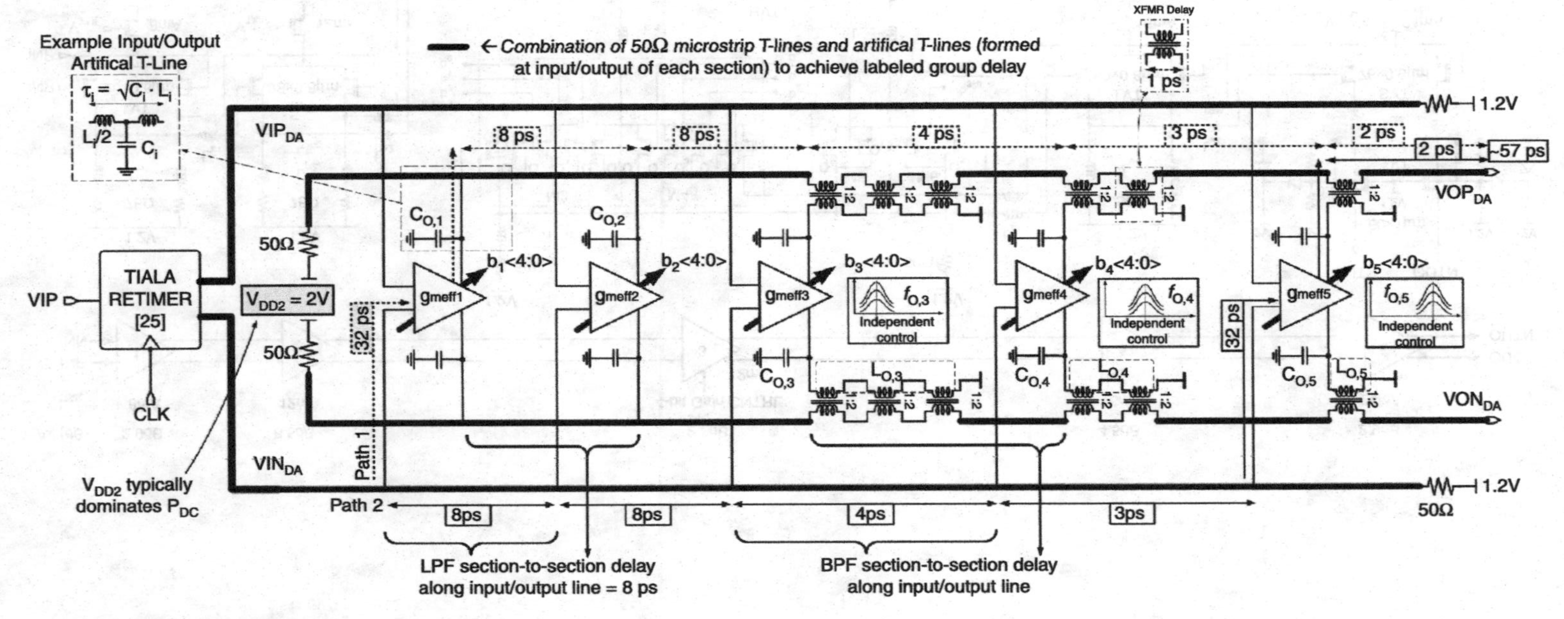

Figure 12.43 Block diagram [14]

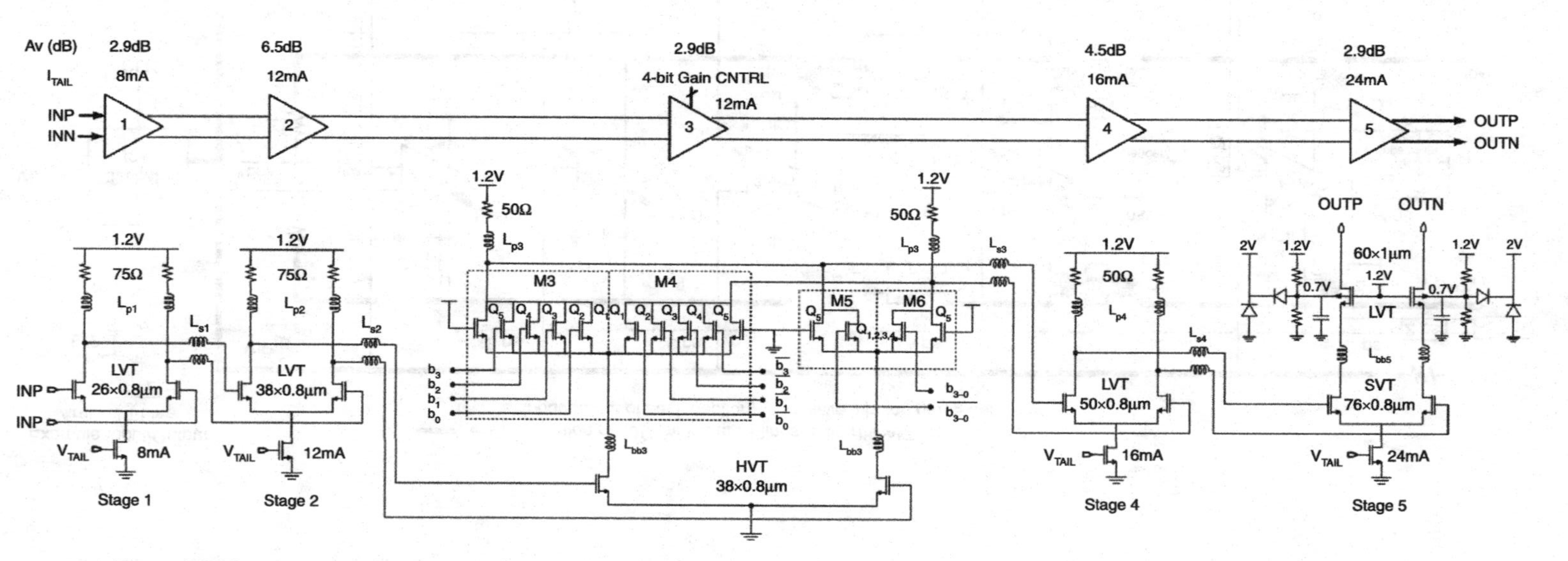

Figure 12.44 Schematic of unit DA section [14]

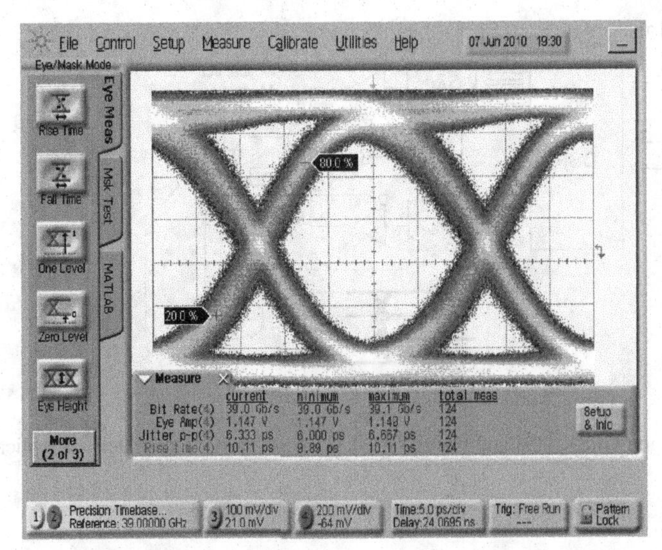

Figure 12.45 39Gb/s eye diagram with $1.2V_{pp}$ per side ($2.4V_{pp}$ differential)

The driver is biased from a 1.2V supply, except for the last stage of the LPF DA sections and the output transmission lines which are connected to a 2V supply. The common-gate MOSFETs in the output cascode stage are placed in deep n-wells such that the voltage between the transistor terminals gate, drain, source and bulk does not exceed 1.1V. The n-well-substrate junction is reverse biased at 2V. However, this is a safe voltage for the deep n-well. Figure 12.45 reproduces a measured 40Gb/s eye diagram with $2.4V_{pp}$ differential swing at the output of this driver.

12.7 HIGH-SPEED DACs

The previous example goes half way to implementing a fully fledged high-speed DAC as a versatile driver capable of synthesizing complex signal waveforms at very high data rates. Such DACs, with lower output swing, have become common place in the latest generation of 40Gb/s and 110Gb/s Ethernet fiber-optic communication systems which employ QPSK, 16QAM and OFDM modulation schemes. Examples of system-level applications of dual-polarization QPSK and of 16 QAM modulation in two different DSP-based 110Gb/s fiber-optic transmitters are shown in Figure 12.46.

These DACs are essentially realized as arrays of (retimed) current-steering stages whose signals are summed up on the output load.

The most common DAC architectures for applications at 40-GS/s or higher are (I) binary-weighted current-steering stages with segmentation [18],[19] and (ii) R-2R ladder [1].

12.7.1 56Gb/s 6-bit DAC in 65nm p-MOS CML

The block diagram of a 56-GS/s 6-bit DAC employing a half-rate, segmented current-steering architecture with inductive peaking is shown in Figure 12.47. It is fabricated in the same 65nm CMOS process as the previous driver. There are a total of 18 output slices, 16 unary slices corresponding to the 4 MSBs (most significant bits), and two binary slices

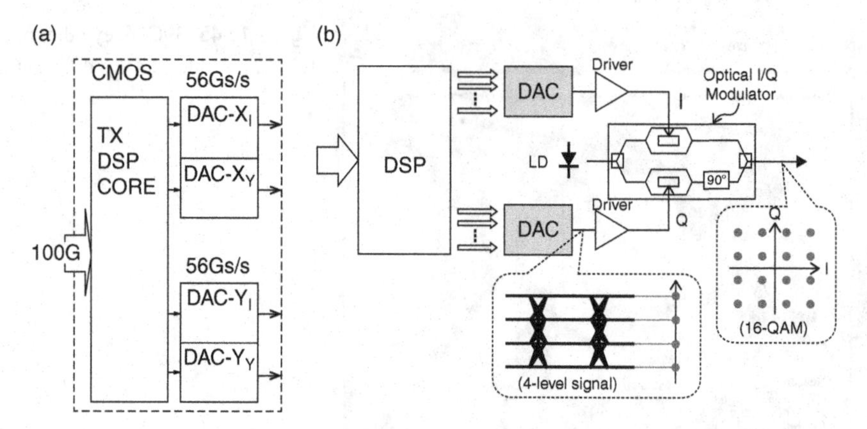

Figure 12.46 (a) Dual-polarization QPSK [19] and (b) 16-QAM [1] DSP-based transmitters for optical fiber communication

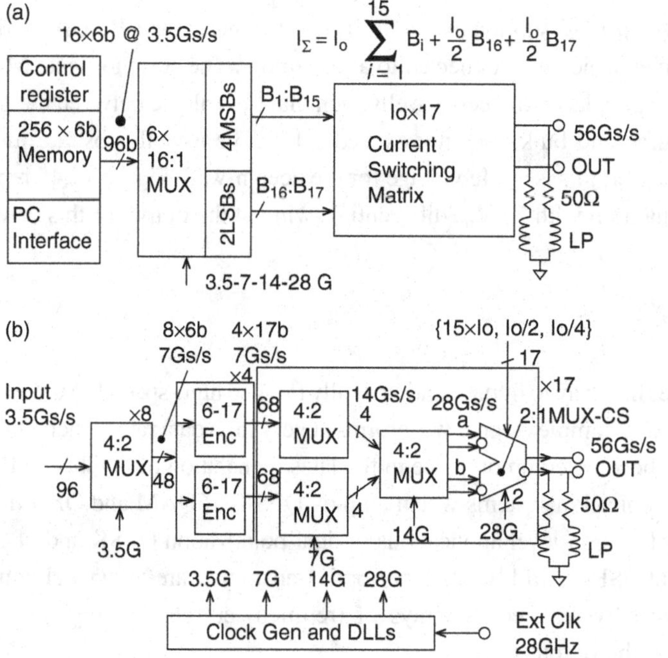

Figure 12.47 (a) Segmented current-steering DAC architecture and (b) MUX and clock generation details [19]

corresponding to the 2 LSBs (least significant bits). In order to save power, the last 2:1 MUX drives the output current-steering pair directly, without retiming, and is realized with conventional CMOS pass gates and a 28GHz clock. As illustrated in Figure 12.48, p-MOSFETs are used in the output current-steering differential pair. Although, compared to n-MOSFETs, the p-MOSFETs degrade the switching speed in this technology, this topology permits referencing to GND and use of thick-oxide cascodes in the current source

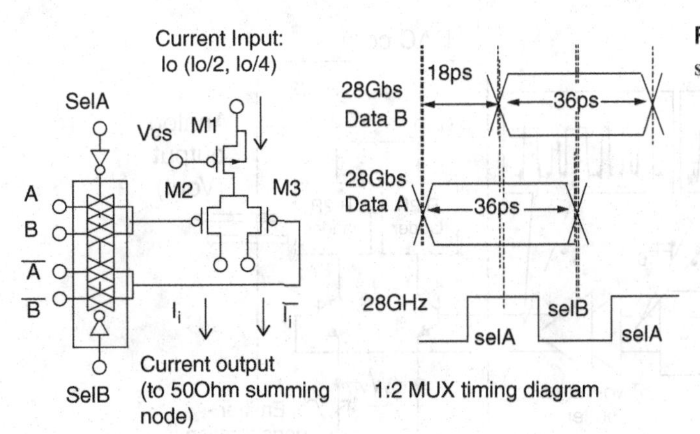

Figure 12.48 Half-rate current steering slice topology [19]

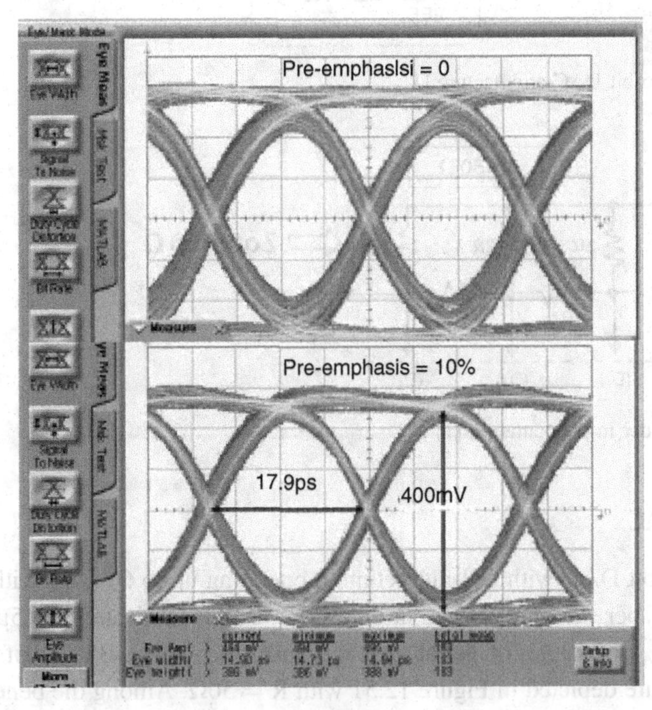

Figure 12.49 Differential 56Gb/s eye diagrams with and without pre-emphasis [19] showing 400mV$_{pp}$ swing, 200mV$_{pp}$ per side

with $V_{DD} = 2.5$V. Well-matched, high-impedance current sources are critical for achieving good SFDR in DACs. In the absence of full-rate retiming, this architecture relies on the DSP engine to compensate for imperfections in signal waveform (DCD, clock and data skew). The measured 56-Gs/s eye diagrams with 400mV$_{pp}$ differential swing are reproduced in Figure 12.49. The p-MOSFET current-steering stage is expected to perform even better in strain-engineered 45nm and 32nm CMOS technologies where the p-MOSFET is just as fast as the n-MOSFET.

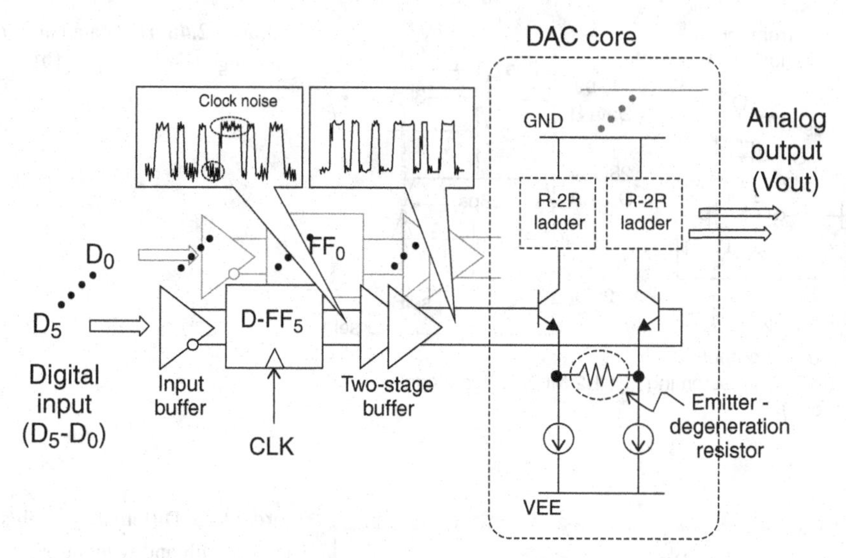

Figure 12.50 60Gb/s InP HBT 6-bit DAC architecture [1]

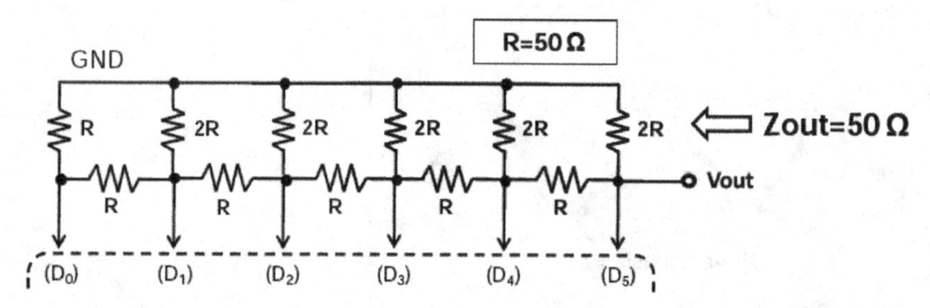

Figure 12.51 R-2R resistive ladder implementation [1]

12.7.2 60GS/s InP HBT 6bit DAC

The second example is a 6-bit DAC with full-rate retiming operating up to 60 GS/s with $1V_{pp}$ differential swing ($500mV_{pp}$ per side) and shown in Figure 12.50. It is fabricated in 0.5μm InP HBT technology with f_T/f_{MAX} of 290/320GHz at $J_{pfT} = 3.5mA/\mu m^2$, and a BV_{CEO} of 4V. It employs the R-2R architecture depicted in Figure 12.51 with R = 50Ω. Among the benefits of the R-2R ladder DAC architecture are

- reduced capacitance at the output node and increased bandwidth because of the minimal number of slices,
- reduced glitches between transitions,
- use of only unit current sources which improves matching between slices, reduces circuit complexity, footprint, and therefore output and interconnect capacitance.

The output stage is realized as an HBT current-steering stage with resistive degeneration for improved bandwidth. Prior to reaching the current-steering stage, the 6-bit wide digital data are first retimed with a full-rate clock and the clock feed-through is filtered by a two-stage buffer.

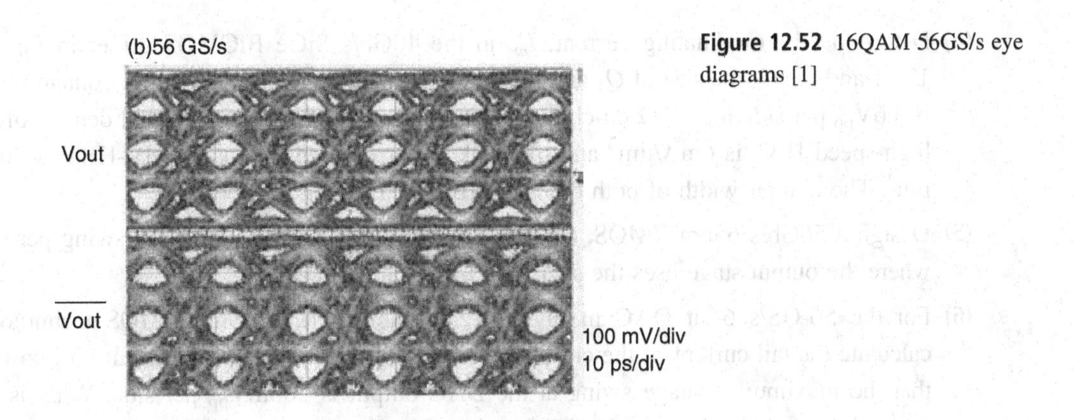

Figure 12.52 16QAM 56GS/s eye diagrams [1]

Figure 12.52 reproduces the 16QAM eye diagrams measured at the differential output at 56GS/sec with an output voltage swing of 500mV$_{pp}$ per side.

Summary

In this chapter we have reviewed the main high-speed digital output driver topologies and the waveshape control functions. Drivers can be classified in backplane/cable drivers, laser drivers and optical modulator drivers. These drivers typically operate in limiting mode and employ a current-steering differential pair architecture. Low-current/voltage swing output stages such as VCSEL and backplane drivers are realized with lumped stages, whereas optical modulator drivers typically employ a distributed output stage. The waveshape control functions implemented in the driver include retiming, pulse-width control, output swing control, pre-emphasis control and DC output level control. There is clear trend in recent years to implement these control functions using high-speed DACs and a DSP engine.

Problems

(1) Design the bias current and transistor size of a VCSEL driver output stage fabricated in 90nm CMOS technology with a modulating current of 10mA. What should be the value of the peaking current source if 25% pre-emphasis is specified?

(2) Design the DCD circuit for the lumped predriver in Figure 12.41 to ensure + /– 20% of pulse-width control around the 50% midpoint.

(3) Design a differential version of the 40Gb/s output driver in Figure 12.15 with 3V$_{pp}$ swing per side using the stacked PA concept in Chapter 6. Implement 4-bit output current control. What is the size of the LSB device? Use the parameters of the 45nm SOI CMOS technology provided in Appendix 9. Assume that each transistor has DC drain–source voltage of $V_{DS} = 1V$.

(4) Determine the modulating current, I_m, in the 40Gb/s SiGe BiCMOS driver in Figures 12.41 and 12.42, the size of Q_1 and Ml, knowing that the maximum output voltage swing is 3.6V$_{pp}$ per side into 75Ω on-chip and off-chip loads. The peak f_T current density of the high-speed HBT is 6mA/μm^2 and the peak f_T current density for the HV-HBT is 3mA/μm^2. The emitter width of both types of HBTs is 180nm.

(5) Design a 56Gb/s 65nm CMOS, 6-bit distributed DAC diver with 1.2V$_{pp}$ swing per side where the output stage uses the segmented-gate Gilbert cell.

(6) For the 56-GS/s 6-bit DAC in Figure 12.48 fabricated in 65nm CMOS technology, calculate the tail current of the current-steering stage for the least significant bit knowing that the maximum voltage swing at the DAC output is 200mV$_{pp}$ per side. What is the optimal gate width of the p-MOSFET in the LSB slice for maximum speed, knowing that the peak f_T current density of the p-MOSFET is 0.2mA/μm? What is the required power gain of the Mach-Zehnder modulator driver that needs to be placed after this DAC if the modulator requires an 8V differential swing?

(7) Calculate the tail current and the p-MOSFET gate width if the 65nm CMOS 6-bit DAC in Problem 5 were to employ an R-2R ladder. Assume that the maximum output swing remains 200V$_{pp}$ per side.

(8) Explain how you could combine the 7-stage DA driver concept, with the 6-bit segmentated DAC architecture and the n-MOSFET cascode in deep n-well current steering stage topology to obtain a segmented large-swing 9-bit DAC with 1.2V$_{pp}$ per side.

(9) Design the DAC topology in [20] with HBTs and RC ladder.

(10) Draw the schematics of an R-2R 6-bit n-MOS DAC with 600mV$_{pp}$ swing per side operating at 56-GS/s. Indicate how you can add two additional bits by replacing the current-steering output stage with a segmented-gate layout Gilbert cell. What is the tail current and n-MOSFET size of the LSB slice knowing that the peak f_T current density is 0.4mA/μm.

(11) Design a 56GS/s, 2V$_{pp}$ per side 6bit DAC with R-2R ladder using stacked 45nm SOI technology. Use technology data from Appendix 9.

(12) Design a 10-stage, 3V$_{pp}$ swing per side, distributed 100Gb/s Mach-Zhender modulator driver using the InP HBT technology from Section 12.7.2. Use the same output stage topology as in Figure 12.16. Find the size of the HBTs in the current steering stage and the interstage inductance value on the output transmission line, knowing that $F = 0.25$ps, $C_{jC}/A_E = 8$fF/μm^2, and $C_{cs}/l_E = 0.6$fF/μm. Calculate the 3dB bandwidth of the output line.

(13) Come up with a possible logic function implementation of the "npeak" and "ppeak" blocks for the circuit in Figure 12.24 which allows for separate rising edge and falling edge pulses.

(14) Draw the block diagram of a 6-bit SiGe HBT distributed DAC driver with 3V$_{pp}$ swing per side assuming that there are 7 DA cells. How would you choose the topology of each cell if you were to implement 3 MSB and 3 LSB segmentation? What are the tail currents of the segmented MSB and LSB cells? Assume the same SiGe BiCMOS process as in Example 12.1.

REFERENCES

[1] M. Nagatani, H. Nosaka, K. Sano, K. Murata, K. Kurishima, and M. Ida, "A 60-GS/s 6-bit DAC in 0.5-μm InP HBT technology for optical communications systems," *IEEE CSICS Digest*, in print, October 2011.

[2] E. Säckinger, *Broadband Circuits for Optical Fiber Communication*, Chapter 8, John Wiley & Sons, 2005.

[3] A. Narashima, B. Analui, Y. Liang, T.J. Sleboda, S. Abdalla, E. Balmater, S. Gloeckner, D. Guckenberger, M. Harrison, R.G.M.P. Koumans, D. Kucharski, A. Mekis, S. Mirsaidi, D. Song, and T. Pinguet, "A fully integrated 4× 10Gb/s DWDM optoelectronic transceiver implemented in a standard 0.13μm CMOS SOI Technology," *IEEE JSSC*, **42**, pp. 2736–2744, December 2007.

[4] C. L. Schow, A. V. Rylyakov, B. G. Lee, F. E. Doany, C. Baks, R. A. John, and J. A. Kash, Transmitter Pre-Distortion for Simultaneous Improvements in Bit-Rate, Sensitivity, Jitter, and Power Efficiency in 20Gb/s CMOS-driven VCSEL Links, OFC, 2011.

[5] C. Kromer, G. Sialm, C. Berger, T. Morf, M. Schmatz, F. Ellinger, and D. Erni G-L. Bona, "A 100mW 4×10Gb/s transceiver in 80nm CMOS for high-density optical interconnects," IEEE ISSCC Digest,pp. 334–335, February 2005.

[6] D. S. McPherson, F. Pera, M. Tazlauanu, and S. P. Voinigescu, "A 3V fully differential distributed limiting driver for 40Gb/s optical transmission systems," *IEEE JSSC*, **38**(9): 1485–1496, 2003.

[7] M. Möller, T.F. Meister, R. Schmid, J. Rupeter, M. Rest, A. Schöpflin and H.-M. Rein, "SiGe retiming high-gain power MUX for directly driving and EAM up to 50 Gbit/s," *Electronics Letts*, **34**(18): 1782–1784, September 1998.

[8] K. Joohwa and J. F. Buckwalter, "A 40Gb/s optical transceiver front-end in 45nm SOI CMOS technology," *Proceedings of the IEEE CICC*, pp. 1–4, San Jose, CA, September 2010.

[9] T. Y. K. Wong, A. P. Freundorfer, B. C. Beggs, and J. E. Sitch, "A 10Gb/s AlGaAs/GaAs HBT high power fully differential limiting distributed amplifier for III-V Mach–Zehnder Modulator," *IEEE JSSC*, **31**(10):1388–1393, October 1996.

[10] J. Cao, M. Green, A. Momtaz, K. Vakilian, D. Chung, K.-C. Jen, M. Caresosa, X. Wang, W.-G. Tan, Y. Cai, I. Fujimori, and A. Hairapetian, "OC-192 transmitter and receiver in standard 0.18-μm CMOS," *IEEE JSSC*, **37**(12):1768–1780, December 2002.

[11] S. P. Voinigescu, D. S. McPherson, F. Pera, S. Szilagyi, M. Tazlauanu, and H. Tran, "A comparison of silicon and III-V technology performance and building block implementations for 10 and 40Gb/s optical networking ICs," *Compound Semiconductor Integrated Circuits*, pp. 27–58, October 2003.

[12] A. Kern, A. Chandrakasan, and I. Young, "18Gb/s optical IO: VCSEL driver and TIA in 90nm CMOS," *Symposium of VLSI Circuits Digest of Technical Papers*, pp. 276–277, June 2007.

[13] D. Kucharski and Y. Kwark, IEEE VLSI Circuits Symposium Digest D. Kuchta, D. Guckenberger, K. Kornegay, M. Tan, C-K. Lin, and A. Tandon, "A 20Gb/s VCSEL Driver with Pre-Emphasis and Regulated Output Impedance in 0.13μm CMOS" *IEEE ISSCC Digest*, pp. 222–223, February 2005.

[14] R. A. Aroca, P. Schvan, and S. P. Voinigescu, "A 2.4Vpp, 60Gb/s, mm-Wave DAC-based CMOS driver with adjustable amplitude and peaking frequency," *IEEE JSSC*, **46**: 2226–2239, October 2011.

[15] M. H. Shakiba "A 2.5Gbps adaptive cable equalizer," *IEEE ISSCC Digest*, pp. 396–397, February 1999.

[16] A. Balteanu and S. P. Voinigescu, "A cable equalizer with 31dB of adjustable peaking at 52GHz," *IEEE BCTM Digest*, Capri, Italy, pp. 154–157, October 2009.

[17] R. A. Aroca and S. P. Voinigescu, "A large swing, 40Gb/s SiGe BiCMOS driver with adjustable pre-emphasis for data transmission over 75Ω coaxial cable," *IEEE JSSC*, **43**(10): 2177–2186, October 2008.

[18] P. Schvan, D. Pollex, and T. Bellingrath, "A 22 GS/s 6b DAC with integrated digital ramp generator," *IEEE ISSCC Digest*, pp. 122–123, February, 2005.

[19] Y. M. Greshishchev, D. Pollex, S.-C. Wang, M. Besson, P. Flemeke, S. Szilagyi, J. Aguirre, C. Falt, N. Ben-Hamida, R. Gibbins, and P. Schvan, "A 56GS/s 6b DAC in 65nm CMOS with 256×6b memory," *IEEE ISSCC Digest*, pp. 194–195, February 2011.

[20] A. Bālteanu, P. Schvan, and S. P. Voinigescu, "6-bit Segmented RZ DAC Architecture with up to 50-GHz Sampling Clock," *IEEE IMS Digest*, June, 2012.

13 SoC examples

This chapter presents a possible design flow, along with biasing, isolation and layout strategies suitable for mm-wave SoCs. Competing transceiver architectures, self-test, and packaging approaches are reviewed next, followed by examples of mm-wave SoCs for a wide range of new applications.

WHAT IS A HIGH-FREQUENCY SoC?

Although a precise definition is difficult to formulate, we define a high-frequency SoC as a single-chip radar, sensor, radio or wireline communication transmitter, receiver or transceiver that includes all high-frequency blocks, sometimes even the antennas, along with digital control and signal-processing circuitry.

Examples include:

- 2GHz cell-phone or 5GHz wireless LAN transceivers
- 40Gb/s or 100Gb/s SERDES
- 60GHz radio transceiver
- 77GHz automotive radar transceiver
- W-, D-, and G-Band active and passive imagers.

13.1 DESIGN METHODOLOGY FOR HIGH-FREQUENCY SoCs

Most foundries have recently decided that the MOSFET and SiGe HBT compact models should capture the parasitic capacitance and resistance of only the first 1–2 metal layers, the minimum required for contacting all the device terminals. The rationale invoked is that this approach allows circuit designers more flexibility in the physical layout of transistors and of commonly encountered transistor groupings such as interdigitated differential pairs, interdigitated Gilbert cell quads, and latches. One (major) impediment is that, since most of the backend parasitics are not accounted for in schematic-level simulations, the burden is passed on to the designer to first lay out and then extract the parasitic impedance of the full wiring stack above the transistor in order to get an accurate simulation of circuit performance. This, in turn, creates a problem since, in most cases, the design starts with schematic-level simulations to find the optimal transistor size. In sub 65nm CMOS technologies, without extracting the parasitics of the transistor layout, the simulated circuit performance can be optimistic by as much as 30–50%. Traditional schematic-level circuit design thus becomes less accurate than "back-of-the-envelope" hand design. A fundamental paradigm shift from traditional analog and RF circuit design is therefore required in the design of mm-wave nanoscale CMOS SoCs. In contrast, schematic level design with dedicated RF models, or schematic level simulations in III-V and SiGe HBT technologies, is still reasonably accurate because the RF model includes RC layout parasitics. A so-called "RF-model," extracted from S-parameter measurements de-embedded only to the edge of the transistor and reflecting a complete wiring stack, is sometimes provided by foundries for a limited set of transistor layouts to facilitate accurate circuit simulation early in the design phase, before layout.

Additionally, even when RF models are available, the design flow for mm-wave ICs is complicated by the need to model every piece of interconnect longer than 5 … 10μm as an inductor or distributed transmission line. An effective method to contain the modeling effort is to include all interconnect leading to and from an inductor in the inductor itself, and to extract the 2-π equivalent circuit of the ensemble from the simulated S-parameters of the inductor using a 2.5D EM field simulator.

At the circuit cell level, the main goal is to minimize the footprint by merging the transistor layouts of differential pairs and mixing quads, and thus to shrink the length and parasitic capacitance of local interconnect. The accurate extraction of RC parasitics at the layout-cell level (i.e. interdigitated transistor or varactor cell, cascode cell, differential-pair cell, switching quad cell, cross-coupled pair cell, etc.) is critical to accurately capture and model the significant gain and noise figure degradation observed in circuits with nanoscale MOSFETs. The MOSFET source, drain, and gate series resistances are further degraded by layout contact and via resistance. Because of the larger transconductance, junction capacitances and smaller C_{bc}/C_{be} ratio (i.e. reduced Miller effect) at the same bias current, circuits realized with HBTs are less sensitive to layout parasitics than those with MOSFETs.

Analog and RF design flows vary from company to company and sometimes from designer to designer, based on individual experience and preference. A top-down-top approach that has been found to be most efficient in the design of mm-wave SoCs through 200GHz at the University of Toronto is described next. In this flow, shown in Figure 13.1, the design hierarchy is first assembled at the symbol (symbol view), layout padframe (layout view), and behavioral (behavioral model view) level from the system specification starting in a top-down fashion, from the SoC symbol and layout padframe, down to the basic circuit-cell level. Symbol and layout placeholders with labelled pins and pads are employed at this stage for blocks whose schematics and layouts are yet to be finalized or even started. Design and optimization of the basic cell schematics, component values, layouts, and extracted layout views of each component and basic circuit cell is performed next. Schematic and layout optimization, and simulation of schematics with extracted layout parasitics proceeds in a bottom-up fashion from the most basic cells to the top level of the SoC. The design flow is summarized below.

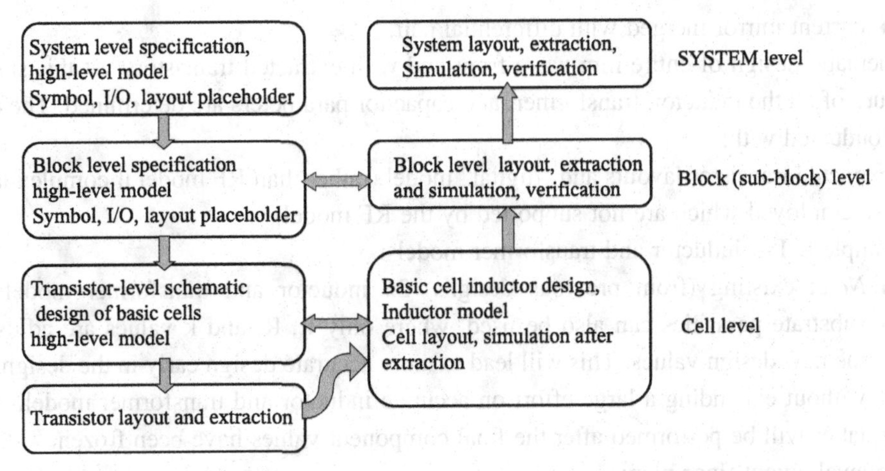

Figure 13.1 Example of top-down design flow employed in mm-wave SoC design

13.1.1 Millimeter-wave design flow

1. Draw the top-level symbol, pad frame and create a high-level model (e.g. VerilogA, AHDL, etc.) which will be used in early-stage verification.
2. Draw the symbols of all the higher-level blocks in the hierarchy in a top-down fashion starting from the top-level and descending step-by-step in the hierarchy. Create a high-level circuit block model to be used in early-stage system verification.
3. Transistor-level schematic design of the lowest (1st-level) cells using the RF models (if available) as proof of concept for architecture. A high-level model for the cell should be created to be used in system-level verification simulations:
   - Transistor sizes and bias currents are set at this stage.
   - Use planar MOSFETs with 2μm to 4μm finger widths in 130nm CMOS/BiCMOS, 0.7μm to 1.5μm finger widths in 65nm CMOS, 0.4μm to 0.7μm figer width in 32nm CMOS, and 0.2μm to 0.4μm finger width in 22nm CMOS. Smaller finger widths dramatically degrade f_T without further improving f_{MAX} of NF_{MIN}. In the case of multi-gate or FinFETs, the finger width is defined by the width, w_{fin}, and height, h_{fin}, of the fin: $w_f = 2h_{fin} + w_{fin}$.
     - Surround all transistors (HBT and MOSFETs) with a ring of substrate contacts (also known as p-taps).
     - In merged-layout Gilbert cells and differential pairs, the substrate contact ring need only be placed around the Gilbert cell or around the differential pair.
   - Transistor layout and RC extraction. Use dedicated optimized layouts for:
     - common-source/common-emitter (CS/CE) topology with inductive degeneration
     - grounded CS/CE topology
     - common-gate (CG) topology
     - latch
     - mixing quad
     - VCO
     - differential pairs

- current mirror merged with differential pair.

4. Schematic design of entire mm-wave front-end with extracted transistors. At this stage the values of all the inductor, transformer, and capacitor parameters are determined. The design is conducted with:
   - extracted transistor layouts and "digital" model (rather than RF model if complex layouts are employed which are not supported by the RF model)
   - simple R-L-k inductor and transformer models:
     - *Note*: existing (from previous designs) 2π inductor and transformer models with substrate parasitics can also be used, where only L, R, and k values are adjusted to the new design values. This will lead to more accurate design early in the design stage, without expending a large effort on accurate inductor and transformer modeling. The latter will be performed after the final component values have been frozen.
5. Top-level layout floor-plan:
   - Start early in the design phase by using existing block layouts from other tapeouts as placeholders for new blocks of comparable complexity
   - Get an initial idea of the length of interconnect between blocks and add rough interconnect models into the individual block schematic, rather than adding the interconnect at a higher hierarchical level. This avoids using basic components such as transmission lines and inductors at higher levels in the design hierarchy, and will simplify hierarchical simulation, extraction and LVS.
6. Inductor and transformer design
   - Use your 2.5-D EM field simulator of choice to simulate the *S*-parameters versus frequency for interconnect, inductors, transformers, coupled-lines, and any other layout sections that require accurate EM-field simulations and models. From the EM-field simulated *Y*- or *S*-parameters, extract and create schematic_sim views of the lumped 2π equivalent circuit of inductors, transformers, etc. This model works with all types of circuit simulations and is the most sensible approach to use in "design" mode, when the component values and layout are not frozen.

     Note: schematic_sim views do not get checked by LVS. The onus is on the designer to verify correctness of schematic_sim view versus schematic view. Schematic_sim views are needed only for custom components such as inductors, transformers, t-lines for which the design kit does not provide accurate models. Also note that the layout of custom inductors and transformers is not typically verified by LVS. An unintended open or short circuit can often occur inside of a transformer or inductor layout and can only be checked visually.
   - From the 3-D or 2.5-D EM-field simulated *S*-parameters of the EM components, also create a "black-box" *S*-parameter file model to be used in analysis mode, in the latter stages of the design process, once the inductor/transformer values become final (or close to final). This "black-box" schematic view of the EM components in your layout can be used for more accurate small signal or noise analysis simulations but may sometimes cause problems in DC-operating point, non linear or time domain analysis. In design mode, the "black-box" model is inefficient since it must be re-simulated and recreated

every time the layout of a component changes. This makes it expensive and time consuming to sweep parameters or to optimize circuit block performance. Moreover, the "black-box" model cannot be used outside the range of frequencies contained in the simulated or measured *S*-parameter file.

7. Simulate the transceiver at top-level with all building blocks extracted.
   - Lay out each cell and include the supply and ground distribution mesh in the cell. Pay attention to where you place the contact pin. It impacts the extracted parasitic resistance value.
   - Extract RC parasitics of the entire cell, without double-counting for inductor/transformer parasitics that have been modeled elsewhere (e.g. in the inductor or transformer model).
8. Re-simulate, re-design, re-model inductors and transformers as necessary.
9. Lay out transceiver top-level completing modeling of interconnect between blocks:
   - model closely spaced interconnect lines as coupled lossy lines with a 2.5D EM simulator.
   - re-design/add passives as necessary to improve inter-block matching, e.g. in tuned circuits, add series MiM cap to resonate out long interconnect between circuit blocks at the frequency of interest.
   - include inter-block interconnect in lower level cells.

Details of some of the more critical aspects of the design flow are discussed next.

13.1.2 Design library organization

Library and cell structure

- Label each library and circuit cell with a meaningful name. A common practice in the industry is for the name to start with the designer's initials. For example, rt_160g_imager stands for the 160GHz imager library and the library designer is Rameses the Third. The initials are important to distinguish designs from each other when a complex chip, including blocks from multiple designers, is assembled in a single gdsii file. The library and cell names should be self-explanatory, as short as possible, and typically not longer than 16 characters.
- Draw the symbol of your chip, including all pins, and label them.
- Draw the layout of your chip pad frame including all pad names. Your symbol and layout should have exactly the same orientation and pin/pad location. For example, if OUTP is the second pin from the left on the NORTH side of the symbol, OUTP should be the second pad from the left on the NORTH side of your layout, etc.
- Draw the symbols, schematics, and layout of each block in your chip hierarchy ensuring that the same signal flow and pin/pad orientation and location exists in the symbol, in the schematic, and in the layout. This is important for you and for other users of your block to understand and verify its correct operation and layout.
- Basic components: transistors, diodes, resistors, inductors, etc. should only appear in the lowest 2–3 levels in the design hierarchy. If basic components are used at higher hierarchical levels alongside more complex blocks, the DRC (design rule checking) and LVS (layout versus schematic checking) of your block will take a significantly longer time to execute.

- Design blocks and cells so that they directly abut in layout, without requiring interconnect. This will simplify parasitic extraction and avoid doubly extracting parasitics.
- Never use a circuit cell that does not have both layout and schematic views, except perhaps, for the mesh. However, a schematic view of the mesh is always recommended since it will simplify verification and will help you identify IR drops in the supply and ground planes.

Schematics

- Draw all physical wires in your schematics. This helps others, and yourself, to understand and verify your circuit. Avoid using net labels instead of physical wires to connect nets in your schematics. If you cannot draw the wires in your schematic you will not be able to draw the layout.
- Do not draw more than 10 components in a schematic.
- Use only size A or B pages for your schematics. Larger page sizes make your schematic too complex, too difficult to draw in layout, and you will not be able to read the component values and operating point information in a print out or on the screen. If your schematic becomes too large, break it up in smaller units and…
- Make sure that all text labels, including component values and bias points, are clearly as visible in the schematics, that they do not overlap, and that no wires cross them. This will help in debugging your circuit later.
- Always annotate bias information, component size, and major performance parameters in the schematic if possible. Later, this will help you and others to understand the operation of the circuit, to verify and debug it, if necessary.
- Label all nets in your schematics. The names should be self-explanatory, as short as possible, and not longer than 16 characters. You should use buses wherever possible to avoid clutter in your schematics. For example, nets b0, b1, b2,…bn can be lumped together in a bus and labelled b[0:n]. The same applies for components (transistors, resistors, capacitors, antenna diodes, varactors, etc.) connected in parallel.
- Make the symbols as small as possible, yet still easy to read, to contain the size of the page.

13.1.3 Mesh layout and dummy generation strategies in nanoscale CMOS

In order to satisfy the strict metal density rules of nanoscale silicon technologies, a mesh cell consisting of all metal, poly, and active layers can be developed which also provides low-inductance, low-resistive ground and supply connections for each circuit cell, Figure 13.2 [1]. Control voltage lines and bias de-coupling MiM capacitors can also be included in this mesh. Ideally, the mesh should have the following features:

- coarse, wide metal ground mesh with grounded substrate taps throughout the SoC;
- "mesh" creates low resistance power and ground planes, like in a printed circuit board, PCB;
- provides distributed de-coupling of the power supply plane over the ground plane;
- substrate contacts, (deep) n-well contacts, or interlaced substrate/n-well contacts to ground and V_{DD}, respectively, or with both p-taps and n-well connected to ground;
- meets metal/poly/active density requirements.

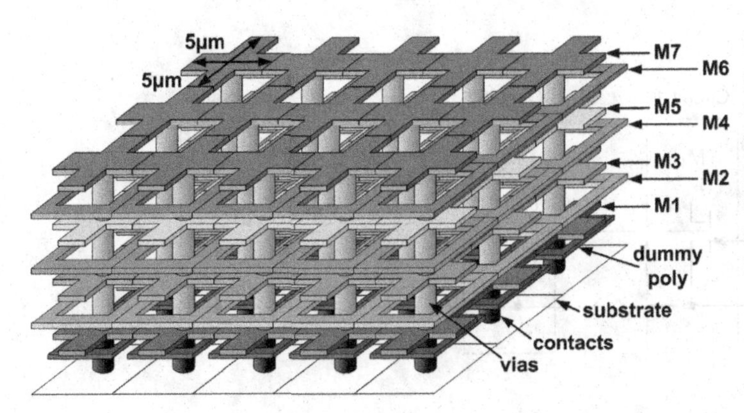

Figure 13.2 3-D view of possible supply, control line, and ground distribution mesh which also satisfies metal density rules and provides distributed supply de-coupling, distributed grounded substrate, low-inductance and low-resistance ground plane [1]

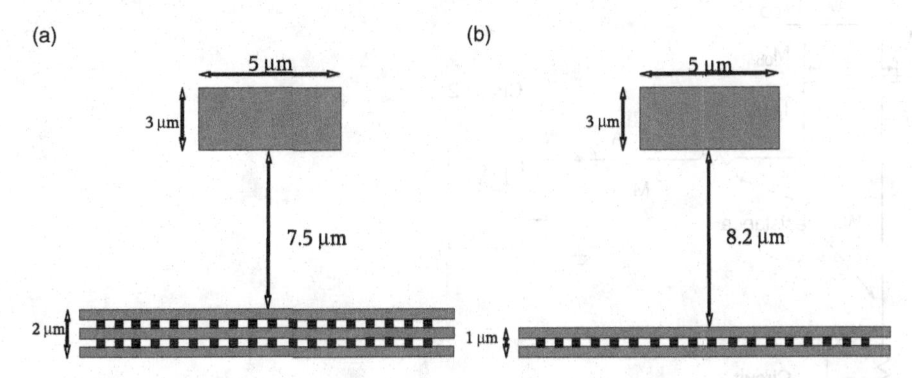

Figure 13.3 Example of using (a) three or (b) two bottom metal layers shunted together to reduce the ground resistance and losses in the ground plane of microstrip or grounded coplanar waveguides on top of a silicon substrate

An example of a multi-metal ground mesh for transmission lines is illustrated in Figure 13.3. In this case, the bottom two to three metal layers (depending on the BEOL) are shunted together to provide a low-resistance, low-inductance ground plane with similar thickness as the signal line. If the bottom three metal layers are employed as ground plane below transmission lines, the Metal 2 plane can be interrupted to allow shielded bias and/or control lines to cross the high-speed signal lines without affecting their loading and without injecting noise into the substrate since the control lines are surrounded by Metal 1 and Metal 3 ground planes.

13.1.4 Local bias supply distribution and supply de-coupling strategies

Power and bias routing is critical in a single-chip high-frequency transceiver. Although the same current-steering circuits are employed for bias current distribution on a large mm-wave chip as for low-frequency analog circuits, the layout parasitics and the broadband signal filtering and power-supply rejection techniques play a central role. Figure 13.4 illustrates two examples of CMOS and SiGe BiCMOS current-steering circuits based on MOSFET and bipolar current mirrors. De-coupling capacitors with values ranging from a few hundred of

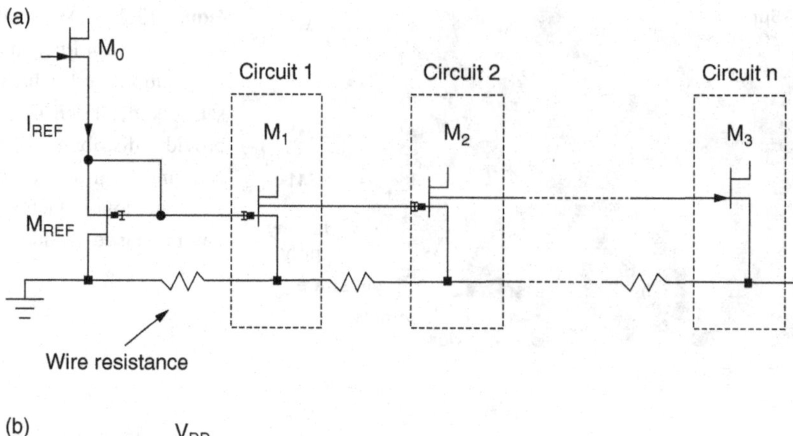

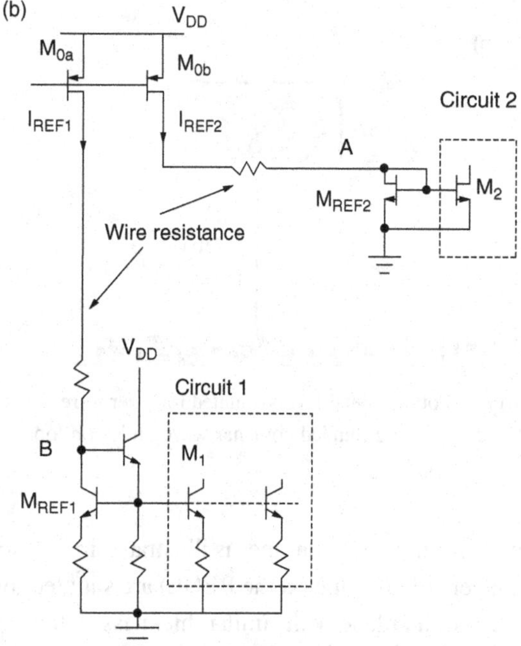

Figure 13.4 Example of bias current distribution circuits: (a) CMOS and (b) BiCMOS

fF to several pF are placed at every current mirror input to prevent noise and high-frequency signal leakage back into the supply currents of adjacent circuit blocks. The ground wire resistance, exacerbated in nanoscale technologies, can affect the accuracy of the current-mirroring ratio and must be minimized, particularly when HBTs are employed without resistive degeneration. Several common-sense guidelines are summarized below:

- Use wide lines (to maximize capacitance to ground, minimize wiring inductance and resistance) for all supply voltage and control voltage distribution planes and lines.
- Provide local power-supply decoupling by placing unit 200fF to 1pF (depending on application frequency) MOM or MIM capacitors, both at supply nodes to GND and at GND nodes to the supply, as shown in Figure 13.5. The value of the de-coupling capacitor unit is chosen

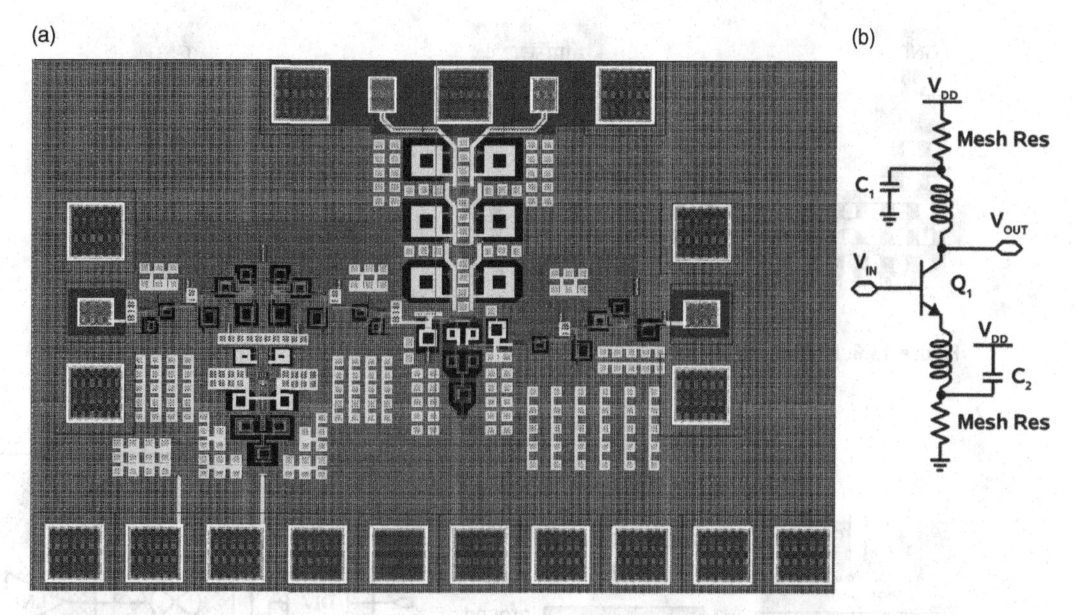

Figure 13.5 (a) Layout of mm-wave transceiver illustrating distributed bias supply de-coupling with MIM or MOM capacitor (b) Local bias supply de-coupling

such that it does not resonate below the highest frequency of operation of the circuit and its layout aspect ratio must be as close to that of a square as possible in order to minimize parasitic inductance.

13.1.5 Block-to-block isolation strategies

- Substrate must be grounded with p-taps to a wide solid ground plane wherever possible.
- Circuit blocks must be surrounded by a "moat" consisting of:
  - narrow, deep n-well connected to a "quiet" V_{DD} pad,
  - contiguous p-diffusion into the substrate, connected to ground and encircling the deep n-well,
  - a Faraday cage formed by the grounded p-tap, Metal 1, Metal 2 … top Metal.
- A grounded metal Faraday cage should also surround transmission lines (microstrip or grounded coplanar) which carry sensitive signals (Figure 13.6).
- Route bias lines below the ground plane when crossing HF signal lines (Figure 13.6).
- Do not route bias lines close to each other. Shield them with a Faraday cage.
- Avoid crossing bias lines by placing ground planes between them (Figure 13.7).
- A ground plane should be placed between bias lines and control lines that cross each other.
- Isolate regions of the chip with different functionality and which operate at different signal levels, Figure 13.7.
- Isolate regions of the chip with reverse-biased pn junctions and transformers. **Warning!** Large area reverse-biased p-n junctions and the SOI buried oxide provide little or no isolation at the upper mm-wave frequencies because of the large capacitance to the substrate/handle

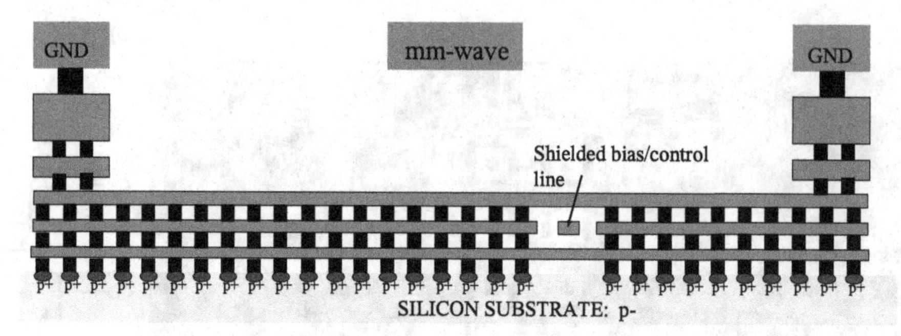

Figure 13.6 Example of Faraday cage isolation

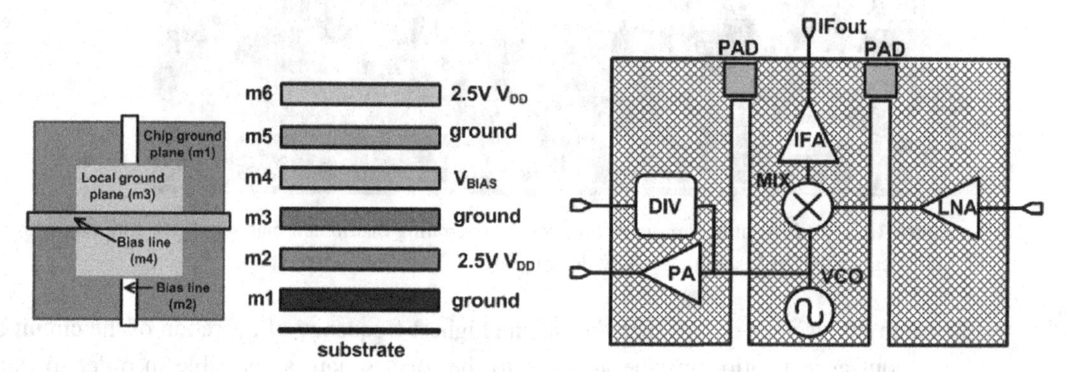

Figure 13.7 Example of (a) bias, control, and supply line shielding with local ground planes, (b) metal layer allocation between supply and ground planes, and (c) separation of supply planes for reduced crosstalk between critical transceiver blocks [2]

wafer. Frequent grounded substrate p-taps are the only effective way to steer all currents away from the substrate at mm-wave frequencies.

- Avoid crossing the bias lines of different circuit blocks which operate under significantly different signal conditions: eg. PA bias with LNA supply, VCO supply with PA supply, as illustrated in Figure 13.7.
- "Quiet" ground and "quiet" V_{DD} pads must be provided for isolation structures.

Examples of these isolation techniques are illustrated in Figures 13.6 and 13.7.

13.2 TRANSCEIVER ARCHITECTURES, PACKAGING, AND SELF-TEST FOR MM-WAVE RADIO, RADAR, AND IMAGING SENSORS

The first industrial silicon mm-wave transceiver systems operating at 60GHz and higher frequencies have been announced recently. These include:

- 60GHz phased array radio transceivers for wireless streaming of uncompressed HDMI signals at up to 5Gb/s data rates, fabricated in SiGe BiCMOS [3] and 65nm CMOS technologies [4],

- SiGe HBT 77GHz automotive cruise-control radar transceiver arrays [5],[6], and
- an 80GHz active imaging array for real-time airport security screening [7],[8] in a SiGe HBT technology.

The automotive radar and active imaging transceivers address potentially large-volume commercial markets which, along with passive imaging, are new to silicon. All these SoCs are characterized by a very high level of functional integration on a single chip. Several other mm-wave SoCs in SiGe BiCMOS and nanoscale CMOS technologies, and operating at 94GHz and in the 110–170GHz range, are currently being developed in academic and industrial research laboratories around the world.

Apart from the obvious circuit design hurdles that must be overcome, some of the most difficult challenges common to all mm-wave systems, and which have a crucial impact on the final system architecture, concern:

- the phase locked, low-phase noise, mm-wave local oscillator signal generation and distribution across the chip to multiple receivers and/or transmitters,
- low-cost, at-speed, on-die self-test for large volume production, and
- low-cost mm-wave packaging and antenna integration.

13.2.1 Transceiver architecture

Different architectures have been utilized to address the PLL and local oscillator (LO) signal generation and distribution at the upper mm-wave frequencies. These include:

1. Fundamental frequency VCO with fractional or integer divider chain locked to a low-frequency crystal reference and fundamental frequency LO distribution tree, as sketched in Figure 13.8,
2. Subharmonic ($f/2, f/3, f/n$) VCO locked to a low-frequency crystal reference and followed by a multiplier chain. The LO signal distribution can be either at the fundamental frequency, f, after the multiplier chain (Figure 13.9), or at the VCO frequency (f/n) followed by local multiplier chains Figure 13.10, and
3. Push–push (2nd harmonic) VCO followed by divider chain locked to a low-frequency crystal reference and fundamental frequency LO distribution network, Figure 13.11.

The first approach is the most common in GHz-range RF systems because it results in the least amount of parasitic spurs and, arguably, the lowest power consumption. However, it places the most stringent demands on the speed, in terms of f_T, f_{MAX}, and on the noise figure of the transistor. If the SoC is designed to operate at frequencies larger than the transistor $f_{MAX}/3$–$f_{MAX}/2$, the phase noise of the VCO may turn out to be worse than in the second architecture. Moreover, for applications that require 10% or higher tuning range, and phase noise better than −80dBc/Hz at 100kHz offset and/or better than −95dBc/Hz at 1MHz offset, a fundamental frequency VCO may prove too challenging. Fabrication of a static or regenerative divider operating at $f_{MAX}/2$may also prove elusive. Nevertheless, a 300GHz fundamental frequency

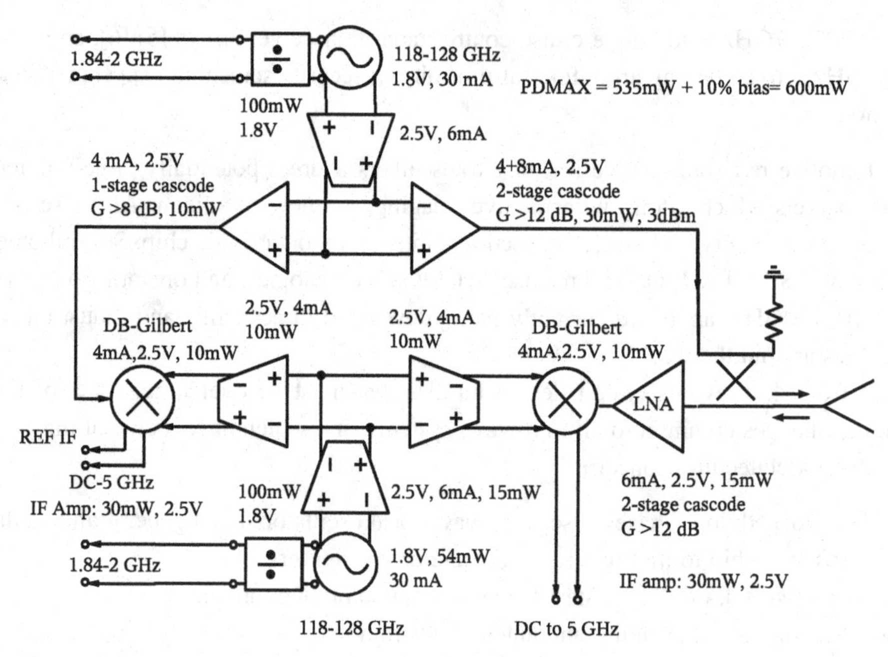

Figure 13.8 Fundamental 120GHz LO distribution architecture with fundamental VCO

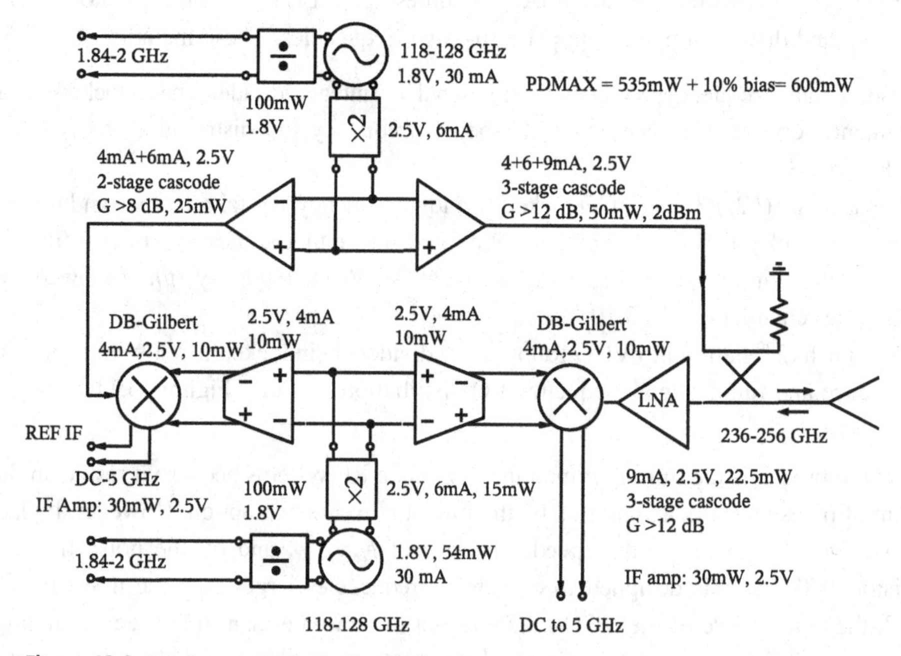

Figure 13.9 Fundamental LO distribution architecture with subharmonic VCO [10]

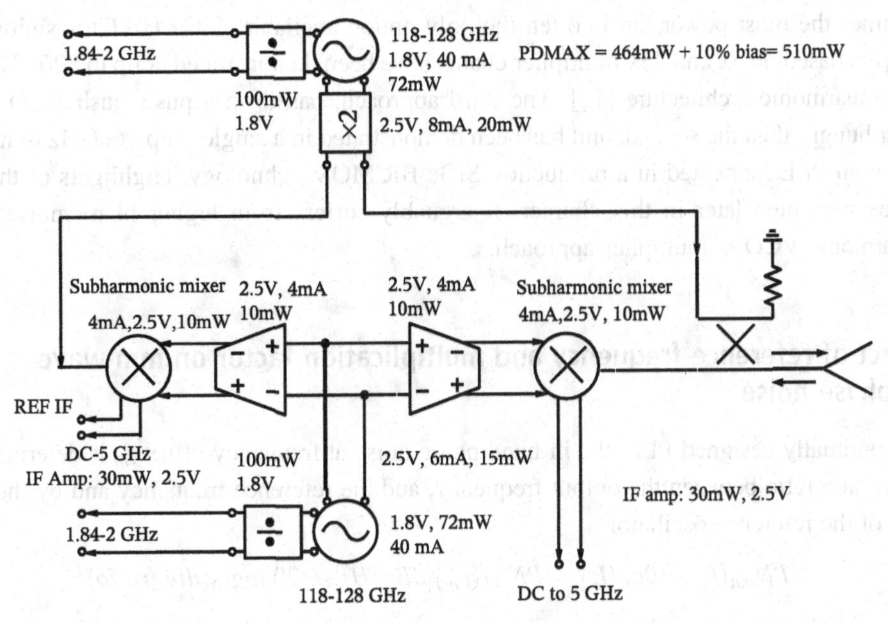

Figure 13.10 240GHz transceiver architecture with subharmonic VCO and subharmonic LO distribution [10]

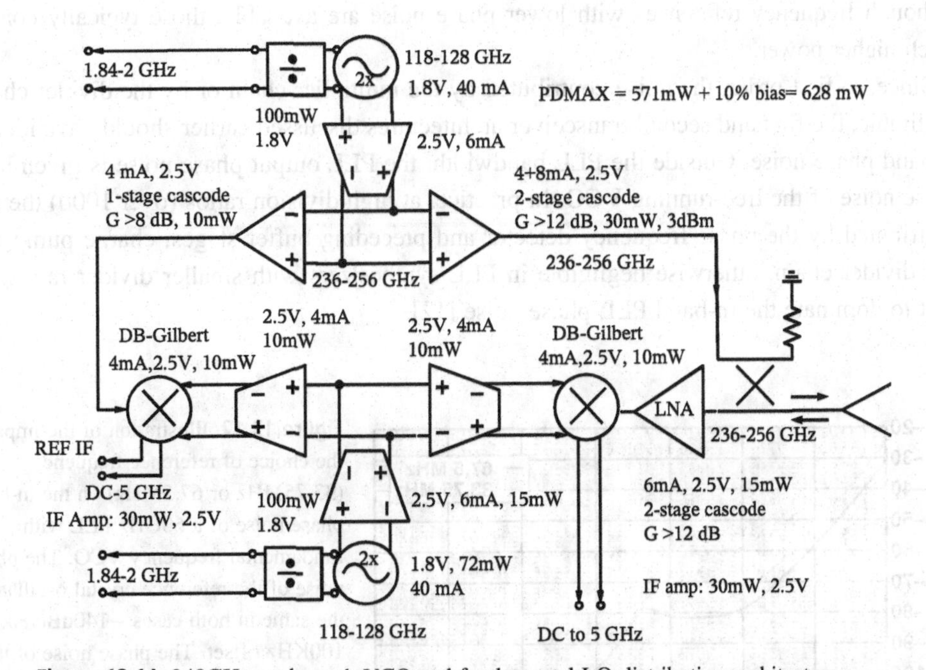

Figure 13.11 240GHz push–push VCO and fundamental LO distribution architecture

PLL with a Miller divider, the highest reported to date, has been demonstrated in an InP HBT technology with f_{MAX} of 1.1THz [9].

The second architecture is recommended when the f_{MAX} of the available transistor technology is less than two times larger than the operation frequency of the transceiver. It generally consumes the most power but is often the only option available. SiGe HBT transmitters and receivers based on 5x and 45x multiplier chains have been demonstrated at up to 820GHz using this subharmonic architecture [11]. The third approach, based on a push–push VCO, is less power hungry than the second, and has been demonstrated in a single-chip 160GHz transceiver array with PLL fabricated in a production SiGe BiCMOS technology. Highlights of that chip will be presented later in this chapter. It arguably suffers from higher phase noise than a subharmonic VCO + multiplier approach.

13.2.2 Impact of reference frequency and multiplication factor on mm-wave PLL phase noise

In an optimally designed PLL, the in-band phase noise at frequency offset f_m is determined by the division ratio between the output frequency, and the reference frequency and by the phase noise of the reference oscillator

$$PN_{out}(f_m)[dBc/Hz] = PN_{ref}(f_m)[dBc/Hz] + 20 \cdot log_{10}(div\_ratio) \tag{13.1}$$

A low-cost, low-power (1–2 mW) crystal oscillator operating at 10s of MHz typically exhibits a phase noise better than −140dBc/Hz at 10kHz offset and a noise floor of −150 to −160dBc/Hz. Although frequency references with lower phase noise are available, those typically consume much higher power.

Since, to first order, the noise contributed by the multiplier chain or by the divider chain is negligible, the first and second transceiver architectures discussed earlier should have identical in-band phase noise. Outside the PLL bandwidth, the PLL output phase noise is given by the phase noise of the free-running VCO. In practice, at high division ratios (over 1000) the noise contributed by the phase frequency detector and preceding buffer stages, charge pump filter, and divider chain, otherwise negligible in PLL synthesizers with smaller divider ratios, may start to dominate the in-band PLL phase noise [12].

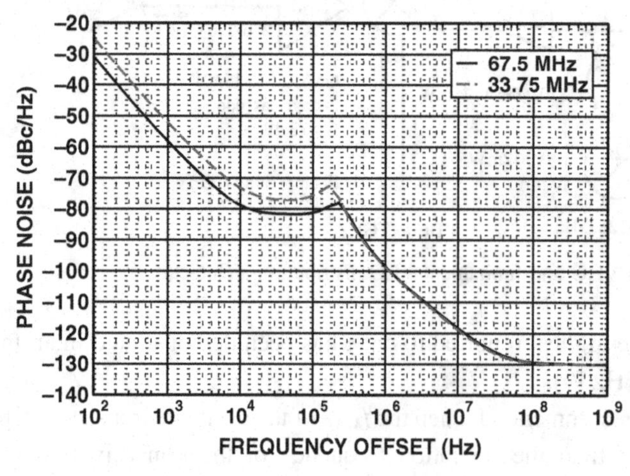

Figure 13.12 Illustration of the impact of the choice of reference frequency (33.75MHz or 67.5MHz) on the in-band phase noise of a 60GHz PLL with fundamental frequency VCO. The phase noise of the reference crystal oscillator is the same in both cases −140dBc/Hz at 100KHz offset. The phase noise of the free-running 60GHz VCO is −100dBc/Hz at 1MHz offset

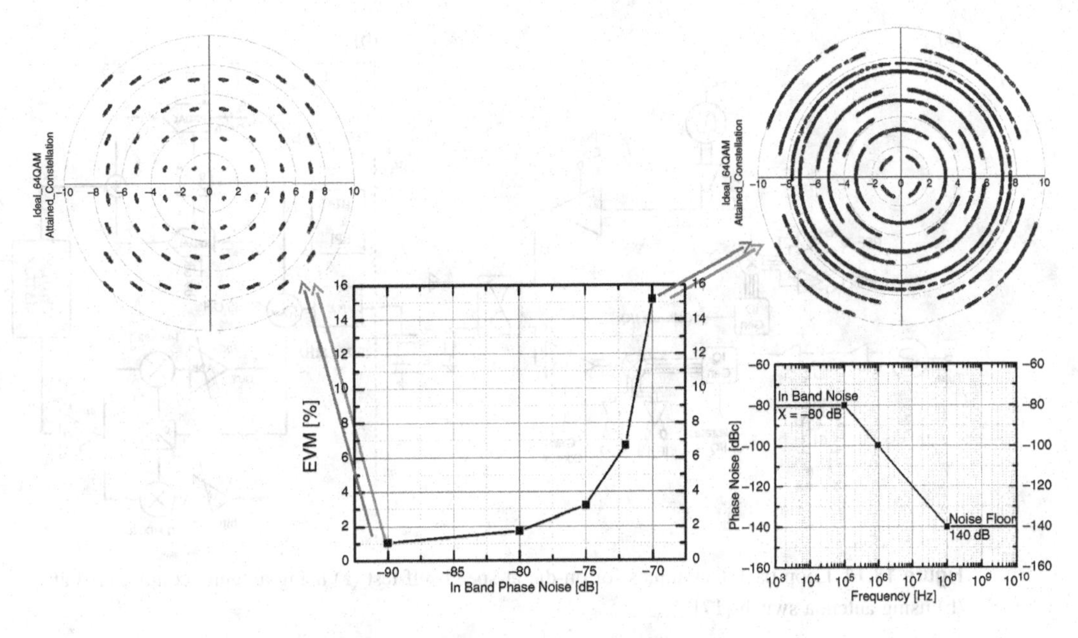

Figure 13.13 Simulated EVM degradation as a function of the in-band and out-of-band PLL phase noise in a directly modulated 64-QAM transmitter [system level simulations by A. Balteanu]

The impact of the in-band and out-of band PLL phase noise on the EVM degradation of a directly modulated 64QAM transmitter is simulated in Figure 13.13. The system-level simulation results indicate that, even at a low PLL bandwidth of 100kHz, the in-band phase noise should be lower than −72dBc/Hz if the EVM target is 10% (or 20dB).

13.2.3 Self-test features (this section is adapted from *IEEE Microwave Magazine* article [13])

Self-test features are needed to verify the functionality of mm-wave SoCs in production, without resorting to expensive mm-wave equipment and relatively time-consuming mm-wave measurements. Varying degrees of functional self-test can be implemented with:

- at-speed loop-back between the transmitter and receiver which allows for the evaluation of the bit error rate, EVM, and IQ mismatch of the entire transmit–receive chain;
- detectors, whose DC voltage tracks the signal levels at various locations in the transmitter and receiver;
- dedicated test receivers and test transmitters which can be coupled to the main transceiver to accomplish full *S*-parameter testing; and
- integrated noise source which can be switched to the receiver input [14] or whose output noise is upconverted using the transmit mixer or a dedicated test transmitter [15].

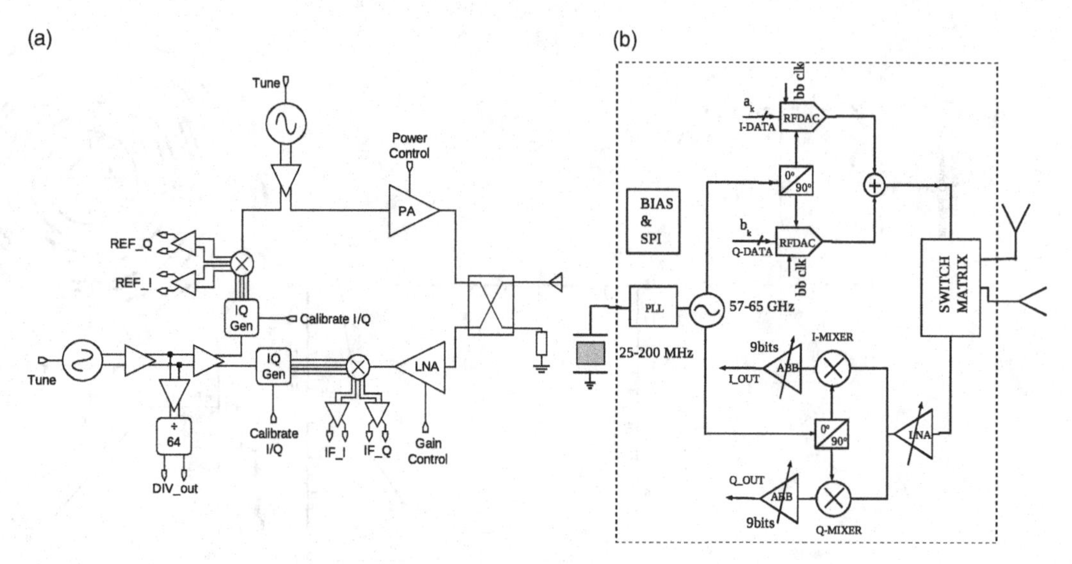

Figure 13.14 Loop-back techniques for on-die at-speed self-test (a) using antenna coupler [16] and (b) using antenna switch [17]

Loop-back testing requires a transceiver chip, rather than a separate transmitter and receiver, and can be implemented using

- antenna couplers [16], Figure 13.14(a)
- antenna switches [17], Figure 13.14(b)
- the test probe itself (Figure 13.15(a)), and
- the packaged chip with in-package or on-board antenna(s) and placing a reflector at a known distance in front of the antenna [18], Figure 13.15(b).

In all these scenarios, only DC and low-frequency signals are being measured and supplied to the chip. Therefore, some form of calibration with mm-wave instrumentation and measurements of the mm-wave inputs and outputs will have to be performed on a small sample of circuits to acquire the absolute values of transmitter output power, transmitter spectral mask, phase noise, receiver linearity and noise figure.

In the case of the 60GHz transceiver in Figure 13.14(b), the receiver noise figure can be estimated by applying a noise signal generated by a DC-to-2GHz noise source at the data inputs of the transmitter, upconverting it to 60GHz and looping it through the antenna switch to the receiver. Because of the small signal nature of the noise, the data path buffers in the transmitter DAC and the mm-wave DAC itself operate in linear mode. This idea has been recently tested in simulation [15]. Note that, in loop-back, the phase noise of the on-die VCO used both in the transmitter and the receiver is significantly reduced due to self-mixing. For example, phase noise as low as −120dBc/Hz at 100KHz offset has been measured in loop-back in this 60GHz transceiver, although the phase noise of the free-running VCO at 100KHz is not better than −90dBc/Hz.

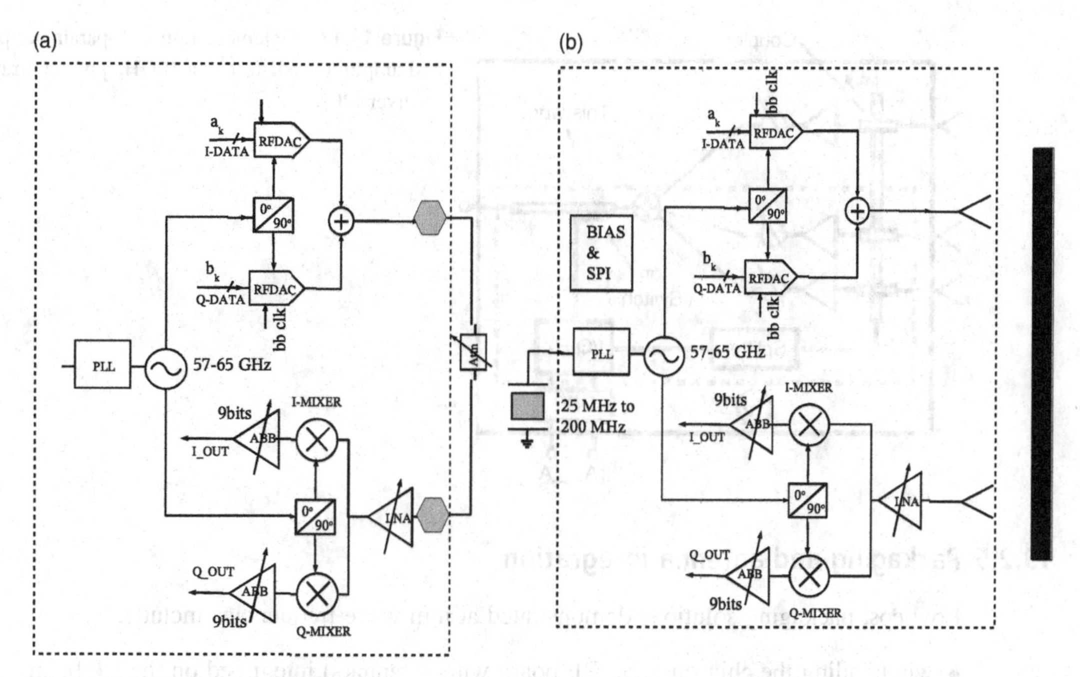

Figure 13.15 Loop-back techniques for (a) in-probe or (b) in-package at-speed self-test [18]

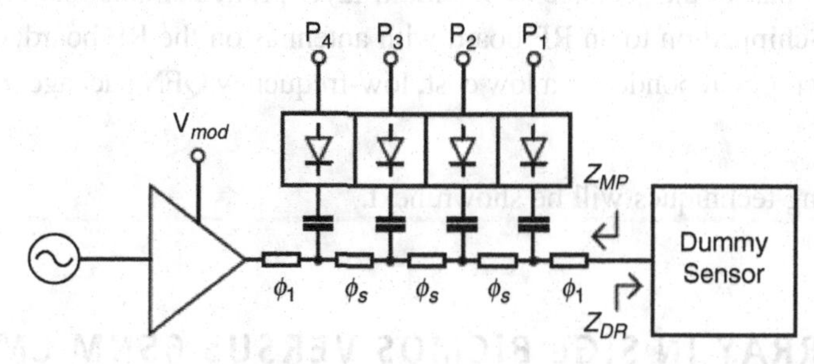

Figure 13.16 On-die single-port S-parameter measurement using a 120GHz reflectormeter [19]

In addition to built-in functionality, recent efforts have been directed at measuring amplitude and phase in self-test mode. One such example is illustrated in Figure 13.16, where a self-calibrating 120GHz reflectormeter, based on four appropriately spaced detectors, was integrated in a SiGe HBT technology to perform single-port impedance measurements [19]. Similarly, the 120GHz sensor transceiver in Figure 13.14(a) featuring separate transmit and receive VCOs, antenna coupler, and two IQ receivers, is capable of measuring and correcting its own IQ amplitude and phase mismatch [16]. Finally, in Figure 13.17, a dedicated test transceiver is selectively coupled to the inputs of each lane of a 10GHz receiver phased array to perform on-chip self-test of the amplitude and phase response of each receiver lane [20].

In all these cases, the coupling of the test signal must be accomplished with minimal degradation of the noise figure, linearity and signal integrity of the circuits under test, CUT.

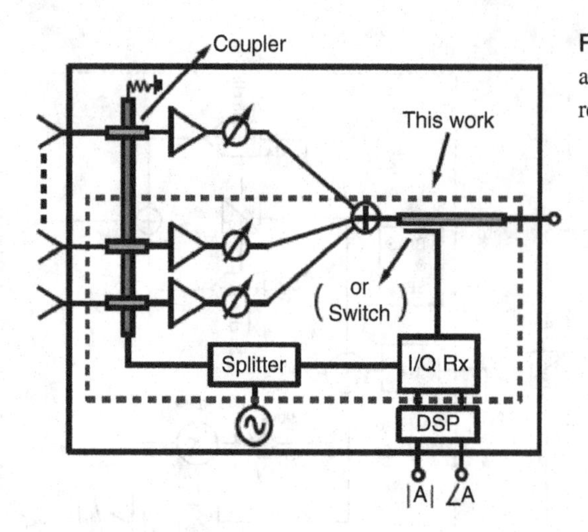

Figure 13.17 Implementation of S-parameter phase and amplitude self-test in a 10GHz phased array receiver [20]

13.2.5 Packaging and antenna integration

Low-cost packaging solutions demonstrated at mm-wave frequencies include:

- wirebonding the chip on to an RF board with antenna(s) integrated on the RF board;
- flip-chipping on to an RF substrate with antenna(s) integrated in the RF substrate, where the RF substrate can be either part of the package or of a multi-layer printed circuit board (PCB);
- wafer-scale package flip-chipped on to an RF board with antennas on the RF board;
- antennas on die with the chip wirebonded to a low cost, low-frequency QFN package (only at W-Band and above).

Examples of these packaging techniques will be shown next.

13.3 60GHZ PHASED ARRAY IN SiGe BiCMOS VERSUS 65NM CMOS

The emergence of 60GHz wireless standards such as IEEE802.15.3c, ECMA, WiGig, and WirelessHD, has enabled the introduction of consumer products which support fixed and portable wireless data transmission of high-definition video content at data rates of 3–7 Gb/s. Because of the very directive nature of 60GHz radiation, and because it is severely attenuated by walls, furniture, pets, and humans, phased array technology is employed in such systems in order to ensure robust non-line-of-sight (NLOS) communication. When an unobstructed direct link between the transmitter and the receiver cannot be established, the beam-steering capability of the phased array ensures that the signals from the transmitter reach the receiver by taking advantage of reflections from the walls, ceiling and surrounding objects. This is illustrated in Figure 13.18 where reflection from the ceiling, which introduces a penalty of about 10dB in the link margin, circumvents a more significant loss of 20dB encountered when a person obstructs the direct path between the transmitter and the receiver [21].

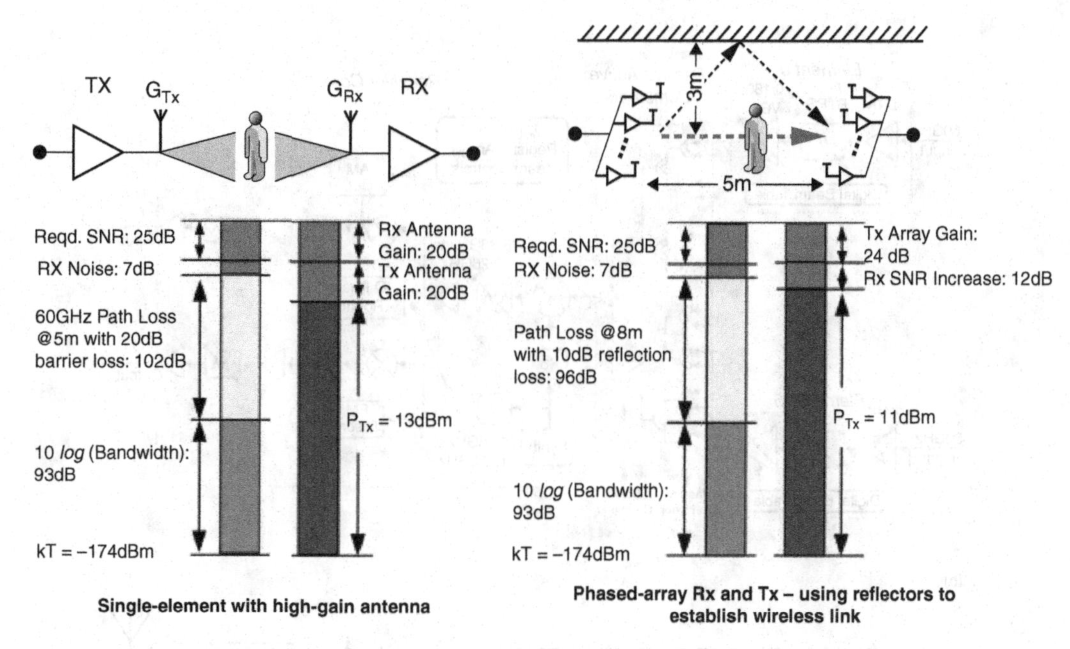

Figure 13.18 Illustration of the benefits of phased array in 60GHz NLOS links [21]

Two such systems, fabricated in 130nm SiGe BiCMOS and 65nm standard digital CMOS technology, are illustrated in Figures 13.19 and 13.20, respectively. They are remarkably similar in terms of architecture, performance, power consumption, overall die area, and packaging solution with antennas embedded in the package. Both rely on a RF phase shifting, phased array architecture, with 16 transmit (TX) and 16 receive (RX) lanes in the first case, and with reconfigurable 32 TX and 4 RX lanes, or 8 TX and 32 RX lanes in the CMOS system. A double-heterodyne receiver architecture with sliding IF is employed in each case. The RF phase shifting architecture was preferred over IF or baseband phase shifting arrays because it minimizes the number of components that have to be duplicated, thus reducing power consumption. At the same time, RF phase shifting poses severe design challenges because of the linearity, noise figure and bandwidth requirements that now have to be satisfied at mm-wave rather than at IF or baseband frequencies.

The SiGe BiCMOS system consists of separate transmitter and receiver chips and covers all 4 channels (58.32GHz, 60.48GHz, 62.64GHz, and 64.8GHz), identical in all the 60GHz standards. In contrast, the CMOS transceiver phased array is integrated as a single chip, in two versions (i) source transceiver with 32 TX and 4 RX lanes, and (ii) sink transceiver with 8 TX and 32 RX lanes. It covers only two of the four 60GHz channels. None of the two transceiver phased arrays integrates the digital baseband processor, the PHY, or the MAC layer with the mm-wave front-end.

In both transceivers, the mm-wave front-end consists of transmitter lanes with phase shifters and PAs, Figure 13.21, and of receiver lanes, Figure 13.22, with a single-ended, variable gain LNA, and phase shifters, Figure 13.23. The SiGe BiCMOS transmit lanes employ fully differential circuits based on cascode topologies operated from a 2.7V supply, whereas the

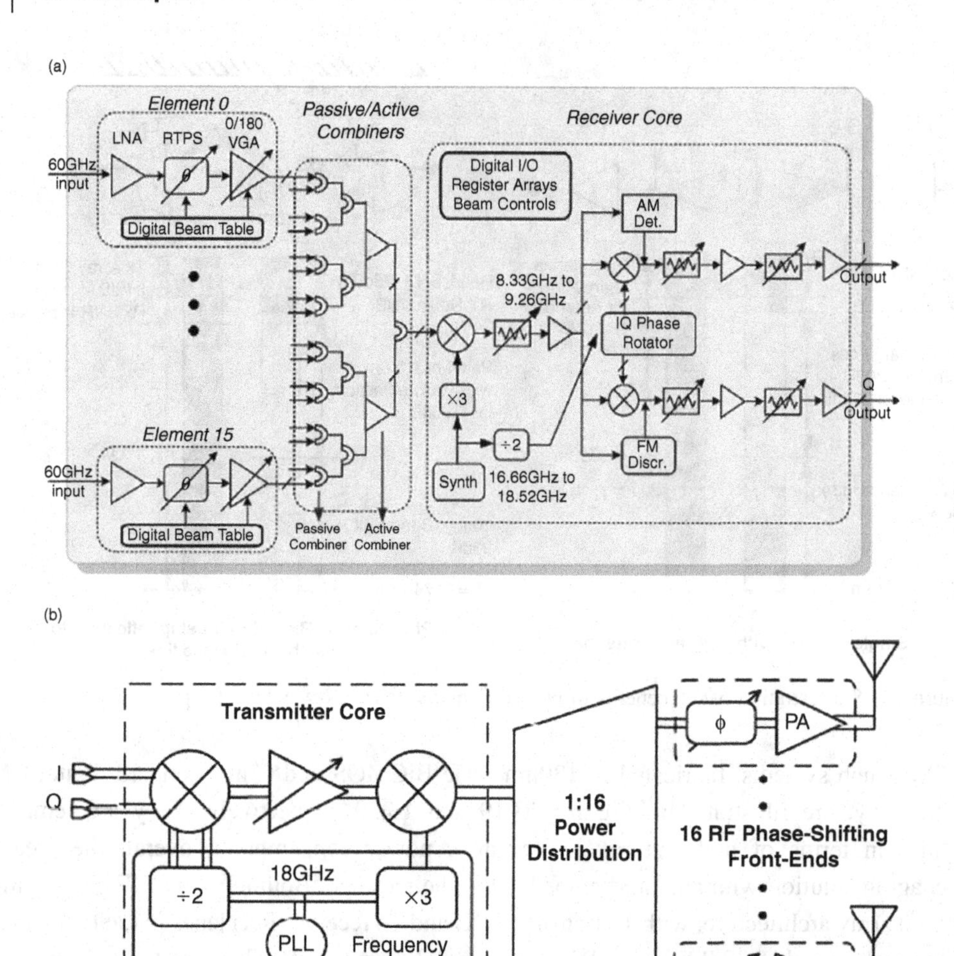

Figure 13.19 60GHz SiGe BiCMOS phased array (a) receiver and (b) transmitter block diagram [22]

CMOS transmit lanes use single-ended topologies and 1V supply. The transmit lane OP_{1dB} is up to 13.5dBm, and 6dBm, in the SiGe BiCMOS and 65nm CMOS chips, respectively, with a peak PAE of 22% claimed for the CMOS PA. The overall *EIRP* of the packaged transmit array, which includes antenna gain, is reported to vary between 33–39 dBmi across the 4 channels for the SiGe BiCMOS chip, and reaches 28dBmi for the CMOS transmitter array. It is important to remind the reader that the *EIRP* numbers should not be confused with the total signal power, P_{TX}, at the transmit array output which cannot exceed 13.5dBm + 10log16 = 25.5dBm and 6 dBm + 10log32 = 21dBm in the SiGe BICMOS and CMOS case, respectively. The difference between *EIRP* and P_{TX} is ideally given by the aggregate gain of the 16 and 32 antennas, respectively. In reality, OP_{1dB} variation across the 4 channels, power-combining losses due to antenna coupling, and antenna gain variations across the array are responsible

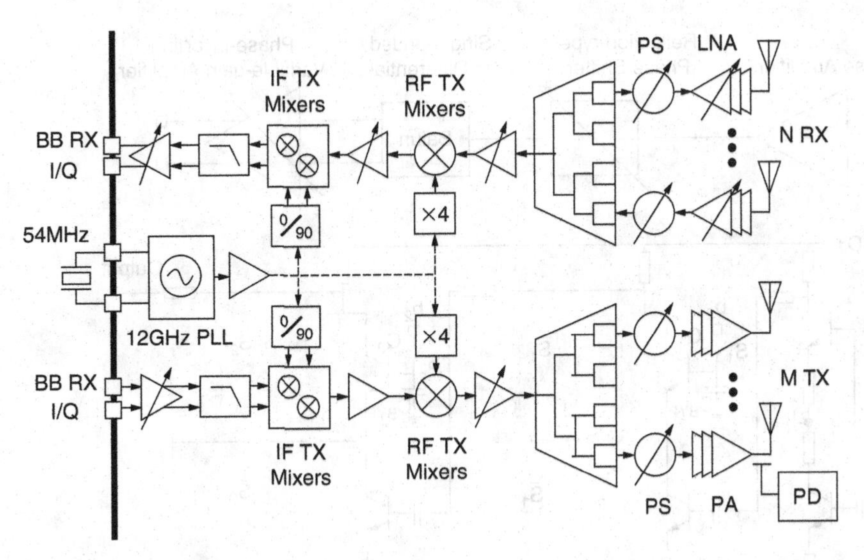

Figure 13.20 CMOS phased array transceiver block diagram [4]

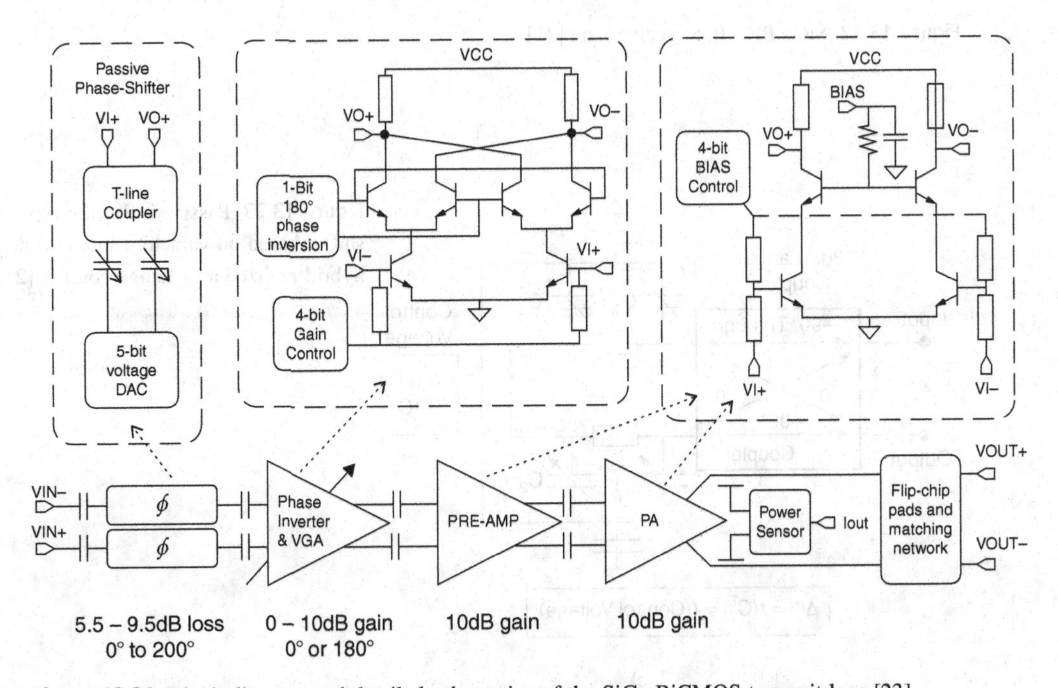

Figure 13.21 Block diagram and detailed schematics of the SiGe BiCMOS transmit lane [23]

for *EIRP* degradation and explain the 6dB (a factor of 4) variation in *EIRP* observed across the four 60GHz channels.

It is instructive to dwell a bit further on this factor of 4 (6dB) in effective *EIRP* to better understand the significant margin for necessary improvement in circuit, antenna, and package design that it implies. With perfect free-space power combining, identical antenna gains, and a TX lane output power of 13.5dBm, 4 lanes rather than 16 lanes, would deliver the same *EIRP*

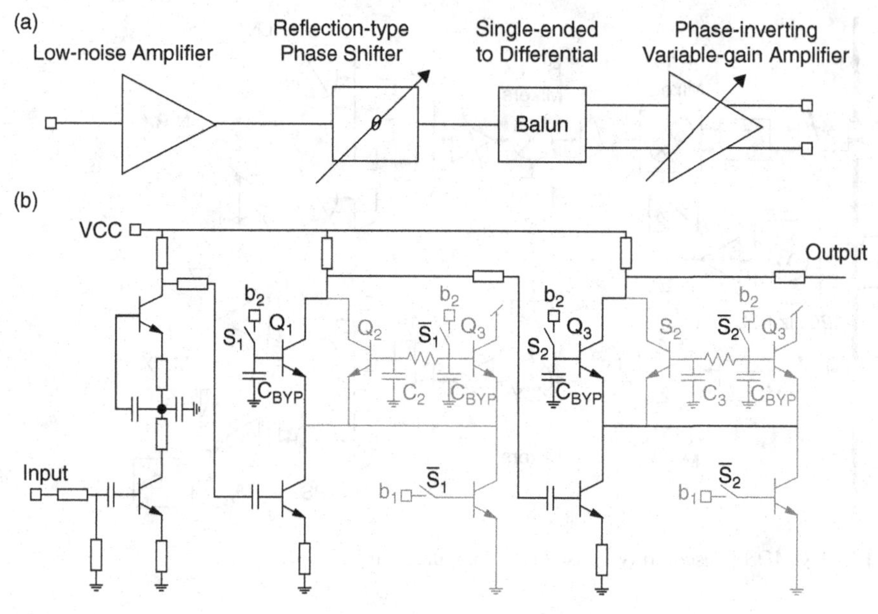

Figure 13.22 SiGe BiCMOS receive lane [21]

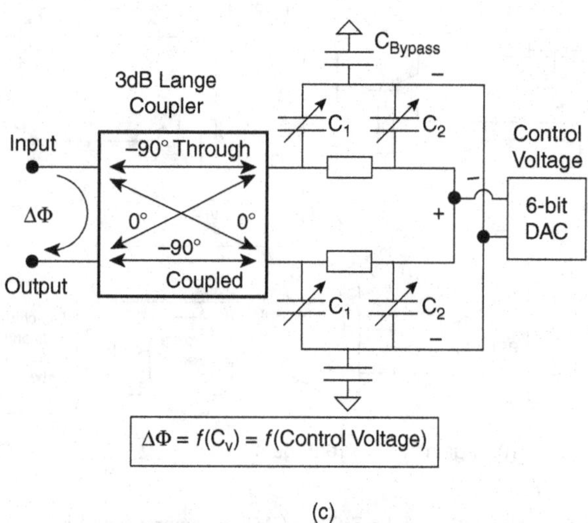

Figure 13.23 Passive reflection-type phase shifter based on varactors and a 90-degree hybrid realized as a Lange coupler [21]

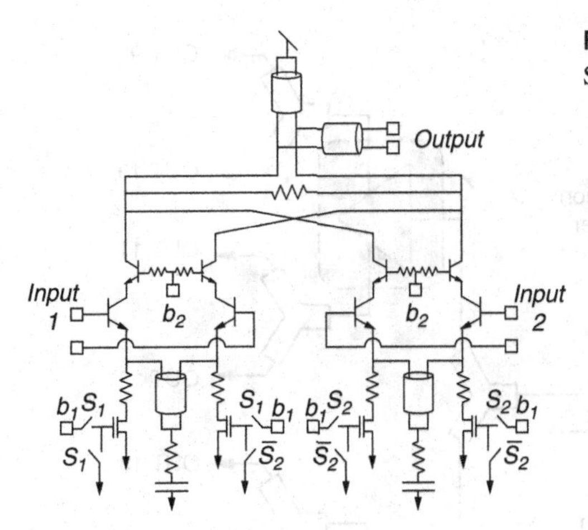

Figure 13.24 Schematics of the power combiner in the SiGe BiCMOS receiver [21]

with a four-fold reduction in TX path power dissipation, significant die area and package size reduction. All of these point to significant system cost reduction, and to the important role that good, optimal circuit and package design play in the overall system cost structure and power consumption, and, ultimately, to the environmental impact of good RF circuit, package, antenna, and system design.

The SiGe BiCMOS receiver lane employs a single-ended cascode LNA with switchable gain, whose schematic is shown in Figure 13.22, and a passive reflection-type phase shifter, Figure 13.23. The 7–8 dB loss of the passive phase shifter has little impact on the overall receiver noise figure (<7dB) because of the large, 22dB gain of the LNA. The main benefit of the passive phase shifter is its linearity and low-power consumption. A variable gain amplifier is inserted after the phase shifter to compensate for the insertion loss variation of the phase shifter at different phase control settings, and to convert the mm-wave signal to differential format before power combining.

In a system with upconversion, like the ones discussed here, the 60GHz signal combining and distribution networks are the most challenging and power hungry because they have to maintain linearity at high-power levels while operating over a bandwidth of 9GHz if all IEEE 80.15.3c channels are to be covered. To accomplish this, both the SiGe BiCMOS and the 65nm CMOS chips employ active power combiners in the receiver, Figure 13.24, and active/passive power splitters for upconverted signal distribution in the transmitter, Figures 13.25–13.26. In the SiGe BiCMOS transmitter and receiver, the 60GHz signal distribution and combining networks are fully differential, whereas, to save power, in the CMOS transceiver they were implemented with single-ended topologies.

Although in both technologies, $f_{MAX} > 3\times60\text{GHz} = 180\text{GHz}$, for arguable reasons (see also the 77GHz radar transceivers in Section 13.4) related to power savings, PLL phase noise and 60GHz fundamental-frequency VCO phase noise and tuning range, the LO-signal generation is subharmonic. The SiGe BiCMOS system employs a fundamental frequency 16.66 to 18.62GHz SiGe HBT cross-coupled VCO followed by a tripler. The VCO is locked to a 308MHz crystal reference by a fractional divider chain and PLL. In the CMOS phased array, the 2.5-V PLL is

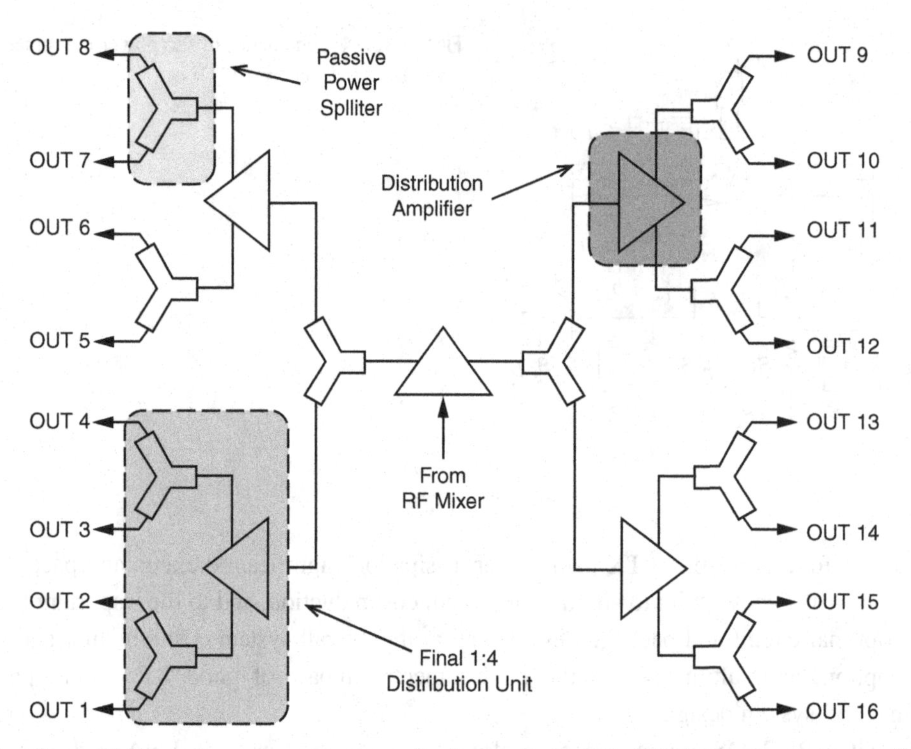

Figure 13.25 Active and passive power distribution in the SiGe BiCMOS transmitter array [23]

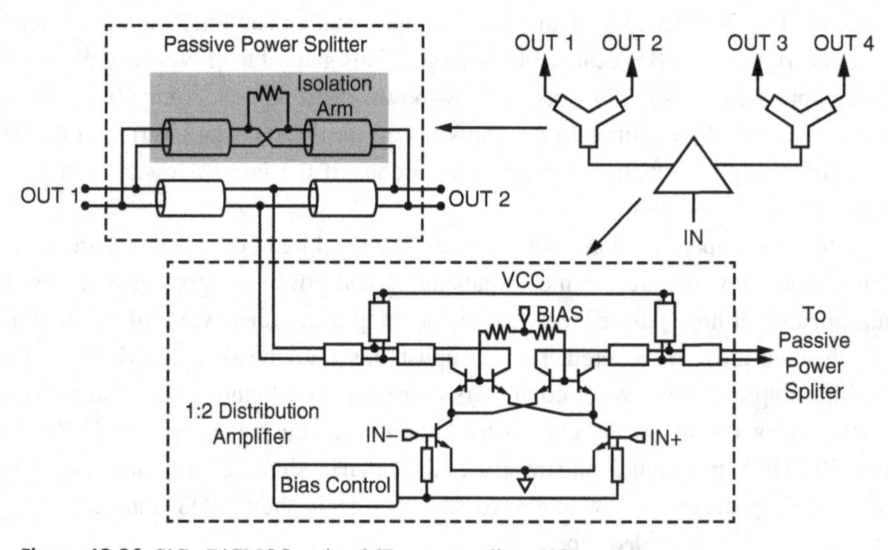

Figure 13.26 SiGe BiCMOS active 3dB power splitter [23]

locked to a 54MHz reference and includes a 12.096–12.528GHz fundamental-frequency VCO followed by a quadrupler. The CMOS quadrupler, shown in Figure 13.27, consists of a differential doubler stage with single-ended output and drawing 1.6mA, followed by a single-ended doubler stage, both biased in the subthreshold region of the MOSFET to maximize

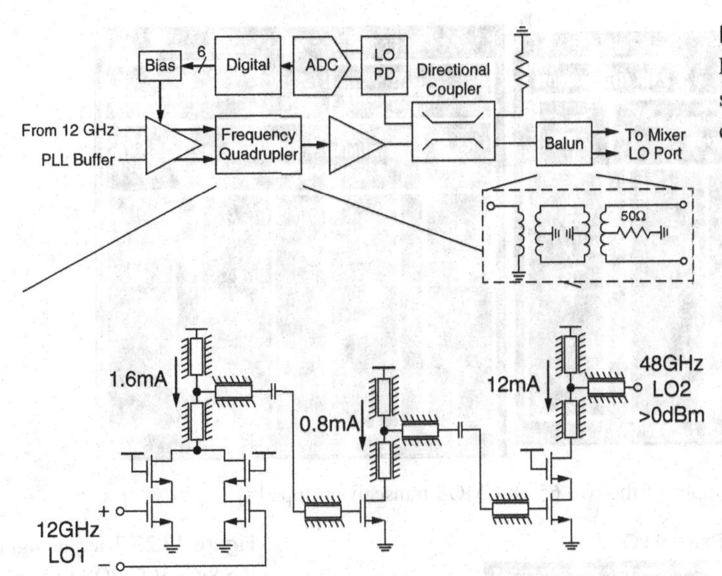

Figure 13.27 Block digram of the LO signal distribution and schematic of the 12–48GHz quadrupler in 65nm CMOS [4]

efficiency. The output stage is formed by a tuned cascode which provides isolation and matching to the 50GHz directional coupler. Two cascaded transformers are employed for single-ended to differential conversion at the mixer port. They ensure excellent common-mode suppression which is otherwise difficult to achieve at mm-wave frequencies with a single transformer. The reported phase noise of the 48GHz LO signal in the CMOS chip is −66dBc/Hz at 75KHz offset in the PLL band, and −96dBc/Hz at 1MHz offset [4]. For comparison, in the SiGe BiCMOS PLL, the measured phase noise is −80dBc/Hz in band, and −90dBc/Hz at 1MHz offset, outside the PLL bandwidth. By pure coincidence (unless the two crystal oscillators have identical phase noise), the in-band phase noise scales almost perfectly with the crystal reference frequencies of the two chips: 20log10 (308MHz/54MHz) = 15dB. In both phased arrays, the phase noise at 1MHz depends primarily on the multiplied (by 3 and by 4, respectively) phase noise of the free-running 18GHz SiGe VCO and 12GHz CMOS VCO, respectively. It is also noteworthy that, while the rest of the 65nm CMOS circuits operate from 1V, the 12GHz section of the CMOS PLL operates from 2.5V, likely to ensure that the demanding VCO phase noise and tuning range specifications are satisfied.

Figure 13.28 reproduces the die photographs of the source and sink 65nm CMOS transceiver chips. For comparison, the die photographs of the SiGe BiCMOS transmitter and receiver arrays are shown in Figure 13.29. Both chips are flip-chipped in ball grid array (BGA) packages with embedded patch antennas. The package for the CMOS chips, Figure 13.30, is ceramic and has 128 pins and 40 embedded patch antennas. In contrast, an open-cavity, organic LCP (liquid crystal polymer) BGA with 288 pins and 16 embedded antennas is employed for each of the SiGe BiCMOS transmitter and receiver chips, Figure 13.31.

Figure 13.32 illustrates the use of a built-in memory table on the SiGe BiCMOS chips to implement fast digital beam-steering without relying on complex beamforming algorithms and frequent communication with the digital baseband chip. A demo link test setup is reproduced in

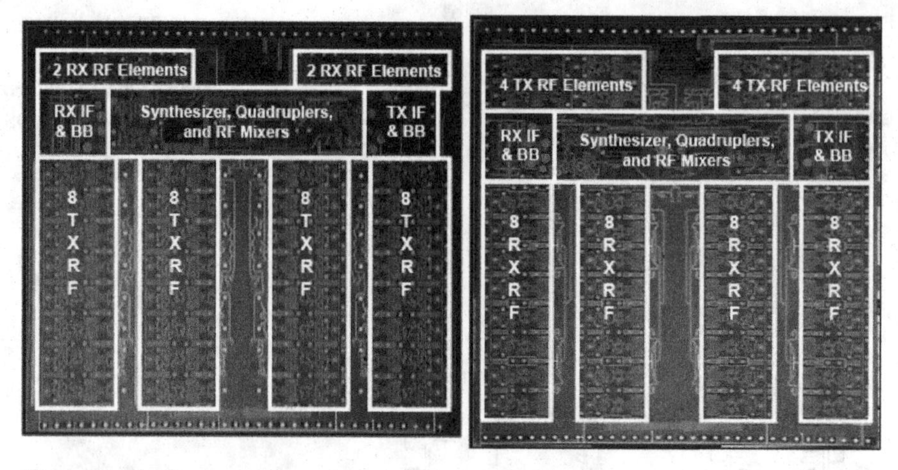

Figure 13.28 Die photographs of the two 65nm CMOS transceiver chips [4]

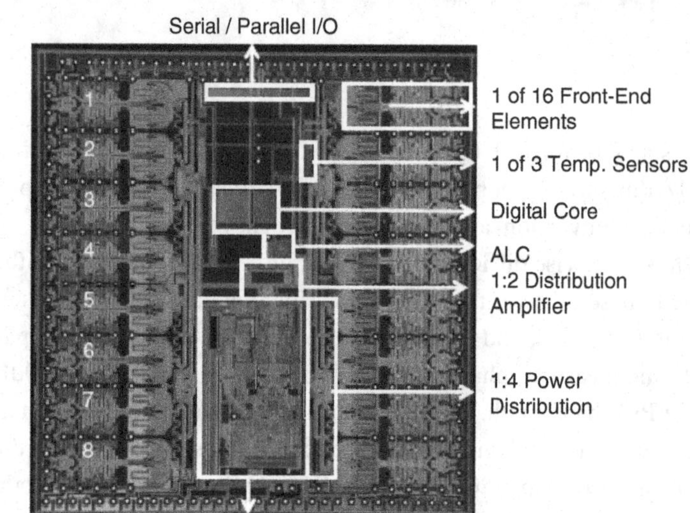

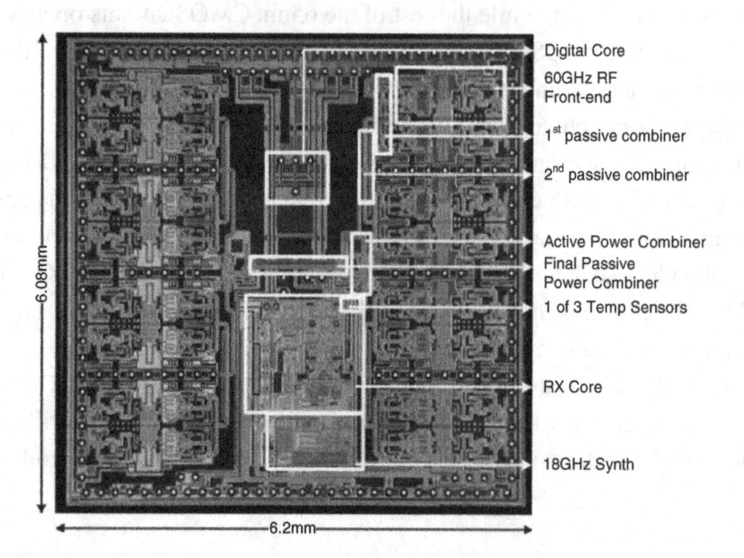

Figure 13.29 Die photographs of the SiGe BiCMOS transmitter and receiver array chips [22]

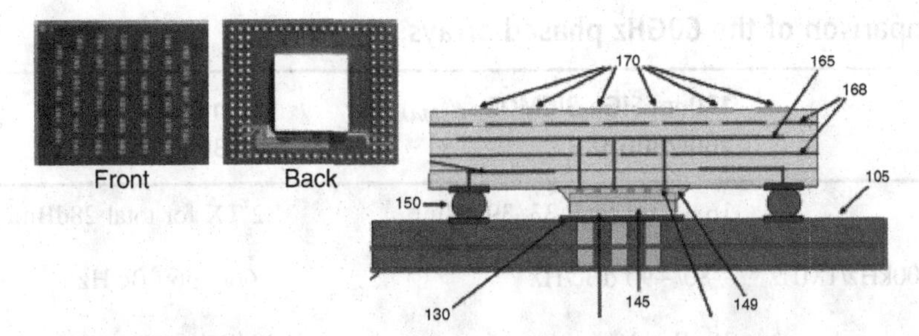

Figure 13.30 128pin, 21mm × 17mm ceramic flip-chip BGA package with 40 embedded antennas for the 60GHz CMOS phased array [4]

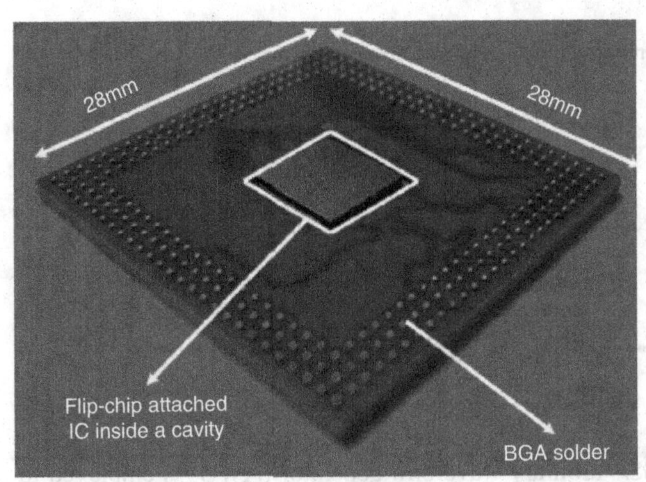

Figure 13.31 288-pin 28mm x 28mm organic (liquid crystal polymer = LCP) open cavity package with 16 embedded antennas for the 60GHz SiGe BiCMOS phased array [22]

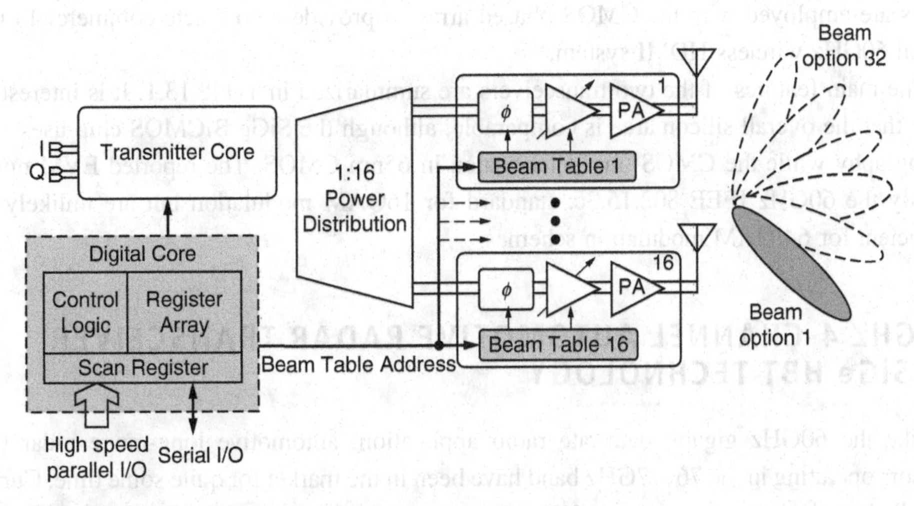

Figure 13.32 TX digital control for fast beamforming [23]

Table 13.1 **Comparison of the 60GHz phased arrays.**

Technology	130nm SiGe BiCMOS f_T/f_{MAX} 200/280GHz	65nm CMOS f_T/f_{MAX} 180/300GHz
TX EIRP	16 Tx for total 33–39 dBmi	32 TX for total 28dBmi
Phase noise at 100kHz/1MHz	−80/−90 dBc/Hz	−66/−96 dBc/Hz
Tx EVM	19dB (11%)	20dB (10%)
Maximum TX/RX P_D, 25°C,	2.7W/1.8W	1.8W/1.25W
Total die size TX + RX	81.571mm^2	77.155mm^2

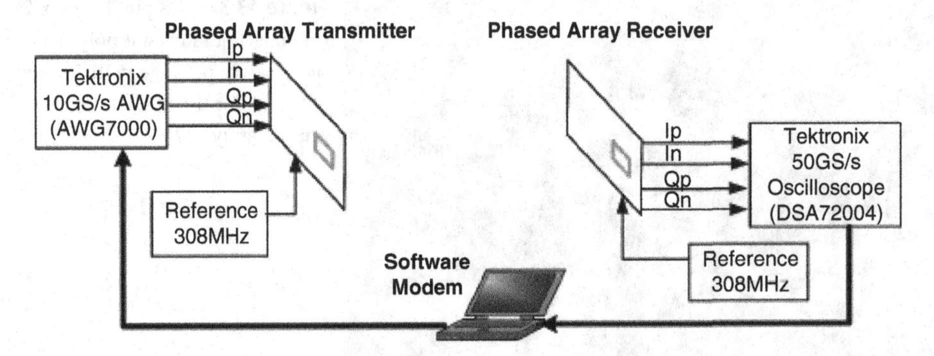

Figure 13.33 60GHz link demo setup [21]

Figure 13.33, where a 10-GS/s arbitrary waveform generator (AWG) is employed to provide the differential IQ analog baseband inputs to the SiGe BiCMOS transmitter and a 4-channel 50-GS/s oscilloscope is used to display the received signal constellation. Dedicated baseband chips are employed with the CMOS phased array to provide a complete commercial bidirectional 60GHz wireless HDMI system.

The main features of the two transceivers are summarized in Table 13.1. It is interesting to note that the overall silicon area is comparable, although the SiGe BiCMOS chip uses 130nm lithography while the CMOS one is fabricated in 65nm CMOS. The reported EVM numbers satisfy the 60GHz IEEE 802.15.3c. standard for 16QAM modulation but are unlikely to be sufficient for 64 QAM modulation schemes.

13.4 77GHZ 4-CHANNEL AUTOMOTIVE RADAR TRANSCEIVER IN SiGe HBT TECHNOLOGY

Unlike the 60GHz gigabit data-rate radio application, automotive long-range radar (LRR) sensors operating in the 76–77GHz band have been in the market for quite some time. Currently installed exclusively in premium luxury cars to provide active comfort functions such as

Table 13.2 **Typical 77GHz LRR, MRR, and SRR specification [24].**

Parameter	LRR Value	MRR Value	SRR Value
Maximum transmit *EIRP*	55dBm	−9 dBm/MHz	−9 dBm/MHz
Frequency band	76–77GHz	77–81GHz	77–81GHz
Bandwidth	600MHz	600MHz	4GHz
Distance or range, R_{MIN}, R_{max}	10–250 m	1–100 m	0.15–30 m
Distance or range resolution, ΔR	0.5m	0.5m	0.1m
Distance or range accuracy, δR	0.1m	0.1m	0.02m
Velocity resolution Δv	0.6m/s	0.6m/s	0.6m/s
Velocity accuracy δv	0.1m/s	0.1m/s	0.1m/s
Angular accuracy $\delta\phi$	0.1°	0.5°	1°
3dB beamwidth in azimuth $\pm\phi_{max}$	±15°	±40°	±80°
3dB beamwidth in elevation $\pm\theta_{max}$	±5°	±5°	±10°

adaptive cruise control (ACC) where the vehicle actively accelerates or brakes to relieve the driver from monotonous tasks, these sensors have been typically realized with discrete III-V components. The recent introduction of 77GHz single-chip ACC transceivers, fabricated in SiGe HBT or SiGe BiCMOS technology and employing a frequency modulated continuous wave, FMCW, radar sensor architecture, promises to significantly reduce cost and improve reliability, making it affordable to equip vehicles in the mid-range and lower-cost segments of the automotive market [24]. A typical specification for a 77GHz LRR sensor is summarized in Table 13.2 [24]. The table also includes specification for future medium (MRR) and short (SRR) range radars. The latter are required for more sophisticated driver assistance systems which implement passive and active safety measures, such as triggering airbag inflation only after a crash, or sudden emergency breaking and deceleration when an obstacle is detected on the road.

Unlike 60GHz radios which are specified for indoor operation in the 0–75°C range, ACC radar transceivers must withstand severe environmental conditions with operating temperatures ranging between −40°C and 125°C, and must satisfy long-term (>10 years) stringent reliability requirements. Furthermore, the system IF frequency is in the 100KHz to 10MHz range, where 1/f noise is large. All of these requirements make SiGe BiCMOS technology, with its sub 1KHz 1/f noise corner and excellent mm-wave and noise performance at elevated temperatures, particularly attractive for ACC LRR.

The block diagram and die photograph of a commercial single-chip SiGe HBT radar transceiver, in mass production since 2009 [24], are reproduced in Figure 13.34. It features

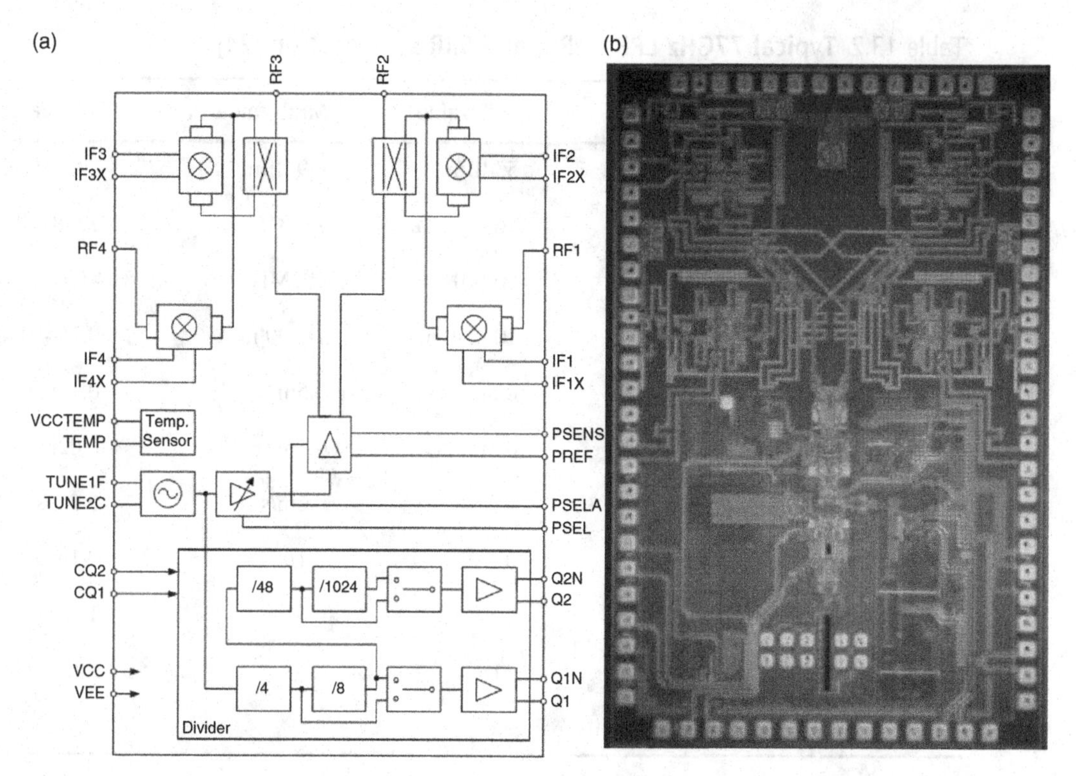

Figure 13.34 77GHz automotive radar transceiver (a) block diagram and (b) die photograph, $3.125 \times 2.1\ \text{mm}^2$ [5]

two transmitters and four receivers in the upper half of the block diagram and die photo, a fundamental 77GHz VCO with associated prescaler (divider chain), and a variety of power and temperature sensors to monitor the correct operation, in the bottom half. A significant architectural difference compared to the 60GHz phased array transceivers, which do not transmit and receive at the same time, is that, in the FMCW Doppler radar transceiver, the receivers must be designed to operate under large reflection from the antenna and large leakage from the transmitter with which they share the antenna port through an antenna coupler whose isolation is at best 25–30dB. The solution adopted here is to eliminate the low-noise amplifier in the receivers and to design a high-linearity downconvert mixer with acceptable noise figure. The downconvert mixer 1/f noise is critical because it is the close-in phase noise of the VCO and the 1/f noise of the mixer at 100kHz IF, rather than receiver noise figure above 10MHz, that limits the sensitivity of the radar. When trying to resolve two closely spaced targets, the transmitter phase noise is also important.

Another noteworthy difference is the use of a large power, fundamental frequency, buffered VCO (as opposed to low-frequency, low-power VCO followed by a multiplier) which simultaneously drives two transmit channels, each with +9dBm output power, four downconvert mixers, and the 77GHz divider chain. To minimize the impact of ground inductance and improve robustness to supply noise and crosstalk, all circuit blocks are realized with differential

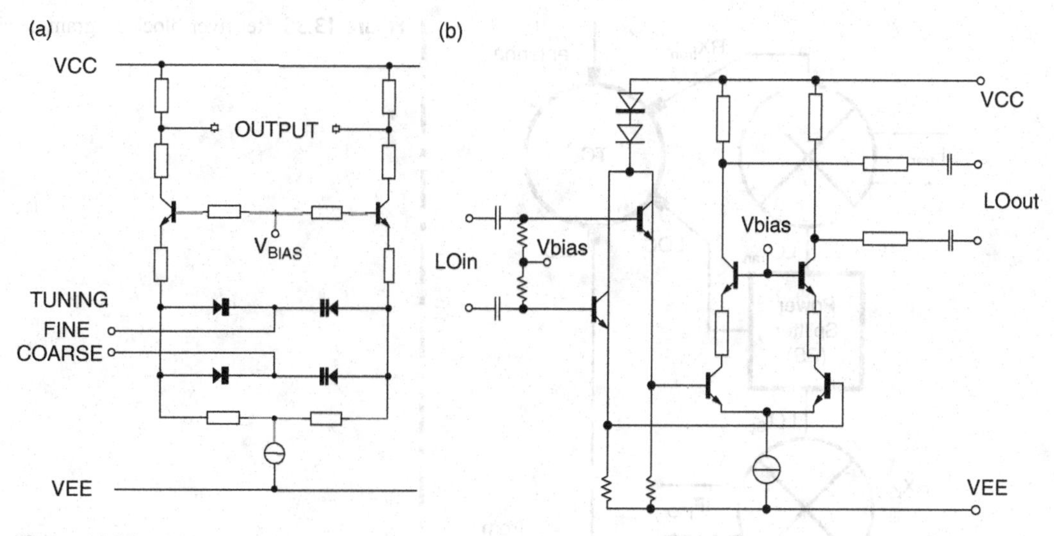

Figure 13.35 77GHz (a) VCO and (b) LO buffer schematics [5]

topologies. This, again, is in contrast with the 60GHz phased arrays discussed earlier, where, to minimize power consumption, at least part of the mm-wave front-end was realized with single-ended topologies.

The radar transceiver technology features SiGe HBTs with 0.18μm emitter width and f_T/f_{MAX} larger than 200GHz, four copper metallization layers, MiM capacitors, two types of polysilicon resistors, and a TaN thin-film resistor used for realizing resistor values below 100Ω. The microstrip transmission lines are represented as narrow, open rectangles.

The schematics of the VCO and LO buffer are shown in Figure 13.35. A Colpitts topology with high-Q varactors is employed in the VCO to reduce phase noise and provide an adequate tuning range over process and temperature variation. Both fine and coarse controls are available to reduce K_{VCO} (important for PLL stability and PLL filter design) and to prevent phase noise degradation. The large power LO buffer features an emitter-follower and a cascode with inductive broadbanding. Damping resistors, rather than current sources, are employed as loads in the emitter-follower stage to avoid parasitic oscillations which might otherwise occur due to capacitively loading the emitter. For the same reason, the interconnect length between the emitter-follower and the input of the cascode is minimized [25].

The frequency divider chain includes a regenerative first stage and static ECL or CML master–slave flip-flop stages. Its high-speed differential output provides a selectable divide ratio of 4 or 32, which facilitates locking to an external 19GHz DRO [26] or to a lower-frequency reference. A second, lower-frequency output, below 50MHz, with selectable divide ratios up to 1572864 is used to monitor the VCO frequency by a microcontroller [5].

The power amplifier following the buffered VCO consists of a differential EF-CE-EF-cascode chain [5]. The output power is adjustable in analog fashion over 8dB and can be reduced by another 20dB in a digital control step to comply with FCC rules when the vehicle is stopped. The transmit channel output power and phase noise are better than + 5.5dBm and −72.6dBc/Hz at 100KHz offset, respectively, over the −40°C to 125°C range. This phase

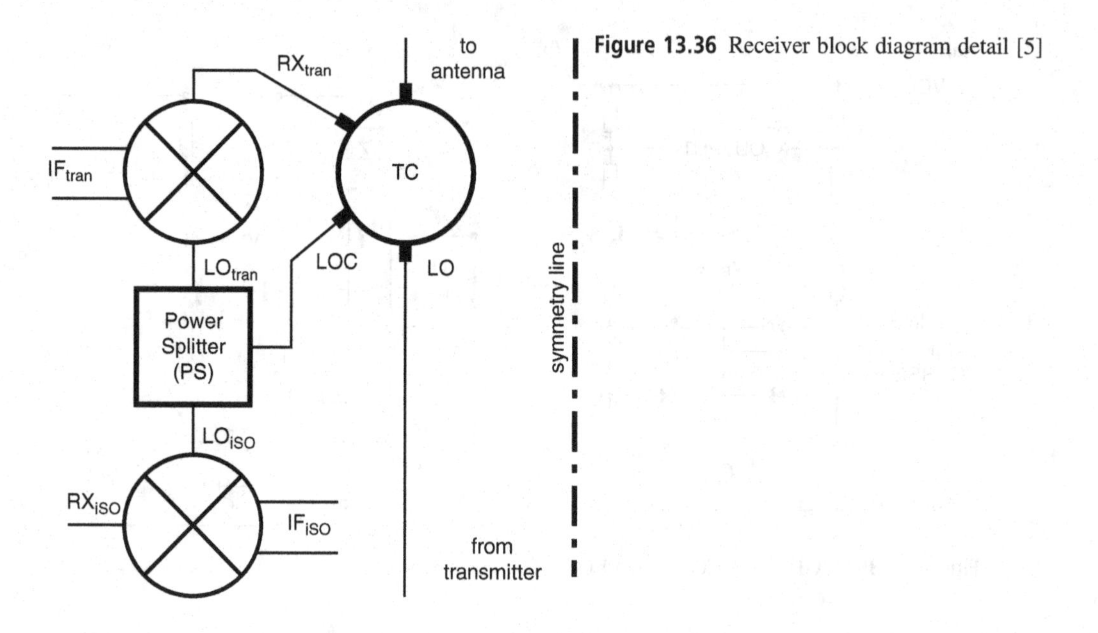

Figure 13.36 Receiver block diagram detail [5]

noise is inadequate for LRR applications but can be improved in an offset PLL which mixes the divided-by-4 LO signal with an external 19GHz DRO, and after additional division, locks it to a lower-frequency PLL [24].

The 4-channel receiver is formed by two Rat-Race transfer couplers, two Wilkinson power dividers (splitters), and four mixers. As illustrated in Figure 13.36, perfect symmetry is maintained between the two receiver halves to minimize coupling from the transmitter. Receivers 2 and 3, at the top of Figure 13.34, and whose performance is degraded by the antenna coupler loss, are used in transmit–receive channels, while receivers 1 and 4, labeled as "iso," operate as isolated receive-only channels. The signal from the power amplifier is applied to the LO arm of the Rat-Race coupler. One half is sent to the antenna port while the other half is sent to the Wilkinson power splitter at the LOC arm of the Rat-Race coupler. The power splitter drives the LO ports of the two Gilbert cell mixers that form the transferring and the isolated receive channels, respectively. Similarly, an incoming receive signal at the antenna port is split equally in two between the power amplifier output, where it is properly terminated, and the transferring mixer input, RXtran, where it is downconverted [5].

The mixer, similar to that shown in Figure 13.37 [27] but with the LO and RF baluns removed, consumes 35mA and employs a Gilbert cell core driven by a common-base differential pair at the RF port. The same LO-buffer topology as in Figure 13.35(b) is employed between the Wilkinson power splitter and the LO mixer input. The common-base stage at the RF input provides higher input linearity (0dBm) than the more traditional common-emitter differential pair with inductive degeneration. The penalty is a higher noise figure which is acceptable in this application.

At room temperature, the isolating mixers have a conversion gain of 14dB and a typical noise figure of 17dB at an IF frequency of 100KHz. The isolation between the four mixers, which operate simultaneously, is better than 35dB. The whole transceiver consumes 3.3W from a

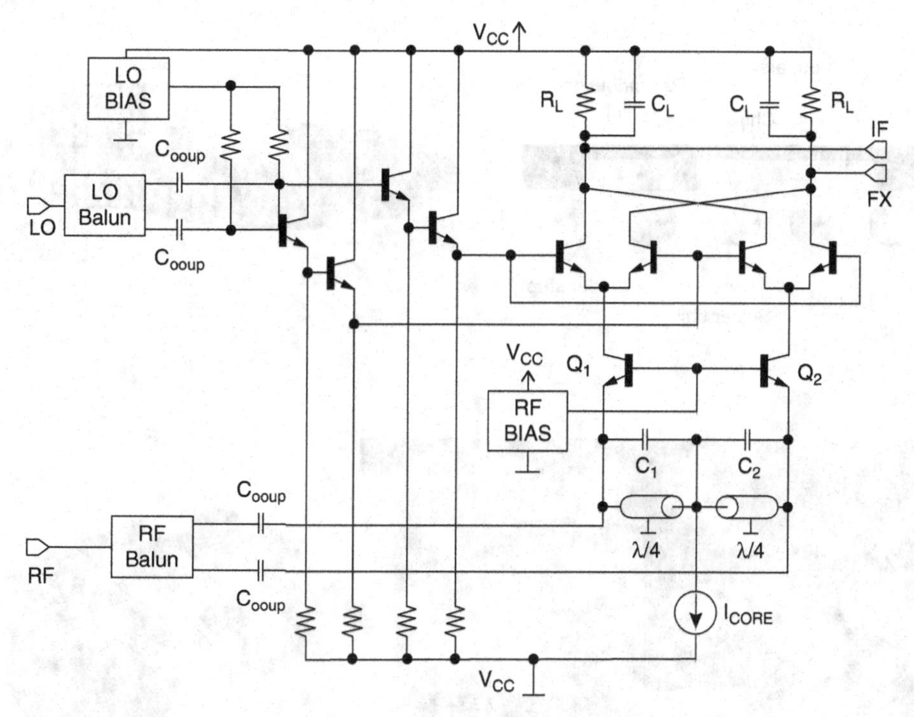

Figure 13.37 77GHz high-linearity downconvert mixer [27]

5.5V supply. One of the packaging solutions is an embedded wafer level BGA (eWLB) package [24], as illustrated in Figure 13.38, where the antenna is formed in the redistribution layer.

13.5 70–80GHZ ACTIVE IMAGER IN SiGe HBT TECHNOLOGY

Another new application for mm-wave transceivers which is receiving considerable attention is in active and passive imaging systems for security, medical, and industrial applications. Figure 13.39 is a conceptual view of a stand-off screening system for security at airports, sports and entertainment venues, large malls, etc. [7]. The system is meant for real-time detection of concealed objects as people pass through, without having to stop.

There are several reasons that make mm-wave frequencies at W-(75–110GHz), D – (110–170GHz), and G-Bands (140–220GHz) particularly attractive for imaging. First, mm-wave radiation easily passes through a variety of materials such as clothes, wood, cardboard, and is strongly reflected by metals, water and skin. Second, the lateral resolution of an image improves with increasing frequency since it is directly proportional to the wavelength, and finally, large bandwidths are available at mm-wave frequencies to apply frequency modulation which improves range resolution.

An active imaging array architecture, consisting of a planar multi-static (i.e. with many separate transmit and receive antennas) sparse array of transmitters and receivers with digital

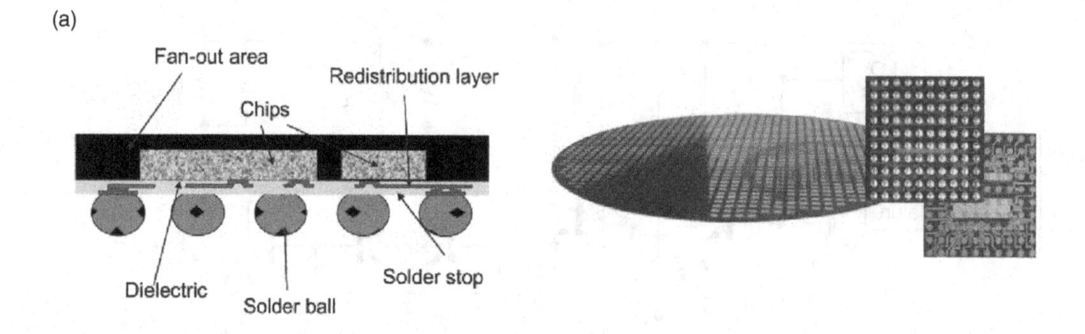

Figure 13.38 77GHz automotive radar transceiver wafer level packaging solution [24]

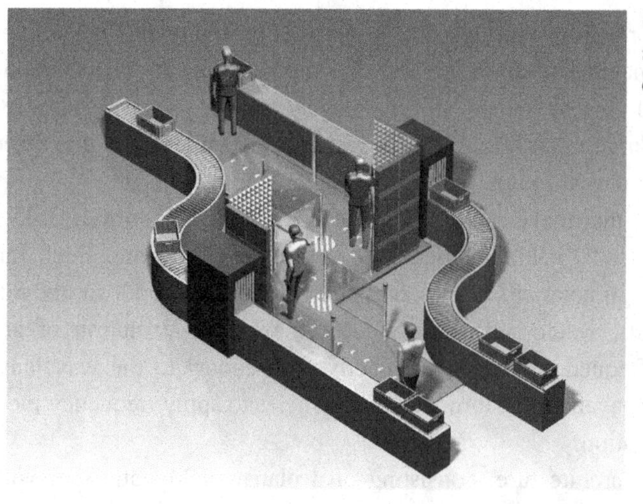

Figure 13.39 Vision of the airport security stand-off screening equipment [7]

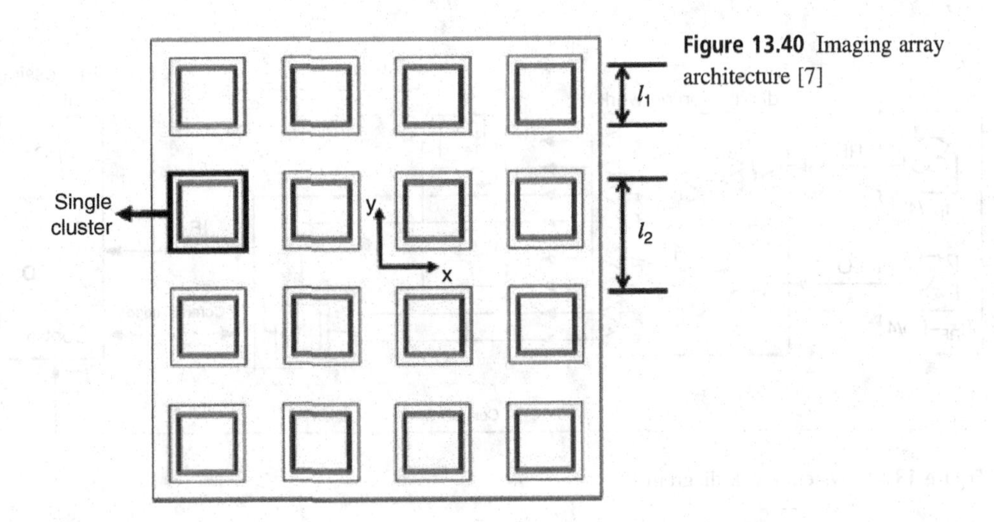

Figure 13.40 Imaging array architecture [7]

(i.e. baseband) beamforming, and which operates in the 72–80GHz range is illustrated in Figure 13.40 [7]. Compared to a conventional array with uniform λ/2 spacing between the antennas, the sparse array significantly reduces the number of required mm-wave transmitters and receivers, system cost, acquisition time, volume of collected data, and the power consumption, without degrading the image resolution.

The imager architecture consists of a uniform 2D array of square antenna clusters of length l_1, with a cluster-to-cluster spacing l_2. The transmit antennas, ideally spaced 0.5–0.75λ apart to avoid image aliasing, are placed in two horizontal lines at the top and bottom edges of each cluster. The receive antennas have identical spacing with the transmit ones, and are placed in two vertical lines at the left and right edge of the cluster, respectively. For example, if there are 24 transmit or 24 receive antennas in a line, spaced 0.75λ apart at 80GHz, this results in l_1 = 67.5mm and l_2 = 135mm.

To achieve the desired array aperture of at least 50 cm, 4×4 uniformly spaced clusters are employed, resulting in a total of 768 transmitters and 768 receivers, each with its own antenna. An array of about 1600 channels is claimed to be adequate for scanning humans [7]. Because of the special arrangement of transmitters and receivers, and the digital signal-processing algorithms employed for image correction [7], the degree of sparsity achieved is approximately 5%. For comparison, a dense uniform 2D array with the same aperture size (50cm×50cm) and having antennas placed at a 0.75λ pitch would result in over 30,000 antennas. The significant reduction of the numbers of transmitters and receivers is critical in achieving the stated goal of containing the power consumption of the entire system to less than 3kW. This translates in a power consumption smaller than 200mW per channel [8].

The image resolution equations are given by

$$\delta_x = \frac{\lambda D}{L_x^t + L_x^r},\ \delta_y = \frac{\lambda D}{L_y^t + L_y^r},\ \text{and}\ \delta_z = \frac{c}{2\Delta f} \tag{13.2}$$

where L'_x and L'_y are the aperture widths of the transmitter array along the x and y directions, respectively, L^r_x and L^r_y are the aperture widths of the receiver array along the x and y

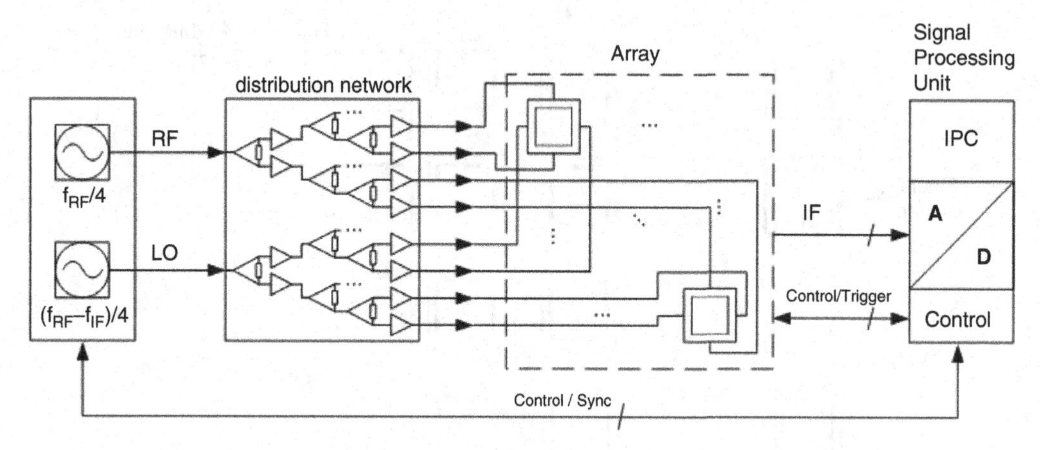

Figure 13.41 System block diagram [7]

directions, λ is the wavelength, D is the distance between the imaging array and the object to be imaged, c is the speed of light and Δf is the modulation bandwidth of the transmitter. In this system [7], the modulation bandwidth is 8GHz and $D = L^t_x = L^t_y = L^r_x = L^r_y = 50$ cm, which results in a range (depth) resolution of 2 cm and a lateral resolution of 2mm. For comparison, the range resolution of the LRR with 600MHz of frequency modulation bandwidth was 25cm.

The system, whose block diagram is illustrated in Figure 13.41, operates by sending a stepped-frequency beam to the target and synchronously collecting the reflected data for each TX-RX pair and for each selected frequency in the 72–80GHz band. The transmitters are turned on sequentially, with only one transmitter operating at a given time, while the receivers are always on, collecting data in parallel. To maximize the dynamic range, which has to be better than 60dB for high-quality imaging [7], and to prevent the transmitter signal from desensitizing the receiver, heterodyne receivers are employed which operate with an IF frequency of 20MHz. The IF signals are digitized to extract the amplitude and phase information of the reflected data collected by each receive channel.

The most challenging part of the system is the synchronous RF and LO signal distribution to the 768 transmitters and 768 receivers, respectively. To save power while ensuring synchronous data collection, the LO and transmit RF signals are generated from the same stable signal source, at one fourth of the intended frequency, and distributed by coaxial cables to each cluster. The isolation between channels is determined by the coupling between antennas on the PCB and between the signal distribution traces. To satisfy image quality requirements, a channel-to-channel isolation better than 40dB was specified [8].

Separate RX (Figure 13.42) and TX (Figure 13.43) SiGe chips consisting of four receiver channels and an LO-signal quadrupler, and four transmit channels and a RF-signal quadrupler, respectively, are employed in each cluster. The TX chips consume 568mW from 3.3V supply and have an output power of 0dBm per lane while occupying a die area of 4mm^2. Power detectors are weakly coupled to the output of each TX lane to monitor the transmit power and their DC voltage is monitored with an analog multiplexer (MUX). Each receive channel has a downconversion gain of 23dB, IP1dB of −30dBm and a single-sideband noise figure of 9.5dB.

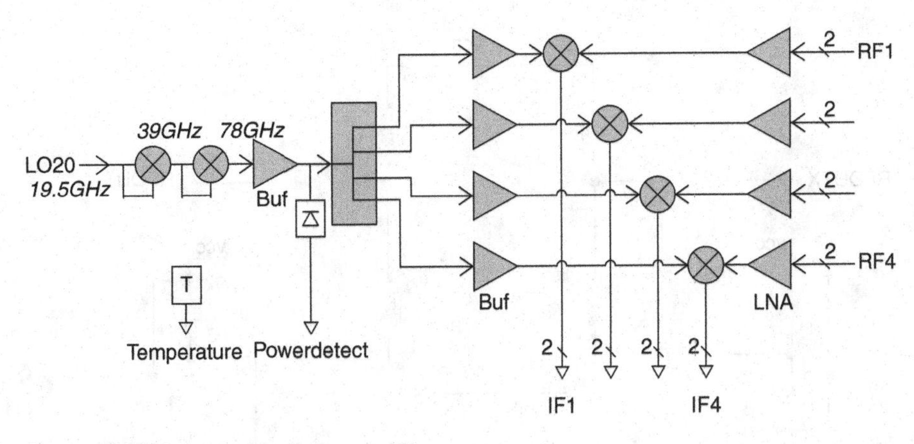

Figure 13.42 Receiver block diagram [8]

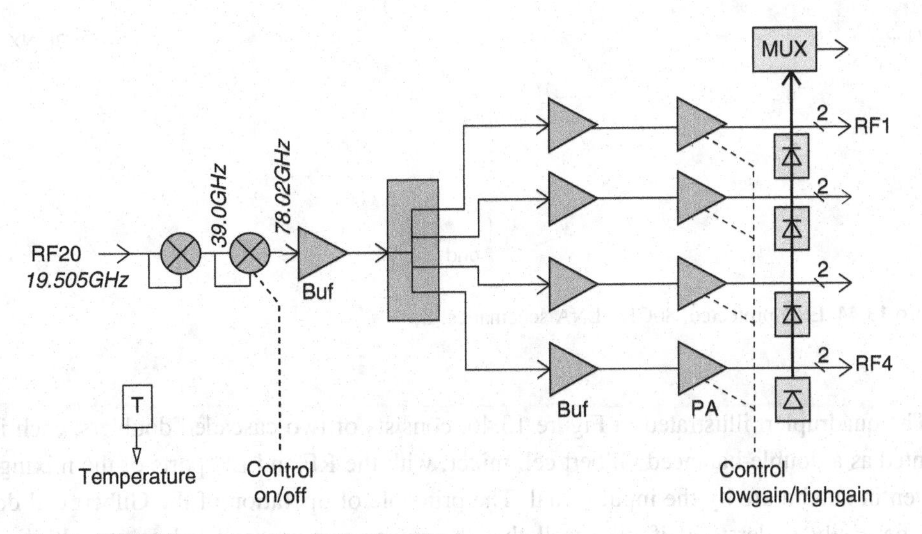

Figure 13.43 Transmitter block diagram [8]

The 4-channel receiver chip is identical in size to the transmitter chip, and consumes 638mW from 3.3V supply. As in the 60GHz radio transceiver arrays, Wilkinson power dividers are employed on both chips to distribute the multiplied LO/RF signal to the 4 receive/transmit channels. Buffers are placed at the LO port of each downconvert mixer to compensate for the loss of the Wilkinson power splitter.

All circuits employ fully differential topologies to reduce crosstalk between channels and to minimize sensitivity to ground inductance. The schematic of the cascode LNA with inductive degeneration and ESD protection based on shorted $\lambda/4$ lines is shown in Figure 13.44. Since the linearity requirements are not as stringent as in the LRR sensor due to the smaller transmit power of only 0dBm, the downconvert mixer (Figure 13.45) is realized as a traditional double-balanced Gilbert cell with emitter-follower as the IF output buffer. The combined power consumption of the LNA, mixer, and IF buffer is 33mA from 3.3V.

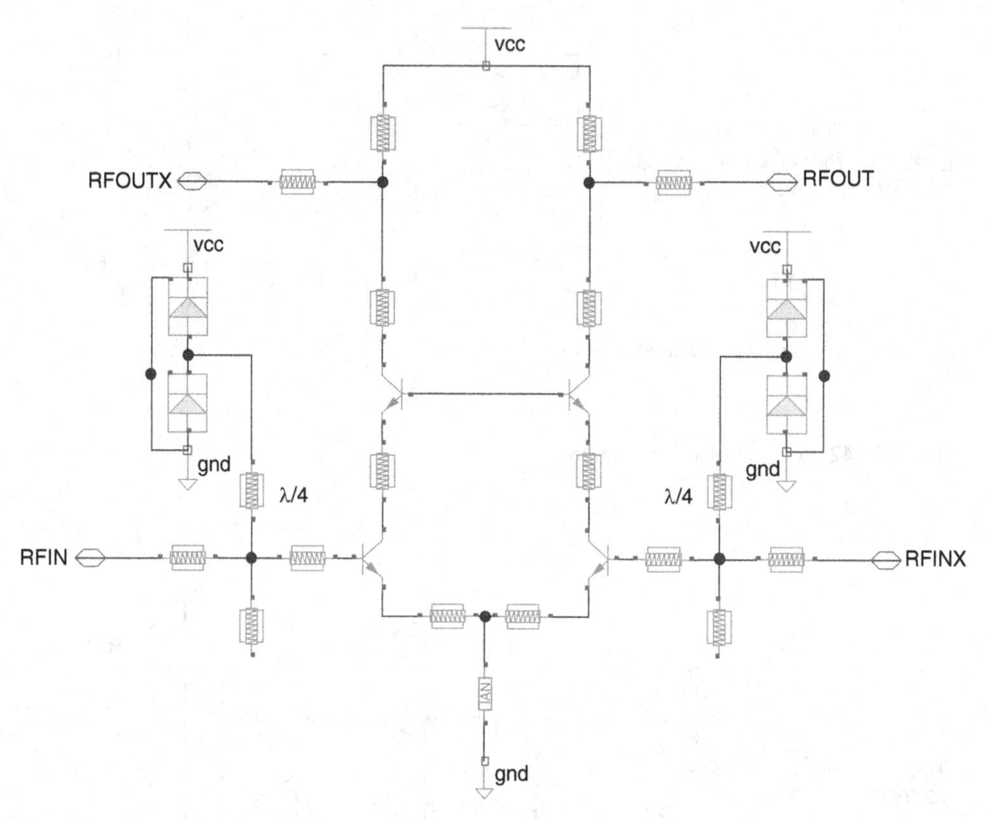

Figure 13.44 ESD-protected, 80GHz LNA schematics [8]

The quadrupler, illustrated in Figure 13.46, consists of two cascaded doublers, each implemented as a double-balanced Gilbert cell mixer, with the RF and LO ports of the mixing quad driven in quadrature by the input signal. The principle of operation of the Gilbert cell doubler can be easily understood if we recall that the mixer can be modeled as a multiplier with A$\cos(\omega t)$ applied to the LO port at the gates of the mixing quad, and $A\sin(\omega t)$ applied to the RF port, at the emitters of the mixing quad

$$s_{out} = \cos(\omega t) \cdot \sin(\omega t) = \frac{1}{2}\sin(2\omega t) \tag{13.3}$$

The power amplifier (Figure 13.47) also employs a differential Gilbert cell topology with a DC control voltage applied at the bases of the mixing-quad transistors. About 30dB of gain control is obtained by steering the output current between the power supply (AC ground) and the output loads. To fully turn off the transmitter, the current in the second multiplier stage can also be turned off. It should be noted that, because of the relatively relaxed specifications, the transmitter and receiver chips could have been also implemented in 65nm CMOS technology.

Photos of the 4×4 cluster demonstrator and of an individual cluster are shown in Figure 13.48 [7]. An absorber sheet is placed between the antenna arrays in the center of each cluster to reduce antenna coupling and to avoid standing wave formation between the imaged object and

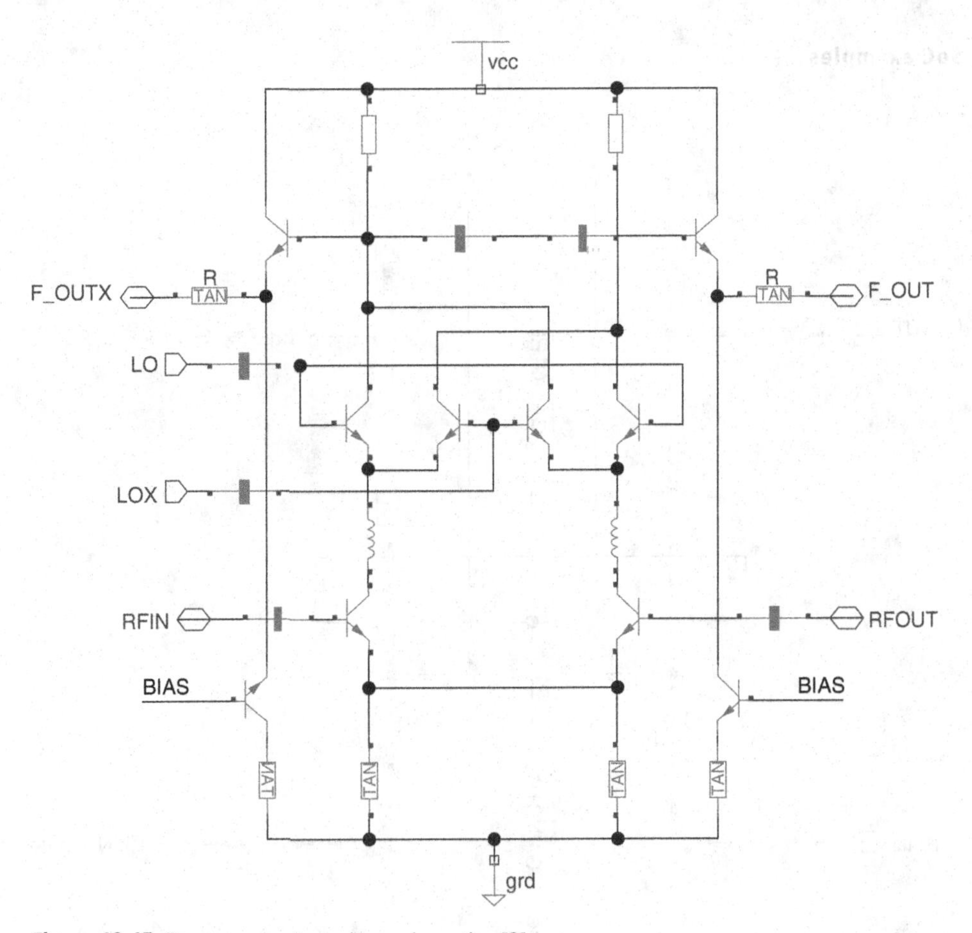

Figure 13.45 Downconversion mixer schematics [8]

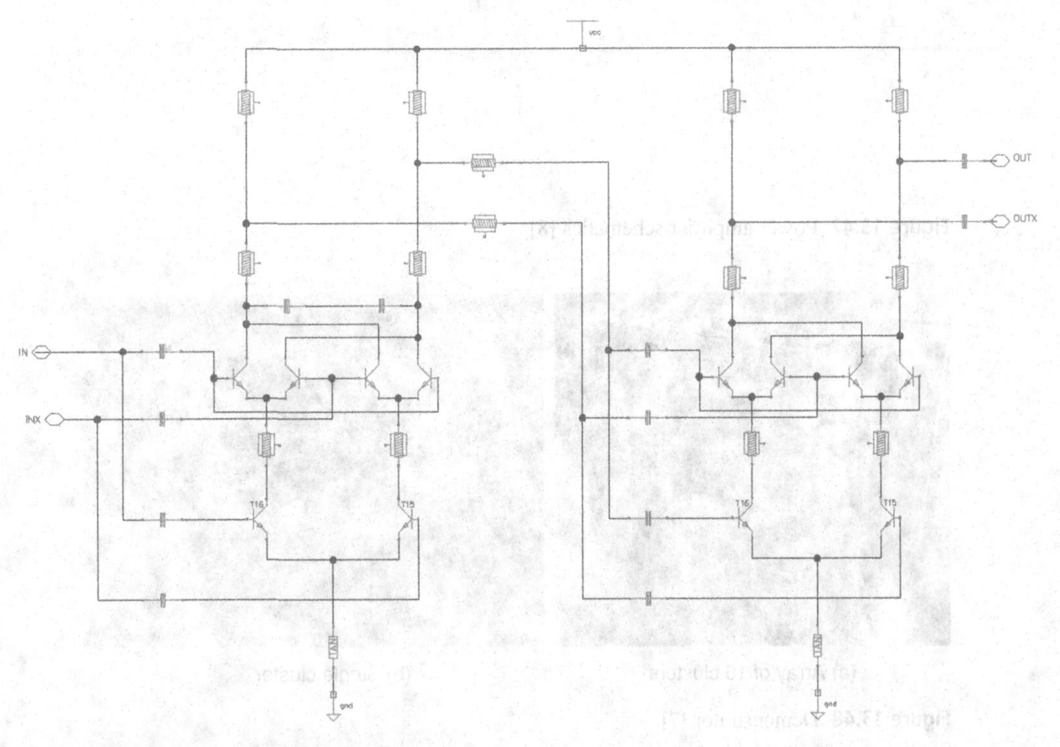

Figure 13.46 Quadrupler schematics [8]

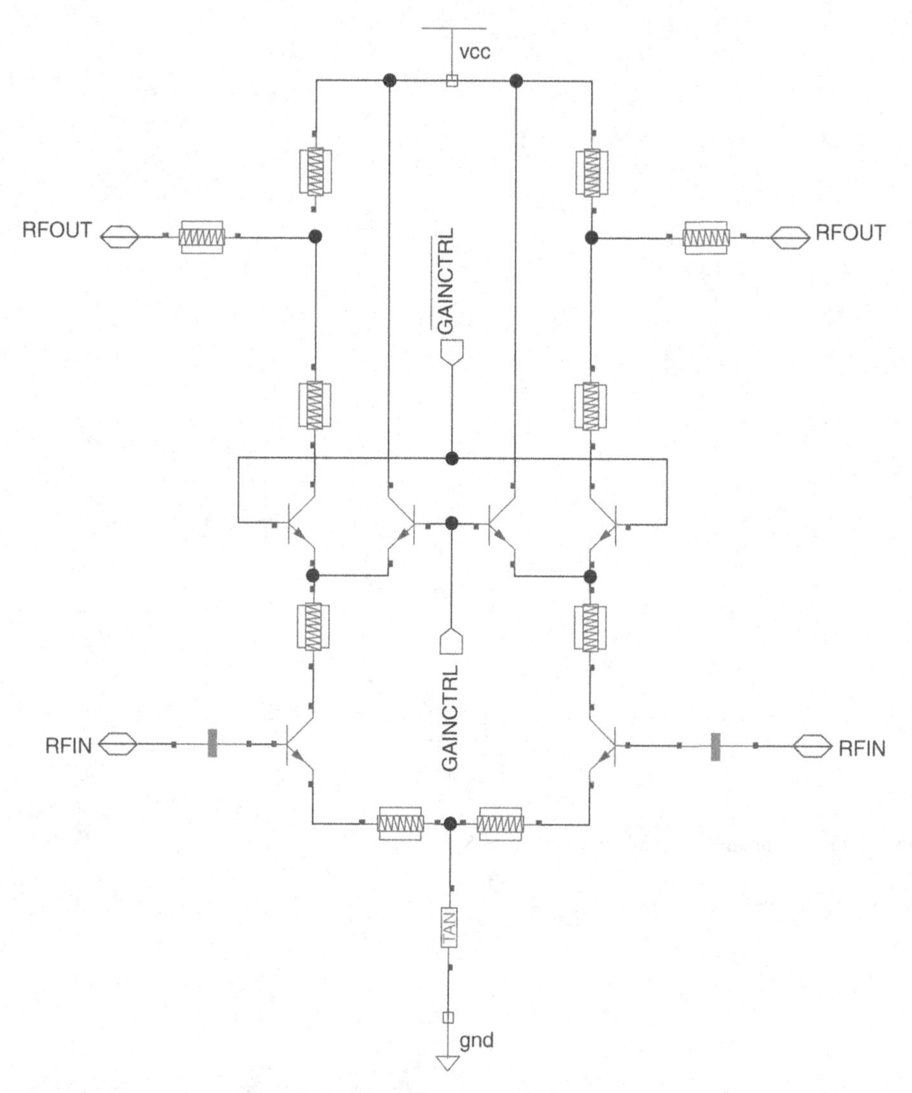

Figure 13.47 Power amplifier schematics [8]

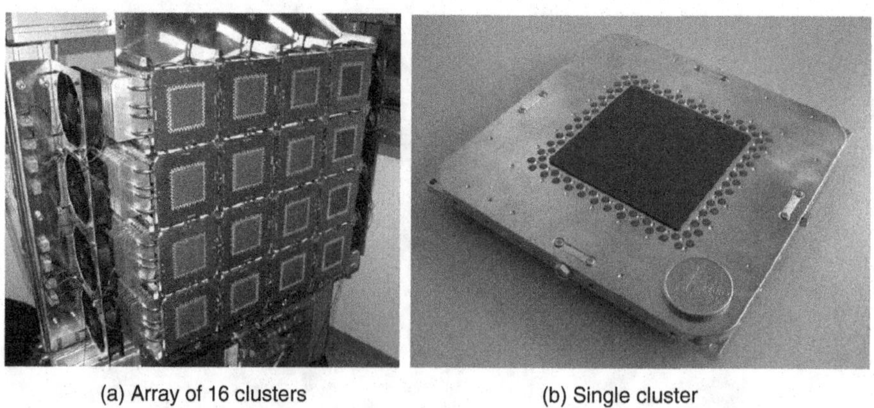

(a) Array of 16 clusters

(b) Single cluster

Figure 13.48 Demonstrator [7]

the metallic surface of the array. The measurement time for collecting data at 32 frequency steps between 72 and 80GHz is about 160ms, adequate for the real-time imaging of humans.

One of the most important aspects of system operation is the image calibration procedure. Two standards are employed for that purpose: a match and an offset short, realized with an absorber placed in front of the array, and a metal plate placed at a known distance, respectively. The match measurement M_m is used to correct for the coupling between the transmit and receive antennas. For example, for closely spaced TX-RX pairs, the transmitter signal drives the receiver in compression and calibration fails. Those (fewer than 0.1%) TX-RX pair combinations are masked out during image collection. The offset short measurement helps to correct the phase of the signals between receivers and to equalize the amplitude of the signals over the relatively large operation frequency band [7].

The imaged object reflection coefficient, Γ, obtained from a collected measurement M at the corresponding TX-RX pair is given by:

$$\Gamma = \frac{M - M_m}{M_s - M_m}\Gamma_s \tag{13.4}$$

where Γ_s is the simulated reflection coefficient from the offset short plate, M_m is the measurement (amplitude and phase) of the match standard for that particular TX-RX pair, and M_s is the measurement (amplitude and phase) of the offset short for the same TX-RX pair. The simulation of Γ_s is conducted based on the known geometry of the imager array and on the known location of the offset short standard with respect to the imaging array.

13.6 150–168GHZ ACTIVE IMAGING TRANSCEIVER WITH ON-DIE ANTENNAS IN SiGe BiCMOS TECHNOLOGY

An example of how the previous active imager concept can be scaled in frequency to double its lateral and range resolution while achieving even higher levels of integration on a single chip is illustrated in Figure 13.49 [28]. At 160GHz, the wavelength becomes smaller than 1.5mm, making it economical for synchronized antenna arrays to be integrated on the die, when a thick

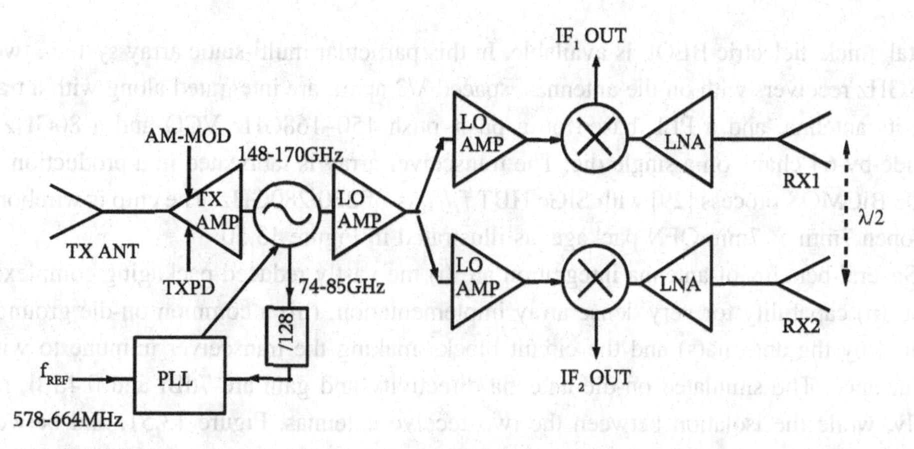

Figure 13.49 Block diagram of 150–168GHz active imaging transceiver [28]

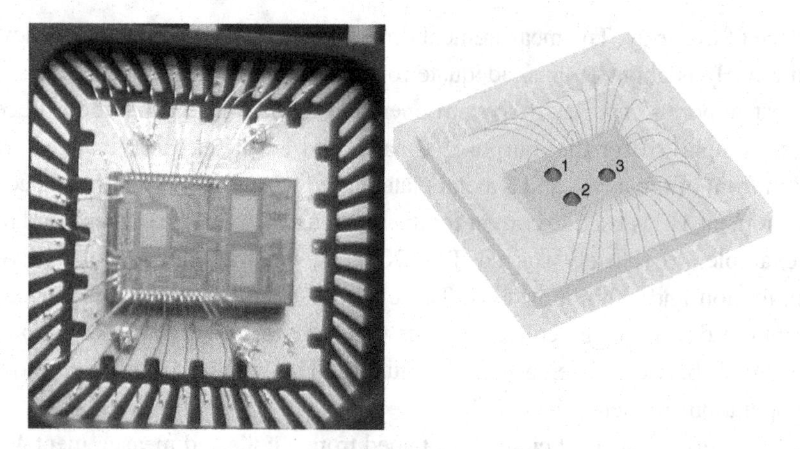

Figure 13.50 (a) Packaging solution and die photo (b) EM simulation port setup [28]

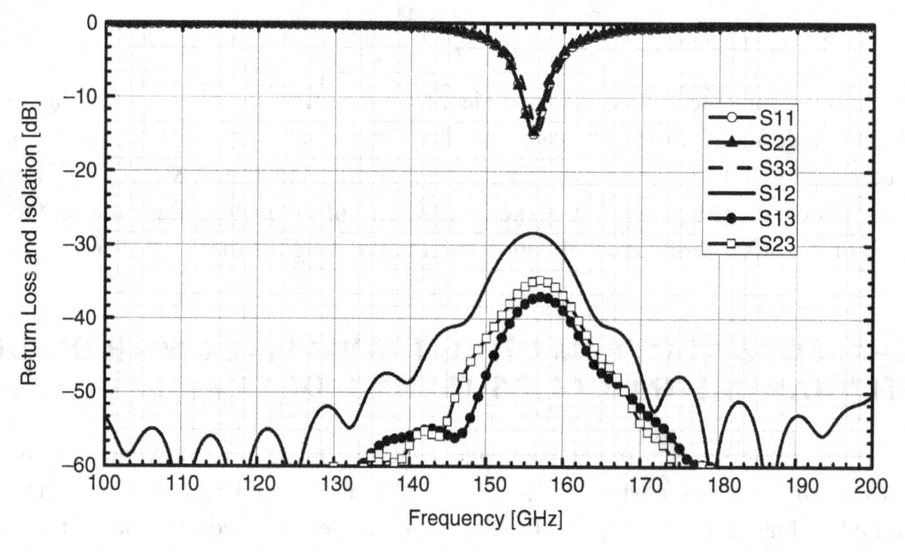

Figure 13.51 Simulated TX-RX antennas isolation and return loss [28]

metal, thick dielectric BEOL is available. In this particular multi-static array system, two 150–168GHz receivers with on-die antennas, spaced λ/2 apart, are integrated along with a transmitter, its antenna, and a PLL based on a push–push 150–168GHz VCO and a 80GHz static-divide-by-64 chain, on a single die. The transceiver array is fabricated in a production 130nm SiGe BiCMOS process [29] with SiGe HBT f_T/f_{MAX} of 230/280GHz. The chip is wirebonded in an open 7mm × 7mm QFN package, as illustrated in Figure 13.50.

Several benefits of antenna integration are (i) the vastly reduced packaging complexity and cost, (ii) capability for very dense array implementation, (iii) a common on-die ground plane shared by the antenna(s) and the circuit blocks making the transceiver immune to wirebond inductance. The simulated on-die antenna directivity and gain are 7dBi and 0.4dBi, respectively, while the isolation between the two receive antennas, Figure 13.51, and between the receive and transmit antennas is better than 28dB and 35dB, respectively.

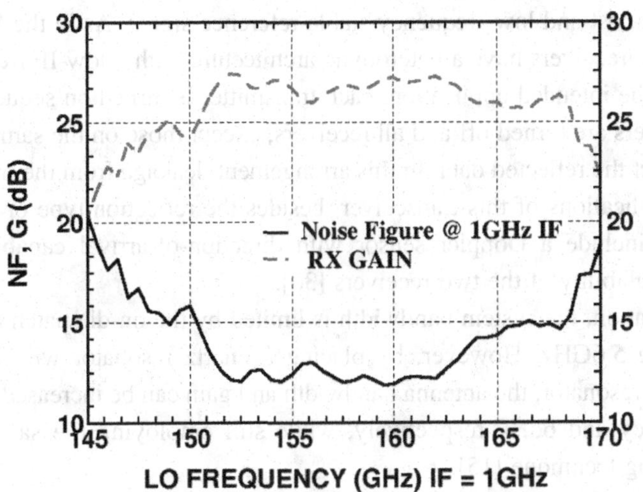

Figure 13.52 Measured receiver double-sideband noise figure and downconversion gain [28]

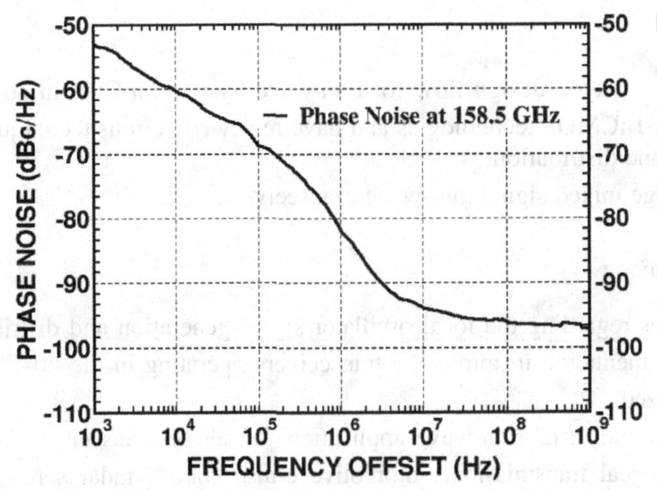

Figure 13.53 Measured phase noise at 158.5GHz [28]

The 3dB bandwidth of the receiver breakout with on-die VCO, but without antenna, shown in Figure 13.52, covers the entire 150–168GHz range. The DSB noise figure and downconversion gain of each receive lane are 12dB and 27dB, respectively. The transmitter output power, before the antenna, is + 1dBm and was estimated from the measured gain and saturated output power of a transmit amplifier breakout, and from the measured output power, in the −10dBm to −8dBm range, of the VCO breakout. As illustrated in Figure 13.53, the PLL phase noise at 158.5GHz is −68 and −82dBc/Hz at 100KHz and 1MHz offset, respectively. It varies by less than 1dB across the frequency band and from 25°C to 100°C. The whole chip, with both receivers and the transmitter turned on, consumes 1.8W from 1.8, 2.5, 3.3V, 3.6V (for VCO) supplies.

The integrated PLL, requiring a stable reference in the 600MHz range, facilitates the synchronization of such multiple transceiver chips in a sparse active imaging array, by distributing a relatively low-cost and low-frequency stable reference signal. As in the W-band imager discussed earlier, the receivers have a heterodyne architecture with a low-IF frequency in the 1–50MHz range. In the intended application, each transmitter is turned on sequentially, while all the other transmitters are turned off and all receivers, except those on the same die as the active transmitter, collect the reflected data. In this arrangement, leakage from the transmitter is minimized. Other applications of this transceiver, besides the reflection-type or inverse scattering active imagers, include a Doppler sensor with direction-of-arrival capability by taking advantage of the availability of the two receivers [30].

In the current implementation, the system bandwidth is limited by the on-die patch antenna bandwidth, simulated to be 5.6GHz. However, by placing a quartz resonator with a patch antenna on-top of an on-die resonator, the antenna bandwidth and gain can be increased to over 10% of the center frequency and 6dBi, respectively, while still employing the same QFN package and die wirebonding technique [15].

Summary

In this chapter we have discussed a design flow for mm-wave System-on-Chip in advanced nanoscale CMOS and SiGe BiCMOS technologies and have reviewed various techniques for:

- supply and ground plane distribution,
- biasing circuits for large mixed-signal mm-wave, transceivers,
- block isolation, and
- bias de-coupling techniques.

Critical architectural choices regarding the local oscillator signal generation and distribution, and at speed self-test implementation in mm-wave transceivers operating in the 60–300GHz range have also been analyzed.

Finally, examples of new industrial mm-wave applications of silicon transceivers in high-data-rate wireless HDMI signal transmission, automotive cruise control radar sensors, and active imagers for stand-off people screening were presented and competing solutions implemented in 65nm CMOS and SiGe BiCMOS technologies were analyzed.

Problems

13.1 Figure 13.54 shows the block diagram of a 76–81GHz automotive radar transceiver array for collision avoidance and automatic cruise control (ACC). The radar array consists of one transmitter with + 15dBm output power and its own antenna, and 4 receivers, each with its own antenna. Each receive path consists of a tuned LNA centered at 78.5GHz with 10% bandwidth and 20dB built-in automatic gain control. The LNA has a 50-Ω noise figure of 4dB at maximum gain, a variable gain of +5 to +15dB, and IIP3 of −5dBm at minimum gain, and −10dBm at maximum gain. The mixer has a noise

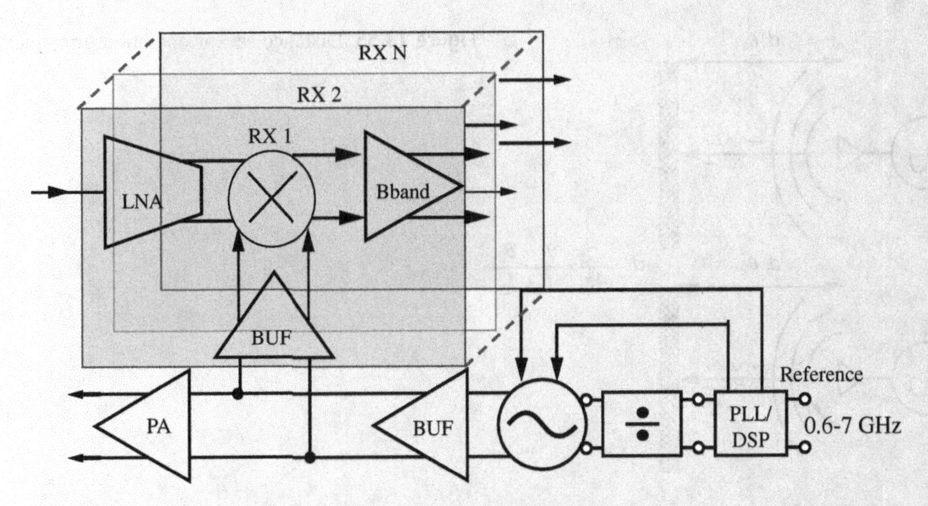

Figure 13.54 80GHz radar transceiver array block diagram

figure of 12dB, an IIP3 of 0dBm and a power gain of 6dB. An amplitude shift keying (ASK) modulator is implemented in the PA allowing for a modulation rate as high as 2Gb/s, useful in collision avoidance applications.

(a) Calculate the sensitivity of each receiver lane in collision avoidance mode at room temperature assuming an ideal baseband bandwidth of 2GHz, and an SNR of 13dB for an error rate of 1E-6. How does the receiver sensitivity change in the ACC mode of operation knowing that the baseband bandwidth is only 200MHz and that the SNR remains 13dB for an error rate of 1E-6?

(b) Calculate the IIP3 of the each receiver lane when the gain of the LNA is set to its minimum value of 5dB.

(c) The baseband outputs of the four receivers are added linearly. Assuming that the signals received at the antennas of each of the four receiver lanes have identical amplitude and phase, how does the SNR of the receiver array compare to that of a single receiver lane? Assume that the noise of each receiver lane is not correlated with that of other lanes. Assume that all receiver lanes are identical. What about the receiver array noise figure. Is it better than the noise figure of a single receive lane?

13.2 For the 150GHz active imager, determine the required frequency modulation range and square array aperture if an object placed 10cm away from the array is to be imaged with a lateral resolution better than 0.1mm and a range resolution of 1cm.

13.3 A D-Band total power radiometer consumes 50mW and has a noise bandwidth of 20GHz centered on 160GHz. It consists of a 5-stage amplifier with 36dB gain and 7dB noise figure, followed by a diode detector with an NEP of 6 pW/$\sqrt{\text{Hz}}$. The 1/f noise corner of the detector is 500Hz and that of the entire radiometers is 100Hz.

(a) If the responsivity of the diode detector is 10kV/W, calculate the responsivity of the entire radiometer.

(b) Calculate the NEP of the radiometer.

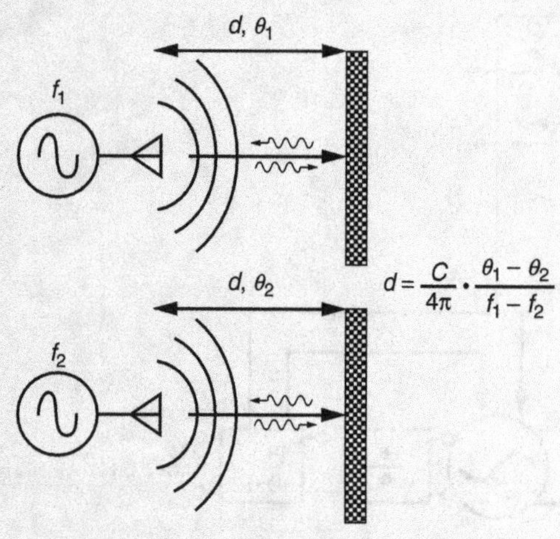

Figure 13.55 Distance sensor operation principle

(c) Calculate the NETD of the total power radiometer assuming an integration time of 2ms.

(d) A 50×50 pixel, 2-dimensional wafer-scale integrated passive imaging video camera with 10Hz frame rate is fabricated based on this total power radiometer unit. Each pixel includes an on-die antenna with 50% efficiency and is spaced $\lambda/2$ apart from the adjacent pixels. What is the area occupied by the array and what is its power consumption if the 2D imaging array is scanned one row at a time? How is the NETD of each pixel degraded by the antenna efficiency?

13.4 A super-heterodyne D-Band SiGe BiCMOS distance sensor with integrated antenna operates in the 120–125GHz band by measuring the time(phase) delay to and from the target, as illustrated in Figure 13.55 [16]. The sensor employs the monostatic, single-chip transceiver architecture in Figure 13.14(a) and features a transmitter and two IQ receivers, one of which acts as the reference channel. Both the transmit and the receive VCOs are phase locked to crystal references with 50 ppm stability. The first IF frequency is in the 1.5–2GHz range. Two transmit frequencies: 121GHz and 122.5GHz are employed to determine the distance to the target. The bandwidth, determined by the integration time, is at 50kHz.

(a) If the transmitter output power is 3dBm, the antenna efficiency and directivity are 50% and 10dBi respectively, and the receiver noise figure is 7dB, calculate the SNR of the received signal if the target is located at a distance of 1m and has a cross-section of 0.1m^2.

(b) Calculate the precision with which a distance of approximately 1m can be measured knowing that phase noise cancelation occurs for all distances of interest in this application.

REFERENCES

[1] E. Laskin, M. Khanpour, R. Aroca, K.W. Tang, P. Garcia, and S.P. Voinigescu, "95GHz receiver with fundamental frequency VCO and static frequency divider in 65nm Digital CMOS," *IEEE ISSCC Digest*, pp. 180–181, February 2008.

[2] S.T. Nicolson, P. Chevalier, B. Sautreuil, and S.P. Voinigescu, "Single-Chip W-Band SiGe HBT Transceivers and Receivers for Doppler Radar and Millimeter-Wave Imaging," *IEEE JSSC*, **43**(10): 2206–2217, October 2008.

[3] A. Valdes-Gacia, S.T. Nicolson, J-W Lai, A. Natarajan, P-Y. Chen, S.K. Reynolds, J-H. C. Zhan, D.K. Kam, D. Liu, and B.A. Floyd, *et al.*, "A SiGe BiCMOS 16-Element Phased array Transmitter for 60GHz Communications," *IEEE ISSCC Digest*, pp. 218–219, February 2010.

[4] S. Emami, R.F. Wiser, E. Ali, M.G. Forbes, M.Q. Gordon, X. Guan, S. Lo, P.T. McElwee, J. Parker, J.R. Tani, J.M. Gilbert, and C.H. Doan, "A 60GHz CMOS phased array transceiver pair for multi Gb/s wireless communications," *ISSCC Digest Tech. Papers*, pp. 163–164, February 2011.

[5] H.P. Forstner, H. Knapp, H. Jager, E. Kolmhofer, J. Platz, F. Starzer, M. Treml, A. Schinko, G. Birschkus, J. Bock, K. Aufinger, R. Lachner, T. Meister, H. Schafer, D. Lukashevich, S. Boguth, A. Fischer, F. Reininger, L. Maurer, J. Minichshofer, and D. Steinbuch, "A 77GHz 4-channel automotive radar transceiver in SiGe," in Proc. *IEEE Radio Frequency Integrated Circuits Symp.* pp. 233–236, June 2008.

[6] S. Trotta, B. Dehlink, A. Ghazinour, D. Morgan, and J. John, "A 77GHz 3.3V 4-channel transceiver in SiGe BiCMOS technology," in Proc. *IEEE Bipolar/BiCMOS Circuits and Technology Meeting BCTM 2008*, pp. 186–189, October 2008

[7] S.S. Ahmed, A. Schiessl, and L.-P. Schmidt, "Novel Fully Electronic Active Real-Time Millimeter-Wave Imaging System based on a Planar Multistatic Sparse Array," in the *IEEE Microwave Symposium Digest*, June 2011.

[8] M. Tiebout, H.-D. Wohlmuth, H. Knapp, R. Salerno, M. Druml, J. Kaeferboeck, M. Rest, J. Wuertele, S.S. Ahmed, A. Schiessl, and R. Juenemann, "Low Power Wideband Receiver and Transmitter Chipset for mm-Wave Imaging in SiGe Bipolar Technology," in Proc. of the 2011 RFIC Conference, June 2011.

[9] M. Seo, M. Urteaga, M. Rodwell, and M. Choe, "A 300GHz PLL in an InP HBT Technology," to be presented at IEEE MTT-S Int. *Microwave Symp.*, *Baltimore*, June 2011.

[10] S.P. Voinigescu, IEEE 2011 BCTM Short Course, October 2011.

[11] E. Ojefors et al., A 820GHz SiGe Chipset for Terahertz Active Imaging Applications," *IEEE ISSCC Digest*, pp. 224–225, February 2011.

[12] S. Shahramian, A. Hart, A. Tomkins, A.C. Carusone, P. Garcia, P. Chevalier, and S.P. Voinigescu, "Design of a Dual W– and D-Band PLL," *IEEE JSSC*, **46**: 1011–1022, May 2011.

[13] K. Yau, E. Dcquay, I. Sarkas, and S.P. Voinigescu, "On-wafer Device, Circuit and SoC Characterization and Self-Test above 100GHz," *IEEE Microwave Magazine*, pp. 30–54. February 2012.

[14] B.A. Floyd, D.R. Greenberg, R.M. Malladi, B.A. Orner, and S.K. Reynolds, "Radio Frequency Integrated Circuit with on-chip noise source for self-test," *US Patent Application 2009/0190640 A1*, Jul. 20, 2009.

[15] R. Agethen and D. Kissinger, "Built-in Test Architectures for Zero-IF Automotive Radar Receivers," *Workshop on Automotive Radar Sensors in the 76–81GHz Frequency Range, EuRAD, European Microwave Week*, Paris, October 1, 2010.

[16] I. Sarkas, J. Hasch, A. Balteanu, and S. Voinigescu, "A Fundamental Frequency, Single-Chip 120GHz SiGe BiCMOS Precision Distance Sensor with Above IC Antenna Operating over Several Meters," *IEEE MTT*, **60**(3), pp. 795–812, March 2012.

[17] E. Laskin, A. Tomkins, A. Balteanu, I. Sarkas, and S.P. Voinigescu, "A 60GHz RF IQ DAC Transceiver with on-Die at-Speed Loopback," *IEEE RFIC Symposium Digest*. pp., June 2011.

[18] E. Laskin, P. Chevalier, B. Sautreuil, and S.P. Voinigescu, "A 140GHz Double-Sideband Transceiver with Amplitude and Frequency Modulation Operating over a few Meters," *IEEE BCTM Digest*, pp. 178–181, October 2009.

[19] B. Laemmle, K. Schmalz, C. Scheytt, D. Kissinger, and R. Weigel, "A 122GHz Multiprobe Reflectometer for Dielectric Sensor Readout in SiGe BiCMOS Technology," 2011 *IEEE CSICS*, October 2011.

[20] O. Inac, D. Shin, and G.M. Rebeiz, "A Phased Array RFIC with Built-In Self-Test using an Integrated Vector Signal Analyzer," 2011 *IEEE CSICS*, October 2011.

[21] A. Natarajan, S.K. Reynolds, S.T. Nicolson, J-H. C. Zhan, D.K. Kam, D. Liu, Y-L.O. Huang, A. Valdes-Garcia, and B.A. Floyd, "A Fully-Integrated 16-Element Phased array Receiver in SiGe BiCMOS for 60GHz Communications," *IEEE JSSC*, **46**(5): 1059–1075, May 2011.

[22] A. Valdes-Garcia, S.K. Reynolds, A. Natarajan, D.K. Kam, D. Liu, J-W Lai, Y-L.O. Huang, P-Y. Chen, M-D. Tsai, J-H. C. Zhan, and S.T. Nicolson, "Single-Element and Phased array Transceiver Chipsets for 60GHz Gb/s Communications," *IEEE Communication Magazine*, pp. 120–131, April 2011.

[23] A. Valdes-Garcia, S.T. Nicolson, J-W Lai, A. Natarajan, P-Y. Chen, S.K. Reynolds, J-H. C. Zhan, D.K. Kam, D. Liu, and B.A. Floyd, "A Fully Integrated 16-Element Phased array Transmitter in SiGe BiCMOS for 60GHz Communications," *IEEE JSSC*, **45**(12): 2757–2773, December 2010.

[24] J. Hasch, E. Topak, T. Zwick, R. Schnabel, R. Weigel, and C. Waldschmidt, "Millimeter Wave Technology for Automotive Radar Sensors in the 77GHz Frequency Band," *IEEE MTT*, March 2012, pp. 845–860.

[25] H. Li, H.-M. Rein, T. Suttorp, and J. Böck, "Fully integrated SiGe VCOs with powerful output buffer for 77GHz automotive radar systems and applications around 100GHz," *IEEE JSSC*, **39** (10): 1650–1658, October 2004.

[26] H. P. Forstner, H. D. Wohlmuth, H. Knapp, C. Gamsjager, J. Bock, T. Meister, and K. Aufinger, "A 19GHz DRO downconverter MMIC for 77GHz automotive radar frontends in a SiGe bipolar production technology," in Proc. *IEEE Bipolar/BiCMOS Circuits and Technology Meeting*, pp. 117–120, October 2008.

[27] B. Dehlink, H.-D. Wohlmuth, H.-P. Forstner, H. Knapp, S. Trotta, K. Aufinger, T. Meister, J. Böck, and A. Scholtz, "A highly linear SiGe double-balanced mixer for 77GHz automotive radar applications," *IEEE RFIC Symposium Digest*, pp. 235–238, June 2006.

[28] I. Sarkas, E. Laskin, A. Tomkins, J. Hasch, A. Balteanu, E. Dacquay, L. Tarnow, P. Chevalier, B. Sautreuil, and S.P. Voinigescu, "Silicon-based radar and imaging sensors operating above 120GHz," *IEEE Mikon 2012*, May 2012, pp. 91–96.

[29] G. Avenier, M. Diop, P. Chevalier, G. Troillard, B. Vandelle, F. Brossard, L. Depoyan, M. Buczko, C. Leyris, S. Boret, S. Montusclat, A. Margain, S. Pruvost, S.T. Nicolson, K.H. K. Yau, N. Revil, D. Gloria, D. Dutartre, S.P. Voinigescu, and A. Chantre, "0.13μm SiGe BiCMOS technology for mm-wave applications," *IEEE JSSC*, **44**: 2312–2321, September 2009.

[30] J. Hasch, U. Wostradowski, R. Hellinger, and D. Mittelstrass, "77GHz radar transceiver with integrated dual antenna elements," *2010 GeMIC Digest*, March 2010.

APPENDIX 1
Trigonometric identities

$$\sin x = \frac{1}{csc x} \quad \cos x = \frac{1}{sec x} \quad \tan x = \frac{\sin x}{\cos x} \quad \cot x = \frac{1}{\tan x}$$

$$\sin^2 x + \cos^2 x = 1 \qquad 1 + \tan^2 x = sec^2 x \qquad 1 + \cot^2 x = csc^2 x$$

$$\sin\left(\frac{\pi}{2} - x\right) = \cos x \qquad \cos\left(\frac{\pi}{2} - x\right) = \sin x \qquad \tan\left(\frac{\pi}{2} - x\right) = \cot x$$

$$\sin(-x) = -\sin x \quad \cos(-x) = \cos(x)$$

$$\sin(x \pm y) = \sin x \cdot \cos y \pm \cos x \cdot \sin y \qquad \cos(x \pm y) = \cos x \cdot \cos y \mp \sin x \cdot \sin y$$

$$\sin(2x) = 2 sin x \cdot \cos x \qquad \cos(2x) = \cos^2 x - \sin^2 x = 2\cos^2 x - 1 = 1 - 2\sin^2 x$$

$$\sin x \cdot \sin y = \frac{1}{2}[\cos(x-y) - \cos(x+y)] \qquad \cos x \cdot \cos y = \frac{1}{2}[\cos(x-y) + \cos(x+y)]$$

$$\sin x \cdot \cos y = \frac{1}{2}[\sin(x+y) + \sin(x-y)] \qquad \cos x \cdot \sin y = \frac{1}{2}[\sin(x+y) - \sin(x-y)]$$

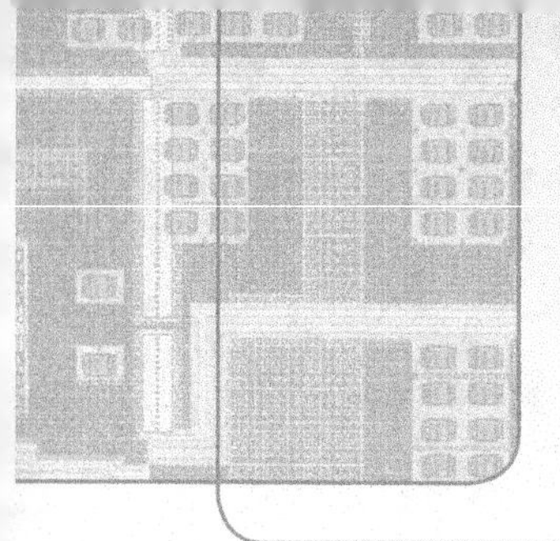

APPENDIX 2

Baseband binary data formats and analysis

In all the transceiver examples discussed previously, binary data is employed at baseband. The remainder of this chapter reviews techniques to encode, generate, and characterize the quality of the baseband data signals.

A2.1 LINE CODES

The spectrum of very long random sequences of data extends to very low frequencies. This can cause detection and IC integration problems because systems with very large DC gain also suffer from DC offsets, wander, and 1/f noise, all of which degrade the SNR of the received data signal. One solution to suppress these effects is to place very large capacitors in series on the signal path. However, large capacitors in the microfarad range are too expensive to integrate monolithically.

As a result, the baseband data are typically encoded such that they exhibit:

- DC balance (i.e. approximately equal numbers of "1"s and "0"s),
- short run lengths of random sequences,
- high transition density (which helps to simplify clock and data recovery circuits).

Data coding solutions include:

- scrambling
- block coding:
  - o 8B10B: no more than 5 "1" or "0" in a row
  - o 64B/66B
- combination of the above.

A2.2 GENERATING PSEUDO-RANDOM DATA (PRBS)

A common technique to characterize baseband circuits is to monitor their behaviour when pseudo-random data are applied at their input. A pseudo-random data sequence (PRBS) can be generated using a linear shift register with feedback, consisting of D-type flip-flops and XOR gates, as shown in Fig.A2.1 for 2^7–1 PRBS.

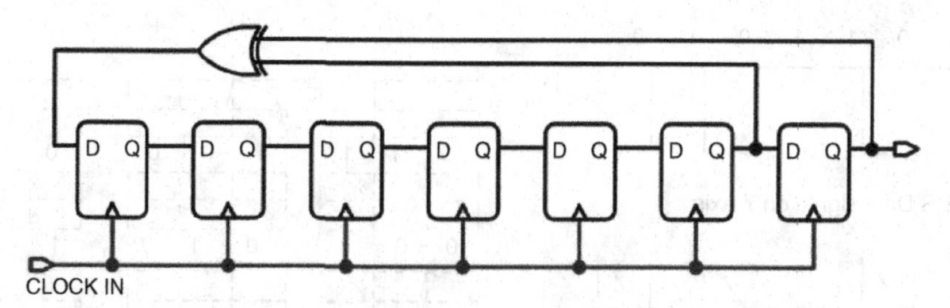

Figure A2.1 Schematic of a linear shift register with feedback for generating a 2^7–1 PRBS

Typical pattern lengths employed in testing are 2^7–1, 2^{15}–1, 2^{23}–1, and 2^{31}–1. Examples of generator polynomials that result in PRBS are shown below.

- for 2^{23}–1: $x^{23} + x^{18} + 1$;
- for 2^7–1: $x^7 + x^6 + 1$

A longer pattern results in a lower minimum frequency of the spectrum. In transistor-level circuit simulations, 2^7–1 patterns are often used to avoid the long simulation times demanded by higher order generator polynomials.

A2.3 CREATING AN EYE DIAGRAM

Fig. A2.2 gives a pictorial view of the technique employed to create and display eye diagrams. A saw-tooth signal is used for the x-axis to fold back each bit or each integer number of consecutive bits (typically 2 or 3) to the time origin.

A2.4 RISE AND FALL TIMES

The delay and rise/fall times of a single-pole linear system can be obtained from the first and second order impulse response of the system.

$$t_d = \frac{\int_{-\infty}^{\infty} t h(t) dt}{\int_{-\infty}^{\infty} h(t) dt} \tag{A2.1}$$

$$\left(\frac{t_R}{2}\right)^2 = \frac{\int_{-\infty}^{\infty} t^2 h(t) dt}{\int_{-\infty}^{\infty} h(t) dt} - t_d^2 \tag{A2.2}$$

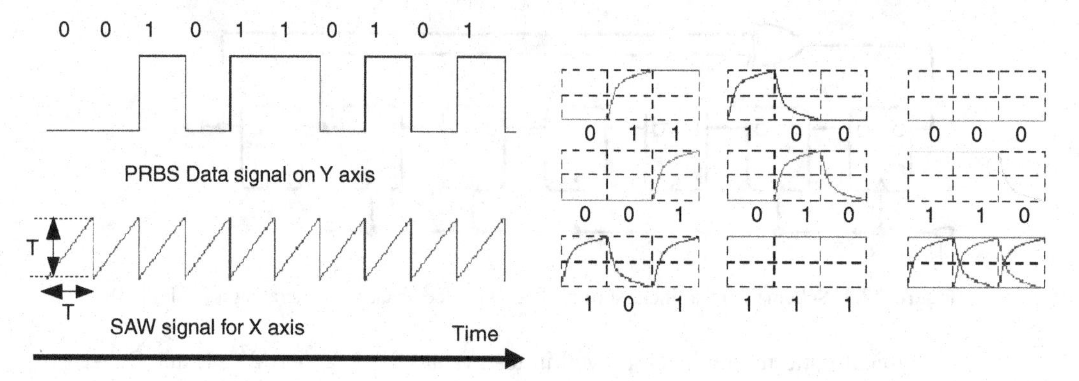

Figure A2.2 Creating an eye diagram using a sawtooth signal for the x-axis

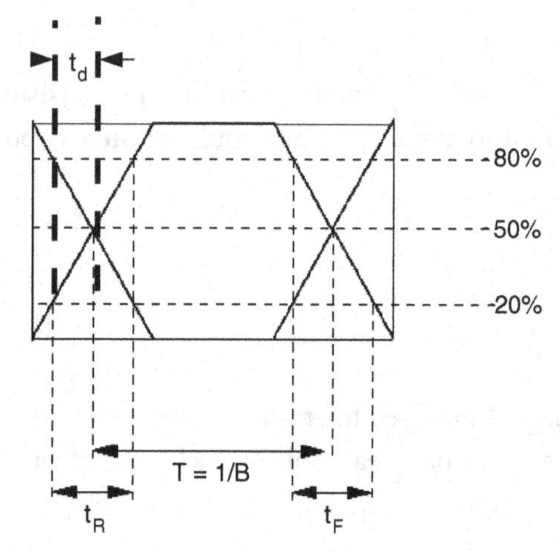

Figure A2.3 Definitions of rise, fall, and delay times

$$t_d = t_{d1} + t_{d2} \ ; \quad t_R = \sqrt{t_{R1}^2 + t_{R1}^2}. \tag{A2.3}$$

Their definition is also depicted on the idealized eye diagram of Fig.A2.3 to illustrate how they are measured in the lab. T is the data signal period. It should be noted that the rise and fall times, t_R and t_F, are obtained from the x-axis coordinates when the rising (or falling) edge of the pulse reaches 20% and 80% of the signal amplitude. In same cases the 10% and 90% points are used instead, resulting in larger rise and fall times. To avoid ambiguity, rise and fall times must be specified in terms of percentages of signal amplitude.

In either definition, the measurement set-up will affect the measurement, increasing the rise and fall times. The impact of the set-up is difficult to de-embed, except in very simple situations where the setup can be approximated by a first order filter response. For a linear system, the relationship between the rise time and the small signal 3-dB bandwidth is given by

$$\omega_{3\mathrm{dB}} \approx \frac{ln(0.9/0.1)}{t_R} = \frac{2.2}{t_R} \tag{A2.4}$$

A2.5 PULSE-WIDTH DISTORTION (PWD) OR DUTY-CYCLE-DISTORTION (DCD)

- *T* remains the same
- Eye crossing <> 50%
- Pulse width is distorted

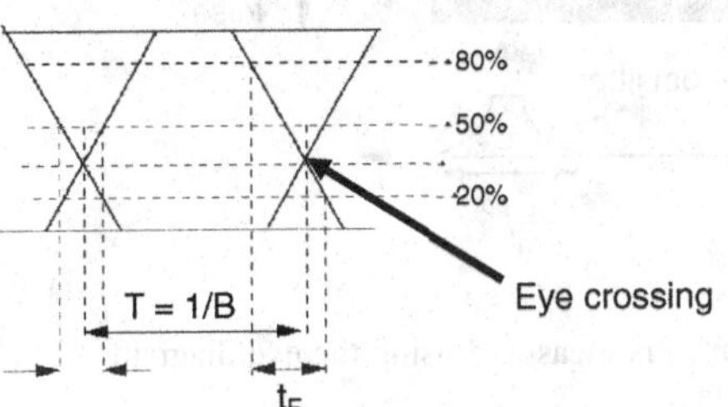

Figure A2.4 Illustration of pulse-width distortion

A2.6. JITTER GENERATION

Deterministic jitter, t_{DJ}

- due to insufficient bandwidth and nonlinear group delay
- shows up as multiple-edges
- specified as peak-to-peak value.

Random jitter, t_J

- due to phase noise in the clock signal
- specified in rms values

$$t_j^{pp} \leq 2 \cdot Q \cdot t_j^{rms} + t_{DJ}^{pp} \tag{A2.5}$$

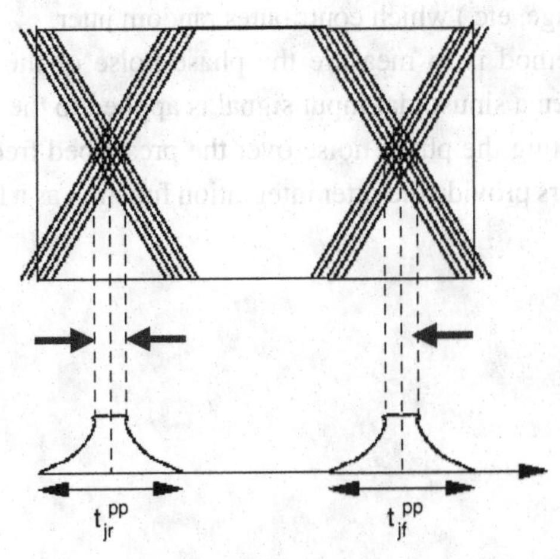

Figure A2.5 Illustration of the impact of random and deterministic jitter on the measured eye diagram

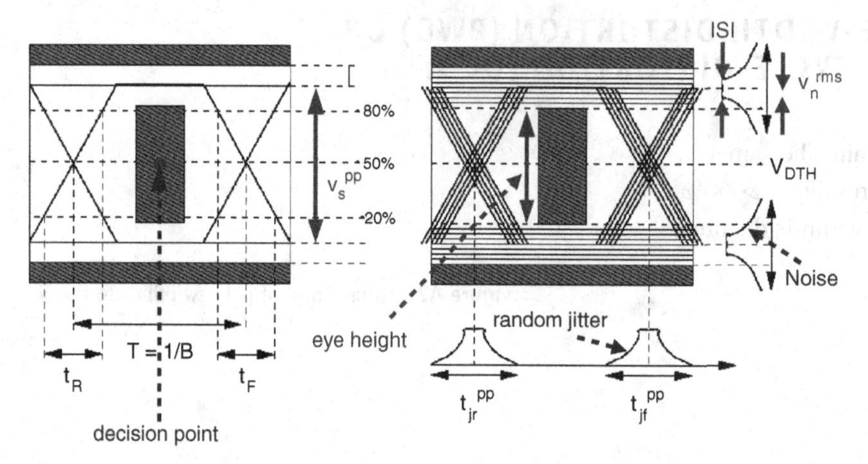

Figure A2.6 Illustration of the main data signal parameters measured using the eye diagram

A2.7 EYE MASK TEST

From the measured eye diagram on the oscilloscope, the quality factor of the eye can be calculated as

$$Q = 1/(2Bt_j{}^{rms}) \tag{A2.6}$$

$$Q = v_s{}^{pp}/(2v_n{}^{rms}) \tag{A2.7}$$

A $Q = 7$ corresponds to a BER of 10^{-12}.

Most modern oscilloscopes provide the eye Q as a feature.

A2.8 THE LINK BETWEEN RANDOM JITTER AND PHASE NOISE

Random jitter is typically measured using an oscilloscope and applying a sinusoidal signal (usually at the highest operation frequency) to the circuit. Such an approach does not de-embed the impact of the measurement setup (probes, cables, package, etc.) which contributes random jitter.

An alternate (arguably more precise) method is to measure the phase noise of the output directly with a power spectrum analyzer when a sinusoidal input signal is applied to the circuit. The jitter can be obtained simply by integrating the phase noise over the prescribed frequency range. Most modern power spectrum analyzers provide the jitter integration function as a feature.

A2.9 FIBER CHARACTERISTICS

Loss and bandwidth

- 2.5 dB/km @ 0.85 μm (short reach data)
- 0.4 dB/km @ 1.3 μm

- 0.25 dB/km @ 1.55 μm (long haul)
- The C+L bands (around 193 THz /1.55 μm) provide more than 10 THz of bandwidth around 193 THz.

Dispersion (D)

Dispersion in the fiber broaden the pulse as the signal propagates through the fiber. Three sources of dispersion can be identified

- modal dispersion (multimode fibre)
- chromatic dispersion:
  - 0 @ 1.3 μm
  - 17 ps/(nm × km) @ 1.55 μm
- polarization mode dispersion: D_{PMD} typically 0.1ps/km$^{0.5}$

$$D = \frac{1}{L} \times \frac{\partial \tau}{\partial \lambda} \tag{A2.8}$$

$$\Delta T = |D| \times L \times \Delta\lambda \tag{A2.9}$$

$$\Delta T = D_{PMD} \times \sqrt{L} \tag{A2.10}$$

APPENDIX 3
Linear matrix transformations

Z to S

$$S_{11} = \frac{(z_{11}-1)(z_{22}+1) - z_{12}z_{21}}{(z_{11}+1)(z_{22}+1) - z_{12}z_{21}}; \quad S_{12} = \frac{2z_{12}}{(z_{11}+1)(z_{22}+1) - z_{12}z_{21}};$$

$$S_{21} = \frac{2z_{21}}{(z_{11}+1)(z_{22}+1) - z_{12}z_{21}}; \quad S_{22} = \frac{(z_{11}+1)(z_{22}-1) - z_{12}z_{21}}{(z_{11}+1)(z_{22}+1) - z_{12}z_{21}}.$$

S to Z

$$z_{11} = \frac{(S_{11}+1)(1-S_{22}) + S_{12}S_{21}}{(1-S_{11})(1-S_{22}) - S_{12}S_{21}}; \quad z_{12} = \frac{2S_{12}}{(1-S_{11})(1-S_{22}) - S_{12}S_{21}}$$

$$z_{21} = \frac{2S_{21}}{(1-S_{11})(1-S_{22}) - S_{12}S_{21}}; \quad z_{22} = \frac{(1-S_{11})(1+S_{22}) + S_{12}S_{21}}{(1-S_{11})(1-S_{22}) - S_{12}S_{21}}.$$

Z to Y

$$y_{11} = \frac{z_{22}}{\Delta z}; \quad y_{12} = \frac{-z_{12}}{\Delta z}; \quad y_{21} = \frac{-z_{21}}{\Delta z}; \quad y_{22} = \frac{z_{11}}{\Delta z}.$$

Y to Z

$$z_{11} = \frac{y_{22}}{\Delta y}; \quad z_{12} = \frac{-y_{12}}{\Delta y}; \quad z_{21} = \frac{-y_{21}}{\Delta y}; \quad z_{22} = \frac{y_{11}}{\Delta y}.$$

Z to H

$$h_{11} = \frac{\Delta z}{z_{22}}; \quad h_{12} = \frac{z_{12}}{z_{22}}; \quad h_{21} = \frac{-z_{21}}{z_{22}}; \quad h_{22} = \frac{1}{z_{22}}.$$

H to Z

$$z_{11} = \frac{\Delta h}{h_{22}}; \quad z_{12} = \frac{h_{12}}{h_{22}}; \quad z_{21} = \frac{-h_{21}}{h_{22}}; \quad z_{22} = \frac{1}{h_{22}}.$$

Z to A

$$A = \frac{z_{11}}{z_{21}}; \quad B = \frac{\Delta z}{z_{21}}; \quad C = \frac{1}{z_{21}}; \quad D = \frac{z_{22}}{z_{21}}.$$

A to *Z*

$$z_{11} = \frac{A}{C}; \quad z_{12} = \frac{\Delta A}{C}; \quad z_{21} = \frac{1}{C}; \quad z_{22} = \frac{D}{C}.$$

S to *Y*

$$y_{11} = \frac{(1 - S_{11})(1 + S_{22}) + S_{12}S_{21}}{(1 + S_{11})(1 + S_{22}) - S_{12}S_{21}}; \quad y_{12} = \frac{-2S_{12}}{(1 + S_{11})(1 + S_{22}) - S_{12}S_{21}};$$

$$y_{21} = \frac{-2S_{21}}{(1 + S_{11})(1 + S_{22}) - S_{12}S_{21}}; \quad y_{22} = \frac{(1 + S_{11})(1 - S_{22}) + S_{12}S_{21}}{(1 + S_{11})(1 + S_{22}) - S_{12}S_{21}}.$$

Y to *S*

$$S_{11} = \frac{(1 - y_{11})(1 + y_{22}) + y_{12}y_{21}}{(1 + y_{11})(1 + y_{22}) - y_{12}y_{21}}; \quad S_{12} = \frac{-2y_{12}}{(1 + y_{11})(1 + y_{22}) - y_{12}y_{21}};$$

$$S_{21} = \frac{-2y_{21}}{(1 + y_{11})(1 + y_{22}) - y_{12}y_{21}}; \quad S_{22} = \frac{(1 + y_{11})(1 - y_{22}) + y_{12}y_{21}}{(1 + y_{11})(1 + y_{22}) - y_{12}y_{21}}.$$

S to *H*

$$h_{11} = \frac{(1 + S_{11})(1 + S_{22}) - S_{12}S_{21}}{(1 - S_{11})(1 + S_{22}) + S_{12}S_{21}}; \quad h_{12} = \frac{2S_{12}}{(1 - S_{11})(1 + S_{22}) + S_{12}S_{21}};$$

$$h_{21} = \frac{-2S_{21}}{(1 - S_{11})(1 + S_{22}) + S_{12}S_{21}}; \quad h_{22} = \frac{(1 - S_{11})(1 - S_{22}) - S_{12}S_{21}}{(1 - S_{11})(1 + S_{22}) + S_{12}S_{21}}.$$

H to *S*

$$S_{11} = \frac{(h_{11} - 1)(1 + h_{22}) - h_{12}h_{21}}{(1 + h_{11})(1 + h_{22}) - h_{12}h_{21}}; \quad S_{12} = \frac{2h_{12}}{(1 + h_{11})(1 + h_{22}) - h_{12}h_{21}};$$

$$S_{21} = \frac{-2h_{21}}{(1 + h_{11})(1 + h_{22}) - h_{12}h_{21}}; \quad S_{22} = \frac{(1 + h_{11})(1 - h_{22}) + h_{12}h_{21}}{(1 + h_{11})(1 + h_{22}) - h_{12}h_{21}}.$$

S to *A*

$$A = \frac{(1 + S_{11})(1 - S_{22}) + S_{12}S_{21}}{2S_{21}}; \quad B = \frac{(1 + S_{11})(1 + S_{22}) - S_{12}S_{21}}{2S_{21}};$$

$$C = \frac{(1 - S_{11})(1 - S_{22}) - S_{12}S_{21}}{2S_{21}}; \quad D = \frac{(1 - S_{11})(1 + S_{22}) + S_{12}S_{21}}{2S_{21}}.$$

A to *S*

$$S_{11} = \frac{A + B - C - D}{A + B + C + D}; \quad S_{12} = \frac{2(AD - BC)}{A + B + C + D}; \quad S_{21} = \frac{2}{A + B + C + D}; \quad S_{11} = \frac{B - A - C + D}{A + B + C + D}.$$

A to *Y*

$$y_{11} = \frac{D}{B}; y_{12} = \frac{-\Delta A}{B}; y_{21} = \frac{-1}{B}; y_{22} = \frac{A}{B}.$$

Y to *A*

$$A = \frac{-y_{22}}{y_{21}}; B = \frac{-1}{y_{21}}; C = \frac{-\Delta y}{y_{21}}; D = \frac{-y_{11}}{y_{21}}.$$

H to *A*

$$A = \frac{-\Delta h}{h_{21}}; B = \frac{-h_{11}}{h_{21}}; C = \frac{-h_{22}}{h_{21}}; D = \frac{-1}{h_{21}}.$$

A to *H*

$$h_{11} = \frac{B}{D}; h_{12} = \frac{\Delta A}{D}; h_{21} = \frac{-1}{D}; h_{22} = \frac{C}{D}.$$

Y to *H*

$$h_{11} = \frac{1}{y_{11}}; h_{12} = \frac{-y_{12}}{y_{11}}; h_{21} = \frac{y_{21}}{y_{11}}; h_{22} = \frac{\Delta y}{y_{11}}.$$

H to *Y*

$$y_{11} = \frac{1}{h_{11}}; y_{12} = \frac{-h_{12}}{h_{11}}; y_{21} = \frac{h_{21}}{h_{11}}; y_{22} = \frac{\Delta h}{h_{11}}.$$

S to *T*

$$T_{11} = \frac{1}{S_{21}}; T_{12} = \frac{-S_{22}}{S_{21}}; T_{21} = \frac{S_{11}}{S_{21}}; T_{22} = S_{12} - \frac{S_{11}S_{22}}{S_{21}}.$$

T to *S*

$$S_{11} = \frac{T_{21}}{T_{11}}; S_{12} = T_{22} - \frac{T_{21}T_{12}}{T_{11}}; S_{21} = \frac{1}{T_{11}}; S_{22} = \frac{-T_{12}}{T_{11}}.$$

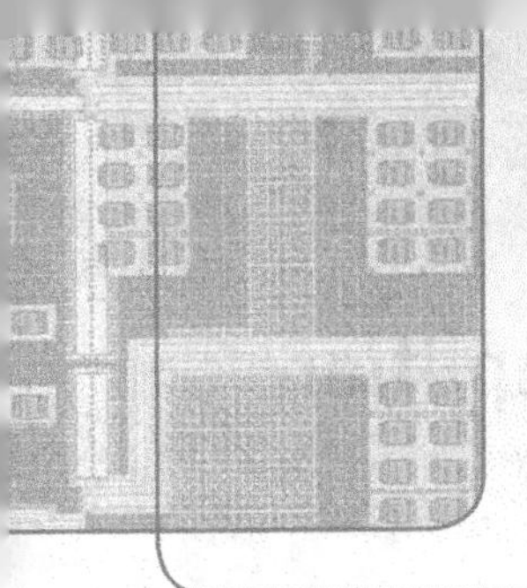

APPENDIX 4
Fourier series

The Fourier series for a function **f**: [−π,π] → **R** is

$$a_o + \sum_{n=1}^{\infty} [a_n \cos(n\omega t) + b_n \sin(n\omega t)]$$

where

$$a_o = \frac{1}{2\pi} \int_{-\pi}^{\pi} f(t)dt,$$

$$a_n = \frac{1}{\pi} \int_{-\pi}^{\pi} f(t)\cos(n\omega t)dt,$$

and

$$b_n = \frac{1}{\pi} \int_{-\pi}^{\pi} f(t)\sin(n\omega t)dt.$$

USEFUL EXPRESSIONS

$$\int_{-\pi}^{\pi} \cos(k\omega t)\cdot\sin(n\omega t)dt = 0; \quad \int_{-\pi}^{\pi} \cos^2(n\omega t)dt = \pi; \quad \int_{-\pi}^{\pi} \cos(k\omega t)\cdot\cos(n\omega t)dt = 0 \text{ if } n \neq k.$$

The Fourier series of a square wave function over period 2π and maximum and minimum values of 1 and −1, respectively is given by

$$f(t) = \frac{4}{\pi} \sum_{n=1,3,5}^{\infty} \frac{1}{n} \sin(nt).$$

APPENDIX 5

Exact noise analysis for a cascode amplifier with inductive degeneration

Let us consider a cascode amplifier with series–series feedback formed by L_G and L_S and with a resonant parallel tank load formed by the output capacitance of the transistor, C_D and L_D, as shown in Figure A5.1. Note that, alternatively, the loss resistances of L_S and L_G, R_{LS}, R_{LG}, respectively, could be accounted for by absorbing them in the source, and gate resistance, respectively, of the transistor/cascode stage.

The equivalent noise sources at the input of the feedback network are expressed as a function of the loss resistance of each inductor

$$v_{nf}^2 = 4kT\Delta f(R_{LG} + R_{LS}) \text{ and } i_{nf} = 0. \tag{A5.1}$$

From them, we can derive the noise parameters of the feedback network in the noise impedance formalism as

$$G'_{nf} = 0;\ R'_{uf} = R_{LG} + R_{LS};\ Z'_{corf} = 0;\ Z_{11f} = R_{LG} + R_{LS} + j\omega(L_S + L_G);\ Z_{12a} = 0,\ Z_{12f} = j\omega L_S. \tag{A5.2}$$

If the Q of inductors L_G and L_S is assumed infinite (lossless feedback) then $R_{uf} = 0$ and Z_{11f} becomes purely imaginary.

The expressions of the Y-parameters of a cascode stage as a function of those of the CS (A) and CG (B) stages are

$$y_{11a} = \frac{y_{11A}(y_{22A} + y_{11B}) - y_{21A}\,y_{12A}}{y_{22A} + y_{11B}};\ y_{12a} = \frac{-y_{12A}\,y_{12B}}{y_{22A} + y_{11B}};\ y_{21a} = \frac{-y_{21A}\,y_{21B}}{y_{22A} + y_{11B}};$$
$$y_{22a} = \frac{y_{22B}\,(y_{22A} + y_{11B}) - y_{21B}\,y_{12B}}{y_{22A} + y_{11B}}. \tag{A5.3}$$

For simplicity but justifiably, we assume that the y_{12a} of the cascode stage is 0.

$$z_{11a} = R_s + R_g + \frac{y_{22a}}{y_{11a}y_{22a} - y_{12a}y_{21a}} \approx R_s + R_g + \frac{1}{j\omega(C_{gs} + 2C_{gd})} = R_s + R_g - j\frac{f_{Ta}}{f g_m} \tag{A5.4}$$

$$z_{11} = R_s + R_g + z_{11\mathrm{a}} + z_{11f} = R_s + R_g + R_{LS} + R_{LG} - j\frac{f_{Ta}}{f g_m} + j\omega(L_S + L_G) \tag{A5.5}$$

$$z_{21a} = \frac{-y_{21a}}{y_{11a}y_{22a} - y_{12a}y_{21a}} \approx \frac{-y_{21a}}{y_{11a}y_{22a}} = jR_P\frac{f_{Ta}}{f} \tag{A5.6}$$

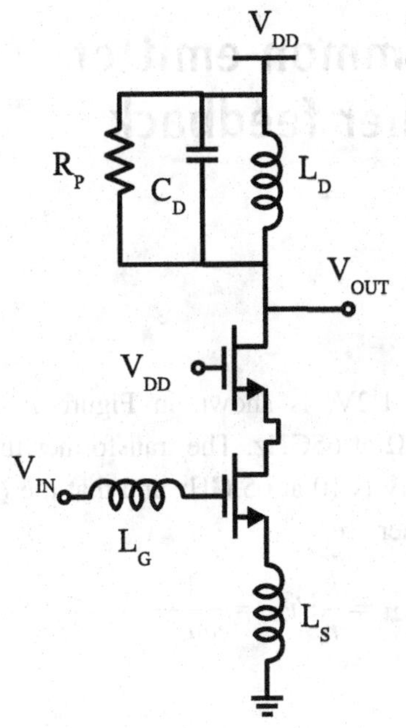

Figure A5.1 Schematics of a cascode LNA stage with inductive degeneration

$$z_{21} = Z_{21a} + Z_{21f} = R_{LS} + jR_P \frac{f_{Ta}}{f} + j\omega L_S \approx jR_P \frac{f_{Ta}}{f} \quad \text{(A5.7)}$$

In a properly designed amplifier $\omega(L_G + L_S) = f_{Ta}/(fg_m)$ and

$$Z_{in} = Z_{11} - \frac{Z_{21}Z_{12}}{Z_{22}} \approx R_s + R_g + R_{LS} + R_{LG} + \frac{2\pi R_P f_{Ta} L_S + \omega^2 L_S^2}{R_P} \approx R_s + R_g + R_{LS} + R_{LG} + \omega_{Ta} L_S \quad \text{(A5.8)}$$

where R_P is the equivalent resistance of the output tank at resonance and includes the impact of the output resistance of the (cascode) stage and the finite Q of L_D.

From (3.177)–(3.179), we can now derive the expressions of the noise parameters of the entire amplifier with feedback.

$$G'_n = \frac{|Z_{21a}|^2}{|Z_{21}|^2} G'_{na} Z'_{cor} = Z'_{cora} + R_{LG} + R_{LS} + j\omega(L_G + L_S); \; R'_u = R_{LG} + R_{LS} + R'_{ua} \quad \text{(A5.9)}$$

$$Z_{sopt} = \sqrt{R^2_{sopta} + \frac{R_{LG} + R_{LS}}{G_{na}} + 2R'_{cora}(R_{LG} + R_{LS}) + (R_{LG} + R_{LS})^2} + j[X_{sopta} - \omega(L_G + L_S)] > R_{sopta} \quad \text{(A5.10)}$$

$$F_{MIN} = 1 + 2G'_{na}\left[R'_{cora} + R'_{sopt} + R_{LG} + R_{LS}\right] > F_{MINa} \quad \text{(A5.11)}$$

Indeed, the real part of the optimum noise impedance and the minimum noise figure increase slightly compared to those of the transistor/cascode due to the losses in the feedback network described by R_{LG} and R_{LS}. However, R_{SOPT} and F_{MIN} will be unchanged from those of the cascode stage if the feedback is lossless.

APPENDIX 6

Noise analysis of the common-emitter amplifier with transformer feedback

A transformer-feedback LNA, biased with $V_{CC} = 1.2\text{V}$, is shown in Figure A6.1. The SiGe HBT is sized for optimal noise matching to 50Ω at 65GHz. The transformer turn ratio is $n = n_s/n_p$. Assume that the Q-factor of the secondary is 10 at 65 GHz and that the Q of the primary is infinite. The H-parameters of the transformer are

$$h_{11} = j\omega(1-k^2)L_{11}; h_{12} = \frac{-k}{n}; h_{21} = \frac{k}{n}; h_{22} = \frac{1}{j\omega L_{22}} \tag{A6.1}$$

where

$$n = \sqrt{\frac{L_{22}}{L_{11}}} \text{ and } k = \frac{M}{\sqrt{L_{11}L_{22}}} \tag{A6.2}$$

$$\begin{bmatrix} V_1 \\ I_2 \end{bmatrix} = \begin{bmatrix} h_{11} & h_{12} \\ h_{21} & h_{22} \end{bmatrix}\begin{bmatrix} I_1 \\ V_2 \end{bmatrix} \text{ and } h_{11} = \frac{1}{y_{11}}; h_{12} = \frac{-y_{12}}{y_{11}}; h_{21} = \frac{y_{21}}{y_{11}}; h_{22} = \frac{\Delta_y}{y_{11}} \tag{A6.3}$$

SOLUTION

In this circuit with series–shunt feedback (voltage feedback) the transformer forms the feedback network and $\beta = h_{12f} = -k/n$ where k is the coupling coefficient and n is the turn ratio of the transformer. The exact solution is obtained using H-matrices and the transformation between Y- and H-matrices of a two-port

$$h_{11} = h_{11a} + h_{11f} = \frac{1}{y_{11a}} + j\omega\left[L_B + (1-k^2)L_{11}\right] \tag{A6.4}$$

$$h_{11} = R_b + r_E + \frac{1}{j\omega(C_\pi + C_\mu)} + j\omega\left[L_B + (1-k^2)L_{11}\right] \tag{A6.5}$$

$$h_{21} = h_{21a} + h_{21f} = j\frac{f_T}{f} + \frac{k}{n} \approx j\frac{f_T}{f}; \quad h_{12} = h_{12a} + h_{12f} = \frac{-y_{12a}}{y_{11a}} - \frac{k}{n} = \frac{1}{1+\frac{C_\pi}{C_\mu}} - \frac{k}{n} \tag{A6.6}$$

$$h_{22} = h_{22a} + h_{22f} = \frac{\Delta_y}{y_{11a}} + \frac{1}{j\omega L_{22}} = g_o + \frac{g_m}{1+\frac{C_P}{C_m}} + j\omega(C_m + C_{cs}) + \frac{1}{jwL_{22}} \ll \frac{1}{R_P}. \tag{A6.7}$$

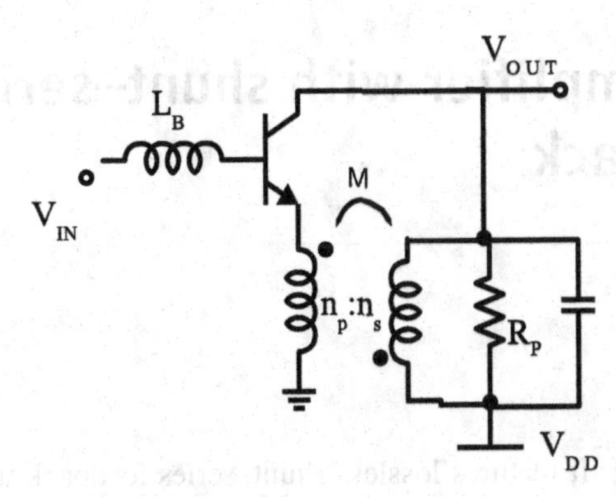

Figure A6.1 Series–shunt transformer feedback amplifier

We notice that the amplifier can be neutralized if $n/k - 1 = C_\pi/C_\mu \cong 10$. In general, this condition, which makes the amplifier with feedback unilateral, is easier to accomplish in MOSFETs than in HBTs because $C_\pi/C_\mu >> C_{gs}/C_{gd} = 2$.

If the imaginary parts of the input and output impedances are tuned out, then

$$Z_{IN} = h_{11} - \frac{h_{12}h_{21}}{h_{22}} \approx R_b + r_E + \frac{1}{j\omega(C_\pi + C_\mu)} + j\omega\left[L_B + (1-k^2)L_{11}\right] + j\frac{k}{n}\frac{f_T}{f}R_P \quad \text{(A6.8)}$$

If we assume an ideal transformer with infinite Q, $v_{nf} = i_{nf} = 0$; $R'_{uf} = 0$; $G'_{nf} = 0$; $Z'_{corf} = 0$ and Real $(h_{11f}) = 0$.

$$z_{sopt} = R_{sopta} + j\left[X_{sopta} - \omega(1-k^2)L_{11}\right]; \quad F_{MIN} = 1 + 2G'_{na}\left[R'_{cora} + R_{sopt}\right] = F_{MINa}. \quad \text{(A6.9)}$$

APPENDIX 7

Common-source amplifier with shunt–series transformer feedback

Consider the wideband LNA in Figure A7.1. It features lossless shunt–series feedback using a transformer. Knowing that the G-parameters of the transformer can be expressed as

$$g_{11f} = \frac{-j}{\omega L_P} + G_P; \quad g_{12f} = \frac{M}{L_P}; \quad g_{21f} = \frac{-M}{L_P}; \quad g_{22f} = j\omega L_S\left(1-k^2\right) + R_{SEC}$$

derive the expressions of the noise parameters of the amplifier with feedback. Assume $k = 0.6$, $L_S = 0.75\text{nH}$, $L_P = 3.2$ nH, $M = 0.93$, $n = 2.3$.

SOLUTION

The G-parameters of the transistor are obtained from the y-parameters of the transistor

$$g_{11a} = y_{11a} - \frac{y_{12a}y_{21a}}{y_{22a}} \approx j\omega\left(2C_{gd} + C_{gs}\right) \approx j\frac{f g_m}{f_{Ta}} \tag{A7.1}$$

$$g_{22a} = \frac{1}{y_{22a}} = \frac{\frac{g_o^2}{g_m} + G_L - j\omega\left(C_{gd} + C_{db}\right)}{\left(\frac{g_o^2}{g_m} + G_L\right)^2 + \omega^2\left(C_{gd} + C_{db}\right)^2} \tag{A7.2}$$

$$g_{12a} = \frac{y_{12a}}{y_{22a}} \approx 0 \tag{A7.3}$$

$$g_{21a} = \frac{-y_{21a}}{y_{22a}} \approx \frac{g_m G_L + g_o^2 - j\omega g_m\left(C_{gd} + C_{db}\right)}{\left(\frac{g_o^2}{g_m} + G_L\right)^2 + \omega^2\left(C_{gd} + C_{db}\right)^2} \tag{A7.4}$$

The g-parameters of the entire amplifier with feedback are obtained by adding the g-parameters of the amplifier and of the feedback networks

$$g_{11} = g_{11f} + g_{11a} \approx G_P - \frac{j}{\omega L_P} + j\frac{f g_m}{f_{Ta}} \tag{A7.5}$$

$$g_{22} = g_{22f} + g_{22a} = \frac{\frac{g_o^2}{g_m} + G_L - j\omega\left(C_{gd} + C_{db}\right)}{\left(\frac{g_o^2}{g_m} + G_L\right)^2 + \omega^2\left(C_{gd} + C_{db}\right)^2} + j\omega L_S\left(1-k^2\right) + R_{SEC} \tag{A7.6}$$

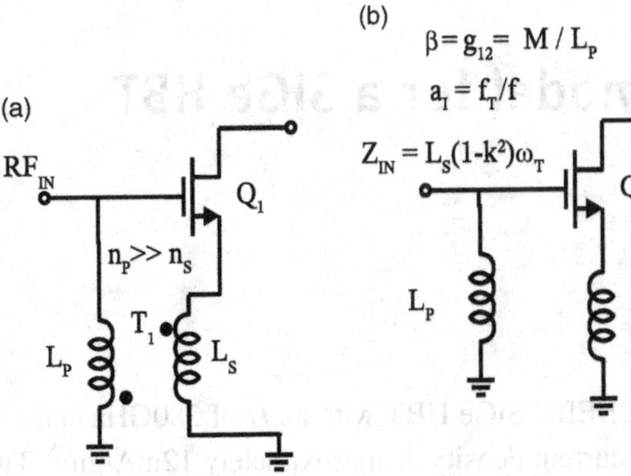

Figure A7.1 (a) CS MOSFET LNA with shunt-series feedback using transformer T_1 (b) Open loop amplifier with loading from the feedback network

$$g_{12} = g_{12f} + g_{12a} \approx \frac{M}{L_P} \tag{A7.7}$$

$$g_{21} = g_{21a} + g_{21f} \approx \frac{-M}{L_P} + \frac{g_m G_L + g_o^2 - j\omega g_m (C_{gd} + C_{db})}{\left(\frac{g_o^2}{g_m} + G_L\right)^2 + \omega^2 (C_{gd} + C_{db})^2} \approx \frac{g_m G_L + g_o^2 - j\omega g_m (C_{gd} + C_{db})}{\left(\frac{g_o^2}{g_m} + G_L\right)^2 + \omega^2 (C_{gd} + C_{db})^2} \tag{A7.8}$$

The noise sources at the input of the transformer-feedback network are:

$$v_{nf} = \frac{g_{12f}}{g_{11f} g_{22f} - g_{12f} g_{21f}}; \quad v_{n2f} \approx \frac{M}{L_s} v_{n2f}; \quad i_{nf} = i_{n1f}$$

$$\overline{v_{n1f}^2} = 4kT\Delta f \frac{M^2}{L_s^2} R_{SEC}; \quad \overline{i_{nf}^2} = 4kT\Delta f G_p$$

$$R_{nf} = \frac{M^2}{L_S^2} R_{SEC}; \quad G_{uf} = G_P; \quad Y_{corf} = 0 \text{ and Real } (g_{11f}) = G_P.$$

Note that, if the transformer is lossless, $G_P = 0$ and $R_{SEC} = 0$ and

$$R_n \approx R_{na}; \quad G_u = G_{ua}; \quad Y_{cor} = Y_{cora}$$

$$Y_{sopt} = G_{sopta} + j\left(B_{sopta} + \frac{1}{\omega L_P}\right)$$

$$F_{MIN} = 1 + 2R_{na}\left[G_{cora} + G_{sopta}\right] = F_{MINa}$$

$$Y_{IN} = g_{11} - \frac{g_{12} g_{21}}{g_{22}} \approx \frac{-j}{\omega L_P} + j\frac{f g_m}{f_{Ta}} + g_m \frac{M}{L_P} + G_P. \tag{A7.9}$$

The feedback is purely reactive and therefore it does not degrade the noise figure. Unfortunately in this case, it also does not change the real part of the optimum noise impedance from that of the transistor alone. As a result, the optimal transistor size and bias current for noise matching are still as large as in the case without feedback.

The output conductance g_o includes the load impedance of the next stage, typically a CG transistor and can therefore be large (as in the g_m of next stage.)

APPENDIX 8
HiCUM level 0 model for a SiGe HBT

This model is for a 120nm × 4.5μm CBEBC SiGe HBT with an f_T of 280GHz and an f_{MAX} of 380GHz. BV_{CEO} = 1.6V. The peak f_T current density is approximately 12mA/μm^2. The model includes self-heating and avalanche breakdown.

LEVEL = 13.00 (for HSPICE)

FLSH = 1.000

CJE0 = 12.86f AJE = 1.100 VDE = 898.1m ZE = 169.9m

IBES = 3.490E-20 IRES = 4.700f IBCS = 0.000 MBE = 1.018 MRE = 2.500 MBC = 1.000

IS = 40.20a MCF = 1.007 VEF = 1.000E + 20

IQF = 6.500m IQR = 1.000E + 20 IQFH = 3.400m TFH = 10.00u

RBI0 = 6.000 VR0C = 1.000E + 20 FGEO = 656.0m

T0 = 215.0f DT0H = 15.00a TBVL = 14.10a TEF0 = 175.0f GTE = 800.0m THCS = 2.900p AHC = 7.000m

TR = 0.000 RCI0 = 15.50 VLIM = 250.0m VPT = 2.500 VCES = 100.0m

CJCI0 = 7.790f VDCI = 700.1m ZCI = 200.0m VPTCI = 100.0

CJCX0 = 0.000 VDCX = 700.0m ZCX = 190.0m VPTCX = 1.000E + 20 FBC = 750.0m

RBX = 8.500 RCX = 11.50 RE = 4.000

VDS = 1.000 ZS = 10.00m VPTS = 100.0 CJS0 = 5.350f

ZETACT = 3.668 ZETABET = 3.350

RTH = 2.777K CTH = 6.979n

VGB = 990.0m VGE = 1.118 VGC = 1.118 VGS = 1.115

F1VG = −102.4u F2VG = 432.1u ALT0 = 2.260m

ZETACI = 418.2m ZETARCX = 458.6m ZETARBX = 420.0m ZETARBI = 458.9m KT0 = 9.872u ZETARE = −2.414

ALVS = 0.000 ALCES = 0.000 ALKAV = −35.46m ALEAV = −1.159m

EAVL = 18.57 KAVL = 7.412

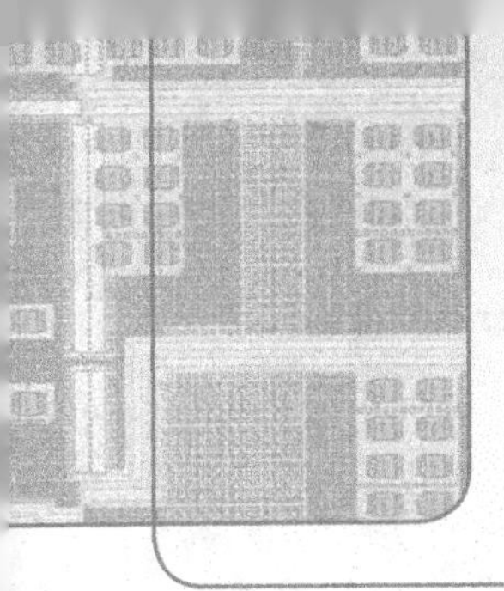

APPENDIX 9

Technology parameters

150NM GaN HEMT PARAMETERS

VT = −6V; g'_{mn} = 0.45mS/μm; the peak f_T of 70GHz occurs at 0.3mA/μm, the peak f_{MAX} is 150GHz, V_{MAX} = 40V, J_{sat} = 1mA/μm, C'_{gs} = 0.9fF/μm; C'_{gd} = 0.1fF/μm; C'_{ds} = 0.15fF/μm;

SiGe HBT PARAMETERS ARE PROVIDED PER EMITTER LENGTH L_E OR PER EMITTER AREA $W_E \times L_E$

Ic@peak f_T = 14mA/μm^2; w_E = 0.13μm; β = 500; V_{CESAT} = 0.3V, V_{BE} = 0.9V, peak f_T = 230GHz.
tau = 0.3ps; C_{je} = 17fF/μm^2; C_μ = 15fF/μm^2; C_{cs} = 1.2fF/μm;
R'_b = 75Ω × μm, R'_E = 2Ω × μm^2, R'_c = 20Ω × μm

180NM THICK OXIDE N-(P)MOSFET PARAMETERS

|VT| = 0.5V; $g'_{mn} = 2g'_{mp}$ = 0.5mS/μm; $g'_{dsn} = 2g'_{dsp}$ = 0.04mS/μm; the peak f_T of 60GHz occurs at 0.3mA/μm for 1V < V_{DS} < 1.8V.
For both n-channel and p-channel devices: C'_{gs} = 1fF/μm; C'_{gd} = 0.3fF/μm; $C'_{sb} = C'_{db}$ = 0.6fF/μm;
R'_s (n-MOS) = R'_d = 200Ω × μm; R'_s (p-MOS) = R'_d = 500Ω × μm;
R_{shg}(n/p-MOS) = 5Ω/sq; poly contact resistance R_{CON} = 5Ω. Total R_g/2μm finger contacted on one side = 50Ω.

130NM N-(P)MOSFET PARAMETERS

|VT| = 0.4V; Idsn at minimum noise = 0.15mA/μm; $g'_{mn} = 2g'_{mp}$ = 0.7, mS/μm; $g'_{dsn} = 2g'_{dsp}$ = 0.07mS/μm; the peak f_T of 80GHz occurs at 0.3mA/μm for 0.4V < V_{DS} < 1.2V.
For both n-channel and p-channel devices: C'_{gs} = 1.0fF/μm; C'_{gd} = 0.4fF/μm; $C'_{sb} = C'_{db}$ = 1.1fF/μm;
R'_s (n-MOS) = R'_d = 200Ω × μm; R'_s (p-MOS) = R'_d = 400Ω × μm;

R_{shg}(n/p-MOS) = 8Ω/sq; poly contact resistance R_{CON} = 10Ω. Total R_g/2μm finger contacted on one side = 70Ω. Noise params: P = 1, C = 0.4, R = 0.25.

65NM N-(P)MOSFET PARAMETERS

L = 45nm, |VT| = 0.3V; Idsn at minimum noise = 0.15mA/μm; $g'_{mn} = 2g'_{mp}$ = 1.3mS/μm; $g'_{dsn} = 2g'_{dsp}$ = 0.18mS/μm; the peak f_T of 170GHz occurs at 0.3mA/μm for 0.3V < V_{DS} < 1V.
For both n-channel and p-channel devices: C'_{gs} = 0.7fF/μm; C'_{gd} = 0.4fF/μm; $C'_{sb} = C'_{db}$ = 0.7fF/μm;
R'_s (n-MOS) = R'_d = 200Ω × μm; R'_s (p-MOS) = R'_d = 500Ω × μm;
R_{shg}(n/p-MOS) = 15Ω/sq; poly contact resistance R_{CON} = 40Ω.
Total R_g/0.7μm finger contacted on one side is 158Ω. Noise params: P = 1, C = 0.4, R = 0.25

45NM N-(P)MOSFET PARAMETERS

L = 30nm, |VT| = 0.3V; Idsn at minimum noise = 0.15mA/μm; $g'_{mn} = 2g'_{mp}$ = 1.8mS/μm; $g'_{dsn} = 2g'_{dsp}$ = 0.18mS/μm; the peak f_T of 270GHz occurs at 0.3mA/μm for 0.3V < V_{DS} < 1V.
For both n-channel and p-channel devices: C'_{gs} = 0.7fF/μm; C'_{gd} = 0.3fF/μm; $C'_{sb} = C'_{db}$ = 0.6fF/μm;
R'_s (n-MOS) = R'_d = 200Ω × μm; R'_s (p-MOS) = R'_d = 400Ω × μm;
R_{shg}(n/p-MOS) = 20Ω/sq; poly contact resistance R_{CON} = 70Ω.
Total R_g/0.7μm finger contacted on one side = 200Ω. Noise params: P = 1, C = 0.4, R = 0.25

32NM N-(P)MOSFET PARAMETERS

L = 22nm, |VT| = 0.25V; Idsn at minimum noise = 0.2mA/μm; $g'_{mn} = g'_{mp}$ = 2.6mS/μm; $g'_{dsn} = g'_{dsp}$ = 0.4mS/μm; the peak f_T of 350GHz occurs at 0.4mA/μm for 0.3V < V_{DS} < 0.9V.
For both n-channel and p-channel devices: C'_{gs} = 0.55fF/μm; C'_{gd} = 0.3fF/μm; $C'_{sb} = C'_{db}$ = 0.55fF/μm;
R'_s (n-MOS) = R'_d = 200Ω × μm; R'_s (p-MOS) = R'_d = 200Ω × μm;
R_{shg}(n/p-MOS) = 15Ω/sq; gate contact resistance R_{CON} = 110Ω.
Total R_g/0.5μm finger contacted on one side is 200Ω. Noise params: P = 1, C = 0.4, R = 0.25.

<chip>		
	chip x =64	; dimensions of the chip in um in x direction
	chip y =64	; dimensions of the chip in um in y direction
	fft x = 256	; x-fft size (must be a power of 2)
	fft y = 256	; y-fft size
	TechFile = 65nm_typ.tek	; the name of this file
	TechPath =.	; the pathname of the data files
<layer>	0	; Bulk Substrate
	rho = 10	; Resistivity: ohm-cm
	t = 315.5	; Substrate Thickness: microns
	eps = 11.7	; Permitivity: relative
<layer>	1	; Oxide Layer (poly layer, eps = avg(4.1,3.9))
	rho = 1e7	; Resistivity: ohm-cm, insulator
	t = 0.61	; thickness (microns)
	eps = 4	; Permitivity: relative
<layer>	2	; Low-k Oxide Layer (M1 to M5 layers)
	rho = 1e7	; Resistivity: ohm-cm, insulator
	t = 1.86	; thickness (microns)
	eps = 3.0	; Permitivity: relative
<layer>	3	; Oxide Layer (M6 and M7 layers)
	rho = 1e7	; Resistivity: ohm-cm, insulator
	t = 3.47	; thickness (microns)
	eps = 4.15	; Permitivity: relative
<layer>	4	; Passivation Layer (Simplified)
	rho = 1e7	; Resistivity: ohm-cm, insulator
	t = 1.9	; thickness (microns)
	eps = 4.9	; Permitivity: relative (weighted average)
<metal> 0		
	layer = 1	; substrate contact layer
	rsh = 16500	; Sheet Resistance Milli-Ohms/Square
	t = 0.105	; Poly Thickness (microns)
	d = 0.275	; Distance from bottom of layer (microns)
	name = poly1	; name used in ASITIC
	color = red	; color in ASITIC
<via>	0	; contact (poly to M1)
	top = 1	; via connects up to this metal layer

	bottom = 0	; via connects down to this metal layer
	r = 30	; resistance per via
	width = 0.09	; width of via
	space = 0.11	; minimum spacing between vias
	overplot1 =.085	; minimum dist to substrate metal
	overplot2 =.18	; minimum dist to metal 1
	name = polyc	; name in ASITIC
	color = white	; color in ASITIC
<metal> 1		
	layer = 2	
	rsh = 78	
	t = 0.22	
	d = 0	
	name = m1	
	color = cadetblue	
<via> 1		; metal 1 to metal 2
	top = 2	
	bottom = 1	
	r = 5	
	width = 0.10	
	space = 0.10	
	overplot1 =.04	
	overplot2 =.04	
	name = via12	
	color = cadetblue	
<metal> 2		
	layer = 2	
	rsh = 78	
	t = 0.22	
	d = 0.41	
	name = m2	
	color = yellow	
<via>	2	; metal 2 to metal 3
	top = 3	
	bottom = 2	
	r = 5	

	width = 0.10	
	space = 0.10	
	overplot1 =.04	
	overplot2 =.04	
	name = via23	
	color = yellow	
<metal> 3		
	layer = 2	
	rsh = 78	
	$t = 0.22$	
	$d = 0.82$	
	name = m3	
	color = brown	
<via>	3	; metal 3 to metal 4
	top = 4	
	bottom = 3	
	$r = 5$	
	width = 0.10	
	space = 0.10	
	overplot1 =.04	
	overplot2 =.04	
	name = via34	
	color = brown	
<metal> 4		
	layer = 2	
	rsh = 78	
	$t = 0.22$	
	$d = 1.23$	
	name = m4	
	color = green	
<via>	4	; metal 4 to metal 5
	top = 5	
	bottom = 4	
	$r = 5$	
	width = 0.1	
	space = 0.1	

	overplot1 =.04	
	overplot2 =.04	
	name = via45	
	color = green	
<metal> 5		
	layer = 2	
	rsh = 78	
	$t = 0.22$	
	$d = 1.64$	
	name = m5	
	color = cyan	
<via>	5	; metal 5 to metal 6
	top = 6	
	bottom = 5	
	$r = 0.6$	
	width = 0.4	
	space = 0.4	
	overplot1 =.1	
	overplot2 =.1	
	name = via56	
	color = cyan	
<metal> 6		
	layer = 3	
	rsh = 19.1	
	$t = 0.90$	
	$d = 0.6$	
	name = m6	
	color = red	
<via>	6	; metal 6 to metal 7
	top = 7	
	bottom = 6	
	$r = 1.5$	
	width = 0.5	
	space = 1.1	
	overplot1 =.1	
	overplot2 =.25	

	name = via67
	color = red
<metal> 7	
	layer = 3
	rsh = 23
	$t = 1.17$
	$d = 2.3$
	name = m7
	color = white

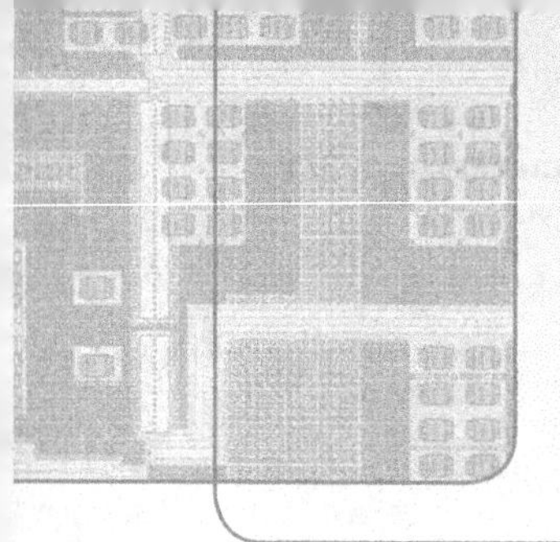

APPENDIX 10

Analytical study of oscillator phase noise

A10.1 A STUDY OF PHASE NOISE

In this section we look into how phase noise arises from noise currents in an oscillator, using an analytic power series model. An analysis is conducted based on this power series model, and this is used to predict the phase noise of an oscillator. This circuit is then made to oscillate with transient simulation and is studied for its phase noise performance with the harmonic balance method using a proprietary simulator (ADS). The results are then compared. The basis of this study will be a half circuit test bench of a Colpitts oscillator as in Figure 10.14(a) and Figure 10.41 of Chapter 10, but with a HBT instead of a MOSFET.

The topic of phase noise was introduced in Section 10.1.4 of Chapter 10. Equation (10.15) simply assumed a phase noise existing at a frequency offset from the fundamental oscillation frequency, but does not explain how this phase noise arises from real physical noise current sources (e.g. resistors, lossy inductors, transistor shot noise, etc.) present inside the oscillator circuit. The latter is studied in more detail in this Appendix using a one-port equivalent circuit for the oscillator, as shown in Figure A10.1.

The inductor, L_{pt}, is assumed in parallel with a resistor R_{pt}, at the base input of the transistor. (An equivalent series L-R representation is also possible.) One end of this inductor is placed at the desired DC base bias voltage of the transistor. The value of R_{pt} is calculated based on the assumed Q of an actual linear inductor, at the oscillation frequency. The resistor R_{pt} is shown inside the "one-port" as illustrated in Figure A10.1. When this assembly is in steady-state oscillation, the input to the one-port must by definition be purely capacitive because the net negative resistance created inside the oscillator must balance any sources of positive resistance. Note that in the Colpitts oscillator the transistor inside the one-port does not have its emitter connected to ground, so the subsequent analysis of the one-port is of the prototype structure of this oscillator, not just of the proprietary device models included in it.

A10.2 USE OF AUTOMATIC OPTIMIZATION IN THE DESIGN OF THE OSCILLATOR

In the oscillator of Figure A10.1 the circuit parameters to be chosen are L_{pt}, R_{pt} (obtained from the specified Q of the inductor), C_1, C_2, the current I_{BIAS}, the transistor model and size, and the load. The load and any following circuitry connected to it are part of the one-port. The desired

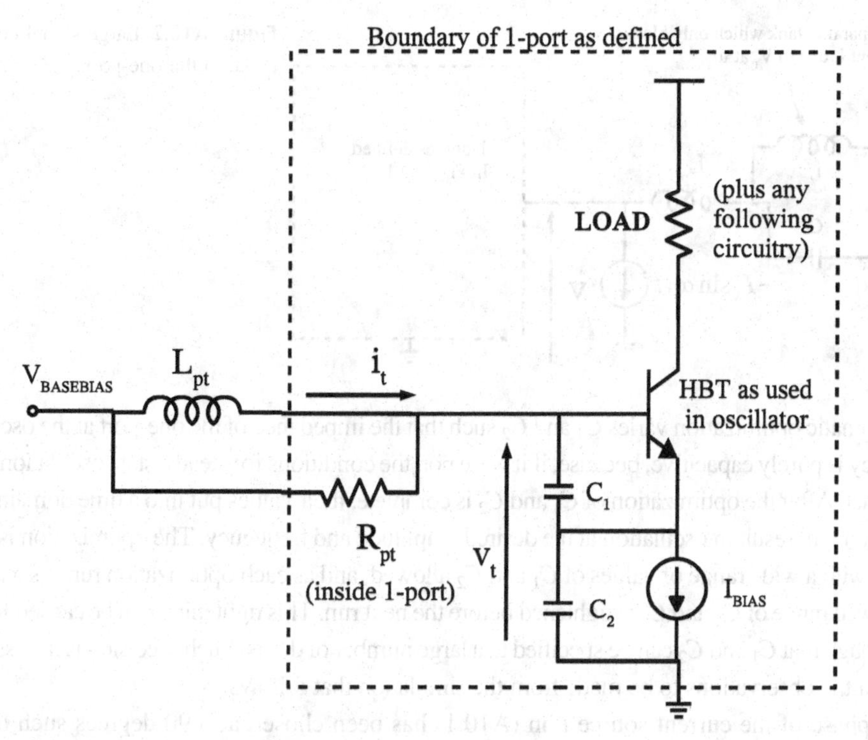

Figure A10.1 One-port equivalent of a Colpitts oscillator

amplitude and frequency are specified at the beginning of the optimization. In contrast, in normal oscillator design by simulation, circuit non-linearities make the achieved amplitude and frequency indefinite, requiring much iteration. However in this Appendix we use automatic optimization in the simulator to obtain the choice of the circuit parameters such that, when simulated, the circuit will achieve accurately the desired amplitude and frequency.

The circuit being optimized is the one-port of Figure A10.1, without the inductor L_{pt}. The aim is to duplicate the conditions that exist in the actual oscillator. From the specified value of L_{pt} and the desired fundamental voltage amplitude, the desired amplitude I_1 of the fundamental current driving the one-port is calculated. The latter is given by

$$i_t = -I_1 \sin(\omega_1 t) \tag{A10.1}$$

and will create a large signal voltage at the fundamental oscillation frequency and various harmonics across the one-port. For these harmonics to see the same impedance as in the real oscillator, L_{pt} must also be seen by them. Yet the current i_t must not be allowed to flow through L_{pt}. This scenario can be realized by adding a parallel LC tank trap at the fundamental frequency in series with L_{pt}, as shown in Figure A10.2, which will block the flow of the fundamental current component through L_{pt}, but which will act as a short circuit at all harmonics of the fundamental. This allows the correct harmonic voltages to develop across L_{pt}, as would be in the case of the real oscillator.

In the one-port optimization method the choices of transistor size and bias current can be so as to satisfy the considerations presented earlier for minimum noise, or whatever the design goal may be.

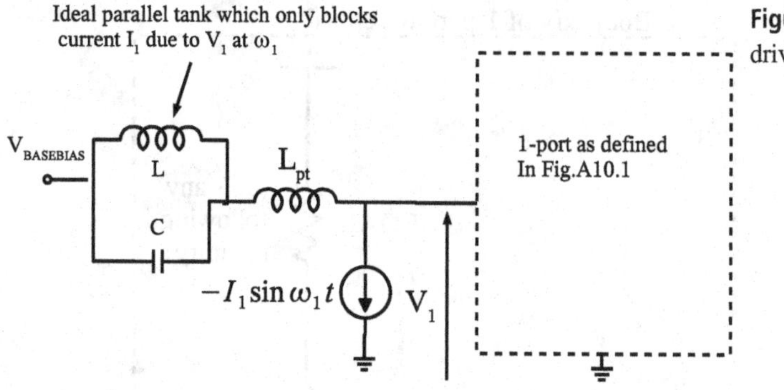

Figure A10.2 Large signal current drive to the one-port

The automatic optimization varies C_1 and C_2 such that the impedance of the one-port at the oscillation frequency is purely capacitive, because, if it were not, the conditions for steady-state oscillation would not be met. After the optimization of C_1 and C_2 is complete, their values put into a time domain or *HB* simulation will result in oscillation at the desired amplitude and frequency. The optimization is begun initially with a wide range of values of C_1 and C_2 allowed, and as each optimization run is successful, the allowed range of C_1 and C_2 is tightened before the next run. This tightening can be carried forward long enough that C_1 and C_2 can be specified to a large number of digits. Such precision is necessary for clarity of the observations to be made from the simulation that follows.

The phase of the current source i_t in (A10.1) has been chosen at +90 degrees such that the large signal fundamental of voltage is precisely along the positive real axis, given that the input of the one-port was constrained to be purely capacitive in the *HB* optimization. This makes the one-port impedance essentially flat near the oscillation frequency. The known operating voltage and frequency of the one-port now allows to study the harmonic behavior of the oscillator, including phase noise, in a way that is not practical in a free-running oscillator simulation.

Because the values of capacitors C_1 and C_2 can be tightened as much as desired during the optimization, these values, when put into normal transient simulation, will result in precisely the desired oscillation frequency and amplitude. As well, the one-port can be used in simulation to predict the behavior as regards phase noise. This can be compared to normal phase noise as predicted in *HB*-based oscillator noise simulation, and certain interesting conclusions can be reached.

The harmonic balance method exercises non-linear components in the time domain and then moves to a truncated Fourier series to express the time waveform in the frequency domain. As a result, it is well suited to dealing with harmonic effects resulting from periodic excitation. In the section that follows, the results of *HB* simulation will be compared to a simple power series model for the internals of the one-port.

A10.3 LARGE SIGNAL BEHAVIOR OF THE ONE-PORT

What is going on inside the detailed transistor models and other elements inside an actual oscillator simulation will be very complex and the details of the algorithm used in the *HB* simulation are not easily available. In this analysis therefore, the behavior of the one-port, i.e. the tank voltage v_t, will be encapsulated analytically by a power series, truncated here at the seventh-order term. This should be a high enough order that the result of the *HB* simulation of the one-port, as will be done here, will look identical to that in the actual time domain or *HB* simulation of the oscillator.

The current i_t of (A10.1) is substituted in

$$v_t = a_1 i_t + a_2 i_t^2 + a_3 i_t^3 + a_4 i_t^4 + a_5 i_t^5 + a_6 i_t^6 + a_7 i_t^7 \tag{A10.2}$$

The coefficients a_n of the power series are complex, allowing for the phase shifts between voltage and current at the various harmonics. The zero-order term, a_0, is omitted[1]. The result after much trigonometric manipulation is

$$v_t = v_{t0} + v_{t1} + v_{t2} + v_{t3} + v_{t4} + v_{t5} + v_{t6} + v_{t7} \tag{A10.3}$$

where the various harmonics of v_t are found to be

$$v_{t0} = \left(\frac{1}{2}a_2 I_1^2 + \frac{3}{8}a_4 I_1^4 + \frac{5}{16}a_6 I_1^6\right)$$

$$v_{t1} = -\left(a_1 I_1 + \frac{3}{4}a_3 I_1^3 + \frac{5}{8}a_5 I_1^5 + \frac{35}{64}a_7 I_1^7\right) \cdot \sin(\omega_1 t)$$

$$v_{t2} = -\left(\frac{1}{2}a_2 I_1^2 + \frac{1}{2}a_4 I_1^4 + \frac{15}{32}a_6 I_1^6\right) \cdot \cos(2\omega_1 t)$$

$$v_{t3} = \left(\frac{1}{4}a_3 I_1^3 + \frac{5}{16}a_5 I_1^5 + \frac{21}{64}a_7 I_7^7\right) \cdot \sin(3\omega_1 t)$$

$$v_{t4} = \left(\frac{1}{8}a_4 I_1^4 + \frac{3}{16}a_6 I_1^6\right) \cdot \cos(4\omega_1 t)$$

$$v_{t5} = -\left(\frac{1}{16}a_5 I_1^5 + \frac{7}{64}a_7 I_1^7\right) \cdot \sin(5\omega_1 t)$$

$$v_{t6} = -\left(\frac{1}{32}a_6 I_1^6\right) \cdot \cos(6\omega_1 t)$$

$$v_{t7} = \left(\frac{1}{64}a_7 I_1^7\right) \cdot \sin(7\omega_1 t).$$

Because the fundamental of the large signal current, (A10.1), is defined on the positive imaginary axis, the pure capacitance of the one-port means that the fundamental large signal voltage, v_{t1} of (A10.3), will be at a phase angle of exactly 0 degrees, i.e. along the positive real axis (after the optimization, of course). This does not mean that the phase angle of a_1 is at exactly −90 degrees because in (A10.3) the higher-order odd numbered power series coefficients are also seen to take part in the fundamental voltage. The complex coefficients in a_2 and higher will also have phase shifts. These will show the voltage harmonics to be delayed at higher frequencies, for example due to RC low-pass action at the base input of the transistor. For low-frequency oscillators, these phase shifts will reduce to a minimum.

A harmonic balance simulation of the one-port was done, limited to order 7. This gave results only to the seventh harmonic, as is consistent with (A10.3). Note that the odd-order coefficients of a_n give only the fundamental and odd-order harmonics of v_t, while the even-order a_n give only even harmonics. Since v_{t6} and v_{t7} contain only a_6 and a_7, the values of a_6 and a_7 can be solved for by directly post-processing the sixth and seventh harmonic results of the one-port *HB* simulation. Then a_6 and a_7 can similarly be used with the v_{t4} and v_{t5} obtained from the *HB* simulation to calculate a_4 and a_5. Working similarly with lower order v_{tn} will finally give all the a_n. The final group of a_n coefficients is particular to that design of the one-port.

The large signal impedance at the fundamental frequency can be obtained from the v_{t1} term of (A10.3) as

$$Z_1 \stackrel{\text{def}}{=} \frac{v_{t1}}{i_t} = \left(a_1 + \frac{3}{4}a_3 I_1^2 + \frac{5}{8}a_5 I_1^4 + \frac{35}{64}a_7 I_1^6\right) \quad \text{(A10.4)}$$

Z_1 will be purely capacitive and of constant value for a specified oscillator inductor L_{pt} and a specified fundamental amplitude. However, depending on frequency, on the DC bias current, and on the transistor size and model chosen, the harmonics of the fundamental change. This shows up as a change in a_n, even though Z_1 is a constant. This behavior would be of interest in phase noise calculations where mixing of noise from harmonics of the fundamental is considered.

A10.4 SMALL SIGNAL BEHAVIOR OF THE ONE-PORT

For the study of small signal behavior, the parallel tank trap of Figure A10.2 was modified to also allow the proper currents and voltages for small offset frequencies very near the fundamental, by making C very large and L very small. A range of one-port prototypes for the Colpitts oscillator of varying current density was explored in the *HB* simulator with a fixed L_{pt}, fixed fundamental tank voltage amplitude, current, and frequency. That is, Z_1 in (A10.4) was fixed. When small signal noise currents were applied to the one-port *HB* simulation at a small positive offset frequency in addition to the large signal fundamental (+1 MHz compared to 40 GHz), an upper frequency sideband of noise voltage at 40.001 GHz was observed, as shown in Figure A10.3(a). In addition, a lower-frequency noise voltage sideband was also observed at 39.999 GHz. The amplitude of the lower sideband noise voltage was typically 5–15% of the amplitude of the upper sideband noise voltage, depending on the transistor current density. The amplitude of both noise voltage sidebands was found to be a linear function of the small signal noise current amplitude. Similar small signal noise sidebands were observed at the voltage harmonics of the fundamental, again of unequal amplitudes for the upper versus the lower sideband. Other sidebands at multiples of the offset frequency were negligible. That is, mixing did not give rise to additional sidebands beyond the first. Since this is not consistent with simple mixing in the presence of a non-linearity, it was decided to investigate if the same effects arose in the one-port power series analysis.

In the power series model of (A10.2), assume two, statistically independent, small signal noise currents, i_{usn} and i_{lsn}, of equal amplitude, added at the upper sideband, $\omega_1 + \delta\omega$, and at the lower sideband, $\omega_1 - \delta\omega$, respectively, defined as

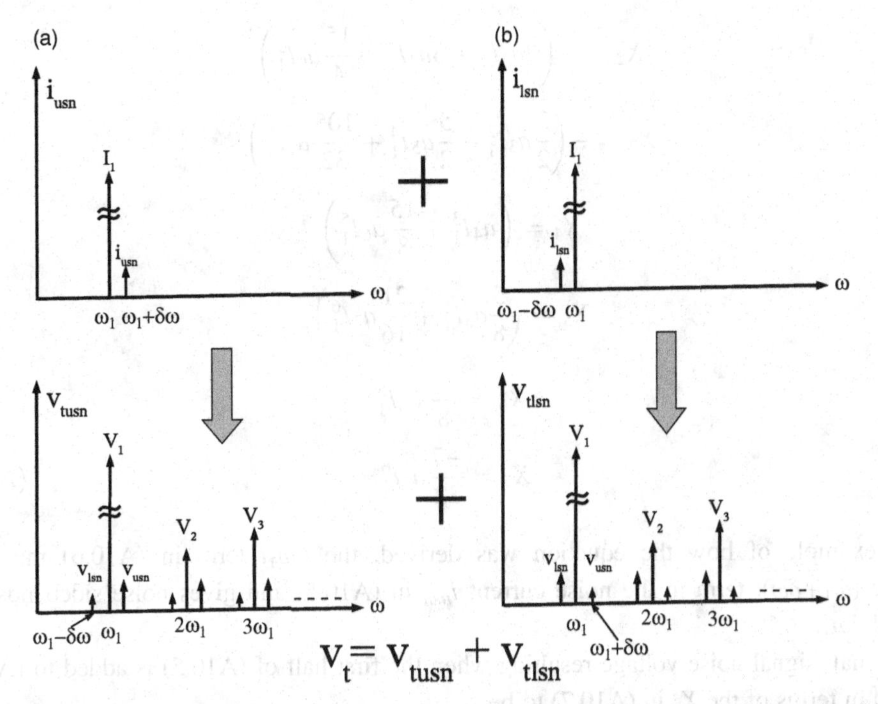

Figure A10.3 Illustration of the formation of asymmetrical voltage noise sidebands when (a) an upper sideband only or (b) lower sideband only, small signal noise current is added to the large signal current excitation at the fundamental

$$i_{usn} = I_n\cos[(\omega_1 + \delta\omega)t] \text{ and } i_{lsn} = I_n\cos[(\omega_1 - \delta\omega)t]. \tag{A10.5}$$

This is related to the simple assumption of a pure phase noise that was done in Section 10.1.4, but physical noise current sources present in the oscillator are now considered. The trigonometric expansion of (A10.2) was repeated with each of these noise currents separately added to the large signal current of (A10.1). All terms of higher order than one in the small signal noise currents were discarded as negligible. The large signal voltage part of the solution was the same as before, in (A10.3).

(a) Upper sideband noise current

The small signal noise voltage due to the upper sideband noise current i_{usn}, v_{tusn}, in Figure A10.3(a), was solved to be

$$\begin{aligned} v_{tusn} = [&X_1 + X_2\sin(\omega_1 t) + X_3\cos(2\omega_1 t) + X_4\sin(3\omega_1 t) + X_5\cos(4\omega_1 t) \\ &+ X_6\sin(5\omega_1 t) + X_7\cos(6\omega_1 t)]I_n\cos[(\omega_1 + \delta\omega)t] \end{aligned} \tag{A10.6}$$

Here the X_n are complex numbers as follows

$$X_1 = \left(a_1 + \frac{3}{2}a_3I_1^2 + \frac{15}{8}a_5I_1^4 + \frac{35}{16}a_7I_1^6\right)$$

$$X_2 = -\left(2a_2I_1 + 3a_4I_1^3 + \frac{15}{4}a_6I_1^5\right)$$

$$X_3 = -\left(\frac{3}{2}a_3I_1^2 + \frac{5}{2}a_5I_1^4 + \frac{105}{32}a_7I_1^6\right)$$

$$X_4 = \left(a_4I_1^3 + \frac{15}{8}a_6I_1^5\right)$$

$$X_5 = \left(\frac{5}{8}a_5I_1^4 + \frac{21}{16}a_7I_1^6\right)$$

$$X_6 = \frac{-3}{8}a_6I_1^5$$

$$X_7 = \frac{-7}{32}a_7I_1^6. \tag{A10.7}$$

As an example of how the equation was derived, the $6\omega_1 t$ term in (A10.6) multiplies with the $(\omega_1+\delta\omega)t$ term in the noise current i_{usn} in (A10.5) and gives noise sidebands near $5\omega_1$ and $7\omega_1$.

The small signal noise voltage resulting when the first half of (A10.5) is added to (A10.1) is found in terms of the X_n in (A10.7) to be

$$v_{tusn} = \frac{I_n}{2}X_2\sin[(-\delta\omega)t] + I_nX_1\cos[(\omega_1+\delta\omega)t] + \frac{I_n}{2}X_3\cos[(\omega_1-\delta\omega)t]$$

$$+\frac{I_n}{2}X_2\sin[(2\omega_1+\delta\omega)t] + \frac{I_n}{2}X_4\sin[(2\omega_1-\delta\omega)t] + \frac{I_n}{2}X_3\cos[(3\omega_1+\delta\omega)t] + \frac{I_n}{2}X_5\cos[(3\omega_1-\delta\omega)t]$$

$$+\frac{I_n}{2}X_4\sin[(4\omega_1+\delta\omega)t] + \frac{I_n}{2}X_6\sin[(4\omega_1-\delta\omega)t] + \frac{I_n}{2}X_5\cos[(5\omega_1+\delta\omega)t] + \frac{I_n}{2}X_7\cos[(5\omega_1-\delta\omega)t]$$

$$+\frac{I_n}{2}X_6\sin[(6\omega_1+\delta\omega)t] + \frac{I_n}{2}X_7\cos[(7\omega_1+\delta\omega)t]. \tag{A10.8}$$

Equation (A10.8) predicts that a lower sideband voltage occurs along with the upper sideband one when an upper sideband noise current is applied, as illustrated in Figure A10.3(a)[2]. The coefficients X_n decrease in magnitude as n increases, so the lower sidebands are smaller in amplitude than the upper sidebands. For example, the upper sideband at the third harmonic of the fundamental depends on X_3, while the lower sideband depends on X_5. It is noteworthy that only the X_n with n odd react back to near the fundamental and odd harmonics, while the X_n with n even react back to near to DC and to even harmonics. Note that adding the small signal noise current did not give rise to additional sidebands beyond the first. This was verified in simulation.

A harmonic balance simulation for the one-port oscillator prototypes was carried out with an upper sideband noise current only, which gave numerical values for the fundamental component of the voltage and its harmonics up to the seventh. These results were fully in accordance with (A10.8).

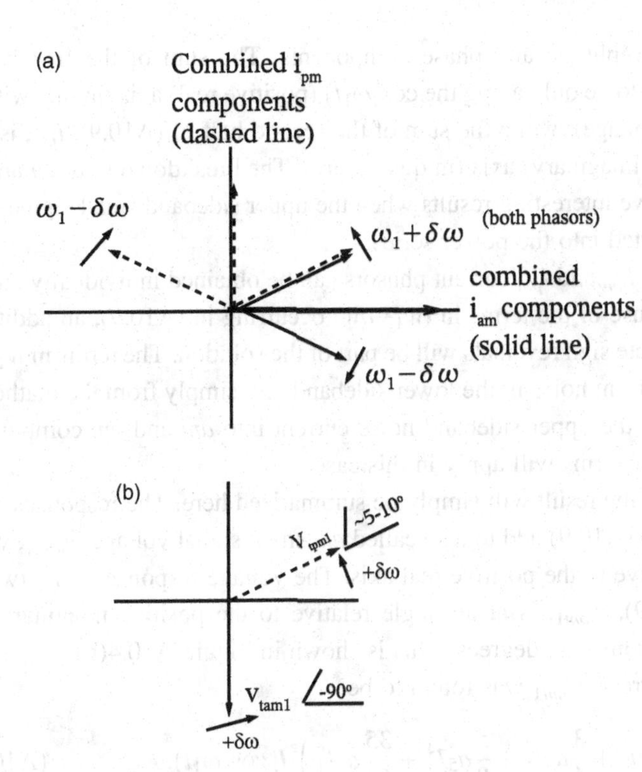

Figure A10.4 (a) Small signal current at $\omega_1 + \delta\omega$ broken down into four components. (b) Voltage response to i_{am} components (solid line) and i_{pm} components (dashed line). *Note*: Frame of reference is rotating counterclockwise at ω_1 and components of small signal current are rotating relative to it

(b) Lower sideband noise current

The exploration was done as in Figure A10.4(a), but this time with just a lower sideband noise current applied. Again, asymmetric noise voltage sidebands were observed, as seen in Figure A10.3(b), with the lower sideband noise voltages being the larger ones. The result of the power series analysis is identical except that in (A10.8) $\delta\omega$ is replaced by $-\delta\omega$. Again, *HB* simulation verified this case.

Therefore, as illustrated in Figure A10.3, if both upper and lower sideband noise currents which are uncorrelated with each other exist in an oscillator, the noise voltage at the upper sideband will be obtained by the *rms* addition of the larger noise voltage caused by the upper sideband noise current, and the smaller noise voltage caused by the uncorrelated noise current in the lower sideband, and vice-versa.

The behavior of the one-port in Figure A10.3(a) will now be explained in terms of the different response of the power series to small signal currents that are phase aligned with the fundamental large signal voltage versus the response to small signal currents whose phase is in quadrature with the fundamental component of the voltage.

Using simple trigonometric manipulation, the small signal upper sideband noise current at frequency offset $+\,\delta\omega$ in (A10.5) can be recast as

$$i_{usn} = \frac{I_n}{2}[(\cos(\omega_1+\delta\omega)t + \cos(\omega_1-\delta\omega)t)] + \frac{I_n}{2}[(\cos(\omega_1+\delta\omega)t - \cos(\omega_1-\delta\omega)t)] = i_{am} + i_{pm}. \tag{A10.9}$$

This splits the current into amplitude and phase components. The sum of the first half of (A10.9),defined as i_{am}, is seen to be only along the $\cos(\omega_1 t)$ (positive real) axis (in line with the large signal fundamental of voltage), while the sum of the second half of (A10.9), i_{pm}, is only along the $-\sin(\omega_1 t)$ (positive imaginary) axis (in quadrature). The breakdown into *am* and *pm* components will be seen to give interesting results when the upper sideband small signal noise current i_{usn} in (A10.9) is inserted into the power series.

The voltage response to the i_{am} and i_{pm} current phasors can be obtained individually and the results added. Note that, because of the terms in (w_1-dw) occurring in (A10.9), an additional form of (A10.8) with appropriate sign reversals will be part of the solution. The terms in w_1-dw in (A10.9) are not from the current noise in the lower sideband, but simply from the mathematical artifact of breaking down the upper sideband noise current into *am* and *pm* components. The sign reversals for the $-dw$ terms will apply in this case.

The solution is lengthy and the result will simply be summarized here. The responses to the two *am* current components in (A10.9) add to a so called *am* small signal voltage, v_{tam1}, which is at a -90 degree angle relative to the positive real axis. The voltage response to the two *pm* current components in (A10.9), v_{tpm1}, is at an angle relative to the positive imaginary axis which, however, is different from -90 degrees. This is shown in Figure A10.4(b).

For the *am* small signal current, v_{tam1} was found to be

$$v_{tam1} = \left(a_1 + \frac{3}{4}a_3 I_1^2 + \frac{5}{8}a_5 I_1^4 + \frac{35}{64}a_7 I_1^6\right) \cdot I_n \cos(\omega_1 t). \qquad \text{(A10.10)}$$

In terms of the power series coefficients this is identical to the response to the large signal fundamental current in (A10.4) except that the small signal current cause is along the positive real axis while the small signal voltage is along the negative imaginary axis.

The resulting *am* small signal impedance becomes

$$Z_{am1} = \left(a_1 + \frac{3}{4}a_3 I_1^2 + \frac{5}{8}a_5 I_1^4 + \frac{35}{64}a_7 I_1^6\right). \qquad \text{(A10.11)}$$

Note that the *am* subscript for Z_{am1} is in terms of the angle of the *am* current cause. The angle of the resulting voltage is at -90 degrees to this.

To test the validity of (A10.11) a large signal current was applied in the *HB* simulator along with a small signal current in phase with it, i.e. *am* in the definition of (A10.9). The small signal a_n coefficients in (A10.10) were obtained in the *HB* simulator by post processing the small signal voltage harmonics that resulted. The small signal a_n coefficients were exactly the same as those obtained from the *HB* post processing of the large signal harmonics. At first glance this is surprising. An intuitive explanation is that the small signal current i_{am} in (A10.9) along the real axis does not seek to change the amplitude of the voltage v_t which is on the imaginary axis and hence does not exercise the amplitude across any non-linearity, only the phase. When a pure phase change of the fundamental of v_t occurs, the phasor joining the old v_t will be governed by the same numerical behavior as that for the large signal of v_{t1} itself. Thus the power series coefficients for the small signal *am* current are identical to those for the large signal fundamental of current.

A quite different behavior is exhibited for the *pm* small signal currents defined in (A10.9). The small signal fundamental frequency solution from the substitution of these currents in the power series is

$$v_{tpm1} = -\left(a_1 + \frac{9}{4}a_3 I_1^2 + \frac{25}{8}a_5 I_1^4 + \frac{245}{64}a_7 I_1^6\right)\cdot I_n\cdot \sin(\omega_1 t) \qquad \text{(A10.12)}$$

and the small signal fundamental frequency impedance to *pm* currents is

$$Z_{pm1} = \left(a_1 + \frac{9}{4}a_3 I_1^2 + \frac{25}{8}a_5 I_1^4 + \frac{245}{64}a_7 I_1^6\right). \qquad \text{(A10.13)}$$

Note that the value of the a_n coefficients seen in (A10.12) by *pm* small signal currents are different from the large signal a_n coefficients. This is because the small signal *am* current performs an incremental swing around the large signal current. Also, the fractions multiplying with the a_n coefficients in (A10.13) are different from those for Z_{am1}. This suggests that Z_{pm1} may not be purely capacitive as Z_{am1} is.

In postprocessor calculation from the results of *simulation* for small signal *pm* current, the small signal a_n coefficients were found to be different from those for i_{am}. This is not surprising since the non-linearities in the transistor model are exercised differently. These non-linearities are exercised by small signal *pm* currents rather than the large signals that set the coefficients of Z_1 in (A10.4). The result from HB simulation is that Z_{pm1} has a phase shift from the *pm* axis which depends on the transistor DC bias current and is typically around minus 80–85 degrees, so v_{t1} is at a small positive angle, of about plus 5–10 degrees. The main component of this small signal voltage is along the *am* axis, but the smaller component along the *pm* axis will therefore be part of the small signal voltage phase response and would affect the phase noise.

Z_{pm1} was found in *HB* simulation to be different in magnitude from Z_{am1}, the difference being smaller at low transistor current densities and increasing to up to 15% higher at large current densities. It might be wondered if the phase component of v_{tpm1} would affect the overall phase noise significantly. However, in the section on oscillator phase noise prediction it is discussed how the infinite-Q resonance of the pure capacitance of Z_{am1} with the tank inductor, L_{pt}, does not apply to Z_{pm1}. That means that Z_{pm1} does not form an ideal resonator with L_{pt}. This is because its shift in magnitude and angle removes the infinite-Q response to the *pm* component of the small signal voltage and shifts its frequency. The phase noise component due to v_{tpm1} would therefore not be significant.

Comparing these predictions from the analysis to the results obtained from *HB* simulation, Part A of Table A10.1 shows the large and small signal impedances for one-port prototypes with four different bias currents, I_{BIAS}. The simulated values of Z_{am1} and Z_1 are identical, which is consistent with the analysis. However, as was discussed in the paragraph following (A10.13), the simulated values of Z_{pm1} are different from those of Z_{am1} and Z_1.

Table A10.1 **Simulations and calculations for a range of prototypes of varying current density.**

The transistor DC current bias I_{BIAS} was varied as in the table for various prototypes, at the fundamental frequency of 40 GHz, voltage magnitude/angle = 1.25/0 volts, tank inductance = 101 pH, desired oscillator fundamental impedance magnitude/angle Z_1 = 25.4/−90 Ω.

Note: For simplicity, various values reported below are truncated in the number of digits, but the number actually used for automatic *HB* optimization was much more precise.

PART A: Large and small signal *HB* impedance simulations of the one-port at the fundamental frequency of oscillation for various transistor bias current, I_{BIAS}.

For transistor bias current	10 mA	12 mA	14 mA	16 mA
Large signal impedance of one-port (magnitude/angle) obtained using automatic optimization.	25.4/ −90.0°	25.4/ −90.0°	25.4/ −90.0°	25.4/ −90.0°
Small signal *am* impedance of one-port, Z_{am1}, (it was found in *HB* simulation to be identical to large signal impedance, Z_1)	25.4/ −90.0°	25.4/ −90.0°	25.4/ −90.0°	25.4/ −90.0°
Small signal *pm* impedance of one-port, Z_{pm1}, (as found from *HB* simulation)	25.7/ −82.1°	23.4/ −80.3°	28.6/ −85.2°	29.1/ −85.1°
Part B: Study of *HB* simulator using R_{pt} theoretical resistor noise only, which was calculated from (A10.14) as 0.8011 × 10^{-11} A/√Hz. Both upper and lower sideband noise is included.				
rms upper sideband voltage of one-port from R_{pt} only, calculated from analysis [10^{-10} V/√Hz].	2.048	1.956	2.160	2.178
rms upper sideband voltage of one-port from R_{pt} only, calculated from HB simulation [10^{-10} V/√Hz].	2.060	1.980	2.177	2.200
Part C: Comparison of oscillator phase noise from R_{pt} only, in normal *HB* simulation versus that predicted in one-port calculation				
Oscillator phase noise due to R_{pt} only, calculated from one-port [dBc/Hz]	−109.8	−109.8	−109.8	−109.8
Oscillator phase noise due to R_{pt} only, calculated from *HB* simulation [dBc/Hz].	−110.3	−109.1	−107.0	−106.8
Part D: Comparison of complete oscillator phase noise using one-port versus that from normal *HB* oscillator simulation				
Equivalent noise current for whole one-port, from *HB* simulation [10^{-11} A/√Hz].	3.067	2.489	5.530	6.300
Oscillator phase noise calculated from one-port, due to all noise sources [dBc/Hz]	−98.1	−99.9	−93.0	−91.8
Oscillator phase noise calculated from actual *HB* simulation, due to all noise sources [dBc/Hz]	−97.8	−98.3	−90.1	−88.4

A10.5 RMS VOLTAGE NOISE PREDICTION IN THE ONE-PORT USING THEORETICAL R_{PT} NOISE ONLY

First, the *rms* voltage noise of a simple RC lowpass filter was simulated to verify the accuracy of the *HB* simulator. Since the RC circuit is linear, there is no difference between the *am* and *pm* responses and the check for an agreement between theory and simulation is straightforward. The theoretical resistor current noise spectral density from parallel resistor R_{pt} is

$$I_n = \sqrt{4kT/R_{pt}} \tag{A10.14}$$

This current was used with the RC filter to predict the filter output voltage spectral density in V/√Hz in the passband. The theoretical *rms* voltage noise showed complete agreement with that predicted in the *HB* simulator, verifying that in the harmonic balance simulator the basic theoretical noise current as in (A10.14) is calculated for linear resistors.

Next, noise in the one-port rather than in the simple RC circuit was studied with only white current noise I_n from the resistor R_{pt} present. Noise sources from the transistor and any other resistors in the one-port were turned off. Note that the *HB* simulation here was only of the one-port, not of the oscillator.

The noise voltage at the upper sideband caused by the upper sideband noise current is

$$v_{t1,usn} = I_n Z_{am1}\cos[(\omega_1 + \delta\omega)t] - I_n Z_{pm1}\sin[(\omega_1 + \delta\omega)t] \tag{A10.15}$$

An additional upper sideband noise voltage, uncorrelated with the one in (A10.15), will occur because of the noise current present in the lower sideband, as (A10.8) shows. Thus, we need only calculate these two contributions and add them in a *rms* fashion to get the total *rms* noise in either sideband. Part B of Table A10.1 compares the *rms* voltage noise in the one-port calculated from the theoretical R_{pt} noise, compared to the noise predicted in the *HB* simulator. The total *rms* noise is somewhat higher at higher bias currents I_{BIAS}, in columns 3 and 4, consistent with the higher Z_{pm1} impedance seen previously in part A of Table A10.1. The agreement between *HB* simulation and *rms* voltage calculation in part B is thus consistent with the finding that the small signal impedance to *am* currents is different from that to *pm* currents.

In the analysis, mixing of white R_{pt} noise from (near to) DC or from (near to) harmonics of the offset frequency is not included as a source of noise at the offset frequency. The same is true for noise from small signal currents at offsets from the harmonics of the fundamental. The close agreement of the *rms* voltage noise obtained via the one-port with actual *HB* simulation suggests that such mixing is not being considered inside the *HB* simulator when the noise source is the linear resistor R_{pt}.

A10.6 OSCILLATOR PHASE NOISE PREDICTION USING THE ONE-PORT WITH R_{PT} NOISE ONLY

Now the effect of adding the inductor L_{pt} in parallel with the one-port is being considered. A high-Q parallel resonance between Z_{am1} and the inductor now occurs. The precise details of how the *HB* simulator performs phase noise calculations must involve a very high-Q resonance, as only this could give the very high gain that noise current experiences in causing phase noise. Since automatic optimization is used in the one-port to make the input impedance purely capacitive at the desired oscillation frequency, adding the ideal linear inductor L_{pt} results in an infinite-Q parallel LC filter. This has infinite impedance at resonance and drops off at $1/f$ as illustrated in the white noise region of Figure 10.11(b) of Chapter 10. Thus the phase noise at small frequency offsets can be calculated and compared to the *HB* simulation of the oscillator noise.

The value of Z_{pm1} in Part A of Table A10.1 departs from the ideal capacitance of Z_{am1}, meaning that the Q seen by i_{pm1} is greatly reduced and the *am* noise voltage along the large signal voltage fundamental is quite low (typically in ratio ~1/4000 in the simulator). In contrast, for the i_{am} component of noise current, the infinite Q tank greatly amplifies the noise caused perpendicular to the large signal voltage. Since only i_{am} sees the ideal tank, the phase noise calculation will use a fixed value for the equivalent one-port capacitance (which was set by the one-port optimization).

Part C of Table A10.1 shows this one-port phase noise prediction for noise current due to R_{pt} only. The *HB* simulator equivalent result is shown for comparison. The results are quite close at the two lower current biases I_{BIAS}, but depart systematically to predict about 3 dBc/Hz poorer phase noise at the two higher current biases. This is ascribed to noise mixing from noise currents near the harmonics of the fundamental. These harmonics are higher at higher current bias.

A10.7 OSCILLATOR PHASE NOISE PREDICTION USING ALL NOISE SOURCES IN THE ONE-PORT

Part D of Table A10.1 looks at the one-port phase noise prediction when other noise sources additional to R_{pt} are considered in the one-port, such as transistor noise. When all the noise sources are turned on it is no longer possible to use theoretical values to calculate the noise as was done for R_{pt} in the previous part. An equivalent noise current was therefore obtained, using the *rms* voltage noise given in preliminary one-port *HB* simulation for this case, as shown in Part D. Note that for low-phase noise the optimum transistor DC bias current for this oscillator would be near 12 mA. The four equivalent noise currents were used in the analysis to predict the oscillator phase noise. The total equivalent noise current in Part D of Table A10.1 should be compared to the noise current from the inductor resistor alone (0.8011e-11 A/√Hz) in Part C. The total equivalent noise current varies as the DC current bias is varied, consistent with noise changing in the transistor as current density changes. Again, at higher transistor bias currents the one-port phase noise prediction is lower than the *HB* simulation result. This is once more ascribed to higher harmonics of the fundamental at higher transistor bias currents.

The fact that the harmonic balance simulator operates on orthogonally different *am* and *pm* small signal impedances is a fundamental outcome of the harmonic balance simulation method, which has been shown in this Appendix to be based on a truncated power series representation. The logical consequence of this is that it is not correct to speak of a single linear small signal circuit model for the noise in the oscillator if power-series representation gives a correct description of reality.

A10.8 CONCLUSIONS

The *HB*-based simulation of oscillator phase noise is seen in the analysis to be strongly affected by the power-series representation of circuit non-linearities. It can be concluded from this Appendix that the power-series model used correctly encapsulates a lot of complexity going on inside the *HB* simulator. The simulation results lend support to the most significant outcome of this analytical model: the appearance of partially correlated upper and lower noise sidebands around the fundamental oscillation frequency and its harmonics caused by the presence of physical, statistically independent noise sources in the oscillator circuit. Whether the real oscillator displays this behavior remains an interesting question. It is not clear at present how to conduct an experiment that would support or refute the existence of the partial correlation of the upper and lower noise sidebands in an oscillator's output spectrum. Another interesting observation coming out of the previous analysis is that the very same physical noise sources that produce the phase noise sidebands cause much smaller amplitude noise in the oscillator.

ENDNOTES

1 It can be shown that a large signal current such as (A10.1) causes a DC shift in the bias point of the transistor, which cancels the DC term in (A10.3).

2 There is also a small DC shift in the solution (A10.8) associated with the $\sin(-\delta\omega t)$ term.

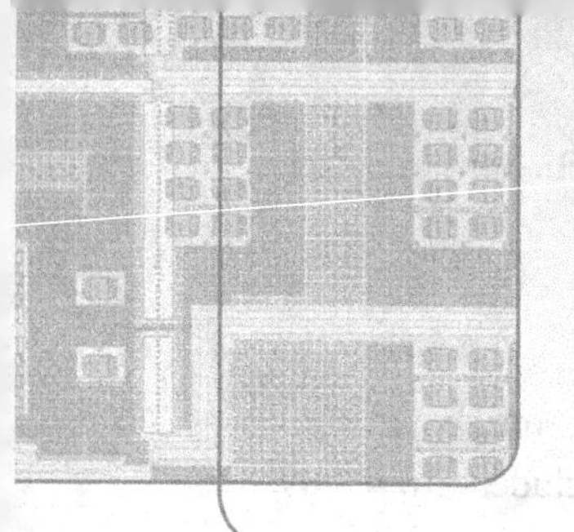

APPENDIX 11

Physical constants

Description	Symbol	Value and unit
Electron charge	q	1.6×10^{-19} C
Boltzmann's constant	k	1.38×10^{-23} J/K
Permittivity of vacuum	ε_0	8.854×10^{-12} F/m
Permittivity of silicon	ε_{Si}	1.04×10^{-10} F/m
Permittivity of SiO_2	ε_{ox}	3.45×10^{-11} F/m
Planck's constant	h	6.63×10^{-34} J–s
Free-electron mass	m_0	9.1×10^{-31} kg
Velocity of light in vaccum	c	3×10^{8} m/s
Thermal voltage (T = 300 K)	kT/q	0.02587 V

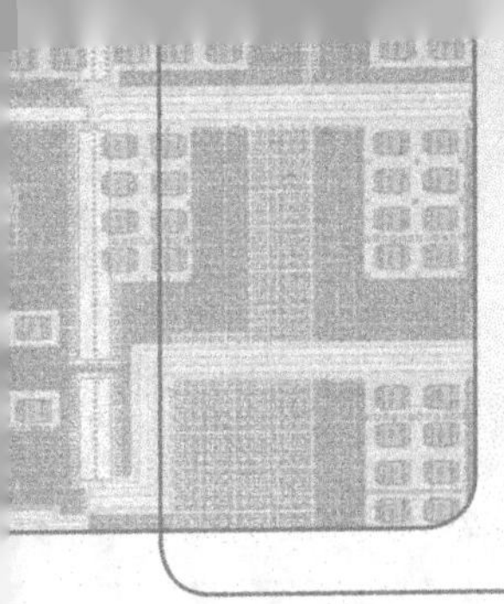

APPENDIX 12
Letter frequency bands

Band designator (waveguide)	Frequency band in GHz
L	1 to 2
S	2 to 4
C	4 to 8
X	8 to 12
Ku	12 to 18
K	18 to 26.5
Ka	26.5 to 40
V (WR−15)	40 to 75
E (WR−12)	60 to 90
W (WR−10)	75 to 110
F (WR−8)	90 to 40
D (WR−6)	110 to 170
G (WR−5)	140 to 220
G (WR−4)	170 to 260
G (WR−3)	220 to 325
Y (WR−2)	325 to 500
Y (WR−1.5)	500 to 750
Y (WR−1)	750 to 1100

INDEX

Printed in the United States
By Bookmasters